# 12th ISIEM (International Seminar on Industrial Engineering & Management): "Industrial Intelligence System on Engineering, Information, and Management"

IOP Conference Series: Materials Science and Engineering
Volume 847

Batu, Indonesia
17-19 March 2020

**Editors:**

**Wahyukaton**
**Rahmi Maulidya**
**Wisnu Sakti Dewobroto**
**Desinta Rahayu Ningtyas**
**Bambang Cahyadi**
**Lina Gozali**
**Sidik Nurjaman**
**Yogi Yogaswara**
**Riana Magdalena**

ISBN: 978-1-7138-4730-4
ISSN: 1757-8981

**Printed from e-media with permission by:**

Curran Associates, Inc.
57 Morehouse Lane
Red Hook, NY 12571

**Some format issues inherent in the e-media version may also appear in this print version.**



Institute of Physics
Dirac House, Temple Back
Bristol BS1 6BE UK

Phone: 44 1 17 929 7481
Fax: 44 1 17 920 0979

techtracking@iop.org

**Additional copies of this publication are available from:**

Curran Associates, Inc.
57 Morehouse Lane
Red Hook, NY 12571 USA
Phone: 845-758-0400
Fax: 845-758-2633
Email: curran@proceedings.com
Web: www.proceedings.com

# TABLE OF CONTENTS

Preface

The Committee

Logo

Peer review statement

Planning the New Factory Layout of PT Hartekprima Listrindo using Systematic Layout Planning (SLP) Method ........ 1
*L Gozali, L Widodo, S.R. Nasution, N Lim*

The use of FMEA for the Quality Control Analysis of Greige Fabrics (case study in the Weaving Department of PT. Unitex, Tbk) ........ 9
*P Fithri, D Jovie Andra, E Wirdianto, Taufik*

What is Effective Use of Big Data? The Consensual Definition of Effective Use of Big Data ........ 17
*Feliks P Sejahtera Surbakti*

Selection strategy implementation of cleaner production using ISM and AHP Method in Chemical Laboratory of Service Industries ........ 24
*D Rimantho, D Ardinia*

Improving Productivity and Quality of Medium Voltage Cable Production ........ 32
*A T Pratama, T Saraswati, F Prianggara, T Savitri*

Analysis of textile dye production scheduling using FCFS, CDS and Heuristic Pour methods ........ 40
*N Y Hidayah, M Syafrizal, M Darmawan*

Greedy Heuristics for the Maximum Covering Location Problem: A case study of Optimal Trashcan Location in Kampung Cipare – Tenjo – West Java ........ 48
*Harummi Sekar Amarilies, A A N Perwira Redi, Ilma Mufidah, Reny Nadlifatin*

Determining optimum eco paving block compositions by using factorial design method ........ 56
*A T Purwandari, A Ratnamirah, N Parwati, W N Tanjung*

Design of Facility Location for new model of Medical Pharmaceutical Refrigerator Production Area on PT. XYZ ........ 64
*R Prasetyani, A Siti, G A Timang*

Managing Internal Bullwhip Effect to Plan Product Distribution in a Garment Factory ........ 72
*S Hidayat, SW Fauzia, N Nurhasanah*

Production quality improvement of Yamalube Bottle with Six Sigma, FMEA, and Data Mining in PT. B ........ 80
*Rina Fitriana, Johnson Saragih, Dea Prameswari Larasati*

Retailer's EOQ model considering demand and holding cost of the defective items under carbon emission tax ........ 88
*Yosef Daryanto, Bellachintya Reira Christata, Ika Murti Kristiyani*

Design of Multicriteria Decision Making Tools for IT Project Selection: A Case from Software House ........ 95
*R P Soesanto, W Tripiawan, I Darmawan*

Prototype Design of Plastic Waste Processing Equipment ........ 103
*N Parwati, R Nurdina, A T Purwandari, W N Tanjung*

Quality Control Analysis of Tube SandwichUsing Six Sigma Method in PTXYZ ........ 111
*A S Sundari, E P Setyawati, D Rimantho*

Allocation of Maltodextrin Raw Material Orders by Fuzzy Analytic Network Process (FANP) and Goal Programming Methods (Study Case: PT. Neopangan Selaras Indonesia) ........ 119
*R Magdalena, S P Dias, A P Ginting*

Integration of conjoint analysis and QFD for new product development in manufacturing small and medium enterprises (case study: a food company) ........ 127
*Wilson Kosasih, Lithrone Laricha Salomon, Alfred Darius Halim*

Integrated model for Multi-criteria Supplier Selection and Order Allocation Problem ........ 136
*Tjutju Tarliah Dimyati*

Web-Based Application of Reliability Availability Maintainability and Cost of Unreliability Method to Analyze Performance of the Machine ........ 144
*Judi Alhilman, Muhammad Fadhil Habibie, Wawan Tripiawan*

Evaluation of Effectiveness and Cost of Machine Losses using Overall Equipment Effectiveness (OEE) and Overall Equipment Cost Loss (OECL) Methods, a case study on Toshiba CNC Machine ........ 152
*Sarastya Dewi, Judi Alhilman, Fransiskus Tatas Dwi Atmaji*

Safety stock analysis of ship fuel in shipping company (Case study: white oil ship PT. Pertamina (Persero)) ........ 160
*Nuzul Fatma Septiana, Iwan Sukarno*

Designing Templates to Support and Monitoring the Activities of Material Requipment Planning (MRP) ........ 168
*Franka Hendra, R Ambarita, K Eko Putranto, Supriyono*

Life Cycle Cost Analysis in Construction of Green Building Concept, A Case Study ........ 176
*MK Kamaralo, J Alhilman, FTD Atmaji*

Productivity Analysis Stevedore a Descriptive Analysis Method with Integration, Importance Performance Analysis, Quality Function Deployment (Case Study: PT. Port Indonesia III (Persero) Branch Gresik) ........ 184
*E Widodo, H Suprayitno, Suparno, Umaiyah*

A Multi-objective model for a Chemical Industry considering economic risk and environment ........ 192
*James Yu, Sri Retno Purwaningsih, Hui Ming Wee*

Development of waiting time predictor based Artificial Neural Network ........ 200
*V Limlawan, P Anussornnitisarn*

Using the Bees Algorithm to solve combinatorial optimisation problems for TSPLIB ........ 208
*A H Ismail, N Hartono, S Zeybek, D T Pham*

Integrated FMEA-MCDM For Prioritizing Operational Disruption in Production Process ........ 217
*Muqimuddin, M. L. Singgih*

Optimizing production layout and capacity via FlexSim– A case study of Y factory ........ 225
*Tai-Jung Chen, Yu-Ching Lee, Chin-Hsin Chiang*

Economic Feasibility Study of Onshore Exploration Oil Field Development using Gross Split Contract ........ 233
*Muhammad Ariyon, Aldo Setiawan, Refiandi Reza*

Preventive Maintenance Scheduling for Shifter Machine in Flour Mills ........ 239
*Wahyukaton*

Performance Evaluation of an EWMA p Chart Based on Improved Square Root Transformation to detect Small Shift Process Variation ........ 246
*Arumsari, J N Rifqi*

Warehouse Improvement Evaluation using Lean Warehousing Approach and Linear Programming ........ 254
*Y. Prasetyawan, N. G. Ibrahim*

The Proposed OEE-SIGMA Prediction for Increased Profits ........ 262
*Y Prasetyawan, F Giffari, B S A Putera*

Analysis of Overall Effectiveness on Hall Separator Punching Machine at PT. DNIA ........ 270
*Iphov Kumala Sriwana, Nadya Syauqillah*

Multi Sensor-Based Failure Diagnosis using the Mahalanobis Taguchi System ........ 278
*B S P Susanto, N Kurniati*

Improved YM Laser Machine Performance with Overall Equipment Effectiveness and Fault Tree Analysis Methods Implementation at PT. XYZ ........ 287
*Arief Suwandi, Zulhilmi Naufal Aulia*

Quality improvement on Common Rail Type-1 Product using Six Sigma Method and Data Mining on Forging Line in PT. ABC ........ 295
*Rina Fitriana, Johnson Saragih, Salma Defina Fauziyah*

Environmental Management System (EMS) within Construction Site: A case study in Kelantan State, Malaysia ........ 303
*A Omran, SHS Yahya*

Quality Improvement for Product Body 2-1 at PT. X ........ 313
*Johnson Saragih, Rina Fitriana, Tri Andriyan*

Forecasting for Steel Production using Artificial Neural Networks and Feasibility Analysis of Plant Regeneration Acid Development in PT. XYZ ........ 320
*A Sasongko, I A Marie, F Arifin*

Analysis of the impacts of Motor Vehicle Exhaust Emissions at Pancasila University on health in order to create A Green Campus ........ 327
*M Y M Sholihin, R Prasetyani, B C Mukti*

Implementation of backpropagation artificial neural network for early detection of vitamin and mineral deficiency ........ 335
*Nina Sevani, Iwan Aang Soenandi, Fajar Saputra*

Optimizing Schedule In Furniture Planning ........ 342
*Abdul Talib Bon*

Book trolley design for Telkom University Library using User Centred Design (UCD) method ........ 350
*Mira Rahayu, Frans Arianto Silalahi, Erna Febrianti*

Designing A Reading Chair using Kansei Engineering Approach ........ 358
*Mira Rahayu, Hilman Ardian Ekananda, Ilma Mufidah*

Designing Information System for Student Practicum Assessment in the Laboratory ........ 365
*Ahmad Arif Nurrahman, Novan Pizary Husen, Otong Rukmana*

Analysis of Choice Shrimp Technology based on Business Process, Productivity, Financial and Risk ........ 373
*Wydzka Tasha Aulia Akbar, Endang Chumaidiyah, Meldi Rendra*

Raw Materials Inventory Planning in Automotive Industries by EOQ Method Consider with the Contract Agreement ........ 381
*Teguh Sri Ngadono, Zulfa Fitri Ikatrinasari*

Three-Level Supply Chain considering Direct and Indirect Transportation Cost and Carbon Emissions ........ 388
*A K S Ong, R A C Robielos, Y. T. Jou, H M Wee*

Designing a Device for Measuring Human Reaction Time ........ 396
*DI Rudolf, V Triyanti*

Risk Mitigation Strategies on Supply Chain PT. X ........ 403
*S P D Kristiana, C W Oktavia, R Magdalena, M A Lilajati*

Integrated ISM-ANP Method for Supplier Selection Criteria Analysis: A Case Study of Construction Company ........ 411
*C Natalia, C W Oktavia, S G Tjhong, T P Hidayat*

Warehouse Layout Designing of Cable Manufacturing Company using Dedicated Storage and Simulation Promodel ........ 419
*W Septiani, G A Divia, S Adisuwiryo*

Performance measurement using Balance Score Card and Analytic Network Process in Elastomer Switch Keypad Manufacturer Indonesia ........ 427
*C O Doaly, L L Salomon, A K J Arta*

A conceptual framework for Servitisation of the manufacturing companies to deliver Product–Service Systems solutions: A study case of the Indonesian Motorcycle Industry ........ 434
*D R S Dewi, S Pittayachawan, E Tait*

Ergonomic Intervention to Improve The Productivity of Brick Press Tool in Small and Medium Entreprise (SME) Akheng Kobar ........ 442
*Lamto Widodo, Silvi Ariyanti, Andreas Jason*

Inventory and Order System Development at PT.X ........ 450
*Steffi Ratanadewi, Marsellinus Bachtiar*

Mitigation of Supply Chain Risk using HOR Model at PT. Sumber Karya Indah ........ 458
*Ahmad, Wilson Kosasih, H J Kristina, Lamto Widodo, Kristina Pasaribu*

Determination of Multi-Product Distribution using Capacitated Vehicle Routing Problem (CVRP) and Product Cubication Dimensions Restriction ........ 466
*Yogi Yogaswara, Neng Resi Andriyani*

Reducing Waste in Manufacturing Industry using Cost Integrated Value Stream Mapping Approach ........ 473
*Wilson Kosasih, C O Doaly, Shabara*

Measurement of Feasibility and Risk Level on Modern Embroidery Kebaya Boutique Establishment in Jakarta ........ 481
*R W Hanaa, E Chumaidiyah*

The Effects of Sleep Quality on Vigilance and Driving Performance in a Train Simulator ........ 489
*D Siswanto, H Iridiastadi, K Muslim*

The NPD Process Design Canvas: Tool for NPD Process Creation ........ 496
*M Iqbal, A Suzianti*

EPR Approach for Better Waste Management System for Mobile Phone Design in Indonesia ........ 504
*A B Junianto, D C Sugandha*

Vehicle Routing Problem with Heterogeneous Fleet, Split Delivery, Multiple Product, Multiple Trip, and Time Windows: A Case study in fuel distribution ........ 512
*S M Nugroho, L Nafisah, M S A Khannan, H Mastrisiswadi, M N Ramdhani*

Analysis of socio-technical approach and Socio User Experience Network Analysis (SNA) to address objections ergonomic loom in the village of Lumban Suhi-Suhi ........ 520
*L N Huda, A F Giffari*

Determinants of Donation Behaviour on Flood Disasters in Indonesia ........ 528
*D. S. Utomo, N. Paoprasert, R. Yousuk*

Designing the Smart Health Function towards Puskesmas (Citizen Health Centre) based on Smart City Concept ........ 536
*Alif Miftahul Jannah, Muharman Lubis, Rd. Rohmat Saedudin, S Fritasya Dwiputri*

Scheduling and Allocation of Airport Service Manpower by considering Time and Work Constraints using M-MAPTWTC Method: a case study ........ 543
*P Moengin, F Darussalam*

Investigating Supply Chain issues in the Food Processing Industry ........ 549
*S Z Khan, MA Ashfaq, M U Awan, H URehman, S A Kamal, N T X Hoa, M Shafiq*

Development of Low-Cost Multi-Input Automated Storage and Retrieval System (AS/RS) for Educational Purposes ........ 558
*Y Theopilus, S S Tjandra, B Sagara*

Workplace Design Process at Indonesian Manufacturing SMEs ........ 566
*D Herwanto, A Suzianti*

A System Performance Analysis of Ship to Shore Operation Considering Crane Availabilities using Simulation Approach ........ 574
*BD Aqsha, N Siswanto, Suparno, M I Tama*

Comparison Analysis of Tajuk's Onion Production Efficiency (Allium Ascalonicum) in Rejoso Sub-District – Nganjuk ........ 581
*D R Ulansari, I N Pujawan*

Assessment of Safety Culture Maturity Level in Production Area of a Steel Manufacturer ........ 589
*A Sudiarno, A A C Sudarni*

ITSM Analysis using ITIL V3 in Service Operation in PT.Inovasi Tjaraka Buana ........ 596
*Muharman Lubis, Rizky Cherthio Annisyah, L Lyvia Winiyanti*

Design and Analysis of Electrical Ergonomic Bionic Grip Wrench ........ 603
*A A Ramadhan, F A Putra, H Wirawan, T R Sahroni*

Risk Assessment User Interface Design's by using Object Orientation Programming Approximation in Wooden Toys Industry ........ 609
*M Mutiasari, W N Tanjung MT, N Parwati, A T Purwandari, U W Islamiah*

Supply Chain Risk Evaluation Information System Design using Fuzzy House of Risk in Wooden Toy Industry ........ 617
*U W Islamiah, W Nurchayanty Tanjung, N Parwati, A Tri Purwandari, M Mutiasari*

ERP Implementation in Crisis Management: A Case Study of Government-Owned Electricity Company ........ 625
*Muharman Lubis, Exa Parmita, Lyvia Winiyanti Lumingkewas*

Discrete Event Simulation Modelling for Classifying the Container Yard Availability Considering Dock Unloading Activity ........ 633
*M I Tama, N Siswanto, Suparno, BD Aqsha*

Lean Startup Approach on Product Design and Manufacture Facility Planning in Uncertain Business Climate ........ 641
*W S Dewobroto, I A Marie*

Design of Mobile and Integrated Tyre Repair Tools for Motorcycle ........ 649
*T Eben Haezar, W Samdan, A Yudha, S Taufik Roni*

Optimasation of Mechanical Cassava Peeling System Parameters ........ 658
*I M Gana, A A Shhu, A Gbabo*

Influence of moisture dependent physical properties of fluted pumpkin vital to development it's processing equipments........ 666
*J J Shiru, I M Gana*

Workload and Fatigue Assessment on Air Traffic Controller ........ 674
*V Triyanti, H A Azis, H Iridiastadi, Yassierli*

Solar Panel-based Wireless Battery Charging System using Fuzzy Control Method........ 681
*Alfarid Hendro Yuwono, Muhammad Rivai, Tri Arief Sardjono*

The Application of Soft System Methodology to design the Conceptual Model for Intelligent Supply Chain Model of Natural Fibre Agroindustry ........ 688
*N Nurhasanah, Machfud, D Mangunwidjaja, M Romli*

Analysing the cause of idle time in loading and unloading operation at Indonesian international port container terminal: Port of Tanjung Priok case study........ 697
*F Suprata, C Natalia, A Sugioko*

Combination of Value Stream Mapping and House of Risk Methods to Eliminate Waste in Productivity Enhancement in Production Area of Fertilizer Company ........ 705
*F Suryanti, A Sudiarno, A A C Sudarni*

Measurement of Physiological and Psychological Workloads of Mechanical Department Operator PT. XYZ ........ 712
*B Cahyadi, A S Maryanti, G A Timang*

Clustering on Small-scale Food Manufacturing Industry in West Jakarta: A Fuzzy Analytical Hierarchy Process Approach ........ 720
*S Sulistiandi, B Marpaung, O Sunardi*

Management of Plastic Waste Recycling by Value Stream Mapping ........ 727
*Tiena Gustina Amran*

Assessing supply chain practices and how they are perceived to impact performance of firms in Sierra Leone: A Case Study in a telecommunication company (Sierratel)........ 736
*Hassan Andrew Fornah, I Nyoman Pujawan*

Investigating the impact of supply chain management on the performance of manufacturing industries in Sierra Leone: case study of Sierra Leone bottling company (SLBC) ........ 744
*Alimamy Kamara, I Nyoman Pujawan*

Determination of Standard Time and Output Production of Spring Frame Mattress Components using Work Sampling Method........ 752
*Taufiqur Rachman, Iphov Kumala Sriwana, Stefanny Liyanawati Zalukhu*

**Author Index**

# PREFACE

This program book is published in line with the Twelfth International Seminar on Industrial Engineering and Management (12$^{th}$ ISIEM). The International Seminars on Industrial Engineering and Management (ISIEM) is an annual seminar to provide an effective forum for invited speakers, academicians, engineers, professionals and practitioners coming from universities, research institutions, government agencies and industries to share or exchange their ideas, experiences and recent progresses in industrial engineering and management and other related fields in dealing with the dynamics and challenges of the 21$^{st}$ century.

This 12$^{th}$ ISIEM is hosted by seven universities, namely Atma Jaya Catholic University of Indonesia, Universitas Trisakti, Universitas Esa Unggul, Universitas Al-Azhar Indonesia, Universitas Tarumanagara, Universitas Pasundan, and Universitas Pancasila. The seminar main theme for this year is **Industrial Intelligence System on Engineering, Information and Management**. Under this theme, we will explore sustainable innovation in industrial technology, information, and management of global issues. The articles cover a broad spectrum of topics in Industrial Engineering and Management, namely Quality Engineering & Management (QM), Decision Analysis & Information System (DAIS), Supply Chain Management (SCM), Production System (PS), Industrial System (IS), Operation Research (OR), and Ergonomics & Product Design (ER&PD).

The articles in this issue provide an overview of critical research issues reflecting on past achievements and future challenges. Those papers were selected from 149 abstracts, and we will send these papers to IOP for publication as an Open Access Proceeding. This is the third time we have had MOU with IOP in United Kingdom to publishing the papers that is indexed by Scopus. This year's seminar become special as more delegates and papers come and received from various universities as well as countries. We are hosting more than 110 delegates both local and from abroad.

I would like to give special commendation to our keynote speakers **Prof. Dr. Abdul Talib Bon** of Universiti Tun Hussein Onn Malaysia and Prof. **Dr Hui-Ming Wee** of Chung Yuan Christian University Taiwan. We are also grateful to our International partners, namely Kasetsart University Thailand, Bright Star University Libya, Chung Yuan Christian University Taiwan, and Universiti Tun Hussein Onn Malaysia, for their contribution to enrich the variety of articles and participants. We are deeply grateful to PT. LEN Industri for sponsoring our seminar. We appreciate all reviewers and editors, for their commitment, effort and dedication in undertaking the task of reviewing all the abstracts and full papers. Examining large number of submissions in a relatively short time frame is always challenging. Highest appreciation is also given to all members of committees for their mutual efforts and invaluable contribution to success of this seminar. Without their help and dedication, it would not be possible to produce this program book in such a short time frame.

Finally, special thanks to all delegates of 12th ISIEM for their contributions. We hope the information in this Program book are useful to all of you. Thank you.

Vivi Triyanti ST., M.Sc.
Chair of Committee

IOP Conf. Series: Materials Science and Engineering **847** (2020) 011002 doi:10.1088/1757-899X/847/1/011002

# THE COMMITTEE

**Steering Committee:**

| | |
|---|---|
| • Dr. Rina Fitriana, S.T, M.M, IPM | Universitas Trisakti |
| • Dr. Iphov Kumala Sriwana, S.T., M.Si., IPM | Universitas Esa Unggul |
| • Trifenaus Prabu Hidayat, S.T., M.T. | Universitas Katolik Indonesia Atma Jaya |
| • Ir. Toto Ramadhan, M.T. | Universitas Pasundan |
| • Ir. Ahmad Chirzun, M.T. | Universitas Al Azhar Indonesia |
| • Wilson Kosasih, S.T., M.T., IPM | Universitas Tarumanagara |
| • Nur Yulianti Hidayah, S.T., M.T. | Universitas Pancasila |

**Organizing Committee:**

| | | |
|---|---|---|
| • Chair: | Vivi Triyanti, S.T., M.Sc. | Universitas Katolik Indonesia Atma Jaya |
| • Vice Chair: | Dr. Winnie Septiani, S.T., M.Si., IPM | Universitas Trisakti |
| • Exchequer: | Dr. Iphov Kumala Sriwana, S.T., M.Si., IPM | Universitas Esa Unggul |
| | • Dr. Winnie Septiani, S.T., M.Si., IPM | Universitas Trisakti |
| • Secretariat: | Emelia Sari, Ph.D. | Universitas Trisakti |
| • Paper Coordinator: | Nunung Nurhasanah, S.T., M.Si. | Universitas Al Azhar Indonesia |
| | • Aprilia Tri Purwandari, S.T., M.T. | Universitas Al Azhar Indonesia |
| | • Dr. Lamto Widodo, S.T., M.T., IPM | Universitas Tarumanagara |
| | • Wawan Tripiawan, S.T., M.T. | |
| | • Nur Yulianti Hidayah, S.T., M.T. | Universitas Pancasila |
| • Event & Accommodation Coordinator: | Stefani Prima Dias Kristiana, S.T., M.Sc | Universitas Katolik Indonesia Atma Jaya |

*The 12th International Seminar on Industrial Engineering and Management (12th ISIEM)*
*Amarta Hills Hotel and Resort, Batu-Malang-Indonesia, March 17-19, 2020*

IOP Conf. Series: Materials Science and Engineering **847** (2020) 011002 doi:10.1088/1757-899X/847/1/011002

| | |
|---|---|
| • Christine Natalia, S.T., M.T. | Universitas Katolik Indonesia Atma Jaya |

**Reviewer:**

Chief Reviewer:

| | |
|---|---|
| Nunung Nurhasanah, S.T., M.Si. | Universitas Al Azhar, Indonesia |

Members:

| | |
|---|---|
| • Prof. Matteo M. Savino | Sanio University, Italy |
| • Prof. Ir. Dr. Sha'ri bin Mohd Yusof | Universiti Teknologi Malaysia, Malaysia |
| • Ir. Dr. Syuhaida Ismail | Universiti Teknologi Malaysia, Malaysia |
| • Dr. Muhammad Shafiq | University of Engineering and Technology Taxila, Pakistan |
| • Fajar Kurniawan, S.T., M.Si. | Saint Mary's University of Hong Kong |
| • Dr. Ir. Hj. Arumsari, M.Sc., IPU | Universitas Pasundan, Indonesia |
| • Dr. Ir. Hj. Tjutju Tarliah Dimyati, MSIE., IPM | Universitas Pasundan, Indonesia |
| • Ir. Wahyukaton, M.T. | Universitas Pasundan, Indonesia |
| • Dr. Ir. Yogi Yogaswara, M.T. | Universitas Pasundan, Indonesia |
| • Dr. Ir. Syarif Hidayat, M.Eng.Sc, M.M. | Universitas Al Azhar Indonesia, Indonesia |
| • Nunung Nurhasanah, S.T., M.Si. | Universitas Al Azhar Indonesia, Indonesia |
| • Dr. Iphov Kumala Sriwana, S.T., M.Si., IPM | Universitas Esa Unggul, Indonesia |
| • Dr. Ir. Nofi Erni, M.M., IPM | Universitas Esa Unggul, Indonesia |
| • Prof. Parwadi Moengin, Ph.D. | Universitas Trisakti, Indonesia |
| • Dr. Winnie Septiani, S.T., M.Si., IPM. | Universitas Trisakti, Indonesia |
| • Ronald Sukwadi, S.T., M.M., Ph.D., IPM. | Universitas Katolik Indonesia Atma Jaya, Indonesia |
| • Vivi Triyanti, S.T., M.Sc. | Universitas Katolik Indonesia Atma Jaya, Indonesia |
| • Dr. Lamto Widodo, S.T., M.T., IPM. | Universitas Tarumanagara, Indonesia |

**International Partnership:**

| | |
|---|---|
| • Prof. Dr. Abdul Talib Bon | Universiti Tun Hussein Onn, Malaysia |
| • Prof. Dr Hui-Ming Wee | Chung Yuan Christian University, Taiwan |
| • Yung-Tsan Jou, Ph.D | |
| • Dr. Abdelnaser Omran Ali | Bright Star University, Libya |
| • Dr. Osama Saad Hammad Salleh | |
| • Mr. Alzaroog Saleh Abdulali Emhamed | |
| • Dr.Peerayuth Charnsethikul | Kasetsart University, Thailand |
| • Dr. Pornthep Anussornnitisarn | |

IOP Conf. Series: Materials Science and Engineering **847** (2020) 011003 doi:10.1088/1757-899X/847/1/011003

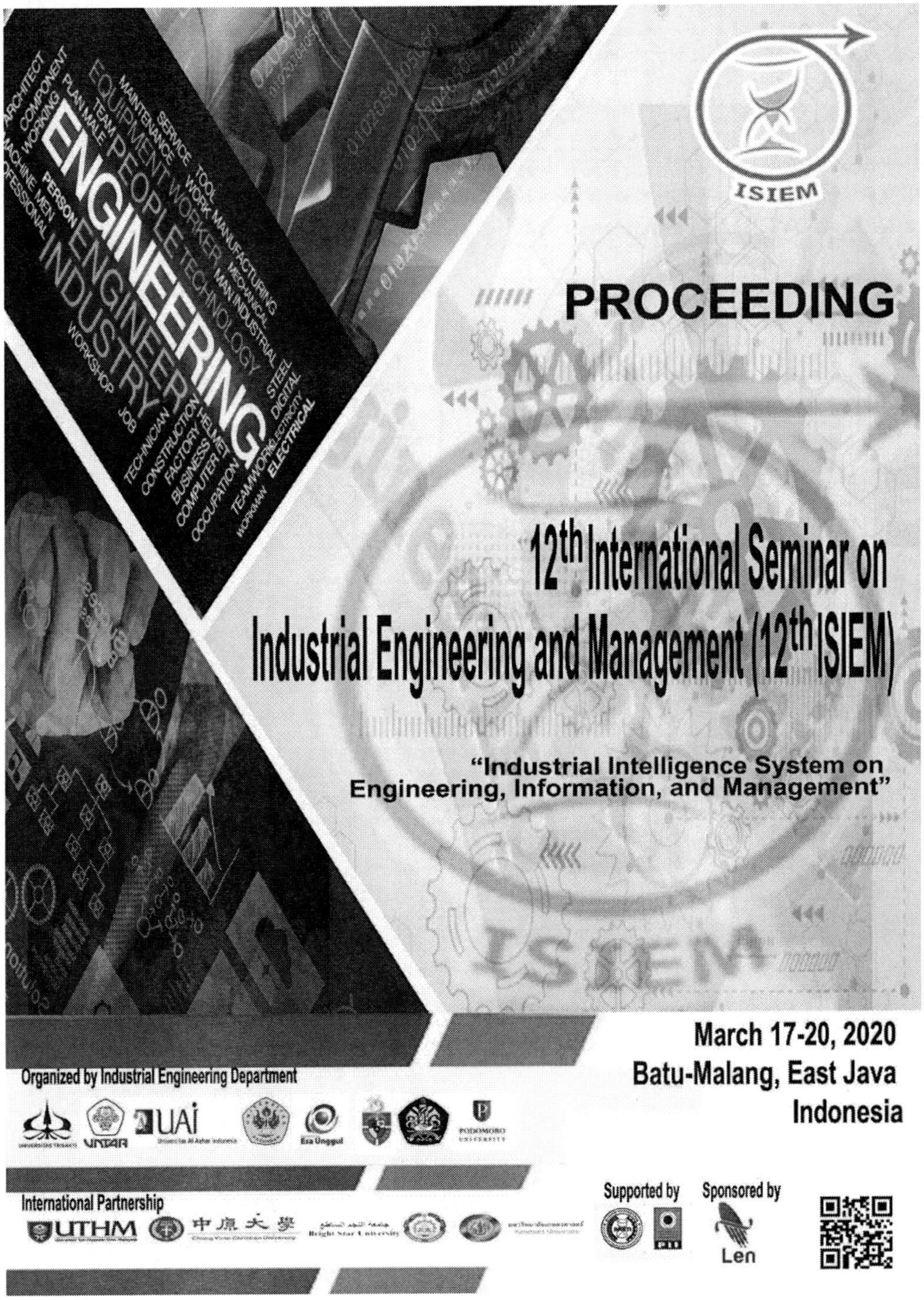

# Peer review statement

All papers published in this volume of *IOP Conference Series: Materials Science and Engineering* have been peer reviewed through processes administered by the proceedings Editors. Reviews were conducted by expert referees to the professional and scientific standards expected of a proceedings journal published by IOP Publishing.

# Planning the New Factory Layout of PT Hartekprima Listrindo using Systematic Layout Planning (SLP) Method

**L Gozali[1], L Widodo[2], S.R. Nasution[3], and N Lim[4]**

[1,2,3,4]Industrial Engineering Department, Faculty of Engineering, Tarumanagara University, Jl. S.Parman no.1, Jakarta, Indonesia
[1]ligoz@ymail.com, [2]lamtow@yahoo.com, [3]sitirohananasution@gmail.com, [4]nicholsonlim28@gmail.com

**Abstract.** PT Hartekprima Listrindo is a manufacturer that produces silent, open and mobile/trailer types of diesel-powered generators, along with AMF/ATS panels and synchronous panels with the HARTECH brand. The generators produce eight Kva power variants up to 3000 Kva power. PT Hartekprima Listrindo plans to increase the production capacity by relocating its factories to a new place. The production capacity of the new plant is planned to achieve 20 generators/year in a 5184 m2 total land area of production. This relocation design uses Systematic Layout Planning (SLP) method, and the layout design is simulated. The SLP method is carried out by describing material flow in the production process using a Multi-Product Process Chart (MPPC), and then continued by using Activity Relationship Chart (ARC) which illustrates the relationship among the activities of the departments. The next stage of layout design is describing the Activity Relationship Diagram (ARD) and Area Allocation Diagram (AAD). The comparison layout design resulted in 2 alternative layouts that are better than the initial layout. The chosen layout design is alternative design 1 because it reduces transportation time and increases movement efficiency.
**Keywords**: Systematic Layout, Activity Relationship Chart, Activity Relationship Diagram, Area Allocation Diagram.

## 1. Introduction

The recently tight competition pushes the companies to increase efficiency and productivity in all production process. These activities are carried out by reducing the production costs, improving the quality of goods or products and on-time delivery. With the increased demand from consumers, the company plans to do a factory facility layout that is more flexible, productive, effective and efficient in order to improve customer service. Re-layout activities also include designing new layout for work facilities and production by arranging the order of the process flow with the existing production facilities.

Facility layout is the design of plant layout as a planning and integration of the production flow for getting the most effective and efficient interrelation between operators, equipment, and the process of material transformation from receiving area to finishing area. [1]. The factory layout is a part of facility design, which consists of the allocation of the factory and the design of the building where the factory layout is closely interrelated with material handling [2]. The purpose of these activities is to arrange the layout of production facilities to be more effective and efficient so that production time is minimal and produces a quantity of products of the highest quality. Facility layout is defined as the process of designing a facility, including analysis, planning, design and arrangement of facilities, physical equipment, and people which aimed for improving the production efficiency and services [3]. Setting the layout of facilities and work areas is a problem found in this industry.

The problems found in the production process of this company are because the layout does not match the order of the production process and many delays occur among the production processes. The purpose of this research is to produce a good factory layout which comply with the production flow with Systematic Layout Planning method to increase productivity.

### *1.1 Systematic Layout Planning (SLP)*

Systematic Layout Planning is a systematic and organized approach to planning a layout [4]. The SLP method is applied because it can minimize the flow of material and consider the relationship between the room, the need for space and available space. In addition, the SLP method is also a simple and an easy method to implement.

Richard Muther developed a layout design method called Systematic Layout Planning (SLP) [4], with factory layout planning procedures such as:

- Conduct initial data collection, namely product design data, process design and production schedule design.
- Determine material flow (Flow Process Chart/FPC).
- Determine the relationship between activities or activities (Activity Relationship Chart/ARC).
- Create a relationship diagram of activity and flow (Activity Relationship Diagram/ARD).
- Determine the number of rooms needed and adapted to the available room size.
- Make a room relationship diagram (Area Allocation Diagram/AAD).
- Make modifications and limits in making alternative layouts.
- Make alternative layouts.
- Evaluate and choose alternative layouts.

The Systematic Layout Planning (SLP) method is included in the conventional analysis techniques. The SLP method is often used in designing layouts because it is done by following a sequence of interrelated (systematic) stages [5]. The SLP method also uses quantitative inputs such as distance and frequency of material movement and qualitative input such as the degree of relationship of activities in the stages of analysis, so that the analysis is done better. In addition, the condition of the company where the research was conducted also supports the application of the SLP method, which is not too large, have relatively simple layouts, and the irregularities in the flow of materials.

In addition, the SLP method also has detailed procedures in arranging layouts based on the sequence of processes, then building block diagrams, and ultimately making detailed layouts of each plant.

## 2. Methods

This research uses the SLP method, starting with Operation Process Chart, then continued with Multi Product Process Chart, From To Chart, Activity Relationship Chart (ARC), Activity Relationship Diagram (ARD), Area Allocation Diagram (AAD). After AAD, the research purpose 2 layout problem solutions. After testing the problem with distance criteria, then the research found the best solution for the problem.

## 3. Result and Discussion

### *3.1. Flow Process Chart (FPC)*

Flow Process Chart is the most commonly used map for layout design which is based on the planning process. FPC is a complete operating diagram/chart that includes an analysis of the material flow of production for generator. In addition, FPC is also a tool for recording all of production activities and showing the distance of equipment from raw materials to finished goods [6]. Below is Figure 1, the FPC of PT Hartekprima Listrindo.

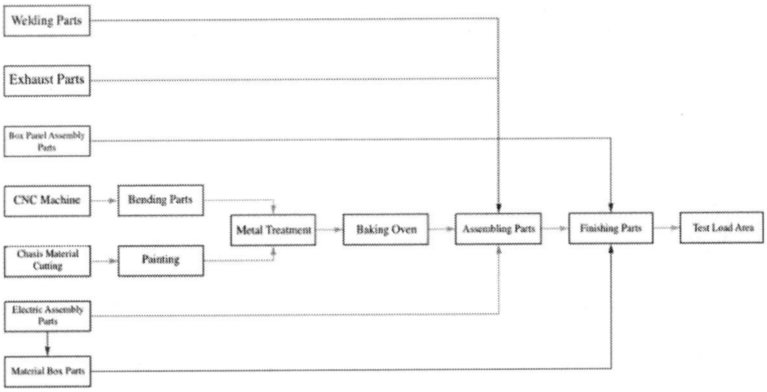


**Figure** 1. FPC of PT Hartekprima Listrindo

### *3.2. Operation Process Chart (OPC)*

OPC is a diagram that describes the steps of the production process experienced from operations and quality checking [6]. Figure 2 shows the Operation Process Chart of Generator HT 80P (silent).

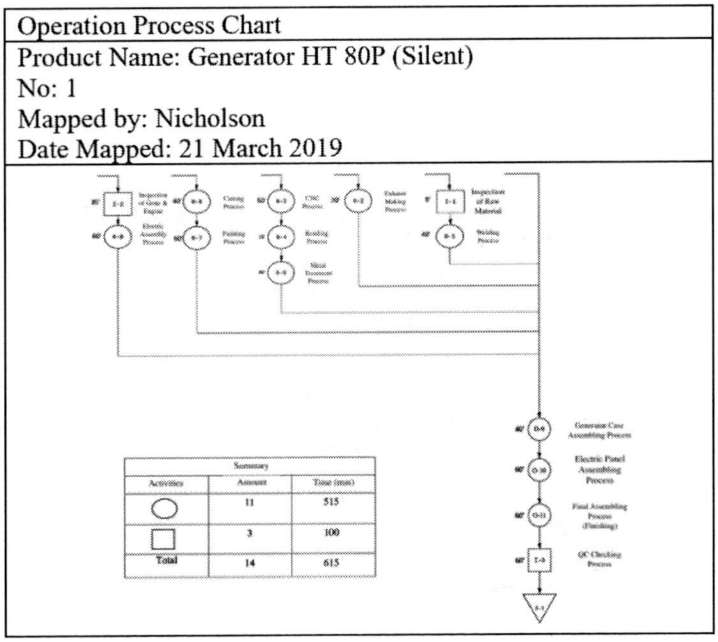


**Figure** 2. OPC Manufacture of Generator HT 80P

### *3.3. Multi Product Process Chart (MPPC)*

Multiproduct process chart (MPPC) is the picture of a simple production process for making products from the beginning of the material to the end products based on the analysis of material handling and the occurrence of backflow [6]. Figure 3 shows the Multi Product Process Chart of generator set HT 80 P.

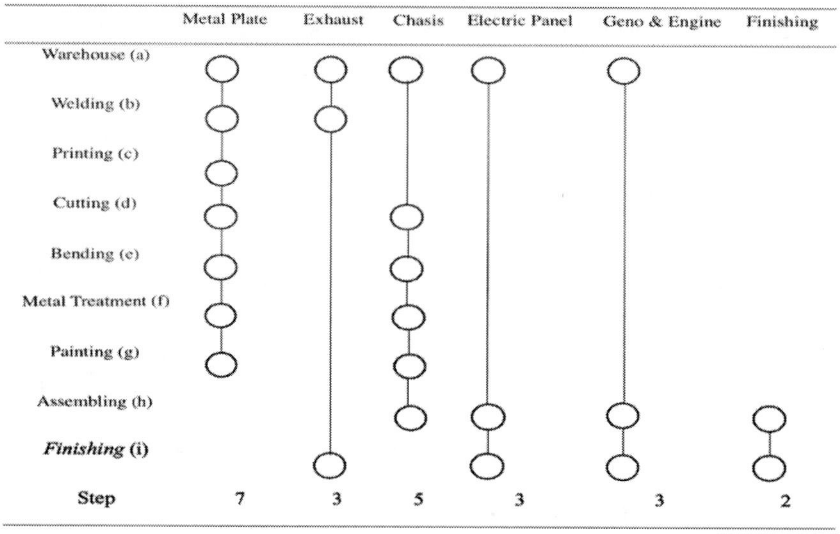


**Figure** 3. MPPC Generator HT 80 P

### *3.4. Activity Relationship Chart (ARC)*

ARC is a simple method or technique in designing facility layout based on the degree of activity relations that are often stated in "qualitative" research and tends to be considered on subjective techniques from each facility or department [6]. Figure 4 shows the Activity Relationship Chart of PT Hartekprima Listrindo.

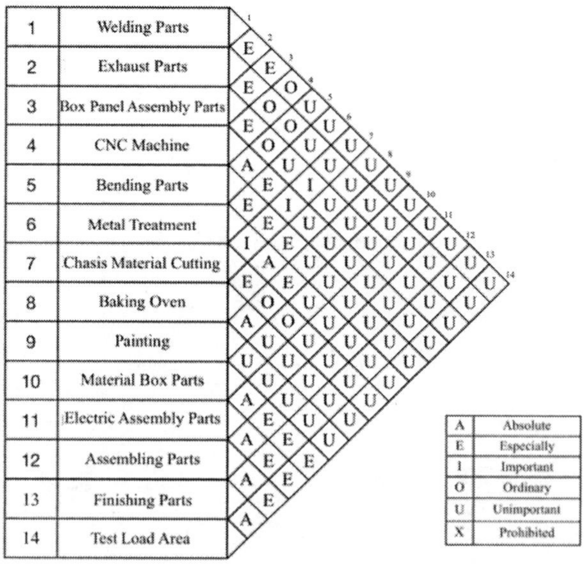


**Figure** 4. Activity Relationship Chart (ARC)

### *3.5. Activity Relationship Diagram (ARD)*

Activity Relationship Diagram basically explains the relationship among material flow patterns, their location and each supporting department of the production [7]. Figure 5 shows Activity Relationship Diagram of alternative layout 1 and Activity Relationship Diagram alternative 2.

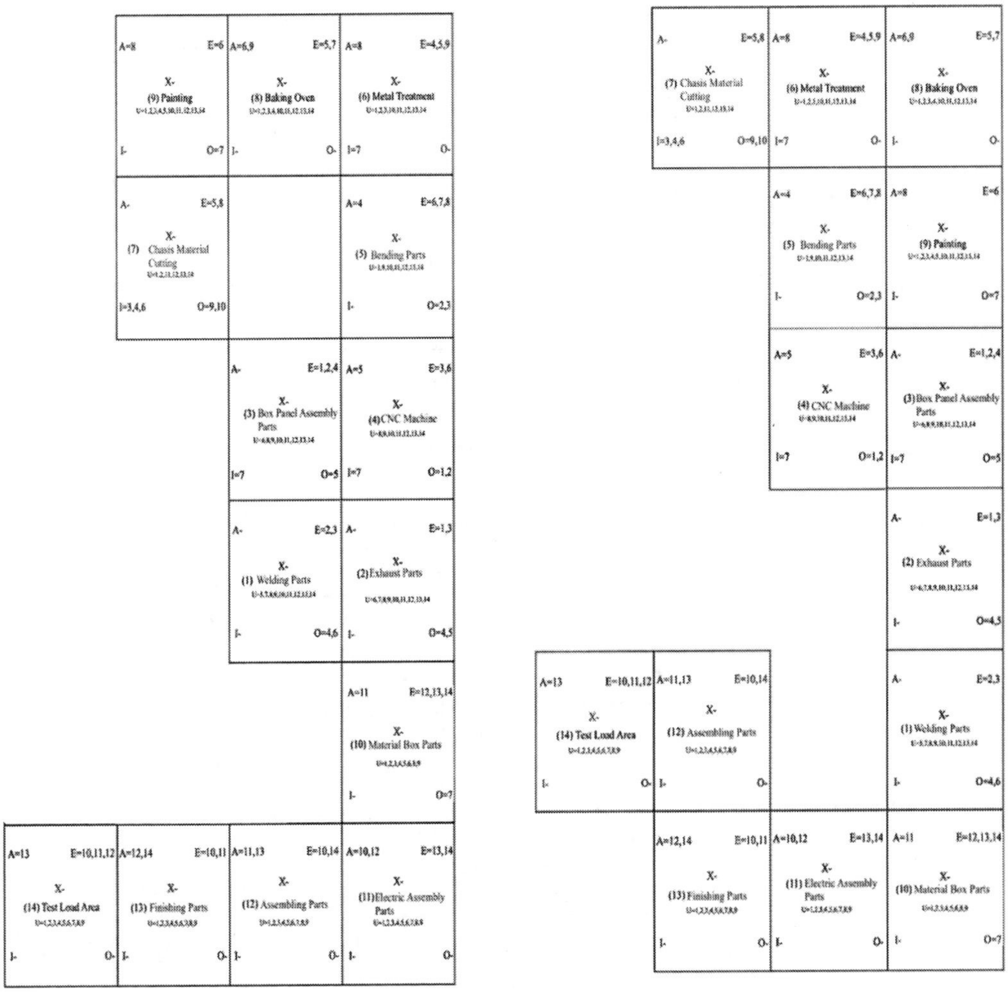


**Figure** 5. ARD Alternative layout 1 (left) and ARD Alternative layout 2 (right)

### *3.6. Area Allocation Diagram (AAD)*

Area Allocation Diagrams (AAD) shows templates globally, the information only shows the placement of area, while complete visualization images can be seen in the final template/layout which is the final result of analyzing and planning factory layout [8]. Below is AAD which is based on the ARD that was designed in the section 4.5. Figure 6 shows the alternative AAD layout 1 and alternative AAD layout 2.

Remarks in Figure 7: 1 = Welding Parts, 2 = Exhaust Parts, 3 = Box Panel Assembly Parts, 4 = CNC Machine, 5 = Bending Parts, 6 = Metal Treatment, 7 = Chasis Material Cutting, 8 = Baking Oven, 9 = Painting, 10 = Material Box Parts, 11 = Electric Asembly Parts, 12 = Assembling Parts, 13 = Finishing Parts, 14 = Test Load Area.

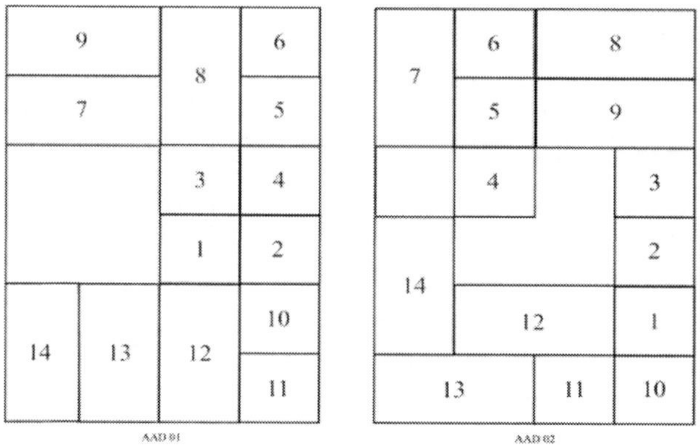


**Figure** 6. Alternative AAD layout 1 and Alternative AAD layout 2

### *3.7. Alternative Layout Designs*

Alternative selection is based on the distance of material transportation and the placement of spaces for each part of production. The new lay out which is based on alternative layouts that have been chosen, alternative layout 1 and alternative layout 2, show different placements to the initial layout of the factory. The Figure 7 shows the initial layout of process flow.

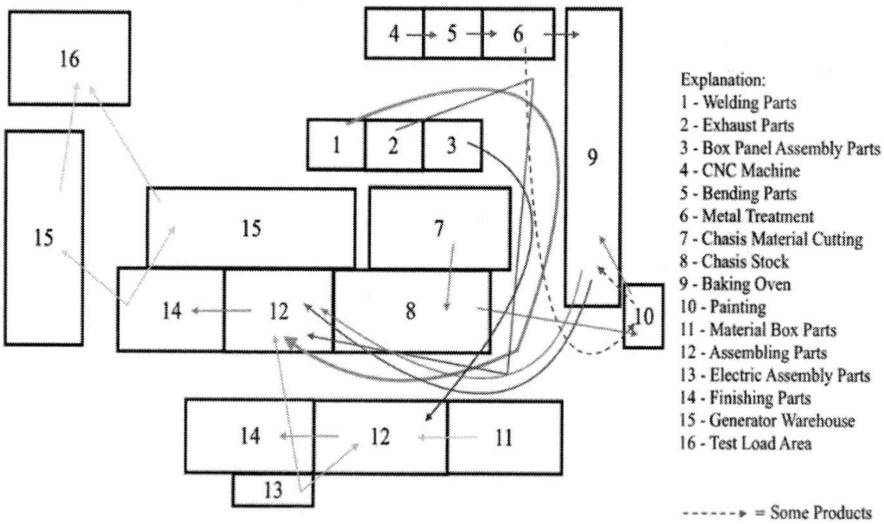


**Figure** 7. Initial Layout Flow Process

Therefore, the Alternative layout 1 and Alternative layout 2 process can be made by using the steps from the previous section. Figure 8 shows the flow production process of alternative layout 1 and layout 2.

IOP Conf. Series: Materials Science and Engineering **847** (2020) 012001 doi:10.1088/1757-899X/847/1/012001

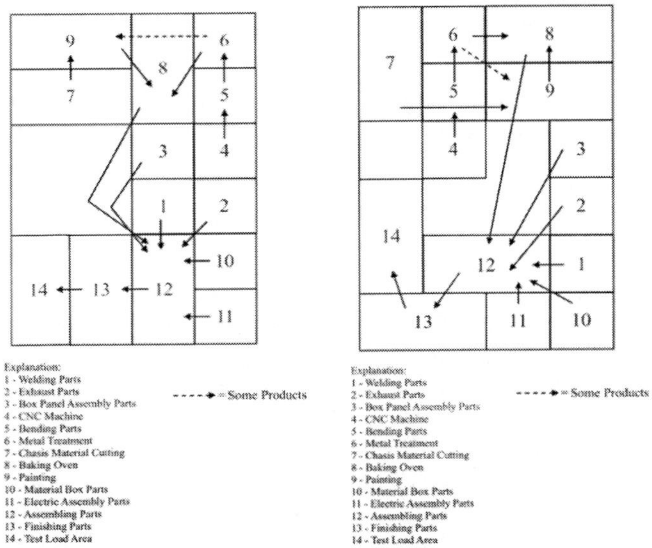


**Figure** 8. Flow Process Layout Alternative 1 (left) and Layout Alternative 2 (right)

Designing new layout for the new factory of PT Hartekprima Listrindo based on the method and techniques produced 2 alternative layouts. The alternative layouts are produced based on the production flow, which can reduce the distance of the excessive transportation and efficient movement. The results of the alternative layouts selected which has better movement and minimize distance. Figure 9 shows the selected alternative layout results.

**LAYOUT PT HARTEKPRIMA LISTRINDO**

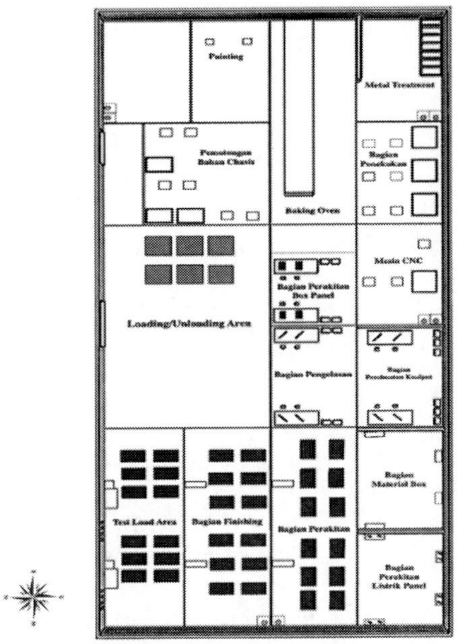


**Figure** 9. Selected Alternative Layout

## 4. CONCLUSION

The research's result at PT Hartekprima Listrindo shows the following conclusion:
This research applied the Systematic Layout Planning (SLP) method that result the better alternative layouts with ARC, ARD and AAD techniques. The recommendation of this research found 2 alternative layouts, alternative layout 1 and alternative layout 2. The best layout chosen is alternative 1 which has only 1 intersecting part and relatively short displacement distance, compared to alternative layout 2 which has 2 intersecting parts.

## 5. REFERENCES

[1] Apple, James M. 1990. *Tata Letak Pabrik dan Pemindahan Bahan.* 3rd Edition, (Bandung: ITB)
[2] Meyers, F.E. 1992. *Plant Layout and Material Handling.* (New Jersey: Prentice Hall)
[3] Wignjosoebroto, Sritomo. 1992. *Pengantar Teknik & Manajemen Industri.* (Surabaya: Guna Widya)
[4] Purnomo, Hari. 2004. *Perencanaan dan Perancangan Fasilitas* 1st Edition, (Yogyakarta: Penerbit Graha Ilmu)
[5] Wignjosoebroto, Sritomo. 2009. *Tata Letak Pabrik dan Pemindahan Bahan.* 3rd Edition, (Surabaya: ITS)
[6] RS Nuranisa, L Widodo, N Erni. 2013. Usulan Perbaikan Rancangan Tata Letak Penyimpanan Bahan Baku Berdasarkan Kriteria Pemakaian Bahan, 69-80.
[7] L Putrajaya, L Gozali, S Ariyanti. 2013. The Proposed Layout Design Using Factory Systematic Layout Planning Method at PT Jasa Laksa Utama.
[8] P Andriani, L Gozali, IA Marie. 2015. Factory Plastic Bag Layout Design in Elite Recycling Indonesia Extension.

IOP Conf. Series: Materials Science and Engineering **847** (2020) 012002 doi:10.1088/1757-899X/847/1/012002

# The use of FMEA for the Quality Control Analysis of *Greige* Fabrics (case study in the Weaving Department of PT. Unitex, Tbk)

**P Fithri[1,2], D Jovie Andra[1,3], E Wirdianto[4], and Taufik[5]**

[1]Industrial Engineering Department, Engineering Faculty, Universitas Andalas, Kampus Unand Limau Manis Pauh Padang, Sumatera Barat, Indonesia
[2]primafithri@eng.unand.ac.id, [3]dedejovieandra@gmail.com, [4]e.wirdianto@gmail.com , [5]taufik@ft.unand.ac.id

**Abstract.** PT Unitex, Tbk, is a textile company in Bogor. The problem that is found in this company is the number of defects that occur in the weaving section, namely reaching. If conditions like this are allowed, it will cause losses to the company. With these conditions, PT Unitex wants to reduce the number of defects to a minimum.For this reason, the main problem raised in this study is how to control the quality of defective fabrics produced by PT Unitex, Tbk, to support and evaluate the quality control policies implemented by PT Unitex, Tbk by using Pareto diagram tools, and fishbone diagrams later in analysis using the Failure Mode And Effect Analysis (FMEA) method. The results of this research are obtained: In the period July - December 2018, 5 types of fabric defects were most dominant, namely OSA TOSHI, NIHON DOSHI, USUDAN, TATE GIRE, and ATSUDAN. Analysis of Failure Mode and Effect Analysis of the causes of failure obtained from the fishbone diagram obtained the highest severity value for human factors, the highest occurrence value for less precision and control, and the highest detection value for engine maintenance.
**Keywords:** Textile company, Defects, Weaving section, Failure Mode And Effect Analysis (FMEA) method

## 1. Introduction

Increased consumer demand for personal needs, such as clothing, makes the level of competition in the textile industry more stringent. Quality is a very powerful tool in maintaining a company's business [1]. It can compete with other products; therefore, every company must improve its efficiency and minimize product defects that occur in every product production process [2], [3]. In Indonesia, demand for clothing products is increasing every year and can be seen in Figure 1.

Based on Figure 1, the demand for clothing products in Indonesia has increased from 2011-2018. With the demand for clothing products in 2018 amounting to 20036 (in millions of USD). Based on these data, each industry must improve the quality of its products to be accepted by consumers and be able to compete with other industries to get consumers. Quality control has an important role in the company. One of them is by showing the quality of its products and not ignoring product failures or product defects. Product quality is the main focus of the company. So that with increasing demand for a product will have an impact on increased sales results. According to Irwan and Haryono (2015), PT. Unitex, Tbk, which is engaged in the textile industry, is a company that manufactures fabrics consisting of several processes. In general, the fabric production process at PT. Unitex, Tbk consists of spinning, weaving, and dyeing processes.

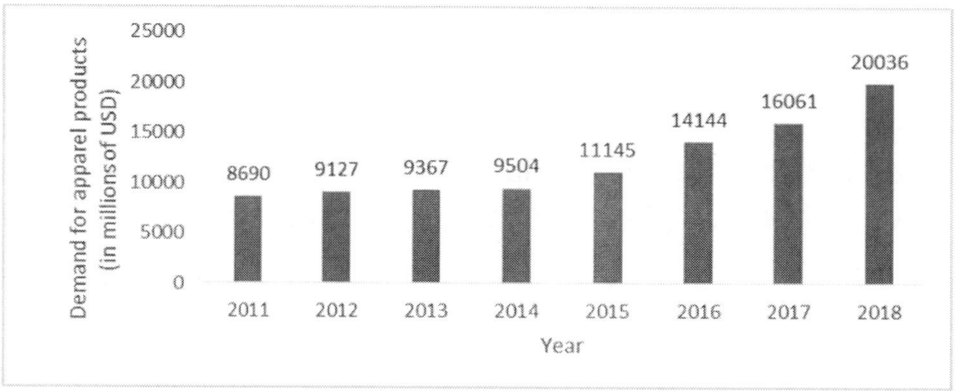


**Figure 1**. Demand for clothing products in Indonesia According to the Economic Intelligence Unit

The Spinning section is processed by cotton or polyester yarn and blended with polyester. In this Weaving section, the fabric is processed from yarn or blended yarns with other materials such as yarn weaving with a process of weaving the yarn so that it becomes a raw cloth (Greige Cloth). Fabric defects occur in the weaving process; this is because several causes make the fabric defective. The weaving process uses Shokki machines; there are seven types of shokki machines, including New DB, AJL DB, DB 810, DB 813, Old AJL 30, Old AJL H4, and New AJL. Based on previous research conducted by Tejaskumar S. Parsana and Mihir T. Patel in 2014 stated that FMEA was used to identify and analyze all failure modes from various parts of the system, the effects of these failure modes on the system and how to avoid failures or moderate the effects from the failure system [4]. Jigor Doshi conducted other research and Darshak Desai in 2016, Failure Mode and Effect Analysis (FMEA) is a quality tool used to identify potential failures and related effects on processes and products, so that continuous improvement in quality can be achieved by reducing it [5]. FMEA also can be used for analysed the risk priority [6]–[8].With so many defects that occur in a process will cause harm to a company. Therefore, research on the FMEA method as a means system for quality control for shokki machines in the Weaving department at PT. Unitex, Tbk needs to be reviewed and efforts made in creating continuous and better product quality improvements going forward. Based on the background description in the previous sub-chapter, the following problem can be summarized: how to control the quality of the product defects in the greige fabric in the Weaving department at PT. Unitex, Tbk by using the Failure Mode and Effect Analysis method to reduce the defect level of the greige fabric products and their causes?

The objectives that can be expected to be achieved in this study are:1. Identify possible causes of product defects. 2. Analyzing the Causes of product defects to reduce the level of fabric product defects in the Weaving department at PT. Unitex, Tbk.3. Provide recommendations for improvement of the causes of fabric defects.

## 2. Method

The initial stage of the research carried out in the form of direct observation to the field to see the process that is taking place. The purpose of the observations made is to find out the condition of the company's real production. Observations were made generally about the quality control system used by the company. Collecting complete information related to all processes and conditions that occur obtained directly through the field supervisor and employees. Based on this information, a picture of the problems associated with quality control is found in the company.

A literature study is a stage of preparing the theoretical foundation that can support research originating from various sources that are considered relevant. So, we get supporting theories that will be used to solve the problems in this study. Literature sources used include those from books, journals, and research from other parties that are considered relevant. The literature used can provide an overview related to the basic theories of quality control using the Failure Mode and Effect Analysis (FMEA) method as well as other supporting theories that help in conducting discussions to resolve this problem.

At this stage, identification of the problems that occur in the company. One of the main problems that occur is the quality of fabric production. The quality of the fabric produced from the production process carried out shows that there are still fabrics that are in grade C due to defects found in the fabric. So that effort is needed to improve the quality control of fabrics.

Sources of data used in this study are primary data and secondary data. Primary data were obtained from direct observations in the field and interviews of the authors with related parties, including the head of the weaving division, field supervisors, and operators who were directly involved in handling the system or process that was taking place while the secondary data obtained by the author from sources that have been collected are production data and the number of defects, production processes and product quality characteristics.

The data processing that will be carried out in this study is based on the stages contained in the Failure Mode and Effect Analysis method which consists of the following steps: Identification of the type of failure that occurred, identification of the cause of the failure, determination of rating severity, occurrence, and detection, Calculation and ranking RPN values, analysis of causes of defects, proposed improvements and controls. Before doing data processing, it is necessary to recapitulate production data, which includes production data of both grade A and grade C products for each type of fabric. Then, the data is processed using the Failure Mode and Effect Analysis (FMEA) method following the stages.

1. Identify the types of failures that occur an initial step in the FMEA method is to look for any possible fabric failures or defects that occur in the system as a whole. Seeing the type of failure that can occur from data obtained from the company or can also be direct observation.
2. Identify the Cause of failure next step is to identify the cause of the failure that has been obtained from the initial step. The tool used in identifying problems using fishbone diagrams and Pareto diagrams.
3. Determination of Rating Severity, Occurrence, and detection The next step is weighing the severity, occurrence, and detection of several factors obtained from the fishbone diagram. Then to get the RPN value, it is done severity, occurrence, and detection multiplication.
4. Failure Mode and Effect Analysis (FMEA) AnalysisAfter identifying the consequences of fabric defects, the next step is to create an Analysis of Failure Mode and Effect Analysis, where the severity, occurrence, and detection values are first searched.

Then the next step is to multiply the rating value to get the RPN value which is then performed an analysis of the severity value due to process failure, an analysis of the occurrence value of the cause of failure, an analysis of the detection value of the control carried out, and an analysis of the RPN value of the failure factor that happened. The last stage of the research conducted is to draw conclusions from the results of practical work research and provide suggestions regarding improvements that can be made by others in future research.

## 3. Result and discussion

Data collection was carried out by collecting data on the amount of production of greige cloth combined two types of fabric, namely yarn dyed and piece dyed. Data on the type of fabric defects with grade C for each type of machine used. The data comes from the production report data and defects in the Weaving section in the December 2018 period. The defect product data collected consists of the number of defects produced, the percentage of defects produced from each fabric produced, and the machine used, and the types of defects that occur. The defect product data collected is defect product data in the period July - December 2018 obtained from the archive of the Weaving section which can be seen in Table 1.

**Table 1**. Production Results of Defective Fabrics by Type of Fabric Type and Machine Type for the Period July - December 2018

| Month | Type of Fabric | Type of Machine NEW DB | DB 810 | DB 813 | AJL 30 | AJL DB | AJL H4 | NEW AJL |
|---|---|---|---|---|---|---|---|---|
| July | Yarn Dyed | 6355.55 | 4276.745 | 9863.16 | 5484.45 | 0 | 0 | 10296.1 |
| | Piece Dyed | 0 | 0 | 328.9 | 329.5 | 3874.27 | 730.57 | 0 |
| August | Yarn Dyed | 7137.96 | 5303.49 | 5603.49 | 973.075 | 6356.13 | 429.5 | 2881.975 |
| | Piece Dyed | 0 | 0 | 973.075 | 429.5 | 2881.975 | 1041.47 | 0 |
| September | Yarn Dyed | 8759.04 | 6870.3 | 8864.83 | 8271.33 | 0 | 0 | 13620.27 |
| | Piece Dyed | 0 | 8759.04 | 0 | 6870.3 | 631.995 | 8854.83 | 299 |
| October | Yarn Dyed | 13135.68 | 9859.11 | 11926.03 | 10871.55 | 0 | 0 | 0 |
| | Piece Dyed | 0 | 0 | 3256.725 | 403 | 6476.235 | 5740.45 | 0 |
| November | Yarn Dyed | 8276.72 | 5518.4 | 7629.89 | 7337.96 | 0 | 0 | 12359.23 |
| | Piece Dyed | 0 | 0 | 269.185 | 383 | 3578.935 | 1310.29 | 0 |
| December | Yarn Dyed | 12489.6145 | 7274.3 | 9595.94 | 9001.96 | 332.5 | 0 | 14292.35 |
| | Piece Dyed | 0 | 0 | 857.79 | 312 | 5300.45 | 1913.49 | 0 |

Based on data on the types of defects that occur in each machine, ranking is obtained based on the type of fabric defects that often occur that will be identified the cause of the defects can be seen in Table 2.

**Table 2.** Ranking of types of fabric defects that often occur

| Type of Fabric Defect | Frequency(meters) July | August | September | October | November | December | Average | Percentage | Cumulative Percentage |
|---|---|---|---|---|---|---|---|---|---|
| OSATOSHI | 8302.49 | 9342.16 | 10502.14 | 12276.97 | 10617.57 | 11939.47 | 10496.80 | 29.89 | 29.89 |
| NIHONDOSHI | 4367.4 | 4955.07 | 5727.4 | 6791.27 | 5756.77 | 6840.19 | 5739.68 | 16.35 | 46.24 |
| USUDAN | 2721.295 | 3129.905 | 3583.62 | 4156.42 | 3474.71 | 4009.21 | 3512.53 | 10.00 | 56.24 |
| TATE GIRE | 2564.75 | 2952.825 | 3467.745 | 4366.025 | 3207.525 | 3916.605 | 3412.58 | 9.72 | 65.96 |
| ATSUDAN | 2171.57 | 2616.735 | 3041.39 | 3757.63 | 2885.72 | 3480.47 | 2992.25 | 8.52 | 74.48 |
| TATE HIKE | 1285.33 | 1250.93 | 1451.18 | 3404.3 | 2829.6 | 3134.22 | 2225.93 | 6.34 | 80.82 |
| SHORT PICK | 1702.045 | 1877.83 | 2373.26 | 2789.09 | 1774.89 | 3104.505 | 2270.27 | 6.47 | 87.29 |
| MURAITO | 1000.31 | 1174.83 | 1690.235 | 2178.885 | 1469.585 | 2537.665 | 1675.25 | 4.77 | 92.06 |
| ITOKOZU | 989.02 | 1122.565 | 1316.065 | 2269.075 | 852.355 | 1859.025 | 1401.35 | 3.99 | 96.05 |
| NOZZLE | 5109.99 | 301.33 | 473.66 | 1162.96 | 521.66 | 756.91 | 1387.75 | 3.95 | 100.00 |
| Total | 30214.2 | 28724.18 | 33626.695 | 43152.63 | 33390.385 | 41578.27 | 35114.39 | | |

The use of Pareto diagrams in identifying the causes of fabric defects to show the problem is based on the order of the many types of fabric defects that occur. The sequence starts with the number of problems that occur the most until the least occur. The graph is indicated by the highest bar graph (far left) to the lowest graph. Pareto diagram of the average fabric defects in the period July - December 2018 can be seen in Figure 2.

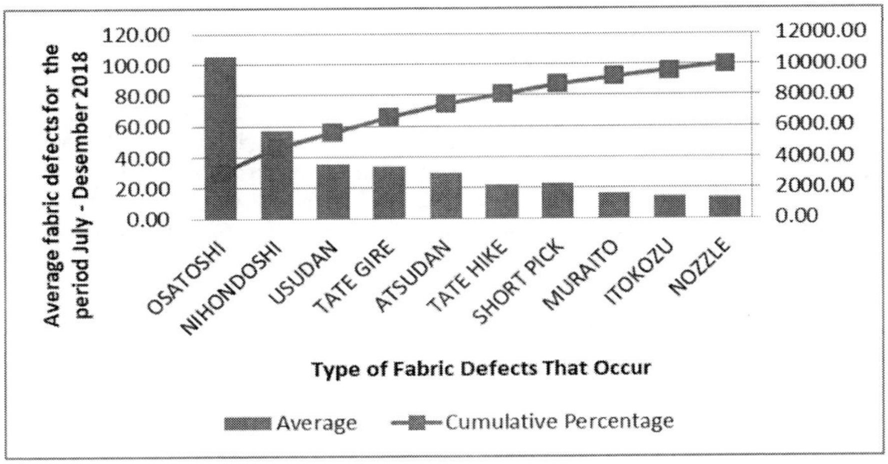


**Figure 2**. Pareto Diagram of Type of Disability July-December 2018

Based on Figure 2, it can be seen that ten types of fabric defects dominate in PT. Unitex, Tbk. OSA TOSHI is the most dominant type of fabric defect that is OSA TOSHI along 10496.80 meters, with a cumulative frequency of 29.75%. Then followed by NIHON DOSHI along 5739.68 meters with a cumulative frequency of 46.02%, and so with other fabric defects.

Based on the causes of fabric defects of each type of fabric defects that occur, there are several potential causes of defective products, including human, machine, method, material, and environmental factors. The following is a description of the potential causes of the problem for each factor that affects product defects:

1. Human Factors
   The human factor is one of the factors causing the dominant type of defect in PT. Unitex, Tbk. This is because one operator supervises many machines, so the operator feels tired while working, which results in decreased accuracy. Then the lack of supervision of the machine that results in negligence of workers, causing the machine to be uncontrolled, and workers make mistakes when setting the machine.
2. Engine Factor
   Machine factors also affect the fabric defects, this is caused by damage to the engine shokki such as wear and tear on the machine, setting the wrong machine, the machine is less sensitive so that when the thread breaks the machine does not die, there is some damage to the machine that makes the thread break, and also possible due to lack of maintenance of these machines.
3. Material Factors
   The material used to produce greige fabric is obtained from spinning. Material from spinning often does not comply with the standards that should be. Because the average fabric defect occurs because the thread is broken when weaving. And some threads have different thicknesses.
4. Environmental Factors
   Environmental factors also affect the production process of greige cloth. This is because, at the time of the production process, there is dust around the machine, which causes the dust to accumulate in the middle so that it is also woven, which results in the fabric being grade C or defective.
5. Method Factors
   The method factor is also one of the causes of the many fabric defects that occur in PT. Unitex, Tbk. Fabric defects are also caused by improper machine setting procedures such as incorrectly setting the speed of the machine which makes the machine not work effectively. Then the fabric defects also often occur because the operator does not do routine checking when woven in the

shokki unit. Then the production control in the sizing unit lacks this is because, during the sizing process, many threads break up in the size box.

After identifying the consequences of fabric defects, the next step is to make a Failure Mode and Effect Analysis Analysis, where the severity, occurrence, and detection values are first searched. Then the next step is to multiply the rating value to get the RPN value which is then performed an analysis of the severity value due to process failure, an analysis of the occurrence value of the cause of failure, an analysis of the detection value of the control carried out, and an analysis of the RPN value of the failure factor that happened. Following is the Analysis of Failure Mode and Effect Analysis, which can be seen in Table 3.

**Table 3**. Failure Mode and Effect Analysis

| No | Factor | Duc to process failure | Severity | Cause of failure | Occurance | Control carried out | Detection | RPN | Total RPN per Factor | Rank |
|---|---|---|---|---|---|---|---|---|---|---|
| 1 | Human | Defects occur | 8 | Not careful | 8 | Operators are given a reprimand | 4 | 256 | 512 | 2 |
| 2 | | Defects occur | 8 | Less control | 8 | Operators are given a reprimand | 4 | 256 | | |
| 3 | Machine | The production process is interrupted | 4 | Engine failure | 7 | Perform periodic engine maintenance | 7 | 196 | 460 | 3 |
| 4 | | The production process is interrupted | 4 | Incorrect engine settings | 6 | Perform periodic engine maintenance | 7 | 168 | | |
| 5 | | The production process is interrupted | 4 | less sensitive engine | 4 | Checking the machine and looking for causes related to the machine are less sensitive | 6 | 96 | | |
| 6 | Material | defects in production result | 7 | low yarn strength | 8 | for longer payroll and process improvement | 4 | 224 | 518 | 1 |
| 7 | | defects in production result | 7 | thead does not meet specifications | 7 | control the material to be used whether it is according to specifications or not | 6 | 294 | | |
| 8 | Enviromental | defects in production result | 7 | cotton dust scattered | 6 | checking around the machine | 5 | 210 | 210 | 5 |
| 9 | Method | Defects occur | 5 | Incorrect engine setting procedure | 6 | conducted training for operators | 6 | 180 | 300 | 4 |
| 10 | | Defects occur | 5 | free weft remains on the fabric | 6 | operators are given a reprimand | 4 | 120 | | |

There are several kinds of assessments in the FMEA method, namely, severity, occurrence, and detection. From Table 4., it can be seen that the greatest severity points are found in the human factor which is worth eight. This is because human factors are the cause of fabric defects. With the inaccuracy of the operator makes the control of the machine becomes less.

Based on Table 4, it can be seen that the highest Occurance value of the writer is given, namely on human and material factors. With a very low level of detail and control will cause the operator to make mistakes that cause damage to the fabric. And also, the low quality of the yarn can cause fabric defects to occur so that when the thread breaks, the operator does not realize that there will be a fabric defect that will cause the fabric to be grade C.

The highest weighting Detection value is given to maintenance with a weight value of 7. The author gives a value of 7 because these factors need to be done because of the many damages to the machine that results in disruption of the production process.

After weighting has been done, then the RPN value will be obtained from the multiplication of severity, occurrence, and detection. Then after that, the RPN score from the largest to the lowest. Based on Table 4, it can be seen that material factors get the first rank. Therefore material factors are a priority to be repaired first because material factors are the biggest contributing factors for fabric defects.

Recommendations for improvement that can be given are related to the problem factors that occur, namely, first making a good SOP that aims to make it easier for each operator to operate the machine correctly and prevent errors in the use of the machine that will later affect the fabric production process. Secondly, making machine control forms. Making this form is intended so that the state of the machine can be controlled easily both by the maintenance. This is one way to reduce the possibility of damage that occurs in the engine, such as worn machines. The controlled maintenance of the machine is expected to reduce fabric defects that occur due to damage to the machine.

Third, the procurement of training for operators. The purpose of this training provision is expected for each operator to prioritize operating the machine by specified standards. Training can be done by outsiders and insiders as needed. It is also hoped that the material in training can change the way of thinking that operators not only carry out tasks by their respective responsibilities but also pay attention to the damage and quality of products produced. Fourth, for material factors. The step that must be done is to communicate the problem to the spinning department. It could also be suggested that the spinning department is done for a long time.

## 4. Conclusion

The results of this research are obtained: In the period July - December 2018, 5 types of fabric defects were most dominant, namely OSA TOSHI, NIHON DOSHI, USUDAN, TATE GIRE, and ATSUDAN which would serve as samples and priority handling. Identification of the causes of fabric defects is made using a fishbone diagram with the causes of defects found including from human factors, materials, machinery, environment, and methods. Analysis of Failure Mode and Effect Analysis of the causes of failure obtained from the fishbone diagram obtained the highest severity value for human factors, the highest occurrence value for less precision and control, and the highest detection value for engine maintenance. Proposed improvements that can be given to the problems above, namely making a good SOP and sticking to each machine, making a form to control the state of the machine components, providing training to the operator, and communicating the problem to the Spinning section.

**Acknowledgment**

Authors would like to thank the Industrial Engineering of Engineering Faculty of Universitas Andalas for providing publication grant FY 2019.

## 5. References

[1] D. Bauer, Thomas; Fiedler, Lars; Jacobs, Jeff; Spillecke, 2017. "The secret to great marketing analytics ? Connecting with decision makers," *McKinsey Co.*

[2] G. Akrani, 2013. "What is Product Quality? Definition Meaning Importance," *Kalyan City Life Blog.*

[3] M. R. Suryoputro, M. Sugarindra, and H. Erfaisalsyah, 2017. "Quality Control System using Simple Implementation of Seven Tools for Batik Textile Manufacturing," in *IOP Conference Series: Materials Science and Engineering*, doi: 10.1088/1757-899X/215/1/012028.

[4] T. S. Parsana and M. T. Patel, 2014. "A Case Study: A Process FMEA Tool to Enhance Quality and Efficiency of Manufacturing Industry," *Bonfring Int. J. Ind. Eng. Manag. Sci.*, doi: 10.9756/bijiems.10350.

IOP Conf. Series: Materials Science and Engineering **847** (2020) 012002 doi:10.1088/1757-899X/847/1/012002


[5] J. Doshi and D. Desai, 2017. "Application of failure mode & effect analysis (FMEA) for continuous quality improvement - multiple case studies in automobile SMEs," *Int. J. Qual. Res.*, doi: 10.18421/IJQR11.02-07.

[6] D. M. Barends, M. T. Oldenhof, M. J. Vredenbregt, and M. J. Nauta, 2012. "Risk analysis of analytical validations by probabilistic modification of FMEA," *J. Pharm. Biomed. Anal.*, doi: 10.1016/j.jpba.2012.02.009.

[7] W. Wang, X. Liu, Y. Qin, and Y. Fu, 2018. "A risk evaluation and prioritization method for FMEA with prospect theory and Choquet integral," *Saf. Sci.*, doi: 10.1016/j.ssci.2018.08.009.

[8] P. Fithri, N. A. Riva, L. Susanti, and B. Yuliandra, 2018. "Safety analysis at weaving department of PT. X Bogor using Failure Mode and Effect Analysis (FMEA) and Fault Tree Analysis (FTA)," in *2018 5th International Conference on Industrial Engineering and Applications, ICIEA 2018*, pp. 382–385, doi: 10.1109/IEA.2018.8387129.

# What is Effective Use of Big Data? The Consensual Definition of Effective Use of Big Data

**Feliks P Sejahtera Surbakti**[1]

[1]Industrial Engineering Department, Atma Jaya Catholic University of Indonesia, Jakarta, 12930, Indonesia
feliks.prasepta@atmajaya.ac.id

**Abstract.** Big Data has attracted significant attention in recent years due to its potential. The use of Big Data is believed to be another source of competitive advantage for organizations. A leading organization in the future is an organization that can use Big Data effectively. However, what constitutes effective use of Big Data remains elusive and evidence suggests that most Big Data projects fail to deliver on their expectations and that executives are slow to adopt Big Data. Existing Engineering Management and Information Systems research have explored effective use in a variety of contexts; however, it is yet to specifically consider the unique characteristics of Big Data. Accordingly, there is a need for rigorous study on what the core of the concept of effective use of Big Data. The aim of the study is to understand better the effective use of Big Data. A multiple case studies approach is followed in the investigation. Eight organizations from private and public organizations were investigated. Data were gathered primarily using semi-structured interviews and through document analysis. The study results show the core of the concept of effective use of Big Data.
**Keywords**: Big Data. Effective Use, Definition, Qualitative Research

## 1. Introduction

The effective use of Big Data has the ability to realize data into actionable insights. This paper builds understanding on what constitute to effective use of Big Data. Study shows that although most organizations have access to Big Data, they do not have the capacity to use it effectively [1, 2]. Indeed, evidence suggests that organizations struggle with how to use Big Data effectively [3, 4]. Moreover, another study show that although executives understand the need for the value realisation of Big Data for increasing profits, less than a quarter of them believe their organization makes value from Big Data use [5]. In spite of the failures as well as slow progress, the opportunities of Big Data is promising and widely acknowledged, for example [6] highlights that Big Data can deliver a new wave of productive growth and change economies.

How to use Big Data effectively is a complex phenomenon. So far, the effective use of Big Data is rarely covered in literature. It is still unclear what effective use of Big Data is. While the study of effective use should give direction in such a context, Theory of Effective Use [7] was established on the foundation of common Information Systems and Management, which may not contain the unique attributes of Big Data. Developing a consensual definition of effective use of Big Data is a necessary step towards providing guidance to organizations such that the worth of Big Data can be better realized. This research is aligned with [8], who states that academicians can make important

IOP Conf. Series: Materials Science and Engineering **847** (2020) 012003 doi:10.1088/1757-899X/847/1/012003

contributions to academic and practice, by *"introducing new constructs"* and *"by better conceptualising existing constructs"*, which is what this study does with how to use Big Data effectively.

The objective of this study is to define the core of the concept of effective use of Big Data in public and private sectors context. Effective use in the Information Systems and Management discipline is defined as the degree to which the use of a system increases achievement of the goals an organization has set concerning the system [7]. While there is agreement on this definition in the discipline, there is a lack of understanding of effective use in the context of Big Data. Consensus, however, is important. Ronda-Pupo and Guerras-Martin [9] argue that the level of consensus on a definition of a concept shown by a scientific community can be used as a measure of the development of a body of knowledge. Big Data has instead developed rapidly and arguably in a disorganized manner, that such a universally accepted formal statement denoting what is the meaning of the effective use of Big Data does not exist. Even though theory and definition of effective use in the Information Systems and Management discipline have been proposed, the author shows that the knowledge developed on effective use of Information Systems and Management needs to be revisited in the Big Data context.

The rest of the paper is organized as follows. In the following section, I explain my method to undertake the formulation of definition and proposing a core of concept and definition of effective use of big in the result and discussion section. Finally, I conclude the paper with a summary of the contribution and discussion of related limitations.

## 2. Methods

In this study, an exploratory case studies method is selected because it allows for a collection of rich descriptive data and is suitable for presenting a unique case in research areas where there is lack of theory [10]. To understand how to use big data effectively, data are gathered primarily using semi-structured interviews and through document and video analysis. The semi-structured interviewing technique was chosen as the main data collection technique in this case study approach to enable the researcher to ask for more detail, probe an issue, and go back and forth among important points and request further explanation. Interviews are suitable to explore topics and explain other findings [11].

It is important to choose the right case organization and have available to the researcher data from the case organization when conducting case studies. As stated earlier private and public sectors as chosen the case organizations. This decision was made as Big Data provides organizations with opportunities that are still largely unexploited. However, without a better understanding of the complex issues involved in real-word settings, the effective use Big Data remains elusive. Therefore, the public and private organizations offered an excellent opportunity to study, in which using Big Data is critical to their competitiveness and success.

A total of 12 participants from the eight organizations were interviewed, as summarized in Table 1. Eisenhardt [12] suggests from 4 to 10 is ideal number of cases. The interviews averaged 60 minutes in duration, with a total time of 781 minutes. Based on the interviews data set, the author found that saturation happened within the twelfth interviews. This empirical evidence is supported by Guest, Bunce [13] who argued within the twelve interviews, new themes emerged infrequently as analysis continued.

**Table 1.** Interview Participants

| Participant | Role | Experience | Industry |
|---|---|---|---|
| Participant 1 | Research Manager | >2 years | Resources |
| Participant 2 | Artificial Intelligence Manager | >7 years | Resources |
| Participant 3 | Risk and Information Management Manager | >3 years | Finance |
| Participant 4 | Engagement Manager | >2 years | Logistics |
| Participant 5 | CEO and Founder | >4 years | Start-Up |
| Participant 6 | Data Analyst | >2 years | Health |

**Table 1.** Interview Participants

| Participant | Role | Experience | Industry |
|---|---|---|---|
| Participant 7 | Director, Health Analytics | >15 years | Health |
| Participant 8 | Project Manager | >10 years | Health |
| Participant 9 | Analytics Architect | > 6 years | Sports |
| Participant 10 | Sport Scientist | >10 years | Sports |
| Participant 11 | Data Scientist | >3 years | Energy |
| Participant 12 | Head of Data & Analytics | >10 years | Finance |

The data analysis process was conducted iteratively as encouraged by the grounded theory methodology. The coding started with a process referred to as microanalysis [14], that is, a line-by-line analysis of each semi-structured interview transcript to identify initial codes was carried out. The author used NVivo 12 to organise and analyse the data to generate and label nodes/codes following the guidelines from Saldaña [15].

## 3. Result and Discussion

This section provides the research study's findings assembled through a general inductive approach to qualitative research. Triangulation of the 12 participants was an important strategy that facilitated any inclination towards researcher bias. The analysis was anchored by following research question: What is the meaning of effective use of Big Data?

Further in-depth analysis produced "units of meaning" used to reconstruct participants narratives into a cohesive yet agile statement about the meaning of effective use of Big Data, without losing the essence of narratives in their entirety. Burton-Jones and Grange [7] put forth the concept of effective use in Information Systems and Management context, which refers to the use of a system "*in a way that helps attain the goals for using the system*" (p.2) and propose adaptation and learning actions as the drivers of effective use. Their definition of use draws on information theory to fully integrate the nature and purpose of an information system, and thus is inclusive of not only the system, but also of the user and the tasks. From this definition, there are three keywords for effective use: "in a way", "using the system" and "attain the goals", and the ultimate measure of effective use is to attain the goals (G).

On the other hand, De Mauro, Greco [16] analysed the most frequent keywords included in Big Data related articles and considering their mutual relationships and identified four top research themes in current literature, namely: 1. Driver 2. Information, 3. Technology, 4. Methods. To define effective use of Big Data, the author classified the definition into 5 themes that derived from Burton-Jones and Grange [7] and De Mauro, Greco [16]: Driver (D), Information (I), Technology (T), Methods (M), and Goal (G).

The appendix summarises the meaning of effective use of Big Data that come from empirical evidence. By looking to the definition that presented by participants and the whole interviews, the author proposes that the core of the concept of effective use of Big Data can be claimed by:

- Define the problem to describe effective Big Data initiatives should starts from clear goals and problem driven instead of fuzzy objectives and data driven and show the value to the organization; (D)
- Using insights from information to describe the transformation from information to insights as the main pathway Big Data is impacting organizations and society; (I)
- Build Big Data infrastructure and data analytics to describe Big Data initiatives should be use driven instead of technology driven, Big Data technology and analytics as support mechanisms not the goals; (T)
- Scalability and sustainability mean that the organizations must be able to choose the scope of the projects and get the data, powered by a sustainable growing process from performance and financial perspectives; (M)

- Make solution to describe effective use of Big Data means solve the problem. (G)

The author argues the definition of effective use of Big Data as follows:
*"Effective use of Big Data refers to harnessing value from Big Data by defining the problem, showing the value to the organization, and making solution by using insights from data and build Big Data infrastructure and data analytics by considering scalability and sustainability."*

## 4. Conclusion

The analysis of twelve participants provides the first empirically derived picture of the emerging organizational construct that is supposed to foster effective use of Big Data in the organization. It has revealed several insights of theme for the definition of effective use of Big Data. Each of theme is described with a comprehensive set of explanation that help to develop an effective use of Big Data and further research such an organizational construct.

The theoretical contribution of this study is the definition of effective use of Big Data, which is both empirically grounded and theory informed. The study provides better understanding of the essential concept of effective use of Big Data. The theory on Information Systems and Management in Big Data context is scarce. Furthermore, effective use is not-well captured in the Engineering Management and Information Systems literature (17) and towards context-specific of Big Data, rather than general theories. The study is one of the first effort to document the effective use of Big Data. We, therefore, contribute to study on effective use of Big Data by providing concrete concept for the organizational effective use of Big Data. The author further adds to effective use in Information Systems and Management research by identifying the core concept and definition of effective use in Big Data context. These contributions build an important foundation for future study.

The selection of the sample and the ways of gathering data exposes the study to certain limitations. Overall, two different types of limitation apply and shall be addressed in future research. First, for restrictions of interviewee availability, only one interview was conducted for most of the organizations. A more diverse perspective from users of the organizations and from different roles within the organizations may reveal a more differentiated picture. Future research should consider a case study approach and analyse the organization from different stakeholder perspectives.

Second, this study focuses on organizations that are early adopters of the Big Data use, the major organizations that have already established Big Data use initiatives and are addressing the issues of effective use. However, the author cannot say with certainty that these options are the only viable or most effective ones at this point. A broader study of organizations that want to realize value from Big Data within the organization using different approaches may reveal a better understanding of the effective use of Big Data.

**Acknowledgements**
I gratefully acknowledge the funding sources that made my PhD work possible; LPDP (Indonesia Endowment Fund for Education).

## 5. References

[1] Günther WA, Mehrizi MHR, Huysman M, Feldberg F. 2017. Debating big data: A literature review on realizing value from big data. *The Journal of Strategic Information Systems*, **12**(3), 191-209.
[2] Sejahtera F, Wang W, Indulska M, Sadiq S, editors. 2018. Enablers and Inhibitors of Effective Use of Big Data: Insights From A Case Study. *Pacific Asia Conference on Information Systems (PACIS)*, Yokohama, Japan.
[3] Partners N. 2019. *Big Data and AI Executive Survey 2019*.
[4] Surbakti FPS, Wang W, Indulska M, Sadiq S. 2019. *Factors influencing effective use of big data: A research framework*. Information & Management In Press.

[5] Smart C, De Meyer-Heydenrych C, editors. 2018. The Use of Big Data in Marketing: An Emerging Market Financial Services Industry Perspective. The 2018 Annual Conference of the Emerging Markets Conference Board, Wits Business School.
[6] Mikalef P, Pappas IO, Krogstie J, Giannakos M. 2018. Big data analytics capabilities: a systematic literature review and research agenda. *Information Systems and e-Business Management*, **16**(3), 547-78.
[7] Burton-Jones A, Grange C. 2012. From use to effective use: a representation theory perspective. *Information Systems Research,* **24**(3), 632-58.
[8] Barki H, Paré G, Sicotte C. 2008. Linking IT implementation and acceptance via the construct of psychological ownership of information technology. *Journal of Information Technology,* **23**(4), 269-80.
[9] Ronda-Pupo GA, Guerras-Martin LÁ. 2012. Dynamics of the evolution of the strategy concept 1962–2008: a co-word analysis. *Strategic Management Journal*, **33**(2), 162-88.
[10] Yin RK. 2017. *Case study research and applications: Design and methods.* (London: Sage publications)
[11] Saunders M, Lewis P, Thornhill A. 2009. *Research methods for business students.* Pearson education.
[12] Eisenhardt KM. 1989. Building theories from case study research. *Academy of management review*, **14**(4), 532-50.
[13] Guest G, Bunce A, Johnson L. 2006. How many interviews are enough? An experiment with data saturation and variability. *Field methods*, **18**(1), 59-82.
[14] Strauss A, Corbin J. 1990. *Basics of qualitative research techniques: Grounded Theory Procedures and Techniques*. (Newbury Park, CA: Sage publications)
[15] Saldaña J. 2015. *The coding manual for qualitative researchers.* (London: Sage).
[16] De Mauro A, Greco M, Grimaldi M, editors. 2015. What is big data? A consensual definition and a review of key research topics. *AIP conference proceedings,* AIP.
[17] Burton-Jones A, Volkoff O. 2017. How can we develop contextualized theories of effective use? A demonstration in the context of community-care electronic health records. *Information Systems Research*, **28**(3), 468-89.

**Appendix**

| Participant | Effective Use of Big Data | D | I | T | M | G |
|---|---|---|---|---|---|---|
| 1 | *I think effective use of Big Data in the field of my work is about* ***making solutions*** *for* ***business problems*** *by completing* ***prediction models*** *and* ***optimizations algorithms*** | ✓ | | | ✓ | ✓ |
| 2 | *From business perspective, it means the* ***scalability*** *at first glance, so first of all, scalability to be able to choose what we were doing globally and to be able to get all the data that you're being provided by the sensory so far and all this new data that is coming from large device and be able to scale all this stuff but also* ***sustainability*** *because so I used to work as database professional so and I know that a lot of figures and row we were already able to scale lots of data using regular databases but the problem is that it was not kind of sustainable because the performance we started to decrease a large and also the cost is increasing the large. So, I think when we talk about Big Data, mainly what comes in my head is scalability, and on this scalability, which is powered by a sustainable growing process from a financial perspective but also for* ***performance perspective*** | | | | ✓ | ✓ |

| Partici pant | Effective Use of Big Data | D | I | T | M | G |
|---|---|---|---|---|---|---|
| 3 | *I think its related to the first answer, basically bottom-up, it's not top-down,* ***bottom-up, what is the value, what is the potential benefit, and start from there,*** *because once you start with that, and then you can gather, okay I need this data, and it is the feasibility okay, I need this data. So, a lot of approaches basically getting the data first, and think about this case later, but I believe in use cases first and then get the data and then* ***build your infrastructure*** *based on whatever you need* | ✓ | | ✓ | | |
| 4 | *I think personally it has to come down to using a Big Data approach gives you some measure of uplift and whatever you're trying to do compared to a different approach, or no approach may be, some level of uplift in some level of measurable success because I think that's you know, this is doing analysis for analysis say, but then there's doing analysis to achieve an outcome, and* ***ultimately you want to achieve an outcome that you can measure compare it to some benchmark.*** | | | | | ✓ |
| 5 | *Always more efficient to* ***start with the end in mind.*** *Then move* ***back to see what data the organization need to collect,*** *and* ***how would the organization remove the noise and do desired transformations*** | | | | ✓ | ✓ |
| 6 | *The mission of the company is to make solar energy much more accessible and much more useful, so do things like initial introduction [...] and helping save the customer money at the same time. So that's the effective use of the data, if* ***we can take all this usage data and take price data,*** *like tariff data, and also our knowledge about where the people live, and we can do things like put the customers together in a big virtual power plant,* ***we can achieve our mission directions, we can save the customers money and it can be a win-win all around*** | | | | | ✓ |
| 7 | *Ultimately in our environment, effective use of Big Data is what we can help to inform the development of health policy and shape the decisions of health of ministry, and he or she can feel that the decision that they take withstanding difficult requirements, sometimes that may override,* ***the decision they make firmly on evidence*** | ✓ | | | | ✓ |
| 8 | *I'm not sure have a 100% answer for that one, I guess, I think effective use of Big Data is no not really that different from effective use of regular data is, making sure that being able* ***to get to help you answer the questions that you want answered*** | ✓ | | | | ✓ |
| 9 | ***Make a difference in the real world.*** *So, the impact on the health service of Australians. It changes how Australians get their health services, how long they live, how they able to maintain their health.* ***That's the context*** *why we are doing* ***data analytics*** *in the health context* | | | | ✓ | ✓ |
| 10 | *Effective use of Big Data, mm, I don't know the answer for that. I think* ***it is making better use,*** *not just, you bringing data, to* ***the purpose of generate more value*** *out of it so it's like an asset so yeah you're gonna have effective use of Big Data it needs to be able to be using way that* ***adds value,*** *otherwise all you're doing is* | | | | | ✓ |

IOP Conf. Series: Materials Science and Engineering **847** (2020) 012003 doi:10.1088/1757-899X/847/1/012003

| Participant | Effective Use of Big Data | D | I | T | M | G |
|---|---|---|---|---|---|---|
| | *creating a storage of information, so if you can't drive any value out of it, that nicely is ineffective* | | | | | |
| 11 | *So effective use of Big Data means* ***influencing change; making decisions****. Well, it's just using* ***visualization tools****.* ***It's using information****, which is updated regularly with minimal input from humans, but it is pulling out the necessary information; live dashboards, tools which people can use without having to look at the data themselves.* ***And that's how you're able to visualise things, and to understand it, and to make change, and to effect change*** | | ✓ | | | ✓ |
| 12 | *It's that making sure that we are* ***solving problems****, making sure that you can* ***show the value to the organization****, and* ***understanding the business need first****, and those* ***problem statements first****.* | ✓ | | | | ✓ |

# Selection strategy implementation of cleaner production using ISM and AHP Method in Chemical Laboratory of Service Industries

**D Rimantho[1,2], and D Ardinia[1]**

[1-2]Industrial Engineering, Pancasila University, Srengsengsawah-Jagakarsa, DKI Jakarta, Indonesia
[2] dino.rimantho@univpancasila.ac.id

**Abstract.** The chemical analysis in the service industry cannot be separated from energy use, water, consumables, and chemicals related to the amount of waste produced. The number of hazardous solid waste generated in the CRS Chemical Laboratory division has increased in certain months, ranging from 400-600 kg. This shows the low implementation of green production in the company. Therefore, the purpose of this study is to analyze the key criteria and determine the best strategies to reduce waste. The method used in this study is Interpretive Structural Modeling to find key criteria and the Analytic Hierarchy Process method to determine the priority selection of clean production implementation strategies. This study uses a questionnaire distributed to four experts. The results showed that there were nineteen sub-criteria analyzed in determining key criteria. The sub-criteria for lack of awareness from both individuals and organizations regarding changes for the better is a key criterion in the application of cleaner production. Furthermore, the best alternative calculation results using the AHP method obtained by the implementation of periodic green production socialization to employees has the greatest valuation of 0.57 with a Consistency Ratio (CR) of 0.09. Further research is needed to evaluate the alternatives chosen.
**Keywords**: Analytic Hierarchy Process (AHP), Interpretive Structural Modelling (ISM), Key Criteria, Priority, Cleaner Production

## 1. Introduction

The development of environmentally friendly industries has become a global demand to support the achievement of sustainable development that is successful currently. In 1992, clean production was emphasized as an important means of achieving sustainable development in Agenda 21, which was adopted at the United Nations Conference on Environment and Development [1]. Cleaner production was introduced by the United Nations Environment Program (UNEP) in May 1989 and was officially proposed in September 1989 at the Promotion on Cleaner Production seminar in Canterbury. Indonesia agreed to adopt the definition conveyed by UNEP that clean production is preventive and integrated. Therefore, the strategy needs to be applied continuously to the production process and product life cycle with the aim of reducing risks to humans and the environment.

Cleaner production is an environmental management strategy that is directed towards prevention and is integrated so that it can be applied to the entire production cycle. It has the aim to increase productivity by providing better levels of efficiency in the use of raw materials, energy, and water, encouraging better environmental performance through reducing sources of waste generation and emissions and reducing the impact of products on the environment from product life with designs environmentally friendly, but effective in terms of cost.

Study conducted by Shah describes that clean production can be applied at all levels of decision making in the industry with its main focus of applying cleaner technologies and techniques, and can be applied to various sectors (for example, industry, infrastructure, housing, and hospitality services) and organizational size (from large petrochemical plants to small company industries) [2]. In other countries, the application of cleaner production has also been widely applied. In the study described by Vukadinovic et al. in Serbia, the application of clean production is carried out in large petrochemical companies by optimizing steam systems and modification [3]. The steps implemented led to a reduction in total energy consumption by 6% and a reduction in greenhouse gas emissions by 9 tons of CO2. In Laos, the application of cleaner production was also applied at the Manoluck Hotel [4]. After the hotel implemented clean production, there was a 65% reduction in specific energy consumption and greenhouse gas emissions, 76% in water consumption, wastewater generation and a significant reduction in the waste generation for disposal by waste sorting and composting for use as a soil conditioner. Based on the description shows that net production can be applied in various industrial sectors both manufacturing and services.

In Indonesia, the application of cleaner production has contributed significantly to the batik industry [5], which can improve environmental performance in terms of reducing the concentration of BOD, COD, and TSS by 85%, 89%, and 98% respectively in waste generated. The clean production option implemented is to substitute a material input strategy that is replacing chemical dyes with natural dyes during the coloring process.

The CRS Chemical Laboratory Division also has issues related to environmental management. This can be known from the activities or activities carried out, such as the use of energy, electricity, water, hazardous chemicals, and other consumables that will be related to the amount of waste produced. Related to the amount of waste generated both solid and liquid waste, the amount of waste should still be more minimized because in the actual situation there is still the use of consumables and chemicals that are not yet efficient.

Cleaner production is an integrated environmental management strategy that, if properly and truly implemented, will become an alternative company in an effort to increase the efficiency and effectiveness of resource use. Although the actual conditions illustrate that the company has not implemented clean production properly. That is because there are several factors that have the potential to influence and hinder the process of implementing clean production. Thus, the application of a cleaner production process requires appropriate decision making in determining strategies in various types of companies. Numerous researches related to decision-making strategies in solving environmental issues have been done in previous studies. For example, some of these studies include research on renewable energy [6], municipal waste management [7], improving the quality of wastewater in the food industry [8], Strategy for enlightening Environmental Performance of Small Scale Cracker Industry [9], Cleaner Production Application at Small Scale Bakery Industry [10]. However, research related to strategies regarding the application of cleaner production in the analysis services industry has never been done.

Therefore, the purpose of this study is to determine the factors that have the potential to contribute to the implementation of clean production in the CRS Chemical Laboratory division. In order to determine the key factors in the implementation of cleaner production using the ISM method and determine the priorities used in the selection of cleaner production implementation strategies based on the key factors of the sub-criteria by using the AHP method.

## 2. Methods

Primary data collection was carried out in several ways such as field observations, brainstorming and interviews with employees in the CRS Chemical Laboratory division. The questionnaire method was carried out by giving a list of questions (questionnaire), both openly and closely, to be filled in by respondents (experts) where the list of questions had been prepared in advance. In this study, the distribution of questionnaires was divided into four stages. The first stage is intended to determine criteria and sub-criteria, the second stage is to determine the contextual relationship between sub-criteria for determining key criteria, the third stage to determine alternative proposals, and the fourth stage to determine pairwise comparisons between alternative strategies in implementing clean production. The number of experts used in this study was as many as four people. The experts used in this study have different backgrounds such as academics, environmental experts, company leaders.

One method that can be applied in strategic policy planning is modeling using ISM. In addition, the ISM method can be applied in the analysis of problems that have high complexity that cannot be solved by operational research methods and descriptive statistics. There are several stages in implementing the ISM method, namely:

1. Identification of factors that can be done through literature studies and brainstorming
2. Making contextual relationships between factors based on the purpose of the model being developed
3. Develop a single SSIM interaction matrix. There are 4 symbols used to describe the type of relationship between two factors in the system under consideration.
    a. V: the relationship of factor Ei to Ej, not vice versa
    b. A: factor Ej relationship with Ei, not vice versa
    c. X: interrelated relationship between Ei and Ej (vice versa)
    d. O: indicates that Ei and Ej are not related
4. Make a reachability matrix that changes the SSIM symbol into a binary matrix, i.e., for V, Ei becomes 1 while Ej will be 0, for X, Ei will be 1 and Ej will be 1, for O Ei is 0 and Ej is 1, for A , Ei is 0 and Ej is also 0.
5. Create partitions from the final reachability matrix to different levels
6. Make a graph based on the final reachability matrix
7. Make digraph as ISM models
8. Check for consistency based on the ISM model that has been built

After the key criteria are obtained, the next priority strategy determination is obtained using the Analytic Hierarchy Process (AHP) method. The steps that must be done in the AHP method for solving a problem begin with defining the problem and making the details of the desired solution then the hierarchical structure is made. The pairwise comparison matrix is made for the contribution or influence of each relevant element for each influential criterion at the level above it. Furthermore, the geometric mean value is calculated. After collecting all the paired comparative data, then the inverse values are entered along with entry number 1 along the main diagonal, then priority is searched and a consistency test is sought.

The implementation of the same steps is carried out for all levels and clusters in the hierarchy. Hierarchical composition or synthesis is used to weight the priority vectors by weighting the criteria, and adding up all the weighted priority entries concerned with priority entries from the next lower level, and so on. The result is an overall priority vector for the lowest level of the hierarchy. The final step is to conduct a consistency evaluation for the entire hierarchy. Calculate the Consistency Ratio using the formula: CR = CI / IR. Where CR = Consistency Ratio, CI = Consistency Index, and IR = Random Consistency Index. Check consistency if the consistency ratio (CI / IR) is less or equal to 0.1, then the calculation results can be declared correct. If not, the calculation must be corrected.

## 3. Result and discussion

Based on the results of discussions with experts who are parties involved in research in the field, various factors that contribute to causing net production have not been implemented in the company. The factors are divided into five namely human, material, method, machine and environment. The five factors then developed and obtained information about several sub-criteria of the five main factors. Initially from the results of brainstorming obtained 25 sub-criteria that contribute.

After distributing the first questionnaire and calculating the validation of each sub-factor, 19 valid sub-factors were obtained. Several of these sub factors are lack of information (MA 1), lack of training (MA 2), inconsistency (MA 3), lack of awareness (MA 4), weak internal communication (MA 5), inappropriate use of materials and electricity (MT 1), the presence of some chemicals that exceed the shelf life (MT 2), the absence of work instruction (WI) about clean production (ME 1), the application of WI about separation waste has not properly (ME 2), the implementation of the engine's internal maintenance schedule has not been carried out regularly (ME 3), there are no appropriate restrictions in the classification of samples (ME 4), the application of the WI regarding the use and placement of machines has not properly (ME 5), the implementation of building maintenance schedules has not regularly (ME 6), old machines (MS 1), machinery often experience problems (MS 2), the existence of improper installation and placement of machines (MS 3), narrow spaces and limited space (LI 1), inappropriate temperatures in several rooms (LI 2), and air polluted from the use of chemicals (LI 3).

**Table 1** Structural Self Interaction Matrix (SSIM)

| | MA1 | MA2 | MA3 | MA4 | MA5 | MT1 | MT2 | ME1 | ME2 | ME3 | ME4 | ME5 | ME6 | MS1 | MS 2 | MS 3 | LI1 | LI2 | LI3 |
|---|---|---|---|---|---|---|---|---|---|---|---|---|---|---|---|---|---|---|---|
| MA1 | | X | V | A | X | V | V | X | V | V | V | V | V | V | V | V | V | V | V |
| MA2 | | | V | A | A | V | V | X | V | V | V | V | V | V | V | V | V | V | V |
| MA3 | | | | A | A | V | V | A | V | V | V | V | V | V | V | V | V | V | V |
| MA4 | | | | | V | V | V | V | V | V | V | V | V | V | V | V | V | V | V |
| MA5 | | | | | | V | V | V | V | V | V | V | V | V | V | V | V | V | V |
| MT1 | | | | | | | A | A | A | A | A | A | A | A | A | A | O | A | O |
| MT2 | | | | | | | | A | A | O | O | A | O | O | O | O | O | A | A |
| ME1 | | | | | | | | | V | V | V | V | V | V | V | V | V | V | V |
| ME2 | | | | | | | | | | V | V | O | O | O | O | O | O | O | V |
| ME3 | | | | | | | | | | | V | A | O | V | V | 0 | 0 | V | V |
| ME4 | | | | | | | | | | | | X | V | O | V | O | O | O | O |
| ME5 | | | | | | | | | | | | | V | V | X | V | V | V | V |
| ME6 | | | | | | | | | | | | | | V | V | O | V | V | O |
| MS1 | | | | | | | | | | | | | | | X | A | O | X | X |
| MS2 | | | | | | | | | | | | | | | | A | A | X | X |
| MS3 | | | | | | | | | | | | | | | | | A | V | V |
| LI1 | | | | | | | | | | | | | | | | | | V | O |
| LI2 | | | | | | | | | | | | | | | | | | | V |
| LI3 | | | | | | | | | | | | | | | | | | | |

Based on the results of the second stage questionnaire that has been filled out by the respondents, the next step is to make a Structural Self Interaction Matrix (SSIM) matrix as in table 1. Furthermore, based on Table 1 about the SSIM matrix then made in the form of a Reachability Matrix (RM) table by replacing V, A, X, O become numbers 1 and 0. Thus, the results obtained in Table 2. Moreover, after the RM table is obtained with a combined opinion, then the transitivity rule is checked and the Driver Power value is calculated by summing an element horizontally. While the Dependent value is obtained by summing factors vertically as in Table 2.

Then, based on the results of the interpretation of the final Reachability Matrix (RM) matrix, a hierarchy of relationships between factors that contribute to the clean production cannot be drawn up, which can be explained in Figure 1.

Figure 1 provides information that the lack of awareness of both individuals and organizations regarding changes in a better direction (MA 4) is a factor that has the highest value of power drivers or is called a key factor. Diagraph aims to classify sub-criteria into four groups. The groups are the result of the calculation of the RM that has been tested for transitivity and arranged by placing on each ordinate (x; y) respectively. Digraph in Figure 2 provides information, such as:

a. Five criteria are in the Sector 1 area: weak driver - weak dependent variables (Autonomous). Sub-criteria included in this sector are not related to the system and may have few relationships, although the relationship can be strong. Criteria in this sector are ME 2, ME 3, MS 3, and LI 1.

**Table 2** Reachability Matrix (RM)

| $E_i$ \ $E_j$ | MA1 | MA2 | MA3 | MA4 | MA5 | MT1 | MT2 | ME1 | ME2 | ME3 | ME4 | ME5 | ME6 | MS 1 | MS 2 | MS 3 | LI1 | LI2 | LI3 | Driver power (DP) |
|---|---|---|---|---|---|---|---|---|---|---|---|---|---|---|---|---|---|---|---|---|
| MA1 | 1 | 1 | 1 | 0 | 1 | 1 | 1 | 1 | 1 | 1 | 1 | 1 | 1 | 1 | 1 | 1 | 1 | 1 | 1 | 18 |
| MA2 | 1 | 1 | 1 | 0 | 0 | 1 | 1 | 1 | 1 | 1 | 1 | 1 | 1 | 1 | 1 | 1 | 1 | 1 | 1 | 17 |
| MA3 | 0 | 0 | 1 | 0 | 0 | 1 | 1 | 0 | 1 | 1 | 1 | 1 | 1 | 1 | 1 | 1 | 1 | 1 | 0 | 11 |
| MA4 | 1 | 1 | 1 | 1 | 1 | 1 | 1 | 1 | 1 | 1 | 1 | 1 | 1 | 1 | 1 | 1 | 1 | 1 | 1 | 19 |
| MA5 | 1 | 1 | 1 | 0 | 1 | 1 | 1 | 1 | 1 | 1 | 1 | 1 | 1 | 1 | 1 | 1 | 1 | 1 | 1 | 18 |
| MT1 | 0 | 0 | 0 | 0 | 0 | 1 | 0 | 0 | 0 | 0 | 0 | 0 | 0 | 0 | 0 | 0 | 0 | 0 | 0 | 1 |
| MT2 | 0 | 0 | 0 | 0 | 0 | 1 | 1 | 0 | 0 | 0 | 0 | 0 | 0 | 0 | 0 | 0 | 0 | 0 | 0 | 2 |
| ME1 | 1 | 1 | 1 | 0 | 0 | 1 | 1 | 1 | 1 | 1 | 1 | 1 | 1 | 1 | 1 | 1 | 1 | 1 | 1 | 17 |
| ME2 | 0 | 0 | 0 | 0 | 0 | 1 | 1 | 0 | 1 | 1 | 1 | 0 | 0 | 0 | 0 | 0 | 0 | 0 | 1 | 6 |
| ME3 | 0 | 0 | 0 | 0 | 0 | 1 | 0 | 0 | 0 | 1 | 1 | 0 | 0 | 1 | 1 | 0 | 0 | 1 | 1 | 8 |
| ME4 | 0 | 0 | 0 | 0 | 0 | 1 | 0 | 0 | 0 | 0 | 1 | 1 | 1 | 0 | 1 | 0 | 0 | 0 | 0 | 5 |
| ME5 | 0 | 0 | 0 | 0 | 0 | 1 | 1 | 0 | 0 | 1 | 1 | 1 | 1 | 1 | 1 | 1 | 1 | 1 | 1 | 12 |
| ME6 | 0 | 0 | 0 | 0 | 0 | 1 | 0 | 0 | 0 | 0 | 0 | 0 | 1 | 1 | 1 | 0 | 1 | 1 | 0 | 6 |
| MS1 | 0 | 0 | 0 | 0 | 0 | 1 | 0 | 0 | 0 | 0 | 0 | 0 | 0 | 1 | 1 | 0 | 0 | 1 | 1 | 5 |
| MS2 | 0 | 0 | 0 | 0 | 0 | 1 | 0 | 0 | 0 | 0 | 0 | 0 | 0 | 1 | 1 | 0 | 0 | 1 | 1 | 5 |
| MS3 | 0 | 0 | 0 | 0 | 0 | 1 | 0 | 0 | 0 | 0 | 0 | 1 | 0 | 1 | 1 | 1 | 0 | 1 | 1 | 7 |
| LI1 | 0 | 0 | 0 | 0 | 0 | 0 | 1 | 0 | 0 | 0 | 0 | 0 | 0 | 1 | 1 | 1 | 1 | 1 | 0 | 4 |
| LI2 | 0 | 0 | 0 | 0 | 0 | 1 | 1 | 0 | 0 | 0 | 0 | 0 | 0 | 1 | 1 | 0 | 0 | 1 | 1 | 6 |
| LI3 | 0 | 0 | 0 | 0 | 0 | 0 | 1 | 0 | 0 | 0 | 0 | 0 | 0 | 1 | 1 | 0 | 0 | 0 | 1 | 4 |
| Dependence | 5 | 5 | 6 | 1 | 3 | 17 | 11 | 5 | 7 | 9 | 10 | 10 | 9 | 13 | 16 | 9 | 8 | 14 | 13 | 171 |

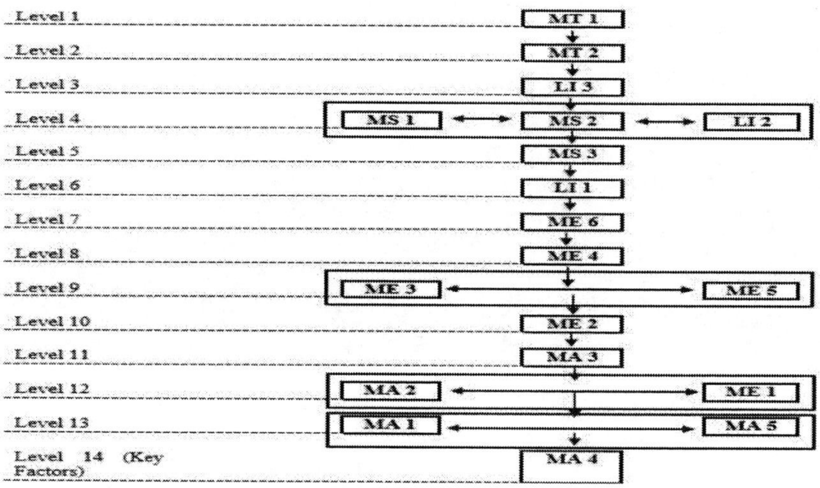


**Figure 1** Hierarchy Structure Application of Cleaner Production

b. Seven criteria are in the Sector 2 area: weak driver - strongly dependent variables (Dependent). Sub criteria within this sector are generally not free variables. The criteria that enter into this sector are MT 1, MT 2, ME 4, MS 1, MS 2, LI 2, and LI 3.
c. One criterion is in Sector 3: Strong driver - Strongly dependent variables (Linkage). The sub-criteria that enter into this sector are ME 5.
d. Six criteria are in the Sector 4 area: strong driver - weak dependent variables (Independent). Sub criteria that are in this sector are free variables. The criteria that enter into this sector are MA 1, MA 2, MA 3, MA 4, MA 5, and ME 1.

The determination of alternative strategies is based on sub-criteria that are key elements like the purpose of the assessment. The previous results showed that the key element was the lack of awareness

IOP Conf. Series: Materials Science and Engineering **847** (2020) 012004 doi:10.1088/1757-899X/847/1/012004

both from individuals and organizations (MA 4). Based on this, several proposed strategies were obtained from the results of brainstorming with company leaders. Some of the proposed strategies are:

a. Conduct training for employees in the CRS Chemical Laboratory division on the implementation of clean production (AS 1).

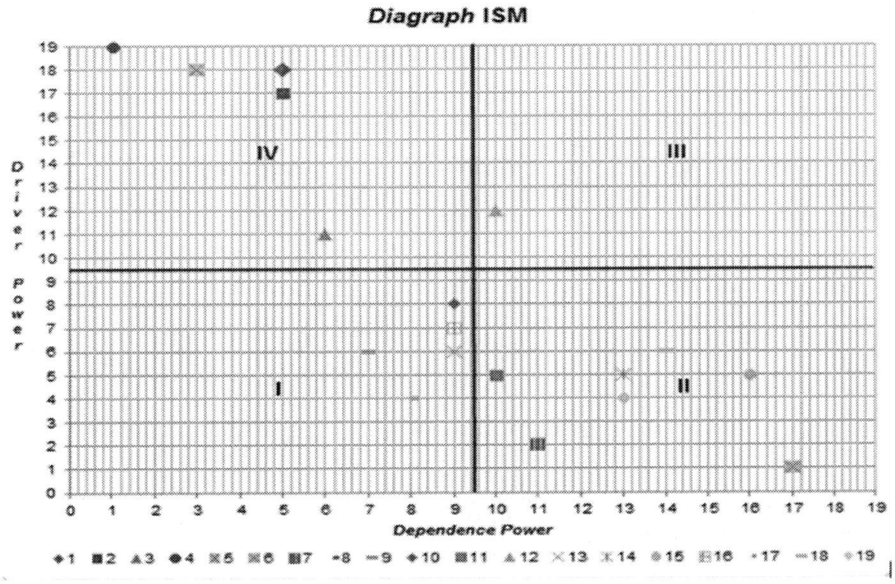


**Figure 2** Diagraph Driver Power – Dependence

b. Conduct regular socialization to employees in the CRS Chemical Laboratory division regarding the importance of implementing clean production (AS 2).
c. Conducting checks or supervision related to activities and work environment on a regular basis to support the application of cleaner production in the CRS Chemical Laboratory division (AS 3).
d. Make guidance in the form of work instructions regarding clean production programs (AS 4).

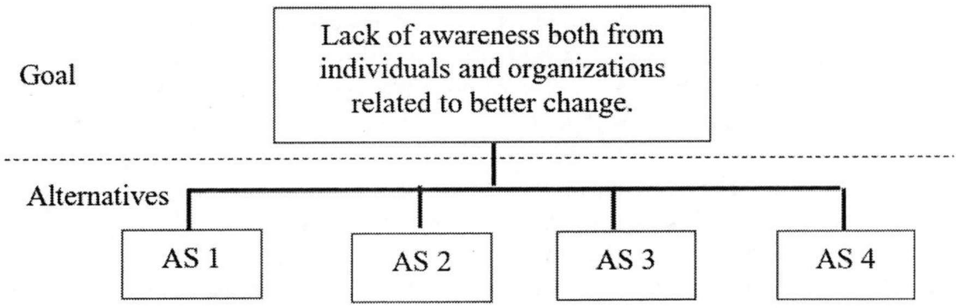


**Figure 3** Alternative selection hierarchy structure

Then an alternative selection hierarchy structure can be made shown in Figure 3. From the proposed strategy, an alternative selection analysis was carried out by implementing the AHP method. Determining the priority of alternatives is obtained from the results of the fourth questionnaire, using pairwise comparisons of the alternatives to be chosen. The results of the pairwise comparison questionnaire on alternatives are shown in Table 3.

**Table 3** The results of the pairwise comparison questionnaire on alternatives

| Alternatives | Respondents | | | | Alternatives |
|---|---|---|---|---|---|
| | 1 | 2 | 3 | 4 | |
| AS 1 | 1/5 | 1/7 | 1/7 | 1 | AS 2 |
| AS 1 | 5 | 7 | 7 | 5 | AS 3 |
| AS 1 | 5 | 5 | 7 | 5 | AS 4 |
| AS 2 | 5 | 7 | 7 | 7 | AS 3 |
| AS 2 | 5 | 7 | 7 | 7 | AS 4 |
| AS 3 | 1 | 1/7 | 1 | 1 | AS 4 |

The results of the calculation of the normalization weighting of pairwise comparison judgments between alternatives are shown in Table 4.

**Table 4** Normalization weighting of pairwise comparison

| Alternative | AS 1 | AS 2 | AS 3 | AS 4 | Σ | X |
|---|---|---|---|---|---|---|
| AS 1 | 0.19 | 0.16 | 0.39 | 0.40 | 1.15 | 0.29 |
| AS 2 | 0.75 | 0.64 | 0.43 | 0.48 | 2.29 | 0.57 |
| AS 3 | 0.03 | 0.10 | 0.07 | 0.05 | 0.24 | 0.06 |
| AS 4 | 0.03 | 0.10 | 0.11 | 0.07 | 0.32 | 0.08 |

The calculations show that the weight of the priority the alternative selected is AS 2 (socialization green production) with a weight of about 0.57, the second alternative is AS 1 (Training) with a weight of around 0.29, Alternative AS 4 with a weight of about 0.08, and alternative AS 3 with a weight of approximately 0.06. The result also shows that the consistency test obtained a CI value of 0.09. Based on the provisions of Saaty that the CI value must be ≤ 10%, then the consistency value can be accepted or in other words, the hierarchy can be stated consistently.

The implementation of cleaner production in the CRS Chemical Laboratory division is still very low. This is due to the lack of awareness from both individuals and organizations to support the implementation of clean production programs. Therefore, to increase awareness, a regular outreach program is needed related to the benefits and importance of clean production both in economic and environmental terms.

Clean production socialization is an initial strategic step that really needs to be done. It aims to increase awareness of both individuals and an organization related to better change. In order to realize the socialization program related to clean production involves several stakeholders to support this program such as managers, supervisors, supervisors, and all employees. Therefore, the first step that will be taken is to form a special team that is responsible for helping to drive the clean production program that will be implemented. The special team will assist in determining the schedule for the implementation of the socialization and the agenda of the activities to be carried out. The form of socialization is to hold a special meeting to review the clean production program and the distribution of leaflets related to the implementation of the clean production program. The meeting will be held at the beginning of each month where all the employees are participants.

The presentation was carried out as a form of implementation of the socialization program. The presentation material consisted of an initial introduction to clean production, an introduction to the benefits of implementing clean production, a discussion of the agenda of activities to be carried out to support the clean production program, to an evaluation of the implementation of clean production. Some things that become notes and findings will be evaluated later to be discussed planning the improvement process.

IOP Conf. Series: Materials Science and Engineering **847** (2020) 012004 doi:10.1088/1757-899X/847/1/012004

## 4. Conclusion

Cleaner production is one of the keys to the success of sustainable development. This study attempts to identify and analyse the key criteria for implementing clean production in a chemical laboratory. In addition, this study also determines strategies for key sub-criteria. There are nineteen factors that contribute to the not yet implemented net production. By applying the ISM method the results of the criteria for lack of awareness from both individuals and organizations regarding better change are key elements. Whereas the alternative strategy that becomes a priority is periodic socialization to employees regarding the importance of applying clean production, the weight of the assessment is around 0.57. This research is the first step in identifying a green production program in the service industry. This research also still has many shortcomings. Thus, further and in-depth analysis and assessment need to be done in subsequent research.

## 5. References

[1] Nalukowe, Barbara Batumbya. 2006. *Sustainable Industrial Development in Uganda trough Cleaner Production: Case Study of Sugar Corporation of Uganda Ltd (SCOUL).* available at: www.gcpcenvis.nic.in, *Thesis,* 2006, (accessed 4 November 2017).

[2] Shah, Jigar. 2012. *Guidance Notes on Tools for Pollution Management: Cleaner Production*, available at: http://siteresources.worldbank.org, (accessed 24 March 2018).

[3] Vukadinovic, B., Aleksandar Subotin, Ivanka Popovic, dan Mirjana Kijevcanin. 2018. *Cleaner Production and Environmental Sustainability: Analysis of The Serbian Petrochemical Plant.* available at: https://www.researchgate.net/publication/322240648_Cleaner_production_and_environmental_sustainability_Analysis_of_the_Serbian_petrochemical_plant, (accessed 24 March 2018).

[4] Gupta, Permod Kumar. 2014. *Resource Efficient Cleaner Production: Case Study Hotel Sector in Hotel Manulock Luang Prabang.* Available at: http://www.laocpc.org, (accessed 24 March 2018).

[5] Sirait, M. 2018. Cleaner Production Options for Reducing Industrial Waste: The Case of Batik Industry in Malang, East Java-Indonesia. Jakarta, *IOP Conf. Series: Earth and Environmental Science* **106,** 012069

[6] Rimantho D., Putra WA., Hidayah NY., 2018. Determining the Key Criteria Development of Renewable Energy in Indonesia Using a Combination ISM and AHP Methods, *Journal of Sustainability Science and Management*, **13**(2), December 2018

[7] Kholil , Eriyatno , Sutjahyo SH., 2008. Pengembangan Model Kelembagaan Pengelola Sampah Kota dengan Metode ISM (Interpretative Structural Modeling) Studi Kasus di Jakarta Selatan, Sodality: *Jurnal Sosiologi Pedesaan*, **2,** 1.

[8] Rimantho D dan Rosdiana H. 2017. Penentuan Faktor Kunci Peningkatan Kualitas Air Limbah Industri Makanan Menggunakan Interpretive Structural Modelling (ISM). *Jurnal Ilmu Lingkungan*, **15**(2), 90-95

[9] Khuriyati N., Wagiman, Kumalasari D., 2015. Cleaner Production Strategy for Improving Environmental Performance of Small-Scale Cracker Industry, *Agriculture and Agricultural Science Procedia* **3**, 102 – 107

[10] Bantacut T., and Zulaikha SK., 2019. Study on Possibility of Cleaner Production Application at Small Scale Bakery Industry in Pekalongan Indonesia, *Chemical and Process Engineering Research*, **59**, DOI: 10.7176/CPER

# Improving Productivity and Quality of Medium Voltage Cable Production

**A T Pratama[1,2], T Saraswati[1,3], F Prianggara[1,4] and T Savitri[1,5]**

[1]Department of Industrial Engineering, Faculty of Engineering and Information Technology, Swiss German University, The Prominance Tower, Alam Sutera, Tangerang, Indonesia
[2]aditya.pratama@sgu.ac.id, [3]triarti.saraswati@sgu.ac.id,
[4]farhan.prianggara@student.sgu.ac.id, [5]theodora.savitri@student.sgu.ac.id

**Abstract.** A cable manufacturer has produced varieties of products, such as low voltage cables, medium voltage cables, telephone cables, and fiber-optic cables. The main goal this company is to achieve ultimate customers satisfaction through timely delivery, consistency in product quality, and best service. The company faces a problem where daily production cannot reach 6,000 meters/day as requested by management. Moreover, defects have been found in certain processes. A method of countering those problems is proposed by integrating OEE, SMED, and HFACS. Based on the analysis results, the OEE from insulation and screening machines are not meet the company target which is below 60%. In addition, visual defect shown as the most frequent type of defect that occurred in the medium voltage cable production process with 56%. SMED has been utilized to separate the internal task and external task, then the daily production can achieve 6,165 meters/day by reducing the setup time based on the simulation result. In addition, HFACS has been used to analyze the main problem due to human factor aspects, then it is recommended to add extra operator, extra material handling for lifting/moving products, and a re-data checking procedure in the business process.
**Keywords**: cable manufacture, OEE, SMED, HFACS, defects, productivity

## 1. Introduction

A cable manufacturer was established in 1973, located in Tangerang, has produced varieties of products, such as low voltage cables, medium voltage cables, telephone cable, and fiber-optic cables. The main goal this company is to achieve ultimate customers satisfaction through timely delivery, consistency in product quality, and best service. In achieving this goal, the company always upholds national and international standards. The division experts regularly carry out accurate examinations and control that supported by state of the art at the technology, product engineering, and quality control. Supervision and quality control are implemented at every product stage. The supervision and quality control starts from customer order, raw material, inspection, production process, final product inspection, product delivery, until after-sales service.

The cable manufacturer encounters production problem for medium voltage cable which has not to be able to work optimally. This company shows the production output cannot meet the demand target, and the production in the current condition has a defect that is only found in certain process. Since there is long time on the setup time, the daily demand of 6,000 meters is hard to be achieved. In addition, the company has reached some defect rates in certain months with a percentage of 1-2%. The defect product obstructs the production process since it needs time to locate the defect area and repair the defect. Then, the estimated time for the customer receives the finish products can be missed from the target that has been set by production planning department.

Several literatures have explained OEE (Overall Equipment Effectives) assessment that has been used to increase productivity of a production line and focused only on the developing maintenance program. OEE is one method that used to measure the success of Total Productive Maintenance (TPM) program. OEE is procedure used to determine the effectives of machine that consists of availability (A), performance (P), and Quality (Q) [1]. OEE has been used to identify maintenance improvement potential at the croissant production line by analysing failure data that cover a period of 15 months. The evaluation of OEE provide a useful a guide to the aspects of production process, which identifies the critical points of the line that require further improvement through effective maintenance strategy [2]. By combining OEE and Fuzzy FMEA (Failure Mode Effect Analysis), the main root cause that often-occurred damage to the machine can be identified, then how to solve the problems to increase machine productivity can be prescribed [3].

The objectives of this paper are to identify the factors and improve the productivity of medium voltage cables by increasing the Overall Equipment Effectiveness rate of each machine to exceed than 60%. Moreover, the cause of the defect in medium voltage cable production will be analyzed to improve the quality of the current production state. This paper provides several considerable impacts to the company that by considering the implementation of OEE method, the higher efficiency of production line can be increase by reducing the setup time. Moreover, the source of defect due to human error can be identified, then the quality of product can be improved.

## 2. Method

Overall Equipment Effectiveness (OEE) concept is very useful for analyzing the availability, performance and quality of machines in production plants. OEE measurement tool was developed from the TPM concept [4]. The goal of TPM is to achieve zero breakdown and zero defects related to equipment. The TPM concepts puts much attention on productivity equipment, since they have a high influence on quality, productivity, cost, inventory, safety and health, and production output [4]. OEE can be defined as a measure of total equipment performance, that is, the degree to which the equipment is doing what is supposed to do [5]. OEE is a three-part analysis tool for equipment performance based on its availability, performance, and quality rate of the output. OEE can points to hidden capacity in a manufacturing process and lead to balanced flow. OEE can be used to track and trace improvements or decline in equipment effectiveness over a period of time [6]. In this paper, data is acquired through direct observation, interviews, and time study. The data is in the form of path description of the production line, production method, production process, and others. Stopwatch is used to retrieve the preparation time, loading time, unloading time, installation time, and other activities that needed to complete the process. In addition, the data is also collected from company record such as: data of medium voltage cable production, defect, quality, and maintenance activities in 2017 and 2018. The value of OEE can be measured based on the company record through calculation of ratio on Availability, Performance, and Quality on each machine.

Single Minutes Exchange of Dies (SMED) is a method that focuses on the recognition of internal and external activities. This relates primarily by converting and internal activity and an external activity. This relates primarily by converting an internal activity to an external amount as much as possible by also minimizing the internal ones [7]. SMED is characterized as a scientific approach for the reduction of setup times, and which can be applied in any industrial unit and for any machine [8]. In SMED, internal setup time and external setup time are two important operations that involved. Internal setup time can be only carried out when the machine or process has stopped, and external setup time could be performed while the machine or process is still in operating. Based on the OEE result and fishbone diagram, the root cause on long setup time can be solved by increasing Availability ratio through SMED. In the SMED, the external activities during setup time may be done, while the process (internal activity) is still operational.

Human Factor Analysis and Classification System (HFACS) was specifically developed to define the latent and active failures implicated in Reason's 'Swiss cheese' [9]. Meanwhile, [10] explains accident causation with the breakdown of interaction between fundamental elements in an organization; decision makers, line management, preconditions, productive activities, and defences. In Swiss cheese model, elements constitute a production system which receive input and produce some outputs. The human contributions to the breakdown of these elements are separated into two categories active and latent failures. Active failures are errors where the effect can be seen almost immediately, while latent failures refer to the errors whose adverse consequences may lie dormant within the system for a long period of time [10]. One limitation of Reason's model that respond by HFACS framework is that it does not defines the window of accident opportunity. Human Factor Analysis and Classification System (HFACS) is a framework tool to assist an investigation process to systematically identify active and latent failures within an organization that culminated in the accident. The goal of HFACS is to underlying casual factors that lead to an accident and understand the process framework. HFACS describes four levels of failure, such as: Unsafe Acts, Preconditions for Unsafe Acts, Unsafe Supervision, and Organizational Operations [9]. Based on the fishbone diagram, most of the product defects is affected by human error. Human error is not only from the Man category, but also can be influenced by Management and Method category. In this paper, the HFACS is used to define the cause of visual defect during the process.

## 3. Result and Discussion

This paper studies in one of the plants owned by the cable manufacturer, which produce medium voltage cable. Medium voltage cable is a power distribution cable with a range from 2,000 to 35,000 volts. The production of medium voltage cable must through eight processes with quality check for each process before continuing to next process. The eight processes flow are Drawing, Stranding, Insulation, Screening, Cabling, Inner Sheath, Armour Sheath, Outer Sheath with Marking, and Final Test. The medium cable production starts from drawing process where the aluminium bought from a producer is stretched by being forcibly pulled to make the diameter from 0.8 mm into the desired dimension. The stretched aluminium then continues to a stranding process, where 35 stretched cables are twisted into one single core, layer by layer. Afterwards, the twisted single core cable is insulated in the insulation process. In insulation process, the previous finished product is coated with 100°C melted Cross-Linked Polyethylene (XLPE). The finished coated cable has three layers of coating, the inner layer called semi-conductive compound, the middle and the thickness layer called XLPE layer, and the outer layer is strippable semi-conductive compound. After being coated, the cable continues to the screening process. In the screening process, there are three stages of screening: semi-conductive water blocking, metallic screen, and non-conductive water blocking. Those process mentioned above are producing single core cable and after the single core has finished the screening process, three finished single cores will be twisted in the cabling process. Then, the twisted cable will be covered with Galvanized Steel Tape in armour process, to protect the cable since medium voltage cables are commonly implemented for underground in streets, so it will not break if passed by heavy load or when exposed to sharp tools during excavation process. The last process is called outer sheath process, where the armour is coated with red extrude 90°C grade PVC and stamped for the product code and company name right after when the PVC still in the hot state. Figure 1 shows the production process flow of medium voltage cable.

Based on data acquisition on processing time of each stations, model of existing condition has been developed by using Tecnomatix Plant Simulation®. The simulation of this model shows that the existing capacity for producing medium voltage cable can achieved 5,753 meters/day or still below 6,000 meters/day.

The company has categorized the defect into Visual, Dimension, Electrical, and Construction defect. Based on the secondary data, the most significant influence defect can be analyzed by using Pareto chart as shown in Figure 2.

IOP Conf. Series: Materials Science and Engineering **847** (2020) 012005 doi:10.1088/1757-899X/847/1/012005

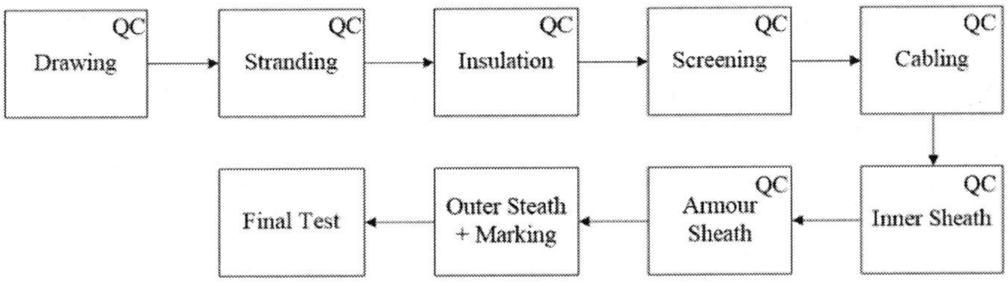


Figure 1. Medium voltage cable production process flow.

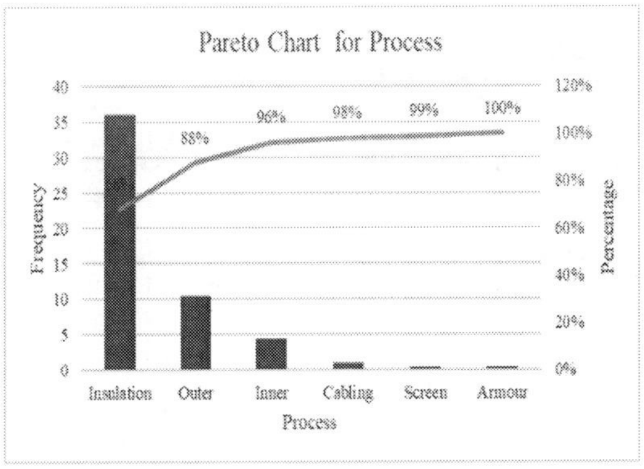


Figure 2. Pareto chart for medium voltage cable production

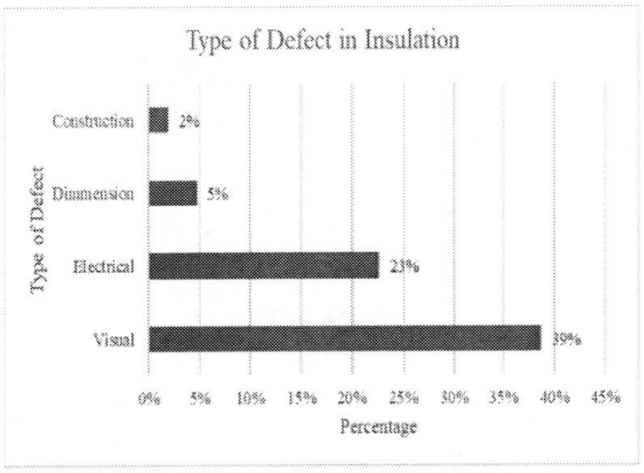


Figure 3. Clustered bar for type of defect in insulation

As can be seen from Figure 2, Insulation and Outer process have the most significant defect influence on the defect percentage by total 88%, where 68% from Insulation process itself. After analyzing and find the most significant influence process to the defect is the Insulation process, the next step is to find which kind of defect that frequently occurs in the Insulation process. From Figure 3, visual defect shown as the most frequent type of defect that occurred in the medium voltage cable production process with 39% and followed by electrical defect with 23%. Electrical defect can be happened because of influenced by visual defect. To explore more in detail the causes of visual defect, this paper uses Fishbone Diagram for further analyzing the root-cause of visual defect. Figure 4 is a Fishbone Diagram that is used to find the root cause of visual defects.

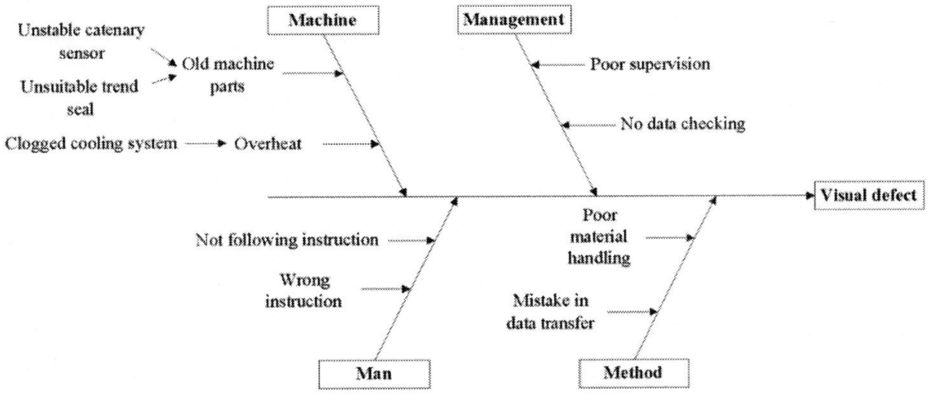


Figure 4. Fishbone diagram for visual defect

IOP Conf. Series: Materials Science and Engineering **847** (2020) 012005 doi:10.1088/1757-899X/847/1/012005

Based on Figure 4, the visual defects are caused by four types of categories, those are: Machine, Man, Method, and Management. In the other hand, from the analysis obtained, most of the damage is affected by human error. Human error is not only from the Man category, but also can be influenced by Management and Method category. Human errors that can cause visual defect in Insulation process can be divided into two categories: Collision and product does not meet the customer request. Collision is a defect caused by errors in material handling because workers ignore procedures. Whereas product does not meet the customer request, are defects that occur because of input data errors or errors in the factory that result in the products produced not in accordance with what is desired by the customer. Table 1 shows the Human Factor Analysis and Classification System (HFACS) to summarize and analyse the human errors that cause collision and product does not meet the customer request.

Table 1. Human Factor Analysis and Classification System

| HFACS Level | Classification | | | Risk |
|---|---|---|---|---|
| | Errors | Violations | | |
| | **Decision Errors**<br>• Worker does not use tools to move the finished products<br>• Worker does not follow instructions<br>• Worker sets the wrong tools<br><br>**Skill-Based Errors**<br>• No skill needed<br><br>**Perceptual Errors**<br>• Worker does not know the effect of not following instructions<br>• Worker fills the wrong form from the office | **Routine**<br>• Moving item without any moving tools<br>• No data checking | **Exceptional**<br>- | |
| **1. Unsafe Acts** | | | | • Collision<br>• Delivered product does not meet the customer request<br>• Unintentional errors/unsafe acts. |
| **2. Pre-conditions for Unsafe Acts** | **Conditions of Operators** | **Personnel Factors** | **Environmental Factors** | • Unintentional errors/data input<br>• Human error<br>• Uncoordinated |
| | **Adverse Mental States**<br>• Careless worker<br>• Loss of situational awareness<br>• Worker experiences fatigue<br>• Mental fatigue<br><br>**Adverse Physiological States**<br>• Medical illness<br>• Unable to lift heavy products<br><br>**Physical/Mental Limitation**<br>• Impossible to lift finished products<br>• Visual limitations | **Crew Resource**<br>• Failed to coordinate<br>• Failed to conduct adequate briefing<br><br>**Personal Readiness**<br>- | **Physical Environment**<br>-<br><br>**Technological Environment**<br>• Manual data transfer | |
| **3. Unsafe Supervision** | **Inadequate Supervision**<br>• Failed to provide oversight<br><br>**Planned of Inappropriate Operations**<br>• No planned operation<br>• Failed to provide correct data | **Failed to Correct the Problem**<br>• Worker continue to violate the instructions | **Supervisory Violations**<br>• Cannot to supervise at any time | • Human errors<br>• Lead to unsafe act<br>• Unprofessional mindset |
| **4. Organizational Operations** | **Resource Management**<br>• Lack of supervision<br>• No daily briefing<br>• Lack of training<br>• Limited material handling tools | **Organizational Climate**<br>• No reward or punishment | **Organizational Process**<br>• No double checking<br>• No standard procedure for moving finished product | • Lead to data error<br>• Lead to unsafe act<br>• Unprofessional mindset |

Table 2 shows the current condition, expected condition, and proposed improvement based on two categories of wrong material handling and product does not meet the customer request. The first defect problem caused by human error is wrong material handling due to the current condition where there is limited moving tools and workers moving the finished products by rolling them. The expected conditions are the company provide a more efficient and safe moving tools and also workers to follow the procedure given for material handling. To solve this problem, there is three improvements to be proposed, first is hoist crane can save more time to moving the item and reduce the possibility of defects caused by collisions. Second, the company should give proper training and a presentation about the importance of material handling. Moreover, third is by assigning a person for a specific supervision task in the factory.

Table 2. HFACS proposed improvement

| Summary | Current Condition | Expectation | Proposed Improvement |
|---|---|---|---|
| Wrong material handling | • Limited material handling tools | • Efficient and safe material handling tools<br>• Worker follows the procedures for material handling | • Install crane<br>• Provide training and importance information related to material handling<br>• Assign person to supervise |
| Product does not meet the customer request | • Move the item by rolling the finished products | • No errors during data transfer<br>• Computer system is used for transferring the data | • Add procedure for data checking |

The second defect is an error due to wrong data transfer resulting in a wrong production. The expectation of solving this case is by eliminating the error in data transfer using a computer system. Due to the current situation where all the data transfer is still using a paper form, implementing computer system to all section of the company can take a huge investment. In result, the possible improvement to be proposed to the company is adding a re-data checking procedure in the business process.

Based on the simulation results, the OEE rate has been analysed by utilizing secondary data of each machine. OEE rate calculation depends on three main ratios, such as Availability, Performance, and Quality. The OEE value does not meet the target since the obtained value of OEE is below 60%. Table 3 shows the OEE ratio on overall machines

Table 3. OEE ratio on overall machines

| Station | Type of Machine | Availability | Performance Rate | Quality Production Rate | OEE |
|---|---|---|---|---|---|
| 1 | Drawing 1 | 84% | 96% | 100% | 80% |
| | Drawing 2 | 84% | 96% | 100% | 81% |
| 2 | Stranding ST 54 A | 87% | 94% | 99% | 82% |
| | Stranding ST 54 C | 86% | 94% | 99% | 81% |
| 3 | Insulation JR | 70% | 86% | 98% | 59% |
| | Insulation SUMITOMO | 75% | 80% | 98% | 59% |
| 4 | Screening TP-2 | 61% | 97% | 100% | 59% |
| | Screening TP-5 | 61% | 95% | 99% | 58% |
| | Screening TP-6 | 61% | 95% | 100% | 58% |
| 5 | Cabling 26 A | 82% | 79% | 100% | 65% |
| | Cabling 26 B | 83% | 80% | 100% | 66% |
| 6 | Inner | 71% | 98% | 100% | 70% |
| 7 | Armoring TP-4 | 73% | 95% | 100% | 70% |
| | Armoring TP-3 | 76% | 95% | 100% | 72% |
| 8 | Other | 77% | 91% | 99% | 69% |

The company has targeted the OEE on each machine must be larger or equal to 60%. The OEE value on the Insulation JR, Insulation SUMITOMO, TP-2 Screening, TP-5 Screening, and TP-6 Screening have not yet reached the targeted of OEE standard values. Based on Table 3, the low availability especially on Insulation and Screening machines is one factor that makes the OEE value does not reach the target of the company. The low availability rate on those machines was caused by long period of setup time due to lack of supporting operator and material handling tool. The operator waits until the machine stops and then he/she prepares new bobbins. If the supporting operator is present, he/she can prepare new bobbins while the machine is in process. Furthermore, the bobbins are heavy weight, then the operator must wait for the other operator to use the material handling tool for installing/moving the bobbins.

Since the long period of setup time has significant impact on low availability rate and OEE rate, then Single Minutes Exchange of Die (SMED) was used as method to reduce the setup time. Table 4 shows the setup activities at Insulation JR machine after applying SMED. The setup consists of two activities, internal and external setup. Internal setup is activities that require an inactive (shut down) process, meaning that operator no orders to do set up until the machine stops. External setup activities may be done while the process is operational. From the SMED result on Insulation JR machine, the setup time decreased to 70.12 minutes/day from 109.71 minutes/day.

Table 4. Setup activities of Insulation JR machine after SMED

| No | Work Element | Task Time (s) | Internal | External |
|---|---|---|---|---|
| 1 | Loading input | 470.30 | | |
| 2 | Unloading input | 461.15 | | |
| 3 | Running | 18850.00 | | |
| 4 | Preparation new loading input | 200.20 | | |
| 5 | Preparation new loading output | 181.30 | | |
| 6 | Unloading output bobbin | 503.70 | | |
| 7 | Unloading input bobbin | 372.75 | | |
| 8 | Checking cable | 402.30 | | |
| 9 | Checking bobbin | 385.60 | | |
| 10 | Taking and putting tools | 1608.45 | | |
| 11 | Cleaning | 1997.10 | | |
| Total Setup Time | | Internal (min) | | External (min) |
| | | 70.12 | | 39.59 |

The similar SMED method has been applied for other Insulation and Screening machine and found that significant decrease on setup time. The setup time decreased to 70.24 minutes/day from 109.81 minutes/day on Insulation Sumitomo machine. While, on Screening TP-2, TP-5, and TP-6, the setup time decrease from 170.80 minutes/day to 101.01 minutes/day, from 161.90 minutes/day to 92.21 minutes/day, and from 161.09 minutes/day to 88.34 minutes/day, respectively.

By using the same production model on Figure 2 and modify the setup time in accordance to SMED analysis, the simulation shows that the number of production output has reached 6,165 meters/day from 5,753 meters/day, then it can fulfil the daily customer request.

## 4. Conclusion and Recommendation

It is necessary to improve productivity so that the output can be maximized. In this case, each operating process must be analyzed for the cause of the problem so that how well the resources are managed and utilized to achieve optimal results. The OEE from insulation and screening machines are not meet the company target which is below 60%. Low availability rate that caused by long period of setup time is the main factor that causes the OEE of the insulation and screening machines does not meet the target of the company. SMED method can be used to reduce the setup time and increase the output daily output of the company.

Meanwhile, the quality of medium voltage cable production in this company gives a significant impact on the product quality, production schedule, waste material, and delivery time. Based on the Pareto chart, Insulation process has the highest defect rate for the past two years. Defects that often occur in Insulation processes are Visual and Electrical defects, where Electrical defects occur due to Visual defects. Fishbone diagram shows that three categories that inflict Visual defects caused by human errors from various aspects. Human Factor Analysis and Classification System analysis to look for potential errors caused by human errors.

The results of the analysis using Human Factor Analysis and Classification System shows that human error that results in collide is workers who do not comply with the rules and tools that use for material handling. While the occurrence of production errors is due to an error entering the data. Both of these errors can occur due to lack of supervision and professionalism when working.

A new material handling is proposed to be installed in intention to minimize the possibility of Visual defect caused by a collide. The installation of a crane also uses to decrease the setup time. For maximizing the use of cranes, an extra operator is needed for helping the main operator in preparing the material. In addition, regular training is also needed for workers in the factory to know the importance and influence of material handling on products, safety and cost to the company.

Business Process Improvement is a new procedure of re-data checking before all data continues to the manufacturing department. The aim of the data checking is to detect an error in data transfer and to correct any errors that exist to prevent production errors caused by input data errors.

Since this paper shows a proper result in improving the production quality, the future study related to the quality improvement in production will be required. In addition, the company want to increase the daily production of medium voltage cable up to 10,000 meters/day. Then, feasibility study related to the additional machines and operators need to be obtained.

## 5. References

[1] Dutta, S., Dutta, A. K., 2016. A Review on the experimental study of Overall Equipment Effectiveness of various machines and its improvement strategies through TPM implementation. *International Journal of Engineering Trends & Technology*, **36**(5), 224-232.

[2] Tsarouhas, P., 2019. Improving operation of the croissant production line through overall equipment effectiveness (OEE) A case study. *International Journal of Productivity and Performance Management*, **68**(1), 88-108.

[3] Hunusalela, Z. F., Perdana, S., and Usman, R. 2019. Analysis of productivity improvement in hard disc spare parts production machines based on OEE, FMEA, and fuzzy value in Batam. *In IOP Conference Series: Materials Science and Engineering*, **508**(1).

[4] Seiichi Nakajima, 1988. *Introduction to TPM: total productive maintenance*. Productivity Press.

[5] Williamson, R.M., 2006. *Using Overall Equipment Effectiveness: The metric and the measure.* Strategic Work System, Inc., Columbus NC 28722, 1-6.

[6] Bulent, D., Tugwell, P., and Greatbanks, R., 2000. Overall equipment effectiveness as a measure of operational improvement - A practical analysis. *International Journal of Operations & Production Management*, **20**(12), 1488-1502.

[7] Joshi, M. R. and Naik, P. G., 2012. Reduction in setup time by SMED. *International Journal of Modern Engineering Research*, **2**(1), 442-444.

[8] S.A. Shingo, 1985. *Revolution in Manufacturing: the SMED System*. Productivity Press, Cambridge, MA.

[9] Wiegmann, D., Shappell S., 2003. *A human error approach to aviation accident analysis*, Aldershot, (England: Ashgate).

[10] Reason, J., 1990. The contribution of latent human failures to the breakdown of complex systems. *Philosophical Transactions of the Royal Society (London), series B*. 327: 475-484.

[11] Wood, D.J. and Logsdon, J.M., 2017. Theorising business citizenship. In *Perspectives on corporate citizenship*: 83-10.

# Analysis of textile dye production scheduling using FCFS, CDS and Heuristic Pour methods

**N Y Hidayah[1,2], M Syafrizal[1,3,] and M Darmawan[1]**

[1]Industrial Engineering Department Pancasila University, Jl. Srengseng Sawah, Jagakarsa, Jakarta, Indonesia
[2]nurhidayah@univpancasila.ac.id, [3]muhammadsyafrizal968@gmail.com

**Abstract.** PT. X is a company engaged in the chemical industry that produces various types of textile dyes for several garment companies in Indonesia and abroad. The manufacture of various dye textiles has a different processing time because it is influenced by the quality and parameters of each product. The production scheduling system used by the company is First Come First Served (FCFS). Companies often experience delays in the timely delivery of orders due to the production scheduling system that is used by the company was not optimal. This research was conducted to schedule a sequence of textile dye processing in PT. X by using the FCFS, Campbell Dudek Smith (CDS) and Heuristic Pour to get shorter makespan and minimize delivery delays. The results of research at PT.X showed that the scheduling method which had the fastest makespan was the Heuristic Pour scheduling method with a total makespan of 562.4 hours (23 days) and could save time by 47.96 hours or 7.86% compared to the FCFS method.
**Keywords**: production scheduling methods, makespan, textile dye, delivery delay

## 1. Introduction

Production activities are the main activities that must be carried out by companies engaged in manufacturing. Companies that implement a make to order system will produce goods according to customer orders. Therefore, companies must try to complete the production process in accordance with the specified deadline in order to minimize delivery delays. Then we need an appropriate sequence of workmanship to produce the fastest time to complete the job. Scheduling is an effort to manage activities or work with the aim of achieving efficient use of facilities, time and costs.

PT. X is a company engaged in the chemical industry that produces various types of textile dyes for several garment companies in Indonesia and abroad. The company produces based on incoming orders (make to order). The company produces 20 families of textile coloring products with more than 300 variants of textile dyes. Each variant has the same sequence of production processes which can result in idle time on materials and machines which greatly affect the product finishing time. The method used by companies to schedule their products was First Come First Served (FCFS) where first-time orders would be prioritized. Based on the data obtained, it was known that in February 2019, PT. X experienced delays in sending orders from three garment companies in Indonesia. The delay could be caused by the scheduling method used which was not quite right because the FCFS method does not pay attention to the production process time and due date of each order. Table 1 shows the data of shipping delays experienced by PT. X.

IOP Conf. Series: Materials Science and Engineering **847** (2020) 012006 doi:10.1088/1757-899X/847/1/012006

**Table 1.** Shipping delay in February 2019

| | Shipping delay | Percentage |
|---|---|---|
| PT A | 3,005 kg | 37% |
| PT B | 3,110 kg | 38% |
| PT C | 2,015 kg | 25% |

This research was conducted to make a production schedule with the right method so that the sequence of workmanship was obtained with the amount of time needed to complete all orders to become shorter. In this research, the scheduling method used was the Heuristic Pour algorithm and Campbell Dudek Smith (CDS).

## 2. Methods

Scheduling can be interpreted as allocating a number of resources to perform a number of tasks or operations in a certain period of time and is a decision making process whose role is very important in the manufacturing and service industries [1]. This study uses 3 production scheduling methods, namely:

- First Come First Served (FCFS) [2]. FCFS is a method where orders that come first will be prioritized to be done. This method was used by PT. X to schedule orders from consumers.
- Campbel Dudek Smith (CDS) [3]. This method solves the problem of n jobs on m machines into two groups, then sequences jobs on the two machines using the Johnson algorithm. After obtaining as many m-1 alternative job sequences, then the order with the smallest processing time will be chosen. The Johnson algorithm rule was developed for n jobs that are done on two machines in sequence. In the calculation of the CDS algorithm, we can use the following formula:

$$t^*_{i,1} = \sum_{i=1}^{k} t_{i,k} \quad (1)$$

$$t^*_{i,2} = \sum_{i=1}^{k} t_{i,m-k+1} \quad (2)$$

  The steps for scheduling with the CDS algorithm are as follows:

  1. Determine the number of iterations, i.e. the number of work center minus 1 or K = m - 1.
  2. Take the first scheduling (K = 1). For all jobs, look for the minimum value of $t^*_{i,2}$ which is the processing time on the first and second machines, where $t^*_{i,1} = t_{i,1}$ and $t^*_{i,2} = t_{i,2}$. Do it also for the next iteration.
  3. Perform Johnson's rules. If the minimum time is obtained on the first machine ($t_{i,1}$) place the task at the beginning of the scheduling sequence and if the minimum time is obtained on the second machine ($t_{i,2}$), the task is placed at the end of the scheduling sequence.
  4. Remove the tasks from the list and arrange them in the form of a scheduling sequence. If there are no jobs left, the scheduling is complete. Thus, the processing time of the two machines namely the first machine ($t^*_{i,1}$) and the second machine ($t^*_{i,2}$) on the K-scheduling has been completed (see formulas 1 and 2).
  5. The calculation is continued based on the next task then calculates the makespan.
  6. Continue to do these steps until we get the order of assignment with the smallest makespan.
- Heuristic Pour Algorithm [4]. This method was developed by Hamid Davoud Pour (2001) in completing flow shop scheduling with the aim of minimizing makespan based on a combination approach. The following notation is used:
  - $P_{ij}$ : the processing time of job *i* on machine *j*.
  - $C_{ij}$ : the time span between when job *i* on machine *j* starts (t = 0) until the job is finished.
  - $\sum C_i$ : sum of completion time for job *i* on all machines.

• $F_{max}$ : the time span between when the work can begin until the work is finished (makespan)
The steps for scheduling the Heuristic Pour algorithm are as follows:

1. Choose job 1 as the first sequence while in the work order so that job 1's processing time on all machines is considered zero.
2. Placing other jobs (other than those that have been selected as the first order, that is job 1) in the next sequence.
3. Select the smallest processing time for each machine.
4. Increase completion time for each $P_{ij}$ by increasing processing time rules, namely by adding cumulative processing time from the smallest to the largest in each $P_{ij}$.
5. Calculates the sum of completion time ($\sum C_i$) for each existing job.
6. Sort $\sum C_i$ with the increasing order rule (i.e. ordering from the smallest to the largest) to be placed in the order after the job that has been selected for the first order (ie job 1).
7. After obtaining a sequence where job 1 is the first order, then calculate the $F_{max}$ from the sequence.
8. Repeat steps 1-7 for each existing job that will be placed first in the sequence of job execution until the minimum $F_{max}$ value is obtained.
9. Repeat steps 1-8 for the job that will occupy the next position that is in the second, third, and so on after the job is selected for the first position with a minimum $F_{max}$ value.

The next step was calculated performance parameters. In the research of production scheduling, there are several performance parameters used to determine better methods, namely Efficiency Index (EI) which is a comparison between the proposed scheduling method and the method used by the company, formulated as follows [2]:

$$\text{EI} = \frac{F_{max} \text{ method used by the company}}{F_{max} \text{ proposed scheduling method}} \quad (3)$$

If EI = 1, then both methods have the same performance, if EI <1, the proposed scheduling method has poor performance compared to the method used by the company.

## 3. Result and Discussion

### 3.1. Customer request data

Data on customer requests for May 2019 along with due dates can be seen in Table 2 while data on machines used for the production of textile dyes can be seen in Table 3.

**Table 2**. Textile dyes demand in May 2019

| | Product Code | Colour Code | Demand (kg) | Due Date (days) | Purchase Order Date |
|---|---|---|---|---|---|
| Terasil Borwn 2RFL | YROH | | 15,500 | 20 | 23/04/2019 |
| Teratop Black HL-BL | YHAL | | 17,000 | 20 | 24/04/2019 |
| Teratop Black LF-01 | YHAL | | 24,000 | 20 | 25/04/2019 |
| Teratop Black HL-NF-01 | YHAL | | 8,500 | 20 | 29/04/2019 |
| Terasil Blue WW-2GS | YROH | | 10,500 | 20 | 29/04/2019 |
| Teratop Blue HL-B | YHAL | | 25,000 | 20 | 29/04/2019 |
| Novasol Blue RS MD | YFER | | 30.000 | 20 | 30/04/2019 |
| Terasil Orange GLN | YROH | | 12.000 | 20 | 02/05/2019 |
| Terasil Orange HLNF | YROH | | 28.000 | 20 | 03/05/2019 |
| Terasil Red W-4BS | YROH | | 49.000 | 20 | 03/05/2019 |

IOP Conf. Series: Materials Science and Engineering **847** (2020) 012006 doi:10.1088/1757-899X/847/1/012006

**Table 3**. Name and capacity of machines in each work station

| | Machine Station | Amount (unit) | Production Capacity/ Machine (kg) | Total Capacity Machine (kg) |
|---|---|---|---|---|
| Pasting Material | R10 (M1) | 1 | 12,000 | 12,000 |
| Milling | R20, R21 (M2) | 3 | 14,000 | 42,000 |
| Adjustment Process | R32, R33 (M3) | 2 | 10,000 | 20,000 |
| Spray Dryer Process | T50 (M4) | 1 | 1,000 | 1,000 |
| Packing Process | R50 (M5) | 1 | 40,000 | 40,000 |

The following is the processing time data of each product obtained from the production department. Each product has a different processing time which is influenced by product quality and parameters.

**Table 4**. Processing time for each product (hours).

| | M1 | M2 | M3 | M4 | M5 |
|---|---|---|---|---|---|
| YROH (J1) | 18 | 50 | 56 | 1 | 9 |
| YHAL (J2) | 21 | 61 | 42 | 1 | 10 |
| YHAL (J3) | 19 | 46 | 63 | 1 | 13 |
| YHAL (J4) | 23 | 44 | 47 | 1 | 15 |
| YROH (J5) | 15 | 55 | 43 | 1 | 10 |
| YHAL (J6) | 19 | 58 | 39 | 1 | 10 |
| YFER (J7) | 31 | 57 | 53 | 1 | 15 |
| YROH (J8) | 17 | 31 | 54 | 1 | 9 |
| YROH (J9) | 16 | 32 | 36 | 1 | 8 |
| YROH (J10) | 19 | 37 | 40 | 1 | 11 |

Following is an example of calculating the total processing time required in M2 on J1:

$$\text{Time in M2 on J1} = \frac{\text{Cycle time} \times \text{number of units}}{\text{production capacity}} = \frac{50 \text{ hours} \times 15{,}500 \text{ kg}}{14{,}000\ kg} = 55.35 \text{ hours}$$

$$\text{Machine Station Time} = \frac{\text{Time required in M2 on J1}}{\text{Machine Amount}} = \frac{55.35}{3} = 18.45 \text{ hours}$$

**Table 5**. The total processing time required based on each order to each machine station (hours).

| | M1 | M2 | M3 | M4 | M5 |
|---|---|---|---|---|---|
| J1 | 23.25 | 18.45 | 43.40 | 15.50 | 3.49 |
| J2 | 29.75 | 24.69 | 35.70 | 17.00 | 4.25 |
| J3 | 38.00 | 26.29 | 75.60 | 24.00 | 7.80 |
| J4 | 16.29 | 8.90 | 19.98 | 8.50 | 3.19 |
| J5 | 13.13 | 13.75 | 22.58 | 10.50 | 2.63 |

| | M1 | M2 | M3 | M4 | M5 |
|---|---|---|---|---|---|
| J6 | 39.58 | 34.52 | 48.75 | 25.00 | 6.25 |
| J7 | 77.50 | 40.71 | 79.50 | 30.00 | 11.25 |
| J8 | 17.00 | 8.86 | 32.40 | 12.00 | 2.70 |
| J9 | 37.33 | 21.33 | 50.40 | 28.00 | 5.60 |
| J10 | 77.58 | 43.17 | 98.00 | 49.00 | 13.48 |

### *3.2. First Come First Served (FCFS)*

Based on the FCFS method, the production order was J1-J2-J3-J4-J5-J6-J7-J8-J9-J10 with a makespan of 610.36 hours or 25 working days, can be seen in table 6.

**Table 6**. Makespan calculations with the FCFS method.

| | M1 | M2 | M3 | M4 | M5 |
|---|---|---|---|---|---|
| J1 | 00.00/23.25 | 23.25/41.70 | 41.70/85.10 | 85.10/100.60 | 100.60/104.09 |
| J2 | 23.25/53.00 | 53.00/77.69 | 85.10/120.80 | 120.80/137.80 | 137.80/142.05 |
| J3 | 53.00/91.00 | 91.00/117.29 | 120.80/196.40 | 196.40/220.40 | 220.40/228.30 |
| J4 | 91.00/107.29 | 117.29/126.19 | 196.40/216.38 | 220.40/228.90 | 228.90/232.09 |
| J5 | 107.29/120.42 | 126.19/139.94 | 216.38/238.83 | 238.83/249.33 | 249.33/259.60 |
| J6 | 120.42/160.00 | 160.00/194.52 | 238.83/287.58 | 287.58/312.58 | 312.58/318.83 |
| J7 | 160.00/237.50 | 237.50/278.21 | 287.58/367.08 | 367.08/397.08 | 397.08/408.33 |
| J8 | 237.50/254.50 | 278.21/287.07 | 367.08/399.48 | 399.48/411.48 | 411.48/414.18 |
| J9 | 254.50/291.83 | 291.83/313.16 | 399.48/449.88 | 449.88/477.88 | 477.88/483.48 |
| J10 | 291.83/369.41 | 369.41/412.58 | 449.88/547.88 | 547.88/596.88 | 596.88/610.36 |

Note: a / b = start time / end time

Production time from start to finish for each product using the FCFS scheduling method can be seen in figure 1.

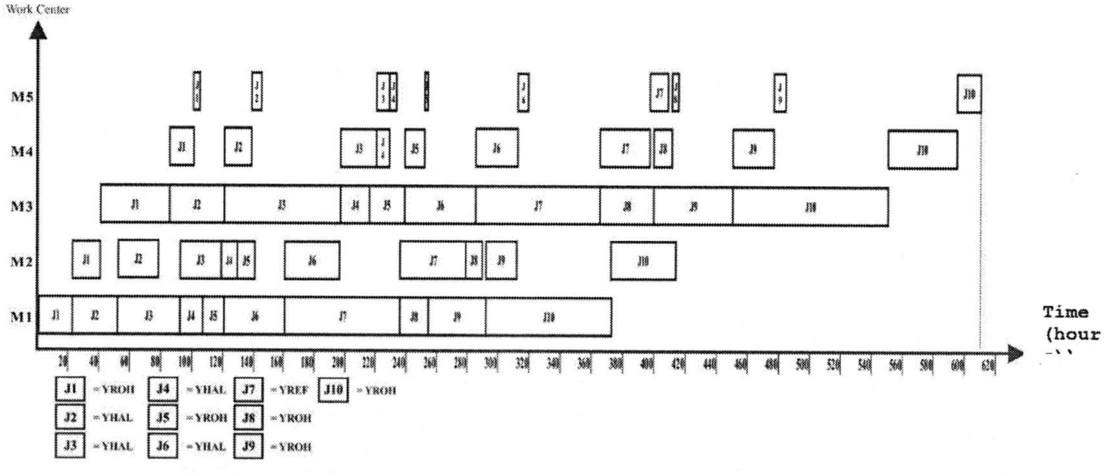


**Figure 1.** Job scheduling for May 2019 with the FCFS method.

### *3.3. Heuristic Pour*

Based on calculations using the Heuristic Pour scheduling method, the optimum of job scheduling was J4-J8-J5-J1-J2-J9-J3-J10-J7-J7-J6 with $F_{max}$ or makespan of 562.4 hours or 23 days. The makespan calculation is presented in Table 7. Production time from start to finish for each product using the Heuristic Pour scheduling method can be seen in Figure 2.

**Table 7**. Makespan calculation using the Heuristic Pour method.

| | M1 | M2 | M3 | M4 | M5 |
|---|---|---|---|---|---|
| J4 | 00.00/16.29 | 16.29/25.19 | 25.19/45.17 | 45.17/53.67 | 53.67/56.86 |
| J8 | 16.29/33.29 | 33.29/42.15 | 45.17/77.57 | 77.57/89.57 | 89.57/92.27 |
| J5 | 33.29/46.42 | 46.42/60.17 | 77.57/100.02 | 100.02/110.52 | 110.52/113.15 |
| J1 | 46.42/69.67 | 69.67/88.12 | 100.02/143.42 | 143.42/158.92 | 158.92/162.41 |
| J2 | 69.67/99.42 | 99.42/124.11 | 143.20/178.90 | 178.90/195.90 | 195.90/200.15 |
| J9 | 99.42/136.75 | 136.75/158.08 | 178.90/229.30 | 229.30/257.30 | 257.30/262.90 |
| J3 | 136.75/174.75 | 174.75/201.04 | 229.30/304.90 | 304.90/328.90 | 328.90/336.70 |
| J10 | 174.75/252.33 | 252.33/295.50 | 304.90/402.90 | 402.90/451.90 | 451.90/465.38 |
| J7 | 252.33/329.83 | 329.83/370,54 | 402.90/482.40 | 482.40/512.40 | 512.40/523.64 |
| J6 | 329.83/369.41 | 370.54/405.06 | 482.40/531.15 | 531.15/556.15 | 556.15/562.40 |

Note: a / b = start time / end time

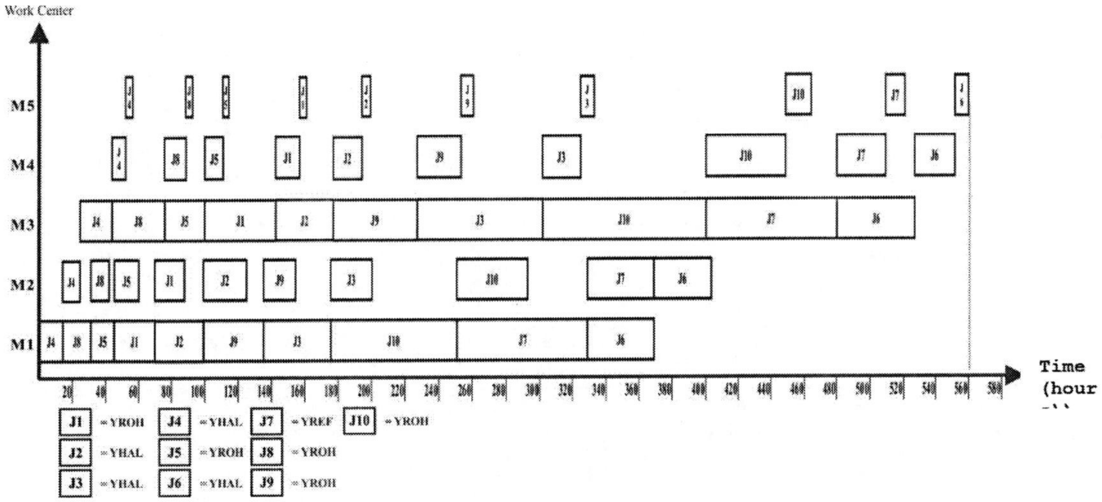


**Figure 2.** Job scheduling for May 2019 with the Heuristic Pour method.

### *3.4. Campbell Dudek Smith (CDS)*

The first step in compiling the CDS algorithm was to determine the number of iterations or working stages.

K (number of iterations) = number of work centre - 1 = 5 - 1 = 4

Thus the process of scheduling using the CDS method was done by 4 iterations. Find the minimum $t^*_i$ value which is the processing time on the first and second machines, where $t^*_{i,1} = t_{i,1}$ and $t^*_{i,2} = t_{i,2}$. According to Johnson's Rules, if the minimum time is obtained on the first machine ($t_{i,1}$), place the work at the beginning of the scheduling sequence and if the minimum time is obtained on the second machine ($t_{i,2}$), the job is placed at the end of the scheduling series. Based on the calculation using the CDS method, the sequence of jobs with the smallest makespan was 638.62 hours or 27 days with the priority sequence of jobs that was J10-J7-J3-J6-J9-J2-J1-J8-J5-J5-J4 obtained from the 4th iteration (K4). The calculation of the makespan can be seen in Table 8. The completion time of the production process can also be seen on the Gantt Chart in Figure 3.

**Table 8**. Makespan calculation using the CDS method.

| | M1 | M2 | M3 | M4 | M5 |
|---|---|---|---|---|---|
| J10 | 00.00/77.58 | 77.58/120.75 | 120.75/218.75 | 218.75/267.75 | 267.75/281.23 |
| J7 | 77.58/155.08 | 155.08/196.79 | 218.75/298.25 | 298.25/328.25 | 328.25/339.50 |
| J3 | 155.08/193.08 | 196.79/223.08 | 298.25/373.85 | 373.85/397.85 | 397.85/405.65 |
| J6 | 193.08/232.66 | 232.66/267.18 | 373.85/422.60 | 422.60/447.60 | 447.60/453.85 |
| J9 | 232.66/269.99 | 269.99/291.32 | 422.60/473.00 | 473.00/501.00 | 501.00/506.60 |
| J2 | 269.99/299.74 | 299.74/324.43 | 473.00/508.70 | 508.70/525.70 | 525.70/529.95 |
| J1 | 299.74/322.99 | 324.43/342.88 | 508.70/552.10 | 552.10/567.60 | 567.60/571.09 |
| J8 | 322.99/339.99 | 342.88/351.74 | 552.10/584.50 | 584.50/596.50 | 596.50/599.20 |
| J5 | 339.99/353.12 | 353.12/366.87 | 584.50/606.95 | 606.95/617.45 | 617.45/620.08 |
| J4 | 353.12/369.41 | 369.41/378.31 | 606.95/626.93 | 626.93/635.43 | 635.43/638.62 |

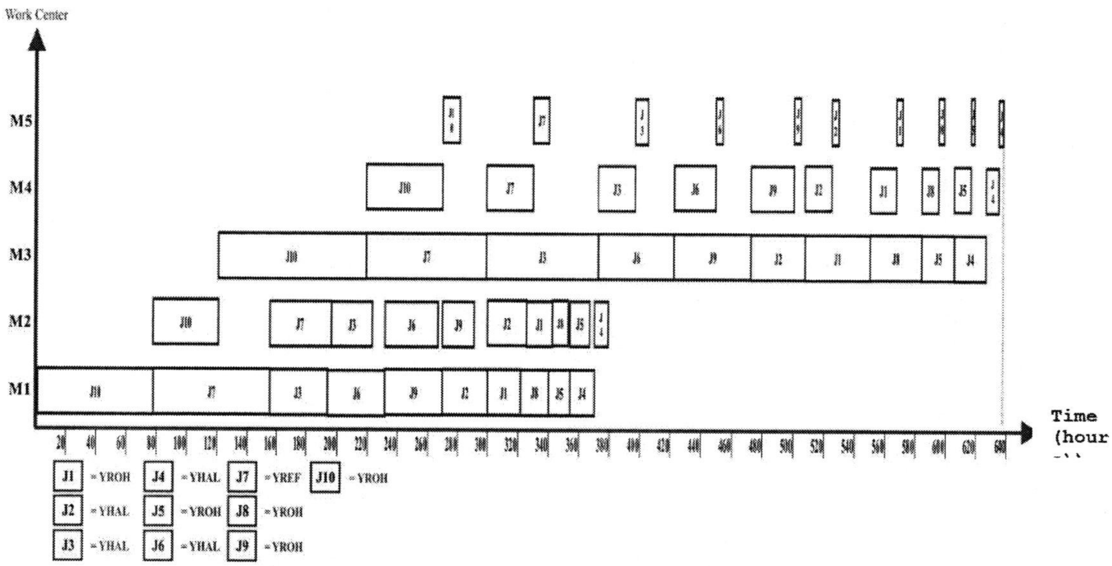


**Figure 3.** Job scheduling for May 2019 with the CDS method.

From the three scheduling methods, the smallest makespan value was found in the Heuristic Pour method. By using equation 3, it could be calculated the value of performance parameters (EI) that showed that the Heuristic Pour method was better than the scheduling method used by the company, namely FCFS.

$$EI_{Heuristic\ Pour} = \frac{610{,}36}{562{,}4} = 1{,}085$$

EI Heuristic Pour value > 1 which means that the performance of the Heuristic Pour scheduling method was better than the FCFS scheduling method. So the best production scheduling method for PT.X was the Heuristic Pour method with a makespan 562.4 hours (23 days) and the company could save time by 47.96 hours or 7.86% compared to the FCFS method.

## 4. Conclusion

The best production scheduling method for PT. XYZ was Heuristic Pour with total makespan of 562.4 hours (23 days) with the order of processing, namely Teratop Black HL-NF-01 (J4) - Orange GLN (J8) - Blue WW-2GS Successful (J5) - Terasiop Brown 2RFL (J1) - Teratop Black HL-BL (J2) - Terrain Orange HLNF (J9) - Teratop Black LF-01 (J3) - Terasil Red W-4BS (J10) - Novasol Blue RSMD (J7) - Teratop Blue HL-B (J6). By using the Heuristic Pour scheduling method the company could save time by 47.96 hours or 7.86%.

## 5. References

[1] Baker K R and Trietsch D 2009 *Principles of Sequencing and Scheduling* (America: John Wiley & Sons)
[2] Nasution A H and Prasetyawan Y 2008 *Perencanaan dan Pengendalian Produksi.* (Yogyakarta: Graha Ilmu).
[3] Ginting Rosnani 2009 *Penjadwalan Mesin.* 1st Edition (Yogyakarta: Graha Ilmu).
[4] Pour Hamid Davoud 2001 A New Heuristic for n_job m_machine Flow Shop Problem *J. Production Planning and Control,* **12**(7), pp. 648-653.
[5] Blazewich J, Ecker K H, Pesch E, Schdimt G and Weglarz J 2001 Second Edition (Berlin: Springer)
[6] N Steven 2009 *Production and Operations Analysis.* Sixth Edition. (Singapore: McGraw-Hill).

# Greedy Heuristics for the Maximum Covering Location Problem: A case study of Optimal Trashcan Location in Kampung Cipare – Tenjo – West Java

**Harummi Sekar Amarilies[1,4], A A N Perwira Redi*[1,5], Ilma Mufidah[2,6], and Reny Nadlifatin[3,7]**

[1]Department of Logistics Engineering, Universitas Pertamina, Kebayoran Lama, Jakarta, Indonesia
[2]Department of Industrial Engineering, Telkom University, Jl. Telekomunikasi Terusan Buah Batu Bandung-40257, West Java, Indonesia
[3]Department of Technology Management, Institut Teknologi Sepuluh Nopember, Surabaya, 60111, Indonesia

[4]sekaramarilies@gmail.com, [5]wira.redi@gmail.com,
[6]ilmamufidah@telkomuniversity.ac.id, [7]reny@its.ac.id
*Corresponding author

**Abstract.** Having a rapidly growing economy without proper waste handling system and infrastructure has made Indonesia one of the top contributors of plastic marine debris in the world: 3.22 million metric tons of mismanaged plastic waste per year. The mismanagement of waste is worse in the rural area where the number of available public trashcan is limited and the people have no idea where their trash actually ends up. In this research, a case study of waste management in Kampung Cipare - Tenjo, a small village 80 kilometers away from Jakarta, is conducted. The biggest purpose of the research is to educate people about sanitation and waste management, initiated by computing the most optimum number and locations of the public trashcans. The problem is addressed as Maximum Covering Location Problem (MCLP) to determine a set of facility locations that maximizes the total demand population served by the facilities within a prespecified maximum service distance. Two greedy heuristics algorithms: Greedy Adding Algorithm (GAA) and Greedy Adding with Substitution Algorithm (GAAS) are utilized to solve the problem. A sensitivity analysis is also done to check the result on the effect of trashcan number and maximum service distance to demand coverage. Out of the 26 location candidates, it is concluded that the final 10 locations are the most optimum in terms of demand coverage and service distance. The result of the research has been implemented in Kampung Cipare-Tenjo.
**Keywords**: Maximum Covering Location Problem, Greedy Heuristics, waste management, trashcan location optimization

## 1. Introduction

As one of the top five most populated countries with rapidly growing economy, Indonesia has become one of the biggest contributors to marine pollution with its plastic waste. The insufficient waste handling system and infrastructure has put Indonesia as the second biggest contributors to plastic debris (Jambeck, 2015). Plastic debris in the ocean endangers the life of marine organism as well as human's existence, since it circulates in the ocean, changes marine ecosystem, absorbs other pollutants and consistently remains in nearly all forms of marine life (McKinsey, 2017) in the food chain, with human in the top of the chain. The biggest supplier to this plastic marine debris is the uncollected plastic waste that ends up in the river and waterways, goes on to the ocean, forming the plastic gyres.

The waste generation rate in Indonesia is 0.52 kg per person per day with 11% of it is the plastic waste, resulting to 3.22 million metric tons mismanaged plastic waste which contributes up to 1.29 million metric tons of plastic marine debris to the world (Jambeck, 2015). Figure 1 displays the data of waste management in Indonesia, based on the Indonesian Statistics Bureau report of the year 2016 (Badan Pusat Statistik, 2016). It is shown that within the year of 2014, in daily basis, only 71.5% waste in major cities in Indonesia is managed correctly. Although the number is actually better compared the previous years, this data shows that the remaining 28.5% waste in major cities is not transported to the landfill. The chance is that the "untreated" waste was independently managed by the citizen. However, only a small number is treated in correct ways such as recycling and composting. Meanwhile, the rest is discarded in the river or other waterways and eventually ends up in the ocean.

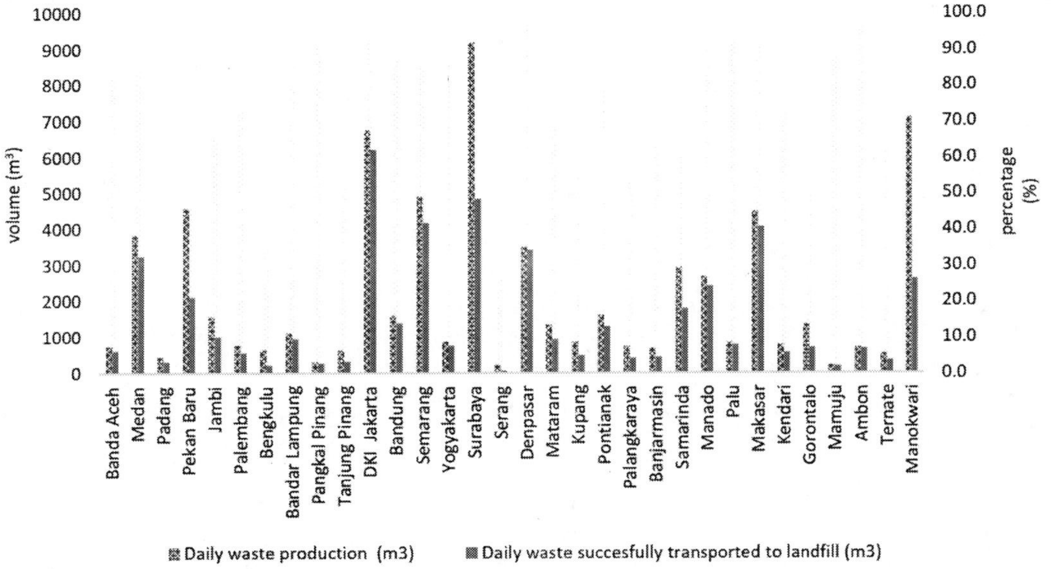


**Figure 1.** Waste production and the volume transported to landfills in major cities of Indonesia

The situation in the city perhaps indicates that the importance of waste management is not yet understood by most Indonesians. However, the situation gets worsen in the rural area where the number of available public trashcans is limited and the incorrect waste management is practiced. Not many people care about sanitation and the effects of trash mismanagement to their health. In this study, we took the case of Kampung Cipare - Tenjo, a small village in Bogor, West Java, that is located only 80 km away from the current capital city of Indonesia: Jakarta. Tenjo consists of around 120 households with average earnings around Rp 2,500,000 or US$180 a month and surely, waste management is not in the top priority yet. The people of Tenjo has no specialized area dedicated as a landfill, no trashcans outside the houses nor on the side of the road. In the public areas of Tenjo, scattered plastic bottles are swept away during rainy days and end up in the waterways, while the household trash is regularly burned. People of Tenjo still have no idea that the plastic trash in the waterway will end up in the ocean and the burned garbage produces dangerous gases harmful to their own health. To alleviate this situation, we are taking initiative to provide several public trashcans that optimally located in the vicinity of the houses in Tenjo. Moreover, a social engineering has been conducted to educate the importance of a proper waste management. However, due to fund limitation, the number of trashcans being donated is limited.

In this problem, there are candidates of trashcan locations that provide coverage to the nearby houses. A demand area is considered covered if it is within a predefined service distance. Thus, the basic nature of the considered problem is that of the maximum covering location problem (MCLP). The objective of MCLP is to determine a set of facilities locations that maximize the total demand population serviced by the facilities within a prespecified maximum service distance. The maximum covering location problem is a relaxation of set covering problem (SCP) when it is not feasible to cover all customers thus it needs to predefined the number of facilities being allocated. The application of MCLP application can be found in several topics such as emergency and military services, locating police stations, schools, plants, bus stops, and fire stations. However, as far as our knowledge, the application of MCLP for trashcan locations optimization is not yet being considered. The situation motivates this study to propose the use of MCLP for trashcan location optimization with a study case of Kampung Cipare – Tenjo. Moreover, due to the nature of the problem complexity is NP-Complete. For an NP-Complete problem, a heuristics or metaheuristics is perhaps the most feasible way to generate a solution. Thus, we proposed two greedy heuristics called Greedy Adding Algorithm (GAA) and Greedy Adding with Substitution Algorithm (GAAS) to solve the problem in a reasonable computational time.

The remaining content of this paper is organized as follows: In Section 2, we describe the methods of research that include the formulation of the maximum covering location problem for the trashcan location optimization and the proposed greedy heuristics for solving the problem. Section 3 is about the computational experiments result and discussion, while section 4 is the conclusion and future research. Section 5 is the Acknowledgement and the References can be found in section 6.

## 2. Methods

### *2.1 Maximum Covering Location Problem for Trashcan Location Optimization*

In this study, the formulation of the MCLP is based on the model proposed by Church and ReVelle (1974). First, there is known demand node and candidate facility location denoted as set $I$ and $J$, respectively. Demand node $i \in I$ associated with demand $h_i$. A binary coefficient $a_{ij}$ takes value of 1 if a facility at candidate $j \in J$ can cover demands at node $i \in I$, otherwise it is 0. The decision variable is a binary decision $X_j$ equal to 1 if we locate a facility at $j \in J$, otherwise 0. Based on this decision variable, the node being covered can be decided and it is noted as $Z_i$. The node being covered shown by the value of $Z_i$ equals to one, otherwise 0. The number of facilities opened is restricted to a number of $P$. The MCLP formulation then adopted with the problem context which is optimally locating trashcan to maximize the demand coverage. The model is formulated as follows.

$$\text{Maximize} \sum_{i \in I} h_i Z_i \tag{1}$$

$$Z_i \leq \sum_{j \in J} a_{ij} X_j \qquad \forall i \in I \tag{2}$$

$$\sum_{j \in J} X_j \leq P \tag{3}$$

The objectives (1) is to maximize the number of demands covered based on the decision variable. Constraints (2) state the demand $i \in I$ cannot be covered unless at least one facility that covered the demand note is selected. Constraints (3) restricts the number of facilities being used is no more than $P$.

*2.2 Greedy Heuristics Algorithms*

Basically, the greedy adding algorithm with substitution procedure (GAAS) is trying to select the best facility being located at each step of the algorithm. The algorithm basically consists of four steps. The first step is to find the candidate site that covers the most uncovered demand, then locating the facility at that site. A substitution procedure is conducted only if more than one facility has been located. After that, the covered demands are being updated. The algorithm terminates after a number of predefined sites have been located or all demands have been covered. A more detail about the substitution algorithm is that it is basically considers removing every selected candidate site and replacing it with every non-selected candidate site. The algorithm then selects the best selected site after the swap procedure. The variants without the substitution procedure is called Greedy Adding Algorithm (GAA). The implementation of these algorithms is based on Daskin and Maass (2019). The flowchart of the algorithm can be seen in Figure 2.

## 3. Computational Experiments Result and Discussion

The proposed greedy algorithm is implemented in C++ programming language using Microsoft Visual Studio 2019 on a PC with specification Intel Core i7 3.6Ghz – 16GB RAM. The following subsection described the dataset and the numerical results.

*3.1. Dataset*

The proposed greedy adding with substitution algorithm was evaluated using a real-case dataset based on the condition of Kampung Cipare – Tenjo. Demands nodes are assumed from the 120 households, with 4-5 people in every house the total demand is 569 people. It is assumed that the service distance coverage is 60 meters. The trashcans can be selected among 26 location candidates that approved by the community. Due to the budget limitation, only 10 trashcans can be donated.

*3.2. Computational Results*

Figure 3 shows the illustration of the trashcan locations optimization result in Kampung Cipare – Tenjo. The results showed that there are at least ten trashcans needed to fully cover the demand. Moreover, a more detailed experiments are conducted to compare the computational results between Greedy Adding Algorithm with Substitution procedure (GAAS) and Greedy Adding Algorithm without substitution procedure (GAA). It is known that the number of facilities being used have impact on the final results. Thus, a sensitivity analysis regarding a different number on parameter P is conducted.

The computational results in Table 1 show the comparison between GAAS and GAA in terms of objective function value (demand coverage) and computational time. The results shown that with the given dataset, GAAS result is able to select the facilities that cover all the demands by using only 10 facilities, on the other hand, GAA was not able to show the similar performance. The average percentage difference between demand coverage obtained by GAAS and GAA is 1.6% that shows GAAS perform better than GAA. However, the computational time of GAA is 0.06s in average, considered smaller compared to GAAS that having average computational time of 0.5s.

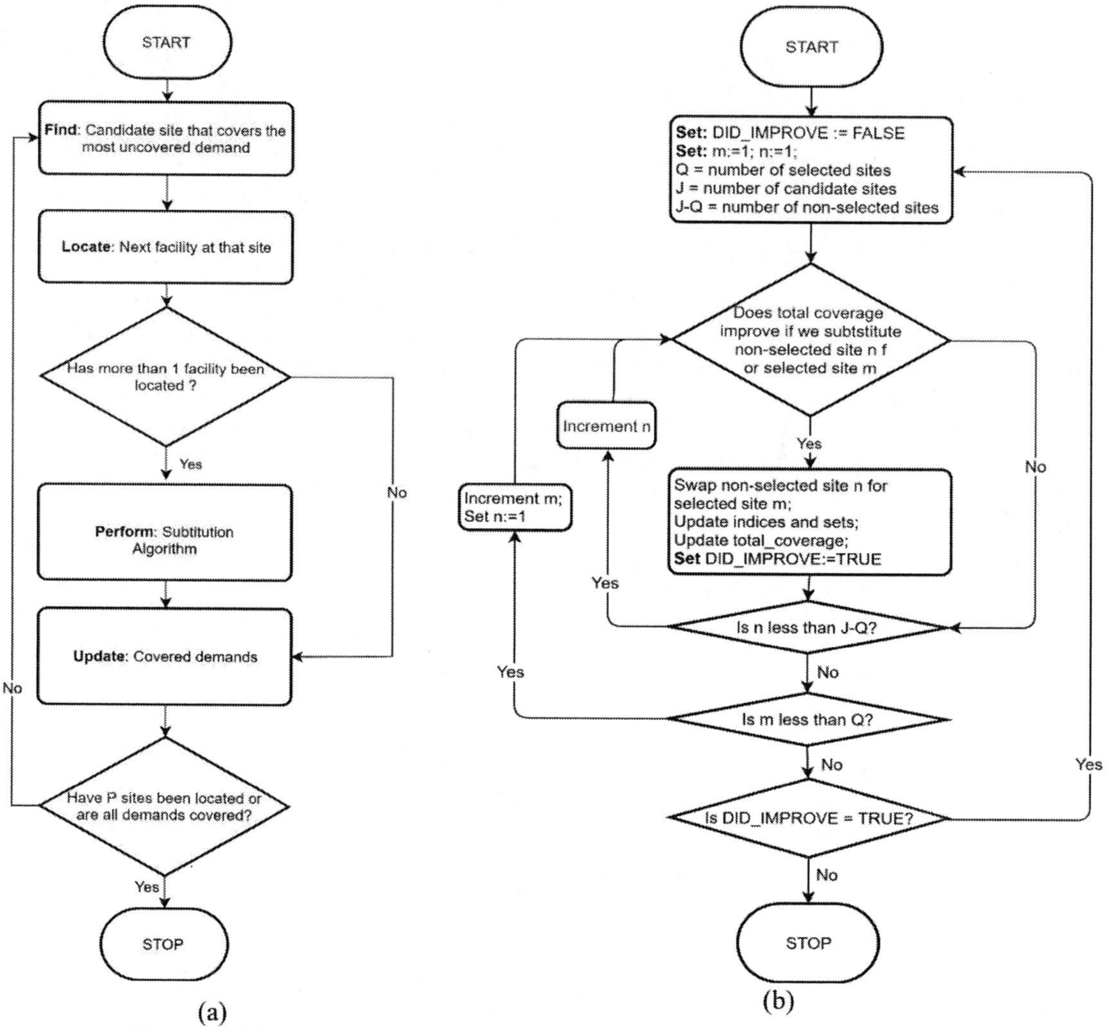


**Figure 2.** (a) Greedy adding with substitute algorithm (b) substitute algorithm

**Table 1**. Results comparison between Greedy Adding Algorithm with Substitution procedure (GAAS) and Greedy Adding Algorithm (GAA)

| Facility | GAAS | | | GAA | | | Diff. (%) |
|---|---|---|---|---|---|---|---|
| | Demand Covered | Percent (%) | Time (s) | Demand Covered | Percent (%) | Time (s) | |
| 1 | 156 | 27.42 | 0.01 | 156 | 27.42 | 0.01 | 0.00 |
| 2 | 303 | 53.25 | 0.29 | 303 | 53.25 | 0.09 | 0.00 |
| 3 | 388 | 68.19 | 0.33 | 388 | 68.19 | 0.03 | 0.00 |
| 4 | 438 | 76.98 | 0.38 | 438 | 76.98 | 0.08 | 0.00 |
| 5 | 481 | 84.53 | 0.43 | 456 | 80.14 | 0.03 | 5.48 |
| 6 | 511 | 89.81 | 0.49 | 500 | 87.87 | 0.09 | 2.20 |
| 7 | 534 | 93.85 | 0.53 | 522 | 91.74 | 0.03 | 2.30 |
| 8 | 554 | 97.36 | 0.76 | 535 | 94.02 | 0.06 | 3.55 |
| 9 | 565 | 99.30 | 0.83 | 560 | 98.42 | 0.09 | 0.89 |
| 10 | 569 | 100.00 | 0.92 | 560 | 98.42 | 0.09 | 1.61 |
| Average | | | 0.50 | | | 0.06 | 1.60 |

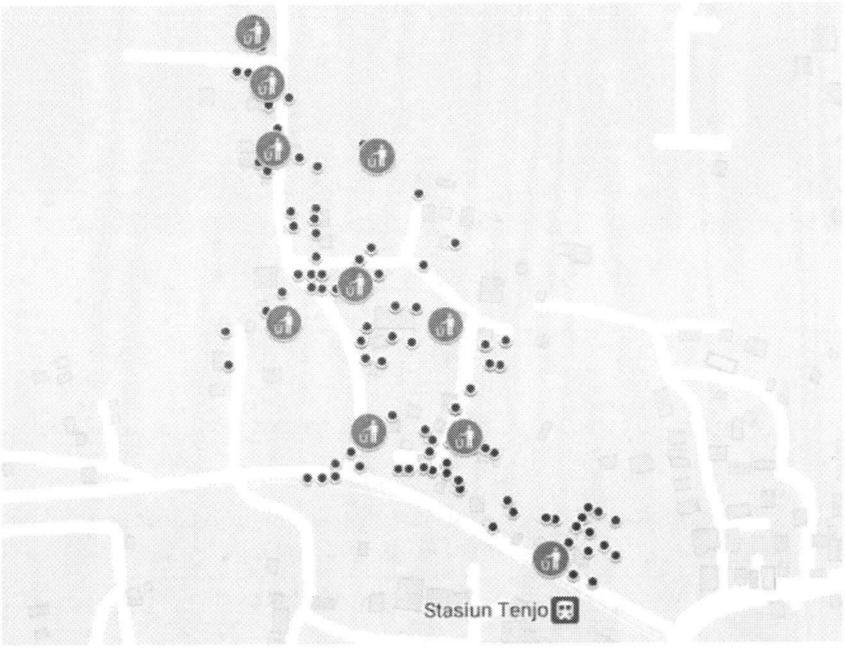


**Figure 3.** Illustration of the trashcan location optimization result in Kampung Cipare – Tenjo

A further analysis regarding the number of selected facilities and coverage service distance is conducted. This sensitivity checking is done by generating the illustrations in Figure 4 and Figure 5 based on the result in Table 1. The figures show that along with the increasing number of demand facility being selected, the percentage of coverage demand is steadily increasing. Meanwhile, with increasing number of facilities, the additional demand being covered is reduced. It shows that there is a possibility of one demand location that might be covered by more than one facility. Furthermore, the effect of increasing coverage service distance to the number of demands covered is shown in Table 2. The results show that increasing coverage distance by twice of the original distance will result to smaller number of facilities that should be provided, only four trashcans are needed to fulfil all demands. This shows that there is a trade-off between the increasing service level of providing more facilities and using efficient budget by reducing the service level of coverage distance.

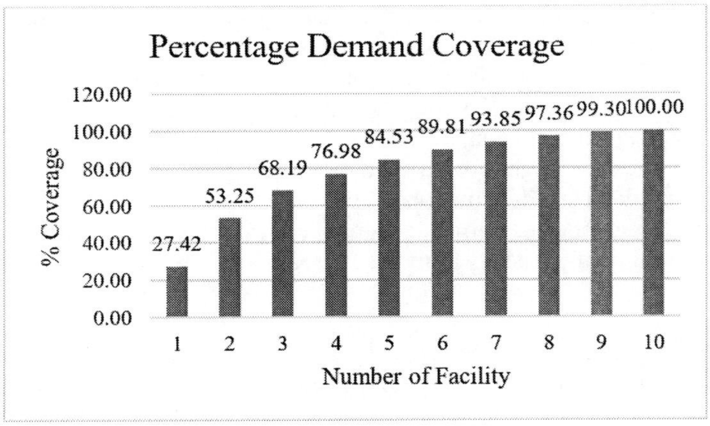


**Figure 4** Analysis on the effect of increasing number of selected facility to the percentage demand coverage

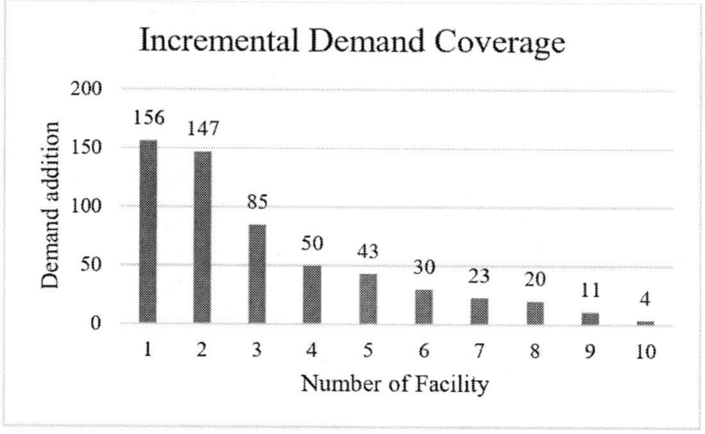


**Figure 5.** Analysis on the effect of increasing number of selected facilities to the incremental demand coverage

**Table 2.** The effect of increasing coverage service distance to the number demand covered in the solution of GAAS with p=4.

| Coverage Distance | Demand Covered | Percentage (%) |
|---|---|---|
| 60 | 438 | 76.98 |
| 80 | 497 | 87.35 |
| 100 | 539 | 94.73 |
| 120 | 569 | 100.00 |

## 4. Conclusions and Future Research

This study proposes the use of maximal covering location problem for optimizing the trashcan locations, with a case study in Kampung Cipare – Tenjo, West Java, Indonesia. Due to its complexities, two greedy heuristics have been proposed, namely Greedy Adding Algorithm (GAA) and Greedy Adding with Substitution Algorithm (GAAS). Numerical experiments are conducted to show the effectiveness of those two algorithms in solving the problem. It is shown that the use of GAAS is able to obtain better demand coverage compared to GAA with the average percentage difference of 1.46%. However, GAA can perform faster with the average of computational time of GAAS and GAA of 0.5s and 0.06s respectively. The result also shown that all ten provided trashcans can cover all the demand locations. Moreover, the analysis on number of selected facilities and coverage service distance shows the increasing number of facilities has impact on the demand coverage. It is also shown that there is a condition where one demand can be coverage by more than one facility. The analysis also shows that increasing distance service coverage can lead to a smaller number of facilities being used. Finally, the application of this result has been successfully implemented in Kampung Cipare – Tenjo.

Further research direction might consider the optimization of the waste collection to the landfill. Moreover, a more complex real-life situation can also be considered such as different cost of locating the facility, forecasting number of demands from the attribute of the family, and constraint of the minimum distance location of trashcans from the houses.

## Acknowledgement

The authors gratefully acknowledge the support that given by Universitas Pertamina that funded the research under UPREACH 2019 collaboration grant. Furthermore, we also gratefully thank our UPREACH 2019 team that helped with data collection and activities conducted for this research.

## 5. References


[1] Badan Pusat Statistik 2016 Statistik lingkungan hidup Indonesia 2015/2016. *Jakarta: Badan Pusat Statistik.*

[2] Chung, Chen-Hua 1986 Recent applications of the maximal covering location planning (MCLP) model. *Journal of the Operational Research Society,* **37**(8), 735-746.

[3] Church, Richard, and ReVelle, Charles 1974 The maximal covering location problem. *Papers in regional science,* **32**(1), 101-118.

[4] Damanhuri, Enri, and Padmi, Tri 2010 Pengelolaan sampah. *Diktat Kuliah Teknik Lingkungan Institut Teknologi Bandung, TL-3104,* 5-10.

[5] Daskin, Mark S, and Maass, Kayse Lee 2019 Location Analysis and Network Design *Operations, Logistics and Supply Chain Management* (pp. 379-398): Springer.

[6] Jambeck, Jenna *et al.* 2015 Marine Pollution: Plastic waste inputs from land into the ocean Science **347** 768-770

[7] McKinsey Center for Business and Environment 2017 Stemming the Tide: Land-based Strategies for a Plastic-free Ocean. *Ocean Conservancy,* 1-17.

[8] Megiddo, Nimrod, Zemel, Eitan, and Hakimi, S Louis 1983 The maximum coverage location problem. *SIAM Journal on Algebraic Discrete Methods,* **4**(2), 253-261.

[9] Murray, Alan T 2016 Maximal coverage location problem: impacts, significance, and evolution. *International Regional Science Review,* **39**(1), 5-27.

[10] Redi, AAN Perwira, Maghfiroh, Meilinda FN, and Vincent, F Yu 2013 *Discrete particle swarm optimization with path-relinking for solving the open vehicle routing problem with time windows.* Paper presented at the Proceedings of the Institute of Industrial Engineers Asian Conference 2013.

IOP Conf. Series: Materials Science and Engineering **847** (2020) 012008 doi:10.1088/1757-899X/847/1/012008

# Determining optimum eco paving block compositions by using factorial design method

**A T Purwandari[1,2], A Ratnamirah [1,3], N Parwati[1,4], and W N Tanjung[1,5]**

[1]Industrial Engineering Department, Faculty of Science and Technology, Universitas Al Azhar Indonesia, Jakarta, Indonesia
[2]aprilia@uai.ac.id, [3]alikaaratnamirah@gmail.com, [4]niken.parwati@uai.ac.id, [5]widya@uai.ac.id

**Abstract.** Some plastic waste processing into economically valuable recycle products has been done. By using a plastic waste processing machine, a variety of products made from plastic waste can be made, one of which is a paving block that is commonly used as building material. This paving block is called as eco paving block to show the ecology side by using plastic waste as raw material. It has been produced with various compositions and type of plastics, but the optimal composition is not yet known to produce good quality. So, this experiments were carried out using the factorial design as one of the Design of Experiment (DOE) methods to determine the effect of type of plastics and compositions ratio to the eco paving block compressive strength and also it can be obtained optimal composition of the eco paving block. The result showed that the ratio of compositon is 2:3 of sand and plastic can produce optimum eco paving block based on compressive strength. This paving block has an average compressive strength 9.91 MPa. It can be classified into D quality classification that can be used for yards, parks, and others.
**Keywords**: factorial design method, eco paving block, plastic, compressive strength

## 1. Introduction

The amount of plastic waste that is not managed properly can potentially damage the ecosystem. Based on World Economic Forum (WEF) data, only 5% of plastic are effectively recycled, while the other 40% ends up in landfills, and the rest ends in ecosystems such as the ocean. Indonesia got second ranks as the largest contributor of plastic waste in the world with 5.4 million tons per year [1]. The principles of waste management are reduction in using of plastic products such as platics bag, platics packaging, etc., optimizing the use of plastic can be reused, producing or processing plastic into a product that is useful and has economic value, replace or avoid the use of disposables with items that can be used repeatedly, and respect to build a sense of caring and loving the environment so that, can get used to and wise before choosing or using a product.

Therefore, from the many methods, technologies, and ways to recycle plastic waste into a new product, the use of plastic waste processing machine is one of the technologies that is very effective and efficient in producing recycle products from plastic waste. The product that can be made from processing plastic waste is paving block. Basically, paving block made of cement and sand. Some communities or companies have produced eco paving block with various quality. Based on SNI 03-

0691-1996, the classification for paving block is measured based on the compressive strength as shown in Table 1.

**Table 1.** Quality standard of paving block classification [2]

| Quality | Utility | Compressive Strength (MPa) | |
|---|---|---|---|
| | | Average | Minimum |
| A | Road Hardener | 40 | 35 |
| B | Parking Lot | 20 | 17 |
| C | Pedestrian | 15 | 12.5 |
| D | City Park | 10 | 8.5 |

Actually, some experiments to determine best composition of paving block by using plastic mixture has conducted with varoius material, parameters, and methods. Previous research used variation of plastic type ratio as parameter. The result showed that the best comparison of ratio is 1 of body of plastic bottle,1 of plastic bag, 4 of cap of plastic bottle [3]. Other research found that optimum compressive strength occured in the addition of 0.5% *polietilena tereftalat* fiber and it increased of 42.23% when compared to normal paving block [4]. While, the other study proved that the value of paving block compressive strength can be increased by using Polypropylene (PP) plastic waste as an aggregate substitution in the manufacture of paving bloc [5]. The paver block with the addition of PP fiber at 0.15% showed better strength performance compared to conventional pavement [6]. Meanwhile in other research showed the utilization of High Density Polyethylene (HDPE) for making paving block. At normal temperatures, it can be very hard and suitable for making paving block [7]. Eco paving block made from HDPE wastes and sand composite generates higher performance relative to the conventional one [8].

Based on this idea, this experiment will be conducted to investigate the effect of composition ratio and type of plastic to the quality of eco paving block and also it can be obtained optimum composition of eco paving block by using experimental method of Design of Experiment (DOE) based on factorial design. Good quality eco paving block are paving block that have high compressive strength value (MPa) and low absorption value (percentage of water absorption).

## 2. Methods

### *2.1. Experimental Design*

Design of Experiment (DOE) is a research conducted to study or discover something about existing processes or compare the effects of several conditions on a phenomenon [9]. Generally, DOE is used in designing an experiment to be able to find out the response and characteristics of a factor and element to a test variable [10]. Factorial design is a kind of DOE method usually that used to produce more objective experiments. In factorial design method, every possible combination of all factors and levels in the experiment to determine the effect of interactions between factors in factorial experiments could be investigated.

Factors and levels in this study obtained by interviewing the experts and did literature review from previous research. In this experiment, it is used 2 factors consist of types of plastic at 2 levels (PP and HDPE) and ratio of sand and plastic at 3 levels (250:250; 200:300; 300:200). There are 3 replications for each of the 6 ($2^1$ x $3^1$) level factor combinations. So, this study has 18 (6 x 3) of total treatments. Then the data tested by using factorial design to determine influence each factor to the response variable. Hypothesis formulation for each factor to be analyzed is as follows :

$H_{01}$ : The types of plastic has no effect significantly on the compressive strength
$H_{11}$ : The types of plastic has effect significantly on the compressive strength
$H_{02}$ : The composition ratios has no effect significantly on the compressive strength
$H_{12}$ : The composition ratios has effect significantly on the compressive strength

$H_{03}$ :The interaction between type of plastic and composition ratio has no effect significantly on the compressive strength

$H_{13}$ :The interaction between type of plastic and composition ratio has effect significantly on the compressive strength

In order to increase accuracy in experiments, usually it is conducted randomization. Randomization is the design technique used to guard againts such a nuisance factor [11]. By using Minitab Statistical Software, experimental design of this study can be seen in Table 2.

**Table 2.** Experimental Design

| Std Order | Run Order | Replica-tions | Types of Plastic | Composi-tion Ratio | Std Order | Run Order | Replica-tions | Types of Plastic | Composi-tion Ratios |
|---|---|---|---|---|---|---|---|---|---|
| 9 | 1 | 2 | HDPE | 3 | 3 | 10 | 1 | HDPE | 3 |
| 8 | 2 | 2 | HDPE | 2 | 5 | 11 | 1 | PP | 2 |
| 11 | 3 | 2 | PP | 2 | 6 | 12 | 1 | PP | 3 |
| 7 | 4 | 2 | HDPE | 1 | 17 | 13 | 3 | PP | 2 |
| 12 | 5 | 2 | PP | 3 | 13 | 14 | 3 | HDPE | 1 |
| 10 | 6 | 2 | PP | 1 | 15 | 15 | 3 | HDPE | 3 |
| 1 | 7 | 1 | HDPE | 1 | 16 | 16 | 3 | PP | 1 |
| 2 | 8 | 1 | HDPE | 2 | 14 | 17 | 3 | HDPE | 2 |
| 4 | 9 | 1 | PP | 1 | 18 | 18 | 3 | PP | 3 |

### *2.2. Performing the Experiment*

This experiment is focused on paving block made by plastic waste and sand. The choice of material is an effort to reduce the amount of plastic waste in the environment that is processed by using an plastic waste machine. The size of paving block for this experiment has length is 13 cm, width is 13 cm, and height is 6 cm with a modified cube base shape, as the shape of the paving block mold design shown in Figure 1.

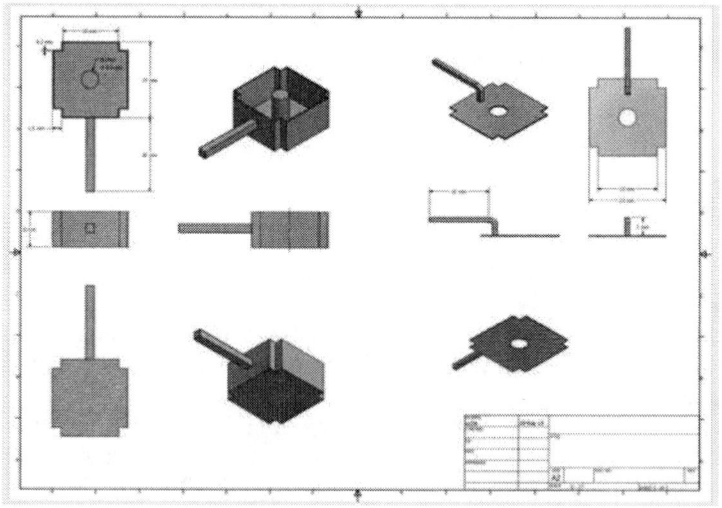

**Figure 1.** Paving block mould design

In conducting experiments, 18 of paving block samples were taken. The making of plastic paving block uses the help of plastic waste processing machine, as shown in Figure 2 and 3.

**Figure 2.** Shredder machine

**Figure 3.** Extrusion machine

Based on Figure 4 and 5, there are two types of plastic waste processing machine used in making eco paving block with different uses and functions. Shredder is used to destroy and refine plastic waste, while extrusion machine is used to melt the results of chopping and as a medium used in the process of mixing between plastic seeds with fine aggregate (sand). Figure 4 showed the eco paving block that produced by using the machines.

**Figure 4**. Plastic paving block

The variable response of this research is compressive strength of the eco paving block. The process compressive strength testing of eco paving block was carried out in the Concrete Laboratory, using CTM (Compression Testing Machine) as in Figure 5.

**Figure 5.** Compression testing machine

### *2.3. Statistical data analysis*
Analysis of variance is important analysis techniques to the factorial design method. Analysis of variance is a technique used as a tool to analyze the average difference between groups in an experiment or research by conducting a hypothesis test to test the interaction between one factor of the

sample taken in an experiment to know the interaction between variables and their effect on a treatment. In this research has been used statistical software to compute p-value. The α-value has been selected is 0.05. If *p-value* is below *α-value*, there are factors or factor interactions are significantly affects the response variable. To be able to conduct the analysis of variance, the data must be normally distributed and homogeneous [12].

#### 2.3.1. Model Adequacy Checking

In the analysis of variance, the uses of normal probability plot of the residuals to check the assumption of normality is usually more effective and straight forward [10]. If the underlying error distribution is normal, this plot will resemble a stright line. The other plot that used for model adequacy checking is plot of residuals versus fitted, plot of residuals versus, and histogram.

## 3. Result and Discussion

### 3.1. Compressive strength result

The results of compressive strength by using the compression testing machine to all eco paving block with all variation of treatment can be seen on Table 3.

**Table 3.** Compressive strength test results

| Treatments | Types of Plastic | Composition Ratio | CompressiveStrength (MPa) |
|---|---|---|---|
| 1 | HDPE | 3 | 7.29 |
| 2 | HDPE | 2 | 9.99 |
| 3 | PP | 2 | 9.65 |
| 4 | HDPE | 1 | 8.42 |
| 5 | PP | 3 | 7.24 |
| 6 | PP | 1 | 8.4 |
| 7 | HDPE | 1 | 8.86 |
| 8 | HDPE | 2 | 9.87 |
| 9 | PP | 1 | 8.25 |
| 10 | HDPE | 3 | 7.51 |
| 11 | PP | 2 | 9.20 |
| 12 | PP | 3 | 7.52 |
| 13 | PP | 2 | 9.99 |
| 14 | HDPE | 1 | 8.25 |
| 15 | HDPE | 3 | 7.41 |
| 16 | PP | 1 | 8.08 |
| 17 | HDPE | 2 | 9.87 |
| 18 | PP | 3 | 7.52 |

From Table 3 can be seen that the highest value of compressive strength is 9.99 MPa. It is obtained from treatment number 2 and 13. Both of treatments used the same compositions, these are 200 gram of sand and 300 gram of plastic. Therefore, the types of plastic are different.

### 3.2. Ststistical analysis

To conduct ANOVA tests, the data must be normally distributed and homogeneous in nature. The normality assumption has conducted by using residual plots as shown in Figure 6. Based on residual plot, the residual value is obtained that follows a linear line, so the residual data can be assumed to be normally distributed and the requirements for the ANOVA test can be fulfilled. For the histogram, a pattern that follows a normal curve is obtained, so that it can be said that the data is normally distributed. Versus fits graph is used to see whether residuals have a constant value or not. The pattern produced on the garfik versus fits does not form a particular pattern and tends to be constant and

IOP Conf. Series: Materials Science and Engineering **847** (2020) 012008 doi:10.1088/1757-899X/847/1/012008

identical to either the top or bottom of the zero line. As for versus order, it is used to describe whether the residuals are free from one another. The graph versus the order obtained shows a random pattern at the center of the line or the zero line on the y axis or does not have a certain pattern, so it can be said that the residual data are mutually independent. From all the graphs on the residual plot, it is obtained that the residual data meet the normal distribution assumption.

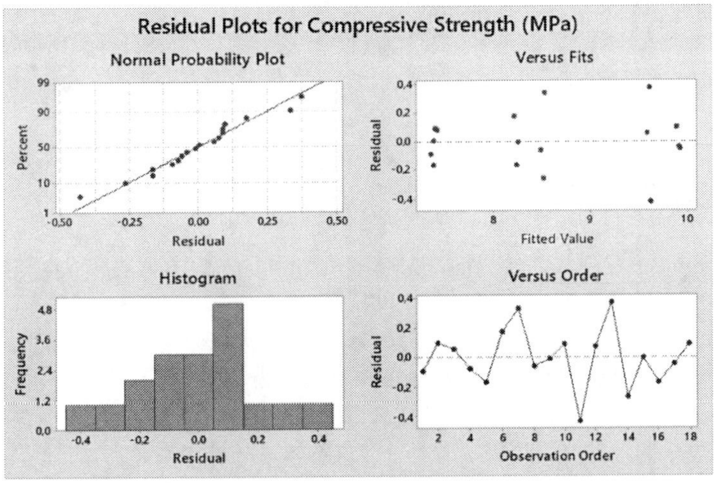


**Figure 6.** Residual plots

Based on homogeneity test, the variance of each data is similar. The *p-value* is 0.168 or greater than the alpha value of 0.05. The data has fulfilled the analysis of variance assumptions. Table 4 shows that there is no significant effect between types of plastic on the compressive strength. Meanwhile, the composition ratio have a significant effect on the compressive strength, because the *p-value* obtained is 0,000 or less than the alpha value, reject $H_{02}$, which means composition ratio has a significant effect on the compressive strength. But the interactions of factor levels has no effect significantly on the compressive strength. It can be showed through the *p-value* is 0.508. It was greater than an alpha value, reject $H_{13}$.

**Table 4.** Analysis of variance result

| Term | p-value |
|---|---|
| Types of Plastic | 0.164 |
| Composition Ratio | 0.000 |
| Types of Plastic* Composition Ratios | 0.508 |

So, the main effect that can influence the compressive strength of the paving block is composition ratio as shown in Figure 7. It is known that the material composition ratio factor has an influence between each level of composition when compared to the type of plastic factor, with a level 2 material composition that has a very significant effect on the compressive strength of eco paving block when compared to the composition of level 1 and 3. The greater slope, the more significant effect on the response variable.

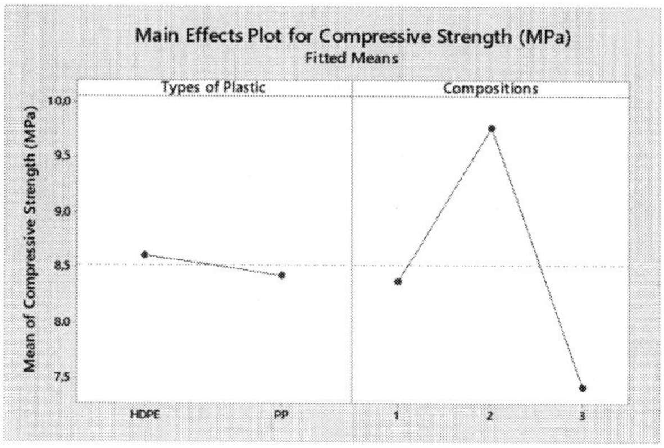


**Figure 7.** Main effects plot

Interactions between factors are also important to identify because interactions between these factors can influence the response variables, as in Figure 8.

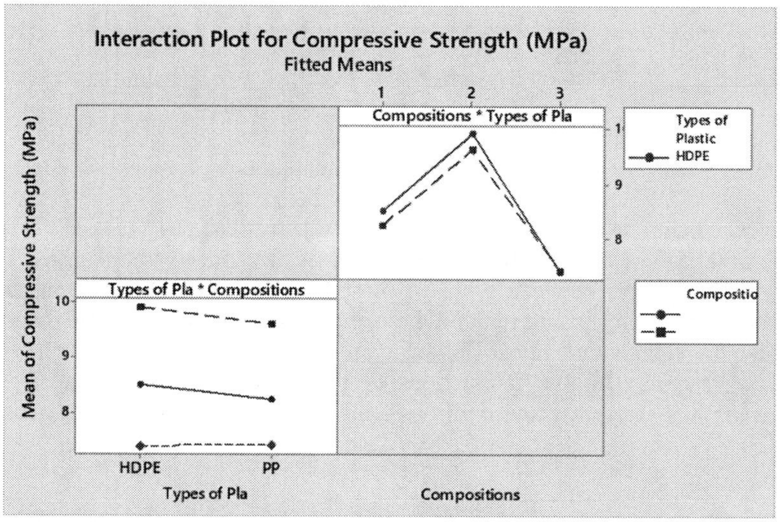


**Figure 8.** Interaction plot

Based on the interaction plot, it is known that to achieve optimal compressive strength on plastic paving block, used plastic with HDPE type with a composition of level 2 materials because the interaction plot for the type of plastic and the material composition ratio is located at the highest value among the other material composition ratios. So, it can be assumed that the optimum eco paving block composition ratio is the eco paving block made by 200 gram of sand and 300 gram of plastic with the optimum average value of compressive strength is 9.91 MPa. It can be determined the eco paving blocks can be categorized into quality D paving blocks based on paving block quality standard as shown in Table 1.

The result of the experiment showed that in the making of eco paving block by using plastic waste mixture, the quantity of plastic must be greater than the sand to produce the better quality of eco paving block. The type of plastic that used in this research only two type of plastic, these are HDPE and PP. It need to be investigated of use the other type of plastic for making eco paving block, esspecially thermoplastic, such as Polyethylene Terephthalate (PET), Low Density Polyethylene (LDPE), and so on.

## 4. Conclusion

Based on data processing and analysis that has been done, the conclusions can be obtained as follows:

1. Factor that has an influence significantly on the eco paving block compressive strength is the composition ratio of material, meanwhile the type of plastic has no effect on the compressive strength.
2. The optimum composition for the eco paving block is 2:3 of sand and plastic which can produce an average compressive strength is 9.91 MPa.

**Acknowledgment**
This work is supported by Greant Research Prime founded by LP2M Universitas Al Azhar Indonesia

## 5. References

[1] Jambeck J R, Geyer R, Wilcox C, Siegler TR, Perryman M, Andrady A, Narayan R and Law KL 2015 Plastic waste inputs from land into the ocean. *Science* **347,** 768–771

[2] The Indonesian Standardization Body 1996 *Paving block SNI 03-0691-1996* (Jakarta: The Indonesian Standardization Body)

[3] Frasman A S and Tarigan J 2013 Utilization of Plastic Bottle Waste as Eco-friendly Material (Economic Plastic Fiber) Paving Block with an Environmental Friendly Concept with Compressive Strength Test, Shock Test and Water Absorption. *Journal of Civil Engineering* **2** (North Sumatra: University of North Sumatra)

[4] Burhanuddin B and Darmanijati MRS 2018 Utilization of Plastic Waste for Raw Material in Making Paving Block. *Journal of Environmental Engineering* **18**

[5] Kusuma G A 2019 Utilization of Poly Propylene Plastic Chopped as Aggregate Substitution on Paving Block Available at: https://dspace.uii.ac.id/ (Yogyakarta: Universitas Islam Indonesia)

[6] Ananthi A, Sulthan M S and Vinitha V 2017 Study on the Effects of Polypropylene Fiber in Concrete Paver Blocks. *International Journal of Construction Engineering and Planning* **3,** 32-38

[7] Sari K I and Nusa A B 2019 Utilization of HDPE (High Density Polyethylene) Plastic Waste as a Materials for Paving Block. *Main Technical Bulletin* **15**

[8] Ingabire D, Ntihemuka F, Mugabo G, Isabane R S and Turatimana T 2018 Recycling High-Density Polyethylene (HDPE) into Construction Materials as a key step in Plastic Waste Reduction: Case of Kigali City. *Rwanda Journal of Engineering* **1**

[9] Coleman D E and Montgomery D C 1993 A Systematic Approach to Planning for a Designed Industrial. *Experiment Technometrics* **35,** 1-12

[10] Coleman D E and Montgomery D C 1997 *Design and Analysis of Experiment.* 4th Edition (New York: John Wiley&Sons, Inc)

[11] Coleman D E and Montgomery D C 1991 *Design and Analysis of Experiment.* 4th Edition (New York: John Wiley&Sons, Inc)

[12] Kirk R E 1995 *Experimental design: Procedures For The Behavioral Sciences.* 3rd Edition (Belmont, CA, US: Thomson Brooks/Cole Publishing Co)

# Design of Facility Location for new model of Medical Pharmaceutical Refrigerator Production Area on PT. XYZ

**R Prasetyani[1,2], A Siti [1,3] and G A Timang[1,4]**

[1]Industrial Engineering Department, Pancasila University, Jl. Srengseng Sawang, Jagakarsa, Jakarta Selatan, Indonesia
[2]rini.prasetyani@univpancasila.ac.id , [3]auliasiti12@gmail.com, [4]gitatimang20@gmail.com

**Abstract.** PT. XYZ is a manufacturing company that produces medical devices. The company has planned to develop a new product called Medical Pharmaceutical Refrigerator (MPR). Based on research, the company has provided an area for MPR production of 648 m2. After analyzing the process, 5 areas are used in conjunction with other products, but one area called Polyurethane Injection produces a product flow mismatch in the PU area. This problem is solved by designing a layout using the Industrial Facilities method and Systematic layout planning with the Product layout approach. Based on the results of the study, proposed two alternative layouts which then analyzed the weight of material transfer using the From-to-chart method. Layout design 1 has a weight transfer material of 9322.09 per day, while the proposed layout design 1 produces a weight transfer material of 5750.5 per day. So, the MPR production layout is determined using the layout design 2. By determining this layout, a biomedical warehouse area was moved at of 49.42 m2.
**Keywords**: Facilities layout, Facilities Industry Method, Systematic Layout Planning, Product Layout, From to Chart.

## 1. Introduction

The design of facility layout is one of the things that need to be considered because it is closely related to the production process. The design is an activity that includes the utilization of the area available for the placement of machinery, the smooth movement of material transfer, material storage, and worker determination. A good layout will provide efficient material flow, shorter transfer distances, minimum material transfer costs, and shorter production processing times. Therefore, to achieve an optimal production process, design and layout of the production floor and facilities are required in a company. PT XYZ is a manufacturing company that manufactures medical devices and plan to add a new product called the Medical Pharmaceutical Refrigerator (MPR). Until now, the company has not planned the facility layout design for MPR products, but the area for production has been provided for 648 m2. After analyzing the process, 5 areas are used in conjunction with other products that are already running, but one area called Polyurethane Injection produces a mismatch in the product flow in the PU area. Based on this description, this research will analyze the determination of the MPR production area requirements with the available area and follow the production flow that has been determined at 5 existing work stations. And how to determine the layout of facilities or work stations to best meet the MPR production capacity.

## 2. Methods

Below are the steps were undertaken for the study:

1. Conduct field studies to understand the actual problems that occur in the field and conduct literature studies to find out how to solve problems that occur in the object of research.
2. Collect data that supports the problem-solving process in research. Primary data is the data of direct observation in the production area and interviews with related parties, and secondary data is data that has been provided by the company including product characteristics, production processes, number of production requests and machine sizes.
3. Designing facility layout to optimize MPR production output using systematic layout planning method. The stages in designing the facility layout can be explained following the sequence of activities developed by Richard Muther, namely through an approach known as Systematic Layout Planning (SLP). [3] The stages in the making the plan are: (a) Activity Relationship Chart, (b) Activity Template Block Diagram, (c) making two proposed alternative layout designs, and (d) evaluating the layout
4. Propose two alternative layouts then determine the best layout using the from-to-chart method. From-to-chart is a conventional method that is often used for layout planning. The result of From to chart is the weight of material transfer between the amount of material moved and distance traveled. The smallest material transfer weight is the best alternative design because the distance and the amount transferred is better and more efficient.[1] The technique used for distance measurement is the Rectilinear method, which is the distance measured following a perpendicular path. Rectilinear distance measurements used the following notation:
$$d_{ij} = |x_i - x_j| + |y_i - y_j| \tag{1}$$
After that, enter the multiplication results of the amount of material moved and the distance of movement at each work station. Then the weight value of the material transfer will be obtained.
5. Inferring the results obtained in this study following the initial purpose of the study, namely to determine the facilities, area requirements and facility layout planning and the best layout obtained based on the overall results of the study. As well as advice to companies or further research development.

## 3. Result and Discussion

### *3.1. Product Characteristics*

Medical Pharmaceutical Refrigerator (MPR) is one of the medical devices that has a size of 800x500x1820 mm, consisting of 233 parts that are purchased and assembled.

### *3.2 Production Process*

The main production processes carried out for the manufacture of MPR consist of several things there are Transfer of raw material into the production area, Assembly of the inner box, outer box, cooling, evaporator, door and electrical, the inspection process, charging refrigerant, vacuum, aging unit, and packing. Based on the production process, a map of the operation process is then made, and the required machine data and machine facilities are obtained, then analyzed according to the size needed to be entered on the required area sheet

### *3.3 Layout Type Analysis*

The best layout type that can be applied is a layout based on product flow (Product Layout), this is seen from the products produced in large quantities and processes that take place repeatedly and similar. Also, it can be seen in the application of previous products that machines are arranged in the order of a process.

*3.4 Determination of Area Needs*

Based on the machine size data, the facilities obtained from the operational process map and the operator's needs are determined by the area needs that have been given 50% leeway for the crossing road requirements as follows:

**Table 1.** Results of Determining the Need for Area

| Work Station | Machine Name | Space Requirements ($m^2$) |
|---|---|---|
| Inner Box Assembly | BM/Rack/Work Table | 15,76785 |
| Outer Box Assembly | Rack/Jig Outer Box | 8,499 |
| Combine Cabinet Assy | CM/Rack/Work Table | 21,7575 |
| Cooling Assembly | BRM/Rack/Work Table | 5,5605 |
| Evaporator Assembly | BRM/Rack/ Work Table | 7,566 |
| Door Assembly | Work Table | 7,00275 |
| Unit Process | BRM/Rack/Lifter unit | 7,44375 |
| Vacuum Unit | VM | 7,425 |
| Refrigerant Charging | RCM | 3,21 |
| Welding Unit | UWM, HLD/Lifter Unit | 3,225 |
| Electrical Assembly | Work Table | 4,3275 |
| Safety Inspection | DPM, ECT, WIM | 3,25575 |

*3.5 Facility Layout Design*

The stages of designing the facility layout using systematic layout planning methods.

*3.5.1 Material Transfer Planning.* The characteristics of the raw material used for MPR assembly have varying sizes with a total number of parts of 233 items, as well as a low material flow rate and displacement distance with a relatively short product design layout type. Then applied material transfer is done manually, with a minimalist special Trolley Picking.

*3.5.2 Making Activity Relationship Chart.* The activity relationship chart is an analysis phase that is made in the form of a chart and is used to show the level of activity relationships between work stations. Each determination of the level of relationship is made by considering several reasons intended for the smooth production process.

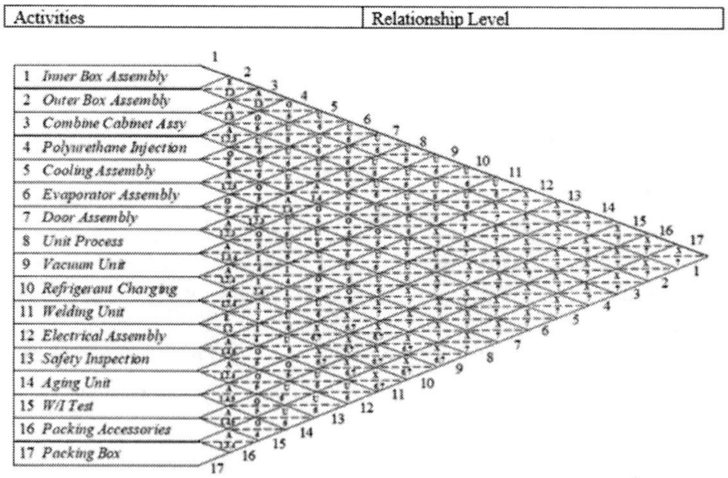


**Figure 1.** Activities-Relationship Chart

**Table 2.** Table of reasons for work station connection

| Code | The Reason |
|---|---|
| 1 | Work flow sequence |
| 2 | Use the same operator |
| 3 | Transfer of material |
| 4 | Facilitates repair |
| 5 | Facilitate supervision |
| 6 | Job interdependence |
| 7 | Noise, dust |

*3.5.3 Creation of a Block Diagram Activity Template.* The activity relationship diagram between work stations is made to show the combination of the level of visual relationship between work station area and the product flow that occurs in each process. Activity template block diagrams are made based on activity related worksheets. Two alternative ATBD designs were made as follows.

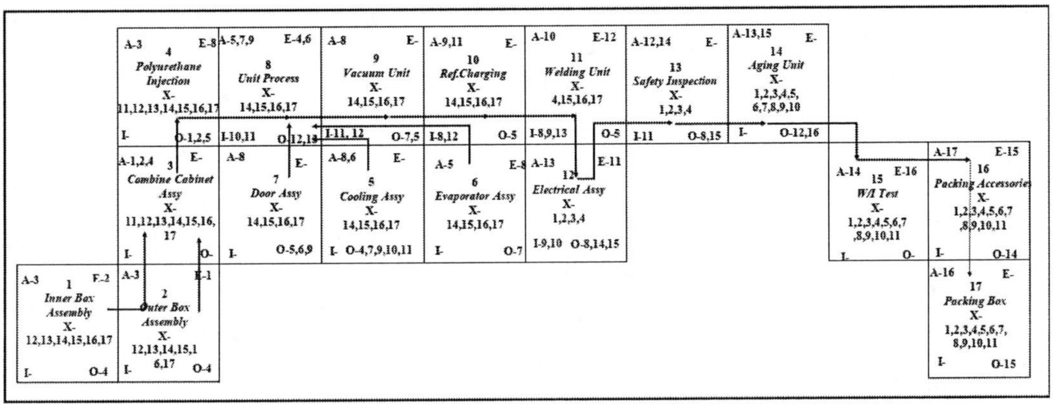


**Figure 2.** Proposal ATBD 2

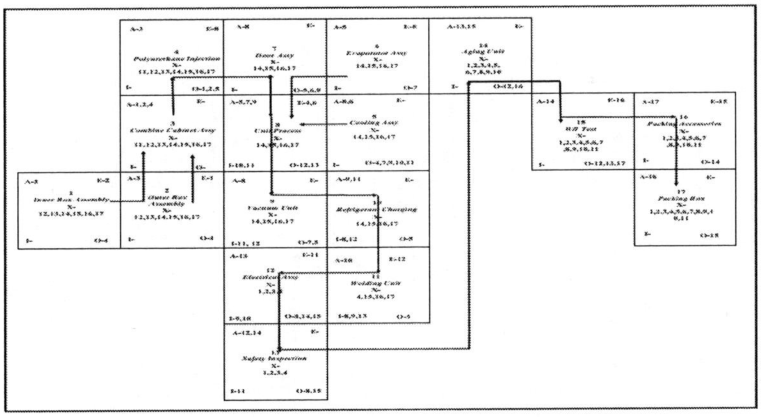

**Figure 3.** Proposal ATBD 1

*3.5.4 Designing Alternative Layouts.* The alternative of layout design is based on a block diagram activity template in the form of a block scale with a certain scale and adjusted to the available area.

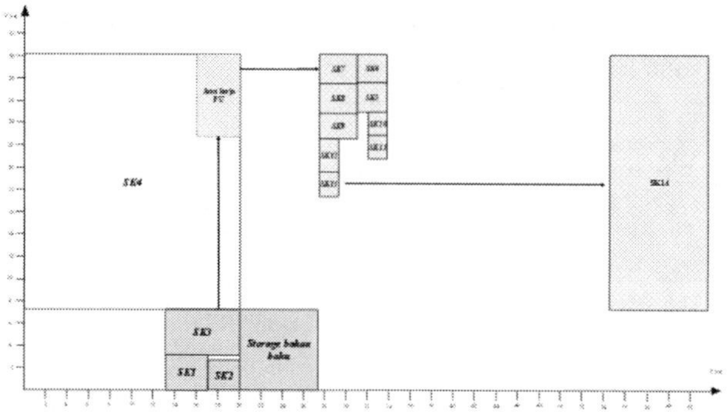

**Figure 4.** Proposal Block Layout 1

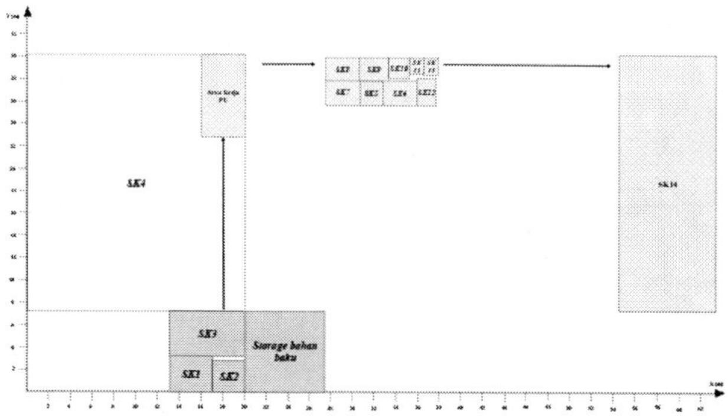

**Figure 5.** Proposal Block Layout 2

*3.5.5 Evaluation of Alternative Layouts.* Two alternative layouts have been made and evaluated, then the best alternative design is chosen. In the layout evaluation used the analysis of the movement of material activity between workstations on both layouts using the From-to-Chart method, used to determine the smallest material transfer weight. The flow of material is drawn based on the activity flow in the Activity template block diagram. Calculation of the distance between activities using the rectilinear distance measurement method.

- Proposed Layout Analysis 1

  Based on the block layout, the first suggestion is to get coordinate points at each workstation that can be seen in Figure 6.

| Dari \ Ke | 0 | 1 | 2 | 3 | 4 | 5 | 6 | 7 | 8 | 9 | 10 | 11 | 12 | 13 | 14 | 15 | 16 | 17 |
|---|---|---|---|---|---|---|---|---|---|---|---|---|---|---|---|---|---|---|
| 0 | | 10,3 | | | | | 32,46 | 26,3 | | | | | 39,54 | | | | 53,93 | |
| 1 | | | 4,06 | | | | | | | | | | | | | | | |
| 2 | | | | 6,08 | | | | | | | | | | | | | | |
| 3 | | | | | 23,81 | | | | | | | | | | | | | |
| 4 | | | | | | | | 14,23 | | | | | | | | | | |
| 5 | | | | | | | | | 5,22 | | | | | | | | | |
| 6 | | | | | | | | | 6,58 | | | | | | | | | |
| 7 | | | | | | | | | 4,74 | | | | | | | | | |
| 8 | | | | | | | | | | 3,01 | | | | | | | | |
| 9 | | | | | | | | | | | 2,26 | | | | | | | |
| 10 | | | | | | | | | | | | 1,51 | | | | | | |
| 11 | | | | | | | | | | | | | 1,57 | | | | | |
| 12 | | | | | | | | | | | | | | 1,09 | | | | |
| 13 | | | | | | | | | | | | | | | 31,32 | | | |
| 14 | | | | | | | | | | | | | | | | 7,07 | | |
| 15 | | | | | | | | | | | | | | | | | 11,03 | |
| 16 | | | | | | | | | | | | | | | | | | 5,21 |
| 17 | | | | | | | | | | | | | | | | | | |

**Figure 6.** FTC Distance Between Centers Analysis 1

After that, make a mapping with from-to-chart with the amount of material transferred between workstations that can be seen in Figure 7.

| Ke / Dari | 0 | 1 | 2 | 3 | 4 | 5 | 6 | 7 | 8 | 9 | 10 | 11 | 12 | 13 | 14 | 15 | 16 | 17 |
|---|---|---|---|---|---|---|---|---|---|---|---|---|---|---|---|---|---|---|
| 0 | | 104 | | | | | 33 | 44 | | | | | 47 | | | | 43 | |
| 1 | | | 64 | | | | | | | | | | | | | | | |
| 2 | | | | 21 | | | | | | | | | | | | | | |
| 3 | | | | | 14 | | | | | | | | | | | | | |
| 4 | | | | | | | | 33 | | | | | | | | | | |
| 5 | | | | | | | | | | | | | | | | | | |
| 6 | | | | | | | | | 8 | | | | | | | | | |
| 7 | | | | | | | | | 8 | | | | | | | | | |
| 8 | | | | | | | | | | 8 | | | | | | | | |
| 9 | | | | | | | | | | | 8 | | | | | | | |
| 10 | | | | | | | | | | | | 8 | | | | | | |
| 11 | | | | | | | | | | | | | 8 | | | | | |
| 12 | | | | | | | | | | | | | | 8 | | | | |
| 13 | | | | | | | | | | | | | | | 8 | | | |
| 14 | | | | | | | | | | | | | | | | 8 | | |
| 15 | | | | | | | | | | | | | | | | | 8 | |
| 16 | | | | | | | | | | | | | | | | | | 23 |
| 17 | | | | | | | | | | | | | | | | | | |

**Figure 7.** FTC Amount of Material Analysis 1

Based on the amount of material and the distance that has been obtained, make a weight transfer matrix which shows the multiplication between the amount of material and the distance as follows.

| Ke / Dari | 0 | 1 | 2 | 3 | 4 | 5 | 6 | 7 | 8 | 9 | 10 | 11 | 12 | 13 | 14 | 15 | 16 | 17 | |
|---|---|---|---|---|---|---|---|---|---|---|---|---|---|---|---|---|---|---|---|
| 0 | | 1071,2 | | | | | 1071,18 | 1157,2 | | | | | 1858,38 | | | | 2588,64 | | 7746,6 |
| 1 | | | 259,84 | | | | | | | | | | | | | | | | 259,84 |
| 2 | | | | 127,68 | | | | | | | | | | | | | | | 127,68 |
| 3 | | | | | 333,34 | | | | | | | | | | | | | | 333,34 |
| 4 | | | | | | | | 469,59 | | | | | | | | | | | 469,59 |
| 5 | | | | | | | | | 41,76 | | | | | | | | | | 41,76 |
| 6 | | | | | | | | | 52,64 | | | | | | | | | | 52,64 |
| 7 | | | | | | | | | 37,92 | | | | | | | | | | 37,92 |
| 8 | | | | | | | | | | 24,08 | | | | | | | | | 24,08 |
| 9 | | | | | | | | | | | 18,08 | | | | | | | | 18,08 |
| 10 | | | | | | | | | | | | 12,08 | | | | | | | 12,08 |
| 11 | | | | | | | | | | | | | 12,56 | | | | | | 12,56 |
| 12 | | | | | | | | | | | | | | 8,72 | | | | | 8,72 |
| 13 | | | | | | | | | | | | | | | 250,56 | | | | 250,56 |
| 14 | | | | | | | | | | | | | | | | 56,56 | | | 56,56 |
| 15 | | | | | | | | | | | | | | | | | 88,24 | | 88,24 |
| 16 | | | | | | | | | | | | | | | | | | 41,68 | 41,68 |
| 17 | | | | | | | | | | | | | | | | | | | |
| | | | | | | | | | | | | | | | | | | | 9322,09 |

**Figure 8.** FTC Transfer Weight Analysis 1

- Proposed Layout Analysis 2
  Based on the block layout, the first suggestion is to get coordinate points at each workstation that can be seen in Figure 9.

IOP Conf. Series: Materials Science and Engineering **847** (2020) 012009 doi:10.1088/1757-899X/847/1/012009

| Ke / Dari | 0 | 1 | 2 | 3 | 4 | 5 | 6 | 7 | 8 | 9 | 10 | 11 | 12 | 13 | 14 | 15 | 16 | 17 |
|---|---|---|---|---|---|---|---|---|---|---|---|---|---|---|---|---|---|---|
| 0 | | 10,3 | | | | 31,9 | | | 28,04 | | | | 21,64 | | | | 53,83 | |
| 1 | | | 4,06 | | | | | | | | | | | | | | | |
| 2 | | | | 6,08 | | | | | | | | | | | | | | |
| 3 | | | | | 23,81 | | | | | | | | | | | | | |
| 4 | | | | | | | | | | | | | | | | | | |
| 5 | | | | | | | 2,38 | | | | | | | | | | | |
| 6 | | | | | | | | | | | | | | | | | | |
| 7 | | | | | | | | | 3,08 | | | | | | | | | |
| 8 | | | | | | 3,86 | 5,74 | | | 3,01 | | | | | | | | |
| 9 | | | | | | | | | | | 4,65 | | | | | | | |
| 10 | | | | | | | | | | | | 2,14 | | | | | | |
| 11 | | | | | | | | | | | | | 11,44 | | | | | |
| 12 | | | | | | | | | | | | | | 3,03 | | | | |
| 13 | | | | | | | | | | | | | | | 33,02 | | | |
| 14 | | | | | | | | | | | | | | | | 7,07 | | |
| 15 | | | | | | | | | | | | | | | | | 11,03 | |
| 16 | | | | | | | | | | | | | | | | | | 5,21 |
| 17 | | | | | | | | | | | | | | | | | | |

**Figure 9.** FTC Distance Between Centers Analysis 2

After that, make a mapping with from-to-chart with the amount of material transferred between workstations.

| Ke / Dari | 0 | 1 | 2 | 3 | 4 | 5 | 6 | 7 | 8 | 9 | 10 | 11 | 12 | 13 | 14 | 15 | 16 | 17 |
|---|---|---|---|---|---|---|---|---|---|---|---|---|---|---|---|---|---|---|
| 0 | | 104 | | | | 33 | | | 44 | | | | 47 | | | | 43 | |
| 1 | | | 64 | | | | | | | | | | | | | | | |
| 2 | | | | 21 | | | | | | | | | | | | | | |
| 3 | | | | | 14 | | | | | | | | | | | | | |
| 4 | | | | | | | | | | | | | | | | | | |
| 5 | | | | | | | 15 | | | | | | | | | | | |
| 6 | | | | | | | | | | | | | | | | | | |
| 7 | | | | | | | | | 11 | | | | | | | | | |
| 8 | | | | | | 8 | 8 | | | 8 | | | | | | | | |
| 9 | | | | | | | | | | | 8 | | | | | | | |
| 10 | | | | | | | | | | | | 8 | | | | | | |
| 11 | | | | | | | | | | | | | 8 | | | | | |
| 12 | | | | | | | | | | | | | | 8 | | | | |
| 13 | | | | | | | | | | | | | | | 8 | | | |
| 14 | | | | | | | | | | | | | | | | 8 | | |
| 15 | | | | | | | | | | | | | | | | | 8 | |
| 16 | | | | | | | | | | | | | | | | | | 8 |
| 17 | | | | | | | | | | | | | | | | | | |

**Figure 10.** FTC Amount of Material Analysis 2

Based on the amount of material and the distance that has been obtained, make a weight transfer matrix which shows the multiplication between the amount of material and the distance as follows:

| Ke / Dari | 0 | 1 | 2 | 3 | 4 | 5 | 6 | 7 | 8 | 9 | 10 | 11 | 12 | 13 | 14 | 15 | 16 | 17 | |
|---|---|---|---|---|---|---|---|---|---|---|---|---|---|---|---|---|---|---|---|
| 0 | | 1071,2 | | | | 1052,7 | | | 1233,76 | | | | 1017,08 | | | | 96,83 | | 4471,57 |
| 1 | | | 259,84 | | | | | | | | | | | | | | | | 259,84 |
| 2 | | | | 127,68 | | | | | | | | | | | | | | | 127,68 |
| 3 | | | | | 333,34 | | | | | | | | | | | | | | 333,34 |
| 4 | | | | | | | | | | | | | | | | | | | |
| 5 | | | | | | | 35,7 | | | | | | | | | | | | 35,7 |
| 6 | | | | | | | | | | | | | | | | | | | |
| 7 | | | | | | | | | 14,08 | | | | | | | | | | 14,08 |
| 8 | | | | | | 30,88 | 45,92 | | | 24,08 | | | | | | | | | 100,88 |
| 9 | | | | | | | | | | | 37,2 | | | | | | | | 37,2 |
| 10 | | | | | | | | | | | | 17,12 | | | | | | | 17,12 |
| 11 | | | | | | | | | | | | | 91,52 | | | | | | 91,52 |
| 12 | | | | | | | | | | | | | | 24,24 | | | | | 24,24 |
| 13 | | | | | | | | | | | | | | | 264,16 | | | | 264,16 |
| 14 | | | | | | | | | | | | | | | | 56,56 | | | 56,56 |
| 15 | | | | | | | | | | | | | | | | | 88,24 | | 88,24 |
| 16 | | | | | | | | | | | | | | | | | | 88,24 | 88,24 |
| 17 | | | | | | | | | | | | | | | | | | | |
| | | | | | | | | | | | | | | | | | | | 5750,5 |

**Figure 11.** FTC Transfer Weight Analysis 1

From the results of the calculations in Figure 11, the determination of the proposed layout 2 produces a displacement weight of 5750.5, which causes a backflow of semi-finished material.

Based on the above calculation, the determination of the proposed layout 2 has a displacement weight of 5750.5 per day, which is smaller than the proposal layout 1, which has a displacement weight of 9322.09 per day. Thus, the MPR production layout was set using the proposed layout 2, although there was a backflow for semi-finished materials because the delivery time for raw materials and semi-finished materials did not occur at the same time, so it would not interfere with the smooth flow of materials.

## 4. Conclusion

Based on the result and the discussion above, the conclusion for this research are:

1. To increase the production area for the new model MPR with a production capacity of 4153 units, an area of 98.2966 m2 is required. By determining the number of work stations as many as 17 areas of which 5 of them are existing work stations, then the addition of operators is needed as well as the addition of several new machines as follows: 1 piece of Bending, Crane, Refrigerant charging, Ultrasonic welding, and Safety inspection machine, 3 piece of Brazing and Vacuum machine.
2. By determining the production layout based on product layout, an analysis of the weight of material transfer in the two alternative layout proposals obtained layout proposal 1 has a displacement weight of 9322.09 per day and proposal layout 2 has a displacement weight of 5750.5 per day. The best MPR production layout to use is the proposed layout 2. By setting the layout, a biomedical warehouse area of 49.42 m2 is required.

## 5. References

[1] J M Apple 1990 *Tata Letak Pabrik dan Pemindahan Bahan*. 3rd Ed. (Bandung: Institut Teknologi Bandung).

[2] H Purnomo 2004 *Perencanaan & Perancangan Fasilitas*. (Yogyakarta: Graha Ilmu).

[3] S Wignjosoebroto 2003 *Tata Letak Pabrik dan Pemindahan Bahan*. 3rd Ed. (Surabaya: Guna Widya).

[4] R P Naganingrum 2012 Perancangan Ulang Tata Letak Fasilitas di PT. Dwi Komala Dengan Metode Systematic Layout Planning, *Final Project* (Solo: Teknik Industri Fakultas Teknik Universitas Sebelas Maret).

[5] S Heragu 1997 *Facilities Design*. (Boston: Pws Publishing Company)

[6] A W Ekarini, Centrausaha.Com 2018. [Online]. Available: Https://Centrausaha.Com/Lokasi-Industri-Pabrik-Manajemen-Operasional/. [Accessed 3 May 2019].

# Managing Internal Bullwhip Effect to Plan Product Distribution in a Garment Factory

**S Hidayat[1], SW Fauzia[2], and N Nurhasanah[1]**

[1]Lecturer, Industrial Engineering Department, Faculty of Science and Technology, Universitas Al Azhar Indonesia
[2]Student, Industrial Engineering Department, Faculty of Science and Technology, Universitas Al Azhar Indonesia
[1]syarif_hidayat@uai.ac.id

**Abstract** - XYZ is a family garment business established in 2004. XYZ produces and sells garments for 6-12 years old boys with brands like Forboys, Wellboy, Nashkid, Yepao, and Kid Dreams. The types of garments are *oblong, setelan* and *wangky*. XYZ has three garment Cut-make-trim (CMT) partners in Rangkas, Serang and Pandeglang. The three CMT partners has been giving XYZ neat and tidy garment products, ready to compete in Tanah Abang wholesale center. Based on the demand forecast calculation XYZ sends its order for production to the CMTs. In return, based on the quantity ordered, the CMTs order the raw materials from XYZ. This study attempts to measure the "internal" bullwhip effect which is observed as the fluctuations of orders sent by the CMTs to XYZ, based on the production orders from XYZ to the CMTs. XYZ experienced high inventories and its mounting costs due to this internal bullwhip effect. Further, the method used in this study investigate the appropriate forecasting method. The best forecasting method is the Double Exponential Smoothing, which also results in the reduction in the bullwhip effect ratio, and improvement of scrap factor (around 10.60%). The reduction of BE ratio is quite significant from 0.344, 0.242 and 0.717 for *oblong*, *setelan* and *wangky* product types respectively.
**Keywords:** Garment business, Cut-make-trim, Internal bullwhip effect, Scrap factor.

## 1. INTRODUCTION

XYZ is a family owned small medium garment industry which produces different type of clothing for boys aged 6-12 years. The brands are Forboys, Wellboy, Nashkid, Yepao, and Kid Dreams. XYZ is very committed to quality and always attempts to satisfy the customers with best product quality. XYZ employs a unique production processes. XYZ buys initial textile raw materials from its suppliers which are sent to some cutting partners in Sukabumi area. XYZ is experiencing fluctuative demand from the garment market, while it has to control the order to its CMT partners, who in return will send orders for raw materials from XYZ. Further, XYZ experiences various reject and low quality products which causes product returns. To minimize this matter XYZ applies a scrap factor (SF) when calculating the orders to its CMTs.

Figure 1 shows how XYZ operates its business between allocating production order to the three CMTs, based on its demand forecast. Based on these production orders the CMTs orders textile raw materials from XYZ. XYZ must carefully control the orders such that it does not develop mounting number of garment products which should have been sent or sold through its market outlets. The internal bullwhip effect (BWE) emerges when production orders were sent to the three CMTs and other partners, CMTs and other partners send requests for raw materials from XYZ, which may be uncontrollably higher than is necessary. Packaged finished garments were received back at XYZ warehouse in Jakarta from all CMTs and partners.

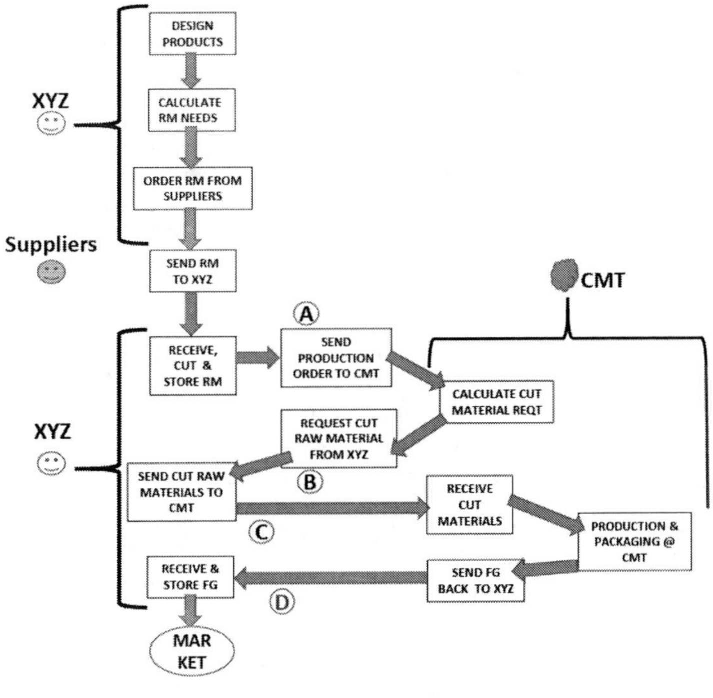


**Figure 1.** XYZ Business Process Flow

The detail business process at XYZ may be explained as follows:

1. XYZ designed the garment to be produced. The owner of XYZ evaluated the design prototype before order for production is released. Required materials to produce the garment are calculated and then ordered from the suppliers. Received raw materials at XYZ Jakarta warehouse are cut following the design pattern, then are stocked in XYZ Jakarta warehouse.
2. XYZ calculates the demand as the basis for assigning production order to the CMTs (point A). CMTs orders the required cut materials from XYZ (point B). This is the point where internal BWE emerges.
3. Cut raw materials are sent to CMTs in Rangkas, Serang and Pandeglang, as well as to other partners for further sewing processes (point C). Finished garments at all the CMTs are packed and sent back to XYZ in Jakarta (point D). This is where the number of finished garments are accumulated from all partners.
4. XYZ made the final finishing touch on the garments before being sent to market/sales outlets at Tanah Abang. XYZ has to control the flow and inventory of its various raw materials and finished products in its Jakarta warehouse.

## 2. METHODS

Due to the informal and intuitive forecasting using previous historical sales data, XYZ does not develop proper and accurate raw material orders from its suppliers, does not develop proper production assignments to its CMT partners, and does not send appropriate amount of cut raw materials to the CMT partners. The result is that XYZ receives "uncontrolled" and "inaccurate" number of finished garments from the CMT partners. All these garments as well as the initial raw materials, and cut raw materials are piling up in XYZ Jakarta warehouse. The Jakarta warehouse is in fact a mansion which was turned to be a garment processing facilities without proper design and layout. The phenomenon of highly uncontrollable fluctuation of raw material and finished garments at XYZ warehouse is what we describe as "internal bullwhip effect". We describe it as such because it is happening inside a company material logistics.

XYZ experienced high number of rejects finished garments from CMTs. From this experience XYZ employed a certain "scrap factor" when calculating number of production assignments to the CMTs. This numbers were determined roughly from past 12 months (returned) sales on each class of product and each CMTs. This practice of applying scrap factor helped reduced the huge level of inventories at XYZ. The handling costs incurred with this huge inventories has significantly eroded its profitability and negatively affect the morale of its Jakarta staff and workers due to the high piles of materials everywhere.

XYZ needs to measure the level of its "internal bullwhip effect" and then find more ways to get rid of the mounting level of its inventories. The idea to help XYZ is by improving the forecasting method with more "scientific" models.

### *2.1 Distribution Requirement Planning (DRP)*

Distribution requirement planning is a system that determine demand for supplies to the center of the distribution of, combining historical demand, and as its input in order to the system of production and material [1]. Another expert stated that DRP is a forecasting based system that plans and determines the right inventory level to ensure satisfactory flow of products distribution. DRP gives future demand visibility pertaining to the need to shipping from source stocking points to destinations selling points. This will help the firm to take corrective measures before undesirable occurrence develops into a crisis [2].

### *2.2 Scrap Factor*

To reduce the effect of unwanted order quantity accumulation, the formula to calculate DRP is merged with the scrap factor. The formula is applied to planned order release, rather than on gross requirement, for scrap estimate lost during the distribution of products. The scrap factor is the percentage that is used in the calculation of the DRP to anticipate the loss in the distribution process [3]. The planned order release quantity which include scrap factor is calculated as follows:

$$PO\ Release\ Quantity = \frac{PO\ Receipt\ Quantity}{(1-\%Scrap)} \quad (1)$$

### *2.3 Forecasting*

Management is continuously faced with a fast-paced flow of business planning and decision-making situations. Forecasting is an engineering analysis done by a quantitative and qualitative approach to estimate the future using past reference data. A sound forecasting system is a necessity in the expanding supply change management world, allowing firms to cope with the ever-changing shifts in demands for their products and resources. The common goal is to have the least amount of inventory to satisfy the customers' demands for its products, and at the same time minimize the cost of buying and holding the inventory [5]. In this study three models of forecasting method are used. (1) Double Moving Average (DMA), (2) Exponential Smoothing Method (DES), and (3) Triple Exponential Smoothing (TES), following the approach by Siregar [5] to analyse the best method in Palm Oil production.

### *2.4 Bullwhip Effect and Bullwhip Effect Measurement*

A phenomenon that is now well known as the BWE is the variability of orders increases as they move up the supply chain from retailers to wholesalers to manufacturers to suppliers. Other researchers describe BWE as the large discrepancy between inventory and demand, or the jump from small order in the downstream to high request in the upstream [4]. Existing approaches that aim at quantifying the BWE neglect the network structure of supply chains. Some major factors said to cause or contribute to BWE are [6, 7, 8]:

- The causes are often interconnected and tracking down the sources of the disruption is difficult. Other than identifying the causes for the BWE, it is difficult to quantify the degree to which each cause affects the supply chain. For this reason, many of the studies in this area either base their empirical results on a narrow range of causes, or study simple supply chains using simulations.
- Rationing and shortage gaming – buyers' strategic ordering behaviour as a possible cause of the bullwhip effect. In an environment where there is supply shortage, buyers tend to overorder to secure resources for themselves, and suppliers tend to correct this over ordering by rationing back to smaller quantities. These two patterns of behaviour are interrelated and mutually reinforce one another.
- Past demands are not used in forecasting future customer demand. up-to-S policy where S is a constant. Other conditions by which these causes were derived include: (a), No fixed order cost, (b) Re-supply comes from an infinite source with a constant fixed lead time (c) Purchase cost for the product is stationary.
- Lack of communication between each link in the supply chain makes it difficult for processes to run smoothly. Managers can perceive a product demand quite differently within different links of the supply chain and therefore order different quantities.
- Demand information – relying on past demand information to estimate current demand information of a product does not take into account any fluctuations that may occur in demand over a period of time.
- When the retail and manufacturing industries do not smooth their production relative to their demand, i.e., they impose more volatility on their suppliers than they face the demand from their customers.

In this study, we attempt to consider the confusion of a manufacturer in managing the flow of its order of raw materials from its suppliers, the production order to its external CMT partners who must use the cut raw materials from the manufacturer, and getting the finished garment back from the CMTs as a problem of internal bullwhip effect.

***Bullwhip Effect Measurement***

BWE needs to be measured and controlled to achieve a better company inventory and cost position. The formula for measuring bullwhip effect ratio (= BE) is shown below [4]:

$$\text{BE} = \frac{CV\ (Order)}{CV\ (Demand)} \tag{2}$$

$$\text{CV Order} = \frac{\sigma\ (Order)}{\mu\ (Order)} \tag{3}$$

$$\text{CV Demand} = \frac{\sigma (Demand)}{\mu\ (Demand)} \tag{4}$$

Where:

| | | |
|---|---|---|
| BE | = | Bullwhip effect ratio |
| $CV_{(Order)}$ | = | Coefficient of Variable for sales (= order) |
| $CV_{(Demand)}$ | = | Coefficient of Variable for demand |
| $\sigma_{(Order)}$ | = | Standard Deviation for sales (= order) |
| $\mu_{(Order)}$ | = | Average Sales (= order) |
| σ (Demand) | = | Standard Deviation for demand |
| μ (Demand) | = | Demand average |

This paper attempts to measure the internal BWE in the garment business of XYZ and the appropriate solution to reduce the BWE. The reason for reducing the effect is obviously to reduce the high cost of garment inventory mounting up due to the error in the demand forecast. Figure 2 shows the methodology applied in this study. Data collection are obtained directly from the management and staff of XYZ. When the data is considered sufficient they are processed using three forecasting methods and analyzed. BWE is then calculated before and after the forecast.

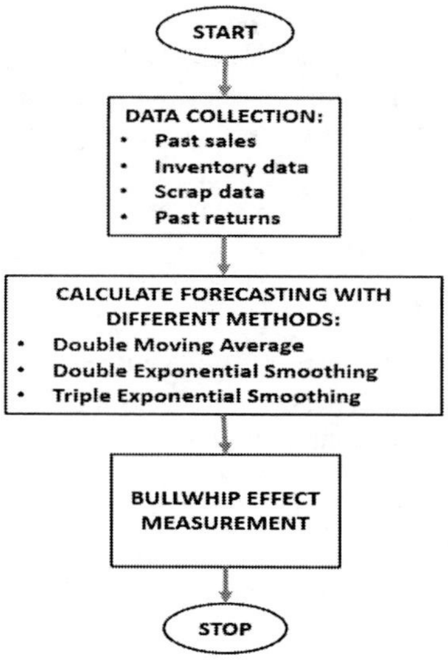


**Figure 2.** Research Flowchart

## 3. RESULTS AND DISCUSSION

*3.1 Data Collection*

*3.1.1 Allocation Order Production Between XYZ and CMT*

Figure 3 describes how the internal BWE occurred in the order of XYZ to CMT concerned and the return to XYZ. This is a horizontal redrawing of Figure 1.

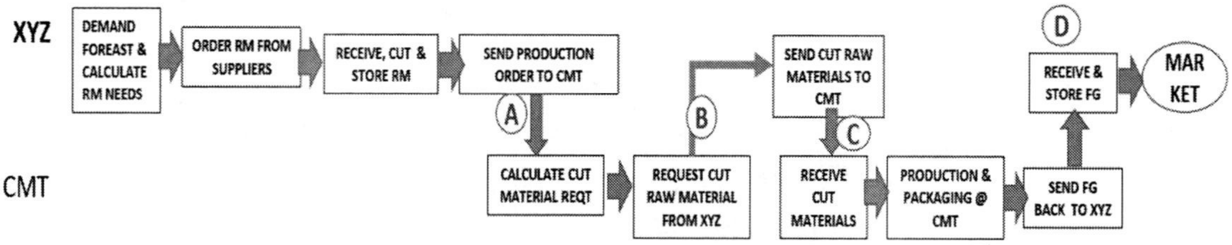


**Figure 3.** Mapping of XYZ Bullwhip effect

The three CMTS in Serang regency are CMT Tarji, CMT Basri and CMT Toro, following the names of the CMT facility owners. Fig. 4 explains the internal distribution network of XYZ which consists of 3 levels of distribution.

- At level 0 is XYZ who assigns production order to CMTs (outsourcing).
- At level 1 is the three CMTs as outsource garment producing; CMT will order cut material or other material required to XYZ as material for making the garment. The three CMTs include the scrap factor in this "support factor" in anticipation of damaged product by the market.
- At level 2 is XYZ who will store the product received (after finishing steps)
- At level 3 is the Marketing Centre that deals directly with customers.

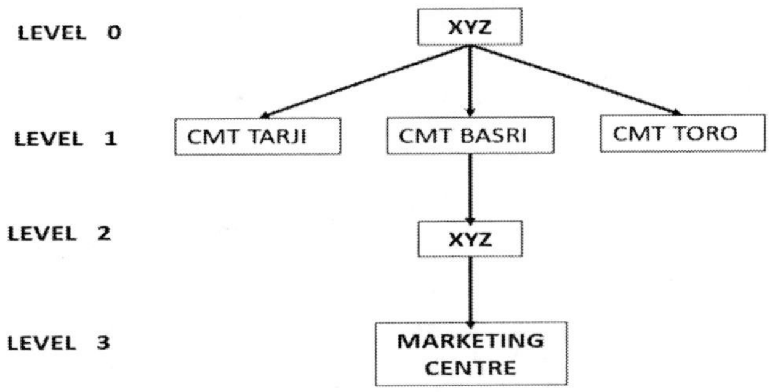


**Figure 4.** Internal Distribution Levels of XYZ

*3.1.2 Returns Data*

Some defective product or products are returned back to XYZ for several factors, namely: (1) The neck is unsymmetrical, (2) Moist, and (3) Untidy stitches. Table 1 shows the product return data for the three CMTs.

**Table 1.** Returns Data of XYZ

| No | Month | Returned product (unit) | | | Total | % of total product |
|---|---|---|---|---|---|---|
| | | CMT Tarji | CMT Basri | CMT Toro | | |
| 1 | January | 699 | 306 | 677 | 1,682 | 3.52 |
| 2 | February | 688 | 730 | 661 | 2,079 | 5.08 |
| 3 | March | 526 | 161 | 678 | 1,365 | 2.92 |
| 4 | April | 459 | 610 | 467 | 1,536 | 2.74 |
| 5 | May | 454 | 833 | 675 | 1,962 | 4.93 |
| 6 | June | 661 | 864 | 345 | 1,870 | 4.61 |
| 7 | July | 678 | 699 | 654 | 2,031 | 4.58 |
| 8 | August | 700 | 1,028 | 786 | 2,514 | 5.32 |
| 9 | September | 651 | 526 | 675 | 1,852 | 4.39 |
| 10 | October | 972 | 972 | 456 | 2,400 | 5.30 |
| 11 | November | 455 | 809 | 432 | 1,696 | 4.71 |
| 12 | December | 467 | 809 | 745 | 2,021 | 5.10 |
| | Total | 7,410 | 8,347 | 7,251 | 23,008 | 4.43 |

*3.2 Forecasting*

After the data from the three CMTs are collected and plotted, they are ready for forecast. Forecasting methods used are the Double Moving Average (DMA), Double Exponential Smoothing (DES) by Brown and Triple Exponential Smoothing (TES) by Brown.

Table 2 shows the result of all three forecasting calculations. The figure for the third product (*Wangky*) shows the smallest value of MAPE (mean absolute percentage). For all CMTs (Tarji, Basri and Toro) the smallest MAPE values is achieved using DES method.

**Table 2.** MAPE Result of CMTs

| No | Method | CMT Tarji | | | CMT Basri | | | CMT Toro | | |
|---|---|---|---|---|---|---|---|---|---|---|
| 1 | Double Moving Average | 0.38 | 0.57 | 0.29 | 0.26 | 0.38 | 0.23 | 0.22 | 0.30 | 0.26 |
| 2 | Double Exponential Smoothing | 0.25 | 0.32 | 0.25 | 0.27 | 0.22 | 0.15 | 0.18 | 0.23 | 0.23 |
| 3 | Triple Exponential Smoothing by Brown | 0.28 | 0.37 | 0.25 | 0.27 | 0.22 | 0.16 | 0.22 | 0.26 | 0.26 |

### *3.2.1 Scrap Factor Calculation*

The next stage is to calculate manually to determine the percentage scrap factor. Table 3 shows the result. Obviously the calculation for the third product (*oblong*) gives the lowest value of the scrap factor. This data are obtained by dividing the numbers returned products from Table 1 by the total production of each CMT.

**Table 3.** Calculation Result Scrap Factor of Three CMTs

| No | CMT | Oblong | Setelan | Wangky |
|---|---|---|---|---|
| 1 | Tarji | 10.69% | 11.88% | 11.41% |
| 2 | Basri | 14.05% | 15.76% | 17.33% |
| 3 | Toro | 12.45% | 13.22% | 12.87% |

### *3.2.2 Bullwhip Effect Measurement*

Table 4 shows the result of BE measurement for the next 3 months. The table shows that the measured BE ratio are better (or lower) after the orders and flows are based on better forecasting method, namely Double Exponential Smoothing (DES). For *oblong* products at CMT Tarji the reduction is 0.49/1.41 = 0.347; for *setelan* product 0.20/0.83 = 0.241 while for *Wangky* product, the ratio is reduced 0.71/0.99 = 0.717.The figures for the other two CMTs are shown in the table.

**Table 4.** Calculation of Bullwhip Effect Product all CMTs

| Garment type | Para-meter | Sales Value Before and After Forecasting | | | | | | | | |
|---|---|---|---|---|---|---|---|---|---|---|
| | | CMT Tarji | | | CMT Basri | | | CMT Toro | | |
| | | Be-fore | After | Im-prove | Be-fore | After | Im-prove | Be-fore | After | Im-prove |
| *Oblong* | Average | 6,621 | 3,797 | | 4,550 | 5,013 | | 4,578 | 4,749 | |
| | Std Dev | 1,104 | 218 | | 289 | 3 | | 991 | 29 | |
| | Coef Var | 0.17 | 0.06 | | 0.06 | 0.00 | | 0.22 | 0.01 | |
| | BE Ratio | 1.41 | 0.49 | 0.344 | 1.28 | 0.01 | 0.010 | 1.80 | 0.05 | 0.028 |
| *Setelan* | Average | 6,151 | 3,519 | | 4,310 | 5181 | | 4,221 | 4,718 | |
| | Std Dev | 1,389 | 192 | | 764 | 80 | | 374 | 18 | |
| | Coef Var | 0.23 | 0.05 | | 0.18 | 0.02 | | 0.09 | 0.00 | |
| | BE Ratio | 0.83 | 0.20 | 0.242 | 0.80 | 0.07 | 0.087 | 1.65 | 0.07 | 0.043 |
| *Wangky* | Average | 6,198 | 3,480 | | 4,378 | 3,707 | | 4,165 | 4,877 | |
| | Std Dev | 540 | 217 | | 245 | 38 | | 626 | 23 | |
| | Coef Var | 0.09 | 0.06 | | 0.06 | 0.01 | | 0.15 | 0.00 | |
| | BE Ratio | 0.99 | 0.71 | 0.717 | 0.78 | 0.14 | 0.186 | 0.87 | 0.03 | 0.031 |

## 4. CONCLUSION

Based on the results of data processing that has been carried out, some conclusions can be drawn as follows:

1. The best forecasting method is the double exponential smoothing (DES) which has the smallest value of MAPE (Mean Absolute Percentage). For *oblong* product from CMT Tarji = 0.25%. For *wangky* Produk from CMT Basri = 0.15%. For *oblong* product from CMT Toro = 0.18%.
2. The reduction of BE ratio is quite significant from 0.344, 0.242 and 0.717 for *oblong*, *setelan,* and *wangky* product types respectively. For the CMT Tarji, the reduction are from 0.241 to 0.717 for CMT Tarji products. The reduction of BE ratio for CMT Basri are 0.009, 0.087, and 0.186; and for CMT Toro are from 0.344, 0.242 and 0.717 for *oblong*, *setelan* and *wangky* product types respectively. In general the BE ratio becomes less when the the right forecasting method is used.

3. The BW value has been decreasing with the use of appropriate forecasting method DES. The smaller the value of BW the better sales level would be achieved because there would be less sales returns.

The recommended future research area is performing the calculations for the rest of the products. The idea of Internal BWE may sound strange or alien. But in Indonesia the practice of production outsourcing is everywhere. Not only in the garment, but also in other kinds of products like food, even for bicycles or automobiles. Definitely the concept and definition need to be smoothen or improved to be acceptable by the research community.

## 5. REFERENCES

[1] Bowersox D J, Closs, D J, Cooper M B, and Bowersox J C 2013 *Supply Chain Management*, 4th edition. (Singapore: McGraw-Hill).

[2] Bozarth C and Handfield R 2008 *Introduction to Operations and Supply Chain Management*. 2nd edition. (New Jersey: Pearson Education Inc.).

[3] Firmansyah D 2010 *Optimasi Pendistribusian Produk Dengan Menggunakan Distribution Requirement Planning Berdasarkan Nilai Scrap Factor dan Bullwhip Effect (Study Kasus Di Phia Deva Sleman-Yogyakarta).* (Yogyakarta: Universitas Islam Indonesia).

[4] Asmono FT 2011 *Analisa Bullwhip Effect Pada Retail Air Minum Dalam Kemasan (Study Kasus CV Tirta Mekar Jaya).* (Surakarta: Universitas Muhammadiyah Surakarta).

[5] Siregar B, et al. 2017 *Comparison of Exponential Smoothing Methods in Forecasting Palm Oil Real Production*. *J. Phys.: Conf. Ser.* **801**, 012004.

[6] Alony I, and Munoz A 2007 The Bullwhip effect in complex supply chains. *International Symposium on Communications and Information Technologies ISCIT 2007*, 17-19 Oct, 1355-1360.

[7] Hassanzadeh A, Jafarian A, and Amiri M 2014 Modeling and analysis of the causes of bullwhip effect in centralized and decentralized supply chain using response surface method. *Applied Mathematical Modelling*. **38**, 2353–2365.

[8] Eric S 2009 The Bullwhip effect in supply chains-An overestimated problem? *International Journal of Production Economics*. **118**(1), March 2009, 311-322.

# Production quality improvement of Yamalube Bottle with Six Sigma, FMEA, and Data Mining in PT. B

**Rina Fitriana**[1], **Johnson Saragih**[1], **Dea Prameswari Larasati**[1]

[1] Quality Engineering Laboratory, Department of Industrial Engineering, Faculty of Industrial Technology, Universitas Trisakti
[2]rinaf@trisakti.ac.id, [3]johnson_saragih@yahoo.com, [4]deaprameswari19@gmail.com

**Abstract.** PT. B is one of the manufacturing company engaged in the field of plastic packaging. One of the products produced by companies that bottle Yamalube to customer PT. Yamaha. Based on daily reports of production companies, found that the percentage of disability have an average of 14% per month in which the company standard is 5% per month. It is necessary for Six Sigma to improve the quality of the bottles with the steps Define-Measure-Analyse-Improve-Control (DMAIC). In the define phase diagram created SIPOC and Quality Plan Yamalube bottle production process. In the measure phase created Sigma level. At this stage the known value of 0.16 DPU, DPMO of 250,000 and the results showed that the conversion DPMO sigma level Yamalube bottle production process is equal to 2.17. In the analyze phase using a decision tree is known that bottles attributes that influence the decision of rejection is the position of the label, the number of labels, colours and perforated. By using FMEA method is known that improvements will be prioritized on the label of disability are not two sides to the root cause of the wear rubber. Repairs carried out by making the inspection SOP rubber, checksheet and proposals in the form of anti-static vacuum tool.
**Keywords**: Six Sigma, DMAIC, Decision Tree, FMEA,

## 1. Introduction

PT. B is one of the manufacturing company in the field of plastic packaging which was founded in 1997 and has one branch in Cikarang. Due to the variety of the products produced, the company received orders from various companies in the form of products and wide range of colours. The number of plastic bottles were booked also relatively high so that the production process runs for 24 hours. In one line of blow moulding machines, namely B-42 engine that produces Yamalube bottle, found the number of defects with an average percentage of 14%, which target the defect percentage set by the company amounted to 5%. It is therefore necessary improvements using Six Sigma methods to minimize the level of defects in the product bottle Yamalube by identifying defects.

Six Sigma is an organizational approach to eliminate distortions and reduce waste in the process by using statistical science approach.[1]

Past research is needed to become a reference in the study to be performed. As previous studies related to this study are shown in Table 1.

IOP Conf. Series: Materials Science and Engineering **847** (2020) 012011 doi:10.1088/1757-899X/847/1/012011

**Table 1**. Related Research and Outstanding Research

| NO | TOPICS | WRITER'S NAME |
|---|---|---|
| 1 | Integration of Six Simadan cause effect matrixdan improvement using DOE for process improvement rolling mill | K Ganguli [2] |
| 2 | Six Sigma Approach to Reduce Defect in Process of Plastic Bottle Blow Molding Machine ASB in 2000 ml | P Adhi, N Widha, F Cheers [3] |
| 3 | Application of Six Sigma, manufacture optimum setting with DOE and manufacturing standardization QC PASS with Decision Tree | R Fitriana, D Sugiarto, J Saragih, A Bagio [4] |
| 4 | FMEA application in production processes to identify risks kegagaan gloves ATM | N Budi, A Martanto [5] |
| 5 | Integration of FMEA and FTA in determining the root cause of disability products cause insulator | D Fitria, D Ardianto, Y Yuniati [6] |
| 6 | DMAIC integration with FMEA to determine the PVC pipe production process improvement | D Caesaron, S John [7] |
| 7 | Integration of Six Sigma, FMEA to make improvements in product quality Boncabe | D Rukmayadi, S Sugiarti [8] |
| 8 | Making the quality control system for the production of bread using Six Sigma methods | Safrizal, Muhajir [9] |
| 9 | Ishikawa application to determine the cause of disability of bottled drinking water and improved use of FMEA | Ni Wayan, Sri Mulyani, I Wayan [10] |
| 10 | Applying DMAIC Six Sigmadengan and the proposed method is based on Decision Tree | RFitriana, J Saragih, S Sarasaty [11] |
| 11 | Application Six Sigma DMAIC method in Metallurgical Operations | L Girmanova, M Solc, J Kliment, A Divokova, V Miklos [12] |
| 12 | Application of Six Sigma with the proposed improvements using the IOE to improve the maintenance process | M Zasadzien [13] |
| 13 | Model design of quality products by using data mining in R Bakery Company | R Fitriana, J Saragih and N Luthfiana [14] |
| 14 | Integration of Six Sigma, FMEA and Data Mining to improving the quality of the production process of plastic bottles | The position of current research |

## 2. Methods

Research had been conducted by collecting data daily reports in November 2018 and interviews with plant managers to determine the flow of the production process and defects that occur in the product.

In conducting the necessary stages of DMAIC certain methods for each phase, as for the activities to be performed and tools that will be used in each phase can be seen in Table 2.

IOP Conf. Series: Materials Science and Engineering **847** (2020) 012011 doi:10.1088/1757-899X/847/1/012011

**Table 2**. Details Tools DMAIC

| Phase | Description | Outputs | Tools and Techniques |
|---|---|---|---|
| *Define* | - Defining production process<br>- Defining quality plan | - Diagram SIPOC<br>- Table Quality Plan | - Interview and discussions |
| *Measure* | - Choosing suitable control chart<br>- Calculating DPMO Value<br>- Calculating Sigma Value | - Data In *Control*<br>- Nilai DPMO<br>- Nilai Sigma | - P control chart<br>- DPMO Value calculation<br>- DPMO to Sigma value conversion |
| *Analyze* | - Analyzing dominant defect<br>- Analyzing cause and effect of defects<br>- Creating root cause diagram<br>- Analyzing decision making process<br>- Choosing prioritized improvement | - *Decision Tree*<br>- Dominant defect<br>- Cause of defect<br>- *Root cause of defect*<br>- Highest RPN value | - *Data Mining*<br>- Pareto Diagram<br>- Ishikawa Diagram<br>- Root cause diagram<br>- FMEA |
| *Improve* | - Proposing improvement | - SOP<br>- *Checksheet*<br>- *Anti Static Vacuum* | - *Brainstorming* |
| *Control* | - Calculating DPMO value after improvement<br>- Calculating Sigma value after improvement | - New DPMO Value<br>- New Sigma Value | - DPMO Value calculation<br>- DPMO to Sigma value conversion |

## 3. Result and Discussion

In the define phase diagram SIPOC made to define the production process. SIPOC diagram is shown in Table 3.

**Table 3.** SIPOC Diagram

| Supplier | Input | Process | Output | Customer |
|---|---|---|---|---|
| Sales Dept | Customer order | Order quantity given to PPIC dept | Customer order | PPIC Dept |
| PPIC Dept | Customer order | Production Planning | Purchase order | Suppliers |
| Incoming material warehouse | Raw material & masterbatch | Transport to production floor | Production floor quantity | Production warehouse |
| Die pin | Mixed liquid material | Shape liquid material | Parison | Mould |
| Magazine | Label | Dipindahkan ke mould | Label | Mould |
| Transit Area | Yamalube bottle | Transport to shipping area | Yamalube bottle | Shipping area |
| Shipping area | Yamalube bottle | Shipping to customer | Yamalube bottle | PT. Yamaha |

In addition, SIPOC diagram, Quality Plan is created to define the quality control is done in the company on product Yamalube bottle. Quality Plan are shown in Table 4.

**Table 4.** Quality Plan

| No. | Process Name | Machine / Device / Department | Methods | | | | Reaction Plan |
|---|---|---|---|---|---|---|---|
| | | | Evaluation Technique | Sample | | Control Method | |
| | | | | Size | Frequency | | |
| 1 | Incoming Material | Magnifier Lamp | visual Test | 100 gr | per incoming material | sampling | Reject / Return to supplier |
| | | Color Test Device | Color Test | 100 gr | per incoming material | sampling | Reject / Return to supplier |
| 2 | Bottle Inspection | censorship | label Check | unit of production | continuous | continuous Inspection | reject |
| | | Leak Test Device | Leak Test | unit of production | continuous | continuous Inspection | reject |
| | | Inspector | visual Inspection | unit of production | continuous | continuous Inspection | reject |

Based on data from the company's daily report of November 2018, the type of defects that occurs on the product label is tilted label, the labels are not two sides, leaked bottle and a non-standard colour.

After that Critical to Quality (CTQ) is made to describe the type of disability that occurs. Critical to Quality are shown in Table 5.

**Table 5.** Critical to Quality

| No. | Defect | Details | Occurs in the process |
|---|---|---|---|
| 1 | Tilted label | Position the label on the bottle is not straight | *In Mold Labeling* |
| 2 | Missing label | Labels are not on the front and back | *In Mold Labeling* |
| 3 | Leaked bottle | There is a hole in the bottle | Mixing raw materials |
| 4 | Non-standard color | The color of the bottle is too thick or thin | Melting raw materials and masterbatches |

Based on the explanation in Table 5, it is known that there are four types of disability that occurs in the bottle Yamalube products. The first disability is slanted label marked with a label that is not a straight position and occur in the process of in mould labelling. Both are not two sides of the label of disability where there is no label on the front or back of the bottle and on the process of in mould labelling. The third was a hole where there is a hole that occurs in the mixing seeds of raw materials and the fourth is not a standard colour where the colour of the bottle is too thin occurring in smelting raw materials and masterbatch (dye).

Furthermore, the map p is made to determine whether the data is within the control or not. In making the map control researcher using Minitab software. The control chart can be seen in Figure 2.

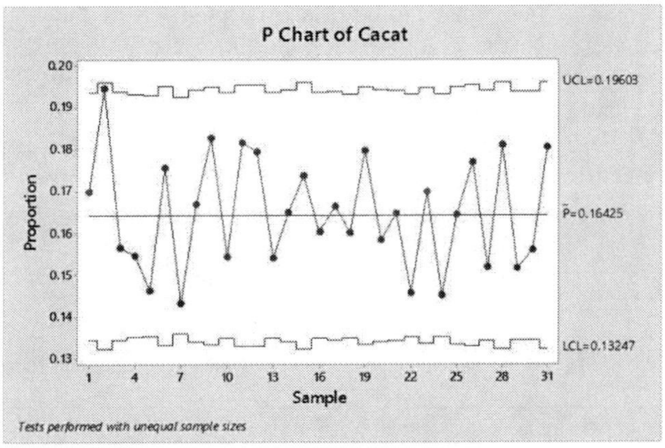


**Figure 1.** Control chart P

From a control chart that has been created, it can be seen that these data are under control. After making sure the data is within the control of the process capability that is calculated by calculating the DPU (defects per unit). Based on values obtained DPU calculation process capability by 0:16. After calculating process capability, then the value will be calculated DPMO. DPMO calculation results show that the possibility of defects in the amount of 1,000,000 units of production is 250,000 units. Having obtained the DPMO values Sigma value will be calculated and the calculation results obtained from Sigma value is 2.17.

The next step is the Analyze phase where it will be made a decision tree disability Yamalube bottle using Weka software 3.8. The results of the decision tree shown no figure 2.

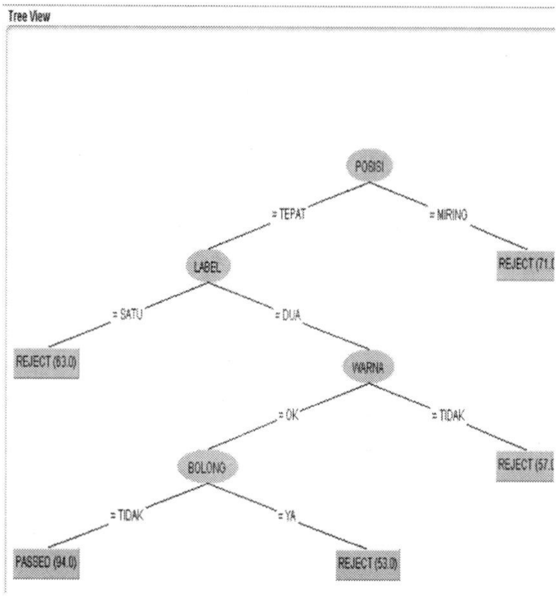


**Figure 2.** Decision Tree Disability Bottle Yamalube

**Table 6.** If Then Rules

| No | Fungsi |
|---|---|
| 1 | If position "MIRING<br>Then category "REJECT" |
| 2 | if Position"TEPAT"<br>LABEL "SATU"<br>Then KATEGORI "REJECT" |
| 3 | If POSISI "TEPAT"<br>LABEL "DUA"<br>WARNA "OK"<br>BOLONG "TIDAK"<br>Then KATEGORI "PASSED" |
| 4 | if POSISI "TEPAT"<br>LABEL "DUA"<br>WARNA "OK"<br>BOLONG "YA"<br>Then KATEGORI "REJECT" |
| 5 | if POSISI "TEPAT"<br>LABEL "DUA"<br>WARNA "TIDAK"<br>Then KATEGORI "REJECT" |

Based on the decision tree, we can conclude that the position of the label discount most impact, followed by a number of labels, the colour of the bottle and a hole in the bottle. To facilitate the understanding of the decision tree is made if then rules. Then there are five If the Rules of the result of making a decision tree. Table if then rules can be seen in Table 6.

Furthermore, the calculation of Risk Priority Number (RPN) with FMEA method by multiplying the value of Severity (S), Occurrence (O) and Detectability (D). The calculations show that disability is not a two-sided label has the highest RPN value is 336. Table FMEA calculation can be seen in Table 7.

**Table 7.** FMEA

| Function Process | Potential Failure Mode | Potential Effect of Failure | Severity | Potential Cause of Failure of process | Occurrence | Current Control | Detection | Recommended Action | RPN |
|---|---|---|---|---|---|---|---|---|---|
| **Resin melting** | Non-standar color | Rejected bottle | 5 | Extruder temperature is not standardized | 6 | Temperature check | 6 | Extruder standarized temperature | 180 |
| **Bottle shaping and in mould labelling** | Missing label | Rejected bottle | 8 | Vacuum rubber is worn out | 6 | Vacuum check | 7 | Vacuum check SOP | 336 |
| | | | | Storage temperature is too high | 5 | Room temperature resetting | 2 | Standardized room temperature | 80 |
| | Tilted label | Rejected bottle | 8 | Label is not stored correctly | 5 | Magazine check | 3 | Magazine check standard | 120 |

The next step is to make improvements Improve stage in accordance with the highest RPN value on the FMEA method. The highest RPN value is on the label is missing label caused by a worn rubber vacuum mouth. In determining the vacuum rubber worn or not, there are some indicators that note. These indicators are the holes in the rubber. If the rubber has holes invisible to the eye, it has been ascertained that the rubber is worn out. The next indicator is a rough rubber surface. If when the rubber tangible feels rough, it can be categorized as the rubber is worn out. One final indicator rubber stiffness. When the rubber is worn out the rubber will become stiff and inflexible,

Therefore made recommendations for improvement is to establish Standard Operating procedure (SOP) inspection rubber vacuum mouth. The SOP is shown in Figure 3.

| bumimulia | **STANDARD OPERATING PROCEDURE**<br>INSPEKSI KARET VAKUM MESIN *IN MOULD LABELLING*<br>*PLANT* I |
|---|---|
| **TUJUAN**<br>Untuk memastikan karet mulut vakum dalam kondisi baik | |
| **ALAT YANG DIGUNAKAN**<br>1. Checksheet | |
| **PENANGGUNG JAWAB**<br>1. *Leader* | |
| **PROSEDUR** | |
| NO | URAIAN PROSEDUR |
| 1 | Mengamati permukaan karet |
| 2 | Memberi tanda centang pada kolom "BOLONG" pada *cheecksheet* apabila terdapat lubang yang kasat pada dan tanda ( - ) apabila tidak terdapat lubang yang kasat mata |
| 3 | Meraba permukaan karet |
| 4 | Memberi tanda centang pada kolom "KASAR" pada *checksheet* apabila permukaan kasar dan tanda ( - ) apabila permukaan halus |
| 5 | Menggerakan karet ke berbagai arah |
| 6 | Memberi tanda centang pada kolom "KAKU" pada *checksheet* apabila karet kaku dan tanda ( - ) apabila karet lentur |
| 7 | Memberi tanda centang pada kolom "YA" untuk penggantian karet atau memberi tanda centang pada kolom "TIDAK" apabila penggantian tidak perlu dilakukan |
| Gambar karet mulut vakum | |

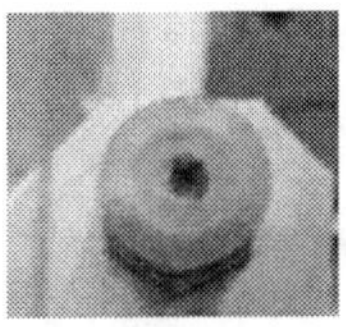

**Figure 3.** SOP Inspection rubber vacuum

| bumimulia | Departemen Quality Control | No. Dokumen | : QC - 135 |
|---|---|---|---|
| | PT. Bumimulia Indah Lestari | Tanggal | : 30-Mar-2019 |
| | Checksheet pengecekan vakum | Halaman | : 1 |
| Mesin | : B-42 | Nama Produk | : Yamalube |
| Customer | : Yamaha | Inspektor | : Dea Prameswari |

Pengecekan Karet Mulut Vakum

| TANGGAL | BOLONG | KASAR | KAKU | GANTI KARET | |
|---|---|---|---|---|---|
| | | | | YA | TIDAK |
| 1/4/2019 | ✓ | ✓ | ✓ | ✓ | |
| 8/4/2019 | - | - | - | | ✓ |
| | | | | | |
| | | | | | |
| | | | | | |
| | | | | | |
| | | | | | |
| | | | | | |
| | | | | | |
| | | | | | |
| | | | | | |
| | | | | | |
| | | | | | |
| | | | | | |
| | | | | | |
| | | | | | |
| | | | | | |
| | | | | | |
| | | | | | |

**Figure 4.** Example check sheet

Check sheet need to be made to ensure that when the rubber is in check, the gum condition and the action taken. Check sheet shown in Figure 4.

Repair using SOP and check sheet are improvements that can be made in the near term, while the proposed improvements in the long term is to replace the vacuum mouth with anti-static vacuum shown in Figure 5.

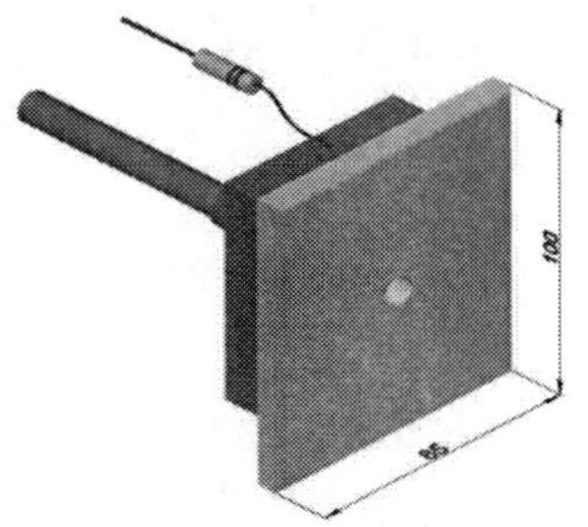


**Figure 5.** Anti-Static Vacuum

*Anti-static vacuum* is a tool used in the process of in mould label to move the label into a mould or mould bottle. As for how the anti-static vacuum is as follows:

1. The arm will take the label by using a vacuum
2. The label will be attached to the bearing surface of the vacuum. At the same time, the voltage of 12kV 7- given for 35 to 60 milliseconds on a plate in a vacuum and the wind stopped
3. The label will be attached to the pads and arm moves towards mold
4. Sleeve label release in the mold

## 4. Conclusion

Based on the results of data processing that has been done, the conclusion of this research as follows:

1. After calculation of DPMO known that DPMO value amounted to 250,000 units, and the result of the conversion value to the value of Sigma obtained DPMO sigma level at 2,17
2. The highest RPN value is 336, on the label is missing label caused by a worn rubber vacuum mouth.
3. Based on *decision tree* who has made it known that most influential attribute bottle to bottle disability is the position followed by the number, colour and perforated.
4. The improvement is given based on the results of data analysis is making SOP, check sheet and anti static vacuum.

## 5. References

[1] MN Nasution 2015 *Integrated Quality Management (Total Quality Management)* Third Edition. (Bogor: Ghalia Indonesia)

[2] G Kunal 2012 Improvement Process for Rolling Mill Through The DMAIC Six Sigma Approach *International J. for Quality Research.* **6**(3): 221-231

[3] PA Krismasurya, Nasir Widha and Farela Cheers 2014 Six Sigma Approach to Reduce Defect in Process of Plastic Bottle Blow Molding Machine ASB in 2000 ml *J of Industrial Engineering and Management Systems* **3**(1), pp. 189-199.

[4] R Fitriana, D Sugiarto, J Saragih and A Bagio 2014 Aplikasi Six Sigma Dan Data Mining Untuk Meningkatkan Kualitas Pada Industri Manufaktur. *Seminar Nasional Teknik Industri BKSTI*, pp. 92-98

[5] NB Puspitasari and A Martanto 2014 The use of FMEA in Production Process Failure Risk Identifying ATM Gloves (Weaving Machine Tools) (Case Study PT. Asaputex Jaya Tegal). *J. TI Undip* **9**(2), pp 93-98

[6] DF Barry, D Adianto and YYuniati 2015 Usulan Pengendalian Kualitas Produk Isolator Dengan Metode Failure Mode Effect Analysis (FMEA) dan Fault Tree Analysis (FTA). *J. online Institut Teknologi Nasional* **3**(2), pp 82-91

[7] D Caesaron and S John 2015 Implementation of DMAIC approach for PVC Pipe Production Process Improvement (Case Study PT. Rusli Vinilon). *J Metris,* pp. 91-96

[8] D Rukmayadi and S Sugiarti 2015 Method Approach Six Sigma (DMAIC) for Product Quality Improvement Boncabe CV Kobe and Lina Food *J. of Industrial Engineering & Management Systems* **8**(1), pp. 1-11

[9] Safrizal and Muhajri 2016 Pengendalian Kualitas dengan Metode Six Sigma. *J. Manajemen Dan Keuangan* **2**(5), pp. 615-626

[10] N Wayan, A S Dewi, S Mulyani and I Wayan Arnata 2016 Attributes of Packaging Quality Control Method Using Failure Mode Effect Analysis (FMEA) in Production Process Bottled Water. *J Agroindustrial Engineering and Management* **4**(3), pp. 149-160.

[11] R Fitriana, J Saragih and S Sarasaty 2014 Implementation of Six Sigma and Data Mining to Improve Die Casting Production Process at PT. AB. *Proceedings of the 7th International Seminar on Industrial Engineering and Management* pp. 64-71.

[12] L Girmanova, MSolc, J Kliment, A Divokova, and Vojtech Miklos 2017 Application of Six Sigma DMAIC Methodology Using in The Process of Product Quality Control in Metallurgical Operation. *Acta Technologica Agriculturae* **4**, Nitra Slovaca Universitas Agriculturae Nitriae pp 104-109.

[13] M Zasadzien 2017 Application of The Six Sigma Method for Improving Maintenance Processes - Case Study *Proceedings of the 6th International Conference on Operations Research and Enterprise Systems*, pp. 314-320.
[14] R Fitriana, J Saragih and N Luthfiana 2017 Model business intelligence system design of quality products by using data mining in R Bakery Company. *IOP Conf. Series: Materials Science and Engineering* **277**, 012005.

# Retailer's EOQ model considering demand and holding cost of the defective items under carbon emission tax

**Yosef Daryanto[1], Bellachintya Reira Christata[2], Ika Murti Kristiyani[3]**

[1,2,3] Department of Industrial Engineering, Universitas Atma Jaya Yogyakarta,
Jl. Babarsari 43, Yogyakarta, Indonesia
yosef.daryanto@uajy.ac.id, 195603031@students.uajy.ac.id, ika.murti@uajy.ac.id

**Abstract.** This paper presents a retailer's inventory model considering imperfect production process and material handling that cause some defective items. The defective items possess a fraction of its original utility. Therefore, after a quality inspection, the retailer holds them in a different area until they are sold entirely. The proposed model considers the demand and holding cost of the defective items. It also incorporates carbon emission costs from transportation and storage activities. The objective function is to maximize the expected total profit, which simultaneously minimizes the total carbon emissions. A numerical example illustrates the model implementation. From this data set, the optimum order quantity is 1770.1 units, and the backorder quantity is 580.3 units, which give an expected total profit of $ 1,227,945 per year and expected total emissions of 0.765 ton$CO_2$/year. Further analysis shows a trade-off between economic and environmental performance. Incorporating carbon emission costs into an EOQ model will reduce the expected total carbon emissions. However, it also causes a reduction in the expected total profit.
**Keywords**: EOQ, defective items, demand, carbon emission

## 1. Introduction

In many manufacturing environments, machine reliability and human error are part of existing production and material handling problems. They may cause some defective items with considerable economic consequences. This situation affects inventory decisions, such as in economic order quantity (EOQ) models. Salameh and Jaber [1] and Cardénas-Bárron [2] constructed a fundamental approach to address this situation by setting a quality inspection. Then, Maddah and Jaber [3] made some improvements concerning the probability function of the defective items. Many researchers assume that the defected items can be sold as a single lot [1], [2], [3], [4], [5], while Teng and Hsu [6] identified a situation in which the defective items are gradually depleted following customer demand. In real business, the demand rate of defective items is lower than the demand rate of a good product. Other researchers were allowing shortages and assuming either full or partial backorder [6], [7], [8], [9], [10].

The green logistics system is an emerging research topic [11]. Recently, studies on inventory models with concern on environmental impact attracted many researchers. Battini et al. [12] identified some sources of emissions in inventory management activity including material handling, transportation, storage, sales, and waste disposal. Therefore, inventory decision models may incorporate carbon tax and carbon cap policies [13], [14]. Previous studies show that carrying more inventory may not be the wrong decision as it can lower the total emissions by reducing energy usage. Chen et al. [15] developed EOQ models to study the effect of various carbon emission policies. They considered emissions from ordering and inventory holding activities and found a possibility of higher emission reduction percentage compared to the increase in cost. Hence, there are possibilities of a trade-off between economic and environmental performance [16]. Wang and Ye [17] proved the benefit of incorporating carbon emissions into decision making in terms of lower carbon emissions and cost. They compared the total carbon emissions resulted from the EOQ and just-in-time (JIT) approach. Recently, Taleizadeh et al. [18] considered carbon emissions, shortages, and sales discounts. The objective is to optimize the order cycle time and selling price.

Recent research simultaneously considers the effect of defective items and carbon emissions in an EOQ model. Kazemi et al. [19] identified a lower total expected profit when an EOQ model incorporates carbon emission costs. Besides, the model will also result in a smaller order size. Wee and Daryanto [20] examined EOQ models when there is a percentage of defective items in a received lot considering carbon emissions costs and a full backorder. Sarkar et al. [21] allowed the repairment of the defective items. In their proposed model, the holding cost and carbon emission from holding the perfect and repaired items are different.

## 2. Methods

Besides maximizing its profit, a retailer is willing to reduce the environmental impact of its logistics activities by optimizing the lot size and shortage backorder. An extended EOQ model is studied, taking into consideration the probability, demand rate, and holding cost of the defective items. Further, carbon emission cost from delivering and holding the items, and backorder cost are counted. Figure 1 illustrates the inventory model for one replenishment cycle. The proposed model is constructed based on the model from Maddah and Jaber [3], Teng and Hsu [6], and Wee and Daryanto [20].

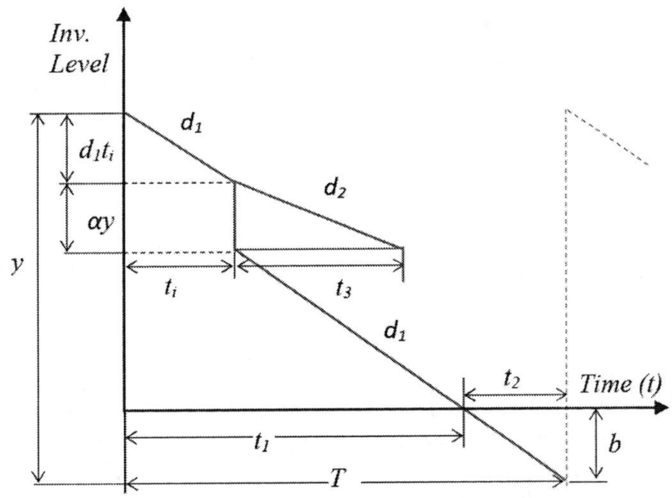


**Figure 1.** Inventory level with defective items sale

The notations for all parameters and variables are listed below:

| | |
|---|---|
| $T$ | order cycle time (year); (decision variable) |
| $y^*$ | optimum order quantity (units); (decision variable) |
| $b^*$ | optimum backorder quantity (units); (decision variable) |
| $d_1$ | rate of customer demand for good quality items (units/year) |
| $d_2$ | rate of customer demand for defective items (units/year); $d_1 > d_2$ |
| $\alpha$ | rate of defective items per lot $y$, |
| $f(\alpha)$ | probability density function (pdf.) of $\alpha$ |

| | |
|---|---|
| $i$ | rate of quality inspection (units/year); $i >> d_1$ |
| $t_i$ | quality inspection period (year) |
| $P_p$ | selling price of good quality items (\$/unit) |
| $P_d$ | selling price of the defective items (\$/unit); $P_p > P_d$ |
| $C_o$ | cost of ordering per cycle (\$/cycle) |
| $C_p$ | per unit purchasing cost (\$/unit) |
| $C_i$ | per unit quality inspection cost (\$/unit) |
| $C_{h1}$ | unit cost for holding good quality items (\$/unit/year) |
| $Ch_2$ | unit cost for holding defective items (\$/unit/year) |
| $C_b$ | penalty cost due to backorder (\$/unit) |
| $C_d$ | fixed cost per order delivery (\$/cycle) |
| $C_f$ | variable cost from fuel consumption (\$/liter) |
| $C_t$ | carbon tax (\$/ton$CO_2$) |
| $l$ | delivery distance (km) |
| $F$ | direct emission from fuel usage (ton$CO_2$/liter) |
| $E$ | indirect emission from electricity usage (ton$CO_2$/kWh) |
| $w$ | product weight (ton/unit) |
| $a$ | fuel efficiency for empty truck (liter/km) |
| $b$ | fuel efficiency from truckload (liter/km/ton) |
| $c$ | average electricity consumption per unit inventory (kWh/unit) |
| $ETR$ | expected total revenue (\$) |
| $ETC$ | expected total cost (\$) |
| $ETPU$ | expected total profit per unit time (\$) |
| $E[.]$ | expected value operator of $\alpha$ |

Further, the proposed model works under some assumptions:
1. The retailer plans an item with constant demand.
2. A particular probability density function of the defective items is known.
3. The inspection rate is much higher than the demand rate so that no shortage during the inspection.
4. After inspection, the retailer keeps the defective items in a different area. They will be sold at a lower price and have a lower demand rate.
5. Shortages are allowed and completely backordered.

Figure 1 shows that the retailer receives $y$ units at the beginning of the cycle. During the period $[0, t_i]$, the inventory level depleted due to customer demand $d_1$. At the same time, the retailer performs a quality inspection.

$$t_i = y/i$$

At time $t_i$, the inspection ends, and the defective items $\alpha y$ being separated into a different area. Inventory level of the perfect items is continuously decreasing due to customer demand $d_1$ and reaches zero at time $t_1$. Meanwhile, the inventory level of the defective items also decreases in a lower rate $d_2$ and reaches zero at time $t_3$. Further, shortages are accumulated during $t_2$ with $d_1$ rate and will be completely backordered at the beginning of the next cycle.

From Fig. 1, $T = t_1 + t_2$ and $T = (1-\alpha)y/d_1$. Due to the probability of defective items, $E[T] = (1-E[\alpha])y/d_1$. The $ETR$ comes from the total sales of the perfect as well as the defective items. Hence,

$$ETR = (1 - E[\alpha])yP_p + E[\alpha]yP_d \quad (1)$$

$ETC$ per cycle = ordering cost ($C_1$) + purchasing cost ($C_2$) + inspection cost ($C_3$) + holding cost of perfect items ($C_4$) + holding cost of defective items ($C_5$) + backorder cost ($C_6$) + transportation cost ($C_7$)

$$ETC = C_1 + C_2 + C_3 + C_4 + C_5 + C_6 + C_7 \tag{2}$$

In which $C_1 = C_o$, $C_2 = yC_p$ , and $C_3 = yC_i$.

The amount of inventory per cycle for perfect items is similar to Wee et al. [8]. The inventory holding cost considers the traditional holding cost for perfect items ($C_{h1}$) and carbon emission cost ($cEC_t$) due to energy consumption in holding the inventory. Therefore, for the perfect items

$$C_4 = (C_{h1} + cEC_t)\left(\frac{1}{2}\frac{(y - E[\alpha]y - b)^2}{d_1} + \frac{E[\alpha]y^2}{i}\right) \tag{3}$$

The amount of inventory per cycle for defective items is similar to Teng and Hsu [6]. Because $t_3 = E[\alpha]y/d_2$, one has

$$C_5 = (C_{h2} + cEC_t)\frac{(E[\alpha]y)^2}{2d_2} \tag{4}$$

The backorder cost is

$$C_6 = \frac{1}{2}\frac{C_b b^2}{d_1} \tag{5}$$

The delivery activity implies transportation costs (e.g., setup, fuel, etc.) and carbon emission costs from fuel combustion. The transportation cost consists of a fixed cost ($C_d + 2alC_f$) and a variable cost ($blC_f wy$), which depends on the delivery quantity. Hence, the carbon emission cost also has a fixed cost ($2alFC_t$) and a variable cost ($blwFC_t y$). Therefore,

$$C_7 = C_d + 2alC_f + blwC_f y + 2alFC_t + blwFC_t y \tag{6}$$

Finally, the *ETC* per cycle becomes

$$\begin{aligned} ETC = {} & C_o + yC_p + yC_i + (C_{h1} + cEC_t)\left(\frac{1}{2}\frac{(y - E[\alpha]y - b)^2}{d_1} + \frac{E[\alpha]y^2}{i}\right) + (h_2 + cEC_t)\frac{(E[\alpha]y)^2}{2d_2} \\ & + \frac{1}{2}\frac{C_b b^2}{d_1} + C_d + 2alC_f + blwC_f y + 2alFC_t + blwFC_t y \end{aligned} \tag{7}$$

From Maddah and Jaber [3],

$$ETPU = \frac{ETR - ETC}{E[T]}$$

Therefore,

$$ETPU = \frac{\begin{array}{c}(1 - E[\alpha])yP_p + E[\alpha]yP_d - C_o - yC_p - yC_i - (C_{h1} + cEC_t)\left(\frac{1}{2}\frac{(y - E[\alpha]y - b)^2}{d_1} + \frac{E[\alpha]y^2}{i}\right) \\ -(h_2 + cEC_t)\frac{(E[\alpha]y)^2}{2d_2} - \frac{1}{2}\frac{C_b b^2}{d_1} - C_d - 2alC_f - blwC_f y - 2alFC_t - blwFC_t y\end{array}}{\frac{(1 - E[\alpha])y}{d_1}} \tag{8}$$

Taking the second derivative of *ETPU* with respect to *b* and *y*, one has

$$\frac{\partial^2 ETPU}{\partial b^2} = -\frac{(C_{h1} + cEC_t + C_b)}{y}\frac{1}{1 - E[\alpha]} \tag{9}$$

$$\frac{\partial^2 ETPU}{\partial y^2} = \left(\frac{b^2}{y^3}(C_{h1} + cEC_t + C_b) + \frac{2d_1}{y^3}(C_o + C_d + 2alFC_t + 2alC_f)\right)\frac{1}{E[\alpha] - 1} \tag{10}$$

$$\frac{\partial^2 ETPU}{\partial b \partial y} = \frac{b}{y^2}(C_{h1} + cEC_t + C_b)\frac{1}{1 - E[\alpha]} \quad (11)$$

$$\left(\frac{\partial^2 ETPU}{\partial b \partial y}\right)^2 - \left(\frac{\partial^2 ETPU}{\partial b^2}\right)\left(\frac{\partial^2 ETPU}{\partial y^2}\right) = -\frac{2d_1(C_{h1} + cEC_t + C_b)(C_o + C_d + 2alFC_t + 2alC_f)}{y^4}\frac{1}{(E[\alpha] - 1)^2} \leq 0 \quad (12)$$

From equations (9), (10), and (12), the profit function *ETPU* is strictly concave. Further, setting the first derivative of *ETPU* with respect to *b* and *y* equal to zero, $b^*$ and $y^*$ can be obtained as follow:

$$b^* = \frac{(C_{h1} + cEC_t)y^*(1 - E[\alpha])}{(C_{h1} + cEC_t + C_b)} \quad (13)$$

$$y^* = \sqrt{\frac{(2d_1C_o + 2d_1C_d + 4d_1alC_f + 4d_1alFC_t + C_bb^2 + C_{h1}b^2 + cEC_tb^2)id_2}{C_{h1}(id_2 - 2d_2iE[\alpha] + 2d_1d_2E[\alpha] + d_2iE[\alpha]^2) + C_{h2}d_1iE[\alpha]^2 + (cEC_t)(d_2i + 2d_1d_2E[\alpha] - 2d_2iE[\alpha] + d_1iE[\alpha]^2 + d_2iE[\alpha]^2)}} \quad (14)$$

Further, the ETE per unit time from inventory holding and delivery activities can be subtracted from Eq. (7) as

$$ETE = \left(cE\left(\frac{1}{2}\frac{(y - E[\alpha]y - b)^2}{d_1} + \frac{E[\alpha]y^2}{i}\right) + cE\frac{(E[\alpha]y)^2}{2d_2} + 2alF + blwFy\right)\left(\frac{d_1}{(1 - E[\alpha])y}\right) \quad (15)$$

## 3. Result and discussion

A set of data is provided to illustrate the research problem and its solution. The data is adapted from Teng and Hsu [6], with some additional transportation and carbon emissions parameters from Wee and Daryanto [20], as follows:

$d_1 = 50{,}000$; $d_2 = 30{,}000$; $P_p = 50$; $P_d = 40$; $i = 175{,}200$; $w = 0.01$ ton/unit;
$C_o = 100$; $C_b = 20$, $C_p = 25$, $C_{h1} = 10$; $C_{h2} = 8$; $C_i = 0.5$; $C_t =$ \$75/ton$CO_2$; $C_d =$ \$100/delivery
$l = 10$ km; $F =$ \$0.75/L; $a = 27$ L/ 100km; $b = 0.57$ L/100km/ton;
$c = 1.44$ kWh/unit; $F = 2.6 \times 10^{-3}$ ton$CO_2$/L; $E = 0.5 \times 10^3$ ton$CO_2$/kWh;

The percentage of defective items $\alpha$ is assumed to have a uniform distribution with:

$$f(\alpha) = \begin{cases} 25, & 0 \leq \alpha \leq 0.04 \\ 0, & otherwise \end{cases}, \qquad E[\alpha] = 0.02$$

Solving the maximizing function of *ETPU* using equations (13) and (14) simultaneously, one has $b^* =$ 580.3 units and $y^* =$ 1770.1 units with the expected total profit \$ 1,227,945 per year. Further, the expected total emissions are 0.765 ton$CO_2$/year.

Suppose that there is no carbon tax regulation (i.e., $C_t = 0$). Solving the problem simultaneously, one has $b^* =$ 577.8 units and $y^* =$ 1768.8 units with the $ETPU =$ \$ 1,228,002 per year and $ETE =$ 0.766 ton$CO_2$/year.

The above results show that the proposed inventory model has successfully reduced the expected total carbon emissions by incorporating carbon emission into an EOQ model. However, it results in a reduction in the expected total profit. A similar lower expected total profit situation was gained by Kazemi et al. [19] in their study.

## 4. Conclusion

This study proposes an EOQ model to solve a profit maximization problem of a retailer that willing to reduce the environmental impact of its inventory decisions in terms of total carbon emissions. The model incorporated a percentage of defective items in a received lot, followed by inspection and separation of the defective items. It considers the sales rate and holding cost for both the perfect and defective items and a complete backorder. Optimal order quantity and maximum backorder level are stated. A numerical example is presented, and the result shows that incorporating carbon emissions into an EOQ model may reduce total carbon emissions although there is a reduction in the expected total profit.

The proposed model can be extended for cases with a partial backorder and items with deterioration [22], [23]. Besides, this study considered a deterministic customer demand for defective items with a known selling price. Therefore, future research could determine the optimal price for the defective items to increase the demand rate.

## Acknowledgments

The authors acknowledge the editor and reviewers for their services, comments, and suggestion. We also thank Universitas Atma Jaya Yogyakarta for supporting this research under Hibah Penelitian Percepatan Guru Besar No. 382/HB-PEN PGB/LPPM/XI/2019.

## 5. References

[1] Salameh M K and Jaber M Y 2000 Economic order quantity model for items with imperfect quality, *International Journal of Production Economics* **64** pp 59-64

[2] Cárdenas-Barrón R 2000 Observation: "Economic production quantity model for items with imperfect quality" [Int. J. Production Economics 64 (2000) 59-64], *International Journal of Production Economics* **67** pp 201

[3] Maddah B and Jaber M Y 2008 Economic order quantity for items with imperfect quality: Revisited, *International Journal of Production Economics* **112** pp 808-815

[4] Khan M, Jaber M Y and Bonney M 2010 An economic order quantity (EOQ) for items with imperfect quality and inspection errors, *International Journal of Production Economies* **133** pp 113-118

[5] Hsu L F 2012 A note on "An economic order quantity (EOQ) for items with imperfect quality inspection errors", *International Journal of Industrial Engineering Computation* **13** pp 695 -702

[6] Teng H M and Hsu P H 2013 Revisit economic ordering strategy for items with imperfect quality and backordering. *International Journal of Management and Enterprise Development* **12**(3) pp 251-262

[7] Rezaei J 2005 Economic order quantity model with backorder for imperfect quality items, *Proc. of the 2005 IEEE International Engineering Management Conference*. DOI: 10.1109/IEMC.2005.1559191

[8] Wee H M, Yu J and Chen M C 2007 Optimal inventory model for items with imperfect quality and shortage backordering. *Omega* **35** pp 7-11

[9] Chang H C and Ho C H 2010 Exact closed-form solutions for "optimal inventory model for items with imperfect quality and shortage backordering. *Omega* **38** pp 233-237

[10] Taleizadeh A A and Zamani-Dehkordi N 2017 A stochastic lot sizing model with partial backordering and imperfect production processes, *International Journal of Inventory Research* **4**(1) pp 75-96

[11] Shaharudin M S, Fernando Y, Jabbour C J C, Sroufe R and Jasmi M F A 2019 Past, present, and future low carbon supply chain management: A content review using social network analysis. *Journal of Cleaner Production* **218** pp 629-643

[12] Battini D, Persona A and Sgarbossa F 2014 A sustainable EOQ model: theoretical formulation and applications. *International Journal of Production Economics* **149** pp. 145-153

[13] Bouchery Y, Ghaffari A, Jemai Z and Dallery Y 2012 Including sustainability criteria into inventory models. *European Journal of Operational Research* **222** pp 229-240

[14] Benjaafar S, Li Y and Daskin M 2013 Carbon footprint and the management of supply chains: insights from simple models *IEEE Transactions on Automation Science and Engineering* **10**(1) pp 99-116

[15] Chen X, Benjaafar S and Elomri A 2013 The carbon-constrained EOQ. *Operations Research Letters* **41** pp 172-179

[16] Bonney M and Jaber M Y 2011 Environmentally responsible inventory models: non-classical models for a non-classical era. *International Journal of Production Economics* **133** pp. 43-53

[17] Wang S and Ye B 2018 A comparison between just-in-time and economic order quantity models with carbon emissions. *Journal of Cleaner Production* **187** pp. 662-671

[18] Taleizadeh A A, Hazarkhani B and Moon I 2018 Joint pricing and inventory decisions with carbon emission considerations, partial backordering and planned discounts. *Annals of Operations Research* https://doi.org/10.1007/s10479-018-2968-y

[19] Kazemi N, Abdul-Rashid S H, Ghazilla R A R, Shekarian E and Zanoni S 2018 Economic order quantity models for items with imperfect quality and emission considerations. *International Journal of Systems Science: Operations & Logistics* **5**(2) pp 99-115

[20] Wee H M and Daryanto Y 2020 Imperfect quality item inventory models considering carbon emissions. In Shah N H and Mittal M *(eds.) Optimization and Inventory Management* (Singapore: Springer Nature) pp 137-159

[21] Sarkar B, Ahmed W, Choi S B and Tayyab M 2018 Sustainable inventory management for environmental impact through partial backordering and multi-trade-credit-period, *Sustainability* **10**(12) Article No 4761

[22] Daryanto Y, Wee H M, Widyadana G A 2019 Low carbon supply chain coordination for imperfect quality deteriorating items, *Mathematics* **7**(3) Article No 234

[23] Kung K Y, Huang Y D, Wee H M, Daryanto Y 2019 Production-inventory system for deteriorating items with machine breakdown, inspection, and partial backordering, *Mathematics* **7**(7) Article No 616

# Design of Multicriteria Decision Making Tools for IT Project Selection: A Case from Software House

**R P Soesanto[1], W Tripiawan[2], and I Darmawan**

School of Industrial Engineering, Telkom University, Bandung - Indonesia
raysoesanto@telkomuniversity.ac.id[1], wawan.tripiawan@gmail.com[2]

**Abstract.** Information Technology as an enabler helps organization to improve business by gaining more competitive advantage, it plays a significant role for organization to compete. Many organizations sense the pressure to leverage their investment towards information system. Selecting the optimal portfolio of IT projects is becoming increasingly important as the dependency on IT for organizational performance increases. The purpose of this research is to design tools for IT project selection tools in the software house. Combination of Delphi, AHP, and Factor rating method is used to gain the prioritize of the projects. Agile method is used to develop the application. The application is built in web platform to help the organization decision. Future research can be done to consider more variables for choosing the right project for the organization.
**Keywords:** Agile Method, AHP, Factor Rating, Project Selection, Software House

## 1. Introduction

The growth of Information Technology (IT) leads the significance of using IT as an enabler to help improve business process from organization. The development of IT has brought changes into the business environment in the organization, in example it can enhance the customer service, reduce distribution cost, etc [1]. Nowadays the utilization of information technology for the organization becomes a requirement to support the business [2]. Information Technology (IT) plays a significant role in an organization's ability to compete in the marketplace. Due to many advantages can be obtained through the implementation of IT, many organizations senses the pressure to leverage their investment towards information system [3]. Selecting the optimal portfolio of IT projects is becoming increasingly important as the dependency on IT for organizational performance increases [4]. The process of project selection is not easy, because there is a lot of criteria that must be taken into consideration, besides there are many issues such as too many projects, aligning business strategy and projects and poor coordination between projects and the resistant from the organization. In a software company, usually there is two type of project that is occurred during the business, which is the internal project that can help the company to gain advantage and the external project which is the main activity of the software company. The research is needed to support start-up company to guide them in taking a decision and avoid a choice that could lead to business failure. Most of the IT project-based companies are dependent on the project revenues as the source of income, therefore the optimal project selection is needed to increase the profitability of the organization. The need and authorization for projects, regardless of the selection methods, can come from one or more sources: marketplace opportunity, business need, customers, advances in technology, legal, and social [5].

The purpose of this research is to design the tools to pick and choose the suitable IT project for software company both internally and externally using multicriteria decision making (MCDM). MCDM methods are commonly used to rank multiple alternatives according to multiple and often conflicting criteria, therefore it is suitable to use in the IT project selection [6]. The object of this study is a start-up company, especially software house. According to Archibald, Thomas, Betts, and Johnston[7], start-up companies are considered to be an important element in a successful economy. In addition to the IT project selection, the Delphi method is also often used as a reference in qualitative determinations. The most important requirements for the use of Delphi are the need for experts' judgment, group consensus to achieve the results, anonymity in data collection, a complex, multidimensional, and interdisciplinary problem, lack of consensus and imperfect knowledge, experienced and capable experts, dispersion of experts, no time limitation, and lack of cost-effective method [8]. The Delphi method is beneficial when the problem at hand can benefit from collective, subjective judgments or decisions and when group dynamics do not allow for effective communication (e.g., time differences, distance, and personality conflicts) [9]. In this research, Analytical Hierarchy Process (AHP) is used. The Analytic Hierarchy Process (AHP) is a theory of measurement through pairwise comparisons and relies on the judgements of experts to derive priority scales. It is these scales that measure intangibles in relative terms. The correlations are made utilizing a size of total decisions that speaks to, the amount increasingly, one component overwhelms another regarding a given property. The decisions might be conflicting, and how to gauge irregularity and improve the decisions, when conceivable to acquire better consistency is a worry of the AHP [9]. Factor rating method is used to determine the highest score of each possible project, the method multiply the value with the weight of each indicators [10]. The advantage of this method it is most widely used in analytical technique and most popular technique because a wide variety of factor can be included in the analysis [11]. In factor rating method, Most Important Factors must be identified in evaluating alternative sites for the new facility. Then a weight between 0 and 100 should be assigned to each of these factors. Agile method is used to develop the application because the methods can break down the projects into small chunk and support the dynamic change in the requirement from the stakeholder [12], the dynamic change can embrace the flexibility and the agility in the development process which something that the traditional approach can't be supported [13].

## 2. Methods

The purpose of this research is to design the tools to pick and choose the suitable IT project for software company both internally and externally using multicriteria decision making (MCDM). The object of the study is one software house company in Bandung, Indonesia. To choose the suitable project, it is needed to create the criteria based on the stakeholder perspective. The criteria are gathered from previous research and validate with the stakeholder. The stakeholder in this research is the CEO, CTO, COO of the company. Delphi method is used to collect the suitable criteria from the stakeholder, according to Grime [9], Delphi is suitable for facing the collective judgments such as group decision where effective communication is limited. After the criteria is defined, the next step is to develop the selection method, Analytical Hierarchy Process (AHP) is used to select the best portfolio by determine the weight of each criteria. AHP is suitable for multicriteria decision because it measured through pairwise comparisons and relies on the judgements of experts to derive priority scales [14]. To determine the weight of each criteria it is needed to check the consistency of the data. To gain which project that is need to be selected, factor rating is used to calculate the weight and the value given by the stakeholder.

After the selection method is developed the next step is to create the tools for calculate the result. Agile method is used to develop the application. Agile have the ability to rapidly adapt to volatile requirements which the plan-driven approach can't accommodate [15], the unique characteristic of agile method is the modularity on development process level, the iterative phase with short cycles enabling fast verifications and corrections [16]. Figure 1 shows the agile method phase. The first phase is the planning, in this phase the project backlog is created. After the product backlog is created, the next step is the sprint process, each sprint consists of iterative method from requirement, develop, and testing phase. The sprint process usually takes one weeks to develop and is done by selecting the most possible product backlog to be developed [12]. In each sprint, daily meeting is essential, after each sprint is finished the next step is the retrospective meeting with the product owner, the product owner in this research is the CEO of the company.

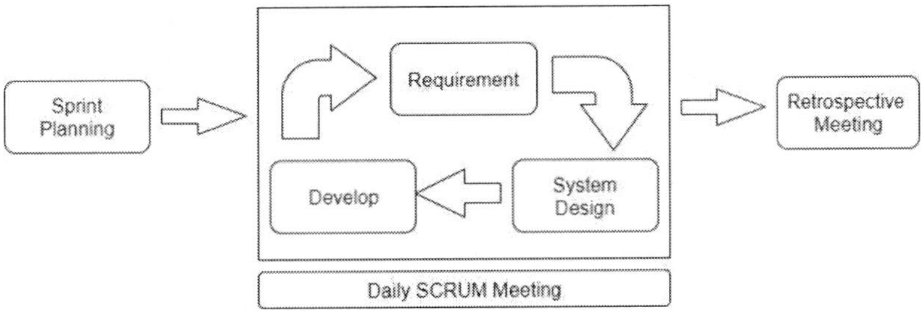


**Figure 1.** Agile Method SCRUM

## 3. Result and Discussions

### *3.1. Criteria Development*

In criteria development, literature study is conducted to gather the suitable criteria for IT Project selection. Table 1 shows the criteria gathered for this research.

**Table 1.** Criteria for IT Project Selection

| No. | Criteria | Author |
|---|---|---|
| 1 | Politics | [4] |
| 2 | Value / Opportunity | [4]; [5] |
| 3 | Execution | [4]; [6] |
| 4 | Time-Pressure | [4]; [6]; [17] |
| 5 | Relevant Methods | [4] |
| 6 | Resource | [4] ; [17]; [18] |
| 7 | Risk | [4]; [6] |
| 8 | Management Commitment | [4] |
| 9 | Strategy Alignment | [4] ; [5]; [6] [18] |
| 10 | Social | [5]; [19] |
| 11 | Legal | [5] |
| 12 | Cost | [17]; [18];[19] |
| 13 | Profitability | [18]; [19] |

After develop the base of criteria the next step is to validate the criteria to the stakeholder. Data gathering is conducted to the CEO, CTO, and COO of the company. All the chief officer is asked to prioritize which criteria is the most important for the company. Delphi method is used in the form of a questionnaire. The consensus of each stage is set to 60%.

The consensus is obtained after 3 iterations, the result of the prioritized information system is shown in Table 2. From the result, it is known that only 10 criteria that is selected by the chief officer. The "Execution" criteria is changed into difficulty because all three stakeholders agree that the execution of the project is depend on the difficulty of the IT Projects.

### *3.2. Weight Calculation*

After the criteria is developed. The next step is calculating the weight of each criteria. The calculation of the weight of the criteria was done by using analytical hierarchy process (AHP). The result of the weight calculation can be seen in Table 2. In order to ensure that the respond from respondents was consistent, the calculation of consistency ratio was performed. The value of consistency ratio obtained from the calculation is 0.097. Since the value of consistency ratio is less than 0.1, the respond from respondents is considered consistent and valid. After the weight of each criteria was determined, the next step was to determine the scale of each criteria. Table 3 shows the scale of each criteria.

Table 2. Criteria Result and Weight Calculation

| Criteria# | Criteria | Weight |
|---|---|---|
| C1 | Value / Opportunity | 5% |
| C2 | Difficulty | 2.9% |
| C3 | Time-Pressure | 25.6% |
| C4 | Resource | 9.4% |
| C5 | Risk | 13.6% |
| C6 | Management Commitment | 3.3% |
| C7 | Strategy Alignment | 2.2% |
| C8 | Social | 1.5% |
| C9 | Cost | 18.8% |
| C10 | Profitability | 17.7% |

Factor rating method was used to calculate the score for each IT project. The score generated from factor rating will be used as based for stakeholder to make decision related IT project that will be selected. The step of factor rating is [11]:

1. Identify the criteria
2. Determine the weight for each factor that is reflecting as the relative importance of each criteria
3. Determine the scale for each criterion
4. Determine the score for each criterion for each alternative based on scale, then normalized the score
5. Multiply the score and the weight for each alternative
6. Make recommendation based on highest score

Therefore, the model to calculate the score for each project is as follow:

$$SAS_i = 0.005(NC1_i) + 0.029(NC2_i) + 0.256(NC3_i) + 0.094(NC4_i) + 0.136(NC5_i) + 0.033(NC6_i) + 0.022(NC7_i) + 0.015(NC8_i) + 0.188(NC9_i) + 0.177(NC10_i) \quad (1)$$

$$NC1_i = \frac{SC1_i}{4} \times 10 \quad (2)$$

$$NC2_i = \frac{Min\{SC2\}}{SC2_i} \times 10 \quad (3)$$

$$NC3_i = \frac{Min\{SC3\}}{SC3_i} \times 10 \quad (4)$$

$$NC4_i = \frac{Min\{SC4\}}{SC4_i} \times 10 \quad (5)$$

$$NC5_i = \frac{Min\{SC5\}}{SC5_i} \times 10 \quad (6)$$

$$NC6_i = \frac{SC6_i}{Max\{SC6\}} \times 10 \quad (7)$$

$$NC7_i = \frac{SC7_i}{Max\{SC7\}} \times 10 \quad (8)$$

$$NC8_i = \frac{SC8_i}{Max\{SC8\}} \times 10 \quad (9)$$

$$NC9_i = \frac{Min\{SC9\}}{SC9_i} \times 10 \quad (10)$$

$$NC10_i = \frac{SC10_i}{Max\{SC10\}} \times 10 \quad (11)$$

Where,

| | | |
|---|---|---|
| SASi | : | Factor rating score for project-i |
| NCij | : | Normalized score for project-i in criteria-j |
| SCij | : | Score for project-I in criteria-j |
| Min {SCnj} | : | Minimum score of criteria-j from all project |
| Min {SCnj} | : | Maximum score of criteria-j from all project |

**Table 3.** Scale of Each Criteria

| Criteria# | Scale | Priority |
|---|---|---|
| C1 | 1-10 | Higher C1, higher priority |
| C2 | 1-10 | Lower C2, higher priority |
| C3 | 1-10 | Lower C3, higher priority |
| C4 | 1-10 | Lower C4, higher priority |
| C5 | 1-10 | Lower C5, higher priority |
| C6 | 1-10 | Higher C6, higher priority |
| C7 | 1-10 | Higher C7, higher priority |
| C8 | 1-10 | Higher C8, higher priority |
| C9 | 1-10 | Lower C9, higher priority |
| C10 | 1-10 | Higher C10, higher priority |

### *3.3. Development*

The development phase start with the product backlog. Product backlog is a list that contain all information needed for developing the product [20]. Product backlog is a master list of all project[21]. In the first phase of product development, the product backlog contains the known requirement from the customer. The product backlog result can be seen in Table 4.

**Table 4.** Product Backlog

| No. | Description | Estimation (Hours) |
|---|---|---|
| 1 | Initial Analysis | 4 |
| 2 | Model Development | 8 |
| 3 | IT Project Selection Support System | 12 |

From the product backlog, it is known there is 3 product backlog and in the actual development process, there is 3 sprint backlog that is equal to the product backlog. In each sprint, the development phase is done by defining the menu map for the application. Figure 2 shows the menu map of the application.

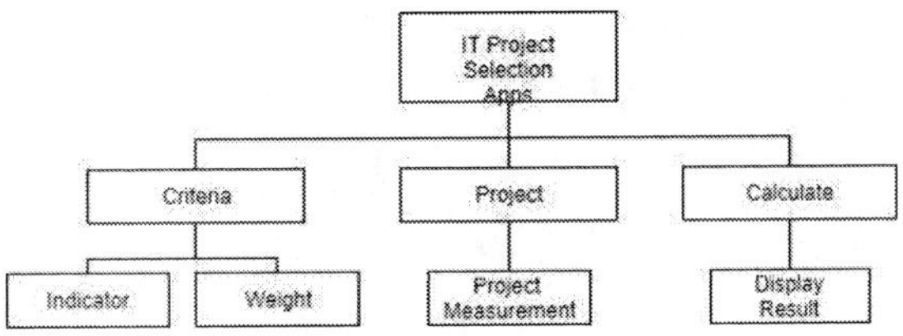


**Figure 2.** Menu Map

IOP Conf. Series: Materials Science and Engineering **847** (2020) 012013 doi:10.1088/1757-899X/847/1/012013

After the menu mapping is done, the next step is creating the Entity Relationship Diagram (ERD) for the design of the database and the Activity Diagram to map the business process into technical view. Figure 3 shows the activity diagram for creating the indicator.

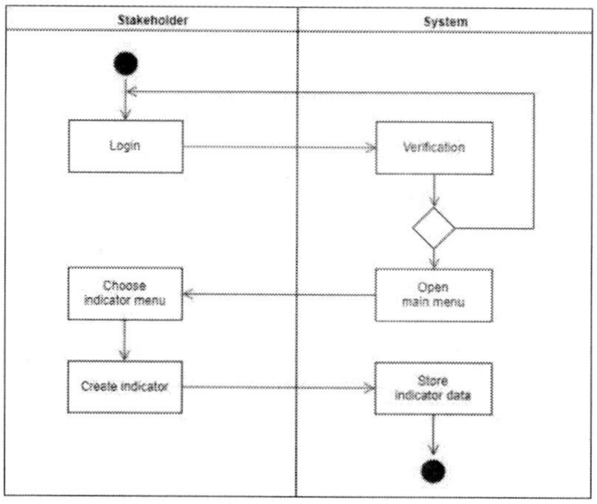


**Figure 3.** Activity Diagram Example

The application is developed based on the activity performed by the stakeholder in selecting the right IT project portfolio. The initial activity is the stakeholder login into the system and choose the criteria menu, in this menu all criteria and the weight are defined. The system will respond by displaying the current criteria that is set by default. After the criteria and weight is defined then the stakeholder can input the current project with the respective value of each indicator. When user choose the calculation menu, the result of the application will be shown. The model form previous step is used to calculate the result. After the design phase is done, the next step is to develop the application, in this research the application platform is web based and using PHP as a programming language.

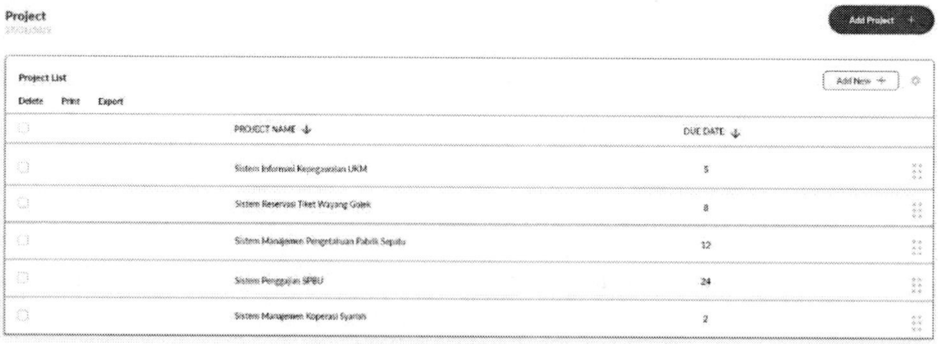


**Figure 4.** Project Menu

Figure 4 shows the project menu, in this menu the project that will be chosen by the stakeholder is shown. In the project menu, when the add project button is clicked, it will prompt the user to input the value of each criteria from 1 to 10 scale.

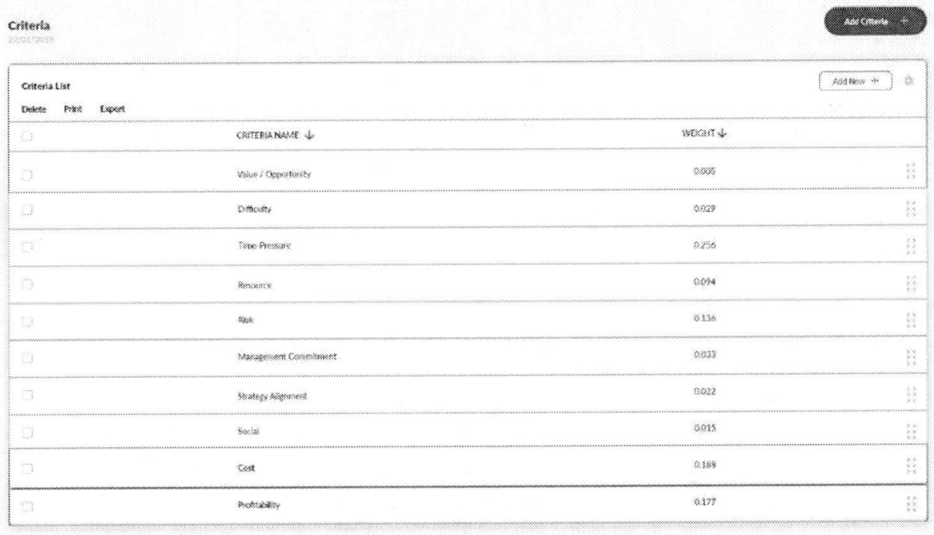


**Figure 5.** Criteria Menu

Figure 5 shows the criteria menu, this menu shows the current criteria set by the stakeholder. When the add criteria is clicked, it will prompt the user to enter the criteria name and the weight of each criteria. In this research, the weight of each criteria is gained from the AHP calculation in the previous phase by giving the AHP questionnaire to the stakeholder.

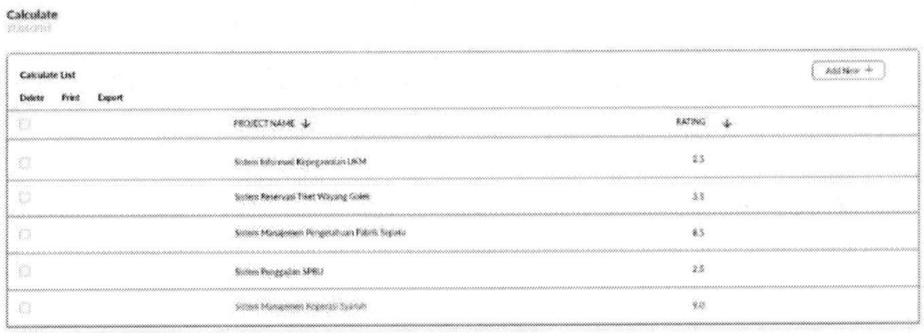


**Figure 6.** Calculation Result

Figure 6 shows the calculation result menu, in this menu each project is scored by the value given by the stakeholder multiply with the weight of each indicator. The highest score will be shown with the green colour in the project name.

## 4. Conclusions

This research tends to guide a start-up IT company in taking better decision by designing the tools to pick and choose the suitable IT project. From the result, we can conclude that in IT Project selection, there are 10 criteria that used to prioritized in IT Project; namely Value/Opportunity, Difficulty, Time-pressure, resource, risk, management commitment, strategy alignment, social, cost, and profitability. From thus 10 criteria, 5 are in higher criteria - higher priority; Value/Opportunity, management commitment, strategy alignment, social, and profitability. And the other 5 have lower criteria - higher priority; Difficulty, Time-pressure, resource, risk, and cost. Future research can be done to test the instrument and add more specific criteria.

## 5. Reference

[1] T. Borshalina and N. Amelia 2016 Internet Based ICT by Indonesian SMEs (Case Study: Batik Trusmi), pp. 231–242.
[2] Y. Fahmi 2018 Analysis of Enterprise Resource Planning (ERP) Systems Implementation. *J. Telemat. Informatics*, **6**(2), pp. 3–8.
[3] R. P. Soesanto, W. Tripiawan, A. F. Rizana, A. Kurniawati, and F. Ramadhan 2019 Information System Planning for Emerging Start-Up Company: A Case from Software House.
[4] K. Pedersen 2016 IT Project Selection: Politics, Experience and good Friends. **19**(1), pp. 55–70.
[5] I. Ahmad, Basheer and ul Haq 2016 Project Selection Techniques, Relevance & Applications in Pakistan Project Selection Techniques, Relevance & Applications in Pakistan. *Int. J. Technol. Res.*, **4**(2), pp. 52–60.
[6] M. Tavana 2019 A hybrid mathematical programming model for optimal project portfolio selection using fuzzy inference system and analytic hierarchy process. *Eval. Program Plann.*, **77**(April), 101703.
[7] T. W. Archibald, L. C. Thomas, J. M. Betts, and R. B. Johnston 2002 Should Start-up Companies Be Cautious? Inventory Policies Which Maximise Survival Probabilities. *Manag. Sci.*, **48**(9), pp. 1161–1174.
[8] A. Habibi, A. Sarafrazi, and S. Izadyar 2015 Delphi Technique Theoretical Framework in Qualitative Research. *Int. J. Eng. Sci.* (December).
[9] B. M. M. Grime and G. Wright 2016 Delphi Method. *Wiley StatsRef Stat. Ref. Online*, (August), pp. 1–6.
[10] A. F. Rizana and R. P. Soesanto 2019 Design of Decision Support System Application for Determining Scholarship Grantee Using Analytical Hierarchy Process and Factor Rating.
[11] P. Sharma, R. K. Phanden, and V. Baser 2012 Analysis for Site Selection Based on Factors Rating," *Int. J. Emerg ing trends Eng. Dev. ment*, **6**(2), pp. 616–622.
[12] Philosophie 2019 *Agile Handbook*. Philosophie.
[13] K. Curcio, T. Navarro, A. Malucelli, and S. Reinehr 2018 Requirements engineering: A systematic mapping study in agile software development," *J. Syst. Softw.*, **139**, pp. 32–50.
[14] T. L. Saaty 2008 Decision making with the Analytic Hierarchy Process," *Int. J. Serv. Sci.*, **1**(1), pp. 83–98.
[15] R. Hoda, N. Salleh, J. Grundy, and H. Mien 2017 Systematic literature reviews in agile software development : A tertiary study. *Inf. Softw. Technol.*, **85**, pp. 60–70.
[16] L. Williams 2002 Agile Software Development. *Comput. Sci. Educ.*, **12**(3), pp. 167–168.
[17] S. Song, F. Yang, and Q. Xia 2019 Computers & Industrial Engineering Multi-criteria project portfolio selection and scheduling problem based on acceptability analysis. *Comput. Ind. Eng.* **135**(June), pp. 793–799.
[18] E. Karasakal and P. Aker 2016 A multicriteria sorting approach based on data envelopment analysis for R & D project selection problem.
[19] Ş. GÜR, M. Hamurucu, and T. Eren 2016 Using Analytic Network Process And Goal Programming Methods For Project Selection In The Public Institution. *Les Cah. du Mecas*, **12**(Mcdm), pp. 36–51.
[20] K. Schwaber and J. Sutherland 2017 *Scrum Guide* **19**(6), p. 504.
[21] M. C. Layton 2015 *Scrum for Dummies* - A Wiley Brand, p. 403.

# Prototype Design of Plastic Waste Processing Equipment

**N Parwati[1,2], R Nurdina[1,3], A T Purwandari[1,4], and W N Tanjung[1,5]**

[1]Industrial Engineering Major, Faculty of Science and Technology, Al Azhar Indonesia University, Komplek Masjid Agung Al Azhar, Jalan Sisingamangaraja, Kebayoran Baru, South Jakarta 12110 Indonesia
[2]niken.parwati@uai.ac.id, [3]dinasantoso08@gmail.com, [4]aprilia@uai.ac.id, [5]widya@uai.ac.id

**Abstract.** Garbage is the remnants of human daily activities or natural processes that are solid or semi-solid, in the form of organic or inorganic substances, and are biodegradable or non-biodegradable which are considered to be no longer useful and thrown into the environment [2]. Therefore, it is necessary to develop a plastic waste processing device in order to produce an effective output and use of time efficiently. In this research, the development of plastic waste processing equipment products is done by integrating the chopper and heating machine into one tool, then can produce various kinds of recycle products. This plastic waste processor can process thermoplastic plastic types. Thermoplastic is a type of plastic that can be recycled or reprinted by reheating processes, such as Polyethylene (PE), Polystyrene (PS), ABS, Policarbonate (PC), High Density Polyethylene (HDPE) and Polypropylene (PP). To make a prototype design of a plastic waste processing device, a product architecture is carried out for the arrangement of the functional parts of a product in the form of a scheme, then an industrial design is carried out to see the level of importance of the product based on the application of aesthetic and ergonomic art. Product design was designed using Autodesk Inventor software. Then the last process, which is making prototypes of plastic waste processing equipment in accordance with consumer needs. By doing this research it is hoped to be one alternative in dealing with the problem of plastic waste through the creativity of its users in producing recycle products in order to be useful and have a sale value.
**Keywords**: plastic waste pollution, plastic recycle process, kano, quality function deployment

## 1. Introduction

Plastic waste is one of the biggest contributors to environmental problems. There are several alternatives that can be done to process plastic waste, namely by 3R (Reuse, Reduce, Recycle). Previous plastic recycling activities have been carried out by changing plastic waste into fuel oil [5] and making a trash can that is able to recycle plastic waste into plastic ore [3]. In this paper, we explain our study about customer preferences in recycling plastic machine, that can produce a variety of recycle products. And using the Kano Method to classify each attribute of consumer needs into the category of Kano. We also use QFD to determine the product technical specifications and priority the needs in building the machine.

## 2. Methods

In this research, three stages are carried out, namely the identification of the attributes of consumers' needs for products into the Kano category based on Blauth's Formula and staging into the Kano quadrant

IOP Conf. Series: Materials Science and Engineering **847** (2020) 012014 doi:10.1088/1757-899X/847/1/012014

diagram. By using the Kano Method, the development team can find out the attributes of consumer needs that must be prioritized and not. So the development team can focus on increasing consumer satisfaction with plastic waste processing equipment products. Then determine product specifications by making the House of Quality (HoQ) metrics.

## 3. Result and Discussion

### *3.1 Data Collection*

*3.1.1 Operation Process Chart (OPC).* The Operation Process Chart (OPC) describes the work elements, work sequence, and division of work elements in the production process of plastic waste processing equipment. The work map on making plastic waste processing equipment provides information on processing time, number of scarp, types of activities carried out, and tools or machines used on each work element. Based on OPC drawings, it can be seen that there are 22 operations and 2 inspection activities with a total total time of 175 minutes [1].

*3.1.2 Bill of Material (BOM).* The Bill of Materials (BOM) describes the definition of the final product which consists of a list of items, materials, or materials needed to assemble, mix or produce the final product. In this BOM consists of various levels of the level of needs of a product section. The plastic waste processing equipment consists of 4 levels [1]. Table 1 is a table of Bill of Material.

**Table 1.** Bill of Material

| Part Number Level | | | | | Information | Amount | Units | Decision |
|---|---|---|---|---|---|---|---|---|
| 0 | 1 | 2 | 3 | 4 | | | | |
| A000 | | | | | Recycling Plastic Machine | 1 | Units | Make |
| | A1100 | | | | Electric Motor | 1 | Units | Buy |
| | B1100 | | | | Gear Box | 2 | Units | Buy |
| | C1100 | | | | Poelly | 2 | Units | Buy |
| | D1100 | | | | V-Belt | 1 | Units | Buy |
| | E1100 | | | | Shedder | 1 | Units | Make |
| | | E100 | | | Hopper | 1 | Units | Make |
| | | | E110 | | Iron plate | 4 | Cm | Buy |
| | | E200 | | | Blade | 1 | Units | Buy |
| | | | E210 | | Bolt | 1 | Units | Buy |
| | F1100 | | | | Extrution | 1 | Units | Make |
| | | F100 | | | Barrel | 1 | Units | Buy |
| | | | F110 | | Screw Barrel | 1 | Units | Buy |
| | | | | F111 | Steel | 1 | Cm | Buy |
| | | | F120 | | Nozzle Heater | 4 | Units | Buy |
| | | | F130 | | Bearing | 3 | Units | Buy |
| | | | | F131 | Iron | 1 | Cm | Buy |
| | | | F140 | | Gear | 2 | Units | Buy |
| | | F200 | | | Controller Temperatur | 1 | Units | Buy |
| | | | F210 | | Solid State Relay | 1 | Units | Buy |
| | G1100 | | | | Iron plate | 1 | Cm | Buy |
| | H11000 | | | | Framework | 1 | Cm | Make |
| | | H100 | | | Galvaniz Ram Wire | 2 | Units | Buy |
| | I1100 | | | | Drawer | 1 | Units | Make |
| | | I110 | | | Drawer Handle | 1 | Units | Buy |

*3.1.3 Assembly Chart*
An assembly chart contains a diagram of the relationship between the components to be assembled into a product. Assembly Chart is useful to show the components of a product and explain the sequence of assembly of these components. Here is a picture of the Assembly Chart of plastic waste processing equipment products.

*3.2 Data Processing*

*3.2.1 Architecture Product*
*3.2.1.1 Product Scheme.* The Product Schema is a diagram that illustrates the team's understanding of the elements that make up the product. The scheme for plastic waste processing equipment products has physical elements as well as energy and material flow. Some elements are related to critical components and some elements are still explained functionally contained in Figure 1.

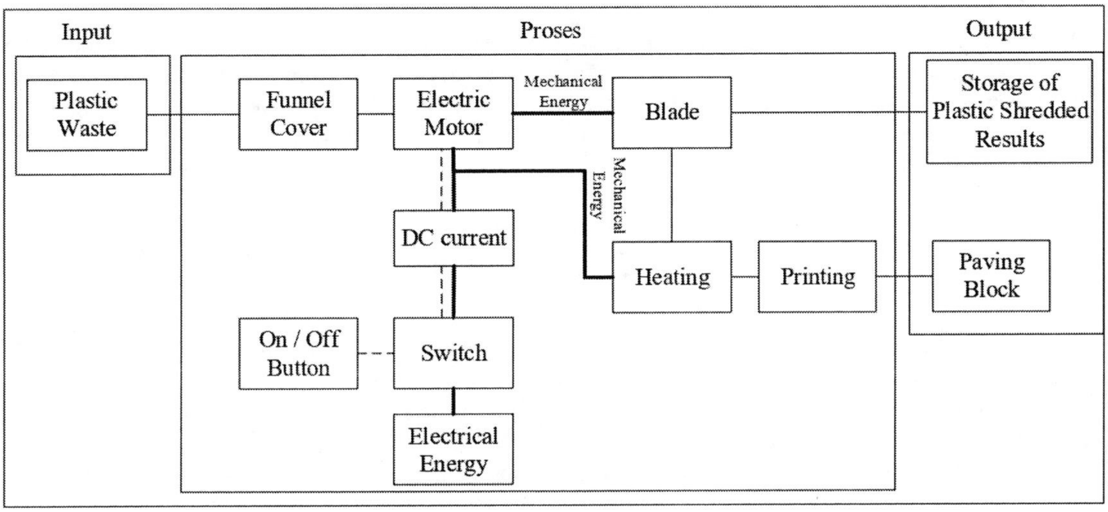


**Figure 1.** Product Scheme

Based on Figure 1 which is the product scheme in the plastic waste processing equipment, it can be seen that there are physical elements, namely in the lid of the chopper, on / off button, switch, storage of the results of plastic chopping (storage), and paving block molds. Then the functional elements are in the electric motor, blade, and heater. In the product scheme above it can be seen that the blade will function when the on button is activated. In the process generates DC current so that it activates an electric motor. Then there is a change in electrical energy into motion energy on the chopper blade. The blade will chop the input of plastic waste that enters through the mouthpiece lid. Electrical energy flowed from an electric motor also activates plastic heaters. Then the resulting plastic chopped can pass through two choices of lanes. The first path is to go directly to the plastic heating area and the second path is stored in a plastic shredder storage box. If the first path is selected, the chopped plastic will be heated to a plastic melt and put into a mold. Then the mold produces a paving block product.

*3.2.1.2 Fundamental and Incidental Interactions.* In the system-level design phase a fundamental and incidental interaction identification process is carried out to better control the coordination of the process. Fundamental interactions are interactions that have been planned and known from the start. This interaction is easy to understand because the process is a basic product. Then incidental interactions are interactions that occur suddenly, unplanned and have risks. This is illustrated in Figure 2.

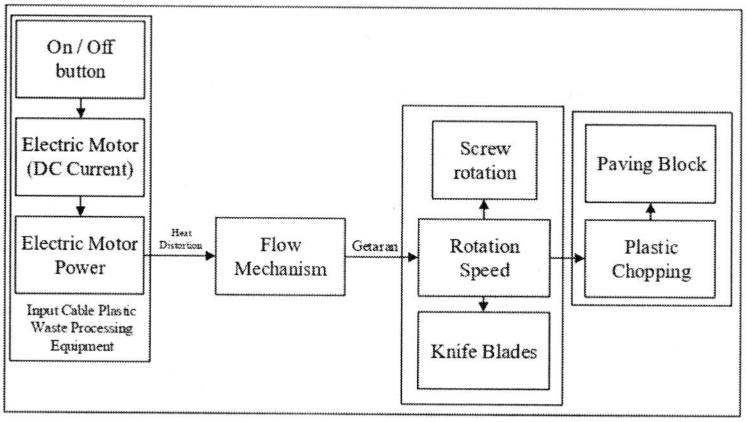


**Figure 2.** Fundamental and Incidental Interactions

Figure 2 is a fundamental interaction that occurs is the on / off button functions to activate the electric motor, then the electrical energy generated will make the blade and screw in the plastic waste processing equipment move. Incidental interactions that occur are the heat distortion generated from the energy released by the electric motor, then the vibration caused by the heating screw rotation and the power supplied by the electric motor.

*3.2.1.3 Concept Drafting.* The product concept is a picture or estimate, the working principle and shape of a product. In addition, the product concept is also a brief description of how the product can satisfy the needs and wants of customers.

- Classification Tree

The classification tree is used as a way to develop product concepts in plastic waste processing equipment. The classification tree is done by making a number of alternative choices that can be chosen to make product concept variations. Alternative choices in the classification tree were made by a team of plastic waste processing equipment developers. Selection of alternatives is done by identifying benchmark products and the results of voice of customers. Then an analysis is performed to determine the parts of the tool that can be used as a differentiator with existing tools. By knowing the advantages and disadvantages that each alternative has. Selection of these variations is feasible to make.

- Combination Table

The combination table results from alternative choices given in the classification tree. So that produced 4 kinds of concepts that are feasible and feasible in making plastic waste processing equipment contained in Figure 3.

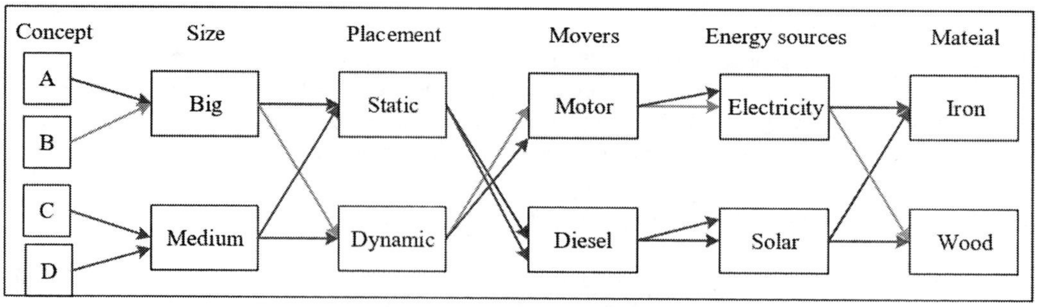


**Figure 3.** Combination Table

IOP Conf. Series: Materials Science and Engineering **847** (2020) 012014 doi:10.1088/1757-899X/847/1/012014

Based on the Figure 3 which is combination table produced 4 kinds of concept variations in making plastic waste processing equipment. Concept A is that the product has a large size that is 2 m long, 8 m wide, and 2 m high, the product placement is static, the engine drive uses diesel, the energy used is diesel, and uses iron-based materials. In concept B the product has a large size of 2 m, width 8 m, and height 2 m, product placement is dynamic, the engine drive uses a motor, the energy used is electricity, and uses wood-based materials. In the C concept, the product has a medium size of 1 m, width 6 m, and height 1 m, the product placement is static, the engine drive uses a motor, the energy used is electricity, and uses iron-based material. In the D concept the product has a medium size of 1 m, width 6 m, and height 1 m, the product placement is static, the engine drive uses diesel, the energy used is diesel, and uses wood-based materials.

*3.2.1.4 Concept Selection.* Concept selection helps the development team in refining and refining the concept. Then it can establish one or more promising concepts and be the focus in further testing and assess the concept by paying attention to customer needs and other criteria by comparing the strengths and weaknesses of a concept.

- Screening Concept

Concept screening is an evaluation process in the form of thoughts aimed at narrowing down the alternatives through 0, positive (+), and negative (-) assessments for each selection criteria. This is due to the development team knowing more about the conditions for selecting variations in each alternative concept contained in Table 2.

**Table 2.** Screening Concept

| **Selection Criteria** | **Concept** | | | |
|---|---|---|---|---|
| | **A** | **B** | **C** | **D** |
| Product Size | 0 | 0 | + | + |
| Product Placement | 0 | + | + | 0 |
| Movers | - | + | + | - |
| Energy Resources | - | + | + | - |
| Material | + | - | + | - |
| Amount + | 1 | 3 | 5 | 1 |
| Amount - | 2 | 1 | 0 | 3 |
| Amount 0 | 2 | 1 | 0 | 1 |
| **Final Score** | -1 | 2 | 5 | -2 |
| **Rank** | 3 | 2 | 1 | 4 |
| **Continue?** | NO | YES | YES | NO |

From Table 2, if the index value possessed by the selection criteria of 0% -33% indicates that the criterion is negative (-), meaning that if a selection criterion is used on the concept of X then it is worse. If the index value owned by the selection criteria is 34% -67%, it indicates that the criteria are neutral (0), meaning that when using the selection criteria on the concept X, the properties are the same or neutral. Then if the index value possessed by the selection criteria of 68% -100% indicates that the criteria are positive (+), meaning that if a selection criterion is used on the concept of X then it is better.

- Scoring Concept

Concept evaluation uses weighting selection criteria and a better rating scale. Concept evaluation can be skipped if concept screening results in a dominant concept. At this stage the rating and burden are given by the development team to produce a score or scoring for each concept contained in Table 3.

**Table 3.** Scoring Concept

| Selection Criteria | Load | Concept | | | |
|---|---|---|---|---|---|
| | | B | | C | |
| | | Rating | Value | Rating | Value |
| Product Size | 15% | 3 | 0,45 | 4 | 0,6 |
| Product Placement | 15% | 5 | 0,75 | 5 | 0,75 |
| Movers | 25% | 4 | 1 | 4 | 1 |
| Energy Resources | 20% | 4 | 0,8 | 4 | 0,8 |
| Material | 25% | 2 | 0,5 | 5 | 1,25 |
| Total Value | 100% | 3,5 | | 4,4 | |
| Rank | | 2 | | 1 | |
| Continue? | | NO | | YES | |

Based on Table 3, results of scoring metrics, it can be seen that there are loads or weights that are used as reference criteria for selection on the attributes of plastic waste processing equipment. The total value of the load given to the selection criteria must be 100%. Rating on the rating uses the Likert scale 1-5, where the value of 1 (strongly disagree), 2 (agree), 3 (neutral), 4 (disagree), and 5 (agree). Each rating value given by an expert is multiplied by the weight value. Resulting in the highest value of 4.4 in concept C. Concept C in plastic waste processing equipment has a medium product size of 1 m long, 6 m wide and 1 m high, dynamic placement, using dynamo or electric motors as penggerknya, energy sources which is generated is electricity, and uses iron material.

*3.2.1 Design Industry*
Industrial design assessment is carried out to determine the ergonomic and aesthetic aspects possessed by plastic waste processing equipment products. Ergonomic aspects are aspects of products that relate to the human or consumer side. Then the aesthetic aspect is the product aspect that is related to the visual form of the product to attract the interest of consumers.

*3.2.1.1 Ergonomic Aspects.* Ease of use in plastic waste processing equipment can be said to be quite easy (medium). Because the placement of every function on the machine can be seen clearly. This product also has guidance or other information through social media, namely Instagram. Then the ease of maintenance on plastic waste processing equipment can be said to be quite easy (medium). Because maintenance on this machine does not need to be done in the short term. The quantity of user interaction with plastic waste processing equipment can be said to be quite frequent (high). Because when using this product, consumers or users have a role in running the function of the tool. Then the level of security possessed by plastic waste processing equipment can be said to be safe (high). Because this product has a lid on the plastic counter and the whole product is covered (full coverage) with an iron plate. Thus allowing users not to be affected by the heater in the plastic melting process contained in Table 4.

**Tabel 4.** Ergonomic Apects

| Ergonomics | | | |
|---|---|---|---|
| Category | Level of Importance | | |
| | Low | Medium | High |
| Ease of Use | | ● | |
| Ease of Care | | ● | |
| Quantity of User Interactions | | ● | |
| Security | | | ● |

*3.2.1.2 Aesthetic Aspects.* Differentiation that is owned by plastic waste processing equipment can be said to be different (moderate) from similar products. Because this product is designed to have 2 main functions, namely the counter and heater that are joined together or integrated in one machine. So that in one machine the same can produce 2 different outputs namely plastic chopping and plastic melting (recycle product). Then the prestige of ownership of this tool can be said to be quite high. Because with products that have an all in one function, consumers who buy these products have a purpose in their use. This product can be used as a commercial product, which means it can provide material profits for its owners. In addition this product can be used as a forum for learning and information on how to process plastic waste properly. By creating a plastic waste processing tool that is integrated between the counter and the heater, the development team feels there is still much that can be developed (quite high) from this tool such as the selection of product materials, machine components used, design contained in Table 5.

**Table 5**. Estectic Aspects

| **Aesthetics** | | | |
|---|---|---|---|
| **Category** | **Level of Importance** | | |
| | **Low** | **Medium** | **High** |
| Product Differentiation | | | ● |
| Prestige of Ownership, Fashion, Impression | | | ● |
| Team Motivation | | | ● |

*3.2.2 Prototype*
The prototype design is done by using Autodesk Inventor software on plastic waste processing equipment based on the results of product concept selection contained in Figure 4 and Figure 5.. The concept chosen was the C concept, which was a medium-sized device that was 1 m long, 6 m wide, and 1 m high, placed a dynamic product, moved using an electric motor, generated energy sources, and the material used was iron contained in Figure 6.

**Figure 4.** Illustration of a 3D Image of a Plastic Waste Processor

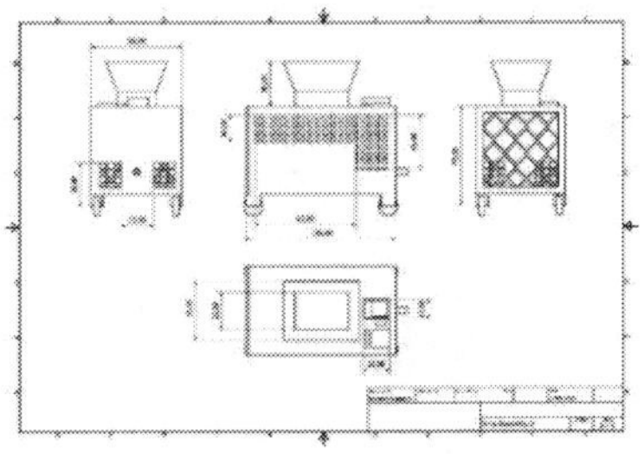

**Figure 5**. Illustration of a 2D Image of a Plastic Waste Processor

**Figure 6.** Prototype of Plastic Waste Processor

## 4. Conclusion

Based on data processing and analysis that has been done, the following conclusions can be obtained: Making prototypes of plastic waste processing equipment capable of functioning as expected. In the process of counting and heating the plastic can work well. In the enumeration process, the blade does not make noise when the machine is used. But in the process of heating the molten plastic produced is still small in number. That is because the placement of fillament in the heating pipe is still not right, so it needs improvement to be able to produce the desired plastic melt. Based on the product concept selection. The concept chosen was the C concept, which was a medium-sized device that was 1 m long,

6 m wide, and 1 m high, placed a dynamic product, moved using an electric motor, generated energy sources, and the material used was iron.

**Acknowledgement**
The author would like to thank *Lembaga Penelitian dan Pengabdian Masyarakat Universitas Al Azhar Indonesia* (LP2M UAI) for funding the paper to participate in this seminar and to be published in the proceeding.

## 5. References

[1] Andriani M 2019 *Implementasi Algoritma CART Untuk Uji Kelayakan Bisnis Alat Pengolah Sampah Plastik.* (Jakarta: Universitas Al Azhar Indonesia).
[2] Anonim. 2008 Undang-undang Republik Indonesia Nomor 18 Tahun 2008 tentang Pengelolaan Sampah. (Jakarta: Sekretariat Negara).
[3] Ichlas N, Nofriadi and Rusmardi 2014 Pengembangan Mesin Pencacah Sampah/Limbah Plastik Dengan Sistem Crusher Dan Silinder Pemotong Tipe Reel. *Seminar Nasional Sains dan Teknologi*. Fakultas Teknik Muhamadiyah. Jakarta. ISSN: 2407-1846.
[4] Ogedebe P M and Jacob B P 2012 Software Prototyping: *A Strategy to Use When User Lacks Data Processing Experience*. ARPN Journal of Systems and Software, II (6).
[5] Sari L 2017 Kajian Potensi Pemanfaatan Sampah Plastik Menjadi Bahan Bakar Cair. *Jurnal Teknik Lingkungan*. **3**(1), Agustus 2017. ISSN: 2549-1652.
[6] Ulrich K T and Steven D E 2001. *Perancangan dan Pengembangan Produk.* (Jakarta: Salemba Teknika).

# Quality Control Analysis of Tube SandwichUsing Six Sigma Method in PTXYZ

**A S Sundari[1,2], E P Setyawati[1,3] and D Rimantho[1,4]**

[1]Department of Industrial Engineering, Universitas Pancasila, Jakarta 12640, Indonesia
[2]anggina.sandy@univpancasila.ac.id, [3]ekaputrisetyawati@gmail.com, [4]dino.rimantho@univpancasila.ac.id

**Abstract.** Quality is one of the key aspects in every production activity. This research aims to improve the quality of the product on tubing machines used Six Sigma method with DMAI phases. At phase of define, the problem is found attubing machines with defects glue, printing, improper size and paper extension. In the measure phase, DPMO value is 3691 and sigma value is 4,18. In the analyzephase, the most dominant problem type of defects are glue and printing, will be analyzed according to machine, material and method factors. In the improve phase, the cause of problem will be sorted from the highest to the lowest uses FMEA method. Causes of glue problems are paper rolls are from many suppliers, the extruder needs time to be stable, the age of machine is old, piece of paper tucked in the knife bearing and not optimal material handling. The priority order of printing problems is not optimal material handling, piece of paper tucked in the knife bearing, the paper stuck in the printing unit, and ink are from several suppliers. Proposed improvements with 5W1H can be given according the order of root problems from the largest to the smallest.
**Keywords**: 5W1H, Defect Per Million Opportunities (DPMO), Failure Mode and Effect Analysis (FMEA), Quality, Six Sigma.

## 1. Introduction

Every manufacturing industry is required to provide the best in all aspects to meet customer needs and maintain customer loyalty. One aspect that is intended is the quality of the product. One method that can be used is the Six Sigma method. Six Sigma is a method for increasing productivity and profitability by applying methods from statistical tools to identify and measure waste and to show the steps for improvement. The Six Sigma concept itself is a continuous improvement effort in the hope of reducing defective products so that they can approach zero defects [1].

PT.XYZ is one of the construction materials companies in Indonesia, especially in the manufacture of cement. ABC Division is one of the supporting divisions at PT.XYZ which has the role to produce and provide all types of bags cement for PT. XYZ. ABC division has three types of machines used to produce cement bags.In December 2018 until February 2019 one of the machines in ABC division which is TM1 had a reject proportion of 0.63%, it means the reject proportion is higher than the reject standard of the company which is 0.40%. With the high proportion of reject products every month in the production of tube sandwiches, it will affect the costs incurred on the production floor.

Therefore, some actions and efforts are needed to reduce the high proportion of rejects on TM1 by first identifying the cause of the most dominant problem then analyzing and providing solutions for improvement of the problem. Therefore, the authors are interested in conducting research using the Six Sigma method to reduce the high proportion of rejects that occur in TM1.

## 2. Methods

Six Sigma is a method for increasing productivity and profitability. In addition, Six Sigma is an application of statistical tools to identify, measure and analyze the defects or wastage of a product and process so it leads to steps for improvement. In the result of Six Sigma application in the company it is hoped that this can overcome the production problems associated with product defects [1]. The phases of implementing quality improvement with Six Sigma consist of five phases, namely using DMAIC (Define, Measure, Analyze, Improve, and Control) [2].

### *2.1. Define*

The first phase-define-serves as the platform for the team to get organized, determine the roles and responsibilities of each member of the team, establish team goals and milestones and review the process steps. The key points to be defined at this phase are the voice of the customer, the scope of the project, the cause and effect prioritization (a list that the team creates for pursuing the specific project based on cause and effect criteria) and project planning. The purpose of this phase is to set the foundations for the work ahead in solving a problem. This means that an excellent understanding of the process must exist for all team members, as well as complete understanding of the CTQ characteristics. After CTQ factors are identified, everyone in the team must agree on developing an operational definition for each CTQ aspect.

### *2.2. Measure*

The second phase of the DMAIC model—measure—is when the team establishes the techniques for collecting data about current performance that highlights project opportunities and provides a structure for monitoring subsequent improvements. Upon completing this phase, we expect to have a plan for collecting data that specifies the data type and collection technique, a validated measurement system that ensures accuracy and consistency, a sufficient sample of data for analysis, a set of preliminary analysis results that provides project direction and baseline measurements of current performance.

### *2.3. Analyze*

The third phase—analyze—serves as an outcome of the measure phase. The team at this phase should begin streamlining its focus on a distinct group of project issues and opportunities. In other words, this phase allows the team to further target improvement opportunities by taking a closer look at the data.Thisphase can be done by formulating some of the biggest causes of problems, so that the next phase (improve) becomes more focused and focused. Statistical analysis-based thinking plays a very important role at this phase. In this phase, it needs a thorough analysis so that in the next phase there is no error.Determine LCL, CL and UCL for control chart with these formula [3]:

$$LCL = \bar{p}_o - 3\sqrt{\frac{\bar{p}_o(1-\bar{p}_o)}{n}} \quad (1)$$

$$CL = \bar{p}_o \quad (2)$$

$$UCL = \bar{p}_o + 3\sqrt{\frac{\bar{p}_o(1-\bar{p}_o)}{n}} \tag{3}$$

Where $\bar{p}_o$ is the proportion of defective products determined by the company and $n$ is sample size. The general form of DPMO fuction is :

$$DPMO = \frac{DPU}{CTQ} \times 1.000.000 \tag{4}$$

Where *DPU* is Defect Per Unit and *CTQ* is Critical to Quality of product. *DPU* is generally chosen as :

$$DPU = \frac{Total\ Defect}{Total\ Production} \tag{5}$$

The sigma value is obtained through the sigma conversion table from DPMO value.

*2.4. Improve*

The fourth phase—improve—aims to generate ideas; design, pilot and implement improvements; and validate the improvements. Perhaps the most important items in this phase are the process of brainstorming, the development of the "should be" process map, the review and/or generation of the current FMEA (failure mode and effect analysis), a preliminary cost/benefit analysis, a pilot of the recommended action and the preliminary implementation process.

*2.5. Control*

The fifth stage—control—is to institutionalize process or product improvements and monitor ongoingperformance.This stage is the place where the transition from improvement to controlling the process and ensuring that the newimprovement takes place. Of course, the transition is the transferring of theprocess from the project team to theoriginal owner. The success of this transfer depends upon aneffective and very detailed control plan.

**3. Result and Discussion**

The discussion in this study uses the Six Sigma method with the stages of DMAI (Define, Measure, Analyze, and Improve) to define, measure, analyze the causes of the high proportion of rejects found in TM1so the improvements can be made in cause of potential failure.

*3.1. Define*

*3.1.1. Identify the types of defects.* The types of defects are determined based on references from the product check sheet as well as interviews with related parties there are six types of defect there are gluing, printing, mismatch size, paper extension, overlapping, and the number of plies are not appropriate.

*3.1.2. Determine CTQ.* CTQ determination is based on sample data taken by observing in the form of random sampling of 25 samples with a sample size of 560 pcs each.Total CTQ found was 4, gluing, mismatch size, printing and paper extension.

*3.1.3. Determine the most dominant type of defect.* From the sample data that has been collected, the most dominant type of defect can be determined by making a Pareto Diagram. Pareto diagram can show the most dominant type of defect so that treatment or repair can be started from the dominant cause first.The Pareto Diagram of defect is given in Figure 1.

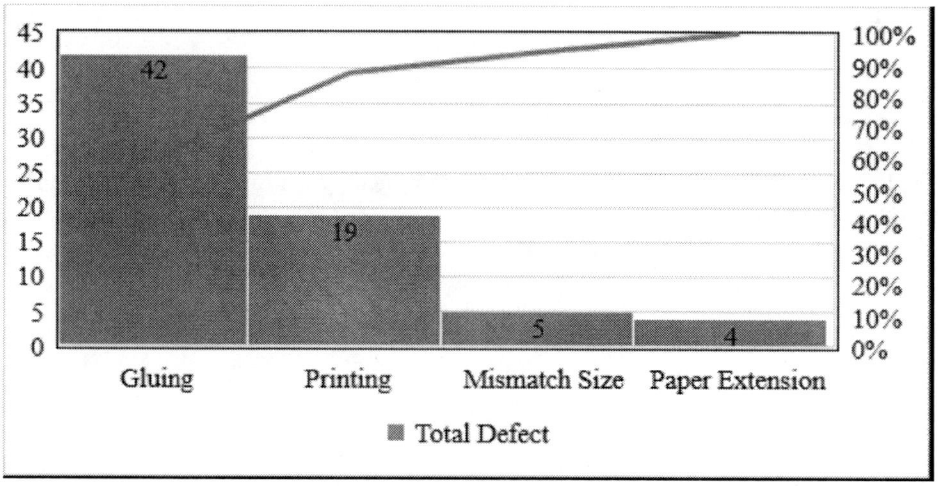


**Figure 1.** Pareto Diagram of defect.

Then the most dominant type of defect is due to gluing problems then followed by problems due to printing, mismatch size and paper extention. According to the 80:20 rule [4], this study will focus more on analyzing the types of defects caused by gluing and printing.

*3.2. Measure*

*3.2.1 P control chart.* By using equations (1), (2), and (3), the following results are LCL equal to 0, CL equal to 0,004 and UCL equal to 0,012. The P control chart is given in Figure 2.

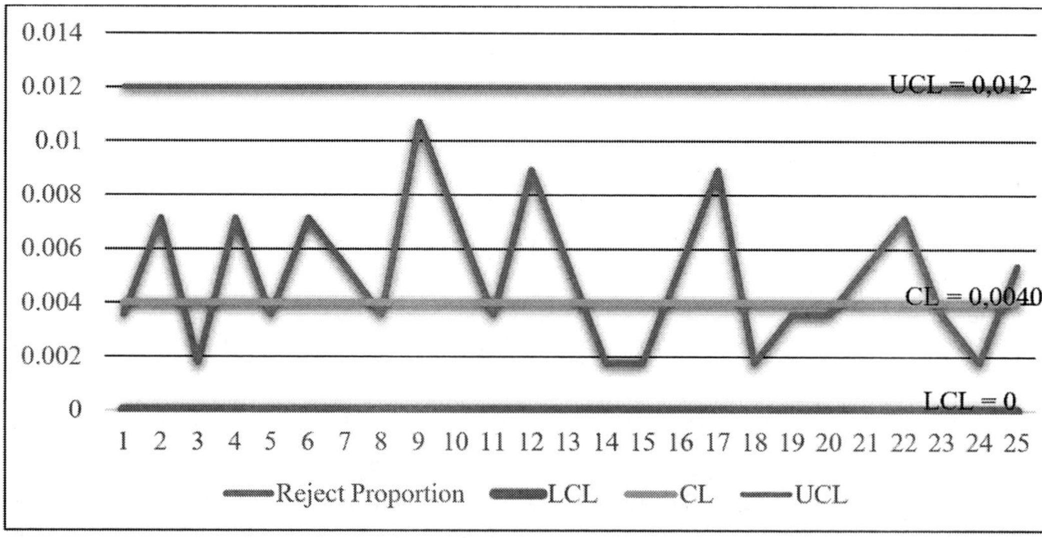


**Figure 2.** P control chart.

All sample points are within the control limits. However, there are 13 sample points above the central line (CL) which in this case is the standard set by the company. Therefore, further analysis is needed.

*3.2.2 Measure DPU, DPMO and Sigma value.* DPU and DPMO values based on equations (4) and (5), sigma values based on conversion tables are given in Table 1.

**Table 1.** DPU, DPMO and Sigma value.

| No. | Number of Defect | CTQ | DPU | DPMO | Sigma |
|---|---|---|---|---|---|
| 1. | 2 | 1 | 0,0036 | 3571 | 4,19 |
| 2. | 4 | 2 | 0,0071 | 3571 | 4,19 |
| 3. | 1 | 1 | 0,0018 | 1786 | 4,41 |
| 4. | 4 | 3 | 0,0071 | 2381 | 4,32 |
| 5. | 2 | 1 | 0,0036 | 3571 | 4,19 |
| 6. | 4 | 3 | 0,0071 | 2381 | 4,32 |
| 7. | 3 | 1 | 0,0054 | 5357 | 4,05 |
| 8. | 2 | 1 | 0,0036 | 3571 | 4,19 |
| 9. | 6 | 3 | 0,0107 | 3571 | 4,19 |
| 10. | 4 | 1 | 0,0071 | 7143 | 3,95 |
| 11. | 2 | 1 | 0,0036 | 3571 | 4,19 |
| 12. | 5 | 1 | 0,0089 | 8929 | 3,87 |
| 13. | 3 | 2 | 0,0054 | 2679 | 4,28 |
| 14. | 1 | 1 | 0,0018 | 1786 | 4,41 |
| 15. | 1 | 1 | 0,0018 | 1786 | 4,41 |
| 16. | 3 | 2 | 0,0054 | 2679 | 4,28 |
| 17. | 5 | 2 | 0,0089 | 4464 | 4,11 |
| 18. | 1 | 1 | 0,0018 | 1786 | 4,41 |
| 19. | 2 | 1 | 0,0036 | 3571 | 4,19 |
| 20. | 2 | 1 | 0,0036 | 3571 | 4,19 |
| 21. | 3 | 1 | 0,0054 | 5357 | 4,05 |
| 22. | 4 | 1 | 0,0071 | 7143 | 3,95 |
| 23. | 2 | 1 | 0,0036 | 3571 | 4,19 |
| 24. | 1 | 1 | 0,0018 | 1786 | 4,41 |
| 25. | 3 | 2 | 0,0054 | 2679 | 4,28 |
| | | Total | | 3691 | 4,18 |

### *3.3. Analyze*

*3.3.1 Analyze the causes of defect with Fishbone Diagram.* Pareto diagrams show the dominant problem caused by the type of defect caused by gluing and printing problems. the further analysis of these two problems using Fishbone Diagrams is given in figure 3 and figure 4.

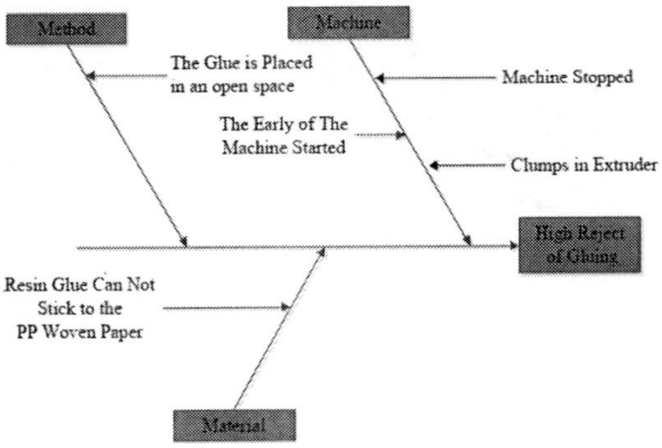


**Figure 3**. High reject of gluing Fish Bone.

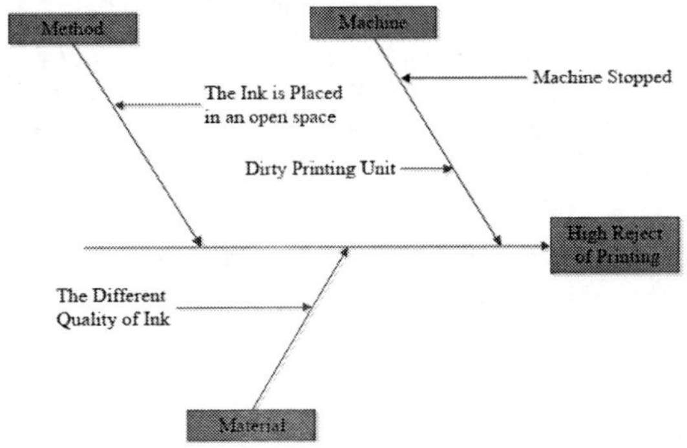


**Figure 4**. High reject of printing Fish Bone.

*3.3.2. Determine failure modes with CFME.* After the problem is found in the fishbone, find the root of the problem until there are no more causes. CFME is used before making FMEA as a development of the fishbone[5]. CFME due to gluing and printing issues are given in Figure 5 and Figure 6.

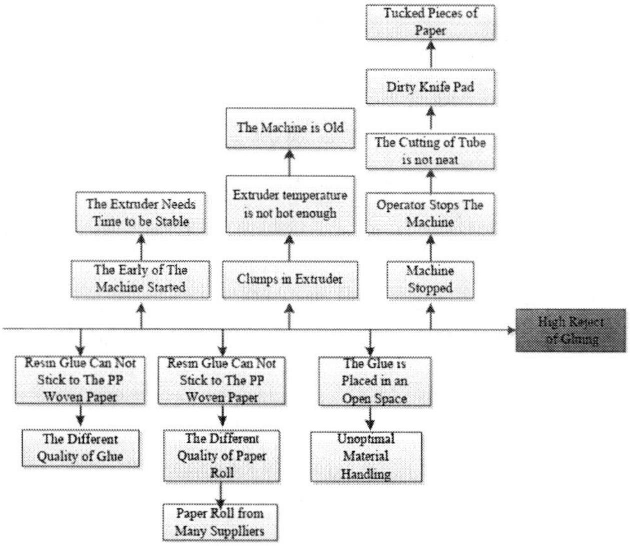


**Figure 5**. High reject of gluing CFME.

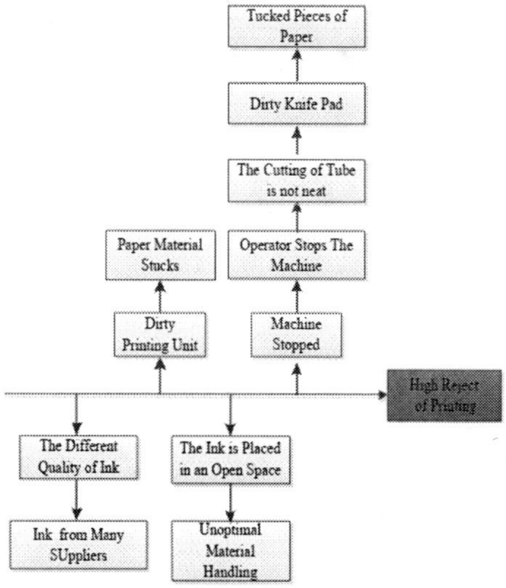


**Figure 6**. High reject of printing CFME.

*3.3.3. Determine the problem priority with FMEA.* The high reject of gluing FMEA is given in Table 2 and the high reject of printing FMEA is given in Table 3.

**Table 2.** High reject of gluing FMEA.

| Factor | Failure Mode | S | Cause of Failure | O | Current Process Control | D | RPN | Rank |
|---|---|---|---|---|---|---|---|---|
| Machine | Tucked pieces of paper | 7 | Cracks in the open knife pad can cause the pieces of paper get inside | 5 | Change periodically the knife pad | 5 | 175 | 4 |
| | The machine is old | 6 | The machine was made in 1975. The machine spesifications have not changed much over time | 7 | Perform regular and scheduled maintenance | 6 | 252 | 3 |
| | *The* extruder need time to be stable | 7 | The extruder takes approximately a minute in order to distribute the glue stably | 6 | Eximining tube product, especially in the resin glue section | 7 | 294 | 2 |
| Material | Paper roll from many supplier | 8 | Paper roll from 4 different suppliers and each suppliers has different quality | 7 | Replace paper roll with specifications | 6 | 336 | 1 |
| | The different quality of glue | 4 | The quality of each shipment of resin glue is different | 5 | Replace glue with specifications | 4 | 80 | 6 |
| Method | Unoptimal Material Handling | 5 | Unoptimal Material Handling | 4 | Stir and filter the glue before use | 6 | 120 | 5 |

IOP Conf. Series: Materials Science and Engineering **847** (2020) 012015 doi:10.1088/1757-899X/847/1/012015

**Table 3.** High reject of printing FMEA.

| Factor | Failure Mode | S | Cause of Failure | O | Current Process Control | D | RPN | Rank |
|---|---|---|---|---|---|---|---|---|
| Machine | Tucked pieces of paper | 7 | Cracks in the open knife pad can cause the pieces of paper get inside | 6 | Change periodically the knife pad | 5 | 210 | 2 |
| | Tucked pieces of paper *in* printing unit | 8 | The paper cover the stamp from *printing unit* | 6 | Stop the machine and clear the printing unit | 3 | 144 | 3 |
| Method | Unoptimal Material Handling | 7 | Unoptimal Material Handling | 6 | Stir the ink before use | 7 | 294 | 1 |
| Material | Ink from many supplier | 6 | Ink from 2 different suppliers and each suppliers has different quality | 4 | Replace ink with specifications | 6 | 144 | 4 |

*3.4. Improve*

At this stage an analysis of an action plan is carried out through the improvement of the sourcesof defective products with the 5W1H method. 5w1h method aims to determine preventive and corrective actions from the cause of the problem [6]. High reject of gluing proposed improvement is given in Table 4 and high reject of printing proposed improvement is given in Table 5.

**Table 4.** High reject of gluing proposed improvement.

| Rank | Failure Mode | Proposed Improvement |
|---|---|---|
| 1. | Paper roll from many supplier | Check samples of the material before it is received and inspect the material before use |
| 2. | The extruder need time to be stable | During the first 1 minute when the machine is running, set the machine at a slow speed, so that the extruder and paper rates are in line |
| 3. | The machine is old | Replace the machine components according to the latest standards |
| 4. | Tucked pieces of paper | Pay attention to the cleanliness of the blade bearings along with the cutting unit and carry out regular checks and changes |
| 5. | Unoptimal Material Handling | Provides cover for resin glue and potato starch glue, as not to be exposed to outside air |

IOP Conf. Series: Materials Science and Engineering **847** (2020) 012015 doi:10.1088/1757-899X/847/1/012015

**Table 5.** High reject of printing proposed improvement

| Rank | Failure Mode | Proposed Improvement |
|---|---|---|
| 1. | Unoptimal Material Handling | Provides cover for ink, as not to be exposed to outside air |
| 2. | Tucked pieces of paper | Pay attention to the cleanliness of the blade bearings for the cutting unit and carry out regular checking and replacement |
| 3. | Tucked pieces of paper in printing unit | Make sure the printing unit is clean and there are no extra materials, especially on the stamp. |
| 4. | Ink from many supplier | Check samples of the material before it is received and inspect the material before use. |

## 4. Conclusion

Based on the research objectives to be achieved, the conclusions obtained from the results of the research are the DPMO value obtained from the production process on tubing machinesequal to 3691. The achievement of sigma value is 4.18. Even though the achievement of the sigma value is above average, there must always be continuous improvement to achieve the Six Sigma target. From this research, types of defects that often occur in tubing machines are gluing, printing, mismatched sizes, and paper Extention. Proposed improvements for gluing problems include check samples of the material before it is received and inspect the material before use, set the machine at a slow speed for the first minute, so that the extruder and paper rates are in line, replace the machine components according to the latest standards, pay attention to the cleanliness of the blade bearings along with the cutting unit and carry out regular checks and changes and provide cover for resin glue and potato starch glue, as not to be exposed to outside air. Proposed improvements to the printing problem areprovide cover for ink, as not to be exposed to outside air, pay attention to the cleanliness of the blade bearings for the cutting unit and carry out regular checking and replacement, make sure the printing unit is clean and there are no extra materials (especially on the stamp) and check samples of the material before it is received and inspect the material before use. For further research, it would be better to continue this research so that the control stage in the DMAIC stage is complete by implementation of the proposed improvement given then conduct an evaluation and comparison with the previous conditions. In addition, research can also be added regarding the cost aspects that have not been discussed in this study.

## 5. References

[1] Brue G 2005 *Six Sigma for Managers: 24 Lesson to Understand and Apply Six Sigma Principles in Any Organization* (New York: McGraw-Hill Education)
[2] Stamatis D H 2004 *Six Sigma Fundamentals: A Complete Guide to The System, Method, and Tools* (New York: Production Press)
[3] Mitra A 2016 *Fundamentals of Quality Control and Improvement* (New Jersey: John Willey and Sons Inc.)
[4] Ma'arifM S, Tanjung H 2003 *Manajemen Operasi* (Jakarta: PT GramediaWidiasarana)
[5] Nasution S R, Respati D 2011 *Perbaikan kualitas dengan metode PDCA untuk Mengurangi Cacat Produk Frame Actuator pada Proses Frame Assembly di PT SCJI. FTUP Engineering Journal* **XXIV**(2), 69 – 77
[6] Martins A C A, Leite J C, Parente R S, Mustafa E V, DinolaI C S, Santos A D J D 2019 Method of Problem Analysis and Solving applied to Quality and Productivity in a Furniture Industry. *International Journal of Advanced Engineering Research and Science (IJAERS)* **6**, 587-610

# Allocation of Maltodextrin Raw Material Orders by *Fuzzy Analytic Network Process* (FANP) and *Goal Programming* Methods (Study Case: PT. Neopangan Selaras Indonesia)

**R Magdalena[1,2], S P Dias[1,3], and A P Ginting[1,4]**

[1]Universitas Katolik Indonesia Atma Jaya, Jakarta, Indonesia
[2] riana.magdalena@atmajaya.ac.id, [3]stefani.primadyas@atmajaya.ac.id,
[4]pricillaginting@gmail.com

**Abstract**..PT. Neopangan Selaras Indonesia is a food company that produces seasoning. One of the raw materials with the most frequent problems with delays and wrong allocation is maltodextrin, which is 72%. To determine the need for evaluation for each maltodextrin supplier using the Fuzzy Analytical Network Process (FANP) and Goal Programming methods. The existence of a method of merging these two studies. Determination of the proportion of the criteria and sub-criteria in the company by minimizing changes in each assessment. The FANP method in this study produced criteria and assessments of each supplier that could be considered in the allocation of the purchase of raw materials. The best supplier of weights was obtained by PT. Indo Asia Tirta Manunggal (0.4421). The Goal Programming method provides a solution by minimizing total costs by considering the price of each supplier, maximizing the purchase value by considering the weight generated. This method produces a solution that produces a total cost of 0.134%, increasing the purchase value of 1.143%.
**Keywords**: Fuzzy-ANP, Goal Programming, Order Allocation, Supplier Selection

## 1. Introduction

Competition in the industrial world, especially for each company is now classified as very tight, so companies need to make their strategies to defend from the competition. Quality, flexibility, product diversity, rapid response, and competition that occur mainly in the industrial field becomes important and the main thing for the company in obtaining customers [4]. Maintaining competition in the industrial world in a company is done by fulfilling all factors, such as obtaining raw materials at the right vendor, the production process following the plan so that it does not cause delays in shipping products to customers.

Ordering of raw materials is seen from suppliers who meet several requirements of the company itself, where suppliers are one of the most important parts of the company for the continuity of the business of a company. Supplier selection is a strategic activity especially if the supplier supplies an item that will be used for the long term as an important supplier [3]. The selection of suppliers in several companies must be carefully considered by the company itself and requires special analytical methods that can make it possible to minimize errors in selecting suppliers who will be used as the main raw material in the production process. Ordering of raw materials is one aspect of the company where careful planning must be done. Errors in allocating orders to suppliers can result in high costs due to the purchase of raw materials.

PT. Neopangan Selaras Indonesia is a company engaged in the food industry whose main focus is on the production of herbs. There are 4 main raw materials used by this company, namely Sugar, Salt, Monosodium Glutamate (MSG) and Maltodextrin. PT. Neopangan Selaras Indonesia has a make to order system where there are requests or orders from customers, the company will carry out the production process. This company applies a value that is the quality of the products produced must be following established standards, which means that the allocation of orders to each supplier must be done appropriately. At present, PT.Neopangan Selaras Indonesia does not yet have definite criteria and sub-criteria in selecting suppliers, where usually this company makes an order to several suppliers based on the availability of goods available at each supplier when the company makes an order. PT. Neopangan Selaras Indonesia also often experiences a budget increase tendency due to errors in the allocation of raw materials to each supplier. This research focuses on raw material Maltodextrin. Maltodextrin is a raw material that has a role as a sweetener or glucose chart with the physical form of white powder, which according to the company this raw material has the most delays for the 5 suppliers. 72% of the delay that often occurs in the raw material Maltodextrin.

This research was conducted by combining two methods, namely the Fuzzy Analytic Network Process (FANP) and Goal Programming. The combination of Fuzzy and ANP is one of the advantages in this study where FANP is used to identify the relationship between criteria by producing weights so that priorities can be identified in the selection of suppliers by minimizing subjective from the company itself. Methods that help produce allocations to each supplier based on the weights obtained and also the prices set by each supplier. This research will be assessed by 3 respondents who are considered to represent the company.

## 2. Methods

### *2.1. Data Collection*

The data used during the research are as follows:

- Primary Data
  Data obtained directly by researchers following the problems of the company itself. The primary data contained in this study are interview data and questionnaire results to 3 respondents. The questionnaire in this study consisted of: criteria questionnaire, questionnaire linkages between sub criteria, paired comparison criteria and related sub criteria questionnaire, and paired supplier assessment questionnaire.
- Secondary Data
  Data obtained from books, research journals, lecture material, as well as historical data that support research. Secondary data used in this study consisted of: Dickson's Vendor Selection Criteria, supplier capacity data, price data for maltodextrin raw materials, delay data on maltodextrin raw materials, and data request for maltodextrin raw materials

### *2.2. Fuzzy Analytical Network Process (FANP) Methods*

The steps in this research are as follows [2]:

- Determine the criteria and sub-criteria following company conditions in the selection of maltodextrin suppliers. Determination of the criteria is based on Dickson's Vendor Selection Criteria [1].

**Tabel 1.** Criteria and Sub-Criteria Used

| Criteria | Sub-Criteria |
|---|---|
| Quality (K) | Age of Material<br>Material Function<br>Material Compliance with Specifications |
| Delivery (P) | Accuracy of Order Quantity<br>Accuracy of Delivery Time<br>Availability of Material |
| Perfomance History (SP) | Past Performance<br>Flexibility to change the number of orders<br>Fulfillment in the cooperation agreement |
| Warranties and Claim Policies (G) | Response Speed<br>Ease of contact<br>Service time<br>Ease of Replacing Reject tems |
| Price (H) | Discounts and Bonuses<br>Payment Method<br>Product Prices |
| Communication system (SK) | Communication Skills<br>The Speed of Information Exchange<br>Flexibility in Sharing |
| Management and Organization (MO) | Ability to Follow Procedures |

- Describe the ANP structure model in which there are several clusters according to the relationship of company conditions in the selection of suppliers obtained from questionnaire data collection 3 and filling in pairwise comparisons with scale [5].
- Convert the resulting value on questionnaires 4 and 5 into the Triangular Fuzzy Number scale.
- Calculate the geometric mean by combining several values of each respondent into one unit in each upper boundary (l), middle boundary (m), and lower limit (u). The geometric average is calculated using the formula:

$$w_i = \sqrt[n]{w_{n-2} \cdot w_{n-1} \cdot w_n} \quad (1)$$

- Calculate defuzzification by combining all of the averages to get a single value from the upper limit (l), middle limit (m), and lower limit (u).

$$F_{ij} = \frac{l_{ij} + m_{ij} + u_{ij}}{3} \quad (2)$$

- Calculate and make a normalization matrix and priority vectors. The normalization matrix is obtained from the previous defuzzification value which will be normalized dividing the cell value by the total column in that cell. Priority Vector Calculations are carried out with the following formula:

$$VP = \frac{Number\ of\ Rows}{n} \quad (3)$$

- Perform consistency test calculations, where this data will be said to be consistent if the value produced is less than 10%, conversely, if the value of this test is produced more than 10%, it is necessary to retrieve the data. There are two calculations of consistency test, namely Consistency Index and Consistency Ratio with the following formula:

$$CI = \frac{(\lambda_{max} - n)}{(n-1)} \quad (4)$$

$$CR = \frac{CI}{RI} \quad (5)$$

- Making Supermatrix in 3 steps.
  Unweighted Supermatrix is a supermatrix made from a recapitulation of priority vector values that have been calculated by existing comparisons. Weighted Supermatrix is a multiplication supermatrix between the unweighted supermatrix and cluster matrix, the final result of each column being 1. Limit Supermatrix is a weighted supermatrix multiplication supermatrix by itself, in which several iterations are carried out until the matrix is stable and the results obtained for each row have the same value and the total column is 1.

### *2.3. Mathematical Model with Goal Programming Method*

The Mathematical Model is made following the conditions of the company, where by formulating the goals to be achieved first. Where the objective function to be considered as a proposal is as follows:

- Minimize the Total Cost of Purchasing Raw Materials
- Maximizing the Value of Purchasing Maltodextrin
- Minimize the Total Cost of Purchasing & Maximizing the Value of Purchasing

**Indeks**
i = Supplier Index, i = 1,2,..,n

**Parameter**
$B_i$ = Raw Material Costs from suppliers –i
$V_i$ = Valuation Weight from supplier–i
D = Demand
$G_i$ = Capacity from supplier –i

**Decision Variabe**
$k_i$ = The order quantity for suppliers–i
$l_i$ = Decision variable whether raw material will be ordered or not at the supplier –i
1, there are ordering raw materials at the supplier –i
0, there is no ordering of raw materials at the supplier -i

**Goals Function**

$$\text{Min } Z_1 = \sum_{i=1}^{n} B_i \cdot k_i \text{ ; with i = 1,2,..., n} \tag{6}$$

$$\text{Max } Z_2 = \sum_{i=1}^{n} V_i \cdot k_i \text{ ; with i = 1,2,..., n} \tag{7}$$

**Limit Function**

- The Total order is equal to the total demand

$$\sum_{i=1}^{n} k_i = D \text{ ; with i = 1,2,...,n} \tag{8}$$

- The Order Allocation for each supplier does not exceed the supply capacity limit of the supplier

$$\sum_{i=1}^{n} k_i \leq G_i \cdot l_i \text{ ; with i = 1,2,...,n} \tag{9}$$

- Minimum suppliers are distributed with 2 suppliers

$$n_{min} \leq \sum_{i=1}^{n} l_i \leq n_{max} \text{ ; with i = 1,2,...,n} \tag{10}$$

- $k_i \geq 0, \ i = 1, 2, 3$
  $k_i$ : *integer*

Mathematical model combining the two conditions above by function number 3, the mathematical model is obtained as follows:

**Parameter**
$T_t$ = The Purchase value obtained from the LINGO software in Max TV calculation.
R = The Purchase Budget limit that will be issued by the company is obtained from the LINGO software in Min TC calculation.

**Decision Variabel**
$m_{i^-}$ = Achievement below the target of the objective function
$m_{i^+}$ = Achievement above the target of the objective function

**Goals Function**
$Min\ m_{i^-} + m_{i^+}$

**Limit Function**

- Minimizing the total purchase cost, the order allocation does not exceed the purchase budget limit for the company, which is R
$$\sum_{i=1}^{n} B_i \,.\, k_i + m_{i^-} - m_{i^+} = \text{R}\text{ ; with } \text{i} = 1,2,..,\text{n} \quad (11)$$
- Maximization of purchase value based on weight, order allocation which gives the highest toal purchase value.
$$\sum_{i=1}^{n} V_i \,.\, k_i + m_{i^-} - m_{i^+} = T_t \quad (12)$$
- The Total order is equal to the total demand
$$\sum_{i=1}^{n} k_i = D\text{ ; with i} = 1,2,\ldots,\text{n} \quad (13)$$
- The Order Allocation for each supplier does not exceed the supply capacity limit of the supplier
$$\sum_{i=1}^{n} k_i \leq G_i \,.\, l_i\text{; with i} = 1,2,\ldots,\text{n} \quad (14)$$
- Minimum suppliers are distributed are 2 suppliers
$$n_{min} \leq \sum_{i=1}^{n} l_i \leq n_{max}\text{ ; with i} = 1,2,\ldots,\text{n} \quad (15)$$
- $k_i \geq 0$; for i = 1,2, ..., n
- $l_i = 0\ atau\ 1\ integer$; for i = 1,2, ..., n

## 3. Result and Discussion

Suppliers used in this study on maltodextrin raw materials consist of 5 suppliers at PT. Neopangan Selaras Indonesia, which is shown in table 2.

**Tabel 2.** Supplier of Maltodextrin Raw Material

| Supplier | Price | Capacity (kg) |
|---|---|---|
| PT. Indo Asia Tirta Manunggal (P1) | IDR 9.400 | 5000 |
| PT. Astamulya Mandiri (P2) | IDR 9.200 | 5000 |
| PT. Satya Anugerah Nutritama (P3) | IDR 9.500 | 5000 |
| PT. Galic Bina Mada (P4) | IDR 9.300 | 5000 |
| PT. Selalu Lancar Maju Karya (P5) | IDR 8.900 | 5000 |

There are 7 criteria and 20 sub-criteria which are considered by PT. Neopangan Selaras Indonesia. The criteria and sub-criteria used can be seen in Table 1. Ratings or rankings obtained are based on the results of the supermatrix limit. The sub-criteria priority weights in local and global can be seen in Table 3 with the main sub-priority being the ability to follow procedures by 0.0815. Supplier weights are shown in Table 4.

**Tabel 3.** Global Rank & Local Rank

| Criteria | Sub-Criteria | Global Weight | Normalized By Cluster | Global Rank | Local Rank |
|---|---|---|---|---|---|
| Quality | Age of Material | 0,0431 | 0,4372 | 7 | 1 |
| | Material Function | 0,0253 | 0,2565 | 15 | 2 |
| | Material Compliance with Specifications | 0,0302 | 0,3063 | 11 | 3 |
| Delivery | Accuracy of Order Quantity | 0,0143 | 0,3132 | 18 | 2 |
| | Accuracy of Delivery Time | 0,0183 | 0,4007 | 17 | 1 |
| | Availability of Material | 0,0131 | 0,2861 | 19 | 3 |
| History | Past Performance | 0,0284 | 0,2665 | 12 | 2 |
| | Flexibility to change the number of orders | 0,0188 | 0,1762 | 16 | 3 |
| | Fulfillment in the cooperation agreement | 0,0594 | 0,5572 | 2 | 1 |
| Garansi | Response Speed | 0,0570 | 0,3450 | 3 | 1 |
| | Ease of contact | 0,0468 | 0,2832 | 6 | 2 |
| | Service time | 0,0268 | 0,1620 | 14 | 4 |
| | Ease of Replacing Reject tems | 0,0347 | 0,2098 | 10 | 3 |
| Price | Discounts and Bonuses | 0,0382 | 0,3017 | 9 | 3 |
| | Payment Method | 0,0482 | 0,3811 | 5 | 1 |
| | Product Prices | 0,0401 | 0,3173 | 8 | 2 |
| Communication system | Communication Skills | 0,0118 | 0,1312 | 20 | 3 |
| | The Speed of Information Exchange | 0,0504 | 0,5622 | 4 | 1 |
| | Flexibility in Sharing | 0,0275 | 0,3066 | 13 | 2 |
| Management & Organization | Ability to Follow Procedures | 0,0815 | 1,0000 | 1 | 1 |

**Tabel 4**. The weight of each supplier

| Alternative | Weight | Rank |
|---|---|---|
| PT. Indo Asia Tirta Manunggal | 0,4421 | 1 |
| PT. Astamulya Mandiri | 0,1352 | 5 |
| PT. Satya Anugerah Nutritama | 0,1434 | 2 |
| PT. Galic Bina Mada | 0,1414 | 3 |
| PT. Selalu Lancar Maju Karya | 0,1379 | 4 |

The results of the goal programming method, obtained values based on 3 conditions namely cost minimization, value maximization and a combination of cost minimization and value maximization. From the results of the Recommendation, it can be used as a basis for allocating other than January 2019. The January 2019 election was due to this month having accumulated demand at PT. Neopagan

IOP Conf. Series: Materials Science and Engineering **847** (2020) 012016 doi:10.1088/1757-899X/847/1/012016

Selaras Indonesia. The results of the Recommendation of Allocation with the help of LINGO software are shown in Table 5.

**Tabel 5**. Recommendation of Allocation Orders in January 2019

| Goals | Allocation Orders in January 2019 | | | | | Total Value | Cost Incurred |
|---|---|---|---|---|---|---|---|
| | P1 | P2 | P3 | P4 | P5 | | |
| **Min TC** | 500 | 5000 | 0 | 5000 | 5000 | 2293,55 | IDR 141.700.000 |
| **Max TV** | 5000 | 0 | 5000 | 5000 | 500 | 3703,45 | IDR 145.450.000 |
| **Min TC & Max TV** | 5000 | 168 | 332 | 5000 | 5000 | 3677,322 | IDR 142.699.600 |

While the actual order allocation in January 2019 is shown in Table 6. It is seen that the costs incurred by January 2019 are said to be incompatible with the budget calculations previously planned by PT. Neopangan Selaras Indonesia.

**Tabel 6**. Actual Allocation Order in January 2019

| Actual Allocation Order in January 2019 | | | | | Total Value | Cost Incurred |
|---|---|---|---|---|---|---|
| P1 | P2 | P3 | P4 | P5 | | |
| 5000 | 4000 | 4500 | 0 | 2000 | 3672,4 | IDR144.350.000 |

From the results of the proposed allocation it can be seen that in the TC minimization, the costs incurred are lower than the actual allocation results but with the existing value below the actual value. In TV maximization conditions it is shown that the proposed value is higher than the actual value, but the resulting costs are greater than the actual costs incurred by the company. In the Combined condition between Min TC and Max TV, it is known that the cost value obtained is lower than the actual value and the value obtained is greater than the actual. The most optimal proposal overall can be seen in the merging condition between Min TC and Max TV from Table 5. Table 7 shows a comparison of the achievement of the combined goal between actual and recommendation in January 2019.

**Tabel 7**. Percentage of Actual and Recommendation Allocation Differences

| Goal Programming | Actual | Recommendation | Differences |
|---|---|---|---|
| **Total Cost** | 3672,4 | 3677,322 | -0,134% |
| **Total** | IDR144.350.000 | IDR142.699.600 | 1,143% |

From the percentage comparison table between actual and recommendation, it can be seen that by using goal programming, the recommendation can save costs by 0.134% and also increase the value by 1.143%.

## 4. Conclusion

The results of the Fuzzy ANP calculation are weights generated for the assessment or evaluation of each supplier. PT. Neopangan Selaras Indonesia considers 7 criteria and 20 sub-criteria in selecting suppliers of maltodextrin raw materials. These criteria include quality, delivery, history and performance of the company, warranty & service complaints, price, communication systems, management & organization.

The resulting weights can determine priority sub-criteria based on PT. Neopangan Selaras Indonesia, represented by 3 respondents who took over in the interaction of raw material sealing activities. The resulting priority sub-criteria is the ability to follow procedures, which is equal to 0.0815 from the global ranking, while the lowest sub-criteria is the ability to communicate. Weights obtained for each supplier is PT. Indo Asia Tirta Manunggal at 0.4421. Weight with sequence 2 is PT. Satya Anugerah Utama in the amount of 0.1434. The number 3 weight is PT. Galic Bina Mada of 0.1414. The 4th weight is PT. Always Current Advanced Work of 0.1379. The weight with the 5th order is PT. Astamulya Mandiri of 0.1352.

Mathematical modelling is done with 3 conditions and is also limited by the actual conditions of the company itself. The 3 conditions are cost minimization, value maximization and a combination of cost minimization and value maximization. Based on the results obtained, it is known that the optimal solution is when using the combined conditions of Min TC and Max TV, where the allocation is carried out as much as 5000 kg to the supplier of PT. Indo Asia Tirta Manunggal, 168 kg to the supplier of PT. Astamulya Mandiri, 322 kg to the supplier of PT. Satya Anugerah Utama, 5000 kg to the supplier of PT. Galic Bina Mada, 5000 kg to the supplier of PT. Always Working Smoothly Forward. This optimal allocation results in a decrease of actual costs of IDR 142.669.600 or as much as 0.134% of the actual and an increase of 4.922 or in other words as much as 1.143% of the actual company. R

## 5. References

[1] Dickson G 1966 An Analysis of Vendor Selection System and Decision. *Journal of Purchasing* **2**(1), 5-17
[2] Kang H, Lee A, and Yang C 2012 A Fuzzy Model for Supplier Selection as Applied to IC Packaging. *Journal of Intelligence Manufacturing* **23,** 1477-1488
[3] Pujawan I and Erawan, M 2010 *Supply Chain Management* (Surabaya: Guna Widya)
[4] Shahroudi K and Rouydel H 2012 Using a Multi-Criteria Decision Making Approach (ANP-TOPSIS) to Evaluate Suppliers in Iran's Industry. *International Journal of Applied Operational Research* **2,** 37-48.
[5] Saaty T 1996 *Decision Making with Dependence and Feedback-The Analytic Network Process* (RWS Publication, Pittsburgh).
[6] Saaty T 2006 *Decision Making with the Analytic Network Process* (Springer: RWS Publication, Pittsburgh).

# Integration of conjoint analysis and QFD for new product development in manufacturing small and medium enterprises (case study: a food company)

**Wilson Kosasih[1], Lithrone Laricha Salomon[2] and Alfred Darius Halim[3]**

Department of Industrial Engineering, Faculty of Engineering Universitas Tarumanagara, Indonesia
Jl. Letjend S. Parman No. 1, Jakarta 11440
wilsonk@ft.untar.ac.id[1], lithrones@ft.untar.ac.id[2], alfreddariushalim@gmail.com[3]

**Abstract.** Basically, small and medium enterprises (SMEs) in Indonesia are weak in innovating product development in accordance with customer's needs, making it difficult for SMEs to penetrate the market. This paper discusses the application of integration between Conjoint Analysis and QFD in the development of a new product from SME. The research was conducted at a medium-sized company, which produces a variety of "dimsum" frozen food products. Finally, the integration of the two provides a double benefit in the research and development stage of the manufacturing SME so that a new product that is developed can really be realized. The final design results obtained two new product segments according to its target market, namely for the economic and premium segments.
**Keywords:** Integration, Conjoint Analysis, QFD, New Product Development, SME.

## 1. Introduction

The frozen food market in Indonesia is growing, for example in 2014 amounting to USD 421.8 million and becoming USD 434.3 million in the next year, and is predicted to continue to increase in the coming years [1]. It shows that frozen food products are increasingly in demand by the Indonesian people. Even in its development, there are currently many SMEs that produce frozen food products. According to the Association of Food and Beverage Entrepreneurs (Gapmmi), in Indonesia the number of SMEs that produce frozen food reaches 10% of the total national food processing SMEs [2].

Similar to SMEs in general, which has weaknesses for the following things; product development innovations including packaging design, marketing strategies, and distribution systems. However, frozen food production SMEs have difficulties with raw materials and their production processes so that a good quality management system is needed, starting from the preparation of raw materials to the processing and adequate food technology.

Along with the challenges mentioned above, innovation becomes a vital aspect in strengthening the competitiveness of food processing SMEs, especially frozen food production. The success of an innovation is largely determined whether an idea or a solution or a product offered must already have early adopters or consumers. Refer to Kosasih et.al. [3], discussing innovation is not just "what and how to design a product," but furthermore it related with "let's hear the voice of the customers"[4].

Therefore, this paper discusses how to integrate conjoint analysis and quality function deployment (QFD) in the development of a new product for frozen food production SMEs. The integration of both can offer more comprehensive product development so that it can penetrate a wider market. Conjoint analysis is an experimental approach to measure consumer preferences for an object (product, service, or idea) [5], [6], [7]. This analysis is based on how consumers evaluate the value of objects (real or hypothetical) by combining each different value from each attribute [7], [8]. Meanwhile, QFD is a quality management that focuses on consumers and product development methodologies that are tangible [9]. According to Jikar et.al. [10], QFD is the process of capturing and translating what consumers need into remedial actions involving many parties from the company, including; research and development, engineering and manufacturing, marketing/sales, and services.

## 2. Methods

### *2.1. Conjoint Analysis Approach*

Conjoint Analysis was first developed by psychological expert Luce and statistician Tukey in 1964 in the field of mathematical psychology [5]. Since the mid-1970s, this method has attracted many researchers' attention as one method that is able to clearly describe the way consumers make decisions in the process of selecting products or services that have multi-attributes [11]. Then in the 1980s, conjoint analysis was widely applied to the industrial field. In the 1990s, the use of conjoint analysis was increasingly widespread in various fields of science [5], [11].

This research was conducted at a medium-sized company, which produces a variety of "dimsum" frozen food products. The initial step in this study began by discussing in depth with the management company regarding the market share of their products as well as their marketing mix strategy. In the next stage, the preliminary questionnaire was made to collect data in the form of what attributes were considered important/priority according to consumer preferences, and other supporting questions. Questionnaires that were made before the deployment were tested trial and error first.

Then fractional factorial design is used to stimulate and generate several profile combinations from a set of attributes with their respective levels arranged to describe the features of a product being developed, or called the full-profile method approach (as seen in Fig 1). An assessment uses Likert scale from value 1 to value 5. Value 1 indicates a preference for "dislike" and a value of 5 indicates a preference for "very like". According to the target consumers, the survey respondents who had consumed a "dimsum" frozen food product. The survey results obtained were then calculated the relative importance and the utility value using conjoint analysis. The final profile combination is determined based on the most positive utility value of a level in certain attribute.

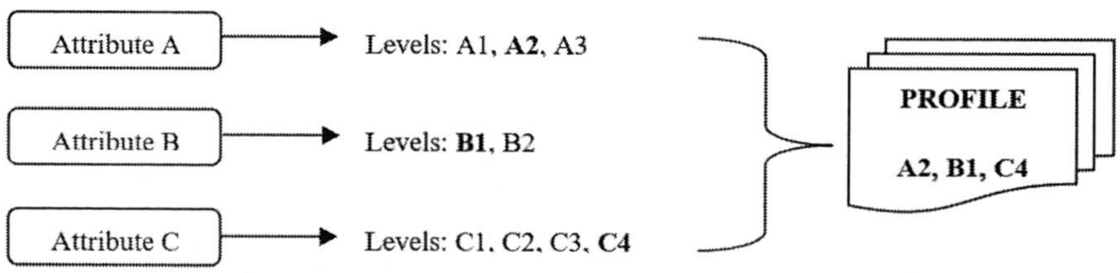


**Figure 1.** Relationships between attributes, attribute levels, and the profile [6],[7]

Conjoint analysis has metric or non-metric dependent variables ($Y$) and has several non-metric independent variables ($X_1, X_2, X_3, ..., X_n$), where the relationship between the variables can be expressed by the following formula [7], [12]:

$$Y = \sum_{i=1}^{n} X_i \quad (1)$$

Mathematical equation that can express the basic model of conjoint analysis is as follows:

$$\mu(x) = \sum_{i=1}^{m} \sum_{j=1}^{k} a_{ij} x_{ij} + \varepsilon_{ij} \quad (2)$$

Where; $\mu(x)$ = total utility of an alternative or stimuli, $a_{ij}$ = utility of the $i^{th}$ attribute ($i$ = 1,2, ..., $m$) and $j^{th}$ level ($j$ = 1,2, ..., $k$), $k$ = numbers of level of each attribute , $m$ = numbers of attribute, $x_{ij}$ = dummy variable of the $i^{th}$ attribute and $j^{th}$ level (value 1 if $j^{th}$ level of $i^{th}$ attribute occurs and 0 if it does not occur), $\varepsilon_{ij}$ = value of stochastic or constant error.
The importance of an attribute that is $I_i$ is defined in the range of part-worth which can be formulated as follows:

$$I_i = \{\max(a_{ij}) - \min(a_{ij})\}, \text{ for every i} \quad (3)$$

Normalizing the importance level of each attribute to ensure its relative importance with other attributes, can be expressed with the following equation:

$$W_i = \frac{I_i}{\sum_{i=1}^{m} I_i} \quad (4)$$

Where; $W_i$ = relative importance weights for each attribute, $I_i$ = range of importance values for each attribute.

2.2. *QFD Application for the Food Industry*
In final step, the product concept would be further developed using the QFD method to plan and design the manufacturing process up to the packaging design to meet consumer needs. Yoji Akao and Katsuyo Ishihara introduced and developed quality function deployment (QFD) from 1965 to 1967, and defined QFD as a method for developing design quality that leads to customer satisfaction and translating customer desires into design targets and critical quality assurance that can be used during development of production or service [13]. In 1972 QFD was first applied by Mitshubishi Heavy Industries Limited in Kobe Shipyard, Japan, and then in the late 1970s, the detailed QFD concept was further developed by Toyota which was then widely applied to the manufacturing industry in the world [14].

Generally, the QFD method consists of four phases of the planning and development matrix known as the four phase model or clausing model [15], [16], namely; a) Product planning matrix (house of quality); b) Design deployment matrix (part deployment); c) Manufacturing planning matrix (process planning); d) Production planning matrix (production operations planning). Refer to Benner, et.al. [16], Hofmeister [17] states that the application of QFD in the food industry has two paths in the product development process based on the needs or voices of customers. These two lines are the packaging deployment path and the food deployment path, each path contains four stages as shown in Figure 2. This method is used because it will be more difficult to integrate the four phase model for food products compared to mechanical products in the industry which are arranged from several components to end products. This case is different for food products where raw materials show a lot of interaction with each other [16].

## 3. Results and Discussion

From the results of the preliminary questionnaire with 214 respondents (see Table 1), it found that there are seven attributes of a "dimsum" frozen food product that were consumers' preferences in choosing the product when ordered from the highest priority level, among others; a) number of contents, b) attractive physical appearance or form, c) flavor, d) opener of packaging, e) packaging form, f) packaging material, and g) form of container. All of these attributes are considered in further product development. Each attribute is developed at a level based on survey results and in-depth discussions with customers, as shown in Table 2.

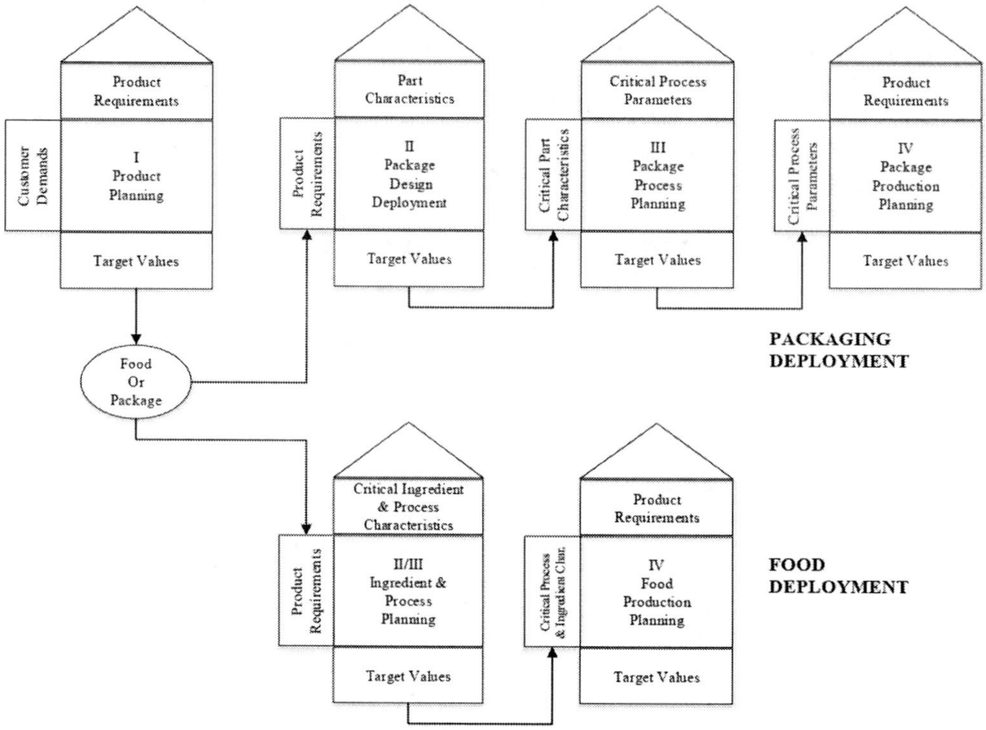


**Figure 2.** A scheme of QFD for the food industry [17]

**Table 1.** Demographic characteristics of respondents

| No. | Variables | Description | Total (N = 214) | |
|---|---|---|---|---|
| | | | Respondents | Percentage |
| 1 | Gender | Male | 127 | 59.35% |
| | | Female | 87 | 40.65% |
| 2 | Age | 15-19 y.o. | 8 | 3.74% |
| | | 20-29 y.o. | 166 | 77.57% |
| | | 30-39 y.o. | 24 | 11.21% |
| | | Equal and greater than 40 y.o. | 16 | 7.48% |
| 3 | Occupation | Student | 109 | 50.93% |
| | | Private employee | 67 | 31.31% |
| | | Entrepreneur | 12 | 5.61% |
| | | Housewife | 14 | 6.54% |
| | | Others: bureaucrat, doctor, teacher, artist, personal trainer, chef, travel guide, bartender, unemployment | 12 | 5.61% |
| 4 | Expenditure per month | IDR 1.000.000,- to IDR 5.000.000,- | 95 | 44.39% |
| | | IDR 5.000.001,- to IDR 10.000.000,- | 69 | 32.24% |
| | | IDR 10.000.001,- to IDR 15.000.000,- | 34 | 15.89% |
| | | Greater than IDR 15.000.000,- | 15 | 7.01% |
| 5 | Region | North Jakarta | 54 | 25.23% |
| | | West Jakarta | 44 | 20.56% |
| | | South Jakarta | 23 | 10.75% |
| | | Central Jakarta | 16 | 7.48% |
| | | East Jakarta | 22 | 10.28% |
| | | Tangerang | 25 | 11.68% |
| | | Bekasi | 11 | 5.14% |
| | | Others | 19 | 8.90% |

**Table 2.** Attributes and its levels

| No | Attributes | Description of Attributes | Levels of Attributes |
|---|---|---|---|
| 1 | Flavor | Variety of stuffing | Salted egg<br>Meat |
| 2 | Physical form | Form of outer appearance | Plain motif<br>Tied motif |
| 3 | Amount of contents | the amount of contents in the packaging | 8 pcs<br>12 pcs |
| 4 | Packaging form | Physical outer appearance of the packaging | Bright full color cover<br>Dark full color cover<br>Partially transparent |
| 5 | Packaging material | Type of packaging material | Plastic<br>Cardboard or paper |
| 6 | Opener of packaging | Mechanism and instructions for opening the packaging | Tear mark<br>Scissors line mark |
| 7 | Container form | Design of the container in the packaging | With partition<br>Without partition<br>Without container |

If the result of the combination is too much, the respondents can be difficult in giving judgment. Therefore, Fractional Factorial Design is used to present and stimulate sixteen profiles of all possible combinations. Then 100 of the 214 respondents were selected as potential early consumers, who would conduct an assessment of all existing profiles. All survey results were tested by Pearson's R correlation. It was found that there were three respondents with Pearson's R value below 0.707 so that the data were not used for further data processing in order to obtain a more accurate the results of utility value and relative importance, as shown in Table 3 and Table 4 respectively.

In this study, "Pao" frozen food products were developed based on market segmentation. Simply stated respondents with expenses below 5 million rupiah per month are categorized as lower class buyers, where 52 respondents and the majority are students. Meanwhile, respondents with expenditure above 5 million rupiah per month are categorized as middle class buyers, where as many as 48 respondents and the majority are employees or entrepreneurs. The products are further divided into two segments, namely economic segment and premium segment. The two segments have different importance levels to the attributes of the product, as seen in detail in Table 4. From the utility of each attribute level, it can be seen that different segments produce different profile combinations as shown in Table 5.

**Table 3.** Average of part-worth utility for economic and premium segments

| Attribute | Levels of Attribute | Economic segment | | Premium segment | |
|---|---|---|---|---|---|
| | | Estimated Utility | Std. Error | Estimated Utility | Std. Error |
| Flavor | Meat | -0.092 | 0.063 | 0.029 | 0.082 |
| | Salted egg | 0.092 | 0.063 | -0.029 | 0.082 |
| Physical form | Tied Motif | -0.030 | 0.063 | -0.003 | 0.082 |
| | Plain Motif | 0.030 | 0.063 | 0.003 | 0.082 |
| Amount of contents | 8 pcs | -0.087 | 0.063 | -0.226 | 0.082 |
| | 12 pcs | 0.087 | 0.063 | 0.226 | 0.082 |
| Packaging form | Bright full color cover | 0.010 | 0.084 | -0.028 | 0.110 |
| | Dark full color cover | -0.105 | 0.099 | 0.145 | 0.129 |
| | Partially transparent | 0.095 | 0.099 | -0.116 | 0.129 |
| Packaging Material | Cardboard or paper | 0.115 | 0.063 | 0.316 | 0.082 |
| | Plastic | -0.115 | 0.063 | -0.316 | 0.082 |
| Opener of packaging | Tear mark | 0.015 | 0.063 | 0.004 | 0.082 |
| | Scissors line mark | -0.015 | 0.063 | -0.004 | 0.082 |
| Container form | With partition | 0.333 | 0.084 | 0.493 | 0.110 |
| | Without partition | -0.044 | 0.099 | -0.167 | 0.129 |
| | Without container | -0.289 | 0.099 | -0.326 | 0.129 |
| | (Constant) | 3.252 | 0.070 | 3.004 | 0.091 |

**Table 4.** Importance level of each attribute for each segment

| Attribute | Importance Level | |
|---|---|---|
| | Premium segment | Economic segment |
| Flavor | 13.857 % | 9.339 % |
| Physical form | 10.012 % | 8.218 % |
| Amount of contents | 9.612 % | 16.465 % |
| Packaging form | 20.798 % | 14.867 % |
| Packaging material | 11.659 % | 18.751 % |
| Opener of packaging | 7.384 % | 6.685 % |
| Container form | 26.677 % | 25.675 % |

**Table 5.** Combined profile results for each segment

| Attribute | Economic segment | Premium segment |
|---|---|---|
| Flavor | Salted egg | Meat |
| Physical form | Plain motif | Plain motif |
| Amount of contents | 12 pcs | 12 pcs |
| Packaging form | Partially transparent | Dark full color cover |
| Packaging material | Cardboard or paper | Cardboard or paper |
| Opener of packaging | Tear mark | Tear mark |
| Container form | With partition | With partition |

*QFD Integration Design*

The results of HOQ design obtained eight technical responses related to consumer needs, including; composition of raw materials, ease of forming the motif, amount of contents in the packaging, the features displayed on packaging, type of packaging used, effort to open the packaging, type of container, size of pao. The importance of each technical response for each segment for details can be seen in Figure 3.

There are ten manufacturing processes that are designed from the development of the food deployment matrix as shown in Figure 4. This process is divided into 2 parts namely "baozi" and its stuffing. The process of making "baozi" considers the following factors; a) the composition of dough or the amount of each raw material used. In this case related to raw materials such as flour, sugar, water, yeast, and several other ingredients; b) Stirring time is related to the duration of kneading all raw materials so that they are evenly mixed. In this case the duration used must not be too long or too fast so that the dough gets time to ferment and expand; c) The type of mold is related to the type of motif produced by the machine. In this case only has 2 types of molds namely, plain and flower motifs; d) The size set on the machine is the size of pao to be printed according to its design specifications; e) Texture related to how a product is felt, such as tender, tough, smooth, creamy, etc.; f) Duration of fermentation is related to how much the dough can expand, where the duration of this process must be in accordance with the composition of the product because if it is not suitable for example too little duration will cause the dough to not expand properly. Conversely, if it takes too long, the texture will become rough and have many holes because the mixture will expand too much causing too much wind in the mixture; g) Heat treatment is related to how to cook the dough, which is used by the oven to steam with the duration of time and temperature that must be considered. The duration will increase with the weight of the dough. The temperature used must be optimal, it should not be too hot or not hot enough.

The manufacturing of product stuffing considers the following factors; a) its composition is directly related to the type of raw material used, wherein the composition of salted egg stuffing consists of salted egg yolk, water sugar, milk, and several other ingredients. Meanwhile, the composition of meat stuffing consists of chicken meat, sugar, salt, and several other ingredients; b) The processing starts from the preparation of raw materials to be processed to become finished products. In salted egg stuffing (economic segment product), the main ingredients are mixed then ground and then steamed. After steaming then add some ingredients and mashed and then steamed again. In meat stuffing (premium segment product), raw chicken meat is seasoned then the meat is grilled. After that, it is ground and then spread a "char-siew" sauce and stirs until blended; c) Heat treatment is related to the process for cooking the product. As mentioned earlier in salted egg stuffing, the mixture is cooked by steaming. In meat stuffing, meat is cooked by roasting.

Besides, the results of the development of the packaging deployment matrix (see Figure 5) identify the six characteristics of this product packaging to meet customer's requirement, among others; a) The container size determines how much pao can be filled onto the container; b) The color of packaging serves to attract customers to buy the product; c) Information and product descriptions are related to the presentation of product composition and nutritional value; d) Carton used is food grade type which tends to be used to pack food. This material is also easier to decompose so it does not pollute our environment; e) The packaging is given a tear mark so that it is possible to open it easily; f) Container with partition that can separate each item so that those do not come into contact with each other.

Then it is planned to be more detailed in developing packaging using the packaging manufacturing deployment matrix (see Figure 6), among others; a) The dimension of packaging is 23 cm x 27 cm x 8 cm; b) The dark color is used as the background of packaging; c) Display product descriptions that aim to provide information about how the product will look; d) Carton used is food grade type; e) a tear mark to make it easier for consumers to open the packaging; f) Containers with 12 partitions according to most interested people from the results of this study.

Figure 7.a illustrates the final design of a new product for each segment developed using the integration of conjoint analysis and QFD methods. Figure 7.b shows an overview of the packaging from various points of view.

## 4. Conclusion

The integration of these two methods makes it possible to plan and develop structured products according to consumer preferences. The QFD used in this study was specifically developed for the food industry so it is appropriate to be applied to food processing SMEs, especially "dimsum" frozen foods. In this study, two new product segments are developed in accordance with the target market, namely for the economic segment and the premium segment. The final specifications of new products obtained are plain motifs, two flavor variants (salted eggs and meat), packaging size 27 cm x 23 cm x 8 cm, dark color, displaying product images on the packaging, food grade cardboard type, there is tear mark, and last it has twelve partitions on the container.

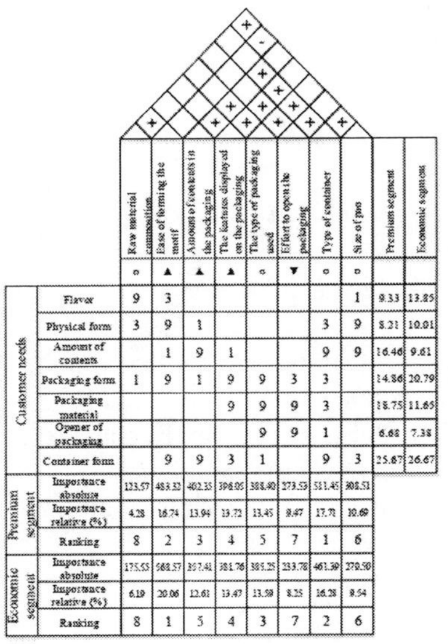


**Figure 3.** Design of HOQ

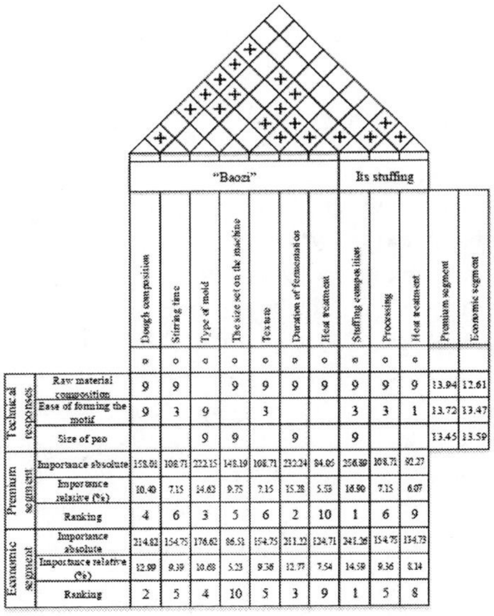


**Figure 4.** Food deployment matrix

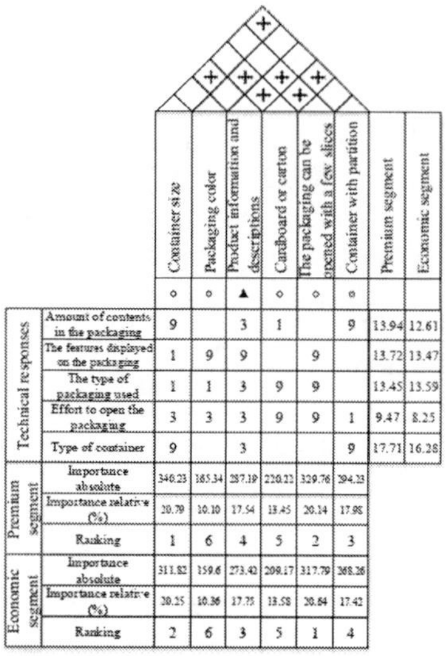


**Figure 5.** Packaging design deployment matrix

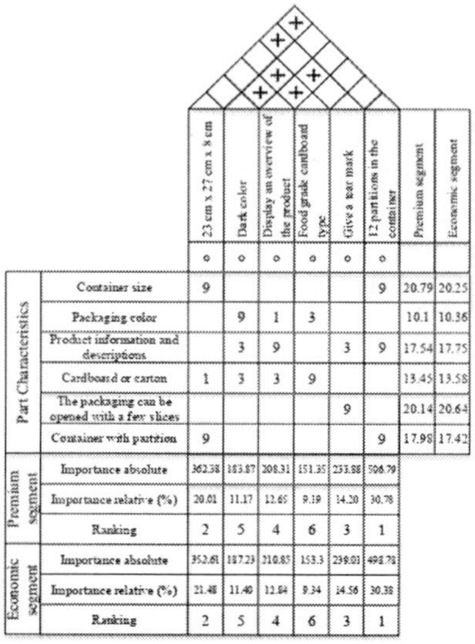


**Figure 6.** Packaging manufacturing planning matrix

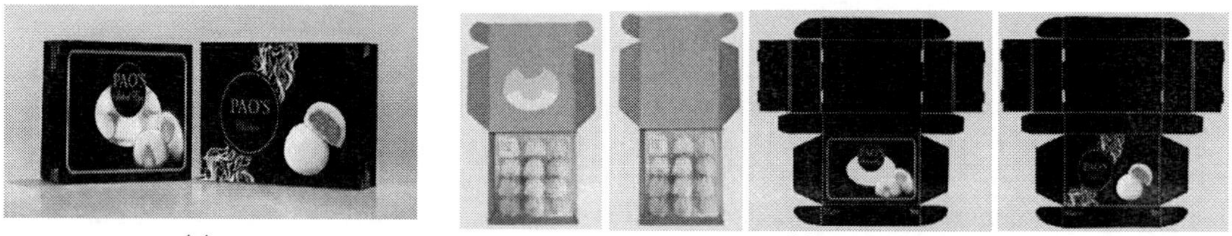


(a) (b)

**Figure 7.** Final design results: (a) new products for each segment, (b) packaging for each segment

**Acknowledgements**
This research recieved a grant from Institute of Research and Society Service, Universitas Tarumanagara. On this occasion, Authors would like to thank to this institution and also to everyone who assisted complete this research.

## 5. References

[1]. See supplementary material at https://www.hexaresearch.com/research-report/indonesia-frozen-food-market.

[2]. See supplementary material at https://ekonomi.bisnis.com/read/20190822/257/1139699/bpom-didesak-tangani-ikm-produsen-frozen-food.

[3]. Kosasih, W., I. Soenandi, and E. Celsia. 2013. Aplikasi QFD untuk Pengembangan Produk Wafer (Studi Kasus: PT. Indo Sari Abadi). *Jurnal Teknik dan Ilmu Komputer*, **2**(7), pp. 258-269.

[4]. Zairi, M. and M. Youssef. 1995. A Review of Key Publications on Benchmarking: part I. *Benchmarking for Quality Management and Technology*, **2**(1), pp. 65-72.

[5]. Kuzmanovic, M., Milan Martić, and Bisera Andrić Gušavac. 2011. Using Conjoint Analysis to Assess Customer Value in The Product Development Process. In *Proceedings of 1st International Symposium Engineering Management and Competitiveness* (Technical Faculty, University of Novi Sad, Zrenjanin, Republic of Serbia), pp. 353-358.

IOP Conf. Series: Materials Science and Engineering **847** (2020) 012017 doi:10.1088/1757-899X/847/1/012017

[6]. Kuzmanovic, M., Bisera Andrić Gušavac, and Milan Martić. 2011. Determining Customer Value Dimensions: a Conjoint Analysis Approach. *Research in Logistic and Production* **1**(3), pp. 139-150.
[7]. Kosasih, W., L.L. Salomon, and R. Hutomo. 2017. Using Conjoint and Cluster Analysis in Developing New Product for Micro, Small, and Medium Enterprises (SMEs) based on Customer Preferences (Case Study: Lampung province's banana chips). *AIP Conference Proceedings 1867*, 020051, https://doi.org/10.1063/1.4994454.
[8]. Hair, J.F., William C. Black, Barry J. Babin, and Rolph E. Anderson. 2009. *Multivariate Data Analysis*. 7th Ed. (New Jersey: Prentice Hall Inc.)
[9]. Chan, L.K. and M.L. Wu. 2002. Quality Function Deployment: A Literature Review. *European Journal of Operational Research* **143**. Pp. 463-497.
[10]. Jikar, V.K., et.al. 2007. Quantitatively Augmented QFD-HOQ. *Asia Pasific Automotive Engineering Conference,* American Society of Mechanical Engineering. Hollywood, Calif.
[11]. Hair, J.F., William C. Black, Barry J. Babin, and Rolph E. Anderson. 2009. *Multivariate Data Analysis*. 7th Ed. (New Jersey: Prentice Hall Inc.)
[12]. Surjandari, Isti. 2010. Conjoint Analysis: Konsep dan Aplikasi, Jakarta: Universitas Trisakti.
[13]. Akao, Y. 1990. *History of Quality Function Deployment in Japan, The Best on Quality*. (New York: Hanser)
[14]. Eldin, N. and V. Hikle. 2003. Pilot Study of Quality Function Deployment in Construction Projects. *Journal of Construction Engineering and Management*, **129**(3), pp. 314-329.
[15]. Cohen, L. 1995. *Quality Function Deployment: How to Make QFD Work for You.* Reading Mass: Addison-Wesley.
[16]. Benner, M., A.R. Linnemann, W.M.F. Jongen, and P. Folstar. 2003. Quality Function Deployment(QFD): Can It Be Used To Develop Food Product? *Food Quality and Preference* **14**, pp. 327-339.
[17]. Hofmeister, K.R. 1991. Quality Function Deployment: Market Success Through Customer Driven Products", in Graf, E. and Saguy, I.S. (Eds), *Food Product Development: From Concept to the Marketplace*, Van Nostrand Reinhold, New York, NY, pp. 189-210.

IOP Conf. Series: Materials Science and Engineering **847** (2020) 012018 doi:10.1088/1757-899X/847/1/012018

# Integrated model for Multi-criteria Supplier Selection and Order Allocation Problem

**Tjutju Tarliah Dimyati**

Industrial Engineering Department, Universitas Pasundan, Jln. Dr. Setiabudi No. 193, Bandung, Indonesia
tjutjutarliah@unpas.ac.id

**Abstract.** This paper discusses an integrated model of selecting suppliers and order allocation for a company that wishes to decide the quantity to be ordered from each supplier on the basis of some qualitative criteria. Since each supplier may have a different performance with respect to these criteria, an integration of analytical hierarchy process and linear program model is proposed to solve the problem in two stages. In the first stage, suppliers are evaluated based on qualitative criteria to consider both tangible and intangible factors in choosing the best suppliers. The output of this stage is the final score of each supplier. In the second stage, a linear programming model is proposed to placing the optimum order quantities among them such that the total final scores of suppliers become maximum. The mathematical programming model is validated through numerical analysis, and the computation result shows that the model is effective and applicable.
**Keywords**: supplier selection, order allocation, multiple criteria decision making, analytical hierarchy process, linear programming.

## 1. Introduction

To be able to survive the current business competition, supplier selection process is a very important issue to be well planned. This is because suppliers will not only determine the price of goods or materials needed, but also the time of delivery. In addition to the price and delivery time, supplier selection is generally based on other criteria, such as goods' quality, after-sales service, response to order changes, supplier's financial condition, supplier location, etc. It is clear that the selection of supplier involves multi-criteria, both quantitative and qualitative. Since there are often conflicts between these criteria, supplier selection should be done by aligning both types of criteria.

Based on the ability of suppliers to meet the needs of buyers, supplier selection issues can be categorized into two types of issues [1]. If all buyers ' needs can be met by one supplier, the issue to be resolved is to choose one best supplier based on predefined criteria. If there is no single supplier can fulfill all of the needs of buyers, then the problem to be solved is to select some suppliers and at the same time specifying the number of items to be ordered from the selected supplier, so that the value of purchase can be maximized. Clearly, the second type of problem is more complex than the first one.

Some models and methods of supplier selection and order quantity allocation problem have been developed by many researchers, where mathematical programming models and approaches have been widely used. Among others is integer linear programming model with the objective of maximizing the preference weight of supplier [2], mixed integer programming (MIP) model with various objectives, such as minimizing both the cost and the number of suppliers [3], minimizing the total of purchasing, inventory, transportation, and supplier management costs [4], minimizing the total of selecting, procuring, holding, and shortage costs [5]. Another methods considering multi-objective have also been developed, such as multi-objective mathematical model which minimize total cost and maximize total social, total environmental, and total economic qualitative scores [6], multi-objective mixed integer non-linear programming (MINLP) model which considered cost, quality and service level [7,8]. and a hybrid of the DEA, fuzzy multi-objective programming and genetic algorithm [9].

The purpose of the study is to create an optimization model to determine the allocation of orders to a number of suppliers, where there are three important issues to be decided. These three issues comprise of determining the ability level of each supplier to meet the criteria set by the buyer, determining the quantities of material to be ordered, and determining the ordering time to fulfil the needs.

This paper is organized as follows. A brief review of the literature is presented in section 2, while section 3 contains the research methodology. Section 4 and 5 contain a numerical example, the results, and the discussions, while the conclusions is presented in section 6.

## 2. Methods

To solve the integrated problem of supplier selection and order allocation, a comprehensive framework is proposed. The proposed approach presents two-stage decision making procedure, where AHP is applied in the first stage to evaluate the weight of each supplier performance, and a linear programming model is applied in the second stage to allocates order quantities between the suppliers. The proposed algorithm consists of the following steps:

1. Constructing a hierarchical system
2. Preparing a pair-wise comparisons matrix and perform the synthesis process.
3. Calculating the weights and testing the consistency ratio.
4. Computing the overall performance score of each supplier.
5. Constructing the linear programming model

### *2.1 Constructing the hierarchical system*

Structuring a hierarchy/decision model, i.e. to construct hierarchies based on the objectives, criteria, sub-criteria and alternatives, so that the overall structure/decision model is visible.

### *2.2 Making pair-wise comparison and perform the synthesis process*

As explained in Saaty [10], after the hierarchy has been structured, the next stage is to do a pair-wise comparison and provide judgement to the pair of criteria, sub-criteria, or alternatives using a basic scale (fundamentals scale) 1 to 9. This assessment is done by comparing a number of combinations of components that exist at each level of the hierarchy, so that the quantitative analysis to determine the magnitude of the weight can be done.

### *2.3 Calculating the weights and testing the consistency ratio*

Perform the normalization process, i.e. making the unit of each criterion, sub-criterion, or alternative to be the same, so that we can compare each criterion, sub-criterion or alternative. Then calculate the Consistency Ratio (CR) to determine the level of consistency of the decision maker in providing judgement. If the CR is < 10% then the decision maker is already consistent in doing their judgement

### *2.4 Computing the total performance score of each supplier*

To determine the performance rating of each supplier, multiply the value of each sub criterion with the weight of that sub criterion, then sum that value to all criteria

### *2.5 Building the linear programming model*

Construct a linear programming model using the suppliers' performance rating as the coefficient of the objective function to be maximized. The goal is to allocate the order quantities subject to some important constraints.

IOP Conf. Series: Materials Science and Engineering **847** (2020) 012018 doi:10.1088/1757-899X/847/1/012018

## 3. Results and Discussions

The following illustration is intended to explain how the proposed model is used. Suppose that the constraints of this linear programming are buyer's demand, supplier's capacity, supplier's defect rate, and supplier's delivery reliability, then the objective function and constraints of this linear programming are as follows.

### *3.1 Notations*

$S_j$ final score of supplier $j$
$V_j$ maximum quantity that can be supplied by supplier $j$
$L_j$ lot-size of supplier $j$
$D$ buyer's demand quantity
$q_j$ percentage of defect rate of supplier $j$
$Q$ buyer's maximum acceptable defect rate
$r_j$ percentage of late delivery of supplier $j$
$R$ buyer's maximum acceptable late delivery
$X_j$ integer decision variable denoting the number of lots ordered from supplier $j$

### *3.2. The Objective Function*

The objective function to maximize the total final scores of all suppliers can be shown as

$$Max \sum S_j X_j L_j \qquad j = 1, 2, ..., n \tag{1}$$

### *3.3. The Constraints*

The important constraints of the problem are buyer's demand quantity, the quality, the delivery reliability, and the supplier's capacity, which are formulated as follows:

*Demand constraint*

The total order quantity of the item from all suppliers must meet the demand quantity of the item. It can be stated as

$$\sum X_j L_j \leq D \qquad j = 1, 2, ..., n \tag{2}$$

*Quality constraint*

The total defect quantity of the item cannot exceed maximum total acceptable defect quantity, therefore,

$$\sum q_j X_j L_j \leq QD \qquad j = 1, 2, ..., n \tag{3}$$

*Delivery constraint*

The total late delivery cannot exceed maximum total acceptable lateness. It can be stated as

$$\sum r_j X_j L_j \leq RD \qquad j = 1, 2, ..., n \tag{4}$$

*Supplier's capacity constraint*

The order quantity of the item cannot exceed the supplier's capacity. It can be stated as

$$X_j L_j \leq V_j \qquad j = 1, 2, ..., n \tag{5}$$

Non-negativity constraint

$$X_j \geq 0, \text{ integer} \qquad j = 1, 2, ..., n \tag{6}$$

### *3.4. Final Model*

The final integrated linear integer programming model can be shown as

$$Max \sum S_j X_j L_j \qquad j = 1, 2, ..., n \tag{7}$$

Subject to:

$$\sum X_j L_j \leq D \qquad j = 1, 2, ..., n \tag{8}$$

$$\sum q_j X_j L_j \leq QD \qquad j = 1, 2, ..., n \tag{9}$$

$$\sum r_j X_j L_j \leq RD \qquad j = 1, 2, ..., n \tag{10}$$

$$X_j L_j \leq V_j \qquad j = 1, 2, ..., n \qquad (11)$$

$$X_j \geq 0, \text{ integer} \qquad j = 1, 2, ..., n \qquad (12)$$

The objective function (7) is to maximize the total final scores of all suppliers. The constraint (8) stating that the total order quantity of the item from all suppliers must meet the demand quantity of the item. The constraint (9) stating that the total defect quantity of the item cannot exceed maximum total acceptable defect quantity. The constraint (10) stating that the total late delivery cannot exceed maximum total acceptable lateness. The constraint (11) stating the supplier's capacity constraints, while constraint (12) stating that the number of lots ordered has to be integer and non-negative value.

*3.5. Numerical Example*

Suppose a company wishes to place an order of 1200 pieces of particular item, for which four suppliers are available. Each supplier accepts orders in lot-sizes that vary from supplier to supplier. The lot-size for supplier A, B, C, and D is 40, 70, 60, and 50, respectively. Each supplier can supply only a maximum quantity given by 800, 500, 600, and 900 units for supplier A, B, C, and D respectively. The previous record shows that the defect rate for supplier A, B, C, and D is 3%, 5%, 1%, and 6% respectively, while the record of on-time delivery for supplier A, B, C, and D is 93%, 98%, 85%, and 92% respectively. The company's maximum acceptable defect rate is 4% and the maximum acceptable lateness is 8%.

*3.5.1 Supplier Selection.* Suppose the company have found five most important criteria and 11 sub-criteria as shown in Table 1, that leads to the hierarchy structure of the problem as shown in Figure 1.

**Table 1.** The main-criteria and sub-criteria for performance evaluation

| Main-criteria | Sub-criteria | Description |
|---|---|---|
| Purchasing Cost | Product price | Price offered by supplier |
| | Condition of payment | Date of maturity of the invoice |
| Product Quality | Defect ratio | Defective percentage rate |
| | Supporting document | All important documents are available |
| Delivery Reliability | On-time delivery | Goods are delivered as per PO date |
| | Delivery accuracy | Goods are delivered in accurate quantities |
| | Delivery time | Total time since order to delivery |
| Service | Responsiveness to request for quotation | Supplier promptly submits quotation after receiving a request for quotation |
| | Responsiveness to claim/problem resolve | Supplier quickly responses back when claim/problem is reported |
| Flexibility | Inventory availability | Supplier' ability to fulfill customer's demand |
| | Negotiability | Supplier' willingness to negotiate on price and delivery schedule |

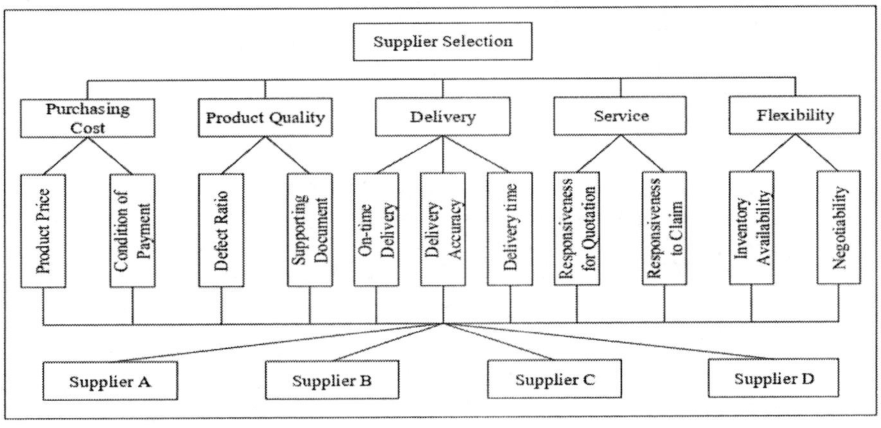


**Figure 1.** The hierarchical structure for supplier selection

The input data consisting of pair-wise comparison matrix of the suppliers for the five main-criteria is given in Table 2.

**Table 2.** The pair-wise comparison for the five main-criteria

| | Purchasing Cost | Product Quality | Delivery | Service | Flexibility |
|---|---|---|---|---|---|
| Purchasing Cost | 1 | 1 | 3/2 | 2 | 2 |
| Product Quality | 1 | 1 | 3/2 | 2 | 2 |
| Delivery | 2/3 | 2/3 | 1 | 3/2 | 3/2 |
| Service | 1/2 | 1/2 | 2/3 | 1 | 3/2 |
| Flexibility | 1/2 | 1/2 | 2/3 | 2/3 | 1 |

For simplicity, let say that the sub-criteria product price is twice more important than the condition of payment, the defect ratio is three times more important than the supporting document, the responsiveness for quotation is 1.5 times more important than the responsiveness to claim resolve, and the inventory availability is twice more important than the negotiability, and the pair-wise comparison matrix for the sub-criteria of delivery is shown in Table 3.

**Table 3.** The pair-wise comparison for the five main-criteria

| | On-time Delivery | Delivery Accuracy | Delivery time |
|---|---|---|---|
| On-time Delivery | 1 | 3/2 | 2 |
| Delivery Accuracy | 2/3 | 1 | 3/2 |
| Delivery time-cost | 1/2 | 2/3 | 1 |

The pair-wise comparison matrices for each of 11 sub-criteria are shown in Table 4 to Table 8.

**Table 4.** Pair-wise comparison for the sub-criteria of purchasing cost

| | Sub-criteria | | | | | | | |
|---|---|---|---|---|---|---|---|---|
| | Product price of supplier | | | | Condition of payment of supplier | | | |
| Supplier | A | B | C | D | A | B | C | D |
| A | 1 | 2 | 3/2 | 3 | 1 | 2/3 | 1/2 | 1/2 |
| B | 1/2 | 1 | 3/2 | 3/2 | 3/2 | 1 | 2/3 | 1/2 |
| C | 2/3 | 2/3 | 1 | 2 | 2 | 3/2 | 1 | 3/2 |
| D | 1/3 | 2/3 | 1/2 | 1 | 2 | 2 | 2/3 | 1 |

**Table 5.** Pair-wise comparison for the sub-criteria of product quality

| | Sub-criteria | | | | | | | |
|---|---|---|---|---|---|---|---|---|
| | Defect ratio of supplier | | | | Supporting document of supplier | | | |
| Supplier | A | B | C | D | A | B | C | D |
| A | 1 | 4/3 | 3/2 | 2 | 1 | 1 | 1 | 1 |
| B | 3/4 | 1 | 2 | 5/2 | 1 | 1 | 1 | 1 |
| C | 2/3 | 1/2 | 1 | 3/2 | 1 | 1 | 1 | 1 |
| D | 1/2 | 2/5 | 2/3 | 1 | 1 | 1 | 1 | 1 |

**Table 6.** Pair-wise comparison for the sub-criteria of delivery

| | Sub-criteria | | | | | | | | | | | |
|---|---|---|---|---|---|---|---|---|---|---|---|---|
| | On-time delivery of supplier | | | | Delivery accuracy of supplier | | | | Delivery time of supplier | | | |
| Supplier | A | B | C | D | A | B | C | D | A | B | C | D |
| A | 1 | 2 | 2 | 3/2 | 1 | 1/2 | 2/3 | 1/3 | 1 | 2 | 3/2 | 4/3 |
| B | 1/2 | 1 | 1 | 2/3 | 2 | 1 | 3/2 | 2/3 | 1/2 | 1 | 2/3 | 2/3 |
| C | 1/2 | 1 | 1 | 2/3 | 3/2 | 2/3 | 1 | 1/2 | 2/3 | 3/2 | 1 | 2/3 |
| D | 2/3 | 3/2 | 3/2 | 1 | 3 | 3/2 | 2 | 1 | 3/4 | 3/2 | 3/2 | 1 |

**Table 7.** Pair-wise comparison for the sub-criteria of service

| | Sub-criteria | | | | | | | |
|---|---|---|---|---|---|---|---|---|
| | Responsiveness for quotation of supplier | | | | Responsiveness to claim resolve of supplier | | | |
| Supplier | A | B | C | D | A | B | C | D |
| A | 1 | 2 | 3 | 3 | 1 | 1/2 | 1/3 | 1/4 |
| B | 1/2 | 1 | 3/2 | 4/3 | 2 | 1 | 1/2 | 1/3 |
| C | 1/3 | 2/3 | 1 | 2/3 | 3 | 2 | 1 | 2/3 |
| D | 1/3 | 3/4 | 3/2 | 1 | 4 | 3 | 3/2 | 1 |

**Table 8.** Pair-wise comparison for the sub-criteria of flexibility

| | Sub-criteria | | | | | | | |
|---|---|---|---|---|---|---|---|---|
| | Inventory availability of supplier | | | | Negotiability of supplier | | | |
| Supplier | A | B | C | D | A | B | C | D |
| A | 1 | 1/3 | 1/4 | 1/5 | 1 | 1/2 | 2/3 | 3/4 |
| B | 3 | 1 | 2/3 | 1/2 | 2 | 1 | 3/2 | 4/3 |
| C | 4 | 3/2 | 1 | 2/3 | 3/2 | 2/3 | 1 | 3/2 |
| D | 5 | 2 | 3/2 | 1 | 4/3 | 3/4 | 2/3 | 1 |

In order to determine the final score of each supplier, the AHP model is performed using *Expert Choice* software, where the result is given in Figure 2.

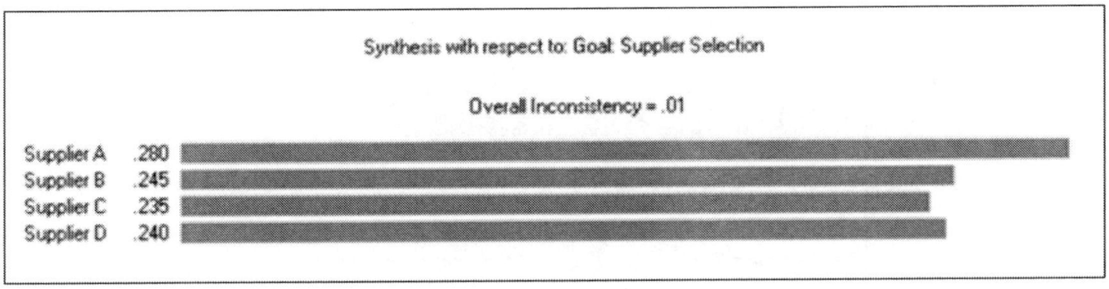


**Figure 2.** The final score of each supplier

*3.5.2 Order Allocation*
In order to find the best order quantities, the total final scores of suppliers should be maximized. The model formulation is as follows.

*3.5.2.1 Objective function*
The objective to maximize the total final scores of all suppliers is expressed as:

$$Max \quad 0.280\,X1 + 0.245\,X2 + 0.235\,X3 + 0.240\,X4$$

*3.5.2.2 The demand constraint*
The total order quantity of the item from all suppliers must meet the demand quantity of the item, therefore:

$$40\,X1 + 70\,X2 + 60\,X3 + 50\,X4 \leq 1200$$

*3.5.2.3 The quality constraint*
The total defect quantity of the item cannot exceed maximum total acceptable defect quantity, therefore:

(0.03 x 40) $X1$ + (0.05 x 70) $X2$ + (0.01 x 60) $X3$ + (0.06 x 50) $X4 \leq$ (0.04 x 1200)
or $1.2\,X1 + 3.5\,X2 + 0.6\,X3 + 3\,X4 \leq 48$

*3.5.2.4 The delivery constraint*
The total late delivery cannot exceed maximum total acceptable lateness. It can be stated as:
(0.07 x 40) X1 + (0.02 x 70) X2 + (0.15 x 60) X3 + (0.08 x 50) X4 $\leq$ (0.08 x 1200)
or $2.8\,X1 + 1.4\,X2 + 9\,X3 + 4\,X4 \leq 96$

*3.5.2.5 The supplier's capacity constraint*
The order quantity of the item cannot exceed the supplier's capacity. It can be stated as

$40\,X1 \leq 800$
$70\,X2 \leq 500$
$60\,X3 \leq 600$
$50\,X4 \leq 900$

The problem is then solved using Linear Interactive Discrete Optimizer (LINDO) software, where the report is shown in Figure 3, where the solution gives an optimal order of 20 lots with supplier A and 8 lots with supplier D, and the total final scores is 7.52. No orders are placed with the supplier B and C.

```
MAX      0.28 X1 + 0.245 X2 + 0.235 X3 + 0.24 X4
 SUBJECT TO
        2)   40 X1 + 70 X2 + 60 X3 + 50 X4 <=   1200
        3)   1.2 X1 + 3.5 X2 + 0.6 X3 + 3 X4 <=   48
        4)   2.8 X1 + 1.4 X2 + 9 X3 + 4 X4 <=   96
        5)   40 X1 <=    800
        6)   70 X2 <=    500
        7)   60 X3 <=    600
        8)   50 X4 <=    900
 END
 GIN       4

LP OPTIMUM FOUND AT STEP       3
 OBJECTIVE VALUE =    7.51999998

        OBJECTIVE FUNCTION VALUE

        1)        7.520000

 VARIABLE        VALUE          REDUCED COST
       X1        20.000000         -0.280000
       X2         0.000000         -0.245000
       X3         0.000000         -0.235000
       X4         8.000000         -0.240000

      ROW   SLACK OR SURPLUS     DUAL PRICES
       2)         0.000000          0.000000
       3)        -0.000001          0.000000
       4)         8.000001          0.000000
       5)         0.000000          0.000000
       6)       500.000000          0.000000
       7)       600.000000          0.000000
       8)       500.000000          0.000000

 NO. ITERATIONS=       3
 BRANCHES=     0 DETERM.=   1.000E    0
```

**Figure 3**. Problem and solution report for the example

The numerical example demonstrates that the model offers a reasonable and realistic solution based on the imposed objective and constraints. The optimum solution shows that to obtain 1200 units of the demanded product, the company has to order 20 lots of product from supplier A which means that the quantity of product ordered from supplier A is (20x40) = 800 units. In addition, the company has to order 8 lots of product from supplier D, which means that the quantity of product ordered from supplier D is (8x50) = 400 units.

Since the company's maximum acceptable defect rate is 4% and the maximum acceptable lateness is 8%, then the total defect product has to be less than (0.04x1200) = 48 units and the total product with late delivery has to be less than (0.08x1200) = 96 units. From the optimum solution it can be resumed that the total number of defect products from both suppliers A and D is (0.03x800) + (0.06x400) = 48 units, which is equals to the company's maximum acceptable defect rate. The total unit of product that will be late delivered both from supplier A and D is (0.07x800) + (0.08x400) = 88 units, which is less than the company's maximum acceptable lateness.

## 4. Conclussion

The overall supplier selection problem of multiple sourcing is not only to select the right suppliers, but also to allocate optimal order quantity among the selected suppliers, based on a number of key criteria such as costs, quality, delivery reliability, supplier's capacity, etc. For these purposes, the two-stage decision making procedure was developed in this paper. Firstly, a hierarchical structure for supplier selection was developed through the analytical hierarchy process, in consideration of both quantitative and qualitative criteria, in order to identify a set of candidate suppliers. Secondly, to help management allocate the optimum order quantities to the candidate suppliers, a linear programming model was formulated such that the total final score of suppliers becomes maximum.

The advantage of this proposed model is that it can deal with multiple criteria, both qualitative or quantitative, such as cost, quality, delivery reliability, customer services, and many other criteria in supplier selection problems. For future research, this model can be extended by considering uncertainties, since many parameters may not be estimated certainly, such as customer demand and supplier's capacity.

## 5. References

[1] Deshmukh AJ and Vasudevan H 2014 Emerging supplier selection criteria in the context of traditional vs green supply chain management. *International Journal of Managing Value and Supply Chains (IJMVSC),* **5** 19-33

[2] Ghosh SK, Naurin Z, Tanzima ZC and Md Sazol A 2018 Supplier selection using integer linear programming model. *Global journal of researches in engineering,* **18** 26-30

[3] Basnet C and Weintraub A 2009 A genetic algorithm for a bicriteria supplier selection problem. *Intl. Trans. in Op. Res.,* **16** 173-87

[4] Hammami R, Frein Y and Hadj-Alouane AB 2012 An international supplier selection model with inventory and transportation management decisions. *Flexible Services and Manufacturing Journal,* **24** 4-27

[5] Zhang JL and Chen J 2013 Supplier selection and procurement decisions with uncertainty demand, fixed selection costs and quantity discount. *Computer and Operation Research,* **40** 2703-10

[6] Azadnia AH 2016 A multi-objective mathematical model for sustainable supplier selection and order lot-sizing under inflation. *International Journal of Engineering (IJE), TRANSACTIONS B: Applications,* **29** 1141-50

[7] Kamali A., Ghomi SF, and Jolai F 2011 A multi-objective quantity discount and joint optimization model for coordination of a single-buyer multi-vendor supply chain. *Computers and Mathematics with applications,* **62** 3251-69

[8] Rezaei J and Davoodi M 2011 Multi-objective models for lot-sizing with supplier selection. *International Journal of Production Economics,* **130** 77-86

[9] Parthiban P, Punniyamoorthy M, Ganesh K and Ranga JG 2009 A hybrid model for sourcing selection with order quantity allocation with multiple objectives under fuzzy environment. *Int. J. Applied Decision Sciences,* **2** 275-98

[10] Saaty TL 1994 How to make a decision: the analytic hierarchy process. *Interface* **24** 19-43

IOP Conf. Series: Materials Science and Engineering **847** (2020) 012019 doi:10.1088/1757-899X/847/1/012019

# Web-Based Application of Reliability Availability Maintainability and Cost of Unreliability Method to Analyze Performance of the Machine

**Judi Alhilman[1,2], Muhammad Fadhil Habibie[1,3], and Wawan Tripiawan[1,4]**

[1]School of Industrial and Systems Engineering Telkom University Bandung, West Java, Indonesia
[2]alhilman@telkomuniversity.ac.id, [3]mfadhilhbb@gmail.com, [4]wawantripiawan@telkomuniversity.ac.id

**Abstract.** Maintenance is important to do in order to maintain or improve the function of the equipment to work optimally. One factor that support the maintenance activities is machine reliability. Higher reliability can reduces process costs. Conversely, engine failure can reduce production output also the business benefits for the community. In business terms, the problem of controlling the Cost of Unreliability (COUR) of equipment and process failure is just a waste of money. Unreliable index costs are simple and practical reliability tools for converting failure data into costs. A long stage in manual calculation of COUR and RAM analysis with more than one applications used as work tool, therefore designed a web-based application with more complete features that can be used to analyze COUR and RAM. This application will simplify the calculation process, analysis, and results management, so it will helps maintenance analysts in doing their work. This application can determine the maintenance policy, predict the performance of machine's reliability, availability and maintenance capabilities. As well as calculating costs from reliability issues.
**Keywords**: Aplication, reliability, COUR, RAM

## 1. Introduction

Maintenance can be defined as a combination of various actions to maintain or repair an item to an acceptable condition. In Journal [1] One method described in determining the cost policy for maintenance is COUR. Research on determining preventive maintenance has been conducted by [2], [3], and [4] with Output Proposed Maintenance Policy in the Form of Maintenance Task and Industrial Application Based Maintenance. Reliability Availability Maintainability (RAM) is a method that can be used to predict and evaluate system performance. The purpose of this method is to increase the ability of the system to operate normally, for a long time, without any failure. Information obtained from the analysis helps management in improving system performance [5]. Therefore, a web-based application is designed to help determine maintenance policies. This application will facilitate the process of calculation, analysis, and storage of the calculated output output for companies that implement maintenance and also record the history of damage for each machine. Then to facilitate analysis using the COUR and RAM methods.

## 2. Methods

### *2.1. Maintenance Management*
According to the journal [1]. Maintenance Management is an activity that is carried out repeatedly with the aim that the equipment is in the same condition as the initial state. Maintenance activities also aim to maintain the equipment in conditions that are acceptable to its use.

### *2.2. Objective Maintenance*
According to [6] the objectives of treatment are to extend the useful life of assets, to guarantee the optimum availability of equipment installed for production, to ensure operational readiness of the equipment needed in an emergency at all times, to ensure the safety of people who use these facilities.

### *2.3. Preventive Maintenance and Predictive Maintenance*
Preventive Maintenance is a scheduled action that aims to maintain the system at a certain level by providing systematic review, detection, and or prevention of impending failure [3]. Corrective maintenance is an unscheduled maintenance caused from failures of component or system with the aims to return the performance back to the initial conditions [7]. Predictive maintenance is a maintenance strategy where the implementation is based on the condition of the machine itself [1].

### *2.4. Cost Of Unreliability (COUR)*
According to [8], COUR means all costs that are caused from all reliability failures, costs associated with bad programs and poor maintenance. COUR gives the results of cost calculations arising from reliability problems, which will show the costs generated by each equipment or machine in the system [8]. Figure 1 explains the costs used in calculating the COUR according to [8].

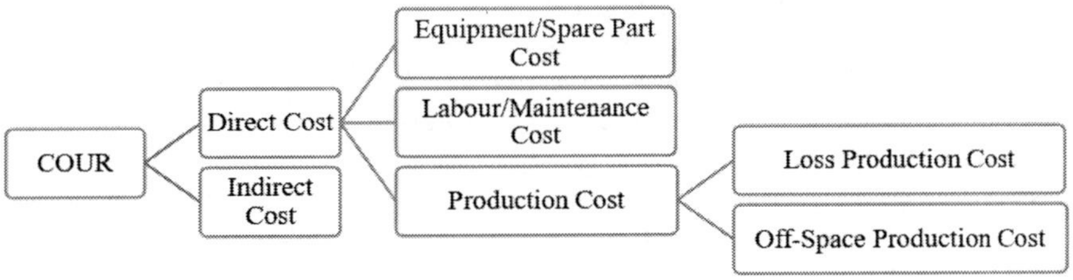


**Figure 1.** COUR Classification

$$COUR = Direct\ Cost + Indirect\ Cost \quad (1)$$

### *2.5. Reliability Availability Maintainability (RAM)*
RAM analysis is a method that can be used to predict the performance of reliability ,availability maintainability of a system or component. The main performance indicator in RAM is availability which is part of the time when the system is fully functional. RAM analysis is used to identify critical and sensitive subsystems in production systems that can have a large effect on system performance. RAM analysis can also be used to assist in the selection of concepts, as well as being able to provide detailed decisions related to front end engineering systems [9].

### *2.6. Related Paper*
Based on research [10] the RAM and COUR methods were used to evaluate the M251 Weaving machine at Buana Intan Gemilang Company. The data that used in the calculation including, system breakdown structure, determining critical subsystems, Reliability Block Diagram (RBD) modeling, determining the representative distribution and distribution plotting (TTF, TTR and DT) data. From the result can be

concluded that the company needs to improve operational conditions. In addition, the lack of lagging indicator value of the company's target is recommended to increase the efficiency of the machine. On [7] research used statistical analysis of damage data to estimate RAM on an automated croissant production line with purpose to determine the critical point on the automated croissant production line that needs improvement through effective maintenance strategies. The methodology used is based on the analysis of damage and repair data with statistical techniques. By using data trend test and serial correlation to determine whether the data is independent and distributed or not. Based on research [11] analyze RAM in wine packaging production lines. In this study, RAM analysis is useful for determining maintenance intervals, planning and organizing adequate maintenance strategies. This study uses the Failure Mode and Effect Analysis (FMEA) method. Based on FMEA analysis the most common factors in wine packaging production lines include low availability of spare parts due to lack of planning and time-consuming delivery. Based on research [9] on cheese production lines over a 17 month period. It is known that the company applies corrective maintenance, ie maintenance when the engine is damaged. Data collection such as TTF and TTR is done when the machine breaks down and begins repairing. The results of the study note the availability value is greater than the actual availability which means the availability of the machine is large enough to be used. Based on research [12] determine the reliability and availability assessment of seabed storage tanks using fault tree analysis. There are four steps in the procedure including determining the system boundary, collecting reliability data, creating a fault tree and estimating the reliability value. The company suggested to do preventive maintenance so that the storage tank remained safe. Based on research [8] oil & gas companies are required to implement long-term strategies to reducing risk in the future. This paper focuses on the economic sustainability aspects, showing how the process must be carried out through the economy with COUR method. The purpose of this paper is to help make the right decisions of RAM improvement. On [1] research discuss about reliability of printing machine. Engine failure can disrupt production target so the company will incur financial losses. The purpose of this method is to calculate the total cost incurred because the printing press cannot be relied on during active repair and downtime. Based on research [5] Efficiency and effectiveness of a maintenance is important for the success and continuity of the organization. Some financial performance indicators for maintenance are presented in the literature but on the contrary, most indicators deal only with direct or preventive maintenance costs as added value and all costs for corrective maintenance as waste. Based on research [5] discusses the influence of the two-dimensional warranty policy to predict the cost of unreliability. This modeling procedure takes into account the observed reduction in the number of warranty claims, making it a more realistic evaluation of the product. On [5] research discusses failures related to the production system, which can result an enormous environmental, health and financial losses. Removing process of critical failures at the design stage results much lower costs than removing the manufacturing stage. The designer must know in advance about the real costs that can not be relied upon, so that it is faster to complete an unsuitable design solution related to the value of a very large loss of failure.

### *2.7. System Design*

*2.7.1. Use Case Diagram.* Use case functions to identify the user and function of a system. Use cases explain the function of the system from the user's perspective and are still within the scope of user understanding [9]. Use cases represent the purpose of a system and explain the sequence of user activities can be seen in Figure 2:

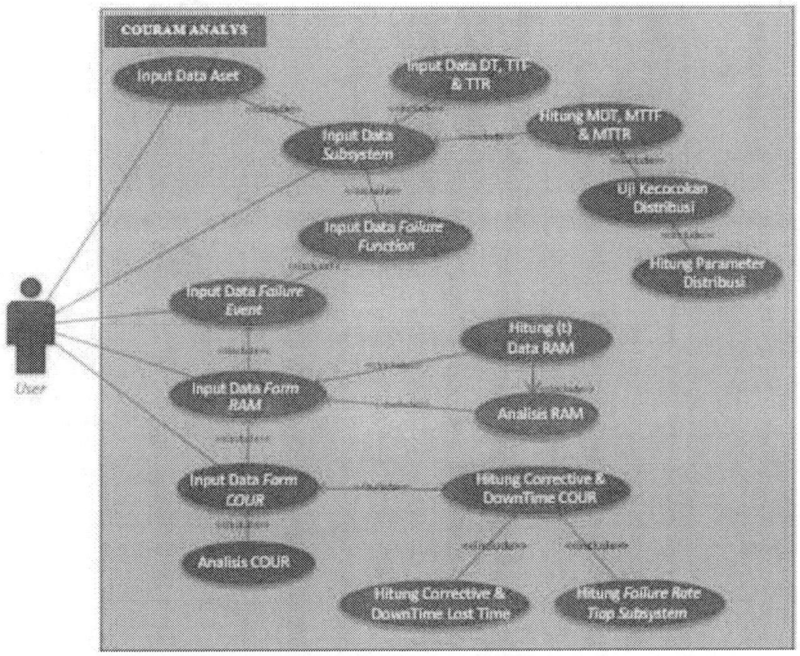


**Figure 2.** Use Case Diagram

### *2.7.2. Interface*

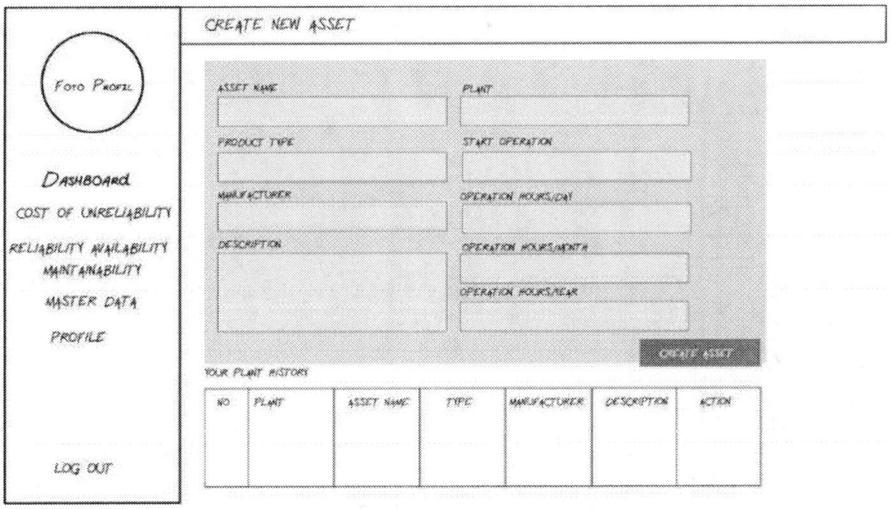


**Figure 3.** Interface

On this page, there is a subsystem information in the form of the asset number and name, the name of the subsystem contained in the asset, failure function, failure mode of the asset. Users enter data in the form of failure causes and effects for each failure mode. Then, the user asks the system to do a COUR analysis with the output of the COUR Corrective & Downtime calculation results. Next, the user enters data assets to calculate RAM. After all data is defined the system can calculate RAM for subsystems.

*2.7.3. Database Design.* The picture below is a database design of the application that has been designed. One of the functions of a database is to store the output of the application.

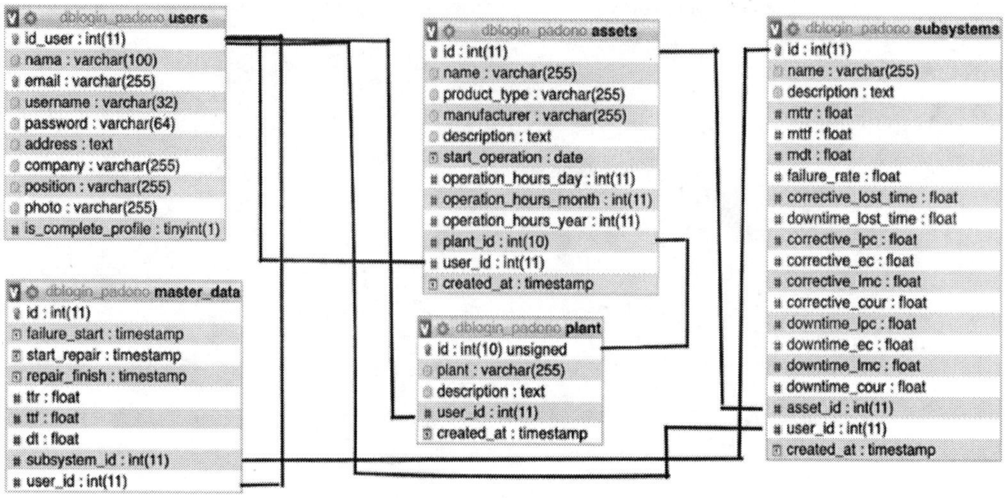


**Figure 4.** Application Database Design

## 3. Result and Discussion

### *3.1. Application Display*

This page displays a list of companies, assets, subsystems, and user accounts that have been stored in the database. The DT, TTF and TTR data page views of the designed application are shown in the following figure. This page displays DT, TTF and TTR data and graphics from each subsystem.

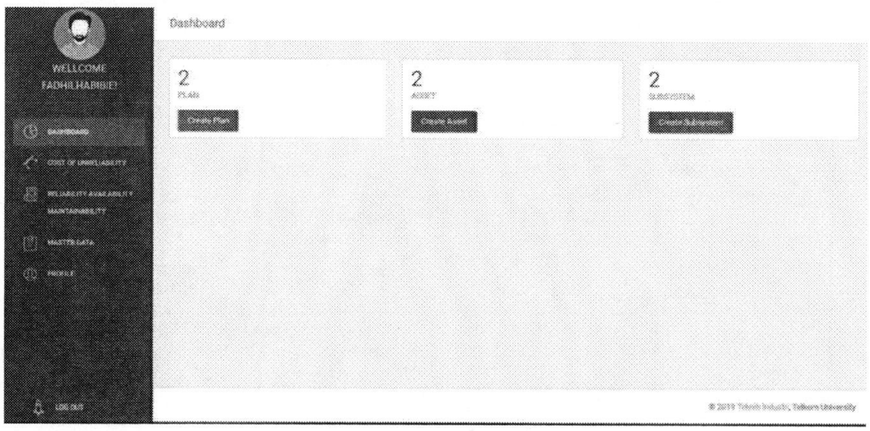


**Figure 5.** Dashboard Page View

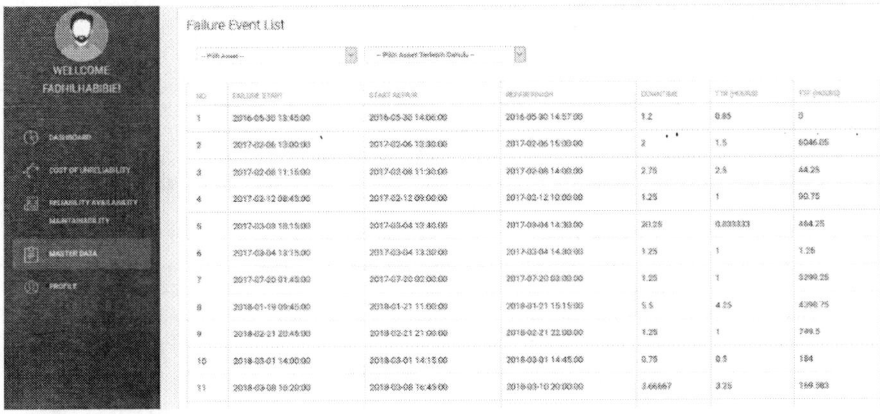


**Figure 6.** Display Data Pages DT, TTF and TTR

### *3.2. Comparison of Manual and Application Calculation Results*

After doing calculations manually and using the application of the case study given, the level of accuracy of the application can be seen in Table 1.

**Table 1.** Manual Calculation and Application of the Tundish Car Subsystem

| Process Name | Manual Result | Application Result | Accurancy |
|---|---|---|---|
| Calculation Parameter TTF | Conveyor<br>$\mu$ = 1271,85 hour $\eta$ = 1081,30 hour<br>$\sigma$ = 1791,34 hour $\beta$ = 0,52 hour<br>$\lambda$ = 0,00079 hour | $\mu$ = 1271,85 hour $\eta$ = 1081,30 hour<br>$\sigma$ = 1791,34 hour $\beta$ = 0,52 hour<br>$\lambda$ = 0,00079 hour | 100% |
| | Capper<br>$\mu$ = 3587 hour $\eta$ = 3777,25 hour<br>$\sigma$ = 3631,36 hour $\beta$ = 0,88 hour<br>$\lambda$ = 0,0003 hour | $\mu$ = 3587 hour $\eta$ = 3777,25 hour<br>$\sigma$ = 3631,36 hour $\beta$ = 0,88 hour<br>$\lambda$ = 0,0003 hour | 100% |
| Calculation Parameter TTR | Conveyor<br>$\mu$ = 4,27 hour $\eta$ = 2,65 hour<br>$\sigma$ = 12,14 hour $\beta$ = 0,83 hour<br>$\lambda$ = 0,234 hour | $\mu$ = 4,27 hour $\eta$ = 2,65 hour<br>$\sigma$ = 12,14 hour $\beta$ = 0,83 hour<br>$\lambda$ = 0,234 hour | 100% |
| | Capper<br>$\mu$ = 2,54 hour $\eta$ = 2,77 hour<br>$\sigma$ = 2,65 hour $\beta$ = 1,14 hour<br>$\lambda$ = 0,393 hour | $\mu$ = 2,54 hour $\eta$ = 2,77 hour<br>$\sigma$ = 2,65 hour $\beta$ = 1,14 hour<br>$\lambda$ = 0,393 hour | 100% |
| Distribution Test TTF | Conveyor<br>Dmax<br>Normal = 0,327<br>Eksponensial = 0,246<br>Weibull = 0,148 | Dmax<br>Normal = 0,327<br>Eksponensial = 0,246<br>Weibull = 0,148 | 100% |
| | Capper<br>Dmax<br>Normal = 0,2496<br>Eksponensial = 0,1167<br>Weibull = 0,1294 | Normal = 0,2496<br>Eksponensial = 0,1167<br>Weibull = 0,1294 | 100% |
| Distribution Test TTR | Conveyor<br>Dmax<br>Normal = 0,4419<br>Eksponensial = 0,4685<br>Weibull = 0,3002 | Dmax<br>Normal = 0,4419<br>Eksponensial = 0,4685<br>Weibull = 0,3002 | 100% |
| | Capper<br>Dmax<br>Normal = 0,2781<br>Eksponensial = 0,1792<br>Weibull = 0,2334 | Dmax<br>Normal = 0,2781<br>Eksponensial = 0,1792<br>Weibull = 0,2334 | 100% |

**Table 1.** Continued

| Process Name | Manual Result | | Application Result | Accurancy |
|---|---|---|---|---|
| Calculation MTTF | | Conveyor | | 100% |
| | MTTF = 2037,50 hour | | MTTF = 2037,50 hour | |
| | | Capper | | 100% |
| | MTTF = 3587,00 hour | | MTTF = 3587,00 hour | |
| Calculation MTTR | | Conveyor | | 100% |
| | MTTR =2,94 hour | | MTTR =2,94 hour | |
| | | Capper | | 100% |
| | MTTR =2,54 hour | | MTTR =2,54 hour | |
| Time Lost | | Conveyor | | 100% |
| | Failure Rate = 0,000491<br>Corrective Lost Time = 50,0291 hour<br>Downtime Lost Time = 129,51 hour | | Failure Rate = 0,000491<br>Corrective Lost Time = 50,0291 hour<br>Downtime Lost Time = 129,51 hour | |
| | | Capper | | 100% |
| | Failure Rate = 0,000278784<br>Corrective Lost Time = 20,33 hour<br>Downtime Lost Time = 111,246 hour | | Failure Rate = 0,000278784<br>Corrective Lost Time = 20,33 hour<br>Downtime Lost Time = 111,246 hour | |
| Money Lost | | Conveyor | | 100% |
| | Corrective Cour = Rp. 4.066. 058. 134,51<br>Downtime Cour = Rp. 10.525.500.281,65 | | Corrective Cour = Rp. 4.066. 058. 134,51<br>Downtime Cour = Rp. 10.525.500.281,65 | |
| | | Capper | | 100% |
| | Corrective Cour =<br>Rp. 1.652. 569. 612,27<br>Downtime Cour =<br>Rp. 9.041.364.517,85 | | Corrective Cour =<br>Rp. 1.652. 569. 612,27<br>Downtime Cour =<br>Rp. 9.041.364.517,85 | |
| RAM Analyst | | Conveyor<br>Reliability | | 100% |
| | Manual | | Application | |

## 4. Conclusion

Based on the test results and analysis discussed in the previous chapter, the following conclusions can be concluded.

1. Applications that have been designed can automatically determine distribution. Calculate and determine the results of MDT, MTTR, and MTTF accurately.
2. Applications that have been designed can accurately calculate and determine COUR maintenance policies.
3. Applications that have been designed can calculate and determine RAM maintenance policies accurately.
4. Applications that have been designed can accurately determine the performance verification indicators of the technical and operational availability of the machine.
5. The application that has been designed has been verified because it has an output with an accuracy level above 95%.

## 5. References

[1] J Alhilman 2017 Cost of unreliability method to estimate loss of revenue based on unreliability data: Case study of Printing Company. *IOP Conf. Ser. Mater. Sci. Eng.*, **277**(1).

[2] FTD Atmaji 2015 Optimasi Jadwal Perawatan Pencegahan Pada Mesin Tenun Unit Satu Di PT KSM, Yogyakarta. *J. Rekayasa Sist. Ind.*, **2**(April), pp. 7–11.

[3] DS Dhamayanti, J Alhilman, and N Athari 2016 Usulan Preventive Maintenance Pada Mesin Komori Ls440 Dengan Menggunakan Metode Reliability Centered Maintenance (RCM II) Dan Risk Based Maintenance (RBM) Di PT ABC," *J. Rekayasa Sist. dan Ind.*, **3**(02), pp. 31–37.

[4] R Ahmad and S Kamaruddin 2012 An overview of time-based and condition-based maintenance in industrial application. *Comput. Ind. Eng.*, **63**(1), pp. 135–149.

[5] A Salonen and M Deleryd 2011 Cost of poor maintenance: A concept for maintenance performance

improvement. *J. Qual. Maint. Eng.*, **17**(1), pp. 63–73.
[6] MT Todinov 2006 Reliability value analysis of complex production systems based on the losses from failures. *Int. J. Qual. Reliab. Manag.*, **23**(6), pp. 696–718.
[7] P Tsarouhas 2019 Statistical analysis of failure data for estimating reliability, availability and maintainability of an automated croissant production line. *J. Qual. Maint. Eng.*
[8] F Vicente 2012 Assessing the cost of unreliability in gas plant to have a sustainable operation," in *Petroleum and Chemical Industry Conference Europe Conference Proceedings, PCIC EUROPE.*
[9] P Tsarouhas, IS Arvanitoyannis, and TH Varzakas 2009 Reliability and maintainability analysis of cheese (feta) production line in a Greek medium-size company: A case study," *J. Food Eng.*, **94**(3–4), pp. 233–240.
[10] DJ Nainggolan, J Alhilman, and NA Supratman 2017 Performance Assessment Based on Reliability of Weaving M251 Machine Using Reliability, Availability & Maintainability (RAM) and Cost of Unreliability (COUR) Methods (Case Study at PT Buana Intan Gemilang)," *Int. J. Innov. Enterp. Syst.*, **1**(01), p. 13.
[11] P Tsarouhas 2018 Reliability, availability, and maintanability (RAM) analysis for Wine Packaging Production Line. *International Journal of Quality & Reliability Management* **35**(3), pp. 821-842.
[12] IH Choi and DChang 2016 Reliability and availability assessment of seabed storage tanks using fault tree analysis. *Ocean Eng.*, **120**, pp. 1–14.

# Evaluation of Effectiveness and Cost of Machine Losses using Overall Equipment Effectiveness (OEE) and Overall Equipment Cost Loss (OECL) Methods, a case study on Toshiba CNC Machine

**Sarastya Dewi[1,3], Judi Alhilman[1,4] and Fransiskus Tatas Dwi Atmaji[2,5]**

[1]School of Industrial and System Engineering Telkom University, Jl. Telekomunikasi No.1, Bandung, Indonesia
[3]rastya.dewi@gmail.com, [4]alhilman@telkomuniversity.ac.id, [5]frans.tatas@gmail.com

**Abstract.** In a manufacturing company, the machine is one of the most important elements in their production process because machine failure can stop the production process. Therefore, the initial step to minimizing losses caused by machine failure can be done by evaluating the machine condition. Evaluation of machine performance is carried out by measuring the effectiveness of the machine with the Overall Equipment Effectiveness (OEE) method. Based on the calculation result, the OEE value of the machine is 68.63% and this value still under the Japanese Institute of Plant Maintenance standard. Six big losses analysis is performed to determine the biggest loss that affects the effectiveness of the machine. The result of six big losses calculation shows that the most influential factor for the low OEE value of the machine is Reduced Speed Loss (39.12%). Causal analysis with a fishbone diagram is done to find out the causes of the highly reduced speed loss. To calculate the equipment cost loss use the Overall Equipment Cost Loss (OECL) methods. The total of the overall equipment cost loss is IDR 849,839,947.53.
**Keywords**: OEE, OECL, six big losses, fishbone diagram, machine performance.

## 1. Introduction

Machine is one of the important elements to support the production process in a manufacturing company. Machine failure that occurs suddenly can disrupt the production process. Based on [1] machine maintenance is inseparable from the problems of efficiency and effectiveness of machinery. Therefore, actions are needed to prevent or overcome the problem of machine downtime. OEE is a quantitative metric used in the manufacturing industry for controlling and monitoring the productivity of production equipment and as an indicator of process drivers and performance improvement [2]. According to Nakajima, OEE is based on three performance aspects: availability rate (A), performance efficiency (P) and Quality rate (Q) [3]. From [4] OEE measurements were carried out on 50 units of machinery in the company that manufactures tractors. OEE values indicate the need to improve machine effectiveness. Then the implementation of Total Productive Maintenance and within 5 years the value of OEE has increased. Based on [3] OEE is also used to help identify issues of availability and efficiency in wire mesh producing companies. This research was conducted at X company, which produces industrial equipment. High demand causes the production process to be carried out for 24 hours and causes machine failure, which disrupts the production process. Evaluations of machine performance is done by measuring the effectiveness of the machine with the OEE method. Then, six big losses identification carried out to know the highest losses from the machine. The highest loss factor will be analyzed with a causal diagram to find out the causes. Machine failures results a loss costs, then the OECL method used to calculate it. This research will focus on Toshiba CNC machine which used to produce arm excavators because it has the highest level of failures.

## 2. Methods

In conducting this research, several steps must be taken as data collection and then calculation to evaluate the state of the machine. Calculations and analysis are carried out in the following steps.

### 2.1. *Overall Equipment Effectiveness*

OEE is used to measure the effectiveness of several equipments or systems in which the measurement process is related to several considerations [5]. According [6] OEE calculation formula is:

$$\text{OEE} = \text{Availability x Performance x Quality} \tag{1}$$

*2.1.1. Availability.* Availability calculation aims to be able to find out how much the level of machine availability. According to [5] the availability value can be calculated by the following formula:

$$\text{Availability (A)} = \frac{\text{loading time} - \text{downtime}}{\text{loading time}} \text{ x } 100\% \tag{2}$$

*2.1.2. Performance Efficiency.* Performance efficiency is the ratio of the total products produced to the total products that could have been produced [6]. performance efficiency can be calculated by the following formula [5]:

$$\text{Performance Efficiency (P)} = \frac{\text{processed amount x theoritical cycle time}}{\text{operation time}} \text{ x } 100\% \tag{3}$$

*2.1.3. Rate of Quality.* The Rate of Quality shows how much success a machine produces in a product that is not defective. According to [5] the value of the rate of quality can be calculated by the following formula:

$$\text{Quality (Q)} = \frac{\text{processed amount} - \text{number of defect}}{\text{processed amount}} \text{ x } 100\% \tag{4}$$

### 2.2. *Six Big Losses*

Six big losses are the six losses that affect the effectiveness of the equipment. The types of losses are as follows [1].

*2.2.1. Equipment Failures.* Equipment Failure is a loss caused by failure of equipments so it takes time to repair. According to [1] equipment failure loss can be calculated by the following formula:

$$\text{Equipment Failure} = \frac{\text{total breakdown time}}{\text{loading time}} \text{ x } 100\% \tag{5}$$

*2.2.2. Setup and Adjustment.* Setup and Adjustment is a loss caused by setup and adjustments such as shifts turnover, changes in operations and products. Setup and adjustment loss can be calculated by the following formula [1]:

$$Setup\ and\ Adjustment = \frac{total\ setup\ and\ adjusment}{loading\ time}\ x\ 100\% \tag{6}$$

*2.2.3. Idling and Minor Stoppages.* Idling and Minor Stoppages are losses caused by the equipment doesn't work for a while due to sensor failure, machine error, and machine idle. This losses can calculate by the following formula [1]:

$$\text{Idling and Minor Stoppages} = \frac{\text{nonproductive time}}{\text{loading time}} \text{ x } 100\% \tag{7}$$

*2.2.4. Reduced Speed Losses.* Reduced Speed is a loss caused by a decrease in production speed. According to [1] the reduced speed value can be calculated by the following formula:

$$Reduced\ Speed = \frac{operation\ time - (theoritical\ cycle\ time\ x\ processed\ amount)}{loading\ time}\ x\ 100\% \tag{8}$$

*2.2.5. Defect Losses.* Defect Loss is a loss caused by a defect in the product produced. According to [1] the value of defect loss can be calculated by the following formula:

$$\text{Defect Loss} = \frac{(\text{total product reject x theoritical cycle time})}{\text{loading time}} \text{ x } 100\% \tag{9}$$

*2.2.6. Reduced Yield.* Reduced Yield is a loss caused by a defective product at the beginning of production time where the machine condition is not stable yet, calculated with following formula [1].

$$Reduced\ Yield = \frac{(theoritical\ cycle\ time\ x\ reject\ production)}{loading\ time}\ x\ 100\% \tag{10}$$

### 2.3. *Pareto Diagram*
Pareto diagram is a diagram that interprets data classification based on the highest order to the lowest order so that the main problem can be identified. The advantage of the Pareto diagram is it can show the most vital problem [7].

### 2.4. *Fishbone Diagram*
Cause and Effect Diagrams (Fishbone Diagrams) developed by Dr. Kaoru Ishikawa in 1943 and often called the Ishikawa Diagram. Cause and Effect Diagrams is a diagram that resembles fish bones where this diagram can show the relationship between the causes and effects of an event [7].

### 2.5. *Overall Equipment Cost Loss*
The OECL method was developed to calculate the overall cost of losses caused by equipment. The cost of loss is calculated using the same three main elements of the OEE method, namely availability, performance, and quality rate. OECL calculation is as follows [8].

$$OECL = Availability\ Losses + Performance\ Losses + Quality\ Losses \tag{11}$$

*2.5.1. Availability Losses.* To find out the cost of losses resulting from availability can be calculated by the following formula [9]:

$$\text{AL} = \text{OL}_\text{A} + \text{PCL}_\text{A} \tag{12}$$

Where,

$$\text{OL}_\text{A} = \text{LT x M}_{\text{cap}} \text{ x PPU} \tag{13}$$

$$\text{LT} = \text{breakdown time} + \text{set up and adjustment time} \tag{14}$$

$$\text{PCL}_\text{A} = \Sigma\left(\frac{\text{LT X EP}_{\text{i,A}}}{\text{Loading time}}\right) \tag{15}$$

where $\text{OL}_\text{A}$ is opportunity loss for availability rate element (Rp), LT is loss time (hour), $\text{M}_{\text{cap}}$ is maximum capacity (units/hour), PPU is profit per unit (IDR/unit), $\text{PCL}_\text{A}$ is production cost loss for availability (IDR), $EP_{i,A}$ expense (IDR/month) of production cost i for availability rate element according to Table I [9].

**Table 1.** Relating Cost with Three Elements of OEE

| | OEE elements | | | |
|---|---|---|---|---|
| Type of cost | A | P | Reject (Qrej) | Rework(Qrew) |
| Opportunity cost | ✓ | ✓ | ✓ | x |
| Direct material cost | x | x | ✓ | x |
| Direct labour cost | ✓ | ✓ | ✓ | ✓ |
| Indirect material cost | x | x | ✓ | ✓ |
| Indirect labour cost | ✓ | ✓ | ✓ | ✓ |
| Facility cost | x | ✓ | ✓ | ✓ |
| Depreciation cost | ✓ | ✓ | ✓ | ✓ |
| Maintenance cost | ✓ | ✓ | ✓ | ✓ |
| Renting cost | ✓ | ✓ | ✓ | ✓ |
| Insurance cost | ✓ | ✓ | ✓ | ✓ |
| Welfare cost | ✓ | ✓ | ✓ | ✓ |
| Rework cost | x | x | x | ✓ |

*2.5.2. Performance Losses.* Performance efficiency losses is calculated to find out the loss caused by the amount of production is under the maximum capacity of the machine. The number of costs resulting from performance loss can be calculated using the following formula [9]:

$$PL = OL_p + PCL_p \quad (16)$$

Where,

$$OL_p = LU \times PPU \quad (17)$$

$$LU = \text{maximum capacity} - \text{actual production} \quad (18)$$

$$PCL_p = \sum\left[\frac{LU}{M_{cap}} x \frac{EP_{i,P}}{\text{Operating time}}\right] \quad (19)$$

where $\mathbf{OL_P}$, opportunity loss for performance efficiency element (IDR), **LU**, loss unit (units); $\mathbf{PCL_P}$, production cost loss for performance efficiency element (IDR), and $\mathbf{EP_{i,P}}$, expense (IDR/month) of production cost **i** for performance efficiency according to Table 1.

*2.5.3. Quality Losses.* From [9] Quality losses are identified from two types of losses namely the reject losses and the rework losses. Therefore, quality losses can be calculated as follows:

$$QL = RejL + RewL \quad (20)$$

Reject loss is a losses that caused by products that not comply with specifications. Losses due to rejects can be calculated by the formula:

$$RejL = OL_{Q-rej} + DML_{Q-rej} + PCL_{Q-rej} \quad (21)$$

Where,

$$OL_{Q-rej} = Rej \times PPU \quad (22)$$

$$DML_{Q-rej} = Rej \times EPDM \quad (23)$$

$$PCL_{Q-rej} = \sum\left[\frac{Rej}{M_{cap}} x \frac{EP_{i,Q-rej}}{\text{Net operating time}}\right] \quad (24)$$

where $\mathbf{OL_{Q_{rej}}}$, opportunity loss for quality rate reject sub element (IDR), **Rej**, number of rejects (units); $\mathbf{DML_{Q-rej}}$ , direct material cost loss for quality rate reject sub element (IDR), **EPDMC**, expense of direct material cost (IDR/unit), $\mathbf{PCL_{Q-rej}}$, production cost loss for quality rate reject sub element (IDR) except direct material cost loss, and $\mathbf{EP_{i,Q-rej}}$, expense (IDR/month) of production cost ***i*** for quality rate reject ($\mathbf{Q_{rej}}$) sub element except direct material cost loss according to Table 1.

IOP Conf. Series: Materials Science and Engineering **847** (2020) 012020 doi:10.1088/1757-899X/847/1/012020

While rework is a product that is not according to specifications but can be fixed. The calculation of rework loss can be written as follows:

$$\mathbf{RewL} = \mathbf{PCL_{Q-rew}} + \mathbf{RwkL_{Q-rew}} \tag{25}$$

Where,

$$\mathbf{PCL_{Q-rew}} = \sum \left[ \frac{\mathbf{Rew}}{\mathbf{M_{cap}}} \mathbf{x} \frac{\mathbf{EP_{i,Q-rew}}}{\mathbf{Net\ operating\ time}} \right] \tag{26}$$

$$\mathbf{RwkL_{Q-rew}} = \mathbf{Rew\ x\ EP_{rew}} \tag{27}$$

where **Rew**, number of reworks (units); $\mathbf{PCL_{Q-rew}}$, production cost loss for quality rate rework sub element (IDR) except direct material cost loss, $\mathbf{EP_{i,Q-rew}}$, expense (IDR/month) of production cost $\boldsymbol{i}$ for quality rate rework ($\mathbf{Q_{rew}}$) sub element according to Table 1, except rework costs, **RewL**, rework loss (IDR), and $\mathbf{EP_{rew}}$, expense of rework (IDR/unit).

## 3. Result and Discussion

### *3.1. Calculation of Overall Equipment Effectiveness*

The OEE calculation of the Toshiba CNC machine is based on the breakdown and production data in 2019. The OEE calculation result can be seen in Table 2.

**Table 2.** Overall Equipment Effectiveness of Toshiba CNC Machine

| Month | A | P | Q | *OEE (%)* |
|---|---|---|---|---|
| 1 | 90.59% | 80.36% | 100.00% | 72.80% |
| 2 | 89.81% | 82.54% | 100.00% | 74.13% |
| 3 | 88.37% | 82.62% | 100.00% | 73.01% |
| 4 | 90.58% | 86.79% | 100.00% | 78.62% |
| 5 | 88.73% | 62.24% | 100.00% | 55.22% |
| 6 | 89.56% | 77.25% | 100.00% | 69.19% |
| 7 | 89.63% | 77.45% | 100.00% | 69.42% |
| 8 | 89.16% | 80.48% | 100.00% | 71.76% |
| 9 | 90.17% | 87.19% | 100.00% | 78.62% |
| 10 | 85.81% | 83.16% | 100.00% | 71.35% |
| 11 | 88.10% | 68.59% | 100.00% | 60.43% |
| 12 | 66.61% | 75.08% | 100.00% | 50.01% |
| Average | 87.26% | 78.64% | 100.00% | 68.63% |

OEE calculations are carried out from January–December 2019 by calculating the machine's Availability, Performance and Rate of Quality values. Based on calculations, the average OEE value of the Toshiba CNC machine is 68.63%. From the OEE values obtained in Table 3, an analysis of whether the OEE value of the Toshiba CNC machine meets the Japan Institute of Plant Maintenance standards.

**Table 3.** Fulfillment OEE of Toshiba CNC Machine with JIPM Standard

| OEE Factor | Calculation Result | JIPM Standard | Standard Fullfilment |
|---|---|---|---|
| Availability | 87.26% | 90% | NO |
| Performance Efficiency | 78.64% | 95% | NO |
| Rate of Quality | 100.00% | 99% | YES |
| OEE | 68.63% | 85% | NO |

In the Table 3, it is known that the Availability and Performance Efficiency values do not meet the JIPM standard. The reason for the low availability is due to the high machine downtime. Downtime can be caused by sudden damage to the machine and a long setup and adjustment time. The low-Performance Efficiency value is caused by the actual production time not as expected. The Rate of Quality meets the standards because in 2019 production no defective products were found. The average OEE value in 2019 is 68.63% and this value does not comply with the JIPM standard, so the action is needed to increase the low OEE value of the Toshiba CNC machine.

### *3.2. Calculation of Six Big Losses*

After calculating OEE, the next calculation of six big losses is performed shown on Table 4.

**Table 4.** Six Big Losses of Toshiba CNC Machine

| No | Six Big Losses | Losses Percentage (%) | Total Losses Percentage (%) |
|---|---|---|---|
| 1 | Reduced Speed Loss | 18.55 | 39.12 |
| 2 | Idling and Minor Stoppages | 13.21 | 27.86 |
| 3 | Equipment Failures | 3.32 | 7.00 |
| 4 | Setup and Adjustment | 9.42 | 19.86 |
| 5 | Reduced Yield | 2.92 | 6.16 |
| 6 | Defect Loss | 0.00 | 0.00 |
| | Total | 47.42 | 100.00 |

According to Table 4, it can be seen that the percentage of the six big losses on a Toshiba CNC machine. The biggest factors that caused losses were Reduced Speed Loss 18.55% or 39.12% from total losses percentage, Idling and Minor Stoppages 13.21% or 27.86% from total losses percentage and Setup and Adjustment 9.42% or 19.86% from total losses percentage. Six big losses greatly affect the machine's OEE value so action must be taken to reduce the six big losses on the machine.

### *3.3. Analysis of Pareto Chart*

The Pareto Chart will represent the six big losses data in the highest to lowest order, so that the main problem can be identified.

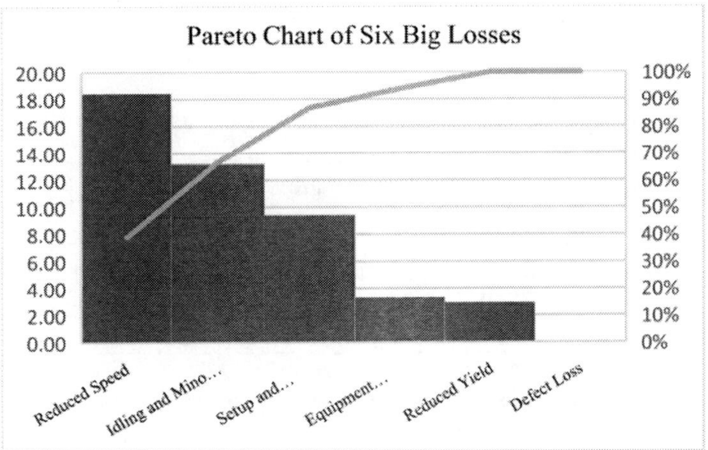


**Figure 1.** Pareto Chart of Six Big Losses from Toshiba CNC Machine

From Figure 1 it can be seen that the six big losses factor with the highest percentage is Reduced Speed 39.12% from all total losses. In pareto theory, 80% of the problems are caused by 20% of the causes. By dealing with 20% of the problems can solve other problems and effectiveness will increase. Then it is necessary to take action to overcome the problem of the high value of reduced speed loss on the CNC machine.

### *3.4. Analysis of Fishbone Diagram*

To identify the cause of the highly reduced speed based on the calculation of six big losses, an analysis was carried out using a fishbone diagram.

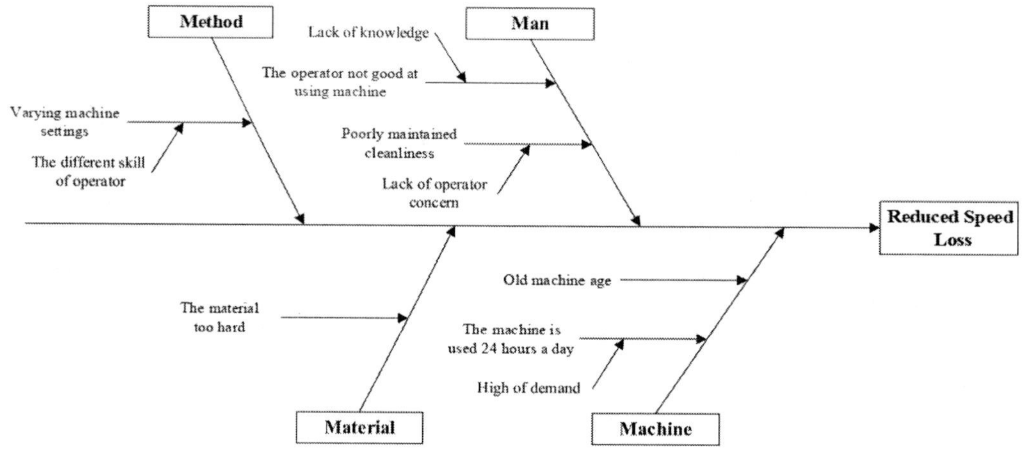


**Figure 2.** Cause Effect Analysis of Reduced Speed Loss Using Fishbone Diagram

Figure 2 shows the causal analysis of the reduced speed loss in the Toshiba CNC machine. The analysis was carried out on four factors namely man, machine, method, and material. In the man factor, operators are often not good at using machines and maintain machine cleanliness due to the lack of operator knowledge. From the machine factor, the use of the machine is almost 24 hours in a day and the old of the machine age. Methods also affect the reduced speed loss, when the production process of the machine is set many times with different speeds, operator skills are needed in the process of setting to avoid failure. Material that is too hard can also cause the machine to work harder when processing material and can cause reduced speed on the machine.

### *3.5. Calculation of Overall Equipment Cost Loss*

In addition to measuring the OEE value on the machine, overall equipment cost loss calculation is also performed. Losses are calculated from three factors which are the same as OEE namely availability, performance and quality. Table 4 is an OECL calculation.

**Table 5.** Amount of Overall Equipment Cost Loss of Toshiba CNC Machine

| Availability Losses | Performance Losses | Quality Losses | Total |
|---|---|---|---|
| IDR 329,182,951.22 | IDR 510,529,680.55 | IDR 10,127,315.77 | IDR 849,839,947.53 |

The OECL calculation is done using the data of the arm excavator production on the Toshiba CNC machine in 2019. In Table 4 the OECL value obtained on the Toshiba CNC machine is IDR 849,839,947.53. The total loss was obtained from Availability Losses of IDR 329,182,951.22, Performance Losses of IDR 510,529,680.55 and Quality Losses of IDR 10,127,315.77.

## 4. Conclusion

Based on research that has been carried out on Toshiba OEE CNC machines, the following conclusions are obtained:

1. In 2019, the Toshiba CNC machine has an OEE value of 68.63%, which is this value under the standard of JIPM
2. From total losses of six big losses, factors that contributed significantly to the low OEE value were 39.12% reduced speed loss, 27.86% idling and minor stoppages and 19.86% setup and adjustment.
3. Based on the pareto chart theory, by dealing with 20% of the problems can solve other problems, so is necessary to take action to overcome the problem of the high value of reduced speed loss.
4. From causal analysis carried out for the high reduced speed losses using a fishbone diagram, the causes are: man, machine, method and material. The main cause of the reduced speed loss is because the operator is not good at using and cleaning the machine, the machine setting technique, the machine age is old and used almost 24 hours a day. It is necessary to increase operator skills and autonomous maintenance.
5. Based on the calculation, the total OECL by the company is IDR 849,839,947.53.
6. For further research should be conducted to improve the effectiveness of the machine.

## 5. References

[1] Alhilman J and Abdillah A F 2019 Analysis of Double indian Ballbreaker Net Sorter Machine Based on Overall Equipment Effectiveness Method Cases in Tea Plantation Plants *IOP Conference Series: Materials Science and Engineering.*

[2] Tsarouhas P H 2013 Evaluation of overall equipment effectiveness in the beverage industry: A case study *Int. J. Prod. Res.*

[3] Wudhikarn R 2011 Implementation of overall equipment effectiveness in wire mesh manufacturing *IEEE International Conference on Industrial Engineering and Engineering Management.*

[4] Gupta P and Vardhan S 2016 Optimizing OEE, productivity and production cost for improving sales volume in an automobile industry through TPM: A case study *Int. J. Prod. Res.*

[5] S Nakajima 1988 Introduction to TPM: Total Productive Maintenance.pdf *Product. Press. Cambridge.*

[6] Borris S 2006 *Total productive maintenance* (McGraw-Hill New York)

[7] Dale B H and others 2003 *Total Quality Management*, Revised 3 rd ed, Anna University.

[8] Wudhikarn R 2012 Improving overall equipment cost loss adding cost of quality *Int. J. Prod. Res.*

[9] Wudhikarn R 2016 Implementation of the overall equipment cost loss (OECL) methodology for comparison with overall equipment effectiveness (OEE) *J. Qual. Maint. Eng.*

# Safety stock analysis of ship fuel in shipping company (Case study: white oil ship PT. Pertamina (Persero))

**Nuzul Fatma Septiana**[1,2], **Iwan Sukarno**[1,3]

[1]Logistics Engineering Department, Universitas Pertamina; Jakarta, Indonesia
[2]nuzulfatma1@gmail.com, [3]iwansukarno@universitaspertamina.ac.id

**Abstract.** Currently, fuel supply of PT. PERTAMINA (PERSERO) SHIPPING for each trip with a predictive safety stock to supply as much as 40% of the total fuel provided. The fuel that is stored for too long will also affect the sediment, thereby reducing the performance of the ship's engine in the long run. Optimization was conducted by using the safety stock formula which influenced by the uncertainty of demand and the uncertainty lead time (sailing time) of the ship. Calculations are based on demand from three distance routes that have the highest frequency assignment. Calculations are performed by comparing the safety stock in the four conditions, namely: (a) the condition of safety stock with combination of three routes (heterogeneous); (b) the condition of the safety stock by successively using the distance 286 NM; (c) 320 NM; and (d) 536 NM. The results obtained by calculating safety stock with the heterogeneous condition was 81.44%; homogeneous conditions within the 286 NM; 320 NM; 536 NM were 15.2%; 7.3%; and 29.2% of the total fuel supplied. Thus, the safety stock application is done by plotting trend charts to get an increase of 0.056% for each NM.
**Keywords:** Fuel Supplies, Safety Stock, Voyage, and Storage.

## 1. Introduction

The new policy applied by the PT. PERTAMINA (PERSERO) SHIPPING is to use a supply system that performed each voyage. Ship activities when loading until the ship unload all cargo (discharge), means that the ship has undertaken a voyage. Fuel consumption used for the one-way trip despite the same service will not spend the same fuel as well. PT. PERTAMINA (PERSERO) SHIPPING had a calculation to control vessel fuel consumption with the Slow Speed and Over Bunker (SSOB) method.

The positive effects of this policy will minimize the occurrence of theft, as well as the use of more fuel-efficient because it prevents sediment caused by nature. On the other hand, the negative effects of this policy are the amount of fuel that is too little allows the vessel ran out of fuel amid the trip. Ship refueling is done at loading port for efficiencies voyage activities. The problems that occur are (1) not all of the port where cargo loading has the willingness of time or (2) lack of fuel stock to supply ships. With a small fuel stock, determining the location when a deviation occurs will be quite difficult. Another issue that will occur is when the ship experiences slow speed which will affect the use of excess fuel. Nowadays, the company implements a safety stock policy for 40% of the supply of fuel used for a single voyage. The determination of the number is the result of a rough prediction of operators in the field. The new problem that will arise is in the determination of safety stock. If the safety stock was too little, it would not fulfill the demand (fuel consumption). By contrast, if the safety stock was too large, it would cause wastage due to sedimentation that would damage the ship's engine performance in the long term.

This issue is the basis for the need to analyze the fuel supply safety policy, whether the amount is sufficiently optimal. This paper will investigate how the computation and application of optimal safety stock to apply to this policy and evaluate the SSOB calculation that is going to be one of the processes to calculate the safety stock.

## 2. Methods

Inventory is a current asset owned by the company as a tool to meet customer demand and is further processed in meeting a particular goal [1]. Meanwhile, inventory is an investment in the form of sales in the future, expecting a return on these sales and minimizing costs [2]. Based on the inventory theory, it can be concluded that inventory is goods or resources that are used for the company's needs to do further processing or immediate resale in the future. Inventory in this paper refers to the amount of fuel that will be consumed during the course of the ship.

In the case of this inventory requires inventory management which can optimize the number and the right time in the store inventory. Inventory management is all the planning, implementation, and supervision to meet the operating needs optimally [3]. The purposes of inventory management are to keep inventory so that the costs can be kept low and to ensure uninterrupted supply during operation (production) [4]. Things to note, inventory control is not able to eliminate all risks due to the amount of inventory [5]. But it also does not mean inventory control or inventory management is done to achieve the lowest inventory. The inventory control processes are centered on the process when investments and stocks are adjusted according to a limit set by company policy [6].

The crucial part of the inventory is the calculation of safety stock. Definition of additional inventory safety stock is used to prevent the situation out of stock [7]. Even so, the safety stock must still be taken into account because of the high safety stock will result in additional costs [8]. safety stock will act as a buffer if there is a surge in demand or normal orders that are not able to close the request [9]. No different from some of the opinions that have been quoted, the authors assume the demand is the amount of fuel consumption during the ship trip.

Calculating safety stock are divided based on the uncertainty of each variable as the uncertainty of demand, the uncertainty of lead time, and a mixture of both uncertainties [10]. Other research also has the same opinion, the size of the safety stock is affected by the variance that is in the required variable [11]. This study takes into account the fuel consumption as demand uncertainty and long lead time of the ship is uncertain. Safety stock calculations taking into account the uncertainty of demand and the lead time is [12];

$$Safety\ Stock = \sqrt{l\ x\ S_d^2 + d^2\ x\ S_l^2} \quad \text{... (2.1)}$$

Information:

d = Demand per voyage (MT/voyage)
$S_d$ = Standard deviation of demand
l = Lead time per voyage (day/voyage)
$S_l$ = Standard deviation of lead time

Calculating safety stock requires preliminary stages of determinations and sample calculations. Sampling is a method used when research is not possible to collect data or process one by one the data of all existing data [13], the population consists of all the data that will be observed, while the sample is a set of existing in a population that is able to represent the entire population to be studied [14], calculation of the sample using the following equation:

$$n = \frac{Z_{\alpha}^2\, p\,(1-p)}{d^2} \quad \text{... (2. 2)}$$

Information:

n = Minimum number of samples required
Z = Degree of confidence
p = the sample proportions in the population
d = Limit of error or absolute precision

Multi-Stage sampling method is used to decide the sample The principle of this method is to select a sample that meets the adjustment of a stage in the sampling design [15]. Then, from this highly specific group, several samples are taken that will represent the population. This method aims to select a sample that is concentrated in a population.

PT. PERTAMINA (PERSERO) SHIPPING is currently implementing these policies consisting of:
1. The vessel will be given the supply of fuel for each voyage.
2. *Safety stock* for ships that will carry out the voyage is 40% of the supply to be provided.

These policies are applied to ships carrying white oil only. This is because these ships generally only have single loading and discharge for a single voyage. The route followed is fairly close and has a certain pattern. In the illustration in Figure 1, it appears that the possibility of water entering the fuel tank so that the performance of the fuel to be reduced. Moreover, the appearance of the sediment if the fuel is in the tank are too high can also reduce the amount of fuel that is still fit for use

.

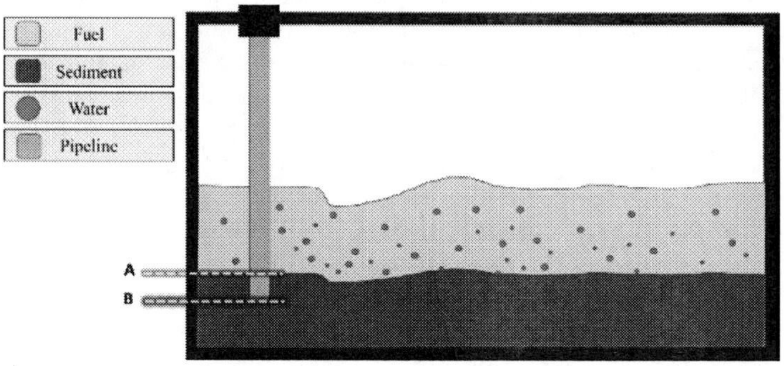


**Figure 1.** Illustration of the fuel tank with the excess sediment

Another possibility is that the engine will consume the sediment of fuel. Figure 1 shows that the ship's engine will consume sediment of fuel, due to the amount of sediment (line A) already exceed the end of the tank pipeline (line B). The sediment, which has a viscosity that is denser, will reduce boat engine performance in the long term.

A voyage is the condition of a journey for single loading and single discharge, and it spends about 50% of the fuel supply. 40% of safety stock policy is applied, the ship ought to be able to do almost three times the activity (twice loading and discharge one or vice versa). Furthermore, if the ship deviated for refueling purposes, it will be heading to the nearest port that probably only spend a little bit of safety stock. Here are some of the potential losses that would be borne by the company if the safety stock is high or low:
1. *Large safety stock* would cause deposits that affect the quantity of fuel (which is still fit for use) and engine performance.
2. *Small safety stock* will affect the ship stopped amid the trip due to run out of fuel.
3. Increase the risk of theft if the safety stock is still left after the ship has deviated.

Therefore, it is necessary to evaluate the stock safety policy currently applied and determine the optimal safety stock level transform and reduce the residue.

## 3. Results and Discussion

Calculations start by taking the sample. SSOB then performed calculations to determine the ideal conditions of a data received. The last step is to calculate the safety stock with four different conditions.

### *3.1. Sample calculation*

Before taking the sample, the authors conducted a proper sampling to determine in advance the target population to be tested samples. Charter ship that still has a contract with PT. PERTAMINA

(PERSERO). Operational of ship should having voyage by only doing single loading and single discharge.Calculating the number of samples needed as in equation (2. 2) with known secondary data up to 0.005 p-values and 95% confident level, the sample that will be used are 2 ships included in the sampling frame. The sample calculation process has not stopped at this stage, still has to do the sorting to choose which is the best ship which, with additional as the two samples. At this stage, sorting will be done based on the Multi Stage Sampling method:

1. Stage 1: Sister ship shorting
   The more similar the ship used as a sample the more accurate. The results showed that further sorting is done by 18 ships or 9 pairs of samples to be processed.
2. Stage 2: Checking the performance speed input data in January 2016 - July 2019
   This inspection function is to see if data on a ship is still available because the ship and sister ship also have a different contract. Thus, there is a possibility that the contract only exists in certain years. The results showed that there were two pairs of ships removed.
3. Stage 3: Checking attribute of the performance speed input data in January 2016 - July 2019
   There are 7 pairs of vessels from the previous sorting. This stage requires the completeness of the attributes of the data input speed performance of each vessel. The results obtained are a pair of ships.
4. Stage 4: Voyage most
   A pair of these ships have the same voyage numbers (the same task time) with a different route. To shorten calculation time, the writer decides to use the data with the highest number of the voyage to avoid duplicate data. This will not affect the results for the two ships are vessels that have the same specifications.

### *3.2. SSOB calculation*

The SSOB calculation is applied to determine the ideal condition of a ship by calculating whether the ship consumes fuel with the quantity that should be and runs at speeds under the agreement. The SSOB calculation will result in :(a) the time it should have while sailing and (b) the amount of excess fuel used. Generally, when a ship experiences a slowdown (the delivery time is longer than it should), it will automatically consume more fuel.

*3.2.1. Slow Speed Calculation.* The purpose of this calculation is to look for differences in actual sailing time and standard sailing time. The process of determining the actual sailing time can be determined by looking at the length of the ship's journey from the ship departing from the port before arriving at the destination port. While checking the sailing time in ideal conditions (Standard sailing time) aims to ensure the speed of the ship following the agreement. The speed of the ship uses the Knot unit(s). 1 Knot is equivalent to 1,852 Km/hour [16]. The speed that is generally used at sea is Knot, except for special cases (powerboat racing and waterskiing), Knot also means Nautical Mile (NM) per hour or Sea Mile per hour [16]. If the actual sailing time is bigger than standard sailing time, it can be concluded that the speed of the ship is slowing down.

*3.2.2. Fuel consumption calculation.* This calculation is done by finding the difference between actual fuel consumption and standard fuel consumption. Actual fuel consumption is the amount of fuel consumed or consumed by the ship to run the engine and crew needs (electricity, heating/cooling, etc.) since the ship departs from the port before arriving at the port of destination. Whereas standard fuel consumption is the maximum fuel consumption that can be consumed by ship. In general, if a ship experiences a slowdown it will consume more fuel than it should.

*3.2.3. Safety stock calculation.* The safety stock calculation requires sailing time variable or the amount of fuel consumption. This calculation has several stages that must be passed before the data can be calculated. The stages among which:

1. Adjust the data of ships experiencing

the slowdown or excess fuel consumption, calculation of fuel consumption, distance, and sailing time per voyage. This adjustment was made to reach the assumption that the ship will get the amount of safety stock in ideal conditions (conditions when the ship does not experience a slowdown or excesssaive fuel consumption).

2. Determine the distance for the calculation of safety stock
Based on the graph in Figure 2, it appears that the three most frequently traveled distances are: 286 NM (39 times) 320 NM (11 times), and 536 NM (19 times). It can be concluded that the three distances can be used as a reference in determining safety stock.
3. Making the dummy data as a real activity representation
The making of the dummy data is intended to get an idea of the ship is in real activity. This is caused by the presence of data to be tested to have an unbalanced amount of data. These conditions are still not able to represent the activities in real activity.

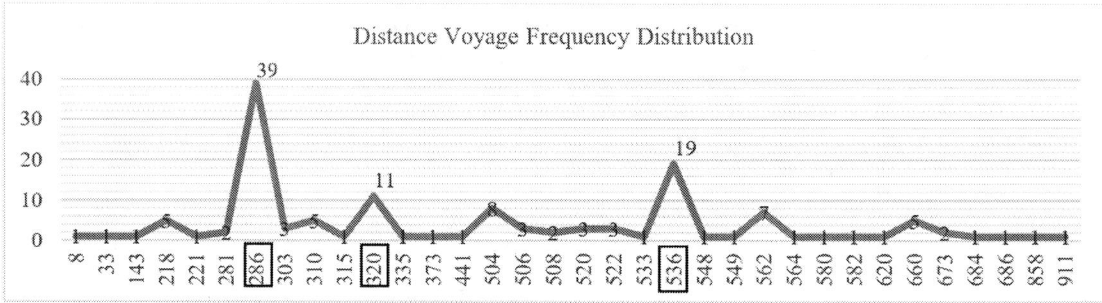


**Figure 2.** Distance Voyage Frequency Distribution Graph

4. Safety stock calculation with 4 conditions
After making a dummy data will be calculated safety stock with four conditions. The first condition is where the calculation condition of safety stock when the distance heterogeneous (containing 3 types of distance) and when the distance is homogeneous (there is only one kind of distance). Determination of dummy data for heterogeneous conditions requires a calculation by comparing if the three types of distance voyage distributed at 136 voyages. Here is a comparison of the distance distribution for the heterogeneous condition. With a total of 3 distance voyages with the highest frequency was 69. The percentage for each distance is:

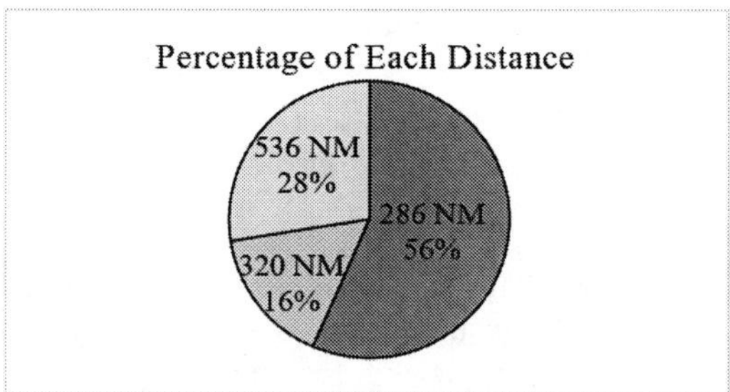


**Figure 3.** Percentage of Each Distance

If there will be 136 voyage numbers for heterogeneous dummy data, then it is necessary:

- *distance* 286 NM = 57% (77 of 136 voyages);
- *distance*320 NM = 16% (22 of 136 voyages); and
- *distance* 536 NM (37 of 136 voyages)

This study will be limited to these 4 conditions. The distance that has been mentioned has represented at least a short distance, a medium distance, and a long distance. Conditions taking into account one type of distance is said to be a homogeneous condition. Whereas, mixing of 3 types of distances is said to be a heterogeneous condition.

a) Heterogeneous condition

Heterogeneous conditions are data compilation conditions with the three types of routes. When the data is inputted by not changing the standard deviation as well as the average of demand and lead times of each route, the new safety stock is 81.44% of the demand.

b) Homogeneous conditions

Homogeneous conditions, means that all demand and lead time is spent adjusting for consumption and time spent on the voyage with that distance only. Based on the dummy data homogeneous safety stock amount of 286 NM, it obtained 15.2% of the demand, 320 NM was 7,3% of the demand, and 536 NM was 29,2% of the demand.

### *3.3. Analysis of safety stock calculation*

The safety stock calculation is carried out by four different conditions. The purpose of grouping is to determine whether the optimal safety stock if carried out with varying distances or distances that do not vary. Other considerations that the authors assume is:

a) The safety stock is determined by applying three distances in one calculation. It is assumed that in real practice activities, the ship will have different routes or travel distances.
b) The safety stock is determined by applying each distance. If the results indicate a more optimal safety stock in the homogeneous conditions, the fuel supply can then be determined by calculating the safety stock at each distance.

Both these assumptions have advantages and disadvantages to each. On the first assumption, the calculation of safety stock will not be repeated. But the high variance causes the result of too much safety stock calculations, even reaching 80% of the supply must be provided. Supply when using heterogeneous dummy data is also relatively small for medium to long-distance routes, but also relatively large for short-distance routes.

In the second assumption, the safety stock calculation must be done repeatedly. Depending on the distance to be traveled. This will not make time-efficient if the ship has a varied route scheduling. The positive side of this calculation is the existence of a fairly rational value because the calculation is certainly more accurate and focused on travel times and fuel consumption are relatively the same.

The calculation is performed by the safety stock formula which is influenced by demand uncertainty and lead time. Based on historical demand data, each ship has a different amount of consumption for each trip. Although the data processed has been normalized by adjusting the data to the ideal conditions, this certainly cannot be used as a reference because the ship may slow down the speed and/or excessive fuel consumption due to travel conditions.

Based on calculations, heterogeneous conditions of safety stock produce more than 80% because of the many variations of the route that must be taken. Each route certainly has a different standard deviation of demand and lead times. So if the calculation combines all these deviations will result in the emergence of a new standard that will be good. Seeing the results of calculations that allocate quite a lot of fuel, this calculation is less than optimal in terms of the amount and risk that occurs if this calculation is applied in real activities. The amount of safety stock that is too much will cause a bad effect (sediment) if the fuel is stored for too long in the tank.

From the above calculation, the safety stock with this homogeneous condition will be more profitable because the amount of additional fuel is also relatively small. This allows the right quantity of fuel to refill the tank. Based on the calculation of the lowest consecutive safety stock when the 320 mileage is 7.3%; mileage 286 by 15.2%; and 536 mileage as much as 29.2% of supply requests. Based on the results of homogeneous calculations, it can be seen that there is no trend line (Figure 4) in easy application to the company.

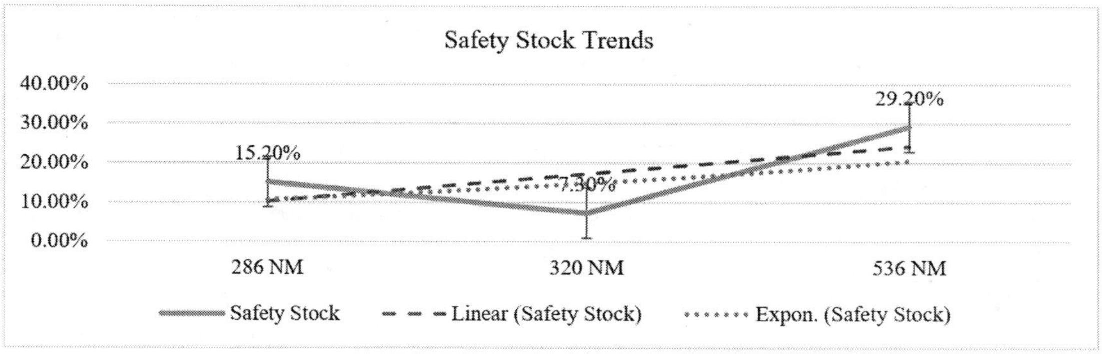


**Figure 4.** Trends of Safety Stock Homogeneous Condition

This difference is also influenced by variations in lead time and demand for each route. This variance is also influenced by field conditions such as waves when sailing ships, changes in direction of travel (to avoid pirates or other disturbances), or other conditions that may occur.

Figure 4 shows the trend line in a linear, exponential, and standard error line to determine the best path as a reference to the percent increase in safety stock per distance voyage. The trend line is used in determining the safety stock approach for various distance voyages without having to do calculations for each distance voyage.

**Table 1.** *Linear and Exponential Plotting Result.*

| Distance Voyage (NM) | Linear | Exponential |
|---|---|---|
| 286 | 10% | 11.5% |
| 320 | 16.5% | 15% |
| 536 | 24% | 20.5% |

The results of plotting the two trend lines can be determined the percent increase in safety stock based on the exponential line is equal to 0.035% per Nautical Mile (NM), while for the linear line has an increase of 0.056% per Nautical Mile (NM). Reviews based on standard errors show linear trend lines better than exponential trend lines. Based on this calculation, companies should use safer percentages because there might be unexpected events that will consume more fuel. also, calculations with linear tendencies give results that are not too far from the error limit illustrated in the graph. the author further recommends using additional fuel by following linear calculations.

## 4. Conclusions

After doing the calculations and discussion of case studies, the conclusions of the previous calculation are:

1. The volume of fuel for each trip has a different volume. But for the three types of distances with the highest frequency, it can be seen based on average fuel consumption.
   *Distance* 286 = 1.786802423 MT
   *Distance* 320 = 1.705224317 MT
   *Distance* 536 = 3.204817322 MT
2. Variables used for calculating safety stock are a demand or fuel consumption \ and lead time or sailing time under ideal circumstances. Furthermore, the calculated average and standard deviation for the calculation of the variable safety stock as well.
3. The amount of safety stock is calculated with 4 conditions. But for accuracy reasons, the safety stock is divided into 3 distance categories (the result of homogeneous condition calculations). *Distance* 286 = 15.2% of demand

*Distance* 320 = 7.3% of demand
*Distance* 536 = 29.2% of demand

Safety stock can only be applied at a specified distance, with an increase of 0.056% for each Nautical Mile (NM) on long-distance shipping. It aims to accurately supply quantities and also avoid the risk of losses arising from the fuel storage period.

## 5. References


[1] N Apriyani and A Muhsin 2017 Pengertian, Fungsi, dan Metode Pengendalian Persediaan (Stock Control) pada Perusahaan Manufaktur. *Jurnal Optimasi Sistem Industri,* **10**, pp. 128-142.

[2] H Budianto and D Ferriswara 2017 Penerapan Metode Pencatatan dan Penilaian Persediaan Barang Menurut Sak Etap pada CV. Tjipto Putra Mandiri Indonesia. *Jurnal Aplikasi Administrasi,* **20**, pp. 124-138.

[3] Indrajit, E Richardus and A R Djokopranoto 2003 *Manajemen Persediaan: Barang Umum dan Suku Cadang untuk Keperluan Pemeliharaan, Perbaikan, dan Operasi*. (Jakarta: PT Gramedia Widiasarana).

[4] M F Mpwanya 2005 Inventory Management as A Determinant for Improvement of Customer Service (Pretoria: University of Pretoria).

[5] J Juventia and L P Hartanti 2016 Analisis Persediaan Bahan Baku PT. BS dengan Metode Economic Order Quantity (EOQ). *Jurnal Gema Aktualita,* **5**, pp. 55-63.

[6] F O Sporta 2018 Effect of Inventory Control Techniques on Organization's," *The International Journal Of Business & Management,* **6**, p. 62.

[7] Constantin 2016 Inventory Management, Service Level And Safety Stock," *Journal of Public Administration, Finance and Law,* **9**, pp. 145-153.

[8] S P Anantadjaya, L Lahindah and C Mekel 2014 Stock Out Analysis: An Empirical Study on Forecasting, Re-Order Point and Safety Stock Level at PT. Combiphar, Indonesia. *Society of Interdisciplinary Business Research,* **3**, pp. 52-64.

[9] U Sharan 2016 Analysis of Safety Stock. *International Research Journal of Engineering and Technology,* **3**, pp. 1063-1065.

[10] M Hudori, 2018 Formulasi Model Safety Stock dan Reorder Point untuk Berbagai Kondisi Persediaan Material. *Jurnal Citra Widya Edukasi,* **10**, p. 221, 3 Desember 2018.

[11] M. E. Nenni and M. M. Schiraldi 2013 Validating Virtual Safety Stock. *International Journal of Engineering Business Management,* **5**, pp. 1-8.

[12] A Ruiz-Torres and F Mahmoodi 2010 Safety stock determination based on parametric lead time and demand. *International Journal of Production Research,* **48**, pp. 2841-2857.

[13] H Taherdoost 2016 Sampling Methods in Research Methodology; How to Choose a Sampling Technique for Research. *International Journal of Academic Research in Management (IJARM),* **5**, pp. 18-27.

[14] R E Walpole, R H Myers, S L Myers and K Ye 2012 *Probability & Statistics for Engineers & Scientists*, 9th edition ed., (Blacksburg: PEARSON).

[15] A Shimizu 2005 Multi Stage Sampling. *Encyclopedia of Biostatistics.*

[16] T Bartlett 2003 *RYA Navigation Handbook*, (Southampton: Royal Yachting Association).

IOP Conf. Series: Materials Science and Engineering **847** (2020) 012022 doi:10.1088/1757-899X/847/1/012022

# Designing Templates to Support and Monitoring the Activities of Material Requipment Planning (MRP)

**Franka Hendra**[1,2,5], **R Ambarita**[3,6], **K Eko Putranto**[4,7] **and Supriyono**[1,8]

[1]Industrial Engineering, Pamulang University,
Jl. Surya Kencana, Pamulang, Tangerang Selatan, Indonesia
[2]Razak Technology and Informatic Faculty of Universiti Teknologi Malaysia
Jl. Sultan Yahya Petra, Kuala Lumpur, Malaysia
[3,4]Industrial Engineering, National Institute of Science and Technology
Jl. Moh. Kahfi II, Srengseng Sawah, Jagakarsa, Jakarta, Indonesia
[4]Agency for The Assesmetn and Application of Technology (BPPT)
Puspiptek Serpong, Tangerang, Indonesia
[5]dosen01508@unpam.ac.id , [6]rahmiambarita.saragih@adidas-group.com ,
[7]kputranto@hotmail.com, [8]dosen01509@unpam.ac.id

**Abstract** The main purpose of this paper is to design a template based on Excel as a support and monitor the operation of the MRP system, as well as optimize inventory value.
The methodology done by identifying the MRP Objective targets that exist in the company and the raw material planning system process flow and comparing the achievement of the company's MRP objects with the results of calculations using templates.
The results of comparison between the achievements of 2015 and 2016, then it can be concluded that the *Excel template* can be used to help the SAP system to delete junk data (*fake*) due toindiscipline *user* in inputting data can provide better performance. From the comparism result with using Template Excel company can achieve CSL > 95%, CSL Infull > 95% and Warehourse Inventory Value reduce until less than IDR 300 billion.
**Keywords**: Material Requipment Planning, Inventory, MRP Objectives, Safety Stock

## 1. Introduction

The level of market demand for personal products from year to year for each country is different and also changes significantly over time. Personal product industry companies must have the flexibility to demand that changes suddenly in order to continue to provide the best service to customers to be able to compete with other companies. Because improving service levels is the most important goal of Supply Chain Management and raw material planning is a key sub-area of Supply Chain Management, the Raw Materials Planning Department in a business is central to achieving the stated goals (Gulyassy, 2010). As such, raw material planning is important and must be able to anticipate changes in demand suddenly.

Excess raw materials outside of safety stock will affect inventory levels which will ultimately impact on warehouse costs (Brian Willcox, 2016). While the unavailability of raw materials at a certain time will result in the cessation of the production process which results in not fulfilling customer orders. Therefore raw material planning needs to be considered so as not to have an impact on the company's inventory level which will result in additional costs.

IOP Conf. Series: Materials Science and Engineering **847** (2020) 012022 doi:10.1088/1757-899X/847/1/012022

If the raw material planning system has been implemented and meets the company's MRP objectives, the raw material planning system can be said to be good. However, what if the raw material planning system cannot meet the objectives of the company's MRP? This research focuses on achieving the MRP objectives from the MRP implementation system.

*1.1. Managementof Raw Materials*

Raw materials are very important resources that make a significant contribution to the costs incurred by the company and the impact on the production environment. During the production phase most of the resource efficiency initiatives and methods have been a major concern in making improvements measured by economic value. At present the need for a resource affinity level has expanded the scope of initiatives to consider a complete manufacturing and industrial system at an economic and environmental level (O Gould et al., 2016).

Study of advanced planning related to various planning processes and methods, such as; sales planning and and operations. Production master scheduling, material planning capacity and control on the production floor (P Jonssong, et al., 2016)

Based on estimated sales and production plans, raw material planning and control can be carried out. Raw material planning can estimate the needs of each piece of raw materials, prepare raw material budgets, predict (estimate) the value of inventory levels, schedule ordering and monitor the performance of production and sales.

*1.1.1. Purchasing.* Included in the scope of purchase is the selection of suppliers, including in the case of purchases, allocation of purchase orders, follow-up to purchases, maintaining good relations with suppliers, evaluating suppliers and assigning values to each supplier.

*1.1.2. Storage Management.* Storage management involves controlling inventory, preserving stores, minimizing obsolete values and damage to raw materials through timely disposal and efficient handling, site maintenance, and storage. Stores are also responsible for verifying inventory values and collecting them in one recorded data. Stores play an important role in company operations.

*1.1.3. Inventory Management and Control.* Inventories generally refer to raw materials stored in warehouses. This is also referred to as an idle product in a company. Inventories are items that are ready to be distributed and also items that will be reused for the production process. The interval between receipt of items purchased and converted to final products will vary from industry to industry depending on the time of the manufacturing cycle (Murray, 2010).

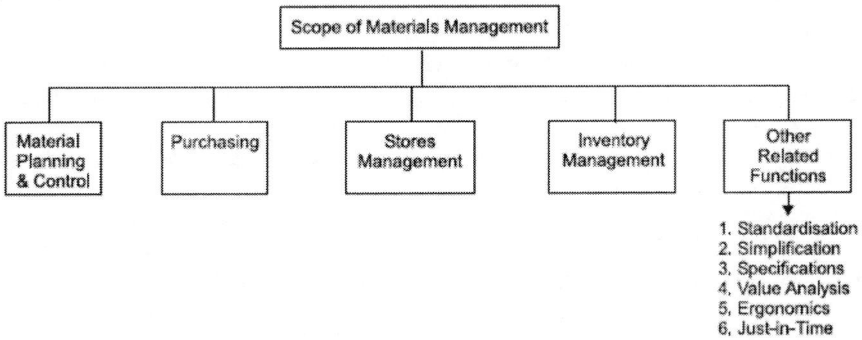


**Figure 1**. Overall Scope of Raw Material Management Raw Material

IOP Conf. Series: Materials Science and Engineering **847** (2020) 012022 doi:10.1088/1757-899X/847/1/012022

*1.2. Requirement Planning (MRP)*

Computer equipment is a part in carrying out raw material planning in more detail as follows Willcox (2016):

- Computer Equipment
- *Bill of Material* (BOM))
- Inventory Notes
- MPS (*Master ProductionSchedule*)

The purpose of the calculation of needs is to determine the number of needs and the delivery date for the finished product. Usually, many of the different raw materials must be planned in one production. In addition, finished products, semi-finished products, raw materials, *operating supplies, trading goods, spare parts,* and other types of items can have a significant influence on characteristics such as the number of sales, number of orders, prices, forms of consumption, cycle conditions, *leadtime* changes , and production conditions (Gulyassy, 2010).

According to him to meet this need it is necessary to automate the planning of many items, while other items require monitoring in more detail, so that the MRP parameters must be studied in more depth to optimize the planning tools (systems) used. The MRP parameters that affect inventory value: a) Production Type, b) Determination of Storage Level (Inventory), c) Consumption, d) ParametersParameters *Lot-Sizing,* e) Procedure *Safety Stock,* f) Forecasting Strategies and g) Item Classification and Product Analysis

According to him MRP (the literature also makes frequency references for program-oriented methods) is a Accurate procedures for determining needs. This procedure follows a demand program where independent needs or planned market demands take the form of warehouse requests or sales requests. Planning procedures based on consumption values are usually used in areas without *in-house production* or in manufacturing companies for the purpose of planning class B or C materials inoperations *supply*. The application of determining the needs process depends on thecontrol class *export* specific. Warehouse is the one that provides value added process and has been recognized as a service, where companies can provide the most tailored services and gain competitive advantage (Arsinijevic. et al, 2018)

The third option, in addition to determining MRP and determining needs based on consumption value, is determining needs based on demand / ordering (often known as raw material planning based on demand / ordering). Based on customer orders / requests in the form of POs, the determination of needs is carried out to fulfill POs of consumers. This determination will lead to the determination of individual needs. This will only occur in 'type production*make-to-order'* where the product will only be produced for consumers.production *Make-to-order* can also be used if a company does not want a lot of excess inventory, out of, or *stocksafety stock* on a material, or if the company wants to convert a fulfillment of demand to a demand that has been ordered (PO).

## 2. Methods

Main step of this research is identified the objective of the company's MRP targets and the flow of the raw material planning process that was already underway. These objective targets will then be compared with the achievements of the current MRP system.

Supporting templates are designed using Microsoft Excel by using formulations in accordance with MRP calculations. From the results of data processing done on the template compared with the MRP output system by SAP. For this reason, testing is carried out at the same time on the current raw material planning system (SAP MRP system) and the excel comparison template.

Tests have been carried out from the beginning of January 2016 and as proof of the end,carried out *screenshoots were* for each stage of the two systems on December 9, 2016 which were carried out on raw material 10165950. Both results of each system will be displayed to describe the parts that constitute weaknesses / deficiencies be identified.

## 3. Result and Discussion

*Template Excel* is designed as a comparison to clean up garbage data (dirty) that exist in the current MRP system as a result ofindiscipline *user* fulfill its function so that the performance of the performance of an MRP system is currently providing much better results against objective MRP companies, especially the value of the inventory warehouse..

### *3.1.Display of Need Value Differences between System and Actua*

The net requirement calculation between SAP system and *Excel template* produces the same results, but in making decisions for the status of raw material values are very different. Figure 10 shows the total MRP requirement displayed from the SAP system for December 9, 2016 for 10165950 raw materials is 104,888.62 KG (663,592 KG + 7,931,845 KG + 4,793,139 KG + 6,402,056 KG + 85,097,983 KG).

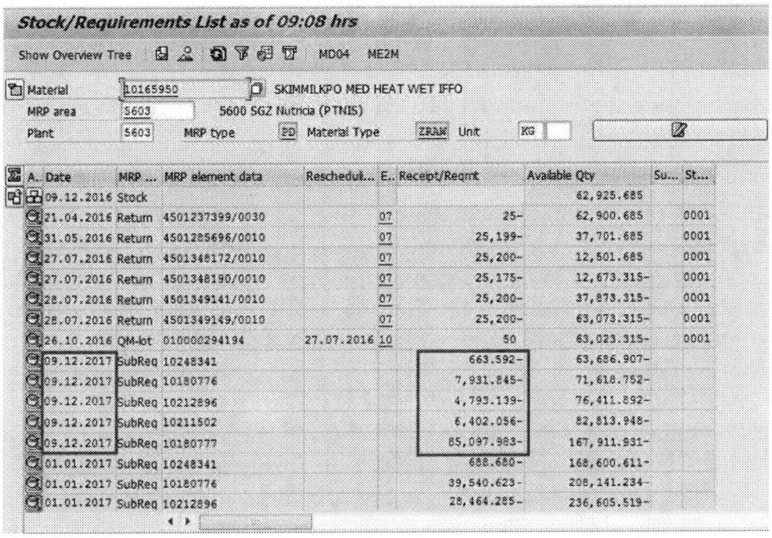
Stock/Requirements List as of 09:08 hrs

Show Overview Tree MD04 ME2M

Material 10165950 SKIMMILKPO MED HEAT WET IFFO
MRP area 5603 5600 SGZ Nutricia (PTNIS)
Plant 5603 MRP type PD Material Type ZRAW Unit KG

| A. | Date | MRP ... | MRP element data | Reschedul... | E.. | Receipt/Reqmt | Available Qty | Su... | St... |
|---|---|---|---|---|---|---|---|---|---|
| | 09.12.2016 | Stock | | | | | 62,925.685 | | |
| | 21.04.2016 | Return | 4501237399/0030 | | 07 | 25- | 62,900.685 | | 0001 |
| | 31.05.2016 | Return | 4501285696/0010 | | 07 | 25,199- | 37,701.685 | | 0001 |
| | 27.07.2016 | Return | 4501348172/0010 | | 07 | 25,200- | 12,501.685 | | 0001 |
| | 27.07.2016 | Return | 4501348190/0010 | | 07 | 25,175- | 12,673.315- | | 0001 |
| | 28.07.2016 | Return | 4501349141/0010 | | 07 | 25,200- | 37,873.315- | | 0001 |
| | 28.07.2016 | Return | 4501349149/0010 | | 07 | 25,200- | 63,073.315- | | 0001 |
| | 26.10.2016 | QM-lot | 010000294194 | 27.07.2016 | 10 | 50 | 63,023.315- | | 0001 |
| | 09.12.2017 | SubReq | 10248341 | | | 663.592- | 63,686.907- | | |
| | 09.12.2017 | SubReq | 10180776 | | | 7,931.845- | 71,618.752- | | |
| | 09.12.2017 | SubReq | 10212896 | | | 4,793.139- | 76,411.892- | | |
| | 09.12.2017 | SubReq | 10211502 | | | 6,402.056- | 82,813.948- | | |
| | 09.12.2017 | SubReq | 10180777 | | | 85,097.983- | 167,911.931- | | |
| | 01.01.2017 | SubReq | 10248341 | | | 688.680- | 168,600.611- | | |
| | 01.01.2017 | SubReq | 10180776 | | | 39,540.623- | 208,141.234- | | |
| | 01.01.2017 | SubReq | 10212896 | | | 28,464.285- | 236,605.519- | | |

**Figure 2**. Display of SAP 10165950 on December 9, 2016
(source: capture from SAP)

Display of needs in the *Excel template* also displays the same basic needs, which are 104,888,616 KG which can be seen in Figure 2. However, the inventory values between the two systems show different statuses. In the SAP system shows that the company experienced a shortage of raw materials on December 9, 2017 to 167,911 KG as indicated by the presence of a negative sign in thecolumn *Available Qty*. Whereas in the *Excel template the* status of raw materials in the warehouse is now sufficient and can be seen in Figure 3 for SOH 1 (values*stock on hand)*.

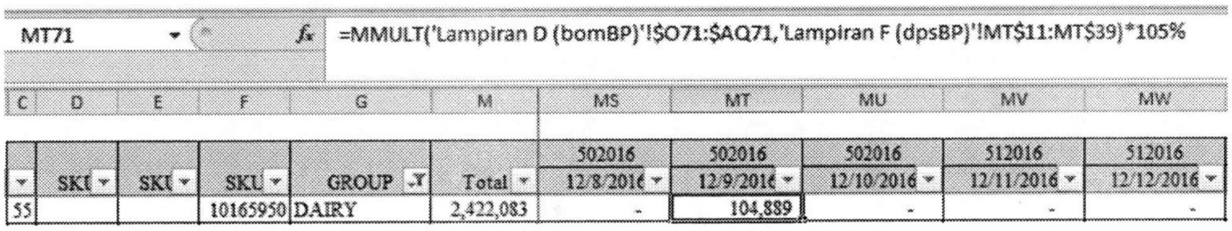
MT71 =MMULT('Lampiran D (bomBP)'!$O71:$AQ71,'Lampiran F (dpsBP)'!MT$11:MT$39)*105%

| C | D | E | F | G | M | MS | MT | MU | MV | MW |
|---|---|---|---|---|---|---|---|---|---|---|
| | | | | | | 502016 | 502016 | 502016 | 512016 | 512016 |
| | SKU | SKU | SKU | GROUP | Total | 12/8/2016 | 12/9/2016 | 12/10/2016 | 12/11/2016 | 12/12/2016 |
| 55 | | | 10165950 | DAIRY | 2,422,083 | - | 104,889 | - | - | - |

Figure 3. SOH Display 1 *Excel Template* 10165950
(source : capture form Template Exel)

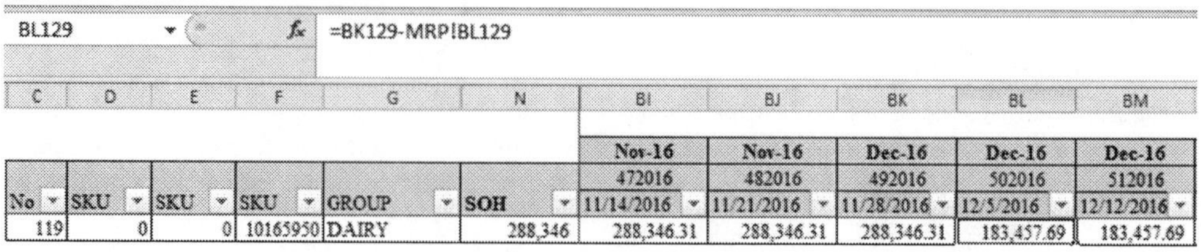

BL129 =BK129-MRP!BL129

| No | SKU | SKU | SKU | GROUP | SOH | Nov-16<br>472016<br>11/14/2016 | Nov-16<br>482016<br>11/21/2016 | Dec-16<br>492016<br>11/28/2016 | Dec-16<br>502016<br>12/5/2016 | Dec-16<br>512016<br>12/12/2016 |
|---|---|---|---|---|---|---|---|---|---|---|
| 119 | 0 | 0 | 10165950 | DAIRY | 288,346 | 288,346.31 | 288,346.31 | 288,346.31 | 183,457.69 | 183,457.69 |

**Figure 4**. Display Basic Requirements for *ExcelTemplate* 10165950
(source: capture form Template Exel)

When checked on the production line and in the warehouse, it turns out that there are no deficiencies with raw materials. Because if the raw material status is experiencing a shortage as shown in SAP's appearance, then on December 9, 2016 there should be a production stop at the company on that day. However, the conditions in the field stated that there were no signs that the factory was experiencing a shortage of raw materials 10165950 at that time.

### *3.2.BOM Exchange with a Bureaucratic Exchange (Approval Request)*

In the SAP system, if you want to condition a BOM exchange, it takes days to get *approval* from various related documents even though the decision has been mutually agreed through the results of the meeting and has been confirmed using *minutes of meetings* that have been shared in electronic mail. Waiting for this process will take a high amount of time which will add up to the *leadtime* of the raw material itself.

Templates Excel can be customized (arranged) as easily as possible to see an overview of the status of raw materials so that efforts to order raw materials can be done more quickly. In this case ordering of raw materials is still carried out using SAP. But the decision making process to order how much is needed can be immediately known by utilizing Excel templatethe existing.

### *3.3.There are many Purchase Orders "Garbage" (Invalid)*

PO *return* is a condition where the PO seems to be returned (withdrawn) even though the actual conditions in the field there is no withdrawal of raw materials in the warehouse. This is usually done for several purposes, one of which is theteam *Finance* to overcome the *gap* that is found in theprocess *stock taking* or it can also be due to the process of price differences between the system and the agreed price of the *supplier*.

In the SAP system, all POvalues *return* will be considered as if the raw materials were actually withdrawn so that they must reduce the value of inventory in the warehouse. In fact, in the case of PO *returns,* there is no need to do anything (ignored) so that the value of raw materials will look good. This is the basis of mistakes so that many mistakes are found in making decisions. When referring to the SAP system, a raw material planner is required to encode the raw material at a negative price as shown in *Available Qty* and can be seen in Figure 11.

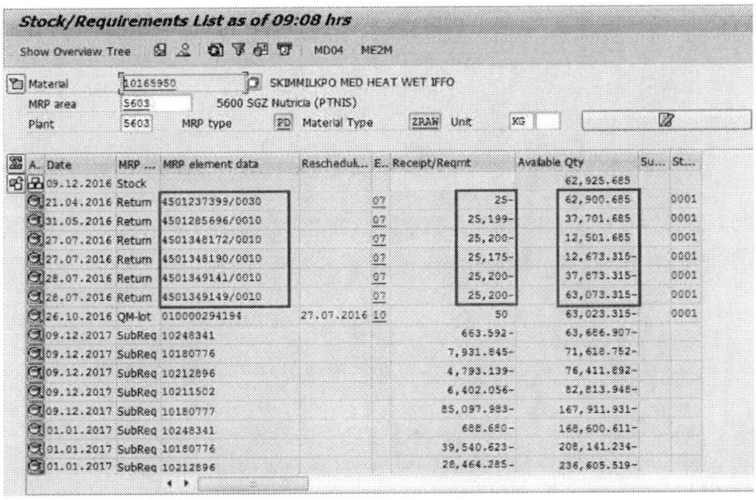


| A.. | Date | MRP ... | MRP element data | Reschedul... | E.. | Receipt/Reqmt | Available Qty | Su... | St... |
|---|---|---|---|---|---|---|---|---|---|
| | 09.12.2016 | Stock | | | | | 62,925.685 | | |
| | 21.04.2016 | Return | 4501237399/0030 | | 07 | 25- | 62,900.685 | | 0001 |
| | 31.05.2016 | Return | 4501285696/0010 | | 07 | 25,199- | 37,701.685 | | 0001 |
| | 27.07.2016 | Return | 4501348172/0010 | | 07 | 25,200- | 12,501.685 | | 0001 |
| | 27.07.2016 | Return | 4501348190/0010 | | 07 | 25,175- | 12,673.315- | | 0001 |
| | 28.07.2016 | Return | 4501349141/0010 | | 07 | 25,200- | 37,873.315- | | 0001 |
| | 28.07.2016 | Return | 4501349149/0010 | | 07 | 25,200- | 63,073.315- | | 0001 |
| | 26.10.2016 | QM-lot | 010000294194 | 27.07.2016 | 10 | 50 | 63,023.315- | | 0001 |
| | 09.12.2017 | SubReq | 10248341 | | | 663.592- | 63,686.907- | | |
| | 09.12.2017 | SubReq | 10180776 | | | 7,931.845- | 71,618.752- | | |
| | 09.12.2017 | SubReq | 10212896 | | | 4,793.139- | 76,411.892- | | |
| | 09.12.2017 | SubReq | 10211502 | | | 6,402.056- | 82,813.948- | | |
| | 09.12.2017 | SubReq | 10180777 | | | 85,097.983- | 167,911.931- | | |
| | 01.01.2017 | SubReq | 10248341 | | | 688.680- | 168,600.611- | | |
| | 01.01.2017 | SubReq | 10180776 | | | 39,540.623- | 208,141.234- | | |
| | 01.01.2017 | SubReq | 10212896 | | | 28,464.285- | 236,605.519- | | |

**Figure 5.** SAP 10165950 display for PO returns
(source: capture from SAP)

In the *excel template,* all POs *return* or PO that is negative will be assumed to be zero so thesystem *excel* will not give results in the calculation process. In this case, of course decision making is easier to manage. The following display PO *returns* in the *excel template* can be seen in

| PO | DAPH | Vendor supplying plant | Item | Document | Material | Pl | Delivery Dat | Received | Remaining | Order |
|---|---|---|---|---|---|---|---|---|---|---|
| 4501237399 | | 10006622 DANONE ASIA PACIFIC HOLDINGS | 30 | 4/20/2016 | 10165950 | 5603 | 21-Apr-16 | (25,175) | 0 | (25,200) |
| 4501285696 | | 10006622 DANONE ASIA PACIFIC HOLDINGS | 10 | 5/31/2016 | 10165950 | 5603 | 31-May-16 | 0 | 0 | (25,199) |
| 4501348172 | | 10006622 DANONE ASIA PACIFIC HOLDINGS | 10 | 7/27/2016 | 10165950 | 5603 | 27-Jul-16 | 0 | 0 | (25,200) |
| 4501348190 | | 10006622 DANONE ASIA PACIFIC HOLDINGS | 10 | 7/27/2016 | 10165950 | 5603 | 27-Jul-16 | 0 | 0 | (25,175) |
| 4501349141 | | 10006622 DANONE ASIA PACIFIC HOLDINGS | 10 | 7/28/2016 | 10165950 | 5603 | 28-Jul-16 | 0 | 0 | (25,200) |
| 4501349149 | | 10006622 DANONE ASIA PACIFIC HOLDINGS | 10 | 7/28/2016 | 10165950 | 5603 | 28-Jul-16 | 0 | 0 | (25,200) |
| 4501349316 | | 10006622 DANONE ASIA PACIFIC HOLDINGS | 10 | 7/28/2016 | 10165950 | 5602 | 28-Jul-16 | 0 | 0 | (304) |
| 4501381578 | 4501382857 | Murrav Goulburn Co-operative | 10 | 8/24/2016 | 10165950 | 5603 | 11-Nov-16 | 0 | 100,800 | 100,800 |
| 4501381582 | 4501382858 | Murrav Goulburn Co-operative | 10 | 8/24/2016 | 10165950 | 5603 | 11-Nov-16 | 0 | 100,800 | 100,800 |
| 4501381586 | 4501382334 | Murrav Goulburn Co-operative | 10 | 8/24/2016 | 10165950 | 5603 | 11-Nov-16 | 0 | 50,400 | 50,400 |

**Figure 6**. Figure 6. Display *Exceltemplate* 10165950for PO returns
(source: capture from Tempate Excel)

*Template Excel template* will only define all *remaining POs* as raw material inventory *on order*. For this reason, all negative values will be defined as the number of zeros to ensure that SOH2 calculations will be maintained properly. The same thing happened to *purchase requisition*.

### *3.4.Not Flexible with New Scenarios The current*

Inflexibility of the MRP system against new scenarios is also identified in one of the MRP's objective factors which cannot be modified, namely the specific raw material consumption factor. *The key user* explained that there were some raw materials that could be used as substitutes for other raw materials if they experienced a *quality issue*. For example, raw material 49991049/10134481 which is a raw material for *skimmed milk powder dry,* if it cannot meet the SMPstandard *dry,* then it can be used as aSMP *wet.* So that the display of raw material inventory in the warehouse can be increased. However, this cannot be done on an SAP system duebureaucracy *approval* to the long. For this reason the *Excel template* can be used to fulfill this request with details can be seen in Figure 6 shown in *highlights* yellow.

IOP Conf. Series: Materials Science and Engineering **847** (2020) 012022 doi:10.1088/1757-899X/847/1/012022

| PLANT | SLOC | MATERIA' | BATCH | Manufacture | Expired Date | *Unrestricted* | Blocke' | QI | Remark |
|---|---|---|---|---|---|---|---|---|---|
| 5602 | 1 | 10165950 | 1611291184 | 8/26/2016 | 8/26/2018 | 25,626 | 0 | 0 | 10165950 |
| 5602 | 1 | 10165950 | 1611291188 | 8/26/2016 | 8/26/2018 | 16,225 | 0 | 0 | 10165950 |
| 5602 | 1 | 10165950 | 1612011190 | 8/26/2016 | 8/26/2018 | 21,075 | 0 | 0 | 10165950 |
| 5602 | 1 | 10165950 | 1611291189 | 8/26/2016 | 8/26/2018 | 0 | 0 | 8,975 | 49991049 |
| 5602 | 1 | 10165950 | 1611281179 | 9/6/2016 | 9/6/2018 | 0 | 0 | 15,775 | 49991049 |
| 5602 | 1 | 10165950 | 1611291186 | 9/6/2016 | 9/6/2018 | 0 | 0 | 25,175 | 49991049 |
| 5602 | 1 | 10165950 | 1612011191 | 9/6/2016 | 9/6/2018 | 0 | 0 | 4,125 | 49991049 |
| 5602 | 1 | 10165950 | 1611281182 | 9/6/2016 | 9/6/2018 | 0 | 0 | 25,000 | 49991049 |
| 5602 | 1 | 10165950 | 1611281180 | 9/8/2016 | 9/8/2018 | 0 | 0 | 9,375 | 49991049 |
| 5602 | 1 | 10165950 | 1610251323 | 7/31/2016 | 7/31/2018 | 0 | 0 | 25,175 | 10134481 |
| 5602 | 1 | 10165950 | 1611201331 | 9/27/2016 | 9/27/2018 | 0 | 0 | 25,200 | 49991049 |
| 5602 | 1 | 10165950 | 1611231332 | 9/27/2016 | 9/27/2018 | 0 | 0 | 25,150 | 49991049 |
| 5602 | 1 | 10165950 | 1611271334 | 9/27/2016 | 9/27/2018 | 0 | 0 | 25,200 | 49991049 |
| 5603 | 1 | 10165950 | 1610261328 | 7/30/2016 | 7/30/2018 | 0 | 0 | 50 | 10134481 |

**Figure 7**. Display *Excel Template* Addition 49991049/10134481
(source: Capture from Template excel)

With all the weaknesses / weaknesses of the SAP system being met due todiscipline *user* that can be helped by the *excel template,* the test is declared successful if the results of *excel template* thisare able to meet MRP objectives to the company. Therefore, the comparison of achievement between Objective MRP in 2015 and Objective MRP in 2016 with details as follows:

a. Level of Customer Service (CSL ontime)
b. Availability of Raw Materials (CSL *infull*

Table 1 shows the second objective of the MRP, the level of customer service and availability of raw materials Both of these MRP objectives increased in 2016, although in 2015 they also met the company's MRP Objective targets, showing that the *excel template* can help the performance of the existing MRP system in SAP in cleaning up waste data

**Table 1**. Achievement 2 Objectives MRP 2015 and 2016

| Total 2015 | | | | | Total 2016 | | | | |
|---|---|---|---|---|---|---|---|---|---|
| PO | Ontime | Infull | Ontime CSL | CSL Infull | PO | ontime | Infull | CSL ontime | CSL Infull |
| 660 387 | 642 021 | 646 270 | 97.2% | 97.9% | 660 387 | 636 492 | 646 270 | 96.4% | 97.9% |
| 1,987,837 | 1,882.619 | 1,893,512 | 94.7% | 95.3% | 1,987,837 | 1,902,268 | 1,923,500 | 95.7% | 96.8% |

**Table 2**. Objective Achieve

| Num | Category | Achieve |
|---|---|---|
| 1 | Customer Service Level (CSL) | >95% |
| 2 | CSL Infull | >95% |
| 3 | Warehouse Inventory Value | < IDR 300 billion |

Inventory value owned by the company has increased from time to time. This time and this are not in accordance with the company's objective targets to minimize the value of inventory. The existence of *an Excel template* gives a good change from time to time. Figure 13 describes the information that the value of inventory owned by the company has shown a better direction.

## 4. Conclusion

*Template excel* designed as a comparison to clean up garbage data (gross) which is in an MRP system is currently a lack of discipline *user* fulfills its function so that the performance of the performance of an MRP system now yield which is better towards the company's MRP objectives, specifically the value of warehouse inventory. For this reason, testing is carried out at the same time on the current raw material planning system (SAP MRP system) and the *excel template* comparison.

IOP Conf. Series: Materials Science and Engineering **847** (2020) 012022 doi:10.1088/1757-899X/847/1/012022

With the results of comparisons between 2015 and 2016 achievements, it can be concluded that the *Excel template* can be used to help SAP systems delete junk data (*fake*) due to indiscipline *user* in inputting data can provide better performance. *The Excel template is* only used as a *tool* supplementary, not a replacement system because the *database* on the SAP system is a special standard *global* that can monitor the company's overall profit and loss.

For smaller scale companies that have not been able to have an MRP calculation system because the price of the MRP system (for example SAP) is not cheap, this *excel template* can be used as an MRP system that can assist planners in making decisions for procurement of raw materials.

## 5. References


[1] Arsenijević, O, Podbregar, I, Šprajc, P, Trivan, D, and Ziegler, Y 2018 Warehouse Management System Implementation: Case Study of TPV Velika Loka. *Organizacija in Negotovosti V Digitalni Dobi Organization and Uncertainty in The Digital Age*, **769**.

[2] Gulyassy, Ferenc et al. 2010 *Materials Planning with SAP.* (Boston: SAP Press).

[3] Gould, O, Simeone, A, Colwill, J, Willey, R, and Rahimifard, S 2016 A Material Flow Modeling Tool for Resource Efficient Production Planning in Multi-Product Manufacturing Systems. *Procedia of 48th Conference on Manufacturing System-CIRP Science Direct*, **41**, 21-26.

[4] Jonsson, P, and Mattsson, SA 2016 Advanced material planning performance: A Contextual Examination And Research Agenda..*International Journal of Physical Distribution & Logistics Management*, **46**(9), 836-858.

[5] Murray, Martin 2010 *SAP Warehouse Management Functionality and Technical Configuration.* (Boston: SAP Press).

[6] Willcox, Brian 2016 *Effective for 2016 Basics of Supply Chain Management.* (United States: APIC)

# Life Cycle Cost Analysis in Construction of Green Building Concept, A Case Study

**MK Kamaralo[1,4], J Alhilman[2,5] and FTD Atmaji [3,6]**

[1] Melviani Karolin Kamaralo, Telkom University, Bandung, West Java, Indonesia
[2] Judi Alhilman, Telkom University, Bandung, West Java, Indonesia
[3] Fransiskus Tatas Dwi Atmaji, Telkom University, Bandung, West Java, Indonesia
[4] karolinmelviani15@gmail.com, [5] alhilman@telkomuniversity.ac.id,
[6]franstatas@telkomuniversity.ac.id.

**Abstract.** Based on data from the Green Building Council of Indonesia, more than one-third of $CO_2$ gas emissions worldwide are generated by buildings, it has an impact on the environment such as global warming, ozone layer depletion, and accumulation of waste. The concept of Green Building is considered very necessary to overcome global warming and improve energy and resource efficiency. In the process of building a building that is Green Building requires a relatively high cost when compared to conventional buildings. Therefore, the Life Cycle Cost (LCC) method is used to determine the total cost needed, the optimal cost of the building, the economic age of the building, the number of crew maintenance and the level of energy efficiency. The analysis using the Life Cycle Cost method requires several related costs such as Initial Costs, Maintenance Costs, Energy Costs, Replacement Costs, and Utility Costs. The analysis was conducted using the Present Worth method within a period of 8 years from the start of building construction. Based on data processing using the Life Cycle Cost method, the optimal cost of a green building concept building is IDR 232,296,615,337 with the economic life of the building being 8 years, the optimal number of maintenance crews is 1 person and the level of energy consumption intensity is very efficient.
**Keywords**: Green Building, Life Cycle Cost, Economical Age, Present Worth Analysis, Energy Efficiency, Global Warming.

## 1. Introduction

The state of the world is currently facing major problems related to the environment such as global warming, ozone depletion, waste buildup and others. In recent years studies have shown that the global climate is changing very rapidly. Based on data from the Green Building Council of Indonesia, all $CO_2$ gas produced by buildings is around 30-40% and therefore buildings have an important role in consuming energy throughout the world [1]. From these problems to minimize the impact that occurs by applying the concept of environmentally friendly buildings or the concept of green building. Building construction with the concept of Green Building is one of the government's efforts in energy efficiency and resource issues that affect the costs required in the Green Building construction process [2]. Indonesia's achievement in implementing development with the Green Building concept can save energy by 853,914 MWh / year, as well as cost savings of $ 68,313,105 / year and CO2 reduction of 605,425 Metric Tons / year. This achievement is expected to increase as development is carried out, especially in Jakarta, Indonesia. The process of building a Green Building requires a relatively high cost when compared to conventional buildings, but in its application that focuses on energy efficiency and resources will be an advantage that affects operational costs, maintenance costs and replacement costs [3]. Therefore, Life Cycle Cost is a tool to optimize buildings with a long-term perspective and to exploit sustainable economic principles [4] and also see how much the total costs are removed from the construction stage up to the economic age of the building and the optimal number of maintenance crews.

## 2. Method

### *2.1. Maintenance Management*
According to the journal [5] Maintenance is defined as an activity so that damaged components/systems will be returned/repaired under a certain condition in a certain period. Maintenance Management is an activity that is carried out repeatedly with the aim that the equipment is in the same condition as the initial state.

### *2.2. Life Cycle Cost*
Based on journal [6] Life Cycle Costs (LCC) are the sum of estimated costs over the entire life cycle from initiation to completion, both equipment and projects. LCC components can be broadly classified into design and development costs, acquisition costs, operating costs and disposal costs. Life Cycle Cost approach emphasizes the consideration of total cost for the operation of the equipment from the initial cost [7].

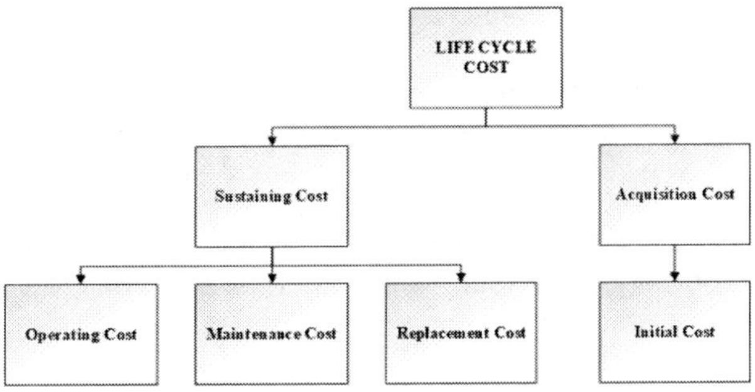


**Figure 1.** Life Cycle Cost Classification

Life Cycle Cost is classified into several parts, there are Sustaining Cost and Acquisition Cost [7]. Sustaining cost is the sum of annual energy or operational costs, annual maintenance costs, and annual replacement costs and for the Acquisition Cost is the sum of annual initial costs including construction costs, initial costs of green building features and administrative costs. In this method, the formula according to [5] :

$$LCC = SC + AC \quad (2.1)$$

Where:
LCC: Life Cycle Cost
SC: Sustaining Cost
AC: Acquisition Cost

### *2.3. LCC Calculation Procedure*
Steps in calculating life cycle cost based on [8], there are :
- Collecting data and categories of costs required
- Input the data of Energy Cost, Operational & Maintenance Cost and Replacement Cost in a year
- Calculate the total cost of energy cost (multiplies the cost of each usage and total usage)
- Calculate the total cost of operational and maintenance cost

- Calculate the total cost of replacement cost (multiplies the cost of each replacement unit and total unit)
- Calculate Sustaining Cost = Total Energy Cost + Total Operational & Maintenance Cost + Total Replacement Cost
- Calculate annual cost of the sustaining cost (using single-payment present worth analysis)
- Single-payment present worth analysis (P= F(P/F,i,n)
- Input the data of Construction Cost, Initial Cost for Feature Green Building, Administrative Cost, and Population Cost
- Calculate the total cost of construction cost
- Calculate the total cost of initial cost (multiplies the cost of each unit and total unit)
- Calculate the total cost of administrative cost (multiplies the construction cost and precentage)
- Calculate Acquisition Cost = Total Construction Cost + Total Initial Cost + Total Administrative Cost + Population Cost
- Calculate annual costs from the acquisition cost category (using single-payment present worth analysis)
- Single-payment present worth analysis (P= F(P/F,i,n)
- Life Cycle Cost = Annual Sustaining Cost + Annual Acquisition Cost

### *2.4. Green Building*
According to DKI Jakarta Provincial Governor Regulation number 38, 2012 that Green Buildings is a building that is responsible for the environment and efficient resources from the planning, implementation of construction, utilization, maintenance, until deconstruction.
The requirements and criteria for developing a Green Building concept that has been determined by the government are:

1. Building Types and Extents, for example, the type of building that is a business, in the field of trade, must have a limit of the entire building floor area of more than 50,000 m2 (fifty thousand square meters), while for the type of building in the health service sector, it must have a floor area of more than 20,000 m2, etc.
2. Building Technical, Technical requirements for green buildings for new buildings include Energy Efficiency, Water efficiency; Indoor air quality, land and waste management; and Implementation of construction activities.

### *2.5. Life Cycle Cost on Green Building*
LCC has a long history in Denmark with sustainable building construction and emphasizes energy saving as a major driver of life cycle costs [9]. Another strong driver of the LCC in Denmark is the construction of buildings with certificates from DGNB for sustainable buildings and urban areas. Strongly supported by industry, the Danish Green Building Council (DK-GBC) was established to manage the DGNB scheme and promote sustainable development more generally. DGNB assesses buildings based on six groups of criteria include (i) Environmental Quality; (ii) Economic Quality; (iii) Social, Cultural and Functional Quality; (iv) Technical Quality; (v) Process Quality; (vi) Site Quality.
Based on research [1] that has been used for environmental evaluations of buildings and related industrial buildings (including construction products, construction systems, buildings, and civil engineering construction) through highly scattered, summarized and regulated literature. The journal also explains that energy use in a building can change to go up or down depending on the consumption of a building at the total cost of construction, material costs, energy costs for materials, operational energy costs, for example, Buildings built with most materials can be used again as a component of new materials and buildings where all materials and components are new. The results show that about 55% of energy can be saved with the material being reused as a new component.
Based on others journal [8] the construction of green building cost a relatively higher cost compared to the cost of conventional building construction. The building with the concept of green building is only

focused on energy efficiency throughout the project life cycle will be a sustainable advantage in its operational costs. From the results of Life Cycle Cost analysis with cost categories consisting of Initial Costs, Energy Costs, Operational Costs, Maintenance, and Replacement Costs, the total cost of the Malaysian Diamond Building cycle is or IDR 759,290,649,000. If you include the Remaining Value in the Life Cycle Cost category, the total cost of living for the Diamond Building will be IDR 559,940,649,000.

## 3. Result and Discussion

### *3.1 Sustaining Cost Calculation*

There are breakdown of the energy cost, operational and maintenance cost and replacement cost estimate which are components in the sustaining cost;

**Table 1** Energy Cost Estimates for Green Building

| Description | Usage | Cost/usage | | Total Cost | |
|---|---|---|---|---|---|
| Day-Lighting Cost | 25.005 kWh | IDR | 1,115 | IDR | 27,881 |
| Garden Lamp ø45 PLC-1x18 W | 0.072 kWh | IDR | 1,115 | IDR | 80 |
| Hanging Lamp REF 23 W | 0.391 kWh | IDR | 1,115 | IDR | 436 |
| Halogen Lamp 50 W | 1.25 kWh | IDR | 1,115 | IDR | 1,394 |
| Hanging Lamp (Decorative) | 0.092 kWh | IDR | 1,115 | IDR | 103 |
| Water Usage Cost/month | 7.57 $m^2$ | IDR | 20,000 | IDR | 151,400 |
| Total Energy Cost | | | | IDR | 181,293 |
| Total Energy Cost/year | | | | IDR | 2,175,517.80 |

Seen from the **Table 1** Energy costs are focused on electricity and water use costs, electricity usage costs in the alerts from PLN are IDR 1,115 per kWh multiplied by total usage per month, while for water use costs are IDR 20,000 per m2 in electricity and water usage utilizing energy-saving lamps and water-saving sanitation technology.

**Table 2** Operational and Maintenance Cost Estimates for Green Building

| Description | Period | Cost/Period | | Total Cost | |
|---|---|---|---|---|---|
| Building Maintenance Cost : | | | | | |
| Cleaning Cost | 12 | IDR | 35,000 | IDR | 420,000 |
| Security and Safety Cost | 2 | IDR | 245,000 | IDR | 490,000 |
| Environmental Control Costs | 1 | IDR | 155,000 | IDR | 155,000 |
| Building and Infrastructure Maintenance Costs | 12 | IDR | 20,000 | IDR | 240,000 |
| Maintenance Feature Green Building Cost : | | | | | |
| Garden Maintenance Costs (Green Area) | 8 | IDR | 25,000 | IDR | 200,000 |
| Rainwater Harvesting Maintenance Costs | 8 | IDR | 25,000 | IDR | 200,000 |
| Maintenance Crew Salary | 1 | IDR | 4,267,349 | IDR | 4,267,349 |
| Total Maintenance | | | | IDR | 22,327,349 |
| Total Maintenance/year | | | | IDR | 267,928,188 |

Based on **Table 2,** Maintenance costs is performed every month on a regular basis and the maintenance costs for each component are different with different maintenance times as well, while the operational costs calculated are maintenance crew salaries obtained from the Jakarta regional minimum of IDR 4,267,349 for one crew member.

**Table 3** Replacement Cost Estimates for Green Building

| Description | Year | | Total Cost |
|---|---|---|---|
| Replacement Cost : | | | |
| Daylighting Replacement Cost | 8 | IDR | 14,811,800 |
| Building Facility Replacement Cost | 5 | IDR | 18,500,000 |
| Rainwater Harvesting Replacement Cost : | | | |
| Pump Replacement | 8 | IDR | 450,000 |
| Water Storage Tank | 8 | IDR | 2,125,000 |
| Water filtration | 5 | IDR | 705,000 |
| Total Replacement Cost | | IDR | 34,466,800 |
| **Total Sustaining Cost** | | IDR | 301,570,506 |

On the **Table 3,** Replacement costs are only done to replace some components that cannot be reused or have been completely damaged and replacement of components is often done about 5 years or more after the age of the component, each component has a different replacement costs with different ages as well.

From all the calculations that have been done, the total obtained from the sustaining cost which is the sum of energy costs, operational & maintenance costs, and replacement costs, the total sustaining cost is IDR 301,570,506.

### *3.2 Acquisition Cost Calculation*

There are breakdown of the construction cost, initial feature green building cost, administrative cost and population cost estimate which are components in the acquisition cost;

**Table 4** Construction Cost Estimates for Green Building

| Description | Volume | Cost/volume | | Total Cost | |
|---|---|---|---|---|---|
| Construction Cost : | | | | | |
| a. Land Cost | 25,677 $m^2$ | IDR | 5,000,000 | IDR | 128,385,000,000 |
| b. Building Cost | | IDR | 14,916,325,790 | IDR | 14,916,325,790 |
| c. Cost of Used Equipment | 36.41% | IDR | 14,916,325,790 | IDR | 5,431,034,220 |
| Total Construction Cost | | | | IDR | 148,732,360,010 |

Based on **Table 4** Estimated construction costs consist of the process of purchasing land, building construction and the cost of consumable equipment. The assumption from the purchase of land in South Jakarta is IDR 5,000,000 for 1$m^2$, for the cost of green building construction costs is IDR 14,916,325,790 while for the cost of consumable equipment is 36.41% of the construction costs.

**Table 5** Initial Cost Feature Green Building Estimates

| Description | Unit | Cost/Unit | | Total Cost | |
|---|---|---|---|---|---|
| Feature Green Building Initial Cost : | | | | | |
| Daylighting : | | | | | |
| a. Lamp TL-5-2x28 W | 78 | IDR | 27,000 | IDR | 2,106,000 |
| b. Downlight PLC-2x26 W | 46 | IDR | 29,800 | IDR | 1,370,800 |
| c. Downlight PLC-2x18 W | 118 | IDR | 27,000 | IDR | 3,186,000 |

IOP Conf. Series: Materials Science and Engineering **847** (2020) 012023 doi:10.1088/1757-899X/847/1/012023

**Table 5** Initial Cost Feature Green Building Estimates

| Description | Unit | Cost/Unit | Total Cost |
|---|---|---|---|
| d. Downlight PLC-1x18 W | 33 | IDR 27,000 | IDR 891,000 |
| e. Downlight PLC-1x13 W | 191 | IDR 38,000 | IDR 7,258,000 |
| Air Conditioning : | | | |
| a. AC Split Duct 1 PK | 3 | IDR 3,300,000 | IDR 9,900,000 |
| b. AC Split Duct 1.5 PK | 15 | IDR 4,100,000 | IDR 61,500,000 |
| c. AC Split Duct 2 PK | 24 | IDR 5,450,000 | IDR 130,800,000 |
| d. AC Split Duct 2.5 PK | 48 | IDR 7,500,000 | IDR 360,000,000 |
| Rainwater Harvesting : | | | |
| a. Recycle Tank | 3 | IDR 3,200,000 | IDR 9,600,000 |
| b. Flush Plants /Lanscape | 3 | IDR 2,500,000 | IDR 7,500,000 |
| Total Initial Feature Green Building Cost | | | IDR 604,011,800 |

As we seen on **Table 5** Initial Cost Feature Green Building is the cost of each component or material used to save energy expended, for example, the use of electricity using energy saving TL-5 LED lamps with several models and types as well as the use of AC Split Duct saving electricity on each floor of the building with some species also save water by utilizing rainwater harvesting. The price of each component is obtained from e-commerce from various lights and air conditioners.

**Table 6** Administrative Cost and others

| Description | Precentage | Construction Cost | Total Cost |
|---|---|---|---|
| Administrative Cost | 5% | IDR 148,732,360,010 | IDR 7,436,618,000.51 |
| Others Cost | 10% | IDR 148,732,360,010 | IDR 14,873,236,001.01 |
| Total | | | IDR 22,309,854,001.52 |
| Total Acquisition Cost | | | IDR 171,646,225,812.00 |

On **Table 6** Administrative costs in construction are required such as the cost of verifying documents, taxes and other government costs, as well as unexpected costs in building construction. Administrative costs and others are assumed to be 5% and 10% of the construction costs for each administration fee and unexpected costs.

From all the calculation, the total of Acquisition Cost is IDR 171,646,225,812.00 before adding up the salvage cost and book value. The total Acquisition Cost will be added to the salvage cost and book value when calculating using present worth analysis.

**Table 7** Population Cost Estimate for Green Building

| Year | Salvage Cost | Construction Cost | Book Value |
|---|---|---|---|
| 1 | IDR 130,884,476,809 | IDR 148,732,360,010 | IDR 16,360,559,601 |
| 2 | IDR 115,178,339,592 | IDR 148,732,360,010 | IDR 14,397,292,449 |
| 3 | IDR 101,356,938,841 | IDR 148,732,360,010 | IDR 12,669,617,355 |
| 4 | IDR 89,194,106,180 | IDR 148,732,360,010 | IDR 11,149,263,272 |
| 5 | IDR 78,490,813,438 | IDR 148,732,360,010 | IDR 9,811,351,680 |
| 6 | IDR 69,071,915,826 | IDR 148,732,360,010 | IDR 8,633,989,478 |
| 7 | IDR 60,783,285,927 | IDR 148,732,360,010 | IDR 7,597,910,741 |
| 8 | IDR 53,489,291,615 | IDR 148,732,360,010 | IDR 6,686,161,452 |

IOP Conf. Series: Materials Science and Engineering **847** (2020) 012023 doi:10.1088/1757-899X/847/1/012023

Population cost calculation consists of book value and salvage costs with a fixed cost construction cost of IDR 148,732,360,010 so that there is a calculation of the remaining cost of a building by multiplying the construction cost and assuming the depreciation 12% per year while for book value calculation is a reduction in construction costs and the remaining costs already divided by the age of the building, the results of salvage cost and book value can be seen from the **Table 7**.

### *3.3 Life Cycle Cost Calculation*

Life Cycle Cost is the sum of annual sustaining costs and annual acquisition costs that previously used a single-payment present analysis or present worth analysis assuming if there were 1 to 8 maintenance crew and the age of the building since it was built 8 years ago, so that the results obtained are age the economic cost of the building (vertical column year in **Table 8**), the number of maintenance crew (horizontal column in **Table 8**) and the optimal total cost of the green building concept, the results of the Life Cycle calculation are as follows:

**Table 8** Life Cycle Cost Calculation

| | Maintenance Crew | | | |
|---|---|---|---|---|
| Year | 1 | 2 | 3 | 4 |
| 1 | IDR 289,717,867,781 | IDR 289,760,305,725 | IDR 289,802,743,669 | IDR 289,845,181,613 |
| 2 | IDR 275,923,724,425 | IDR 275,967,316,681 | IDR 276,010,908,937 | IDR 276,054,501,193 |
| 3 | IDR 264,355,316,693 | IDR 264,400,094,658 | IDR 264,444,872,624 | IDR 264,489,650,589 |
| 4 | IDR 254,761,072,235 | IDR 254,807,068,161 | IDR 254,853,064,087 | IDR 254,899,060,013 |
| 5 | IDR 246,920,029,417 | IDR 246,967,276,432 | IDR 247,014,523,447 | IDR 247,061,770,462 |
| 6 | IDR 240,638,175,511 | IDR 240,686,707,645 | IDR 240,735,239,779 | IDR 240,783,771,913 |
| 7 | IDR 235,745,224,625 | IDR 235,795,076,832 | IDR 235,844,929,040 | IDR 235,894,781,248 |
| 8 | IDR 232,091,782,585 | IDR 232,142,990,773 | IDR 232,194,198,961 | IDR 232,245,407,149 |

| | Maintenance Crew | | | |
|---|---|---|---|---|
| Year | 5 | 6 | 7 | 8 |
| 1 | IDR 289,887,619,556 | IDR 289,930,057,500 | IDR 289,972,495,444 | IDR 290,014,933,388 |
| 2 | IDR 276,098,093,449 | IDR 276,141,685,705 | IDR 276,185,277,961 | IDR 276,228,870,216 |
| 3 | IDR 264,534,428,554 | IDR 264,579,206,519 | IDR 264,623,984,485 | IDR 264,668,762,450 |
| 4 | IDR 254,945,055,939 | IDR 254,991,051,865 | IDR 255,037,047,791 | IDR 255,083,043,717 |
| 5 | IDR 247,109,017,477 | IDR 247,156,264,492 | IDR 247,203,511,507 | IDR 247,250,758,522 |
| 6 | IDR 240,832,304,047 | IDR 240,880,836,181 | IDR 240,929,368,315 | IDR 240,977,900,449 |
| 7 | IDR 235,944,633,456 | IDR 235,994,485,664 | IDR 236,044,337,872 | IDR 236,094,190,080 |
| 8 | **IDR 232,296,615,337** | IDR 232,347,823,525 | IDR 232,399,031,713 | IDR 232,450,239,901 |

Based on **Table 8** the output from the Lifecycle Cost method with an estimated maintenance crew of 1 to 8 people and an estimated building of around 8 years due to construction starting in 2013 is IDR 232,296,615,337 for an optimal Lifecycle Cost with 1 maintenance crew and an economic age of 8 years.

### *3.4 Efficiency Energy Calculation*

**Table 9** Emission $CO_2$ Calculation

| Emission $CO_2$ | Total Energy Consumption | Emission $CO_2$ |
|---|---|---|
| Baseline | 827,337.92 kWh | 737,158.17 kg |
| Design | 659,022.56 kWh | 587,189.10 kg |
| Deviation | 168,315.36 kWh | 149,969.07 kg |

From **Table 9** Carbon dioxide emissions per kWh of annual energy consumption are 0.891 kg. The reduction of carbon dioxide emissions is calculated from the difference in energy consumption between the baseline building and the design multiplied by the annual energy consumption of 0.891 kg so that the reduction of carbon dioxide emissions is 149,968.99 kg.

**Table 10** Energy Consumption Intensity

| No | Floor | Total Area | Total Watt |
|---|---|---|---|
| 1 | Basement Floor | 779.68 $m^2$ | 5990 kWh |
| 2 | 1st Floor | 779.68 $m^2$ | 5284 kWh |
| 3 | 3rd Floor | 889.98 $m^2$ | 5239 kWh |
| 4 | 4th Floor | 890.04 $m^2$ | 7436 kWh |
| 5 | 5th Floor | 689.07 $m^2$ | 2961 kWh |
| | Total | 4028.45 $m^2$ | 26910 kWh |

Seen from **Table 10** that the intensity of energy consumption is said to be very efficient if the intensity value of 4.17-7.92 kWh /$m^2$ so that the result of the division between Total Watts and Total Area is 6.68 then, the intensity of energy consumption in the Green Building is very efficient.

## 4. Conclusion

Based on the analysis and discussion in the previous chapter, the cost categories contained in the Green Building, consist of Initial Costs, Energy Costs, Maintenance Costs, and Replacement Costs, using the Life Cycle Cost method, the optimal cost of a Green Building concept is IDR 232,296. 615,337 with the economic life of the building is 8 years because the construction process starts in 2013, so that the economic age is still followed and the optimal maintenance crew is 1 maintenance crew. The intensity of energy consumption in the Green Building case study is 6.68 kWh / m2, which can be said that the intensity of energy consumption in the green building concept is very efficient than conventional buildings. For future research, it is expected to be able to compare costs used in buildings that are Green Building concepts and conventional buildings, and calculate overall energy efficiency and compare between Green Buildings and conventional buildings.

## 5. References

[1] L. F. Cabeza, L. Rincón, V. Vilariño, G. Pérez, and A. Castell 2014 Life cycle assessment (LCA) and life cycle energy analysis (LCEA) of buildings and the building sector: A review. *Renew. Sustain. Energy Rev.* **29**, pp. 394–416.

[2] F. Salvado, N. M. de Almeida, and A. Vale e Azevedo 2018 Toward improved LCC-informed decisions in building management. *Built Environ. Proj. Asset Manag.* **8**(2), pp. 114–133.

[3] H. Islam, M. Jollands, and S. Setunge 2015 Life cycle assessment and life cycle cost implication of residential buildings - A review. *Renewable and Sustainable Energy Reviews*.

[4] A. Pelzeter 2007 Building optimisation with life cycle costs – the influence of calculation methods. *J. Facil. Manag.* **5**(2), pp. 115–128.

[5] A. R. Eliyus and J. Alhilman 2014 Estimasi Biaya Maintenance Yang Optimal Dengan ( Studi Kasus : Pt Toa Galva ). *J. Rekayasa Sist. Ind.*, pp. 48–54.

[6] C. Ghosh, J. Maiti, M. Shafiee, and K. G. Kumaraswamy 2018 Reduction of life cycle costs for a contemporary helicopter through improvement of reliability and maintainability parameters. *Int. J. Qual. Reliab. Manag.* **35**(2), pp. 545–567.

[7] A. Anggriawan and A. Kurniawati 2015 Biaya Maintenance Dengan (Studi Kasus : PT Telkomsel Indonesia). *J. Rekayasa Sist. Ind.* **2**(3), pp. 33–38.

[8] T. Firsani and C. Utomo 2012 Analisa Life Cycle Cost pada Green Building Diamond Building Malaysia. *Tek. ITS* **1**(September), pp. D34–D38.

[9] K. Haugbølle and L. M. Raffnsøe 2019 Rethinking life cycle cost drivers for sustainable office buildings in Denmark. *Facilities* **37**(9–10), pp. 624–638.

IOP Conf. Series: Materials Science and Engineering **847** (2020) 012024 doi:10.1088/1757-899X/847/1/012024

# Productivity Analysis Stevedore a Descriptive Analysis Method with Integration, Importance Performance Analysis, Quality Function Deployment (Case Study: PT. Port Indonesia III (Persero) Branch Gresik)

**E Widodo[1], H Suprayitno.[2], Suparno[3], and Umaiyah[4]**

Department of Industrial Engineering, Institut Teknologi Sepuluh Nopember, ITS, Surabaya, Indonesia
[1]erwin@ie.its.ac.id,
[2]heri3prayitno@gmail.com,[3]suparno@ie.its.ac.id,[4]najashafa12@gmail.com

**Abstract.** Gresik port is a port with a very strategic location in the region Gresik industrial area, productivity can be achieved with the implementation of a better way of working, organized, effective, efficient, creating something better. the productivity of loading and unloading the target of service to satisfaction of stakeholders, how to determine the factors that influence and by improving the productivity of loading and unloading at the port of Gresik. With descriptive analysis method to know the level of EF: BT is still yet to be achieved was based stevedoring productivity director general of sea transportation standards already achieved. Methods importance performance analysis is used to determine the perception of the services to be improved the speed in the process of loading and unloading, the availability of facilities and port equipment, fleet readiness ocean going ship expeditions, circumstances and facilities yard, competency officers loading and unloading, the speed of completing complain, system lighting port. Methods of quality function deployment used to determine priority in handling an increase in productivity of loading and unloading at the Port General Gresik.
**Key words:** Productivity Stevedoring, Descriptive Analysis, IPA and QFD

## 1. Introduction

Loading and unloading are the activities of the movement of goods from sea transport mode for inland transport modes or vice versa, which includes [1]:

1. Ship operation is loading activities (loading cargo) and unloading (unloading cargo / discharge) from or to the ship.
2. Quay transfer operation is a displacement activity on the dock charge of the side of the hull to the location of the buildup or otherwise, the term cargo which is the second phase of a series of operations dock.
3. Storage operation is the buildup of cargo at the port which is an important step in the transfer of cargo between sea and land freight transport. Which is the third stage of a series of operations quay (berth operation).
4. Receiving / Delivery Operation is the activity and delivery of goods cargo that goes on the side of the hull / dock, warehouse or yard.

Stevedoring productivity is the level of ability and speed of execution of demolition activities handling goods from the ship to the warehouse or yard or vice versa for loading activities of goods from the warehouse / yard up to the vessel. Service performance of port operations is measurable work results achieved in implementing port service vessels, goods and utilization of facilities and tools, in a period of time and a certain unit [2]. Loading and unloading performance measured in tones / Gang / hour is the amount of goods in tons or M3 unloaded / loaded within a period of one hour of work by a gang:

$$\text{TGH} = \frac{\Sigma \text{ Goods unloaded or loaded each ship}}{\Sigma \text{ Aisles each shift each ship x effective hours each ship}} \quad (1)$$

The analytical method of Quality Function Deployment [3] commonly used to determine the priority of the measures to be taken to make improvement. QFD is a matrix diagram tool, which is very useful for organizing the data collected, helping to facilitate the repair process. QFD structure used to organize information known as the House of Quality. In a broad sense, QFD House of Quality to show the relationship between the dependent (WHATS) and independent (HOWS) variable (Woods, 1994. In order to effectively use the QFD requires the participation of a team and discipline inherent in the practice of QFD, which has proven to be a highly experience team building well. In this study, after unloading performance was analyzed by IPA then the variables are in quadrant I, which is a variable with a low performance level but has a high level of interest will be sought and priority handling solutions with QFD to build a house of quality.

Previous research has introduced a new approach that was done by Agus Triyono, Achmad Wicaksono, M. Ruslin Anwar with the title Operational Performance Assessment and Public Port Development Strategy Gresik by using IPA, SWOT IFAS EFAS methods in 2015. M. Rum Raehan, Ludfi Djakfar, Alwafi Puji Raharjo with the title Performance Evaluation General Stevedoring at the Port of Gresik by using IPA, QFD methods in 2017. In this study combines with the integration of descriptive method of analysis, science and QFD to obtain the priority measures to be taken by the management of PT. Port Indonesia III (Persero) branch Gresik to enhance productivity and achievement stevedoring better performance. The Limitations of this study is in the process of loading and unloading of ships in the Port Gresik the public pier and dock pier 78 dry bulk and logs.

The process of loading and unloading at the port is one of the dominant factors that affect the performance of a port. The duration of the process of loading and unloading at the port also has implications for the duration of the turn round time a ship that would make ships more expensive operating costs [4]. From the results of previous studies showing too long ship in the dock because many are waiting for a load so often happens hanging slings, truck loading and unloading systems losing, cargo to be loaded on the ship was still in the warehouse line II. Things like this affect the quality of logistics services. Optimization of time in dealing with the loading and unloading process is expected to improve the performance of loading and unloading. loading and unloading of the performance review and identify the factors that cause the length of the loading and unloading process [5].

### *1.1. Problem formulation*
Based on the background described, the formulation of the problem in this research are:

1. How to determine the factors that affect the performance of loading and unloading of goods at PT. port Indonesia III (Persero) Branch Gresik?
2. How to know stevedoring productivity at PT. Port Indonesia III (Persero) Branch Gresik?
3. How to improve the productivity of loading and unloading of goods at PT. port Indonesia III (Persero) Branch Gresik?

### *1.2. Research Objectives*
Based on the formulation of the problem and the assumptions above, the purpose of this study are as follows:

1. How to determine the existing condition of the performance of loading and unloading in the port of Gresik.
2. How to determine the factors that affect the performance of loading and unloading in the port of Gresik general.
3. What efforts should be made to increase the productivity of public loading and unloading in the port of Gresik.

## 2. Methods

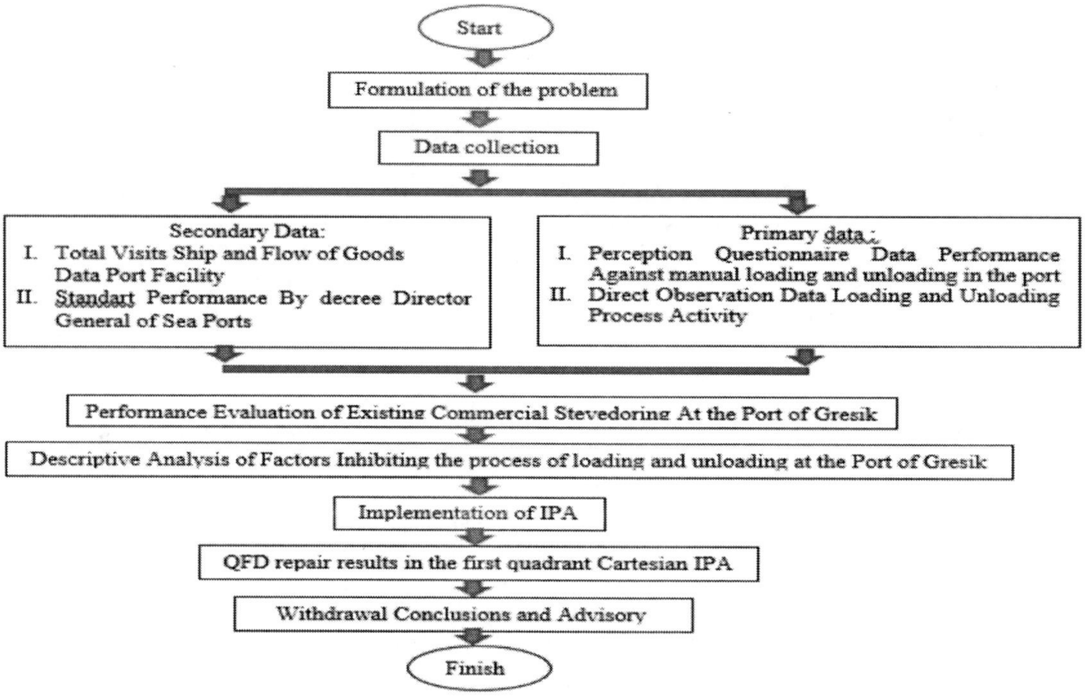


**Figure 1**. Flowchart of Research Methodology

Figure 1 shows that this study takes a place at the pier and dock dry bulk log Gresik 78 public ports, while the flow of this research is the show role in the diagram as follows: Collecting data using secondary data is data from the report Gresik port operations, while primary data obtained by giving IPA questionnaires and interviews as well as suggestions from stakeholders that ship agency, stevedoring companies and owners of goods that have activities in the port of Gresik.

The population used in this study is the number of vessels that carry out loading and unloading in the port of Gresik General during January s / d in March 2019, especially the Dry Bulk Terminal Pier and Log, Pier 78. The number of samples based on the formula [6] as follows:

$$\mathbf{n} = \frac{N}{1+Ne2} \tag{2}$$

n = sample size
N = population size
e = fault tolerance (error) 5%

IOP Conf. Series: Materials Science and Engineering **847** (2020) 012024 doi:10.1088/1757-899X/847/1/012024

## 3. Result and Discussion

### *3.1. The realization of performance-based Stevedoring Descriptive Analysis*

Guidelines used is the Decree of the Director General of Sea Transportation UM.002 numbers / 38/18 / DGLT-11 and Regulation Director General of Sea number: HK.103 / 2/18 / DGLT-16. The standard value of each aspect associated with loading and unloading can be seen in Table 1.

**Table 1**. the standard value of each aspect.

| No. | Service | Unit | 2011 | 2016 |
|---|---|---|---|---|
| A | **Boats services** | | | |
| 1 | Waiting Time | Hour | 1 | 1 |
| 2 | Approach Time | Hour | 2 | 2 |
| 3 | ET: BT | % | 70 | 70 |
| B | **Goods Services** | | | |
| 1 | General Cargo | T / G / H | 30 | 35 |
| 2 | Cargo Bag | T / G / H | 35 | 35 |
| 3 | Liquid Bulk | T / H | 100 | 100 |
| 4 | Dry bulk | T / H | 100 | 150 |
| C | **Utilization Facilities & Operations Equipment** | | | |
| 1 | Berth Occupancy Ratio | % | 70 | 70 |
| 2 | Shed Occupancy Ratio | % | 65 | - |
| 3 | Yard Occupancy Ratio | % | 80 | 65 |
| 4 | Equipment operations | % | 70 | 70 |

Source: Standard Performance Based Sea Relations Port Authority Office Gresik

Based on the Table 1. To determine the effective time: berthing time, the productivity of loading and unloading as well as loading and unloading equipment in the utility set minimum standards, achievement of operational service performance of each indicator ET: BT, performance and readiness for operation of loading and unloading equipment is determined as follows:

a. If the value of the achievement of the above values of operational service performance standards specified, otherwise good;
b. If the value of the achievement of above 90% up to 100% of the value of the operational service performance standards set, considered quite good.
c. If the value of the achievement of less than 90% of the standard values of performance, still inadequate.

**Table 2.** Value Achievement Performance Dry Bulk Terminal Pier & Log January S / D March 2019

| Commodity | Number of Ships | Performance indicators | Standard | Unit | Achievement | Information |
|---|---|---|---|---|---|---|
| General Cargo | 19 | Productivity loading unloading | 35 | T / G / H | 62.3 | Reached |
| | | ET: BT | 70 | % | 53 | Not achieved |
| Dry bulk Logs | 73 | Productivity loading unloading | 150 | T / G / H | 228 | Reached |
| | | ET: BT | 70 | % | 61.5 | Not achieved |
| General Cargo | 40 | Productivity loading unloading | 35 | T / G / H | 62.25 | Reached |
| | | ET: BT | 70 | % | 53 | Not achieved |
| Cargo bag | 7 | Productivity BM | 35 | T / G / H | 56.5 | Reached |
| | | ET: BT | 70 | % | 51.5 | Not achieved |

Source: Port Authority Office Gresik

IOP Conf. Series: Materials Science and Engineering **847** (2020) 012024 doi:10.1088/1757-899X/847/1/012024

Table 2 is influenced by factors that idle time waiting truck unloading coal, equipment trouble on cranes, cargo waiting and waiting labour and also influenced by factors that idle time waiting truck for loading and unloading trucks used also for delivery in the field as well, waiting for hours TKBM labour has not been 24 hours, Trouble equipment, and bed weather.

### *3.2. Importance Performance Analysis Results Analysis Methods*

Measuring the level of interest of the facility and operational performance levels of loading and unloading facilities using a Likert scale of 5 levels. Degree of importance is given a score of 1 (very important / very good to 5 (very important / very good) [7].

By taking the average of the Y-axis (level of importance) and the X-axis (level of performance) are obtained Cartesian diagram as follows:

**Table 3.** The average rate of interest & Performance Levels

| No. | Commentary | Interest | Performance |
|---|---|---|---|
| 1 | Speed in doing process unloading and loading | 3.53 | 3.97 |
| 2 | Availability of facilities and equipment unloading and loading | 3.53 | 3.92 |
| 3 | Readiness ocean going ship expeditions (trucking) | 3.56 | 3.86 |
| 4 | Circumstances and capacities Field cumulation | 3.31 | 3.84 |
| 5 | Pier with Field cumulation | 4.17 | 3.78 |
| 6 | Stacking Field Access to Dock | 3.75 | 3.92 |
| 7 | Warehouse Capacity Stacking | 3.44 | 3.53 |
| 8 | Stacking Warehouse access to the Wharf | 3.72 | 3.75 |
| 9 | Stacking Warehouse Access to the Road great | 3.44 | 3.47 |
| 10 | Warehouse facilities and field cumulation | 2,67 | 3.53 |
| 11 | Availability Labor Unloading and loading | 3.72 | 4.25 |
| 12 | The accuracy of the operation planning | 3.69 | 3.92 |
| 13 | Harbor Area Cleanliness | 3.61 | 3.69 |
| 14 | Speed document administration unloading and loading | 4.00 | 4.28 |
| 15 | Rates unloading fit | 3.67 | 3.83 |
| 16 | Type a commodity goods which in Unloading / unloading | 4.14 | 3.83 |
| 17 | competency officer unloading fit | 3.53 | 3.97 |
| 18 | Speed complaint settlement | 2,94 | 4.11 |
| 19 | Environmental Safety Harbor | 3.89 | 3.58 |
| 20 | Port Facility Condition | 3.58 | 4 |
| 21 | Validity data unloading fit | 3.61 | 3.81 |
| 22 | Port lighting system | 3.5 | 4.25 |
| 23 | Easy Access Information Systems / IT Technology updated | 3.64 | 4.03 |
| 24 | Availability of parking of vehicles in the port area | 3.19 | 3.72 |

Source: The data processing

Based on table 3 we can do in a Cartesian diagram IPA analysis, there are four quadrants divided by two axes. The axis is the abscissa (X) and ordinate (Y), where the axis is based on the average value the overall level of performance (X) and interest (Y). IPA Cartesian diagram analysis can be seen in figure below:

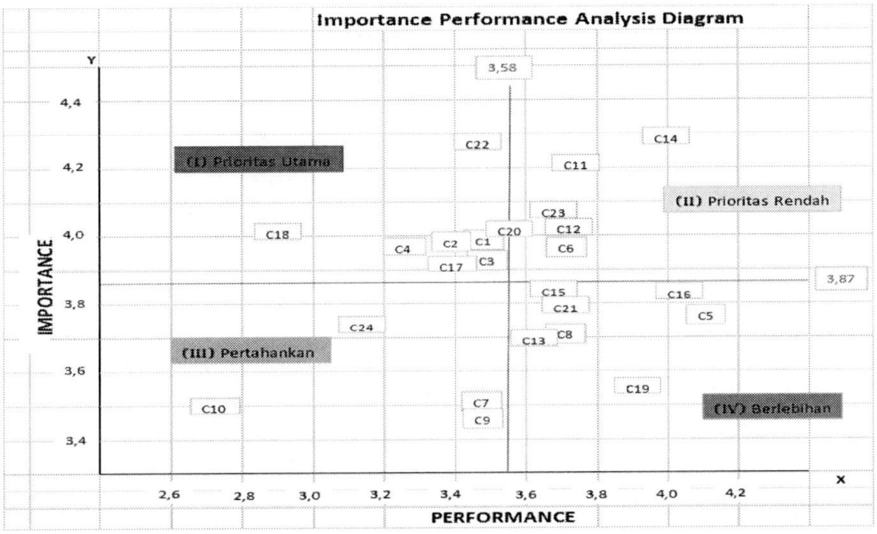


**Figure 2.** Cartesian diagram Importance Performance Analysis

From Figure 2 above results Cartesian diagram IPA analysis is divided into four quadrants, for the first quadrant (top priority) are: (1) C1. The speed in the process of loading and unloading. ; (2) C2. The availability of facilities and port equipment; (3) C3. EMKL fleet readiness; (4) C4. State and yard facilities; (5) C17. Competency Unloading Officer; (6) C18. Speed Complaint Resolution; (7) C22. Lighting system at the Port.

### *3.3. Unloading Productivity Performance Improvement, Quality Function Deployment*

Priority handling productivity improvement unloading at the Port General Gresik used QFD analysis based on the variables contained in the first quadrant analysis results IPA. The next step is to determine the relationship "item what" to "how the item" [8]. It shows in table below:

**Table 4.** The ways how to handle "Item How"

| No. | "What Item" | "Item How" |
|---|---|---|
| 1. | The speed in the process of loading and unloading | 1. Training of Crane Fixed operators whose expertise is uneven in each group harbor loading unloading logs<br>2. At unloading the goods from the manufacturer should be stacked goods used in the field / warehouse |
| 2. | The availability of facilities and port equipment | 3. That for unloading the logs need addition both the number and capacity forklift |
| 3. | Fleet readiness EMKL | 4. Since many goods are loosing then the third coordinate ocean going ship expeditions road traffic<br>5. Adding to the fleet in the event of delays on the road and crowded in the Warehouse destination |
| 4. | State and yard facilities | 6. Need to restructure yard in the given zone according to the type of commodity goods.<br>7. Expanding yard |
| 5. | Officers Competency Unloading | 8. Loading unloading need training to officers on the use of the Tab / for real time activity |
| 6. | Complaint Resolution Speed | 9. Service users in order to coordinate the time of closing time ship out with the port authorities. |
| 7. | Lighting system at the Port | 10. Need additional lighting at the point in the yard. |

Source: the data processing

Based on table 4 we can do in a house of quality analysis, there are two item: (1) item what, (2) item how. Item "what" is divided to be seven sub items namely: loading and unloading until port lighting system as like shown in the figure 3. Item "How" is divided to be ten sub items namely: item how 1 until item how 10 as like shown in the figure 3.

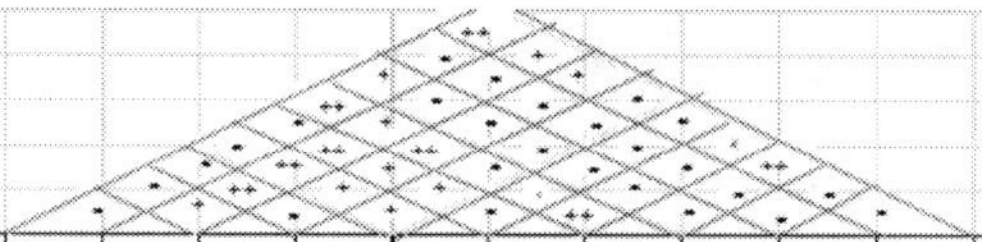

| Average level of performance | Item What \ Item How | Item How 1 | Item How 2 | Item How 3 | Item How 4 | Item How 5 | Item How 6 | Item How 7 | Item How 8 | Item How 9 | Item How 10 | Rata-rata tingkat kinerja | Goal | Improvement Ratio | Raw Weight | Normalized Raw Weight |
|---|---|---|---|---|---|---|---|---|---|---|---|---|---|---|---|---|
| 3.94 | Loading and unloading speed | ◎ | ○ | ○ | △ | ◎ | ○ | △ | ○ | △ | ○ | 3.53 | 3.94 | 1.116 | 4.398 | 0.134 |
| 3.92 | Availability of facilities and equipment | △ | △ | ◎ | △ | △ | ○ | ◎ | △ | △ | ○ | 3.53 | 3.92 | 1.11 | 4.353 | 0.132 |
| 3.86 | Truck fleet readiness | △ | △ | △ | ◎ | ◎ | ○ | △ | △ | △ | △ | 3.56 | 3.86 | 1.084 | 4.185 | 0.127 |
| 3.89 | State of the field buildup facility | △ | ○ | △ | △ | △ | ◎ | ◎ | △ | △ | ○ | 3.31 | 3.89 | 1.175 | 4.572 | 0.139 |
| 3.97 | Loading unload officer competencies | ○ | △ | △ | △ | △ | △ | △ | ◎ | △ | △ | 3.53 | 3.97 | 1.125 | 4.465 | 0.136 |
| 4.11 | Speed of resolution of complaints | △ | △ | △ | △ | △ | △ | △ | ○ | ◎ | △ | 2.94 | 4.11 | 1.398 | 5.746 | 0.175 |
| 4.25 | Port Lighting system | △ | △ | △ | △ | △ | ○ | ○ | △ | △ | ◎ | 3.50 | 4.25 | 1.214 | 5.161 | 0.157 |
| | Contribution | 17 | 11 | 17 | 15 | 23 | 23 | 25 | 19 | 15 | 21 | | | | | |
| | Normalized Contribution | 2.274 | 1.529 | 2.251 | 1.909 | 3.076 | 3.198 | 3.476 | 2.58 | 2.621 | 3.296 | | | | | |
| | Priority | 7 | 10 | 8 | 9 | 4 | 3 | 1 | 6 | 5 | 2 | | | | | |

**Figure 3.** House of Quality

Based on figure 3 To repair the problem based on the analysis of the IPA on the priority measures to be carried out in accordance analysis QFD is expanding its container yard, additional lighting, zoning yard, adding a fleet of trucks when crowded, service users coordinate first with the port authorities when closing time ship, training the use of tan on the loading and unloading personnel, training of crane operators at each loading and unloading labour group, the addition of forklift when unloading logs, goods are loosing to coordinate the road traffic, goods to be stacked first in the yard or warehouse harbour.

## 4. Conclusion

1. According to standard of decree of the Director General of Sea Transportation UM.002 numbers / 38/18 / 11and DGLT-Regulation Director General of Sea number: HK.103 / 2/18 / DGLT-16, then for good productivity unloading dry bulk terminal and the terminal log 78 has been reached, whereas ET: BT is not achieved at both terminals.
2. Descriptive analysis of the results of loading and unloading speed is influenced by several factors including idle time waiting truck, cargo waiting, waiting labor equipment and bed weather trouble.
3. IPA analysis of the results obtained some things that should be improved to increase the productivity of loading and unloading at the port of Gresik namely speed in the process of loading and unloading facilities and port equipment availability, readiness ocean going ship expeditions fleet, State and yard facilities, Competency Unloading Officer, Complaints Resolution Speed, Lighting System in the Port.
4. As suggestions to management of PT. Port Indonesia III (Persero) branch Gresik are need to restructure yard layout with giving appropriate zoning commodity goods to be unloaded or loaded. Establish a commitment to cooperative loading and unloading labor Gresik port so bad to work for 3 shifts 24 hours due to the loading of logs to 3 am alone.

IOP Conf. Series: Materials Science and Engineering **847** (2020) 012024 doi:10.1088/1757-899X/847/1/012024

## 5. Reference


[1] Ashlyzan Razik, Md, Max Tahar, R, Hasrulni Wan, Wan Mahmood and Mohd Rozar N 2015 Integrated Quality Function Deployment (QFD) Model for Dry Bulk Terminal Improvement (DBTI) in Malaysian Ports. *Journal of Economies, Business and Management,* **3**(4), April 2015

[2] Handajani M 2004 Analysis of the operational performance of loading and unloading of containers Port of Tanjung gold Semarang. *Journal of Transportation*. **4**(June 1st), 1-12.

[3] Irawati Buomona Y 2016 Analisis faktor-faktor yang mempengaruhi jumlah bongkar muat di lima pelabuhan barang terbesar di Indonesia. *Thesis* (Surabaya: University Airlangga) 040911174.

[4] Min, HH, Zaili Y, Jasmine SL 2018 Port performance in container transport logistics: A multi-stakeholder perspective. Journal 29 September 2018 - 0967-070X/Elsevier.

[5] Mokhtar K, and Zaly Shah M 2006 *Agregassion models for vessel turnaround time.* Tokyo academic, industry and cultural integration tour in 2006, Shibaura Institute of Technology Japan.

[6] Siddique G and Basak A 2018 Importance-Performance Analysis (IPA) of service quality in public transport of Asansol-durgapur development authority. *The International Journal of Computational engineering research (IJCER)*.

[7] Rum Raekhan M, Djakfar L, and Pujiharjo A 2017 Evaluation of the performance of loading and unloading in the port of Gresik. *Journal vol transportation* (August 2017), 133-44.

[8] Subagio H 2010 Optimization of service performance loading and unloading of containers at the Port of Tanjung Emas. *Thesis*. (Semarang: Diponegoro University).

# A Multi-objective model for a Chemical Industry considering economic risk and environment

**James Yu[1,2], Sri Retno Purwaningsih[1,3], and Hui Ming Wee[1,4]**

[1]Department of Industrial & Systems Engineering, Chung Yuan Christian University, Taiwan
[2]sowhat1130@yahoo.com, [3]p.sriretno@gmail.com, [4]weehm@cycu.edu.tw

**Abstract.** The research proposes a multiple objective model considering economic risk and environment in the chemical industry. The economic order quantity (EOQ) model is used for the economic aspect which considers the transportation modes and all unit quantity discount. The objective of this research is to optimize economic, safety and green performance in terms of cost minimization, risk control, and carbon emission reduction. A solution approach to obtain the optimal solution is proposed. A numerical example is performed to demonstrate the applicability and efficacy of the procedures.
**Keywords:** EOQ model, transportation, quantity discount, multi-objective optimization, chemical industry.

## 1. INTRODUCTION

According to Cefic chemical data, the annual sale of global chemicals in 2015 increased by €434 billion from 2014 – a substantial difference for the chemical industry. Chinese sales were the main reason for this change, with a value of €1,409 billion earned in 2014 when compared to €1,084 in 2014 (a 30% value increase).

Furthermore, transportation generated 14% of global Green House Gas (GHG) emissions in 2010. The main contributors to these transportation emissions were road vehicles, trains, airplanes, and maritime transport ships, which mostly used diesel and gasoline fuels (EPA, 2016). Road and marine transportation methods were the main avenues of transportation used by chemical logistics, in terms of ton-kilometer usage, and also the main contributors of $CO_2$ emissions. Meanwhile, freight transportation accounts for 1/3$^{rd}$ of all transportation emissions, out of which heavy long-distance trucking is a main contributor for $CO_2$ emission (Cefic, 2015). Heavy trucking transportation consists of two categories: the truckload mode (TL) and less-than-truckload mode (LTL).

TL transportation occurs when a firm utilizes capacities that may be less than the full capacity offered by a truck, while paying the cost of a full truck. Thus, a partial fixed cost per load can be calculated by the costs incurred by the truck, the driver, and by other operating expenses. In contrast, LTL transportation is calculated at a constant cost per unit due to standardized rates [1]. Risk management is also an important logistical component due to its ability to prevent interruptions, such as transportation hazards and labor disputes. The transport of dangerous chemicals involves a factor of risk for the driver and passengers, and also to other drivers. In the event of a chemical spill, the materials may be toxic or inflammable, and if spilled on the roadways, may incur cleanup costs and jeopardize traffic. The scale of risk may be anywhere from physical harm to the involved persons to fatalities. Thus, risk can be calculated as the product of the sum of accident probabilities and their corresponding consequences of failure [2]. Modern economic competition requires members in the chemical industry to enter into cooperative relationships between suppliers, allies, and customers in order to overcome a variety of challenges, such as cost, quality, lead time, flexibility, customer service, and accident prevention. In addition, the participants must counter environmental change by minimizing carbon emissions. However, producers often hold a preference for optimization methods that create financial value over risk management and emissions reduction. Thus, risk reduction and carbon emissions control methods that do not increase economic costs present a challenge for supply chain managers.

A traditional and well-known method of supply chain optimization is the economic order quantity (EOQ) model, which was popularized due to its simple design, adaptability, and straightforward application [3]. Harris (1913) first created the model to calculate optimal order quantity based on manufacturing parameters and several cost considerations. Over time, the model underwent various adaptations as additional considerations were used to extend its formula. Mousavi et al. (2016) created a seasonal multi-product, multi-period control model with an all unit discount policy and inflation costs [4]. Tersine et al. (1995) extended the original model by considering fully back-ordered shortages with all unit discounts, incremental order volume discounts, and transportation discounts offered by the seller [5]. Bai et al. (2011) adapted an inventory model using multiple-supplier dynamic demand considerations for order cost, incremental order volume discounts, multiple setups, and all-unit quantity discount systems [6]. Taleizadeh (2014) developed an EOQ model allowing for partial backorders and all-unit discounts [7].

Mendoza et al. (2008) created a single-echelon EOQ model over a finite planning timeframe using two transportation methods [8]. Later, the authors extended the model to an infinite planning horizon [1]. The models, however, did not consider additional aspects such as risk planning and environmental impact minimization. Attempts were made by researchers to extend the model by considering variables such as transportation methods [1], risk management [2,9], and carbon emissions [10]. However, there is no integrated model considering these factors with EOQ.

Multi-objective considerations have seen widespread use in many research fields [11]. One such optimization solution is the Pareto optimal solution, which was applied by several researchers 94,11,12]. Wang et al. (2011) used Pareto optimal solutions to calculate a Pareto frontier distribution which was then presented as a set of optimized solutions [11]. The stated model used only two objective functions. Zhao et al. (2017) adapted a green multi-optimization supply chain model that minimizes hazardous accident risks, carbon emissions, and cost [12]. The model, however, does not consider transportation modes, nor does it consider all unit quantity discounts. This research contributes by developing a multi-objective, two-echelon carbon emissions-minimizing supply chain model for the chemical industry. The model takes into account issues of cost, risk, and environmental concerns by introducing them as decision variables. These modified variables also provide considerations for economic cost decisions.

This paper proposes a multi-objective optimization model which considers economic costs, risk management, and carbon emissions control. In this research, an EOQ model was applied to calculate the system's total cost, which directly impacts the cost aspect. Next, performance of the proposed model is tested via numerical analysis. The aspects of cost minimization, risk control, and carbon emissions reduction can be clarified through this analysis, allowing for an appropriate strategy. The purpose of this research is to analyze and study the two-echelon network while considering economic costs, risk management, and environmental impacts in the chemical industry. The goal is to arrive at an EOQ system which can reduce costs, risks, and carbon emissions, so the research scope matches these aspects accordingly. Lastly, truckload and less than truckload (TL and LTL) transport methods are considered as specific costs along with all-unit quantity discounts as an extension to the model

## 2. METHODS

The proposed two-echelon network has two members: the manufacturer and the distribution center. The research scheme description is portrayed below in Figure 1.

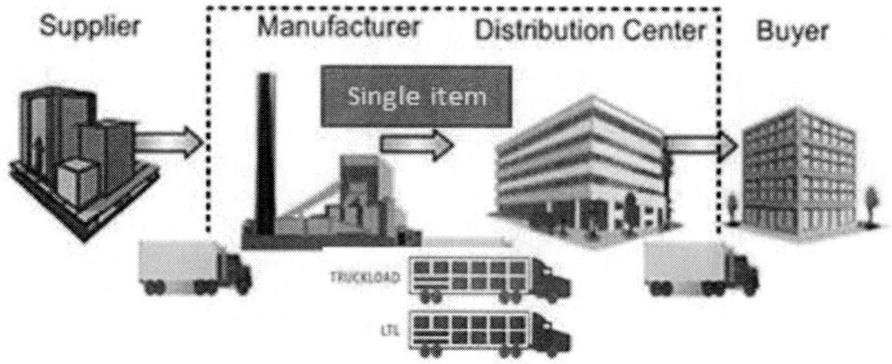


Figure 1. The proposed two-echelon network

*2.1 Notations and parameters*

The corresponding indices, decision variables, and criteria for this EOQ extension are shown below:

Indices

m: manufacturer

d: distribution center

k: the index set for the breakpoints {0,…, B}

Parameters

| | |
|---|---|
| K | individual setup cost, \$ |
| h | holding cost per time unit, \$/unit/month |
| d | demand rate, \$/unit/month |
| Q | order quantity, units |
| CT | fixed cost, \$/truck |
| B | number of price breakpoints |
| TQ | demand time, months/truck |
| QT | truck volume for a full load, units/truck |
| j | quantity of truckloads, units |
| s | LTL cost per unit delivered, \$/units |
| n+1 | number of breakpoints for changes in purchasing cost per unit |
| β | the i-th breakpoint (for the order interval T), month |
| ci | unit cost for item i, \$ |
| l | constant lead time, month |
| r | reorder point, r*=dl |
| demk | main manufacturer demand for raw materials, kg |
| demc | consumer demand c, kg |
| Qij | quantity of transported products |
| caf | capacity of facility, f∈ {m,d} |
| cat | capacity of transportation |
| Invm | risk control investment at risk level m |
| Invl | carbon emissions reduction investment at emissions level l |
| Rf | Probability f that a failure will occur in facility i |
| Rt | goods delivery transportation risk |
| Probm | risk probability at the level m |
| Probf | probability of risk in facility f |
| Probff | 'risk for transport from f to f' |
| Nfp | monetary consequence of a type-p risk failure in facility f, million dollarsNff 'p monetary consequence of an accident resulting from a type-p risk failure during transportation from f to |

f', p ∈ {1,2,3}, million dollars

Lij distance between facility f∈ {m,d} and f' ∈ {m,d}, km

Arij vehicle transport accident probability

CEf total carbon emissions in all network facilities, kgCO2/kg

CEt transportation carbon emissions, calculated from transport distances, kgCO2/kg

EMl emission factor at emission level l, kgCO2/kg

EMf facility f emission factor, kgCO2/kg

EMff' emission factor for transport from f to f', kgCO2/kg

xmd product quantity traveling from the manufacturer to distribution center d

Network decision variables are shown below:

xk 1, if the item is purchased
0, otherwise

xd 1, if distribution center d is chosen
0, otherwise

ym 1, if risk level m is chosen during risk control measures
0, otherwise

zl 1, if carbon emission level is chosen during emission reduction measures
0, otherwise

Assumptions for the network are:

- Distribution center demand is deterministic and known.
- The quantity delivered is known.
- Lead time is constant.
- Shortages are not allowed.
- Order quantities are made only when the inventory runs out.
- The truckload and less-than-truckload transport methods are always available.
- Each truckload delivers a quantity matching the time unit demand.
- Filling a truck to full capacity incurs a cost is lower than shipping via LTL, since equal costs would allow LTL transport benefits at no additional expense.
- The order transportation cost depends on the transportation method and the order quantity.
- Short-term risks are considered for the delivery item and in the facility.

Mathematical formulation

The model was made with three functional objectives: the minimization of risk, emission quantities, and total cost. Minimizing risk

Three risk types exist: fatalities, environmental harm and property damage, and are calculated as the product of the risk probability multiplied by the results of their occurrence. Environmental harm can be further expanded as water, air and soil pollution, and property damage can be classified as any accident that incurs losses on supply chain operations. The resulting function is a sum of these probabilities, shown below [12]:

$$\textit{Minimizing risk} \qquad R_i = \mathrm{Pr}ob_i \cdot \sum_{p=1}^{3} N_{ip} \tag{1}$$

The sum of transportation risks and their corresponding events is shown below [12]:

$$R_{ij} = \mathrm{Pr}ob_{ij} \cdot \sum_{p=1}^{3} N_{ip} \tag{2}$$

$$\mathrm{Pr}ob_{ij} = Ar_{ij} \cdot L_{ij} \cdot Q_{ij} \tag{3}$$

Minimizing these inherent risks can be expressed by the following function:

$$\text{Min } Z_1 = R_f + R_t \tag{4}$$

$$R_f = \ Prob_m \cdot y_m \cdot \sum_{p=1}^{3} N_{mp} + Prob_d \cdot \sum_{p=1}^{3} N_{dp} \tag{5}$$

$$R_t = \ x_{md} \cdot Ar_{md} \cdot L_{md} \cdot \sum_{p=1}^{3} N_{mp} \tag{6}$$

Minimizing carbon emissions

Supply chain carbon emissions for the facility and for transportation can be calculated as follows:

$$\text{Min } Z = CEf + CEt \tag{7}$$

$$CEf = \ EM_l \cdot z_l \cdot dem_k \tag{8}$$

$$CEt = \ EM_l \cdot z_l \cdot dem_k + x_{md} \cdot EM_{md} \cdot L_{md} \tag{9}$$

*2.2 Minimizing total economic cost*

The EOQ model is applied as a part of the total cost. This model considers both truckload and less-than-truckload transportation, in addition to an all-unit quantity discount consideration [1].

The transportation cost C(T) denotes a cost for the order interval T which exists during the period jTQ≤T<(j+1)TQ as shown below:

$$C(T) = \begin{cases} sd(T - jT_Q) + jC_T & \text{for } \ T \in I_j^1, \\ (j+1)C_T & \text{for } \ T \in I_j^2, \end{cases}$$

where $I_j^1 = \{T : jT_Q \leqslant T < (C_T/sd) + jT_Q\}$ and $I_j^2 = \{T : (C_T/sd) + jT_Q \leqslant T < (j+1)T_Q\}$. (10)

The purchasing cost depends on the order interval T:

$$F(T) = \begin{cases} c_o, & \text{if } 0 \leq T < \beta_1, \\ c_1, & \text{if } \beta_1 \leq T < \beta_2, \\ \dots \\ c_n, & \text{if } \beta_n \leq T < \infty. \end{cases} \tag{11}$$

It is then assumed that c0 > c1 > … > … > cn > 0

Adding the all unit quantity discount creates this extension to the EOQ average total system cost:

$$W(T) = \frac{K}{T} + \frac{hdT}{2} + \frac{C(T)}{T} + d \cdot F(T), \tag{12}$$

$$W(T) = \begin{cases} \frac{K - jsdT_Q + jC_T}{T} + \frac{hdT}{2} + sd + d \cdot F(T), & \text{for } T \in I_j^1, \\ \frac{K + (j+1)C_T}{T} + \frac{hdT}{2} + d \cdot F(T), & \text{for } T \in I_j^2. \end{cases} \tag{13}$$

The algorithms used to calculate the optimal order interval values (To and T*) are then applied [1].

The total cost consists of setup, holding, transportation, risk reduction investment costs, and carbon emission reduction investment costs, which are shown with the following expression:

$$\text{Min } Z = FC + HC + TC + RC + CE \tag{14}$$

where FC is the fixed setup cost based on the operation of the facilities. HC is the holding cost, TC, the transportation cost. RC contains all investments made towards risk reduction, and CE the investments made towards carbon emissions reduction.

IOP Conf. Series: Materials Science and Engineering **847** (2020) 012025 doi:10.1088/1757-899X/847/1/012025

## 3. RESULT AND DISCUSSION

Some constraints are considered in the model, including capacity limit constraints and the decision variables. These constraints are as follows:

### 3.1. *Capacity limit constraints*

$$x_{kd} \leq ca_k \quad (15)$$
$$x_{ff'} \leq ca_{ff'} \quad (16)$$

### 3.2. *Constraints of decision variables*

$$x_d \leq x_{dc} \quad (17)$$
$$x_d \in \{0,1\} \ \forall d \quad (18)$$
$$x_{ff'} \geq 0 \quad (19)$$
$$y_m, z_l \in \{0,1\} \text{ and } m \in \{1, \ldots, a\}, l \in \{1, \ldots, b\}, a,b \in Z \quad (20)$$

Solution Approach

The solution approach in this section is meant to determine a Pareto frontier, out of which a set of optimal solutions may be inferred. The decision maker is then able to choose the most beneficial tradeoff for their own purposes. Risk reduction is an additional constraint that must be considered while calculating minimum carbon emissions. The solution values to these goals were added to the total cost function in order to calculate the overall minimal cost. Lingo programming, or ILOG CPLEX, is used to solve the model.

Here, the model is analyzed using numerical analysis. Data was taken from previous studies [1, 12] with several modifications. The intent of these calculations is to find the optimal configurations which minimize all three objective functions.

Table 1. Data of The Parameters

| Parameter | Value | Unit |
|---|---|---|
| d | 7,000 | units/month |
| K | 6,000 | $ |
| h | 0.5 | $/unit/month |
| $C_T$ | 2,000 | $/truck |
| $Q_T$ | 5,000 | units/truck |
| s | 2.5 | $/unit |
| $T_Q$ | 0.57 | months/truck |
| $c_0$ | 7 | $/unit (< 10,000 units) |
| $c_1$ | 6 | $/unit (10,000 ≤ units < 30,000 units) |
| $c_2$ | 5.5 | $/unit (≥30,000 units) |
| $\beta_1$ | 1.43 | months |
| $\beta_2$ | 4.29 | months |
| $dem_k$ | 3652.3958 | kg/d |
| $N_{kp}$ | 1,439 | million $ |
| $N_{kdp}$ | 215.911 | million $ |
| $EM_m$ | 247.64 | kg $CO_2$/ton of product |
| $EM_{kd}$ | 0.14 | kg $CO_2$/ton of product |
| $Ar_{kd}$ | 2.4 x $10^{-4}$ | frequent accidents |
| | 1.61 x $10^{-5}$ | the expected number of accidents |

The optimal results for these three functions, in addition to the order quantity, are shown below in Table 2.

Table 2. Optimal Objective Result

| | |
|---|---|
| The optimal reorder interval, $T_0$ | 1.71 months |
| The optimal reorder interval, $T^*$ (TL and LTL transportation and all unit quantity discount) | 4.29 months |
| Total cost | $ 51,135.71 |
| Optimal order quantity | 30,000 units |
| Inherent Risk | $5,892,937.13 |
| Carbon emissions | 3540.01 kg $CO_2$ |
| Total economic cost | $168,410.80 |

## 4. CONCLUSION

In recent years, due to the rapid development of the logistics industry, the shipping efficiency of goods has significantly improved compared to the past, but the losses in the shipping process will lead to a decline in profit. This study was designed to develop an EOQ cost model that minimizes cost, accident risk , and carbon emissions. The problem was separated into three objective functions, which were then employed to find the optimal Pareto solution. The model's economic relevance results from its use of the EOQ formula. Lastly, numerical analysis was used to test the model's applicability. It is expected that this model can provide users with consideration of cost, risk and other factors as a reference basis for making the best decisions, thereby achieving the goal of improving overall profitability. The factors that affect the development and profitability of the chemical industry are not only the three items mentioned in this article. It is expected that subsequent researchers should consider changes in production and transportation modes and incorporate new considerations in a timely manner to make the model and the results obtained more realistic need.

## 5. REFERENCES

[1] Mendoza, A. and J.A. Ventura 2014 *EOQ Models with Two Modes of Freight Transportation and All-Units Quantity Discounts*, in *Handbook of EOQ Inventory Problems: Stochastic and Deterministic Models and Applications*, T.-M. Choi, Editor. Springer US: Boston, MA. pp. 139-153.

[2] Zhao, R., G. Neighbour, and P. Deutz 2011 Preliminary Study of Hazardous Materials Management: An Optimization Model in the Context of Green Supply Chain Management. in *2011 5th International Conference on Bioinformatics and Biomedical Engineering*.

[3] Teng, J.-T. 2009 A simple method to compute economic order quantities. *European Journal of Operational Research*. **198**(1), pp. 351-353.

[4] Mousavi, S.M., et al., 2016 A bi-objective inventory optimization model under inflation and discount using tuned Pareto-based algorithms: NSGA-II, NRGA, and MOPSO. *Applied Soft Computing*, **43**, pp. 57-72.

[5] Tersine, R.J., S. Barman, and R.A. Toelle 1995 Composite lot sizing with quantity and freight discounts. *Computers & Industrial Engineering*. **28**(1), pp. 107-122.

[6] Bai, Q.-G. and J.-T. Xu 2011 Optimal solutions for the economic lot-sizing problem with multiple suppliers and cost structures. *Journal of Applied Mathematics and Computing*, **37**(1), pp. 331-345.

[7] Taleizadeh, A.A. and D.W. Pentico 2014 An Economic Order Quantity model with partial backordering and all-units discount. *International Journal of Production Economics*. **155**, pp. 172-184.

[8] Mendoza, A. and J.A. Ventura 2008 Incorporating quantity discounts to the EOQ model with transportation costs. *International Journal of Production Economics*. **113**(2), pp. 754-765.

[9] Zhao, R., et al. 2012 Using game theory to describe strategy selection for environmental risk and carbon emissions reduction in the green supply chain. Journal of Loss Prevention in the Process Industries. **25**(6), pp. 927-936.

[10] Battini, D., A. Persona, and F. Sgarbossa 2014 A sustainable EOQ model: Theoretical formulation and applications. *International Journal of Production Economics*. **149**, pp. 145-153.

[11] Wang, F., X. Lai, and N. Shi 2011 A multi-objective optimization for green supply chain network design. *Decision Support Systems*. **51**(2), pp. 262-269.

[12] Zhao, R., et al. 2017 An optimization model for green supply chain management by using a big data analytic approach. *Journal of Cleaner Production*. **142(Part 2)**, pp. 1085-1097.

# Development of waiting time predictor based Artificial Neural Network

**V Limlawan**[1,2] **and P Anussornnitisarn**[1,3]

[1]Faculty of Engineering, Kasetsart University,Bangkok, Thailand
[2]varis.lw@gmail.com, [3]fengpta@ku.ac.th

**Abstract.** Queue is a part of our daily life. Everyday people have to wait in queue for services such as waiting for train, buying a cup of coffee. Unfortunately, it is difficult to eliminate waiting time in queue because it is very costly to do. Some business provides queue-length to each customer for reducing customer anxiety, but the given queue-length is often larger than the actual queue-length. Customers estimated the waiting time from the given queue-length and decide to leaves the queue, so the business may lose customer to their competitor. This paper aims to propose the design of the system that can provide the accurate estimated waiting time to each customer. The proposed system constantly updates queue-length and service rate. The updated queue-length and the updated service rate are used as the input of Artificial Neural Network (ANN) which is the waiting time predictor in the system. ANN with the proposed system is compared to the other predictors such as linear regression, the historical based predictor, Queue-Theory predictor. The result shows that ANN with the proposed system outperforms the others, and more than 95% of predicted waiting time by ANN is accurate within 5 minutes tolerances.
**Keywords**: queuing system, queuing theory, waiting time prediction, artificial neural network, delay estimation

## 1. Introduction

Queue is everywhere in daily activities and service such as bank, hospital, and canteen. People cannot avoid waiting in queue for services. Many studies show that long waiting in queue has a negative effect on service satisfaction although service quality is not quite bad [1, 2]. However, it is difficult to eliminate waiting time in queue because the costs and resources in service systems are limited.

To reduce dissatisfactions, many businesses provide entertainment stuffs for customers while waiting such as television, WiFi and coffee are provided for customers in a waiting area. Moreover, some business provides queue-length information to customers for reducing anxiety, but the given queue-length is often larger than the actual queue-length. Customers estimated waiting time from the queue-length information and decide to leave from the queue. Business loss revenue from leaving customers and may lose the customers to competitors. To solve the problem, providing accurate waiting time for each customer is another way to increase customer satisfaction. With accurate waiting time, customers can plan their activities while waiting for services.

The paper proposes a queue system that can provide the accurate waiting time to each customer. The remaining of this paper is organized as follows. Method of this study is described in section 2.

Experiment result and discussion are presented in section 3. Finally, this work is concluded in section 4.

## 2. Methods

### *2.1. Waiting time prediction*

The goal of our proposed system is to propose the accurate waiting time estimation for each customer, so waiting time predictors are reviewed and discussed as follows.

*2.1.1. Queue-theory based predictor.* Queue-Theory based Predictor (QT) is based on a G/M/n queue system [3]. Predicting waiting time by QT is not complex and use only queue-length, service rate, and number of server in the prediction. Let $\hat{W}_{QT}$ be the waiting time predicted by QT, $q(t)$ be queue-length at time *t*, $u$ be service rate trained by the historical data, $s(t)$ be number of service at time *t*. $\hat{W}_{QT}$ is

$$\hat{W}_{QT} = \frac{q(t)+1}{s(t)\mathrm{u}} \quad (1)$$

Although QT is easy to implement, QT may not accurate in queue systems that has customer abandonment or reneging. The system does not update queue-length when customer leaves from the queue before receiving services. Queue-length used in QT is often larger than the actual queue-length.

*2.1.2. Historical based predictor.* Last customer to enter service (LES) is the historical based predictor which use the waiting time of the customer who recently enter service [4, 5]. LES is not based on queue-length, so incorrect queue-length from customer abandonment does not affect LES. However, queue environment such as arrival rate, service rate may affect the performance of LES. The queue environment of the last customer who enters service might not be the same as the environment of the new customer. For example, the last customer waits for 10 customers in the queue before receiving service, but the new customer has only 2 people in front of him in the queue. Common sense we know that the waiting time of new customer shorter than the last customer.

*2.1.3. Linear regression.* Linear Regression (REG) is applied in many applications [6]. In waiting time prediction, variants of linear regression such as regression spline, quartile regression were applied to predict waiting time in a call centre and an emergency room [7, 8]. Queue-length and number of server are used as the main independent variables in those predictors.

In this paper, the standard linear regression is used in the experiment. The independent variables are queue-length divided by number of server, and the response is waiting time. The linear regression for waiting time prediction is

$$\hat{W}_{REG} = \beta_0 + \beta_1 \frac{Q(t)}{s(t)} \quad (2)$$

where $\hat{W}_{REG}$ is waiting time predicted by the linear regression, $\beta_0$ ,$\beta_1$ are unknown regression coefficients, $q(t)$ is Queue length at time t and $s(t)$ is the number of server at time *t*.

Naturally, linear regression performs well in the problem that response and independent variables have linear relationship. LREG may not accurate if queue-length and number of server do not have linear relationship with waiting time. Moreover, incorrect queue-length may affect the performance of LREG as QT.

*2.1.4. Artificial neural network.* Artificial Neural Network (ANN) is inspired by nervous system of human brain. ANN is a powerful tool that can solve multi-dimensional inputs and outputs problems [9,

10]. ANN has been applied in many applications such as stock prediction, wind forecasting [11]. In waiting time prediction, ANN has been applied to predict waiting time in call-center and bank [7, 12].

In those works, the architecture of ANN is a single hidden layer feedforward, and the main inputs are queue-length and number of server. The results of ANN are slightly better than QT and REG in the queue system with no abandonment, but the results of ANN is somewhat in the queue system with abandonment.

Based on those works, there are no works which consider incorrect queue-length problem from customer abandonment. Incorrect queue-length may affect the performance of waiting time predictor which has queue-length in the predictive models such as QT, REG, ANN.

*2.2. The proposed queue system*

Nowadays, ticket queue system is commonly installed to managing a queue system. Customers arrive and get a ticket which contains queue information such as queue-length, estimated waiting time. After customer got the ticket, customers wait for calling at waiting area.

The ticket queue system make customers feel more relax than the physical queue. Instead of standing in queue, customer can do anything during waiting. However, the disadvantage of the ticket queue system is inaccurate queue information [13]. The ticket queue system does not update queue-length when customer left the system before receiving service, so queue-length and estimated waiting time is often larger than the actual waiting time.

To solve the problem, the proposed system is designed to provide the accurate estimated waiting time to each customer. The design of the system is illustrated in figure 1. Unlike the ticket queue, the proposed system constantly updated queue-length. All cases of arrival and departure are counted by the system. Moreover, service time is also monitored and used to adjust service rate which is the main variable in several waiting time predictors. Every time the system updated queue-length and service rate, the system will send them to waiting time predictor. The predictor will use them to estimate waiting time for a new customer.

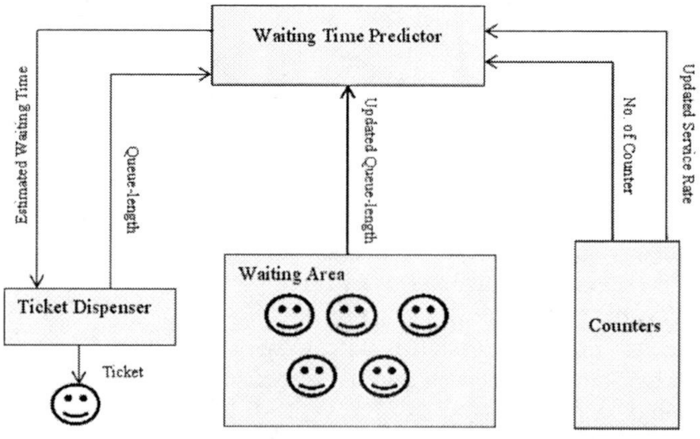


**Figure 1.** Diagram of the proposed queue system.

To monitor queue-length and collect service time, we roughly plan to use the video surveillance and computer vision technologies. Recently, video surveillance has been commonly found in many places, and its cost is inexpensive for every organization. With computer vision technologies, video surveillance can be used for many purposes such as car detection, people counting. The detail of using the technology is beyond the scope of this paper, so we will not discuss in detail.

Service time of each customer is sent to Exponential Weight Average (EWMA) control chart for adjusted service rate in waiting time predictor. EWMA control chart is a tool for detecting mean shift [14]. Let $u_0$ be an initial service rate from the historical data, $u$ be the current service rate, $\sigma$ be the standard deviation of service rate from the historical data, $z_i$ be service rate of customer $i$, $\lambda$ and $L$ be EWMA parameters. The step of EWMA for controlling service rate is as follows

Step 1 Initialize parameters $u = u_0, z_0 = u_0$, $\lambda$ and $L$

Step 2 Calculate $z_i$ from EWMA statistics as follows.

$$z_i = \lambda x_i + (1-\lambda) z_{i-1} \tag{3}$$

Step 3: Calculate lower bound (LCL) and upper bound (UCL) as follows.

$$UCL = u_0 + L\sigma\sqrt{\frac{\lambda}{2-\lambda}\left(1-(1-\lambda)^{2i}\right)} \tag{4}$$

$$LCL = u_0 - L\sigma\sqrt{\frac{\lambda}{2-\lambda}\left(1-(1-\lambda)^{2i}\right)} \tag{5}$$

Step 4: If $z_i > UCL$ or $z_i < LCL$, then the service rate is updated by $u = z_i$

Step 5: Repeat step 2

Once the service rate and queue-length is updated, they will be sent to waiting time predictor. Based on the advantage of ANN, the main predictor of our system is ANN. The architecture of ANN in our system is the single hidden layer feedforward which has three layers as shown in figure 2. The inputs of ANN are queue-length, service rate and number of server. Based on [15], the single hidden layer feedforward is expressed as follow.

$$p = \begin{bmatrix} u(t) \\ Q(t) \\ S(t) \end{bmatrix} \quad \text{, for } t = 1, 2, 3, \ldots \tag{6}$$

$$a^0 = \mathbf{p} \tag{7}$$

$$a^{i+1} = f^{i+1}\left(\mathbf{w}^{i+1} a^i + \mathbf{b}^{i+1}\right), \text{ for } i = 1, 2, 3, \ldots \tag{8}$$

$$\mathbf{a}^2 = \left[\hat{W}_{ANN}\right] \tag{9}$$

where $u(t)$ is the updated service rate at time $t$, $Q(t)$ is the updated queue-length at time $t$ and $S(t)$ is the number of server at time $t$, $t$ is discrete time, $\mathbf{p}$ is the inputs from the external source, $a^i$ is the input vector of layer $i$, $\mathbf{b}^i$ is the biased vector of layer $i$, $\mathbf{w}^i$ is the weights vector of layer $i$, $f^i$ is the activate function of layer $i$.

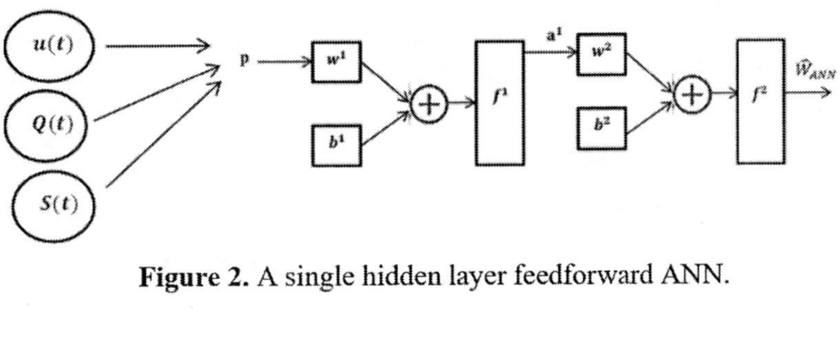


**Figure 2.** A single hidden layer feedforward ANN.

### *2.3. Numerical experiment*

To evaluate the proposed queue system, ANN with the proposed system is compared to QT without the system, LES, REG with the proposed system. The test system is described as follow.

*2.3.1. Test system.* The proposed system is evaluated by the dynamic queue system which was conducted by SIMAN ARENA® Rockwell software. The detail of the tested system is as follows.

- The system opens at 8.00 a.m. and closes at 16.00 p.m.
- The service time ($u$) is normal distribution with a mean of 1 and a standard deviation of 0.75
- The load of the system ($p$) is dynamically changed in a replication.
- The arrival rate is Poisson distribution with a mean of $\lambda$ which is adjusted based on the load $\rho$ and service rate $\mu$.
- The reneging time of customers is 2-Erlang distribution with a mean of 15.
- The summary of test system is shown in table 1.

In the numerical experiment, there are 3 cases for testing ANN with the proposed system. The cases are divided by the shift of service rate. There is no mean service time shift in case 1, and there is a negative shift and positive shift on the case 2 and the case 3 respectively. The shift size is +0.5 and -0.25 in case 2 and case 3. For all cases, the shift in service time begins at 11.00 a.m.

**Table 1.** Summary of test systems.

| Time (hr) | Arrival Rate | Average Service Time (u) | Load ($p$) | Renege Rate |
|---|---|---|---|---|
| 8.00-9.00 | POISSION(λ) λ depends on load (p) and avg. service time (μ) | Norm(1,0.75) | 0.85 | 2-Erlang distribution with a mean of 15. |
| 9.00-10.00 | | Norm(1,0.75) | 0.85 | |
| 10.00-11.00 | | Norm(1,0.75) | 0.95 | |
| 11.00-12.00 | | Norm (1±Shift size,0.75) | 0.99 | |
| 12.00-13.00 | | Norm (1±Shift size,0.75) | 0.99 | |
| 13.00-14.00 | | Norm (1±Shift size,0.75) | 0.95 | |
| 14.00-15.00 | | Norm (1±Shift size,0.75) | 0.85 | |
| 15.00-16.00 | | Norm (1±Shift size,0.75) | 0.85 | |

*2.3.2. Parameter setting.* For training ANN and REG, we simulated 30 replications of the mixed case which combined all cases together, and we simulated 10 replications for each test case. For ANN, the architecture of ANN is a single hidden layer feedforward. The number of neuron in the hidden layer is 5. The activated function of the hidden layer is tan-sigmoid function, and the activated function of the output layer is pure linear function. ANN is trained by backpropagation algorithm (traingd) in MATLAB R2016a. The learning rate of the backpropagation algorithm is 0.05. For other predictors, QT, REG and LES are coded by Excel VBA. Moreover, λ and L of EWMA control chart are set at 0.1 and 2.7 respectively.

*2.3.3. Performance measurement.* To compare the performance of ANN with the proposed system with the other predictors, Mean Square Error (MSE) is used as the performance measurement in the experiment. The error and MSE are as follows.

$$e_i = W_i - \hat{W}_i \tag{10}$$

IOP Conf. Series: Materials Science and Engineering **847** (2020) 012026 doi:10.1088/1757-899X/847/1/012026

$$MSE = \frac{1}{n}\sum_{i=1}^{n} e_i^2 \tag{11}$$

where $e_i$ is the prediction error, $W_i$ is the actual waiting time of customer $i$, $\hat{W}_i$ is the predicted waiting time of customer $i$ and $n$ is the number of the test sample.

Moreover, percentage of accuracy within an acceptable tolerance PA is used to measure the performance of the best predictor. Let $l$ be the acceptance tolerance, $n$ be number of test samples, $e_i$ be an error of $i$ sample, $c_i$ is the accuracy indicator of $i$ sample. The percentage of accuracy is defined as follow.

$$PA \pm l = \frac{1}{n}\sum_{i=1}^{n} c_i \times 100 \tag{12}$$

where, $c_i = \begin{cases} 1 & , e_i \in [-l, l] \\ 0 & , e_i \notin [-l, l] \end{cases}$

## 3. Result and Discussion

The comparison of MSE for each predictor is summarized in table 2. QT is the best predictor in case 1. MSE of ANN with the system is less than QT in case 2 and case 3, by 5.7% and 25.6% respectively. MSE of LES is approximate 4.3 for all cases. REG is the worse predictor in all cases. The comparison of predicted waiting time of QT and ANN in case 3 is illustrated in figure 3. As shown in the comparison, ANN with the system is slightly better than QT in case 3.

**Table 2.** MSE of each predictor in the test system.

| | QT | LES | | REG (with system) | | ANN (with system) | |
|---|---|---|---|---|---|---|---|
| Case | MSE | MSE | %Diff* | MSE | %Diff* | MSE | %Diff* |
| Case 1: No shift | 2.10 | 4.17 | -98.9 | 6.52 | -210.7 | 2.23 | -6.30 |
| Case 2: Positive shift | 2.11 | 4.50 | -113.4 | 6.31 | -199.2 | 1.99 | 5.70 |
| Case 3: Negative shift | 6.48 | 4.47 | 31.1 | 6.28 | 3.1 | 4.82 | 25.6 |

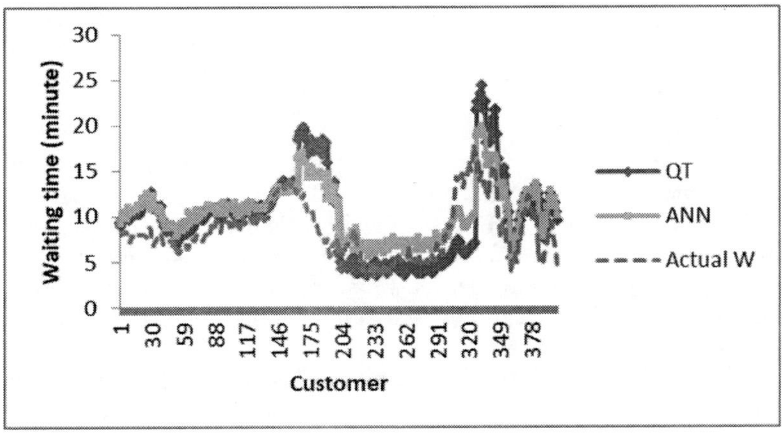


**Figure 3.** Comparison of actual waiting time and predicted waiting time of QT and ANN plot in Case 3.

Table 3 summarized the accuracy percentage of ANN with the system which is the best predictor in MSE comparison. The result shows that more than 95% of predicted waiting times are accurate within 5 minutes, and more than 99% of predicted waiting times are accurate within 10 minutes.

Based on the result, ANN with the system outperforms other predictors in the case that has the shift in service time, but QT is the best predictor in the system that does not have the shift in the service rate. On the other hand, REG with the system is the worse predictor in all cases because the training set is not specific for any cases, and REG has no ability to adapt the model when the service time was shifted.

**Table 3.** Percentage of accuracy within the acceptance tolerance of ANN with the system.

| Case | Accuracy (%) within ±1 min | Accuracy (%) within ±3 min | Accuracy (%) within ±5 min | Accuracy (%) within ±10 min |
|---|---|---|---|---|
| Case 1: No shift | 62.00% | 94.90% | 98.80% | 99.90% |
| Case 2: Positive shift | 61.10% | 95.00% | 99.50% | 100.00% |
| Case 3: Negative shift | 41.30% | 83.30% | 96.70% | 99.90% |

## 4. Conclusion

In this paper, we proposed the design of the queue system that can improve the accuracy of waiting time prediction for each customer. The proposed system solves the losing queue information in the traditional ticket queue system by constantly updating queue-length and service rate. Due to the availability of camera devices and computer vision technology, we roughly plan to use cameras as a sensor to monitor the queue-length and other queue information, so all cases of customer abandonment are counted and update in the system. For service rate, we use Exponential Weight Average (EWMA) control chart to monitor and adjust service rate of the system. Both service rate and queue-length updated from the system are sent to waiting time predictor which is Artificial Neural Network (ANN).

ANN with the proposed system is compared to Queue Theory predictor without the system, the historical based predictor, and linear regression predictor with the proposed system. The result shows that ANN with the system outperforms other predictors in the case that has customer abandonment and a shift in service rate. Moreover, more than 95% of predicted waiting time by ANN is accurate within 5 minutes.

The proposed system solved the losing queue information and can provide the accurate waiting time estimation for each customer. With the accurate waiting time, customers can manage their time and activities while they wait for a service. For future work, more cases on service time and more cases customer abandonment will be studied for investigating the effect of them.

## 5. Reference

[1] Tarloy S 1994 Waiting for Service: The Relationship between Delays and Evaluations of Service. *Journal of Marketing*. **52,** 56-69.

[2] Pruyn, A T H and Smidts A 1998 Effects of waiting on the satisfaction with the service: Beyond objective time measures. *International Journal of Research in Marketing*. **15(4)**, 321-334.

[3] Whitt W 1999 Predicting queueing delays. *Management Science*. **45,** 880-888.

[4] Ibrahim R and Whitt W 2008 Real-time delay estimation in call centres. *Proceedings of the 2008 Winter Simulation Conf* pp. 2876-2883.

[5] Ibrahim R and Whitt W 2009 Real-Time Delay Estimation Based on Delay History *Manufacturing & Service Operations Management Manufacturing & Service Operations Management*. **11(3),** 397-415.

[6] Montgomery D C and Runger G C 2010 *Applied statistics and probability for engineers*. 5th ed (Hoboken, NJ, Wiley).

IOP Conf. Series: Materials Science and Engineering **847** (2020) 012026 doi:10.1088/1757-899X/847/1/012026


[7] Thiongane M, Chan W and Ecuyer P L 2015 Waiting time predictors for multi-skill call centres. *Proceedings of the 2015 Winter Simulation Conference* pp. 3073-3084.

[8] Sun Y, Teow K L, Heng, B H, Oni C K and Tay S Y 2012 Real-Time Prediction of Waiting Time in the Emergency Department, Using Quantile Regression. *Annals of Emergency Medicine.* **60(3),** 299 – 308.

[9] Abiodun O I, Jantan A, Omolara A E, Dada K V, Mohamed N A and Arshad H 2018 State-of-the-art in artificial neural network applications: A survey. *Heliyon* **4(1).**

[10] Moghaddam A H, Moghaddam M H and Esfandyari M 2016 Stock market index prediction using artificial neural network. *Journal of Economics, Finance and Administrative Science* **21,** 89-93.

[11] Ahmed A and Khalid M. 2017 Multi-step Ahead Wind Forecasting Using Nonlinear Autoregressive Neural Networks. *Energy Procedia* **134,** 192-204.

[12] Satya Hermanto R P, Suharjito D and Nugroho A 2018 Waiting-Time Estimation in Bank Customer Queues using RPROP Neural Networks. *Procedia Computer Science.* **135,** 35-42.

[13] Xu S H, Gao L and Ou J 2007 Service Performance Analysis and Improvement for a Ticket Queue with Balking Customers. *Management Science.* **53(6),** 971-990.

[14] Montgomery D C 2012 *Introduction to Statistical Quality Control* 7th ed (John Wiley & Sons, Inc).

[15] Hegan M T, Demuth H B, Beale M H and Jess O D 2014 *Neural Network Design* 2nd eds. [online].

# Using the Bees Algorithm to solve combinatorial optimisation problems for TSPLIB

**A H Ismail**[1,2], **N Hartono**[1,3], **S Zeybek**[1,4], **D T Pham**[1]

[1]Department of Mechanical Engineering, University of Birmingham, Edgbaston, Birmingham B15 2TT, UK
[2]ahi729@student.bham.ac.uk, [3]nxh886@student.bham.ac.uk,
[4]sxz879@student.bham.ac.uk

**Abstract.** The Bees Algorithm (BA) is a metaheuristic algorithm to find good solutions to optimisation problems in reasonable computing times. This paper is the first to report on the use of the BA to solve 9 combinatorial optimisation problems (COP) with more than 100 cities from TSPLIB rigorously to test the performance of the algorithm. The work employed a basic version of the BA for COP and TSPLIB datasets involving between 100 and 200 cities. The results obtained show that deviations from the best-found tour lengths for the datasets with 100 cities and 200 cities were approximately 2.5% and 7.5%. The reason for this jump in deviations was that the number of iterations was kept constant for all experiments while the solution space increased factorially with the number of cities. This research can be replicated and modified through Google Colab.
**Keywords**: bees algorithm, combinatorial optimisation problem, metaheuristics algorithm, travelling salesman problem.

## 1. Introduction

One of the most intensive fields studied in mathematics, engineering, and computer science is the optimisation which involves finding the best solution or a feasible solution from candidates in the solution space. An important class of practical optimisation problems is that of combinatorial problems. Many COPs are NP-hard and thus challenging to solve because the number of combinations exponentially increases with the size of the problem [1]. The need to find acceptable near-optimal solutions in reasonable computing times has motivated researchers to develop heuristic and meta-heuristic algorithms. According to Ouaraab, an advantage of metaheuristic algorithms is that they are simple and flexible [1]. Nature-inspired metaheuristic algorithms including those based on the behaviour of social insects such as honey bees have attracted the attention of many researchers [2].

As stated by Lones, there are 15 nature-inspired metaheuristic algorithms that have each received more than 1,000 citations [3]. One of those algorithms is the Bees Algorithm (BA) proposed by Pham et al. in 2005 [4], [5]. The BA is inspired by the natural honey bees foraging behaviour and can be used to solve continuous and combinatorial optimisation problems. A comparative study of the BA, Particle Swam Optimisation (PSO), and Evolutionary Algorithm (EA) show that the BA outperforms PSO and EA [6]. The BA also outperforms the deterministic simplex method, genetic algorithm, ant colony system, and stochastic annealing optimisation procedure [5]. The BA has been very popular since its development and it has many applications in multi-discipline area from work scheduling, image processing, pattern recognition, manufacturing, arc routing, PCB manufacturing, matrices structure, data partitioning, data transmission in a network, power distribution networks, and robot navigation [7],[8].

The BA has already found more than 400 different applications covering a variety of continuous or combinatorial problems, such as pattern recognition, scheduling, engineering design, neural network training, manufacturing cell formation, control system tuning, dynamic control, data clustering, mechanical design and supply chain design [7], [8], [9], [10]. There have been at least 56 endeavours to develop new versions of the BA. However, only 4 out of those 56 attempts have focused on combinatorial optimisation although COP formed approximately 33% of the total number of applications of the algorithm and the problem of combinatorial optimisation arises in many fields of engineering and other scientific disciplines that use computational approaches, such as artificial intelligence, operations research, electronic commerce, and bioinformatics. The authors believe that the reason for the lack of activity on combinatorial versions of the BA is that there are no available results on the performance of the BA on standard data-sets (test functions) that can be used for comparison in developing the algorithm.

Although there are many publications addressing real-world applications in the combinatorial domain such as machine scheduling [11],[12], cellular manufacturing [13], Printed Circuit Board (PCB) manufacturing [14],[15],[16], vehicle routing [17], it is important to use standard data sets when assessing the performance of an algorithm. The results for one or more standard data sets could be employed to compare different algorithms.

The basic combinatorial version developed by Koc [18] was adopted in this work as the basic version of the BA for combinatorial optimisation ("the Combinatorial BA"). This study extends Koc's previous work that used simple problems up to 51 cities. The performance of many algorithms is satisfactory for simple problems. However, the problem becomes challenging when the number of cities increases. The Google Colab link for this work is provided in the appendix to give a clear starting point to develop the Combinatorial BA.

One of the most common problems in combinatorial optimisation is the travelling salesman problem (TSP). The problem has many variants such as PCB manufacturing, machine scheduling, cellular manufacturing, arc routing, vehicle routing, job shop scheduling, data routing, machine sequencing, computer wiring, DNA sequencing, etc. TSP is a NP-hard optimisation problem because the complexity of the computation increases exponentially with the number of cities [1]. The TSP was initially introduced in 1930 and is one of the most studied COP as it can be used as a benchmark for evaluating the performance of many combinatorial optimisation algorithms [19].

The most widely used dataset library for COP is TSPLIB, which was published by Reinelt in 1991 [20]. The library itself has been cited more than 2,400 times, according to Google Scholar (October 2019). It is a very popular library and has been used mostly by researchers to evaluate optimisation algorithms such as the GA, PSO and ACO. However, no work has been done to measure the performance of the BA on TSPLIB problems involving more than 100 cities. This paper will present the results of testing the basic version of the BA on TSPLIB datasets with 100 to 200 cities. A reason for choosing those datasets was that most metaheuristic algorithms, including the BA, give good results for datasets with fewer than 100 cities.

The basic combinatorial version of the BA adopted in this research employed a permutation random number generator as the global search operator. The basic local search operator was a combination of swap, insertion, and reversion. This local search operator was chosen as the algorithm performed poorly when swap, insertion or reversion alone was used.

### *1.1. Bees Algorithm*

Pham et al. introduced the BA as a nature-inspired metaheuristic optimisation method inspired by the food foraging behaviour of the swarm honey bees [4], [5]. The BA classified as one of the swarm intelligence algorithms. Figure 1 illustrates the flow chart of The BA.

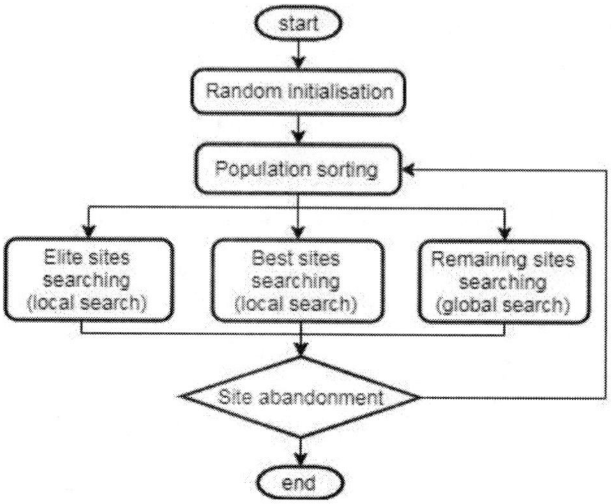


**Figure 1.** Flowchart of Bees Algorithm (basic version)[5]

The BA for the continuous domain has become more advanced and better structured compared to the version for the combinatorial domain. Work on the algorithm can be classified into four major sections: parameter tuning and setting, initialisation, local search (exploitation), and global search (exploration). Most of the developments primarily concentrate on the local search procedure of the BA [7]. All of the previous work on the BA in the continuous domain had different characteristics from the combinatorial version, such as the solution space is real rather than discrete. Therefore, the BA needs to be modified to perform exploitation and exploration in the combinatorial domain without changing the principal procedure of the algorithm.

*1.2. Bees Algorithm in Combinatorial Optimisation Problem*

The first COP solved by BA was reported in 2007 for finding the solution of the machine scheduling [11]. In the same year, BA also had solved PCB assembly sequence as an early TSP based COP [14]. After those publication in 2007, other important works began to emerge in the combinatorial BA algorithm such as timetabling [21], [22], [23], [24], PCB assembly planning [15], [16], circuit designing [25], scheduling [12], [26], [27], [28], [29], cellular manufacturing [13], vehicle routing [17], disassembly sequence planning [30], [31], and path planning [32], [33].

Combinatorial and continuous problems have completely different search principles. The task of searching as the main feature in the combinatorial problem is distinctive from the continuous problem where there is no real distance value [18]. The combinatorial domain has main characteristics [34]: the search space is discrete; the constraints are finite; the solution has ordered sequence and a cost function which related to the combination. Since the BA was originally developed to solve continuous domains, it is important to change the searching section in global and local search with a discrete search operator.

In essence, the BA for COP is almost identical to the algorithm for continuous one. The main difference between continuous and combinatorial version is lying on their (global and local search) operators. The combinatorial version, of course, has to use a discrete random generator replacing the real number generator on the continuous version.

*1.3. Travelling Salesman Problem*

TSP can be classified as symmetrical or asymmetrical TSP. In symmetrical TSPs, the distances between the two cities do not rely on the trajectory direction. For example, if the distance between two cities i and j when travelling from i to j is denoted by *di,j* and if $d_{i,j} = d_{j,i}$ then the TSP is symmetrical and vice versa.

TSP's graph represented the Graph $G = (V,E)$, where V is a set of vertices representing the cities, and all the connecting lines between the cities is *E*. Every edge indicates a possible route between two connected vertices or cities. The variable $d_{i,j}$ is associated with an edge *(i,j)* and represents the Euclidean distance from vertex $(x_i,y_i)$ to $(x_j,y_j)$ as Equation (1). Before executing the BA, these distances of all edges have calculated and store as a distance matrix.

$$d_{i,j} = ((x_i - x_j)^2 - (y_i - y_j)^2)^{0.5} \quad (1)$$

The TSP objective is to find the minimal total tour length of the final closed Hamilton cycle (only visiting once of all cities) tour as defined in Equation (2).

$$tour\ length = \sum_{i=1}^{n} d_{i,i+1} + d_{n,1} \quad (2)$$

Koc, Otri and Zeybek were three of the researchers developing the BA for solving TSPs taken from the TSPLIB dataset up to 100 cities [18],[34],[35].

## 2. Methods

The first step before running the BA is setting up the parameters. There are 5 parameters that have to be initially set up as number of scout bees *(n)*, elite bees *(nep)*, best bees *(nsp)*, elite sites *(e)*, and best sites *(m)*. All the parameters dependently set to the n. The parameter *nsp* and m are 50% of *n*. The e is 40% of *m*. The *nep* has a double number of *nsp*. The n set equal to 40 bees (see Table 1).

After setting up the parameters, random initialisation (see Figure 1) will generate *n* initial solutions. The best e out of *n* initial solutions will be the elite sites which best seeds to be improved. The *m* sites excluding the e sites are sufficiently good seeds as sites to be exploited as well.

**Table 1.** Parameter value of the running experiments

| Parameter | Value |
|---|---|
| Number of runs | 10 |
| Number of iterations | 3,000 |
| Number of scout bees *(n)* | 40 |
| Number of elite bees *(nep)* | 40 |
| Number of best bees *(nsp)* | 20 |
| Number of elite sites *(e)* | 16 |
| Number of best sites *(m)* | 20 |

The elite and best sites are the places needed to be exploited by worker bees (*nep* and *nsp*) using neighbourhood or local search (see Figure 2 to 5). The remaining $(n - e - m)$ bees as insufficiently seeds will explore the other solution spaces using the global search operator. All worker bees will continuously renew the patch a better position for every iteration and abandoned the worn-out one. The position that a worker bees had is representing a solution (complete tour). The best solution is generated by sorting all the best positions from all sites. The code is written in Python and could be retrieved in Google Colab.

```
Input: Cities: List,selection for Swap, Insertion and Reversion: integer
Output: Changed List of cities indexes
procedure LOCAL SEARCH(List, LS)
  begin
    N := sizeOfList(List)
    i_th := generateRandomInteger[0, N − 1]
    j_th := generateRandomInteger[0, N − 1]
    if LS == 0 then
      LS ← generateRandomInteger[0, 3]
    if LS == 1 then
      List ← SWAP(List, i, j)
    if LS == 2 then
      List ← INSERTION(List, i, j)
    if LS == 3 then
      List ← REVERSION(List, i, j)
    return List
  end
```

**Figure 2.** Pseudo-code of the basic local search operator of the combinatorial BA

```
Input: Cities indexes: List, i: integer, j: integer
Output: Changed List of cities indexes
procedure SWAP(List, i, j)
  begin
    Temp = List[i]
    List[i] ← List[j]
    List[j] ← Temp
    return List
  end
```

**Figure 3.** Pseudo-code of swap operator

```
Input: Cities indexes:List, i: integer, j: integer
Output: Changed List of cities indexes
procedure INSERTION(List, i, j)
  begin
    while i < j do
      Temp = List[i + 1]
      List[i + 1] ← List[j]
      i ← i + 1
      j ← j − 1
    return List
  end
```

**Figure 4.** Pseudo-code of insertion operator

```
Input: Cities indexes:List, i: integer, j: integer
Output: Changed List of cities indexes
procedure REVERSION(List, i, j)
  begin
    while i < j do
      Temp = List[i]
      List[i] ← List[j]
      List[j] = Temp
      i ← i + 1
      j ← j − 1
    return List
  end
```

**Figure 5.** Pseudo-code of reversion operator

## 3. Result and Discussion

Nine benchmark problems from TSPLIB [20] were used in this work to evaluate the performance of the combinatorial BA with the basic local search operator. The experimental results are summarised in Table 2. The first column of Table 2 gives the TSPLIB dataset names and the tour length for the best-known solution (*BKS*). The second column lists the tour length for the best solution of 10 runs, each of 3,000 iterations (*Best*). The third column gives the average tour length of the solutions found by the ten runs (*Avg*). The fourth column gives the deviation of the Avg from BKS (*Davg*). The last column lists the deviation of the Best from BKS (*Dbest*). *Davg* and *Dbest* as the accuracy indicators are defined in Equation (3) and (4).

$$Davg = (Avg - BKS).(BKS)^{-1}.100\% \quad (3)$$

$$Dbest = (Best - BKS).(BKS)^{-1}.100\% \quad (4)$$

**Table 2.** Results for ten runs each of 3000 iterations for 100-200 cities data-sets from TSPLIB using the combinatorial BA

| Dataset names (BKS) | Best | Avg | Davg | Dbest |
|---|---|---|---|---|
| KroA100 (21,282) | 21,469 | 21,712.4 | 2.02% | 0.88% |
| KroB100 (22,141) | 22,618 | 22,879.0 | 3.33% | 2.15% |
| KroC100 (20,749) | 20,969 | 21,339.8 | 2.85% | 1.06% |
| KroD100 (21,389) | 21,392 | 21,896.7 | 2.37% | 0.01% |
| KroE100 (22,068) | 22,266 | 22,496.1 | 1.94% | 0.90% |
| KroA150 (26,524) | 27,527 | 27,871.2 | 5.08% | 3.78% |
| KroB150 (26,130) | 27,095 | 27,458.5 | 5.08% | 3.69% |
| KroA200 (29,368) | 31,550 | 32,214.0 | 9.69% | 7.43% |
| KroB200 (29,437) | 32,024 | 32,519.7 | 10.47% | 8.79% |

The overall result shows that the *Davg* starts from 1.94% up to 10.47% and for the *Dbest* from 0.01% up to 8.79%. The results of 100 cities' datasets show that the *Davg* have range between 1.94% and 3.33% while the *Dbest* range between 0.01% and 2.15%. For the 150-200 cities results, the highest *Davg* is 10.47% and the lowest is 5.08%, while the *Dbest* has a range from 3.69% to 8.79%. The performance for 150-200 cities' shows that the deviation is around three times larger than performance of 100 cities' results due to the solution spaces were factorially increasing while the number of iteration remains the same. The authors believe that the accuracy performance of the combinatorial BA will increase if the number of iteration of the algorithm were adjusted according to the problem complexity (number of cities). The poor convergence has been found as other problem in this work. ACO's previous work [36] in 1999 had proved that ACO convergence performance would increase when ACO utilising a constructive heuristic (nearest neighbour) on the candidate solution generator. Later, It provokes the other researchers to be more intense to develop the constructive heuristic [37],[35],[30] as initialisation or global search operator for the algorithm development.

Overall, the basic combinatorial BA has the ability to provide good solutions with average error around 2.5% for 100 cities datasets and 7.5% for 150-200 cities datasets. Figure 6, Figure 7,and Figure 8 show the best-found tour length results for every TSPLIB's datasets in this research.

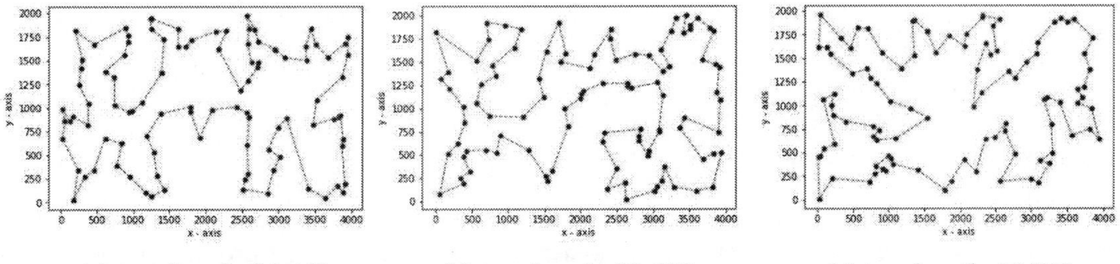


**(a)** tour length: 21,469 **(b)** tour length: 22,618 **(c)** tour length: 20,969

**Figure 6.** Best solution of (a) KroA100, (b) KroB100, (c) KroC100

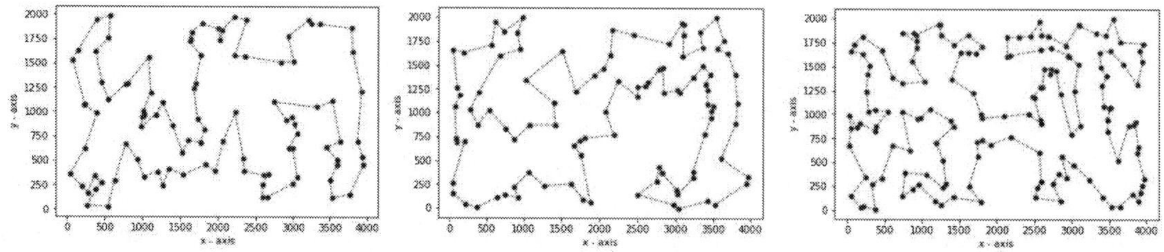

**(a)** tour length: 21,392 **(b)** tour length: 22,266 **(c)** tour length: 27,527

**Figure 7.** Best solution of (a) KroD100, (b) KroE100, (c) KroA150

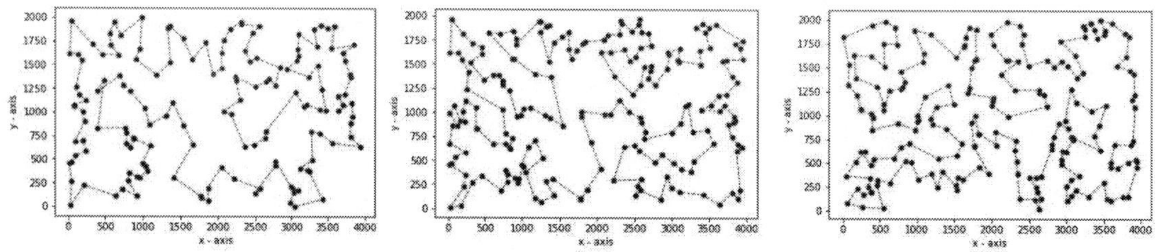

**(a)** tour length: 27,095 **(b)** tour length: 31,550 **(c)** tour length: 32,024

**Figure 8.** Best solution of (a) KroB150, (b) KroA200, (c) KroB200

## 4. Conclusion

This paper has shown that the BA can find near optimal solutions to the COP. The algorithm uses the permutation random number generator as global search operator and a combination of swap, insertion and reversion as the basic local search operator. Experiments using TSPLIB datasets with 100 to 200 cities show that the deviation increases around three times from the 100 cities dataset to the 200 cities dataset. The reason for this surge was that the number of iterations was kept constant for all experiments while the solution space increased factorially with the number of cities. This poor convergence for larger problems is a weakness common to many swarm intelligence or nature-inspired algorithms. The convergence problem is a challenge to address in developing these nature-inspired algorithms including the BA.

For further research, the other constructive and improved heuristic algorithms potentially hybridised to the BA either as a global or local search operator to improve the performance of the algorithm for solving the COP.

## Appendix A. Google Colab link

Google Colab link for this research: https://bit.ly/2MaJTyG or http://bit.ly/2LR7rIO or https://bit.ly/2YYXQVo

## Appendix B. Authors' ORCID

https://orcid.org/0000-0002-6580-4654, https://orcid.org/0000-0003-2314-1394,
https://orcid.org/0000-0002-1298-9499

## Acknowledgments

The authors would like to acknowledge the generous support from the Newton Fund-Ristekdikti, Indonesia Endowment Fund for Education (Lembaga Pengelola Dana Pendidikan Republik Indonesia), University of Birmingham, University of Pancasila, Industrial Engineering University of Pelita Harapan, The Scientific and Technological Research Council of Turkey (TÜBİTAK) and The Fatih Sultan Mehmet Vakif University (FSMVU).

## 5. References

[1] Ouaarab A, Ahiod B and Yang X S 2014 Discrete cuckoo search algorithm for the travelling salesman problem *Neural Computing and Applications* **24** (Springer) pp 1659–69

[2] Bitam S, Batouche M and Talbi E g 2010 A survey on bee colony algorithms *2010 IEEE International Symposium on Parallel & Distributed Processing, Workshops and Phd Forum (IPDPSW) (IEEE)* pp 1–8

[3] Lones M A 2020 Mitigating metaphors: A comprehensible guide to recent natureinspired algorithms *SN Computer Science* **1** (Springer) p 49

[4] Pham D, Ghanbarzadeh A, Koc E, Otri S, Rahim S and Zaidi M 2005 The bees algorithm *Technical Note,* Manufacturing Engineering Centre, Cardiff University, UK

[5] Pham D T, Ghanbarzadeh A, Koc¸ E, Otri S, Rahim S and Zaidi M 2006 The bees algorithm—a novel tool for complex optimisation problems *Intelligent Production Machines and Systems* (Elsevier) pp 454–59

[6] Pham D T, Castellani M and Le Thi H A 2014 Nature-inspired intelligent optimisation using the bees algorithm *Transactions on Computational Intelligence XIII* (Springer) pp 38–69

[7] Hussein W A, Sahran S and Abdullah S N H S 2017 The variants of the bees algorithm (ba): *A survey Artificial Intelligence Review* **47** (Springer) pp 67– 121

[8] Rajasekhar A, Lynn N, Das S and Suganthan P N 2017 Computing with the collective intelligence of honey bees–a survey *Swarm and Evolutionary Computation* **32** (Elsevier) pp 25–48

[9] Yuce B, Packianather M, Mastrocinque E, Pham D and Lambiase A 2013 Honey bees inspired optimization method: the bees algorithm *Insects* **4** (Multidisciplinary Digital Publishing Institute) pp 646–62

[10] Pham D T and Castellani M 2015 A comparative study of the bees algorithm as a tool for function optimisation *Cogent Engineering* **2** (Taylor & Francis) p 1091540

[11] Pham D, Koc E, Lee J and Phrueksanant J 2007 Using the bees algorithm to schedule jobs for a machine *Proceedings of Eighth International Conference on Laser Metrology, CMM and Machine Tool Performance* pp 430–39

[12] Packianather M S, Yuce B, Mastrocinque E, Fruggiero F, Pham D T and Lambiase A 2014 Novel genetic bees algorithm applied to single machine scheduling problem *2014 World Automation Congress (WAC) (IEEE)* pp 906–11

[13] Pham D T, Afify A and Koc E 2007 Manufacturing cell formation using the bees algorithm *Innovative Production Machines and Systems Virtual Conference*, Cardiff, UK

[14] Pham D, Otri S and Darwish A H 2007 Application of the bees algorithm to pcb assembly optimisation *Proceedings 3rd International Virtual Conference on Intelligent Production Machines and Systems (IPROMS 2007)* pp 511–16

[15] Mei C A, Pham D T, Anthony J S and Kok W N 2010 Pcb assembly optimisation using the bees algorithm enhanced with triz operators *36th Annual Conference on IEEE Industrial Electronics Society (IEEE)* pp 2708–13

[16] Ang M C, Ng K W, Pham D T and Soroka A 2013 Simulations of pcb assembly optimisation based on the bees algorithm with triz-inspired operators *International Visual Informatics Conference* (Springer) pp 335–46

[17] Fenton A 2016 The bees algorithm for the vehicle routing problem *Thesis* (University of Auckland, NZ) arXiv preprint arXiv:1605.05448

[18] Koc E 2010 Bees algorithm: theory, improvements and applications *Thesis* (Cardiff University, UK)

[19] Gutin G and Punnen A P 2006 The traveling salesman problem and its variations (Springer Science & Business Media)

[20] Reinelt G 1991 Tsplib—a traveling salesman problem library *ORSA journal on computing* **3** (INFORMS) pp 376–84

[21] Lara C, Flores J J and Caldero´n F 2008 Solving a school timetabling problem using a bee algorithm *Mexican International Conference on Artificial Intelligence* (Springer) pp 664–74

[22] Phuc N B, Khang N T T M and Nuong T T H 2011 A new hybrid ga-bees algorithm for a real-world university timetabling problem *2011 International Conference on Intelligent Computation and Bio-Medical Instrumentation (IEEE)* pp 321–326

[23] Khang N T T M, Phuc N B and Nuong T T H 2011 The bees algorithm for a practical university timetabling problem in vietnam *2011 IEEE International Conference on Computer Science and Automation Engineering* **4** (IEEE) pp 42–7

[24] Abdullah S and Alzaqebah M 2013 A hybrid self-adaptive bees algorithm for examination timetabling problems *Applied Soft Computing* **13** (Elsevier) pp 3608–20

[25] Mollabakhshi N and Eshghi M 2013 Combinational circuit design using bees algorithm *IEEE Conference Anthology (IEEE)* pp 1–4

[26] Yuce B, Pham D, Packianather M S and Mastrocinque E 2015 An enhancement to the bees algorithm with slope angle computation and hill climbing algorithm and its applications on scheduling and continuous-type optimisation problem *Production & Manufacturing Research* **3** (Taylor & Francis) pp 3–19

[27] Mayteekrieangkrai N and Wongthatsanekorn W 2015 Optimized ready mixed concrete truck scheduling for uncertain factors using bee algorithm Songklanakarin *J. Sci. Techn.* **37** 221–30

[28] Ghasemi B, Sadeghi A, Roghani M and Branch S T 2015 The solution of multiobjective multimode resource-constrained project scheduling problem (rcpsp) with partial precedence relations by multi-objective bees algorithm *Silvae Genetica* (ISSN: 0037-5349) **57**

[29] Yuce B, Fruggiero F, Packianather M S, Pham D T, Mastrocinque E, Lambiase A and Fera M 2017 Hybrid genetic bees algorithm applied to single machine scheduling with earliness and tardiness penalties *Computers & Industrial Engineering* **113** (Elsevier) pp 842–58

[30] Liu J, Zhou Z, Pham D T, Xu W, Ji C and Liu Q 2018 Robotic disassembly sequence planning using enhanced discrete bees algorithm in remanufacturing *International Journal of Production Research* **56** (Taylor & Francis) pp 3134–51

[31] Laili Y, Tao F, Pham D T, Wang Y and Zhang L 2019 Robotic disassembly replanning using a two-pointer detection strategy and a super-fast bees algorithm *Robotics and Computer-Integrated Manufacturing* **59** (Elsevier) pp 130–42

[32] Darwish A H, Joukhadar A and Kashkash M 2018 Using the bees algorithm for wheeled mobile robot path planning in an indoor dynamic environment *Cogent Engineering* **5** (Taylor & Francis) p 1426539

[33] Sabri A N, Radzi N H M and Samah A A 2018 A study on bee algorithm and a algorithm for pathfinding in games *2018 IEEE Symposium on Computer Applications & Industrial Electronics (ISCAIE) (IEEE)* pp 224–29

[34] Otri S 2011 Improving the bees algorithm for complex optimisation problems *Thesis* (Cardiff University, UK)

[35] Zeybek S and Ko¸c E 2015 The vantage point bees algorithm *7th International Joint Conference on Computational Intelligence (IJCCI)* **1** pp 340–45

[36] Stützle T, Dorigo M et al. 1999 Aco algorithms for the traveling salesman problem *Evolutionary algorithms in engineering and computer science* **4** 163–83

[37] Ismail A H 2019 Domino algorithm: a novel constructive heuristics for traveling salesman problem *IOP Conference Series: Materials Science and Engineering* **528** (IOP Publishing) p 012043

# Integrated FMEA-MCDM For Prioritizing Operational Disruption in Production Process

**Muqimuddin**[1,3], **M. L. Singgih**[2,4]

[1,2]Department of Industrial Engineering Institut Teknologi Sepuluh Nopember, Surabaya, Indonesia
[3]Muqimuddin03@gmail.com, [4]Moseslsinggih@ie.its.ac.id

**Abstract.** The more complex the production system of a company, definitely makes it sensitive of change or known as disruption. Disruptions always occur in production process and also cause productivity decrease. If there are any kind of disruptions occur, so the desicion maker have to prioritize it appropriately. The risk is going to be greater as a consequance when the priority made is false. FMEA is one method that is often used for evaluation and prioritization, especially failure as one type of disruption to minimize risk which can be adopted to assess and evaluate disruptions. The FMEA has some shortcomings including it is not considering the importance of each factor, not considering uncertain information from expert, not considering the cost as one of the criterion considered, and the level of severity is hard to measure, because on different objects can be measured by different aspect. Therefore, this research aims to develop FMEA model for prioritizing disruptions in production process. The model can improve some shortcomings of conventional FMEA. In achieving this goal, several methods are integrated including Analytical Hierarchy Process (AHP) which aims to determine the importance weighting of RPN element factors. In addition, this study uses the Technique for Order Preference by Similarity to an Ideal Solution for prioritizing disruption. Rough set also uses as a method for aggregating opinion of experts, and also uses for improving uncertain information. The model also considers loss revenue as aspect to measure severity level of disruption in determining prioritization.
**Keywords:** FMEA, Loss Revenue, Uncertain Information, Operational Disruption

## 1. Introduction

Production process is sensitive of change called disruption. Therefore, the disruptions always occur in production process and create loss for manufacturing company. Loss as an risk of diruptions on the company are difference because it is influenced by duration of disruption [1]. If the disruption occurs, the amount of production produced will decrease. The risk is going to be greater when prioritize disruption is false. If there is an error in prioritizing the disruption, it will certainly have consequences for the manufacturing company. The consequences is the risk of another disruption unselected will occur more severe in the production process. As an illustration where prioritizing disruptions without considering loss revenue, then the disruption with low loss revenue is selected as priority than others disruption. However, each diruptions have a different amounts of loss revenue per unit time. And so, it makes the loss revenue of the disruption can't be reduced significantly high. In the other hand, unselected disruption with high loss revenue, can increase greater than the loss revenue of disruption selected. If the priority handling is done on unselected disruption, the reduction of loss revenue would significantly be high. Therefore, in determining priorities it is necessary to consider the correct RPN so that they do not create a greater risk.

Failure Mode and Effect Analysis (FMEA) is one of the methods often used by many researchers, especially in evaluating failure as a type of diruption for risk management. This method can also be used to determine failure priorities. However, some researchers criticize this method because it still has some shortcomings including:

1. In determining the value ofRPN, it doesn't consider the weight of importance of RPN elements [2] [3]
2. In determining the RPN by multiplying the RPN element will be very possible to produce the same RPN value and it is not reliable for determining the priority of failure [4].
3. RPN elements with likert scoring will be possibly contained uncertain information from experts [4].
4. If only considering 3 elements of RPN as RPN including severity, occurance and detection, it has not provided a good desicion result in prioritizing, and needs others RPN for instance cost as a RPN to consider [5].

Some researchers have done to improve FMEA shortcomings to produce better prioritization such as adding cost aspects [6] and use the DEA method to determine risk priorities [7]. In addition to adding cost aspects, it also acomodate uncertain information from expert opinion and has considered the importance weight of each RPN by integrating BWM and GRA to FMEA [4]. Another research have accommodated uncertain information by aggregating expert assesment using the rough set and integrating BWM and TOPSIS for prioritizing risk in grup decision-making [8]. These previous research, the researcher assume each experts have no importaces in determining the most importance of RPN to consider. In fact, they actually have different opinion regarding it. This situation can be shown which each experts have different opinion regarding the most important and the less important RPN to consider. Therefore, to improve this situation, this research aims to propose FMEA method by integrating some method to acomodate FMEA shortcomings. The proposed method is hopely able to prioritize disruption in production process. the hierarchy of diruption priority is shown as Figure 1.

## 2. Methods

The proposed architectural model is shown in Figure 2. The proposed model consists of 4 stages. The first stage is identifying disruptions which occur in production process. Second stage is determining the evaluation scale each RPN (occurrence and detection), while severity is measured by calculating loss revenue each disruptions occured. Third stage is determining the optimal importance weight of each RPN. In this paper, AHP is chosen as method to determine the weight. This method is also able to accomodate difference of expert's opinion regarding the most importance RPN to be considered in prioritizing disruptions without ignore uncertain information from expert. Fourth stage is determining disruption priorities. In proposed model, TOPSIS is also used as method, because it is effective and suitable to determine all priorities. In additional, uncertain information accomodation of expert's opinion, in this model is also used rough set to aggregate expert's assessment of each input of method.

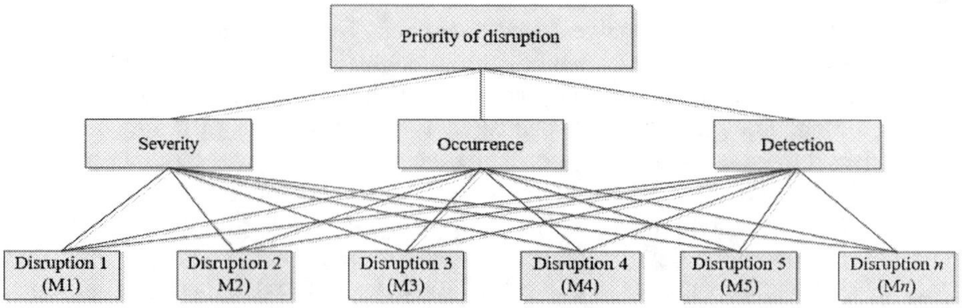


**Figure 1.** Hierarchy of operational disruption priority

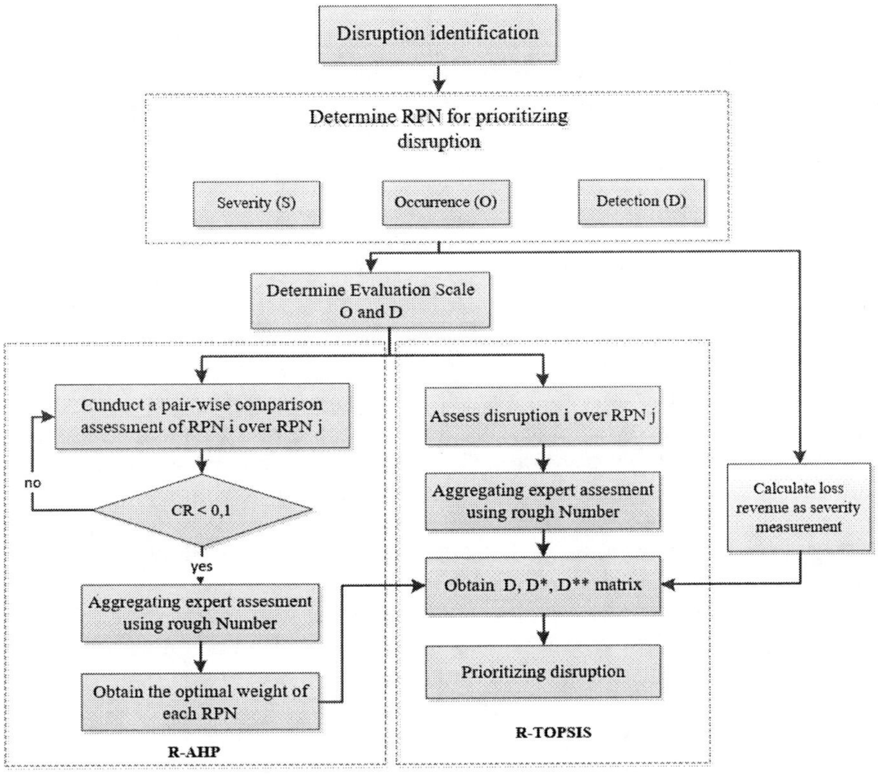


**Figure 2.** Architectural model

*2.1 Loss revenue calculation as severity measurement*

In determining the amount of revenue experienced by a company can be calculated based on multiplying the selling price of the product by amount of demand [9]. Meanwhile, if the company produces the same amount of production (Q) with amount of demand (D), the equation becomes as follows:

$$Revenue = S_x \times Q \quad (1)$$

In calculating Q is by multiplying the average production ($P_y$) and production time ($T_p$) [10]. Whereas if there is disruption time during $T_d$ then Q 'is produced with the following equation:

$$P_y \times (Tp - Td) = Q' \quad (2)$$

Loss revenue is a decrease in the amount of revenue generated by a company, so that it can be generated by multiplying the selling price of type x products ($S_x$) and amount of decreasing production (Q - Q ') so equation 2 can be simplified to:

$$Loss\ revenue = S_x \times P_y \times Td \quad (3)$$

*2.2 Rough Number*

Proposed by Professor Pawlak in 1982, rough set theory is a mathematical method that can acomodate imprecise, incosistent, and incomplete information or knowledge [17]. The steps of determining the rough number value are as follows [8]:

1) Determine lower dan upper approximations

   There is a U that represents all sets containing all objects and R represents a set of t class $(e_1, e_2, \dots, e_t)$ which covers all the objects inside U. R = $\{e_1, e_2, \dots, e_t)$, $e_1 < e_{2<}, \dots, < e_t$. Whereas Y is an arbitrary object of U. $\forall Y \in U, e_v \in R$ dan $1 \leq r \leq t$. Lower and upper bound approximation value $[(\underline{apr}(e_v), \overline{apr}(e_v)]$ of rough number can be obtained using the following equation:

$$Lower\ approximation : \underline{apr}(e_v) = \cup \{Y \in U/R(Y) \leq e_v\} \quad (4)$$

$$Upper\ approximation: \overline{apr}(e_v) = \cup\{Y \in U/R(Y) \geq e_v\} \quad (5)$$

Based on equation (4) will produce element $a_t$ dan and equation (5) will produce element $b_t$. While the number of elements $a_t$ shown with $N_L$ and number of elements $b_t$ shown with $N_U$ Thus it can be calculated Lower and upper limit ($(\underline{lim}\,(e_v), \overline{lim}\,(e_v))$) based on following equations:

$$\underline{lim}\,(e_v) = \frac{\sum_{i=1}^{N_L} a_t}{N_L} \quad (6)$$

$$\overline{lim}\,(e_v) = \frac{\sum_{i=1}^{N_U} b_t}{N_U} \quad (7)$$

2) Determine interval value of rough number
   Rough Number (RN) is represented by $[\underline{e}_v, \overline{e}_v]$ or $[\underline{lim}\,(e_v), \overline{lim}\,(e_v)]$.

*2.3 Analitical Hierarchy Process (AHP)*

AHP is one of the popular methods used by many decision makers for various problems, one of them, for determining the importance weight of RPN. This method is also able to measure the consistency of preferences, manipulate multiple decision makers, handle tangible and intangible RPN, and manage decision making that involves subjective judgment [11] and steps as follows:

1) Determine RPN
   At this step, researchers determine the RPN that will be assessed by the expert.
2) Conduct assessment by comparison of each RPN.
   At this stage one or several decision makers determine the value of the importance of each criterion over other RPN in pairs. The assessment is conducted by expert based on a rating scale 1 - 9. The results of expert assessment then it sorted into following matrix B:

$$B^k = \begin{bmatrix} 1 & B_{12}^k & \cdots & B_{1m}^k \\ B_{21}^k & 1 & \cdots & B_{2m}^k \\ \cdots & \cdots & \ddots & \vdots \\ B_{n1}^k & B_{n2}^k & \cdots & 1 \end{bmatrix} \quad (8)$$

   Where $B_{ij}^k$ shows the comparison value of criterion i over j by respondent k.
3) Calculate the consistency ratio value (CR)
   Before calculate the CR value, it is necessary to determine the maximum value of the eigenvalue ($\lambda_{max}^k$) from the assessment results of each expert. The value is used for calculating concistency index (CI) with the following equation:

$$CI = (\lambda_{max}^k - a)/(a - 1) \quad (9)$$

   Where $a$ is the number of RPN assessed by the expert. After the CI value is obtain, the CR value can be obtained based on calculating the ratio of the CI value and the random concistency index (RI)

$$CR = CI/RI \quad (10)$$

   The RI value is shown in Table 1 below:

**Table 1.** Random concistency index (RI)

| a | 3 | 4 | 5 | 6 | 7 | 8 | 9 | 10 |
|---|---|---|---|---|---|---|---|---|
| RI | 0,52 | 0,89 | 1.11 | 1,25 | 1,35 | 1,4 | 1,45 | 1,49 |

4) Obtain the initial rough matrix.
   The determination of the value of this rough number aims to aggregate the results of the assessment of experts. The aggregation uses equation (1) – (4) and the results are sorted into the following matrix C:

$$\mathbf{C} = \begin{bmatrix} [\mathbf{1},\mathbf{1}] & \tilde{C}_{12} & \dots & \tilde{C}_{1n} \\ \tilde{C}_{21} & [\mathbf{1},\mathbf{1}] & \dots & \tilde{C}_{2n} \\ \dots & \dots & \ddots & \vdots \\ \tilde{C}_{m1} & \tilde{C}_{m1} & \dots & [\mathbf{1},\mathbf{1}] \end{bmatrix} \quad (11)$$

If there are more than 1 rough number $\left[\underline{C}^{e}_{ij}, \overline{C}^{e}_{ij}\right]$I can be aggregated using the following equation:

$$\underline{C}_{ij} = \frac{\underline{C}^{1}_{ij}+\underline{C}^{2}_{ij}+\dots+\underline{C}^{S}_{ij}}{S}; \overline{C}_{ij} = \frac{\overline{C}^{1}_{ij}+\overline{C}^{2}_{ij}+\dots+\overline{C}^{T}_{ij}}{T} \quad (12)$$

Where $\tilde{C}_{ij}$ is a obtained rough number value from the rough set method which is including lower bound value ($\underline{C}_{ij}$and upper bond value ($\overline{C}_{ij}$).

5) Obtanin the optimal weight each RPN
   As for determining the optimal weight, you can use the following equation:

$$\widetilde{W}_j = \left[\sqrt[m]{\prod_{j=1}^{m} \underline{C}_{ij}}, \sqrt[m]{\prod_{j=1}^{m} \overline{C}_{ij}}\right] \quad (13)$$

then the optimal weight is normalized to produce 0-1 using the following equation:

$$\widetilde{W}_j = \widetilde{W}_j / \max \overline{C}_{ij} \quad (14)$$

*2.4. Technique for Order Preference by Similarity to an Ideal Solution (TOPSIS)*

TOPSIS was first proposed by Hwang and Yoon in 1981. The main concept of this method is to determine the best alternative closest to the positive ideal solution (optimal solution) and the farthest from the negative ideal solution (inferior solution). [12]. TOPSIS used by some researchers in decision making because the concept is simple, easy to understand, efficient calculation, and has ability to measure the relative performance of alternative decisions. [13]. The TOPSIS steps are as follows [4]:

1) Determine initial matrix
   The initial matrix contains the results of k expert assessment for each failure mode i over the RPN element j. Assessments are based on scale 1 - 10 evaluation scale used in research. The initial matrix is represented into Y matrix as follows:

$$\mathrm{Y} = \begin{bmatrix} Y_{111}, Y_{112,}, \dots, Y_{11t,} & Y_{121}, Y_{122,}, \dots, Y_{12t,} & \dots & Y_{1m1}, Y_{1m2,}, \dots, Y_{1mt,} \\ Y_{211}, Y_{212,}, \dots, Y_{21t,} & Y_{221}, Y_{222,}, \dots, Y_{22t,} & \dots & Y_{2m1}, Y_{2m2,}, \dots, Y_{2mt,} \\ \dots & \dots & \ddots & \vdots \\ Y_{n11}, Y_{n12,}, \dots, Y_{n1t,} & Y_{n1t}, Y_{n22,}, \dots, Y_{n2t,} & \dots & Y_{nm1}, Y_{nm1,}, \dots, Y_{nmt,} \end{bmatrix} \quad (15)$$

Where $Y_{ijk}$ is the assessment result of failure mode i over RPN j by expert k, i = 1,2,..n ; j = 1,2,..m ; k = 1,2,..t.

2) Determine the initial rough number matrix (D)
   This step aims to aggregate the assessment of experts using the rough set method. the aggregation uses equation (1) – (4) and the results are sorted into the following matrix D:

$$.\,D = \begin{bmatrix} \tilde{G}_{11} & \tilde{G}_{12} & \dots & \tilde{G}_{1n} \\ \tilde{G}_{21} & \tilde{G}_{22} & \dots & \tilde{G}_{2n} \\ \dots & \dots & \ddots & \vdots \\ \tilde{G}_{m1} & \tilde{G}_{m1} & \dots & \tilde{G}_{mn} \end{bmatrix} \quad (16)$$

if there are more than 1 rough number $\left[\underline{G}^{e}_{ij}, \overline{G}^{e}_{ij}\right]$ I can be aggregated using the following equation [11]:

$$\underline{G}_{ij} = \frac{\underline{G}^{1}_{ij}+\underline{G}^{2}_{ij}+\dots+\underline{G}^{S}_{ij}}{S}; \overline{G}_{ij} = \frac{\overline{G}^{1}_{ij}+\overline{G}^{2}_{ij}+\dots+\overline{G}^{T}_{ij}}{T} \quad (17)$$

Where $\tilde{G}_{ij}$ is a obtained rough number value from the rough set method which is including lower bound value ($\underline{G}_{ij}$) and upper bound value ($\overline{G}_{ij}$).

3) Obtain the normalized matrix (D*)

This matrix contains the normalized matrix D in order to produce the value 0 - 1. The normalization with a minimum value can be calculated by the following equation:

$$\tilde{G}_{ij}^{*} = \left[\frac{\underline{G}_{ij}}{G_j^{max}}, \frac{\overline{G}_{ij}}{G_j^{max}}\right], \text{where } G_j^{max} = \max_{1\le i\le m}\{\overline{G}_{ij}\} \tag{18}$$

Then obtained data can be sorted into the following matrix D*:

$$D^{*} = \begin{bmatrix} \tilde{G}_{11}^{*} & \tilde{G}_{12}^{*} & \cdots & \tilde{G}_{1n}^{*} \\ \tilde{G}_{21}^{*} & \tilde{G}_{22}^{*} & \cdots & \tilde{G}_{2n}^{*} \\ \cdots & \cdots & \ddots & \vdots \\ \tilde{G}_{m1}^{*} & \tilde{G}_{m2}^{*} & \cdots & \tilde{G}_{mn}^{*} \end{bmatrix} \tag{19}$$

Where $\tilde{G}_{ij}^{*}$ is normalized preference value of failure mode $i$ over RPN $j$ which is including upper bound value ($\overline{G}_{ij}^{*}$) and lower bound value ($\underline{G}_{ij}^{*}$). $G_j^{max}$ is represented the maximum preference value of the upper preference value of RPN j.

4) Determine weighted normalization matrix (D**)
Matrix D is then normalized again into a matrix D** by multiplying ($\tilde{w}_j$) and normalized matrix and denoted by ($\tilde{G}_{ij}^{*}$) and it can be calculated by the following equation:

$$\tilde{V}_{ij} = [(\tilde{G}_{ij}^{*}) \times (\tilde{w}_j)] \tag{20}$$

And the result can be sorted into matrix D** as follows:

$$D^{*} = \begin{bmatrix} \tilde{V}_{11} & \tilde{V}_{12} & \cdots & \tilde{V}_{1n} \\ \tilde{V}_{21} & \tilde{V}_{22} & \cdots & \tilde{V}_{2n} \\ \cdots & \cdots & \ddots & \vdots \\ \tilde{V}_{m1} & \tilde{V}_{m2} & \cdots & \tilde{V}_{mn} \end{bmatrix} \tag{21}$$

where $\tilde{V}_{ij}$ are value of weighted normalization matrix of failure mode $i$ over RPN $j$ which is including lower and upper bound value [ $\underline{V}_{ij}$, $\overline{V}_{ij}$].

5) Obtain PIS and NIS
The next step is determining PIS and NIS. PIS is a positive ideal solution, while NIS is a negative ideal solution. For determining PIS and NIS can be obtained using the following equation:

$$PIS = (r_1^{*}, r_2^{*}, \ldots, r_m^{*}); where\ r_j^{*} = \max_i \{\overline{V}_{ij}\} \tag{22}$$

$$NIS = (r_1^{-}, r_2^{-}, \ldots, r_m^{-}); where\ r_j^{-} = \max_i \{\underline{V}_{ij}\} \tag{23}$$

6) Calculate the separation distance of PIS and NIS
These distance is calculated based on the following equation:

$$d_i^{*} = \sum_{j=1}^{m} \sqrt{\left(\frac{\left(r_j^{*}-\underline{V}_{ij}\right)^2 + (r_j^{*}-\underline{V}_{ij})^2}{2}\right)} \quad ; d_i^{-} = \sum_{j=1}^{m} \sqrt{\left(\frac{\left(\underline{V}_{ij}-r_j^{-}\right)^2 + (\underline{V}_{ij}-r_j^{-})^2}{2}\right)} \tag{24}$$

7) Calculate closeness coefficient
The last step in this mathod is calculating the value of closeness coefficient (CCi) which is the index value as a reference in determining the priority of failure mode. For calculating CCi,can be obtained using the following equation:

$$CC_i = w^{+}(d_i^{-}/\textstyle\sum_{i=1}^{n} d_i^{-}) - w^{-}(d_i^{*}/\sum_{i=1}^{n} d_i^{*}) \tag{25}$$

Where $w^{+}$ is a given weight for PIS. And $w^{-}$ is a given weight for NIS. When CCi value is higher than 0, it indicate the disruption is close to positive ideal solution (PIS). The priority will selected based on the CCi value is close to 1.

## 3. Discussion

FMEA model is modified by integrating R-AHP and R-TOPSIS for qualitative research. As explanation in first section, FMEA model has 4 shortcomings. It make the desicion result is not relible in prioritization. Hence, these method is intagrated so that the prioritization is more relible than FMEA traditional. Each decision maker certainly has a different view of the RPN that is important to consider. Therefore, this model also uses the AHP method which aims to produce the importance of each criterion. Whereas TOPSIS is used for prioritizing disruption where the determination of the ideal solution is based on the PIS and NIS values which then calculate the CCi value to determine the disruption rating. As for the types of disruption that have the highest positive CCI values are the priority in handling. each expert as a decision holder certainly has a different insight about the type of disorder, so the results of the assessment will be different. Therefore, this model also aggregates the results of the assessment using the rough set method to accommodate uncertain information from the results of expert assessments. Another advantage of developed FMEA model can be used for evaluating risk which the number of experts have a different opinion regarding the most importance RPN to consider although number of expert equal to number of citeria. And it definitely still accomodates uncertain information of experts assessment. This modified FMEA is able to use for prioritizing disruption as one of risk management generally in production process.

## 4. Conclusion

The FMEA model has been carried out by integrating any methods such as R-AHP and R-TOPSIS and has accommodated the level of uncertain information from experts, the level of importance of RPN, and considering the cost aspects in determining priorities. Our proposes model is effective to prioritize disruption in production process and able to make decision better than FMEA traditional. Although the proposed model can improve the shortcoming of FMEA traditional method, it still have a lot of limitation. The proposed will take a long time to implement if there are many RPN to consider. The optimal weight of each RPN will be obtained by much pair-wise comparison using AHP method.

## Appendix A

Table A1

Linguistic Scale for AHP

| Scale | Linguistic | Description |
|---|---|---|
| 1 | *Equal* | *The importance of RPN i is equal to RPN j* |
| 3 | *Moderate* | *RPN i is moderately important to RPN j* |
| 5 | *Important* | *RPN i is important to RPN* |
| 7 | *Very Important* | *RPN i is very important to RPN* |
| 9 | *Extreamly Important* | *RPN i is extreamely important to RPN* |
| 2,4,6,8 | *Value between scale* | *The value if there is vague between scale* |

Table A2

Preference Scale for TOPSIS

| Value | Occurrence (O) | Detection (D) |
|---|---|---|
| 10 | *Disruption Almost Inevitable (FI)* | *Absolute Uncertainty (AU)* |
| 9 | *Very High (VH)* | *Very Remote (VR)* |
| 8 | *Repeated Failure (RF)* | *Remote (R* |
| 7 | *High (H)* | *Very Low (VL)* |
| 6 | *Moderately High (MH)* | *Low (L)* |
| 5 | *Moderate (M)* | *Moderately (M)* |
| 4 | *Relatively Low (RL)* | *Moderately High (MH)* |
| 3 | *Low (L)* | *High (H)* |
| 2 | *Remote (R)* | *Very High (VH)* |
| 1 | *Nearly Impossible (NI)* | *Almost Certain (AC)* |

## 5. References

[1] H. Hishamuddin, R. A. Sarker, and D. Essam 2010 A recovery model for an economic production quantity problem with disruption, *IEEM2010 - IEEE Int. Conf. Ind. Eng. Eng. Manag.*, pp. 198–202.

[2] A. C. Kutlu, 2012 Expert Systems with Applications Fuzzy failure modes and effects analysis by using fuzzy TOPSIS-based fuzzy AHP. *Expert Syst. Appl.*, vol. **39**, pp. 61–67.

[3] K. Chang, Y. Chang, and I. Tsai 2013 Enhancing FMEA assessment by integrating grey relational analysis and the decision making trial and evaluation laboratory approach. *Eng. Fail. Anal.* **31**, pp. 211–224.

[4] H. Lo and J. J. H. Liou, 2018 A novel multiple-RPN decision-making-based FMEA model for risk assessment. *Appl. Soft Comput. J.*, **73**, pp. 684–696.

[5] K. Chang, 2016 Generalized multi-attribute failure mode analysis. *Neurocomputing* **175**, pp. 90–100.

[6] A. Von Ahsen, 2008 Cost-oriented failure mode and effects analysis. *Int. J. Qual. Reliab. Manag.*, 25, pp. 466–476.

[7] M. J. Rezaee, A. Salimi, and S. Yousefi 2017 Identifying and managing failures in stone processing industry using cost-based FMEA. *Adv. Manuf. Technol.*, pp. 3329–3342.

[8] H. Lo and J. J. H. Liou 2019 A novel failure mode and effect analysis model for machine tool risk analysis. *Reliab. Eng. Syst. Saf.*, **183**(November 2018), pp. 173–183.

[9] S. K. Paul, R. Sarker, and D. Essam 2015 A disruption recovery plan in a three-stage production-inventory system, *Comput. Oper. Res.* 57, pp. 60–72.

[10] M. P. Grover, 2015 *Automation Production System, and Computer-Integrated Manufacturing*, 4th ed. (United States of America: Pearson Higher Education, Inc.)

[11] G. Zhu, J. Hu, J. Qi, C. Gu, and Y. Peng, 2015 Advanced Engineering Informatics An integrated AHP and VIKOR for design concept evaluation based on rough number, *Adv. Eng. Informatics,* **29**(3), pp. 408–418.

[12] G.-H. Tzeng and J.-J. Huang, 2011 *Multiple Attribute Desicion Making Method And Application*, 1st ed. (United State of america: Taylor & Francis Group, an informa business).

[13] A. Łatuszyńska, 2014 *Multiple-Rpn Decision Analysis Using Topsis Method For Interval Data In Research Into The Level Of Information Society Development.* Folia Oeconomica Stetin.

IOP Conf. Series: Materials Science and Engineering **847** (2020) 012029 doi:10.1088/1757-899X/847/1/012029

# Optimizing production layout and capacity via FlexSim– A case study of Y factory

**Tai-Jung Chen[1,2], Yu-Ching Lee[1,3], and Chin-Hsin Chiang[1,4]**

[1]Department of Industrial Engineering and Engineering Management, National Tsing Hua University, Hsinchu, Taiwan
[2] abao1126@gapp.nthu.edu.tw, [3] yclee@ie.nthu.edu.tw, [4] ct0806449@gmail.com

**Abstract.** Production capacity maximization and cycle time reduction are considered two of the most important issues for general factories. The paper applies Systematic Layout Planning (SLP), process balance, and system simulation to a small condense units manufacturer to enhance its production capacity and reduce cycle time. The developed solution shortens the total traveling distance by 81.25% and reduces the original travel time by 61.76%. In addition, we calculate the number of machines needed by process balancing and decrease the original cycle time by 64%. It is found that the proposed approach improved both cycle time and travel distance, and free up considerable capacity for manufacturing of other condensing units product types.
**Keywords**: FlexSim; system simulation; Systematic Layout Planning; cycle time reduction; process balance

## 1. Introduction

Globalization has put considerable pressure on small manufacturers to optimize production processes, e.g., by redesigning production facility layouts to reduce transfer time and effort, and by eliminating bottlenecks in production lines [1]. This research aims to reduce the cycle time in a small manufacturer of condensing units (Y factory). The factory produces different types of products including customized products. Each order is at a 50-unit scale. Observation of Y factory's production line Discovered efficiency flaws and bottlenecks, such as non-value travel from operators carrying heavy Work In Process (WIP).

We applied SLP to rearrange optimize factory floor efficiency, identified production bottlenecks, calculated the potential production capacity, and adjusted the number of machines at every station through process balancing. Finally, we used FlexSim, a powerful system simulation tool, to verify our modification and performance. [2]

The constraints of the present study include the following: (1) Station areas do not overlap; (2) Only equipment and product flow are considered; (3) The process ends when the product is finished, without consideration of shipping; and (4) The lead product by output volume is adopted as the test product. Other product types are not considered.

## 2. Methods

The project uses SLP, process balancing, and system simulations. An operation flow chart was developed to identify the manufacturing process. The principles used to draw the operational flow chart are listed in Table 1. SLP is a theory used to arrange production facilities by locating areas with high

IOP Conf. Series: Materials Science and Engineering **847** (2020) 012029 doi:10.1088/1757-899X/847/1/012029

frequency and logical relationships close to each other, thus optimizing speed and efficiency in material flows by minimizing transport, cost, and travel time, while maximizing safety. SLP requires an activity relationship chart, activity relationship diagram, space relationship diagram, and block layout diagram to arrange the layout steps by steps. [3-5] An activity relationship chart is a tabular means of displaying the closeness rating among all pairs of activities or departments. An activity relationship chart includes six closeness ratings which may be assigned to each pair of departments, as well as nine reasons for those ratings (each denoted by a reason code). Examples are shown in Figures 1 and 2. The space relationship diagram shown in Figure 3 shows the activity relationship after considering the actual space of every workstation. The block layout diagram, as shown in Figure 4, is used to adjust the spatial relationship diagram to fit the actual space of the workshop.

Process balancing is a technique used to minimize the Effective Station Cycle Time, which is the average cycle time needed for a station to finish processing a product, and is used to identify production bottlenecks. Figure 5 shows the steps for process balancing.

**Table 1.** Legends for operation process chart.

| No. | Chart | Illustration |
|---|---|---|
| 1 | → ↑ | Horizontal line represents the entrance of the material. Vertical line means the steps of manufacturing |
| 2 | —○—□— | Typical process flow chart |
| 3 | ←○—□— (re-entrant) | A re-entrant process flow |

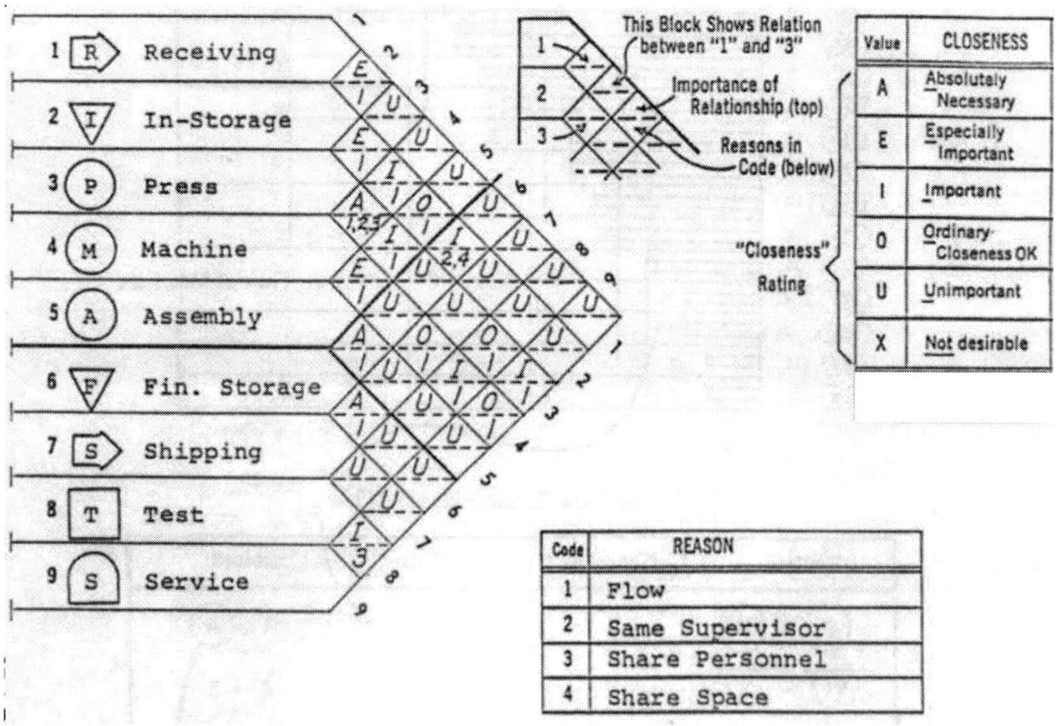


**Figure 1.** Activity relationship chart

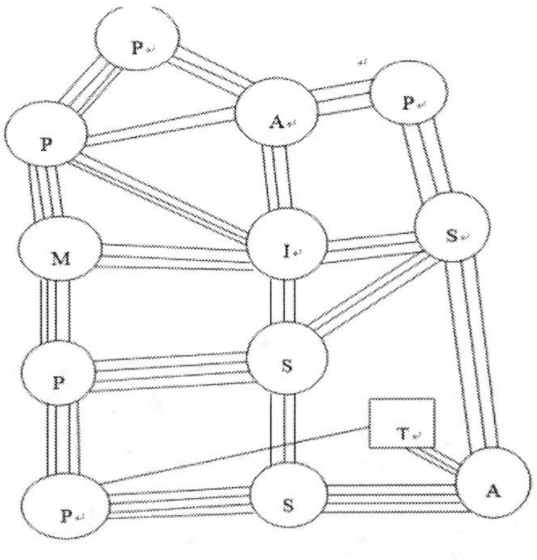


**Figure 2.** Activity relationship diagram.

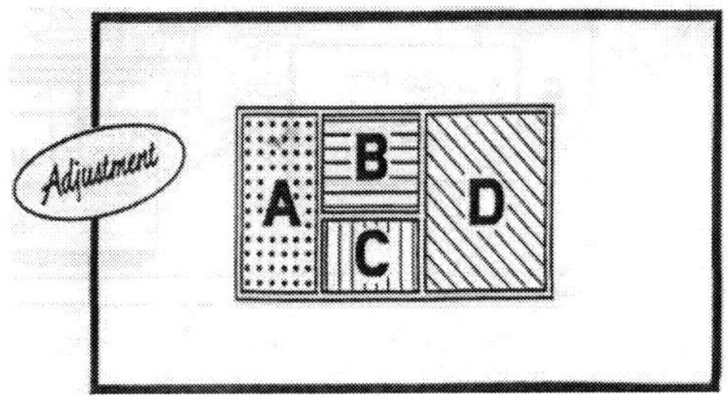


**Figure 3.** Space Relationship diagram.

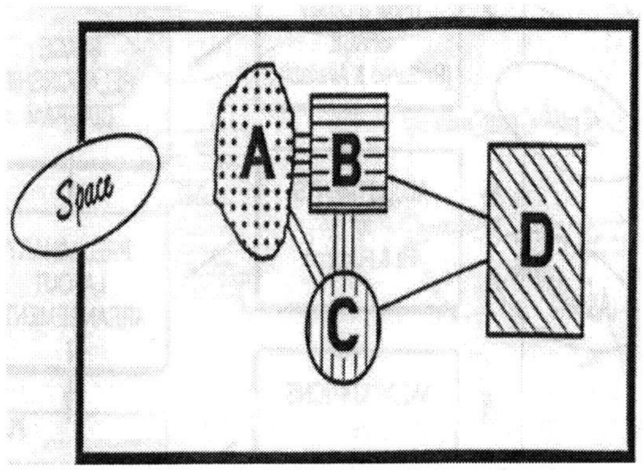


**Figure 4.** Block layout diagram.

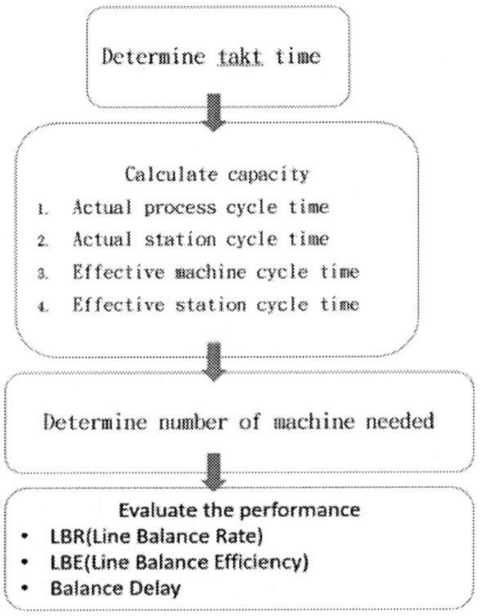


**Figure 5.** Steps to Process Balance.

Terminology related to process balance is listed below. Figure 6 integrates the formula for process balancing.

- Takt Time: Takt time is the average time between the start of production of one unit and the start of production of the next unit, when these production starts are set to match the rate of customer demand.
- Actual Process Cycle Time (APCT): The actual time needed for a process or machine to finish processing a product.
- Actual Station Cycle Time (ASCT): The actual time needed for a workstation to finish processing a product.
- Effective Process Cycle Time (EPCT): The average time needed for a process or machine to finish processing a product.
- Effective Station Cycle Time (ESCT): The average time needed for a workstation to finish processing a product.
- Number of machine needed: The number of machines needed to ensure bottleneck-free production.
- Line Balance Rate (LBR): It is a performance index utilized in process balancing and to evaluate the average utilization rate of each workstation, where a higher LBR indicates a smaller deviation of the effective cycle time between workstations and reduced production workstations, thus greater efficiency.

IOP Conf. Series: Materials Science and Engineering **847** (2020) 012029 doi:10.1088/1757-899X/847/1/012029

- Line Balance Efficiency (LBE): LBE is a performance index for process balancing. It evaluates the average utilization rate of each workstation. A higher LBE indicates a smaller deviation of the effective cycle time between each workstation and reduced production bottlenecks, thus greater efficiency.

A powerful system simulation software, FlexSim, plays an important role as a verification tool in this research. We built two models via FlexSim. The first model represents current status of Y factory. The other model represents the situation after applying the modifications. Thus, we have the opportunity to examine the performance of our approaches [6, 7]. Figure 7 revealed an exemplar of FlexSim model.

$$\text{Takt time} = \frac{\text{daily operation time}}{\textit{daily demand}}$$

$$\text{Number of machine needed} = \frac{\textit{effective machine cycle time}}{\textit{takt time}}$$

$$\text{LBE} = \sum_{i=1}^{N} t_i / N^* \max t_i$$

$t_i$ : $i^{th}$ station's effective cycle time

$\max t_i$ :Maximum of all effective station cycle time

$$\text{LBR} = \left(\prod_{i=1}^{N} t_i\right)^{1/N} / \max t_i$$

N : number of the stations

**Figure 6.** Formulas for process balance.

**Figure 7.** FlexSim model.

## 3. Results dan Discussion

### *3.1 SLP and simulation*

The Bill of Material (BOM) of Y factory was used to identify the relationship between material and final product. A condensing unit is composed of brass tubing, aluminium flake, and an iron shell. An operation process chart was constructed to provide a systematic understanding of product processing. Shown in Figure 8, the product needs to go through seven processes in a particular order. The travel distance of the operator was measured and set as a Key Performance Indicator (KPI) for optimization.

Figure 9 shows the activity relationship diagram based on the processing activity sequence in the storage and production areas, measuring the close relationship between the two areas. Workstations with priority in the product processing sequence should be in close proximity, and the degree of closeness between the two workstations is converted into relationships of descending closeness denoted by A, E, I, O and U, where U denotes nearly no communication between two workstations. Figure 10 shows a spatial relationship diagram based on the activity relationship diagram. We use the spatial relationship diagram to develop the block relationship diagram (Figure 11). Finally, we adjust the spatial relationship diagram to fit the actual area of Y factory and come up with an improved layout. Figures 12 and 13 show the floor plan before and after modification, respectively. Table 2 shows station names by code.

After completing SLP, we construct the layout, and design the flow by coding into our FlexSim model to emulate the travel distance and time. The revised layout shortens the distance between the brass bending workstation and pipe insertion workstation from 7 meters to 1 meter. Reduced travel distance between workstations also improves material handling efficiency. The tables below show that the overall cycle time and travel distance are significantly reduced during the process (Total benefit increased 81.25% by distance reduced and 61.76% by travel time reduced). Figures 14 and 15 respectively show the FlexSim model before and after modification, while Tables 3 and 4 respectively summarize travel distance and travel time after SLP.

### *3.2 Process balance and simulation*

The improved workstation layout shortens the distance between the workstations and transportation time, but we still seek a greater reduction in cycle time using the process balancing method to adjust the production capacity. The data needed for process balancing is displayed in Table 5.

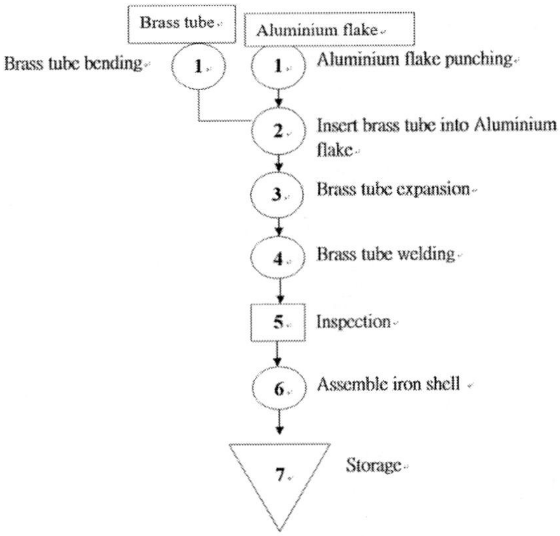


**Figure 8.** Operation process chart of Y factory.

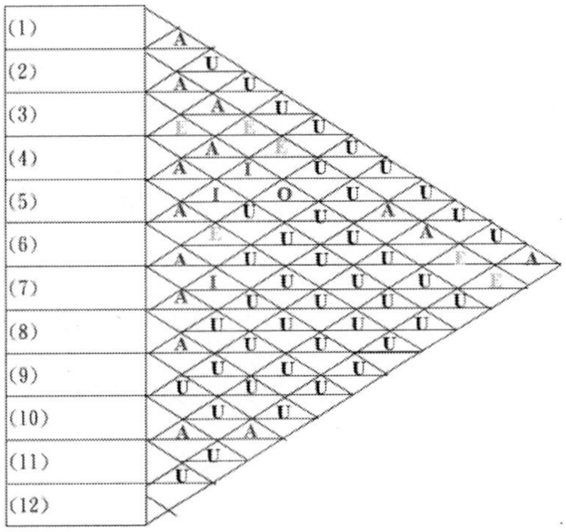


**Figure 9.** Activity relationship chart of Y factory.

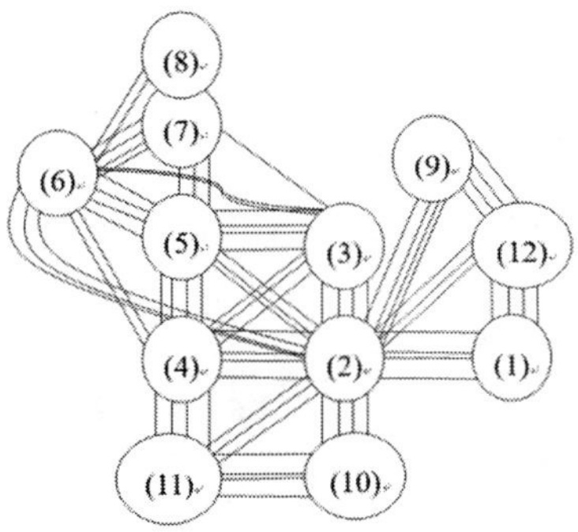


**Figure 10.** Spatial relationship diagram of Y factory.

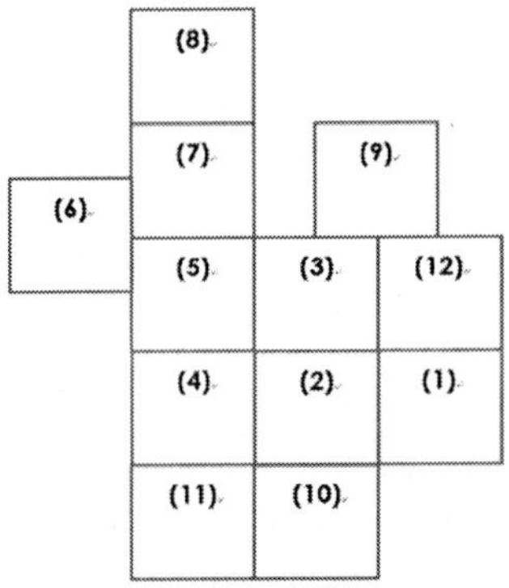


**Figure 11.** Block relationship diagram of Y factory.

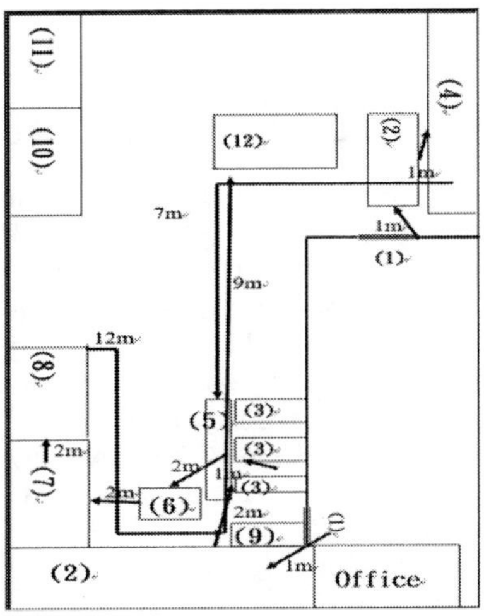


**Figure 12.** Floor plan before SLP.

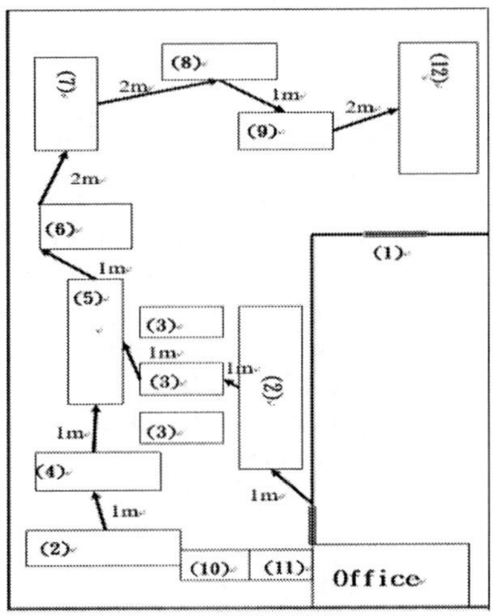


**Figure 13.** Floor plan before SLP.

**Table 2.** Codes of each station.

| Code | Station name | Code | Station name |
|---|---|---|---|
| (1) | Gate | (7) | Welding |
| (2) | Material storage | (8) | Leak detect |
| (3) | Flake punching | (9) | Assembly |
| (4) | Elbowing | (10) | Cutting |
| (5) | Pipe piercing | (11) | Bending |
| (6) | Pipe expansion | (12) | Final product storage |

**Table 3.** Comparisons of distances after applying SLP.

| Distance | Before (m) | After (m) | Benefit (%) |
|---|---|---|---|
| (2) – (3) | 2 | 1 | 50.00% |
| (4) – (5) | 7 | 1 | 85.71% |
| (5) – (6) | 2 | 1 | 50.00% |
| (8) – (9) | 12 | 1 | 91.67% |
| (9) – (12) | 9 | 2 | 77.78% |
| Total benefit | 32 | 6 | 81.25% |

**Table 4.** Comparisons of travel times after applying SLP.

| Travel time | Before (s) | After (s) | Benefit (%) |
|---|---|---|---|
| (2) – (3) | 5 | 2.5 | 50.00% |
| (4) – (5) | 25 | 14 | 44.00% |
| (5) – (6) | 10 | 8 | 20.00% |
| (8) – (9) | 41 | 8 | 80.49% |
| (9) – (12) | 38 | 13 | 65.79% |
| Total benefit | 119 | 45.5 | 61.76% |

**Table 5.** The data needed for process balancing.

| Machine name | APCT | EPCT | Batch size | Takt time | No. of machine needed |
|---|---|---|---|---|---|
| Elbowing machine | 506 | 506 | 1 | 288 | 1.8 |
| Flake punching machine | 82 | 82 | 1 | 288 | 0.3 |
| Pipe piercing machine | 30 | 30 | 1 | 288 | 0.1 |
| Pipe expansion machine | 42 | 42 | 1 | 288 | 0.1 |
| Welding machine | 234 | 33 | 7 | 288 | 0.8 |
| Leak detector | 95 | 95 | 1 | 288 | 0.3 |
| Assembly machine | 120 | 120 | 1 | 288 | 0.4 |

Process balancing reveals the need for an additional brass-bending machine, while the number of other machines remains static. Furthermore, we apply the revised workstation layout in Section B, and then simulate this scenario through FlexSim. We find that the cycle time for finishing the first product of an order is significantly reduced by 46.06% from 4220 s to 2640 s. The cycle time for producing a batch of 50 (i.e., the minimum order size) is shortened by 64.40% from 29,054 s to 14,900 s. Table 6 briefly shows the cycle time improvement after applying process balancing. Among the important indicators of process balance, LBR increased from 15.69% to 28.43%, and LBE increased from 23.17% to 37.01%. Despite these improvements, the LBR and LBE absolute values are still below 50%. This indicates that the Takt Time of the product can be shorter, meaning that there is still plenty of room to increase order sizes to make better use of capacity. Table 7 compares LBR, LBE, and the balance rate after applying process balancing. Figure 16 shows the FlexSim model after process balancing and SLP.

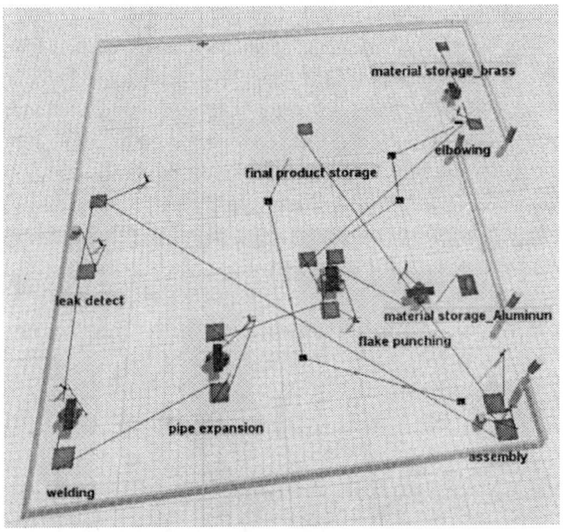


**Figure 14.** FlexSim model (before SLP).

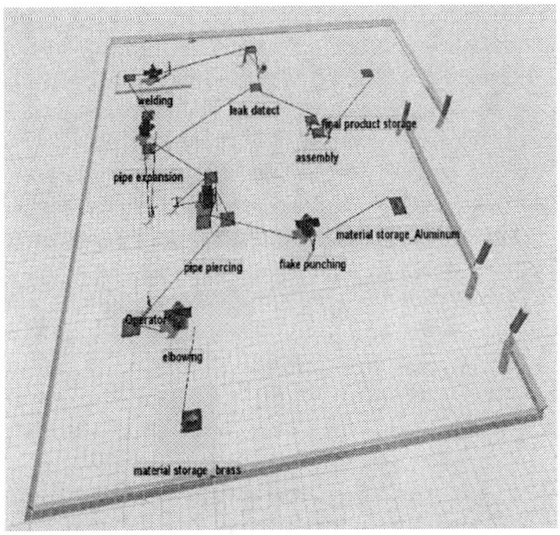


**Figure 15.** FlexSim model (after SLP).

**Table 6.** Cycle time improvements for products.

| Cycle time (s) | Before | After | Benefit |
|---|---|---|---|
| One product | 4220 | 2640 | Decrease by 46.06% |
| 50 products | 29054.48 | 14900 | Decrease by 64.40% |
| Average | 581.0896 | 298 | Decrease by 64.41% |

**Table 7.** Comparisons of performance after process balancing.

| KPI | Before | After | Benefit |
|---|---|---|---|
| LBR | 15.69% | 28.43% | Increase by 12.75% |
| LBE | 23.17% | 37.01% | Increase by 13.84% |
| Balance Delay | 76.83% | 62.99% | Decrease by 13.84% |

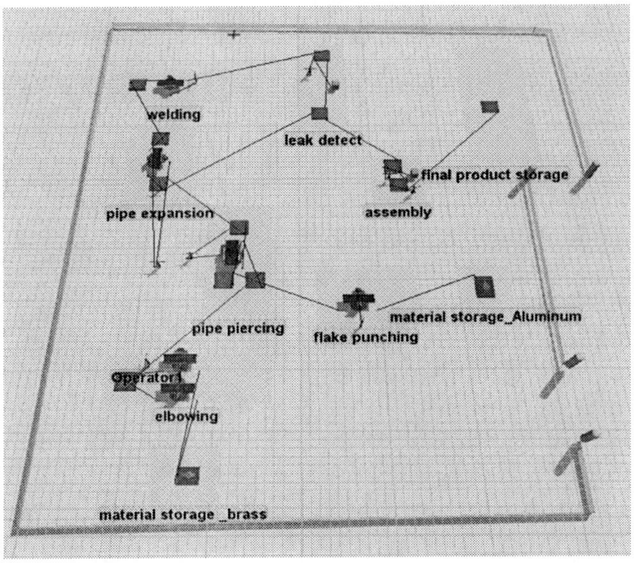


**Figure 16.** FlexSim model after process balancing.

## 4. Conclusions

Production capacity maximization and cycle time reduction are considered two of the most important issues for general factories. This study uses SLP, process balancing, and FlexSim simulation to developed a solution which improved both cycle time and travel distance. The developed solution shortens the total traveling distance by 81.25% and reduces the original travel time by 61.76%. Additionally, the number of machines needed by process balancing and decrease the original cycle time by 64%, freeing up considerable capacity for manufacturing of other condensing unit product types. The approaches applied in Y factory are suitable for other small manufacturers as well.

The layout of Y factory needs to be rearranged to maximum production capacity. Adding one brass-bending machine will simultaneously reduce the travel distance and cycle time. It should be noted that there are limitations of this study including: (1) Ignoring cost analysis for more comprehensive insights and adjustments; (2) The scope of research to facilities doesn't consider multiple production floors.

## 5. References

[1] Sheu D D 2014 Production System Design and Diagnosis 2nd ed. *Taipei: Agitek.* pp. 137-200 (in Chinese)

[2] Lin J T 2001 System simulation theory and application *Tsing Hai Publishing* pp. 1-29 (in Chinese)

[3] Bozer Y A and Miller R D 2007 A re-examination of the distance-based facility layout problem *IIE Transactions* **29**(7), pp. 549-560

[4] Rekiek B De Lit P Pellichero F Falkenauer E and Delchambre A 1999 Applying the equal piles problem to balance assembly lines *International Symposium on Assembly and Task Planning* pp. 399-404

[5] Chiang W C 1998 The application of a tabu search metaheuristic to the assembly line balancing problem *Annals of Operations Research*, **77**, pp. 209-227

[6] Zhu X, Zhang R Chu F He Z and Li J 2014 A Flexsim-based optimization for the operation process of cold-chain logistics distribution centre *J. of Applied Research and Technology* **12**(2), pp. 270-278

[7] Nie X and Wang L 2019 Simulation process design for scheduling mode of railway container terminals based on FlexSim *J. Phys. Conf. Ser.* 1176052012

IOP Conf. Series: Materials Science and Engineering **847** (2020) 012030 doi:10.1088/1757-899X/847/1/012030

# Economic Feasibility Study of Onshore Exploration Oil Field Development using Gross Split Contract

**Muhammad Ariyon[1,2], Aldo Setiawan[1,3] and Refiandi Reza[1,4]**

[1]Universitas Islam Riau, Pekanbaru, Indonesia
[2]aryonmuhammad@eng.uir.ac.id, [3]aldo.setiawan.as@gmail.com,
[4]reza20refiandi@gmail.com

**Abstract.** The condition of the oil and gas world economy is not good and the cost recovery is too large and the results obtained are not comparable to those obtained by the contractor. Thus, the change in the new contract system with the addition of split rules for contractors considers in managing the oil and gas fields. To restore the situation and restore the spirit of the contractor to continue investing in Indonesia, the Indonesian government provides contractors with incentives and other alternatives to change the existing PSC Contract with the Gross Split PSC based on the Minister of Energy and Mineral Resources Regulation No. 52 of 2017. Field X is one of the onshore Exploration fields in the South Sumatra Basin. The field is a commercial field that has potential oil reserves to be managed. To conduct field X development, it is necessary to conduct an economic study of field X development using a Gross Split contract to determine the feasibility level. Based on calculations using the Gross Split Contract field development X has a contractor NPV value of $ M 192,63, Government NPV = 415,235 and Contractor IRR = 41.04%, Government Take = 72.3% and Contractor Take = 27.7%. Based on this, the Government and Contractors will benefit from using the Gross Split PSC Contract because it is economical to develop.
**Keyword:** oil field, gross split, petroleum economic, exploration

## 1. Introduction

Nowadays Indonesia is undergoing various revisions and improvements to regulations, along with the activities of the oil and gas industry, such a situation will certainly have an impact on the investment climate in the upstream oil and gas sector [1]. Oil and gas investment is made based on profit considerations, so the quantities that affect it, namely reserves that produce production, costs, prices, and taxes need to be known [2]. International oil companies and the Government of Indonesia have different points of view regarding the Production Sharing Contract (PSC) [3].

Looking at the historical development of the Indonesian PSC generation from the first generation of 1966 to the current generation, it is always undergoing a policy change that has occurred along with the emergence of new regulations to attract investors to invest in Indonesia. Oil and gas policy is a government attitude towards the form of management of oil and gas resources. Thus the production sharing cost recovery contract formulating the policy depends on the situation in the development of the oil and gas industry [4].

In January 2017, the Indonesian government issued regulation of the Minister of Energy and Mineral Resources No. 8 year 2017 concerning the Gross Split PSC but in August 2017 a regulation was changed from the previous by the Minister of Energy and Mineral Resources to Number 52 year 2017 from some changes to the provisions. The Gross Split Contract is a production sharing contract based on the principle of gross production sharing without a cost recovery mechanism so that all costs incurred for block operations are fully borne by the contractor [5] . Unlike previous production sharing contracts that use fixed net split after tax for contractors and the government, in this new fiscal period, contractor splits are determined using a basic split that is adjusted based on progressive variables and components [6].

This research was conducted to examine the economic feasibility of developing onshore oil exploration fields in the southern Sumatra basin using a gross split contract.

## 2. Methods

### 2.1. *Data collection*

Collection of research data from the field. Field review where the research was conducted. From the data obtained can be known reservoir characteristics, production profile, and costs required for the Onshore Exploration X Field development project. Data obtained for reserves calculation onshore exploration X field is from Geological data. From the field data, it is assumed that the shape of the reservoir is Ellipse then calculated for oil reserves using the volumetric method. Furthermore, the forecasting of the X field production profile is assumed up to 30 years.

### 2.2. *The economic development model of X field based on gross split contracts*

In January 2017, the Government of Indonesia issued Minister of Energy and Mineral Resources (ESDM) Regulation No. 08 of 2017 concerning the Gross Split PSC. In accordance with regulations, the Gross Split PSC is a production sharing contract based on the principle of sharing production without a cost recovery mechanism so that all costs incurred for block operations are fully borne by the contractor [7]. Then in August, the Government of Indonesia revised from the previous law so that it became law No. 52 of 2017 different are some of the split values or from each of the respective components.

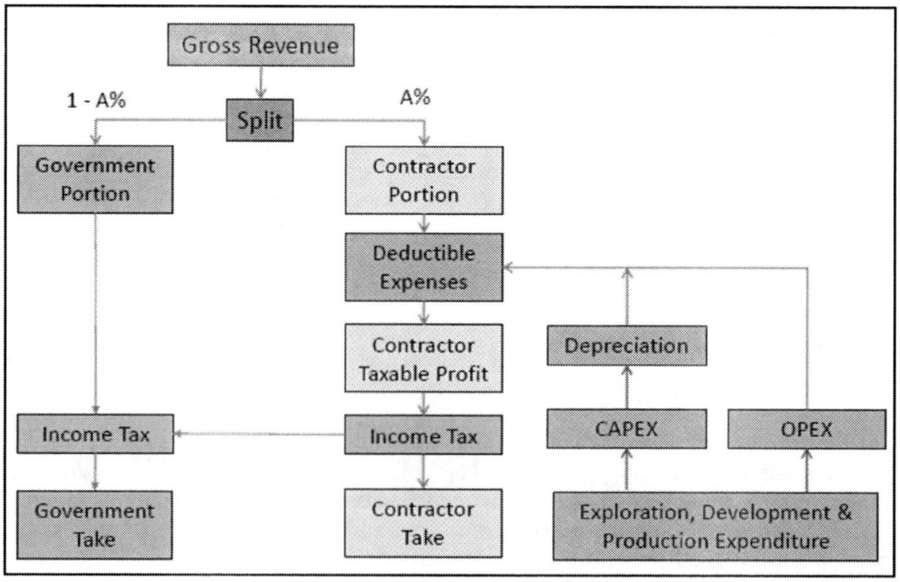


**Figure 1.** PSC management with gross split [8]

The gross split PSC does not have a cost recovery mechanism, therefore PSC contractor income comes solely from its gross production sources (and also has to pay income tax to the government in connection with this income). Government revenue will consist of the government's gross share of production, bonuses, PSC contractors' income taxes and indirect taxes paid by PSC Contractors. Initial base split each PSC gross split will have an initial base split percentage for each field developed within the contract area, which is then adjusted to the specific field factors listed in the Regulations. According to Law No. 52 of 2017, The basic division for PSC Contractors is 43% for oil and 48% for gas. The factors that change the initial base split is because this new scheme leads to PSC contractors who bear greater risk. The regulation provides certain incentives that offer increased production based on regulations which are described as variable and progressive factors. Indicators of project feasibility can be used quantitatively as a guide or evaluation tool in making decisions regarding project feasibility. The project feasibility indicators are: NPV and IRR. Then, sensitivity analysis is a way to see the effect of changing parameters on economic indicators. The parameters used for the sensitivity analysis include: production, price, investment and operating costs.

## 3. Result and Discussion

### *3.1 Field production projection*
X field is an onshore exploration field that is projected to produce oil starting in the 11th year with a total reserve of 170 MMSTB. Table 1 is a table of forecasting results of X field production profiles for 30 years using the decline curve analysis method. In the 16th year onwards production is expected to decline due to decreased reservoir pressure.

**Table 1.** Forecasting result of X field production profiles

| Year | Oil Production (Bbl) | Year | Oil Production (Bbl) | Year | Oil Production (Bbl) |
|---|---|---|---|---|---|
| 0 – 10 | - | 17 | 4.238.548 | 24 | 1.546.850 |
| 11 | 1.776.057 | 18 | 3.670.107 | 25 | 1.339.399 |
| 12 | 2.843.525 | 19 | 3.177.901 | 26 | 1.159.769 |
| 13 | 4.552.575 | 20 | 2.751.705 | 27 | 1.004.230 |
| 14 | 5.821.041 | 21 | 2.382.668 | 28 | 869.550 |
| 15 | 5.623.846 | 22 | 2.063.123 | 29 | 752.933 |
| 16 | 4.895.032 | 23 | 1.786.433 | 30 | 651.955 |

### *3.2 Investment value planning*
Planning for funding needs in X field until the contract period is completed 30 years as shown in Table 2.

**Table 2.** Investment planning

| Year | Investment Cost MUS$ | | | OPEX |
|---|---|---|---|---|
| | Capital | Non Capital | Total | US$/Bbl |
| 0 | - | - | - | 9 |
| 1 | - | 1.94 | 1.94 | 9 |
| 2 | - | 8.01 | 8.01 | 9 |
| 3 | - | 208.01 | 208.01 | 9 |
| 4 | - | 572.03 | 572.03 | 9 |
| 5 | 393.71 | 2627.96 | 3021.67 | 9 |
| 6 | 787.41 | 12932.13 | 13719.54 | 9 |
| 7 | 2850.59 | 17854.15 | 20704.74 | 9 |
| 8 | 2618.39 | 16287.7 | 18906.09 | 9 |
| 9 | 2508.06 | 9437.7 | 11945.76 | 9 |
| 10 | 393.71 | 1901.19 | 2294.9 | 9 |
| 11 to 30 | - | - | - | 9 |

The price of oil used is the average price from January 2018 to October 2019, which is \$ 60 / bbl, while the Operating Cost (Opec) used in the calculation will be \$ 8 / bbl, an escalation factor of 2% per year.

*3.3 The determination of revenue sharing between the government and contractor*
The split between the contractor and the government for the gross split contract scheme has the addition of a split by the provisions of the applicable parameters, namely progressive split, and variable split. At the base split contractor, the share for oil is 57% for the government and 43% for the contractor. The magnitude of progressive split and variables split for the development of X field can be seen in Table 3.

**Table 3** Improvement split field "X"

| Component | Parameter | Split Correction |
|---|---|---|
| Variabel Split | | |
| Field Status | POD | 5% |
| Field Location | Onshore | 0% |
| Reservoir Depth | <2500m | 0% |
| Availability of Supporting Infrastructure | New Frontier | 4% |
| Reservoir Type | Conventional | 0% |
| $CO_2$ Content (%) | - | 0% |
| $H_2S$ Content (%) | - | 0% |
| Oil Specific Gravity | >25 | 0% |
| Component Level in the Country | 30-50 | 2% |
| Production Stages | Primary | 0% |
| Progressive Split | | |
| Oil Price | 65 | 5% |
| Cumulative Total Production | <30 | 10% |
| Contractor Split | Base split+Variabel+progressive | 69% |

*3.4 Economic review using gross split contract*

*3.4.1 Net cash flow.* The net cash flow development amount of X field as shown in Figure 2.

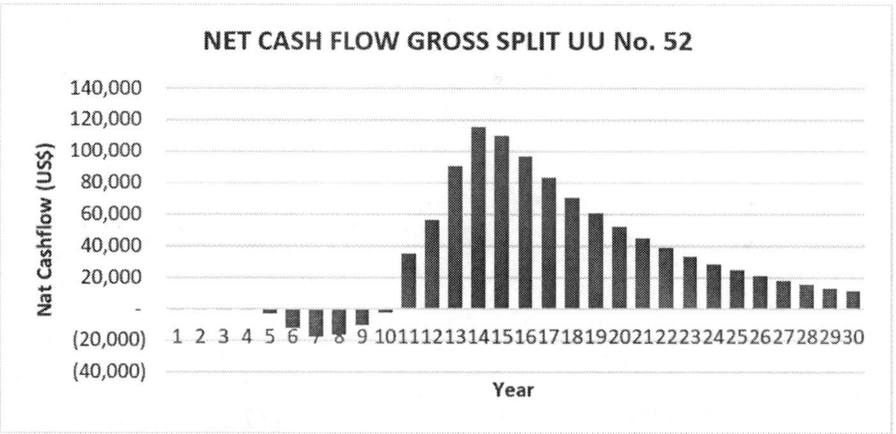


**Figure 2.** Net cash flow development of X field

Based on Figure 2 it can be seen that the contractor's net cash flow has a negative value from the first ten years because the x field has not produced yet, but after producing the cash flow chart has a positive value from year 11 to year 30.

*3.4.2 Oil production sharing between the government and contractor.* Based on the cash flow using the gross split contract system, the distribution of petroleum revenue sharing for the contractor is 27.7% and the government gets the oil sharing amounted to 72.3%. Oil production sharing chart between the government and the contractor as shown in Figure 3.

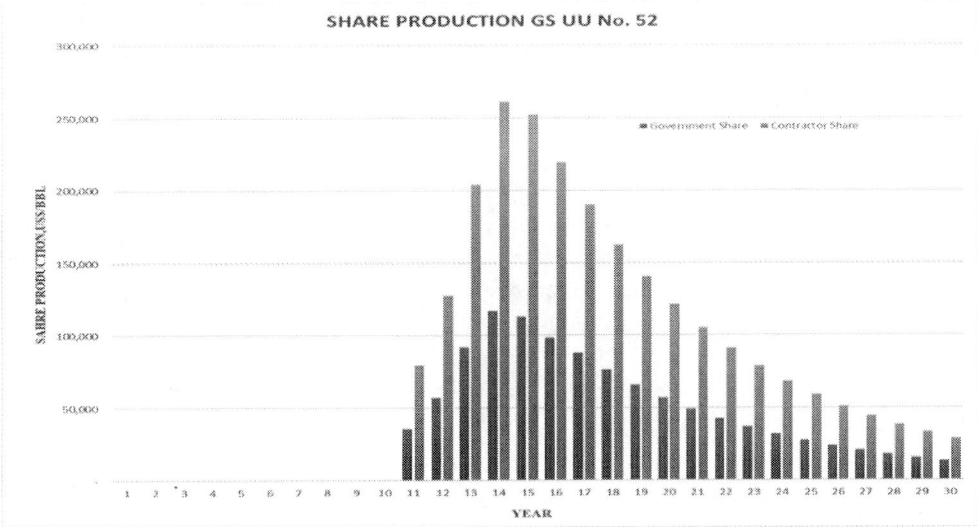


**Figure 3**. Oil production revenue sharing between the government and contractors

*3.4.3 Calculation of economic feasibility indicators.* To find out the economic value and project feasibility development of X field using the gross split contract sharing system is to calculate profit indicators in the form of Net Present Value (NPV) and Internal Rate of Return (IRR). The results of profit indicators calculation for development of oil field X using the gross split contract revenue sharing system are Contraktor NPV 15% = US$ M 192,63; Government NPV 15% = US$ M 415,235 and Contractor IRR = 41.04%. The results of profit indicators on the project above indicate that the NPV is Positive, an IRR greater than the desired MARR (i = 15%)

*3.5 Sensitivity analysis*

Sensitivity to NPV and IRR is economic indicators in field X using Gross Split contracts as shown in Figure 4. Sensitivity for some field conditions with the Gross split contract model is clear that production provides the largest graph slope, then the total price, investment and production costs.

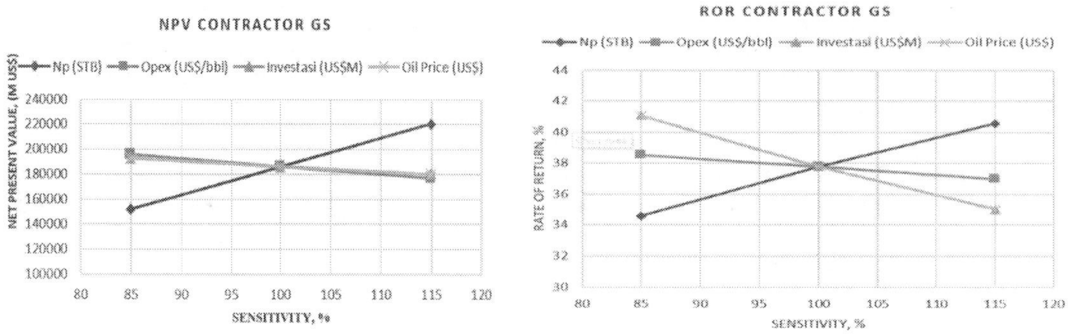


a. Contractor NPV b. Contractor ROR

**Figure 4**. Sensitiviy analysis

## 4. Conclusion

Based on the calculation results it was found that the development of onshore exploration X oil fields using Gross Split contracts is economically valuable and feasible for development. The Gross Split contract provides better and more attractive results to the contractor by providing additional intensive splits if the contractor does so according to specified requirements. Also, if in a field that is to be managed does not reach a certain economy, then based on Article 7 paragraph 1 the gross split ministerial regulation No. 52 of 2017 the Minister of Energy and Mineral Resources will provide additional split regardless of the split until the field is feasible to be managed. For further research, it is recommended to conduct an economic review based on the application of taxes issued by the government in managing the field using a gross split contract.

**Acknowledgement**

Thank you very much for supported by Research Institute and Community Services Universitas Islam Riau.

## 5. References

[1] Pinsent M 2017 *Indonesia's Gross Split PSC: Right structure, wrong split? Wood Mackenzie.*

[2] Michael K, Charles M and Philip D 2010 *The Taxation of Petroleum and Minerals: Principles, Problems and Practice.* Routledge/Internatioanl Monetary Fund.

[3] Widjajono P 1999 *Incentives for Marginal Field Development in Indoesia*. Petrominer.

[4] Muhammad A 2012 Studi Kebijakan Migas di Indonesia. *Journal of Earth Energy Engineering*, **1**, pp. 12-24.

[5] Mohamad J G and Ariel B 2018 Indonesia's New Gross Split PSC: Is It More Superior Than the Previous Standard PSC? in *International Conference on Petroleum and Petrochemical Engineering*.

[6] Surya T K and Jaenudin J 2017 *Proposed Modification of Abandonment and Site Restoration Mechanism in Gross Slit PSC for Marginal Field in Indonesia. SPE 186378.*

[7] Budhi R A, Imam B 2018 *Comparative analysis of financial Production Sharing Contract (PSC) cost recovery with PSC gross split; Case studiy in one of the contractor SKK Migas. Journal of Administrative Business Studies,* **4**, pp. 65-80.

[8] Muhammad A 2018 *Studi Perbandingan Keekonomian Pengembangan Lapangan Minyak Marginal menggunakan Production Sharing Contract dan Gross Split.* In Proceeding Seminar Nasional Teknologi dan Rekayasa.

IOP Conf. Series: Materials Science and Engineering **847** (2020) 012031 doi:10.1088/1757-899X/847/1/012031

# Preventive Maintenance Scheduling for Shifter Machine in Flour Mills

**Wahyukaton**

Industrial Engineering Department, Engineering Faculty, Universitas Pasundan
Jl. Setiabudi 193 Bandung, West Java, Indonesia
wahyukaton@unpas.ac.id

**Abstract.** PT. Indofood Sukses Makmur Tbk, Divisi Bogasari Flour Mills is one of biggest food factory especially for producing flour. In this flour mills division has several machines which are composed from several components, so that the maintenance division has to keep all the machines in a good performance. The factory has three shifter machines which were composed 3 critical components which are motor, bearing, and shifter and they were experienced 133 breakdowns in the last 6 months in 2019. The maintenance was conducted by determining a Mean Time to Failure (MTTF) value and conducted a preventive maintenance. The calculations result showed for 3 critical components were motor has a replacement interval time at 781 hours (day 98$^{th}$) with the availability value was 0,99914, bearing has a replacement interval time at 994 hours (day 124$^{th}$) with the availability value was 0,999255, and shifter has a replacement interval time at 575 hours (day 72$^{nd}$) with the availability value was 0,998729.
**Keywords**: preventive maintenance, MTTF, breakdown, replacement, interval, availability

## 1. Introduction

Maintenance concept is a set of various maintenance interventions (corrective, preventive, condition-based, etc.) and the general structure in which these interventions are brought together. Developing an appropriate maintenance concept is important because of the high direct and indirect costs and because of the operational impact maintenance [1]. Preventive maintenance is an activity to maintain facilities and to fix it, to do a need adjustments or replacements to achieve production process to meet the plan [2]. Preventive maintenance is a combination of all technique, administration actions including monitoring to make sure the systems are properly functioned [3].

PT. ISM Tbk., Bogasari Flour Mills Division has three shifter machines which were composed 3 critical components which are motor, bearing, and shifter and they were experienced 133 breakdowns in the last 6 months in 2019. Figure 1 shows that the critical component from the shifter machine can be determined that there are 3 critical components which are sitter, motor, and bearing. The objectives of this paper are to find the optimum replacement interval time for critical components, and to find the availability value.

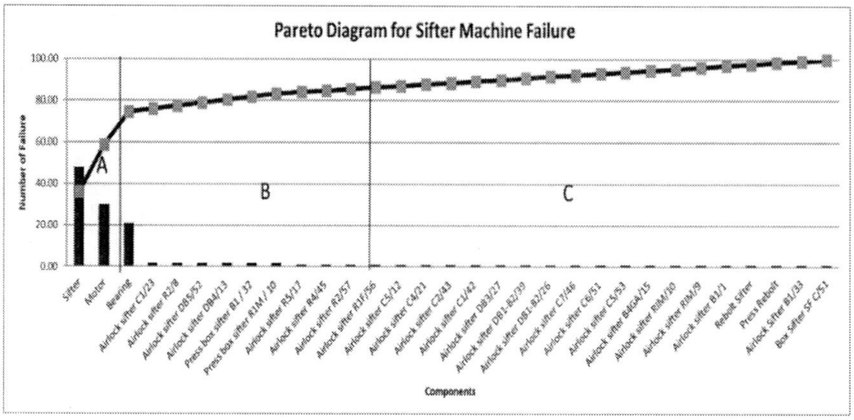


**Figure 1**. Pareto Diagram for Shifter Machine Failure

## 2. Methods

Preventive maintenance is an important element of a maintenance activity and within a maintenance department it normally accounts for a significant proportion of the overall maintenance activity. Preventive maintenance is the care and servicing by maintenance personnel to keep facilities in a satisfactory operational state by providing for systematic inspection, detection, and correction of incipient failures either before their development into major failures or before their occurrence [4,5].

There are many objectives of performing preventive maintenance including improving capital equipment's productive life, reducing production losses caused by equipment failure, minimizing critical equipment breakdowns, and improving the health and safety of maintenance personnel [6].
This research was constructed as follow,

*2.1 Determining Critical Component.*
This step has been done by constructing a pareto diagram as shown in Figure 1, that there are 3 critical components.

*2.2 Calculating Time to Failure (TTF).*
Time to Failure can be calculated by the time component has been repaired until that component will experiencing a failure.

**Table 1**. Motor Failure Data

| No | Date | Machine No | TTF (Hrs) | No | Date | Machine No | TTF (Hrs) |
|---|---|---|---|---|---|---|---|
| 1 | 21/01/2010 | 1252-KLKO-SF | 0.00 | 16 | 28/05/2015 | 1252-KLKO-SF | 1528.00 |
| 2 | 01/02/2010 | 1252-KLKO-SF | 88.00 | 17 | 24/11/2015 | 1252-KLKO-SF | 1440.00 |
| 3 | 08/02/2010 | 1252-KLKO-SF | 56.00 | 18 | 08/01/2016 | 1252-KLKO-SF | 360.00 |
| 4 | 03/01/2012 | 1252-KLKO-SF | 5552.00 | 19 | 08/03/2016 | 1252-KLKO-SF | 480.00 |
| 5 | 19/04/2012 | 1252-KLKO-SF | 856.00 | 20 | 16/02/2017 | 1252-KLKO-SF | 2760.00 |
| 6 | 27/06/2012 | 1252-KLKO-SF | 552.00 | 21 | 20/03/2017 | 1252-KLKO-SF | 256.00 |
| 7 | 17/09/2012 | 1252-KLKO-SF | 656.00 | 22 | 14/07/2017 | 1252-KLKO-SF | 928.00 |
| 8 | 05/11/2012 | 1252-KLKO-SF | 392.00 | 23 | 18/09/2017 | 1252-KLKO-SF | 528.00 |
| 9 | 27/05/2013 | 1252-KLKO-SF | 1624.00 | 24 | 16/10/2017 | 1252-KLKO-SF | 224.00 |
| 10 | 09/09/2013 | 1252-KLKO-SF | 840.00 | 25 | 29/01/2018 | 1252-KLKO-SF | 840.00 |
| 11 | 28/10/2013 | 1252-KLKO-SF | 392.00 | 26 | 27/02/2018 | 1252-KLKO-SF | 232.00 |
| 12 | 24/11/2013 | 1252-KLKO-SF | 216.00 | 27 | 27/05/2018 | 1252-KLKO-SF | 712.00 |
| 13 | 03/03/2014 | 1252-KLKO-SF | 792.00 | 28 | 14/10/2018 | 1252-KLKO-SF | 1120.00 |
| 14 | 06/05/2014 | 1252-KLKO-SF | 512.00 | 29 | 04/03/2019 | 1252-KLKO-SF | 1128.00 |
| 15 | 18/11/2014 | 1252-KLKO-SF | 1568.00 | 30 | 24/05/2019 | 1252-KLKO-SF | 648.00 |

IOP Conf. Series: Materials Science and Engineering **847** (2020) 012031 doi:10.1088/1757-899X/847/1/012031

**Table 2**. Bearing Failure Data

| No | Date | Machine No | TTF (Hrs) |
|---|---|---|---|
| 1 | 14/05/2010 | 1252-KLK0-SF | 0.00 |
| 2 | 30/04/2011 | 1252-KLK0-SF | 2808.00 |
| 3 | 09/06/2011 | 1252-KLK0-SF | 320.00 |
| 4 | 20/06/2011 | 1252-KLK0-SF | 88.00 |
| 5 | 04/09/2013 | 1252-KLK0-SF | 6456.00 |
| 6 | 22/11/2013 | 1252-KLK0-SF | 632.00 |
| 7 | 18/07/2014 | 1252-KLK0-SF | 1904.00 |
| 8 | 30/10/2014 | 1252-KLK0-SF | 832.00 |
| 9 | 25/01/2015 | 1252-KLK0-SF | 696.00 |
| 10 | 30/01/2015 | 1252-KLK0-SF | 40.00 |
| 11 | 02/04/2015 | 1252-KLK0-SF | 496.00 |
| 12 | 27/05/2015 | 1252-KLK0-SF | 440.00 |
| 13 | 09/08/2015 | 1252-KLK0-SF | 592.00 |
| 14 | 27/10/2015 | 1252-KLK0-SF | 632.00 |
| 15 | 07/04/2016 | 1252-KLK0-SF | 1304.00 |
| 16 | 22/07/2016 | 1252-KLK0-SF | 848.00 |
| 17 | 20/10/2016 | 1252-KLK0-SF | 720.00 |
| 18 | 03/12/2016 | 1252-KLK0-SF | 352.00 |
| 19 | 24/02/2017 | 1252-KLK0-SF | 664.00 |
| 20 | 19/07/2017 | 1252-KLK0-SF | 1160.00 |
| 21 | 05/05/2019 | 1252-KLK0-SF | 5240.00 |

**Table 3**. Shifter Failure Data

| No | Date | Machine No | TTF (Hrs) |
|---|---|---|---|
| 1 | 22/02/2010 | 1252-KLK0-SF | 0 |
| 2 | 02/04/2010 | 1252-KLK0-SF | 312 |
| 3 | 17/04/2010 | 1252-KLK0-SF | 120 |
| 4 | 03/01/2012 | 1252-KLK0-SF | 5008 |
| 5 | 17/09/2012 | 1252-KLK0-SF | 2064 |
| 6 | 14/01/2013 | 1252-KLK0-SF | 952 |
| 7 | 28/02/2013 | 1252-KLK0-SF | 360 |
| 8 | 01/06/2013 | 1252-KLK0-SF | 744 |
| 9 | 21/06/2013 | 1252-KLK0-SF | 160 |
| 10 | 15/01/2014 | 1252-KLK0-SF | 1664 |
| 11 | 11/04/2014 | 1252-KLK0-SF | 688 |
| 12 | 22/05/2014 | 1252-KLK0-SF | 328 |
| 13 | 08/06/2014 | 1252-KLK0-SF | 136 |
| 14 | 14/06/2014 | 1252-KLK0-SF | 48 |
| 15 | 02/09/2014 | 1252-KLK0-SF | 640 |
| 16 | 29/12/2014 | 1252-KLK0-SF | 944 |
| 17 | 21/02/2015 | 1252-KLK0-SF | 432 |
| 18 | 29/03/2015 | 1252-KLK0-SF | 288 |
| 19 | 30/04/2015 | 1252-KLK0-SF | 256 |
| 20 | 28/05/2015 | 1252-KLK0-SF | 224 |
| 21 | 09/07/2015 | 1252-KLK0-SF | 336 |
| 22 | 12/09/2015 | 1252-KLK0-SF | 520 |
| 23 | 08/10/2015 | 1252-KLK0-SF | 208 |
| 24 | 12/02/2016 | 1252-KLK0-SF | 1016 |
| 25 | 24/03/2016 | 1252-KLK0-SF | 328 |
| 26 | 16/04/2016 | 1252-KLK0-SF | 184 |
| 27 | 17/05/2016 | 1252-KLK0-SF | 248 |
| 28 | 19/09/2016 | 1252-KLK0-SF | 1000 |
| 29 | 14/11/2016 | 1252-KLK0-SF | 448 |
| 30 | 05/08/2017 | 1252-KLK0-SF | 2112 |
| 31 | 31/08/2017 | 1252-KLK0-SF | 208 |
| 32 | 04/11/2017 | 1252-KLK0-SF | 520 |
| 33 | 11/11/2017 | 1252-KLK0-SF | 56 |
| 34 | 04/12/2017 | 1252-KLK0-SF | 184 |
| 35 | 29/01/2018 | 1252-KLK0-SF | 448 |
| 36 | 19/02/2018 | 1252-KLK0-SF | 168 |
| 37 | 22/02/2018 | 1252-KLK0-SF | 24 |
| 38 | 27/02/2018 | 1252-KLK0-SF | 40 |
| 39 | 13/03/2018 | 1252-KLK0-SF | 112 |
| 40 | 27/05/2018 | 1252-KLK0-SF | 600 |
| 41 | 06/06/2018 | 1252-KLK0-SF | 80 |
| 42 | 11/06/2018 | 1252-KLK0-SF | 40 |
| 43 | 11/08/2018 | 1252-KLK0-SF | 488 |
| 44 | 13/08/2018 | 1252-KLK0-SF | 16 |
| 45 | 12/09/2018 | 1252-KLK0-SF | 240 |
| 46 | 30/09/2018 | 1252-KLK0-SF | 144 |
| 47 | 02/11/2018 | 1252-KLK0-SF | 264 |
| 48 | 21/05/2019 | 1252-KLK0-SF | 1600 |

### *2.3 Goodness of Fit Test.*

A simulation software ARENA was used in this test. The smallest error will determine the fit of the distribution. After several tests using ARENA, the result is all failures are fit to Weibull Distribution.

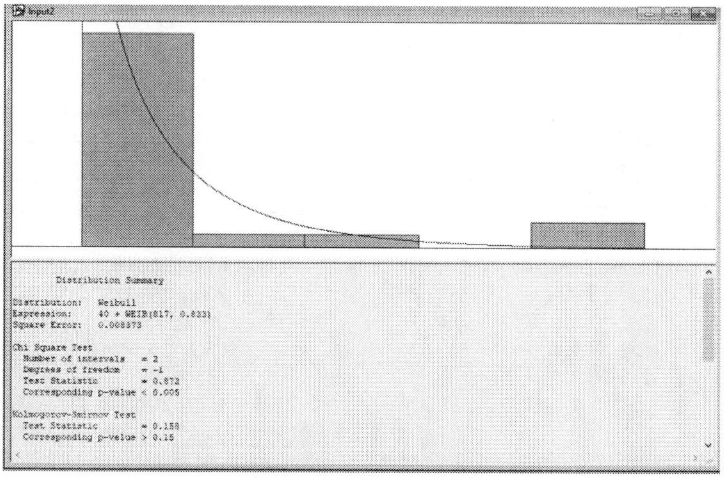


**Figure 2**. Goodness of Fit test Result for Motor Component

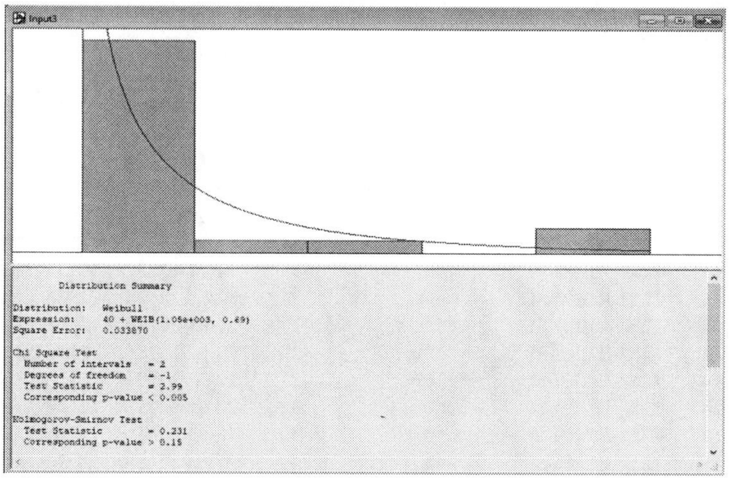


**Figure 3**. Goodness of Fit test Result for Bearing Component

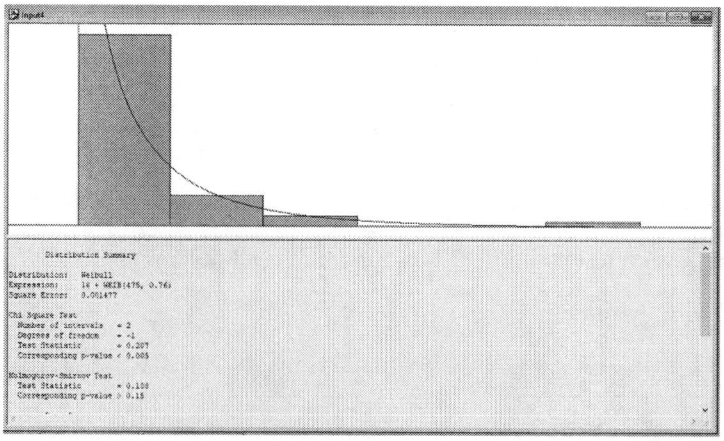


**Figure 4**. Goodness of Fit test Result for Shifter Component

IOP Conf. Series: Materials Science and Engineering **847** (2020) 012031 doi:10.1088/1757-899X/847/1/012031

*2.4 Calculating Mean Time to Failure (MTTF).*

To do this calculation, first it has to find the parameters for the distribution which was done by goodness of fit test, then using the equation of Weibull Maximum Likelihood Estimator (MLE) as follows.
The first derivation of likelihood function for Weibull distribution is to find a parameter β [6]

$$g\left(\hat{\beta}_j\right) = \frac{\left(\sum_{i=1}^{r} t_i^{\beta} \ln t_i\right)+(n-r)t_s^{\beta} \ln t_s}{\left(\sum_{i=1}^{r} t_i^{\beta}\right)+(n-r)t_s^{\beta}} - \frac{1}{\beta} - \frac{1}{r}\sum_{i=1}^{r} \ln t_i = 0 \quad (1)$$

To solve the equation, another method is used, Newton-Rhapson method to solve the equation

$$\hat{\beta}_{j+1} = \hat{\beta}_j - \frac{g(\hat{\beta}_j)}{g'(\hat{\beta}_j)} \text{ where } g'(x) = \frac{dg(x)}{dx} \quad (2)$$

The equation can be solved by integration method to find $\hat{\beta}_j$ value that is maximum to closing 0. The first derivation is

$$g'\left(\hat{\beta}_j\right) = \frac{\left(\sum_{i=1}^{r} t_i^{\beta} \ln^2 t_i\right)\left(\sum_{i=1}^{r} t_i^{\beta}\right)-\left(\sum_{i=1}^{r} t_i^{\beta} \ln t_i\right)}{\left(\sum_{i=1}^{r} t_i^{\beta}\right)} - \frac{1}{\beta} \quad (3)$$

The $\hat{\beta}_j$ value is β = b, can be found by least square curve fitting. The parameter is:

$$\theta = \left[\frac{1}{r}\left\{\sum_{i=1}^{r} t_i^{\beta} + (n-r)t_s^{\beta}\right\}\right]^{\frac{1}{\beta}} \quad (4)$$

**Table 4**. Result of MTTF

| No | Component | Distribution | Parameter | MTTF |
|---|---|---|---|---|
| 1 | Motor | Weibull | β = 940.6896<br>θ = 1.218252 | 781.5047 |
| 2 | Bearing | Weibull | β = 0.952320<br>θ = 953.3532 | 974.5278 |
| 3 | Shifter | Weibull | β = 0.997097<br>θ = 574.4680 | 575.1772 |

*2.5 Determining Component Replacement Interval, Optimum Replacement, Availability Value.*

The Age-Replacement Strategy was used in this step. Calculating the component age from the first time it is installed until this component will having breakdown [6]. If within the determined interval there is no failure, the component is still need to be replaced, this will be the objective of Preventive Maintenance. The availability value can be calculated by these equations,

$$F(t_p) = 1 - e\left[-\left(\frac{t_p}{\theta}\right)\right]^{\beta} \quad (5)$$

$$R(t_p) = e\left[-\left(\frac{t_p}{\theta}\right)\right]^{\beta} \quad (6)$$

$$M(t_p) = \frac{t_p}{F(t_p)} \quad (7)$$

$$D(t_p) = \frac{R(t_p)+F(t_p)}{t_p.R(t_p)+M(t_p)+F(t_p)} \quad (8)$$

$$A(t_p) = 1 - D(t_p) \quad (9)$$

**Table 5**. Values of $F(t_p)$, $D(t_p)$min, $A(t_p)$

| Component | $F(t_p)$ | $D(t_p)$min | $A(t_p)$ |
|---|---|---|---|
| Motor | 0.636306 | 0.00085796 | 0.999142 |
| Bearing | 0.629508 | 0.00074471 | 0.999255 |
| Shifter | 0.631392 | 0.00127044 | 0.998729 |

IOP Conf. Series: Materials Science and Engineering **847** (2020) 012031 doi:10.1088/1757-899X/847/1/012031

## 3. Result and Discussion

After calculation of all parameters of preventive maintenance were done, then the reliability of 3 critical components were compared for before and after preventive maintenance was conducted, and the results are as shown on these figures,

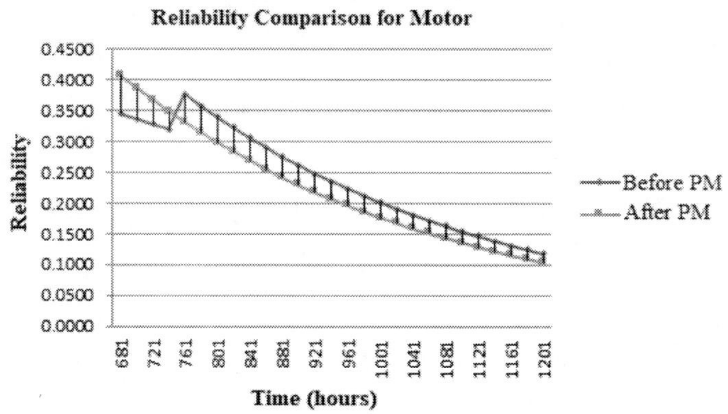


**Figure 5**. Reliability Comparison for Motor Before and After Preventive Maintenance

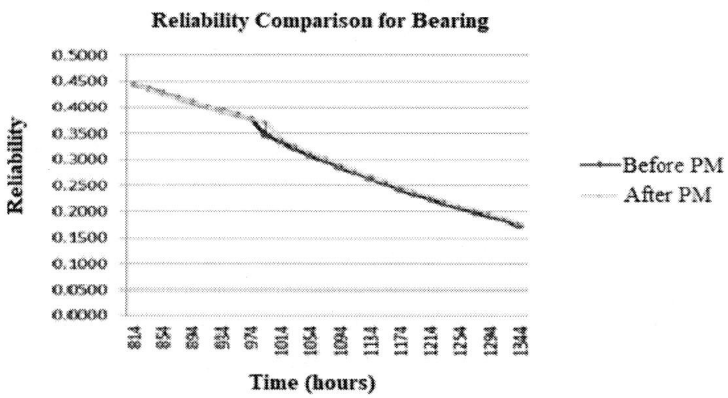


**Figure 6**. Reliability Comparison for Bearing Before and After Preventive Maintenance

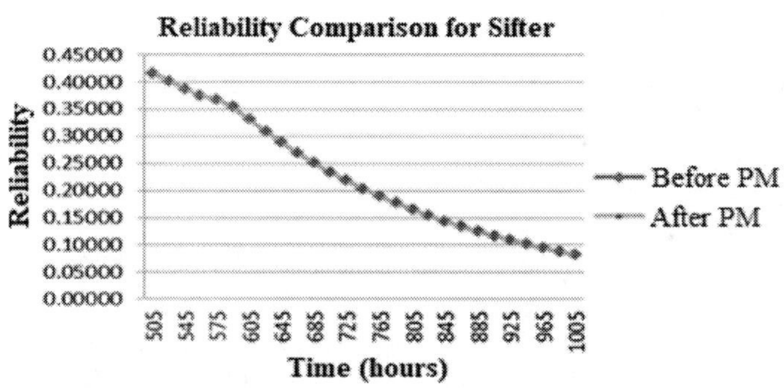


**Figure 7**. Reliability Comparison for Shifter Before and After Preventive Maintenance

Figure 3 shows that reliability value after preventive maintenance is less than reliability value before maintenance. This means that motor component has to be replaced immediately in order to maintain the good performance of the machine.

Figure 4 and Figure 5 show that the value of reliability before and after preventive maintenance are relatively equal. This means that the components are in good performance.

From all preventive maintenance parameters which have been calculated, it is possible to scheduling the components replacement interval, for motor component is in 781 hours or on the day of $98^{th}$, the optimum replacement interval for bearing component is in 994 hours or on the day of $124^{th}$, and the optimum replacement interval for shifter component is in 557 hours or on the day of $72^{nd}$.

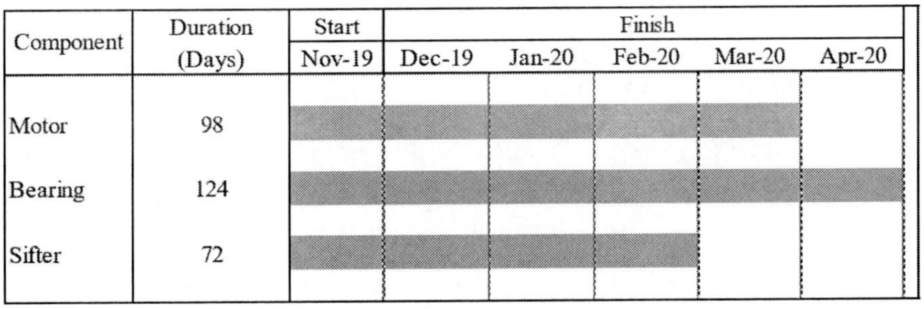


**Figure 8.** Gantt Chart for Scheduling Optimum Replacement Interval for 3 Critical Components

## 4. Conclusion

Based on the values of preventive maintenance parameters, the optimum replacement interval for motor component is in 781 hours or on the day of $98^{th}$, with availability 0.999142. Suppose the component starts on November 2019 then it will be scheduled to be replaced on March 2020. The optimum replacement interval for bearing component is in 994 hours or on the day of $124^{th}$, with availability 0.999255. Suppose the component starts on November 2019 then it will be scheduled to be replaced on April 2020. The optimum replacement interval for shifter component is in 557 hours or on the day of $72^{nd}$, with availability 0.998729. Suppose the component starts on November 2019 then it will be scheduled to be replaced on February 2020.

## 5. References

[1] Pophaley, Mahesh, Krishna Vyas, Ram, 2010. Plant maintenance management practices in automobile industries: A retrospective and literature review. *Journal of Industrial Engineering and Management*, pp. 512-541

[2] O'Connor, P. D. T., and Kleyner, A. 2011. *Practical Reliability Engineering, Design for Reliability, 5th Edition*. (New York: John Wiley & Sons).

[3] Basri, E. I. Abdul Razak, I.H., Ab-Samat H., and Komaruddin S., 2017. Preventive Maintenance Planning: A Review. *Journal of Quality in Maintenance Engineering*, **23**(2), 114-143, doi: 10.1108/JQME-04-2016-0014

[4] Dhillon, B. S. 2006. *Maintainability, Maintenance, and Reliability for Engineers*. Taylor & Francis Group. Boca Raton FL.

[5] Campbell, J., Jardine, A. K. S., and McGlynn, J. 2010. *Asset management excellence: optimizing equipment life-cycle decisions.* In Dekker Mechanical Engineering.

[6] Ebeling, Charles, 1996. Charles Ebeling - *An Introduction To Reliability and Maintainability Engineering*, McGraw-Hill.

# Performance Evaluation of an EWMA p Chart Based on Improved Square Root Transformation to detect Small Shift Process Variation

**Arumsari** [1,2], **and J N Rifqi** [1,3]

[1]Industrial Engeneering Departemen, Pasundan University, Bandung, Indonesia
[2]aarum310355@yahoo.com, [3]rifqijanuar23@gmail.com

**Abstract.** A control chart was used to detect whether the unexpected variation due to special causes was happening. Generally, Shewhart control chart was used, while this chart was not a good one to control the process with small variation. EWMA p chart was one of the alternatives used to monitor small shifts in the process mean. Traditionally, attribute Shewhart p chart has been developed for charting the binomial data or using the normal approximation when percent defective was low and sample size was small. In 100% inspection, these assumptions were not valid. For this reason, EWMA p control chart was modified by improving square root transformation. In this research, the implementation of the modified EWMA p control chart based on Improved square root transformation (ISRT p EWMA chart) was studied. This chart was used to monitor the actual production process where the process has shifted. The research found that. The ISRT p EWMA chart detects the shift twice, while Shewhart p chart does not detect any shift. The research also found that at $\lambda = 0.25$, the shift is detected 2 times, at $\lambda = 0.5$, the shift is detected once while at $\lambda = 0.75$, no shift is detected. The larger the value of the $\lambda$ parameter, the worse the ISRT p EWMA control chart performed.
**Keywords:** Shewhart Control Chart, EWMA p Chart, Small Shift, Square Root Transformation

## 1. Introduction
Statistical Process Control (SPC) is applied for inspecting production or service processes. Typically, it was carried out by control charts. Shewhart developed a control chart for monitoring the defective of a manufacturing process called p chart in early 20th century. The general idea of a control chart consists of signalling an alarm if a specified control limits was exceeded. An alarm will lead to stopping the process in order to prevent from excessive defective unit and restore the process back [3]. Shewhart control charts is characterized by taking only the most recent sample into account. Therefore, Shewhart control charts perform very well to detect a large change in the process parameters but it is not a good one when it used for small shift [4][6]. Nowadays the Shewhart control chart can also be used to attain a sustained continuous quality improvement [6]. The further development of the control chart fundamentals has progressed rapidly. Numerous enhancements of the p chart exist to monitor other process parameters. One of them was developed from Shewhart concept where it takes the previous samples into account. This idea leads to a better performance to detect a small change in the process parameters. Exponential weighted moving average (EWMA) chart and cumulative sum (CUSUM) chart are approaches that considers not only the last sample but it also considers the information about the process contained in the rest of the samples. The CUSUM chart was developed by Page [4]. It is an

alternative type of control chart used to monitor small shifts in the process mean which uses the cumulative sum of deviations from a target. The CUSUM chart plots the cumulative sum of deviations from the target for individual measurements or subgroup means. CUSUM charts improved the ability to detect small shifts (i.e. less than 1.5σ) by charting a statistic that incorporates current and previous data values from the process [4]. CUSUM charts, while not as simple to operate as Shewhart charts, it shown to be more efficient in detecting small shifts in the mean of a process [1]. The EWMA chart is also an alternative to the Shehwart chart in detecting small shifts in the mean process. It introduced by Roberts [7] and was originally called a geometric moving average chart. The name was changed to reflect the fact that exponential smoothing serves as the basis of EWMA charts. Specifically, EWMA was also used on individual observations. It was assumed that observation of the process is normally distributed with mean $\mu$ and variance $\sigma^2$.

Most research proved that the EWMA charts, performed better in detecting the small shift in the manufacturing process compared to Shewhart chart [5][6][7][9] while EWMA chart was more effective than CUCUM chart [2]. EWMA chart is preferable because it is easier to use [7]. EWMA chart is developed for variable and attribute data. Study on EWMA chart for variable was easily found. In this paper the research will focus on EWMA chart for attribute data. EWMA p chart for attribute was introduced by Page [4]. Generally, the traditional EWMA p chart has been developed for charting attribute data characterized by binomial data. This chart can also be developed using the normal approximation with condition as low defect level and the small to moderate sample size where $np < 5$ [8]. In situation where the process is monitored under 100% inspection, these assumptions was not valid. Although the defect level was small, but the sample size was large so that normal assumption is not matched. For this reason, EWMA p chart for detecting a change is modified by improving square root transformation. The application of the modified EWMA p chart based on Improved square root transformation namely ISRT p EWMA is studied. A comparative study of the performance of ISRT p EWMA and the performance of Shewhart p chart is analysed

The objective of the research is how to monitor a small shift process variation using ISRT p EWMA chart and whether ISRT p EWMA chart performed better to detect small shift of process variation compare to Shewhart p chart?

The performance of ISRT p EWMA chart will be studied under the assumptions that monitoring process is done on production line of shoes assembly at PT. Primarindo Asia Infrastruktur Tbk. during March-May 2019. The shift of the process mean is less than 2σ. Parameter L and λ is set at best recommended value from previous research and no extra cost when using the ISRT p EWMA chart.

## 2. Methods

To evaluate the performance of the ISRT p EWMA chart, the chart will be used to monitor on the actual production process of shoes assembly at PT Primarindo Asia Infrastruktur Tbk where all product is 100% inspected. The average process is assumed to be shifting. Usually it was monitored using Shewhart p chart. As a result, when the process was actually shifting, the shift was undetected, and the process continues to run without any action to prevent from excessive defective unit. Therefore, defective unit cannot be reduced. An alternative control chart is needed to replace Shewhart p chart. The ISRT p EWMA chart was chosen to monitor the process under 100% inspection where normality assumption was not fulfilled ($np < 5$) due to large sample size.

The performance of ISRT p EWMA chart was evaluated using the following step:

*First Step. Estimate the $\mu$ and $\sigma$ of the population:*

1. Set initial observation

   Parameter was estimated through initial observation during March 2019. The detail observation presented in Table 1. Total 13.817unit production was inspected during March resulting 535 unit defective or $\bar{p} = \frac{\sum \bar{p_i}}{m} = \frac{535}{13817} = 0.0387$

2. Set parameter $\lambda$ and $L$ for ISRT p EWMA chart.
   $\lambda$ is the weight of past information where $0 < \lambda < 1$, while $L$ is coefficient of control limit of ISRT p EWMA. From previous research, these parameters were set at $\lambda$=0.25 and $L$=0.2998.
3. Convert the initial observation data into the $Z$ Value.
   The ISRT p EWMA chart will plot the value of the weighted moving average ($Z_i$). For this purpose, each data observation was converted into $Z_i$ value using equation (1).
   $$Z_i = \lambda \bar{p}_i + (1-\lambda) Z_{i-1} \quad (1)$$
   Substitute $Z_0 = \bar{p} = 0{,}0387$ using parameter $\lambda$=0.25 and $L$=0.2998 into equation (1) resulting
   $Z_1$ = (0.25 x 0,020) + (1 – 0.25) 0.0387 = 0.0343 and
   $Z_2$= (0.25 x 0,052) + (1 – 0.25) 0.0343 = 0.0389
   Recalculate $Z_i$ for all of the observation and the overall value of $Z_i$ is presented in the 5th column of Table 1.
4. Set up trial control limit UCL and LCL of ISRT p EWMA chart using equation (2) and (3)
   $$UCL_{\text{ISRT p EWMA}} = Z_0 + L\left[\frac{\lambda}{2-\lambda}\left(\frac{1}{2}\sqrt{\frac{1-p}{n}} - \frac{1}{6}\left(\frac{1-p}{n\sqrt{p}}\right)\right)\right] \quad (2)$$
   $$LCL_{\text{ISRT p EWMA}} = Z_0 - L\left[\frac{\lambda}{2-\lambda}\left(\frac{1}{2}\sqrt{\frac{1-p}{n}} - \frac{3}{8}\left(\frac{1-p}{n\sqrt{p}}\right)\right)\right] \quad (3)$$
   Substitute $Z_0$ = 0.0387 and n=572 for the first data using parameter L=2,998 and $\lambda$=0.25 into equation
   (2) and (3) resulting UCL and LCL for ISRT p EWMA as follow:
   $$UCL_{\text{ISRT p EWMA}} = 0.0387 + 2{,}998\left[\frac{0{,}25}{2-0{,}25}\left(\frac{1}{2}\sqrt{\frac{1-0{,}0387}{572}} - \frac{1}{6}\left(\frac{1-0{,}0387}{572\sqrt{0{,}0387}}\right)\right)\right] = 0.0466$$
   $$UCL_{\text{ISRT p EWMA}} = 0.0387 + 2{,}998\left[\frac{0{,}25}{2-0{,}25}\left(\frac{1}{2}\sqrt{\frac{1-0{,}0387}{572}} - \frac{3}{8}\left(\frac{1-0{,}0387}{572\sqrt{0{,}0387}}\right)\right)\right] = 0.0311$$
   The control limit of ISRT p EWMA chart for each initial observation was displayed in Table 1.
5. Decide whether the parameter and trial control limit represent the incontrol situation so that the ISRT p EWMA chart can be used as a bias estimate of mean and variance and used it to monitor the production process. Table 1 indicated that the process was in control so that the parameter can be used to monitor the production process for April. The parameters of ISRT p EWMA chart were $\mu$ = 0,0387, $\sigma$ = 0,0016, $Z_0$ = 0,0385, and the control limits were $UCL_{\text{ISRT p EWMA}}$ = 0,0476 and $LCL_{\text{ISRT p EWMA}}$ = 0,0303
6. To evaluate the performance of the ISRT p EWMA chart, the same observation data was then controlled using Shewhart p chart. The above step is done for the same observation data in Table 1. Each initial observation data was then compiled with control limits UCL and LCL of Shewhart p chart which were calculated using equation (4) and (5) at $\bar{p}$ = 0.0387 as follow:
   $$UCL_{\text{p chart}} = \bar{p} + 3\sqrt{\frac{\bar{p}(1-\bar{p})}{n_i}} \quad (4)$$
   $$LCL_{\text{p chart}} = \bar{p} - 3\sqrt{\frac{\bar{p}(1-\bar{p})}{n_i}} \quad (5)$$
7. Substitute $\bar{p} = 0{,}0387$ and n=575 for 1st observation into equation (4) and (5) resulted the following $UCL_{\text{p chart}}$ and $LCL_{\text{p chart}}$
   $$UCL_{\text{p chart}} = 0.0387 + 3\sqrt{\frac{0{,}0387(1-0{,}0387)}{575}} = 0.0629 \text{ and}$$
   $$LCL_{\text{p chart}} = 0.0387 - 3\sqrt{\frac{0{,}0387(1-0{,}0387)}{575}} = 0.0146$$
   Recalculate all observation data in Table 1 resulted the $UCL_{\text{p chart}}$ and $LCL_{\text{p chart}}$ as presented in 9th and 10th column of Table 1.

**Table 1**. Initial observation to estimate parameter of ISRT p EWMA Chart and Shewhart p Chart.

| No | N | Np | p | Control Limit for ISRT p EWMA | | | | Control Limit for p Chart | | |
|---|---|---|---|---|---|---|---|---|---|---|
| | | | | Zi | UCL | LCL | CL | UCL | LCL | CL |
| 1 | 575 | 12 | 0,020 | 0,0343 | 0,0466 | 0,0311 | 0,0385 | 0,0629 | 0,0146 | 0,0387 |
| 2 | 512 | 27 | 0,052 | 0,0389 | 0,0471 | 0,0308 | 0,0385 | 0,0643 | 0,0131 | 0,0387 |
| 3 | 309 | 10 | 0,032 | 0,0372 | 0,0493 | 0,0291 | 0,0385 | 0,0716 | 0,0058 | 0,0387 |
| 4 | 381 | 13 | 0,034 | 0,0365 | 0,0483 | 0,0298 | 0,0385 | 0,0684 | 0,0091 | 0,0387 |
| 5 | 454 | 15 | 0,033 | 0,0356 | 0,0476 | 0,0304 | 0,0385 | 0,0659 | 0,0116 | 0,0387 |
| 6 | 314 | 17 | 0,054 | 0,0402 | 0,0492 | 0,0292 | 0,0385 | 0,0714 | 0,0061 | 0,0387 |
| 7 | 487 | 18 | 0,037 | 0,0394 | 0,0473 | 0,0306 | 0,0385 | 0,0649 | 0,0125 | 0,0387 |
| 8 | 423 | 16 | 0,037 | 0,0390 | 0,0479 | 0,0301 | 0,0385 | 0,0669 | 0,0106 | 0,0387 |
| 9 | 476 | 24 | 0,050 | 0,0419 | 0,0474 | 0,0305 | 0,0385 | 0,0652 | 0,0122 | 0,0387 |
| 10 | 601 | 24 | 0,039 | 0,0414 | 0,0465 | 0,0312 | 0,0385 | 0,0623 | 0,0151 | 0,0387 |
| 11 | 596 | 20 | 0,033 | 0,0394 | 0,0465 | 0,0312 | 0,0385 | 0,0624 | 0,0150 | 0,0387 |
| 12 | 412 | 14 | 0,034 | 0,0381 | 0,0480 | 0,0301 | 0,0385 | 0,0672 | 0,0102 | 0,0387 |
| 13 | 434 | 14 | 0,032 | 0,0366 | 0,0478 | 0,0302 | 0,0385 | 0,0665 | 0,0109 | 0,0387 |
| 14 | 478 | 15 | 0,031 | 0,0353 | 0,0474 | 0,0305 | 0,0385 | 0,0652 | 0,0122 | 0,0387 |
| 15 | 372 | 19 | 0,051 | 0,0392 | 0,0484 | 0,0297 | 0,0385 | 0,0687 | 0,0087 | 0,0387 |
| 16 | 495 | 16 | 0,032 | 0,0375 | 0,0472 | 0,0306 | 0,0385 | 0,0647 | 0,0127 | 0,0387 |
| 17 | 367 | 12 | 0,032 | 0,0363 | 0,0485 | 0,0297 | 0,0385 | 0,0689 | 0,0085 | 0,0387 |
| 18 | 551 | 21 | 0,038 | 0,0368 | 0,0468 | 0,0310 | 0,0385 | 0,0634 | 0,0141 | 0,0387 |
| 19 | 611 | 30 | 0,049 | 0,0398 | 0,0464 | 0,0313 | 0,0385 | 0,0621 | 0,0153 | 0,0387 |
| 20 | 430 | 23 | 0,053 | 0,0433 | 0,0478 | 0,0302 | 0,0385 | 0,0666 | 0,0108 | 0,0387 |
| 21 | 521 | 18 | 0,034 | 0,0411 | 0,0470 | 0,0308 | 0,0385 | 0,0641 | 0,0134 | 0,0387 |
| 22 | 320 | 10 | 0,031 | 0,0386 | 0,0491 | 0,0292 | 0,0385 | 0,0711 | 0,0064 | 0,0387 |
| 23 | 510 | 21 | 0,041 | 0,0393 | 0,0471 | 0,0307 | 0,0385 | 0,0643 | 0,0131 | 0,0387 |
| 24 | 552 | 27 | 0,048 | 0,0417 | 0,0468 | 0,0310 | 0,0385 | 0,0634 | 0,0141 | 0,0387 |
| 25 | 377 | 12 | 0,031 | 0,0392 | 0,0484 | 0,0298 | 0,0385 | 0,0685 | 0,0089 | 0,0387 |
| 26 | 490 | 16 | 0,032 | 0,0376 | 0,0473 | 0,0306 | 0,0385 | 0,0649 | 0,0126 | 0,0387 |
| 27 | 322 | 10 | 0,031 | 0,0359 | 0,0491 | 0,0292 | 0,0385 | 0,0710 | 0,0065 | 0,0387 |
| 28 | 414 | 17 | 0,041 | 0,0372 | 0,0480 | 0,0301 | 0,0385 | 0,0672 | 0,0103 | 0,0387 |
| 29 | 577 | 26 | 0,045 | 0,0392 | 0,0466 | 0,0311 | 0,0385 | 0,0628 | 0,0146 | 0,0387 |
| 30 | 456 | 18 | 0,039 | 0,0393 | 0,0476 | 0,0304 | 0,0385 | 0,0658 | 0,0116 | 0,0387 |
| Total | 13817 | 535 | - | 1,1558 | - | - | - | - | - | - |
| Mean | | 0,0387 | - | 0,0385 | - | - | - | - | - | - |

Table 1 indicated that the Shewhart p chart was in control, so that the parameters of Shewhart p control Chart were $\bar{p} = 0.0387$, $UCL_{\text{Shewhart p}} = 0.0629$ and $LCL_{\text{Shewhart p}} = 0.146$ and the chart can be used to monitor the production process for April.

**Table 2**. Process Monitoring on April using ISRT p EWMA Chart and Shewhart p Chart.

| $i^{th}$ | N | Np | p | Control Limit for ISRT EWMA p ISRT | | | | Control Limit for p Chart | | |
|---|---|---|---|---|---|---|---|---|---|---|
| | | | | Zi | $UCL_i$ | $LCL_i$ | $CL_i$ | $UCL_i$ | LCL | $CL_i$ |
| 1 | 482 | 14 | 0,0290 | 0,0361 | 0,0473 | 0,0306 | 0,0385 | 0,0651 | 0,0123 | 0,0387 |
| 2 | 402 | 19 | 0,0473 | 0,0389 | 0,0481 | 0,0300 | 0,0385 | 0,0676 | 0,0098 | 0,0387 |
| 3 | 525 | 16 | 0,0305 | 0,0368 | 0,0470 | 0,0308 | 0,0385 | 0,0640 | 0,0134 | 0,0387 |
| 4 | 507 | 25 | 0,0493 | 0,0399 | 0,0471 | 0,0307 | 0,0385 | 0,0644 | 0,0130 | 0,0387 |
| 5 | 490 | 16 | 0,0327 | 0,0381 | 0,0473 | 0,0306 | 0,0385 | 0,0648 | 0,0126 | 0,0387 |
| 6 | 493 | 24 | 0,0487 | 0,0408 | 0,0473 | 0,0306 | 0,0385 | 0,0648 | 0,0126 | 0,0387 |
| 7 | 523 | 13 | 0,0249 | 0,0368 | 0,0470 | 0,0308 | 0,0385 | 0,0640 | 0,0134 | 0,0387 |
| 8 | 531 | 15 | 0,0282 | 0,0346 | 0,0470 | 0,0309 | 0,0385 | 0,0638 | 0,0136 | 0,0387 |
| 9 | 511 | 22 | 0,0431 | 0,0367 | 0,0471 | 0,0307 | 0,0385 | 0,0643 | 0,0131 | 0,0387 |
| 10 | 521 | 15 | 0,0288 | 0,0348 | 0,0470 | 0,0308 | 0,0385 | 0,0641 | 0,0133 | 0,0387 |
| 11 | 475 | 23 | 0,0484 | 0,0382 | 0,0474 | 0,0305 | 0,0385 | 0,0652 | 0,0122 | 0,0387 |
| 12 | 443 | 23 | 0,0519 | 0,0416 | 0,0477 | 0,0303 | 0,0385 | 0,0662 | 0,0112 | 0,0387 |
| 13 | 396 | 18 | 0,0455 | 0,0426 | 0,0482 | 0,0299 | 0,0385 | 0,0678 | 0,0096 | 0,0387 |
| 14 | 374 | 22 | 0,0588 | 0,0466 | 0,0484 | 0,0297 | 0,0385 | 0,0686 | 0,0088 | 0,0387 |
| 15 | 563 | 14 | 0,0249 | 0,0412 | 0,0467 | 0,0310 | 0,0385 | 0,0631 | 0,0143 | 0,0387 |
| 16 | 545 | 21 | 0,0385 | 0,0405 | 0,0469 | 0,0309 | 0,0385 | 0,0635 | 0,0139 | 0,0387 |
| 17 | 559 | 11 | 0,0197 | 0,0353 | 0,0468 | 0,0310 | 0,0385 | 0,0632 | 0,0142 | 0,0387 |
| 18 | 424 | 24 | 0,0566 | 0,0406 | 0,0479 | 0,0301 | 0,0385 | 0,0668 | 0,0106 | 0,0387 |
| 19 | 382 | 14 | 0,0366 | 0,0396 | 0,0483 | 0,0298 | 0,0385 | 0,0683 | 0,0091 | 0,0387 |
| 20 | 363 | 18 | 0,0496 | 0,0421 | 0,0486 | 0,0296 | 0,0385 | 0,0691 | 0,0083 | 0,0387 |
| 21 | 499 | 17 | 0,0341 | 0,0401 | 0,0472 | 0,0307 | 0,0385 | 0,0646 | 0,0128 | 0,0387 |
| 22 | 298 | 16 | 0,0537 | 0,0435 | 0,0495 | 0,0289 | 0,0385 | 0,0722 | 0,0052 | 0,0387 |
| 23 | 400 | 17 | 0,0425 | 0,0433 | 0,0481 | 0,0299 | 0,0385 | 0,0676 | 0,0098 | 0,0387 |
| 24 | 380 | 24 | 0,0632 | 0,0482 | 0,0484 | 0,0298 | 0,0385 | 0,0684 | 0,0090 | 0,0387 |
| 25 | 551 | 29 | 0,0526 | 0,0493 | 0,0468 | 0,0310 | 0,0385 | 0,0634 | 0,0140 | 0,0387 |
| 26 | 441 | 11 | 0,0249 | 0,0432 | 0,0477 | 0,0303 | 0,0385 | 0,0663 | 0,0111 | 0,0387 |
| 27 | 487 | 11 | 0,0226 | 0,0381 | 0,0473 | 0,0306 | 0,0385 | 0,0649 | 0,0125 | 0,0387 |
| 28 | 512 | 15 | 0,0293 | 0,0359 | 0,0471 | 0,0307 | 0,0385 | 0,0643 | 0,0131 | 0,0387 |
| 29 | 498 | 26 | 0,0522 | 0,0400 | 0,0472 | 0,0307 | 0,0385 | 0,0646 | 0,0128 | 0,0387 |
| 30 | 336 | 17 | 0,0506 | 0,0426 | 0,0489 | 0,0294 | 0,0385 | 0,0703 | 0,0071 | 0,0387 |
| Total | 13911 | 550 | - | 1,2062 | - | - | - | - | 0,3470 | - |
| Mean | - | 0,0395 | - | 0,0402 | - | - | - | | | |

*2nd Step: Process monitoring using ISRT p EWMA chart*

The ISRT p EWMA chart is now ready to monitor the production process on April at PT Primarindo Asia Infrastuktur Tbk using the above parameter. All the unit produced on April was inspected and the number of defective units found was presented in Table 2. Using equation (1) and (2) each data during April was converted to Z value and UCL/LCL for ISRT p EWMA chart was calculated. The plot of the data into the ISRT p EWMA chart presented in Figure 1.

Every single data in table 2 was plotted on ISRT p EWMA chart. Overall monitoring result using ISRT p EWMA chart was displayed in Figure 1.

To study the performance of ISRT p EWMA chart, the Shewhart p chart was also developed for the same data observation and the UCL/LCL for Shewhart p chart also presented in Table 2.

The plotted data on Shewhart p Chart and the overall monitoring result was presented in Figure 2.

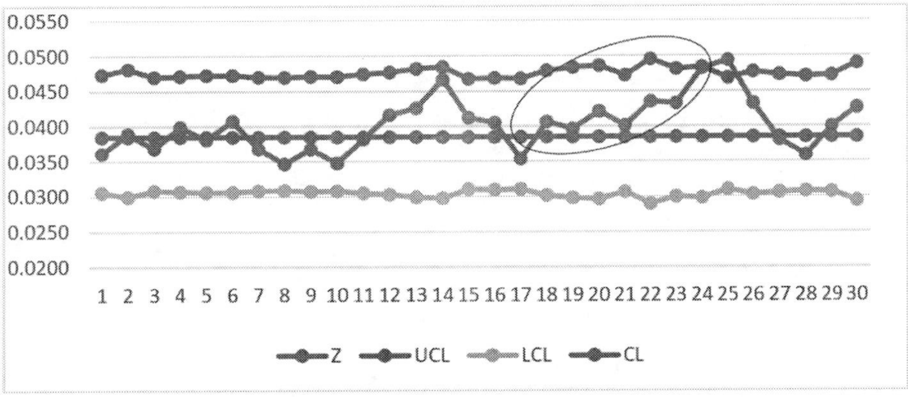


**Figure 1.** Monitoring process variation on April using ISRT p EWMA chart.

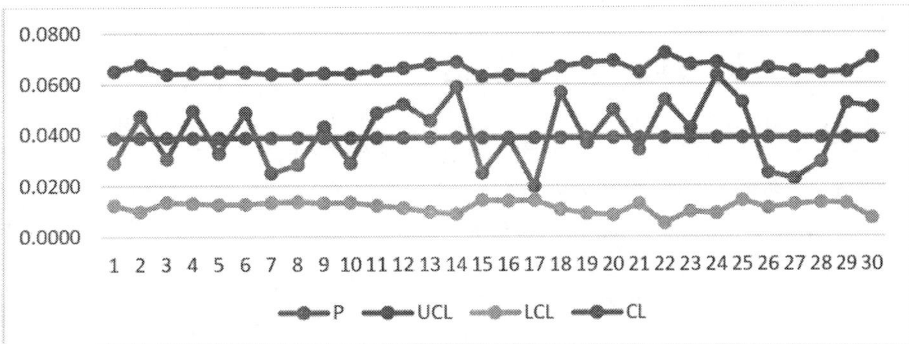


**Figure 2.** Monitoring process variation on April using Shewhart p chart.

Based on ISRT p EWMA chart on Figure 1, the process was found out of control on 18th of April indicated by 7 consecutive point was on the same side of the central line. Also, on 25$^{th}$ of April the chart indicated the out of control condition. It means, the ISRT p EWMA chart detected a shift twice. An action was taken to restore the process back to in control condition. On the other hand, the shift was not detected when Shewhart p chart was used as displayed in Figure 2. In other word, no action was taken, while the process has actually shifted.

## 3. Result ad Discussion

The study resulted that the ISRT p EWMA performed better than traditional Shewhart p chart. Using ISRT p EWMA chart, the shift was detected twice at 18$^{th}$ and 25$^{th}$ observation. Meanwhile, the shift was not detected when traditional Shewhart p chart is used to control the process. In the case of PT Primarindo Asia Infrastruktur Tbk, the Shewhart p chart failed to detect the shift while the process actually has shifted. As a result, the process run in the out of control condition, and automatically produced defectives units. That is the reason why the fraction defective at PT. Primarindo Asia Infrastruktur Tbk was excessive. On the other hand, when ISRT p EWMA chart was used, the shift was

IOP Conf. Series: Materials Science and Engineering **847** (2020) 012032 doi:10.1088/1757-899X/847/1/012032

detected and the process was restored so that defectives unit could be avoided. Therefore, the fraction defective of the process could be reduced.
The result was obtained at a specified parameter L and λ. As a preliminary investigation, in this paper the effect of λ was analysed. To see the effect of λ parameters, the same procedure for the same observation data presented in Table 1 was repeated at λ= 0.25, λ =0.5 and λ = 0.75. The summary results presented in Figure 3a for λ = 0.25, Figure 3b for λ = 0.50 and Figure 3c for λ = 0.75.

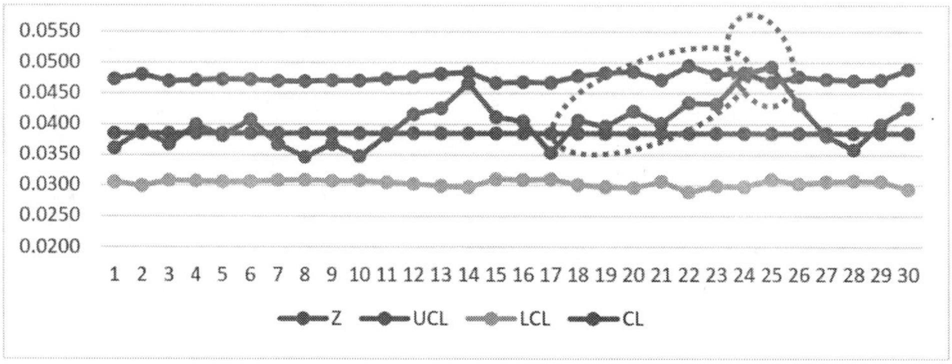


**Figure 3a.** Performance of ISRT p EWMA chart for λ = 0.25.

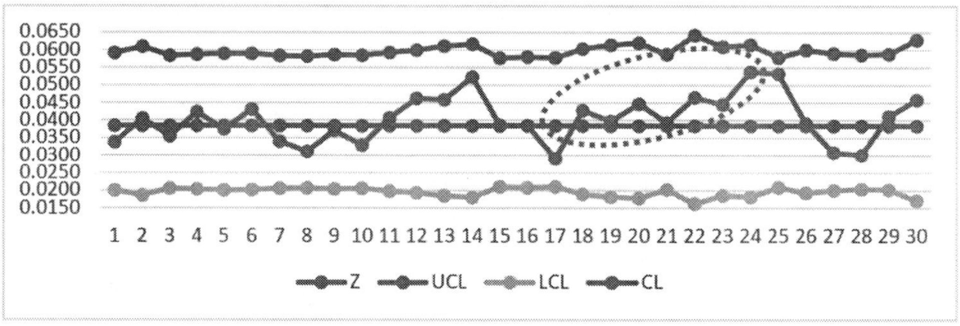


**Figure 3b.** Performance of ISRT p EWMA chart for λ = 0.5.

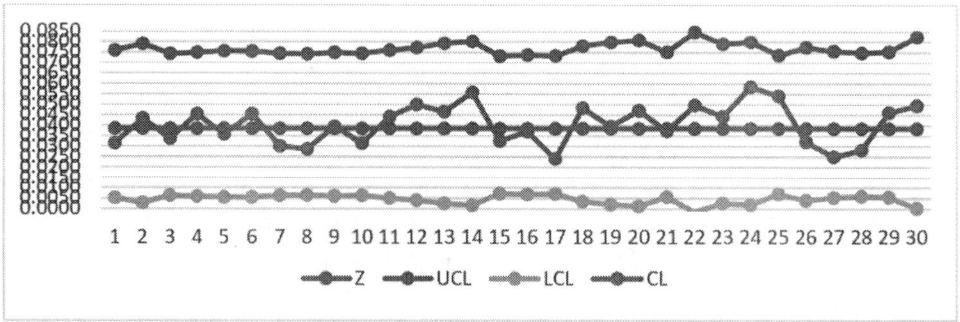


**Figure 3c.** Performance of ISRT p EWMA chart for λ = 0.75.

Figure 3a, 3b, and 3c explained that at λ = 0.25 the shift was detected twice at 18th and 25th observation, while at λ = 0.5 the shift was detected only once at 18th observation and at λ =0.75 the ISRT p EWMA chart could not detect the shift while actually the process has shifted. It means that the greater the value of λ, the worse the performance of ISRT p EWMA chart to detect small shift.

Time limitation leaving L parameter could not be investigated. To get a comprehensive result, it is recommended for further research:

1. Instead of parameter λ, the effect of L parameter should also be investigated.
2. For accurate result, the performance of ISRT p EWMA chart should be measured in term of Average Run Length (ARL).
3. To get a comprehensive result, the ISRT p EWMA chart must be compared against Shewhart p chart as well as p EWMA chart.

## 4. Conclusion

This paper investigates the performance of ISRT p EWMA chart on detecting small shift variation with the following conclusion:

1. The ISRT p EWMA chart detected small shift better than Shewhart p chart.
   The ISRT p EWMA chart detected the shift twice, while Shewhart p chart did not not detect any shift.
2. The value of λ parameter has a significant effect on the performance of ISRT p EWMA chart. The larger the value of λ parameter, the worse the performance of the ISRT p EWMA chart
   At $\lambda = 0.25$, the shift was detected twice, at $\lambda = 0.5$, the shift was detected once while at $\lambda = 0.75$, the shift was not detected.
3. For further research, the performance of the ISRT p EWMA chart, should not only be compared with Shewhart p chart, but also with p EWMA chart as well, in term of Average Run Length for different value of L parameter

## 5. References

[1] Abbas A, Riaz M and Arden M 2010 Enhancing the Performance of Cusum Scale Chart. *Computers and Industrial Engineering*, **63,** pp 400-409.

[3] Erna TH 2018 Modified Exponentially Weighted Moving Average (EWMA) Control Chart on Autocorrelation Data *Journal of Physic:Conf. Ser. 979012097.*

[2] Grant E and Leavenworth R 1996 *Statistical Quality Control* (7th Edition). (Singapore: McGraw Hill)

[4] Knoth S and Steinmetz S 2013 EWMA p charts under sampling by variables. *International Journal of Production Research* **51**(13).

[5] Lucas J and Saccucci M 1990 Exponentially Weighted Moving Average Control Schemes. Properties and Enhancements *Technometrics*, **32(1)**, pp 1-12.

[6] Montgomery D C 2012 *Introduction To statistical quality control.* (USA: John Wiley & Son), pp 120-121.

[7] Robert S W 1959 Control Chart test based on Geometric Moving Average. *Technometric,* **1,** pp 239 – 250.

[8] Sukparungsee S 2014 EWMA p Chart Based on Improved Square Root Transformation, *International Journal of Mathematical, Computational, Physical, Electrical and Computer Engineering* **8**(7).

[9] Vera D C et all 2004 Comparative study of the performance of the CuSum and EWMA control charts *Computer & Industrial Engineering* **46**(4), pp 707-72

IOP Conf. Series: Materials Science and Engineering **847** (2020) 012033 doi:10.1088/1757-899X/847/1/012033

# Warehouse Improvement Evaluation using Lean Warehousing Approach and Linear Programming

**Y. Prasetyawan[1, 2], N. G. Ibrahim [1, 3]**

[1]Department of Industrial Engineering, Institut Teknologi Sepuluh Nopember, Kampus ITS Sukolilo-Surabaya 60111, Indonesia
[2] yudhaprase@gmail.com, [3] naufal.ghani@gmail.com

**Abstract.** Lean manufacturing tools have been applied for several years to improve company's internal logistic. Furthermore, a lot of factors such as globalization, competition and shorter life-cycle product force companies to create production process more efficient and cheaper. The acceptance of lean philosophy in the company means not only respecting the lean principles in the manufacturing area, but also in all the process that performed inside the company. The implementation of lean principles in the warehouse area is a certain step of improvement warehouse process and performance, but also the whole company. However, the implementation of lean principles in warehouse area is relatively new subject in logistics. Most of the research that has been done before only focused on eliminate waste and didn't count the cost that can be saved after lean implementation. This paper's objective is to identify waste that might be able to appear in warehousing process using value stream mapping. After the waste successfully identified, the next step is doing improvement using lean tool and count the implementation cost using linear programming. The result of this paper is choose which tool that make lower implementation cost and higher time reduction
**Keywords**: Lean warehousing, value stream mapping, linear programming

## 1. Introduction

In this modern era, factors such as digitalization, competitiveness, shorter product life cycle, advanced technology and shorter time-to-market have a greater impact than in previous eras on the logistics activities of a company [1]. In fact, to meet the customer needs, warehousing activities must be more optimal in reducing inefficiencies and making warehouses more reliable in terms of cost [2]. Warehouse is a place where there is a movement of goods in which there are activities of receiving, storing, orders picking, sorting and delivering goods [3]. Warehouse is one of the main elements in the supply chain because it involved in every company's business processes such as planning, purchasing, production and distribution processes [4]. The trend in the current era is how to change the warehouse to be more flexible to face demand changing, increase inventory transparency and efficiency of each process in the warehouse [5]. The most popular approach of continuous improvement is lean. When the company decides to implement lean, the company's main focus will shift to how to reduce waste and become more efficient. According to the principles of lean management, there are eight types of waste: waiting, transportation, over-processing, over-production, inventory, motion, defect or errors and underutilised employees [6] [7].

The implementation of lean management not only in manufacturing area but also in another area in company such as supply chain, warehousing and even organization itself. The lean principle must be made a priority when a company wants to maintain and improve its competitive advantages. However, it isn't true to believe that lean management and company's competitiveness can be possible by only achieving lean production. All business processes involved in the company must have the same understanding so that the main goal of lean is achieved by reducing unnecessary activities that do not provide added value. Warehousing is one of the many aspects that are important in achieving lean implementation goals. Lean warehousing is a concept that requires stability, systematic continuity and measurable improvement that involving all employees [8]. Lean warehousing can also be defined as the concept of reducing waste in the upstream or downstream parts of the supply chain to direct customers to the right time and place [9]. In lean warehousing area, a lot of researches have been done with different combination of lean tools and results. Studies that have been done by Swank, 2003 have an effect of reduction in response time 60%, 28% labour cost reduction and 40% reduction of reissues due to errors, Cook et al, 2005 (71% decrease in inbound cycle time, 76% decrease in inventory levels, required storage space decreased by 51%), Pan et,al. 2010 (Used value stream mapping in shorting delivery time), Jaca et al. 2012 (9,34% improvement on the whole warehouse productivity), Bozer, 2012 (Development 'house of lean warehousing' and identified fundamental operational or asset-based changes required to implement lean principles in a warehouse), Dehdari et al. 2013 (Increase warehouse productivity by at least 5%) and Dotoli et al. 2013 using VSM to identify non value added activities and used Gemba Shikumi technique to choose the best response actions [10] [11].

Implementation of lean concepts in warehouses can increase value and reduce costs because warehousing activities are not only a source of cost, but also a source of competitive advantages to achieve higher service levels [12]. The implementation of lean warehousing can also have an impact on internal and external improvements, as indicated by McKinsey study in 2010. The improvement in internal warehouse area could be people (Increase labours' efficiency and flexibility by 1-5%), process (2-9% Reduce handling movement for each product between receiving and shipping), performance management (2-5% Assure continuous improvement of efficiency) and layout (3-4% optimize layout structure and align with local requirements and restrictions). Area of improvement for external warehouse could be interaction with third party (Matching supplier and customer requirement to warehouse restriction) and ownership (Optimise contract of service provider) [13].

According to that fact, most of the research in lean warehousing only focussed on reducing waste. None of them calculate the impact of lean improvement from cost's point of view. It might be happen when a company try to implement lean tool but it takes costly and even the benefit that the company got was less than the cost that company's spent. Therefore, this paper will try to calculate the cost of investment that will be spent in implementing lean tool using linear programming. Linear programming (LP) is a tool for solving optimization problem [14]. Linear programming consist of three component; decision variables, objective function and constraint. Decision variable is a decision variable is a variable whose value will be sought; objective functions a quantitative statement of an optimization case such as maximizing profit or minimizing operating costs while constraint is a number that limited the value of decision variable. This paper will introduce some variable that related on lean warehousing which will be taken into consideration in the selection of lean implementation tools. The overall variables that will be used in considering alternative choices will lead to implementation costs and the tool that resulting the lowest implementation/investment cost to be chosen. The objective function of this problem is minimizing total implementation cost and the decision variables will be in areas such as man, machine, material, method and time.

The aim for the research is to create leaner warehousing system using lean warehousing approach and choose the alternative that given the lowest investment cost using linear programming. In order to deliver the purpose of the research project, the following objectives have been established: Identify the current activities/processes that appeared in warehouse, develop current value stream mapping, calculate current cost that spent for warehousing activities, analysis and eliminate waste based on current value stream mapping using fishbone diagram and 5whys , develop future value stream mapping that have been implementing lean tool such as kanban ,5S and re-layouting and then calculate total implementation cost after applying improvement and choose the tool that resulting lowest implementation cost.

## 2. Methods

There are two main methods that used to make this paper, value stream mapping and linear programming. Both of the method will be described in subsection below.

### *2.1. Value Stream Mapping*

Value stream mapping (VSM) is one of the lean tools that can be used to find out the latest conditions of the company's business processes, in which there is a flow of information and material flow that serves as a consideration for solving problems and helping to make improvements for the future [15]. VSM application in warehouse operation can be used to underline the waste activities (value/non-value added) by concentrating on present state of value stream map. As we know, implementation of lean techniques (both information flow and material flow). There are several step of draw VSM: 1) Draw current state map; 2) Analyse existing conditions. Existing conditions are used to determine waste, bottlenecks and other obstacles; 3) Remove waste. How to eliminate waste in the existing conditions depends on its priority; 4) Draw future state map. After being eliminated, a future map was made and 5) Implementation.

### *2.2. Linear Programming*

After creating VSM and eliminate waste that found on it, the next step is implementing lean tool for improvement. The decision to choose which lean tool will be used must be consider the implementation/investment cost. In this paper, linear programming used to calculate the implementation/investment cost by optimizing on variable that related on lean warehousing such as man, machine and materials. The notation used in this simple mathematical model is as follows.

| | | | |
|---|---|---|---|
| LT | : lean tool | SLI | : salary of indirect labour |
| TCi | : total investment cost of tool i | LI | : number of indirect labour |
| Ii | : implementation cost of tool i | Ei | : material handling i |
| TCPi | : total cost of man power of tool i | HDi | : total handling cost of material handling i |
| TCMi | : total cost of machine of tool i | Mt | : maintenance rate/time |
| TCIi | : total cost of inventory of tool i | Mr | : movement rate per day |
| W1 | : warehouse raw material | Er | : energy consumed by material handling i |
| W2 | : warehouse finished good | HC | : holding cost |
| SLD | : salary of direct labour | To | : turnover rate |
| LD | : number of direct labour | SS | : safety stock |

*2.2.1. Choose lean tool.* After waste successfully identified, the next step is choose lean tool that will implement in warehouse. The tool that will be chosen must consider the total cost of implementation. The consideration can be described as follow:

$$LT = \min\{TC_1, TC_2, TC_3 \ldots TC_n\} \quad (1)$$

The total investment cost can be described as follow:

$$TC_i = I_i + TCP_i + TCM_i + TCI_i \quad (2)$$

*2.2.2. Cost of man power (man).* This mathematical model used to determine number of direct labour (warehouse operator) and indirect labour (administration officer). The calculation can be described as follow:

Objective function:

$$min\ TCP_i = (SLD \times LD_{W1}) + (SLD \times LD_{W2}) + (SLI \times LI_{W1}) + (SLI \times LI_{W2}) \quad (3)$$

Subject to:

$$LD_{W1} + LD_{W2} \leq TLD \quad (4)$$

$$LI_{W1} + LI_{W2} \leq TLI \quad (5)$$

$$LD_{W1}, LD_{W2} \geq 1 \quad (6)$$

$$LI_{W1}, LI_{W2} \geq 0 \quad (7)$$

$$LD_{W1}, LD_{W2}, LI_{W1}, LI_{W2}\ integer \quad (8)$$

Constraint (4) describe the maximum number of direct labour in warehouse 1 and warehouse 2 while constraint (5) describe the maximum number of indirect labour in warehouse 1 and warehouse 2

*2.2.3. Cost of material handling (machine).* This mathematical model used to determine the number and type of material handling that will be used in both warehouses. The calculation can be described as follow:
Objective function:

$$min\, TCM_i = \sum_{i=1}^{n}(HD_1 \times E_1) + (HD_2 \times E_2) + \ldots + (HD_n \times E_n) \quad (9)$$

Subject to:

$$E_1 + E_2 + \cdots + E_n \leq Mt \quad (10)$$

$$E_1 + E_2 + \cdots + E_n \leq Mr \quad (11)$$

$$E_1 + E_2 + \cdots + E_n \leq Er \quad (12)$$

$$E_1, E_2, \ldots, E_n \geq 1 \quad (13)$$

$$E_1, E_2, \ldots, E_n\ integer \quad (14)$$

Constraint (10) describe the maximum number of maintenance rate/time for each type of material handling, constraint number (11) describe the maximum movement rate per day for each type of material handling in a day and constraint (12) describe the maximum of energy consumed for each type of material handling.

*2.2.4. Cost of inventory (material).* This mathematical model used to determine the maximum number of inventory that kept for each warehouse. The calculation can be described as follow:
Objective function:

$$min\, TCI_i = (HC \times W1) + (HC \times W2) \quad (14)$$

Subject to:

$$W1 + W2 \leq To \quad (15)$$

$$W1 + W2 \leq SS \quad (16)$$

$$W1, W2 \geq 1 \quad (17)$$

$$W1, W2\ integer \quad (18)$$

Constraint (15) describe the maximum number of turnover rate for each warehouse while constraint (16) describe the maximum inventory level based on safety stock for each warehouse.

## 3. Result And Discussion

IOP Conf. Series: Materials Science and Engineering **847** (2020) 012033 doi:10.1088/1757-899X/847/1/012033

Before creating VSM, first thing that have to do is identify all the process in warehouse. In this case, warehouse that the process will be observed divided into two warehouse types; raw material warehouse (warehouse A) and finished good warehouse (warehouse B). Both the warehouses consist of same activities; receiving, storing, picking and shipping. From all the process, there are some differences activities in the process of receiving and shipping. In warehouse 1, the goods that received by the warehouse comes from suppliers (raw material) while in warehouse 2 received finished good from production floor. Shipping activity for warehouse 1 is deliver raw material into production area while in warehouse 2 deliver finished product to the customer. Warehouse 1 consist of 27 activities (receiving with 11 activities, storing 5, picking 8 and shipping 3) while warehouse 2 consist of 26 activities (receiving with 2 activities, storing 7, picking 9 and shipping 8). Here is the VSM for warehouse 1 and warehouse 2

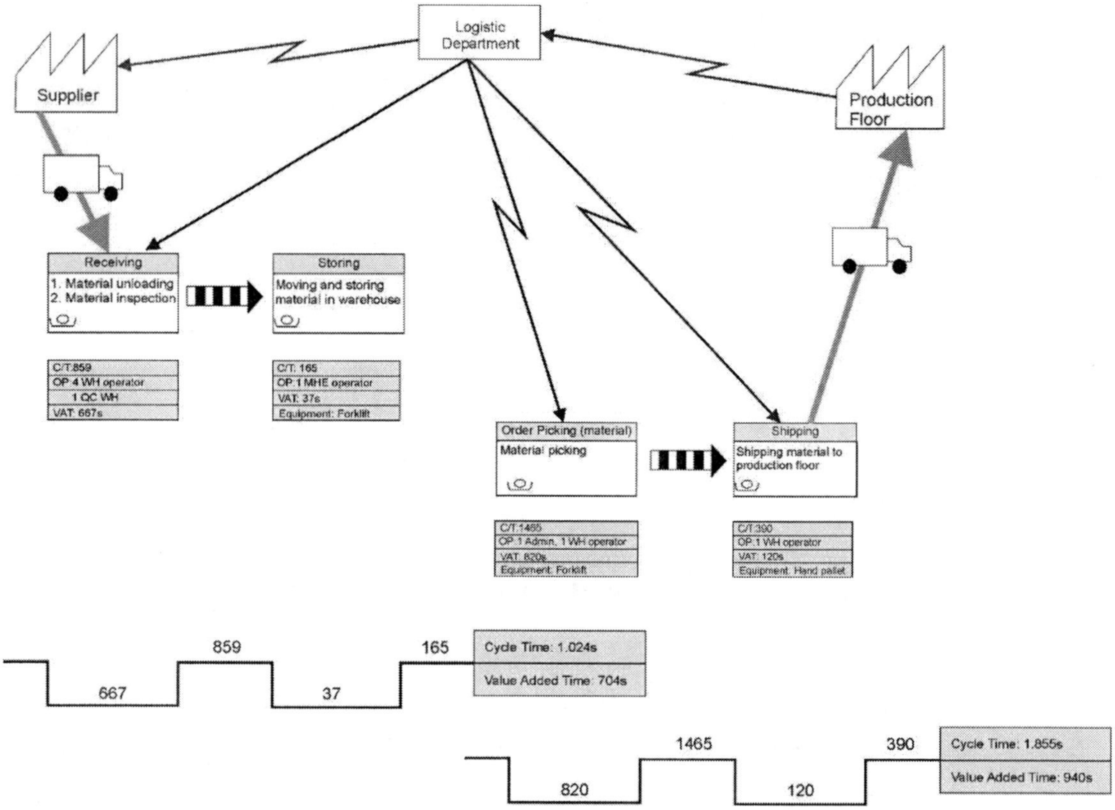


**Figure 1** Current VSM for Warehouse A

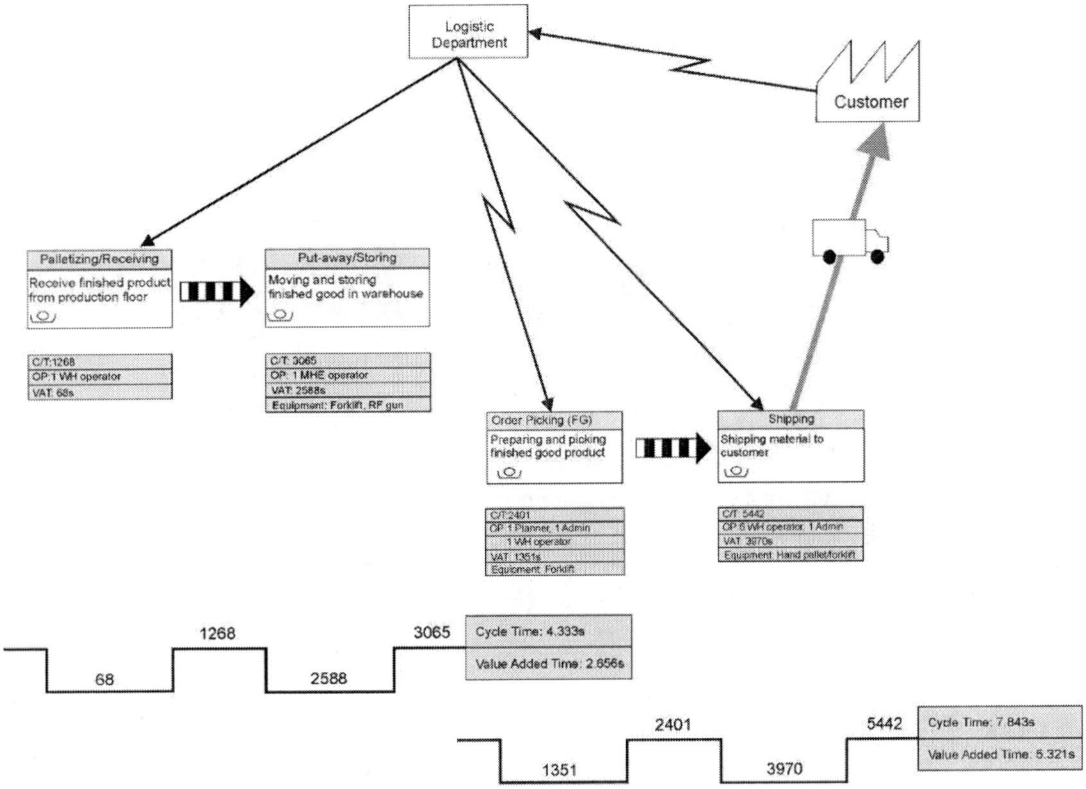


**Figure 2** Current VSM for Warehouse B

After create the current vsm, the next step is identify waste that came up from current vsm and find lean tool for improvement based on linear programming consideration. The proposed lean tools for this case are: re-layouting, invest new material handling, e-kanban, poka-yoke and 5S. Here is the cost and time comparison for the entire proposed lean tool

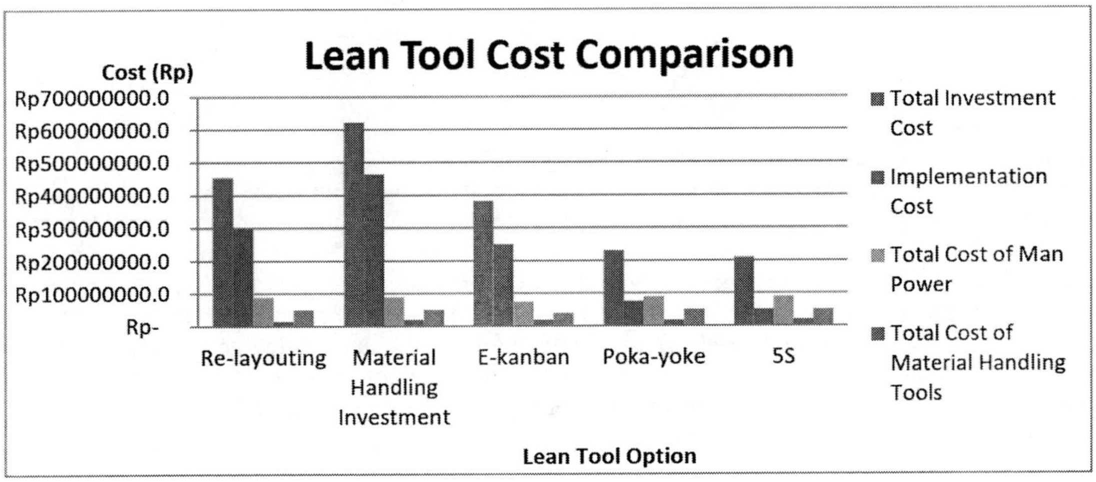


**Figure 3** Lean Tool Cost Comparison

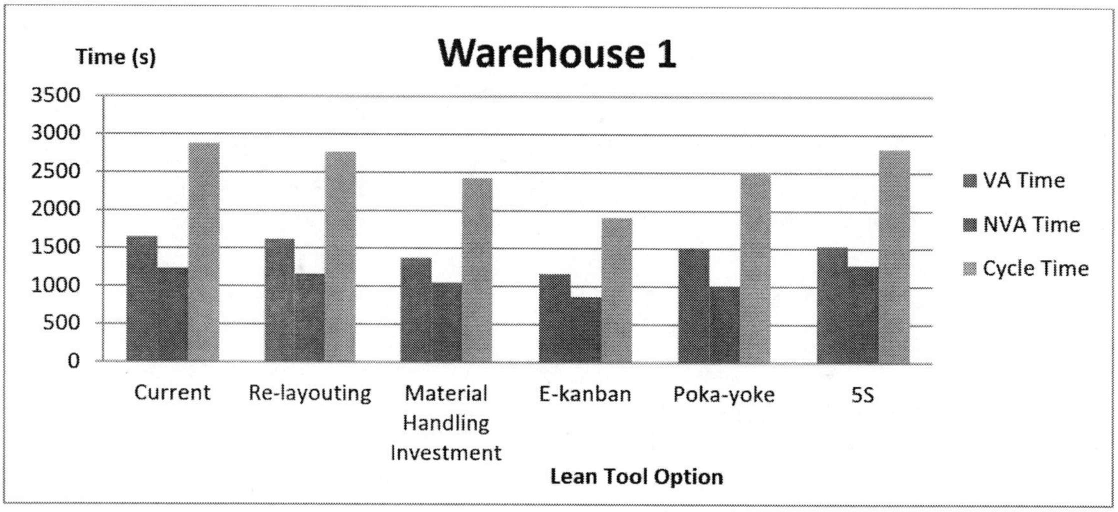


**Figure 4** Time Consumed for Each Lean Tool in Warehouse 1

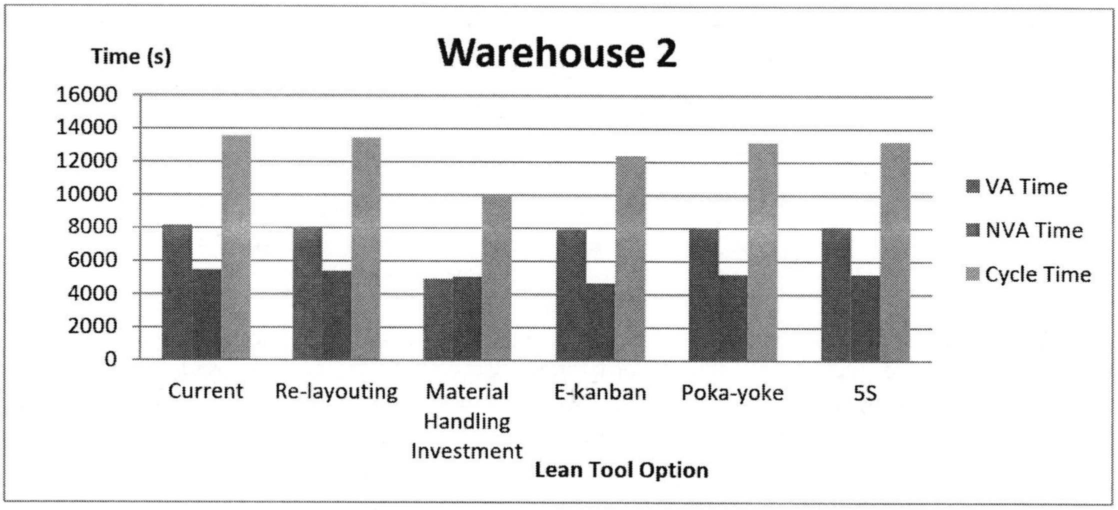


**Figure 5** Time Consumed for Each Lean Tool in Warehouse 2

## 4. Conclusion

Lean warehousing can be used as one of the powerful tools for improves warehouse operation. The combination between lean warehousing and linear programming could make lean warehousing more powerful by considering cost of implementation for each lean tool. Figure 3 showed that the lowest lean implementation cost is 5S, but figure 4 and 5 showed that there is no significant time improvement for implementing 5S. The most significant time improvements are using E-kanban for warehouse 1 and invest new material handling for warehouse 2. Invest new material handling is the most expensive investment cost. Top level management must give the best solution for this problem to find the best solution. This study could be done better by doing simulation for each tool to make sure that the chosen tool gives the best improvement for warehouse performance and using multi criteria decision making to prioritise between cost and time saving.

## 5. References

[1] Dotoli M, Epicoco N, Falagario M, Costantino N. 2015 An integrated approach for warehouse analysis and optimization: A case study. *Computers in Industry* 70. http://dx.doi.org/10.1016/j.compind.2014.12.004

[2] Dotoli M, Epicoco N, Falagario M, Costantino N. 2013 A lean warehousing integrated approach: A case study. DOI: 10.1109/ETFA.2013.6648030

[3] van den Berg JP, Zijm WHM. 1999 Models for warehouse management: Classification and examples. *International Journal of Production and Economic*. 59. https://doi.org/10.1016/S0925-5273(98)00114-5

[4] Rushton AP, C, Baker P. 2010 The Handbook of Logistics and Distribution Management. In Chapter 15: Principles of warehousing: Kogan Page Limited.

[5] Buonamico N, Muller L, Camargo M. 2017 A new fuzzy logic-based metric to measure lean warehousing performance. *Supply Chain Forum: An International Journal*. DOI: 10.1080/16258312.2017.1293466

[6] Braglia M, Carmignani G, Zammori F. 2006 A new Value Stream Mapping approach for complex production system. *International Journal of Production and Research*. 44. DOI: 10.1080/00207540600690545

[7] Shabeena Begam M, Swamynathan R, Sekkizhar J. 2013 Current trends on lean management: a review. *International Journal of Lean Thinking*. 4(2)

[8] Dehdari P. 2014. Measuring the Impact of Lean Techniques on Performance Indicators in Logistics Operations. Doctoral dissertation. Karlsruhe: Institute of Technology Karlsruhe, Faculty of Mechanical Engineering; DOI: 10.5445/KSP/1000036424

[9] Reichart A, Holweg M. 2007 Lean distribution: Concepts, contributions and conflicts. *International Journal of Production Research*. 45(16). DOI: 10.1080/00207540701223576

[10] Visser JJ. 2014 Lean in the warehouse: Measuring lean maturity and performance within a warehouse environment. Rotterdam: Erasmus University Rotterdam, Rotterdam School of Management.

[11] Oey E, Nofrimurti M. 2018 Lean implementation in traditional distributor warehouse - a case study in an FMCg company in Indonesia. *International Journal Process Management and Benchmarking*. 8(1). DOI: 10.1504/IJPMB.2018.088654

[12] Andjelkovic A, Radosavljevic M, Panic DS. 2017 Effects of lean tools in achieving lean warehousing. *Article in Economic Themes*. 54(4). DOI: 10.1515/ethemes-2016-0026

[13] Alicke K, Losch M. 2010 Lean and mean: How does your supply chain shape up? McKinsey. https://www.mckinsey.com/~/media/mckinsey/dotcom/client_service/operations/pdfs/lean_and_mean-how_does_your_supply_chain_shape_up.ashx (Accessed: 27/12/2019)

[14] Winston WL. 2003 Operations Research: Applications and Algorithms. 4th ed.: Duxbury Press.

[15] Masuti PM, Dabade UA. 2019 Lean manufacturing implementation using value stream mapping at excavator manufacturing company. *Materials Today: Proceedings*. http://doi.org/10.1016/j.matpr.2019.07.740

# The Proposed OEE-SIGMA Prediction for Increased Profits

**Y Prasetyawan**[1,2], **F Giffari**[1,3] **and B S A Putera**[1,4]

[1]Department of Industrial Engineering, Institut Teknologi Sepuluh Nopember, Kampus ITS Sukolilo-Surabaya 60111, Indonesia
[2]yudhaprase@gmail.com, [3]arigiffari29@gmail.com, [4]saestubagas@gmail.com

**Abstract.** Most of the organizations adopt methods to measure their operational performance based on timeframe and requirements. One of the most commonly used measures of performance against capability of the equipment is the Overall Equipment Effectiveness (OEE). Overall Equipment Effectiveness (OEE) metric application in industry has been facing a challenge of continuing advancement in industrial operations. A smart OEE is required for a quick response to the dynamism in industrial system. In this study, OEE metric is made smart by integrating sigma continuous improvement tool into it for the enhancement of dynamism required of the traditional OEE model measured basically on three factors-availability, performance and quality. System productivity dynamism is measured and predicted through sigma statistical variation of the production process defined as a ratio of delivered output (supply) to the expected (planned) output. In this research, the model was applied to made dynamics simulation for increased profits of many manufacturing organizations. The result obtained are increased elements in OEE in line with increased of profits. This indicated that smart OEE-sigma model is a better tool for enhancing continuous improvement which can accurately measuring the financial benefits of proposed improvement projects.
**Keywords**: overall equipment effectiveness, sigma metric, continuous improvement, production process

## 1. Introduction

Many manufacturing organizations have trouble accurately measuring the financial benefits of proposed improvement projects. Important projects are often overlooked or not properly prioritized relative to average projects. As a result, a manufacturing area can get mired in driving false improvement projects, never making significant gains in reliability and productivity. And the one proposed of improvement projects in every plant is improve the efficiency. Manufacturing plant is required to embrace a continuous innovation in modern manufacturing systems in order to become productive and being competitive. One of the best efficiency measurement method that can achieve the goal is Overall Equipment Effectiveness (OEE). Overall Equipment Effectiveness (OEE) was launched by Nakajima in the 1980s for measuring productivity of individual equipment in a factory. OEE tool exists to identifies and measures losses from important aspects of manufacturing such as availability, performance and quality rate. This tool can support the improvement of equipment effectiveness and there by its productivity [1]. The OEE concept is becoming increasingly popular and has been widely used as a quantitative tool essential for measurement of productivity especially in the semiconductor manufacturing operations [2].

IOP Conf. Series: Materials Science and Engineering **847** (2020) 012034 doi:10.1088/1757-899X/847/1/012034

Financial statements are the scorecards used to communicate and benchmark a company's business. Understanding the link between OEE and financial statements is of paramount importance in ranking reliability and improvement projects.

It is beyond the scope of this paper to investigate financial accounting. This paper uses a generic simplified approach that can fit most specific cases and give you a relevant first look at financial benefits. Our methodology proposes a new OEE approach for measuring effectiveness in order to enrich the solution to this classic financial problem. This study provides an interesting opportunity to know the financial impact of the best projects and can convey the information to your company's decision-makers in terms they understand and fully support.

## 2. Methods

As stated previously, this paper aimed to optimize the effectiveness and increase profit of company. The OEE has strength to evaluate important aspects of manufacturing through the single measurement tool. The idea is how much impact of changing the factors of OEE element in terms of financial aspects. OEE elements basically can be illustrated as below in figure 1.

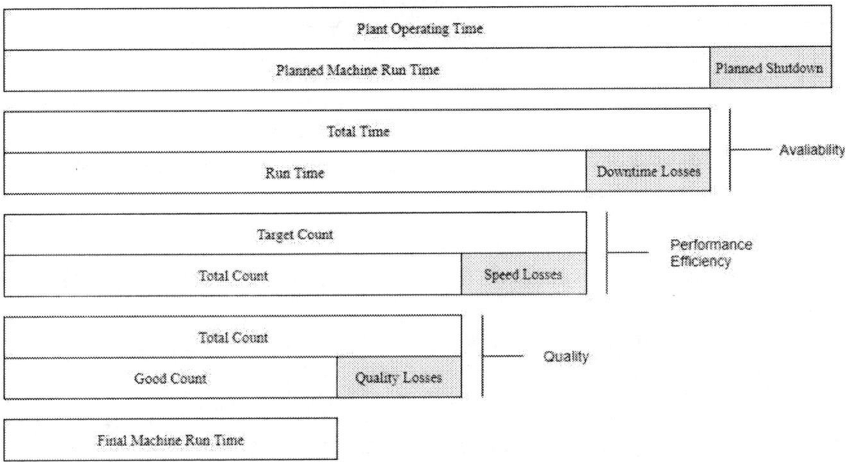


**Figure 1. OEE Elements**

Source : EXOR / DataVisor Marquees (2009). The Complete Guide to Simple OEE

Based on the figure above, effectiveness reduction occurred because losses. There are downtime losses, speed losses, and quality losses. Downtime losses is an event that stop planned production for a length of time such as equipment reparation (planned or unplanned). Speed losses have an affect on manufacturing process to run less than maximum speed or minimum cycle time while production is running. Quality losses is a loss when finished goods can not meet the qualification of quality standard. Each element can be measured by using factors from OEE function.

OEE function that given are as follow.

$$OEE=\alpha\times\beta\times\mu \quad (1)$$

Where the factors are,

$$\alpha \text{ (Availability)}= \frac{\text{delivered production volume}}{\text{expected production volume}} \quad (2)$$

$$\beta \text{ (Performance efficiency)}= \frac{\text{System performance delivered}}{\text{System performance expected}} \quad (3)$$

$$\mu \text{ (Quaility Rate)}= \frac{\text{quality produced delivered}}{\text{quantity of production}} \quad (4)$$

IOP Conf. Series: Materials Science and Engineering **847** (2020) 012034 doi:10.1088/1757-899X/847/1/012034

In the calculation by using heuristic method, there are some scenarios of solution for increasing the effectiveness. There are three scenarios as follow:

1) *Scenario 1*, by control the number of expected volume production or delivered volume production, which means the availability factor will increased.
2) *Scenario 2*, after the first scenario has applied then increase value of speed rate of equipment, which means the availability and performance factor will increased
3) *Scenario 3*, after two previous scenarios has applied then increase the number of good product that meet the standard qualification, which means the availability, performance, and quality factor will increased.

Each scenario have to deal with two cases, where 1) Case 1, the production volume after the scenario given still constant. 2) Case 2, the production volume increased.

In addition to OEE, measuring the performance of equipment can be determined by allocating costs associated with OEE factors. According to Robert C. Hansen [6], the costs associated with OEE consist of direct material costs, direct labor costs, factory overhead costs, and operation expenses. Each cost component is shown at table 2 and table 3.

**Table 1.** Cost Component and Formula

| Cost component | Formula |
|---|---|
| Net sales | $Q_{sales} \times P_{unit}$ |
| Costs | |
| Direct Material | $C_{material}=Q_{material}\times$cost material per unit |
| Direct Labor | $C_{labor}=T_{work}\times$cost of labor per hour |
| Overhead Cost | $C_{overhead}= D+U+I+T+IL+F$ |
| Selling & administration Cost | (Assumption) |
| Operating Income | Income=Net sales-Costs |

**Table 2.** Notation List

| Notation list | |
|---|---|
| $Q_{sales}$ | Amount of finished goods |
| $P_{unit}$ | Price of product per unit |
| $C_{material}$ | Direct Material Cost |
| $Q_{material}$ | Amount of production volume |
| $C_{labor}$ | Direct labor cost |
| $T_{work}$ | Work duration of labor (hour) |
| $C_{overhead}$ | Overhead Cost |
| D | Depreciation |
| U | Utilities |
| I | Insurance |
| T | Property tax |
| IL | Indirect labor |
| F | Factory supplies |

All these components are accumulated into total cost as a parameter of OEE performance. The change of factor can impact the saving cost or extra cost. By using the Heuristic method, the cost performance can be determined.

## 3. Result and discussion

The value of allocation expected and delivered each factor are shown below. The calculation of each factor also determined refers to Equation (1)-(4). The result of calculation is shown below.

**Table 3.** The Data Reference for Analysis

| | Availability | | | Performance | | | Quality | | |
|---|---|---|---|---|---|---|---|---|---|
| Year | Expected Production Volume (million) | Production Volume Delivered (million) | Ratio (α) | Performance Expected (%) | Performance Delivered (%) | Ratio (β) | Quality Product Expected (%) | Quality Product Delivered (%) | Ratio (μ) |
| 2019 | 1.215.000 | 1.150.000 | 0,95 | 93,50 | 80,00 | 0,86 | 80,00 | 73,00 | 0,91 |
| 2020 | 1.421.000 | 1.321.000 | 0,93 | 91,50 | 82,00 | 0,90 | 72,00 | 65,00 | 0,90 |
| 2021 | 1.552.000 | 1.365.000 | 0,88 | 92,00 | 81,50 | 0,89 | 75,50 | 70,50 | 0,93 |

**Table 4.** Recapitulation of OEE Value (Baseline)

| Year | Ratio (α) | Ratio (β) | Ratio (μ) | OEE |
|---|---|---|---|---|
| 2019 | 0,95 | 0,86 | 0,91 | **0,74** |
| 2020 | 0,93 | 0,90 | 0,90 | **0,75** |
| 2021 | 0,88 | 0,89 | 0,93 | **0,73** |

Assume that the schedule running 320 days in a year. Then determine the speed rate by using the equation below

$$R = \frac{\textit{Production volume delivered}}{\textit{Scheduled Time} \times OEE} \quad (5)$$

**Table 5.** Speed Rate Calculation

| Year | Production Volume Delivered (million) | Scheduled Time (hour) | Ideal Speed Rate (Unit / hrs) |
|---|---|---|---|
| 2019 | 1,150,000 | 7,680 | 203 |
| 2020 | 1,321,000 | 7,680 | 229 |
| 2021 | 1,365,000 | 7,680 | 244 |

*Scenario 1*

In this scenario, we calculate the OEE and cost performance after controlling the value of availability in order by 2%, 3%, and 4%. For the first case, the calculation for 2% increasing factor are shown at table 9. Then for case 2, we calculate the new number of goods for new OEE. The result are shown at table below.

**Table 6.** Number of Good Unit Calculation for New OEE (Scenario 1)

| Year | Number of Good Units Made with New OEE | Different Good Units | Different Good Units (%) |
|---|---|---|---|
| 2019 | 1,174,300 | 24,300 | +2,07% |
| 2020 | 1,349,420 | 28,420 | +2,11% |
| 2021 | 1,396,040 | 31,040 | +2,22% |

*Scenario 2*

In this scenario, we calculate the OEE and cost performance after controlling the value of availability and performance in order by 2%, 3%, and 4%. For the first case, the calculation for 2% increasing factor are shown at table 10. Then for case 2, we calculate the new number of goods for new OEE. The result are shown at table below.

**Table 7.** Number of Good Unit Calculation for New OEE (Scenario 2)

| Year | Number of Good Units Made with New OEE | Different Good Units | Different Good Units (%) |
|---|---|---|---|
| 2019 | 1,201,749 | 51,749 | +4,31% |
| 2020 | 1,379,535 | 58,535 | +4,24% |
| 2021 | 1,427,557 | 62,558 | +4,38% |

*Scenario 3*

In this scenario, we calculate the OEE and cost performance after controlling the value of availability, performance, and quality in order by 2%, 3%, and 4%. For the first case, the calculation for 2% increasing factor are shown at table 11. Then for case 2, we calculate the new number of goods for new OEE. The result are shown at table below.

**Table 8.** Number of Good Unit Calculation for New OEE (Scenario 3)

| Year | Number of Good Units Made with New OEE | Different Good Units | Different Good Units (%) |
|---|---|---|---|
| 2019 | 1,228,088 | 78,089 | +6,36% |
| 2020 | 1,410,097 | 89,097 | +6,32% |
| 2021 | 1,458,134 | 93,134 | +6,39% |

**Table 9.** Saving Direct Labor Cost Calculation (Scenario 1 in Case 1)

| Year | Scheduled Time (hr) | Ideal Speed rate (Unit / hrs) | Ratio (α) Increase 2% | New OEE | New Scheduled Time (hr) | Savings Hours | Net saving hour (-96 hour) | New Direct Labor Expenses | Saving cash (Million) |
|---|---|---|---|---|---|---|---|---|---|
| 2019 | 7,680 | 203 | 0,97 | 0,75 | 7,521 | 159 | 63 | $23,80 | $ 0,20 |
| 2020 | 7,680 | 229 | 0,95 | 0,77 | 7,518 | 162 | 66 | $23,79 | $ 0,21 |
| 2021 | 7,680 | 244 | 0,90 | 0,74 | 7,509 | 171 | 75 | $23,77 | $ 0,23 |

**Table 10.** Saving Direct Labor Cost Calculation (Scenario 2 in Case 1)

| Year | Scheduled Time (hr) | Ideal Speed rate (Unit / hrs) | Ratio (α) Increase | Ratio (β) Increase | New OEE | New Scheduled Time (hr) | Savings Hours | Net saving hour (-96 hour) | New Direct Labor Expenses | Saving cash (Million) |
|---|---|---|---|---|---|---|---|---|---|---|
| | | | 2% | 2% | | | | | | |
| 2019 | 7,680 | 203 | 0,97 | 88% | 0,77 | 7,349 | 331 | 235 | $23,27 | $0,73 |
| 2020 | 7,680 | 229 | 0,95 | 92% | 0,79 | 7,354 | 326 | 230 | $23,28 | $0,72 |
| 2021 | 7,680 | 244 | 0,90 | 91% | 0,76 | 7,343 | 337 | 241 | $23,25 | $0,75 |

**Table 11.** Saving Direct Labor Cost Calculation (Scenario 3 in Case 1)

| Year | Scheduled Time (hr) | Ideal Speed rate (Unit / hrs) | Ratio (α) Increase | Ratio (β) Increase | Ratio (μ) Increase | New OEE | New Scheduled Time (hr) | Savings Hours | Net saving hour (-96 hour) | New Direct Labor Expenses (million) | Saving cash (Million) |
|---|---|---|---|---|---|---|---|---|---|---|---|
| | | | 2% | 2% | 2% | | | | | | |
| 2019 | 7,680 | 203 | 0,97 | 88% | 93% | 0,79 | 7,192 | 488 | 392 | $22,77 | $ 1,23 |
| 2020 | 7,680 | 229 | 0,95 | 92% | 92% | 0,80 | 7,195 | 485 | 389 | $22,78 | $ 1,22 |
| 2021 | 7,680 | 244 | 0,90 | 91% | 95% | 0,78 | 7,189 | 491 | 395 | $22,77 | $ 1,23 |

By using equation in Table 1, the calculation of income and relationship between changing factors and income are shown in Table 12. By the same way, we calculate the other scenario and shown in Figure 2,3 and 4.

**Table 12.** Financial Calculation

| Case 1 (Scenario 1) | | Case 2 (Scenario 2) | |
|---|---|---|---|
| Net Sales | $ 115,00 | Net Sales | $ 117,43 |
| Direct Materials used | $ 26,00 | Direct Materials used | $ 26,52 |
| Direct Labor | $ 23,80 | Direct Labor | $ 24,30 |
| Total Factory Overhead | $ 53,00 | Total Factory Overhead | $ 53,00 |
| Operating Income Earnings Before Interest and Taxes (EBITS) | $ 12,20 | Operating Income Earnings Before Interest and Taxes (EBITS) | $ 13,61 |

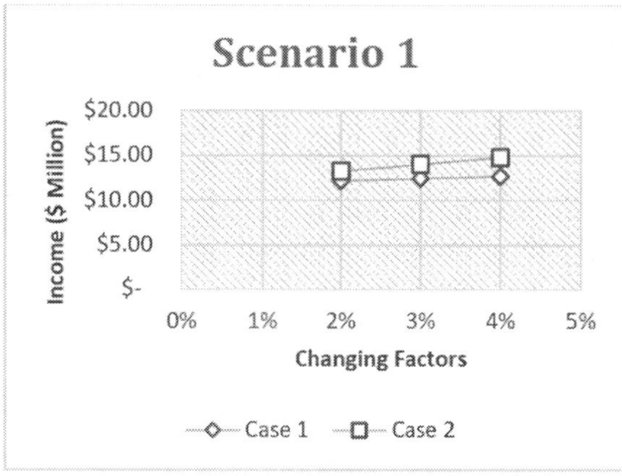


**Figure 2.** Changing Factors VS Income (Scenario 1)

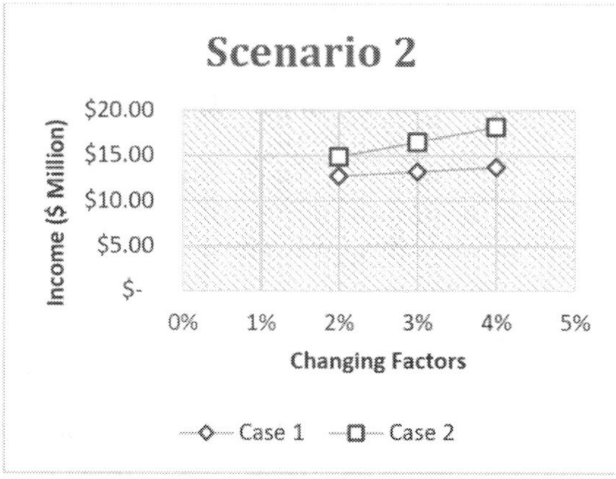


**Figure 3.** Changing Factors VS Income (Scenario 2)

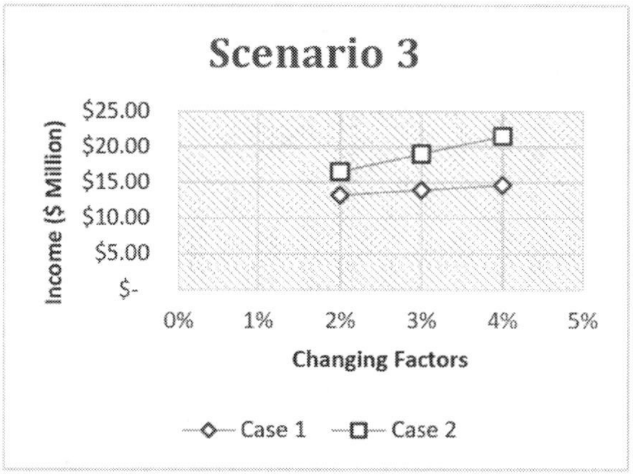


**Figure 4.** Changing Factors VS Income (Scenario 3)

## 4. Conclusion

Based on the results of OEE calculations and measurements of the financial aspects, it is seen that increasing the factors will increase direct material costs, reduce direct labor costs and increase profits from product sales. Case 1 can affect the direct labor cost while in case 2 can affect the direct labor and material cost. All OEE elements need to be considered in order to significantly increase profits. Because the data used is still in the form of historical data, this study needs to examine more patterns of tradeoff between costs and OEE needs in order to get more accurate results.

## 5. References

[1] Nakajima, S., 1988 *Introduction to TPM: Total Productive Maintenance*, Cambridge, MA, Productivity Press.
[2] Huang S.H., Dismukes J.P, Mousalam A., Razzak R.B., Robinson D.E, 2003. Manufacturing Productivity improvement using effectiveness metrics and simulation analysis, *International Journal of Production Research*, **41**(3), 513 – 527.
[3] Fleischer, J., Weismann, U. and Niggeschmidt, S., 2006. Calculation and optimisation model for costs and effects of availability relevant service elements. *Proceedings of LCE*, pp.675-680.

IOP Conf. Series: Materials Science and Engineering **847** (2020) 012034 doi:10.1088/1757-899X/847/1/012034

[4] Williamson R.M., 2006 Using Overall Equipement Effectiveness: the Metric and the Measures, Strategic Work Systems, Inc. Columbus, Available online at: www.swspitcrew.com, (accessed June, 2006)
[5] Bonavia, T. and Marin-Garcia, J. A. 2006. An empirical study of lean production in Ceramic tile industries in Spain. *International Journal of Operations and Production Management*, **26**(5), 505-531. http://dx.doi.org/10.1108/01443570610659883.
[6] Hansen, Robert C., 2001 Overall equipment effectiveness: a powerful production/maintenance tool for increased profits. Industrial Press, Inc. 200 Madison Avenue New York, NY 10016-4078,
[7] Ahuja, I.P.S., dan Khamba, J.S., 2008, Total productive maintenance: literature review and directions, *International Journal of Quality & Reliability Management*, **25**(7), pp.709-756.
[8] Almeanazel, O.T.R. 2010. Total Productive Maintenance Review and Overall Equipment Effectiveness Measurement. *Jordan Journal of Mechanical and Industrial Engineering*, **4**(4). ISSN 1995-6665, Department of Industrial Engineering, Hashemite University, Zarqa, Jordan.
[9] Davis, Roy. 1995. *Productivity Improvements Through TP: The Philosophy and Application of Total Productive Maintenance*. Englewood Cliffs: Prentice Hall Inc.
[10] Almeanazel, O.T.R. 2010. Total Productive Maintenance Review and Overall Equipment Effectiveness Measurement. *Jordan Journal of Mechanical and Industrial Engineering*, **4**(4). ISSN 1995-6665, Department of Industrial Engineering, Hashemite University, Zarqa, Jordan.
[11] Blanchard, S dan Benjamin, 1997, An enchanced approach for implementing total productive maintenance in the manufacturing environment, *Journal of Quality in Maintenance Engineering*, **3**.
[12] Chan, F.T.S., Lau, H.C.W., Ip, R.W.L., Chan, H.K., dan Kong, S., 2005, Implementation of total productive maintenance: a case study, *International Journal Production Economics* **95**, 71-94.
[13] Jeong, K. Y., dan Phillips, D.T., 2001, Operational efficiency and effectiveness measurement. *International Journal of Operations and Production Management*, **21**(11), page 1404-1416.
[14] Muchiri, P. and Pintelon, L., 2008. Performance measurement using overall equipment effectiveness (OEE): literature review and practical application discussion. *International journal of production research*, **46**(13), pp.3517-3535
[15] Shah, M.K., Deshpande, V.A. and Patil, R.M., 2017, February. Case study: Application of Lean tools for Improving Overall Equipment Effectiveness (OEE) & Productivity in panel shop of heavy Fabrication Industry. In *Proceedings of 2nd International Conference on Emerging Trends in Mechanical Engineering*.

IOP Conf. Series: Materials Science and Engineering **847** (2020) 012035 doi:10.1088/1757-899X/847/1/012035

# Analysis of Overall Effectiveness on Hall Separator Punching Machine at PT. DNIA

**Iphov Kumala Sriwana[1], Nadya Syauqillah [2]**

[1] Department of Industrial Engineering, Faculty of Engineering, Universitas Esa Unggul, Jl. Arjuna Utara No 9. Jakarta 11510 Indonesia
[2] Bachelor degree of Industrial engineering Universitas Esa Unggul, Jakarta 11510 Indonesia
iphov.kumala@esaunggul.ac.id

**Abstract.** PT. DNIA is a company engaged in manufacture of automotive parts, one of which is condenser. Manufacture of condenser components requires small parts produced using a hole separator punching machine. However, it deals with high downtime of the machine, resulting in low production performance. This research aimed to identify the extent of hole separator punching machine performance using analysis of Overall Equipment Effectiveness (OEE) and to analyse six big loses which impact on machine downtime. Calculation results show that OEE value obtained, 48.54%, was still below the standard, and therefore continuous improvement attempt is essential to perform. The low OEE value was a result of low performance efficiency which was caused by idling and minor stoppages of 24.54%. In order to improve the performance and carry out idling and minor stoppages loss, it is important to perform improvement attempt in a number of aspects, such as man aspect by training operators to carry machine-related works, machine aspect by repairing abnormal ups and downs of dies, and material aspect by fixing inappropriate position of header tank (material).
**Keywords**: Overall Equipment Effectiveness (OEE), six big losses, down time, productivity

## 1. Introduction

PT. DNIA is engaged in manufacture of automotive parts, i.e. powertrain thermal, one of which is condenser. Manufacture of some components for condenser requires small parts produced with a hole separator punching machine. Historically, downtime frequency of machines used for condenser production is a major cause for high losses man hour. Based on observation in previous studies, it has been noticed that hole separator punching machine shows downtime frequency of 30 times per day. In addition, absence of production data record in previous period results in more difficulty during downtime identification of hole separator punching machine. High downtime frequency of hole separator punching machine leads to delays of condenser production which impacts on a decreased performance, eventually.

In the present study, performance measurement was conducted using an Overall Equipment Effectiveness (OEE) method. OEE is frequently employed to measure TMP performance as reported by [1] or is employed for determining critical machine as reported by [2]. [3] used OEE in performance measurement, while [4] claimed that OEE is the main performance measurement used in mass production environment. Similarly, [5] argued that system and equipment-related performance can be analysed using OEE. The above explanation on OEE becomes a consideration of OEE implementation for minimizing occurrence of machine downtime in order to improve performance, which contributes to an increased productivity, eventually. Once OEE value is measured, identification of six big losses is carried out in order to find out causes for machine downtime.

This research aimed to identify Overall Equipment Effectiveness (OEE) of hole punching separator machine in manufacture of small parts at condenser lines, identify major problems occurring using Six Big Losses in order to discover the most predominantly problem which impacts on a decreased effectiveness of hole separator punching machine during manufacture of small parts at condenser lines, as well as to propose improvement in issues related to effectivity of hole separator punching machine in the manufacture of small parts at condenser lines.

## 2. Methods

Methods used to solve the above mentioned problems consist of three stages, i.e. stage 1 in which calculation of Overall Equipment Effectiveness (OEE) was performed, stage 2 in which six big losses analysis was performed, and stage 3 in which analysis using a number of tools and seven tools in forms of pareto and fishbone diagram was performed.

*Stage 1. OEE Calculation*

[6] reported that Overall Equipment Effectiveness (OEE) is a tool for measuring an efficiency level. It is affected by several factors, i.e. Availability Rate, Performance Rate, and Quality Yield, of which equation is expressed as follows.

$$\text{OEE} = \text{Availability x Performance x Quality}$$

Availability in machine/equipment is a ratio of operation time to loading time of machine, of which equation is expressed as follows.

$$Availability = \frac{Availabprodcle\ Time\ - Downtime}{Loading\ Time}\ x\ 100$$

Performance is a ratio of product quality multiplied by ideal cycle time, to operation time, of which equation is expressed as follows

$$Performance = \frac{Output\ x\ Ideal\ Cycle\ Time}{Operating\ Time}\ x\ 100$$

Quality is a ratio that describes machine capability in manufacturing products complying with specified standard, of which equation is expressed as follows.

$$Quality = \frac{Output\ x\ Defect\ Amount}{Processed\ Amount}\ x\ 100$$

Measurement of OEE value should be carried out with considering standard value that has been specified, and according to [7] World Class OEE value is 85%.

*Stage 2. Analysis of Six Big Losses*

The objective of Six Big Losses is zero breakdown. They are a part of TPM, of which system assists in eliminating six big losses from equipment and processes of production. The whole process of TPM is focussed on eliminating wastes, which are categorized into 6 types of losses [8], namely:

a. Downtime Losses
   - Setup and Adjustment Losses are losses due to setup and adjustment at the beginning and end of engine operation.

$$Set\ Up\ and\ Adjustment\ Losses\ = \frac{Setup\ and\ Adjusment}{Loading\ Time}\ x\ 100$$

   - Equipment Failures Losses are losses due to engine failure which results in shut off.

$$Equipment\ Failures\ Loss = \frac{Breakdown}{Loading\ Time}\ x\ 100$$

b. Speed Losses

- Idling and Minor Stoppages Losses are losses due to an engine momentary stop, stuck engine, and engine idle time.

$$Idle\ and\ Minor\ stoppages\ Looses = \frac{(jumlah\ target - jumlah\ produksi) x ideal\ cycle\ time}{Loading\ Time} x\ 100$$

- Reduced Speed Losses are losses due to un-optimum run of engine (reduced speed of engine operation)

$$Reduced\ Speed\ Losses = \frac{(actual\ cycle\ time - ideal\ cycle\ time)\ x\ output}{Loading\ Time} x\ 100$$

c. Quality Losses

- Defect Losses are losses that occur due to defective or reworked products.

$$Defect\ Losses = \frac{output\ cacat\ x\ ideal\ cycle\ time}{Loading\ Time} x\ 100$$

- Reduced Yield is a material loss required by a machine to produce new products at expected quality.

*Stage 3. Analysis using Tools Contained in 7 Tools*
In this stage, there were only 2 tools employed, i.e. pareto and fishbone diagram. Pareto diagram constitutes a bar chart and a line chart that describe comparison between each type of data and overall data. This diagrams seems to be very simple, however it is very beneficial in control of factory quality [9].

The fundamental function of Fishbone/Cause-effect diagram is to identify and organize possible causes for specific effects, followed by separation of root causes. Identification was carried out using a structured approach that enabled incorporation of a detailed analysis. There were 5 major factors analysed using Fish bone, i.e. man, work method, machine/equipment, and environment [10].

## 3. Results and Discussion

Data used in this research was obtained based on direct observation at the company and company's data source related to production of hole separator punching. Data related to this research such as planned downtime, set up and adjustment, output, breakdown machine, etc. are presented in Table 1.

Table 1. Production Hole Separator Punching Machine

| No | Date | *Operating Time (second)* | *Loading Time (second)* | *Ideal Cycle Time (second)* | *Output* | *Defect* |
|---|---|---|---|---|---|---|
| 1 | 1 | 54212 | 72000 | 14.27 | 2860 | 4 |
| 2 | 2 | 52200 | 72000 | 12.62 | 3000 | 0 |
| 3 | 3 | 55834.4 | 72000 | 15.04 | 2878 | 4 |
| 4 | 4 | 27813 | 43200 | 7.98 | 2740 | 0 |
| 5 | 5 | 51170 | 72000 | 12.04 | 3020 | 0 |
| 6 | 6 | 55757 | 72000 | 14.06 | 3070 | 1 |
| 7 | 7 | 57097.5 | 72000 | 14.87 | 3045 | 22 |
| 8 | 8 | 51225 | 72000 | 13.96 | 2610 | 0 |
| 9 | 9 | 41600 | 72000 | 8.23 | 2920 | 0 |
| 10 | 10 | 40415 | 61199 | 10.57 | 2720 | 3 |
| 11 | 11 | 52116 | 72000 | 10.90 | 3460 | 0 |
| 12 | 12 | 53846 | 72000 | 11.64 | 3460 | 18 |
| 13 | 13 | 54894 | 72000 | 11.58 | 3615 | 1 |
| 14 | 14 | 37793 | 72000 | 8.11 | 2445 | 18 |
| 15 | 15 | 18132 | 43200 | 3.86 | 3060 | 16 |
| 16 | 16 | 52274 | 72000 | 12.09 | 3140 | 0 |
| 17 | 18 | 41043 | 72000 | 7.15 | 3270 | 0 |
| 18 | 19 | 51984 | 72000 | 11.17 | 3360 | 18 |
| 19 | 20 | 43725 | 72000 | 9.32 | 2850 | 6 |

Table 1. Production Hole Separator Punching Machine

| No | Date | *Operating Time (second)* | *Loading Time (second)* | *Ideal Cycle Time (second)* | *Output* | *Defect* |
|---|---|---|---|---|---|---|
| 20 | 21 | 52100 | 72000 | 11.09 | 3400 | 0 |
| 21 | 22 | 42324 | 72000 | 9.87 | 2520 | 0 |
| 22 | 23 | 49391 | 68399 | 11.43 | 3180 | 9 |
| 23 | 24 | 58590 | 72000 | 14.45 | 3300 | 6 |
| 24 | 25 | 45730 | 72000 | 12.63 | 2300 | 0 |
| 25 | 26 | 50451 | 72000 | 10.94 | 3230 | 5 |

**Stage 1. Calculation of Overall Equipment Effectiveness (OEE)**
OEE calculation was initiated with identifying three values of OEE factors, i.e. availability, performance, and quality, as shown in Table 2.

Tabel 2. OEE calculation at Hole Separator Punching Machine

| Date | *Availability* | *Performance* | *Quality* | OEE |
|---|---|---|---|---|
| 1 | 75.29% | 75.29% | 99.86% | 56.61% |
| 2 | 72.50% | 72.50% | 100.00% | 52.56% |
| 3 | 77.55% | 77.55% | 99.86% | 60.05% |
| 4 | 64.38% | 78.63% | 100.00% | 50.62% |
| 5 | 71.07% | 71.07% | 100.00% | 50.51% |
| 6 | 77.44% | 77.44% | 99.97% | 59.95% |
| 7 | 79.30% | 79.30% | 99.28% | 62.43% |
| 8 | 71.15% | 71.15% | 100.00% | 50.62% |
| 9 | 57.78% | 57.78% | 100.00% | 33.38% |
| 10 | 66.04% | 71.13% | 99.89% | 46.92% |
| 11 | 72.38% | 72.38% | 100.00% | 52.39% |
| 12 | 74.79% | 74.79% | 99.48% | 55.64% |
| 13 | 76.24% | 76.24% | 99.97% | 58.11% |
| 14 | 52.49% | 52.49% | 99.26% | 27.35% |
| 15 | 41.97% | 65.18% | 99.48% | 27.22% |
| 16 | 72.60% | 72.60% | 100.00% | 52.71% |
| 18 | 57.00% | 57.00% | 100.00% | 32.49% |
| 19 | 72.20% | 72.20% | 99.46% | 51.85% |
| 20 | 60.73% | 60.73% | 99.79% | 36.80% |
| 21 | 72.36% | 72.36% | 100.00% | 52.36% |
| 22 | 58.78% | 58.78% | 100.00% | 34.55% |
| 23 | 72.21% | 73.60% | 99.72% | 53.00% |
| 24 | 81.38% | 81.38% | 99.82% | 66.10% |
| 25 | 63.51% | 63.51% | 100.00% | 40.34% |
| 26 | 70.07% | 70.07% | 99.85% | 49.02% |

Averaged Overall Equipment Effectiveness (OEE) results obtained was 48.54%. According to Japan Institute of Plant Maintenance (JIPM), 40% ≤ OEE values < 59% are categorized low, and therefore company should improve the existing system performance to prevent it from economic loss and low competitiveness (Nakajima. 1998). Figure 1 shows a comparison graph of OEE at hole separator punching machine.

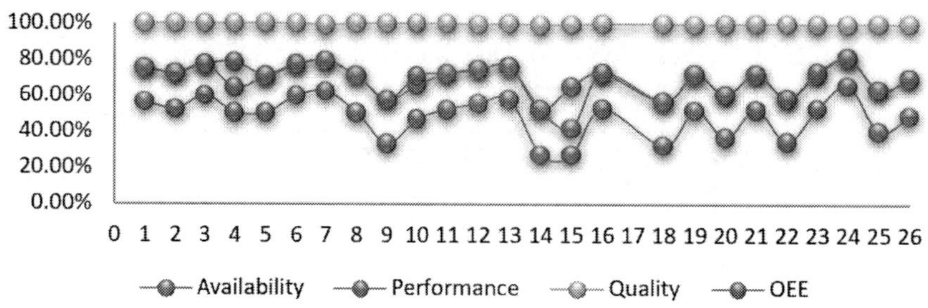


Figure 1. OEE calculation at hole separator punching machine

As seen in Figure 1, performance percentage was slightly lower as compared to availability percentage which was a major factor in performance scope. Calculation of losses in factors that result in low performance percentage was carried out using a six big losses approach.

**Stage 2. Calculation of Six Big Losses**
Calculation of six big losses is beneficial for the company to discover factors producing the largest losses and contributing to low Overall Equipment Effectiveness (OEE). Those causative factors will be considered as the main priority to address.

Table 3. *Six Big Losses* Calculation at Hole Separator Punching Machine

| Date | *Set-up and Adjustment Losses* | *Breakdwon Losses* | *Idle and Minor Stoppages Loss* | *Reduced Speed Loss* | *Rework Losses* | *Reduced Yeild Losses* |
|---|---|---|---|---|---|---|
| 1 | 7.63% | 17.08% | 31.99% | 18.60% | 0.11% | 0.00% |
| 2 | 8.75% | 18.75% | 25.98% | 19.94% | 0.00% | 0.00% |
| 3 | 7.66% | 14.79% | 32.25% | 17.41% | 0.11% | 0.00% |
| 4 | 11.35% | 24.26% | 31.37% | 13.76% | 0.00% | 0.00% |
| 5 | 8.80% | 20.13% | 24.83% | 20.56% | 0.00% | 0.00% |
| 6 | 9.34% | 13.22% | 25.35% | 17.47% | 0.03% | 0.00% |
| 7 | 9.28% | 11.42% | 26.82% | 16.41% | 0.57% | 0.00% |
| 8 | 7.47% | 21.39% | 39.93% | 20.53% | 0.00% | 0.00% |
| 9 | 9.78% | 32.44% | 22.85% | 24.40% | 0.00% | 0.00% |
| 10 | 9.52% | 24.44% | 32.90% | 19.06% | 0.07% | 0.00% |
| 11 | 9.36% | 18.26% | 12.87% | 19.99% | 0.00% | 0.00% |
| 12 | 10.80% | 14.42% | 13.29% | 18.86% | 0.39% | 0.00% |
| 13 | 10.70% | 13.05% | 9.70% | 18.11% | 0.02% | 0.00% |
| 14 | 7.44% | 40.07% | 34.99% | 24.94% | 0.39% | 0.00% |
| 15 | 15.53% | 42.50% | 13.92% | 14.61% | 0.22% | 0.00% |
| 16 | 7.34% | 20.06% | 21.62% | 19.89% | 0.00% | 0.00% |
| 18 | 7.57% | 35.43% | 14.03% | 24.51% | 0.00% | 0.00% |
| 19 | 9.13% | 18.67% | 15.36% | 20.07% | 0.39% | 0.00% |
| 20 | 8.40% | 30.88% | 26.10% | 23.85% | 0.13% | 0.00% |
| 21 | 9.22% | 18.42% | 14.37% | 20.00% | 0.00% | 0.00% |
| 22 | 7.27% | 33.95% | 36.27% | 24.23% | 0.00% | 0.00% |
| 23 | 6.87% | 20.92% | 20.32% | 19.06% | 0.20% | 0.00% |
| 24 | 8.08% | 10.54% | 19.11% | 15.16% | 0.15% | 0.00% |
| 25 | 6.46% | 30.03% | 49.02% | 23.17% | 0.00% | 0.00% |
| 26 | 6.15% | 23.78% | 18.33% | 20.97% | 0.11% | 0.00% |

It is noticed from Table 3 that averaged values of setup and adjustment was 8.80%, equipment failure / breakdown losses was 22.76%, idle and minor stoppages was 24.54%, reduced speed loss was 19.82%, defect loss was 0.12%, and reduced yield loss was 0%.

Table 4 *Six Big Losses* at Hole Separator Punching machine

| *Six Big Losses* | Time Losses (second) | Cumulative | Cumulative Percentage (%) |
|---|---|---|---|
| *Idling and Minor Stoppages* | 0.25 | 0.25 | 32.28% |
| Breakdwon Looses | 0.23 | 0.47 | 62.21% |
| *Reduced Speed Losses* | 0.20 | 0.67 | 88.28% |
| *Set-up and Adjustment Looses* | 0.09 | 0.76 | 99.85% |
| *Defect Losses* | 0.0012 | 0.76 | 100.00% |
| *Reduced Yield Losses* | 0.00 | 0.76 | 100.00% |

**Stage 3. Analysis using Pareto and Fish Bone Diagram**
Results obtained in Table 4 was further analysed using a pareto diagram in order to obtain actual root causes for problems occurring at hole separator punching machine, as given in Figure 2.

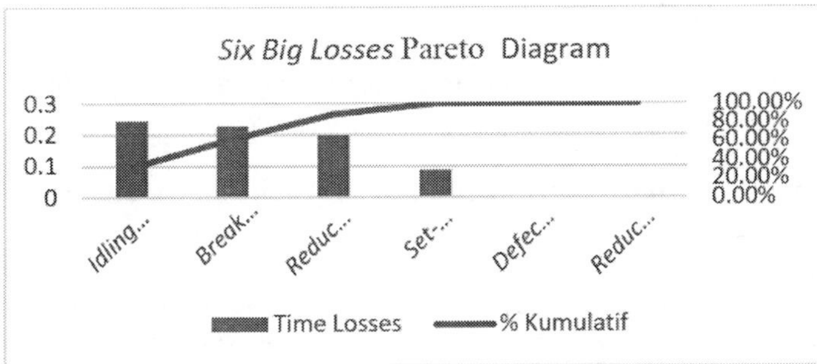


Figure 2. Six big Losses Pareto Diagram

It is noticed from Figure 2 that idling and minor stoppages loss was at 32.28%, which allowed it to be a major cause for low OEE value, and therefore a further analysis was required. The next analysis was carried out using a cause-and-effect diagram (fishbone diagram) as presented in Figure 3.

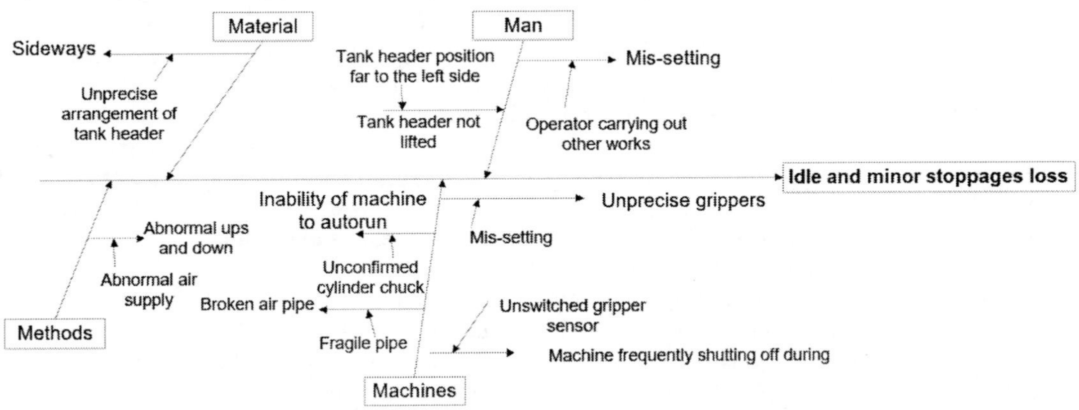


Figure 3. Diagram of Fishbone Idle and Minor Stoppages Loss

As seen in Figure 3, there were several aspects which affected high idle and minor stoppages of hole separator punching machine, which are further described as follows.

1. Man
   Man or operator assigned at the hole separator punching machine often carries double jobs, allowing to carelessness during machine setting and operation.

2. Method
   The obstacle encountered is abnormal ups and downs of dies which are caused by abnormal air supply.
3. Machine
   A number of obstacles encountered include inability of autorun due to unconfirmed cylinder chuck, unprecise or unaligned gripper due to mis-setting at the initial production process, broken air pipe due to fragility, as well as frequently shut-off machine during production process due to unswitched gripper sensor.
4. Material
   Obstacles encountered in the material factor are tank header (material) that is not lifted by gripper due to unprecise or unaligned arrangement as well as tank header position is far to the left side allowing it to crash the slider when being dropped by gripper and resulting in an unprecise position.

Based on the above fishbone idle and minor stoppages losses on man, method, and machine, some recommendations are proposed as follows.

1. Man
   Operators assigned at hole separator punching machine are not burdened by additional works which are not related to the machine.
2. Method
   Periodic calibration of the dies can be performed in the beginning and end of every week (once a week at the least).
3. Machine
   Improvement include periodically preventive maintenance (pm), re-setting in the beginning and end of production, as well as replacement of machine parts that do not working properly.
4. Material
   Positioning of tank header arrangement is carried out using a device. In addition, slope and arrangement of tank header should be measured to prevent it from being dropped or lifted during arrangement.

## 4. Conclusion

According to results as well as data processing and analysis of the present study, it can be concluded that

1. Averaged OEE value of Hole Separator Punching machine of 48.54% was still below the specified standard, and therefore a continuous improvement attempt has to be conducted.
2. Low OEE value was a consequence of low efficiency performance largely contributed by idling and minor stoppages, 24.54%.
3. According to fishbone diagram analysis, factors which impact on idling and minor stoppages loss are as follows.
   a. Man : Operators carrying out works that are not related to machine
   b. Method : Abnormal ups and downs of dies
   c. Machine : Problematic gripper, machine unable to autorun
   d. Material : Inappropriate position of tank header (material)

Some recommendation proposed to improve performance of hole separator punching machine include requirement to implement scheduled and recorded autonomous maintenance in order to facilitate evaluation when problems raise. In addition, OEE calculation should be performed continuously using an information system recorded in database, followed by maintenance, replacement, and re-design of gripper parts.

## 5. References

[1] N. Erni and A. F. Maulana, 2012 Pengukuran Kinerja Mesin Produksi Dengan Metode Overall Equipment Effectiveness Pada PT. Cahaya Biru Sakti Abadi, *J. Inovisi*, **8**(2), pp. 80–91.

[2] I. Latib and O. Doaly, 2018 Usulan Penjadwalan Penggantian dan Pemeriksaan Komponen Kritis Mesin Feeder dan Fanblower (Studi Kasus di PT. Petnesia Resindo), **8**(2), pp. 93–103.

[3] N. Setiawan and M. Annisa, 2017 Pengukuran kinerja produksi hot strip mill dengan metode Overall Equipment Effectiveness (OEE) (Studi Kasus pada Industri Baja di Cilegon, Banten), *teknoin*, **23**(2), pp. 153–160.

[4] A. J. De Ron and J. E. Rooda, 2006 OEE and equipment effectiveness: An evaluation, *Int. J. Prod. Res.*, **44**(23), pp. 4987–5003.

[5] K. M. Cheh, 2014 Analysis of Overall Equipment Effectiveness (OEE) within different sectors in different Swedish industries, Mälardalen University.

[6] S. Boris, 2006 *Total Productive Maintenance.* (United State of Amerika: The McGrawHill Companies, Inc.)

[7] S. Perdana, R. Usman, and Z. F. Hunusalela, 2018 Produktivitas Mesin Produksi Hard Disc Drive Pada PT. XYZ, in *Seminar Nasional dan Diskusi Panel Multidisiplin Hasil Penelitian & Pengabdian kepada Masyaraka*, no. August, pp. 134–139.

[8] S. Nakajima, 1998 *Introduction to Total Productive Maintenance.* (Cambridge: Productivity Press, Inc.)

[9] K. Ishikawa, 1989 *Teknik Penuntun Pengendalian Mutu.* (Jakarta: PT. Mediyatama Sarana Perkasa).

[10] H. A. Prabowo and D. I. R, 2014 Improve the Work Effectiveness With OEE (Overall Equipment Effectiveness) As the Basis for Optimizing Production, *Pasti*, vol. IX, no. 3, pp. 286–299.

# Multi Sensor-Based Failure Diagnosis using the Mahalanobis Taguchi System

**B S P Susanto[1,3] and N Kurniati[2,4]**

[1] Industrial Engineering Department, Institut Teknologi Sepuluh Nopember, Surabaya, Indonesia.
[2]Industrial Engineering Department, Institut Teknologi Sepuluh Nopember, Surabaya. Indonesia.
[3]bagus57susilo@gmail.com, [4]nanikur@gmail.com

**Abstract.** Heavy equipment in the mining industry is an essential instrument in achieving the company's production targets. But on the other hand, heavy equipment is complicated and expensive equipment. These problems can be overcome by leasing the equipment to an external agent. In the lease contract between the lessee and lessor, it is discussed about the maintenance strategy of the equipment that will be rented by the lessee. This research presents a real-time monitoring scheme based on multi-sensor data installed on the machine. Case studies from this research were carried out on excavator units in heavy equipment rental companies. The Mahalanobis Taguchi System (MTS) method is used to handle multi-sensor data. Multi-sensor data is grouped into normal condition groups and abnormal condition groups. The variables used to monitor excavator conditions include vibration, pressure and temperature sensors. Based on the results of the calculation of the excavator threshold conditions using the Mahalanobis Distance (MD) measurement technique, the threshold of normal conditions is (-2.137 to 4.121). In contrast, in abnormal accumulator conditions (4,331 to 39,458), pump failure (40,956 to 138,048), valve failure (2708,104 to 3404,187) and the threshold in the condition of the cooler failure is (10736.160 to 11434.151). This research shows that the MD-based CBM scheme produced can detect, identify, and isolate the failure of excavator components under study.
**Keywords**: Condition Based Maintenance. Mahalanobis Distance, Mahalanobis Taguchi System, Leasing Equipment

## 1. Introduction

In the mining industry, heavy equipment such as excavators is used to load and transport mining material from the mining site to the processing unit. Heavy equipment plays an essential role in achieving the company's production targets. When the device is deteriorating due to its age and usage, an effective way to deal with the incident is to carry out maintenance activities. However, heavy equipment such as excavators is complicated and expensive equipment. The best economical way to overcome this is by leasing equipment to an external agent [1].

Leasing is a cooperation contract between the lessee (lessee) using the equipment owned by the lessor (owner), both of which are separate parties. The lessee routinely pays a certain amount of fees to the lessor [2]. In the lease contract between the lessee and lessor, it is discussed about the maintenance policy for the equipment to be rented by the lessee. This is done so that the lessor can ensure that they guarantee the satisfaction and reliability of the equipment during the rental period.

In general, there are two types of maintenance policies that can be carried out during the contract period, namely Corrective Maintenance (CM) and Preventive Maintenance (PM). CM measures are used to correct equipment failures and return them to operational conditions. While PM is a policy that is more directed to actions to maintain the functional state of equipment in avoiding failure [3]. But lately construction companies, aerospace and automotive are in a trend to develop their company's ability to use the Condition Based Maintenance (CBM) strategy as an effort to meet the demands of their customers [4].

Condition Based Maintenance (CBM) is a maintenance measure carried out following the current status of the equipment. Compared with other maintenance strategies, CBM is considered more effective in overcoming the advantages or disadvantages of the amount of maintenance. If the CBM program is implemented correctly, it is possible for the company to significantly reduce maintenance costs by reducing the amount of maintenance activity. CBM activities consist of diagnostic and prognostic processes. The diagnostic itself consists of detection, isolation, and identification of failures, whereas prognostic is a prediction activity before the failure occurs [5]. The success of implementing CBM depends very much on the company's ability to extract information from a data set that comes from monitoring the condition of the equipment.

Some methods that have been used as tools in implementing CBM include Principal Component Analysis (PCA) and Fuzzy C-Means (FCM) in their research entitled "Application of PCA method and FCM clustering to the fault diagnosis of excavator's hydraulic system" PCA and FCM used as a fault diagnostic approach to improve the reliability of excavator hydraulic systems. FCM clustering is done as a failure classifier to determine failure testing. Fault diagnostics using PCA and FCM have proven to be feasible and effective on excavator hydraulic systems. [6]. Furthermore, other multivariate methods are also used by Yang, et al. [7] in his research entitled "The use of the Mahalanobis-Taguchi System to improve flip-chip bumping height inspection efficiency" overcomes the problem of bumps height inspection efficiency. The Mahalanobis - Taguchi System (MTS) method is used to reduce the number of bump height measurements while maintaining inspection with a high degree of accuracy. Successfully reduced the number of bump height inspections significantly from 10 to 6, and the inspection time could be reduced by 40%. Then Felix et al. [4] have researched excavators to improve the hydraulic performance of excavators through monitoring contamination of hydraulic oil. The study assessed the suitability of particle contamination sensors in carrying out diagnostic and prognostic requirements in implementing CBM. While the research of Soylemezoglu et al. [8] relates to the implementation of MTS based on prognostic decision making for the failure of centrifugal pumps. The main objective of the study is to apply the concept of the MTS model to demonstrate its prognostic ability to detect pump failures and determine or predict pump failure times. This study uses the mean (mean) and three times the standard deviation approach to determine the threshold for each failure condition studied. Helwig et al. [9] also re-used the Multivariate Statistics technique, namely Linear Discriminant Analysis (LDA), which is combined with Mahalanobis distance clustering in hydraulic system monitoring. The purpose of this study is to develop and evaluate a systematic approach to automating the hydraulic system condition monitoring system by analyzing various failure scenarios obtained from experiments on hydraulic systems. The results of the study conclude that the method chosen has proven suitable for detecting different types of errors that show typical symptoms at a predetermined scale. Recent research on the application of CBM in the concept of leasing was carried out by Zhang et al. [10] on this study regarding the satisfaction of lessees and the optimal CBM policy for leasing systems. The purpose of this research is to develop CBM policies for leasing systems, model lessee satisfaction, and make decision models that take into account the total benefits of the leasing system. The conclusion obtained is the CBM policy developed shows that the policy can increase the total lessor profit. In their research conducted a discrete CBM monitoring based on the implementation of inspections expressed by the deterioration rate ($\Omega$) for each time epoch (tk) which then resulted in an Optimal maintenance policy that minimizes cost of maintenance from Preventive

IOP Conf. Series: Materials Science and Engineering **847** (2020) 012036 doi:10.1088/1757-899X/847/1/012036

activities Maintenance (PM) and Corrective Maintenance (CM) after increasing revenue (revenue) and total profit.

Based on various explanations about previous studies, the author will raise several research gaps. The gaps are the use of the Mahalanobis Taguchi System (MTS) to handle the extensive multi-sensor data set in applying CBM in heavy equipment leasing companies. Then in this study, the CBM strategy is based on multi-sensor data installed in the equipment or other words, monitoring is carried out continuously or in real-time. So that applying CBM will make it easier for the lessor to monitor and control the tools that are being used by the lessee during the contract period.

Case Study of this research itself was conducted at PT X, which is a company engaged in heavy equipment leasing. The type of heavy equipment that will be examined in this study is the excavator unit. This is based on the availability of multi-sensor data which is only available in the group. In this study, the application of the condition-based maintenance strategy was applied with the Mahalanobis Taguchi System approach to the excavator hydraulic system at PT X. The choice of the hydraulic system as the object of this study was based on the results of research conducted by Kamil and Ulka [3], which obtained results that the hydraulic system excavators are a vital component of excavator shutdown.

## 2. Methods

This research was conducted at PT X, which is a company engaged in heavy equipment leasing. The heavy equipment unit, which is the object of this research is the excavator unit. The components studied in excavators are hydraulic pumps, accumulators, valves and coolers from the excavator hydraulic system. Multi-sensor data in this study were collected into normal condition data group and abnormal condition data group. An explanation of each of the data conditions is as follows:

**Table 1.** MD transformation and calculation for normal data

| Condition | Information |
|---|---|
| **Normal** | Multi-sensor data in normal conditions is collected when the excavator is in its operating condition. |
| **Failure of Pump** | Multi-sensor data on this condition is collected when a hydraulic pump fails or performance decreases, which can be caused by a leak in the hydraulic pump. This condition is characterized by a decrease in the pressure flow in the hydraulic system. |
| **Failure of Accumulator** | Multi-sensor data on this condition is collected when the hydraulic accumulator is experiencing a failure which is characterized by a decrease in pressure and the emergence of excessive vibration in the hydraulic system. |
| **Failure of Relief Valve** | Multi-sensor data on this condition is collected when the excavator hydraulic valve experiences lag or malfunction, causing an increase in fluid pressure in the hydraulic system. |
| **Failure of hydraulic cooler** | Multi-sensor data on this condition is collected when the hydraulic cooler fails or decreases performance which is marked by an increase in temperature in the hydraulic system. |

The variables in this study consisted of 4 sensor units .the sensor is vibration, temperature, pressure 1, and pressure 2. The following is an explanation of the units and supporting information of the sensors used.

**Table 2.** Research variable information

| Code | Sensor Type | Unit | Information |
|---|---|---|---|
| **X1** | Vibration | mm/s | This component is located in the hydraulic pump casing section. |
| **X2** | Temperature | °C | This component is located in the hydraulic oil tank section |
| **X3** | Pressure1 | Bar | This component is located on the boom cylinder. |
| **X4** | Pressure2 | Bar | This component is located on the travel motor. |

After the data collection process is complete, the next data will be processed using the MTS method. The stages in MTS consist of the following 4 stages [8]:

### *2.1. Construction of Mahalanobis Space (MS)*

The first stage in this process is the transformation of data with z-scores. This process is done because there is variability in the unit of data set that will be processed. The details of this process are as follows:

1) Calculate the mean ($\overline{X}$) of each variable (sensor) on normal data

$$\bar{X}_i = \frac{\sum_{j=1}^{N} X_{ij}}{N} \quad , \; i = 1,2,\dots,p \tag{1}$$

Where :
$X_{ij}$ = *i*th variable in the *j*th observation
N = Number of observations
p = Number of variables

2) Calculates the standard deviation ($\sigma X_i$) on each variable (sensor) from normal data

$$\sigma X_i = \sqrt{\frac{\sum_{j=1}^{N} (X_{ij} - \bar{X}_i)}{N-1}}, i = 1,2,\dots,p \tag{2}$$

3) Furthermore, the transformation of data from each variable (i) on observation to (j) by making a $Z_{ij}$ matrix with the following equation:

$$Z_{ij} = \frac{X_{ij} - \bar{X}_i}{\sigma X_i}, \quad i = 1,2,\dots,p \, , \; j = 1,2,\dots,N \tag{3}$$

4) Then the next step is to verify the average and standard deviations of $Z_{ij}$. after transforming the data, the average value must be 0 :

$$\bar{Z}_i = \frac{\sum_{j=1}^{N} Z_{ij}}{N} = 0, \quad i = 1,2,\dots,p \tag{4}$$

and the standard deviation ($\sigma Z_i$) must be 1.

$$\sigma Z_i = \sqrt{\frac{\sum_{j=1}^{N} (Z_{ij} - \bar{Z}_i)}{N-1}} = 1 \, , \quad i = 1,2,\dots,p \tag{5}$$

Where :
$\overline{Z}_i$ = Mean of the *i*th variable transformation
$\sigma Z_i$ = Standard deviation of the *i*th variable transformation

5) Make a transpose matrix ( $Z_{ij}^T$)
6) Calculates the C correlation matrix from $Z_{ij}$.

$$C = \begin{bmatrix} 1 & r_{12} & \cdots & r_{1p} \\ r_{21} & 1 & \cdots & r_{2p} \\ \vdots & \vdots & \vdots & \vdots \\ r_{p1} & r_{p2} & \cdots & 1 \end{bmatrix} \tag{6}$$

with $r_{ij}$

$$r_{ij} = \frac{\sum_{m=1}^{n} (Z_{im} Z_{jm})}{n-1} \, , i, m = 1,2,\dots,p \tag{7}$$

7) Calculates the inverse of the correlation matrix ($C^{-1}$).
After obtaining the correlation matrix the inverse matrix calculation is then performed with the following formula [2]:

$$C^{-1} = \begin{bmatrix} a_{11} & a_{12} & \cdots & a_{1p} \\ a_{21} & 1 & \cdots & a_{2p} \\ \vdots & \vdots & \vdots & \vdots \\ a_{p1} & a_{p2} & \cdots & a_{pp} \end{bmatrix} = \begin{bmatrix} 1 & r_{12} & \cdots & r_{1p} \\ r_{21} & 1 & \cdots & r_{2p} \\ \vdots & \vdots & \vdots & \vdots \\ r_{p1} & r_{p2} & \cdots & 1 \end{bmatrix}^{-1} \tag{8}$$

8) The last step is to calculate the Mahalanobis Distance ($MD_j$).

$$MD_j = \frac{1}{p} Z_{ij}^T C^{-1} Z_{ij} \tag{9}$$

*2.2. Validation of Mahalanobis Space (MS)*
This stage begins by calculating the abnormal MD of the components of the cooler, valve, pump and accumulator based on sensors Pressure 1, Pressure 2, Temperature, and Vibration. The calculation of abnormal MD uses the mean, standard deviation and correlation matrix of the normal MD calculation in the previous stage. Furthermore, after an abnormal MD value is obtained, validation is performed between normal MD and every abnormal MD. The requirements for normal and abnormal MD distances must exceed more than 1. If the abnormal MD scale does not reach a value greater than 1 than the normal MD, then a new variable will be added.

*2.3. Optimization of Mahalanobis Space(MS)*
This stage purpose of selecting the critical variables used in monitoring for each condition (normal, cooler failure, valve failure, pump failure, and accumulator failure) using the Taguchi orthogonal array (OA) and signal to noise ratio (SNR). On OA calculations process there are 2 level numbers, namely numbers 1 and 2, where level 1 is inclusion means that the variable is included in the MD calculation process, and level 2 is exclusion means that the variable is not included in the MD calculation process while an empty column means no there is a variable placement in that column. This study will use the L8 array (27) to accommodate the variables studied. The calculation process for this array is counted eight times, with a combination of existing variable levels and the SNR type used is larger-the-better. This type of SNR is calculated by the following equation [8]:

$$\eta_q = -10\log\left(\frac{1}{p}\sum_{j=1}^{p}\frac{1}{MD_j}\right) \tag{10}$$

Where :

$\eta_q$ = Signal-to-noise ratio for the $q$th run of the orthogonal array (OA)

Furthermore, the average is calculated for each variable based on the variable level. As well as the difference calculated, using the following equation [8]:

$$\text{Gain} = (\text{Average. SNR})_{\text{Level}-1} - (\text{Average. SNR})_{\text{Level}-2} \tag{11}$$

Semakin tinggi nilai perbedaan, semakin besar pengaruh variabel-variabel ini pada objek. Begitu pula sebaliknya, semakin kecil perbedaan nilainya, semakin rendah dampak variabel-variabel ini pada objek. Sekalipun nilai selisihnya <0, dapat dikatakan bahwa variabel tersebut tidak mempengaruhi tujuan sehingga dapat dihilangkan untuk pemantauan kondisi lebih lanjut.

*2.4. Failure diagnosis with variables selected based on criteria for classification determined by the threshold*
This stage begins with the creation of the MD-Based Diagnostic Scheme. Monitoring data is then performed, where the data will be used to monitor the movement of MD values using MS that has been formed. If MD is from data monitor > T, where T is the threshold of the most expensive distance from normal conditions, then it can be concluded that the status of the tool is displaying abnormal behaviour, and appropriate corrective action is needed. And if MD <T, the condition of the equipment is normal.

## 3. Result and Discussion

### *3.1. The results of MD Calculation from normal condition data*

**Table 3.** MD transformation and calculation for normal data

| Observation | Sensor Data | | | | Transformation Data | | | | MD normal condition |
|---|---|---|---|---|---|---|---|---|---|
| | X1 | X2 | X3 | X4 | Z1 | Z2 | Z3 | Z4 | |
| **1** | 0.562 | 35.582 | 147.450 | 147.340 | 0.311 | 0.488 | -0.092 | -0.428 | 0.292 |
| **2** | 0.557 | 35.500 | 146.960 | 147.170 | 0.107 | -0.028 | -1.355 | -0.944 | 0.640 |
| **3** | 0.555 | 35.422 | 147.880 | 147.990 | 0.025 | -0.519 | 1.017 | 1.546 | 0.968 |
| **⋮** | .... | .... | .... | .... | .... | .... | .... | .... | .... |
| **120** | 0.591 | 35.574 | 146.990 | 147.170 | 1.499 | 0.438 | -1.278 | -0.944 | 1.012 |
| $\bar{X}$ | 0.554 | 35.504 | 147.486 | 147.481 | 0.000 | 0.000 | 0.000 | 0.000 | 0.992 |
| **SD** | 0.024 | 0.159 | 0.388 | 0.329 | 1.000 | 1.000 | 1.000 | 1.000 | 1.043 |

The last step is the MD calculation for each observation from the normal data using the equation (9). The following is an example of MD calculation for observation 1:

$$\bar{X}_1 = \frac{66.528}{120} = 0.554 \qquad \text{and } \sigma X_1 = \sqrt{\frac{0.071}{120-1}} = 0.024$$

Based on the results of verification of the success of the effects of normal data transformation using equations (4) and (5). can be seen that the results of the transformation has addressed the correct results, it is based on the average for all variables has a value of 0 and the standard deviation has also been valued at 1.

### *3.2. The results of the ms validation process*

After obtaining MD on normal condition data, then MD is calculated for abnormal condition data by using mean, standard deviation and correlation matrix from normal MD calculations.

**Table 4.** Validation of normal MD data vs. abnormal cooler data, relief valve, pump and accumulator

| Observation | MD Of Normal Condition | Value MD of a failure condition | | | |
|---|---|---|---|---|---|
| | | Cooler | Valve | Pump | Accumulator |
| **1** | 0.292 | 8578.906 | 2324.393 | 72.208 | 16.851 |
| **2** | 0.640 | 8558.780 | 2173.396 | 77.128 | 13.353 |
| **3** | 0.968 | 8505.433 | 2380.389 | 70.673 | 15.358 |
| **⋮** | .... | .... | .... | .... | .... |
| **120** | 1.012 | 8234.223 | 2368.699 | 66.002 | 13.622 |
| $\bar{X}$ | 0.992 | 8314.401 | 2292.756 | 67.543 | 17.106 |
| **SD** | 1.043 | 87.127 | 86.955 | 12.104 | 4.202 |

Based on the results of the normal and abnormal MD calculations in table 2, the result is that the distance between normal and abnormal MD conditions has been> 1 so that the MS is validated and can proceed to the next stage.

**Table 5.** The threshold for each condition

| Condition | Mean | Std. Deviation | Threshold (T) | |
|---|---|---|---|---|
| | | | Minimum | Maximum |
| **Normal** | 0.992 | 1.043 | -2.137 | 4.121 |
| **Failure of Accumulator** | 17.106 | 4.202 | 4.499 | 29.713 |
| **Failure of Pump** | 67.543 | 12.104 | 31.231 | 103.855 |
| **Failure of Relief Valve** | 2292.756 | 86.955 | 2031.890 | 2553.622 |
| **Failure of hydraulic cooler** | 8314.401 | 87.127 | 8053.019 | 8575.782 |

IOP Conf. Series: Materials Science and Engineering **847** (2020) 012036 doi:10.1088/1757-899X/847/1/012036

### *3.3. The results of the Variable Optimization process*

**Table 6.** Selected variables based on OA and SNR

| **Condition** | Variable (Sensor) | | | |
|---|---|---|---|---|
| | X1 | X2 | X3 | X4 |
| **Failure of Accumulator** | √ | x | √ | √ |
| **Failure of Pump** | √ | x | √ | √ |
| **Failure of Relief Valve** | x | √ | √ | √ |
| **Failure of hydraulic cooler** | x | √ | √ | √ |

Where (√) is the selected variable and (x) is a variable that is not used in monitoring the condition of the equipment. Selected variables will be used in monitoring and diagnosing the condition of the excavator using the scheme created in section 3.4.

### *3.4. Failure Diagnosis Scheme*

This study presents an MD-based failure monitoring scheme for excavator components. Failure detection is the first step in the diagnosis, which indicates the occurrence of errors in the system being monitored or monitored. In this study, MD-based data clustering techniques are used to classify excavator failure data to different failure groups.

In this study, it does not enter into the scope of determining optimal policies for the maintenance of leasing equipment systems but focuses on determining narrative maintenance policies in the application of CBM strategies in heavy equipment leasing companies using MTS. The narrative maintenance policy produced in this study is in the form of an MD threshold-based monitoring scheme using multi-sensor data. The proposed CBM Scheme begins by inputting monitoring data and setting T = 60 minutes. The error detection process starts when the average MD monitoring (μMD) exceeds T1 (threshold from normal conditions). However, if μMD is less than T1, the monitoring process is continued with the addition of testing data by setting T = T + 60 minutes. After the detection process is complete then proceed with the excavator failure identification process, the type of failure is determined by filtering the type of failure based on the threshold value of the Taccumulator, Tpompa, TValve, and Tcooler after the identification process is complete then proceed with the process of isolating the failure which will then proceed with the determination of the policy maintenance tools based on the type of failure studied.

## 4. Conclusion

In this research, a diagnostic condition monitoring scheme is presented using the Mahalanobis Taguchi System method. The proposed scheme is based on the Mahalanobis distance threshold. The conclusion of the proposed approach can be summarized as follows:

1) The threshold of normal conditions is (-2.137 to 4.121) while in abnormal accumulator conditions (4,331 to 39,458), pump failure (40,956 to 138,048), valve failure (2708,104 to 3404,187) and the threshold in the condition cooler failure is (10736.160 to 11434.151).
2) The built scheme has succeeded in distinguishing normal and abnormal conditions from the excavator components under study.
3) The future of research can develop a multi-sensor data-based CBM monitoring scheme in heavy equipment leasing companies by focusing on heavy equipment components that have a series circuit system.

IOP Conf. Series: Materials Science and Engineering **847** (2020) 012036 doi:10.1088/1757-899X/847/1/012036

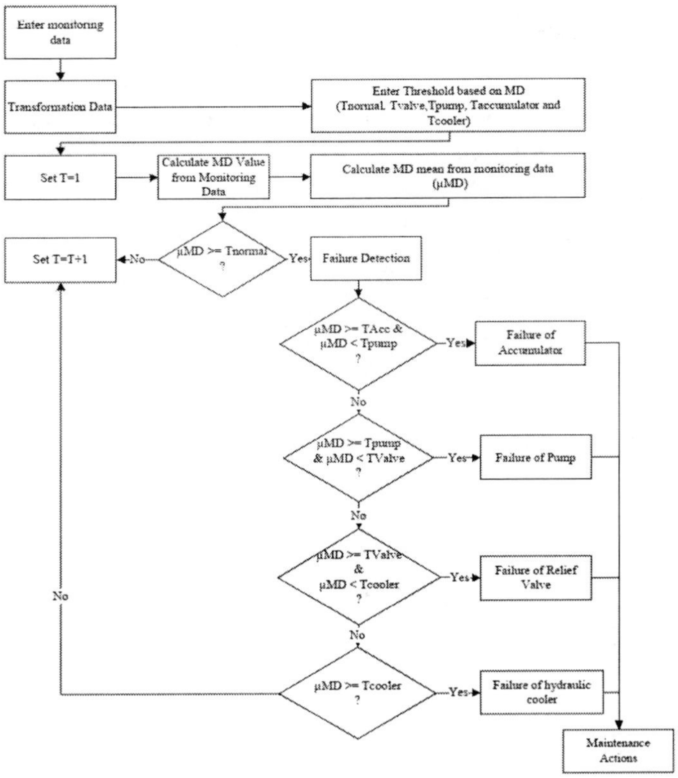


**Figure 1.** Failure Diagnosis scheme

## 5. References

[1]. Husniah, H., Pasaribu, U., Cakravastia, A., and Iskandar, B. 2014. *Performance-based maintenance contract for equipments used in mining industry*. 2014 IEEE International Conference on Management of Innovation and Technology (pp. 543-548). IEEE.

[2]. Deelen, L., Dupleich, M., Othieno, L., Wakelin, O., and Berold, R. 2003. *Leasing for Small and Micro Enterprises*. A guide for designing and managing leasing schemes in developing countries.

[3]. Kamil, I., & Ulka, M. 2006. Penjadwalan Pemeliharaan Komponen Kritis Pada Sistem Hidrolik Dan Engine Unit Excavator Di Departemen Tambang PT. Semen Padang. *Optimasi Sistem Industri*, **6**(1), 31-39.

[4]. Felix, Ng., Harding, J. A., and Glass, J. 2017. Improving hydraulic excavator performance hrough in line hydraulic oil contamination monitoring. *Mechanical Systems and Signal Processing*, **83**, 176-193.

[5]. Jardine, A. K., Lin, D., and D. B. 2006. A review on machinery diagnostics and prognostics implementing condition-based maintenance. *Mechanical systems and signal processing*, 1483-1510

[6]. He, X., and He, Q. 2007. *Application of PCA method and FCM clustering to the fault diagnosis of excavator's hydraulic system*. 2007 IEEE International Conference on Automation and Logistics (pp. 1635-1639). IEEE.

[7]. Yang, T., and Cheng, Y.-T. 2010. The use of Mahalanobis–Taguchi System to improve flip-chip bumping height inspection efficiency. *Microelectronics Reliability*, **50**(3), 407-414

[8]. Soylemezoglu, A., Jagannathan, S., and Saygin, C. 2011. Mahalanobis-Taguchi system as a multi-sensor based decision making prognostics tool for centrifugal pump failures. *IEEE Transactions on Reliability*, **60**(4), 864-878.

[9]. Helwig, N., Pignanelli, E., and Schütze, A. 2015. *Condition Monitoring of a Complex Hydraulic System Using Multivariate Statistics*. IEEE Instrumentation and Measurement Technology Conference, (pp. 210-215).

[10]. Zhang, Y., Zhang, X., Zeng, J., Wang, J., and Xue, S. 2019. Lessees' satisfaction and optimal condition-based maintenance policy for leased system. *Reliability Engineering and System Safety*, **191**, 106532.

# Improved YM Laser Machine Performance with Overall Equipment Effectiveness and Fault Tree Analysis Methods Implementation at PT. XYZ

**Arief Suwandi**[1] **and Zulhilmi Naufal Aulia**[1]

[1]Esa_Unggul_University, west_Jakarta, Indonesia
arief.suwandi@esaunggul.ac.id, zulhilmi949494@gmail.com

**Abstract.** In the development of the shoe-making industry, demanding companies continue to improve the quality and quality of production. PT. XYZ responded by improving the performance of the YM Laser machine which is the main engine in the sports shoe production line. The approach is carried out using the entire Equipment Effectiveness Method and Error Tree Analysis. The purpose of this study is to measure the ratio of value availability, performance ratio, quality ratio, total value of OEE on YM Laser machines, identify the root causes of problems and make improvement plans to improve machine performance. YM Laser Machines have a Overall Average Equipment Effectiveness Value of 75.89%. After researching with the OEE method, it was found that the root of the problem was to reduce speed. Furthermore, the investigation using a cause and effect diagram, obtained the main cause is a machine factor. All problems found were solved using Fault Tree Analysis on engine factors and their improvement efforts to make standard lens replacement so that the lens is always in optimal condition, changing the quality of the switch so that it is not easily damaged again, and regularly updating software and replacing standard bolts.
**Keywords**: Overall Equipment Effectiveness, Performance Ratio, YM Laser Machines, Fault Tree Analysis

## 1. Introduction

In the development of industrial world in the era of globalization, Indonesia's increasing economy has built a climate of increasingly fierce competition between companies. Every company is required to be able to survive and have the ability to compete [1].

Along with the times, the industrial world is also growing rapidly. Productivity is a benchmark to improve company performance. To maintain and improved productivity, the maintenance of production facilities needs to be considered. The intended facility is machine's component that must be maintained so that conditions are always optimal. Machine is one of important component in production process, if the machine stops due to damage, it will produce defect product and energy waste will occurs because the machine operates but doesn't produce product.

The purpose of this research is to measure the performance of the production line, including:

- Calculating the value of the availability ratio, performance ratio, quality ratio and OEE (Overall Equipment Effectiveness) YM Laser machines in the production line at PT. XYZ.
- Look for factors that cause a decrease in overall equipment effectiveness performance.

- Identify the root cause of the problem of lower OEE values compared to its ideal value and cause that dominate the problem and provide improvement suggestion to increase the machine's performance.

*1.1 Overall Equipment Effectiveness*

OEE value is obtained from the multiplication of the three OEE factors, namely availability, performance and quality.

Availability Ratio (AR) aims to measure overall time with unoperated system due to equipment damage, production preparation and adjustment of the machine or component.

$$\text{Availability} = \frac{\textit{Operating Time}}{\textit{Loading Time}}\ x\ 100\% \qquad (1)$$

Performance Ratio is measured with actual operation time ratio from the machine or component with ideal cycle time based on its capacity. The ratio of comparation between the current production level with the planned production level or the ratio of machine capability that obtained from multiplication of the production result with the theoretical cycle time that divided by operation time or multiplication result from operating speed rate and net operating time. The calculation for PR is as follows:

$$\text{Performance} = \frac{\textit{Production total x Ideal Cycle Time}}{\textit{Operation Time}}\ x\ 100\% \qquad (2)$$

Quality Ratio (QR) is a quality loss in the form of number of products that have experienced damage that is related to the equipment. Which is then converted to time by how much time is spent to produce the product that is damaged. The calculation for QR is as follows :

$$\text{Quality} = \frac{\textit{Production total} - \textit{Total Defect}}{\textit{Production Total}}\ x\ 100\% \qquad (3)$$

$$\text{OEE (\%)} = (\text{Availability x Performance x Quality}) \qquad (4)$$

*1.2 Six Big Losses*

In the world of machine's maintenance, known as the six big losses, this is something that every company should avoid. Six big losses are six losses that must be avoided by every company that can reduce the level of effectiveness of a machine. Six big losses are usually categorized into 3 main based on aspects of loss, namely downtime, speed losses and defects. What it meant by downtime is time wasted, where the production process doesn't run as usual due to machine's damage [2]

Downtime consists of two types of losses, namely breakdown and setup and adjustment. Speed losses is a condition where the speed of the production process is interrupted, so the production doesn't reach the expected level. Speed losses consists of two types of losses, namely idling and minor stoppages and reduced speed. Defect is a condition where the product that produced doesn't meet the requested spesifications (nonconformance to standards) [3]. Six big losses in question, among others is :

1. Breakdown

   Machine's or equipment damaged will cause time wasted resulting in losses for the company due to lack of production volume or material loss due to defective product.

$$\text{Breakdown losses} = \frac{\textit{Total Trouble}}{\textit{Loading Time}}\ x\ 100\% \qquad (5)$$

2. Setup and adjustment

   The time that spent on preparing equipment that can't be spent on producing and this is considered a loss. Unnecessary adjustments are also included below.

$$\text{Setup losses} = \frac{\textit{Total Setup}}{\textit{Loading Time}}\ x\ 100\% \qquad (6)$$

3. Idling and Minor Stoppages

   Small stops usually refer to the length of the stop and are a means of classifying stops that are often not noticed (until measured) but the cumulative effect has a very large effect and directly reduces operating time (and increases worker frustration).

$$\text{Idling and minor stoppage} = \frac{(\textit{Loading timg} - \textit{Net operating time})}{\textit{Loading time}}\ x\ 100\% \qquad (7)$$

4. Reduce speed

There is often a heated debate about how fast the process can run and what are the standards to calculate the performance.

$$\textit{Reduce Speed Losses} = \begin{pmatrix} average\ actual\ cycle\ time\ - \\ average\ ideal\ cycle\ time \end{pmatrix} x \left(\frac{total\ production}{loading\ time}\right) x100\% \quad (8)$$

5. Quality Defect and Rework

The most part of reduction in organizational quality defect is a great opportunity to increase productivity and reduce costs. The true cost of poor quality is usually large although most defects can be prevented by using focuse techniques and operators [4]. The aim of each repair team must therefore include zero defects.

$$\text{Defect Losses} = \frac{Total\ defect\ x\ rata\text{-}rata\ actual\ cycle\ time}{Loading\ time} \text{x100\%} \quad (9)$$

6. Reduced Yield or Other Losses

Reduced Yield accounts for defective parts produced from startup until stable (steady-state) production is reached. Reduced Yield can occur after any equipment startup, however, it is most commonly tracked after changeovers. Reduced Yield is a Quality Loss.

$$\text{Yield atau scrap loss} = \frac{ideal\ cycle\ time\ c\ scrap}{Loading\ time}\ x\ 100\%$$

## 2. Methods

The research method that used is overall equipment effectiveness and fault tree analysis.

### *2.1 Overall Equipment Effectiveness*

OEE is a main calculation to measure the success of the implementation of Total Productive Maintenance Program. Valuations relating to OEE follow the global standard, namely 90% for the availability rate value, 95% for the performance rate value and 99% for the quality rate value. The total for OEE value must be 85% form a machine or component. This OEE measurement is very important to identify the machine or equipment that need to be improved the level of productivity or efficiency [5].

Availability Ratio (AR) aims to measure the overall time if there is damage to equipment, production preparation and adjustment of machines or equipments. So, the availability ratio requires value from operating time, loading time and downtime.

Performance Ratio (PR) is measured by the ratio of actual the actual operating speed of a machine or component to an ideal speed based on it's capacity. The ratio is obtained form comparison between current production level with planned production level or machine's capability ratio that obtained from multiplication of produced product with theorytical cycle time that divided by operating time or multiplication from operating speed rate and net operating time.

Quality Ratio is the quality loss in the form of the number that have experienced damage that is related to the equipment, which is then converted to time by how much time is spent to produce the product that is damaged.

### *2.2 Fault Tree Analysis*

Fault tree analysis is a structured diagram analysis that identifies the elements that can cause system failure. This technique is based on deductive logic and can be adjusted to risk identification to analyze the impact of risks arises. Effective application of this technique requires a detailed explanation of the area under discussion. Unwanted results are first identified and then all possible conditions / failures that lead to the event can be identified too. [6].

## 3. Result and Discussion

The research was conducted in the production section PT. XYZ, located in Pakualam, Tangerang. The data used in this research are loading time, downtime, total production, total defects and cycle time. The research was carried out in the cutting line.

### *3.1 OEE Calculation*

All calculations to produce OEE values can be seen in table 1. The formula that used to calculate OEE is : OEE = availability ratio x performance ratio x quality ratio.

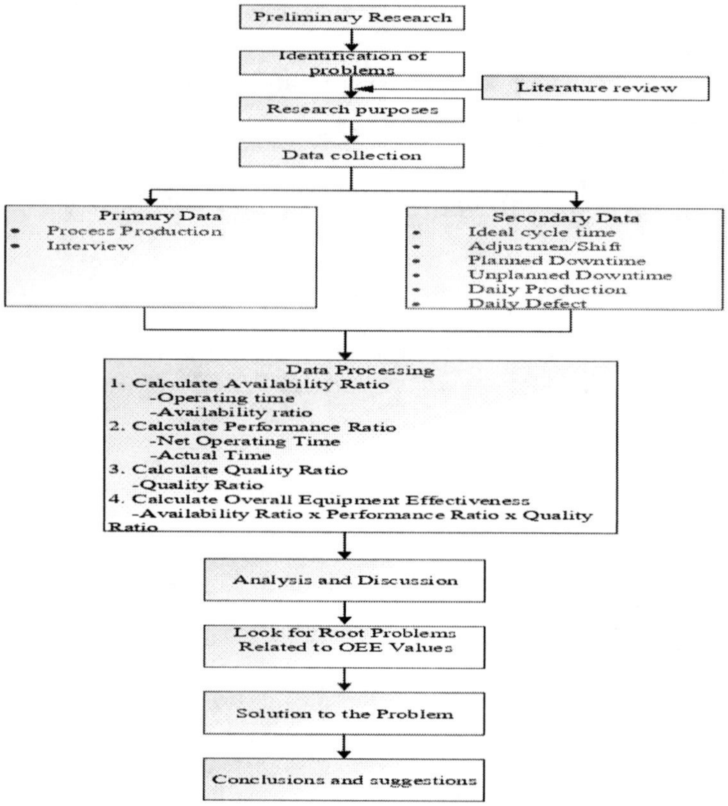


Figure 1. Research flow chart

### *3.2 Six Big Losses Calculation*

The calculation of the value of six big losses is useful to identify how much losses caused by machine's damage. Losses in adjusting time, defect as well as losses resulting from reduced speed and idle time & stoppage minor. The loss calculation is done because the YM Laser machine has an OEE value in March of 75,64% and in April of 76,13% meaning it must be traced and identified because the OEE value was low. There are six big losses on the YM Laser Machine. All Six Big losses calculations can be seen in table 22.

Breakdown or equipment failure is a sudden or unwanted damage to the machine or equipment that causes loss. Breakdown or equipment failure is calculated by dividing the total trouble time with loading time with loading time.

*March calculation :*

$$Breakdown\ Losses = \frac{Total\ Trouble}{Loading\ Time}\ x\ 100\% = \frac{269}{19200}\ x\ 100\%$$
$$= 1.40\ \%$$

Setup losses are losses due to installation, adjustments and also the time that required to substitute types of products to other types of products as well as time that used to change the material and do some additional adjustment. To calculate setup losses, the following are setup losses calculations in March.

$$Setup\ Losses = \frac{Total\ Setup}{Loading\ Time}\ x\ 100\% = \frac{400}{19200}\ x\ 100\% = 2.08\ \%$$

Table 1. Overall Equipment Effectiveness Calculation

| Overall Equipment Effectiveness Calculation in March 2019 | | | | |
|---|---|---|---|---|
| Date | Availability Ratio(%) | Performance Ratio(%) | Quality Ratio(%) | OEE(%) |
| March 1, 2019 | 96.15% | 67.96% | 99.02% | 64.71% |
| March 4, 2019 | 96.04% | 74.65% | 98.90% | 70.90% |
| March 5, 2019 | 97.92% | 86.48% | 99.61% | 84.35% |
| March 6, 2019 | 97.92% | 81.57% | 99.55% | 79.51% |
| March 8, 2019 | 97.92% | 86.72% | 99.61% | 84.59% |
| March 11, 2019 | 97.92% | 93.49% | 99.45% | 91.03% |
| March 12, 2019 | 95.94% | 82.80% | 98.96% | 78.61% |
| March 13, 2019 | 94.48% | 66.08% | 98.46% | 61.47% |
| March 14, 2019 | 96.46% | 72.44% | 98.87% | 69.09% |
| March 15, 2019 | 97.29% | 82.93% | 99.22% | 80.05% |
| March 18, 2019 | 95.83% | 72.09% | 98.94% | 68.35% |
| March 19, 2019 | 96.98% | 77.57% | 99.12% | 74.57% |
| March 20, 2019 | 93.23% | 59.78% | 98.98% | 55.17% |
| March 21, 2019 | 96.98% | 80.96% | 99.39% | 78.03% |
| March 22, 2019 | 97.92% | 91.05% | 99.60% | 88.79% |
| March 25, 2019 | 97.08% | 90.33% | 99.04% | 86.86% |
| March 26, 2019 | 95.00% | 73.33% | 98.54% | 68.64% |
| March 27, 2019 | 93.75% | 65.81% | 98.02% | 60.48% |
| March 28, 2019 | 97.60% | 84.87% | 99.06% | 82.06% |
| March 29, 2019 | 97.92% | 87.82% | 99.41% | 85.48% |
| Total | 1930.31% | 1578.72% | 1981.75% | |
| Average | | | | 75.64% |

The average of the OEE in March was 75,64%.

Table 2. Six Big Losses Calculation

| Month | Breakdown losses | setup losses | other losses | deffect losses | reduce speed | idle & stoppage minor | Total |
|---|---|---|---|---|---|---|---|
| March | 1.40% | 2.08% | 0 | 0.86% | 24.82% | 3.48% | 32.65% |
| April | 1.71% | 2.08% | 0 | 0.88% | 25.06% | 3.79% | 33.53% |
| Total | 3.11% | 4.17% | 0 | 1.74% | 49.89% | 7.27% | 66.18% |
| Average | 1.55% | 2.08% | 0 | 0.87% | 24.94% | 3.64% | 33.09% |

Defect losses is a matter that must be minimized and even must be eliminated from the production process because in addition to providing losses to the material is also detrimental in terms of time. Defect losses can be calculated using the formula.

Defect losses calculation in March,

$$Defect\ Losses = \frac{Total\ defect\ x\ actual\ cycle\ time\ average}{Loading\ time} \text{x}100\% = \frac{438\ x\ 0.37}{19200} \text{x}100\% = 0.86\ \%$$

IOP Conf. Series: Materials Science and Engineering **847** (2020) 012037 doi:10.1088/1757-899X/847/1/012037

Reduce speed losses is an activity caused by a machine that is less than optimal, a decrease in machine's speed occurs the actual cycle time is longer than the ideal cycle time that has been set as a standard. Reduce speed losses can be calculated using the formula.

$$\textit{Reduce Speed Losses} = (actual\ cycle\ time\ average - ideal\ cycle\ time\ average)\ x\ (total\ production) / (loading\ time)\ x\ 100\%$$

$$= \frac{(0.37-0.28) x\ 50456}{19200}\ x\ 100\% = 24.82\ \%$$

Idle and minor stoppage are losses caused by stopping the machine without damage, machine's hour, unfocused operators and machine's idle time. Idling and minor stoppage can be measured using formulas.

Calculation in March,

$$\textit{Idling and minor stoppage} = \frac{(Loading\ timg - Net\ operating\ time)}{Loading\ time}\ x\ 100\%$$

$$= \frac{(19200-17862}{19200}\ x\ 100\% = 3.48\ \%$$

*3.3 Analysis*

Analysis of the results of data processing is done on the calculation of the value of availability performance, quality and overal equipment effectiveness. Based on reference source used in every calculation in OEE has different ideal values. This ideal value is used to find out whether the value is sufficient or not. If not, there must be a repair while sufficient value must only be maintained or increased.

*3.3.1 Cause-Effect Analysis.* Based on the data calculation of losses, the biggest loss in table 2 is on reduce speed. The results of interviews with related parties, the losses were caused by the speed of the machine is below the specified standards, lack of maintenance, lack of preventive maintenance implementation and the schedule changes of components caused the machine can't run optimally as specified. To determine the root of the problem is to use a pareto diagram.

Pareto diagram is useful for determining the priority of solving existing problems. Based on figure 2, the highest priority is on reduce speed with a value of 75,4%, it can be concluded that there is a top priority to be made, namely reduce speed. These priorities are the root of the problems that occur in the YM Laser machine which are the analyzed for the root causes.

*3.3.2 Cause-Effect Diagram Analysis.* To make it easier to identify six big losses, relevant tools are created, that is a cause-effect diagram that formulates an improvement plan to address the root causes. Referring to figure 2, a cause-effect diagram analysis in carried out only on the six big losses that are the priority, namely reduce speed. So that the analysis is more focused on the factors that have high priority which is the cause of the low OEE score. The following is the reduce speed analysis.

Based on figure 3, the reduce speed's cause and effect diagram that causes low productivity there are four factors that affect, among others man, method, machine and environtment.

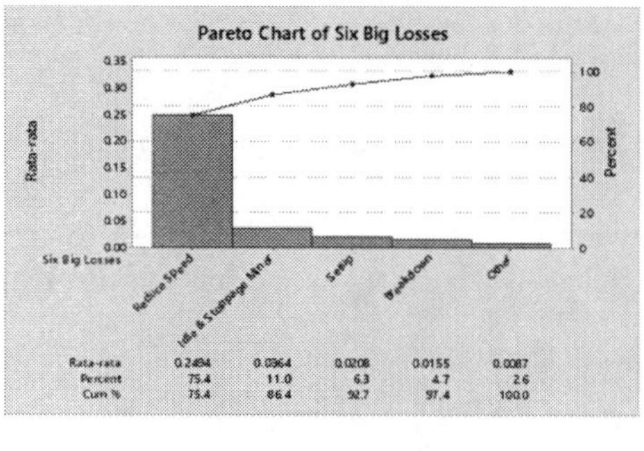


Figure 2. Six Big Losses Average in Pareto Chart

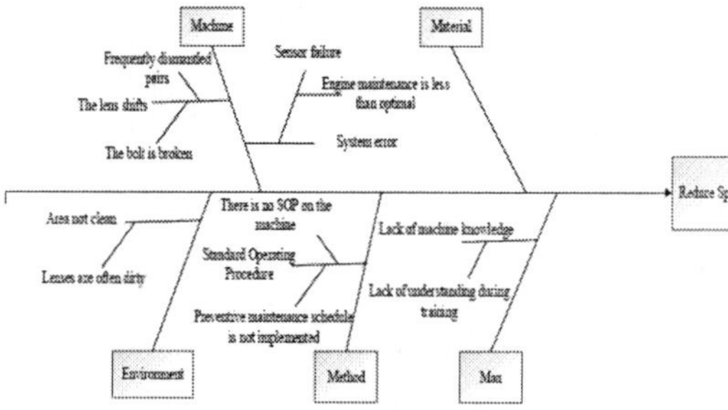


*Figure 3 Reduce Speed's Cause and Effect Diagram*

IOP Conf. Series: Materials Science and Engineering **847** (2020) 012037 doi:10.1088/1757-899X/847/1/012037

1. Machine
   Machine system problems caused by sensors damage due to less optimal maintenance schedules resulting in losses. The machine factor also has several constraints ranging from the lens shift to lack of couplers, slack couplers that are caused by overused. Blower suction power decreases due to compressor leakage which is neglected by the operator.
2. Method
   Standard Operating Procedures (SOPs) that are less socialized due to the absence of SOP sheets on several production lines that result in technicians working not in accordance with company SOPs and the lack of a preventive maintenance schedule.
3. Man
   The ability of employees or operators must be in accordance with the work performed. Lack of operator knowledge of the machine can result in mishandling of small and large things. Sometimes small problems can become big because of the wrong handling.
4. Environment
   Cleanliness of the work area is not get enough attention. This can cause the lens is often dirty thus slowing down machine's performance.

From the analysis of reduce speed cause and effect diagram, there are 3 dominant factors that cause losses, namely blower suction power, lens shift and problematic system.

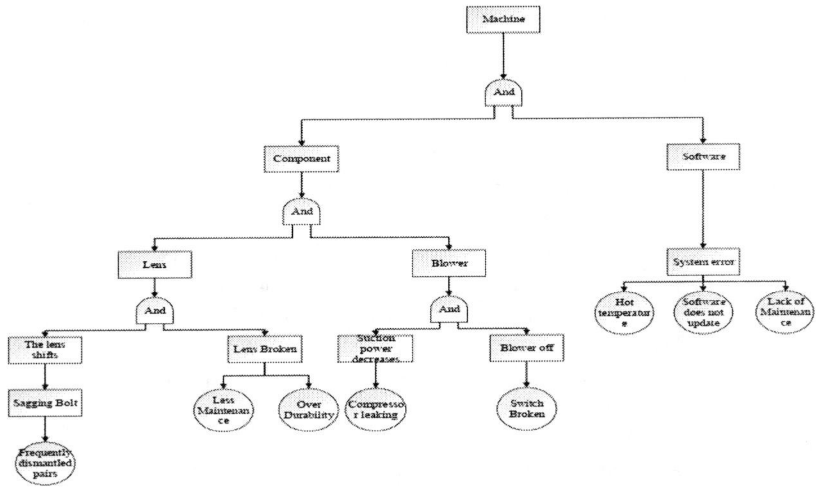


Figure 4. Fault Tree Analysis of Machine Factor

Table 3. Improvement plan to increase OEE value

| Problems | Improvement Plan |
|---|---|
| The lens is often disassembled | Make a standard bolt change so it is not too often disassembled |
| Lack of maintenance on the lens | Improved the supervision so that the preventive maintenance schedule can be wll implemented and makes some work instructions. |
| Excessive lens life | Make a standard lens change so the lens will always in optimal condition |
| Compressor cleaning errors | Provide compressor cleaning procedures so the damage on the compressor will not occur again |
| Sdwitch blower is often damaged | Change the switch blower to a better quality one so it is not too often damaged. Material must be check first before used |
| Software is not updated | Do regular software updates |
| Software is not calibrate | Make a calibration schedule on software so it can be run optimally |

## 4. Conclusions

Based on the results of data processing and analysis of problems can be concluded as follows:

1. 1. The results of the average value of the YM Laser machine availability ratio is said to be ideal with a value of 96.36%, the value of the ideal availability ratio is 90%. The value of the YM Laser machine performance ratio is 79.31% or not ideal, because the ideal value of the performance ratio is 95%. The value of the quality ratio on the YM Laser machine is 99.06%, it is ideal because the ideal quality ratio value is 99%. The result of the overall equipment effectiveness value on YM Laser machine is 75.89%, it has not been said to be ideal because the ideal value of overall equipment effectiveness is 85%, so an increase is needed to increase the OEE value.
2. Performance, availability, and quality factors are sequences of factors that cause OEE values from high to low. The cause of the large losses is the reduced speed caused by suboptimal maintenance.
3. The main cause of the problem of low OEE values on YM Laser machines is the low value of the performance ratio caused by the high reduce speed on YM Laser machines. Then the proposed improvements to the engine factor are as follows :
   - Repair plans for lenses that are often disassembled by changing standard bolts with special bolts so that they are not disassembled frequently.
   - Increase supervision so that preventive maintenance schedules for lenses can be applied and make some work instructions.
   - Make changes to the standard lens so that the lens is always in optimal condition.
   - Provide a compressor cleaning procedure so that compressor damage can be prevented.
   - Change the blower switch to a better one so it doesn't get damaged often and the material must be inspected before use.
   - Perform regular software updates and schedule calibrations in the software so that it can run optimally.

## 5. References

[1] A. Suwandi, T. Y. Zagloel, and A. Hidayatno, 2018 Conceptual model of failure risk control on raw materials inventory system, *IOP Conf. Ser. Mater. Sci. Eng.*, **453**(1), pp. 0–7.

[2] D. Triwardani, A. Rahman, and C. Tantrika, 2012 Analisis Overall Equipment Effectiveness (OEE) dalam Meminimalisi Six Big Losses Pada Mesin Produksi Dual Filters DD07 (studi kasus : PT. Filtrona Indonesia, Surabaya, Jawa Timur, *Jur. Tek. Ind. Univ. Brawijaya*, **7**, pp. 379–391.

[3] A. B. Abdallah, 2013 The Influence of 'Soft' and 'Hard' Total Quality Management (TQM) Practices on Total Productive Maintenance (TPM) in Jordanian Manufacturing Companies, *Int. J. Bus. Manag.*, **8**(21), pp. 1–13.

[4] A. Suwandi, T. Y. Zagloel, and A. Hidayatno, 2017 Enhancing the Quality of Industrial Output in Indonesia – An Analysis of Factors Causing Defective Bolt Stud Cylinder Products, **1**(3), pp. 62–72.

[5] A. R. Utomo, N. Athari, and F. T. D. Atmaji, 2017 Perancangan Maintenance Pada Mesin Rotorvane Dengan Menggunakan Metode Life Cycle Cost (Lcc) Dan Cost of Unreliability (Cour) Di Perkebunan Nusantara VIII Ciater (Lcc) and Cost of Unreliability (Cour) Method in Perkebunan Nusantara Viii Ciater Tot, **4**(2), pp. 3012–3019.

[6] Adinda Febby Mustika, 2014 Analisa Keterlambatan Proyek Menggunakan Fault Tree Analysis ( FTA) (Studi Kasus Pada Proyek Pembangunan Gedung Program Studi Teknik Industri Tahap II Universitas Brawijaya Malang ), pp. 0–9.

# Quality improvement on Common Rail Type-1 Product using Six Sigma Method and Data Mining on Forging Line in PT. ABC

**Rina Fitriana**[1], **Johnson Saragih**[2], **Salma Defina Fauziyah**[3]

Quality Engineering Laboratory Industrial Engineering Department, Faculty of Industrial Technology, Universitas Trisakti, Jl. Kyai Tapa No.1 Grogol Jakarta 11440, Indonesia
rinaf@trisakti.ac.id, johnson_saragih@yahoo.com, salma.defina@gmail.com

**Abstract.** PT. ABC is a manufacturing company engaged in the automotive sector that produces components of cars and motorcycles. This study aims to improve quality by using the six sigma method with the DMAIC stage (Determine, Measure, Analyze, Improve and Control) and analyze it using data mining techniques. At the definition stage 7 defects were identified Critical To Quality namely pockmarked, burry, dent, rust, lapping, marking and NG marking. The measure DPMO value is 28,810 and the sigma value is 3,399. In the analysis phase using data mining with the classification method with the decision tree algorithm, the company's decision in decision-making status of Type 1 Common Rail products agreed on the 10 most interesting parameters. Based on FMEA, it is known as the main cause of the Common Rail Type 1, which is the crust on the dies, marking the product that appears and the trajectory of the heating machine to the forging machine that is too far. The improve phase is using Lifetime Card Dies, the marking is concave inward, reducing the running distance of the heating-forging machine and Controling Check Sheet Product. The DPMO value obtained after improvement is equal to 24095 and the sigma level is 3.476.
**Keywords**: Six Sigma, Data Mining, Classification, Decision Tree.

## 1. Introduction

The rapid development of the manufacturing industry gave rise to intense competition among business actors. This gives encouragement for each company to always improve the products produced in terms of quality. PT. ABC is a company engaged in the automotive sector and was founded in 1986 and located in Cikarang, West Bekasi. One problem that is often faced by the manufacturing industry is that the products produced are not always perfect / defective, defective products are parts that cannot be separated from production. Broadly speaking, this company has 3 main lines, namely forging lines, machining and heat treatment. The study focused on the forging line, because it has a high percentage of defects of 11.24% with a defect tolerance of 6.5%. This research is focused on Type 1 Common Rail parts, because the parts have an average percentage of defects each month of 20%. Defective products in this company are categorized into 2, namely hold product and reject product. Hold product is a defective product that can be repaired so that it becomes a good product, reject products are defective products that cannot be repaired or become scrap. This study aims to improve quality by reducing the number of defects in order to reduce losses such as costs and time incurred from the disability by using the six sigma method through the DMAIC stage and analysis using data mining techniques.

Quality is everything that satisfies customers or according to customer requirements and requirements. To find out whether the quality control performed is good or not, it is done by control chart analysis and to determine the priority of solving the existing problem is done by pareto diagram analysis [1]. Six sigma is a vision of quality improvement towards the target of 3.4 failure per million opportunities (DPMO) for each product transaction (goods / services) and is a new breakthrough in the field of quality management in the form of a control method and quality improvement towards failure rate 0 / zero defect. There are five stages of DMAIC as a characteristic of six sigma, among others, define - measure - analyze - improve - control [2].

Data mining is the process of extracting useful information from data in a large database that has not been known to find meaningful patterns by means of automatic and semi-automatic exploration and analysis. The purpose of data mining techniques is to try to find benefits from a set of data [3].

Classification is the process of finding a model or function that describes and distinguishes classes or concepts of data. The purpose of classification is to find a model from a training set that distinguishes attributes into appropriate categories or classes [4]. Decision tree functions to convert facts into decision trees that represent rules that can be easily understood by natural language [5]. Entropy is the amount of data that is irrelevant to information from a data set. The higher the entropy value means the more symbols appear in the information. Gain is information obtained from changes in entropy in a data set, either through observation or can also be concluded by participating in a data set [6].

Previous research was used as a guideline in conducting research, the following are excerpts of previous research and current research positions can be seen in Table 1.

**Table 1.** Research Quotations and Research Position

| No | Topic | References |
|---|---|---|
| 1 | Use of Decision Tree for Determining the Period of Student Study Program in Informatics Engineering | [4] |
| 2 | Application of C4.5 Algorithm to Predict the Acceptance of Prospective New Employees at PT WISE | [5] |
| 3 | Six Sigma and Data Mining Applications to Improve Quality in Manufacturing Industry | [6] |
| 4 | Process Performance Analysis and Identification of Dominant Disabilities in Bag Making with Statistical Process Control Methods | [7] |
| 5 | Implementation of the DMAIC Approach to Improve PVC Pipe Production Process | [8] |
| 6 | Application of Statistical Process Control (SPC) in Cocoa Beans Processing | [9] |
| 7 | Six Sigma Method Approach (DMAIC) and Audit Process (CPPP) for Quality Improvement at PT. IGP | [10] |
| 8 | Use of FMEA and FTA in Isolator Product Quality Control | [11] |
| 9 | Comparison of Naive Bayes Classification Algorithm, Nearest Neighbour, and Decision Tree in Case Study of Decision Making in Clothing Pattern | [12] |
| 10 | Model design of quality products by using data mining in R Bakery Company | [13] |
| 11 | Integration Between Six Sigma, Data Mining-Association Rules, Statistical Process Control (SPC) Methods in Quality Improvement Efforts in Type 1 Common Rail Part at PT. A B C | Position of this research |

## 2. Methods

The research methodos is using Define-Measure-Analyze-Improve-Control. More detailed stages can be seen in Figure 1.

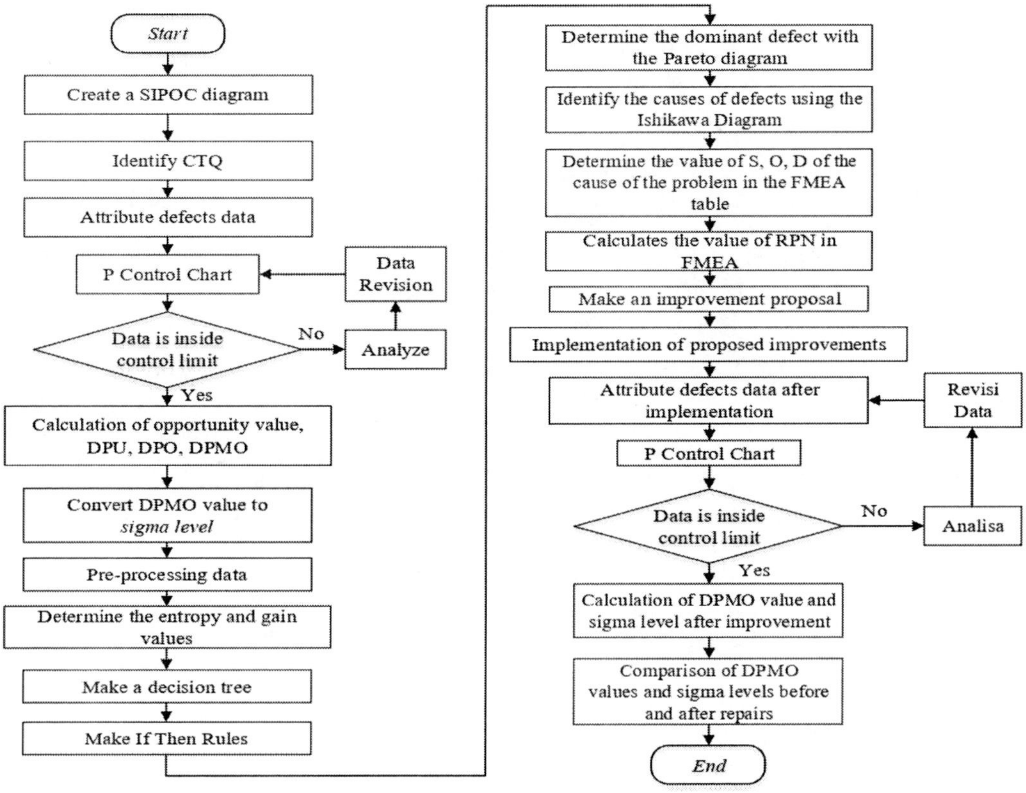


**Figure 1.** Data Processing Flowchart

## 3. Results and Discussion

### *3.1. Define*

Define phase is the process of defining problems, setting goals, and finding opportunities to make improvements. At this stage identification of the process and matters relating to the Type 1 Common Rail part using the SIPOC diagram is carried out. The SIPOC diagram can be seen in the Table 2.

**Table 2.** SIPOC Part Common Rail Type 1

| Supplier | Input | Process | Output | Customer |
|---|---|---|---|---|
| PT. SUMITOMO | Steel Bar | Cutting<br>Heating<br>Forging<br>Trimming | Common Rail Tipe 1 | PT. OTICS Indonesia |
| PT. SUMITOMO | Steel Bar | Cooling Control<br>Grinding<br>Shot Blasting<br>Magnetic Particle Inspection<br>Rust Protection | Common Rail Tipe 1 | PT. OTICS Indonesia |

Critical To Quality (CTQ) is a very important attribute because it is directly related to customer needs and satisfaction with a product and service. CTQ is the requirements desired by the customer. Critical to Quality on Common Rail Type 1 can be seen in Table 3:

**Table 3.** Critical To Quality Part Common Rail Type 1

| No. | Defects | Details | Occur In Process |
|---|---|---|---|
| 1 | Pockmark | Product surface roughness due to powder / dirt attached to dies | Forging |
| 2 | Burry | There are pieces remaining from the trimming process | Trimming |
| 3 | Dent | Product surface damage due to impact | Forging |
| 4 | Rust | There is rust on the surface | Magnetic Particle Inspection |
| 5 | Lapping | Natural folds on the surface of the product | Forging |
| 6 | No Marking | Marking Damaged: There is no / faint print part code | Forging |
| 7 | Marking NG | There is a crust and dent on the marking | Forging |

### *3.2 Measure*

Based on these observations, the percentage of Type 1 Common Rail part defects was 20.17% with the number of defects as many as 605 units from 3000 units. P control chart is used to find out whether the data used is in the control limit or not by using Minitab Software 17.

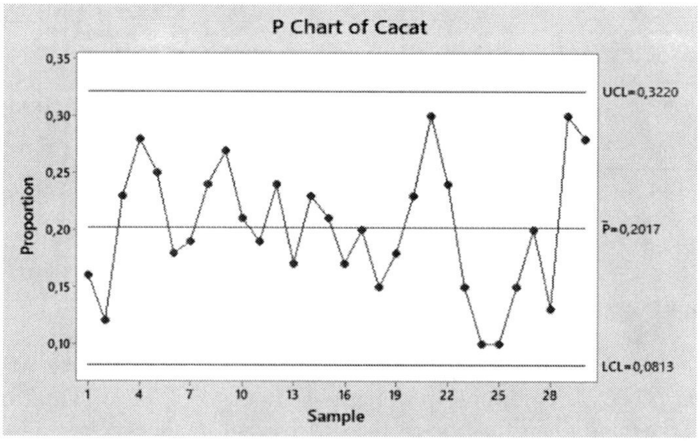


**Figure 2.** P Control Chart

Based on the graph in Figure 2, it can be seen that all data in control or are within the control limit, with an average value of proportions of 0.2017, lower limit values of 0.0813 and an upper limit value of 0.3220. Furthermore, DPMO and sigma level calculations are performed.

$$\text{DPO} = \frac{\text{The number of defects found}}{\text{The number of units examined x the number of CTQ}} = \frac{605}{3000 \text{ x } 7} = 0{,}02881$$

$$\text{DPMO} = \text{DPO X } 1.000.000 = 28810$$

Defect per Million Opportunities (DPMO) in Type 1 Common Rail parts is 28,810 defects in one million opportunities and after conversion, the Common Rail Type 1 sigma part rate is 3.399. Therefore, improvements are needed to increase the sigma level.

*3.3 Analyze*

This stage aims to determine the potential causes of disability that occur so that the proposed improvements are given on target.

*3.4 Classification methods with decision tree algorithms.*

This stage aims to determine whether the sample product is in the hold, NG or QC Passed class. The product is categorized as hold if the product is found defective and the defect can be repaired / rework / repair. The product is classified as NG if the product cannot be repaired to become scrap. The product is categorized as QC Passed if the product is not found defective or has passed the inspection. Based on observational data totaling 3000 product samples, it can be seen that as many as 45 units of products were stated NG, 560 units were declared hold and 2395 units were declared QC Passed. The decision tree results for Type 1 Common Rail product observation data can be seen in Figure 3.

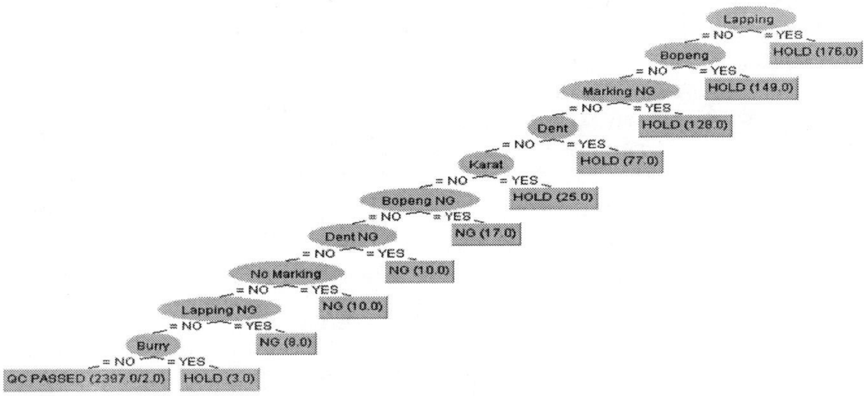


**Figure 3.** Decision Tree

There are ten parameters that are most influential in the production process of Type 1 Common Rail parts. The most influential parameters are lapping, pockmarking, marking NG, dent, rust, NG pock, NG dent, marking, NG and burry lapping. 11 functions are formed which are formed from the decision tree in Figure 3. The first function is that the product is declared to hold if there are lapping defects. The second function of the product is said to hold if there are no lapping defects but found pockmarked defects, and so on. The example of if then rules table can be seen in Table 4.

**Table 4.** *If Then Rules*

| *No.* | *Function* | *No.* | *Function* |
|---|---|---|---|
| 1 | If Lapping "YES" | 5 | IF Lapping "NO" |
| 1 | Then Keputusan "HOLD" | 5 | Bopeng "NO" |
| 2 | IF Lapping "NO" | 5 | Marking NG "NO" |
| 2 | Bopeng "YES" | 5 | DENT "NO" |
| 2 | Then Keputusan "HOLD" | 5 | Karat "YES" |
| 3 | IF Lapping "NO" | 5 | Then Keputusan "HOLD" |
| 3 | Bopeng "NO" | 5 | IF Lapping "NO" |
| 3 | Marking NG "YES" | 6 | Bopeng "NO" |
| 3 | Then Keputusan "HOLD" | 6 | Marking NG "NO" |
| 4 | IF Lapping "NO" | 6 | DENT "NO" |
| 4 | Bopeng "NO" | 6 | Karat "NO" |
| 4 | Marking NG "NO" | 6 | Bopeng NG "YES" |
| 4 | DENT "YES" | 6 | Then Keputusan "NG" |
| 4 | Then Keputusan "HOLD" | | |

IOP Conf. Series: Materials Science and Engineering **847** (2020) 012038 doi:10.1088/1757-899X/847/1/012038

There are 10 parameters that have the most influence on the quality of Type 1 Common Rail products, namely lapping, pockmarking, marking NG, dent, rust, NG pocker, NG dent, NG marking, NG, burry. Determination of the types of dominant defects that appear in the production process of Type 1 Common Rail parts is done by making a pareto diagram. Based on the Pareto diagram 80% of the most dominant types of defects are lapping (30.4%), pockmarked (27.6%), marking NG (21.3%) and dent (14.4%).

In analyzing the causes of defects in products used the FMEA method (Failure Mode and Effect Analysis). FMEA considers the value of Risk Priority Number (RPN) which is the result of multiplication between the values of Severity (Sev), Occurence (Occ), and Detectablity (Det). The summary of FMEA table can be seen in Table 5.

**Table 5**. Summary of FMEA

| Process | Failure Mode | Potential Effects of Failure | Sev | Potential Cause of Failure | Occ | Current Control of Failure | Det | RPN |
|---|---|---|---|---|---|---|---|---|
| Cutting | Lapping | Initial Crack | 7 | At the end of the material is burry | 4 | Product inspection | 6 | 168 |
| Heating | Marking NG | Customer has difficulty to claim products | 6 | Product code appears out | 6 | Product inspection | 5 | **180** |
| Forging | Lapping | Initial Crack | 7 | Material placement with inappropriate or precise dies | 4 | Making stopper | 6 | 168 |
| | | | 7 | Crust on dies | 7 | Surface Treatment Dies | 6 | **294** |
| | | | 5 | The trajectory of the heating machine to the forging machine is too far | 7 | Cooling control | 5 | **175** |
| | Marking NG | Customer has difficulty to claim products | 6 | Product code appears out | 6 | Product inspection | 5 | **180** |

Based on the calculation of the value of the RPN with FMEA, it is known that 3 causes of potential failure have the highest RPN value so that the proposed improvements are focused on the 3 causes of these potential failures. The highest RPN value is 294 with a potential crust on the dies which gives rise to lapping defects and the second highest RPN value that is equal to 180 with the cause of potential product marking failure that arises causing NG marking defects then the next highest RPN value 175 with other potential failure that is, the trajectory of the heating machine to the forging machine is too far to cause pockmarked defects.

### *3.5 Improve*

Improving phase aims to make efforts to improve the existing problems based on data processing that has been done. Based on the analysis using FMEA, the following are suggested improvements:

**Table 6.** Proposed improvements based on FMEA

| Defects | Potential Causes | Proposed Improvement | Implementation |
|---|---|---|---|
| Lapping | There is crust on dies | Lifetime Card Dies | No |
| | | Check Sheet up grade | Yes |
| Marking NG | Product code appears out | Concave product code inward | Yes |
| Pockmarked | The distance between heating and forging is too far, causing crust on the product | Reducing the running distance of heating-forging machines. | No |

*3.6 Concave product code inward*

The product code is intended as the product code. The initial two letters of the code indicate the material code such as the material arrival schedule and production schedule. Whereas, the numbers indicate the dies code, so that it is known the production time of the dies and can know the age of the dies used. Product code makes it easy to track events from materials and dies used, so that if there is a defect or claim from the customer, the company can easily find out the cause of the defect. In fact, the product code that protrudes out has a higher risk of defects because it is susceptible to impact so it can cause defects in the product code or NG marking. Usually NG marking occurs due to a collision between the product and the collision between the product and the pallet. Therefore, if the product code is made concave inward it can minimize the risk of NG marking or defects in the product code.

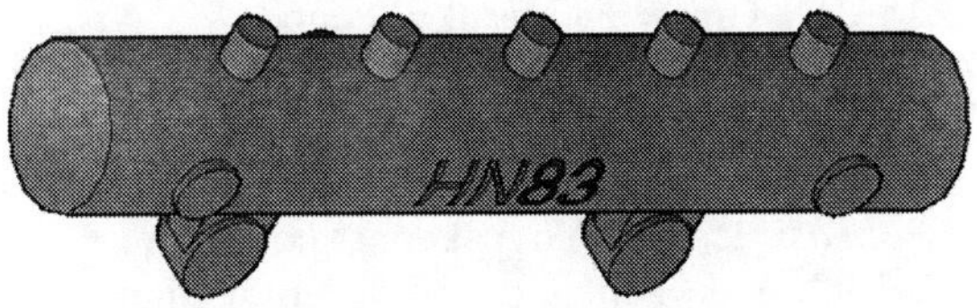


**Figure 4**. Concave Product Code Inward

*3.7 Reducing the running distance of heating-forging machines.*

It can be seen that the layout of heating and forging machines is currently at an altitude of 4.4 meters from the production floor and has a sloping length of 5 m. Machine layout heating and current forging and proposals can be seen in Figure 5.

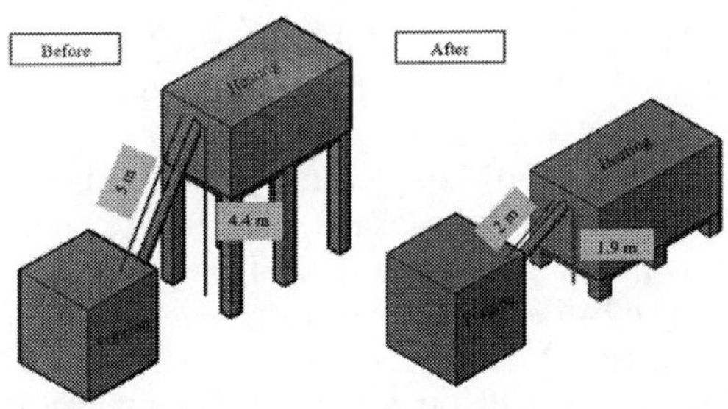


**Figure 5.** Engine Heating and Forging Layout

Proposed repairs with a height of 1.9 meters and a sloping side of 2 meters are designed to cut the long distance. The shorter the distance of the track can reduce the temperature of the material, which means a decrease in electricity consumption, can also reduce the crust arising from a decrease in temperature.

*3.8 Control*

The results of the implementation of the proposed improvements given to determine the effectiveness of the proposed improvements and to find out whether or not the effect of the proposed improvements is significant.

Defect per Million Opportunities (DPMO) in Type 1 Common Rail parts amounted to 24,095 defects in one million opportunities and after conversion, the Common Rail Type 1 sigma part level was 3.476. This shows that the sigma level of the Type 1 Common Rail part production process after implementation of improvement increases by 0.077 sigma.

## 4. Conclusion

Based on data processing and analysis of the results that have been done, it can be concluded from this study that:

1. In the Type 1 Common Rail part production process, the DPMO value before implementation is 28,810 defects in one million opportunities and the sigma value is 3,399.
2. In the analysis phase using data mining with the decision tree algorithm classification method, the company's decision in decision making on the status of Common Rail Type 1 products is based on the 10 most influential parameters, namely lapping, pockmarking, marking NG, dent, rust, pocked NG, NG dent , NG, burry, marking and lapping.
3. Proposed improvements given based on the FMEA table are using markings made concave inward and reducing the heating-forging machine track distance.
4. The DPMO value after the implementation of the proposed improvement is 24095 and the sigma value is 3.476 so that it experiences a sigma increase of 0.077 sigma.

## 5. References

[1] V. Gaspersz, 1997 *Manajemen Kualitas: Penerapan Konsep-Konsep Kualitas Dalam Manajemen Bisnis Total.* (Jakarta: PT Gramedia Pustaka Utama).

[2] V. Gasperz, 2002 *Pedoman Implementasi Program Six Sigma Terintegrasi dengan ISO 9001:2000, MBNQA and HACPP.* (Jakarta: PT Gramedia Pustaka Utama).

[3] B. Santosa, 2007 Aturan Asosiasi. *Data Mining: Teknik Pemanfaatan Data Untuk Keperluan Bisnis Teori and Aplikasi,* 1st Ed. (Yogyakarta: Graha Ilmu).

[4] Rismayanti, 2018 Decision Tree Penentuan Masa Studi Mahasiswa Prodi Teknik Informatika (Studi Kasus: Fakultas Teknik and Komputer Universitas Harapan Meand), *Jurnal Sistem Informasi,* Meand, **2**(01), pp. 16-17.

[5] F.F. Harryanto and S. Hansun, 2017 Penerapan Algoritma C4.5 untuk Memprediksi Penerimaan Calon Pegawai Baru di PT WISE, *Jurnal Jatisi,* Tangerang, **3**(2), pp. 97-98.

[6] R. Fitriana, D. Sugiarto, J. Saragih, A. Bagio, 2014 Aplikasi Six Sigma and Data Mining Untuk Meningkatkan Kualitas Pada Industri Manufaktur, in *Seminar Nasional Teknik Industri BKSTI,* pp. 92-93.

[7] C.V. Gunawan and H. Tannady, 2016 Analisis Kinerja Proses and Identifikasi Cacat Dominan pada Pembuatan Bag dengan Metode Statistical Process Control, *Jurnal Teknik Industri,* Jakarta, **XI**(1), pp. 91-93.

[8] D. Caesaron, S.Y.P Simatupang, 2015 Implementasi Pendekatan DMAIC untuk Perbaikan Proses Produksi Pipa PVC", *Jurnal Metris,* Jakarta, **16**, pp. 9-11.

[9] I.B.Suryaningrat, N. Novijianto, N. Faidah, 2015 Penerapan Metode Statistical Process Control (SPC) Pada Pengolahan Biji Kakao, *Jurnal Agreteknologi,* Jember, **09**(01), pp. 46-49.

[10] S.F. Irawan, 2014 Pendekatan Metode Six Sixma (DMAIC) And Proses Audit (CPPP) Untuk Peningkatan Kualitas Di PT. IGP, on *Jurnal Pasti,* Surabaya, **VIII**(3), pp. 413-420.

[11] D.F. Mayangsari, H. Adianto, Y. Yuniati, 2015 Usulan Pengendalian Kualitas Produk Isolator Dengan Metode Failure Mode and Effect Analysis (FMEA) And Fault Tree Analysis (FTA), on *Jurnal Online Institut Teknologi Nasional,* Bandung, **03**(02), pp. 83-84.

[12] D. Sartika and D.I. Sensue, 2017 Perbandingan Algoritma Klasifikasi Naive Bayes, Nearest Neighbour, and Decision Tree pada Studi Kasus Pengambilan Keputusan Pemilihan Pola Pakaian. *Jurnal Jatisi,* Palembang, **1**(2), pp. 155-156.

[13] R Fitriana, J Saragih and N Luthfiana 2017 *Model business intelligence system design of quality products by using data mining in R Bakery Company* IOP Conf. Series: Materials Science and Engineering **277** 012005

IOP Conf. Series: Materials Science and Engineering **847** (2020) 012039 doi:10.1088/1757-899X/847/1/012039

# Environmental Management System (EMS) within Construction Site: A case study in Kelantan State, Malaysia

**A Omran[1,3] and SHS Yahya[2,4]**

[1]Faculty of Engineering, Bright Star University, El-Breqa, Libya
[2]Faculty of Management, Bright Star University, El-Breqa, Libya
[3]naser_elamroni@yahoo.co.uk

**Abstract.** An Environmental Management System (EMS) was introduced to meet the environmental needs and an organization should have an understanding of what those environmental needs are. In a construction industry, the awareness about the environment performance is improving and the need to protect the environment quality is growing good in order to maintain the sustainability of the development in a country.Spirit of sustainable development embodied and endorsed in the medium- and long-term development strategies such as Third Malaysian Plan (1976-1980) and thereafter, e.g. Seventh Malaysian Plan (1996-2000). Thus, the construction players that involved in the construction projects have to take the responsibility to accept all the regulation related to the environment protection during construction processes. This study concluded that the awareness of EMS among the construction players in Kelantan State is very low. Most of the construction players understand an EMS. However, although they understood what EMS is, yet they did not implement and apply the system within the construction sites. They understand an EMS and its obligation, but some of them, indicated that it is not important to be implemented, as they are doing projects just to get the profit. Although, the construction players understand an EMS, but they do not have any skills and experiences in implementing EMS within construction site.
**Keywords**: Environmental Management System, ISO14000, Construction industry, Malaysia

## 1. Introduction

An Environmental Management System (EMS) was introduced to meet the environmental needs and an organization should have an understanding of what those environmental needs are. Thus, according to [1], an EMS is a useful tool in ensuring that environmental improvements are met. Environment Management System (EMS) is the part of the overall management system that includes organizational structure, planning activities, responsibilities, practices, procedures, processes and resources for developing, implementing, achieving, reviewing and maintaining the environmental policy. The use of this system aims to manage and to produce more sustainable projects that are more efficient in terms of resource and energy use, reducing the impact on the environment, providing better value of living and work conditions in improving economic productivity and competitiveness. EMS is also to produce a corporate environmental plan which will lead the organization to improve environmental performance. The use of EMS approach has challenged the construction players in the construction industry to work carefully and to think the impact on the environment and the quality of life resulting from their construction's activities. It is also providing the perfect platform for the type of systematic approach needed to achieve sustainability. Thus, the EMS has to be implementing by construction players in any development project in order to create a sustainable development. Sustainable construction or 'green construction' describes the responsibility of the construction industry in attaining the sustainability. Sustainable construction is a process whereby, over time, sustainability is achieved. So, the environment management system must be applied by the construction players within construction site in order to strike a balance between conserving the environment and maintaining prosperity in development. The negative effects to the environment caused by construction sites will give very significant impact on the development if the environmental aspects are not properly managed by the management of every construction sites. Although Malaysia has developed expertise and tools such as environmental auditing and environmental impact assessment to help quantify impacts on the cost of environmental regulation, it is hardly keeping up with the pace of development that we are currently experiencing. Several strategies can be employed to avoid this, such as the development and adherence to guidelines of best practices provided by codes of practice and local and national laws. The EMS in Malaysia can be said as growing in a positive way because most of the organization in Malaysia aware and have gained the knowledge about the correct EMS within construction site in order to create the sustainable development. Therefore, in this study aimed at identifying the EMS in a specific area in Malaysia to check the level of awareness of certain companies about the environment impact especially in the construction industry.

### *1.1. Problem Statement*

In a construction industry, the awareness about the environment performance is improving and the need to protect the environment quality is growing good in order to maintain the sustainability of the development in a country. Spirit of sustainable development embodied and endorsed in the medium and long-term development strategies such as Third Malaysia Plan (1976-1980) and thereafter, e.g. Seventh Malaysia Plan (1996-2000). Thus, the construction players that involved in the construction projects have to take the responsibility to accept all the regulation related to the environment protection during construction processes. The image of the modern country can be shown through the variety of the abundant development in the country. In Malaysia, the construction industry is growing up and contributes various developments such as Kuala Lumpur International Airport, Penang Bridge, Kuala Lumpur Twin Tower (KLCC) and others and as the result, it helps to jump up the name of Malaysia. Buildings and structures enabled mankind to meet their social needs for shelter, to meet economic needs for investment and to satisfy corporate objectives. However, the satisfaction of these needs usually comes with a high price for example an irreversible damage to our environment. As the people that involved in the construction industry, they need to alert and aware about the environment condition within construction site. The construction industry plays a major role in improving the quality of the built environment, but it is also impacting on the wider environment in a number of ways.

Everything that consumers, companies, institutions and authorities do has a positive or negative impact upon the environment. To prevent future violations of environmental conservation, regulatory bodies have enacted many laws pertaining to environmental protection including Malaysia but the effectiveness of these regulations is yet to be seen. According to Department of Standard Malaysia (DSM), 2004 only seven (7) organizations out of 310 are the certified organization in EMS ISO 14001 related to construction works, which mean that is only 2.3% out of the figure implement EMS within construction sites. But, in 2009, according to DSM, the number of certified organizations in EMS ISO 14001 are increasing as they are now realized that the important to implement an EMS in creating sustainable development. Some organization whether small or big, private or public organization still did not put the EMS as their top management system in order to put the environment issues as a priority. It seems that some of them put the profit as a priority in their business strategies above other things. Some of them are lack of awareness on the nature and the important of environmental management system practices within construction site and the lack of understanding about the environmental issues. Thus, the construction players in an organization whether private or public sector must be able to find a balance between development with the environment protection and begin to take an action in order to maintain sustainability in doing their construction's activities.

*1.2. EMS Standardization: ISO 14001*

Global environmental awareness grew from its embryonic in the mid-80s, and in this respect the called Brundtland report published by the World Commission on Environment and Development was highly influential and it provided for the environmental activities a global framework [2]. Morrow and Rondinelli were explained that in 1993, it was felt that an international standard was required for environmental management while in 1996, from an idea based on BS 7750, the ISO 14001 standard was born[3]. It was argued to be a step forward given the successes of the quality standard ISO 9000 and intended to replace the numerous and often conflicting sets of criteria found in various countries. ISO 14001 focuses on the processes involved in the creation, management and elimination of pollution, rather than solely on directly reducing pollution [4]. As BS 5750 had been withdrawn with the appearance of the ISO 9000 series, the outcome of emergence of the international standard ISO 14001 meant that the national standard BS7750 together with the national standards in other EU countries were, with common consent also withdrawn. Since the introduction of ISO 14001 many other ISO 14000 standards have also come into operation. The list is shown in Table 1.

**Table 1.** ISO 14000 Standards

| Standard | Description |
|---|---|
| 14000 | Guide to environmental management principle techniques. |
| 14001 | Environmental management systems – Specification with guidance for use. |
| 14010 | Guidelines for environmental auditing – General principles of environmental |
| | auditing. |
| 14011 | Guidelines for environmental auditing – Audit programmes, reviews and |
| 14012 | assessments. |
| | Guidelines for environmental auditing Qualification criteria for environmental |
| 14013/15 | auditors. |
| | Guidelines for environmental auditing – Audit programmes, reviews and |
| 14020/23 | assessments. |
| | Environmental labeling |
| 14024 | Environmental labeling – Practitioner programmes – Guiding principles, practices |
| | and certification Procedures of multiple criteria programmes. |
| 14031/32 | Guidelines on environmental performance evaluation. |
| 14040/43 | Life cycle assessment – General principles and practices. |
| 14050 | Glossary |
| 14060 | Guide for the inclusion of environmental aspects in product standards. |

**Source:** [2]

Melnyk described that the ISO 14001 standard is a process, not performance standard that describes a system to help an organization achieve its own environmental objectives [5]. It is assumed that by helping a firm focus on its manufacturing process, the firm will improve its environmental performance. Thus, the ISO Standard 14001 is a standard that used by organization to achieve their objectives.ISO14001 internationally recognized standard for EMS. This standard already achieves about 120 000 organizations registration.

*1.3. Local Agenda 21*

*1.3.1. About Agenda.* Agenda 21 is a programme run by the United Nations (UN) related to sustainable development and was the planet's first summit to discuss global warming related issues. It is a comprehensive blueprint of action to be taken globally, nationally and locally by organizations of the UN, governments and major groups in every area in which humans directly affect the environment.

Agenda 21 consists of 40 chapters which is all related to the sustainable development. Through the Agenda 21, the community can involve themselves in order to ensure and identify the issues of sustainable development, create an action plan to solve the issues and implement the plan. The basic approach of this program is bottom-up approach which means the community themselves can involve in determining the action plan at their areas. For example, the community that involve directly about the discussion, workshop or any activities related to the local issues. The issues that can be discussed are environment pollution, transportation issues, housing and so on.

*1.3.2. Characteristic of implementing Local Agenda 21(LA21).* To implement LA21, the local authority needs to include the characteristic below:

1. Together to take account about the need of economic, social and environment.
2. To conclude the sustain future aims or target based on consensus.
3. Involving of local community.
4. To embrace a board or forum that consists of important parties to manage the process.
5. To conclude an action plan with certain target.
6. To conclude the sustainable development indicator.
7. To implement the activities and programs to achieve the action plan that has been concluded.

Thus, LA21 is not only implemented in Malaysia. According to a research by the International Council of Local Environmental Initiatives (ICLEI),there are more than 1,800 local authority from 64 countries involved in the activities by LA21 ( until the end of year 1997) that was focused on 11 countries such as Australia, Bolivia, China, Denmark, Finland, Japan, Netherlands, Norway, Republic of Korea, Sweden and United Kingdom [6]. Nowadays, the LA21 is being expanding in the word in order to overcome the global warming problem. All over the world are taking part in LA21 programme together to save the earth. For example, of the programme that recently did was Sustainable Global Warming. Some of the country was taking an action in implementing LA21 by reorganized their objectives of LA21 and prepare more action plan.

*1.4. EMS within Construction Sites*

A study by Tse et. al explained that many natural areas are irreversibly damaged by construction activities which alter their ecological integrity [7]. Thus, there is a growing awareness of the effect of construction activities on the environment. These effects include land use and land deterioration, resource depletion, waste generation, and various forms of pollution [8]. A major environmental impact of construction stems from its consumption of materials, many of which are non-renewable. Site construction produces many atmospheric pollutants. The construction industry produces large volumes of waste and consumes large amounts of energy [9]. In the construction industry, recycling of material waste is a typical example of showing the financial benefits of ISO 14000 EMS. In 1999, about 37,100 tonnes of construction waste are produced each day and while about 80 per cent of it is re-used for reclamation, 7,900 tonnes still go to landfills. This accounted for about 44 per cent of the total waste disposed of at landfills [10]. Besides that, firms who are willing to take a more strategic approach to EMS should be able to integrate pollution prevention throughout the firm's practices and processes and use them to create long-term advantages [11]. Moreover, if firms are not measuring the wastes associated with their production processes and do not have an integrated approach to managing these wastes, then many of the costs are not captured and subsequently placed into overhead [12]. According to the Robert et al, firms which react to new environmental regulations with end-of-pipe (e.g., scrubbers on smokestacks is at the end of the pollution proverbial pipe) solutions to pollution problems are the first to say that environmental regulations have only added to the cost of doing business [12]. Contrary to this often-expressed view, EMS and environmental regulations may do more than just add to the costs of operations. Environmental regulations that place a heavier burden on new entrants confer an advantage on existing firms by increasing the barriers to entry in industries in which pollution abatement is important [13]. This implies that incumbents may be able to use environmental regulations strategically to enhance competitive advantage [12].

IOP Conf. Series: Materials Science and Engineering **847** (2020) 012039 doi:10.1088/1757-899X/847/1/012039

## 2. Methods

A questionnaire survey was distributed to the involved parties in the construction industry which were contractors, developers, and consultants. The questions included in the survey, were focused about the respondent's awareness, knowledge, skill, experiences and their contribution in EMS within construction sites. Also, identifying their awareness about Local Agenda 21 (LA21) and the respondents' involvement in LA21. About 40 questionnaires were distributed among these construction players but only 33 of respondents answered and return back the questionnaires, yielding a response rate of 83%.

## 3. Result and Discussion

### *3.1. Respondents' background*

As presented in Table (2), most of the construction companies participated in this survey came from G07 classification (64%) even though this question were included all the classification grades available by CIDB in the Malaysian construction industry (G01 until G07). However, only companies with 2 classification grades namely G07 and G03 were participated in thus conducted survey. Concerning the company's role, it can be seen from Table (2) that main contractor (70%) are the most important person in playing a role in the company followed by developer (24%). The other company role includes of specialist/sub-contractor and supplier. With regard to the duration of the company/respondent involved in the construction industry. The analysis showed that the highest percentage that the company involved in the construction industry was 64% which shows more than 9 years of involvement. Then, followed by 27% which shows 7 to 9 years of involvement.

**Table 2.** Shows the Respondents' Background

| Items | Percentage |
|---|---|
| **Classification Companybased on CIDB** | |
| G1 | - |
| G2 | - |
| G3 | 36% |
| G4 | - |
| G5 | - |
| G6 | - |
| G7 | 64% |
| **Company role** | |
| Main contractor | 70% |
| Developer | 24% |
| Others | 6% |
| **Duration of the company involved in the construction industry** | |
| 1-4 years | - |
| 4-6 years | 9% |
| 7-9 years | 27% |
| More than 9 years old | 64% |

### *3.2. Awareness to the Environmental Management System (EMS)*

*3.2.1. Understanding of EMS.* The respondents were asked to whether they understand what is Environmental Management System (EMS) or not. However, most of the respondents (36%) are clearly understood it, followed by the thirty three percent who are not sure about it.

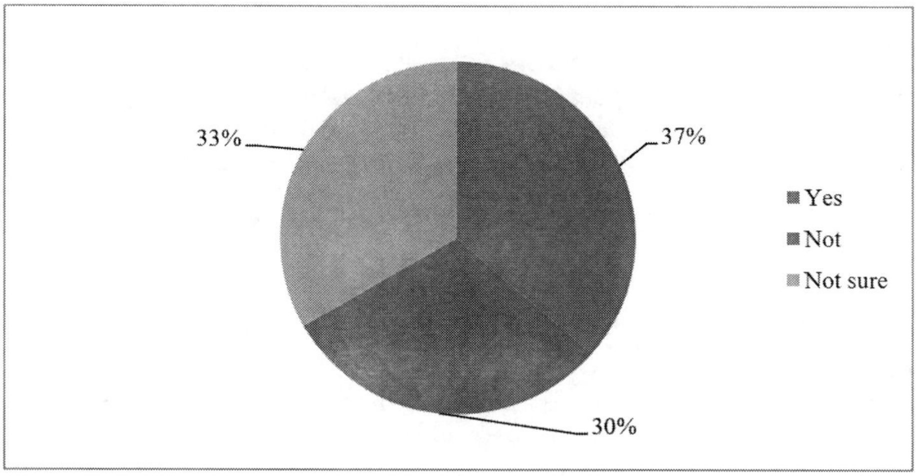


**Figure 1.** Understanding of EMS

*3.2.2. Implementation of environmental policy.* The participants were asked on whether the company has any environmental and sustainability policy or not. An environmental or sustainability policy endorsed by senior management is often a first step taken in EMS implementation. Thus, as presented in Figure (4), most of the companies (82%) did not have an environmental and sustainability policy that endorsed by senior management. However, only 18% of these companies indicated that they have environmental and sustainability policy endorsed by senior management.

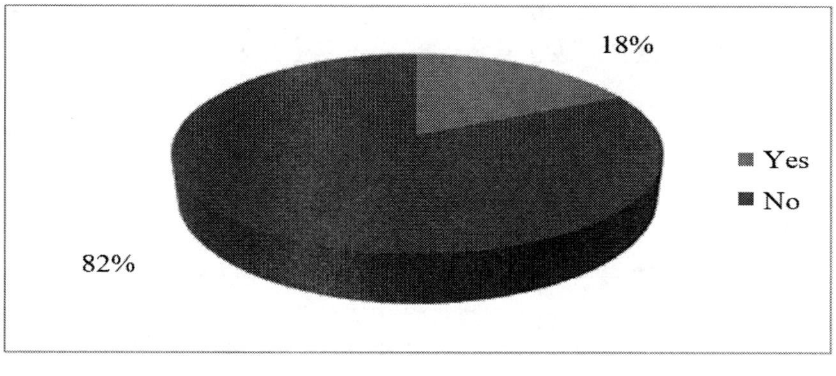


**Figure 2.** Implementation of environmental policy

*3.2.3. EMS-based framework.* Based on the analysis made in Figure (3) concerning the respondents' opinion on EMS framework, It can be seen that most of the respondents (70%) are implementing their EMS based on OHSAS18001 modelfollowed by ISO14001 (21%).

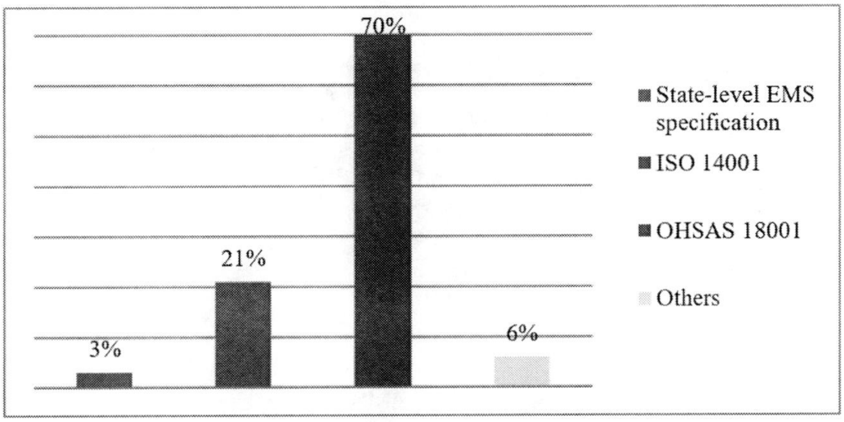


**Figure 3.** EMS-based frameworks

*3.2.4. EMS of the company.* Based on the results obtained with regard to the respondents' view on the situation of EMS. It can be stated that 52% of the participants do not have any current intention of developing an EMS. Although, most of the respondent understand what EMS is, but they did not have any intention to implement and develop an EMS within construction site. Other respondents' views are depicted in Figure (4).

*3.2.5.*

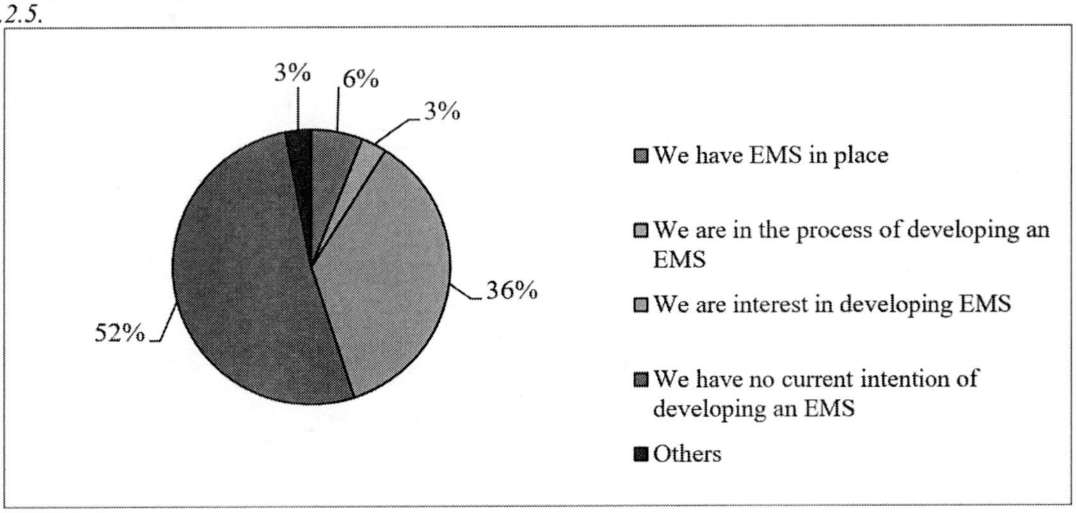


**Figure 4.** EMS of the company

*3.2.6. Implementation of EMS management committee.* Figure (5) shows that 88% of respondents do not have an EMS steering committee or management team, to lead the EMS.

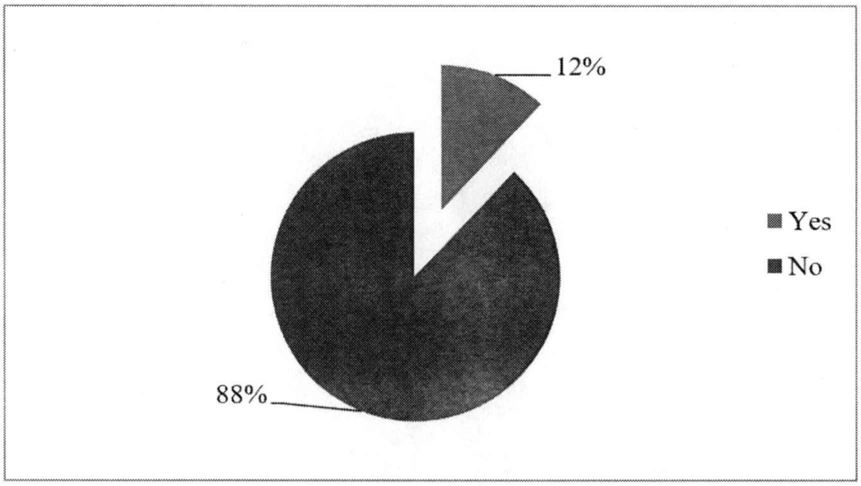


**Figure 5.** Implementation of EMS management committee

*3.3. Local Agenda 21 (LA21)*

*3.3.1. Awareness of LA21.* In this section, respondents were asked on their level of knowledge about LA21.LA21 is a programme towards sustainable development and emphasized on the cooperation Local Authority and participation from the public. Based on the results, it is found that 100% of the respondents never heard about LA21.

*3.3.2. Contribution programme by LA21.* The question attempted to identify whether those who are participated in the survey contributed in any programme by LA21 or not. The analysis has shown that 100% of them never heard about LA21.

*3.3.3. Programme involved with LA21.* The respondents were questioned on their involvement and cooperation with the local authorities programme in managing and planning the environment to create the sustainable development. Ninety seven percent (97%) of those responded to then conducted survey did indicate that they had not involved with local authorities in such programme (See Figure 6).

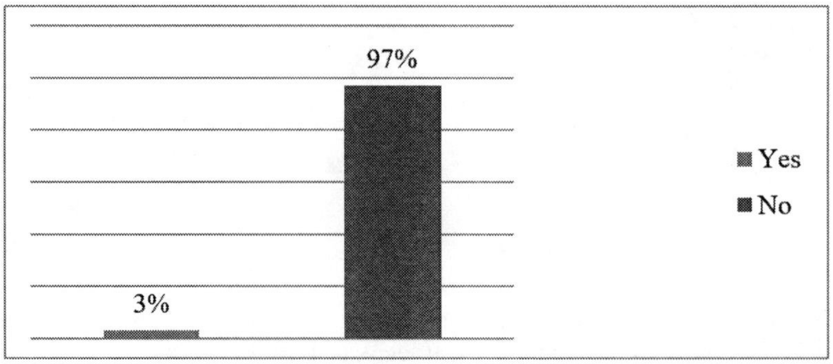


**Figure 6.** Programme involved with LA21

*3.3.4. Awareness of programmes done by LA21.* Referring the participants in this conducted survey on the awareness programmes done by LA121. Surprisingly, it is found that 97% of them are not aware about it (refer to Figure 7).

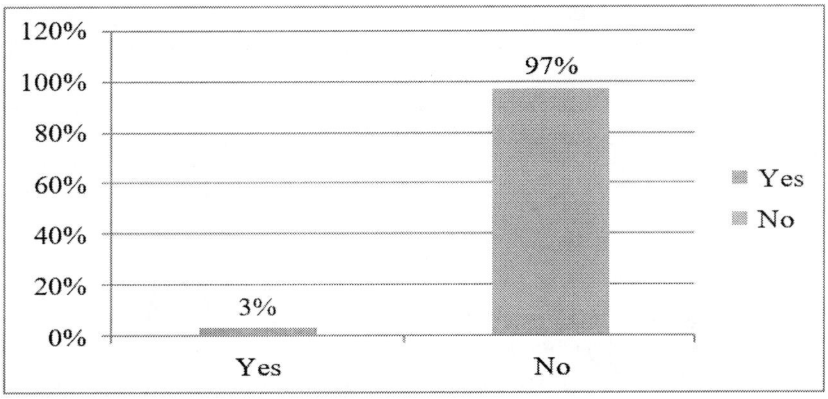


**Figure 7.** Awareness of programme done by LA21

*3.3.5. Implemented activities by LA21.* The respondents were asked to identify whether they are agreedon the action plan and activities implemented by LA21 in their state or not. However, based on the obtained results, it is found 82% of them disagreed with the action plan and activities implemented by LA21. These results is shown in Figure (8).

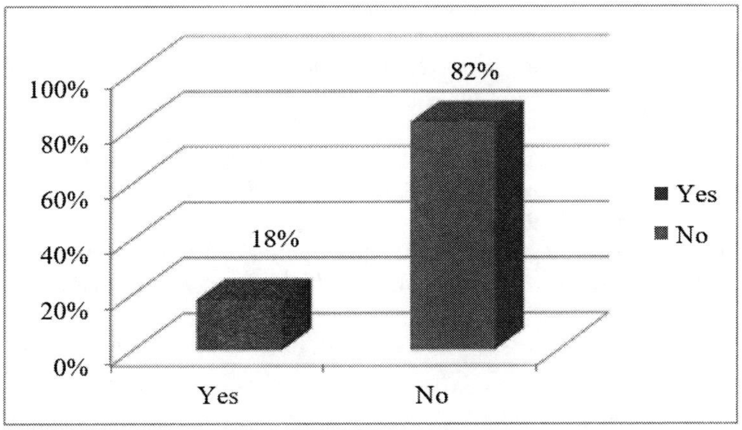


**Figure 8.** Implemented activities by LA21

## 4. Conclusion

The main aim of the study was to check for the existence of environmental management skill and experience among the construction players within construction site. It was also aimed at knowing their awareness and understanding of Environment Management System (EMS). In addition, it was aimed at seeing the responsibilities and acceptance of EMS among the construction players in their construction activities and their understanding to the implementation of EMS in order to create a sustainable development. However, based on the obtained results, it can be concluded that the awareness of EMS among the construction players in Kelantan State is very low. Most of the construction players understand an EMS. However, although they understood what EMS is, yet they did not implement and apply the system within the construction sites. They understand an EMS and its obligation, but some of them, indicated that it is not important to be implemented, as they are doing projects just to get the profit. Although, the construction players understand an EMS, but they do not have any skills and experiences in implementing EMS within construction site. In addition, they do not really know and understand which of the EMS framework, standards and specification they should follow and based on. It was observed that the construction players did not really aware and their level of knowledge of an EMS was very poor. Most of the construction companies do not have an environmental or sustainability policy, they do not have intention to develop an EMS and also do not have EMS steering committee or management team to lead the EMS. Thus, it can be recommended that the construction players should have more environmental training and they should know the EMS so that in the future can create sustainable environment and avoid any negative impacts on the environment.

## 5. References

[1] Tinsley and Pillai (2006)
[2] Wikipedia (2019). *Environmental Management System*, https://en.wikipedia.org/wiki/ISO_14000
[3] Morrow D and Rondinelli D 2002 *Adopting corporate environmental management systems: motivations and results of ISO 14001 and EMAS certification*. European Management Journal, 20(2):159-171.
[4] Melnyk, SA Saroufe, RP and Calantone, R 2013 *Assessing the impact of environmental management system on corporate and environmental performance*, Journal of Operations Management, 21, 329-351.
[5] Melnyk et al. (2003)
[6] DewiAdharina, 2006
[7] Tse, RYC Chiang, YH Hui, ECM Raftery, J and Chan, T 2000 *Review of construction industry in Asia*, Journal of Construction Research, 1(1), 77–86
[8] *People, Settlements, Environment and Development. Nairobi: United Nations Centre for Human Settlements*
[9] UNCHS 1993 *Development of National Capacity for Environmentally Sound Construction. Nairobi: United Nations Centre for Human Settlements.*
[10] (Tse, 1998)
[11] Rondinelli, DA Berry, MA and Vastag, G 1997 *Strategic Programming for Environmental Management: Sonoco's Take-Back Policy*, Business Horizons, 40, (3), 23-32.
[12] (Robert, Steven and Vastag,1998)
[13] Dean, TJ and Brown RL 1995 '*Pollution Regulation as a Barrier to New Firm Entry: Initial Evidence and Implications for Future Research*', Academy of Management Journal, 38(1): 288-303.

# Quality Improvement for Product Body 2-1 at PT. X

**Johnson Saragih[12], Rina Fitriana [13], and Tri Andriyan[14]**

[1] Quality Engineering Laboratory, Department of Industrial Engineering, Faculty of Industrial Engineering, Trisakti Univerity, Indonesia
[2]johnson_saragih@yahoo.com,[3]rinaf@trisakti.ac.id,[4]tri.andryan@ymail.com

**Abstract.** Customer satisfaction is an important indicator for companies to improve the quality of products or services. And if the customer feels satisfied it will automatically increase sales so that it also impacts on the company profits.PT Y is one of the customers who are not satisfied with the product produced PT X, because based on data from January to December 2018, the average monthly defect is 2.34 % whereas the company target is 0.2 %.Therefore in this research it is necessary to minimize the level of defects with the Six Sigma. With five stage, *Define, Measure, Analyse, Improve and Control.* The proposed improvement provided are installing pins on the stopper, making checklist form of dies check, updating the SOP on the bending process, the results *Defect per Million Opportunities (DPMO)* before implementing is 19.268 after implementing is 18.229 or decrease (5.3 %) and Sigma level before implementing is 3.57, after implementing is 3.59 increase (0.5 %)
**Keywords**: Six Sigma, DMAIC, DPMO, Sigma Level

## 1. Introduction

The development of the manufacturing industry is currently very rapidly. Companies must be able to compete to meet the consumer demand for increasingly high. An industry definitely wants the production process can continue to run well so the viability of the company can be assured. Companies are required to competitive and able to survive in the face of competition that goes with running a business strategy.

PT.X is a company engaged in the field of automotive manufacturing for the manufacture of dies and stamping products using raw materials such as sheet / coil steel. Product placement strategies applied by the company is the make to order, in which production is carried out in accordance with customer demand in the form of numbers and specifications. Every day the company can produce hundreds to thousands of products stamping, from the number of products produced, there is the output of a process that does not comply with the company's expectations that the product NG (Not Good). PT.X has eight fixed customer who always contributed stamping products in certain periods with different specifications. This research is devoted to products made for one customer, namely PT.Y. It is always to order stamping products each month with varying amounts. According to historical data the company since January to September 2018, stamping the resulting products to customers PT. Y are contributing to the biggest problem of the seven other customers. Based on historical data the company, from January to September 2018, defect amounting to 2453 unit of products or equivalent with 2.34 %, in the other side the target of company is 0.2 %.

This study focused on a 2-1 body products which the product is produced to customer PT.Y because, it is reach about 24% defects, the biggest defects from the other .The expected goals of their

IOP Conf. Series: Materials Science and Engineering **847** (2020) 012040 doi:10.1088/1757-899X/847/1/012040

improvement in the quality of the product body 2-1 with six sigma methodology through the stages of DMAIC is generated defects in the product can be minimized, so that the company can reduce the losses caused by a defect of the product in the form of fees, timing and effectiveness of production.

**Table 1**. Research Position

| No | TOPIC | Source |
|---|---|---|
| 1 | Application of the method through the stages of the DMAIC Six Sigma for quality control of raw water in food production | [4] |
| 2 | Proposed Improvements To Reduce Number of Defect Product on Sandal Eiger-101 Lightspeed with Method Using Six Sigma | [5] |
| 3 | Implementation of Six Sigma Method To Reduce Paint Bucket in PT.X | [6] |
| 4 | Implementation of DMAIC approach for PVC Pipe Production Process Improvement (Case Study PT. Rusli Vinilon) | [7] |
| 5 | Implementation of Six Sigma and Data Mining to Improve Die Casting Production Process at PT.B | [8] |
| 6 | Quality Improvement for Product Body 2-1 at PT.X | This research position |

Table 1 is a research table on Six Sigma with DMAIC, This position of this research among other studies around the world, can be seen from various aspects above improvement of Product Body 2-1 can be seen from Quality Control methodology is combination with data mining, Failure Mode and Effect Analysis and Fault Tree Analysis.

## 2. Method

The preliminary study carried out by observing the production process for a 2-1 body products, as well as observe the conditions of the working environment, machinery, working methods, raw materials and labor as well as direct interviews with the head of the quality control related to the issues raised in the field of quality. Data used in the study consisted of secondary data and primary data. Secondary data is data that the company has been given, such as product defects historical data, production data and other general data. While the primary data is data collected directly, such as interviews and observations on the production floor. In data processing, analysis of the results.

### *2.1 Define*

Define phase is the initial phase in DMAIC phases. In this stage, how we done to identification

### *2.2 Measure*

The second phase in DMAIC phases is the measure phase. In this stage, the determination of the key characteristics of the product quality (CTQ), the calculation of the proportion control map (Map P), the calculation of the value of DPMO (Defect per Million opportunities) and sigma level.

### *2.3 Analyze*

Analyze stage is the third stage in the DMAIC phases. At this stage begin to analyze the underlying causes of problems that occur are defects in the product attribute body 2-1. Tools used in this stage is a decision tree-data mining, Pareto diagrams, fishbone diagram, FMEA (Failure Mode and Effect Analysis) and FTA (Fault Tree Analysis).

IOP Conf. Series: Materials Science and Engineering **847** (2020) 012040 doi:10.1088/1757-899X/847/1/012040

*2.4 Improve*
Phase repair or improve a fourth phase in DMAIC phases. At this stage, given the proposed improvements in the quality of the product body 2-1 by basic event that has the highest value on the probability of failure of the FTA.

*2.5 Control*
Phase control is the latest stage in a phase of DMAIC. At this stage the proposed improvements that have been given in the previous stage, namely the improve phase will be implemented and measured return value DPMO and sigma level, to determine whether the improvements made a significant impact on product quality.

## 3. Result and discussion

*3.1. Define*
According to methodology in define stage, we done how to indentification defects of product. In this research defects of products is Bland Not Good (Figure 2), Appearance Not Good (Figure 3) and Blank Minus (Figure 4)

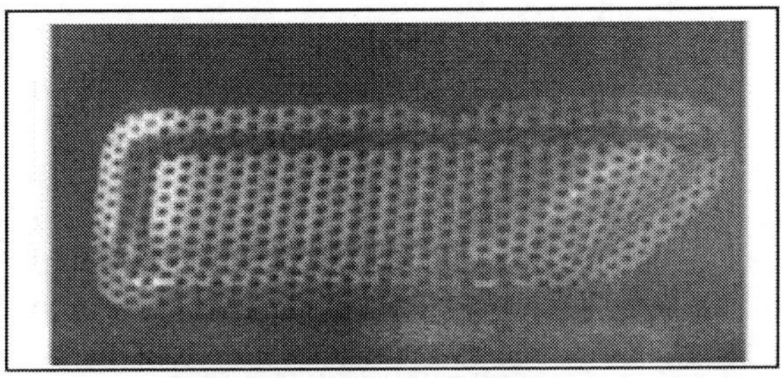

**Figure 1:** Product Body 2-1

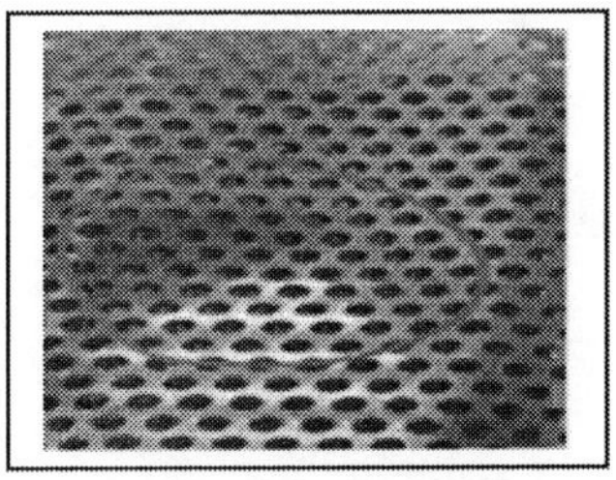

**Figure 2:** Bend NG

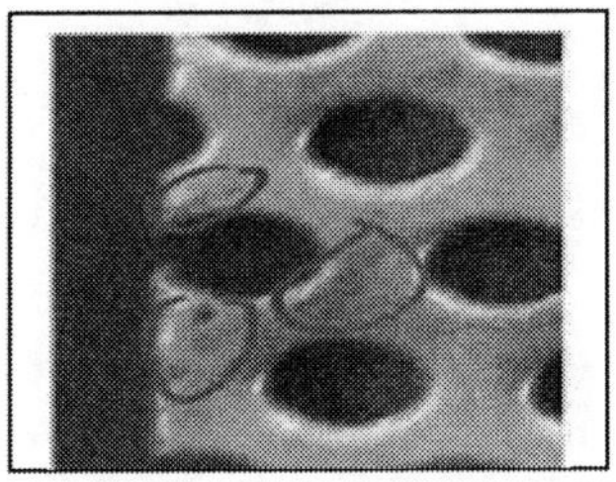

**Figure 3:** Appearance NG

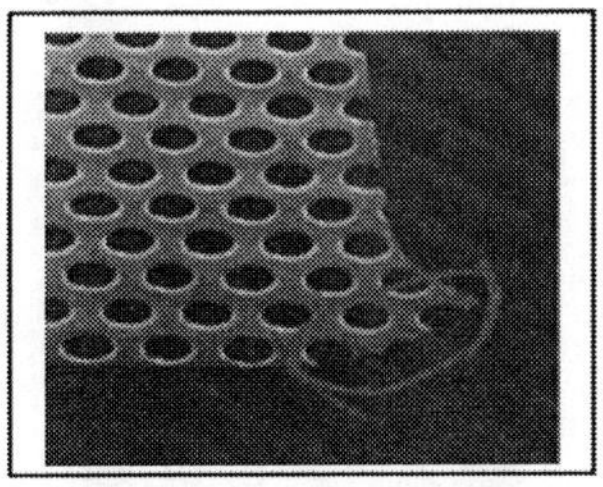

**Figure 4:** Blank Minus

*3.2. Measure*
Measuring is one of the stages used in six sigma, at this stage, control charts are usually in accordance with the data patterns in the production process which is useful for knowing whether the production process is stable or not [1]

Based on the graph p control chart in Figure 5 below, it can be seen that all the data is within the control limit (incontrol) because all the data is between the upper control limit (UCL) and lower control limit (LCL). These results indicate that the process runs steadily. Therefore, there is need for revision of the data.

Value *Defect per Million Opportunities* (DPMO) in the production process body2-1 products amounted to 19 268 defects in one million opportunities. DPMO value is converted to the sigma level to 3.57 sigma. Results value DPMO and sigma level is still far from the desired target of DPMO value that is equivalent to 3.4 DPMO and 6 sigma. The Company has not yet reached the desired target is 6 sigma. Therefore, it is necessary revisions to the production process of the 2-1 body products enable companies to improve the achievement levels as well as lowering the value of DPMO sigma.

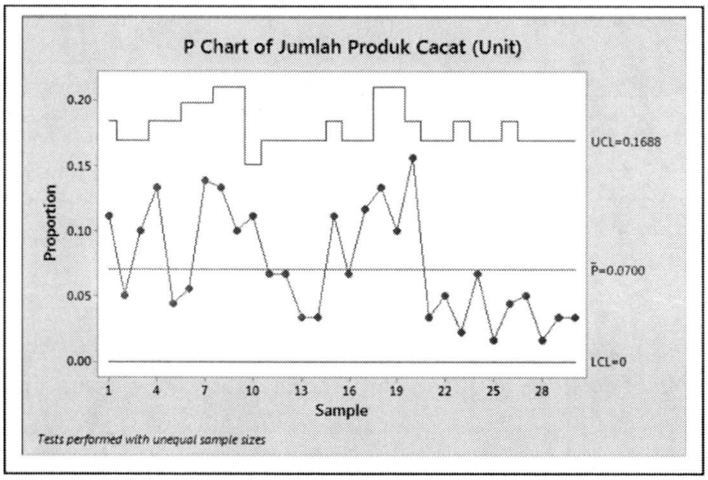


**Figure 5**: P Chart

*3.3. Analyse*

Analyse conducted is to use data mining, which aims to extract information from a collection of data. Information collected is usually the hidden nature, then carried out to predict. [2]

Based on the processing of data mining has been done, the quality characteristics of the most influential are the attributes that include appearance NG (performance of the product is not good), blank minus (Results pieces are not intact), bend NG (Bending is not perfect) and hole NG (The shape of the hole is not well). Type of disability are prioritized for repair is based on 80% of the most dominant disability. 80.20 Pareto diagram is shown in Figure 6.

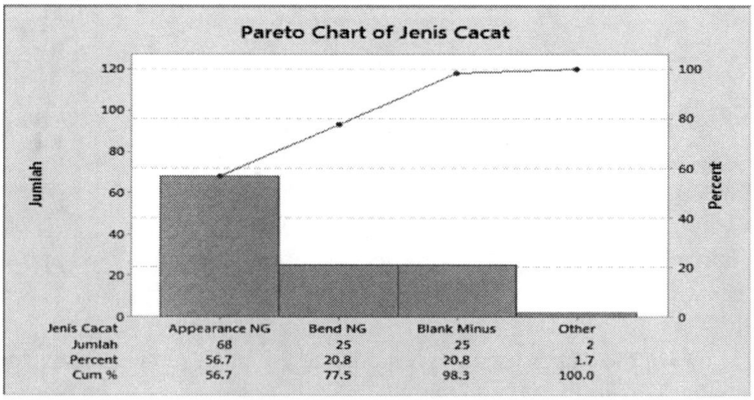


**Figure 6**: Pareto Chart

Based on the Pareto diagram in Figure 8, 80% of disabilities most dominating is the appearance NG (Form no good) with a percentage of 56.7%, then bend NG (Bending imperfect) with a percentage of 20.8%, and the blank minus (Results pieces are not intact) with a percentage of 20.8%. The three types of disability will be analyzed the cause using the fishbone diagram (fish bone).

After all cause problems for the three types of disability described by using ishikawa diagram, then made tables FMEA (Failure Mode and Effect Analysis) is a one the method to analyze the process production before the product reach to the customer [3].The severity, frequency and focus on the detection of failures in the production process of the body 2-1. Determining the value of the criteria of severity, occurrence and detection based on interviews with Mr. Julian as the head of the quality control. The main objective of FMEA table is to determine the cause of problems that must be prioritized to be handled by the largest RPN value. FMEA table can be seen in Table 2.

**Table 2**: Process Failure Mode and Effect Analysis

| No | process functions | Failure `Occurs | Effect of Failure | S | Cause of Failure Process | O | This time controls Do | D | RPN |
|---|---|---|---|---|---|---|---|---|---|
| 1 | blanking | Blank Minus (Results Pieces Not Intact) | *part* can not proceed to the next process and should be rejected | 8 | Operators are not careful and do not complete the first cycle of the process | 5 | Supervision of the performance of the operator | 2 | 80 |
| | | | | 8 | One of reversing material | 4 | Supervision | 2 | 64 |
| | | | | 8 | One uses stopper | 3 | Supervision | 2 | 48 |
| | | | | 8 | blunt dies | 6 | repair dies | 3 | 144 |
| 2 | bending | Appearance NG (No Good Appearance) | Aesthetics products are not good and not in accordance with the wishes of the customer, so that part should be repaired even rejected if conditions are too severe | 7 | Raw material quality is not good | 2 | Checks were tightened when the incoming material | 9 | 126 |
| | | | | 7 | Slamming or throwing the material carriers | 5 | Supervision of the performance of the operator | 2 | 70 |
| | | | | 7 | dies dirty | 4 | *Repair* and replace dies when part results found NG | 5 | 140 |
| | | | | 7 | Lack of lubrication during the process | 8 | Supervision | 3 | 168 |
| | | | | 7 | The laying of the material is not in accordance with the stopper | 6 | Supervision | 2 | 84 |
| | | | | 7 | *punch* and dies not center | 3 | Justifying the location of punch and dies in order to become a center | 7 | 147 |
| | | | | 7 | *dies* rude | 7 | Repair dies when found part NG | 3 | 147 |
| | | Bend NG (Profile Bending Imperfect) | *part reject* because it does not comply with the specifications | 8 | Been lazy and imprecise | 4 | Supervision of the performance of the operator | 2 | 64 |
| | | | | 8 | stopper shifted | 6 | Lay off the process and strengthen the bolt | 3 | 144 |
| | | | | 8 | *dies height* unlocked | 4 | Lay off and lock the height dies | 3 | 96 |
| | | | | 8 | Errors put the material on stopper | 6 | Supervision | 2 | 96 |

### *3.4 Improve the Installation of the Stopper Pin*

Based on discussions with the head of the quality control is a pack of Julian, the need for tools / instruments in addition to fasteners that also serves as a binder stopper on the dies. Additional tools are pin, that is objects that also serves to put a stopper on the dies with directional positions and sturdiness. With the pin mounted on a stopper, be expected if the fasteners loose stopper during the process do not move, because pin as the locking pin and still firmly bind stopper on the dies. Proposal improvement of the stopper placement on the stopper can be seen in Figure 7.

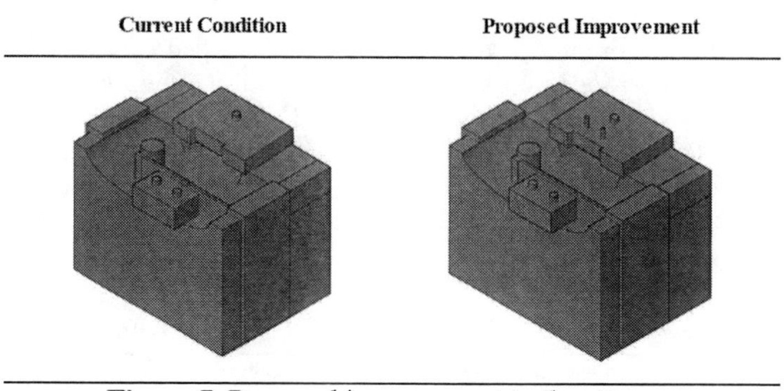


**Figure 7**. Proposed improvement to the stopper

*3.5 Control*

After the proposed improvements are implemented in the company, the results of defects per Million Opportunities (DPMO) is 18.229 and sigma level is 3.59, if we compare with before implementation there will be an increase and decrease as can be seen in Table 3.

**Table 3**. Six Sigma success indicator

| No. | Indicator | Before | After | ( %) |
|---|---|---|---|---|
| 1 | DPMO | 19.268 | 18.229 | Decrease: (5.3%) |
| 2 | Sigma Level | 3.57 | 3.59 | Increase: (0,5%) |

## 4. Conclusion

Based on the results and discussion that has been presented, the conclusion that can be drawn are:

1. Types of defects found in production process 2-1 body composed of defect types of attributes For this type of defect attributes include appearance NG (56,7%),bend NG (20,8%), blank minus (20,8%).
2. Value DPMO (Defect per Million Opportunities) of the production process of body products amounted to 19 268 disability 2-1 in a million chance, while for sigma level is at 3.57 sigma.
3. After the implementation of the proposed improvements, disability percentage obtained by 7.3%, amounting to 18229 DPMO value defects in one million opportunities and sigma level of 3.59. Value percentage disability and DPMO on after the implementation of the proposed improvements decreased and increased sigma level. This shows that the proposed improvements given a significant impact on the improvement of product quality body 2-1.

## 5. References

[1] DC Montgomery 2009 *Introduction to Statistical Quality Control* 6th ed. (New York: John Wiley and Sons, Inc.).

[2] S Adinugroho, YES Sari 2018 *Implementation Data Mining Using Weka* 1st (Malang: UB Press)

[3] DH Stamatis 2003 *Failure Mode and Effect Analysis* 2ed (Milwaukee: American Society for Quality Quality Press).

[4] D Rimantho, DM Mariani 2017 Application of Six Sigma Method on Raw Water Quality Control on Food Production, *J. of Industrial Engineering* **16**, p.1.

[5] Nailah, A Harsono, GP Liansari 2014 Proposed Improvements to Reduce Number of Defect Product on Sandal Eiger-101 Lightspeed with Method Using Six Sigma. *Online J of National Institute of Technology* **2**, pp.257-258.

[6] H Fransiscus, CP Sudarsono IS Astari 2014 Implementation of Six Sigma Method to Reduce Paint Bucket in PT. X. *J. of Industrial Systems Engineering* 3, p.54.

[7] D Caesaron, SYP Simatupang 2015 Implementation of DMAIC approach for PVC Pipe Production Process Improvement (Case Study PT. Rusli Vinilon), *J. Metris* **3,** p.91.
[8] R Fitriana, J Saragih, S Sarasaty 2012 *Implementation of Six Sigma and Data Mining to Improve Die Casting Production Process at PT. B Proceeding 7th Proceeding International Seminar on Industrial Engineering and Management*, pp.64-67.

IOP Conf. Series: Materials Science and Engineering **847** (2020) 012041 doi:10.1088/1757-899X/847/1/012041

# Forecasting for Steel Production using Artificial Neural Networks and Feasibility Analysis of Plant Regeneration Acid Development in PT. XYZ

**A Sasongko[1,2], I A Marie[1,3] and F Arifin[1,4]**

[1]Industrial Engineering Department, Trisakti University, Jakarta, Indonesia
[2]agung.sasongko@trisakti.ac.id,[3]iveline.annemarie@trisakti.ac.id,
[4]fakhrularfn211@gmail.com

**Abstract.** PT. XYZ is a steel production industry company. The Acid Regeneration Plant (ARP) facility has now shown a decrease in plant availability. This study aims to determine steel demand forecasting and also conduct a feasibility study for the development and replacement of ARP technology. Artificial Neural Networks, Linear Regression and Decomposition are used for forecasting experiments. The choice of forecasting method is taken by looking at the smallest MAPE value of each method. Artificial Neural Networks have the smallest MAPE value of 2.58. Then the Artificial Neural Network is used in forecasting requests for 2019 to 2020. The chosen network architecture is 24-12-1 with the traingdx training function. The results obtained are an increase in demand for steel products from 2019 to 2020. The increase in demand is 1,329,398 tons, an increase of 4.59% from the previous 2 years. Then a feasibility study is conducted to assess the construction of new ARP facilities. The feasibility study covers technology selection and investment appraisal. After making a comparison, the technology used is Spray Roaster. Based on the results of the investment appraisal using the NPV, IRR PBP, the new ARP development investment is decent and can meet future steel demand in the period until 2020.
**Keywords**: Forecasting, Artificial Neural Network, Linear Regression, Decomposition, Feasibility Study

## 1. Introduction

PT. XYZ produces various types of steel for various industrial uses in Indonesia. To produce various types of steel with good quality, the production at PT. XYZ must run smoothly. But now the Acid Regeneration Plant (ARP) facility has a problem. The ARP facility operates to regenerate hydrochloric acid (HCl) used for steel cleaning. The HCl will turn into Waste Pickle Liquid (WPL), which is an acidic liquid that has been mixed with Fe ions. WPL will be regenerated into Regenerated Acid (RA), which is HCl that has been regenerated and reused as steel cleaning liquid.

The ARP facility can actually regenerate 48 $m^3$ WPL within 8 hours or per shift, but currently it can only regenerate 35 $m^3$ WPL per shift. This ARP facility becomes mandatory in the production process. In addition to providing HCl for the cleaning process, the ARP facility functions as a waste processor. If the ARP facility does not operate properly, then PT. XYZ must pay the cost to buy Fresh Acid (FA) to increase the shortage of HCl in the production process. Besides that, PT. XYZ has to pay for other companies to process WPL. This made PT. XYZ consider creating a new ARP facility to support production. The feasibility study can be an appropriate method of decision making for a new ARP facility. The feasibility analysis conducts an assessment based on several aspects such as legal aspects, market aspects, technical and production aspects, operational management aspects, environmental aspects and also financial aspects [1].

## 2. Methods

Observation on the production floor as preliminary research has completed with literature study to get problem statement. As result of that, data collection are production historical data, ARP existing analysis and feasibility study indicator. It is used three approach of forecasting of demand which is Artificial Neural Network, Linear Regression and Decomposition and based on the lowest MAPE score from three methods it would used for feasibility study. Research steps carried out in accordance with figure 1.

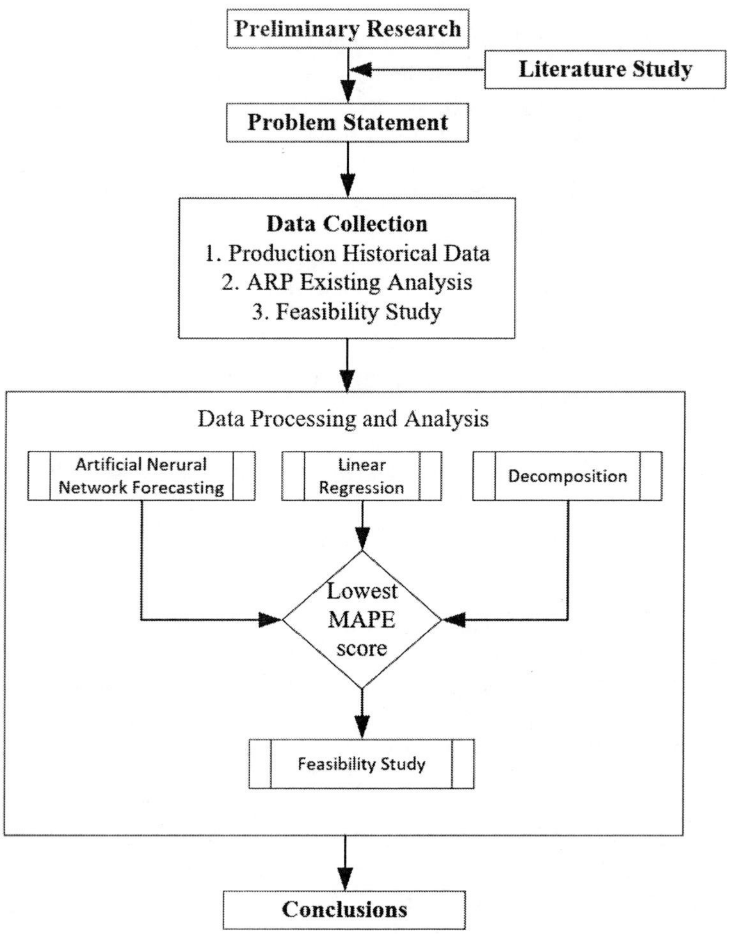


**Figure 1.** Flowchart of research methodology

## 3. Result and Discussion

### *3.1 Forecasting Method of Artificial Neural Networks*

In the artificial neural network forecasting method, the data to be processed must first be normalized [2]. Sales data are transformed at intervals of 0.1 to 0.9. This transformation is a function of sigmoid activation. The result is show in figure 2.

IOP Conf. Series: Materials Science and Engineering **847** (2020) 012041 doi:10.1088/1757-899X/847/1/012041

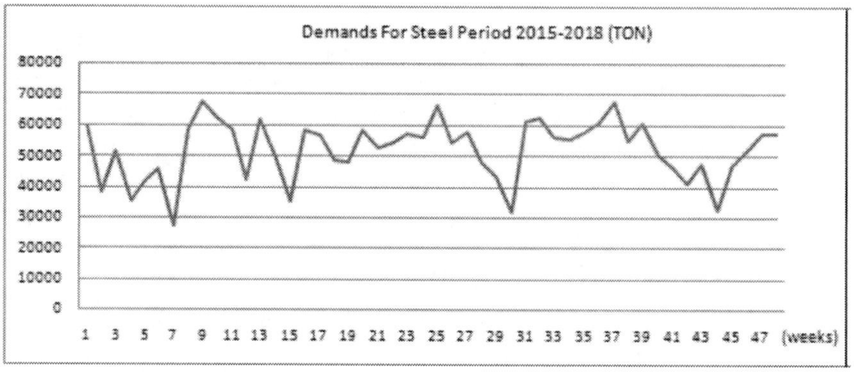


**Figure 2.** Demands for Steel Period 2015-2018 (Ton)

*3.1.1 Artificial Neural Network Training.* Artificial Neural Networks work to recognize a pattern [2]. Artificial neural networks conduct training using these patterns in order to recognize other patterns to be further tested. The training input data pattern used is the CRC production normalization data from 2015 to 2018. While the training target is the CRC sales normalization data for 2017 and 2018. The pattern is created by taking 24 normalization data. Example pattern 1 is the normalization data for CRC sales for the first 24 periods, while the target of pattern 1 is period 25. Pattern 2 starts from period 2 to period 25 and the pattern target 2 is period 26. The pattern is made to get 24 patterns.

*3.1.2 Election of Artificial Neural Network Architecture.* Based on the results of trial and error, network architecture 24-12-1 is the best architecture that gets an MSE value of 0,0009 with 816 iterations / epochs. This architecture uses the traingdx training function with a learning rate value of 0.1. This network architecture also gets optimal weights obtained from the results of trial and error iterations in the 2017 MATLAB software

*3.1.3 Results of Artificial Neural Network Training.* The results of training on artificial neural network models are forecasting for the 2017-2018 period. This value is still in the form of sigmoid activation function. To return these values to normal values, it is necessary to calculate denormalization.

*3.1.4 Best Architecture Testing.* The best architecture is an artificial neural network that can recognize training data patterns. This network architecture must first be tested with other data patterns so that it can be used for forecasting. The test data pattern used is the odd data pattern on the training data pattern. While the test target used is also the target of training in odd periods, other tests are performed by calculating the MAPE value from training in artificial neural network models. And the result of MAPE for Articial Neural Network is 2,58.

### *3.2 Forecasting Method of Linear Regression*
In processing data to forecast steel demand using historical data on steel sales. Historical sales data used are sales from 2017 to 2018. Forecasting with the linear regression method needs to calculate the values (b) and (a) to calculate the forecast value. After calculating the values of (b) and (a), we can calculate the forecast using the linear regression method [3]. And the result of MAPE for Linear Regression is 15,23.

### *3.3 Forecasting Method of Decomposition*
Forecasting using the decomposition method begins by calculating moving averages from existing data [4], namely historical data on steel sales. The moving average calculates an average value of 12 months. Placement of moving average results starts from the 7th period. After calculating the 12-

month moving average value, then looking for a centralized moving average value. Centralized moving average calculates the average value of 2 12 month moving average values. Next is calculating the seasonal factor by dividing the value of demand by the centralized moving average. Next look for seasonal index values. Because of the calculated seasonal factor values, there are 12 values, which are the values for each month of the year, the seasonal index is the same as the seasonal factor. And the result of MAPE for Decomposition is 10,03.

### *3.4 Selection of Forecasting Methods*

The selection of methods that will be used to forecast the 2019-2020 period is done by comparing the accuracy of forecasting. The accuracy of forecasting is done using MAPE values. Table 1 is the MAPE value of the three forecasting methods;

**Table 1.** MAPE Value Comparison

| No | Methods | MAPE Value |
|---|---|---|
| 1 | Artificial Neural Network | 2,58 |
| 2 | Linear Regression | 15,23 |
| 3 | Decomposition | 10,03 |

Based on table 1. shows that the artificial neural network method has the smallest MAPE value. Therefore, the method to be used for forecasting the 2019-2020 period is an artificial neural network.

### *3.5 Current ARP Facility Analysis*

The condition of the ARP facility which has been operating since 1985 is arguably very outdated. Based on data on operating hours from 2015 to 2018, the plant availability of the ARP facility only reached a value of 51%. Ideally, a facility / machine has a plant availability value of around 90%. Based on the calculation of Regenerated Acid processing data and the purchase of Fresh Acid from 2015 to 2018, the CPL line requires 18% HCL of 33,925,168 m3 for each year or 31,412 m3 for each shift. The ARP facility can only supply 18% HCl of 23,851 m3 annually. When the ARP facility does not operate processing WPL and cannot supply 18% HCl to the CPL line, the shortfall will be covered by the purchase of Fresh Acid. With the purchase of Fresh Acid, the 18% HCl requirement will be fulfilled for use in the CPL line as a steel cleaner. Based on 2015 until 2018 demand data, the average steel demand at PT. XYZ is 624,483 tons per year or 578,225 tons per shift. So we can know the value of using HCl 18% for every 1 Ton of steel. This value is 0.0543 m3 / Ton.

### *3.6 ARP Analysis with Results of Forecasting Neural Networks*

Results of forecasting demand for products show an increase in steel demand to 1,329,398 tons, up 4.59% from 2017-2018. If the requirement for HCl 18% during production in the 2015-2018 period is 33,925,168 m3 per year, the acid demand when demand increases is 33,419 $m^3$.

From the calculation above, it shows that the HCl requirement of 18% is still in the design capacity of the ARP facility which is 48m3 / Shift. In fact, the ARP facility was unable to continue to supply the 18% HCl requirement every day. This was supported by the purchase of Fresh Acid in 2015 to 2018 to cover the 18% HCl deficiency originating from the ARP facility. The value of plant availability from this facility is only around 51%. With the addition of the high maintenance cost value every year, it becomes a driving force to construct new ARP facilities.

### *3.7 Feasibility Study of ARP Development*

Some aspects that can be used as an evaluation are legal aspects, market aspects, technical and production aspects, management aspects, environmental aspects and also financial aspects.

*3.7.1 Legal Aspects.* PT.XYZ is a steel company that produces steel from upstream to downstream. PT.XYZ already has permission operating license is also called an Industrial Business License (IUI). In the ARP production unit technology replacement project, no external permission is required based

IOP Conf. Series: Materials Science and Engineering **847** (2020) 012041 doi:10.1088/1757-899X/847/1/012041

on this regulation. The ARP production unit technology replacement project only requires approval from PT.XYZ's top officials because this project is a corporate action.

*3.7.2 Market Aspects.* The demand for steel products is predicted to continue to increase. This increase was followed by the development of infrastructure and automotive industry in domestic. Based on research results from the Market Research & Development Division of PT. XYZ as well as from the POSCO Research Institute, the demand for Cold Rolled Coil (CRC) steel products from 2019 to 2025 will increase as shown in Figure 3.

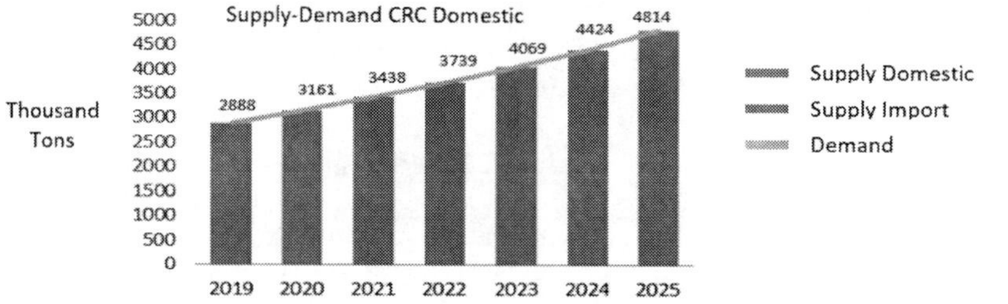


**Figure 3.** Supply-Demand Domestic

*3.7.3 Technical and Production Aspects.* The ARP facility currently uses fluidized bed technology. One of the technology options to choose from is fluidized bed and spray roaster technology. With almost the same investment value, the technology chosen is based on aspects of energy consumption efficiency, price aspects and the number of technology users is a spray roaster.

*3.7.4 Management Aspects.* The project to build a new ARP facility will be carried out by the winning contractor, followed by several ARP technology vendors. The contracting company will handle all construction.

*3.7.5 Environment Aspects.* ARP is a production unit in the Cold Rolling Mill plant. The ARP production unit functions as a waste processor at the CRM plant. The resulting waste is Waste Pickle Liquid (WPL). WPL is a waste from the Contious Pickling Line (CPL) production line. WPL is a concentrated HCl acid containing iron. The WPL will be regenerated to produce Regenerated Acid (RA) used in the CPL production line.

PT.XYZ already has an AMDAL document (government regulation) regarding environmental Feasibility [5]. According to the head of the Health, Safety and Environment Department, project the replacement of the ARP production unit does not require a government permit because PT. XYZ already has the AMDAL documents that have been mentioned.

*3.7.6 Financial Aspects.* Investment valuation criteria used in this aspect are Net Present Value (NPV), Internal Rate of Return (IRR) and Payback Period (PBP) [6]. From the results of the calculation, the project for Spray Roaster is DECENT with the consideration of the value of the payback period <economic age, ie 2 Years 9 Months <10 years. Table 2 show the summary of Spray Roaster Investment Valuations.

**Table 2.** Summary of Spray Roaster Investment Valuations

| Criteria | Value | Status |
|---|---|---|
| NPV | 5.292.000 | Decent |
| IRR | 51,25% | Decent |
| PBP | 2 Tahun 9 Bulan | Decent |

For the calculation of the Fluidized Bed Investment Rating, this project is DECENT with the consideration of the payback period value <economic age, ie 3 Years 7 Months <10 years. Table 3 show the summary of Spray Roaster Investment Valuations.

**Table 3.** Summary of Fluidized Bed Investment Valuations

| Criteria | Value | Status |
|---|---|---|
| NPV | 3.363.072 | Decent |
| IRR | 40,21% | Decent |
| PBP | 3 Tahun 7 Bulan | Decent |

*3.7.7 Sensitivity Analysis.* This sensitivity analysis will analyze three parameters, namely investment, annual benefit and also annual cost [7]. These three parameters are used because all three parameters are included in the cash flow calculation. Investment is the value of capital spent to finance the project. Annual cost is the cost that must be spent in each year. While annual benefit is a benefit that is obtained every year. The following table summarizes the sensitivity values for all parameters for Spray Roaster and Fluidized Bed.

**Table 4.** Summary of Sensitivity Analysis of Spray Roaster and Fluidized Bed

| Parameter | Spray Roaster | | | Fluidized Bed | | |
|---|---|---|---|---|---|---|
| | Cashflow | Sensitive Value | Deviation | Cashflow | Sensitive Value | Deviation |
| Investment | $ 2.546.876 | $ 5.919.402 | 132 % | $ 2.437.264 | $ 4.580.508 | 88 % |
| Annual Benefit | $ 2.714.477 | $ 1.725.817 | 36 % | $ 2.581.477 | $ 1.953.182 | 24 % |
| Annual lCost | $ 979.197 | $ 1.967.857 | 101 % | $ 1.238.695 | $ 1.866.990 | 51 % |

Based on the calculation results of the sensitivity analysis of the two technologies, it can be seen that the fluidized bed technology has a greater sensitivity level compared to the spray roaster technology. The smaller the value of the percentage difference, the more sensitive the parameter. Both technologies have the same pattern in the order of the smallest percentage of difference value. The percentage value of the difference from the smallest to the largest is the annual benefit, annual cost and investment parameters.

Based on the results that have been described, it shows that between the two technologies, fluidized bed technology is the most sensitive to changes. Very sensitive changes are in the annual benefit parameter with a change of 24% has reached the point of sensitivity.

*3.7.8 ARP Projection Analysis After Development.* The construction of new ARP facilities is intended to replace the current ARP facility. The ARP facility currently has a design capacity of 48m3 per shift. In fact, the ARP facility can currently supply 18% HCl of 31.412 m3 per shift for the CPL line, only the 18% HCl supply must be supported by additional supply from the purchase of Fresh Acid. If the ARP facility continues to operate, the costs incurred to operate it will be fairly high. That was caused by the purchase of Fresh Acid and also the cost to carry out maintenance.

After the construction of this new ARP facility, the ARP can supply as much as 48m3 per shift. This new ARP facility can meet the design capacity of existing ARPs. With the new ARP facility, it can also supply 18% HCl for projected demand in 2019-2020 with an estimated HCl requirement of 18% of 33.419 m3 per shift.

## 4. Conclusions

Demand forecasting is done by 3 methods, namely Artificial Neural Networks, Linear Regression and Decomposition. The chosen method based lowest value of MAPE is the Artificial Neural Network method with a value of 2.58. The results of forecasting steel demand for 2019-2020 is 1,329,398 tons, increased by 4.59% of the total demand for 2017-2018.

ARP development is declared DECENT based on Legal Aspects, Market Aspects, Technical and Production Aspects, Management Aspects, Environmental Aspects and also Financial Aspects. The selected ARP technology, Spray Roaster, is based on Technical and Production Aspects. The investment appraisal of this development project was also DECENT with a NPV value of $ 5,292,000, an IRR of 51% and a payback period of 2 Years 9 Months.

The study state that the construction of a new ARP facility with Spray Roaster technology can meet the demand for steel based on forecasting results.

## 5. References

[1] Momin M and Sahadaev R 2017 Feasibility Studies and Important Aspect of Project Management. *International J. of Advanced Engineering and Management*. 2 4 pp 98-100

[2] Kamruzzaman J, et al. 2006 *Artificial Neural Networks in Finance and Manufacturing* (USA: Idea Group Publishing)

[3] Box G E P, et al. 2016 *Time Series Analysis Forecasting and Control* Vol 5 (New Jersey: Wiley)

[4] Makridakis S, et al 1999 *Metode dan Aplikasi Peramalan* Vol 1 (Jakarta: Erlangga)

[5] Murwadji T and Imamulhadi 2018 *Green Banking: The Model and Its Implementation* (Amsterdam: IOS Press)

[6] Gul S, Gul H and Haider M 2018 The Review and Use of Capital Budgeting Investment Techniques In Evaluating Investment Projects: Evidence From Manufacturing Companies Listed On Pakistan Stock Exchange (PSE). *City University Research J.* **8**(2), pp 247-260

[7] Giatman M 2011 *Ekonomi Teknik* Vol 3 (Jakarta: PT RajaGrafindo Persada).

IOP Conf. Series: Materials Science and Engineering **847** (2020) 012042 doi:10.1088/1757-899X/847/1/012042

# Analysis of the impacts of Motor Vehicle Exhaust Emissions at Pancasila University on health in order to create A Green Campus

**M Y M Sholihin[1,2], R Prasetyani[1,3], B C Mukti[1,4]**

[1]Fakultas Teknik Universitas Pancasila, Jl. Srengseng Sawang, Jagakarsa, Jakarta Selatan
[2]yudi_m_s@univpancasila.ac.id, [3]rini.prasetyani@univpancasila.ac.id, [4]bintangcatur10@yahoo.com

**Abstract :** The current carbon monoxide (CO) emissions are the main component of diseases, such as cough, dizziness, nausea, and shortness of breath for those who inhale them. Pancasila University is one of the colleges in the city of Jakarta that cares about the poor quality of air in the city of Jakarta; therefore, Pancasila University wants to apply the concept of green campus as a form of concern for the environment. This research was conducted at Pancasila University to discover the relationship between air quality in the campus, especially carbon monoxide (CO) substance, and its impact on the health of the university's academic community members. The methods used in this study are simple linear regression, HIRARC, and hierarchy of control. The results of this study show a relationship between CO substance variables and complaints about diseases suffered by members of the university's academic community due to the substance, with an R count value of 0.9519. There are substantial risks that the students can experience, including student activities that cause nausea and outdoor employee activities that cause dizziness and nausea. These big risks can be prevented by issuing a decision letter to reduce air pollution at Pancasila University.
**Keywords:** Air Quality, Motor Vehicle Exhaust Gas, Pancasila University, Green Campus

## 1. Introduction

Air is definitely a vital component of life for the survival of humans and other living beings, such as plants and animals. Humans often consider dusty, cloudy, smoky, and smelly air as a normal thing, although bad quality air really affects health [1].

Bad air can also be produced from motor vehicles. Indonesia is a very potential market for selling motor vehicles of any types and brands. This can be seen clearly from the significant increase of the number of motor vehicles. This high growth of motor vehicles, on the one hand, can improve the economic growth rate; on the other hand, it can cause a very serious environmental impact. One of the environmental impacts occurring is air pollution due to exhaust emissions produced by motor vehicle machines. On 4 August 2019, Jakarta ranked second as the most air polluted city in the world according to the data of *AirVisual*. *Air Quality Index* (AQI) of Jakarta is 160, meaning that the Jakarta's air quality is not healthy. AQI is an index describing the air quality severity level in a region [4]. Due to the hazard of the air pollution risks, particularly the health impacts of CO pollutant produced by motor vehicles, and the inexistence of a management program towards the risks that are caused by the pollution from motor vehicles, the campus of Pancasila University located in the capital city of Indonesia, namely DKI Jakarta, has become the research location applying the green campus concept [5].

Currently, Pancasila University is implementing the green campus concept, but there are still several obstacles occurring, such as the lack of the studies of the green campus concept implementation or the inexistence of the tools that can assist students or teachers in conducting research. This present research discusses the impacts of CO (carbon monoxide) pollutant from motor vehicles coming to and going from the area of Pancasila University on the health of the academic community members. It is known that CO pollutant has impacts on health if inhaled in a high amount, and some of the impacts are fainting, fatigue, nausea, dizziness, and also shortness of breath; it might also lead to death [6].

## 2. Methods

### *2.1. Research Object*

The research was conducted in the campus area of Pancasila University on the CO substance as a vehicle exhaust gas entering the area of Pancasila University, and the study focused on its health impact in the academic community members [2].

### *2.2. Pilot Study*

A pilot study was conducted as the basis for obtaining a clear picture on the problems examined and theories applied in this research. This pilot study was divided into two forms as follows.

Field Study A field study is an activity to search for accurate data, explanation, and information suitable for the problems that are brought up in this research by doing direct observation in the field.

Literature Study. A literature study is the phase that has to be conducted to obtain references and the theoretical framework that can support the research. This theoretical framework becomes the reference for solving problems while conducting the research. The sources used in this theoretical framework are, among others, journals, undergraduate theses, and other data [3].

### *2.3. Data Collection*

**Ambient Air Quality Standards of the DKI Jakarta Province Governor's Decision**

The decision of the DKI Jakarta Province Governor number 551/2001 states that the high growth of industry, transportation, and other development in the DKI Jakarta Province can cause ambient air pollution.

**Table 1** Ambient Air Quality Standards Based on the DKI Jakarta Province Governor's Decision

| No | Parameter | Measurement time | Quality Book |
|---|---|---|---|
| **1** | Sulfur Dioxide ($SO_2$) | 1 Hour | 900 μg/Nm3 ( 0,34 ppm) |
| | | 24 Hour | 260 μg/Nm3 ( 0,1 ppm) |
| | | 1 year | 60 μg/Nm3 ( 0,02 ppm) |
| **2** | Carbon Monoxide (CO) | 1 Hour | 26000 μg/Nm3 (23 ppm) |
| | | 24 Hour | 9000 μg/Nm3 (8 ppm) |
| **3** | Nitrogin Dioxide ($NO_2$) | 1 Hour | 400 μg/Nm3 (0,2 ppm) |
| | | 24 Hour | 92,5 μg/Nm3 (0,05 ppm) |

**Table 1** Ambient Air Quality Standards Based on the DKI Jakarta Province Governor's Decision (continued)

| No | Parameter | Measurement time | Quality Book |
|---|---|---|---|
| | | 1 year | 60 µg/$Nm_3$ ( 0,03 ppm) |
| 4 | Oxidant ($O_3$ ) | 1 Hour | 200 µg/$Nm_3$ (0,05 ppm) |
| | | 1 year | 30 µg/$Nm_3$ ( 0,03 ppm) |
| No | Parameter | Measurement time | Quality Book |
| 5 | Hydrocarbon (HC) | 3 Hour | 160 µg/$Nm_3$ ( 0,04 ppm) |
| 6 | Particlet < 10 um ($PM_{2,5}$) | 24 Hour | 150 µg /$m^3$ |
| 7 | Particlet < 2,5 um ($PM_{2,5}$) | 24 Hour | 65 µg /$m^3$ |
| | | 1 year | 15 µg /$m^3$ |
| 8 | Dust (TPS) | 24 Hour | 230 µg /$m^3$ |
| | | 1 Year | 90 µg /$m^3$ |
| 9 | Lead (Pb) | 24 Hour | 2 µg /$m^3$ |
| | | 1 Year | 1 µg /$m^3$ |

*2.4. Number Of Samples*
The number of academic community members at Pancasila University is 13,921. To determine the minimum number of the samples, a calculation was conducted by using the Slovin formula with the error margin of 10% as follows [4].

$$n = \frac{N}{1 + Ne^2} \quad (1)$$

$$n = \frac{13921}{1 + 13921.(0,1)^2}$$

$n = 99.2866 \approx 99$

There were 99 academic community members at Pancasila University that were researched as samples. Primary data are the data that were acquired directly through observation and recorded for the first time. There are several ways to acquire primary data, some of which are as follows.

1. Direct data measurement
   Data were directly acquired through the measurement of air quality in the campus area of Pancasila University and the number of academic community members.
2. Observation
   Observation is data collection by viewing and conducting direct observation of the research object.
3. Questionnaire

The questionnaire was aimed at 99 members of the academic community of Pancasila University.
Secondary data are the data acquired by searching for the literature from various sources, such as journals, books, and reports of the research conducted and intended to support our research.
The data that were obtained were managed using the following method.

1. Questionnaire Validation Test
   The questionnaire that was obtained was validated to discover whether the questions have been understood by the respondents or not.

$$r_{xy} = \frac{n\sum xy - \sum x \sum y}{\sqrt{(n\sum x^2-(\sum x)^2)(n\sum y^2-(\sum y)^2)}} \quad (2)$$

$$r_{xy} = \frac{99\,(5373) - 343\,(3111)}{\sqrt{(99(1267) - (343)^2)(99(23662) - (3111)^2)}}$$

$r_{xy} = 0{,}6353$

The result of r count was compared with the result of r table, where r table for df = n – 2 (99-2 = 97) is 0.1663, so r count is > r table, so that the question number 1 is considered valid. The same calculation was conducted to discover the validity of every question as follows.

**Table 2** Validity of Each Question in Research Questionnaire

| No | Question | r Count | r Table | Valid / Invalid |
|---|---|---|---|---|
| 1 | Do you feel shortness of breath due to inhaling exhaust motor vehicles | 0,6353 | 0,1663 | Valid |
| 2 | Do you feel nauseous due to inhaling exhaust gas from a motorized vehicle | 0,5943 | 0,1663 | Valid |
| 3 | As a result of exhaust gas from motorized vehicles at Pancasila University, there was a feeling of dizziness | 0,4467 | 0,1663 | Valid |
| 4 | Did you experience respiratory symptoms during your activities at Pancasila University? | 0,1784 | 0,1663 | Valid |
| 5 | Do you think air quality at Pancasila University is classified as poor | 0,6733 | 0,1663 | Valid |

2. Simple Linear Regression

**Table 3** Linear Regression between the CO Substance as a Motor Vehicle Gas and the Diseases It Caused

| NO | X | Y | $X^2$ | $Y^2$ | XY |
|---|---|---|---|---|---|
| 1 | 2291 | 73 | 5248681 | 5329 | 167243 |
| 2 | 1146 | 63 | 1313316 | 3969 | 72198 |
| 3 | 1145 | 61 | 1311025 | 3721 | 69845 |
| 4 | 1146 | 60 | 1313316 | 3600 | 68760 |
| 5 | 1145 | 64 | 1311025 | 4096 | 73280 |
| ∑ | 6873 | 321 | 10497363 | 20715 | 451326 |

After acquiring the value of variables x and y, the variables x and y are as follows.

1. Point 1 of the CO substance measurement was located in front of the rectorate building. This activity obtained as much as 2291 μg/m³ with the number 1 question index around 73% prioritizing shortness of breath due to CO. This is because the front part of the rectorate building is the gateway where vehicles come to and go out from Pancasila University. This front part of the building has the largest number of vehicles and pollutants due to heavy traffic as there is a zebra cross for crossing the street from the bus stop of Pancasila University to the train station of Pancasila University.
2. Point 2 of the CO substance measurement was located at the FEB motorcycle parking place. This activity obtained as much as 1146 μg/m³ with the number 2 question index around 63%. There was not so many pollutants in the area because the parking place alone did not produce a high amount of CO. Because of this, the potential disease here is only nausea.
3. Point 3 of the CO substance measurement was located at the FEB yard. This activity obtained as much as 1145 μg/m³ with the number 3 question index around 61%. There were not many pollutants in the area because its location is close to the parking place so that it only causes dizziness.
4. Point 4 of the CO substance measurement was located in front of the Faculty of Pharmacy. This activity obtained as much as 1146 μg/m³ with the number 4 question index around 60%. There were

not many pollutants in the area because its location is far from the main street, but in that area the amount of CO is quite high, so it can cause symptoms of a disease related to breathing.

5. Point 5 of the CO substance measurement was located in front of LIA (English language course) building. This activity obtained as much as 1145 $\mu g/m^3$ with the number 5 question index around 64%. Based on the perception of the academic community members, the air quality in this part of Pancasila University is considered bad.

The next step to determine the constant value (a) and the regression coefficient value (b) is as follows.

$$a = \frac{(\Sigma x^2)(\Sigma y)-(\Sigma x)(\Sigma xy)}{n\,\Sigma x^2-(\Sigma x)^2} \quad (3)$$

$$= \frac{(10497363)(321)-(6873)(451326)}{5 \times 10497363-(6873)^2}$$

$$= 51{,}001$$

$$b = \frac{n\,\Sigma xy-(\Sigma x)(\Sigma y)}{n\,\Sigma x^2-(\Sigma x)^2} \quad (4)$$

$$= \frac{5 \times 451326 - (6873)(321)}{5 \times 10497363- (6873)^2}$$

$$= 0{,}0096$$

The simple linear regression equation can be made as follows.

$Y = 51.001 + 0.0096(X)$

The value of a (constant) is 51.001, showing the amount of the CO substance as a motor vehicle gas that does not affect the disease complaint in a long run and can be defined when the value of CO is 0, so the value of the health complaint of the academic community is 51.001. The regression coefficient (slope) of b is 0.0096, meaning that CO as the motor vehicle gas has a positive correlation with the health complaint of the academic community of Pancasila University, such as dizziness, cough, shortness of breath, and nausea.

3. Simple Correlation

The correlation is used to discover whether variables x and y are mutually related, so they will be tested with the formula as follows.

$$r = \frac{n\sum xy-\sum x\sum y}{\sqrt{(n\sum x^2-(\sum x)^2)(n\sum y^2-(\sum y)^2)}} \quad (5)$$

$$= \frac{5\,(451326) - 6873\,(321)}{\sqrt{(5\,(10497363) - (6873)^2)(5(20715)-(321)^2)}}$$

$$= 0{,}9519$$

The value of r (correlation coefficient) has been known to be as much as 0.9519, where the value of r count is > the value of r table. The value of r table is 0.8054, so it can be assured that there is a correlation between CO produced by motor vehicles and the disease complaints suffered by the academic community members of Pancasila University.

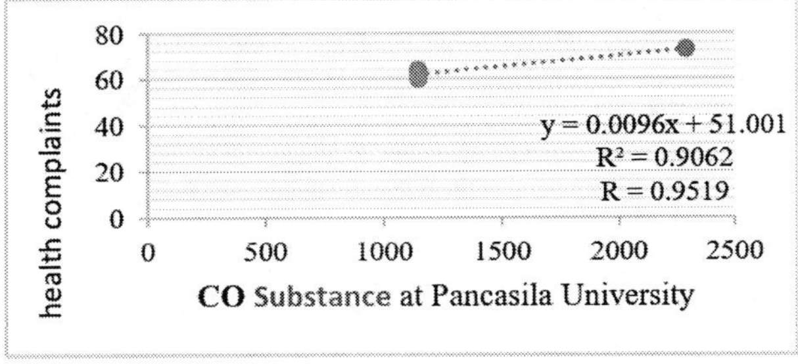


**Figure 1** (Source: Data Management, 2019)

$R^2$ (determinant coefficient) is 0.9062, showing that the variable of CO, the motor vehicle gas affecting the variable of the disease complaint, is 90.62%, while 9.38% is affected by smoking, dust, and others.

## 3. Results And Discussion

Identification of Academic Community Members' Activities at Pancasila University and the Health Impacts of Air Pollution on Health. The identification in this study observed the activities of academic community members (lecturers, employees, and students) and the health impacts of CO as a motor vehicle exhaust gas at the campus area of Pancasila University.

### *3.1. Likelihood*

After discovering the activities of the academic community members, the next step was to determine the likelihood or frequency assessment.

**Table 4** Likelihood Score

| Number of Outdoor Employee Illness Complaints | | | |
|---|---|---|---|
| Type of Complaint | Number of Complaints | Frequency | Likelihood |
| Coughing | 21 | 84 | 4 |
| Shortness of breath | 18 | 72 | 4 |
| Dizziness | 23 | 92 | 5 |
| Nausea | 12 | 48 | 3 |
| The Number of Complaints from Employee in the room | | | |
| Type of Complaint | Number of Complaints | Frequency | Likelihood |
| Coughing | 13 | 54,1 | 3 |
| Shortness of breath | 16 | 66,6 | 4 |
| Dizziness | 12 | 50 | 3 |
| Nausea | 9 | 37,5 | 2 |
| The Number of Complaints from Lecturer Diseases | | | |
| Type of Complaint | Number of Complaints | Frequency | Likelihood |
| Coughing | 15 | 60 | 3 |
| Shortness of breath | 8 | 32 | 2 |
| Dizziness | 11 | 44 | 3 |
| Nausea | 6 | 24 | 2 |
| The Number of Student Illness Complaints | | | |
| Type of Complaint | Number of Complaints | Frequency | Likelihood |
| Coughing | 16 | 64 | 4 |
| Shortness of breath | 13 | 52 | 3 |
| Dizziness | 19 | 76 | 4 |
| Nausea | 10 | 40 | 2 |

### *3.2. Severity*

Severity is the level of danger/graveness incurred from an activity.

**Table 5** Severity Risks

| Level | Desription | Desease Complaint |
|---|---|---|
| 1 | Insignificant | Cough |
| 3 | Moderate | Shortness of breath |
| 1 | Insignificant | Dizziness |
| 2 | Minor | Nausea |

### 3.3. *Hazard Identification Risk Assessment and Risk Control (HIRARC)*

After the scores of likelihood and severity were obtained, the next step was to do the assessment and control of risks by using the HIRARC method:

**Table 6** HIRARC due to CO as a Motor Vehicle Exhaust Gas

| No. | Activity | Risk/impact | L | S | RPN | Hazard Prevention |
|---|---|---|---|---|---|---|
| 1. | Student activity | Cough (1v) | 4 | 1 | 4 | Using personal protective equipment such as a protection mask |
| | | Shortness of breath (1x) | 4 | 1 | 4 | Using personal protective equipment such as a protection mask |
| | | Dizziness (1y) | 2 | 2 | 4 | Using personal protective equipment such as a protection mask |
| | | Nausea (1z) | 3 | 3 | 9 | Issuing a decision letter to reduce air pollution on campus |
| 2. | Lecture activity | Cough (2v) | 3 | 1 | 3 | Using personal protective equipment such as a protection mask |
| | | Shortness of breath (2x) | 2 | 1 | 2 | Using personal protective equipment such as a protection mask |
| | | Dizziness (2y) | 3 | 2 | 6 | Using personal protective equipment such as a protection mask |
| | | Nausea (2z) | 2 | 3 | 6 | Issuing a decision letter to reduce air pollution on campus |
| 3. | Staff activity (inside the building) | Cough (3v) | 3 | 1 | 3 | Using personal protective equipment such as a protection mask |
| | | Shortness of breath (3x) | 4 | 1 | 4 | Using personal protective equipment such as a protection mask |
| | | Dizziness (3y) | 3 | 2 | 6 | Using personal protective equipment such as a protection mask |
| | | Nausea (3z) | 2 | 3 | 6 | Using personal protective equipment such as a protection mask |
| 4. | Staff activity (outside the building) | Cough (4v) | 4 | 1 | 4 | Using personal protective equipment such as a protection mask |
| | | Shortness of breath (4x) | 4 | 1 | 4 | Using personal protective equipment such as a protection mask |
| | | Dizziness (4y) | 5 | 2 | 10 | Issuing a decision letter to reduce air pollution on campus |
| | | Nausea (4z) | 3 | 3 | 9 | Issuing a decision letter to reduce air pollution on campus |

### 3.4. *Risk Matrix*

The risk mapping or risk matrix was done based on the value obtained in the Risk Priority Number (RPN) resulting from likelihood multiplied by the existing severity. The risk level is represented in colors to discover how big the hazard potential that can be caused is. Below is the risk matrix.

| Possibility | The consequences | | | | |
|---|---|---|---|---|---|
| | 1 | 2 | 3 | 4 | 5 |
| 5 | H | H (4y) | E | E | E |
| 4 | M (1v, 1x, 3x, 4v, 4x) | H | E | E | E |
| 3 | L (2v, 3v) | M (2y, 3y) | H (1z, 4z) | E | E |
| 2 | L (2x) | L (1y) | M (2z, 3z) | H | E |
| 1 | L | L | M | H | H |

Figure 2 Risk Matrix

There are four colors in the risk matrix table, namely red, blue, green, and yellow. Those colors mark the criteria of how risky a hazard found in an activity is. The red color shows the extreme hazard level, so we really must pay attention to the risk. The blue color shows a high hazard level, and there are three activities found in this position. Those are student activities causing nausea and staff activities outside the building causing dizziness and nausea. The yellow color shows a medium hazard level, and there are nine activities leading to risks. The green color shows a low hazard level, and there are four activities leading to risks. After doing the risk matrix on CO as a motor vehicle exhaust gas at Pancasila University with the health impact suffered by the academic community members, it is necessary to control or measure the hazards incurred.

## 4. Conclusion

1. The influence of CO as a motor vehicle exhaust gas having the value of R (correlation coefficient) as much as 0.9519 > 0.8054 (r count > r table) shows that there is a correlation between CO as a motor vehicle exhaust gas and its health impacts on the academic community members of Pancasila University. CO as a motor vehicle exhaust gas causes cough, dizziness, nausea, and shortness of breath.
2. CO as a motor vehicle exhaust gas has an impact on the academic community members of Pancasila University who inhale it so that it causes cough, dizziness, nausea, and shortness of breath. The HIRARC table shows two activities leading to a high risk level. First, the staff activities outside the building causes dizziness and nausea; second, the student activities cause nausea. The effort to prevent these diseases is by using administrative controls in the form of issuing a decision letter to reduce the air pollution at Pancasila University and by using personal protective equipment in the form of wearing a protective mask.

## 5. References

[1] P. Insyira, 2016 Evaluasi Kinerja Badan Lingkungan Hidup dalam Uji Emisi Gas Buangan Kendaraan Bermotor Di Kota Pekanbaru, *ejurnal Universitas Riau* **3**.

[2] I. W. Prasetya, Galih Budi, Meidiana, Christia, and Agustin, 2017 Pengurangan Emisi CO2, N2O, dan CH4 dari Kendaraan Bermotor Universitas Negeri Malang, *Jurnal Sosial Ekonomi Pekerjaan Umum* **9**(1).

[3] N. D. Santoso, E. Akmalah, I. Irawati, 2017 Implementasi Konsep *Green Campus* di Kampus Itenas Bandung Berdasarkan Kategori Tata Letak dan Infrastruktur, *Ejurnal Institut Teknologi Nasional* **3**(4).

[4] D. Gusnita, 2010 Berita Dirgantara Green Transport: Transportasi Ramah Lingkungan Dan Kontribusinya Dalam Mengurangi Polusi Udara 2010 *Berita Dirgantara* **11**(2)

[5] S. Supriyadi, A. Nalhadi, and A. Rizaal, 2015 Identifikasi Bahaya dan Penilaian Risiko K3 pada Tindakan Perawatan dan Perbaikan Menggunakan Metode HIRARC (Hazard Identification and Risk Assesment Risk Control) 2015. *Prosiding Seminar NasionalRiset Terapan SENASSET*, ISBN: **978-602-73672-0-3**.

[6] Anonim 2001 Keputusan Gubernur Provinsi DKI Jakarta Nomor **551**/2001, pp. **1–4**.

[7] T. Mayasari, Andhika; Mardyanika, Yus A Prasetya; Sundari, Studi Perencanaan Pengembangan Universitas Hasyim Asy'ari Sebagai Green Campus, 2016 *Jurnal Reaktom*, **01**(01).

[8] A. Budiyono, 2001 Dampak Pencemaran Udara Pada Lingkungan. *Jurnal Berita Dirgantara* **2**(1).

[9] M Y M Solihin, M Arafat 2018 Analisa Pengarung Perubahan Pacing pada bejana silinder terhadap kompresi mesin motor 100cc *simrestek* 2018, proseding e-ISSN : **2621-5934**, p-ISSN : **2621-7112**

[10] M Y M Solihin, 2018 Optimalisasi sistem perawatan dan perbaikan terencana berdasarkan analisis keandalan pada mesin Diesel pengerak generator C10, *Teknobiz jurnal ilmiah* program studi Magister Teknik Mesin.

# Implementation of backpropagation artificial neural network for early detection of vitamin and mineral deficiency

**Nina Sevani**[1,2], **Iwan Aang Soenandi**[1,3] **and Fajar Saputra** [1,4]

[1]Faculty of Engineering and Computer Science, Krida Wacana Christian University, Indonesia
[2]nina.sevani@ukrida.ac.id, [3]iwan.as@ukrida.ac.id,
[4]fajar.2016tin009@civitas.ukrida.ac.id

**Abstract**. Basic Health Research data in 2018 show that around 95.5 percent of Indonesian people less consumption of fruit and vegetables. This condition then leads to the suspicion of vitamin and mineral deficiency in Indonesia people. Several methods have been used to detect vitamin and mineral deficiency, such as convolutional rule-based and certainty factor method. However, these methods are less adaptive to adapt to the changes in symptoms when detecting vitamin and mineral deficiencies. This paper proposes an artificial neural network (ANN) using backpropagation (BPN) to detect the vitamin and mineral deficiencies in the human body. Using 107 input of physical symptoms and 17 output of the type of vitamin and mineral, the architecture of the ANN consist of 107-50-17 neurons for the input layer, hidden layer, and output layer respectively. Based on some trial and error experiments, can be determined the epoch, the learning rate, and the error rate to produce the optimal result of the detection. This experiment using 623 epochs, 0.0517 error rate, and 0.1 for the learning rate. The performance measurement conducted using precision, recall, and F-score, for each class output. The experiment shows the proposed ANN using BPN reaches an accuracy level of 73%.
**Keywords:** Artificial Neural Network; Backpropagation; Deficiency Vitamin and mineral; Health Informatics.

## 1. Introduction

Vitamins and minerals are organic nutrients that are needed in the human body in small amounts. Usually vitamins and minerals can be obtained from outside human bodies by eating vegetables and fruit. However, based on Basic Health Research data 2018 [1], around 95.5 percent of Indonesian people with more than 5 years of age eat fewer vegetables and fruit, so Indonesian people were susceptible to vitamin and mineral deficiency which at the end can cause various diseases.

Some methods have been used for detection deficiencies vitamin and mineral such as rule-based and Certainty Factor methods. Rule-based methods can be used for the detection of vitamin deficiency [2] and detection of mineral deficiencies [3]. The Certainty Factor method can also be used for the detection of vitamin and mineral deficiencies [4,5]. These studies are included in the AI branch called the expert system. However, the expert system is an AI branch that doesn't have the learning process so it is less adaptive to change of symptoms in detecting vitamin and mineral deficiencies [6]. Due to it less adaptive to changes in the symptoms, then the problem here is difficult to determine the most dominant type of deficiency suffered by the patients.

One of the adaptive methods that have the learning process is an artificial neural network [6]. Backpropagation is one of the learning methods that can be used on artificial neural network and has better accuracy compared to other learning methods [7,8]. Based on several studies that has been published, accuracy of backpropagation can reach above 90 percent [9,10,11,12,13,14,15,16,17]. The backpropagation learning process is done through some iterations to adjust the weight in all layers to produce the output with a minimum error value. The goal of this research is to detect vitamin and mineral deficiencies with backpropagation artificial neural network using the proposed architecture in this study.

## 2. Methods

Vitamin and mineral datasets will be used to detect vitamin and mineral deficiencies with backpropagation. The dataset is composed of 107 features and 17 classes. 107 features are 107 symptoms of vitamin and mineral deficiency and 17 classes are 17 types of vitamins and minerals that can be detected. Examples of the 107 symptoms as the input variables for the ANN and the 17 types of the deficiency as the output variables can be seen in table 1. The dataset was obtained based on previous studies [2,3,4,5].

**Table 1.** Examples of the input and output variables

| Examples of Input Variables | Output Variables |
|---|---|
| Double vision; Panting; Lack of concentration; Brittle nails; Hair loss; Decreased endurance; Lack of appetite; Muscle and joint pain; Rough and dull skin; Dental caries; Easily tired; Unstable emotions; Diarrhea; Often cramps in the legs; Pain like rheumatism | Fe; F; I; Zn; A; B1; B2; B3; B5; B6; B7; B9; B12; C; D; E; K |

The backpropagation has three phases in the learning process which are the feedforward phase, the backpropagation of error phase, and the weight adjustment phase and can be seen in figure 1 [18].

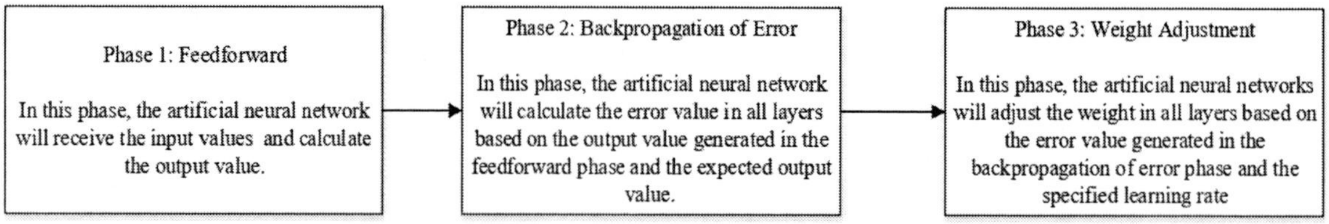


**Figure 1.** Phases in backpropagation

### 2.1. *Feedforward Phases*

First, each neuron on the input layer (*Xi*) receives an input value and passes it to all neurons on the hidden layer (*Yj*). Then, each neuron on the hidden layer (*Yj*) calculates the output value of the hidden layer using equation (1) and using activation function such as in equation (2).

$$Y\_in\,j = \left(\sum_{i=1}^{p} X_i * V_{ij}\right) + V_{0j} \quad (1)$$

$$Y_j = f(Y\_in\,j) \quad (2)$$

Then, each neuron on the output layer (*Zk*) calculates the output value of the output layer using equation (3) and using activation function such as in equation (4).

$$Z\_in\,k = \left(\sum_{j=1}^{q} Y_j * W_{jk}\right) + W_{0k} \quad (3)$$

$$Z_k = f(Z\_in\ k) \tag{4}$$

### *2.2. Backpropagation of Error Phases*

First, each neuron on the output layer (*Zk*) receives an expected target value (*Tk*) and calculates the error value in output layer using equation (5) and calculates the weight correction value for the weight that connects the output layer and the hidden layer using equation (6).

$$\delta_k = (T_k - Z_k) * f'(Z\_in\ k) \tag{5}$$
$$\Delta W_{jk} = \alpha * \delta_k * Y_j \tag{6}$$

Then, each neuron on the hidden layer calculates the error value in the hidden layer using equation (7) and equation (8). Then, calculates the weight correction value for the weight that connects the hidden layer and the input layer using equation (9).

$$\delta\_in\ j = \sum_{k=1}^{r} \delta_k * W_{jk} \tag{7}$$
$$\delta_j = \delta\_in\ j * f'(Y\_in\ j) \tag{8}$$
$$\Delta V_{ij} = \alpha * \delta_j * X_i \tag{9}$$

### *2.3. Weight Adjustment Phases*

Each weight that connects the input layer and the hidden layer adjusts their value using equation (10) and each weight that connects the hidden layer and the output layer adjusts their value using equation (11).

$$V_{ij}(new) = V_{ij}(old) + \Delta V_{ij} \tag{10}$$
$$W_{jk}(new) = W_{jk}(old) + \Delta W_{jk} \tag{11}$$

Then, after reading all data in the dataset, the learning process will stop if the MSE calculated using equation (12) is smaller than the expected error value or the number of epoch has been reached the epoch limit.

$$MSE = \sum_{k=1}^{n} \frac{(Tk - Yk)^2}{n} \tag{12}$$

## 3. Result and Discussion

The architecture of artificial neural networks used in detecting vitamin and mineral deficiencies is 107-50-17 which the architecture can be seen in Figure 2 and obtained the smallest MSE on the trial and error listed in Table 2, marked with red square. The parameters of artificial neural networks are also determined by the trial and error process. The result of the trial and error process for each parameter can be seen in Table 3 to Table 6. In Table 3 is the result for the learning rate, Table 4 for the error limit, Table 5 for the epoch limit, and Table 6 for the activation function. Based on the trial and error process, the parameter of artificial neural networks used in detecting vitamin and mineral deficiencies is 10% for learning rate as seen in red square in Table 3, 0.0517 as error limit and 623 as epoch limit as seen in the last row in Table 4 and Table 5. The activation function is the Hyperbolic Tangent has smallest MSE as seen in red square in Table 6.

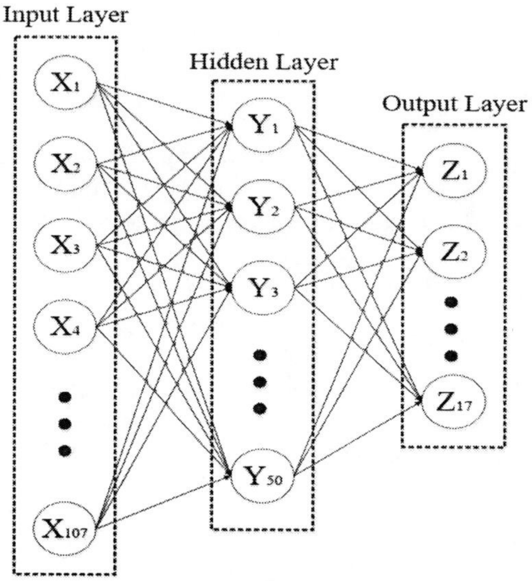


**Figure 2.** Architecture Artificial Neural Network

**Table 2.** Trial and error for architecture ANN

| | Architecture | Epoch | MSE |
|---|---|---|---|
| 1 | 107-10-17 | 1000 | 0.16004 |
| 2 | 107-20-17 | 601 | 0.060881 |
| 3 | 107-30-17 | 331 | 0.16206 |
| 4 | 107-40-17 | 529 | 0.062284 |
| 5 | 107-50-17 | 1000 | 0.046877 |
| 6 | 107-60-17 | | 0.052332 |
| 7 | 107-70-17 | 153 | 0.05469 |
| 8 | 107-80-17 | 1000 | 0.050939 |
| 9 | 107-90-17 | 317 | 0.064988 |
| 10 | 107-100-17 | 165 | 0.061005 |
| 11 | 107-10-10-17 | 1000 | 0.12501 |
| 12 | 107-20-20-17 | 1000 | 0.1157 |
| 13 | 107-30-30-17 | 1000 | 0.059958 |
| 14 | 107-40-40-17 | 1000 | 0.089868 |
| 15 | 107-50-50-17 | 1000 | 0.05816 |
| 16 | 107-60-60-17 | 1000 | 0.05934 |
| 17 | 107-70-70-17 | 1000 | 0.057084 |
| 18 | 107-80-80-17 | 1000 | 0.055597 |
| 19 | 107-90-90-17 | 1000 | 0.058377 |
| 20 | 107-100-100-17 | 1000 | 0.056508 |
| 21 | 107-10-10-10-17 | 1000 | 0.17711 |

**Table 2.** Trial and error for architecture ANN

| | Architecture | Epoch | MSE |
|---|---|---|---|
| 22 | 107-20-20-20-17 | 1000 | 0.15109 |
| 23 | 107-30-30-30-17 | 1000 | 0.054816 |
| 24 | 107-40-40-40-17 | 1000 | 0.054335 |
| 25 | 107-50-50-50-17 | 1000 | 0.05629 |
| 26 | 107-60-60-60-17 | 1000 | 0.060677 |
| 27 | 107-70-70-70-17 | 1000 | 0.058659 |
| 28 | 107-80-80-80-17 | 1000 | 0.054402 |
| 29 | 107-90-90-90-17 | 1000 | 0.053259 |
| 30 | 107-100-100-100-17 | 1000 | 0.05239 |

**Table 3.** Trial and error for learning rate

| | Learning Rate | Epoch | MSE |
|---|---|---|---|
| 1 | 1% | 1000 | 0.11893 |
| 2 | 2% | 1000 | 0.056012 |
| 3 | 3% | 1000 | 0.059579 |
| 4 | 4% | 1000 | 0.16595 |
| 5 | 5% | 1000 | 0.051815 |
| 6 | 6% | 1000 | 0.061316 |
| 7 | 7% | 1000 | 0.060127 |
| 8 | 8% | 1000 | 0.051365 |
| 9 | 9% | 1000 | 0.057172 |
| 10 | 10% | 1000 | 0.049706 |

**Table 4.** Trial and error for error limit

| | Epoch | MSE | | Epoch | MSE |
|---|---|---|---|---|---|
| 1 | 1000 | 0.058264 | 6 | 1000 | 0.052847 |
| 2 | 1000 | 0.059016 | 7 | 1000 | 0.044894 |
| 3 | 1000 | 0.04634 | 8 | 1000 | 0.050041 |
| 4 | 1000 | 0.051016 | 9 | 1000 | 0.054787 |
| 5 | 1000 | 0.051475 | 10 | 1000 | 0.049212 |
| | | | | Average | 0.0517 |

**Table 5.** Trial and error for epoch limit

| | MSE | Epoch | | MSE | Epoch |
|---|---|---|---|---|---|
| 1 | 0.0517 | 764 | 6 | 0.0517 | 502 |
| 2 | 0.0517 | 647 | 7 | 0.0517 | 687 |
| 3 | 0.0517 | 659 | 8 | 0.0517 | 563 |
| 4 | 0.0517 | 703 | 9 | 0.0517 | 627 |
| 5 | 0.0517 | 514 | 10 | 0.0517 | 568 |
| | | | | Average | 623 |

**Table 6.** Trial and error for activation function

| | Activation Function | Epoch | MSE |
|---|---|---|---|
| 1 | Hyperbolic Tangent | 284 | 0.064902 |
| 2 | Sigmoid Biner | 1000 | 0.1578 |
| 3 | Linear Transfer Function | 1000 | 0.204128 |

Then, to determine the performance of the ANN that has been produced, a confusion matrix is used to obtain the value of precision, recall, f-score and accuracy in detecting vitamin and mineral deficiencies. The values of precision, recall, scores for each class and accuracy values in detecting vitamin and mineral deficiencies can be seen in table 7 as follows. The accuracy is 73% obtained from comparing the output of the ANN with the data testing. In this research, the dataset divided into 88 data training that used to produce the output and 30 data testing to test the performance. The accuracy measurement will counted based on the match and mismatch classification between the output of the ANN and the annotation data in the data testing. Table 7 also describes the value of precision, recall, and f-score for each class output as the result of the comparison between output of the ANN and the data testing. For recommendation in the next research, it is interesting to compare the selected input variables that related to the accuracy of output using feature selection method.

**Table 7.** Performance ANN

| Class | Precision | Recall | F Score |
|---|---|---|---|
| Fe | 1.00 | 0.75 | 0.85 |
| F | 0.00 | 0.00 | 0.00 |
| I | 0.00 | 0.00 | 0.00 |
| Zn | 0.00 | 0.00 | 0.00 |
| A | 1.00 | 1.00 | 1.00 |
| B1 | 1.00 | 0.75 | 0.85 |
| B2 | 0.67 | 0.67 | 0.67 |
| B3 | 0.67 | 0.50 | 0.57 |
| B5 | 0.50 | 1.00 | 0.67 |
| B6 | 0.00 | 0.00 | 0.00 |
| B7 | 0.20 | 1.00 | 0.40 |
| B9 | 0.00 | 0.00 | 0.00 |
| B12 | 0.50 | 1.00 | 0.67 |
| C | 1.00 | 1.00 | 1.00 |
| D | 1.00 | 1.00 | 1.00 |
| E | 0.00 | 0.00 | 0.00 |
| K | 0.00 | 0.00 | 0.00 |
| | | **Accuracy** | 73% |

## 4. Conclusion

Based on the analysis and testing that has been done, it can be concluded that backpropagation in artificial neural networks with architecture 107 neurons in the input layer, 50 neurons in the hidden layer, and 17 neurons in the output layer with 10% learning rate, 0.0517 error limit, 622 epoch limit, and Hyperbolic Tangent as an activation function can be implemented in detecting vitamin and mineral deficiencies and produce 73% accuracy rate. For the next research, the implementation of feature selection is needed to increase the accuracy rate and add more training data in the learning process in hopes of improving the error rate.

## 5. References

[1] Badan Penelitian dan Pengembangan Kesehatan Kementerian Kesehatan 2018 *Hasil Utama Riskesdas*. Jakarta: Kementerian Kesehatan Republik Indonesia.

[2] Sevani N and Joshua M 2015 Implementasi forward chaining untuk diagnosa defisiensi vitamin larut dalam lemak berbasiskan web. *Jurnal Informatika*, **10**(2).

[3] Sevani N and Unwaru R 2014 Aplikasi deteksi dini defisiensi mineral mikro pada manusia berbasis web. *Indonesian Journal of Computing and Cybernetics Systems*, **8**(2), 213-222.

[4] Marzuqi T A 2017 *Perbandingan Hasil Deteksi Defisiensi Vitamin dan Mineral dengan Metode Sistem Pakar Fuzzy Mamdani dan Certainty Factor*. Universitas Kristen Krida Wacana. *Final Project*.

[5] Sevani N and Chandra Y J 2016 Web Based Application for Early Detection of Vitamin and Mineral Deficiency. *Communication and Information Technology Journal*, **10**(2), 53-58.

[6] Sutojo T, Mulyanto E and Suhartono V 2010 *Konsep Kecerdasan Buatan*. Penerbit Andi., Palembang.

[7] Adhitya R Y, Ramadhan M. A, Kautsar S, Rinanto N, Sarena S T, Munadhif I and Soeprijanto A 2016 Comparison methods of Fuzzy Logic Control and Feed Forward Neural Network in automatic operating temperature and humidity control system (Oyster Mushroom Farm House) using microcontroller. *International Symposium on Electronics and Smart Devices* (pp. 168-173). IEEE.

[8] Olaniyi E O, Oyedotun O K, Helwan A and Adnan K 2015 Neural network diagnosis of heart disease. *International Conference on Advances in Biomedical Engineering* (pp. 21-24). IEEE.

[9] Afroge S, Ahmed B and Mahmud F 2016 Optical character recognition using back propagation neural network. 2nd *International Conference on Electrical Computer and Telecommunication Engineering* (pp. 1-4). IEEE.

[10] Arulmurugan R. and Anandakumar H 2018 Early detection of lung cancer using wavelet feature descriptor and feed forward back propagation neural networks classifier. *Computational Vision and Bio Inspired Computing* (pp. 103-110). Springer, Cham.

[11] Feshki M G and Shijani O S 2016 Improving the heart disease diagnosis by evolutionary algorithm of PSO and Feed Forward Neural Network. *Artificial Intelligence and Robotics* (pp. 48-53). IEEE.

[12] Hodo E, Bellekens X, Hamilton A, Dubouilh P L, Iorkyase E, Tachtatzis C, and Atkinson R 2016 Threat analysis of IoT networks using artificial neural network intrusion detection system. *International Symposium on Networks Computers and Communications* (pp. 1-6). IEEE.

[13] Nahato K B, Harichandran K N, and Arputharaj K 2015 Knowledge mining from clinical datasets using rough sets and backpropagation neural network. *Computational and mathematical methods in medicine*.

[14] Sarkar A and Pandey P 2015 River water quality modeling using artificial neural network technique. *Aquatic Procedia*, **4**, 1070-1077.

[15] Setti S and Wanto A 2019 Analysis of Backpropagation Algorithm in Predicting the Most Number of Internet Users in the World. *Jurnal Online Informatika* **3**(2), 110-115.

[16] Shinde A, Kale S, Samant R, Naik A and Ghorpade S 2017 Heart Disease Prediction System using Multilayered Feed Forward Neural Network and Back Propagation Neural Network. *International Journal of Computer Applications*, **166**(7), 32-36.

[17] Zhang Y, Phillips P, Wang S, Ji G, Yang J and Wu J 2016 Fruit classification by biogeography-based optimization and feedforward neural network. *Expert Systems*, **33**(3), 239-253.

[18] Jatmiko W, Mursanto P, Hardian B, Bowolaksono A, Wiweko B, Akbar M A, Satwika I P, Immadudin Z, Alvissalim M S, Habibie I, Ma'sum M A, Kurniawan M N 2013 *Teknik Biomedis: Teori dan Aplikasi*. Depok: Fakultas Ilmu Komputer Universitas Indonesia.

IOP Conf. Series: Materials Science and Engineering **847** (2020) 012044 doi:10.1088/1757-899X/847/1/012044

# Optimizing Schedule In Furniture Planning

**Abdul Talib Bon**

Technology Management – Universiti Tun Hussein Onn Malaysia
talibon@gmail.com

**Abstract.** The methodology for developing the production system schedules from a project management perspective is being investigated. System orders are considered to be potentially parallel projects that share the same resources. The mechanics of manufacturing controls are discussed in an existing project management package. Order input to the network is envisaged via the web and, after generation or re-generation of schedules, the estimated fulfilment period and delivery date for each order should be accessible via the internet again. Both ideas were described on the basis of real data from a security door supplier operating a make-from-stock, process-oriented production program. The system provides, on the one hand, the approximate (calculated) completion time of each task and, on the other, the assignment of daily work to assets. Nonetheless, it would have trouble performing more than 3000 activities with concurrent asset scaling.
**Keywords**: Scheduling, Flow Shop, Levelling, Flexibility

## 1. Introduction

The main research hypothesis of this work is that the Project Management (PM) methods, based on critical path principles, with a focus on resource balancing, are an efficient and effective production scheduling method when the form of output is acceptable, i.e. the product does not vary much, the pattern of material flow does not differ too much, and lead times are also predictable. Japanese have discovered that PM software is well suited to production control — especially when performance and delivery dates are significant. For two companies, these systems as complements to MRP systems are provided. [1]

Onwubolu describes an integrated visual flow-shop scheduling system using scheduling techniques such as lower bound, automated generation of near-optimal process sequences and schedule optimization[2]. Outputs are in the form of a Gantt graph. Material requirements preparation can be achieved more efficiently in a laboratory setting using a resource-restricted project scheduling method. This model improves MRP models by adding power constraints and using adjustable lead time lengths[3]. A full overview of the issues and perspectives of flow shop scheduling. [4]

Martinez suggests a technique to integrate a performance perspective into production planning[5]. The definition of project-oriented development is used to demonstrate the hierarchical structure of research and decision-making. Based on a new form of resource workload model, the technique is used to estimate the speed-up of each project for the different levels of resources allocated to it. Leverage utilization allows an optimal level of resources to be identified for each task, which minimizes peak demands by varying aggregate operation intensities.

Thierry points out that make-to-order development is often structured into projects with their own goals[6]. These programs use strategic and essential tools that must be allocated at the level of strategic or tactical planning. Following the development of the initial medium-term plan, the scheduling process consists simply of reviewing the current project schedule when unexpected events occur.

Aurich is implementing the idea of development programs[7]. Under this definition, either a single production line or an entire production site can be planned and run as a company. The Master Schedule of the Production Project consists of uniform phases for the life-cycle-oriented development and function of the production system.

Baek suggests an immersive human-computer scheduling solution in which the system provides a computerized Gantt map comprising a variety of assistance features in static laboratory environments [8]. The optimization behavior of computer-aided human schedulers in scheduling tasks is also observed in the experiment.

PM-based scheduling is simple and intuitive compared to other scheduling strategies, but when the number of tasks / jobs to be scheduled is large software support is needed. The importance of scheduling in the operation of manufacturing systems is illustrated by the sheer number of references which have dealt with the subject in the last 40 years [9]. Apart from PM, the most common scheduling techniques studied are [10]: integer programming for small problems, heuristics-based on major problems and, more recently, meta-heuristics that direct the application of small problems such as simulated Annealing, Taboo Search and Genetic Algorithms.

Evergreen Fibreboard Berhad is engaged in the manufacture and sale of engineered wood products. It provides medium density fiberboard for use in a variety of applications, including furniture, speaker boxes, windows, moldings, flooring and PCB back-up panels, as well as for high-humidity applications in bathrooms, kitchens, flooring, doors and window frames. The company also manufactures particle board for use in veneer, high or low-pressure laminates, melamine and paper cover applications, as well as for use as a core material for the construction of homes and office furniture, shelving and kitchen cabinets and industrial and institutional fixtures. It also sells furniture items for living room, bedroom, work area and storage. In addition, the company manufactures and sells glue, resin, urea formaldehyde concentrates and adhesive products; trades and maintains plantations; and specializes in manufacturing, trading and selling wood products. It exports it services in Asia, Australia, North America, Europe and Africa. The business was founded in 1972 and is based in Batu Pahat, Malaysia.

## 2. Methods

**Figure 1**. Typical door structure without metal cover, wooden panels and some accessories.

The manufacturing that my group going to optimize the scheduling is a security door with its complementary casing process. The door consists of the following parts, see Figure 1: door frame, door covers, door panels, vertical reinforcements, horizontal reinforcements, windshield, lock plate, lock bars, lock bars, safety locks, tubes, hinges, aluminium profiled bars, rubber sealing pads, door handles, glass panels (if required), door-eye, small screws, large screws, plastic caps for hinges and tubes. The door housing consists of the following parts, see Figure 1: the left and the right beams, the horizontal beam, the casing bolts, the splays and the sills.

The type of manufacturing system used for the production of doors and housings is essentially a flow shop consisting of a stable series of operations for all product variants, modifications of the process plan relating, to a large part, to the duration of the operation. The test case used is a manufacturing system that produces safety doors. The original production system consists of five departments, namely metal construction, welding and assembly, painting, final assembly and casing production.

In the metal construction factory, the door frame, the steel board, the housing and the reinforcement are made. Wind blockers, safety boxes, lock plates, etc. are manufactured or purchased separately and stored until necessary at the assembly stage. Both the wooden panels and the metal structure are processed in the paint shop. The assembly shop deals with the welding of the door structure and the housing as well as the assembly of all the auxiliary parts, e.g. glass, handles, rubber insulators, etc. The door housing is assembled on the customer's premises.

The metal construction department consists of a CNC punching station, which mainly makes all the openings necessary on the metal sheet for locks, hinges, rams, etc., a shearing station for cutting to size all the sheets that need to be bent and welded to make the door frame, and a press brake, which forms the door frame elements, including the reinforcement struts.

The welding and assembly unit consists of three welding stations. In the first example, welding is used for initial mounting of reinforcements and hinge bases as well as for welding of door cover to door frame and glass frame to door cover as required. The second one-point welding and the Gas Metal Arc Welding (GMAW) are used for the seam welding of reinforcements and for the welding of lock bars on the lock as well as for the seam welding of the door cover on the door frame. In the sixth, GMAW is used to weld the bolt, the hinges and the wind blocker. There are also three intermediate storage stations for locks / reinforcements / wind blockers / hinges (first station), door frames shaped prior to welding (second station) and ready doors (third station). Painting department cuts, sands down and paints and varnishes wooden door panels in three phases as needed. Varnished panels are processed intermediately before being fed to the assembly unit.

Accessories as well as wooden beams and wooden panels were mounted in the assembly unit. The gate is taken to the finished goods shops. The development department of the casing consists of a calibre-bench on which the casing pieces are cut and positioned in their final position before welding on site. The casing is then sent to an external paint shop for electrostatic painting. To maintain the output of the casing. Regardless of the production of doors, a large inventory of them is retained, which led this analysis to focus on the production of doors alone. Preparation of accessories such as locks, hinges, glass frames for special doors is carried out in auxiliary workstations which are not considered to be necessary for production planning. Remember that the brand variants come in variations of two categories: first, single reinforcement, double reinforcement, glazed.

## 3. Result and Discussion

For each material model, a production flow diagram is drawn and the corresponding Program Evaluation Review Technique (PERT) analysis shows serial and parallel processing, as well as priority constraints, as required, and offers an understanding of critical pathways using the Critical Path Method (CPM) and related graphical tools [9] provided that all appropriate resources are available. This phase is directly comparable to infinite capacity planning.

IOP Conf. Series: Materials Science and Engineering **847** (2020) 012044 doi:10.1088/1757-899X/847/1/012044

Accurate processing times as well as auxiliary process times at each workstation should be required for the creation of PERT diagrams. This is best achieved by doing an evaluation of the stages of research and then conducting time measurements accordingly. According to PERT diagrams for each product version, the design model is initially built and then completed at a later stage by assigning the necessary resources. This is an' instanced' project model further improved later with order-specific information to become an Order Input File (OIF).

In order to construct a full schedule for all current and expected future orders, the models of the project to progress for all applicable orders are compiled into an aggregate plan. Constraints resulting from relative priorities between orders are entered at this point. In addition, existing restrictions, such as set release phases or completion time, may also be entered. The last step that accounts for a practical schedule is the so-called resource levelling, which allows use of the corresponding algorithm that inevitably occurs in computer scheduling systems. This step is directly comparable to that of finite capacity planning.

**Table 1**. Part of an Order Input File. Grey cells represent user input. White cells represent standard template values

| TYPE | ANALYSIS | ID | NAME | TIME | PREDECESSORS | OUTLINE_LEVEL |
|---|---|---|---|---|---|---|
| ARMOUR | DOUBLE | 1 | CNC | 2.3 | | 1 |
| CLOSING | 17 | 2 | CNC sk | 2.3 | | 2 |
| EXT.MATERIAL | ANIGRE | 3 | CNC kap | 2.3 | | 2 |
| EXT.DESIGN | NAUSIKA | 4 | CUT | 3.6 | | 1 |
| INT. MATERIAL | ANIGRE | 5 | Koph sk | 1.2 | 2 | 2 |
| INT. DESIGN | NAUSIKA | 6 | Koph kap | 1.3 | 3 | 2 |
| HANDLE | TWIN | 7 | Stranza | 2.2 | | 1 |
| DATE | 26/8/2018 | 8 | Stranza sk | 2.2 | 5 | 2 |
| REPRESENTATIVE CODE | 1 | 9 | Syg-1-a | 9.8 | | 1 |
| DOOR NUMBER | 19 | 10 | Syg base | 9.8 | 8 | 2 |
| DOOR HEIGHT | 2,10 | 11 | Syg-2 | 22.6 | | 1 |
| DOOR WIDTH | 1,00 | 12 | Syg medesa | 22.6 | 10 | 2 |

The delivery times for each order and for each workstation in the process can now be exported to the shop floor via the intranet, while the estimated fulfillment date of the order can be exported to the customer via the internet. The steps outlined above are further analysed, and descriptions of their implementation are given.

All data relating to the construction of PERT diagrams leading to the construction of project templates are collected for each of the product variants flowing through the system. A production flow diagram is constructed for each product variant. All-in-all nine flowcharts were created in the test case used. A corresponding PERT diagram was constructed on the basis of each flow diagram, the duration of each production or auxiliary stage shall be measured by means of a standard stopwatch-based continuous time measurement and a table shall be constructed containing all such processes and corresponding time durations.

| ID | Stage | Duration (min) | Starting conditions | | | Process | Transport | Control | Delay | Store |
|---|---|---|---|---|---|---|---|---|---|---|
| 1 | Frame punching | 2,29 | - | | | ○ | ▷ | □ | D | ▽ |
| 2 | Frame cutting | 1,21 | after | 1 | | ○ | ▷ | □ | D | ▽ |
| 3 | Frame bending | 1,24 | after | 2 | | ○ | ▷ | □ | D | ▽ |
| 4 | Transporting | 0,60 | after | 3 | | ○ | ▷ | □ | D | ▽ |
| 5 | Cover punching I | 4,46 | after | 1 | | ○ | ▷ | □ | D | ▽ |
| 6 | Cover cutting 1 (3 point) | 0,80 | after | 2 | 5 | ○ | ▷ | □ | D | ▽ |
| 7 | Cover bending I | 1,05 | after | 3 | 6 | ○ | ▷ | □ | D | ▽ |
| 8 | Hinge base welding | 5,78 | after | 4 | | ○ | ▷ | □ | D | ▽ |
| .. | ... | ... | ... | .. | .. | | | | | |

**Figure 2**. Excerpt from flow diagram for variant with single reinforcement without panel

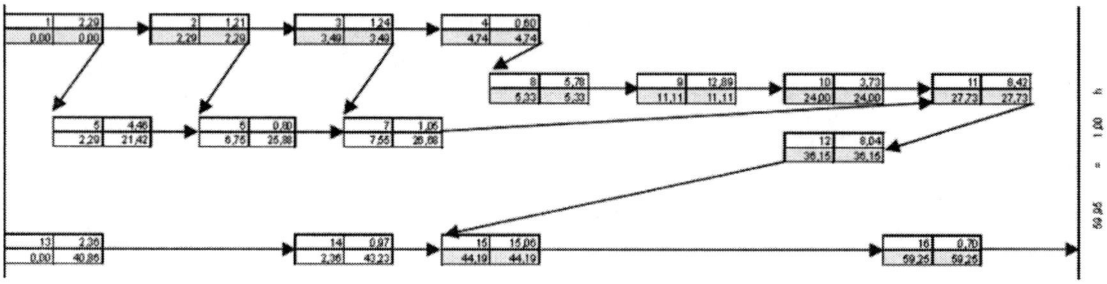

**Figure 3**. PERT diagram corresponding to variant presented in Figure 2.

PERT diagrams contribute to the identification of essential pathways for the processes of each product type, i.e. the time of completion of the product can be determined, assuming that all the assets necessary to carry out each process are available when appropriate, which is obviously a fictitious situation.

Note, too, that, in some cases, the constraints deemed to apply were not correct. For example, painting activities have been observed to occur in a particular order, i.e. panel followed by an event, although this is not appropriate. Therefore, the constraints of the project have been re-examined and mitigated as required. In contrast, certain duties, which were evidently unnecessary to keep separate from others, were adequately represented in them. Usually, this occurs when a part is transferred by a network after the same part has been processed. Such an embodiment simplifies data and saves on the need for computing power.

Each of the output flow diagrams is represented in a spreadsheet format and is complemented by all the information required for the full order processing to become an Order Input File (OIF). This additional information relates to the name of the order, the type of reinforcement, the number of keys, the type of door, the type of handle, the date, the customer code, the brand identification and the measurements (width X height) and the time of issue of the order.

IOP Conf. Series: Materials Science and Engineering **847** (2020) 012044 doi:10.1088/1757-899X/847/1/012044

| ID | Task Name | Duration | Predecessors | Resource Names | Start | Finish |
|---|---|---|---|---|---|---|
| **1** | **⊟ CNC** | **211,5 m** | | | **30/8/05 09:20** | **30/8/05 13:21** |
| 2 | CNC SKAFHS | 2,3 m | | CNC PUNCHING | 30/8/05 09:20 | 30/8/05 09:22 |
| 3 | CNC KAPAKIOY | 2,3 m | | CNC PUNCHING | 30/8/05 13:19 | 30/8/05 13:21 |
| **4** | **⊟ CUT** | **210,5 m** | | | **30/8/05 09:22** | **30/8/05 13:23** |
| 5 | KOPH SKAFHS | 1,2 m | 2 | ΚΟΠΗ | 30/8/05 09:22 | 30/8/05 09:23 |
| 6 | KOPH KAPAKIOY | 1,3 m | 3 | ΚΟΠΗ | 30/8/05 13:21 | 30/8/05 13:23 |
| **7** | **⊟ STRANTZA** | **2,2 m** | | | **30/8/05 09:23** | **30/8/05 09:26** |
| 8 | STRANZA SKAFHS | 2,2 m | 5 | ΣΤΡΑΝΤΖΑΡΙΣΜΑ | 30/8/05 09:23 | 30/8/05 09:26 |
| **9** | **⊟ SYG-1-A** | **9,8 m** | | | **30/8/05 14:14** | **30/8/05 14:24** |
| 10 | SYG BASE | 9,8 m | 8 | ΣΥΓΚ 1 (ΠΟΝΤΑ ΣΤΑΘ) | 30/8/05 14:14 | 30/8/05 14:24 |
| **11** | **⊟ SYG-2** | **22,6 m** | | | **1/9/05 11:28** | **1/9/05 12:21** |
| 12 | SYG MEDESE | 22,6 m | 10 | ΣΥΓΚ 2 (ARGON) | 1/9/05 11:28 | 1/9/05 12:21 |
| **13** | **⊟ SYG-3-A** | **3,7 m** | | | **1/9/05 12:21** | **1/9/05 12:25** |
| 14 | SYG KLEIDI | 3,7 m | 12 | ΣΥΓΚ 3 (ΠΟΝΤΑ ΧΕΙΡΟΣ) | 1/9/05 12:21 | 1/9/05 12:25 |
| **15** | **⊟ SYG-3-B** | **12,1 m** | | | **1/9/05 12:25** | **1/9/05 12:37** |
| 16 | SYG KAPAKI | 12,1 m | 14;6 | ΣΥΓΚ 3 (ΠΟΝΤΑ ΧΕΙΡΟΣ) | 1/9/05 12:25 | 1/9/05 12:37 |

**Figure 4**. Task information in PM system as imported from the OIF

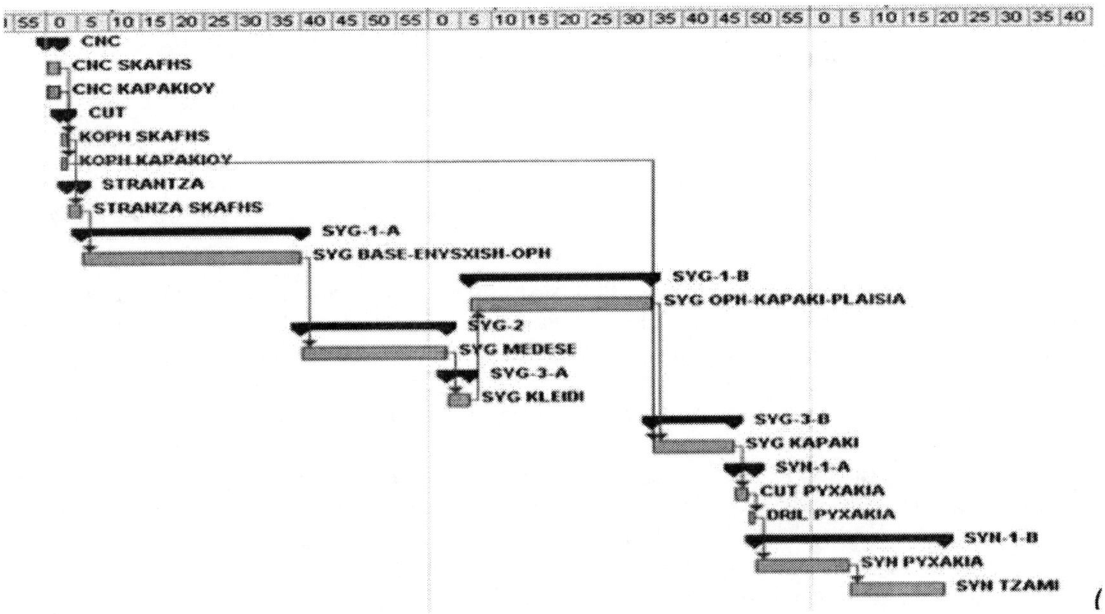


**Figure 5.** Gantt chart corresponding to OIF.

Many other macros have been written to encourage the creation and display of Gantt and CPM maps, to write results in a correct order in the resource job allocation output document, etc.
The PM system requires two types of production information: the approximate (calculated) completion period of each order and the allocation of daily work to assets. The first type is classified as a standard document in the PM system and can be used as a ready-to-use report, see Figure 7. The file representing the second type of information is compiled by exporting asset names, jobs and start-up

IOP Conf. Series: Materials Science and Engineering **847** (2020) 012044 doi:10.1088/1757-899X/847/1/012044

and completion times from the unsorted PM system to the spreadsheet in which form it is then sorted as appropriate, see Table 2.

**0100500000XM-.mpp** as of 12/9/05 08:00

| Dates | | | |
|---|---|---|---|
| Start: | 9/9/05 08:00 | Finish: | 9/9/05 09:01 |
| Baseline Start: | NA | Baseline Finish: | NA |
| Actual Start: | NA | Actual Finish: | NA |
| Start Variance: | 0 m | Finish Variance: | 0 m |

| Duration | | | |
|---|---|---|---|
| Scheduled: | 60,6 m | Remaining: | 60,6 m |
| Baseline: | 0 m | Actual: | 0 m |
| Variance: | 60,6 m | Percent Complete: | 0% |

| Task Status | | Resource Status | |
|---|---|---|---|
| Tasks not yet started: | 24 | Work Resources: | 0 |
| Tasks in progress: | 0 | Overallocated Work Resources: | 0 |
| Tasks completed: | 0 | Material Resources: | 0 |
| Total Tasks: | 24 | Total Resources: | 0 |

**Figure 6.** Sample report excerpt referring to a particular order

**Table 2.** Extract from resource work allocation report

| | ID | Project | Name | Initials | Scheduled work | Start date | Finish Date |
|---|---|---|---|---|---|---|---|
| 1361 | 20 | C:\gd\2900500046JD.mpp | PAINT 1 PERBAZ | BAF1 | 9 mins | 29/8/2005 08:00 | 29/8/2005 08:09 |
| 1337 | 22 | C:\gd\2900500041JT.mpp | PAINT 1 PERBAZ | BAF1 | 9 mins | 29/8/2005 09:48 | 29/8/2005 09:57 |
| 1241 | 23 | C:\gd\2900500001JM.mpp | PAINT 1 PERBAZ | BAF1 | 9 mins | 29/8/2005 09:57 | 29/8/2005 10:06 |
| 1289 | 23 | C:\gd\2900500026JM.mpp | PAINT 1 PERBAZ | BAF1 | 9 mins | 29/8/2005 10:06 | 29/8/2005 10:15 |
| ... | ... | ... | | | | | |

A complete 120 orders (approximately 3000 tasks) exercise was carried out on the real-world data system sponsored by the organization. The results show some idle periods, especially in sheet metal cutting and sheet metal bending stations, which typically achieve lower utilization compared to other stations where the utilization rate is 100 per cent.

The completion time report for each order is intended as a response to the customer, forecasting reasonable delivery dates. The resource loading document (daily) is intended to be used for both allocation and control purposes. Resource loading may be compressed (aggregated) or extended according to perceived needs, as all information provided by the PM system is at the lowest resolution (minutes) level, see example in Figure 8. Remember that although Gantt charts help to visualize the schedules, the documents also provide more accurate information.

The concept of creating a master plan consisting of individual projects corresponding to orders was appropriate as long as the number of projects was relatively small. The system did not seem to manage levelling correctly for a large number of projects, leaving lengthy idle time intervals between many tasks. The notion of master plans was therefore discarded. It has been verified that the individual sub-project files corresponding to the respective order leveling procedure have been well performed by keeping them open. This system also allowed for better control of orders to be scheduled.

| Resource | Work | Start | Finish | Details | 29 Aug '05 | | | | | |
|---|---|---|---|---|---|---|---|---|---|---|
| | | | | | M | T | W | T | F | S |
| CNC1 | 708 mins | 29/8/05 07:48 | 31/8/05 10:12 | Work | 283m | 294m | 132m | | | |
| CUT | 282 mins | 29/8/05 07:51 | 31/8/05 10:13 | Work | 118m | 113m | 51m | | | |
| STR | 338 mins | 29/8/05 07:52 | 31/8/05 10:10 | Work | 134m | 141m | 63m | | | |
| SYG1 | 1.649 mins | 29/8/05 08:06 | 2/9/05 12:46 | Work | 380m | 450m | 418m | 253m | 148m | |
| SYG2 | 2.065 mins | 29/8/05 08:15 | 2/9/05 13:14 | Work | 371m | 450m | 450m | 450m | 344m | |
| SYG3 | 1.631 mins | 29/8/05 08:38 | 2/9/05 13:30 | Work | 286m | 336m | 358m | 356m | 295m | |
| BAF1 | 2.338 mins | 29/8/05 08:00 | 6/9/05 10:48 | Work | 363m | 453m | 450m | 472m | 287m | 100m |
| ST2 | 73.020 mins | 31/8/05 07:35 | 7/9/05 15:55 | Work | | | 12.992m | 11.068m | 14.400m | 6.240m |
| TR1 | 5.994 mins | 30/8/05 07:00 | 7/9/05 07:29 | Work | | 900m | 900m | 900m | 900m | 892m |
| BAF2 | 2.116 mins | 31/8/05 07:25 | 7/9/05 07:55 | Work | | | 426m | 231m | 429m | 254m |
| ST1 | 213.120 mins | 29/8/05 08:09 | 6/9/05 18:48 | Work | 33.120m | 36.960m | 42.720m | 40.800m | 26.400m | 13.440m |
| KTR | 546 mins | 29/8/05 08:00 | 8/9/05 07:21 | Work | 87m | 40m | 45m | 26m | 90m | 74m |
| SYN1 | 2.357 mins | 29/8/05 07:53 | 3/9/05 09:41 | Work | 394m | 451m | 450m | 451m | 450m | 161m |
| SYN2A | 3.561 mins | 29/8/05 09:52 | 8/9/05 08:23 | Work | 411m | 170m | 277m | 169m | 845m | 510m |
| PAK | 1.190 mins | 29/8/05 08:00 | 8/9/05 08:50 | Work | 80m | 20m | | 85m | 240m | 225m |

**Figure 7.** Sample resource loading results

## 4. Conclusion

Based on the experience gained by developing a PM-based scheduling system for a real-world production environment, it has become clear that the sheer volume of data to be handled brings the system to its computational power limits, particularly as regards resource levelling. This is especially true when the product process plan requires repeated storage at the same workstation.

One approach is to divide tasks into larger chunks in order to reduce the number of tasks that the PM system handles simultaneously. This would certainly involve some loss of precision, but it could be tolerable if aggregation parameters were properly established. Order import into the system could be further automated by allowing the customer to fill in product data directly into forms that would be transparently transformed into database tables that would be automatically fed into the PM system.

## 5. References

[1] (Aurich & Barbian, 2004; Martínez, Duje, & Pérez, 1997; Reza Hejazi & Saghafian, 2005; Rom, Tukel, & Muscatello, 2002; Thierry, Lamothe, & Galvagnon, 2004)

[2] Aurich, J. C., and Barbian, P. 2004. Production projects - Designing and operating lifecycle-oriented and flexibility-optimized production systems as a project. *International Journal of Production Research*. https://doi.org/10.1080/00207540410001696348

[3] Martínez, E. C., Duje, D., and Pérez, G. A. 1997. On performance modeling of project-oriented production. *Computers and Industrial Engineering*, *32*(3), 509–527. https://doi.org/10.1016/S0360-8352(97)00005-3

[4] Reza Hejazi, S., and Saghafian, S. 2005. Flowshop-scheduling problems with makespan criterion: A review. *International Journal of Production Research*. https://doi.org/10.1080/0020754050056417

[5] Rom, W. O., Tukel, O. I., and Muscatello, J. R. 2002. MRP in a job shop environment using a resource constrained project scheduling model. *Omega*. https://doi.org/10.1016/S0305-0483(02)00033-6

[6] Thierry, C., Lamothe, J., and Galvagnon, V. 2004. Re-planning support system for make-to-order production with reserved resources. *International Journal of Production Research*. https://doi.org/10.1080/00207540412331282042

# Book trolley design for Telkom University Library using User Centred Design (UCD) method

**Mira Rahayu**[1,2], **Frans Arianto Silalahi**[1,3], **and Erna Febrianti**[1,4]

[1]Telkom University, Bandung Indonesia
[2]mirarahayu@telkomuniversity.ac.id, [3]fansariantono@student.telkomuniversity.ac.id, [4]efebriyati@telkomuniversity.ac.id

**Abstract.** Telkom's library book trolley is a tool to help visitors and librarians returning and shelving books. There are still many complaints from the visitors about the library environment which is untidy and some discrepancies between the information about book's availability on the website and on the bookshelf. In addition, there are complaints from librarians about the visitor's habit which is not follow the guidance in the process of returning books and existing trolley books are small and difficult to control. The purpose of this study is to develop the design of Telkom library book trolley in accordance with the wishes of users using User Centred Design method as a concept to determine the design of Telkom's library book trolley parameters. Survey was conducted to identify user criteria and statistical analysis was used to test the developed hypotheses. The results of this study are the design of Telkom library book trolley parameters that can represent the wishes of users and present Effective, Safe, Health, Comfortable, and Efficient designs. Based on the validation test using marginal homogeneity test, the design is valid in fulfilling the design criteria that the user want with the significance of 5%
**Keywords**: User Centred Design, Effective, Safe, Healthy, Comfortable, Efficient

## 1. Introduction

Telkom University has a library called Open Library. Located in Manterawu Building on the 5th floor has more than 80,500 collection titles with a number of copies of more than 123,937 and has been accredited "A" from the National Library of Indonesia in 2015 (Library, 2011). Telkom University Library has an area of 3200 m 2 and students and outside parties can take advantage of several services provided including book return areas, luggage storage lockers, discussion rooms, mini theatre areas, café areas, multimedia areas, and free areas. Telkom Open Library provides rules that visitors who do reading activities are not permitted to return the book to the shelf but to the place provided (Library, 2018). This was done to maintain the orderliness of the location of the book. But in practice the book return activity still has several obstacles, many visitors do not know the position of returning the book provided so that visitors simply put the reading book on the table or elsewhere, visitors who do not know the rules will likely return the book to an inappropriate rack location, visitors who know the location of the book return feel lazy to go to the location because of the long distance of return, and lack of socialization to return the reading book to the place provided and the regulations in the Telkom Open Library. The Librarian's problem is that librarians are less efficient in the process of returning books because librarians need additional time to stack books that are scattered and bring them to where the time needed for the process requires more time and effort and can occur. Musculoskeletal disorders

(MSDs). Librarians put all the books they want to return at an open table so that visitors take back the tidied book even though there is an appeal not to be taken back during the return process. Librarians use ordinary book trolleys that can only accommodate books relatively little in a matter of amount, so they require a long operating time to return, and librarians will be troubled to rearrange books that are not in accordance with their location because visitors put back the books that are read into the shelves that do not match the location.

Figure 1. Shelving Table

Referring to the background of the existing problems, this study focuses on the study of the design of book search tools that are in accordance with user requests to provide comfort and convenience in the process of returning books. Therefore, the method used to obtain information and analyse all user needs is considered better using a multidisciplinary approach in which engineering competition is combined with quality tools that are centred on the desires of users or specifically for librarians and visitors of the Open Library.

## 2. Methods

### *2.1. User Centred Design Definition*

User Centred Design (UCD) is an approach in designing that places the user as the centre in the design process (Vredenburg, 2002). UCD as one of the design methods is a philosophy based on the needs and interests of users, there is a large influence by users who actively determine the desired design and process that is done repeatedly so that the product desired by users can be fulfilled and understood by all users (Norman, 2013). UCD is a subset or part of Human centred Design (HCD), but it leads to interactions that will occur to users when using products or services that have been designed. UCD and HCD use the same methods and stages but are slightly different in the final concept. If UCD focuses on reciprocal users of a product, HCD focuses on how the product interacts with the overall senses of the user starting from the impact on psychology and other things. The most important part of the UCD method is divided into two. UCD always involves users directly during the design development process and the process is carried out in a repetitive manner in the hope that the products needed by the user can be fulfilled.

### *2.2. Stage in User Centred Design*

a) Understand and specify the context of use at this stage, the process of identifying subjects will use the system or product. This will explain the purpose in certain environmental conditions when consumers will use the system or product. Analysis of existing systems or products such as providing information about problems, including deficiencies and levels of performance and basic satisfaction even though obstacles may be ignored but must be met by the system or product in the future. In this study data collection will be carried out by interviews, observations as well as questionnaires on consumer needs to librarians and visitors to find out the constraints, inputs, and additional information to improve the quality of products to be made.

b) Specify the user requirements, this stage is the process of identifying user needs and establishing functional and other requirements for product or system design. These requirements are intended to be in the context of use for changing new work styles or opportunities to combine products with existing services. In this study data collection will be carried out in the form of attribute validation to be given to respondents.

c) Produce design solutions, this stage is done to bring together the needs of users with design designs that have been developed. In this study, the Axiomatic Design approach is used which starts with mapping of customer attributes to design parameters. This stage aims to validate a design according to the needs and desires of the user. Users will be involved in assessing the design of the design that has been made so that the needs and desires of the user can be clearly described.

d) Evaluate the design, in this stage, the design that has been shown to the user for the next assessment process will be evaluated. Evaluation has the purpose of looking for deficiencies of designs that have been made to suit the needs of users.

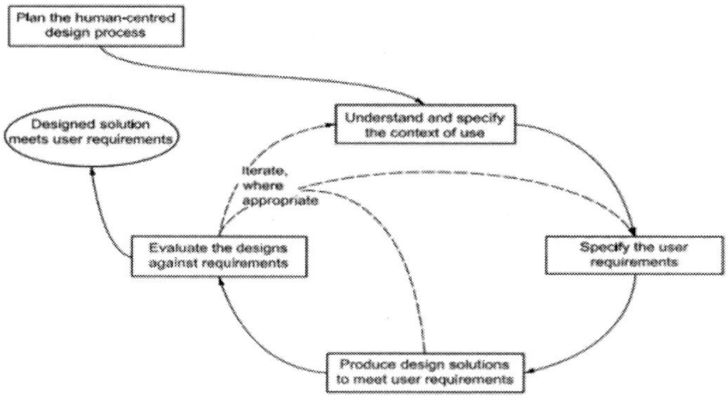


**Figure 2**. Designing stages of User Centred Design based on ISO 9241-210: 2010 (Standardization, 2010)

The advantage of using the User Centred Design method is it lies in full user involvement starting from the design process to the development process. The general product design focuses on the system level design to complete alternative designs, while using the User Centred Design method will continue to implement a simultaneous system between user feedback and alternative concept design processes.

## 3. Result & Discussion

### *3.1 Shelving Process*

The shelving process in the Telkom Open Library is done twice / day at 8:00 and 15:30. The shelving process is carried out by librarians and other staff and lasts 45 minutes or 1 hour 30 minutes. The shelving process starts collecting all the books in the administrative services and reading books used in the library to the shelving table, then the librarians and staff do book solving according to the book classification, then return the book to the shelf according to the classification, and finally check the books which is on the shelf according to the classification or not. If it turns out that there is a book that is not in place according to its classification, the Librarian and staff will return the book to its proper place.

### *3.2 Recapitulation of Number of Books Returned*

The number of books returned to the shelves is 6 days every hour the shelving activities are carried out. This is done to get the average number of books returned to determine the capacity of products that can be accommodated.

**Table 1** Recapitulation of Number of Books Returned

| Day/Date | Time | | Total |
|---|---|---|---|
| | **08.00** | **15.30** | |
| Monday | 476 | 266 | 742 |
| Tuesday | 396 | 306 | 702 |
| Wednesday | 385 | 288 | 673 |
| Thursday | 462 | 372 | 834 |
| Friday | 322 | 286 | 608 |
| Saturday | 258 | - | 258 |
| **Total** | 2299 | 1518 | 3817 |
| **Average** | 383,16 | 253 | 636,16 |

### *3.3 Respondent Profile*

Based on the proportion, the number of samples in this study is unknown, then testing the sample in this study has a minimum limit of 68 respondents. But in this study we used 120 respondents to represent the population of users of the means to return books in the Telkom Open Library and 5 Librarians.

### *3.4 Needs Statement Identification*

Ergonomics requirements can be reduced according to ergonomic aspects, namely EASNE (Effective, Safe, Healthy, Comfortable and Efficient). The collection of data regarding the identification of user needs is done using an open questionnaire and attributes derived from the representation of consumer statements.

**Table 2** Needs Statement List

| Requirements Ergonomics | Description of Needs Statement |
|---|---|
| Effective | The shelving process time will be shorter |
| Safe | Strong product construction |
| | Made of strong material |
| Healthy | Minimize the energy released by the user when using the product |
| Comfortable | Product design that seems neat |
| | There is a sign to inform the procedure and rules for returning the book |
| Efficient | Book return mileage |
| | Number of interactions with the product |
| | Number of products for reading location |
| | Travel time needed for product transfer |
| | The quantity of books that can be accommodated |

### *3.5 Customer Attribute (CA)*

**Table 3** Translation of Customer Attribute

| No | Customer Attribute | Code |
|---|---|---|
| 1 | Effective | CA1 |
| 2 | Safe | CA2 |
| 3 | Healthy | CA3 |
| 4 | Comfortable | CA4 |
| 5 | Efficient | CA5 |

### *3.6 Mapping Functional Requirement (FR) to Design Parameter (DP)*

The produce design solutions process in the stages of the User centred Design method is to obtain attributes from the book trolley design and shelving activities by librarians and staff and the visitor's needs can represent the user's overall needs, then the mapping process is carried out using the Axiomatic

IOP Conf. Series: Materials Science and Engineering **847** (2020) 012045 doi:10.1088/1757-899X/847/1/012045

Design approach. This stage is carried out to provide a deeper explanation of the mapping process. When the Customer Attribute (CA) has been identified, the next step is mapping the CA into a function (FR) which will then be answered by the Parameter Design (DP) as a solution.

**Table 4** Customer Attributes "Effective"

| Code | Code (FR) | Functional Requirement | Code (DP) | Design Parameter |
|---|---|---|---|---|
| CA1 | 1 | Able to achieve predetermined goals | 1 | Effective design |
| | 1.1 | Can be used optimally | 1.1 | Reduced time shelving process |
| | 1.1.1 | Can accommodate extreme size books | 1.1.1 | The dimensions between book slots are adjusted |
| | 1.1.1.1 | The biggest book specifications | 1.1.1.1 | Book height > 32 cm |
| | 1.1.1.2 | Rack numbers that have extreme size books | 1.1.1.2 | Bookshelf number 01,02,13,23,24 |
| | 1.1.1.3 | Distance between book slots | 1.1.1.3 | 35 cm |
| | 1.1.2 | Can divide books by book category | 1.1.2 | It has the coloring of each book category |
| | 1.1.2.1 | Number of colors in the product category | 1.1.2.1 | 24 colours |
| | 1.2 | User interaction with the product | 1.2 | Number of user interactions with the product |
| | 1.2.1 | Number of selected concept interactions | 1.2.1 | 3-5 times |

**Table 5** Customer Attributes "Safe"

| Code | Code (FR) | Functional Requirement | Code (DP) | Design Parameter |
|---|---|---|---|---|
| CA2 | 2 | Design that does not endanger the user | 2 | Strong design |
| | 2.1 | Safe construction | 2.1 | Strong construction |
| | 2.1.1 | Provides a strong connection | 2.1.1 | The connection adjusts the diameter of the construction |
| | 2.1.1.1 | Connection in book slot and buffer | 2.1.1.1 | Slot connection |
| | 2.1.1.2 | Pedestal connection with wheels | 2.1.1.2 | Bolt connection |
| | 2.1.1.3 | Base connection with frame | 2.1.1.3 | Weld connection |
| | 2.1.1.4 | Skeleton connection with support | 2.1.1.4 | Bolt connection |
| | 2.2 | Provide safe material | 2.2 | Has material that is not harmful to the body |
| | 2.2.1 | Strong component material | 2.2.1 | Type of component material |
| | 2.2.1.1 | Material type skeleton | 2.2.1.1 | Stainless Steel |
| | 2.2.1.2 | Type of book slot frame material | 2.2.1.2 | Stainless Steel |
| | 2.2.1.3 | The type of tool material is neat | 2.2.1.3 | Stainless Steel |
| | 2.2.1.4 | Type of buffer material | 2.2.1.4 | Stainless Steel |
| | 2.2.1.5 | Material type of book holding aid | 2.2.1.5 | Canvas |
| | 2.3 | Expenses that can be borne | 2.3 | Product strength |
| | 2.3.1 | Material strength in holding heavy | 2.3.1 | 200 kg |

**Table 6** Customer Attributes "Healthy"

| Code | Code (FR) | Functional Requirement | Code (DP) | Design Parameter |
|---|---|---|---|---|
| CA3 | 3 | Provides minimum workload for users | 3 | Workload and body position |
| | 3.1 | Frequency of transfer of work location | 3.1 | The number of frequencies the operator makes a move |
| | 3.1.1 | Amount of displacement frequency | 3.1.1 | 1x transfer / schedule |
| | 3.2 | Dimensions in accordance with Indonesian anthropometry | 3.2 | Dimensions according to percentiles |
| | 3.2.1 | Height dimension | 3.2.1 | Percentile height |
| | 3.2.1.1 | P50 | 3.2.1.1 | 172 cm |
| | 3.2.2 | The dimensions of the length of the palm | 3.2.2 | Percentile length of palm |
| | 3.2.2.1 | P50 | 3.2.2.1 | 19 cm |
| | 3.2.3 | Elbow height dimensions when standing | 3.2.3 | Percentile elbow height when standing |
| | 3.2.3.1 | P50 | 3.2.3.1 | 107 cm |

**Table 7** Customer Attributes "Efficient"

| Code | Code (FR) | Functional Requirement | Code (DP) | Design Parameter |
|---|---|---|---|---|
| CA5 | 5 | Able to reach goals with little effort | 5 | Easy and fast process |
| | 5.1 | Provide information as visible instructions | 5.1 | Information that can be read clearly |
| | 5.1.1 | Instructions for returning reading books | 5.1.1 | Directions that will be placed on each trolley |
| | 5.1.1.1 | Type of instructions | 5.1.1.1 | Banner |
| | 5.1.1.2 | Banner size | 5.1.1.2 | 85x200 cm |
| | 5.2 | Provides ease of operation | 5.2 | Quantity and capacity |
| | 5.2.1 | The number of trolleys needed | 5.2.1 | 4 pieces |
| | 5.2.2 | Product distance to reading location | 5.2.2 | 3-5 meter |

**Table 8** Customer Attributes "Comfortable"

| Code | Code (FR) | Functional Requirement | Code (DP) | Design Parameter |
|---|---|---|---|---|
| CA4 | 4 | Provides a comfortable impression | 4 | Comfort of use |
| | 4.1 | Coloration in product components | 4.1 | Type of component color |
| | 4.1.1 | Color frame type | 4.1.1 | Silver |
| | 4.1.2 | The color frame of the book slot frame | 4.1.2 | Red, dark red, blue, dark blue, light blue, yellow, orange, green, dark green, light green, white, black, brown, pink, gray, cream, purple, orange red, yellow orange, greenish yellow , turquoise, kevioletic blue, red-black, bluish brown |
| | 4.1.3 | The type of color of the tool is neat | 4.1.3 | Adjusted to the color side of the categorization |
| | 4.2 | Has a tool to tidy up books on a shelf | 4.2 | The tool specifications are neat books |
| | 4.2.1 | The dimensions of the tool tidy up the book on the shelf | 4.2.1 | 27x1x21 cm |
| | 4.3 | Able to provide good control | 4.3 | Wheel and handle design |
| | 4.3.1 | Comfortable wheel design | 4.3.1 | Wheel specifications |
| | 4.3.1.1 | Wheel type | 4.3.1.1 | Caster wheel 80 |
| | 4.3.1.2 | Wheel size | 4.3.1.2 | 4 inch |
| | 4.3.2 | Comfortable handle design | 4.3.2 | Handle shape and specifications |
| | 4.3.2.1 | Shape handle | 4.3.2.1 | Overall connected handle |
| | 4.3.2.2 | Handle size | 4.3.2.2 | Diameter 3 cm |
| | 4.4 | Has a book holding aid | 4.4 | Book anchoring tool specifications |
| | 4.4.1 | Dimensions of book holding aids | 4.4.1 | 150x1x3 cm |

### *3.7 Virtual Design and Alternative choice*

To get a library book trolley that suitable with the wishes of the user, an alternative choice is given to the library book trolley user. In this study 2 alternative choices of virtual design were given to respondents. After selecting from the two alternatives given to users of library book trolleys, the results obtained and selected were designs that truly desired the users and the chosen designs correctly represented the desires of users. In this study the design chosen was design 1 with a very agreeable level of 58% greater with design 2 which was only 34%. Figure 3 is the design of the selected proposal.

### *3.8 Proposed Design Validation*

Design validation is done by comparing the design that the user wants in this case is Effective, Safe, Healthy, Comfortable, and efficient with the chosen design measured at a significance level of 5% (0.05). The effective attribute with product description can be used optimally z values of 0.66, on safe attributes with safe product description when using z values of 0.112, on healthy attributes with product description can reduce workload of workers and reduce the risk of accidental z values of work 0.159, the convenient attribute with product description can provide a comfortable reading location environment of z values of 0.561 and z values of interesting product information of 0.655, in the efficient attribute the product description can inform the return location of the z values reading 0.482, the product information can inform book placement in accordance with book z values classification of 0.261, in the product description can inform the details of the process of returning the reading book z values of 0.506, in the product description it is easy to reach z values of 0.453, on product description easy to use z values of 0.383, on product description has a large amount of z values of 0.452, in the product mobility statement high z values are 0.141, and in the product description it can contain many books having z values of 0.401. Based on the overall z values that have been obtained from each attribute, it shows that the value of $Z > 0.05$ means that there has been a match between the designs made with the design desired by users of the Telkom University library book trolley.

### *3.9 Evaluation*

The evaluation in this study is the final process of the design that has been chosen. Selected designs are evaluated to get improvements in selected designs so that selected designs are expected to be better. There are several evaluations obtained from librarians and respondents, among others, giving evaluations of product construction, color adjustments to products, giving numerical information for categorization in addition to coloring, product placement in strategic areas, material evaluation, and evaluation of RULA & OWAS. Figure 4 are the results of the proposed design evaluation

Figure 3. Design proposal for Telkom University library book trolley

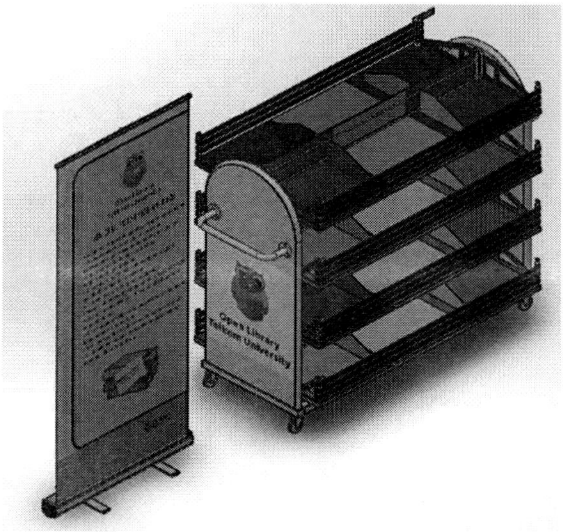

Figure 4. Result in Selected Design

## 4. Conclusion

The use of sections to divide the text of the paper is optional and left as a decision for the author. Where the author wishes to divide the paper into sections the formatting shown in table 2 should be used. The results of the identification of needs statement has 5 attributes (Effective, Safe, Healthy, Comfortable and Efficient) that needed to be developed towards the design of the Telkom University library book trolley. The design of parameters generated through the mapping process is a trolley design has a size of 35 cm (DP1.1.1.3) distance between slots because on the bookshelves in the Telkom University library there is a book that has a height that fits 32 cm so that 1 cm is added so that the user will more easily insert the book into the cart. Book trolleys are categorized by color and number of books so that in the trolley design there are 24 colors (DP1.1.2.1) including red, dark red, blue, dark blue, light blue, yellow, orange, green, dark green, light green, white, black, brown, pink, gray, beige, purple, orange red, orange yellow, greenish yellow, greenish blue, blue, red and blue brown (DP4.1.2). In the case of a book trolley construction made by paying attention to the strength of the product with the connection in

IOP Conf. Series: Materials Science and Engineering **847** (2020) 012045 doi:10.1088/1757-899X/847/1/012045

the book slot and the support using the slot connection (DP2.1.1.1), on the base connection with the wheel using a bolt connection (DP2.1.1.2), on the base connection with a frame using welded joints (DP2.1.1.3), on skeleton joints with supports using bolt joints (DP2.1.1.4). In addition to the construction of products also made by paying attention to the product material with the selected material on the frame is stainless steel material (DP2.2.1.1), on the frame of the selected material book slot is stainless steel (DP2.2.1.2), in the neat tool selected material is stainless steel (DP2.2.1.3), and the supporting tool for the book material used is canvas (DP2.2.1.4). With the material and construction the book trolley can withstand a load of 200 kg (DP2.3.1). Product size adjustments are used to make products that are convenient to use by operators, products are made based on Indonesian anthropometric references with the reference used is eye height when standing using the P50 percentile with a size limit of 172 cm (DP3.2.1.1) and palm length using percentiles P50 size limit of 19 cm (DP3.2.2.1). In the product, there are several features including a book cleaning tool with a dimension of 27x1x21 cm (DP4.2.1) and a book holding aid with a size of 150x1x3 cm (DP4.4.1). In addition, the wheel selection is obtained by the wheel type Caster wheel 80 (DP4.3.1. 1) with a 4 inch wheel size (DP4.3.1.2), an overall handle design (DP4.3.2.1) with a diameter of 3 cm (DP4.3.2.2). Coloration in addition to the book slot frame but also on a colored or fixed frame (DP4.1.1) and it is expected that the book trolley design can reduce the number of operator transfers to only 1x displacement / schedule (DP3.1.1) and user interaction is only 3x- 5x interaction (DP1.2.1). The selected trolley book will be placed at several strategic points with a number of trolleys of 4 (DP5.2.1). To provide information about the procedure for returning books, the location of returning books, etc. regarding the return of books, information media in the form of banners will be placed next to each product (DP5.1.1.1) with a banner size of 85x200 cm (DP5.1.1.2) and the distance of the trolley and tire to the reading location only ranges from 3-5 meters (DP5.2.2). Based on the validation test using the Marginal Homogeneity Test, the design is valid in fulfilling the design criteria that the user want with the significance of 5%.

## 5. Reference


[1] Cochran D.S, E. W., Kubin G., Sesterhenn L.M. 2000. The Application of Axiomatic Design and Lean Management Principles in the Scope of Production System Segmentation. *The International Journal of Production Research*, **38**.

[2] Library, T. O. 2011. *About Us*. Retrieved from https://openlibrary.telkomuniversity.ac.id/

[3] Library, T. O. 2018. *Peraturan Tel-U Open Library*. Retrieved from https://openlibrary.telkomuniversity.ac.id/

[4] Norman, D. A. 2013. *The design of everyday things*, Basic Books, 250 West 57th Street, 15th Floor, New York.

[5] Park, G. J. 2007. *Analytic Methods for Design Practice*. (London: Springer).

[6] Standardization, I. O. f. 2010. Ergonomics of human-system interaction-. In Part 210: *Human-centred design for interactive systems* (Vol. ISO 9241-210:2010(E)). (Switzerland: IHS)

[7] Vredenburg, K., Isensee, S., Righi, C. 2002. *User-centered Design: An Integrated Approach*. Prentice Hall PTR.

# Designing A Reading Chair using Kansei Engineering Approach

**Mira Rahayu[1,2], Hilman Ardian Ekananda[1,3] and Ilma Mufidah[1,4]**

[1]Telkom University, Bandung Indonesia
[2]mirarahayu@telkomuniversity.ac.id, [3]hilmanardian@student.telkomuniversity.ac.id, [4]ilmamufidah@telkomuniversity.ac.id

**Abstract.** Supporting facilities such as reading chair are often used as a scene to read books. However, the existing reading chair has not satisfied the user. This Research uses Kansei Engineering which purpose to design a reading chair that fulfil user needs. Kansei Engineering was chosen in this research because it can translate customer's impression, feeling, and demands on existing products or concepts to design concrete solutions and parameters into product design. This research is done on students in the Bandung area who use reading chair. Using a questionnaire as a tool to collect data that distributed online with google form to 347 respondents. Questionnaire data was processed using KMO statistical test and Barlett test, so from 23 Kansei word that had been obtained there was a reduction to 15 Kansei word would be used into designing reading chair in this research. The results of this research states, the implementation of Kansei Engineering could be done in the design of reading chairs, and there are innovations to meet user needs such as, armrest, headrest and footrest. Also USB port, lights, and book storage area. All of that can be used by users.
**Keywords**: Kansei Engineering, Kansei word, reading chair

## 1. Introduction

Nation is a group of people who share the same identity, language, religion, ideology, culture and history. In Indonesia the culture of fondness for reading is still very low, even though reading will determine the quality of the nation. By reading humans can know everything they don't know. Reading is a skill that can be possessed by every human being, no special skills are needed to be able to read. According to Periyeti (2017) reading can determine the quality of a nation or society, especially for the community with the status of students because with low reading interest will have an impact on the nation's underdevelopment. According to Yassin (2019) one of the factors that influence a person's lack of interest in reading is inadequate means. Inadequate reading facilities will have a negative effect on someone's interest in reading. Whereas adequate reading facilities will encourage someone to read. Supporting facilities such as reading chairs are often used as a place to read books. However, the existing reading chair has not satisfied the user and has not attracted interest in reading. The reading chair is a chair that has a backrest that allows people to sit facing a shelf and can promote relaxation, security, and can be used for hours when reading novels, newspapers and others (*GEAR PATROL*, 2019).

From the results of interviews and questionnaires that have been conducted from a total of 40 respondents reading chair users, the factors that influence reading interest when using a reading chair are features (95%), reading chair design (92.5%), sitting position when using (90%) , sitting position with reading time (87.5%), reading chair size (50%), and reading chair colour (45%). In the current reading chair there are also complaints that are often experienced by users when using a reading chair. From these results it was found that the complaints experienced by users when using the reading chair in general were aches in the back (67.5%), neck (60%), buttocks (40%), waist (27.5%), shoulders (25%), feet (12.5%), thighs (7.5%), and no complaints in any part (7.5%). This research uses Kansei Engineering to produce a product concept in the form of a reading chair that is comfortable and in accordance with user needs by involving feelings and emotions that exist in consumers.

## 2. Methods

Based on the above explanation to improve facilities in terms of reading, the Kansei Engineering approach is used. Kansei Engineering was chosen because it translates customers' impressions, feelings, and demands on existing products or concepts to design concrete design solutions and parameters into product design (Schütte, 2002). According to Schütte (2002) Kansei is the impression one gets from an artefact, environment, or certain situation using all the senses of sight, smell, hearing, feeling, feeling, and recognition of the user. While Kansei Engineering is a product development methodology that translates the impressions, feelings, and demands of customers towards existing products or concepts to design concrete design solutions and parameters into product design.

At Kansei Engineering there are methods that are very structured and can be modified based on the objectives to be achieved. The following is the structure of Kansei Engineering.

- Choice of participant groups, namely choosing group members who have good knowledge about market segmentation.
- Collection of Kansei words namely kansei words that describe the product.
- Data Reduction Methods for Selection of Kansei Words is to reduce data to find the right number of Kansei words.
- Types used scales are different rating scales in the Kansei Engineering context.
- Computerized data collection, which is statistical Kansei Engineering data, requires a large number of different opinions to obtain statistically reliable results.

Connecting the Kansai words to product properties, namely Kansei word, which provides the properties of the product.

## 3. Result & Discussion

The population that will be used in this study are students who have used a reading chair when reading. And Bandung area as the object of sampling. In this study the number of samples needed using a population-based sample size determination table developed by Isaac and Michael. So the number of respondents needed is 347 respondents with an error rate of 5%. In this study using a questionnaire as a data collection tool consisting of 23 questions that have been adjusted to the number of words obtained. Primary data collection is done by distributing questionnaires online to users using google form with a link that is divided into two parts, namely http://bit.ly/KionerHilmanArdian and http://bit.ly/KionerHilmanArdian2. The first questionnaire link was used to test the validity and reliability of each questionnaire obtained and the second questionnaire link was used to process data in factor analysis and KJ method

## 4. Result and Discussion

### *4.1 Identification of Target Markets*

Before making improvements to the product, it is necessary to identify the target market for users using reading chairs. In this study the target market is students who have used a reading chair when reading because students are required to broaden their horizons and enrich themselves by reading a lot of books as well as working on assignments from various literature as a reference.

*4.2 Kansei Word*

Kansei words obtained in number can vary but generally use 50 to 600 words (Schütte, 2002). After collecting the Kansei word, a reduction in the Kansei word is done to reduce the words that have similarities with other words. This is done to reduce ambiguous words between Kansei words to minimize the use of words that are in accordance with the needs of reading chair users. The following are 23 Kansei words that have been reduced as follows:

**Table 1.** Kansei word reduction

| No | Kansei Word | No | Kansei Word |
|---|---|---|---|
| 1 | Smooth | 13 | Appropriate |
| 2 | Stable | 14 | Waterproof |
| 3 | Feature | 15 | Ergonomic |
| 4 | Backrest | 16 | Minimalist |
| 5 | Headrests | 17 | Easy to use |
| 6 | Armrest | 18 | Attractive colours |
| 7 | Premium | 19 | Elegant |
| 8 | Safe | 20 | Futuristic |
| 9 | Comfortable | 21 | Unique |
| 10 | Relax | 22 | Slope |
| 11 | Functional | 23 | Artistic |
| 12 | Ideal | - | - |

The word from the word that has been successfully reduced will then be used to make a question in the form of a questionnaire that will be distributed to reading chair users.

*4.3 Test Validity and Reliability*

- Test Validity

  Validity test can be said to be valid if the correlation coefficient value is 3 0.3 (Indrawati, 2015). Based on the results of the validity test using Spearman shows the results of each Kansei word more than 0.3 which states that the data from the post-test questionnaire is said to be valid.
- Reliability Test

  According to Indrawati (2015) reliability concerns the level of trust, reliability, consistency or stability of the results of a measurement. Cronbach alpha coefficient value $\geq 0.7$ indicates that the questionnaire has a fairly good level of reliability. Based on the reliability test using Cronbach alpha shows that the results obtained are above 0.7, which means the questionnaire data is reliable.

*4.4 Factor Analysis*

Figures KMO and Barlett Test according to Santoso (2012) can be analysed by factor analysis if the value of KMO and Barlett Test is above 0.5 and the significance value is below 0.05. The hypothesis for testing is as follows.

- H0: There is no significant relationship between variables that affect the concept of the reading chair.
- H1: There is a significant relationship between variables that affect the concept of the reading chair.

Based on the results of the KMO and Barlett Test values obtained 0.818 or> 0.5 with a significance value of 0,000 or <0.05 which means accepting the H1 hypothesis or there is a significant relationship between the variables that affect the concept of the reading chair. In the KMO and Barlett Test there is a PCA that can be used to find out the classification of each word word from each grouping result that can be seen in the table 2.

**Table 2.** Grouping of kansei word

| No | *Kansei Word* | Component | Cluster |
|---|---|---|---|
| 1 | Smooth | 1 | |
| 2 | Feature | 1 | |
| 3 | Backrest | 1 | |
| 4 | Headrests | 1 | |
| 5 | Armrest | 1 | Basic |
| 6 | Comfortable | 1 | |
| 7 | Relax | 1 | |
| 8 | Functional | 1 | |
| 9 | Minimalist | 1 | |
| 10 | Attractive colours | 2 | |
| 11 | Futuristic | 2 | Positive Physical |
| 12 | Unique | 2 | |
| 13 | Slope | 2 | |
| 14 | Waterproof | 5 | Adventurous |
| 15 | Ergonomic | 6 | Compromising |

*4.5 Product Concept Description*

After doing a factor analysis and getting four group components from each kansei word, the next is to classify the four groups using the KJ method. According to Nagamachi and Lokman (2011) the KJ method uses affinity diagrams in grouping kansei words called affinity clusters. In the cluster there are several levels of levels, namely zero level, 1st level, and 2nd level, which means that the higher the level, the more detailed it will be.

According to Nagamachi and Lokman (2011) when it can deepen the concept of a product that is initially very general and vague so that it becomes the lowest level concept, we can decide on physical characteristics such as size, design, colour and function. Physical characteristics also need to be translated into technical specifications. The following Figure 1 are the physical characteristics and technical specifications for the reading chair concept.

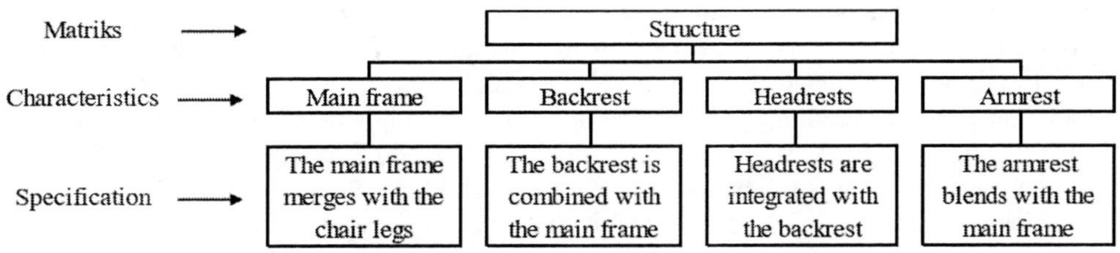


**Figure 1.** Characteristics and specifications of the structure

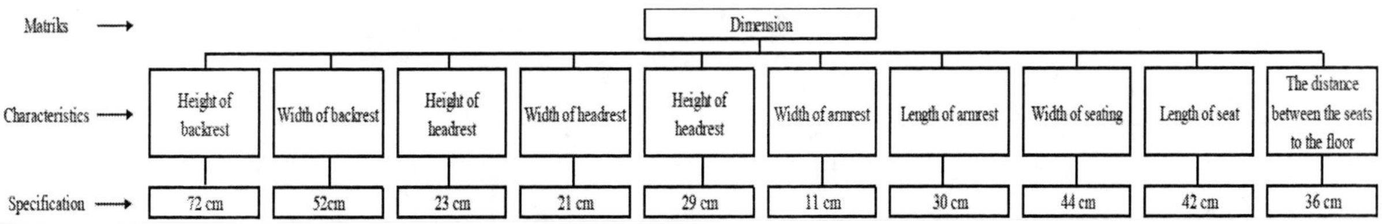


**Figure 2.** Characteristics and specifications of dimension

IOP Conf. Series: Materials Science and Engineering **847** (2020) 012046 doi:10.1088/1757-899X/847/1/012046

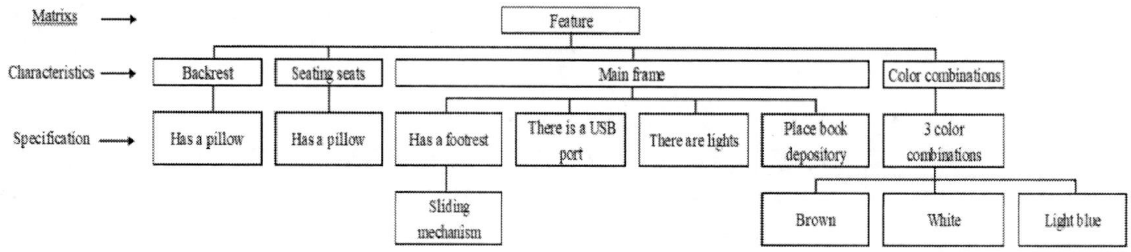


**Figure 3.** Characteristics and specifications of features

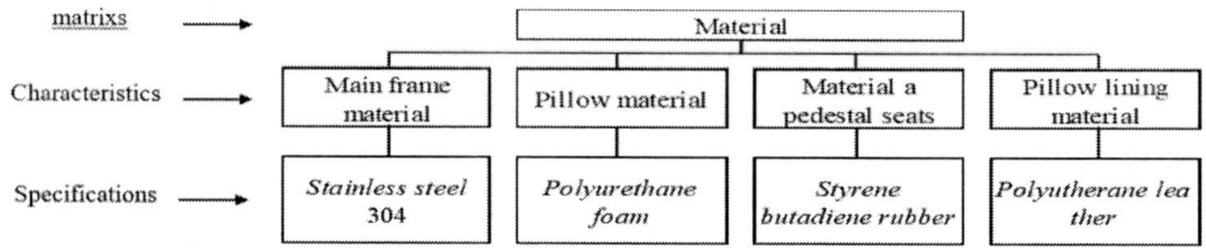


**Figure 4.** Characteristics and specifications of material

### *4.6 Design Idea*
Before visualizing using CAD, it is necessary to design a reading chair design first. Design is done by benchmarking to find references or ideas that are in accordance with the design that will be made. The selection was chosen based on the suitability between the kansei word and the technical specifications obtained on the product design to be made.

### *4.7 Visualization of Design Concept*
After getting the technical characteristics and specifications using the KJ method, the next step is to create a 3D design using solidworks software. The visualization of this design is derived from the benchmarking process on read-read products on the market. So that gives an overview to researchers to make visual designs. The design of this design is also adjusted to the process that the researcher has done in the previous stages so as to get the following visual design innovations.

When the design of the proposal was completed using solidworks software, the researcher conducted a trial to find out the user's response regarding the design of the reading chair that the researcher had designed to 40 respondents and an expert in manufacturing. As a result, users have no problems regarding the form of design that the researchers design, but there are only a few adjustments and additional features in the reading chair needed by users. The adjustments are like the lights that can be adjusted by the user, the backrest which can be adjusted tilt and the brown colour for the frame and the black for the cushion more chosen by the user. As well as additional features needed in the reading chair that is designed is the presence of a folding table that can be used as a place to put books while reading. Then the researcher got expert input regarding the structure of the skeleton used so that there was an improvement in the skeletal structure that the researcher designed. So from that the researchers conducted a design evaluation in the hope that the design evaluated by researchers can be better than the previous design to meet user needs. The following is an evaluation design carried out by the researcher.

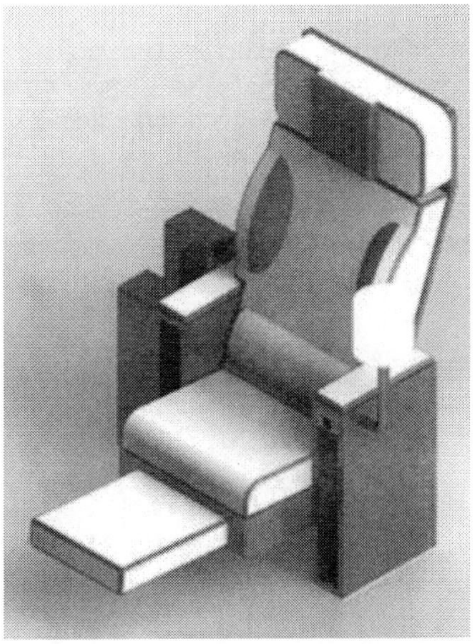

**Figure 5.** Visualization of the proposed design

**Figure 6.** Visualization of evaluation design

## 5. Conclusion

Based on the results of the research that has been done in designing reading chairs that are in accordance with user needs by using Kansei Engineering, conclusions can be drawn as follows.

1. Implementation of Kansei Engineering in making the design of a reading chair can be done because of the design of the view of the user's feelings and needs.
2. The Kansei word test uses processing statistical data to test words that match the user's needs in the design of the reading chair. The result of the 23 words that have been reduced there are 8 words that cannot be used, namely smooth, premium, safe, ideal, appropriate, easy to use, elegant and artistic so that the design of the reading chair only uses 15 Kansei words namely stable, features, backrest, headrests, armrests, comfortable, relax, functional, waterproof, ergonomic, minimalist, attractive colors, futuristic, unique and slope.
3. The use of Kansei Engineering in the manufacture of reading chair design results in several innovations to meet user needs, including the armrests and headrests that are expected by users in the designed reading chair, there are footrests that users can use to give a more relaxed impression when reading, then there is a USB port that functions as an electric charger for users who use gadgets when reading, there are lights that can be used by users when the lighting when reading is felt lacking, there is a book storage area that users can use when they have finished reading. Backrest that can be adjusted by the user. And a folding table that can be used to put books or gadgets while reading.

## 6. References

[1] *GEAR PATROL* 2019 'The 10 Best Reading Chairs of 2019'. Available at: https://gearpatrol.com/2019/02/12/best-reading-chairs/ (Accessed: 22 April 2019).

[2] Indrawati, P. D. 2015 *Metode Penelitian Manajemen dan Bisnis Konvergensi Teknologi Komunikasi dan Informasi*. 1st edn. Edited by D. Sumayyah. (Bandung: PT Refika Aditama).

[3] Nagamachi, M. and Lokman, A. M. (2011) *Innovations of Kansei Engineering*. 1st ed. (Tokyo: Taylor & Francis Group).

[4] Periyeti 2017 'Usaha Meningkatkan Minat Baca Mahasiswa', *Jurnal Pustaka Budaya*, **4**(1), pp. 55–67.

[5] Santoso, S. 2012 *Aplikasi SPSS pada Statistik Multivariat*. 1st ed. (Jakarta: PT Elex Media Komputindo).
[6] Schütte, S. 2002 *Designing Feelings into Products*. (Linköping: Linköpings Universitet).
[7] Yassin, B. A. 2019 *Faktor-Faktor yang Mempengaruhi Minat Membaca, Perpustakaan Universitas Andalas*. Available at: https://pustaka.unand.ac.id/component/k2/item/193-faktor-faktor-yang-mempengaruhi-minat-membaca (Accessed: 1 March 2019).

IOP Conf. Series: Materials Science and Engineering **847** (2020) 012047 doi:10.1088/1757-899X/847/1/012047

# Designing Information System for Student Practicum Assessment in the Laboratory

**Ahmad Arif Nurrahman, Novan Pizary Husen, and Otong Rukmana**

Industrial Engineering, Faculty of Engineering, Universitas Islam Bandung,
Jl. Tamansari No. 1 Bandung 40116 Indonesia
nurrahman18@yahoo.com, npizary@gmail.com, otongrukmana@yahoo.com

**Abstract.** Practicum assessment process at the Laboratory in some university still uses conventional methods ranging from processing practicum value data to presenting assessment information. This certainly can cause various problems that can hamper activities and services in the laboratory. One of the efforts that can be done to improve the practicum assessment process is by designing a practicum assessment information system. The purpose of designing a practicum assessment information system is to identify the information system of current practicum assessments, analyze information system needs, design and build a practicum assessment information system in the Information Systems and Decision Laboratory. The method used in the design of information systems is the Framework for the Application of Systems Thinking (FAST) with the Model Driven Development Strategy model. The FAST method consists of several stages ranging from scope definition, problem analysis, requirements analysis, logical design, decision analysis, physical design, development and testing, implementation and delivery, system operation and maintenance. The design of the proposed improvement of the practicum assessment system in the process block using Data Flow Diagrams (DFD) to get the architecture processes in the practicum assessment system application. The data block uses the Entity Relationship Diagram to design a practicum assessment database system. The communication block uses the Use-Case Diagram design model to get the interface design from the practicum assessment system. The construction of a practicum assessment information system is carried out using XAMPP tools to access the Apache server and the MySQL database and Sublime Text is used to build application interfaces using HTML and PHP programming languages. So that the results of designing the information system for practicum assessment using information technology can be obtained which can be a medium for processing and presenting practical values in the laboratory, which are easily accessible to users and improve aspects of the PIECES framework including performance, information, economy, control, efficiency and service.
**Keywords:** FAST, DFD, Entity relationship diagram, XAMPP

## 1. Introduction

Customer satisfaction is one of the important factors for an organization, especially those engaged in education or academics. There are many advantages to be gained if an organization can manage good relationships with customers, one of which is that it can improve operational efficiency in activities carried out in the organization. The field of education, especially at universities, is one area that is closely related to customer relations because customer satisfaction can be one of the determinants of student graduation at a university. Factors that can influence customer satisfaction are divided into several aspects according to the PIECES framework, including performance, information, economy, control, efficiency, and service. So it takes an effort to be able to manage each of these aspects in order to increase customer satisfaction. At present, efforts have been made to increase customer satisfaction at various universities, one of which will be implemented in the Laboratory of Information Systems and the Decision of Bandung Islamic University.

The Information System and Decision Laboratory is one of 4 laboratories in the Industrial Engineering of Bandung Islamic University. The laboratory is engaged in information systems that provide information about matters related to information technology. Activities carried out in the Information Systems and Decisions Laboratory related to customer relations. The term customer in the laboratory is defined as an entity that is interconnected with each other, namely between student entities, regulatory bodies and laboratory assistant entities. The service used with one process is practicum.

Practicum assessment process in the Information Systems and the Decision Laboratory still uses conventional methods ranging from processing practicum value data to presentation of assessment information. This certainly can cause various problems that can hamper activities and services in the laboratory. One of the efforts that can be done to improve the practicum assessment process is by designing a practicum assessment information system. The application of information technology as a means to design an assessment information system can assist in the various processes that exist in the assessment of practicums such as value input by practicum assistants, database of storing practicum values, presentation of practicum values and also the reporting process of the value of practicum activities and reports of large assignments to lecturers. So that the output of this study is designing an information system for practicum assessment using information technology that can be a medium for presenting practical values in the laboratory, which is easily accessible to users so that it speeds up decision making and increases customer satisfaction.

The research objectives to be achieved are based on the formulation of the problems described above, namely:

1. Identify the practicum assessment information system that is running in the Information Systems and Decision Laboratory
2. Analyzing the needs of the practicum assessment information system in the Information Systems and Decision Laboratory.
3. Design an information system for practicum assessment in the Information Systems and Decision Laboratory.
4. Establish a practicum assessment system in the Information and Decision System Laboratory

## 2. Method

The method used in information system design is the Framework for the Application of System Thinking (FAST) method, using a Model Driven Development Strategy. The stages of the FAST method can be seen in Figure 1

As for the explanation of the stages of the FAST method, namely:

1. Phase 1 - Scope Definition
   Determine the limits of system development by identifying problems, initiatives and objectives. There are two objectives of the scope stage, which is answering the question of whether this problem is sufficiently valuable and setting the size, vision, constraints or limitations of what is needed, as well as the budget and schedule of the project.
2. Stage 2 - Problem Analysis
   Problem analysis is the stage of studying the existing system and analyzing the findings in order to find a deeper understanding of the problems that triggered this project.
3. Stage 4 - Logic Design (Logical Design)
   In the logical design stage is translating business needs into the system model. The term logical design is defined as independent technology

IOP Conf. Series: Materials Science and Engineering **847** (2020) 012047 doi:10.1088/1757-899X/847/1/012047

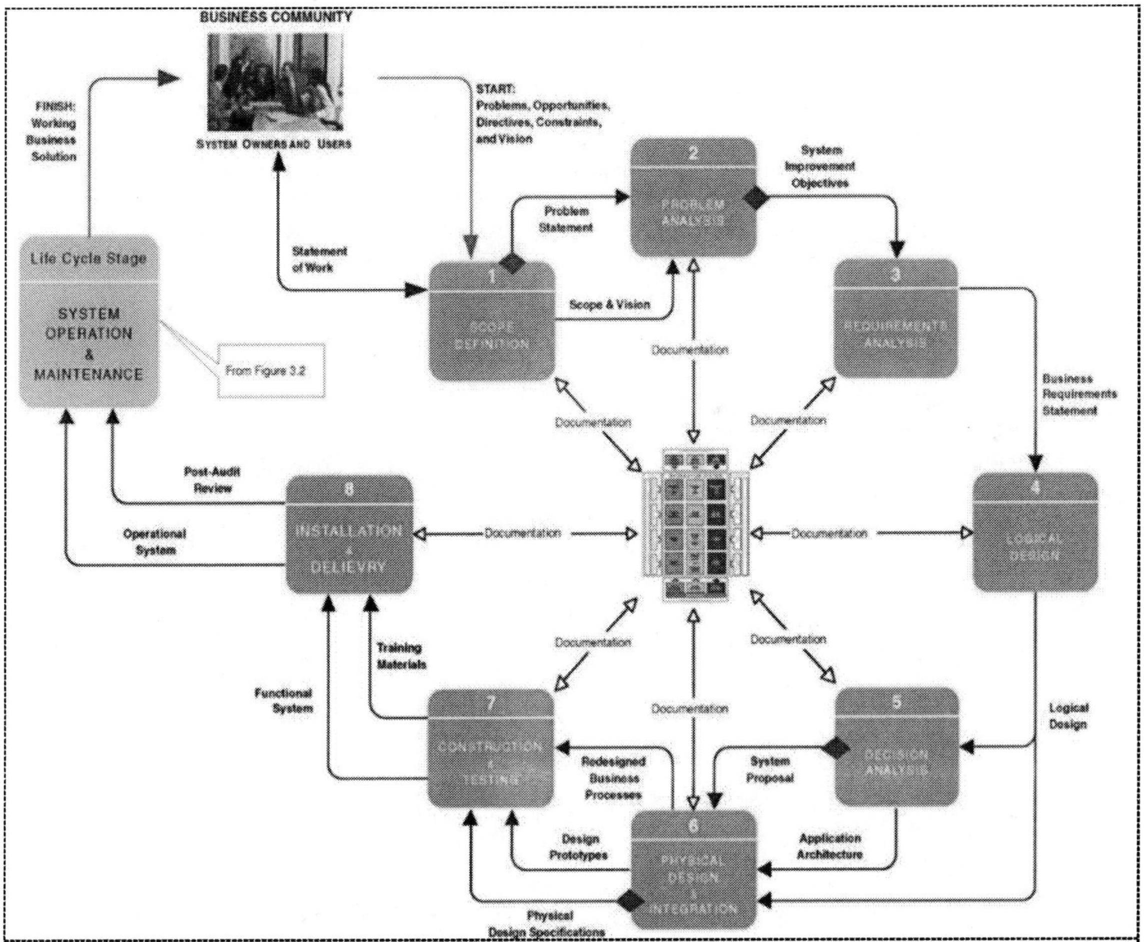


**Figure 1.** FAST Methodology Phase [1]

Analysis of where the meaning of the image describes the system independently of every possible technical solution, the needs of the desired business model must be met by the technical solutions that want to be considered.

4. Stage 5 - Decision Analysis (Decision Analysis)
   Given business needs and logical system models, there are usually many alternatives for designing a new information system to meet these needs. The purpose of this phase is to identify the choice of technical solutions, analyze the solution to the feasibility of the choice, and recommend the choice of system to be designed.
5. Stage 6 - Physical System Design (Physical Design and Integration)
   After the solution is chosen by management, the next step is to transform the business needs in the form of physical specifications that will guide system design.
6. Stage 7 - Development and Testing (Construction and Testing)
   The purpose of this phase is to test the systems that have been developed to determine the level of acceptance of the business processes and system users.
7. Stage 8 - Installation and Delivery (Installation and Delivery)
   This activity is the process of installing software and hardware, socialization and documentation creation.

8. Stage 9 - System Operation and Maintenance
   The final step in the system development cycle is maintenance activities, and also follow up if there are new needs.

## 3. Result

### *3.1. Modeling Interfaces Using Usecase Diagrams*

The system communication model is described using a use-case diagram. Use-case is an interaction that is carried out or received by an actor and relationship that connects the actor and use-case so that the functions in the system will be known. Use-case diagram consists of several components, namely use-case, actor, and relationship (relationship). As for data modeling with the use-case diagram can be seen in Figure 2

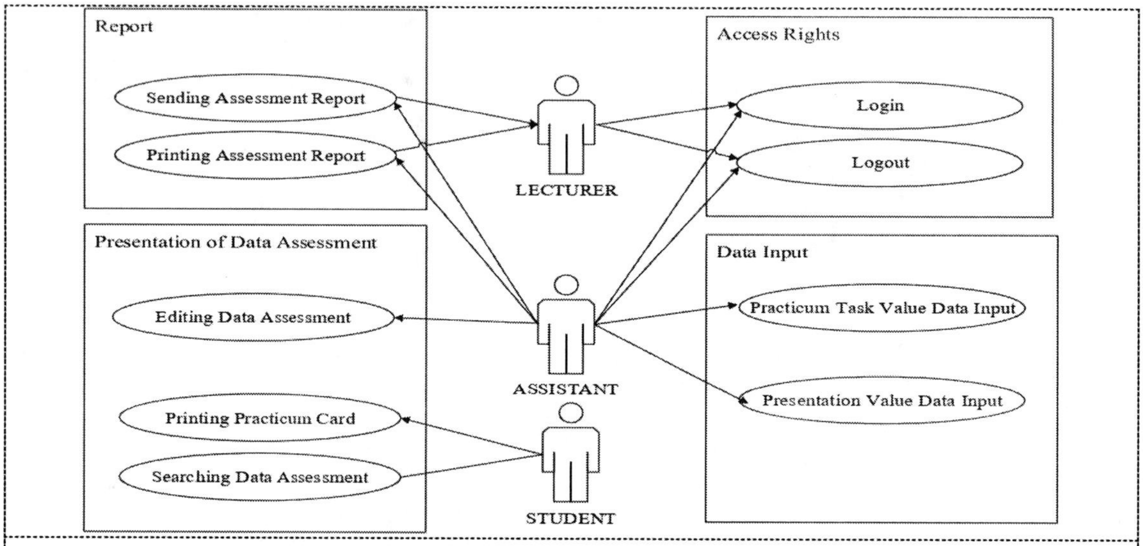


**Figure 2.** Use-Case System Functional Needs Diagram

### *3.2. Modeling the Process Using Data Flow Diagrams*

Context diagram describes the relationship between 3 entities, including assistants, practitioners and lecturers. The system context diagram can be seen in Figure 3

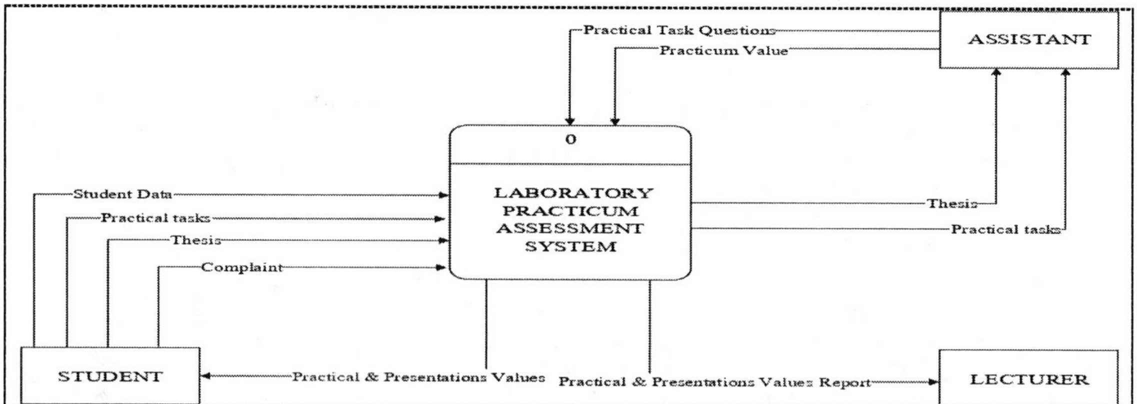


**Figure 3.** Context Diagram

DFD level 1 is a subsystem from the context diagram of the practicum assessment system. There are 8 processes of flow and data storage. The level 1 DFD can be seen in Figure 4.

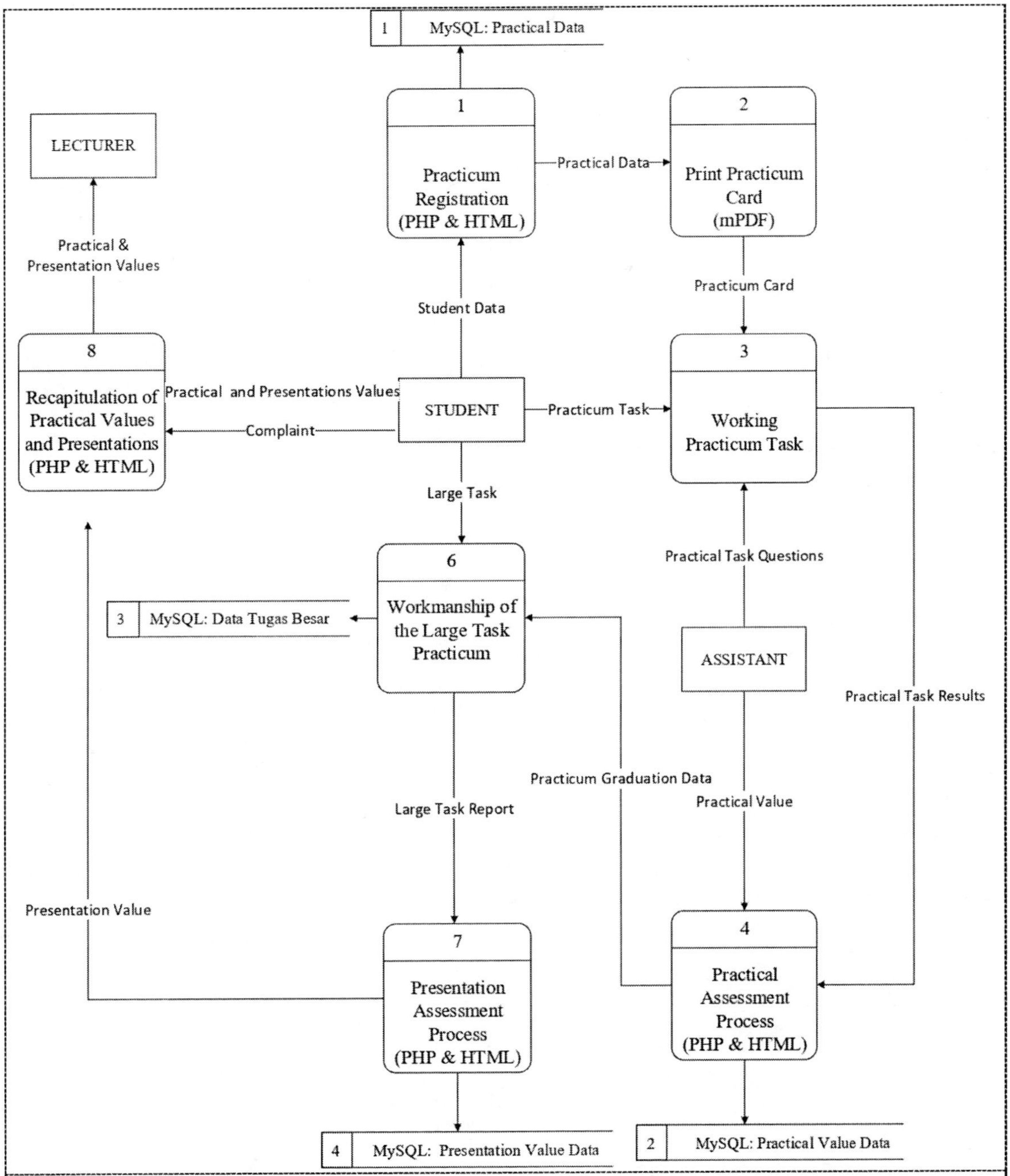


**Figure 4.** Data Flow Diagram Level 1

Level 2 DFD for practicum assessment process there are 5 sub-processes. The level 2 DFD for the practicum assessment process can be seen in Figure 5

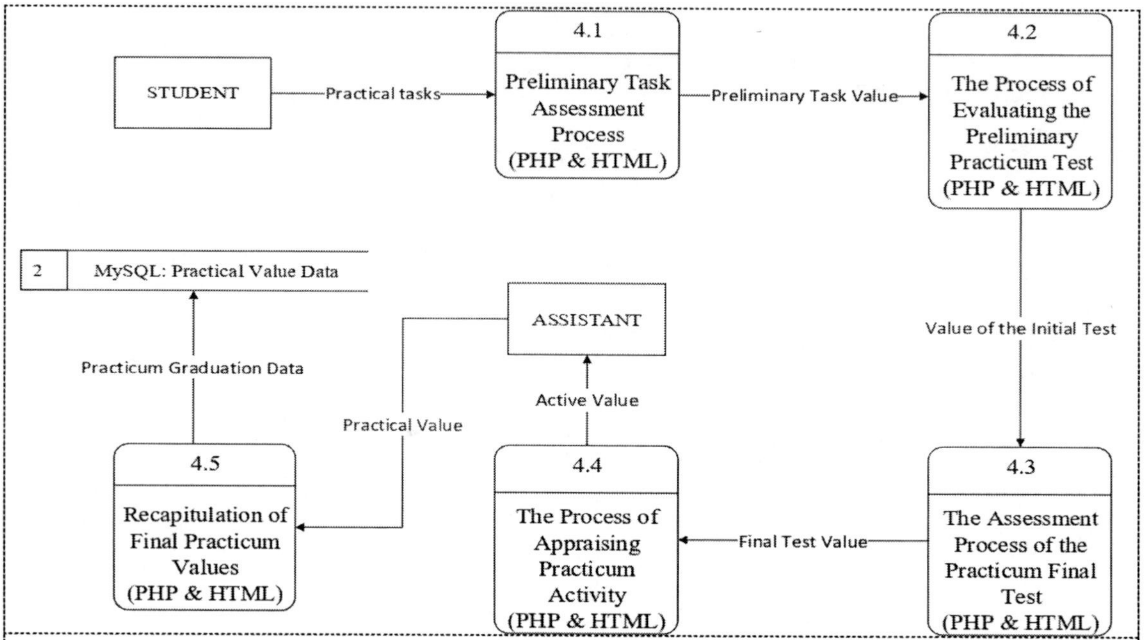


**Figure 5.** Data Flow Diagram Level 2 for Practicum Assessment

DFD level 2 for the presentation assessment process there are 5 sub-processes. The level 2 DFD for the presentation assessment process can be seen in Figure 6.

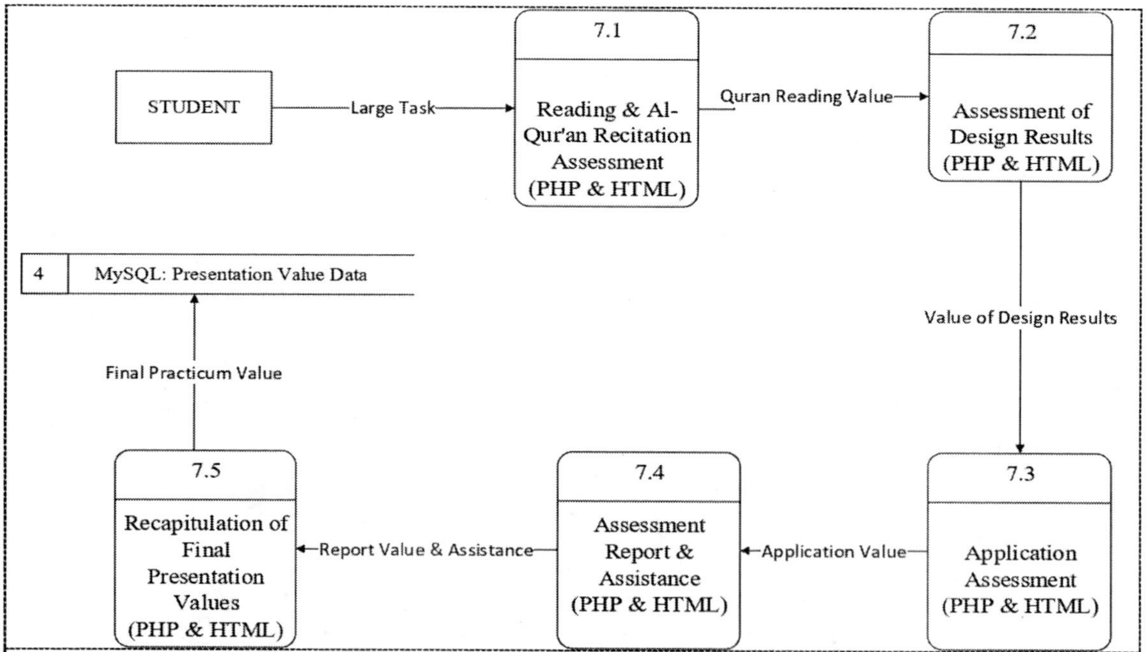


**Figure 6.** Data Flow Diagram Level 2 for Presentation Assessment

### *3.3. The System Interface*

The system interface in program creation refers to the interface design that has been designed beforehand. The description of the system interface for assistant users can be seen in Figure 7.

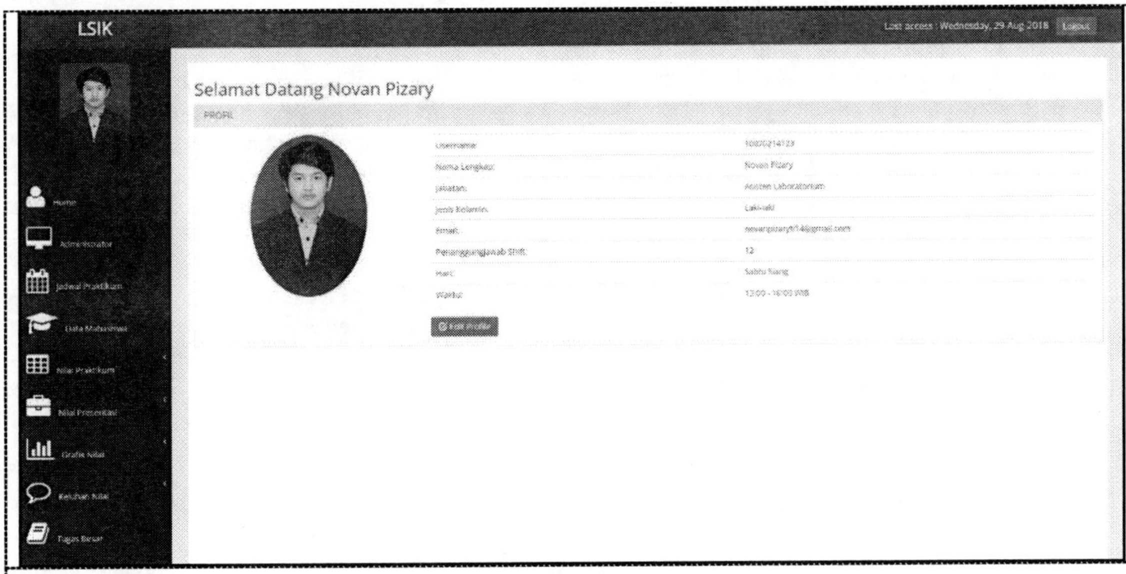


**Figure 7.** User Assistant System Interface

## 4. Analysis

The description of the problem based on PIECES analysis on the previous system has changed and improved on each framework when compared to the proposed system designed. The PIECES analysis for the proposed system is based on performance, information, economy, control, efficiency, and service. The PIECES table is shown in Table 1.

**Table 1** Proposed PIECES System Analysis

| Framework | Analysis |
|---|---|
| Performance | The time needed to manage practicum assessment data is faster than before |
| Information | Data is stored safely because of the database system and avoid data manipulation |
| Economy | Paper expenses can be reduced because the practicum card printing process is done by the practitioner |
| Control | There are security systems such as passwords and value controls being more accurate because they are witnessed by all entities |
| Efficiency | Making practicum cards automatically and directly printing |
| Service | Practical values can be accessed easily anywhere and anytime as long as connected to the internet |

## 5. Conclusion

Based on the results of the research that has been done, then some things that can be concluded are:

1. The practicum assessment system that runs in the Information System and Decision Laboratory is currently still using a conventional system, starting from the practicum card manufacturing process, the process of inputting practicum values on practicum cards and on computers, up to the making of practicum assessment reports. So that the problems faced by the laboratory include the time needed to manage practicum appraisal data relatively long, there is a risk of data loss, there

are paper expenses for printing practicum cards, and there is no security system so anyone can access the system easily.

2. The need for an information system for practicum assessment in the Information and Decision System Laboratory, which is a system that can facilitate the processing of practicum value data, provide storage and data security values that can only be accessed by certain parties, and facilitate the presentation of practicum value information.
3. Designing proposals for improvement of the practicum assessment system on the block process using Data Flow Diagrams (DFD) obtained the results of architectural processes in the application of practicum assessment system. Data blocks using Entity Relationship Diagrams obtained the results of the practicum assessment database system consisting of assistant tables, students, lecturers, practicum assessment, presentation assessment, registration and major assignments. The communication block uses the Use-Case design model diagram obtained by the interface design from the practicum assessment system.
4. Development of practicum assessment information system is carried out using XAMPP tools, MySQL database and Sublime Text generated web-based practicum assessment system consisting of features such as login, input practicum values, input presentation values, large task data input, import data to excel, print cards practicum, assessment information and grading chart. Testing of the practicum assessment system that has been built can work well and is ready to be implemented.

## 6. References

[1] Bentley, L. D., and Whitten, J. L 2007 *System Analysis Design for the Global Enterprise.* 7th. Ed. (New York: McGraw-Hill)

[2] Christy, Novy., 2016 Pengembangan Sistem Informasi Penilaian Praktikum Berbasis Web di STIKI Malang (Sekolah Tinggi Informatika dan Komputer Indonesia. Malang: *Journal of Information and Technology* ISSN 2303-1425)

[3] Connolly, T., Begg, C 2010 *Database Systems: a practical approach to design implementation, and management.* 5th Edition. (America: Pearson Education.)

[4] Faridl, Miftah 2015 *Fitur Dahsyat Sublime Text 3.* (Lug STIKOM.)

[5] Garside, Annisa Kesy 2016 *Perancangan Sistem Informasi Laboratorium Teknik Industri Universitas Muhammadiyah Malang* (Universitas Muhammadiyah Malang. Malang: Seminar Nasional Teknologi dan Rekayasa ISSN 2527-6042)

[6] Indrajit, R. 2000 *Manajemen Sistem Informasi dan Teknologi Informasi.* (Jakarta: P.T Gramedia.)

[7] Jogiyanto, H. 1999 *Analisis dan Desain Sistem Informasi* (Yogyakarta: Andi Offset.)

[8] Kendall, K. K. 2013 *System Analysis and Design. 8th ed.* (USA: Pearson Education, Upper Saddle River, New Jersey)

[9] O'Brien, R. 1998 *Formulating and Processing for Applications* (New York: Technomic)

[10] Sutabri, I. O. R. 2004 *Manajemen Sistem Informasi Dan Teknologi Informasi* (Jakarta: PT. Elex Media Komputindo)

[11] Syafii, M. 2004 *Membangun Aplikasi Berbasis PHP dan MySQL.* (Yogyakarta: Andi Offset.)

[12] Widiastuti, dkk. 2016 Perancangan Sistem Informasi Praktikum Komputer Berbasis Web. Universitas Gunadarma. (Depok: *Prosiding SNATIF* ISBN:978-602-1180-33-4)

IOP Conf. Series: Materials Science and Engineering **847** (2020) 012048 doi:10.1088/1757-899X/847/1/012048

# Analysis of Choice Shrimp Technology based on Business Process, Productivity, Financial and Risk

**Wydzka Tasha Aulia Akbar[1], Endang Chumaidiyah*[1,2], and Meldi Rendra[1,3]**

Industrial Engineering Program, Telkom University, Bandung, Indonesia
[1] wydzkaakbar@gmail.com, [2*]endangcn@gmail.com

**Abstract.** The application of new technology in shrimp ponds, autofeeder, has many advantages, but carries a greater risk compared to conventional shrimp ponds, therefore a more structured analysis is needed to decide the best shrimp pond technology. The comparison is seen from business processes, productivity, financial and risk. The results of business process efficiency level with conventional worth 67.19% while for the autofeeder level worth 88.71%. The productivity results show FCR 1.34 (conventional) and 1.34 (autofeeder); SR 79% (conventional) and 90% (autofeeder); and productivity of 13 tons / ha (conventional) and 25 tons / ha (autofeeder). The financial shows an NPV of Rp 1,515,178.503 (conventional) and Rp 7,721,596,229 (autofeeder); IRR 38.24% (conventional) and 51.23% (autofeeder); payback period 2.68 years (conventional) and 2.17 years (autofeeder); BCR 1,696 (conventional) and 2,065 (autofeeder). Furthermore, for the calculation of risk consisting of production risk and revenue risk with a total risk of 6% for conventional and 31% for autofeeder. In the results of technology selection with an assessment of the 15 criteria above, the results obtained 4 criteria are better for conventional technology and 11 criteria are better for autofeeder technology. So between the two technologies, the selected shrimp pond system is the Autofeeder technology.
**Keywords**: technology selection, shrimp farms.

## 1. Introduction

In line with the shrimp production trend that continued to rise for 5 years from 2009 to 2015, the volume of shrimp exports over the past five years also showed a positive trend despite a decline in 2015. The increase in shrimp production volume is in line with the increase in export volumes (directorate general of aquaculture, 2017). The potential for the development of the Indonesian shrimp industry in shrimp ponds with untapped potential of ponds is 830,900 ha and 70% of conventional shrimp ponds in Indonesia are individual MSME companies [1].

The problem in conventional shrimp culture systems is that mortality increases with decreasing water quality during the maintenance period. The source of the decline in water quality is an increase in metabolic products resulting from the decomposition of feed residues and shrimp feces that are not controlled by water quality management measures. The increase in metabolic products occurs due to inaccurate feeding with an indication of the value of the ratio of feed conversion (feed conversion ratio, FCR) and the ratio of the addition of biomass to the addition of feed is irrelevant resulting in low productivity.

Autofeeder is a structured feeding technology using time performance indicator technology that provides feed automatically using the time program that has been inputted. The emergence of this new technology in shrimp ponds has a positive impact on shrimp farmers because it gives the impact of shrimp growth so quickly and effective and efficient feeding. But the use of autofeeder adds new problems such as the emergence of new diseases. This outbreak is called Hepatopancreatic Necrosis Syndrome (AHPNS), or better known as Early Mortality Syndrome (EMS). All shrimp affected by this disease will experience death in a short time.

## 2. Methods

### *2.1 Technology Management*
Technology management is a study or discussion that links the disciplines of engineering/engineering, science, and management in placing planning, development, and implementation of the ability to form and complete operational and strategic objectives of the company. [2] technology management and several fields within it are very relevant and needed to assist in government decision making, industry leaders and business management studies, including: technology strategy, technology capability development, innovation management, forecasting technology, management technology, manufacturing strategies, business competition, barriers to adopting technology, manufacturing technology and e-business flexibility.

### *2.2 Business Process*
A business process can be explained simply as a flow of activities [3]. A business process is a collection of structured tasks or activities that can produce certain services or products for one or many consumers. Business process modeling is very important in the life of the Business Process Reengineering (BPR) cycle.

### *2.3 Producivity*
Productivity is a measure that states how well resources are managed and utilized to achieve optimal results. Productivity can be used as a measure of the success of an industry in producing goods or services, the higher the ratio, means the higher the product produced. The results obtained by farmers at harvest are called production, and the costs incurred are called production costs. Good farming business is productive or efficient farming. Productive farming means farming that has high productivity. This understanding of productivity is a combination of the conception of business efficiency (physical) and land capacity [4].

- Productivity

$$\frac{Total\ Harvest(tons)}{Land\ Area\ (ha)} \tag{1}$$

- FCR

$$\frac{Total\ Feeds}{Total\ Harvest} \tag{2}$$

- SR

$$\frac{Size * Total\ Harvest}{Total\ Seeds} * 100\% \tag{3}$$

### *2.4 Financial*
The financial aspects of project preparation and analysis explain the financial effects of a proposed project on the parties involved in it. The main purpose of financial analysis is to determine projections about the budget that will be used efficiently by estimating revenue and expenditure during project implementation and in the years to come each year [5].

### *2.5 Risk of Shrimp Culture*
Production Risk Analysis that occurs can be caused due to pests and diseases both sudden and widespread. So that it can result in a decrease in the yield of up to 65% and can even lead to crop failure. Details of production risk analysis in shrimp farming can be seen in Table 6 below: [6]

IOP Conf. Series: Materials Science and Engineering **847** (2020) 012048 doi:10.1088/1757-899X/847/1/012048

**Table 1.** *Risk of Shrimp Culture*

| No | Period | Production (Kg/Ha) |
|---|---|---|
| 1 | cultivation cycle 1 | 158,88 |
| 2 | cultivation cycle 2 | 155,81 |
| 3 | cultivation cycle 3 | 147,72 |
| | a. Average | 154,14 |
| | b. Variance | 33,26 |
| | c. Standard Deviation (V) | 5,77 |
| | d. Coefficient Variance (CV) | 0,04 |

Source: Achmad Rizal et al. (2018)

*2.6 Shrimp Culture System*

Shrimp farming in ponds is the business of raising or enlarging shrimp from seed size (fry) to a size that is fit for consumption. Naturally, shrimp seeds enter the pond along with the tide that irrigates the pond. Shrimp production obtained is uncertain because it only depends on the amount and small amount of shrimp seeds that occur naturally in the sea around the pond. Shrimp seeds can be chosen which is quickly grown and the species is much-loved. (economically important). Life pass (SR), feed conversion ratio and production per hectare. Quality data taken include temperature, dissolved oxygen, pH, salinity and ammonia. Farm fertility can be improved by greater fertilization and water management so that the carrying capacity to maintain shrimp is better. Pest control is more intensified. The construction of ponds, embankment construction and irrigation canals are improved so that the quality of the tabak water can be better controlled and suitable for the life of the shrimp to be maintained. Shrimp cultivation systems in developing ponds are now known to have three levels according to the technology application category, namely the level of simple cultivation (traditional, extensive), middle cultivation level (semi-intensive), and advanced cultivation level (intensive) [7]

*2.7 Conceptual Model*

Depicting a diagram of a set of relationships (Figure 1) between certain factors that are believed to have an impact on or lead to a target condition. In this research process aims to choose the optimal technology alternative using the zero one method. This research can provide a good impact on pond farmers who have not yet decided to use autofeeder technology or conventional systems. The concept of the process is then processed into a research method and data collection needed to describe the problem discussed.

In this study conducted in the village of Brondong, Lamongan, East Java, with the research period 18 February 2019 to 10 June 2019. With direct data retrieval with historical data owned by the farmers for conventional shrimp pond observation objects and autofeeders, where the shrimp pond system with Conventional technology is studied in 2 ponds each with an area of 4,600m2. As for the autofeeder pond system that was studied as many as 5 ponds and 1 treatment pond each with an area of 2500 m2.

## 3. Result and Discussion

*3.1 Business Process*

The business process aspect is used to choose alternative considerations between conventional shrimp ponds and autofeeder shrimp ponds. business process data processing is done by taking the time data used in one cultivation cycle. Interms of deciding this consideration will be based on the results of changes in efficiency through the Business Process whether the results show a better chance or not between the two.

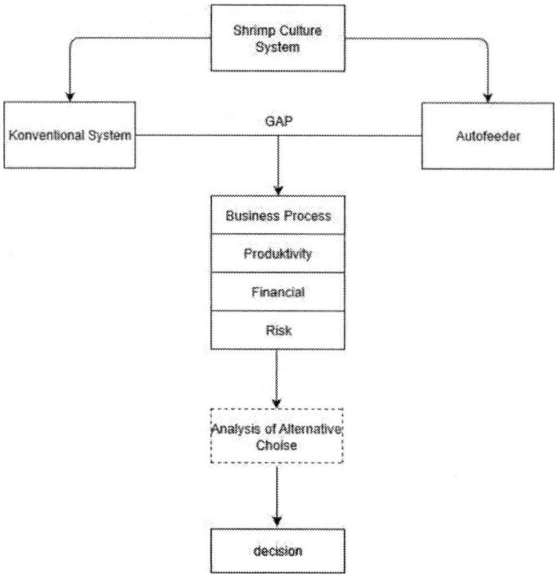


**Figure 1.** Conceptual Model

**Table 2.** Comparison of shrimp farming business processes

| Activity Category | Conventional Shrimp Farming Business Process (hours) | Business Process of Autofeeder Shrimp Farming (hour) |
|---|---|---|
| RVA | 577,4 | 721,3 |
| BVA | 231,85 | 87,9 |
| NVA | 50,05 | 3,9 |
| Total Time (hours) | 859,3 | 813,1 |
| Efficiency | 67,19% | 88,71% |

In the results of the business, process calculations give that the total business process time on conventional shrimp ponds is 859.3 hours higher than the autofeder pond business process with a total time of 813.1 hours. The efficiency level in conventional ponds is 67.19% while in autofeeder ponds it gets an efficiency level of 88.71%. This shows that a good autofeeder farm is better than the business process aspect.

### *3.2 Productivity*

In the aspect of productivity is used to choose an alternative consideration between conventional shrimp ponds and Autofeeder shrimp ponds. In terms of deciding this consideration will be based on the results of the FCR (feed conversion ratio). SR (survival rate) and productivity per hectare through productivity calculation methods in the shrimp culture world. Whether the results show a better chance or not between the two. Therefore. The following is the average productivity of the last 5 cycles in each pond.

**Table 3.** The results of conventional pond productivity calculations

| No | Pond | Pond Area (m2) | ∑ Seed (ekor) | Density (Seed/m2) | DOC (Hr) | Harvest (Kg) | Size (ekor/kg) | ∑ Feed (kg) | FCR | SR (%) | Productivity (ton/ha) |
|---|---|---|---|---|---|---|---|---|---|---|---|
| 1 | West | 4.600 | 329.600 | 72 | 115 | 5.874,70 | 44.4 | 11100,1 | 1.89 | 79 | 12,8 |
| 2 | East | 4.600 | 329.600 | 72 | 115 | 5.838,90 | 44.8 | 10886,2 | 1.86 | 79 | 12,7 |
| TOTAL | | 9.200 | 659.200 | | | 11.714 | | 21986.3 | | | 25 |
| AVERAGE | | 4.600 | 329.600 | 72 | 115 | 5.857 | 45 | 10.993 | 1.88 | 79 | 13 |

In the results of calculations in Table 3, conventional technology productivity which shows the ratio of total feed to shrimp yield (FCR) is an average of 1.88 while the SR which shows the level of shrimp life during cultivation shows an average SR yield of 79% and aquaculture productivity amounting to 13 tons per hectare.

**Table 4.** Calculation results for the Autofeeder pond productivity

| No | Pond | Pond Area (m2) | ∑ Seed () | Density (seed/m2) | DOC (Hr) | harvest (Kg) | Size (Shrimp /kg) | Feed (kg) | FCR | SR (%) | Productivity (ton/ha) |
|---|---|---|---|---|---|---|---|---|---|---|---|
| 1 | A | 2.500 | 356.400 | 143 | 87 | 6.014,68 | 51.9 | 8240,6 | 1.37 | 88 | 24,1 |
| 2 | B | 2.500 | 356.400 | 143 | 90 | 6.460,78 | 49.9 | 8083,2 | 1.25 | 90 | 25,8 |
| 3 | C | 2.500 | 356.400 | 143 | 90 | 6.509,33 | 50.1 | 8209,2 | 1.26 | 91 | 26,0 |
| 4 | D | 2.500 | 355.000 | 142 | 87 | 5.870,32 | 55.5 | 8222,6 | 1.40 | 92 | 23,5 |
| 5 | E | 2.500 | 355.000 | 142 | 86 | 5.771,90 | 54.3 | 8220,9 | 1.42 | 88 | 23,1 |
| TOTAL | | 12.500 | 1.779.200 | | | 30.627 | | 40976,5 | | | 123 |
| AVERAGE | | 2.500 | 355.840 | 142 | 88 | 6.125 | 52 | 8.195 | 1.34 | 90 | 25 |

Whereas in the calculation of Table 4 which shows the productivity of the pond system autofeeder technology shows an average FCR of 1.34 while the SR is 90%, the cultivation productivity is 25 tons/ha. For this productivity assessment, the smallest FCR is the best, the highest SR is the best and the highest productivity is the best.

Based on the comparison of FCR, SR and Productivity tons/ha between conventional technology and autofeeder technology, the following results are obtained:

1. Conventional pond FCR criteria produce FCR 1.88 while autofeeder technology produces FCR 1.34, so the best FCR uses Autofeeder technology because it produces a smaller FCR.
2. SR criteria show that conventional pond SR is 79% while Autofeeder technology is 90%, so SR is best to use Autofeeder technology because it produces higher SR.
3. Productivity criteria indicate that conventional farms produce 13 tons/ha while Autofeeder technology is 25 tons/ha, so the best productivity is to use Autofeeder technology because it produces higher productivity.

Then the assessment of the 3 productivity criteria is the best shrimp farming method is Autofeeder technology.

### *3.3 Financial*

In the financial calculation of the two alternative shrimp farming systems between conventional technology and autofeeder technology, the results are shown in Table 5. The assessment is based on 9 criteria, namely investment, income, operational costs, COGS (Cost of Production) / Kg, profit and loss, NPV (Net Present Value), IRR (Internal Rate of Return), payback period and BCR (Benefit Cost Ratio):

**Table 5.** Financial calculations of conventional technology vs. auto feeder technology

| Caculation | Conventional | Autofeeder | Analysis |
|---|---|---|---|
| 1. Invesment/Ha | Rp 2,008,271,546 | Rp 3,521,385,600 | In the calculation of conventional technology investment is better with a smaller investment of Rp 2,008,271,546 |
| 2. Revenue/Ha (*) | Rp 3.419.141.854 | Rp 4.741.266.282 | In the calculation of Autofeeder technology revenue is better with greater income, amounting to Rp 4,741,266,282 |

**Table 5**. Financial calculations of conventional technology vs. auto feeder technology

| Caculation | Conventional | Autofeeder | Analysis |
|---|---|---|---|
| 3. Operational Costl/Ha (*) | Rp 2.434.005.206 | Rp 2.446.326.459 | In the calculation of conventional technology operational costs is better with a smaller operational costs of Rp 2,434,005,206 |
| 4. COGS/Kg (*) | Rp 64.240 | Rp 46.594 | In the calculation of HPP / Kg autofeeder technology is better with a smaller HPP / kg of Rp. 46,594 |
| 5. Profit & loss/Ha (*) | Rp 886.622.983 | Rp 1.869.861.592 | In the calculation of profit and loss technology, Autofeeder is better with a bigger profit of Rp 1,869,861,592 |
| 6. NPV | Rp 1.515.178.503 | Rp 7.721.596.229 | In the NPV calculation the Autofeeder technology is better with a greater NPV of 7,721,596,229 |
| 7. IRR | 38,24% | 51,23% | In the calculation of IRR, Autofeeder technology is better with IRR greater than 51.23 |
| 8. Payback Period | 2.68 | 2.17 | In PBP calculation, Autofeeder technology is better with smaller PBP of 2.17 |
| 9. BCR | 1.696 | 2.065 | In BCR calculations Autofeeder technology is better with a bigger BCR of 2,065 |

Based on a comparison of the 9 financial criteria above, the results show that conventional technology is better on 2 criteria, namely investment and lower operational costs when compared with autofeeder technology. Whereas for the other 7 criteria, Autofeeder technology is better than conventional technology, which is assessed from the criteria of income, COGS / Kg, profit and loss, NPV, IRR, payback period and BCR. So from the results of financial analysis on the shrimp pond system, the best method is to use autofeeder technology.

### *3.4 Risk*

The risk calculation results consist of production risk and income risk between conventional technology shrimp ponds and autofeeder technology shrimp ponds as shown in Table 6.

**Table 6**. Risks of conventional technology vs. autofeeder technology

| Calculation | Conventional | Autofeeder | Analysis |
|---|---|---|---|
| 1. Production Risk | 3% | 14% | Production risk on conventional technology is better with a smaller risk of 3% |
| 2. Revenue Risk | 3% | 17% | The risk of revenue on conventional technology is better with a smaller risk of 3% |

In the calculation of the total risk in conventional technology calculated from the calculation of the coefficient of variation is 6%. Whereas the autofeeder technology has a coefficient of variation or a total risk of 31%. The results of the coefficient of variation become a reference to determine how high the risk of failure in shrimp farming production.

In the delta ratio calculation (risk difference between conventional shrimp ponds and autofeeder) compared (IRR difference between conventional shrimp ponds and autofeeders).

1. The difference in risk between conventional technology and autofeeder technology which has a delta ratio of 25%.
2. The IRR difference between conventional technology and autofeeder is 12.99%.
3. The risk difference is 25%> greater than the IRR difference of 12.99%, so Autofeeder technology has a high risk with a lower IRR / return.

Based on the comparison between the risk difference and the IRR difference, conventional technology is better when compared with the autofeeder technology in terms of the risk of the shrimp pond system.

*3.5 Analysis of Alternative Choise*

This alternative selection analysis uses the zero one method that is for the selected technology (better) is rated 1 while for those who are not selected given a value of 0. The selection is seen from the largest number of scores between the two alternative shrimp pond systems. The assessment is based on the 15 criteria described above as shown in Table 7 from the calculation results for each criterion in the two alternative technologies.

**Table 7.** Rating of 15 criteria

| Criteria | Result | | Better or Beneficial | |
|---|---|---|---|---|
| | Conventional | Autofeeder | Conventional | Autofeeder |
| Business Process | | | | |
| 1. Efficiency | 67,19% | 88,71% | 0 | 1 |
| Produktivity | | | | |
| 2. FCR | 1,88 | 1,34 | 0 | 1 |
| 3. Survival Rate | 79% | 90% | 0 | 1 |
| 4. Ton/Ha | 13 Ton/Ha | 25 Ton/Ha | 0 | 1 |
| Feasibility Financial | | | | |
| 5. Invement/Ha | Rp 2,008,271,546 | Rp 3,521,385,600 | 1 | 0 |
| 6. Revenue/Ha (*) | Rp 3.419.141.854 | Rp 4.741.266.282 | 0 | 1 |
| 7. Cost of production/Ha (*) | Rp 2.434.005.206 | Rp 2.446.326.459 | 1 | |
| 8. COGS/Kg (*) | Rp 64.240 | Rp 46.594 | 0 | 1 |
| 9. Profit&loss/Ha (*) | Rp 886.622.983 | Rp 1.869.861.592 | 0 | 1 |
| 10. NPV | Rp 1.515.178.503 | Rp 7.721.596.229 | 0 | 1 |
| 11. IRR | 38,24% | 51,23% | 0 | 1 |
| 12. Payback Period | 2.68 | 2.17 | 0 | 1 |
| 13. BCR | 1.696 | 2.065 | 0 | 1 |
| Risk | | | | |
| 14. Production Risk | 3% | 14% | 1 | 0 |
| 15. Revenue Risk | 3% | 17% | 1 | 0 |
| Total Value | | | 4 | 11 |

Description (*): Average for the next 5 years

Based on the analysis of alternative selection in table 6 between conventional technology and autofeeder technology, it was found that conventional technology got a value of 4, while autofeeder technology got a value of 11.

## 4. Conclusion

Calculation of business processes obtained results in the efficiency of conventional ponds worth 67.19% while the autofeeder ponds get an efficiency level of 88.71%. The business process criteria show that the technology of autofeeder ponds is more efficient than conventional ponds.

IOP Conf. Series: Materials Science and Engineering **847** (2020) 012048 doi:10.1088/1757-899X/847/1/012048

Calculation of productivity criteria seen from FCR, SR, and ton/ha productivity shows the results for conventional ponds obtained FCR results of 1.88; SR 79%; productivity of 13 tons/ha. While the autofeeder pond technology obtained FCR results 1.34; SR 90%; productivity of 25 tons/ha. The productivity criteria show that the autofeeder pond technology is higher in productivity compared to conventional ponds.

The calculation of financial criteria calculated from the NPV, IRR, payback period and BCR shows that the autofeeder pond technology is higher than the financial ratio when compared to conventional pond systems. The calculation results for the conventional pond system are NPV Rp 1,515,178.503; IRR 38.24%; payback period 2.68 years; and BCR 1,696. As for the autofeeder pond technology, the results are Rp. 7,721,596,229; IRR 51.23%; payback period 2.17 years; and BCR 2,065.

The results of the calculation of risk criteria consisting of production risk and income risk obtained results of 6% for conventional ponds and 31% for autofeeder ponds. From these risk criteria, it is known that the risk of conventional ponds is smaller than that of an autofeeder pond, so the conventional pond system is better when compared to the autofeeder pond technology.

Based on the overall results of the assessment on the alternative selection between conventional shrimp ponds and autofeeder shrimp ponds it is known that conventional farms get a value of 4, while autofeeder ponds get a value of 11. Based on these assessments, the best alternative of the two shrimp farming systems is the shrimp pond technology autofeeder system.

## 5. References

[1] Chumaidiyah, Endang 2018 Strategy for Capability Development of Knitting Small Medium Enterprises Using SWOT Analysis. *Proceedings of the International Conference on Industrial Engineering and Operations Management Bandung*, Indonesia, March 6-8, 2018.

[2] Samson, D 2002 *Technology Management Text and International Cases.* (New York: Mc Graw Hill)

[3] Yun Chank, Byeung 2011 Business Process Management of Telecommunication Companies: Fulfillment and Operations Support and Readiness Cases, *International Journal of Future Generation Communication and Networking*, **4**(3), pp. 73-74.

[4] Coelli, TJ., DSP Rao, and G.E. Battese 1998 *Efficiency and Productivity Analysis*. (London: Kluwer Academic Publisher)

[5] Rahardja, A Z, Endang Chumaidiyah and W Tripiawan 2018 Feasibility Analysis on the Development of Steel Sheet Zinc Plated and Galvalum Production Factory PT. S Steel. *IOP Conference Series: Materials Science and Engineering* **528**(1), 2019-01-01.

[6] Rizal, Achmad 2018 Analisis Resiko Produksi Dan Pendapatan Budidaya Tambak Udang Rakyat Di Kelurahan Labuhan Deli, Kecamatan Medan Marelan, Kota Medan. *Jurnal Perikanan*, **8**(1), 65-70

[7] Rangka, Nur Ansari 2008 *Potensi dan Kendala Pengembangan Budidaya Udang Vanamei*, Badan Riset dan sumber Daya Manusia Kelautan dan Perikanan, **3**(1).

IOP Conf. Series: Materials Science and Engineering **847** (2020) 012049 doi:10.1088/1757-899X/847/1/012049

# Raw Materials Inventory Planning in Automotive Industries by EOQ Method Consider with the Contract Agreement

**Teguh Sri Ngadono[1,2] and Zulfa Fitri Ikatrinasari[1,3]**

[1]Universitas Mercu Buana, Jl. Meruya Selatan, Jakarta, Indonesia
[2]teguh.ngadono@mercubuana.ac.id, [3]zulfa.fitri@mercubuana.ac.id

**Abstract.** One of industrial sector that currently continuous growing up is automotive industry. Safety glass is a kind of automotive component that always has improvement and development. To support production running well, raw materials supply inventory has to be considered and managed as well. Poly Vinyl Butyral (PVB) as raw material is purchased from 2 suppliers and both of them have a contract agreement. Inventory shall be set up to fulfill the demand and follow the contract. By use Economic Order Quantity (EOQ) Method, inventory is calculated to minimize the cost. So the research objective is to determine inventory system of PVB and inventory cost by optimum. Then researcher got the result is EOQ Method can reduce the inventory cost around 52% than last period by keep consideration with the contact agreement.
**Keywords**: Inventory, Planning, EOQ, Automotive

## 1. Introduction

Automotive industries are continue growing up and become strategic industries because have relation with another economic sectors, absorb bigger employee, and drive middle and micro industrial sector, and utilize simple technologies until complex technology. Indonesia becomes one of countries in South East Asia that has big quantity car production after Thailand. It is proved by some car manufactures from Japan like Toyota, Honda, Suzuki, Nissan, Mitsubishi, Isuzu, Hino built their plants in Indonesia.

Automotive safety glass is important automotive component that shall follow international standard. Two kind of automotive safety glasses are tempered safety glass and laminated safety glass. To produce it, raw materials required are basic glass and polyvinyl butyral (PVB). PVB is used for laminated safety glass only.

AMF is international safety glass group which produces for safety glass car maker manufacture since 1985. Basic glass they can get it from internal plant and buy from their group in overseas, but for PVB they should buy from the suppliers due to no plant in their group and material special handling which is PVB should be kept in cold storage by temperature below than $10^0$ C. To fulfil demand from customer AFG buy PVB from two suppliers with contract agreement as term condition. There for, needs a good inventory planning to minimize their cost. From both suppliers, only PVB Standard Clear type that is managed in the contract.

Zhao and You (2016) studied an EOQ inventory models by pre-sale policy to reduce items where demand depends on the inventory on-hand and items price. Yang, et al. (2015) considered dynamic inventory control and optimize problem at price arrangement on inventory system by adjustment price. Ghorbel, et al. (2014) compared from previews about research works dealing with replenishment

IOP Conf. Series: Materials Science and Engineering **847** (2020) 012049 doi:10.1088/1757-899X/847/1/012049

policies and concluded that inventory should consider the uncertainty on the parameters such as demand, availability, lead time and etc. Nishard and Arunkumar (2018) studied EOQ to be proposed to the company due to current inventory management is not proper which cause overstock and sometimes stock out.

According to Ristono (2013) inventory can be interpreted as goods that are stored for use or sale in the future or period to come. Inventories consist of raw material inventory, semi-finished material inventory and finished goods inventory. Inventories of raw materials and semi-finished materials are stored before used or put into the production process, while inventories of finished goods or merchandise are stored before being sold or marketed. Inventory or inventory is a technique for material management related to inventory. Material management in inventory carried out with several inputs used, namely: the demand that occurs (demand) and the costs associated with storage, and costs if there is a shortage of inventory (shortage).

Inventory control according to Assauri (2010) is one of the activities of the sequence of activities that are closely linked to each other in all of the company's production operations in accordance with what has been planned in advance both in time, quantity, quality and cost.

$$EOQ = \sqrt{\frac{2SD}{IC}} \quad (1)$$

Where is:
EOQ = Economic Order Quantity
S = Ordering Cost
D = Demand QTy per period
C = Holding Cost

Next determine the total cost of supplies (TIC) by adding up the cost of ordering and holding costs. The formula is as follows.

$$TIC = \left[\frac{D}{Q} x\, S\right] + \left[\frac{Q}{2} x\, H\right] \quad (2)$$

Where is:
TIC = Total inventory cost
D = Demand, units per year
Q = Number of items each time ordered
S = Order Cost every time order
H = Storage costs, units per year

A safety stock is an additional inventory held to protect or protect the possibility of a material shortage (stock out). According to Assauri (2010) the possibility of stock out can be caused by the use of raw materials that are greater than originally estimated or the delay in the arrival of raw materials ordered. To determine the cost of safety stock, statistical analysis is used, which is to consider the deviations that have occurred between the estimated use of raw materials and actual use so that the standard deviation is known. The standard deviation formula is as follows.

$$SD = \sqrt{\frac{\Sigma(x-\bar{x})^2}{n}} \quad (3)$$

Where is:
SD = Standard Deviation
x = Demand Qty
$\bar{x}$ = Demand Average
n = Data Qty

While the formula used to calculate the safety stock is as follows.

$$SS = SD \times Z \quad (4)$$

Where is
SS = Safety Stock
SD = Standard Deviation
Z = Safety factor used by the company

In this case, the safety factor in question is the large probability used by the company for the occurrence of stock out. For example, a company uses a probability of 1% for a stock out, then by using a normal frequency distribution table, the value of Z0.01 = 2.33 (Heizer and Render, 2013).

According to Heizer and Render (2013) ROP is an inventory level where when inventory has reached that level, orders must be made immediately and ROP can be calculated with the following formula:

$$ROP = (D \times L) + SS \quad (5)$$

Where is:
ROP = Reorder Point
D = Needs Qty per unit time
L = Lead time
SS = Safety Stock

Then, how to plan inventory system of PVB by optimum to fulfill customer demand, so material shortage can be avoided if there is increasing demand? And how much inventory cost of PVB materials?

The objective this research is to determine inventory system of PVB and inventory cost by optimum.

## 2. Methods

This research is done by quantitative method. To do a measurement, every phenomenon is explained to the some variables and indicators. Each decided variables measured by giving numeric symbols that are different according to the related information. So, calculation technique by mathematical numeric can be done and can get result a conclusion that is generally in a parameter.

The research is exploratory and interview is used by opened question and answer than systematically question. The purpose of this study is to obtain empirical data that can be used in formulating, expanding, and verifying theories and solving problems from phenomena that are obtained theoretically and scientifically. In addition, this research is expected to be a material for further research to develop several methods besides the methods that have been done.

Population and sample is collected only for PVB material and focus to the actual consumption in the last period and the forecasting in the next period that was calculated by Production Planning Department.

Analysis will be done by use Economic Order Quantity. Sensitivity analysis in the EOQ model has significance for management, because after all the results of the EOQ calculation are not the final decision. What the EOQ model shows is an input for management in establishing the final inventory policy decision. Even though EOQ recommends a certain amount of purchases economical in every order, but EOQ may not have considered all aspects of the supply situation. Therefore also, the decision maker must have the freedom to modify the purchase amount recommended by EOQ to be able to meet the environmental characteristics of the inventory problem faced.

According to Tampubolon (2018) sensitivity analysis is very important because the results of the analysis can provide clues to the size error both in terms of cost calculations and in the quantity of inventory. To calculate the analysis of excess quantities, it can be calculated by the following equation (Tampubolon, 2018):

$$\frac{EOQ}{Q} = \frac{1}{2}\left[\frac{EOQ}{Q} + \frac{EOQQ}{EOQ}\right] \quad (6)$$

Where is:
EOQ = EOQ calculation
Q = Q number of orders specified by the company is not based on EOQ

If Q is not the same as EOQ, it will result in marginal costs that will be borne by the company. While marginal cost is the sensitivity ratio obtained minus the lowest sensitivity ratio multiplied by the total inventory cost. If it is made in the form of an equation is as follows:

Marginal cost = (sensitivity ratio - 1) x total inventory cost (7)

## 3. Result and Discussion

AMF inform that previous PVB Standard Clear has inventory cost around US $ 3,733,355.105. Based on information that was collected Ordering Cost (S): US $ 15,000 per one time order, PVB plan consumption (D) as informed in Table 1. and PVB unit price (I) is adjusted according to the contract price with both of suppliers as informed on Table 2. And Holding cost (C) is 20%

Table. 1 Production Planning PVB (unit in Square meter (Sqm))

| Month | *Standard Clear* |
|---|---|
| Jan-19 | 117.191 |
| Feb-19 | 117.191 |
| Mar-19 | 117.191 |
| Apr-19 | 117.191 |
| May-19 | 117.191 |
| Jun-19 | 117.191 |
| Jul-19 | 117.191 |
| Aug-19 | 117.191 |
| Sep-19 | 117.191 |
| Oct-19 | 117.191 |
| Nov-19 | 117.191 |
| Dec-19 | 117.191 |
| TOTAL | 1.406.286 |

Table 2. PVB Contract Information

| Vendor | Contract Qty | Rebate | Unit Price | Order Amount | Rebate |
|---|---|---|---|---|---|
| Vendor A | 1.100.000 | 1% | $ 4.00 | $ 4.400.000 | $ 4,400 |
| Vendor B | 600.000 | #N/A | $ 4.40 | $ 2.640.000 | |
| TOTAL Contract | 1.700.000 | | TOTAL Amount | $ 7.040.000 | |

From the table above, the price of PVB standard clear films used by using an average of the total value purchased from the two suppliers divided by the amount of PVB purchased so that the price per unit (I) is US $ 4.14. From the results of the forecasting, the need for Standard Clear PVB for the next period (D) is 1,406,286 m2 with the number of contracts agreed upon with both suppliers (Q) of 1,700,000 m2. If the number of orders to Vendor A meets the agreed contract, the company will get a rebate of 1% of the purchase value of US $ 4,400 which can reduce inventory costs. Assuming that in one month there are 4 weeks. Then EOQ is calculated as formula (1) and total cost (TC) in Table 3.

Table 3. EOQ and TC PVB Standard Clear

| Month | Demand (m2) | Unit Price ($US) I | Order Cost ($US) S | Demand per Week (m2) D | Holding Cost C | EOQ | TC with EOQ ($US) |
|---|---|---|---|---|---|---|---|
| Jan-19 | 117.191 | 4,14 | 15.000 | 29.297,63 | 20% | 32.581 | 148.269 |
| Feb-19 | 117.191 | 4,14 | 15.000 | 29.297,63 | 20% | 32.581 | 148.269 |
| Mar-19 | 117.191 | 4,14 | 15.000 | 29.297,63 | 20% | 32.581 | 148.269 |
| Apr-19 | 117.191 | 4,14 | 15.000 | 29.297,63 | 20% | 32.581 | 148.269 |
| May-19 | 117.191 | 4,14 | 15.000 | 29.297,63 | 20% | 32.581 | 148.269 |
| Jun-19 | 117.191 | 4,14 | 15.000 | 29.297,63 | 20% | 32.581 | 148.269 |
| Jul-19 | 117.191 | 4,14 | 15.000 | 29.297,63 | 20% | 32.581 | 148.269 |
| Aug-19 | 117.191 | 4,14 | 15.000 | 29.297,63 | 20% | 32.581 | 148.269 |
| Sep-19 | 117.191 | 4,14 | 15.000 | 29.297,63 | 20% | 32.581 | 148.269 |
| Oct-19 | 117.191 | 4,14 | 15.000 | 29.297,63 | 20% | 32.581 | 148.269 |
| Nov-19 | 117.191 | 4,14 | 15.000 | 29.297,63 | 20% | 32.581 | 148.269 |
| Dec-19 | 117.191 | 4,14 | 15.000 | 29.297,63 | 20% | 32.581 | 148.269 |
| TOTAL | 1.406.286 | | | | | 390.970 | 1.779.229 |

Because the company already has a contract with an agreed supplier of 1,700,000 sqm, then assuming there are 52 weeks in one year the number of orders that must be done in one week is 1,700,000 / 52 = 32,692 m2

$$\frac{EOQ}{Q} = \frac{1}{2}\left[\frac{EOQ}{Q} + \frac{Q}{EOQ}\right] = \frac{1}{2}\left[\frac{32581}{32692} + \frac{32692}{32581}\right] = 1.000005836977161$$

This means that Q is equivalent to EOQ so there is no marginal cost borne by the company. So that a rebate of US $ 4,400 can be obtained as purchase profit.

After calculating the EOQ, the next step is the calculation of Safety Stock and ROP. Standard deviation is calculated by using equation (3) in a theoretical study, safety stock using equation (4) and ROP using equation (5). In this calculation it is assumed that in 1 week there are 4 weeks to calculate the number of requests in 1 month, while the average weekly demand is obtained from the number of requests in one year divided by 52 weeks with the assumption in one year there are 52 weeks. The inventory safety factor used by the company is 1.8 weeks to avoid fluctuations demand from customers and in the ROP calculation the specified delivery time / lead time of delivery is 6/7 days or 0.86 weeks assuming 4 days of shipping on shipping and 2 days for complete the Customs and Clarance process at the Port. The company has established a 2-day Customs Clearance process because it has obtained the Priority Facility from Customs. So that the calculation is obtained as in Table 4. Below

Table 4. Safety Stock and ROP PVB Standard Clear

| Month | Qty Demand (m2) | Pland Demand per Week (m2) | Average Demand per Week (m2) | EOQ (m2) | Safety Stock (m2) | ROP (m2) |
|---|---|---|---|---|---|---|
| Jan-19 | 117.190,5 | 29.297,6 | 27.044.0 | 32.581 | 2.028,3 | 25.208,8 |
| Feb-19 | 117.190,5 | 29.297,6 | 27.044.0 | 32.581 | 2.028,3 | 25.208,8 |
| Mar-19 | 117.190,5 | 29.297,6 | 27.044.0 | 32.581 | 2.028,3 | 25.208,8 |
| Apr-19 | 117.190,5 | 29.297,6 | 27.044.0 | 32.581 | 2.028,3 | 25.208,8 |
| May-19 | 117.190,5 | 29.297,6 | 27.044.0 | 32.581 | 2.028,3 | 25.208,8 |
| Jun-19 | 117.190,5 | 29.297,6 | 27.044.0 | 32.581 | 2.028,3 | 25.208,8 |
| Jul-19 | 117.190,5 | 29.297,6 | 27.044.0 | 32.581 | 2.028,3 | 25.208,8 |
| Aug-19 | 117.190,5 | 29.297,6 | 27.044.0 | 32.581 | 2.028,3 | 25.208,8 |
| Sep-19 | 117.190,5 | 29.297,6 | 27.044.0 | 32.581 | 2.028,3 | 25.208,8 |
| Oct-19 | 117.190,5 | 29.297,6 | 27.044.0 | 32.581 | 2.028,3 | 25.208,8 |
| Nov-19 | 117.190,5 | 29.297,6 | 27.044.0 | 32.581 | 2.028,3 | 25.208,8 |
| Dec-19 | 117.190,5 | 29.297,6 | 27.044.0 | 32.581 | 2.028,3 | 25.208,8 |

IOP Conf. Series: Materials Science and Engineering **847** (2020) 012049 doi:10.1088/1757-899X/847/1/012049

From Table 4 can be seen the figure of Safety Stocks and ROP (Figure 1) as below.

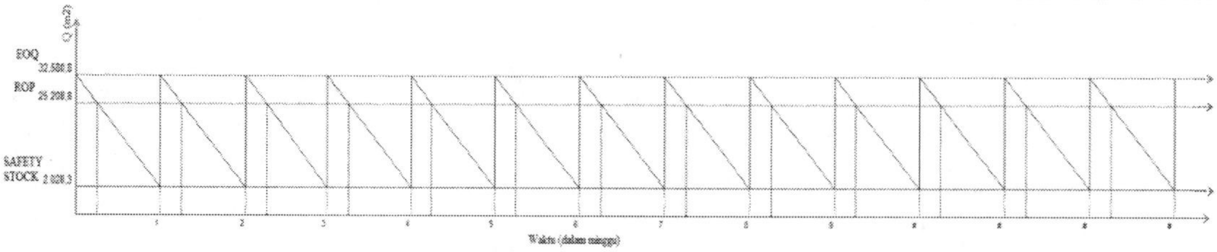


Figure 1. Safety Stock and ROP PVB Standard Clear

In calculating the total inventory cost, a greater result is obtained because the number of contracts with suppliers is greater than the number of demand forecasts in the coming period. So that there is an excess that must be borne by the company resulting in the total inventory costs being greater when compared to using the total cost calculation of the EOQ method. So that if marginal cost is formulated due to excesses that must be borne as follows:

Marginal cost = (sensitivity ratio - 1) x total inventory cost (8)

Marginal costs are increases in total costs that come from the production of one unit of output. Marginal cost measures the additional input costs needed to produce each unit of subsequent output. Because fixed costs do not change when there are output costs, marginal costs reflect changes in variable costs. The cause of the marginal cost to the company is the excess stock that must be borne.

Determination of the economic order Quantity (EOQ) is done if the supply of raw materials depends on several suppliers, so it is necessary to consider the amount of material inventory purchases as needed. This is consistent with the data obtained that the company already has a purchase cooperation contract with the supplier of PVB material.

## 4. Conclusion

Standard Clear PVB inventory costs decreased 52% from US $ 3,733,355.105 to US $ 1,779,225.156. From the number of contracts agreed with the supplier for PVB Standard Clear did not result a marginal cost and the sales rebate could be given from the suppliers as purchase profit around US $ 4,400

EOQ calculation should be reviewed for the next period if there is alteration of contract agreement related the term condition about volume and unit price. And it can be consideration for AMF to check the inventory planning for another materials.

## 5. References


[1] Assauri, Sofjan 2010 *Manajemen Pemasaran: Dasar, Konsep & Strategi* (Jakarta: Raja Grafindo Persada)

[2] Ghorbel, N, Addouche, A, El-Mhamedi, Boujelbene, Y 2014 Replenishment policies in static and dynamic spareparts inventory control: A survey *International Journal of Engineering Development and Research* **2**(4), (Madurai:www.ijedar.org) p 3636

[3] Heizer, J., and Barry, R. 2013 *Manajemen Operasi*. (Jakarta: Salemba Empat)

[4] Nishard, I and Arunkumar 2018 Analysis of Inventory Management by Using Economic Order Quantity Model - A Case Study. *International Journal for Research in Applied Science & Engineering Technology (IJRASET)* **6**(VI), (India/www.ijraset.com) p. 309-315

[5] Ristono, A 2013 *Manajemen Persediaan* (Yogyakarta: Penerbit Graha Ilmu)

[6] Tampubolon, Manahan P. 2018 *Manajemen Operasional (Operation Management)* (Jakarta: Ghalia Indonesia)

[7] Yang, B., Gao, C., Liu N., and Xu L 2015 Dynamic Inventory and Pricing Policy in a Periodic-Review Inventory System with Finite Ordering Capacity and Price Adjustment Cost. *Mathematical Problems in Engineering* (London: Hindawi Publishing Corporation) Article ID 269695

[8] Zhao, L, You, J, 2016 Optimal Pricing and Ordering Policy for Deteriorating Items with Stock and Price Dependent Demand and Presale Rebate *Scientific Programing* (London: Hindawi Publishing Corporation) Article ID 1507285

IOP Conf. Series: Materials Science and Engineering **847** (2020) 012050 doi:10.1088/1757-899X/847/1/012050

# Three-Level Supply Chain considering Direct and Indirect Transportation Cost and Carbon Emissions

**A K S Ong[1,3], R A C Robielos[1,4], Y. T. Jou[2,5] and H M Wee[2,6]**

[1]Department of Industrial Engineering and Engineering Management, Mapua University, Intramuros, Manila
[2]Department of Industrial & Systems Engineering, Chung Yuan Christian University, Taiwan
[3]ardvinkester@yahoo.com; [4]racrobielos@mapua.edu.ph, [5]ytjou@cycu.edu.tw
[6]weehm@cycu.edu.tw

**Abstract.** Managing sustainable supply chains has become a challenge for most industries primarily because environmental, social, and economic aspects must be taken into account, and the difficulty in satisfying customer demands for sustainability and environmental protection has been increasing globally, making sustainability the focus of numerous studies on green supply chains. Previous studies are usually only applicable to a specific product supply chain. Therefore, this paper review model that focuses on a general three-level supply chain considering direct and indirect transport and industrial carbon emissions and costs. Specifically, the model involves joint economic lot sizing considering a single set-up-multiple-delivery (SSMD) policy.
**Keywords**: carbon tax, single set-up-multiple-delivery, supply chain, sustainability

## 1. Introduction

Current research trends deal with optimizing sustainable supply chains (Tancrez et al., 2011), which include minimizing the total costs for supply chains, and addressing environmental factors such as minimizing total carbon emission as a form of competitive advantage to other industries (Sana, 2010). This optimization is a challenge due to the consideration of the three dimensions: environmental, social, and economic aspects of the supply chain (Zadjafar and Gholamian, 2018). In addition, Zadjafar and Gholamian (2018) stated that customer awareness regarding sustainability and environmental conservation has been rising globally, making it the focus of most studies, and especially in global supply chains. Improved sustainability optimizes environmental factors, products, and cost. In line with this, carbon emissions are classified to better affix the costs incurred in an industry. These carbon emission costs are considered as indirect and direct emission costs. Indirect carbon emission costs are the consumptions of electricity, heating, cooling, steam, and the like for industrial use. Direct carbon emission costs cover the production of materials. In addition, transportation costs are divided into indirect and direct carbon emission costs as well. Indirect transportation emission costs result from business-related travel, pick-up and distribution (considering the weight of the material), and other outsourced activities (Wangsa, 2017). Direct transportation emission costs cover company-owned vehicles, equipment, fuel combustions, etc. According to Miyamoto and Takeuchi (2019), to reduce carbon emissions, implementation of the 1995 Kyoto Protocol is required. The parties involved in sustainable supply chains should also consider additional factors such as a carbon tax and carbon capping in annual costs (Daryanto et al., 2019). A carbon tax is a cost implementation to a certain amount of carbon emission produced corresponding to an amount of currency which varies by each country (Ding et al., 2019). On the other hand, a carbon cap is define as having a minimum amount of allowable carbon emission through a production process, for which any emission in excess would incur a corresponding monetary cost (Yi et al., 2019).

Numerous studies regarding sustainable supply chains have been done, including studies that dealt with a three-echelon supply chain considering deteriorating items (Daryanto et al., 2019). In addition, imperfect quality product-inventory modelling (Sana, 2010), sustainable inventory modelling for pulp and paper mills (Zadjafar and Gholamian, 2018), location-inventory (Tancrez et al., 2011), permissible delay in payment of a three-echelon supply chain (Sarkar, et al., 2018), and a two-echelon supply chain with permissible delay in payment for multi-constraint products (Priyan and Uthayakumar, 2014). Permissible delay in payment describes a scenario wherein the supplier can give ample time for the manufacturer to pay for ordered products, and similarly for the retailer with the manufacturer. The results of these studies showed great promise in producing sustainable supply chains for environmental emission factors and total supply chain costings. This study focused on the paper of Sarkar et al. (2018) wherein permissible payment delay in credit was considered in a three-echelon supply chain. It also considered a single supplier, single manufacturer, with multiple retailers following a single-set-up multiple delivery (SSMD) policy for the inventory system of a total supply chain. In the paper of Sarkar et al. (2018), the model considered transportation and variable transportation costs together. Moreover, the carbon emission cost was set to be constant with respect to the transportation cost of a player of the supply chain. In a real-world scenario, the carbon emission results from different factors (*i.e.* industrial and transportation) that comply with the Kyoto Protocol. The main goal of the Kyoto Protocol is reducing air pollution from industrial waste including through production, transportation, or inventory, with the goal to reduce at least 5% percent of the annual $CO_2$ emissions in the European Union and 37 other countries (Miyamoto and Takeuchi, 2019). Miyamoto and Takeuchi (2019) summarized studies on shifting to alternative sources of renewable energy and its effect on Kyoto Protocol implementation. The result of their study showed that reductions of carbon emissions would be beneficial to countries applying the protocol. However, shifting resources may not be applicable to every sector of a supply chain due to costing constraints. Moreover, Maamoun (2019) showed how the Kyoto Protocol reveals hidden carbon emission reduction and optimization of costs to be utilized by different industries. Because of that, studies are being conducted to minimize the production of greenhouse gas (GHG) emissions by parties of a supply chain (Daryanto, et al., 2019; Hovelaque and Bironneau, 2014; Jauhari, et al., 2014; Sarkar et al., 2015; Sarkar, et al., 2018; Wangsa, 2017).

This study discusses Sarkar et al. (2018) and the direct and indirect industrial and transportation carbon emissions. The paper therefore proposes a model that focuses on a three-level supply chain considering direct and indirect transport and industrial carbon emissions. Specifically, the model involves a joint economic lot sizing considering a single set-up-multiple-delivery (SSMD) policy with a multi-delay payment, aiming to minimize the total cost and the total carbon emission of the three-level supply chain.

The objective of the model is to minimize the total cost of the sustainable supply chain by deriving the model using an algebraic approach. In addition, revenue earned, and alternative cost will be considered because of the multi-level in credit payment scenarios to determine which delay-in-payment scenario would be beneficial to the whole supply chain. The optimum result was obtained among the four scenarios by using numerical examples. This study will also address how the penalty and incentive policies of carbon capping will affect total cost. Lastly, sensitivity analysis was made to check the optimality of the findings.

The conclusions from this paper may benefit managers in different sectors of a three-level supply chain mainly because it provides a general model for the total costing incurred is provided, and because the optimal result will minimize total cost and carbon emissions. Lastly, under the Kyoto Protocol, the model will serve as a platform to determine which part of the supply chain or costing to manipulate to reduce cost and emissions.

The model utilized numerical examples and sensitivity analysis to determine the optimal result of the proposed mathematical model. The study developed the model with Maple Software for the derivation and computation. The rest of the paper is organized as follows: the review of literature for the second section, model definition, notation, and then assumptions, followed by the mathematical model without the permissible payment for credit for the third section, the model having permissible payment for credit policy and incentive and penalty for carbon tax for the fourth section, and conclusion and recommendation for the fifth section.

## 2. Methods

This study discusses related literature to find the gap in the studies relating to supply chain dealing with carbon emissions. Several studies (summarized in Table 1) have examined reducing carbon emission, but they still have a long way of improvement for international application because incorporating all carbon emission factors is difficult in a small scale. The study of Wangsa (2017) considered indirect and direct carbon emission from transportation and industrial sources which could be used when extending a study. The focus on the specified emissions should be taken seriously for increase in the improvement of the reduction of carbon emission (Almer and Winkler, 2017).

To reduce carbon emissions, governments have implemented carbon cap-and-trade and carbon taxes for major carbon emitters (He et al., 2014). A carbon cap is the amount of permitted carbon emissions that may be produce by a certain system, for which emission in excess would result in a penalty payment. The cap-and-trade system trades an amount of carbon emission that was unused to other supply chains through a specific carbon market. A carbon tax is a constant rate in payment given to a level of carbon emission produced.

One model that is usually created deals with a Single-Set-up-Multiple-Delivery (SSMD) Policy. This is because the multiple delivery incorporated in the supply chain deals greatly on the decision-making factor in the total cost reduction (Chan et al., 2018). Transportation as one factor causes not only cost but carbon emission as well. The study of Chen et al. (2018) deals on how the combination of the multiple delivery helps in the reduction of total cost of the supply chain system in a real-life scenario. The coordination of the single supplier to multiple buyer helps in the manipulation of the total cost. In the study of Beck et al. (2017), the conditions set for the different shipment set-up for multiple buyers were considered. The shipment policies considered showed promise for future researches to adapt. Both studies of Chen et al. (2018) and Beck et al. (2017) covered SSMD policy but did not consider carbon emission.

Business industries are currently dealing with much when it comes to sustainability. This is mainly because of the knowledge the market gains due to the customer demand of having greener production processes and commodities (Zadjafar and Gholamian, 2018). For example, Daryanto et al. (2019) considered carbon emission focusing on deteriorating items of a three-echelon supply chain. The study dealt with a single supplier to a single buyer. Moreover, Daryanto et al. also did a study in 2018 covering sustainable inventory management in deteriorating items and imperfect quality items that considered carbon emission without affecting the profitability of the supply chain. In addition, the study of Zadjafar and Gholamian (2018) also dealt with sustainable inventory model but focused on items such as pulps and mills.

The global competitive environment is high when it comes to supply chains, which are dealing greatly on minimizing their total costs. Different studies were formulated in-order to minimize costing, which discovered that considering the total supply chain would be beneficial for cost manipulation rather than individual systems of the supply chain (Priyan and Uthayakumar, 2014). Like the study done on multi-commodity supply chain of a two-echelon supply chain by Sadjady and Davoudpour (2011), they considered lead time and inventory costs. They also dealt with the lot-size and location and planning both tactical and strategic considered. The study covered mainly the inventory and lead times to minimize the costs. Their strategy involved a heuristic solution algorithm but needs further research when it comes to the capacity limitations and transportation modes.

Table 1 lists the related literature summary for the proposed model considering transportation, inventory, carbon emission, carbon tax, and multi-delay in payment.

**Table 1.** Literature review summary

| Author | Specific Item | Transportation | Inventory | Multi-delay in payment | Carbon Emission | Carbon Tax |
|---|---|---|---|---|---|---|
| Sarkar et al. (2016) | | √ | √ | | √ | |
| Ghosh et al. (2017) | | √ | √ | | √ | |
| Beck et al. (2017) | | √ | √ | | | |
| Qi et al. (2017) | √ | √ | √ | | √ | |
| Tang et al. (2017) | | √ | √ | | √ | √ |
| Wangsa (2017) | | √ | √ | | √ | √ |
| Chan et al. (2018) | | √ | √ | | | |
| Manupati et al. (2018) | | √ | √ | | √ | √ |
| Daryanto et al. (2018) | √ | √ | √ | | √ | √ |
| Sarkar et al. (2018) | √ | √ | √ | √ | √ | |
| Zadjafar and Gholamian, (2018) | √ | √ | √ | | √ | √ |
| Marchi et al. (2019) | √ | √ | √ | | √ | |
| Daryanto et al. (2019) | √ | √ | √ | | √ | √ |
| Our 4 model | | √ | √ | √ | √ | √ |

## 3. Result and Discussion

The cost summation of the players of the supply chain, as generally seen in equation (1), will be optimized to determine the least cost that may be incurred of the whole system.

$$Total\ cost\ =\ Supplier\ Costs + \ Manufacturer\ Costs\ +\ Retailer\ Costs \tag{1}$$

The cumulative cost of the three-level supply chain is simplified in equation (2).

$$TC(C_T, M_{CT}, R_{CT}, S_R, M_R) = \left(\frac{C_T}{2}\left(R_{CT}\left(\frac{h_{sr}M_{CT}D^2}{S_RP_s} + \frac{2h_sD^2}{M_RP_s} - \frac{h_sM_{CT}D^2}{P_s} + \frac{h_sD^2}{P_m} + h_sM_{CT}D - h_sD - \frac{h_mD^2}{P_m} + h_mD\right)\right) + \frac{2h_mD^2}{P_m} - h_mD\right) + \left(\frac{1}{C_T}\left(\frac{1}{R_{CT}}\left(\frac{A_s+O_sS_R}{M_{CT}} + A_m + O_mM_R\right)\right) + O_r\right) +$$

$$\left(C_T\left(M_{CT}\left(\frac{D\Delta_{I1}(H_{em}+C_{em}+S_{em}+E_{em})E_IC_{tax}}{S_s} + \frac{D\Delta_{I2}C_{tax}}{S_s} +\right.\right.\right.$$

$$\frac{\gamma\delta(2D_s+D_m)}{S_s} + \frac{C_{tax}[D\Delta_{T1}W+\gamma(2D_s+D_m)]}{S_s}\Big)\Big) +$$

$$\left(R_{CT}\left(\frac{D\Delta_{I1}(H_{em}+C_{em}+S_{em}+E_{em})E_I C_{tax}}{S_m} + \frac{D\Delta_{I2}C_{tax}}{S_m} + \right.\right.$$

$$\frac{\gamma\delta(2D_m+D_r)}{S_m} + \frac{C_{tax}[D\Delta_{T1}W+\gamma(2D_m+D_r)]}{S_m}\Big)\Big) +$$

$$\frac{D\Delta_{I1}(H_{em}+C_{em}+S_{em}+E_{em})E_I C_{tax}}{S_r} + \frac{C_T\Delta_{T1}\gamma(2D_r)C_{tax}}{S_r}\Big) \quad (2)$$

The decision variables were derived algebraically to get the least value. (Appendix A shows all variable notations)

$$TC\ (C_T, M_{CT}, R_{CT}, S_R, M_R) = C_T\left(K_1 + K_2 + \frac{K_3}{2} + \frac{K_4}{2}\right) + \left(\frac{A}{T}\right) \quad (3)$$

$$= \frac{\left(T\sqrt{K_1+K_2+\frac{K_3}{2}+\frac{K_4}{2}} - \sqrt{A}\right)^2}{T} + 2\sqrt{(K_1 + K_2 + \frac{K_3}{2} + \frac{K_4}{2})(A)} \quad (4)$$

The total cost equation reaches minimum value when the decision variable with respect to $C_T$ is

$$C_T^* = \sqrt{\frac{A}{K_1+K_2+\frac{K_3}{2}+\frac{K_4}{2}}} \quad (5)$$

The minimum value of the total cost would become

$$TC_1(M_{CT}, R_{CT}, S_R, M_R) = 2\sqrt{(K_1 + K_2 + \frac{K_3}{2} + \frac{K_4}{2})(A)} \quad (6)$$

Since the optimal value for $C_T$ depends on the other decision variables, $M_{CT}, R_{CT}, S_R, M_R$, these would be derived from the equation containing the specific variable as seen in equation (46) for the $R_{CT}$ variable.

$$TC_1(M_{CT}, R_{CT}, S_R, M_R) = 2\sqrt{(K_1 + K_2 + \frac{K_3}{2} + \frac{K_4}{2})(A)} \quad (7)$$

Which translates to

$$TC_1(M_{CT}, R_{CT}, S_R, M_R) = 2\sqrt{\left(\frac{\emptyset_2}{R_{CT}} + O_r\right)\left(\frac{\varphi_2 R_{CT}+\alpha_2}{2} + \frac{2R_{CT}\varphi_3}{2} + \frac{M_{CT}\varphi_4}{2} + \frac{\varphi_5}{2}\right)} \quad (8)$$

Simplifying,

$$= 2\sqrt{\left(\frac{\emptyset_2}{R_{CT}} + O_r\right)\left[R_{CT}\frac{J_1}{2} + \frac{J_2}{2} + \frac{J_3}{2}\right]} \quad (9)$$

Solving for the decision variable $R_{CT}$, the optimal value would be equation (46).

$$= 2\sqrt{\left(\sqrt{J_1 R_{CT} O_r} - \sqrt{\frac{J_2\emptyset_2+J_3\emptyset_2}{R_{CT}}}\right)^2 + 2\sqrt{J_1 O_r}\sqrt{J_2\emptyset_2 + J_3\emptyset_2} + J_1\emptyset_2 + J_2 O_r + J_3 O_r} \quad (10)$$

$$R_{CT}^* = \sqrt{\frac{J_2\emptyset_2+J_3\emptyset_2}{O_r J_1}} \quad (11)$$

The resulting cost function after deriving $R_{CT}$ with respect to the other decision variables $M_{CT}, S_R, M_R$, would be equation (12).

$$TC_1(M_{CT}, S_R, M_R) = \sqrt{2}\left(\sqrt{\emptyset_9} + \sqrt{O_r J_2}\right) \quad (12)$$

The portion with the decision variable $M_{CT}$ was obtained to get equation (13).

$$\sqrt{\emptyset_9} = \sqrt{O_r(\varphi_4 M_{CT} + \varphi_5) + M_{CT}J_5(\varphi_1) + + \frac{\emptyset_1 \alpha_1}{M_{CT}} + J_6} \quad (13)$$

The optimal value for $M_{CT}$ would therefore be

$$M_{CT}^{*} = \sqrt{\frac{\emptyset_1(\alpha_1)}{O_r\varphi_4 + O_r\varphi_5 + J_5\varphi_1}} \quad (14)$$

The total cost becomes equation (15) with respect to the other decision variables, $S_R, M_R$.

$$TC_1(S_R, M_R) = \sqrt{J_6 + 2\sqrt{(\emptyset_1\alpha_1)(J_5\varphi_1 + O_r\varphi_4 + O_r\varphi_5)}} \quad (15)$$

By simplifying and solving for the equation containing the decision variable $S_R$, the optimal variable is,

$$= \sqrt{J_5(\alpha_1 J_7)\emptyset_1(\varphi_1)} \quad (16)$$

$$= \sqrt{\left(\sqrt{J_5(\alpha_1 J_7)} + \sqrt{\emptyset_1\varphi_1}\right)^2} \quad (17)$$

$$S_R^{*} = \sqrt{\frac{A_s h_{sr} D^2}{P_s(\varphi_0 O_s)}} \quad (18)$$

The remaining variable in the equation is $M_R$ and the total cost function becomes,

$$TC_1(M_R) = \sqrt{(J_5\alpha_1 J_7)} + 2\sqrt{\frac{A_s h_{sr} D^2}{P_s}(\varphi_0 O_s)} \quad (19)$$

Taking the equation containing $M_R$, as equation (20),

$$J_5(\alpha_1 J_7) = (A_m + O_m M_R)(O_r\varphi_4 + O_r\varphi_5)(J_5 J_7 \alpha_1)\left[\frac{2h_s D^2}{M_R P_s} + \frac{h_s D^2}{P_m} - h_s D - \frac{h_m D^2}{P_m} + h_m D\right] \quad (20)$$

Simplifying,

$$= \left(\sqrt{M_R(J_9 O_m)} - \sqrt{\frac{2J_7 A_m h_s D^2}{P_s M_R}}\right)^2 + 2\sqrt{(J_7 O_m)(\frac{2J_7 A_m h_s D^2}{P_s})} \quad (21)$$

The optimum result for variable $M_R$ is

$$M_R^{*} = \sqrt{\frac{2J_7 A_m h_s D^2}{P_s(J_9 O_m)}} \quad (22)$$

The final total cost equation for the supply chain becomes equation (59).

$$\boldsymbol{ETC} = 2\sqrt{\frac{A_s h_{sr} D^2}{P_s}(\varphi_0 O_s)} + J_8 + J_{10} + 2\sqrt{O_m J_9 \frac{2J_7 A_m h_s D^2}{P_s}} \quad (23)$$

## 4. Conclusion

This study formulated a three-level supply chain that considers multi-level credit payments. This model would be considered as a platform for interim financing wherein the costs would be monitored in order to boost total profit from the sales.

The results of the model show great differences when applying multi-level-credit payment policy. From the numerical example, giving ample time for the payment would result in an increase in the profitability of the system due to the revenue, but the time-period should not be longer than the cycle time. Moreover, the allowance of delays in payment period would result in a good partnership after a positive outcome. The result showed a better profitability from 17.05% to 43.72%. It could then be seen from the result that giving enough time for the payment per player of the supply chain would give rise to a higher profitability because of the revenue income but not longer than the cycle time. In addition, the delay in payment period would result in a good partnership amongst players of the supply chain after a positive result (Priyan and Uthayakumar, 2014). The result also shows that the eco-friendly model and economic costing will help not only the industry, but the consumers and the environment as well. Managers would be able to utilize the model to create a better trade-off on running the system to have minimum cost for all players of the supply chain. Lastly, having to manipulate the period of payment delay in credit would result in a better profitability, noting that the higher holding cost for the retailer compared to the permissible allotted time for payment would not result to the optimal solution. Considering the cycle time and the allowed time in credit payment would be the key strategy in creating an optimized total cost that would be beneficial to all the players of the supply chain.

This would be beneficial to the parties of the supply chain as a benchmark in deciding on the best outcome for their system regarding the optimality of the costing based from the different illustrated (*i.e.* holding, transport, industrial, carbon emission, carbon tax, interest payment). Optimum minimization would be beneficial for applying the SSMD policy if the industrial cost considering the carbon emission would be less than the holding cost. In addition, the algebraic approach done for the paper presents a positive outcome and would be easier to manipulate in real-life scenarios. Compared to integrated derivations where results are sometimes difficult to manipulate for the application of the different supply chains in the industry, this model is flexible for adaptation to a supply chain's needs. The presented numerical example and sensitivity analysis for the main parameters would be easily manipulated for application purposes of different SSMD policy-applied industries. The study could be extended if the demand will be made in a stochastic approach to simulate a real-life scenario when it comes to inventory.

## 5. References

[1] Almer C., Winkler, R. 2015 Analyzing the effectiveness of international environmental policies: The case of the Kyoto Protocol. *Journal of Environmental Economics and Management*, **82**, 125-151.

[2] Beck, F., Glock, C., Kim, T. 2017 Coordination of a production network with a single buyer multiple vendor with geometrically increasing batch shipments. *International Journal of Production Economics*, **193**, 633-646.

[3] Chan, C., Fang, F., Langevin, A. 2018 Single-vendor multi-buyer supply chain coordination with stochastic demand. *International Journal of Production Economics*, **206**, 110-133.

[4] Daryanto, Y., Wee, H., Astanti, R. 2019 Three-echelon supply chain model considering carbon emission and item deterioration. *Transportation Research Part E*, **122**, 368-383.

[5] Ding, S., Zhang, M., Song, Y. 2019 Exploring China's carbon emissions peak for different carbon tax scenarios. *Energy Policy*, **129**, 1245-1252.

[6] Maamoun, N., 2019. The Kyoto Protocol: Empirical evidence of a hidden success. *Journal of Environmental Economics and Management*, **95**, 227-256.

[7] Marchi, B., Zanoni, S., Zavanella, L., Jaber, M. 2019 Supply chain models with greenhouse gases emissions, energy usage, imperfect process under different coordination decisions. *International Journal of Production Economics*, **211**, 145-153.

[8] Miyamoto, M., Takeuchi, K. 2019 Climate agreement and technology diffusion: Impact of the Kyoto Protocol in international patent applications for renewable energy technologies. *Energy Policy*, **129**, 1331-1338.

[9] Priyan, S., Uthayakumar, R. 2014 Two-echelon multi-product multi-constraint product returns inventory model with permissible delay in payments and variable lead time. *Journal of Manufacturing Systems*, **36**, 244-262.

[10] Qi, Q., Wang, J., Bai, Q. 2017 Pricing decision of a two-echelon supply chain with one supplier and two retailers under a carbon cap regulation. *Journal of Cleaner Production*, **151**, 286-302.

[11] Sana, S. 2010 A production-inventory model of imperfect quality products in a three-layer supply chain. *Decision Support Systems*, **50**, 539-547.

[12 Sarkar, B., Ganguly B., Sarkar, M., Pareek, S. 2016 Effect of a variable transportation and carbon emission in a three-echelon supply chain model. *Transportation Research Part E*, **91**, 112-128.

[13] Sarkar, B., Ahmed, W., Kim, N. 2018 Joint effects of variable carbon emission cost and multi-delay-in-payments under single-setup-multiple-delivery policy in a global sustainable supply chain. *Journal of Cleaner Production*, **185**, 421-445.
[14] Tiwari, S., Daryanto, Y., Wee, H. 2018 Sustainable inventory management with deteriorating and imperfect quality items considering carbon emission. *Journal of Cleaner Production*, **192**, 281-292.
[15] Wangsa, I. 2017 Greenhouse gas penalty and inventive policies for a joint economic lot size model with industrial and transport emissions. *International Journal of Industrial Engineering Computations*, **8**, 453-480.
[16] Yi, B., Xu, J., Fan, Y. 2019 Coordination of policy goals between renewable portfolio standards and carbon caps: A quantitative assessment in China. *Applied Energy*, **237**, 23-35.
[17] Zadjafar, M., Gholamian, M. 2018 A sustainable inventory model by considering environmental ergonomics and environmental pollution, case study: Pulp and paper mills. *Journal of Cleaner Production*, **199**, 444-458.
[18] Zhou, W., Chen, L., Ge, H. 2012 A multi-product inventory control model with joint replenishment strategy. *Applied Mathematical Modelling*, **37**, 2039-2050.

IOP Conf. Series: Materials Science and Engineering **847** (2020) 012051 doi:10.1088/1757-899X/847/1/012051

# Designing a Device for Measuring Human Reaction Time

**DI Rudolf** [1] **and V Triyanti** [1,2]

[1]Atma Jaya Catholic University of Indonesia Jalan Jenderal Sudirman 51, Jakarta 12930, Indonesia
[2] vivi.triyanti@atmajaya.ac.id

**Abstract.** Human reaction time is the time needed for someone to take action shortly after the person receives information. Not many measuring devices that can measure the time or speed of human reactions in detail, which are regarding the cognitive (thinking) and the motoric (action) aspects. Therefore, this research aims to design a reaction time measuring device for a simple activity. The device should calculate the speed of human reactions from the cognitive and motoric aspects. A series of product development phases was conducted, i.e user requirements identification, concept development, embodiment design, construction, and testing. As a result, the selected concept used stimulus from light to measure reaction time, considering the difference of light color, flash rate, and intensity. A series of trials proved that those characteristics could be used to measure and distinguish reaction time ($p$-value $< 0.05$), considering cognitive and motoric time. The device was also proven can differ reaction time normal and fatigue condition.
**Keywords**: Reaction time, a cognitive, motoric, product development phase

## 1. Introduction

Humans have 3 aspects of ability, namely: cognitive, affective, and psychomotor abilities [1]. Of the three aspects, psychomotor is an interesting aspect to study because psychomotor is related to the skill (ability) or the ability to act after someone receives certain stimuli or knowledge. Psychomotor represents a person's ability to receive knowledge and then actuate it into the form of movement following the commands received by the brain. According to [2] Psychomotor is the ability that prioritizes physical skills consisting of perception, readiness, guided movements, accustomed movements, complex movements, adjustment of movement patterns, and creativity. Psychomotor skills related to actions that require coordination between nerves and the brain [3] and to the workings of muscles so that they cause body movements [4].

A common method to measure psychomotor ability is by measuring the reaction time. The reaction time comes from the word "reaction" and "time". Reaction means the activity or action that arises because of an order or an event. Reaction ability can be determined by measuring the time interval between the receipt of stimuli with the answer (response). The Reaction time is the time spent between the appearance of a stimulus or stimulation from the start of the reaction until a certain reaction is done. Reaction time or speed can be identified by measuring the shortest time needed to give a kinetic answer after receiving a stimulus.

Considering the human information system model [5], humans must process the information or stimuli they receive in certain sequences before executing an action, which are perceptual encoding, central processing, and responding. Within the steps, the human process the stimulus received by the sensory, make perception based on long term memory, select response based on working memory, and finally execute the responses. The processes are measured from the moment stimulus is accepted by the human sensory until the reaction move start. Perception and response selection activity need more thinking skills than physical skills [5]. After the steps, responses will be actuated by executing some motions. The execution time is carried out until the end of the movement or until the task is completely fulfilled, depending on the complexity of the task.

A famous method of psychomotor performance measurement by measuring time reaction is the Psychomotor Vigilance Test (PVT) [6][7]. PVT measures the time needed to give a response for a simple stimulus (dot on a screen). The method usually used to detect fatigue or sleepiness [7]. In this method, the reaction is measured as a total time between stimulus task is given, until the task is responded completely. Considering the human information system model by Wickens [5], there are no established tools or approaches available to measure reaction time, which take into consideration the thinking time and the execution time. Therefore, the problem to be discussed in this research is how to design a tool to measure the ability of human psychomotor to perform a simple task, considering the cognitive time and the execution time.

This study aims to design a device to measure the psychomotor aspects of humans, especially in its relations with reaction time. In this study, the reaction time is defined as the ability of individuals to decide on the time stimulus accepted to the time responses finished in the shortest possible time. A product design methodology [8][9][10] was performed to design the device. The device is planned to be used in the university laboratory; therefore, user requirements were identified to know what aspect of psychomotor ability that interested to be explored further. Based on the requirement, a concept of measurement device was developed and constructed.

## 2. Methods

At this stage, the researcher developed alternatives to the reaction time measuring device concepts according to the user requirement. The selected concept was built into a real product. The design phase was carried out as follows:

### *2.1 Identification of Requirement*

The collection of user identification was done in two ways, which were literature and interview. The device is intended to be used in a university laboratory. Therefore, interviews with potential users (university students) were done to obtain user needs or wants. By translating the needs or wants of users to clearer sentences, the user requirements list was built. The list consisted of requirements that were organized into several hierarchies, which are primary, secondary needs, and tertiary needs. For each tertiary need, a quantitative technical need was defined. However, the technical needs and target was not presented in this paper. Based on the user requirement list, a functional breakdown was conducted as a basis for morphological charts and concepts generation

### *2.2. Design Phase: concept development, embodiment design, detail design*

Based on the functional breakdown, a morphological chart is built. The morphological chart is used to generate concepts for each function in the functional breakdown. By elaborating ideas from the morphological chart, some alternatives concepts of the device were generated. The Pugh's Selection Method [8][9] was used to select and evaluate each concept that most fulfill user requirements. As a result, one concept was selected and continued to further stage. In this manuscript, the morphological chart and the Pugh's Selection Method is not presented.

The second stage is the embodiment design stage. At this stage, the chosen concept was formed physically. The embodiment design phase was divided into 3 phases, namely: product architecture design, configuration design, and parametric design [9] the final stage was the detailed design phase. In this phase, the finalization of the design before product realization is carried out. In this paper, the second and third stages is not discussed in detail. Only the final product design is displayed.

### *2.3. Testing*
After the device was constructed, it was tested to the potential user. Testing was done to test that all prescribed functions were run well and to test that the tool can differentiate certain user's conditions. In this research, the prescribed condition was at normal and fatigue condition. Fatigue can cause a decrease in psychomotor abilities [11]. The condition of fatigue can reduce both capacities to work and the efficiency of achievement [12]. Therefore, testing was done two times on the same user, in a normal and fatigue condition. User in normal condition when she/he is in good health, fit and focused. On the other hand, the condition of fatigue in this study was defined as a condition where users are in a state of exhaustion due to lack of sleep. In this condition, the test is carried out when the participant was not sleeping (staying up late). This manuscript only presented the final results. The detailed test was not presented in this manuscript.

## 3. Result and Discussion

### *3.1. Identification of Requirements*
Based on the user requirement identification phase, 12 tertiary requirements were expected to be carried out by this tool (Table 1). In addition to needs related to features such as easy to operate and ergonomics, there are several performance needs related to the expected function. If the existing speed measurement tool (PVT) can only measure the total reaction time, then this tool is expected to be able to separate thinking time (cognitive time) and execution time. Besides, this tool is also expected to consider several factors that may be related to the reaction time, namely the color of light, flash, and light intensity. For the three factors considered, the division is as follows: color of the light (dark and bright), flash (flashing and steady), light intensity (high and low). Based on these needs, technical needs and target values are determined.

**Table 1** User Requirements List

| Primary | Secondary | Tertiary |
|---|---|---|
| **Features** | Easy to operate | Can be operated by 1 person |
| | | Easy to carry |
| | | Light |
| | Ergonomics | The display can be read easily |
| | | The size of part fit to human anthropometry |
| **Performance** | Measure information processing steps | Measure thinking time |
| | | Measure execution time |
| | Considering stimulus movement | Measure reaction in steady light |
| | | Measure reaction to flashing light |
| | Considering stimulus color | Measure reaction to light color |
| | | Measure reaction to bright color |
| | Considering the intensity of light | Measure reaction in high intensity |
| | | Measure reaction in low intensity |

Every need is broken down into functions and sub-functions. The establishment of functional breakdown has the purpose of knowing what functions should appear on the product according to the user's wishes. Table 2 presented examples of functional breakdown related to the functions of measuring reaction time, considering color difference (hard and easy to see color).

### *3.2. Design phase*
From several concepts that were built, one concept was chosen to be developed (Figure 1). The concept uses three factors that can be manipulated to measure reaction time. These factors are the color difference factor, the number of lights flashing (flash rate), and light intensity. In the concept that was built, these three factors serve as rounds of play on the tool.

**Table 2.** Functional Breakdown

| User requirement | Function | Subfunction |
|---|---|---|
| Measure reaction to light color | Measure reaction time considering color differences | 6.1 Build starting mechanism (to start the game) |
| | | 6.2 Giving the order (as input) |
| | | 6.3 Display output |
| Measure reaction to bright color | | 6.4 Measure time to think (cognitive time) |
| | | 6.5 Measure actuation time |
| | | 6.6. Receive user's respond |
| | | 6.7. Display the outputs |

a. Round 1: Color factor
Red and white lights represent the color difference. Red represents bright colors that are easier to see than white. Red and white lights were randomly changing. To measure the reaction time, the user must choose a lamp that lights up as quickly as possible. This round is done twice. The final value displayed is the average of the two trials

b. Round 2: The number of blinking lights
Flash is often used to signify a special condition, such as danger or state of emergency. This steady and flash condition is used as a factor in measuring reaction speed. Flash rate is set into three levels, namely low flash rate (300 blinks/minute, medium (700 blinks/minute), and difficult (2000 blinks/minute). This round is done twice. The final value shown is the average of the two trials

c. Round 3: Light intensity
The level of light intensity consists of low and bright light levels. There is a button (rotary switch) that is used to set the level of light intensity, namely the level of low light with the symbol Low (R) 170 lux or high light level with the symbol High (T) 225 Lux. This round is done twice. The final value displayed is the average of the two trials

The cognitive response is measured the first time the response was given, until just before the hand moves to respond. The motor response is measured starting from the hand moving until the hand turns off the signal. To measure both times, a sensor was used to detect motion. The hand is placed as close as possible to the motion sensor. After the signal is turned on and the cognitive time is over, the hand reacts by approaching the button. When moving, the hand activates the motion sensor. This time will be recorded as the beginning of movement time and ended when the hand has turned off the signal. Cognitive time is calculated by calculating the total response time to motor response time. In the chosen concept, the user must press the button located at the bottom of the lamp that is considered to have the correct answer. When the respondent has completed all three rounds, the screen will display the reaction time along with the number of errors made on the screen.

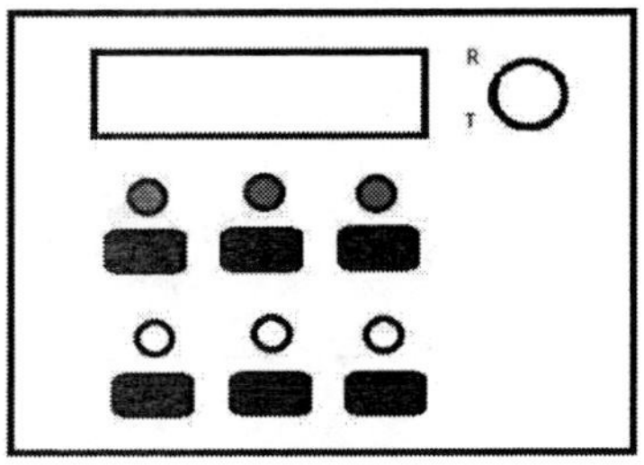


**Figure 1** Selected Concept

### *3.3. Construction*

The final design and component of the selected concept are presented in Figure 2 and Figure 3. The device is portable and has total dimension of the casing is only 15x9x10 $cm^3$ to make the device easy to carry. To start the time calculation, the user must press the big green push button. This push button will trigger the program to start the time and display the game. At round-1, reaction time was tested against 2 different colors, red and white. This round consists of 2 sub-rounds. At each sub round, the program will turn on the white or red lights randomly. The user must press the small push button on the bottom of the active lamp to stop the time calculation. The finger must pass the infrared sensor to press the push button located at the bottom of the light. The sensor stops the cognitive time calculation. When the user pushes the small push button at the bottom of the lamp, the program automatically stops the total time calculation. The cognitive and motoric time is displayed on the screen. To continue the game to the next sub round, the user must press the big green push button again.

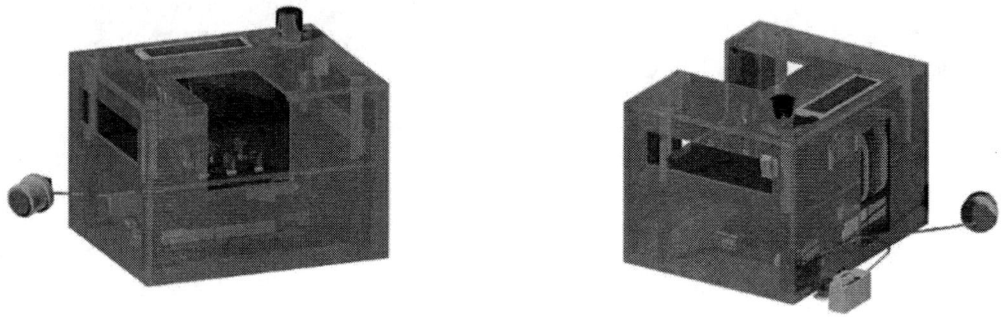

**Figure 2** Developed Concept

Broadly speaking, the measurement tool for the speed of human reaction is divided into several sections, namely as follows:

1. Casing
2. Adaptor
3. Arduino Board
4. Fishbone
5. LED lamp
6. Infra-red sensor
7. LCD
8. PCB
9. Green push button
10. Small push button
11. Switch
12. Battery's holder
13. Frame
14. Intensity switch

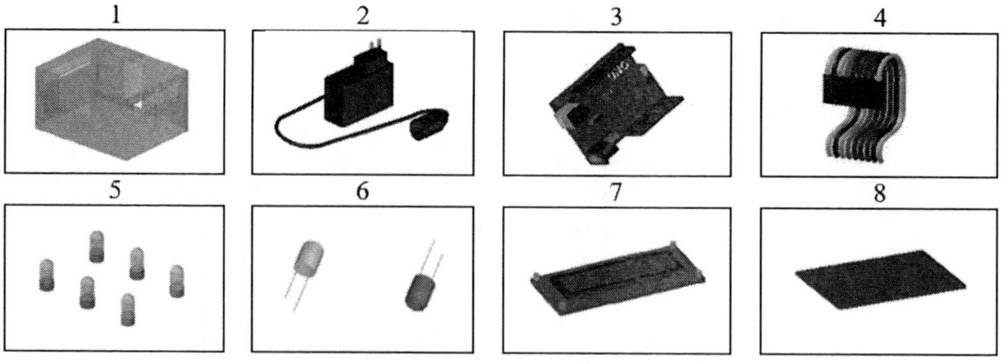


**Figure 3** Components of the developed device

IOP Conf. Series: Materials Science and Engineering **847** (2020) 012051 doi:10.1088/1757-899X/847/1/012051

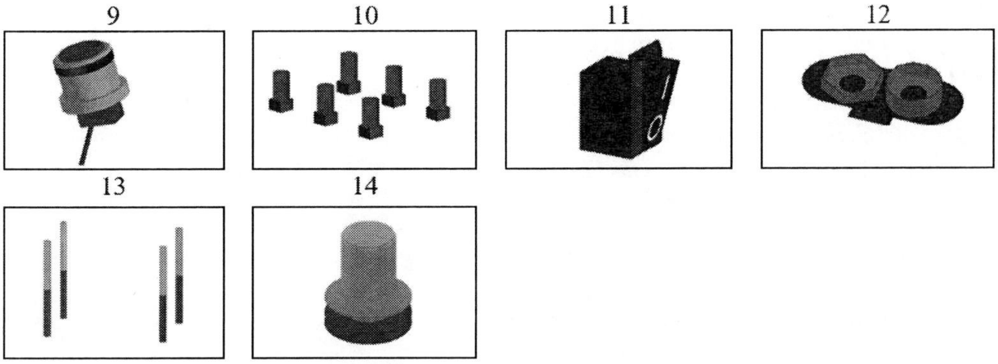


**Figure 3** Components of the developed device (cont.)

After completing round 1, the respondent presses the large green push button to start the next round. This round using flash rate to examine the difference in reaction time. This round consists of three rounds. The sub round 1, 2, and 3 represent slow, medium, and high flash rate sequentially. Similar to round one, the user must press the small push button under the flashing lights and automatically activate the infrared sensor. The cognitive and motoric time then is displayed on the screen. After finishing round 2, the user starts round 3 by pressing again the green push button. In this round the respondent will test the light intensity factor. Before starting the round (by pressing big green button), user must turn the rotary switch to set the intensity of light. To have a bright or dim light, the switch must be switch upward or downward. For this round, there are two sub rounds for each intensity level.

*3.4. Testing*

After construction, the tool was tested on users. A series of experiments were conducted to prove that the device is sensitive to distinguish conditions of the user's reaction time, especially when fatigued. Based on the results of trials on 280 samples, it was concluded that the device could differentiate normal conditions with fatigue significantly (p-value < 0.05). The value of reaction time when fatigue is greater than normal.

*3.5. Limitation and Recommendation*

The developed tool has shown its performance to measure cognitive and motoric reaction time. However, during the construction, some limitations were also found. Of the six pushbuttons used on the lights, two buttons are rather difficult to press. The difficulty sometimes causes it is necessary to press the button repeatedly until the button responds. The problem can be caused by PCB lines that are too small or push buttons not soldered properly. The second limitation is the Arduino engine which gets hot easily so that it sometimes causes errors in the running of the program. For example, the push button does not respond properly or an abnormal display. Arduino itself requires output between 9v - 12V. It turns out that the adapter used has an output of more than 12v, which is 16v. The unfit specification makes the tool hot quickly when used. It is also recommended to use an industrial component as a push button. The industrial component button has a larger field than the push button currently in use so that it can receive responses better. Besides, the adapter used must be adjusted to the needs of the program, so it does not cause the tool to heat up quickly.

## 4. Conclusion

The reaction time device was designed and constructed to distinguish cognitive time and motor time. The device has the ability to consider the factors of light color difference, flash rate, and light intensity. With dimensions that are quite small and portable, these reaction time gauges are quite easy to carry. The device is proven to be able to distinguish between reaction time in normal and fatigue conditions. The current weakness is in the tool components that have limited capabilities, especially in the sensitivity of the buttons and component durability. For further research, it is expected to develop a device with more reliable components.

## 5. References

[1] Hoque M E 2016 Three Domains of Learning: Cognitive, Affective and Psychomotor. *The Journal of EFL Education and Research* **2** 45-49
[2] Sagala S 2013 *Konsep dan Makna Pembelajaran* (Bandung: Alfabeta)
[3] Djemari M 2004 *Penyusunan Tes Hasil Belajar* (Yogyakarta: UNY)
[4] Arikunto S and Jabar C S A 2009 *Evaluasi program pendidikan: pedoman teoritis praktis bagi mahasiswa dan praktisi pendidikan* (Jakarta: Bumi Aksara)
[5] White E J, McMahon M, Walsh M T, Coffey J C, Sullivan L O 2017 Toward a Model of Human Information Processing for Decision-Making and Skill Acquisition in Laparoscopic Colorectal Surgery. *Journal of Surgical Education* **75** 1-5
[6] Khitrov M Y, Laxminarayan S, Thorsley D, Ramakrishnan S, Rajaraman S, Wesensten N J, and Jaques Reifman 2014 PC-PVT: A platform for psychomotor vigilance task testing, analysis, and prediction. *Behav Res Methods* **46** 140-47.
[7] Basner M, Mollicone D, Dinges D 2011 Validity and Sensitivity of a Brief Psychomotor Vigilance Test (PVT-B) to Total and Partial Sleep Deprivation. *Acta Astronautica* **69** 949-959
[8] Ulrich K T and Eppinger S D 1995 *Product Design and Development* (Singapore: McGraw-Hill)
[9] Dieter G E 2009 *Engineering Design.* 4th Edition. McGraw-Hill Internationals Editions.
[10] Widodo I D 2003 *Perencanaan dan Pengembangan Produk.* (Yogyakarta: UII Press Indonesia)
[11] Singer 1975 *Psychomotor Domain.* (US: Springer US).
[12] Budianto I 2015 *Sumbangan Kecerdasan IQ, Keseimbangan, dan Kecepatan Reaksi Terhadap Kemampuan Kiper Dalam Mengantisipasi Arah Bola.* (Semarang: UNS)

# Risk Mitigation Strategies on *Supply Chain* PT. X

**S P D Kristiana[1,2], C W Oktavia[1,3], R Magdalena[1,4], and M A Lilajati[1,5]**

[1]Industrial Engineering Study Program, Faculty of Engineering, Atma Jaya Catholic University of Indonesia, Jakarta, Indonesia
[2]stefani.kristiana90@gmail.com,[3]chendrasari@gmail.com,
[4]riana.magdalena@atmajaya.ac.id, [5]anindyalilajati@gmail.com

**Abstract.** Every supply chain activity in a company is not separated from risk, therefore risk management in a company is strongly required for risk handling. PT. X is a manufacturing company that produces food flavors. Dependence with suppliers and consumers causes uncertainty in the company's supply chain flow. This uncertainty can lead to adverse effects called risk. Risk can be minimized by carrying out risk management. This study aimed to design a risk management by identifying risk events from the company's business processes based on Supply Chain Operation Reference (SCOR), to determine the causes of priority risks and then identifying risk mitigation strategies using House of Risk stage 1 (HOR1), and ranking risk mitigation strategies by considering company criteria in determining risk mitigation strategies using the Analytical Hierarchy Process (AHP). From the results of the study, 21 risk events dan 13 priority risk causes' identified, and 9 risk mitigation strategies' determined. Risk mitigation strategies' rank sequentially were conducting briefings and evaluations every day, coordinating with suppliers continuously, making signboards at the warehouse, tightening the implementation of Standard Operating Procedures (SOP), providing ongoing training, evaluating supplier performance, monitoring more frequently, demand management with considering safety stock and service levels, and had supplier reserves. So, this study suggests PT.X to implement the risk mitigation strategies for minimizing the risks in the company.
**Key word**: Analytical Hierarchy Process (AHP), House of Risk (HOR), Mitigation, Risk Management, Supply Chain

## 1. Introduction

The level of competition in the food industry is increasing, along with the growth of the food industry so it is necessary to create a strategy to increase competitive advantage. One competitive advantage that can be achieved by implementing supply chain management [1]. The challenges of uncertainty, complexity, and the flow of a very wide supply chain network are the main sources in the difficulty of supply chain management [2]. This uncertainty can produce adverse consequences or can be called a risk [3]. Various types of risks and causes of risk can arise in the company's business processes that have the effect of disrupting the smooth running of business processes [4]. Therefore, the risks that occur in the supply chain must be managed appropriately to reduce or eliminate company losses [5]. This approach is known as supply chain risk management. Risk management in the supply chain aims to reduce the probability of risk events and increase the ability to recover from an interruption [6]. Risk management in the supply chain involves several processes namely risk identification, risk assessment, risk evaluation, and risk mitigation [7]. PT. X is a manufacturing company engaged in the processing of food flavors. The process of procuring raw materials at PT. X is very dependent on the supplier and the production process related to the quantity of production is very dependent on consumer demand. This dependency makes PT. X is faced with uncertainties that can produce risks. Currently, PT. X does not yet have risk management that discusses the proposed risk management and risk mitigation strategies that must be carried out to deal with risk. Therefore, companies need to identify risks, identify the causes of risk and mitigate actions against the causes of risk to minimize or eliminate risks that occur in the company's supply chain so that an effective and efficient production process is formed. This research will identify the risk events of the company's business processes based on 5 core Supply Chain Operation Reference (SCOR) processes, namely plan, source, make, deliver, and return. Furthermore, the identified risk events will be processed using the House of Risk phase 1 method. The HOR method emphasizes the risk causes that affect risk events occur. The risk causes are source that drives the occurrence of risk events. Phase 1 helps for identifying and prioritizing the causes of risk that need to be done in advance of a mitigation strategy based on an Aggregate Risk Potential (ARP) assessment that calculates the severity, the frequency of occurrence of the cause of the risk, and the correlation between the risk and the cause of the risk. The highest ARP value is the basis for decision making for risk incident mitigation strategies. Next, develop a risk mitigation strategy for the risk causes by considering the selection of criteria determined by the company using the Analytical Hierarchy Process (AHP). With the use of AHP method, the ranking of risk mitigation strategies can be done by reviewing the criteria based on level of interest.

## 2. Methods

### *2.1. Data Collection*

This study discusses the problem of risk events that occur in companies by applying the HOR and AHP method. The HOR method emphasized the risk causes that affect risk events occur, while AHP method was used to develop the ranking of risk mitigation strategies. The initial stage is data collection, where the data collection is done by interviewing and distributing questionnaires to respondents namely factory manager, purchasing manager, quality control manager. The interview stage is needed to collect data relating to problems experienced by companies that are closely related to risk events, the company's business processes, and solutions provided by the company to the problem of risk events and causes of risk. Distributing questionnaires is needed for risk event confirmation questionnaire, risk identification questionnaire, risk assessment questionnaire (severity, occurrence, and relationship), questionnaire determining risk mitigation strategy, paired comparison questionnaire (between criteria and between alternative risk mitigation strategies in terms of each criterion). The five questionnaires were filled out by respondents following their duties and responsibilities in the company's business processes.

### *2.2. House of Risk Phase 1 Method*

The steps to do HOR 1 are as follows:

1. Classifying business processes that occur in the company that can help identify potential risk events (Ei). Identification of business processes can be done using the SCOR model (plan, source, make, deliver and return).
2. Assess the magnitude of the impact (severity) of risk events if they occur (Si). Assessment can use the Shahin scale. The Shahin scale uses a scale of 1-10 where the severity of a risk scale of 1 presents no impact and the scale of 10 represents the extreme impact.
3. Identifying the cause of risk (Aj) from each risk event that has been identified and assessing the occurrence of the cause of risk (occurrence) against each cause of risk (Oi). Assessment can use the Shahin scale. The Shahin scale for the causes of risk uses a scale of 1-10 where the scale 1 presents almost never happens and the scale of 10 shows almost certainly occurs.
4. Develop a matrix of relationships/correlations between risk events and the causes of risk symbolized by Rij. This Rij can be interpreted as how likely the cause of risk j can cause the risk

event to occur. The scale used is 0, 1, 3 9. Scale 0 presents the absence of correlation and 1, 3.9, respectively present low correlation, medium correlation, and high correlation.

5. Calculate the Aggregate Risk Potential j (ARP j) which is the result of the likelihood of the occurrence of the causes of risk j and a collection of the causal impacts of each risk event caused by the causes of risk j by using the calculation formula:
$$ARPj = Oj \Sigma Si Rij \quad (1)$$
6. Make ranking based on ARP value from the largest ARP value to the smallest ARP value.
7. Selecting some of the causes of ARP risk value with the highest ranking, can use the 80:20 pareto principle of ARPj
8. Identify actions deemed relevant to prevent the cause of risk. One cause of risk can be handled by more than 1 preventive measure and 1 preventive measure can reduce several causes of risk.

The next step is to develop a risk mitigation strategy for the risk causes by considering the selection of criteria determined by the company using the Analytical Hierarchy Process (AHP).

### *2.3. Analytical Hierarchy Process (AHP) Method*

AHP method in this study is used to rank risk mitigation strategies based on the weights of the results of pairwise comparisons between criteria and pairwise comparisons between alternative risk mitigation strategies in terms of each criterion. The steps in processing data using the AHP method carried out in this study are as follows:

1. Arrange the hierarchical structure starting from the objectives at level 1, criteria at level 2, to alternatives at level 3. In this study, the objectives used are risk mitigation strategies. The criteria used are the criteria considered by PT. X to carry out risk mitigation strategies. Alternatives in this hierarchical structure are several risk mitigation strategies.
2. Make AHP modelling on Super Decisions software according to the hierarchical structure that has been created.
3. Fill in the pairwise comparison matrix on the Super Decisions software found on the Pairwise Comparison toolbar. This paired matrix value is obtained from the pairwise comparison questionnaire.
4. The ranking of risk mitigation strategies obtained when all paired comparison matrices have been filled and have been consistently shown in the inconsistency value $< 0.1$ in the Super Decisions software.

## 3. Result and Discussion

### *3.1. Identifying Risk Events and Assessing Risk Events*

Risk events in a company's supply chain activities are identified by understanding and mapping the company's core processes based on plan, source, make, deliver, and return into its sub-processes. The results of risk identification are 3 risk events in the scope of the plan, 5 risk events in the source scope, 7 risk events in the make scope, 4 risk events in the deliver scope, and 2 risk events in the scope of return. According to the assessment by respondents, the risk event that has the highest severity value is in the scope of make, that is, some products do not meet specifications. In the scope of deliveries, product damage during travel, and in the scope of returns, there are complaints to suppliers and complaints from customers. Risk events that have been identified have varying severity values, in the scope of deliveries, product damage during the trip, and in the scope of returns for complaints to suppliers and complaints from customers. Where, each risk event is worth 9, which means it has a serious impact on the company. The results of this recapitalization can be seen in Table 1.

**Table 1.** Identification of risk events and assessment of risk events

| Core Process | Sub-Process | Risk Event | $E_i$ | $S_i$ |
|---|---|---|---|---|
| *Plan* | Material planning | The difference in quantity in the system with the actual | E1 | 3 |
| | | Incorrect estimation of orders | E2 | 5 |
| | Production planning | A sudden change in production plans | E3 | 5 |
| *Source* | Scheduling material reception | *Due date of company acceptance not fulfilled* | E4 | 7 |
| | Purchasing of material | Incompatibility of demand needs with financial planning | E5 | 3 |
| | Material inspection & reception | Delay in checking material specifications | E6 | 2 |
| | | Material received does not match the order | E7 | 7 |
| | Storing material | Material placement does not match the type of goods | E8 | 4 |
| *Make* | Production preparation | Error providing product formula | E9 | 7 |
| | Production process | The production process was forced to stop | E10 | 5 |
| | | Postponement of production | E11 | 5 |
| | | Existing material cannot be used | E12 | 7 |
| | Inspection of production results | There are products that do not meet specifications | E13 | 9 |
| | Product packaging | Product packaging results are imperfect | E14 | 8 |
| | | The product label does not match the product type | E15 | 9 |
| | Storing product | Products accumulate in the transit area | E16 | 4 |
| *Deliv-er* | Delivery Scheduling | Late delivery of products to customers | E17 | 4 |
| | Delivering preparation | *Certificate of Analyst is not attached* | E18 | 4 |
| | Delivering product | Product damage during travel | E19 | 9 |
| *Return* | Returning of material to the supplier | There is a complaint to the supplier | E20 | 9 |
| | Receipiont of rejected product | There are complaints from customers | E21 | 9 |

### *3.2. Identification of Risk Agents and Assessment of Risk Agents*
Causes of risk are identified based on risk events that have been identified previously. One cause of risk can trigger several risk events. Based on the results of data collection, 34 causes of risk were successfully identified and 2 causes of risk that have a high frequency of occurrence are indicated by a value of 10. The results of identifying the causes of risk and assessing the causes of risk can be seen in Table 2.

**Table 2.** Identification of risk agents and an assessment of risk agents

| Risk Agent | $A_j$ | $O_j$ |
|---|---|---|
| PPIC did not immediately update the system | A1 | 2 |
| Lack of monitoring of actual inventory quantities | A2 | 2 |
| Warehouse inaccuracy calculates the quantity of material remaining on a card stock | A3 | 4 |
| Fluctuation of demand | A4 | 10 |
| Seasonal Factors | A5 | 10 |
| Changes in the quantity of consumer demand | A6 | 6 |
| Some requests must be prioritized | A7 | 6 |
| Sudden material demand | A8 | 8 |
| Stunted on the way | A9 | 1 |
| Natural disasters | A10 | 1 |
| Rely on a single supplier | A11 | 8 |
| Supplier is late sending material | A12 | 8 |
| Inaccurate price reference | A13 | 3 |
| Fluctuations in the exchange rate of the rupiah against the dollar | A14 | 5 |
| Documents (COAs and Halal certificates) are incomplete | A15 | 3 |
| Material quality does not meet specifications | A16 | 9 |
| The quantity received does not match the order | A17 | 3 |
| Lack of knowledge of warehouse operators in laying materials | A18 | 8 |
| Irregularity in material placement | A19 | 8 |
| Material grouping information is not listed | A20 | 8 |
| Storage of product formulas is not well ordered | A21 | 7 |
| Electricity and / water supply has stopped | A22 | 2 |
| Engine failure | A23 | 2 |
| There is a WIP product that must be followed up on | A24 | 4 |
| Material shortage | A25 | 9 |
| The careless operator in handling material | A26 | 9 |
| Material is stored for too long in storage | A27 | 3 |
| The operator is negligent in carrying out Work Instructions | A28 | 3 |
| Operators are less skilled at using work tools | A29 | 6 |
| The inaccuracy of production operators in packaging | A30 | 7 |
| Request for delivery delay by the customer | A31 | 4 |
| Limited fleet capacity | A32 | 4 |
| Inaccuracy of the QC | A33 | 4 |
| Inaccurate handling of products by shipping services | A34 | 3 |

### *3.3. Risk Evaluation*

At the risk evaluation stage an Aggregate Risk Potential (ARP) calculation is obtained using equation (1), from the formula there are 3 assessments that affect the risk assessment, namely the severity level of the risk event, the value of the occurrence of the cause of risk, and the correlation value between the risk event and the cause of the risk. ARP calculation results for the causes of A12 risk have the highest value of 864. This shows that it is necessary to prioritize the causes of A12 risk for mitigation measures. The causes of priority risks are presented in Table 3 based on the 80:20 principle from the Pareto diagram.

**Table 3**. Selected risk agent.

| *Rank* | Aj | Risk agent | ARP | % | %KUM |
|---|---|---|---|---|---|
| 1 | A12 | Supplier is late sending material | 864 | 12% | 12% |
| 2 | A16 | Material quality does not meet specifications | 810 | 11% | 23% |
| 3 | A26 | The careless operator in handling material | 567 | 8% | 30% |
| 4 | A11 | Rely on a single supplier | 504 | 7% | 37% |
| 5 | A4 | Demand is very volatile | 450 | 6% | 43% |
| 6 | A5 | Seasonal Factors | 450 | 6% | 49% |
| 7 | A29 | Operators are less skilled at using work tools | 432 | 6% | 55% |
| 8 | A25 | Material shortage | 405 | 5% | 61% |
| 9 | A30 | The inaccuracy of production operators in packaging | 378 | 5% | 66% |
| 10 | A18 | Lack of knowledge of warehouse operators in laying materials | 288 | 4% | 70% |
| 11 | A19 | Irregularity in material placement | 288 | 4% | 74% |
| 12 | A20 | Material grouping information is not listed | 288 | 4% | 78% |
| 13 | A28 | The operator is negligent in carrying out Work Instructions | 243 | 3% | 81% |

### *3.4. Ranking of Risk Mitigation Strategies*

The cause of the selected risk is then determined by its mitigation strategy then the risk mitigation strategy is ranked in terms of the criteria that the company considers to determining the risk mitigation strategy using the AHP method. The criteria that companies consider to determining risk mitigation strategies are obtained from interviews, namely preventing the recurrence of problems, increasing company efficiency, improving the quality of human resources, easy to do and not creating new problems. The initial step taken to rank the risk mitigation strategy is to arrange a hierarchy. In this study, the application of the AHP method was used with the help of Super Decision software. Figure 1. presents the AHP hierarchical structure modelling in the Super Decision software.

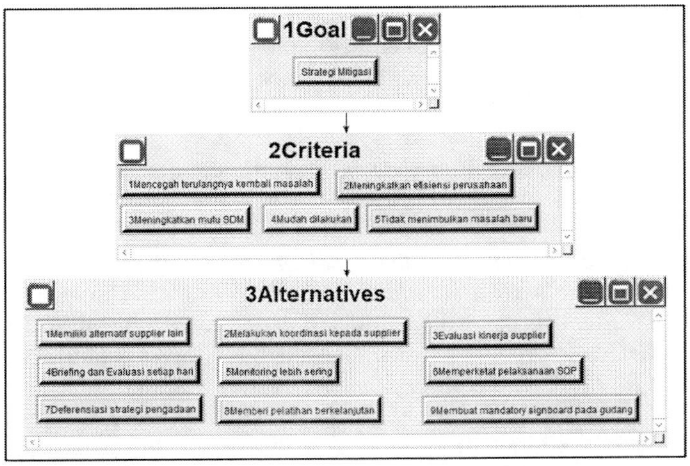


**Figure 1.** AHP hierarchy structure in Super Decision software.

Figure 2. Presents ranking of risk mitigation strategies using the Super Decisions software. Ranking based on total, normal and ideal values. So that, the recommendation for this company is to implement alternatives which is resulted from this study such as to conduct briefings and evaluations every day, coordinate with suppliers continuously, create sign boards in warehouses, tighten the implementation of Standard Operating Procedures (SOP), provide ongoing training, evaluate supplier performance, monitoring more frequently, controlling inventory taking into account safety stock and service levels, and having supplier reserves.

**Alternative Rankings**

| Graphic | Alternatives | Total | Normal | Ideal | Ranking |
|---|---|---|---|---|---|
| | 1Memiliki alternatif supplier lain | 0.0181 | 0.0362 | 0.1931 | 9 |
| | 2Melakukan koordinasi kepada supplier | 0.0792 | 0.1584 | 0.8451 | 2 |
| | 3Evaluasi kinerja supplier | 0.0462 | 0.0923 | 0.4925 | 6 |
| | 4Briefing dan Evaluasi setiap hari | 0.0937 | 0.1875 | 1.0000 | 1 |
| | 5Monitoring lebih sering | 0.0328 | 0.0657 | 0.3503 | 7 |
| | 6Memperketat pelaksanaan SOP | 0.0699 | 0.1399 | 0.7461 | 4 |
| | 7Deferensiasi strategi pengadaan | 0.0216 | 0.0432 | 0.2304 | 8 |
| | 8Memberi pelatihan berkelanjutan | 0.0639 | 0.1278 | 0.6815 | 5 |
| | 9Membuat mandatory signboard pada gudang | 0.0746 | 0.1491 | 0.7956 | 3 |

**Figure 2.** Result of mitigation risk strategy rank

## 4. Conclusion

The conclusion of this study are as follows:

1. House of Risk phase 1 is carried out to map business processes, identify risk events, risk causes, and risk assess. There are 21 risk events and 34 risk causes that potentially occur in the supply chain flow at PT NFI.
2. There are 13 priority risk causes selected to determine the mitigation strategy based on the ARP value that falls into the 80:20 principle of the Pareto diagram.
3. Based on the Analytical Hierarchy Process (AHP) method, the ranking of risk mitigation strategies in sequence is to conduct briefings and evaluations every day, coordinate with suppliers continuously, create sign boards in warehouses, tighten the implementation of Standard Operating Procedures (SOP), provide ongoing training, evaluate supplier performance, monitoring more

frequently, controlling inventory taking into account safety stock and service levels, and having supplier reserves.

Suggestion for the future research is to use another method to find the relationship between risk events and risk causes for example with ANP (Analytic Network Process) method.

## 5. References

[1] Indrajit E R 2016 *Supply Chain Management.* (Jakarta: Preinexus)

[2] Pujawan I N dan Mahendrawathi 2017 *Supply Chain Management.* 3rd edition (Yogyakarta: ANDI Yogyakarta)

[3] Hadiguna R A 2016 *Manajemen Rantai Pasok Argoindustri: Pendekatan Berkelanjutan untuk Pengukuran Kinerja dan Penilaian Risiko.* (Padang: Andalas University Press)

[4] Oktavia C W 2013 Analisis dan Mitigasi Risiko pada Proses Pengadaan Barang dan Jasa dengan Pendekatan Metode Interpretive Structural Modelling (ISM), Analitic Network Process (ANP), dan House of Risk (HOR). *Prosiding Seminar Nasional Manajemen Teknologi.* **19.**

[5] Waters D 2011 *Supply Chain Risk Management Vulnerability and Resilience in Logistics.* 2nd ed (London: Kogan Page Limited)

[6] Pujawan I N dan Geraldin L H 2009 House of Risk: a Model for Proactive Supply Chain Risk *Management. Business Process Management Journal.* **15**(6), 953-967

[7] Kayis B dan Karningsih P D 2010 SCRIS: A knowledge based system tool for assisting organizations in managing supply chain risks. *Journal of Manufacturing Technology Management.* **23**(7), 834-852.

# Integrated ISM-ANP Method for Supplier Selection Criteria Analysis: A Case Study of Construction Company

**C Natalia[1,2], C W Oktavia [1,3], S G Tjhong [1,4] and T P Hidayat[1,5]**

[1]Industrial Engineering Department, Engineering Faculty Atma Jaya Catholic University of Indonesia, Jl. Jend. Sudirman No.51, Jakarta, Indonesia
[2]chrisnatalia@atmajaya.ac.id, [3]chendrasari@gmail.com, [4]sherlygunawant@gmail.com, [5]trifenausprabuhidayat@gmail.com

**Abstract.** Supplier is one of the most important parts in supply chain. Supplier ensures the availability of materials for company's continuity. Supplier selection using several criteria is needed to get a reliable supplier. Therefore, selecting key criteria for supplier selection is essential. This study discussed the relationships between criteria to obtain supplier selection key criteria in a construction company with interpretive structural modeling (ISM) and selecting supplier with analytic network process (ANP). Data used in this study are initial criteria, the relationships between criteria, the grouping of criteria, and pairwise comparison are obtained from interviews and questionnaires. Using ISM method, 19 key criteria are identified, such as availability and integrity. The best supplier based on key criteria is obtained from ANP method. The result of integrated ISM and ANP method shows that the key criterion with the highest weight is availability, and supplier 1 is the best supplier based on key criteria. Availability refers to the availability of materials in supplier, and this key criterion is considered as very important thus having spare suppliers must be considered.
**Keywords**: Supplier Selection, Interpretive Structural Modeling, Key Criteria, Analytic Network Process, Construction Company

## 1. Introduction

Raw material supplier is a part of supply chain management that greatly influence the production process [1]. Supplier selection has an essential part in business because the performance and efficiency of the supplier directly affect the success of the company [2]. Supplier selection is a multi-criteria decision-making problem that consider many criteria to deliver a solution [3]. Thus to get a maximum supply chain performance, suplier selection must consider criteria that relevant to company goals [4]. The supplier selection list by [5] is the most common criteria used in supplier selection. [6] then updated the list by comparing supplier selection criteria in the year 1966-1990 and 1990-2001 and some criteria, such as operational control, packaging ability, and training aids, are not used anymore. But there are also new criteria, such as reliability, flexibility, and long term relationship. [7] compares studies in the year 2005 to 2009 and get that traditional criteria which are price, quality, delivery, and service are still in use. However, the study also finds new criteria such as supplier's profile and availability. Therefore it can be seen that there is a change in supplier selection criteria tren.

Every company uses different supplier selection criteria. [8] conducted a study in an automobile company and used criteria such as cost, quality, and long term relationship to select the supplier. A study by [4] uses 9 criteria, for example, price, material specification, and quality consistency. [9] conducted supplier selection in a construction company using 8 criteria, such as quality, delivery, location, and warranties and claim policies. The supplier selection criteria in the three studies mentioned above differ between one another. Location and warranties and claim policies are used in the construction project but not taken into consideration in other studies. The difference in criteria is caused by the construction project supply chain that is different from the conventional supply chain.

A large number of parties involved in the construction project supply chain makes the project unique, and the process is temporary. The nature of the construction project causes it has triple constraint that is cost, quality, and construction time; thus, optimization of time and resources must be done [10]. One of the stages in the construction project is selecting suppliers. The supplier selection process can be done much more effective and efficient when focusing on key criteria [11].

Key criteria hold a vital role in supplier selection because of the considerable influence it can cause to the supplier selection criteria model. The interpretive structural modeling (ISM) method can produce key criteria. This method identified the direct and indirect relationships from criteria that define the problem to obtain key criteria. Key criteria is criteria that have high driving power that can affect many criteria in the model [12]. A study by [13] uses ISM to identifies the mutual relationships among supplier selection process enablers and to develop a hierarchical model of the enablers. However, the study is limited to only determining the key enablers and has not implemented them in the supplier selection process.

ISM method can be used as a basis for weighting supplier selection criteria as done by [11] that used ISM with the technique for others preference by similarity to ideal solution (TOPSIS). The study uses ISM to determine the key criteria then followed by supplier selection using TOPSIS with criteria weighting from experts. [14] integrate ISM with ANP to get third-party logistics (3PL). The ISM method produces a model that take into consideration in evaluating the criteria. Then the ANP method is used to get the best third-party logistics providers. The identification of key criteria has not been done in this study as conducted in research by [11] whereas it does not consider the interrelationship between criteria when selecting a supplier. ANP method uses network model that has feedback structures and is not rigid to illustrate the relationships between criteria. Thus the model from ISM can be used as a basis in ANP method.

Supplier selection criteria from previous studies, including the one conducted in the construction company, are still determined with a literature study with the approval from the company and key criteria identification for supplier selection has not been done. Therefore, this study uses an integrated ISM and ANP methods to get the best supplier focusing on key criteria. ISM is used to identified key criteria that are used for supplier selection with ANP, which also consider the relationships between criteria. This study is conducted in a construction company in Jakarta and the result of this study helps the construction company to conduct a more effective and efficient supplier selection with key criteria.

## 2. Methods

The study is divided into two sections, the ISM method to get the key criteria for supplier selection and the ANP method to weigh the criteria and supplier. ISM method consists of 6 steps, namely initial criteria, structural self-interaction matrix (SSIM), reachability matrix, level partition and conical matrix, ISM model, and MICMAC analysis. ANP method consists of 4 steps: key criteria grouping, ANP modeling, pairwise comparison, and ANP data processing.

### *2.1 Initial Criteria*

The criteria used in this study are 38 supplier selection criteria collected from several studies in table 1. The criteria are selected by an expert, a procurement manager from the construction company using a questionnaire. At the initial stage, The company chooses criteria that considered as crucial in supplier selection. the obtained 19 criteria : quality, delivery, price, repair service, reliability, flexibility, attitude, communication system, performance history, consistency, long term relationship, reciprocal arrangement, quality standard, integrity, professionalism, reputation and position in industry, operational control, amount of past business, and availability.

**Table 1.** Supplier Selection Criteria

| No. | Criteria | Sources | No. | Criteria | Sources |
|---|---|---|---|---|---|
| 1 | Quality | [5], [6], [7], [15] | 20 | Process improvement | [6] |
| 2 | Delivery | [5], [6], [7], [15] | 21 | Product development | [6], [7], [16] |
| 3 | Price | [5], [6], [7], [15], [16] | 22 | Inventory cost | [6] |
| 4 | Repair service | [5], [6], [7] | 23 | JIT | [6] |
| 5 | Technical capability | [5], [6], [7], [15] | 24 | Quality standard | [6] |
| 6 | Production facilities and capacity | [5], [6], [7], [15], [16] | 25 | Integrity | [6], [7] |
| 7 | Financial position | [5], [6], [7], [15], [16] | 26 | Professionalism | [6] |
| 8 | Management and organization | [5], [6], [7], [15], [16] | 27 | Research | [6] |
| 9 | Reliability | [6], [7] | 28 | Cultural | [6] |
| 10 | Flexibility | [6], [7] | 29 | Reputation and position in industry | [5], [6], [15] |
| 11 | Attitude | [5], [6] | 30 | Labor relations record | [5] |
| 12 | Communication system | [5], [6], [7] | 31 | Operational control | [5] |
| 13 | Performance history | [5], [6], [7], [15] | 32 | Packaging ability | [5] |
| 14 | Geographical location | [5], [6], [7], [15] | 33 | Training aids | [5] |
| 15 | Consistency | [6] | 34 | Desire for business | [5] |
| 16 | Long term relationship | [6] | 35 | Amount of past business | [5], [7] |
| 17 | Procedural compliance | [5], [6] | 36 | Warranties and claim policies | [6] |
| 18 | Impression | [5], [6] | 37 | Supplier's profile | [7] |
| 19 | Reciprocal arrangement | [5], [6] | 38 | Availability | [7] |

### *2.2 Structural Self-Interaction Matrix (SSIM)*
SSIM consists of supplier selection criteria that have been selected in the previous step. The interconnections between criteria determined by the expert's opinions are shown in this matrix and it used four symbols. V means there is a relation from i to j, A for relation from j to i, X for there are connections from both i and j, and O means there is no connection between i and j. The matrix can be seen in table 2.

### *2.3 Reachability Matrix*
Reachability matrix is developed from SSIM by changing the VAXO notation into biner with the following rules:

- V – the entries in the reachability matrix are 1 for (i, j) and 0 for (j, i).
- A – the entries in the reachability matrix are 0 for (i, j) and 1 for (j, i).
- X – the entry for both directions is 1.
- O – the entry for both directions is 0.

Then the transitivity or the indirect relationships in the matrix are checked. The transitivity follows these rules; if criteria i can affect criteria j, and criteria j can affect criteria k, then criteria i can affect criteria k indirectly. The cell that undergoes transitivity is highlighted. The reachability matrix after transitivity can be seen in table 3.

### *2.4 Level Partition and Conical Matrix and Model ISM*
The reachability matrix after transitivity is divided into several levels by eliminating criteria that have the same reachability set and intersection. The reachability set is a set of criteria it can affect. On the contrary, the antecedent set is a set of criteria that can affect the criteria. While the intersection is the criteria listed in both reachability and antecedent set. The level partition can be seen in table 4. Then a conical matrix is made by sorting the reachability matrix by level partition to help the making of the ISM model. Next, The structural model or digraph is made from level partitioning and the relationships

IOP Conf. Series: Materials Science and Engineering **847** (2020) 012053 doi:10.1088/1757-899X/847/1/012053

between criteria in the reachability matrix. Next, the digraph is transferred into ISM model by changing the element node with criteria. The ISM model can be seen in Figure 1.

**Table 2.** SSIM

| Criteria | Criteria j ( Coloumn) | | | | | | | | | | | | | | | | | | | | |
|---|---|---|---|---|---|---|---|---|---|---|---|---|---|---|---|---|---|---|---|---|---|
| Criteria i (row) | 21 | 20 | 19 | 18 | 17 | 16 | 15 | 14 | 13 | 12 | 11 | 10 | 9 | 8 | 7 | 6 | 5 | 4 | 3 | 2 | 1 |
| 1 Quality | O | O | O | V | A | V | O | A | A | V | V | A | V | V | O | O | X | V | X | O | |
| 2 Delivery | A | O | V | V | A | V | A | A | O | V | V | O | V | A | O | A | V | X | O | | |
| 3 Price | A | O | X | V | A | V | A | A | O | O | X | V | O | O | O | O | O | O | | | |
| 4 Repair service | O | O | V | V | O | O | O | V | X | X | X | O | V | X | X | O | X | | | | |
| 5 Reliability | A | V | X | V | A | X | A | O | O | X | V | A | V | X | A | X | | | | | |
| 6 Flexibility | O | O | X | V | O | O | A | X | X | X | X | O | V | V | A | | | | | | |
| 7 Attitude | O | O | X | V | V | V | X | X | V | X | X | O | V | X | | | | | | | |
| 8 Communication system | O | O | V | V | O | O | X | O | O | V | X | O | V | | | | | | | | |
| 9 Performance history | A | O | V | V | O | O | X | A | A | X | X | A | | | | | | | | | |
| 10 Consistency | O | V | V | V | X | X | X | A | V | V | V | | | | | | | | | | |
| 11 Long term relationship | A | O | X | X | O | A | A | A | A | X | | | | | | | | | | | |
| 12 Reciprocal arrangement | O | O | X | V | O | O | X | X | A | | | | | | | | | | | | |
| 13 Quality standard | O | O | V | V | X | V | O | A | | | | | | | | | | | | | |
| 14 Integrity | O | O | V | V | O | V | X | | | | | | | | | | | | | | |
| 15 Professionalism | O | O | V | V | V | V | | | | | | | | | | | | | | | |
| 16 Reputation and position in industry | O | O | O | O | V | | | | | | | | | | | | | | | | |
| 17 Operational control | O | O | V | V | | | | | | | | | | | | | | | | | |
| 18 Desire for business | A | O | A | | | | | | | | | | | | | | | | | | |
| 19 Amount of past business | A | O | | | | | | | | | | | | | | | | | | | |
| 20 Warranties and claim policies | O | | | | | | | | | | | | | | | | | | | | |
| 21 Availability | | | | | | | | | | | | | | | | | | | | | |

**Table 3.** Reachability matrix after transitivity

| CRITERIA | Criteria j (Coulumn) | | | | | | | | | | | | | | | | | | | | | *DrivingPower* |
|---|---|---|---|---|---|---|---|---|---|---|---|---|---|---|---|---|---|---|---|---|---|---|
| Criteria i (row) | 1 | 2 | 3 | 4 | 5 | 6 | 7 | 8 | 9 | 10 | 11 | 12 | 13 | 14 | 15 | 16 | 17 | 18 | 19 | 20 | 21 | |
| 1 | 1 | 1 | 1 | 1 | 1 | 1 | 1 | 1 | 1 | 1 | 1 | 1 | 1 | 1 | 1 | 1 | 1 | 1 | 1 | 0 | 0 | 19 |
| 2 | 1 | 1 | 1 | 1 | 1 | 1 | 1 | 1 | 1 | 1 | 1 | 1 | 1 | 1 | 1 | 1 | 1 | 1 | 1 | 1 | 0 | 20 |
| 3 | 1 | 0 | 1 | 1 | 1 | 1 | 1 | 1 | 1 | 1 | 1 | 1 | 1 | 0 | 1 | 1 | 1 | 1 | 1 | 1 | 0 | 18 |
| 4 | 1 | 1 | 1 | 1 | 1 | 1 | 1 | 1 | 1 | 1 | 1 | 1 | 1 | 1 | 1 | 1 | 1 | 1 | 1 | 1 | 0 | 20 |
| 5 | 1 | 1 | 1 | 1 | 1 | 1 | 1 | 1 | 1 | 1 | 1 | 1 | 1 | 1 | 1 | 1 | 1 | 1 | 1 | 1 | 0 | 20 |
| 6 | 1 | 1 | 1 | 1 | 1 | 1 | 1 | 1 | 1 | 1 | 1 | 1 | 1 | 1 | 1 | 1 | 1 | 1 | 1 | 0 | 0 | 19 |
| 7 | 1 | 1 | 1 | 1 | 1 | 1 | 1 | 1 | 1 | 1 | 1 | 1 | 1 | 1 | 1 | 1 | 1 | 1 | 1 | 1 | 0 | 20 |
| 8 | 1 | 1 | 1 | 1 | 1 | 1 | 1 | 1 | 1 | 1 | 1 | 1 | 1 | 1 | 1 | 1 | 1 | 1 | 1 | 0 | 0 | 19 |
| 9 | 0 | 1 | 1 | 1 | 1 | 1 | 1 | 1 | 1 | 1 | 1 | 1 | 0 | 1 | 1 | 1 | 1 | 1 | 1 | 0 | 0 | 17 |
| 10 | 1 | 1 | 1 | 1 | 1 | 1 | 1 | 1 | 1 | 1 | 1 | 1 | 1 | 1 | 1 | 1 | 1 | 1 | 1 | 1 | 0 | 20 |
| 11 | 1 | 1 | 1 | 1 | 1 | 1 | 1 | 1 | 1 | 1 | 1 | 1 | 1 | 1 | 1 | 1 | 1 | 1 | 1 | 0 | 0 | 19 |
| 12 | 1 | 1 | 1 | 1 | 1 | 1 | 1 | 1 | 1 | 1 | 1 | 1 | 1 | 1 | 1 | 1 | 1 | 1 | 1 | 1 | 0 | 20 |
| 13 | 1 | 1 | 1 | 1 | 1 | 1 | 1 | 1 | 1 | 1 | 1 | 1 | 1 | 1 | 1 | 1 | 1 | 1 | 1 | 0 | 0 | 19 |
| 14 | 1 | 1 | 1 | 1 | 1 | 1 | 1 | 1 | 1 | 1 | 1 | 1 | 1 | 1 | 1 | 1 | 1 | 1 | 1 | 1 | 0 | 20 |
| 15 | 1 | 1 | 1 | 1 | 1 | 1 | 1 | 1 | 1 | 1 | 1 | 1 | 1 | 1 | 1 | 1 | 1 | 1 | 1 | 1 | 0 | 20 |
| 16 | 1 | 1 | 1 | 1 | 1 | 1 | 1 | 1 | 1 | 1 | 1 | 1 | 1 | 0 | 1 | 1 | 1 | 1 | 1 | 1 | 0 | 19 |
| 17 | 1 | 1 | 1 | 1 | 1 | 1 | 1 | 1 | 1 | 1 | 1 | 1 | 1 | 0 | 1 | 1 | 1 | 1 | 1 | 1 | 0 | 19 |
| 18 | 0 | 0 | 1 | 1 | 0 | 1 | 1 | 1 | 1 | 0 | 1 | 1 | 0 | 0 | 0 | 0 | 0 | 1 | 1 | 0 | 0 | 10 |
| 19 | 1 | 1 | 1 | 1 | 1 | 1 | 1 | 1 | 1 | 1 | 1 | 1 | 1 | 1 | 1 | 1 | 1 | 1 | 1 | 0 | 0 | 19 |
| 20 | 0 | 0 | 0 | 0 | 0 | 0 | 0 | 0 | 0 | 0 | 0 | 0 | 0 | 0 | 0 | 0 | 0 | 0 | 0 | 1 | 0 | 1 |
| 21 | 1 | 1 | 1 | 1 | 1 | 1 | 1 | 1 | 1 | 1 | 1 | 1 | 0 | 0 | 1 | 1 | 0 | 1 | 1 | 0 | 1 | 17 |
| *Dependence Power* | 18 | 18 | 20 | 20 | 19 | 20 | 20 | 20 | 20 | 19 | 20 | 20 | 17 | 15 | 19 | 19 | 18 | 20 | 20 | 12 | 1 | 375 |

### *2.5 MICMAC Analysis*

The value of driving power and dependence that are obtained from the reachability matrix after transitivity in table 1. These values are inputted to driving power-dependence diagram for MICMAC analysis. The chart is divided into four categories after drawing lines at half of the maximum values from driving power and dependence. The four categories are:

- *Autonomous factors*. Because of the low driving power and dependence in this factor, the criteria in this category can be separated from the system. No criterion is in this category
- *Linkage factors*. The criteria in this factor are high in both driving power and dependence, making it unstable. Because it has high driving power, criteria in this category are considered as key criteria.

The criteria in this category are quality, delivery, price, repair service, reliability, flexibility, attitude, communication system, performance history, consistency, long term relationship, reciprocal arrangement, quality standard, integrity, professionalism, reputation and position in industry, operational control, and amount of past business.

- *Dependent factors*. This factor has low driving power and high dependence power that makes it vulnerable to change. The criteria in this category are desire for business and warranties and claim policies (criteria 20)
- *Independent factors*. This factor has high driving power, but low dependence; thus, it can be the key criteria. The criteria in this category is availability (criteria 21).

**Table 4.** Level partition

| Criteria | *Reachability Set* | *Antecedent Set* | *Intersection* | Level |
|---|---|---|---|---|
| 1 | 1,2,3,4,5,6,7,8,9,10,11,12,13,14,15,16,17,18,19 | 1,2,3,4,5,6,7,8,10,11,12,13,14,15,16,17,19,21 | 1,2,3,4,5,6,7,8,10,11,12,13,14,15,16,17,19 | II |
| 2 | 1,2,3,4,5,6,7,8,9,10,11,12,13,14,15,16,17,18,19,20 | 1,2,4,5,6,7,8,9,10,11,12,13,14,15,16,17,19,21 | 1,2,4,5,6,7,8,9,10,11,12,13,14,15,16,17,19 | III |
| 3 | 1,3,4,5,6,7,8,9,10,11,12,13,15,16,17,18,19,20 | 1,2,3,4,5,6,7,8,9,10,11,12,13,14,15,16,17,18,19,21 | 1,3,4,5,6,7,8,9,10,11,12,13,15,16,17,18,19 | II |
| 4 | 1,2,3,4,5,6,7,8,9,10,11,12,13,14,15,16,17,18,19,20 | 1,2,3,4,5,6,7,8,9,10,11,12,13,14,15,16,17,18,19,21 | 1,2,3,4,5,6,7,8,9,10,11,12,13,14,15,16,17,18,19 | II |
| 5 | 1,2,3,4,5,6,7,8,9,10,11,12,13,14,15,16,17,18,19,20 | 1,2,3,4,5,6,7,8,9,10,11,12,13,14,15,16,17,19,21 | 1,2,3,4,5,6,7,8,9,10,11,12,13,14,15,16,17,19 | II |
| 6 | 1,2,3,4,5,6,7,8,9,10,11,12,13,14,15,16,17,18,19 | 1,2,3,4,5,6,7,8,9,10,11,12,13,14,15,16,17,18,19,21 | 1,2,3,4,5,6,7,8,9,10,11,12,13,14,15,16,17,18,19 | I |
| 7 | 1,2,3,4,5,6,7,8,9,10,11,12,13,14,15,16,17,18,19,20 | 1,2,3,4,5,6,7,8,9,10,11,12,13,14,15,16,17,18,19,21 | 1,2,3,4,5,6,7,8,9,10,11,12,13,14,15,16,17,18,19 | II |
| 8 | 1,2,3,4,5,6,7,8,9,10,11,12,13,14,15,16,17,18,19 | 1,2,3,4,5,6,7,8,9,10,11,12,13,14,15,16,17,18,19,21 | 1,2,3,4,5,6,7,8,9,10,11,12,13,14,15,16,17,18,19 | I |
| 9 | 2,3,4,5,6,7,8,9,10,11,12,14,15,16,17,18,19 | 1,2,3,4,5,6,7,8,9,10,11,12,13,14,15,16,17,18,19,21 | 2,3,4,5,6,7,8,9,10,11,12,14,15,16,17,18,19 | I |
| 10 | 1,2,3,4,5,6,7,8,9,10,11,12,13,14,15,16,17,18,19,20 | 1,2,3,4,5,6,7,8,9,10,11,12,13,14,15,16,17,19,21 | 1,2,3,4,5,6,7,8,9,10,11,12,13,14,15,16,17,19 | II |
| 11 | 1,2,3,4,5,6,7,8,9,10,11,12,13,14,15,16,17,18,19 | 1,2,3,4,5,6,7,8,9,10,11,12,13,14,15,16,17,18,19,21 | 1,2,3,4,5,6,7,8,9,10,11,12,13,14,15,16,17,18,19 | I |
| 12 | 1,2,3,4,5,6,7,8,9,10,11,12,13,14,15,16,17,18,19,20 | 1,2,3,4,5,6,7,8,9,10,11,12,13,14,15,16,17,18,19,21 | 1,2,3,4,5,6,7,8,9,10,11,12,13,14,15,16,17,18,19 | II |
| 13 | 1,2,3,4,5,6,7,8,9,10,11,12,13,14,15,16,17,18,19 | 1,2,3,4,5,6,7,8,10,11,12,13,14,15,16,17,19 | 1,2,3,4,5,6,7,8,10,11,12,13,14,15,16,17,19 | II |
| 14 | 1,2,3,4,5,6,7,8,9,10,11,12,13,14,15,16,17,18,19,20 | 1,2,4,5,6,7,8,9,10,11,12,13,14,15,19 | 1,2,4,5,6,7,8,9,10,11,12,13,14,15,19 | III |
| 15 | 1,2,3,4,5,6,7,8,9,10,11,12,13,14,15,16,17,18,19,20 | 1,2,3,4,5,6,7,8,9,10,11,12,13,14,15,16,17,19,21 | 1,2,3,4,5,6,7,8,9,10,11,12,13,14,15,16,17,19 | II |
| 16 | 1,2,3,4,5,6,7,8,9,10,11,12,13,15,16,17,18,19,20 | 1,2,3,4,5,6,7,8,9,10,11,12,13,14,15,16,17,19,21 | 1,2,3,4,5,6,7,8,9,10,11,12,13,15,16,17,19 | II |
| 17 | 1,2,3,4,5,6,7,8,9,10,11,12,13,15,16,17,18,19,20 | 1,2,3,4,5,6,7,8,9,10,11,12,13,14,15,16,17,19 | 1,2,3,4,5,6,7,8,9,10,11,12,13,15,16,17,19 | II |
| 18 | 3,4,6,7,8,9,11,12,18,19 | 1,2,3,4,5,6,7,8,9,10,11,12,13,14,15,16,17,18,19,21 | 3,4,6,7,8,9,11,12,18,19 | I |
| 19 | 1,2,3,4,5,6,7,8,9,10,11,12,13,14,15,16,17,18,19 | 1,2,3,4,5,6,7,8,9,10,11,12,13,14,15,16,17,18,19,21 | 1,2,3,4,5,6,7,8,9,10,11,12,13,14,15,16,17,18,19 | I |
| 20 | 20 | 2,3,4,5,7,10,12,14,15,16,17,20 | 20 | I |
| 21 | 1,2,3,4,5,6,7,8,9,10,11,12,15,16,18,19,21 | 21 | 21 | IV |

### *2.6 Key Criteria Grouping as part ANP Model*

The key criteria from ISM method as an output from ISM method is grouped in this step as an input for ANP method. This is the integration step between the two methods and it is done by brainstorming with expert from company. The grouping is based on the similarity of key criteria. The key criteria is 19 from

ISM method are grouped into 4 cluster, namely: relationship, performance, service, and product. The relationship cluster is a set of key criteria that has the same characteristic that is the relationship between the supplier and company up until now. The key criteria in this cluster are long term relationship, reciprocal arrangement, amount of past business, and performance history. Performance cluster is a set of key criteria that denote the supplier success rate that can be assessed by the company itself or from outside of the company. The key criteria in this cluster are reputation and position in industry, delivery, operational control, and reliability. Service cluster is a set of key criteria that is related to the supplier's attitude in serving customers and the easiness of service given by the supplier. Repair service, flexibility, attitude, communication system, integrity, and professionalism are in this cluster. The product cluster is a set of key criteria that are related to the materials. The criteria in this cluster are quality, price, consistency, quality standard, and availability.

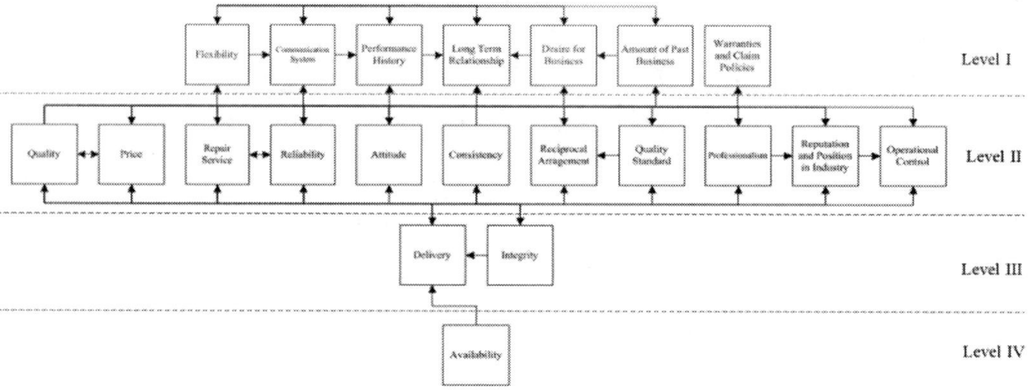


**Figure 1.** ISM Model

**Table 5.** Criteria Weights

| Cluster | Cluster Weight | Criteria | Local Weight | Global Weight |
|---|---|---|---|---|
| Relationship | 0,088953 | Long term relationship | 0.11910 | 0,015288 |
| | | Amount of past business | 0.20052 | 0,02574 |
| | | Performance history | 0.31801 | 0,040821 |
| | | Reciprocal arrangement | 0.36237 | 0,046516 |
| Performance | 0,215626 | Reputation and position in industry | 0.18105 | 0,030364 |
| | | Delivery | 0.21780 | 0,036527 |
| | | Operational control | 0.25126 | 0,042139 |
| | | Reliability | 0.34988 | 0,058678 |
| Service | 0,135443 | Communication system | 0.04251 | 0,008315 |
| | | Repair service | 0.04832 | 0,009453 |
| | | Flexibility | 0.12375 | 0,024208 |
| | | Attitude | 0.18175 | 0,035555 |
| | | Professionalism | 0.22300 | 0,043623 |
| | | Integrity | 0.38068 | 0,074469 |
| Product | 0,559979 | Quality | 0.10021 | 0,026361 |
| | | Quality standard | 0.10442 | 0,027468 |
| | | Consistency | 0.23023 | 0,060566 |
| | | Price | 0.25025 | 0,06583 |
| | | Availability | 0.31490 | 0,082837 |

## 3. Result and Discussion

### *3.1 ISM Results Discussion*

This study uses 19 key criteria by ISM. The key criteria are quality, delivery, price, repair service, reliability, flexibility, attitude, communication system, performance history, consistency, long term relationship, reciprocal arrangement, quality standard, integrity, professionalism, reputation and position in industry, operational control, amount of past business, and availability.

Based On MICMAC Analysis has obtained, availability criteria becomes the most important criterion for selecting a supplier in this study because This factor has high driving power, but low dependence; thus, it can be the key criteria. Constantly changed schedule caused by the number of parties involved in projects makes a change in material requirements. Thus the company may need to do urgent orders, and the suppliers are expected to have high availability in time of urgent orders.

### *3.2 ANP Results Discussion*

The processing for the ANP method is done with Super Decision software. The result from the software, global and local weights for each criterion, is shown in table 5. The highest global weight is availability criteria. This criteria also has an vital position in the ISM model, which is the lowest level, making it the base for the model. Both results support that availability is the most important criteria for selecting a supplier in this study. Construction project needs high availability of materials in supplier because the constantly changing schedule. High availability caused a need for supplier flexibility in order change.

**Table 6.** Supplier Weights

| No. | Alternatives | Value |
|---|---|---|
| 1 | Supplier 1 | 0,372335 |
| 2 | Supplier 2 | 0,131659 |
| 3 | Supplier 3 | 0,260464 |
| 4 | Supplier 4 | 0,128839 |
| 5 | Supplier 5 | 0,106683 |

From table 6, supplier 1 has the highest weight, 0,372335 followed by supplier 3, supplier 2, supplier 4, and supplier 5. Supplier 1 is significantly higher than the others because it has the best performance in all criteria. Supplier 3 has the same weight in reputation and position in industry with supplier 1, the attitude and professionalism performance of this supplier is also high. On the other hand, supplier 2, supplier 4, and supplier 5 have similar values. Supplier 4 excels in reputation and position in industry, but the availability performance is the lowest. While supplier 2 has a better performance in availability between the three, so the overall weight is better. A construction project needs a high availability of materials in the supplier, so the company can consider having spare suppliers.

### *3.3 Managerial Implications*

ISM model depicts interconnections between supplier selection criteria in hierarchic structure, thus giving the order of importance of criteria. Criteria in the lower level can affect the criteria above them; thus the effect of performance change can be estimated. For example, price fluctuations can affect the customer's desire to establish a long term relationship with the supplier. Focusing on key criteria can simplify the supplier selection process. Before identifying the key criteria, the company considers the desire for business and claim and warranties policies criteria. These two criteria are eliminated because they do not give significant impact on supplier selection decision. The supplier selection process can be simplified by omitting these criteria.

From ANP results, availability criteria has the highest weights. Availability refers to the availability of materials in suppliers for urgent orders. Because one supplier may not be able to provide the materials in all urgent orders, having a spare supplier must be considered.

## 4. Conclusion

The supplier selection problem is one of the most essential parts in supply chain operation. The contribution of this study is on the implemetation of ISM method to identify key criteria for supplier selection and integrates with ANP method to get the best supplier focusing on key criteria.The key criteria are identified to help the decision-maker to focus on criteria that has a more significant influence on the supplier selection model. From this study 19 key criteria are identified with criteria availability has the most influence on the overall ISM model and has the highest weight from ANP, thus making it the most critical criterion. This is caused by the ever-changing construction project schedule. A construction project needs a high availability of materials in the supplier, therefore it is suggested that the company can consider having spare suppliers.

However, this study has limitation that still focused on one specific construction company. Thus a further study using judgment from several experts can be done. Expert's opinions can differ between one another, making a change in the criteria. This change can trigger an increase or decrease of transitivity in the ISM method that affects the final solution. Moreover, a pairwise comparison questionnaire for ANP is considered wearying by the expert filling the questionnaire because a large number of assessments has to be done. Thus an error in filling the questionnaire could occur. Therefore, TOPSIS can be used to reduce the pairwise comparison assessment to just interconnections between the criteria.

## 5. References

[1] Limansantoso M F 2013 Pemilihan Supplier Produk Calista dengan Metode Analytical Hierarchy Process (AHP) Pada PT. Buana Tirta Utama – Gresik. *Calyptra: Jurnal Ilmiah Mahasiswa Universitas Surabaya* **2**(1), 1-20

[2] Özbek A 2015 Supplier Selection with Fuzzy TOPSIS. *Journal of Economics and Sustainable Development* **6**(18), 114-125

[3] Kargi V S A 2016 Supplier Selection for A Textile Company Using Fuzzy TOPSIS Method. *Yoneim Ve Ekonomi* **23**(3), 789-803

[4] Kurniawati D, Yuliando H and Widodo K H 2013 Kriteria Pemilihan Pemasok Menggunakan Analytical Network Process. *Jurnal Teknik Industri* **15**(1), 25-32

[5] Dickson G W 1966 *Analysis Of Vendor Selection Systems And Decisions.* Wiley Periodicals, Inc.

[6] Cheragi S H, Dadashzadeh M and Subramanian M 2011 Critical Success Factors For Supplier Selection: An Update. *Journal of Applied Business Research* **20**(2), 91-108

[7] Mukherjee K 2016 Supplier selection criteria and methods: past, present and future. *International Journal of Operational Research* **27**(1/2), 356-373

[8] Gupta A K, Singh O P and Garg R K 2015 Analytic Network Process (ANP): An Approach for Supplier Selection in an Automobile Organization. *European Journal of Advances in Engineering and Technology* **2**(9), 83-89

[9] Wicaksono A, Rahman A, and Tantrika C F M 2015 Pemilihan *Supplier* Baja H-Beam dengan Integrasi Metode Analytical Hierarchy Process and Technique For Order Preference by Similarity to Ideal Solution (Study Kasus: CV. Dharma. Kencana). *Jurnal Rekayasa and Manajemen Sistem Industri* **3**(1), 65-74

[10] Rani H A 2016 *Manajemen Proyek Konstruksi* (Sleman: Penerbit Deepublish)

[11] Verma 2014 Implementation of Interpretive Structural Model and Topsis in Manufacturing Industries for Supplier Selection. *Industrial Engineering Letters* **4**(5), 1-8

[12] Attri R, Dev N and Sharma V 2013 Interpretive Structural Modeling (ISM) approach: An Overview. *Research Journal of Management Sciences* **2**(2), 3-8

[13] Kumar S and Kant R 2013 Supplier Selection Process Enablers: An Interpretive Structural Modeling Approach. *International Journal of Mechanical and Industrial Engineering (IJMIE)* **3**(1), 89-95

[14] Thakkar J, Deshmukh S G, Gupta A D and Shankar R 2005 Selection of Third-Party Logistics (3PL): A Hybrid Approach Using Interpretive Structural Modeling (ISM) and Analytic Network Process (ANP). *Supply Chain Forum* **6**(1), 32-46

[15] Weber C A, Current J R and Benton W C 1991 Vendor Selection Criteria and Methods. *European Journal of Operational Research* **50**, 2-18.

[16] Ellram L and Cooper M 1990 Supply Chain Management, Partnership, and the Shipper - Third Party Relationship. *The International Journal of Logistics Management* **1**(2), 1-10

IOP Conf. Series: Materials Science and Engineering **847** (2020) 012054 doi:10.1088/1757-899X/847/1/012054

# Warehouse Layout Designing of Cable Manufacturing Company using Dedicated Storage and Simulation Promodel

**W Septiani [1,2], G A Divia[1,3] and S Adisuwiryo[1,4]**

[1]Industrial Engineering Faculty of Industrial Technology Trisakti University, Jl. Kyai Tapa No.1 Grogol, Jakarta, Indonesia
[2]winnie.septiani@trisakti.ac.id, [3]gebbyg.aqiilah@gmail.com, [4]sc.adisuwiryo@trisakti.ac.id

**Abstract.** PT. XYZ is a cable manufacturing company that has several types of cables produced. Goods produced are stored in finished goods warehouses with two types of packaging, namely haspel and roll. Laying of goods in open goods warehouses in the form of randomly arranged haspels is mixed between each type of haspel. The purpose of this study was to evaluate the layout of finished goods warehouses for haspel types so that they can be grouped and minimize displacement time. This study uses a simulation design that is implementing a real system using a ProModel simulation and provides improved layout evaluation with a dedicated storage method approach, i.e. goods or materials placed in a fixed location. The first step is to collect and identify the real system and then implement using a ProModel simulation resulting in a total displacement time of 140.27 hours. Improvements to the layout of the finished goods warehouse using a dedicated storage method to group haspels. From the calculation results, there are two scenarios, the first is the application of haspel grouping from the calculation of a dedicated method with displacement time of 139.21 hours and the second scenario adds material handling from the first scenario, the displacement time is 128.08 hours. The results of scenario two resulted in a decrease in displacement time of 8.69% by grouping the haspel and adding material handling.
**Keywords**: warehouse layout, simulation, dedicated storage, displacement time

## 1. Introduction

PT. XYZ is a company engaged in the field of cable production which produces various types of cables such as electric cables, communication cables, and others. The cables are stored in a warehouse in an open or closed area. The area of finished goods warehouse is 15.456 m2 while the area of finished goods warehouse is 3.228 m2. Large cables are wound in a haspel and stored in a finished goods warehouse in an open area. While small sized cables are rolled using a roll and stored in a finished goods warehouse in an enclosed area Haspel used was haspel of wood, namely haspel a construction made of wood and iron as reinforcement.

The warehouse is a supporting facility that has an important role in the production system [1], [2] [3]. There are many factors that influence warehouse design, such as order picking methods, size, and layout of the storage system, material handling systems, product characteristics, demand trends, turnover rates and space requirements [4]. Based on the results of a literature study [5] warehouse performance measurement can be measured from 5 size dimensions namely time, quality, cost and productivity.

The focus of this research is on finished goods warehouses in open areas. The finished goods warehouse in an open area is divided into six areas namely A1, A2, A3, A4, A5 and A6. The type of haspel that is stored in the warehouse area consists of 27 types of haspels with different types of sizes (H50, H55, H60 etc.). However, haspel storage in that area has not been classified by type. Laying of items that are still mixed, not grouped. This causes a long time to search for goods in the warehouse. In addition, the location of laying the warehouse is far from the location of packing or shipping. Haspel transfer activity is carried out by using a forklift.

Based on these problems, it is necessary to arrange the product storage location in the finished product warehouse. This study aims to design a simulation model of finished goods warehouse with a dedicated storage method. This method can provide solutions in the preparation of products in the warehouse to optimize the use of the area available in the warehouse [2]. Class-based storage methods can also increase productivity in processes in the warehouse, such as reducing the time of material uptake and the distance of uptake [4]. Thus it can be known how much it actually needs the area required by the finished product warehouse.

Evaluation of layout improvement is simulated with Promodel discrete event simulation software. Modeling is a windows based system with an intuitive graphical interface and object oriented modeling construct [6]. Promodel is also powerful as a simulation tool for solving manufacturing and service system problems [7], [8] and [6]. Previous studies that have been conducted related to simulations [9] and [10]. [11] Succeeded in using the dedicated storage method to replace items that were mixed in a place and arranged according to the category and consideration of the activities of the goods. This method is used to classify goods in a fixed place and not change again. This research will produce the usual warehouse layout design for open areas and the evaluation of proposals will be simulated using a promodel.

## 2. Methods

Data collection was done through interviews with the operator and the head of the warehouse of finished goods to find out which activities in the warehouse, the method of placement haspel and retrieval haspel in warehouse. Direct observation to observe activity in finished goods warehouse and measure the time displacement. Historical Data the required data include the type of haspel, the data of exit and entry haspel, and the frequency of the displacement haspel. The instruments used in this study include 1) interview guidelines; 2) Observation sheet; 3) field notes and stationery to record interview results; 4) recording device was used to record all interactions and communications that were not recorded in the study. In this study there were 5 stages in conducting, namely:

1. Analysis of existing system. The system analyzed is the layout system of finished product warehouses in the outside area.
2. Designing the simulation model layout of the warehouse at this time.
   a. Simulation models consist of structural elements, operational and numerical data.
   b. Testing the distribution of simulation input data using Stat fit.
   c. Running simulation using Promodel.
   d. Verification and validation.
3. Improved layout of the warehouse of finished goods used methods dedicated storage.
   Improvement of warehouse layout used dedicated storage method:
   a. Calculate space requirement for every haspel.
   b. Calculate Floor Space Requirements for every haspel.
   c. Rank the biggest T/S value to the smallest.
   d. Calculate total distance traveled.
   e. Haspel placement.
4. Evaluation of scenario improvement of warehouse layout using simulation Promodel.

IOP Conf. Series: Materials Science and Engineering **847** (2020) 012054 doi:10.1088/1757-899X/847/1/012054

5. In selecting the warehouse layout proposal ANOVA and Least Significant Different (LSD) calculations are performed to compare the proposed model with the initial system model conducted in this study the analysis of the results of the improved layout of the warehouse of finished goods.

## 3. Result and Discusion

### *3.1. Analysis System*

Warehouse of finished goods on the open area is divided into 6 storage areas, namely A1, A2, A3, A4, A5 and A6. This warehouse saves the 27 types of haspel with a variety of sizes. Distribution of types of haspel on each area is done randomly. The laying of the haspel is still often mixed with each other, not grouped by type. This has an impact on the search time items long. Warehouse layout can currently be seen in Figure 1. In Figure 1 it can be seen that the finished goods warehouse is divided into several areas which are given a different colour and each area of the contents by some kind of haspel. The arrow explains the flow of the movement of goods in the warehouse. Arrow coloured orange explaining the transfer of goods from packing to storage area and green from the storage area to the shipping.

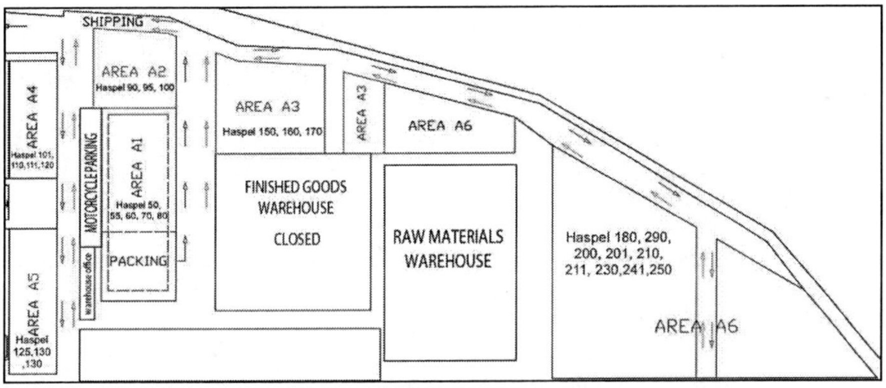


**Figure 1**. Current Layout of the Goods Warehouse

The sequence of flow of activities that occur on the warehouse of finished goods PT. XYZ can be seen in Figure 3.

**Figure 2.** Haspel stack

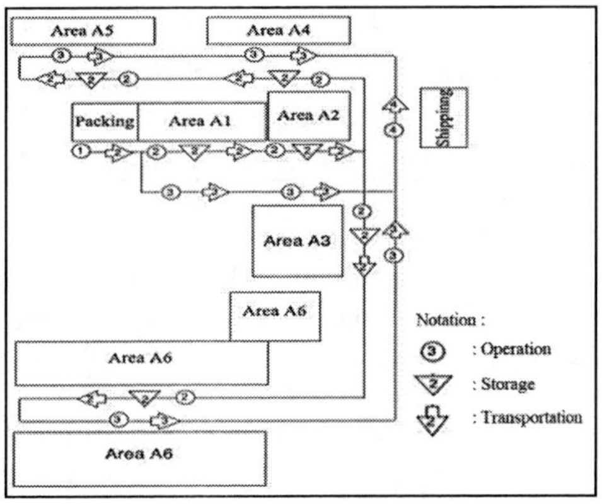


**Figure 3.** The flow of the Transfer Process of Finished Goods in the Warehouse of PT XYZ

IOP Conf. Series: Materials Science and Engineering **847** (2020) 012054 doi:10.1088/1757-899X/847/1/012054

### *3.2. Designing a Simulation Model with Promodel*

Building a credible and valid simulation model requires specific steps to be followed [12]. The division of the elements simulation based on the perspective of simulation using Promodel can be seen in Table 1 (Structural element) and Table 2 (Operational element).

**Table 1**. Structural Elements Simulating a Warehouse Layout System

| Element | Definition | Element in system |
|---|---|---|
| Entities | All this that participate in the system. | Consist of 27 types of Haspel (Haspel 50, Haspel 55, etc.) |
| Location | The place or location where the entity is processed | Area A1, A2, A3, A4, A5, A6, Shipping, Packing |
| Resource | Resources used to carry out the process | Material handling equipment (3 forklift) |
| Path Network | The path through which entities in the system | The route used material handling in warehouse of finished Goods |

**Table 2**. Operational Elements Simulating a Warehouse Layout System

| Element | Definition | Element in system |
|---|---|---|
| Arrivals | The arrival of entities in the system | The arrival of the goods into the warehouse, with input form the number of items in the Qty Each |
| Processing | The process carried out in the system | The process observed is the movement of finished goods from packing to area and area to shipping |

Output variables in this research consist of two variables, namely the total time of displacement and the utility of material handling. The total time displacement in this study was defined as the total time of retrieval of the goods and the transport that occurs in the warehouse of finished goods for one month.

Distribution testing in this study used the Stat: fit application. Distribution testing was carried out to determine the suitability of the data with the type of distribution. The results of processing stat: fit data for Area A1 can be seen in Table 3.

**Table 3.** Summary of Probability Distribution for Area 1

| # | Name | Distribution | Expression |
|---|---|---|---|
| 1 | Receiving time from packing to A1 | Lognormal | 0.765+L(0.563, 0.179) |
| 2 | Order picking time Haspel 50 | Lognormal | 0.765+L(0.563, 0.179) |
| 3 | Order picking time Haspel 55 | Uniform | U(1.33,0.3) |
| 4 | Order picking time Haspel 60 | Lognormal | 0.284+L(1.07, 0.178) |
| 5 | Order picking time Haspel 70 | Uniform | U(1.36, 0.31) |
| 6 | Order picking time Haspel 80 | Uniform | U(1.39, 0.3) |
| 7 | Shipping time from A1 to Shipping | Lognormal | 0.714+L(0.376,5.e-002) |

Verification of the model used two methods, namely the Verification of the model with the Bottom-Up and verification of the model with the facility trace at Pro Model. The results of the verification showed that the model simulation of the same system actually. Replication was performed five times with the value of *half width* which is desired by 0.1 and $Z_{0.025} = 1.96$.

Model validation based on the comparison of the displacement time taken from the multiplication of % in use of material handling with the total time of the simulation. The next stage was to compare the simulation results with a time displacement using material handling in a real system (Table 4).

IOP Conf. Series: Materials Science and Engineering **847** (2020) 012054 doi:10.1088/1757-899X/847/1/012054

**Table 4.** System Validation

| Use of Material Handling | Replacement Time of the Haspel (minutes) | | Time difference |
|---|---|---|---|
| | Real System | Simulation | |
| Forklift 1 | 3213.41 | 3587.83 | -374.41 |
| Forklift 2 | 3272.05 | 3298.31 | -26.25 |
| Forklift 3 | 3253.09 | 4202.21 | -949.13 |

Based on the results and the calculation of *half width*, obtained interval value as follows:

$$-1607.47 \leq \mu\,((1\text{-}2)) \leq 707.61 \quad (1)$$

From the results of the interval it can be seen that there is a hose between the interval that has been calculated and the interval includes the numbers 0. Under these conditions it can be concluded that $H_0$ is accepted, then the results of the real system and the simulation system is otherwise the same.

### *3.3. Improvement of Warehouse Layout using the Method Dedicated Storage*

The results of the improvement of warehouse layout used the method of dedicated storage was as follows:

1. **Calculate Space Requirement for every Haspel**
   In this calculation it was assumed one Block can fit at 3.5 m$^2$ taken from the largest dimensions of haspel, namely haspel 50.

$$\text{Space requirement Haspel 50} = \frac{\text{Maximum storage requirements of each product}}{\text{storage capacity}} \quad (2)$$

$$= \frac{11}{19} \approx 1 \text{ Slot}$$

   An example of the calculation of the floor area haspel 50 :
   Floor area requirements haspel 50 = Space requirements x product dimensions
   = (1 x 3.5) = 3.5 m$^2$

2. **Calculate Throughput for every Haspel**
   The calculation of the throughput to determine the amount of activity the storage and delivery of finished goods in the period of one month.

$$T = \frac{\text{average reception activity / day}}{\text{number of transfers at one time}} + \frac{\text{average shipping activity per day}}{\text{number of transfers at one time}} \quad (3)$$

$$= \frac{0{,}8}{1} + \frac{0{,}87}{1} = 1{,}67 \text{activity}$$

3. **Rank the biggest T/S Value to the smallest**
   To determine the laying of the goods, it used the comparison between the throughput and storage. The calculation of the placement of goods so use the following formula:

$$\text{Placement of finished goods: haspel 50} = \frac{\text{throughput}}{\text{space requirement}} = \frac{1{,}67}{1} \approx 1 \quad (4)$$

4. **Calculate Distance ever Block to I/O**
   The calculation of the distance of travel of the material handling is done by using the method of rectilinear distance. This calculation is taken from the center point of the storage area and a middle area packing and shipping and for the input and the output. Calculation Results with the Dedicated Storage Method can be seen in Table 5.
5. **Product Placement**
   Product placement in Table 6 is done by putting a product that has a high T/S on the location with the closest distance to the input-output, and next put the smallest T/S on the location with the next closest distance to the input-output, and so on.

**Table 5.** Calculation Results with the Dedicated Storage Method

| Calculation Method Dedicated Storage | | | | | | | | | | | | | |
|---|---|---|---|---|---|---|---|---|---|---|---|---|---|
| | | Space requirement | | | | Needs of area | | Throughput calculation | | | | | |
| No | Haspel type | Maximum receipt of finished goods (unit) | Haspel dimension ($m^2$) | Capacity / block (unit) | Space requirement calculation (block) | Dimension / block ($m^2$) | The calculation of floor space requirements ($m^2$) | The average receipt per month (unit) | The average delivery per month (unit) | Throughput (T) | T/S comparison | Distance calculation |
| 1 | 50 | 11 | 0.18 | 19 | 1 | 3.5 | 3.5 | 0.80 | 0.87 | 1.67 | 1.67 | 46.15 |
| 2 | 55 | 13 | 0.21 | 16 | 1 | 3.5 | 3.5 | 1.10 | 0.90 | 2.00 | 2 | 46.15 |
| 3 | 60 | 44 | 0.25 | 14 | 4 | 3.5 | 1.4 | 4.97 | 3.23 | 8.20 | 2.05 | 46.15 |
| 4 | 70 | 8 | 0.30 | 11 | 1 | 3.5 | 3.5 | 1.27 | 0.60 | 1.87 | 1.87 | 46.15 |
| 5 | 80 | 15 | 0.39 | 9 | 2 | 3.5 | 7 | 2.27 | 1.90 | 4.17 | 2.09 | 46.15 |

**Table 6.** Product placement

| Area | Haspel type | Distance (m) |
|---|---|---|
| A1 | H50, H55, H60, H70, H80 | 46.15 |
| A2 | H90, H95, H100 | 16.48 |
| A3 | H101, H110, H111, H120 | 62.57 |
| A4 | H125, H130, H140 | 118.37 |
| A5 | H150, H160, H170 | 12.24 |
| A6 | H180, H190, H200, H201, H210, H211, H230, H241, H250 | 76.41 |

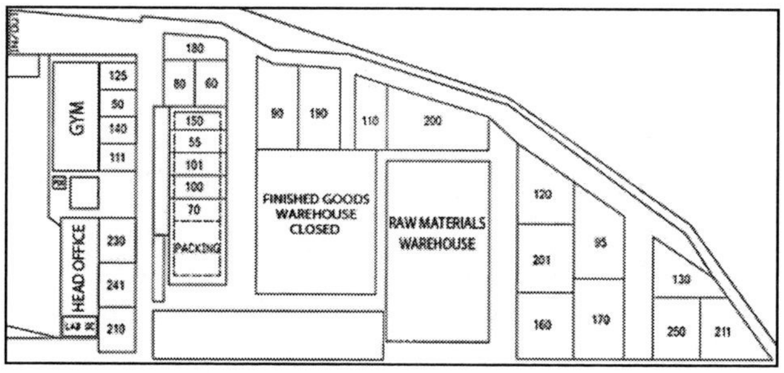


**Figure 4.** Proposed Improvement of Warehouse Layout with Dedicated Storage Method

With the calculation of the dedicated storage obtained some changes in the location of the laying of the goods in the warehouse. By applying such methods, there are some changes the location of the laying of the goods in the warehouse like Figure 3.

### *3.4. Scenarios for Improvement of Layout of Product Warehouse*

The design of the scenarios is made to determine the possibilities of repair solutions warehouse layout. In this research produced two improvement scenarios layout.

1. **Scenario 1**

   Scenario 1 is the result of an improved warehouse layout based on the dedicated storage method. From the calculation results of the method for each type of haspel has a specific area and is not mixed by other haspels. The application of the results of these methods results in a time of goods movement of 139.394 Hours.

2. **Scenario 2**

   The second scenario uses item laying calculations like scenario one with the addition of material handling units. The addition of material handling is done on material handling which has the highest% in use. The material handling that was added was found in forklift 1 which became 2 units. Simulation results show the time of moving goods in the warehouse of 128.145 hours.

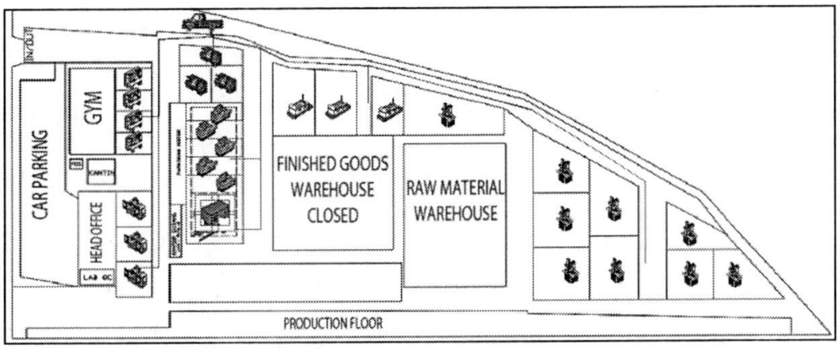


Figure 5. Warehouse Layout Simulation Model Scenario 1

### *3.5. Evaluation of The Proposed Improvement Warehouse Layout*

The key performance indicator used to evaluate the results of warehouse layout improvements is the time of movement. The results obtained from the simulation of taking goods are different for each proposed improvement and decreased which can be seen in Table 7.

**Table 7.** The Evaluation of scenario improved Layout

| Replication | Displacement time of the Haspel (minutes) | | |
|---|---|---|---|
| | Existing | Scenario 1 | Scenario 2 |
| 1 | 140.27 | 139.21 | 128.08 |
| 2 | 140.28 | 139.33 | 128.21 |
| 3 | 140.57 | 139.19 | 127.93 |
| 4 | 140.41 | 139.54 | 128.04 |
| 5 | 140.60 | 139.54 | 128.82 |
| Mean | 140.49 | 139.39 | 128.22 |

Based on the results of the calculation of the ANOVA test obtained value of F count equal to 3484,29 and F table of 3.89. If F count > F table then H0 is rejected. H0 used is there a significant difference between the average system output current simulation and proposal. F count is greater than the F table so that H0 is rejected which means that there is a significant difference against the average output of the simulation system at this time and the proposal. After doing the ANOVA test carried out advanced test that test the methods of LSD to see the differences between each model, the calculations can be seen in Table 8.

**Table 8.** Testing the simulation output using the LSD Method

| Treatment | mean | Treatment | mean | Score different | Result |
|---|---|---|---|---|---|
| Existing System | 140.49 | Scenario 1 | 139.39 | 0.35 | significantly different |
| Scenario 1 | 139.39 | Scenario 2 | 128.15 | 0.35 | significantly different |

**Table 9.** Comparison Results of Initial Models and Proposed Models

| Model | Displacement time (hour) | percentage decrease in time |
|---|---|---|
| Initial layout | 140.27 | |
| Scenario 1 | 139.21 | 0.75% |
| Scenario 2 | 128.08 | 8.69% |

The results of comparison of the proposed model with the initial model can be seen in Table 9. From the results of the comparison table of the initial model and the proposed model it can be seen that for the proposed model 1 decreased by 0.75% and the proposed model 2 decreased by 8.69%. So the proposed model 2 is the best proposed model.

## 4. Conclusion

In this paper compiled several conclusions, namely:

The simulation results using Promodel on the current warehouse layout produce a product transfer time of 140.27 hours in a month. This research has produced two proposed improvements. The first scenario results in an improved warehouse layout using the dedicated storage method. 27 Types of Haspel in warehouses are grouped into 6 areas based on their level of activity. The second scenario results in an improvement in the first scenario plus the addition of material handling tools based on the highest percentage of utilities.

The results of the evaluation with the simulation of the first scenario produce a displacement time of 139.21 hours per month while the second scenario produces a simulation time of 128.08 hours per month. The simulation output has been tested with ANOVA and LSD, the results show a significant difference between the two scenarios, so the output of the two simulations can be compared. Scenario one results in a time decrease of 0.78% and scenario two is 8.73%. The selected improvement scenario is scenario two.

The Key performance indicator used in this study is still limited on time and the utility of material handling. This research can be developed using the key performance indicator of the other, such as quality, cost and productivity. In addition, this research also has yet to calculate storage space efficiency.

## 5. References

[1] Muharni Y, Kulsum and Khoirunnisa M 2019 Warehouse layout designing of slab using dedicated storage and particle swarm optimization. *IOP Conf. Ser.: Mater. Sci. Eng.* **532,** 1-11.

[2] Fumi A, Scarabotti L and Schiraldi MM 2013 Minimizing warehouse space with a dedicated storage policy. *Int. J. Eng. Bus. Manag.* **5,** 1-8.

[3] De Koster RBM, Johnson AL and Roy D 2017 Warehouse design and management. *Int. J. Prod. Res.* **55** 6327-30.

[4] Chan FTS and Chan HK 2011 Improving the productivity of order picking of a manual-pick and multi-level rack distribution warehouse through the implementation of class-based storage. *Expert. Syst. Appl.* **38,** 2686-700.

[5] Staudt FH, Alpan G, Di Mascolo M and Rodriguez CMT 2015 Warehouse performance measurement: A literature review. *Int. J. Prod. Res.* **53,** 5524-44.

[6] Benson D 1998 Simulation modeling and optimization using ProModel. *Proc. 1997 Winter Simul. Conf.,* 587-93.

[7] Khalili MH and Zahedi F 2013 Modeling and simulation of a mattress production line using Promodel. *Proc. 2013 Winter Simul. Conf.,* 2598-609.

[8] Lee CKM, Zhang S and Ng KKH 2019 In-plant logistics simulation model for the catering service industry towards sustainable development: A case study. *Sustain.* **11,** 1-11.

[9] Suhardini D, Septiani W and Fauziah S 2017 Design and simulation plant layout using systematic layout planning. *IOP Conf. Ser.: Mater. Sci. Eng.* **277,** 1-8.

[10] Moengin P, Septiani W and Herviana S 2014 A discrete-event simulation methodology to optimize the number of beds in hospital. *Proc. World Congr. Eng. Comput. Sci.* **2,** 902–7.

[11] Valinda C and Puspitasari NB 2016 Penataan fasilitas rak untuk optimasi inventory menggunakan metode dedicated storage pada klinik Ananda. *Ind. Eng. Online J.* **5,** 1–11.

[12] Clark R and Krahl D 2011 Roadmap to success: your first simulation model. *Proc. 2011 Winter Simul. Conf.,* 1470-80.

# Performance measurement using Balance Score Card and Analytic Network Process in Elastomer Switch Keypad Manufacturer Indonesia

**C O Doaly[1,2], L L Salomon[1,3] and K J Arta A[1,4]**

[1]Universitas Tarumanagara, Jl. S. Parman No. 1 Jakarta Barat, Indonesia
[2]carlaol@ft.untar.ac.id, [3]lithrones@ft.untar.ac.id, [4]kholidarta@gmail.co.id

**Abstract.** Elastomer Switch or rubber keypad are manufactured from silicon rubber properties to make button on cars, cellphone seals and TV remote buttons. From the sales analysis report show that trends that occur in company's sales volume are declining in recent years with an average of 204 complaints per year, so the measurement of company performance needs to be done. In this study, company performance is measured by applying the Balance Scorecard which provides a multi-dimensional evaluation framework and weighting calculations based on the Analitic Network Process (ANP). A KPI Performance Aplication Design is generated from this research to help companies know what perspectives affect the company. The results of performance measurement using the Balance Scorecard Method show that the total performance of the company has a value of 2.741 which means that the company's performance is still not good, but it is almost close to good enough value. So it needs some improvement for the future
**Keywords**: Performance Measurement, Balanced Scorecard, ANP

## 1. Introduction

Performance measurement using the Balanced Scorecard method based on Analytical Network Process has been widely used, such as at PT. X under the auspices of PT.PLN [1] and PT. Telemedia Network Horizon [2]. Performance measurement can also be done using Human Resource Scorecard to measure the performance of human resources [3] and use the Integrated Performance Measurement System (IPMS) method to measure performance related to stakeholder's requirements [4]. From previous studies it was found that this performance measurement produces information in the form of weights from each of the existing perspectives and can provide levels from the perspective that is not too influential to the most influential. Judging from the company's financial data that has decreased profits and increasingly fierce competition, PT.Polymatech Indonesia feels the need to re-arrange the strategy by reviewing the strategy objectives in the competition and evaluating the company's internal capabilities because currently the company's performance measurements are only using the system performance measurement that emphasizes financial aspects. This can occur due to many things such as lack of consumers, the existence of defective goods in shipping, and poor employee performance. In previous studies, the Balanced Scorecard system has a role for management as a vehicle for planning the company's long-term performance, helping management carry out and carry out its mission [5]. So this research wants to help solve company problems by using the Balanced Scorecard Method and weighting using the Analytic Network Process.

IOP Conf. Series: Materials Science and Engineering **847** (2020) 012055 doi:10.1088/1757-899X/847/1/012055

## 2. Methods

### *2.1 Balanced Sorecard*
The Balanced Scorecard (BSC) is one of the performance measurement system models with strategy as a starting point translating into performance measures. Developed by Robert S. Kaplan and David P. Norton of Harvard Business School [6]. Basically the Balanced Scorecard is a management system for companies to invest in the long term for customers (Customer), Learning and employee growth, including management (Learning and growth), internal business processes (Systems) in order to obtain financial results that enable the development of business organizations rather than just managing your bottom line for short-term results [7].

### *2.2 Analytical Network Process*
The Analytical Network Process (ANP) approach is largely ignored compared to the Analytical Hierarchy Process (AHP) approach which is linear in structure and does not accommodate feed-back. This is because AHP is relatively simpler and easier to implement, whereas ANP is deeper and more suitable for complex, complex decision making and requires a variety of interactions and dependencies. [8]. Analytical Network Process (ANP) method is the development of the AHP method. ANP allows for interaction and feedback from elements in the cluster (inner dependence) and between clusters (outer dependence) [9].

### *2.3 Key performance Indicator (KPI)*
Key Performance Indicators (KPI) are a series of key indicators that are measurable and provide strategic target information that has been imposed on an organization successfully achieved [10].

## 3. Result and Discussion
Strategy maps can provide a framework that shows the causal relationships, vision and mission, and the company's strategy, and strategic objectives for each perspective. In this study the strategy map for PT. Polymatech Indonesia can be seen in Figure 1.

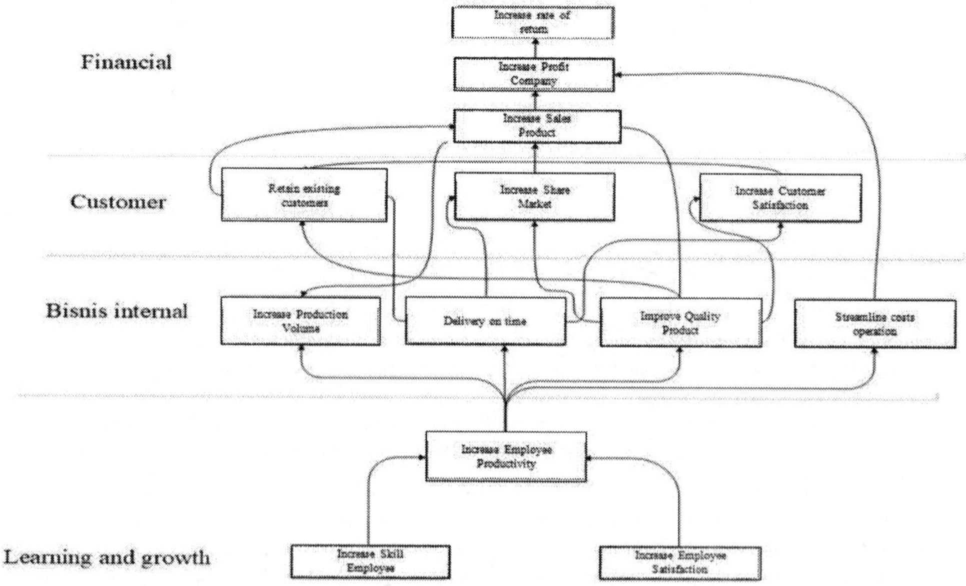


**Figure 1**. Strategy map

IOP Conf. Series: Materials Science and Engineering **847** (2020) 012055 doi:10.1088/1757-899X/847/1/012055

To measure performance at elastomer switch keypad manufacturing with the Balanced Scorecard method it is necessary to first assess the performance of each strategic goal from every perspective that exists in the Balanced Scorecard method. Table 1 shows an assessment of the company's current performance.

**Table 1**. Current company performance

| Perspective | Strategic Target | KPI | Target | Assessment | Score | Current Achievements | Results |
|---|---|---|---|---|---|---|---|
| Finance | Increase Return Rate | ROA | 10 % | ROA > 10 % | 5 | 7 % | 3 |
| | | | | 8% ≤ ROA < 10 % | 4 | | |
| | | | | 6% ≤ ROA < 8 % | 3 | | |
| | | | | 4% ≤ ROA < 6 % | 2 | | |
| | | | | ROA < 4% | 1 | | |
| | Increase Company Profit | NPM | 8 % | NPM > 8 % | 5 | 5 % | 4 |
| | | | | 5% ≤ NPM < 8 % | 4 | | |
| | | | | 2% ≤ NPM < 5 % | 3 | | |
| | | | | 0% ≤ NPM < 2 % | 2 | | |
| | | | | NPM < 0% | 1 | | |
| | Increase Product Sales | Increase in Sales | 5% | PP > 5 % | 5 | -15% | 1 |
| | | | | 3% ≤ PP < 5 % | 4 | | |
| | | | | 1% ≤ PP < 3 % | 3 | | |
| | | | | -2% ≤ PP < 1 % | 2 | | |
| | | | | PP < -2% | 1 | | |
| Customer | Retain existing customers | Customer Reference | 95 % | RP ≥ 95 % | 5 | 92% | 4 |
| | | | | 90% ≤ RP < 95 % | 4 | | |
| | | | | 85% ≤ RP < 90 % | 3 | | |
| | | | | 80% ≤ RP < 85 % | 2 | | |
| | | | | RP < -2% | 1 | | |
| | Increase Market Share | Market Growth Supply | 25 % | NPP ≥ 25 % | 5 | -8 % | 1 |
| | | | | 20% ≤ NPP < 25 % | 4 | | |
| | | | | 15% ≤ NPP < 20 % | 3 | | |
| | | | | 10% ≤ NPP < 15 % | 2 | | |
| | | | | NPP < 10 % | 1 | | |
| | Increase Customer Satisfaction | Number of Customer Complaints | <100 | JKT ≤ 100 % | 5 | 204 | 3 |
| | | | | 100% ≥ JKT > 200 % | 4 | | |
| | | | | 200% ≥ JKT > 300 % | 3 | | |
| | | | | 300% ≥ JKT > 400 % | 2 | | |
| | | | | JKT > 400 % | 1 | | |
| | Delivery of goods on time | % Delays in delivery | 5% | KP ≤ 5 % | 5 | 20 % | 2 |
| | | | | 5% ≥KP > 10 % | 4 | | |
| | | | | 10% ≥ KP > 15 % | 3 | | |
| | | | | 15% ≥ KP > 20 % | 2 | | |
| | | | | KP > 20 % | 1 | | |
| Internal business | Make all Operations Cost effective | Operating costs | Rp 1,3 B | BO ≤ 1,3 B | 5 | Rp 1,7 B | 3 |
| | | | | 1,3 B ≥BO > 1,6 B | 4 | | |
| | | | | 1,6 B ≥ BO > 1,9 B | 3 | | |
| | | | | 1,9 B ≥ BO > 2,2 B | 2 | | |
| | | | | BO > 2,2 M | 1 | | |
| | Increase the volume of the Company's production | Production Volume | 50000 /day | VP ≥ 50000 | 5 | 36000/day | 3 |
| | | | | 40000 ≤ VP < 50000 | 4 | | |
| | | | | 30000 ≤ VP < 40000 | 3 | | |
| | | | | 20000 ≤ VP < 30000 | 2 | | |
| | | | | VP < 20000 | 1 | | |
| | Improve Product Quality | % Number of Product Defects | 5 % | CP ≤ 5 % | 5 | 20 % | 3 |
| | | | | 5% ≥CP > 15 % | 4 | | |
| | | | | 15% ≥ CP > 25 % | 3 | | |
| | | | | 25% ≥ CP >35 % | 2 | | |

**Table 1**. Current company performance

| Perspective | Strategic Target | KPI | Target | Assessment | Score | Current Achievements | Results |
|---|---|---|---|---|---|---|---|
| Growth and learning | Increase Employee Productivity | Revenue / Employee | Rp 50 M | CP > 35 % | 1 | Rp 32.917.351,- | 3 |
| | | | | RE > 50 M | 5 | | |
| | | | | 40 M ≤ RE < 50 M | 4 | | |
| | | | | 30 M ≤ RE < 40 M | 3 | | |
| | | | | 20 M ≤ RE < 30 M | 2 | | |
| | | | | RE < 20 M | 1 | | |
| | Increase employee quality | Learning days | 4 days | LD ≥ 4 day | 5 | 2 days | 3 |
| | | | | 3 day ≤ NPP < 4 day | 4 | | |
| | | | | 2 day ≤ NPP < 3 day | 3 | | |
| | | | | 1 day ≤ NPP < 2 day | 2 | | |
| | | | | NPP < 1 day | 1 | | |
| | Increase employee satisfaction | Employee turn over | 0 % | TOK ≤ 0 | 5 | 0,15 % | 4 |
| | | | | 0% ≥TOK > 0,25 % | 4 | | |
| | | | | 0,25% ≥ TOK > 0,5 % | 3 | | |
| | | | | 0,5 % ≥ TOK > 0,75 % | 2 | | |
| | | | | TOK > 0,75 % | 1 | | |

Each perspective in the balanced scorecard is related to one another in accordance with the concept of the Analytical Network Process. ANP functions to see the amount of contribution made by each perspective and each of these strategic objectives.

### *3.1 Weighting using ANP Method*
This weighting is done by forming a supermatrix, where the supermatrix consists of sub-matrices which are composed of a set of relationships between the two levels contained in the model. There are 3 stages of supermatrix in ANP.

#### *3.1.1 Unweighted Supermatrix*
This supermatrix contains an eigenvector generated from the whole paired comparison matrix in the network. The results of the Unweighted Supermatrix in this study can be seen in Table 2.

#### *3.1.2 Weighted supermatrix*
This supermatrix is obtained by multiplying all the weights in the unweighted supermatrix by the weight of each cluster. Weighted Supermatrix calculation results in this study can be seen in Table 2.

#### *3.1.3 Limitting Supermatrix*
After the Weighted Supermatrix calculation process is complete, a Limitting Supermatrix calculation is performed to obtain a stable priority value. This is done by normalizing the weights on each strategic goal to get the contribution of each strategic goal. The results of Limitting Supermatrix in this study can be seen in Table 2.

### *3.2 Performance measurement of the Balanced Scorecard Method*
Performance measurement from each perspective, namely financial, customer, internal business, as well as learning and growth, is done by multiplying the results of the assessment of company performance in each perspective with the weights obtained from the analysis of current company performance. Performance measurements from four perspectives can be seen in Table 3.

After the performance value of each perspective is known then overall company performance is measured. This measurement is done by multiplying the results of the performance evaluation of each perspective with the weight of each perspective itself, then after adding them up, the company performance results are seen on a Likert scale, namely (1-5). The results of measurements of the overall performance of the company can be seen in Table 4.

**Table 2**. Weighting of ANP

| Perspective | Strategic Target | Weight *Unweighted Supermatrix* | Weight *Weighted supermatrix* | Weight *Limitting Supermatri* |
|---|---|---|---|---|
| Finance | Increase rate of return | 1 | 0,563 | 0,255 |
| | Increase company profits | 1 | 0,563 | 0,255 |
| | Increase product sales | 0,648 | 0,365 | 0,165 |
| Customer | Retain existing customers | 0,268 | 0,062 | 0,028 |
| | Increase market share | 0,794 | 0,183 | 0,083 |
| | Increase Customer Satisfaction | 0,353 | 0,081 | 0,037 |
| Internal business | Increase production volume | 1 | 0,132 | 0,060 |
| | Delivery on time | 0,293 | 0,039 | 0,018 |
| | Improve product quality | 0,567 | 0,075 | 0,034 |
| | Streamline the operating costs | 0,373 | 0,049 | 0,022 |
| Learning and Growth | Increase employee productivity | 0,330 | 0,025 | 0,011 |
| | Improve employee quality | 0,679 | 0,051 | 0,023 |
| | Increase employee satisfaction | 0,321 | 0,024 | 0,011 |
| | TOTAL | | 2,211 | 1 |

**Table 3**. Performance measurement results

| Perspective | Strategic target | KPI | Results | Weight | Rx W |
|---|---|---|---|---|---|
| Finance | Increase Return Rate | ROA | 3 | 0,25 | 0,76 |
| | Increase Company Profit | NPM | 4 | 0,25 | 1,02 |
| | Increase product sales | Increase in Sales | 1 | 0,16 | 0,16 |
| | Total | | | 0,67 | 1,95 |
| | Value of Financial Perspective Performance | | | | 2,88812 |
| Customer | Retain existing customers | Customer Retention | 4 | 0,028 | 0,112 |
| | Increase Market Share | Market Growth Value | 1 | 0,083 | 0,083 |
| | Increase Customer Satisfaction | Number of Customer Complaints | 3 | 0,037 | 0,110 |
| | Total | | | 0,15 | 0,30 |
| | Value of Customer Perspective Performance | | | | 2,06802 |
| Internal business | Delivery of goods on time | % Lateness | 2 | 0,018 | 0,035 |
| | Make effective all operating costs | Operating costs | 3 | 0,022 | 0,067 |
| | Increase Company Production Volume | Production Volume | 3 | 0,060 | 0,179 |
| | Improve Product Quality | % Number of Defects | 3 | 0,034 | 0,102 |
| | Total | | | 0,12 | 0,35 |
| | Performance Value of Internal Business | | | | 3 |
| Learning and Growth | Increase Employee Productivity | Revenue / Employee | 3 | 0,011 | 0,034 |
| | Improve Employee Quality | Learning Days | 3 | 0,023 | 0,069 |
| | Increase Employee Satisfaction | Turn over | 4 | 0,011 | 0,044 |
| | Total | | | 0,05 | 0,15 |
| | Value of Learning and Growth Perspective Performance | | | | 3,24112 |

**Table 4.** Company performance results

| Perspective | Score | Weight | Score x Weight |
|---|---|---|---|
| Finance | 2,888 | 0,563 | 1,625 |
| Customer | 2,068 | 0,230 | 0,476 |
| Business | 3,000 | 0,132 | 0,397 |
| Growth | 3,241 | 0,075 | 0,243 |
| Total | | | 2,741 |

Based on the calculation, the total value of the company's performance is 2.741, which means that the company's performance has not been good, but it is almost close to good enough value, by making some improvements it is expected that the performance will be better. To be able to find out how the company's performance results are designed an application Key Performance Indicator (KPI) by considering the output, financial, Process and Resource aspects that can provide information on achieving strategic goals and perspectives of what is most influential in improving company performance. In Figure 2 shows the output of the KPI application output, Figure 3 for the financial perspective view, Figure 4 for the process perspective and Figure 5 for the Perspective Resources display

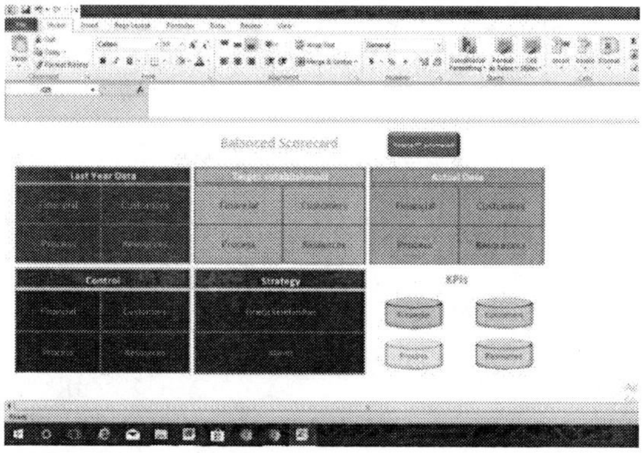

**Figure 2.** KPI output display

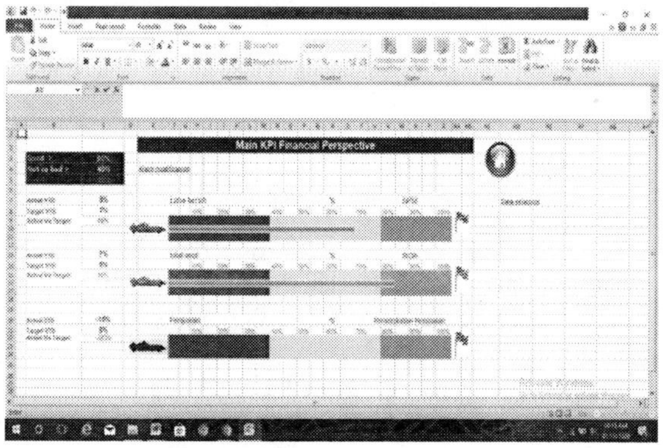


**Figure 3.** Perspective financial display

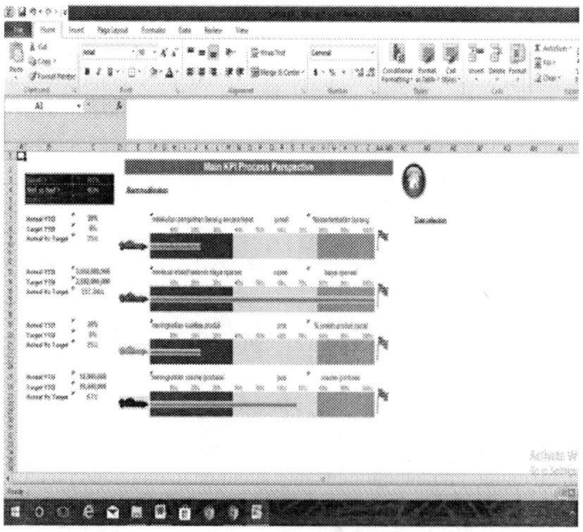

**Figure 4.** Perspective process display

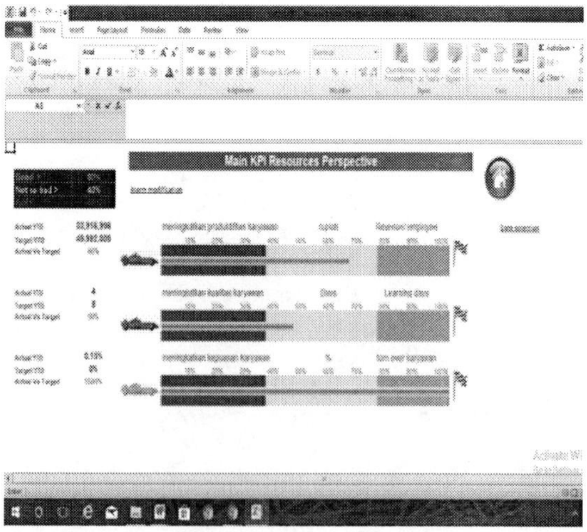


**Figure 5.** Display of perspective resources

## 4. Conclusion

Performance measurement using the Balance Scorecard Method in Elastomer Switch Keypad Manufacturer Indonesia is very useful. The results of the study show that the total performance of the company has a value of 2.741, which means the company's performance is still not good, but it is almost close to good enough value. So it needs some improvement for the future. Based on calculations made from 4 perspectives in the Balanced Scorecard method, namely finance, customers, internal business, and learning and growth, the weighting result for financial perspective is 0.566, Customer is 0.22, Internal Business Process is 0.13 and the weight value is Learning & Growth is 0.075. With this measurement, it is found that the strategic targets are not in accordance with the company's targets, namely increasing product sales, increasing market share, delivering goods on time, making all operating costs effective, and increasing employee productivity. strategy for PT. Polymatech Indonesia so that it can improve the performance of the company's KPIs to the maximum.

## 5. References

[1] Vanany, Iwan 2003 Analytic Network Process. *Jurnal Teknik Industri.* (Surabaya: Universitas Kristen Petra)
[2] Riza, Bob Subhan 2015 *Pengukuran Kinerja Perusahaan Menggunakan Analytic Network Process.* (Medan: Universitas Potensi Utama)
[3] Lorisa C, and Doaly CO 2017 Pengukuran Kinerja Sumber Daya Manusia Dengan Human Resource Scorecard di PT Trio Jaya Steel. *Jurnal Teknik Industri,* **7**(3), pp. 132-146.
[4] Doaly C., Salomon L.L., and Steven C. 2018 Pengukuran Kinerja dengan Metode IPMS. *Proc. Seminar Nasional Teknologi dan Sains III 2018*, pp. 230-240.
[5] Gaol, CHR. Jimmy 2014 *A to Z Human Capital Manajemen Sumber Daya Manusia.* (Jakarta: PT.Grasindo)
[6] Kaplan, R. and D. Norton 1996 *Translating strategic Into Action – The Balanced Scorecard.* (Boston: Harvard Business School Press)
[7] Gaspersz, Vincent 2006 *Sistem Manajemen Kinerja Terintegrasi Balanced Scorecard dengan Six Sigma*. (Jakarta: PT. Gramedia Pustaka).
[8] Rusydiana, Aam S and Devi, Abrista 2013 *Analytic Network Process* Pengantar Teori Dan Aplikasi. (Bogor: Smart Publishing)
[9] Saaty, TL and Vargas, Luis G. 2006 *Decision Making with The Analytic Network Process.* (United States of America: Springer Science).
[10] Soemahadiwidjojo, Arini T 2015 *Panduan Praktis Menyusun KPI.* (Jakarta: RAS).

IOP Conf. Series: Materials Science and Engineering **847** (2020) 012056 doi:10.1088/1757-899X/847/1/012056

# A conceptual framework for Servitisation of the manufacturing companies to deliver Product–Service Systems solutions: A study case of the Indonesian Motorcycle Industry

**D R S Dewi[1,2,3], S Pittayachawan[2,4] and E Tait[2,5]**

[1]Widya Mandala Surabaya Catholic University, Kalijudan 37, Surabaya, Indonesia
[2]Royal Melbourne Institute of Technology, 445 Swanston, Melbourne, Australia
[3]dian.dewi@rmit.edu.au, [4]siddhi.pittayachawan@rmit.edu,au,
[5]elizabeth.tait@rmit.edu.au

**Abstract.** This study develops a framework for the servitisation of manufacturing that covers area of upstream and downstream supply chain to provide a bundling of product and service. An integration of upstream and downstream supply chain is essential to support the Product–Service Systems. However, a research of integrating the upstream and downstream supply chain is still nascent. A Dynamic capability is used as underpinning theoretical framework for this research. Six hypotheses are developed to build the conceptual framework particularly to investigate the relationship between the supply chain capabilities and sustainability performance of Product–service systems in the Indonesian motorcycle industry.
**Keywords**: Product–Service systems, supply chain capabilities, servitisation, sustainability performance.

## 1. Introduction

Product–service systems (PSS) can be thought of as the act of integrating products and services or as a market proposition that extends the traditional functionality of the product by embedding these services [1]. PSS is defined as "a business model focused toward the provision of a marketable set of product and service designed to be economically, socially and environmentally sustainable, with the final aim of fulfilling customer's need" [2]. PSS offers many benefits to companies. It provides an integrated solution by delivering a marketable bundling of product–service to customers [3]. This offer creates a differentiation strategy that increases companies' competitiveness [2, 4]. Effective bundling of product–service also increases customer satisfaction [5]. As a result, companies' profitability increases by having the advantage of bundling of product–service offering [2]. Moreover, creating an integrated system for PSS could increase overall resource productivity, and efficient use of resources [4]. This leads to decreasing companies' operational cost [2, 5]. Consequently, PSS has attracted big companies such as Apple, Xerox, Hyundai, LG, KIA, Toyota, Kone, Electrolux, Caterpillar and Panasonic to adopting PSS [5, 6].

Due to the complexity of the motorcycle as product, manufacturers should sell their product as bundling of product and service. Maintenance and servicing are compulsory for automotive product to maintain products' performance. Such service should be handled by authorised service centres that belong to the manufacturer. To improve the performance of manufacturer as producer of the product and service centres as producer of the service, a collaboration in the supply chain including suppliers, manufacturers and service partners is paramount [7]. An integration of upstream and downstream supply chain (SC) is essential for providing a bundling of product–service [8].

Several dynamic capabilities needed for supporting the integration of upstream and downstream SC in the organisations including: knowledge assessment, partner development, co-evolving, re-conceptualisation and reflexive control from upstream SC and innovative service delivery from downstream SC [9] that directly affect the sustainability performance of the PSS. PSS studies have been explored; however, this area needs further investigation for integrating the upstream and downstream SC. A few studies have explored the model integrating of upstream and downstream SC perspectives in delivering a bundling of product–service [10, 11]. However, their studies have not covered the complete aspects of upstream SC such as logistic integration, information technology integration and collaboration. Therefore, this study develops a framework for the servitisation of manufacturing to provide PSS by investigating the relationship between SC capabilities (upstream and downstream SC) and the sustainability performance of PSS in the Indonesian motorcycle industry.

## 2. Methods

Theoretical foundation and hypotheses development will be developed to investigating the relationship between SC capabilities and the sustainability performance of PSS. Six hypotheses will be developed to discuss the relationships among seven constructs including sustainability performance, innovative service delivery, reflexive control, re-conceptualisation, co-evolving, partner development and knowledge assessment.

### *2.1. Sustainability performance*

Sustainability performance is defined as the performance that contributes to balancing the profit (economic) and the planet (environment) [12]. Some studies have only used environment aspect in their sustainability performance definitions [13]. However, these have been criticized for not including all three aspects of sustainability[14]. Hassini, Surti and Searcy [15] definition of sustainability performance is described as "the capability to conduct business with a long-term goal of maintaining the well-being of the economy, environment, and society", which is compatible with Elkington [16] that proposes three dimensions of the triple bottom line including profit, planet, and people. The economic performance refers to financial and marketing measurement resulting from sustainability activities that improve the company's current condition [17]. It can be acquired by elevating customer satisfactions with providing bundling of product– service [18]. This added service creates a company's revenue and profitability [19]. Environmental performance refers to the capability of the company to reduce waste and minimise the use of hazardous materials [20]. Social performance refers to the impact of sustainability practices on a social aspect such as an image of the company from the stakeholders view point including suppliers, customers, employees, and communities [21]. It can be attained by the provision of more employability in the community [22], improvement of product–service image [23] thus leading to a rising of the company image [2].

### *2.2. Innovative service delivery*

Innovative service delivery is "an inherently dynamic process, which is more dependent for its success than traditional product-centric marketing on continuous adaptation to the evolving nature of customer needs and the technological aspects of provisions" [9]. Kindström, Kowalkowski and Sandberg [9] propose three capabilities related to innovative service delivery including customer-linked service, service delivery process, and orchestrating the service systems. Customer-linked service is the capability of knowing the customers' need [9]. Such capabilities create a close relationship with customers [24]. They include memorable customer service experience [11], differentiation [25], adjustment of service to the product [26], and innovation [27]. Service delivery process is the capability to deliver an integrated bundling of product–service to customers [9]. Arranging service systems is the capability of extending the firm resource, particularly with external parties collaboration, into a new market by proposing a

continuous innovation bundling of product–service [9]. Collaboration with service partners is crucial to maintaining a close relationship with customers [28]. In the motorcycle industry, service means delivering technical expertise that could prolong product's life cycle, for example, service maintenance, technical consulting and extended warranty of the product [29-31]. It decreases the negative impact on the environment [4]. Furthermore, many job vacancies to the social community will be offered as the service need human skills [32]. Accordingly, we can assume that innovative service delivery is positively associated with sustainability in economic, environmental and social perspectives. The above arguments lead to the hypothesis:
*H6: Innovative service delivery positively affects the sustainability performance*

### *2.3. Re-conceptualisation*
Re-conceptualisation is defined as the capability to improve the supply chain to be more sustainable [33]. It is the key component to achieve a sustainable SC [34]. A general policy by the government such as polution charges, extended producer responsibility on product's end of life, eco-labelling informative policy stimulates the environmentally better products and services [35]. Closed-loop SC is a way of re-conceptualising the SC [34], to deliver social well-being and economic wealth while utilising within the limit of our planet [36]. Closed-loop SC is defined as activities related to refurbishing and remanufacturing a product [37], involving activities identified as reduce, reuse, and recycle [38]. A product's take-back program, maintenance, and advice on efficient use are included as closed-loop SC activities [39]. Closed-loop SC has been empirically tested and found to have a positive relationship with sustainability performance [38]. Frequently, incentives from the government including subsidy, tax exemption and loan encourage a company to implement closed-loop practices [37]. The other way to reduce the burden on the environment through maintenance and repair [40]. The pressure from communities, non-government organisations and government/policy makers lead to more sustainable goals. The above arguments lead to the hypothesis:
*H5: Re-conceptualisation positively affects the sustainability performance*

### *2.4. Reflexive Control*
Reflexive control is defined as the capability of gathering information, evaluating, and sharing, and consequently aims to control supply chain functionality [41]. The capability to control will ensure the continuity of the companies to achieve their goal by doing continuous improvement of their operational capabilities [42]. Therefore, reflexive control requires more than just the collection of historical financial data but also the persistent monitoring of the functionality of the SC using evaluation of key performance indicators [43]. Moreover, partners' activities are monitored and audited through standards and certification by third parties such as the ISO 14001 or European Union Eco-Management and Audit Scheme [44]. By partners monitoring the system regularly, the total performance can be monitored hence the long-term development can be maintained effectively [45]. The capability of mitigating risk will increase with transparency and monitoring [46]. Mandal et al. [46] confirm that SC controlling activity such as monitoring is associated positively with an environmental objective. Accordingly, reflexive control should be positively associated with sustainability. The above arguments lead to the hypothesis:
*H4: Reflexive control positively affects the sustainability performance*

### *2.5. Co-evolving*
Co-evolving is defined as the capability of creating new resources consistently by enhancing collaboration within the single supply chain [47]. For example, a new form of resources can be created by the dynamic learning routines based on information sharing among the stakeholders including suppliers, manufacturers and service partners in the SC [47]. Each cooperation and collaboration within SC should be treated as a way to attain continuous learning among the SC to develop new capabilities and performances [48]. Then, co-evolving can be described as the improved collaborations among stakeholders in the SC [33]. Partner-based strategies are new resources as a result of SC collaboration/ co-evolution [49]. The integration of product–service can be acquired through alliances and partnerships

[50]. Long-term collaboration is focused on the capability of the SC to provide spare parts, warranty services, knowledge-intensive services and delivery of PSS to customers [30]. Due to the product–service bundling complex offering, collaboration among stakeholders in the SC becomes the backbone for achieving the SC sustainability goal [51, 52].Accordingly, we hypothesise that:
*H3: Co-evolving positively affects the sustainability performance*

### *2.6. Partner Development*
Partner development refers to the capability of the strongest partner in the SC to improve the capability of the entire network of SC [53]. Partner development programs help the weakest partner in the SC to achieve the sustainability performance in the SC [43]. PSS is an innovation business proposition that offers an innovative bundling of product–service [54]. Specific capabilities including service development processes [55], capability development [56], and learning [42] are crucial. An enthusiasm to enhance SC overall performance should be demonstrated by all stakeholders in the SC [33]. Partner development programs assist in achieving that goal [57] by knowledge sharing development and partners training [43], for example, educating the service partners in technical expertise related to service and maintenance [19]. Agi and Nishant [58] confirmed in the study of green supply chain management that partner development assistance is needed to achieve sustainability. Correspondingly, partner development implementation is required to promote sustainability. The above arguments lead to the hypothesis:
*H2: Partner development positively affects the sustainability performance*

### *2.7. Knowledge assessment*
Knowledge assessment is defined as the capability to access and understand the knowledge from SC partners [43]. Defee and Fugate [47] describe knowledge accessing as "a capability held by two or more parties that fosters an understanding of the current knowledge resources possessed by each party". Each partner in the SC should develop its capabilities by accessing and understanding the capabilities from other partners and use them for the benefit of the entire SC [47], which opposes to the traditional learning orientation that emphasises acquiring and absorbing knowledge [33]. Knowledge assessment is essential element to the Triple Bottom Line concept [37], as well as knowledge to deliver a bundling of product–service[59]. Nevertheless, knowledge spread among SC stakeholders including suppliers, manufacturers and service partners would not be possible without the help of technology to deliver the information. Members of the SC must agree on technology integration such as common IT interfaces and shared licensing [44]. Furthermore, knowledge assessment about market and customer knowledge is also beneficial to obtain an insight into customer needs [60] to respond appropriately to market changes [61]. Collaboration by building knowledge with service partners is preferred rather than just delegating the service activities [10]. Kumar, Subramanian and Arputham [38] confirmed that knowledge sharing, learning and acquisition are important capabilities for sustainability. The above arguments lead to the hypothesis:
*H1: Knowledge assessment positively affects the sustainability performance*

## 3. Result and Discussion
Working from compendious literature reviews have guided to the development of proposed conceptual model, purposely to answer the main objective of this study. Beske, Land and Seuring [33] represent the upstream SC including knowledge assessment, partner development, co-evolving, reflexive control and re-conceptualisation. Kindström, Kowalkowski and Sandberg [9] represent the downstream SC (innovative service delivery). This study proposed to integrate the upstream and downstream SC by extending the work of Beske, Land and Seuring [33] and Kindström, Kowalkowski and Sandberg [9] as shown in Figure 1.

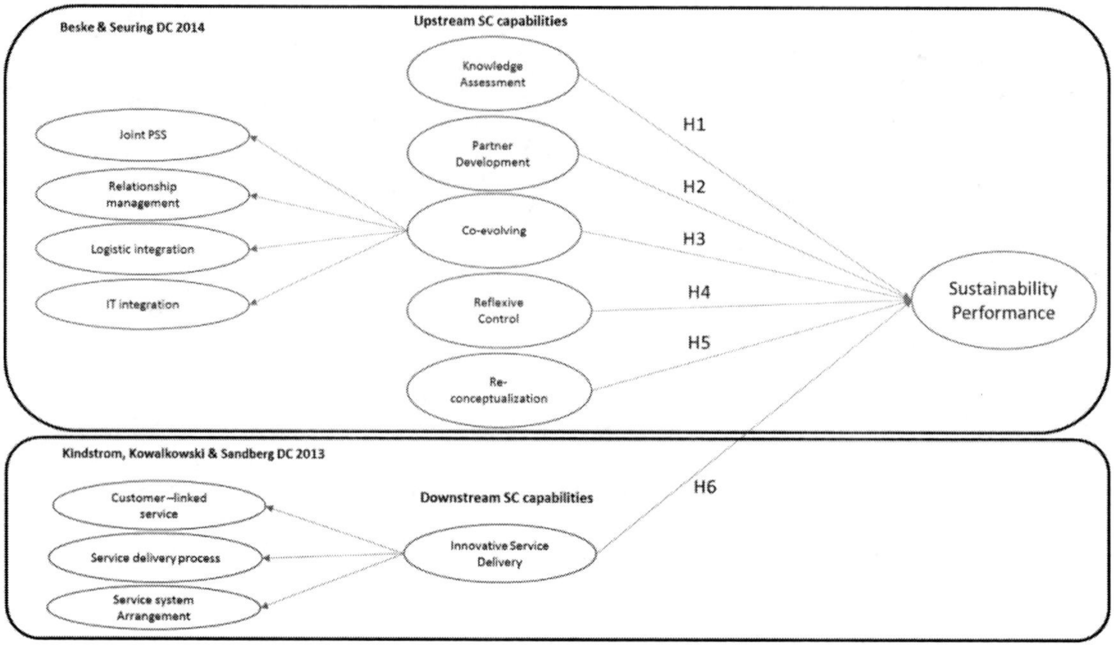


**Figure 1**. Proposed conceptual model

This study contributes to academic knowledge by enriching the theoretical knowledge of building a conceptual framework for the servitisation of manufacturing companies by investigating the relationship between SC capabilities and sustainability performance of PSS. Current literatures have nascent research on the perspective of both upstream and downstream SC. Therefore, this research aims to examine the integration of upstream and downstream SC with their relationship to the sustainability performance of PSS in the Indonesian motorcycle industry. Filling the gap of knowledge by integrating upstream and downstream SC contributes to PSS organisation knowledge. The result of this study will also be of value to practitioners to have a better understanding of upstream SC, downstream SC capabilities and sustainability performance of PSS in the Indonesian motorcycle industry. This will offer practical guidance for the manager to develop SC capabilities to enhance the sustainability performance of PSS in the Indonesian motorcycle industry.

## 4. Conclusion

We contribute to develop a comprehensive conceptual framework for servitisation of the manufacturing companies by investigating the relationship between SC capabilities and the sustainability performance of PSS. In this paper, we identified several SC capabilities from upstream and downstream SC. These are knowledge assessment, partner development, co-evolving, reflexive control, re-conceptualisation, innovative service delivery. These SC capabilities are hypothesised positively affect the sustainability performance. However, further research is needed to expand the operationalisation of the proposed conceptual framework.

## Acknowledgment

Author acknowledge support from the Indonesia Endowment Fund for Education (LPDP) for its grant to conduct this study

## 5. References

[1] Baines, et al. 2007 State-of-the-art in product-service systems. *Proceedings of the Institution of Mechanical Engineers, Part B: Journal of Engineering Manufacture* **221**(10): 1543-1552.

[2] Annarelli, Alessandro, Cinzia Battistella, and Fabio Nonino 2016 Product service system: A conceptual framework from a systematic review. *Journal of Cleaner Production* **139**: 1011-1032.
[3] Baines and Lightfoot 2013 Servitization of the manufacturing firm: Exploring the operations practices and technologies that deliver advanced services. *International Journal of Operations & Production Management* **34**(1): 2-35.
[4] Mont, Oksana 2002 Clarifying the concept of product–service system. *Journal of Cleaner Production* **10**(3): 237-245.
[5] Beuren, Fernanda Hänsch, Marcelo Gitirana Gomes Ferreira, and Paulo A. Cauchick Miguel 2013 Product-service systems: a literature review on integrated products and services. *Journal of Cleaner Production* **47**: 222-231.
[6] Ryu, Hokyoung, et al. 2018 Servicizing Solutions for Manufacturing Firms: Categorizing Service Ideas from Product-Service Integrated Examples. *The Design Journal* **21**(2): 267-302.
[7] Nudurupati, Sai S, et al. 2016 Eight challenges of servitisation for the configuration, measurement and management of organisations. *Journal of Service Theory and Practice* **26** (6): 745-763.
[8] Vural, Ceren Altuntas 2017 Service-dominant logic and supply chain management: a systematic literature review. *Journal of Business & Industrial Marketing* **32**(8): 1109-1124.
[9] Kindström, Daniel, Christian Kowalkowski, and Erik Sandberg 2013 Enabling service innovation: A dynamic capabilities approach. *Journal of Business Research*.
[10] Ayala, Wolfgang Gerstlberger, and Alejandro G Frank 2018 Managing servitization in product companies: the moderating role of service suppliers. *International Journal of Operations & Production Management*.
[11] Agarwal, Renu and Willem Selen 2009 Dynamic Capability Building in Service Value Networks for Achieving Service Innovation. *Decision Sciences* **40**(3): 431-475.
[12] Neto, J Quariguasi Frota, et al. 2008 Designing and evaluating sustainable logistics networks. *International Journal of Production Economics* **111**(2): 195-208.
[13] Sarkis, Joseph, Qinghua Zhu, and Kee-hung Lai 2011 An organizational theoretic review of green supply chain management literature. *International journal of production economics* **130**(1): 1-15.
[14] Seuring, Stefan 2013 A review of modeling approaches for sustainable supply chain management. *Decision support systems* **54**(4): 1513-1520.
[15] Hassini, Elkafi, Chirag Surti, and Cory Searcy 2012 A literature review and a case study of sustainable supply chains with a focus on metrics. *International Journal of Production Economics* **140**(1): 69-82.
[16] Elkington, John 1997 Cannibals with forks: The triple bottom line of twentieth century business. *Capstone: Oxford*.
[17] Zhu, Qinghua, Joseph Sarkis, and Yong Geng 2005 Green supply chain management in China: pressures, practices and performance. *International Journal of Operations & Production Management* **25**(5): 449-468.
[18] Baines, Tim, et al. 2009 Towards an operations strategy for product-centric servitization. *International Journal of Operations & Production Management* **29**(5): 494-519.
[19] Gebauer, Heiko, Marco Paiola, and Nicola Saccani 2012 Characterizing service networks for moving from products to solutions. *Industrial Marketing Management*.
[20] Zhu, Qinghua, Joseph Sarkis, and Kee-Hung Lai 2008 Confirmation of a measurement model for green supply chain management practices implementation. *International Journal of Production Economics* **111**(2): 261-273.
[21] Newman, Alexander, et al. 2016 The impact of socially responsible human resource management on employees' organizational citizenship behaviour: the mediating role of organizational identification. *The International Journal of Human Resource Management* **27**(4): 440-455.

[22] Hong, Jiangtao, Yibin Zhang, and Minqiu Ding 2018 Sustainable supply chain management practices, supply chain dynamic capabilities, and enterprise performance. *Journal of Cleaner Production* **172**: 3508-3519.
[23] Wang and Jun Dai 2018 Sustainable supply chain management practices and performance. *Industrial Management & Data Systems* **118**(1): 2-21.
[24] Gebauer, Heiko, Anders Gustafsson, and Lars Witell 2011 Competitive advantage through service differentiation by manufacturing companies. *Journal of Business Research* **64**(12): 1270-1280.
[25] Story, Vicky M, et al. 2017 Capabilities for advanced services: A multi-actor perspective. *Industrial Marketing Management* **60**: 54-68.
[26] Bustinza, et al. 2015 Servitization and competitive advantage: the importance of organizational structure and value chain position. *Research-Technology Management* **58**(5): 53-60.
[27] Raddats, Chris, et al. 2017 Interactively developed capabilities: evidence from dyadic servitization relationships. *International Journal of Operations & Production Management* **37**(3): 382-400.
[28] Boon-itt, Sakun, Chee Yew Wong, and Christina WY Wong 2017 Service supply chain management process capabilities: Measurement development. *International Journal of Production Economics* **193**: 1-11.
[29] Williams 2007 Product service systems in the automobile industry: contribution to system innovation? *Journal of Cleaner Production* **15**(11): 1093-1103.
[30] Resta, Barbara, et al. 2017 Enhancing the Design and Management of the Product-Service System Supply Chain: An Application to the Automotive Sector. *Service Science* **9**(4): 302-314.
[31] Paiola, Marco, et al. 2013 Moving from products to solutions: Strategic approaches for developing capabilities. *European Management Journal* **31**(4): 390-409.
[32] Halme, Minna, et al. 2006 Sustainability evaluation of European household services. *Journal of Cleaner Production* **14**(17): 1529-1540.
[33] Beske, Philip, Anna Land, and Stefan Seuring 2014 Sustainable supply chain management practices and dynamic capabilities in the food industry: A critical analysis of the literature. *International Journal of Production Economics* **152**: 131-143.
[34] Pagell, Mark and Zhaohui Wu 2009 Building a More Complete Theory of Sustainable Supply Chain Management Using Case Studies of 10 exemplars. *Journal of Supply Chain Management* **45**(2): 37-56.
[35] Ceschin, Fabrizio and Carlo Vezzoli 2010 The role of public policy in stimulating radical environmental impact reduction in the automotive sector: the need to focus on product-service system innovation.
[36] Vezzoli, Carlo, Fabrizio Ceschin, and Jan Carel Diehl 2015 Sustainable Product-Service System Design applied to Distributed Renewable Energy fostering the goal of sustainable energy for all. *Journal of Cleaner Production* **97**: 134-136.
[37] Raut, Rakesh D, Balkrishna Narkhede, and Bhaskar B Gardas 2017 To identify the critical success factors of sustainable supply chain management practices in the context of oil and gas industries: ISM approach. *Renewable and Sustainable Energy Reviews* **68**: 33-47.
[38] Kumar, Nachiappan Subramanian, and Ramkumar Arputham 2018 Missing link between sustainability collaborative strategy and supply chain performance: Role of dynamic capability. *International Journal of Production Economics* **203**: 96-109.
[39] Coenen, Jannie, Rob E. C. M. van Der Heijden, and Allard C. R. van Riel 2018 Understanding approaches to complexity and uncertainty in closed-loop supply chain management: Past findings and future directions. *Journal of Cleaner Production* **201**: 1-13.
[40] Williams, Andrew 2007 Product service systems in the automobile industry: contribution to system innovation? *Journal of Cleaner Production* **15**(11): 1093-1103.
[41] Gruchmann, Tim and Stefan Seuring 2018 Explaining logistics social responsibility from a dynamic capabilities perspective. *The International Journal of Logistics Management*: 08-2017-0200.
[42] Sandberg, Erik and Mats Abrahamsson 2011 Logistics capabilities for sustainable competitive advantage. *International Journal of Logistics Research and Applications* **14**(1): 61-75.

[43] Beske, Philip, Anna Land, and Stefan Seuring 2012 Sustainable supply chain management practices and dynamic capabilities in the food industry: A critical analysis of the literature. *International Journal of Production Economics* **152**: 131-143.
[44] Beske, Philip and Stefan Seuring 2014 Putting sustainability into supply chain management. *Supply Chain Management: an international journal* **19**(3): 322-331.
[45] Parmigiani, Anne, Robert D Klassen, and Michael V Russo 2011 Efficiency meets accountability: Performance implications of supply chain configuration, control, and capabilities. *Journal of operations management* **29**(3): 212-223.
[46] Mandal, Santanu, et al. 2016 Achieving supply chain resilience: the contribution of logistics and supply chain capabilities. *International Journal of Disaster Resilience in the Built Environment* **7** (5): 544-562.
[47] Defee, Clifford and Brian S. Fugate 2010 Changing perspective of capabilities in the dynamic supply chain era. *The International Journal of Logistics Management* **21**(2): 180-206.
[48] Eisenhardt and D. Galunic 2000 Coevolving: At last, a way to make synergies work. *Harvard Business Review* **78**(1): 91-101.
[49] Eisenhardt and Jeffrey A Martin 2000 Dynamic capabilities: what are they? *Strategic management journal* **21**(10-11): 1105-1121.
[50] Bustinza, et al. 2017 Product–service innovation and performance: the role of collaborative partnerships and R&D intensity. *R&D Management.*
[51] Ayala, et al. 2017 Knowledge sharing dynamics in service suppliers' involvement for servitization of manufacturing companies. *International Journal of Production Economics* **193**: 538-553.
[52] Eloranta, Ville and Taija Turunen 2015 Seeking competitive advantage with service infusion: a systematic literature review. *Journal of Service Management* **26**(3): 394-425.
[53] Seuring, Stefan and Martin Müller 2008 From a literature review to a conceptual framework for sustainable supply chain management. *Journal of Cleaner Production* **16**(15): 1699-1710.
[54] Manzini and Vezzoli 2003 A strategic design approach to develop sustainable product service systems: examples taken from the 'environmentally friendly innovation' Italian prize. *Journal of Cleaner Production* **11**(8): 851-857.
[55] Kindström, Daniel and Christian Kowalkowski 2009 Development of industrial service offerings: a process framework. *Journal of service Management* **20**(2): 156-172.
[56] Fischer, Thomas, et al. 2010 Exploitation or exploration in service business development? *Journal of Service Management* **21**(5): 591-624.
[57] Zanoni, Simone and Lucio Zavanella 2012 Chilled or frozen? Decision strategies for sustainable food supply chains. *International Journal of Production Economics* **140**(2): 731-736.
[58] Agi, Maher AN and Rohit Nishant 2017 Understanding influential factors on implementing green supply chain management practices: An interpretive structural modelling analysis. *Journal of environmental management* **188**: 351-363.
[59] Böhm, Eva, Andreas Eggert, and Christoph Thiesbrummel 2017 Service transition: A viable option for manufacturing companies with deteriorating financial performance? *Industrial Marketing Management* **60**: 101-111.
[60] Reim, Wiebke, Vinit Parida, and Daniel Örtqvist 2015 Product–Service Systems (PSS) business models and tactics – a systematic literature review. *Journal of Cleaner Production* **97**: 61-75.
[61] Ngai, Eric WT, Dorothy CK Chau, and TLA Chan 2011 Information technology, operational, and management competencies for supply chain agility: Findings from case studies. *The Journal of Strategic Information Systems* **20**(3): 232-249.

# Ergonomic Intervention to Improve The Productivity of Brick Press Tool in Small and Medium Entreprise (SME) Akheng Kobar

**Lamto Widodo[1,2,4], Silvi Ariyanti[3,4], and Andreas Jason[1]**

[1]Industrial Engineering Department, Tarumanagara University, Indonesia
[2]Work System Design and Ergonomic Laboratory, Tarumanagara University, Indonesia
[3]Industrial Engineering Department, Mercu Buana University, Indonesia
[4]lamtow@ft.untar.ac.id ; ariyantisilvi41@gmail.com

**Abstract.** Akheng Kobar is one of the small and medium enterprises (SMEs) that is growing, which produces concrete blocks manually. The brick making process has an ergonomic problem, because the posture of workers in a squatting and bending position for a long time during work. Based on interviews with workers, there were complaints of pain in the back, arms, wrists, calves and thighs. Other complaints include the neck, hands and ankles. In this study, an analysis of initial working conditions, tool design, implementation and results analysis were carried out. Analysis of current conditions using REBA (Rapid Entire Body Assessment), obtained a score of 11, and the JSI (Job Strain Index) analysis score was 39. The results indicate that the process of bricks production must be changed and improved. Furthermore, it was carried out to design a concrete brick press, with the aim of reducing the level of ergonomic risk and increasing productivity. Tool design based on operator complaints, body posture analysis and anthropometric data. Of the several alternatives, obtained the best design alternatives and then made a prototype. Implementation results show a decrease in REBA Score from 11 to 7. Ergonomic Risk Factor decreased from 9 to 5, Job Strain Index scores decreased from 39 to 6.8, and standart time of process decreased by 44,38%.
**Keywords:** ergonomic, intervention, productivity, REBA, JSI, brick

## 1. Introduction

Ergonomics is the study of the nature, abilities, and limitations of humans to design a work system so that people can live and work on that system properly [1][2]. Thus the desired goals of the work can be achieved effectively, safely and comfortably. The development of ergonomic tools is needed to reduce musculoskeletal disorders related to work and take into account factory needs [3]. Ergonomics talks about identifying work risks and how to reduce those risks. Applications can be either simple analysis up to modern technology [4], even to the use of expert systems [5]. Previous research was carried out on the work systems of the pressing industry [6], furniture industry [7], welding work stations [8], and public facilities [9] by utilizing the principle of ergonomics to increase work productivity. In this research, a case study was carried out on 'Akheng Kobar' Small and Medium Enterprises, a brick industry located on Jalan Tanjung Raya 2, RT 004, RW 8, Banjar Serasan Sub-district, East Pontianak District, Pontianak City, Kalimantan. This company produces bricks manually, with the type of bricks

produced are 3 cavity bricks. From the observations of UKM Akheng Kobar, the position of the worker when performing activities is not ergonomic, ie the worker works with the body position squatting and bending. This is because the tools and materials are placed on the floor. The process is done in a long period of time and repeatedly. In this case, it is seen that working conditions are not ergonomic and can be at risk to the physical condition of workers.
This research uses REBA (Rapid Entire Body Assessment) analysis, a method developed in the field of ergonomics and can be used quickly to assess the work position or posture of an operator's neck, back, arms, wrists and feet [10] [11] [12]. Besides this method is also influenced by coupling factors, external loads supported by the body and the activities of workers. Risk categories according to REBA analysis can be seen in Table 1. The results of the REBA assessment show that the work processes are the most risky, and the most dangerous are the pressing and laying processes.

**Table 1.** REBA score and level of MSDs Risk [10]

| **REBA Score** | **Risk Level** | ***Action Level*** | **Action** |
|---|---|---|---|
| 1 | Negligible risk | 0 | No action reqiured |
| 2-3 | Low risk | 1 | Change may be needed |
| 4-7 | Medium risk | 2 | Further investigation, change soon |
| 8-10 | High risk | 3 | Investigate and implement change |
| 11-15 | Very high risk | 4 | Implement change |

Other methods are also used to assess work risks, namely JSI (Job Strain Index) [13] and ERF (Ergonomic Risk Factor) [14] . JSI is a method used to evaluate work against the risk of musculoskeletal disorders in the Distal Upper Extremity (DUE) section including the elbows, forearms, wrists, and hands. While the ERF assesses any ergonomic factors that can cause work risks which consist of awkward postures, forcing, repetition, vibration, static loading, stress, and extreme temperatures. The results of the JSI and ERF methods also show that pressing is the most dangerous process.

From the results of the preliminary analysis, it was found that the largest score indicating un ergonomic conditions was found in the pressing and laying process. There were worker's complaints of pain in the back, arms, wrists, calves and thighs. Other complaints include the neck, hands and ankles. To solve this problem, an ergonomic intervention is carried out, starting with an ergonomic risk analysis of the initial conditions, designing work tools, implementing the tools and analyzing the results. By using this tool, it is expected to reduce the pain complaints felt by workers as well as to reduce ergonomic work risks.

## 2. Methods

This research was conducted at Akheng Kobar UKM which produces a brick located in Pontianak with a number of workers is 2 person. These workers carry out activities alternately in the brick press (press) and laying section. The study began with direct observations and interviews with workers. This is to find out physical complaints, work completion time, assessment of body posture scores using REBA, JSI and ERP methods. After that, designing tools, making prototypes and implementing them are carried out. The results of the implementation are the final evaluations which are the conclusions of the research and as a basis for suggestions for improvement in the future.

## 3. Result and Discussion

The following are the results of direct observations in the field of brick building SME workers. Ergonomics risk factors are most common in the brick pressing process, can be seen in Table 2.

IOP Conf. Series: Materials Science and Engineering **847** (2020) 012057 doi:10.1088/1757-899X/847/1/012057

**Table 2.** Analysis of 'Ergonomic Risk Factor'

| No | Activity | Awkward Posture | Force | Reps | Vibration | Static loading | Contact Stress | Extreme Temperature |
|---|---|---|---|---|---|---|---|---|
| 1 | Brick Pressing Process | ● | ● | ● | | ● | ● | ● |
| 2 | Brick Laying Process | | ● | ● | | | | ● |

From the results of the calculation of the brick pressing process REBA obtained a score of 11 where the meaning of the value required immediate action because of the very high risk. Meanwhile, the process of brick laying obtained a score of 7. The following are the results of REBA calculations in the three processes that can be seen in Table 3.

**Table 3.** Calculation results from REBA analysis

| Activity | Posture Position | REBA Score |
|---|---|---|
| Brick Pressing Process |  | 11 |
| Brick Laying Process |  | 7 |

| Risk Factor | Rating Criterion | Observation | | | Multiplier | Left | Right |
|---|---|---|---|---|---|---|---|
| Intensity of Exertion (Borg Scale - BS) | Light | Barely noticeable or relaxed effort (BS: 0-2) | | | 1 | 13 | 13 |
| | Somewhat Hard | Noticeable or definite effort (BS: 3) | | | 3 | | |
| | Hard | Obvious effort; Unchanged facial expression (BS: 4-5) | | | 6 | | |
| | Very Hard | Substantial effort; Changes expression (BS: 6-7) | | | 9 | | |
| | Near Maximal | Uses shoulder or trunk for force (BS: 8-10) | | | 13 | | |
| Duration of Exertion (% of Cycle) | < 10% | **Calculated Duration of Exertion** (from inputs below) | | | 0.5 | 3 | 3 |
| | 10-29% | **User Inputs** | **Left** | **Right** | 1.0 | | |
| | 30-49% | Total observation time (sec. | 1800 | 1800 | 1.5 | | |
| | 50-79% | Single exertion time (sec.) | 40 | 40 | 2.0 | | |
| | ≥ 80% | Number of exertions during observation time | 37.5 | 37.5 | 3.0 | | |
| | **lculated Duration of Exertion (%)** | | 83.3 % | 83.3 % | | | |
| Efforts Per Minute | < 4 | **Calculated Efforts Per Minute** (from inputs above) | | | 0.5 | 0.5 | 0.5 |
| | 4 - 8 | | **Left** | **Right** | 1.0 | | |
| | 9 - 14 | | 1.25 | 1.25 | 1.5 | | |
| | 15 - 19 | | | | 2.0 | | |
| | ≥ 20 | | | | 3.0 | | |
| Hand/Wrist Posture | Very Good | Perfectly Neutral | | | 1.0 | 2 | 2 |
| | Good | Near Neutral | | | 1.0 | | |
| | Fair | Non-Neutral | | | 1.5 | | |
| | Bad | Marked Deviation | | | 2.0 | | |
| | Very Bad | Near Extreme | | | 3.0 | | |
| Speed of Work | Very Slow | Extremely relaxed pace | | | 1.0 | 1 | 1 |
| | Slow | Taking one's own time | | | 1.0 | | |
| | Fair | Normal speed of motion | | | 1.0 | | |
| | Fast | Rushed, but able to keep up | | | 1.5 | | |
| | Very Fast | Rushed and barely/unable to keep up | | | 2.0 | | |
| Duration of Task Per Day (hours) | <1 | | | | 0.25 | 1 | 1 |
| | 1< 2 | | | | 0.50 | | |
| | 2 < 4 | | | | 0.75 | | |
| | 4 ≤ 8 | | | | 1.00 | | |
| | > 8 | | | | 1.50 | | |
| Results Key | | SI ≤ 3 | Job is probably safe | | | 39 | 39 |
| | | 3 < SI < 7 | Job may place individual at increased risk for distal upper extremity disorders | | | | |
| | | 7 ≤ SI | Job is probably hazardous | | | | |

**Figure 1.** JSI evaluation results on the Pressing Process

Job Strain Index (JSI) is done in the brick pressing process, and the brick laying process on the rack. The results obtained in the pressing process are the most dangerous. Calculation of JSI pressing process can be seen in Figure 1.

**Brick Press Tool Design**

Based on the analysis of REBA, Ergonomic Risk Factor and Job Strain Index that has been done, the highest score obtained in the pressing process. Therefore, product design is focused on reducing the level of risk of physical hazards in the pressing process. The first stage in design is to determine the needs matrix which is obtained from the identification of customer needs to find out what the customer's desires for the product. The need matrix is seen in Table 4.

**Tabel 4.** Need Matrix

| No | Need | level of importance |
|---|---|---|
| 1 | Convenient Ways of Work | 5 |
| 2 | Speed up the Production Process Time | 4 |
| 3 | No Need To Squat For Pressing | 5 |
| 4 | Easy-to-Use | 5 |
| 5 | Reducing Manpower Expenditures when Pressing | 4 |
| 6 | Durability | 3 |

The next step is to establish technical specifications and target value specifications. The aim is to reveal precise and measurable details about what the product must be able to fulfill. Finally obtained 3 concept design tools as in Figure 2, Figure 3 and Figure 4.

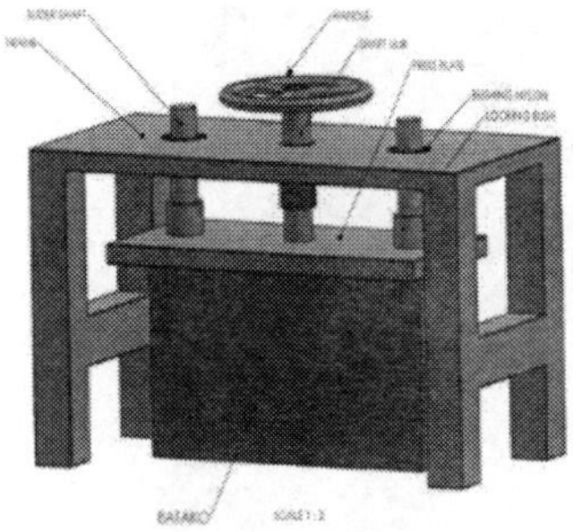

**Figure 2.** 1st Concept

**Figure 3.** 2nd Concept

**Figure 4.** 3rd Concept

From the 3 concepts, conceptual assessment is carried out to choose 1 final concept that can meet expectations and solutions to overcome existing problems. Based on the results of the concept assessment found that concept 1 is the chosen concept. Conceptual size is based on Indonesian Antopometry data [7]. Following are the dimensions of concept 1 which can be seen in Table 5.

**Table 5.** Final dimension of tool

| No | Description | Dimension (cm) |
|---|---|---|
| 1 | Height | 25 cm |
| 2 | Width | 22 cm |
| 3 | Length | 49 cm |

Analysis of REBA by using 3ds Max software aims to simulate a tool that has already passed the design stage. Based on these simulations, a REBA score of 5 is obtained, which indicates a reduction in the value of risk at work. The 3ds Max simulation can be seen in Table 6.

**Table 6.** REBA Analisis by using *Software* 3ds Max

| No | 3ds Max Simulation | Activity | REBA Score |
|---|---|---|---|
| 1. | 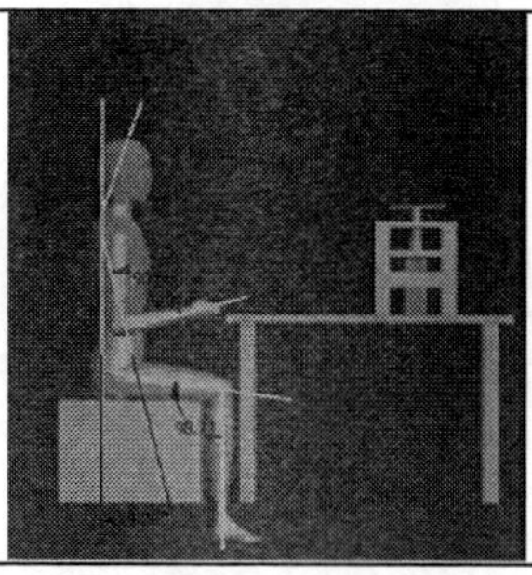 | Pressing Process | 5 |

**Analysis of Implementation Results**

After implementing the work station, Ergonomic Risk Factor analysis is done. The results can be seen in Table 7, where there is a decrease from 9 to 5. The REBA score decreased from 11 to 7. The analysis of the REBA score after implementation can be seen in Table 87. The Job Strain Index (JSI) score decreased from 39 to 6.8, which indicates that the work process is safe. Analysis of JSI scores after implementation can be seen in Figure 5.

**Table 7.** Ergonomic Risk Factor Analysis After Implementation

| No | Activity | Awkward Posture | Force | Reps | Vibration | Static loading | Contact Stress | Extreme Temperature |
|---|---|---|---|---|---|---|---|---|
| 1 | Brick Pressing Process | | | ● | | | | ● |
| 2 | Brick Laying Process | | ● | ● | | | | ● |

**Table 8.** REBA Score after implementation

| Activity | Body Position | REBA Score |
|---|---|---|
| Pressing Process |  | 7 |

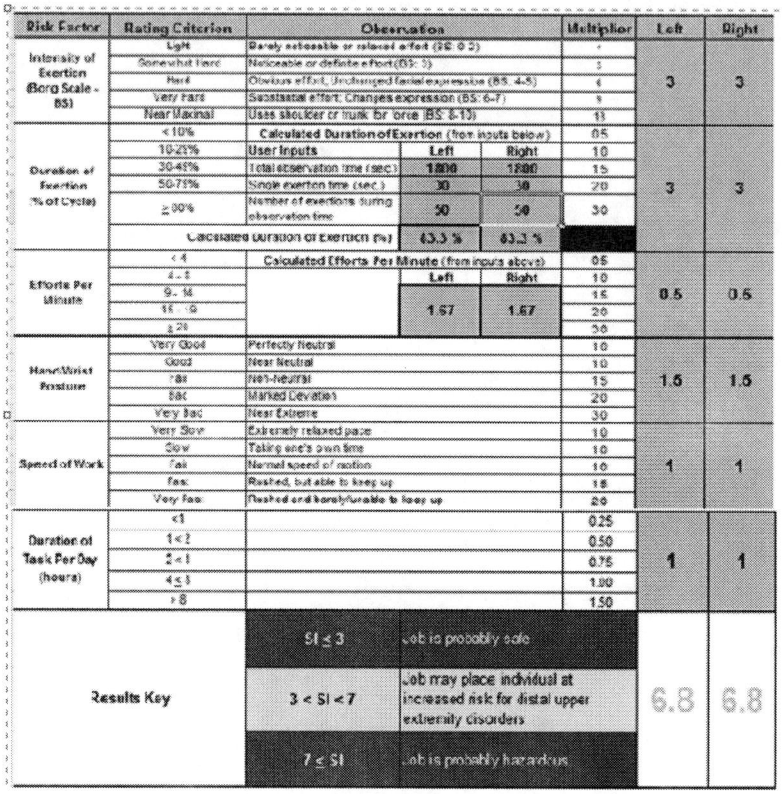

| Risk Factor | Rating Criterion | Observation | | | Multiplier | Left | Right |
|---|---|---|---|---|---|---|---|
| Intensity of Exertion (Borg Scale - BS) | Light | Barely noticeable or relaxed effort (BS: 0-2) | | | 1 | 3 | 3 |
| | Somewhat Hard | Noticeable or definite effort (BS: 3) | | | 3 | | |
| | Hard | Obvious effort; Unchanged facial expression (BS: 4-5) | | | 6 | | |
| | Very Hard | Substantial effort; Changes expression (BS: 6-7) | | | 9 | | |
| | Near Maximal | Uses shoulder or trunk for force (BS: 8-10) | | | 13 | | |
| Duration of Exertion (% of Cycle) | < 10% | Calculated Duration of Exertion (from inputs below) | | | 0.5 | 3 | 3 |
| | 10-29% | User Inputs | Left | Right | 1.0 | | |
| | 30-49% | Total observation time (sec) | 1800 | 1800 | 1.5 | | |
| | 50-79% | Single exertion time (sec) | 30 | 30 | 2.0 | | |
| | ≥ 80% | Number of exertions during observation time | 50 | 50 | 3.0 | | |
| | Calculated Duration of Exertion (%) | | 83.3 % | 83.3 % | | | |
| Efforts Per Minute | < 4 | Calculated Efforts Per Minute (from inputs above) | | | 0.5 | 0.5 | 0.5 |
| | 4 - 8 | | Left | Right | 1.0 | | |
| | 9 - 14 | | 1.67 | 1.67 | 1.5 | | |
| | 15 - 19 | | | | 2.0 | | |
| | ≥ 20 | | | | 3.0 | | |
| Hand/Wrist Posture | Very Good | Perfectly Neutral | | | 1.0 | 1.5 | 1.5 |
| | Good | Near Neutral | | | 1.0 | | |
| | Fair | Non-Neutral | | | 1.5 | | |
| | Bad | Marked Deviation | | | 2.0 | | |
| | Very Bad | Near Extreme | | | 3.0 | | |
| Speed of Work | Very Slow | Extremely relaxed pace | | | 1.0 | 1 | 1 |
| | Slow | Taking one's own time | | | 1.0 | | |
| | Fair | Normal speed of motion | | | 1.0 | | |
| | Fast | Rushed, but able to keep up | | | 1.5 | | |
| | Very Fast | Rushed and barely/unable to keep up | | | 2.0 | | |
| Duration of Task Per Day (hours) | < 1 | | | | 0.25 | 1 | 1 |
| | 1 < 2 | | | | 0.50 | | |
| | 2 < 4 | | | | 0.75 | | |
| | 4 ≤ 8 | | | | 1.00 | | |
| | > 8 | | | | 1.50 | | |
| Results Key | SI ≤ 3 | Job is probably safe | | | | 6.8 | 6.8 |
| | 3 < SI < 7 | Job may place individual at increased risk for distal upper extremity disorders | | | | | |
| | 7 ≤ SI | Job is probably hazardous | | | | | |

**Figure 5.** *Job Strain Index* Score after implementation

Evaluation of the results of implementation is also carried out on the production process time. The results show a decrease in working time by 43.38% which can be seen in Table 9. This means that by implementing the use of design tools can increase productivity.

**Table 9.** Comparison of Production Process Time

| No | Time Criteria | Before Implementation (sec) | After Implementation (sec) | Percentage of Time Reduction (%) |
|---|---|---|---|---|
| 1 | Cycle Time | 47.94/ 1 product | 56.80/2 products | 40.75 |
| 2 | Normal Time | 49.85/ 1 product | 59.072/2 Products | 40.75 |
| 3 | Standard Time | 64.71/ 1 product | 73.272/ 2 Products | 43.38 |

## 4. Conclusion

In the brick production process, work stations that have the highest ergonomic risk are the pressing process, which is indicated by high REBA, JSI and ERF scores. The design of the tool is carried out to mitigate this risk based on operator complaints, posture analysis and anthropometric data. Implementation results show a decrease in REBA Score from 11 to 7. Ergonomic Risk Factor decreased from 9 to 5, and Job Strain Index scores decreased from 39 to 6.8. Before carrying out the implementation, it is known that the standard time for the pressing process and brick laying is 64.71 seconds for 1 product. After implementing the product, it is known that the standard time for the pressing process and laying of concrete blocks is 73.272 seconds for 2 products. So there was a decrease in working time by 44.38%.

IOP Conf. Series: Materials Science and Engineering **847** (2020) 012057 doi:10.1088/1757-899X/847/1/012057

In addition, this study focuses on operator complaints and operator posture, not paying attention to other aspects such as the environment and the detail of the energy required. For further research, it is recommended that these factors be considered, so that the results are more comprehensive.

## 5. References

[1] IEA 2016 *What is Ergonomic*. International Ergonomic Association, Thônex, Canton of Geneva, Switzerland.

[2] R. S. Bridger 2009 *Introduction to Ergonomic*, 3rd ed. Boca Raton, [Florida] London, [England] New York: CRC Press.

[3] M. Aptel, L. Claudon, and J. Marsot 2002 Integration of ergonomics into hand tool design: Principle and presentation of an example. *Int. J. Occup. Saf. Ergon.*, **8**(1), pp. 107–115.

[4] M. Gašová, M. Gašo, and A. Štefánik 2017 Advanced Industrial Tools of Ergonomics Based on Industry 4.0 Concept. in *Procedia Engineering*.

[5] S. Pavlovic-Veselinovic, A. Hedge, and M. Veselinovic 2016 An ergonomic expert system for risk assessment of work-related musculo-skeletal disorders. *Int. J. Ind. Ergon.*

[6] L. Widodo, S. Ariyanti, and F. A. Kurniawan 2019 Perancangan Stasiun Kerja Ergonomis Pada Stasiun Kerja Pressing CV. Karyamitra Lestari. *J. Ilm. Tek. Ind.*.

[7] L. Widodo, I. W. Sukania, and R. Sugiono 2017 Rancangan Furniture Dan Tata Ruang Dengan Dimensi Terbatas Secara Ergonomis. *J. Ilm. Tek. Ind.*

[8] S. Ariyanti, L. Widodo, M. Zulkarnain, and K. Timotius 2019 Design Work Station of Pipe Welding with Ergonomic Approach. *SINERGI*, 2019.

[9] L. Widodo, A. Adianto, D. Debby, and S. Rohananasution 2019 Ergonomic portable toilet for women in public facilities. in *IOP Conference Series: Materials Science and Engineering*.

[10] S. Hignett and L. McAtamney 2000 Rapid Entire Body Assessment (REBA). *Applied Ergonomics*.

[11] N. A. Ansari and D. M. J. Sheikh 2014 Evaluation of work Posture by RULA and REBA: A Case Study," *IOSR J. Mech. Civ. Eng.*, **11**(4), pp. 18–23.

[12] M. Middlesworth 2014 A Step-by-Step Guide Rapid Entire Body Assessment (REBA). *Ergon. Plus Inc*.

[13] M. ève Chiasson, D. Imbeau, K. Aubry, and A. Delisle 2012 Comparing the results of eight methods used to evaluate risk factors associated with musculoskeletal disorders. *Int. J. Ind. Ergon.*

[14] N. Jaffar, A. H. Abdul-Tharim, I. F. Mohd-Kamar, and N. S. Lop 2011 A literature review of ergonomics risk factors in construction industry. *Procedia Engineering*.

IOP Conf. Series: Materials Science and Engineering **847** (2020) 012058 doi:10.1088/1757-899X/847/1/012058

# Inventory and Order System Development at PT.X

**Steffi Ratanadewi[1,2] and Marsellinus Bachtiar[1,3]**

[1]Industrial Engineering Program Atmajaya Catholic University, Campus 3 BSD, Serpong, Indonesia
[3]STEFFI.2015043010@student.atmajaya.ac.id, [4]marsellinus.bachtiar@atmajaya.ac.id

**Abstract.** PT X is Indonesian cutting tools importer, which sells various kinds of cutting tools, such as Carbide End Mill and Reamer. This company is facing two problems. The first is inefficient order process. The second problem is inefficient inventory update, which takes about eight minutes. Logistic Manager double checks the product quantity by counting physical product, to make sure that the quantity is updated. This process affects both order and procurement process. The objective of this study is to propose a new system, which will be able to perform simpler business process and reduce completion time. Modified Waterfall Approach of System Development Life Cycle (SDLC) was chosen, as it is suitable for well – understood processes, where doing overlapping phases is permitted. This study develops UML Diagrams, also medium and high-fidelity prototypes. User Interface Testing, (consists of learnability, error, memorability, and satisfaction measurement) was done to examine system performance. Result shows that the proposed system performs better than current system. The completion time of create order is reduced 27.51% to 3.11 minutes, availability check is reduced 75.99% to 2.3 minutes, incoming inventory update is reduced 99.67% to 7.69 seconds, and outgoing inventory update is reduced 94.95% to 12.46 seconds.
**Keywords**: inventory, order management, system, SDLC

## 1. Introduction

We are living in the middle of Industry 4.0. It is a representation of the fourth revolution in manufacturing that makes it possible to gather and analyse data across machines, also enabling faster, more flexible, and more efficient processes to produce high quality goods at lower costs.

The technological transformations can lead a country into its economic development. Administrative and manufacturing tasks such as manual data inputting, quality checking, and product packaging would be declined as they could be helped with a network of machines that connects and shares information to each other.

PT X, which was established in 1996, is a cutting tools importer. The business is considered as trading, which buys different kind of products from another country and sells them to the consumer afterwards. PT X sells various kinds of cutting tools, such as Carbide End Mill, Carbide Ball Nose, Drill, Reamer, Tap, and HSS End Mill. Cutting tool is used to do material removal on a work piece.This company has two suppliers, one in Taiwan and the other in Czech Republic. It sells its products to ten manufacturing companies, which are located in Jabodetabek area. Each customer usually buys 30 up to 40 pieces of cutting tools per month. The selling price ranges IDR 80,000 up to IDR 1,700,000 per piece. Thus, the business model is B2B. Note: The time delay from importing is excluded under this paper.

PT X has two main problems: first is about customer order and the second one is about inventory update. Customer order process starts when a customer sends its Purchase Order (PO) to company e-mail. After receiving the PO, Sales Manager checks the product types and prices on the catalogue. This process is performed to make sure that the customer didn't write wrong product information on PO. PT X sells more than 100 product items, whose price cannot be memorized one by one. Moreover, the prices change often. As an importer, PT X needs to adjust the price according to the exchange rate. Sales Manager sends the updated catalogue to customers by e-mail, but sometimes the customer still wrote the older version of price on the PO. It makes Sales Manager had to check each PO detail of every Customer. This activity, which takes about four minutes, is considered to be inefficient because it is actually a non-value-added task. The completion time extends if customer writes the wrong information, because Sales Manager needs to inform the customer, then customer needs to fix the PO before the order is processed.

The second problem is inventory update. Sales Manager needs to wait for another eight minutes to get the product availability information. The process starts when Sales Manager asks Logistic Manager to check product quantity. Then Logistic Manager checks the quantity on the inventory report, which is still manually written. Logistic Manager usually continues checking the physical inventory to make sure that the products are available. This happens because sometimes the quantity on report does not match with physical inventory. This process is considered to be inefficient either, because it is a non-value-added activity.

It does not only affect the order process, but also procurement. This process is done by Procurement Manager. Procurement Manager is responsible to order product from supplier when DOI (Days of Inventory) number has reached safety stock, so company will be able to provide sufficient product items whenever needed by customer. Procurement Manager updates the DOI daily, thus Procurement Manager needs to wait for Logistic Manager to give product quantity information. Procurement Manager then calculates the DOI of each product item. This usually takes more than two hours, because there are so many items, and the calculation is still done manually. Another problem arises when Procurement Manager calculates the number incorrectly, hence it affects the order process.

These problems are addressed by using one of Industry 4.0's building blocks: cloud computing. Using an integrated system, PT X will get benefit of shorter, easier, and more efficient process. A business process analysis will be performed to identify the inefficient activities, so the process can be simplified and human error can be reduced. Therefore, the completion time will be reduced. Company may use the spare time to do value-added or more important activities.

The main objective of this research is to propose an integrated system model that connects customer order and inventory update process. Special feature such as order tracking will be built to help Customer see whether the order has been dispatched or completed. This research will be conducted using System Development Life Cycle (SDLC) methodology. It provides detailed, step by step plans to follow, which are needed for making an organized and goal-oriented project. The Modified Waterfall Approach of SDLC is chosen as it is suitable for creating a simple, predictable system, but flexible enough to enable revision of the previous phases.

The proposed system will be created using UML Diagrams. The diagrams will help researcher to define the business process, scenarios, and information flow. A basic user interface prototype will be created using Axure RP to show the display and basic information flow. Then an updated version of prototype will be built to test the functional and usability factors. The system must work well and easy to use, so the process can give shorter completion time

## 2. Methods

### *2.1. Research Object*

The research object is PT X, which is located in West Jakarta. This company is chosen using non-probability sampling (convenience sampling), because of its willingness to provide information and improve business process by developing information system.

*2.2. Preliminary Study*

Preliminary study was conducted in three days (22 October 2018, 23 October 2018, and 25 October 20 18). Researcher came to the company for the first time on 22 October 2018 to ask for permission, gather company profile information, main problems, overall business process, and conduct feasibility study. On 23 October 2018, researcher collected information about company organizational structure with responsibilities and authorities of each employee. On 25 October 2018, researcher collected information about customers, suppliers, and products.

*2.3. Scope of the Research*

This research defines research scopes and limitation) as follows:

1. This study focuses on improving order and inventory process.
2. This study doesn't cover financial aspects of PT X.
3. This study only gathers the estimated completion time of the current system.
4. This quantity used for DOI calculation is taken from April 29th, 2019.
5. This study ends in the first phase of usability testing. It doesn't cover system installation.
6. The diagrams and basic prototype were built from January to April 2019, while the updated prototype and testing were conducted from May to June 2019.

*2.4. Research Design*

This research is considered as an exploratory case study. An exploratory study is one kind of studies, which needs some preliminary works to be done in advance. Thus, the solution ideas or models could be developed based on the preliminary work. This research could be considered as a case study as its objective is to solve company problems.

This research collects data using both of unstructured and structured interviews. Unstructured interview was conducted on the first day of preliminary study, while structured interview was conducted on the second and third day of preliminary study and during data collection phase.

*2.5. Modified Waterfall Method*

The Modified Waterfall Approach provides more flexible steps as it contains overlapping phases and enables going back to the previous phases [1]. The overlapping phases also increases the process efficiency, because some phases could be done at a time. This study is more well – understood as it has clear processes, so most of the requirements can be predicted on the beginning phases. Moreover, this study has time and budget limitation. The limitation of this approach will be overcome by creating programmer schedule. The Modified Waterfall approach will be used for this study. Its four phases are explained as follows:

*2.5.1. Project Planning.* Project planning is identifying the scope of a new system, ensuring the project's feasibility, developing schedule, planning resources, and budgeting for the whole project. The activities and tasks shown in a schedule, see Table 1.

**Table 1.** Project Schedule

| Activities | Day |
|---|---|
| Identify Current Business Process | D1-D8 |
| Design New System | D1-D24 |
| Create Prototype | D24-D57 |
| Review Planning, Analysis and Design Phases | D24-D57 |

*2.5.2. Analysis.* Analysis is made by understanding current business processes of PT X, which consists of order and procurement process. The business processes are gathered by doing face to face interview with Director and observation. This phase was done on 9 January and 10 January 2019. Researcher created basic flowchart, which consists of start, end, arrow, process, and decision symbols, to visualize the business processes. Then researcher analysed the shortcomings of current processes, inefficient and non-value-added activities which can be improved. Later, system needs are defined based on the shortcomings, and detail requirements are defined based on discussion with Director and all managers, on 14 January 2019.

*2.5.3. Design.* Design is the phase of developing the solution system algorithms. The first thing to do is defining system features. UML was created through activity diagram [2] , use case diagram , use case description , domain model class diagram , multilayer diagram and design class diagram.

*2.5.4. Implementation.* This study creates the basic and updated prototype and conduct just one phase of functional and usability testing. Researcher created basic prototype from 15 April until 29 April 2019 using Axure RP. The updated prototype is built using PHP programming language, Laravel framework, and MySQL database constructor. Prototype was built by programmer from 10 May 2019 to 4 June 2019. Testing uses the updated prototype as object.

Usability testing [3] [4] [5] consists of learnability, error, memorability, and satisfaction - is conducted once by Director and managers, to ensure that the system is easy to use. First, researcher prepares testing scenarios, then users are given short demo of how the system works and what they have to do. Then learnability test is done by following the scenarios. Researcher needs to write down the completion time. At the same time, researcher needs to see if users make any mistake. Memorability testing is conducted seven days after doing learnability testing, by doing the tasks on the same scenarios. Satisfaction is conducted using System Usability Scale (SUS) questionnaire after completing memorability testing. SUS is a personal administered questionnaire, which is filled by the user oneself.

## 3. Result and Discussion

### *3.1. Current Business Process*

PT X's business process can be divided into two big parts: order activity and procurement activity. Both are still done manually, without using an integrated system. Figure 1 shows two general business process flows, explaining both activities. Order activity starts when customer sends PO [6]. Customer will need to confirm the order. Later on, company will ship the order, and customer will have to pay in 30 days after accepting the products. Procurement activity starts when Procurement Manager needs to order from supplier. Procurement Manager will have to send PO and confirm the order afterwards. The difference between order and procurement activity lies in the payment. Use Case diagram (figure 2) shows the relation of activities and actor within the inventory cycle.

### *3.2. System Requirement*

This step defines system requirements and the users [1] [2][7] . Table 2 summarizes three main activities which needs improvement, each with stakeholder, actor, time, current performance, and consequence .

IOP Conf. Series: Materials Science and Engineering **847** (2020) 012058 doi:10.1088/1757-899X/847/1/012058

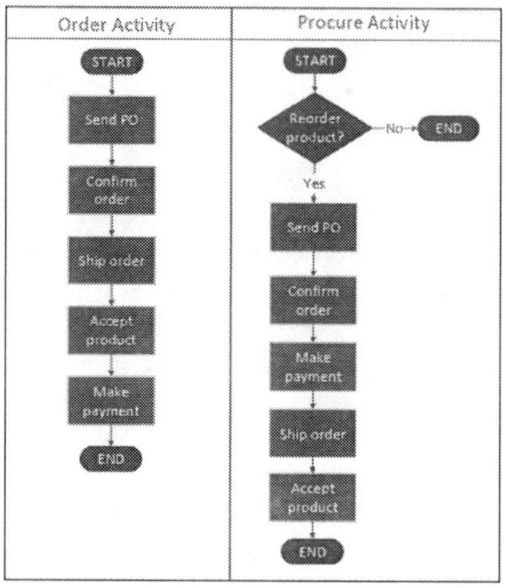


**Figure 1**. Current Business Process

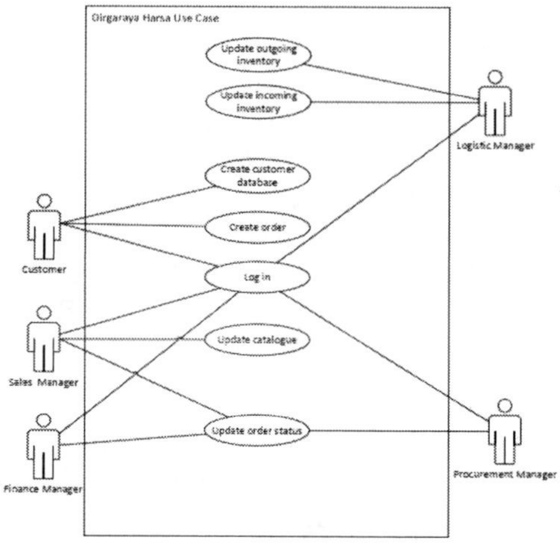


**Figure 2**. Use Case Diagram

**Table 2.** Current System Requirement.

| Activity | Create PO | Inventory Update | DOI Calculation |
|---|---|---|---|
| **Stakeholder** | Sales Manager<br>Customer | Sales Manager<br>Procurement Manager | Procurement Manager |
| **Actor** | Customer | Logistic Manager | Procurement Manager<br>Logistic Manager |
| **Time** | Anytime customer needs to order | Anytime product enters and exits the warehouse | Every day |
| **Current Performance** | Customer is able to create PO on their own, based on given catalogue. But the catalogue is often updated, and it can't be updated real time. There are also various kinds of products with similar name. | Logistic Manager updates inventory regularly, but the calculation was done manually using calculator. | Company has a method already, but Procurement Manager needs to wait for Logistic Manager, then calculates the DOI manually. |
| **Consequence** | PO details often mismatch with updated catalogue. Sales Manager needs to phone the customer to make sure, and customer needs to revise the PO. | Logistic Manager isn't sure if the calculation is correct, so it takes time to check physical products whenever the information is needed. | Procurement Manager needs to take hours for doing this activity, and sometimes DOI was wrongly calculated. |

### *3.3. Proposed System*

The proposed system will provide four features: online catalogue, online order, inventory update, and order tracking. The features will perform real time business process updates [6] .

- **Online catalogue.** Customers will be able to open online catalogue on the website to see product list, descriptions, and price. The catalogue will be updated by Sales Manager. Shortly after the update has been saved. Customer will be able to access it.
- **Online order**. Customer will be able to create order on the website, by choosing the product on catalogue, entering the quantity and also payment method. The order will be generated into PO, delivery order, and invoice.
- **Inventory monitoring**. Logistic Manager can get automatic inventory updates from the system, by confirming every incoming and outgoing order notification. The notification is created by the system

IOP Conf. Series: Materials Science and Engineering **847** (2020) 012058 doi:10.1088/1757-899X/847/1/012058

after Customer or Procurement Manager placed the order. This feature will also have Procurement Manager get automatic DOI updates. Procurement Manager will be informed if the DOI has reached safety stock, so he will be able to create decisions faster.

- **Order tracking**. Customer will can see whether the order has been placed, dispatched, cancelled, or finished.

*3.4. Prototype Design*

The user interface is be divided into six parts, based on different roles:

- Customer. Customer is able to sign up, login, create order, and view profile on the website.
- Sales Manager. Sales Manager is able to login, access order documents, update order status, update catalogue, view inventory, and view profile.
- Finance Manager. Finance Manager is able to login, access order documents, update order status, view inventory, view catalogue, and view profile.
- Procurement Manager. Procurement Manager is able to login, update order from supplier, view inventory, view catalogue, and view profile.
- Logistic Manager. Logistic Manager is able to login, update outgoing and incoming inventory, view catalogue, view inventory, and view profile.
- Director. Director is able to login, view customer order, view order from supplier, view inventory, view catalogue, and view profile.

Figure 4 – 7 show the user interface of updated prototype.

**Figure 3**. Catalogue

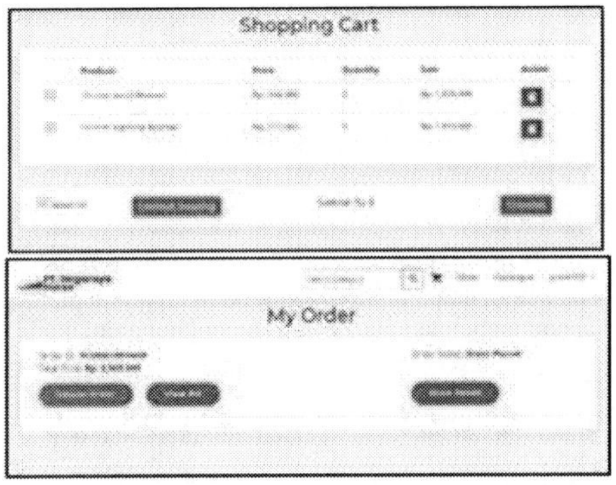


**Figure 4**. Shopping Cart and History

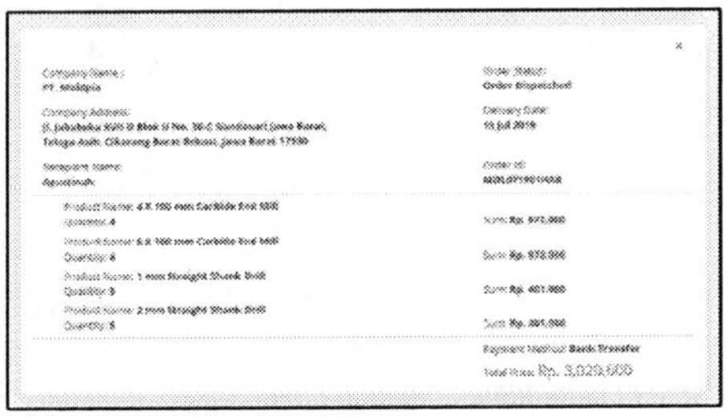

**Figure 5**. Customer Order

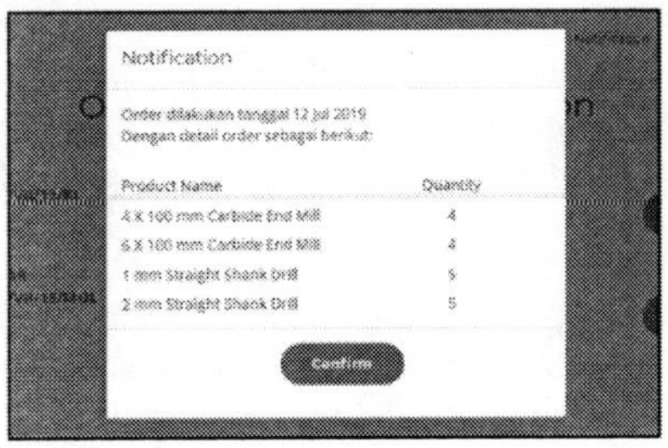


**Figure 6**. Notification

*3.5. Comparison of Current and Proposed System*

Three main activities including create order, availability check, and inventory update, are used to compare current and proposed system.

A. Create order

- **Task**. Create order in current system consists of three tasks. On the other hand, it consists of six tasks in the proposed system. All six tasks of the proposed system can be summarized into one task, which is equal to the first task of current system. Customer doesn't need to send the PO while performing the proposed system, because the PO is automatically generated and accessed by Sales Manager. Sales Manager doesn't need to check product name and price while performing the proposed system either, because the PO is automatically generated from the updated catalogue. The proposed system provides one task only, and therefore it is simpler than the current system.
- **Time**. The first two tasks of current system are not able to be measured or estimated, because researcher does not contact the Customers. On the other hand, Sales Manager usually takes four minutes to check product name and price (assumed that Customer orders five product items). The overall completion time of current system is more than four minutes, because definitely Customer takes seconds to create and send the PO. The completion time of proposed system is taken from the average of memorability testing results. The completion time is 162.63 seconds (3.11 minutes), and therefore it is shorter than current system

B. Availability Check

- **Task**. The availability check in current system consists of three tasks: Sales Manager gives PO to Logistic Manager, Logistic Manager checks product availability, and Logistic Manager informs Sales Manager about the availability. On the other hand, the proposed system consist of the second task.
- **Time**. The overall estimated time of current system is 8 minutes (assumed that Customer orders five product items). On the other hand, estimated time of the proposed system is 138.07 seconds (2.3 minutes), which is the average of memorability testing results. The proposed system has shorter completion time, because it has only one actor and there is no physical inventory check. Logistic Manager usually performs physical inventory check, so it takes longer

C. Inventory Update.

- **Task**. This activity consists of the same actor and task in both current and proposed system. The difference is only Logistic Manager updates the inventory on book and counts the quantity manually. On the other hand, Logistic Manager updates the inventory by confirming the outgoing and incoming notifications, in the proposed system. The proposed system has a formula which adds and subtracts product quantity after a notification has been confirmed. Therefore, the proposed system provides easier task.
- **Time**. It takes about five minutes to update outgoing product (assumed that Customer orders five product items), according to the Logistic Manager. The incoming product update takes about two minutes (assumed that only one item received). On the other hand, the estimated completion time of updating outgoing product is 12.46 seconds, while the incoming product update takes 7.69 seconds. Therefore, both activities performed using the proposed system take shorter completion time than current system.

### *3.6. Limitation of the Prototype*

Both functional and user interface testing show good results, however, this proposed system still has some limitations. First is about the notification display. Notifications and orders are set in list view, which shows the oldest to the newest list. It will make the users take longer and put more effort to find the newest order or notification. However, this has not been fixed because of time limitation.

Second is about user interface testing, which has not been performed by customers, but Sales Manager instead. Having the customers test the system will give detailed perspective of the proposed system, as they are the main stakeholders of this system. Customers have different background, culture, and skill, which will be able to share rich insight before going further with the system. Testing can be done by gathering the customers or by conducting separate tests in each place.

IOP Conf. Series: Materials Science and Engineering **847** (2020) 012058 doi:10.1088/1757-899X/847/1/012058

The third one is about web hosting. The proposed system is built using local hosting, whose program and database are stored internally on desktop. Not only having to download the installer and program, users also have to perform several steps every time. These steps take time to be performed, compared with web hosting.

## 4. Conclusion

The problems regarding the order and inventory process are inefficient processes, such as PO detail check, inventory update, and DOI calculation. These problems are addressed by creating an integrated information system. UML Diagrams are built to identify the new business process [8] , user interface design, and information flow. The basic prototype is created using Axure RP before developing the updated prototype, which represents the final user interface. Functional and user interface testing show that all features of the system work properly and system is easy to use.

## 5. References

[1] Munassa NM and Govardhan A 2010 A Comparison Between Five Models Of Software Engineering. *International Journal of Computer Science* **7**(5) p 94-101.

[2] Satzinger JW, Jackson RB and Burd SD 2005 *Object - Oriented Analysis and Design with the Unified Process* (Boston: Course Technology).

[3] Hornbaek K 2006 Current Practice in Measuring Usability *International Journal of Human-Computer Studies* **64**(2), pp 79-102.

[4] Maramba AC 2019 Methods of Usability Testing in the Development of eHealth Applications: A Scoping Review *International Journal of Medical Informatics* **126** p 95-104.

[5] Marien S, Legranda D, Ramdoyald R, Nsengad J, Ospinad G, Ramond V and Spinewinea A 2019 A User-Centered design and usability testing of a web-based medication reconciliation application integrated in an eHhealth network *International Journal of Medical Informatics* **126** pp. 138-146.

[6] Croxton K 2003 The Order Fulfillment Process. *The International Journal of Logistics Management*, **14**(1), 19-32.

[7] Dennis A , Wixom BH and Roth RM 2012 *Systems Analysis & Design* (New Jersey: John Wiley & Sons, Inc.)

[8] Birkmeier DQ, Klöckner S and Overhage S 2010 An Empirical Comparison of the Usability of BPMN and UML Activity Diagrams for Business Users. *European Conference on Information Systems* (Pretoria: Scholar One), pp. 1-12

# Mitigation of Supply Chain Risk using HOR Model at PT. Sumber Karya Indah

**Ahmad[1], Wilson Kosasih[1], H J Kristina[1], Lamto Widodo[1], and Kristina Pasaribu[1]**

[1]Departement of Industrial Engineering, Universitas Tarumanagara, Jakarta, Indonesia
Ahmad@ft.Untar.ac.id

**Abstract**, In supply chain activities has always been a potential risk arising, therefore risk management is very necessary for the handling of risk. In the company that produces women handbags such as PT. Sumber Karya Indah, in its supply chain activities, has the opportunity to arise risk. Therefore, it is necessary to do risk analysis and design of mitigation action, to reduce the risk of interference that is likely to arise in its the supply chain. The research was conducted using a house of risk model consisting of 2 phases. The first phase is risk identification and risk agent, which is then done a measurement of severity and occurrence level as well as a calculation of aggregate risk priority (ARP) value. The second phase is risk mitigation planning. Data retrieval methods are performed by interviews, observations, and historical data as supporting data. After the research has been obtained results that there are 28 risk events and 32 risk agents. There is 7 mitigation action that can be used and expected to mitigate risk in its supply chain. The result is the House of Risk Assessments Management Template using VBA Macro Excel.
**Keywords:** Supply Chain, ARP, HOR, Risk Mitigation

## 1. Introduction

Supply Chain Management is a complete cycle chain management that starts from raw materials, to operational activities in the company, continues to distribution to consumers who use the product. SCM is an approach or method that integrates with collaborative foundations. Companies that are in a supply chain, in essence, want to satisfy the same consumers, they must work together to make products that can be received by customers, both in terms of price, quality, and accuracy of the send time. Cooperation between elements in the supply chain must be done well so that the objectives can be achieved. Managing the supply chain is not easy. Some of the challenges that can pose risk to the supply chain are the complexity of supply chain structure and uncertainty.[1]

Risk is often interpreted as uncertainty that can cause problems and also bring profitable opportunities for the company and people per person. The risk is associated with possible (probability) losses especially those that cause problems. Risk becomes an important problem if the losses it causes are not known for sure.[2]

Risk management is a field of science that discusses how an organization applies measures in mapping the various problems that exist by placing various management approaches comprehensively and systematically.[3] Supply Chain Risk Management (SCRM) is a combination of Supply Chain Management and Risk Management concept, where Supply Chain Risk Management collaborate with an existing company in the supply chain to apply Risk process Management.[4]

PT. Sumber Karya Indah has problems in supply chain activities, such as suppliers and consumers who are erratic and do not have contracts. The risks that occur from these problems are the uncertainty of the number of stock raw materials, the change in the quality of raw materials and the quality of different products, as well as demand uncertainty. The risks arising from supply chain problems hurt the company both in terms of cost and time. From the above problems, it is necessary to analyze existing risks and draft the risk mitigation strategy to prevent and mitigate the risks that may arise in the supply chain at PT. Sumber Karya Indah. To identify and measure the potential risks involved in the supply chain at PT. Sumber Karya Indah uses the House of Risk (HOR) method.

The method of HOR developed by Pujawan and Geraldin (2009) is an innovative model for supply chain risk management. HOR1 is used to rank each risk agent based on their aggregate risk potentials. HOR2 is intended to prioritize the proactive actions that the company should pursue to maximize the cost-effectiveness of the effort in dealing with the selected risk agents in HOR1. For illustrative purposes, a case study is presented.[5] This method is used because of the focus on preventive actions such as reducing the likelihood of possible risk agents for more proactive supply chain risk management.[6]

Based on the explanation above, the purpose of this research is to mapping supply chain activity at PT. Sumber Karya Indah to determine risk agent and risk event prioritized to be handled using HOR1 and the erection of risk mitigation actions that can minimize risk occurrence by using HOR2.

## 2. Methods

The identification stage is the first step in this research that is observing directly the problem at the research location. After identification the problem formula is created and then set research objectives. Furthermore, library studies and field studies to support research to run well and correctly.

The second stage is the mapping stage of supply chain activity, where supply chain activity is divided into three parts of procurement activity, production process activities, and distribution activities. The mapping of supply chain activities is obtained by interviews and observations.

The data processing stage is the next stage after the data has been collected. Data processing aims to determine the severity of a risk event and the event of a risk agent. This risk analysis is then mapped to the House of Risk 1 (HOR1). Identification of risk events and risk agents is conducted by an interview with the General Manager who has worked from the beginning of the company. In addition to business processes are also identified parts that are responsible for the business process for each business process.

In such models, the risk event and the source of risk are given the value of its correlate, and then calculated the aggregate risk priority (ARP) value of each risk source. From these results, it is then recorded by using the principle 80/20 of the Pareto diagram to determine selected risk agent. After HOR1 is mapped, mitigation actions are identified that are mapped on HOR2. in this phase the total value of the effectiveness of mitigation actions (TEk) is calculated, the level of difficulty in mitigation actions (Dk) and the total effectiveness of the level of difficulty in carrying out mitigation actions (ETDk). [7] [8]

## 3. Results and Discussion

### 3.1. Supply Chain and Mapping supply chain

Supply chain PT. Sumber Karya Indah consists of suppliers, company, retail, and customer. The following can be seen supply chain as a whole from PT. Sumber Karya Indah works in Figure 1.

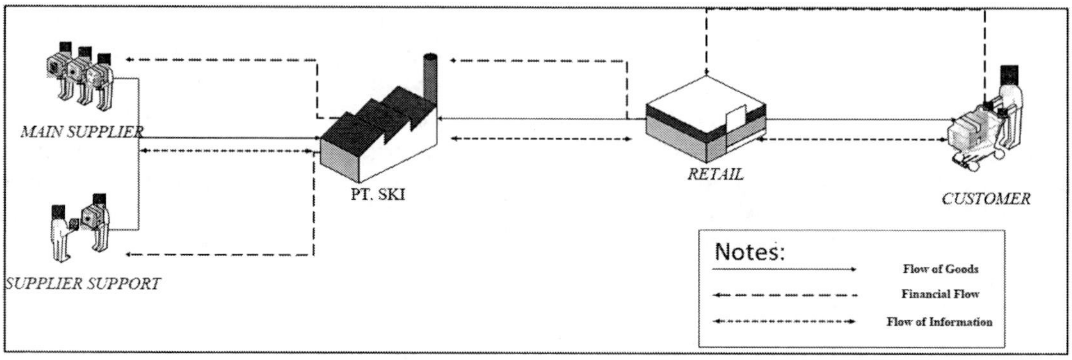


**Figure 1**. Supply Chain PT. Sumber Karya Indah

The procurement activity of this company is influenced by its production type. The make to stock production type begins with the stock checking in the PT. Sumber Karya Indah retail warehouse. When the remaining stock has reached the minimum limit (approximately five to six bags), then the retail part will contact the general administration of the factory will be known by the general manager. General Manager will convey to the head of the production type of goods to be produced. Retail PT. Sumber Karya Indah will deliver the order to the factory General Administration. The administration conveyed to the general manager of product ordering specifications. General Manager considers the model and quality of materials. Then the general manager will inform the sales price range in retail to be delivered to the customer.If the price has been agreed, then the retail part will contact the administration of the factory will be known by the general manager. General Manager will convey to the head of the production type of goods to be produced. Then, the head of the production will divide the task into the design supervisor, and production supervisor. If the raw material is available, the production process can be started, but if the stock of raw material is less or unavailable, then the general Administration will order the raw material on the supplier.

Based on the classification the company's supply chain activity description is shown in Table 1.

**Table 1**. Mapping supply chain activity in five business processes

| Process | Sub Process |
|---|---|
| ***Plan*** | 1. Demand forecasting<br>2. Calculation of Material needs<br>3. Production Planning<br>4. Material Planning and control |
| ***Source*** | 1. Delivery scheduling from Supplier<br>2. Accepting, checking and storing raw materials<br>3. Choosing a Supplier |
| ***Make*** | 1. Production Scheduling<br>2. Production process<br>a. Draw a pattern<br>b. *Cutting*<br>c. *Assembly*<br>3. Product Inspection |
| ***Deliver*** | 1. Delivery selection<br>2. Sending bills to customers |
| ***Return*** | 1. Return scheduling<br>2. Making a return |

### *3.2. Risk Event and Risk Agent*

From the process of mapping the supply chain activity in Table 1, further identifying and measuring risk events and risk agents. These measurements are performed to determine the severity of a risk event and determine the occurrence (level of probability occurring) of the risk agent. This measurement was done by an interview with a General manager who has worked in the company from the beginning of the company and the General Administration that has been working for more than 10 years. The results of the measurements are shown in Table 2 and Table 3.

**Table 2.** Risk Event Measurement Results

| $E_i$ | Risk Event | $S_i$ |
|---|---|---|
| E1 | Mistake the magnitude of forecasting | 5 |
| E2 | Error requirement calculation | 4 |
| E3 | Sudden changes in the production plan | 4 |
| E4 | Production scheduling does not match | 6 |
| E5 | Gaps between recorded and available stock | 7 |
| E6 | Incorrect/accurate inventory parameters | 5 |
| E7 | Raw material delays from suppliers | 7 |
| E8 | Specifications of raw materials received are not by booking | 6 |
| E9 | Received raw materials damaged | 6 |
| E10 | The raw materials submitted are not inspected by the goods | 5 |
| E11 | Changes in the quality of raw materials | 6 |
| E12 | Fault choosing supplier | 7 |
| E13 | Delayed production schedule | 7 |
| E14 | Mechanical malfunction | 6 |
| E15 | Error drawing master pattern | 8 |
| E16 | Cutting is not perfect | 9 |
| E17 | Assembly is not perfect | 7 |
| E18 | Product quality degradation during the process | 6 |
| E19 | Quantity of repair products exceed standard | 6 |
| E20 | Not able to meet all requests | 7 |
| E21 | Fault delivery service selection | 7 |
| E22 | Delayed delivery of products | 5 |
| E23 | Product shipped to the wrong destination | 7 |
| E24 | Cancellation of shipment | 6 |
| E25 | The difference in the bills recorded by customers and companies | 6 |
| E26 | System error | 6 |
| E27 | Return product late | 6 |
| E28 | Product returns canceled | 7 |

The severity value states how much disruption/impact a risk event can cause. The scale used in severity is a scale level of 1-10 with the sense that the value of 1 means that no failure/disruption effect occurs and the value of 10 means almost certain failure/disturbance effects.

**Table 3**. Risk Agent Measurement Results

| $A_j$ | Risk Agent | $O_j$ |
|---|---|---|
| A1 | Significant demand increase | 4 |
| A2 | Seasonal factors | 6 |
| A3 | Sudden request from a customer | 4 |
| A4 | Less coordination in the raw material ordering section | 3 |
| A5 | Transport interference | 4 |
| A6 | External factors | 4 |
| A7 | Inspection of unexamined raw materials receiver | 5 |
| A8 | Interference on raw materials during travel | 5 |
| A9 | Absence of contracts established with suppliers | 8 |
| A10 | Supplier can't fulfill orders | 2 |
| A11 | Technical disruption to the negotiation process with the supplier/customer | 4 |
| A12 | Enterprise internal factors | 4 |
| A13 | Limited human resources | 6 |
| A14 | No maintenance | 8 |
| A15 | Buildup of raw materials for too long | 6 |
| A16 | Working procedure production process is less obvious | 7 |
| A17 | Waiver of work procedure by the production operator | 7 |
| A18 | Negligence of Labor | 6 |
| A19 | An Inscrutiny product inspection process | 5 |
| A20 | Power supply disrupted | 2 |
| A21 | Overworking Hours | 6 |
| A22 | Production Target is relatively high | 6 |
| A23 | The difficulty level of product manufacture | 6 |
| A24 | Large Product variations | 5 |
| A25 | Error of identity granting product | 4 |
| A26 | No Standardized products | 7 |
| A27 | Distance factor and communication between company and customer | 5 |
| A28 | Less coordination in the shipping Section | 5 |
| A29 | Inadequate transportation | 5 |
| A30 | Natural disasters | 2 |
| A31 | It system interference | 2 |
| A32 | Return of returned product does not match specifications | 4 |

After the risk event and the risk agent have been identified, the next step is to determine the value of the inspection between the two. if there is no examination, the value of 0. low examination is represented by the number 1, moderate examination by 3, and 9 represents the high. both data were collected using interviews with managers who had worked at the company since its inception.

### *3.3. House of Risk 1 (HOR 1)*

Activity mapping in this model is done by entering the severity of the measurement of risk events as shown in Table 2. While the events of the risk agent and inspection value can be seen in Table 3. Determination of the ARP value in the following formula can be:

$$ARP_j = O_j \sum_i S_i R_{ij} \tag{1}$$

with ARPj= Aggregate risk potential value from j risk agent, Oj= occurrence from j risk agent, Si= severity of risk event i, and Rij = correlation value between j risk agent and risk event i. The result of the mapping of the HOR 1 was then rendered by using the Pareto diagram shown in Figure 2.

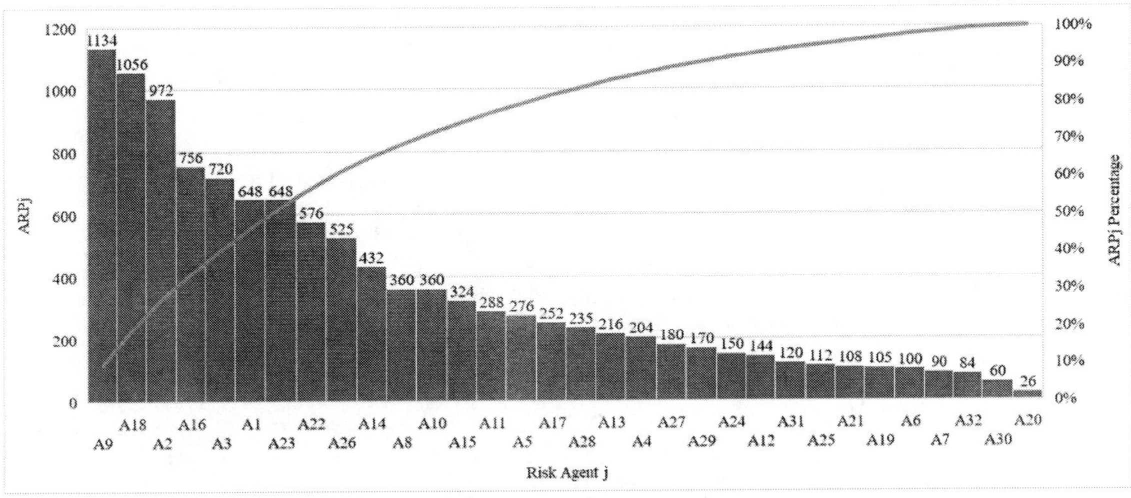


**Figure 2**. ARP Diagram

Based on the ARP diagram in Figure 2 and using the 80/20 principle from Pareto, risk agents will be chosen as a basis for consideration in choosing risk mitigation actions. From Figure 2 it can be seen that the risk agents A9, A18 and A2 are the highest with each ARP value shown in Table 4.

**Table 4**. Risk agent and ARP value

| $A_j$ | Risk Agent | ARP | ARP Percentage | Cumulative ARP Percentage |
|---|---|---|---|---|
| A9 | Absence of contracts established with suppliers | 1134 | 9,9% | 9,9% |
| A18 | Negligence of Labor | 1056 | 9,2% | 19,2% |
| A2 | Seasonal factors | 972 | 8,5% | 27,7% |

These risk agents will be included in the HOR 2 model to determine mitigation actions. Mitigation actions are taken to reduce the impact of risk agents. Recommended mitigation actions are shown in Table 5

**Table 5**. Recommended Mitigation Actions

| Aj | Risk Agent | Mitigation Action | $MA_k$ |
|---|---|---|---|
| A9 | Absence of contracts established with suppliers | Planning procedures for procurement of raw materials | MA1 |
| | | Perform performance evaluation with supplier | MA2 |
| A18 | Negligence of Labor | Conducting personal training in production section | MA3 |
| | | Assess work results every month | MA4 |
| | | Improve coordination between sections | MA5 |
| A2 | Seasonal factors | Looking for a supplier that has different characteristics | MA6 |
| | | Coordination with similar industries for collaborative work | MA7 |

### *3.4. House of Risk 2 (HOR 2)*

Measuring the effectiveness of the level of difficulty is a step that must be determined in the next stage by dividing the total value of effectiveness (TEk) by the level of difficulty through action. The level of effectiveness of the difficulty is aimed at ranking the priority of all mitigation actions, using the following formulas. the results of the calculation can be seen in table 6.

$$TE_k = \sum ARP_j E_{jk} \quad (2)$$
$$ETD_k = TE_k / D_k \quad (3)$$

with TEk = Total effectiveness of the mitigation action k, ARPj = Aggregate Risk Potential value of risk agent j, Ejk = the value of the collation between j risk agent and mitigation action k, ETDk = ratio of effectiveness to the difficulty of implementing a mitigation action k, and Dk = value difficulty in the application of k.

The degree of difficulty (Dk) is 3 (easy-to-implement mitigation action), 4 (rather easy to implement mitigation actions), or 5 (hard-to-implement mitigation actions). The priority of all mitigations can be seen in Table 6 and Figure 3. The priority based on the largest ETDk value.

**Table 6**. House of Risk 2 (HOR2)

| Aj | Mitigation Action | | | | | | | ARPj |
|---|---|---|---|---|---|---|---|---|
| | MA1 | MA2 | MA3 | MA4 | MA5 | MA6 | MA7 | |
| A9 | | | | | | 3 | 3 | 1134 |
| A18 | 9 | 1 | 1 | | | | | 1056 |
| A2 | | | | 9 | 3 | 3 | | 972 |
| TEk | 9504 | 1056 | 1056 | 8748 | 2916 | 6318 | 3402 | |
| Dk | 5 | 5 | 3 | 4 | 4 | 4 | 5 | |
| ETDk | 1900,8 | 211,2 | 352 | 2187 | 729 | 1579,5 | 680,4 | |
| Rk | 2 | 7 | 6 | 1 | 4 | 3 | 5 | |

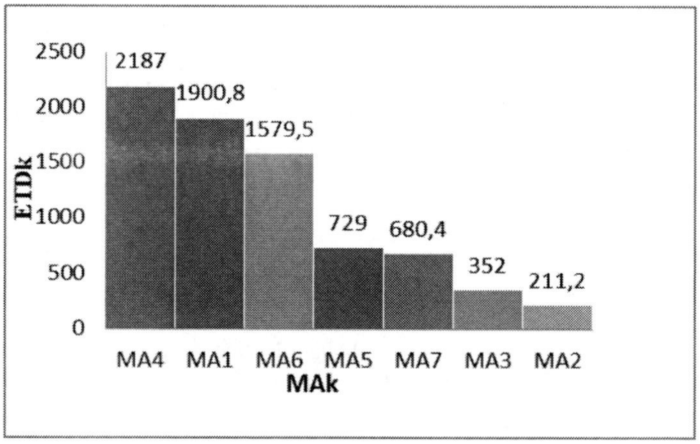


**Figure 3**. Priority of Mitigation Action

Table 3 shows mitigation actions carried out in sequence in accordance with ETDk values. it can be seen that the highest ETD value is MA4 (evaluating work results every month) of 2187, where mitigation actions are most likely to be done first to prevent A8 (labor negligence). The most difficult mitigation action to take is MA2 (evaluating performance with suppliers) which has an ETD value of 211,2. This mitigation action is carried out to overcome A9 (no contracts are formed with suppliers).

### *3.5. HOR Assessment Management Template*

Once the HOR model is calculated, the next step is to provide accessibility for PT. SKI by performing risk recording. This entry will help PT. SKI to make future decisions. Researchers use VBA Macro Excel to create the House of Risk Assessments Management Template. The result of this template refers to the result of a HOR model that is mitigation action. Using this template automates the calculation of HOR, making it easier for PT. SKI to identify and analyze the risk agent occurring in the company. This template is based on the risk events that occur in the company and the risk agents causing the risk events to occur. both of these can change from time to time, because of several factors that influence. These two factors are the main variables in making templates.

## 4. Conclusion

From the supply chain activities of PT. Sumber Karya Indah has identified 28 risk events and 32 risk agents. From the results of the mapping in HOR 1, it is known that three selected risk agents are considered in mitigation actions, namely: A9 (absence of contracts formed with suppliers) with an ARP value of 1134, A18 (negligence of Labor) with an ARP value of 1056, and A2 (seasonal/seasonal factor) with an ARP value of 972.

From the mapping result of HOR 2, obtained 7 draft risk mitigation action, namely: MA1 (planning of raw material procurement activity procedure) with an ETD value of 1900.8; MA2 (conducting supplier performance evaluation) with an ETD value of 211.2; MA3 (conducting personal training) with an ETD value of 352; MA4 (Assess work every month) with ETD value of 2187; MA5 (improves coordination between parts) with an ETD value of 729; MA6 (looking for a supplier that has different characteristics) with an ETD value of 1579.5; and MA7 (coordination with similar industries to collaborate on work) with an ETD value of 680.4.

## 5. References

[1] Pujawan, IN and ER Mahendrawathi 2018 *Supply Chain Management*, 3rd Ed. (Yogyakarta: Andi).

[2] Siahaan, Hinsa 2009 *Manajemen Risiko pada Perusahaan & Birokrasi.* (Jakarta: Elex Media Komputindo).

[3] Kurniawan, Dwi Cahaya 2018 Analisis dan Mitigasi Risiko Proses Make, Deliver, Return dengan Pendekatan Model Green Supply Chain Operation Reference (Green SCOR) dan Metode House of Risk (HOR) pada PT. Globalindo Intimates. *Final Project.* (Yogyakarta: Universitas Islam Indonesia).

[4] Handayani, Dwi Iryaning 2016 A Review: Potensi Risiko Supply Chain Risk Management. *Spektrum Industri.* https://doi.org/10.12928/si.vl4il.3701 (accessed on August 27, 2019).

[5] Pujawan, IN and Geraldin, LH 2009 House of Risk: A Model for Proactive Supply Chain Risk Management. *Business Process Management Journal*, **15**(6): 953–67. https://doi.org/10.1108/14637150911003801 (Accesed on November 27, 2019)

[6] Sasongko, Didi Adji 2018 Analisis Risiko dan Strategi Aksi Mitigasi pada Usaha Penjualan Mesin Teknologi Tepat Guna dengan Metode House of Risk (Studi Kasus: Toko Sedia Mesin). *Seminar Nasional IENACO 2018,* ISSN 2337-4349. https://publikasiilmiah.ums.ac.id/xmlui/ handle/11617/9837 (Accesed on September 3, 2019)

[7] Kristanto, Bayu Rizki, and Hariastuti, LP 2014 Aplikasi Model House of Risk (HOR) untuk Mitigasi Risiko pada Supply Chain Bahan Baku Kulit. *Jurnal Ilmiah Teknik Industri*, **13**(2), 149-157. http://jurnal.ums.ac.id/index.php/ jiti/article/ view/633 (Accesed on August 27, 2019)

[8] Ulfah, Maria, Mohamad SM, Sukardi, and Sapta R 2016 Analisis dan Perbaikan Manajemen Risiko Rantai Pasok Gula Rafinasi dengan Pendekatan House of Risk. *Jurnal Teknologi Industri Pertanian*, **26**(1): 87-103. http://jai.ipb.ac.id/index.php/jurnaltin/article/view/13129 (Accesed on August 24, 2019)

IOP Conf. Series: Materials Science and Engineering **847** (2020) 012060 doi:10.1088/1757-899X/847/1/012060

# Determination of Multi-Product Distribution using Capacitated Vehicle Routing Problem (CVRP) and Product Cubication Dimensions Restriction

**Yogi Yogaswara[1] and Neng Resi Andriyani[2]**

[1,2] Department of Industrial Engineering, Faculty of Engineering Universitas Pasundan, Indonesia
[1] yogiyoga@unpas.ac.id

**Abstract.** CVRP is one of the transportation problems with the characteristic of the limited capacity of the vehicle. PT. Trikarsa Mitra Utama experienced transportation problems in the distribution of 3 types of paint products to 12 consumers, with 2 different units of vehicles. This research discuss how to determine the distribution route of paint products by considering the capacity of the vehicle based on weight and different cubication dimensions of the product. The method used for this problem is the Saving Clarke and Wright method besides that the Cube IQ software is used to loading products on the vehicle based on cubication dimensions. Besides being able to influence the process of preparing transportation routes, the results of the optimum loading process will also be used as a plan for product preparation in vehicle. The routes formed from the calculation results are classified into 3 clusters. Cluster 1 with route 0-H-K-D-I-L-0, cluster 2 with route 0-G-F-J-E-C-0, cluster 3 with route 0-A-B-0 gives a total mileage of 288 km and a travel time of 695 minutes using the same vehicle by *colt diesel engkel (CDE)*.
**Keywords:** CVRP, Saving Clarke and Wright, cubication dimension, loading process, Cube IQ.

## 1. Introduction

Transportation activities have a big role in the level of customer satisfaction and business interests of the company. Transportation problems are commercial problems that must be resolved by companies that carry out logistical activities. A case study conducted at a company that makes the production of various types of paints with various uses such as protective coatings, floor coatings, marine coatings, industrial metal coatings. The problem that occurs in this company lies in the logistics system, the cost of transportation out of logistics companies that are considered high enough to distribute goods from depots to 12 different locations of consumers. This is indicated as a result of the suboptimal determination of transportation routes for the distribution of paint products to 12 consumers resulting in high transportation costs.

Different vehicle transport capacities make transportation problems classified into Capacitated Vehicle Routing Problems (CVRP) [1]. The choice of transportation equipment of the company must be in accordance with the number of consumers concerned, so that the demand can be fulfilled exactly once and there is no double shipping [2]. The choice of the vehicle will not only be considered from the weight of the product but also from its dimensions, so that the feasibility of planning is determined by both the weight and dimension. The addition of the cubication dimension restriction was made due to the fact that the process of distributing various types of paint products at this company was not carried out in bulk, but was packaged in a different package.

The problem formulation of this research is how to determine the distribution route of paint products by considering the capacity of the vehicle based on different cubication dimensions of the product. The objective to be achieved in this research is to get the results that are close to optimum from a Capacitated Vehicle Routing Problem (CVRP) model using the Saving-Clarke & Wright (saving) method by considering the transport capacity based on cubication dimensions through the product loading process. So that it is expected to provide minimum mileage accompanied by a product preparation plan on the vehicle.

## 2. Methods

In the transportation system determining the route and scheduling of vehicles is an important component that can affect the high transportation costs. Vehicle route problems are often referred to as Vehicle Routing Problems (VRP). Vehicle routing problem (VRP) is an extension of the basic traveling salesman problem (TSP) [2].

Along with the development of VRP that has been done and hence the limited capacity of the term vehicle VRP called the Capacity Vehicle Routing Problem (CVRP). The CVRP problem model has the following constraints [3]:

1. Each route will start and end at one depot.
2. Each consumer will only be visited by one vehicle.
3. The total demand for each route must not exceed the vehicle capacity.

The purpose of VRP is to deliver the product to a group of consumers whose request is known by only spending a minimum cost and starting and ending at one or more depots. The output of this problem is a route that produces a low cost and is feasible for each vehicle.

The process of moving goods / materials from the the polling area to a delivery vehicle is called loading. This process is a form of warehouse services to users and consumers. The loading process is largely determined by the type, classification and characteristics of the product. The Framework of methods can be seen in Figure 1.

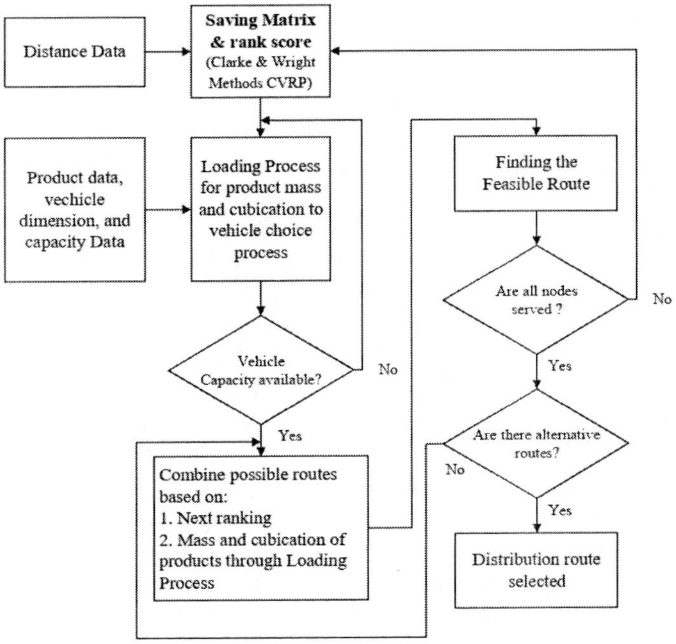


**Figure 1.** Methods Framework

Loading activity is raising the product (raw material, parts, goods-in-process, finished goods) to the truck that has gone through checking, then it will be sent according to the destination of each consumer [4]. Activities that include in the loading process are checking products (raw materials, parts, goods-in-process, finished goods), whether they are in accordance with the number of requests, the type of product and the quantity in accordance with consumer demand.

Cube IQ is a software to help in planning loading activities properly and regularly, which will result in the most efficient compilation. Cube IQ has operating modes for container loading, truck loading, palletization and cartonization. There are various possibilities for Cube IQ in loading activities, are [5]:

1. Create a 3D load diagram of the plan that uses complex loading rules that include loading sequences, partial loads, and load distribution.
2. Optimizing the load under the stack as a whole, in accordance with the rules.
3. Keep loading cases in full.
4. Distribute loading instructions in a 3-D diagram.

## 3. Result and Discussion

### *3.1. Data of Product Size, Vehicle and Saving Ranking Matrix*

Distance Measurement Tools TM application Google Maps is used for collecting some data includes the vehicle's mileage and travel time.

It is known that products X, Y and Z have cylinder shape so the dimensions are owned in the form of diameter and height. For product X lot contains 12 units of product X packed in cartons with arrangement 3 elongated units, 2 widened units, and stacked 2 stacks for product height with a carton thickness of 0.3cm each its side. For more details can be seen in Table 1 below.

**Table 1.** Data of Product Size

| Dimension | Product Size (cm) | | | |
|---|---|---|---|---|
| | X (1 kg) | Y (10 kg) | Z (18 kg) | X (Lot) |
| Length | 13.05 | 27.30 | 30.00 | 39.75 |
| Width | 13.05 | 27.30 | 30.00 | 26.70 |
| Height | 12.00 | 26.30 | 33.80 | 24.60 |

In addition to company-owned product data, transport capacity data is also a main data used in research. Vehicle data used consists of two types of vehicles, the CDE standard and Suzuki Pick-Up. For more details can be seen in Table 2 below.

**Table 2.** Data of Vehicles

| Vehicle Type | Capacity (Maximum) | | | | |
|---|---|---|---|---|---|
| | L(cm) | W(cm) | H(cm) | Vol(m3) | Weight(kg) |
| *CDE Standard* | 237 | 155 | 129 | 4.739 | 2000 |
| Suzuki Pick-Up | 220 | 148 | 100 | 3.256 | 800 |

Technological advancements in the field of Intelligent Transport System (ITS) provide sophisticated equipment that can be used as a consideration to assist the operation of the transportation system [6]. One of the ITS technologies that is integrated with software as a transportation system aids is the Global Positioning System (GPS). Related to transportation problems in this study, the use of ITS technology, the Global Positioning System (GPS) in the Distance Measurement Tools Google Maps application, is used to collect some data which includes mileage and vehicle travel time. Distance data between nodes taken using the Google Maps application are as follows in Table 3.

**Tabel 3.** Matrix Distance Between Nodes (km)

| | 0 | A | B | C | D | E | F | G | H | I | J | K | L |
|---|---|---|---|---|---|---|---|---|---|---|---|---|---|
| 0 | | 24,8 | 37,4 | 5,3 | 33,3 | 15,4 | 25,1 | 3,1 | 57,3 | 30,4 | 29,7 | 55,9 | 30,4 |
| A | | | 22,5 | 21,6 | 53.4 | 23,2 | 44,7 | 27,7 | 76,7 | 51 | 49,1 | 79,2 | 51 |
| B | | | | 30,9 | 62.8 | 34,1 | 55,5 | 37,4 | 87,6 | 61 | 60,3 | 86,1 | 61 |
| C | | | | | 37.9 | 11,8 | 27,6 | 7,1 | 59 | 32,1 | 33,5 | 60,2 | 32,1 |
| D | | | | | | 41,6 | 10,3 | 38,9 | 18,9 | 2,5 | 3,1 | 16,1 | 2,5 |
| E | | | | | | | 35,1 | 17,3 | 67,2 | 38,9 | 38,3 | 65,3 | 38,9 |
| F | | | | | | | | 26,9 | 36,5 | 7,9 | 7,3 | 35,1 | 7,9 |
| G | | | | | | | | | 58,6 | 27,8 | 27,1 | 57,2 | 27,8 |
| H | | | | | | | | | | 24,9 | 25,6 | 3,8 | 24,9 |
| I | | | | | | | | | | | 1,4 | 19,5 | 0,0 |
| J | | | | | | | | | | | | 19,7 | 0,7 |
| K | | | | | | | | | | | | | 18,2 |
| L | | | | | | | | | | | | | |

The saving matrix is obtained from distance matrix data has been calculated using the equation saving that is $sij = ci0 + c0j - cij$ with $cij$ is the distance from customer $i$ to customers $j$ and 0 are depots [7]. Score of saving that has been obtained from the results calculations using these equations presented in the form of ranking in Table 4.

**Table 4.** Saving Score Ranking

| No | Route | *Saving* | No | Route | *Saving* | No | Route | *Saving* |
|---|---|---|---|---|---|---|---|---|
| 1 | H-K | 109,4 | 23 | B-E | 18,7 | 45 | A-H | 5,4 |
| 2 | D-K | 73,1 | 24 | A-E | 17,0 | 46 | A-J | 5,4 |
| 3 | D-H | 71,7 | 25 | B-C | 11,8 | 47 | A-F | 5,2 |
| 4 | K-L | 68,1 | 26 | C-E | 8,9 | 48 | A-D | 4,7 |
| 5 | I-K | 66,8 | 27 | A-C | 8,5 | 49 | A-I | 4,2 |
| 6 | J-K | 65,9 | 28 | B-D | 7,9 | 50 | A-L | 4,2 |
| 7 | H-I | 62,8 | 29 | B-K | 7,2 | 51 | C-H | 3,6 |
| 8 | H-L | 62,8 | 30 | D-E | 7,1 | 52 | C-I | 3,6 |
| 9 | H-J | 61,4 | 31 | B-H | 7,1 | 53 | C-L | 3,6 |
| 10 | D-I | 61,2 | 32 | B-F | 7,0 | 54 | B-G | 3,1 |
| 11 | D-L | 61,2 | 33 | E-I | 6,9 | 55 | C-F | 2,8 |
| 12 | I-L | 60,8 | 34 | E-L | 6,9 | 56 | G-H | 1,8 |
| 13 | D-J | 59,9 | 35 | B-I | 6,8 | 57 | G-K | 1,8 |
| 14 | J-L | 59,4 | 36 | B-J | 6,8 | 58 | C-J | 1,5 |
| 15 | I-J | 58,7 | 37 | E-J | 6,8 | 59 | A-K | 1,5 |
| 16 | D-F | 48,1 | 38 | B-L | 6,8 | 60 | C-G | 1,3 |
| 17 | F-I | 47,6 | 39 | E-K | 6,0 | 61 | F-G | 1,3 |
| 18 | F-L | 47,6 | 40 | G-I | 5,7 | 62 | E-G | 1,2 |
| 19 | F-J | 47,5 | 41 | G-J | 5,7 | 63 | C-K | 1,0 |
| 20 | F-H | 45,9 | 42 | G-L | 5,7 | 64 | C-D | 0,7 |
| 21 | F-K | 45,9 | 43 | E-H | 5,5 | 65 | A-G | 0,2 |
| 22 | A-B | 39,7 | 44 | E-F | 5,4 | 66 | D-G | 0,0 |

### *3.2. Transportation Routes Calculation and Arrangement*

The experiment was carried out into three proposed transportation routes presented in experiment 1 (P1), experiment 2 (P2), and experiment 3 (P3), the results were obtained in the form of total mileage and total time distribution of paint to 12 consumer points is 306 km and 744 minutes for the model Experiment 1 (P1). The experimental model 2 (P2) is 307 km and 744 minutes, and experiment model 3 (P3) are 288 km and 695 minutes, respectively.

From the results of calculations, experiment 3 is the route chosen in this study, because it provides a minimum total distance and travel time compared to experiments 1 and 2. The routes are formed from experiment 3 is 0-K-H-D-I-L-0 joined in cluster 1, route 0-G-F-J-E-C-0 incorporated in cluster 2, and routes 0-A-B-0 incorporated in cluster 3. The final output through capacity loading process which can be seen in sequence in Figure 2, 3, and 4.

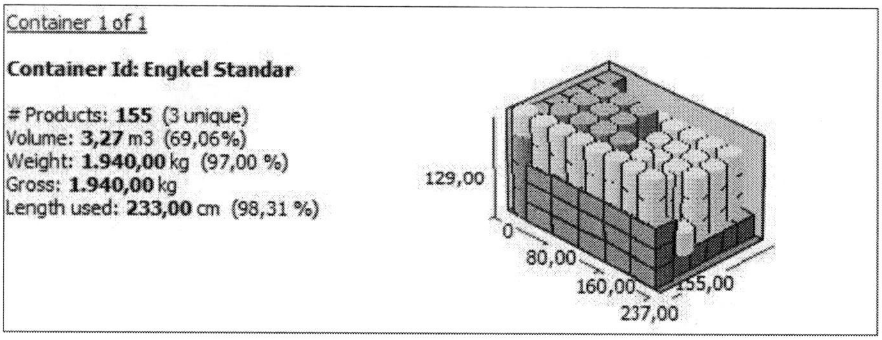


**Figure 2.** Vehicle capacity for node H and K, D, I and L, through the loading process using Cube IQ software after manual merging for Cluster 1.

In Figure 2 the merging of points I and L into Cluster 1 is the remaining transport capacity as described is the optimum decision (feasible route) because it is no longer possible to merge. The feasible route formed is the 0-H-K-D-I-L-0 route or referred to as Cluster 1.

There are 12 consumers who will be designated as product distribution nodes, determined 5 nodes that already have distribution routes namely H, K, D, I, and L which are incorporated in a cluster, namely cluster 1. While 7 points that do not have a distribution route, will be formed in the next cluster with the same process. The formation of cluster 2 is based on the ranking of ranking that is not related to the previous cluster (H, K, D, I, and L).

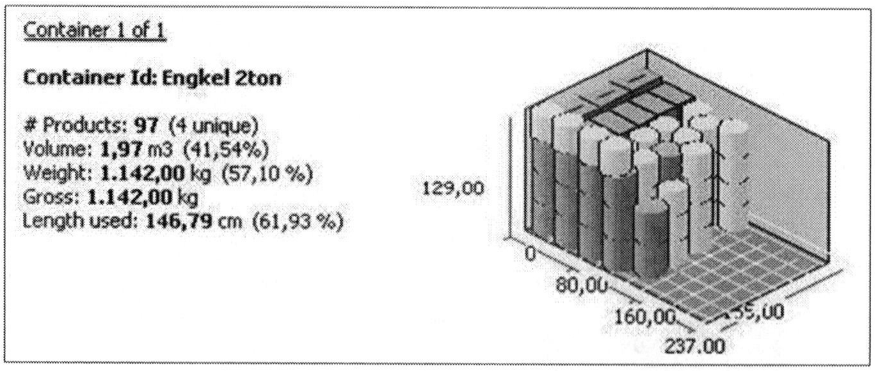


**Figure 3.** Vehicle capacity for node G, F, J, E and C, through the loading process using Cube IQ software for Cluster 2.

IOP Conf. Series: Materials Science and Engineering **847** (2020) 012060 doi:10.1088/1757-899X/847/1/012060

From Figure 3, it can be seen that the volume capacity of vehicles is still available at 58.46%, but the optimum route in Cluster 2 has been formed because there are no more nodes that need to be merged. Then the feasible route formed is 0-G-F-J-E-C-0 or referred to as Cluster 2.

The merging point conducted in experiment 3 is point B with cluster 3, so that the main ledge conducted in this experiment is to determine the feasibility of loading the merge point B with cluster 3 using Cube IQ software. Point A has a demand of 1907 kg (52 units of product X lot, 1 unit of product X, 40 units of product Y, and 49 units of product Z) and point B has demand of 54 kg (3 units of product X lot, and 1 unit of product Z ) then the loading process can be seen in Figure 4.

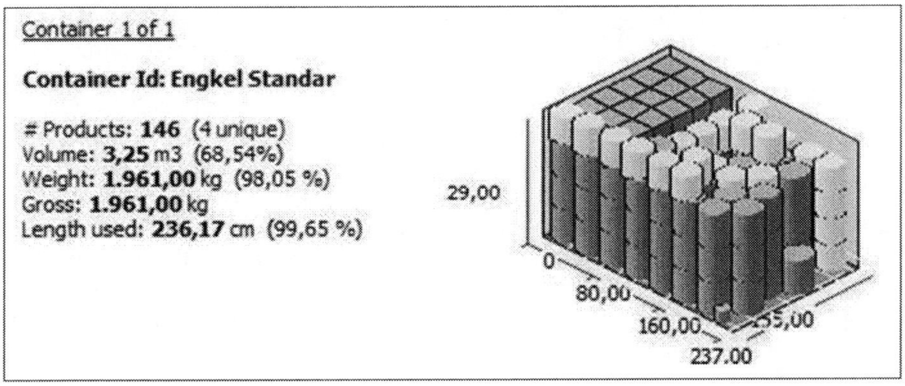


**Figure 4.** Vehicle capacity for node A and B, through the loading process using Cube IQ software for Cluster 3.

The results of experiments that determine the chosen distribution route consisting of distance and travel time can be seen in Table 5.

**Table 5.** Transportation Route for Cluster 1, 2 and 3

| Cluster | Route | Route Between | Distance (km) | Total Distance (km) | Speed (Km/hr) | Route Time (minute) | Total Time (minute) |
|---|---|---|---|---|---|---|---|
| 1 | 0-H-K-D I-L-0 | 0-H | 57.3 | 57.3 | 24.91 | 138 | 138 |
| | | H-K | 3.8 | 61.1 | 19.00 | 12 | 150 |
| | | K-D | 16.1 | 77.2 | 27.60 | 35 | 185 |
| | | D-I | 2.5 | 79.7 | 18.75 | 8 | 193 |
| | | I-L | 0 | 79.7 | 0 | 0 | 193 |
| | | L-0 | 30.4 | 110.1 | 16 | 114 | 307 |
| 2 | 0-G-F-J-E-C-0 | 0-G | 3.1 | 3.1 | 31.00 | 6 | 6 |
| | | G-F | 26.9 | 30 | 28.82 | 56 | 62 |
| | | F-J | 7.3 | 37.3 | 21.90 | 20 | 82 |
| | | J-E | 38.3 | 75.6 | 20.70 | 111 | 193 |
| | | E-C | 11.8 | 87.4 | 33.71 | 21 | 214 |
| | | C-0 | 5.3 | 92.7 | 28.91 | 11 | 225 |
| 3 | 0-A-B-0 | 0-A | 24.8 | 24.8 | 29.76 | 50 | 50 |
| | | A-B | 22.5 | 47.3 | 32.93 | 41 | 91 |
| | | B-0 | 37.4 | 84.7 | 31.17 | 72 | 163 |
| Total Distance | | | | 288 | Total Time | | 695 |

There is an operating time that is owned by the company as well as all 12 consumers, all of which have the same operating time from 08:00 to 17:00. In principle, the distribution time should not exceed the company's operating hours. In this case both the sender and receiver have operating hours because consumers are manufacturing companies, so the operating time is not a time window.

The feasibility of delivery time can be seen in a scheduling. Scheduling product delivery to 12 consumer points involves data such as travel time, loading and unloading time, rest time, start time, finish time, and operational time. From the results of the study it was found that the distribution of products divided into 3 clusters distributed within 2 working days. Cluster 1 is distributed on day 1 with distribution time starting at 08:00 in the morning until 14:17 in the afternoon. Whereas cluster 2 and cluster 3 are distributed on the 2nd day with the distribution starting at 08:00 am to 04:28 pm. With the distribution schedule as follows, it can be said that the distribution process is still in the safe category because it does not come out of the operating time of the company's product delivery, and does not come out of the operating time of 12 consumers as the time of receipt of the product.

## 4. Conclusion

Cluster 1 which contains route 0-H-K-D-I-L-0 with a total distance of 110.1 km and total distribution lead time is 387 minutes. Cluster 2 which loads route 0-G-F-J-E-C-0 with a total distance amounting to 92.7 km and total travel time distribution of 300 minutes. Cluster 3 contains the route 0-A-B-0 with a total distance of 84.7 km and total distribution travel time in the amount of 208 minutes.

Future research needs to be developed with add limits First In First Out (FIFO) into the loading process can provide more result representative because it can make it easy operator at the time of product decline (unloading). The compiling of loads using the Cube IQ software deliver the most efficient process, may also be used for joint shipment.

## 5. References

[1] Ballou R H 2004 *Business Logistics/Supply Chain Management* 5th Edition (USA: pearson Prentice Hall)

[2] Yogaswara Y 2015 *Proceeding 8th International Seminar on Industrial Engineering and Management* **1** (Malang: East Java Indonesia), p. 137

[3] Toth P and Vigo D 2002 *And Overview of Vehicle Routing Problem* (Society of Industrial and Applied Mathematics), pp. 1-26.

[4] Heragu S 2008 *Facilities Design* 3rd edition (CRC Press)

[5] Olsson J 2017 *Solving a Highly Constrained Multi-level Container Loading Problem from Practice* (Linkoping University: Division of Optimization/Dept. of Mathematics)

[6] Pop P C, Sitar C P, Zelina I, Lupse V and Chira C 2011 Heuristic Algorithms for Solving the Generalized Vehicle Routing Problem. *International Journal Computers, Comunication, and Control*, **11**(1), pp. 158-165.

[7] Altinel I K and Öncan T 2005 A new enhancement of the Clarke and Wright savings heuristic for the capacitated vehicle routing problem. *The Journal of the Operational Research Society* **56**(8), pp. 954–961.

IOP Conf. Series: Materials Science and Engineering **847** (2020) 012061 doi:10.1088/1757-899X/847/1/012061

# Reducing Waste in Manufacturing Industry using Cost Integrated Value Stream Mapping Approach

**Wilson Kosasih**[1,a], **C O Doaly**[1,b] **and Shabara**[1,c]

[1]Department of Industrial Engineering, Universitas Tarumanagara, jl. Letjend S. Parman No. 1 West of Jakarta, Indonesia
[a]wilsonk@ft.untar.ac.id, [b]carlaol@ft.untar.ac.id, [c]sabhara.rw@gmail.com

**Abstract.** In recent years, every manufacturing industry has tried to eliminate various wastes in its operations. The effect of wastes will have a direct impact on increasing costs. Therefore, the cost approach is integrated into the value stream mapping to obtain an illustration of the cost losses incurred. This paper describes how to apply integrated value stream mapping with the cost approach. Finally, this approach is considered effective in mapping the amount of cost losses incurred due to the wastes that occur along the value stream.
**Keywords**: waste, integration, value stream mapping, cost approach

## 1. Introduction

In the manufacturing industry, a company's productivity can be seen from its ability to carry out the production process effectively and efficiently. Since the mass production system is widely introduced, all manufacturers or manufacturing industries strive to produce products that have a high degree of conformance to standards (high level of homogeneity), so that they are free from zero defect. By being aware of all these things, the company is demanded more so that it can meet customer expectations and can further increase productivity by reducing existing waste.

With a view to avoiding waste and inefficiency so that the cost of production per unit will be low which in turn will make product prices more competitive. From the customer's point of view, waste is any activity that the customer does not want to pay, if they know that it happened. Today in Indonesia many manufacturing industries have applied the concept of lean, which strives continuously to transform all business activities / processes into something that is beneficial to customers [1]. Lean manufacturing is a production practice that considers all the expenditure of available resources to obtain economic value without waste [2]. The lean approach focuses on efficiency without reducing the effectiveness of the process including increasing value-added operations, reducing waste, and meeting customer expectations [3]. The lean production concept introduces a very effective and efficient production system, which uses less resources to produce higher quality at a lower cost as a result [1], [4].

One tool that can be used to describe the entire business process in implementing lean manufacturing is Value Stream Mapping (VSM) [2], [3], [4], [5], [6], [7] To further optimize the retrieval then the VSM can be integrated with cost analysis using the concept of Acitivity Based Costing (ABC) in the value stream [8], [9]. The basic concept of ABC is that a product will consume activities, activities consume resources, and all of these resources require costs. Therefore, this paper comprehensively discusses the application of VSM integrated with the perspective of ABC concepts.

## 2. Methods

### *2.1. Value Stream Mapping Approach*

Value Stream Mapping (VSM) is a tool that combines lean manufacturing concepts and techniques to visualize the entire flow of material and information needed when the product runs throughout the manufacturing process [5], [6]. The material flow path of a product is traced back from the final operation and its journey to the raw material storage location. VSM helps detect waste and its source in the manufacturing process. VSM was originally developed in 1995 as a rationale for the collection and use of a set of tools that can help researchers or practitioners identify waste along the value stream and, therefore, need to find the right way to eliminate it. This tool is able to identify activities that are value added and non-value added in the manufacturing industry, making it easier to find the root causes of problems in the process [7]. VSM is also useful so that improvements made can be more focused on the overall waste in the system. VSM makes it easier for company management to make improvements in the manufacturing process.

VSM consists of two types, including; a) Current state map is the configuration of the current production value stream; b) Future state map is a blueprint for lean transformation desired in the future. The steps in making VSM, including [5]; a) First, determine the value stream that needs to be improved; b) Understand how the current process is as a baseline; c) Design a flow that fits the lean principle as a vision for the future; d) Develop a detailed implementation plan so that objectives can be achieved; e) Implementation and check whether the VSM goals are achieved.

### *2.2. Cost Integrated Value Stream Mapping*

The concept of cost integrated value streams is to map and measure the costs contained in value streams, which means that cost integrated value stream mapping is a combination of cost aspects and value stream maps. Costs calculated in the form of value added costs and non value added costs. Value added costs are generated by calculating the direct costs of each process or activity while non value added costs are generated by calculating the holding cost per inventory [8], [9].

According to Garrison et.al. [10], activity based costing is a costing method designed to provide manager with cost information for strategic decisions and other decisions that may affect capacity as well as fixed costs. Basically the ABC principle is that resources are allocated to activities and then activities are allocated to cost objects based on their use. There are several advantages of ABC systems in determining product costs, namely [11]; a) More realistic product costs, especially in the manufacturing industry; b) More and more overhead costs can be traced to the product; c) ABC states that activities cause costs, not products, and products consume activities; d) ABC helps reduce costs (cost reduction), and identifies non-value added activities.

This study was conducted at one of the national manufacturing companies that produce valves. The problem that occurs is the delay in production results and the mismatch of production output results with the specified production targets. The types of products discussed further in this study are gate valve types, because they have the highest number of daily requests. Table 1 shows demand data for several types of products in January 2019. This research was conducted by taking data directly obtained from the results of direct observation and interviews. Data collected in the form of production flow, cycle time, production amount, valve reject, SIPOC and costs incurred based on the ABC concept. The costs collected include; machine usage costs, material usage costs, labor costs, and inventory holding costs. Retrieval of time data using stop-watch with repetition of 30 observations for each process.The cycle time obtained is tested for data adequacy. If the data is sufficient, a data uniformity test can be done by determining the upper control limit (UCL) and the lower control limit (LCL). If the data is in the range of LCL and UCL then data normality tests can be performed.

The next step is to create a current integrated cost stream value map based on the identified value added and non value added costs. The next stage, an analysis is carried out to identify existing waste in the current state map and then make an analysis of improvements based on the causes of waste using the 5W & 1H method. The steps that can be used to implement a cost integrated value stream are basically the same as creating a value stream mapping in general, but a cost analysis is added by integrating the cost-line into the VSM along with the time-line. To calculate the total value added cost (TVAC) and the total value added non cost (TVNAC), you can use the following formula [8], [9]:

Value added activity cost = mi + CTi $\left(\frac{Mi+Li}{3600}\right)$ (1)

Non value added activity cost = hi x Ii (2)

TVAC = $\sum_{i=1}^{n}$ mi + CTi $\left(\frac{Mi+Li}{3600}\right)$ (3)

TVNAC = $\sum_{i=1}^{n+1}$ hi x Ii (4)

Where; CT = cycle time, M = rate machine per hour; L = rate hourly operator; m = material costs (mi = 0, if there is no additional material in the activity); I = inventory (for raw materials, WIP, finished goods); h = holding cost per inventory.

**Table 1**. Data of demand

| No. | Type of product | Monthly demand for January | Daily demand |
|---|---|---|---|
| 1 | Gate valve | 441 | 20 |
| 2 | Check valve | 410 | 19 |
| 3 | Globe valve | 272 | 13 |
| 4 | Ball valve | 441 | 20 |
| 5 | Butterfly valve | 118 | 6 |

## 3. Results and Discussion

The valve manufacturing process from the beginning to the shipment consists of casting, machining, assembling, pressure testing, painting, and finally a final inspection, as shown in Figure 1. The casting process has several sub-processes, including; injection wax, dipping, lost wax, melting, sintering, pouring, and blasting. The machining process consists of several sub-processes such as; turning, drilling, welding and lapping. Then each process is carried out a time study to obtain the standard time and setup time, as shown in Table 2 and Table 3. Table 4, respectively, shows the cost of using the machine per hour for each process or sub-process.

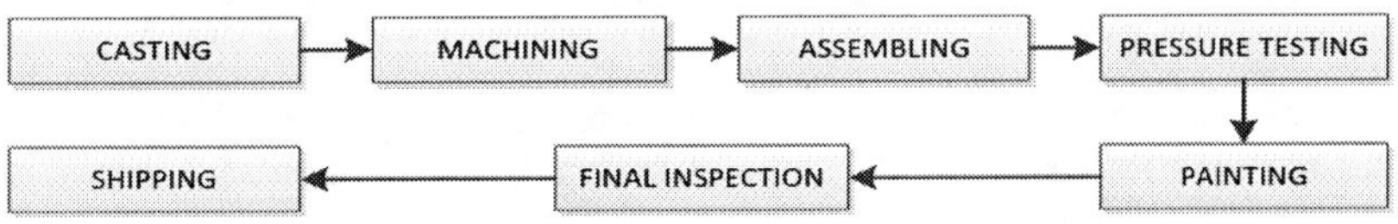


**Figure 1**. Valve production process

**Table 2**. Standard time measurement results*

| No. | Process or sub-process | Standard time (seconds) | No. | Process or sub-process | Standard time (seconds) |
|---|---|---|---|---|---|
| 1 | Injection Wax | 855.28 | 9 | Drilling | 623.48 |
| 2 | Dipping | 755.52 | 10 | Welding | 686.73 |
| 3 | Lost wax | 870.47 | 11 | Lapping | 371.98 |
| 4 | Sintering | 928.37 | 12 | Assembling | 1944.16 |
| 5 | Melting | 1854.64 | 13 | Pressure testing | 915.36 |
| 6 | Pouring | 753.34 | 14 | Painting | 1106.75 |
| 7 | Blasting | 809.45 | 15 | Final inspection | 16.73 |
| 8 | Turning | 3147.42 | 16 | Shipping | 148.29 |

*) Each repetition time measurement 30 times per process or sub-process. Then the adjustment factor is considered using the Westinghouse method.

IOP Conf. Series: Materials Science and Engineering **847** (2020) 012061 doi:10.1088/1757-899X/847/1/012061

**Table 3**. Machine setup time

| Process | Production equipment | Setup time (minute) |
|---|---|---|
| Casting | Auto wax injection | 10 |
| | electric auto clave | 5 |
| | Electric Furnace HT | 15 |
| | Induction Furnace | 15 |
| | Shot blasting hanger | 5 |
| Machining | CNC Lathe | 15 |
| | Radial drilling | 10 |
| | Welding SMAW | 2 |
| | Lapping Polish | 5 |
| Pressure testing | Hydrostatic Pressure | 10 |
| Painting | Spray Gun | 5 |

**Table 4**. Machine usage costs per hour

| Process | Production equipment | Total Rate per hour |
|---|---|---|
| Casting | Auto wax injection | IDR 42,551.12 |
| | Dipping (Manual with tools) | IDR 0.00 |
| | Electric auto clave | IDR 12,471.88 |
| | Electric Furnace HT | IDR 264,110.40 |
| | Induction Furnace | IDR 190,746.40 |
| | Pouring (Manual with tools) | IDR 0.00 |
| | Shot blasting hanger | IDR 50,621.16 |
| Machining | Lathe | IDR 24,210.12 |
| | Radial drilling | IDR 3,228.02 |
| | Welding SMAW | IDR 1,320.55 |
| | Lapping Polish | IDR 16,140.08 |
| Assembling | Manual | IDR 0.00 |
| Pressure Testing | Hydrostatic Pressure | IDR 44,018.40 |
| Painting | Spray Gun | IDR 322.80 |
| Final Inspection | Manual | IDR 0.00 |
| Shipping | Manual | IDR 0.00 |

To count labor hour rate data from the district minimum wage is needed, work hours data, and data on the number of workdays per month. The following is an example calculation labor hour rate; assumption, the district minimum wage = Rp. 3,555,834.- ; working days = 22 days; working hours = 8 Jam, then labor hour rate = 3,555,834 : (22/8) = IDR. 20,203.60. Meanwhile, Table 5 shows the amount of material costs required to manufacture valve products.

**Table 5**. Material costs

| Material | Unit | Quantity | Price per unit (IDR) | Total Price (IDR) |
|---|---|---|---|---|
| HS25051000 | kg | 4 | 4,000 | 16,000 |
| DCMT | pcs | 1 | 5,000 | 5,000 |
| RD260 | stem | 1 | 500 | 500 |
| Paint | liter | 2 | 1,500 | 3,000 |
| Scrap iron | kg | 1 | 4,500 | 4,500 |
| Sandpaper | sheet | 2 | 1,000 | 2,000 |

Determination of holding costs is based on average inventory and is expressed in rupiah currency. Costs included in holding costs include; warehouse usage / rental fees, maintenance fees, insurance fees, capital fees, obsolescence fees and taxes from inventory. Table 6 shows the holding costs per inventory for WIP that occur in each process or sub-process.

*Current vs Future Cost Integrated Value Stream Mapping*

Cost value added and non-value added can be calculated using Eq (1) and Eq (2). Example calculation total value added cost to the process injection wax are as follows:

$$\sum_{i=1}^{n} 16{,}000 + 855.28\left(\frac{42551.12+60610.8}{3600}\right) = \text{IDR. } 40{,}508.97$$

Example calculation total non value added cost on the process injection wax are as follows: $\sum_{i=1}^{n+1} 110 \times 103.16 = \text{IDR. } 11{,}347.60$

**Table 6**. Holding cost per inventory

| No. | Inventory | Holding Cost | No. | Inventory | Holding Cost |
|---|---|---|---|---|---|
| 1 | Raw material warehouse | IDR 103.16 | 9 | WIP turning | IDR 103.33 |
| 2 | WIP injection wax | IDR 103.16 | 10 | WIP radial drilling | IDR 103.33 |
| 3 | WIP dipping | IDR 103.16 | 11 | WIP welding | IDR 103.33 |
| 4 | WIP lost wax | IDR 103.16 | 12 | WIP lapping | IDR 103.33 |
| 5 | WIP sintering | IDR 103.16 | 13 | WIP assembling | IDR 103.37 |
| 6 | WIP melting | IDR 103.16 | 14 | WIP pressure testing | IDR 103.37 |
| 7 | WIP pouring | IDR 103.16 | 15 | WIP painting | IDR 103.37 |
| 8 | WIP blasting | IDR 103.16 | 16 | Finished good | IDR 103.37 |

By calculating using this formula as mentioned earlier, we obtain the total value added cost and the total non value added cost for entire process, can be seen respectively in Table 7 and Table 8.

**Table 7**. Value added cost

| No. | Process | Value Added Cost | |
|---|---|---|---|
| 1 | Injection Wax | IDR | 40,508.97 |
| 2 | Dipping | IDR | 4,240.06 |
| 3 | Lost Wax | IDR | 12,851.30 |
| 4 | Sintering | IDR | 83,739.28 |
| 5 | Melting | IDR | 123,587.05 |
| 6 | Pouring | IDR | 4,227.83 |
| 7 | Blasting | IDR | 20,468.28 |
| 8 | Turning | IDR | 79,157.52 |
| 9 | Drilling | IDR | 4,058.09 |
| 10 | Welding | IDR | 8,459.92 |
| 11 | Lapping | IDR | 5,755.31 |
| 12 | Assembling | IDR | 21,821.68 |
| 13 | Pressure testing | IDR | 22,091.83 |
| 14 | Painting | IDR | 6,310.44 |
| 15 | Final inspection | IDR | 93.89 |
| 16 | Shipping | IDR | 832.22 |
| | Total | IDR | 438,203.67 |

**Table 8**. Non-value added cost

| No. | Inventory | Non Value Added Cost | |
|---|---|---|---|
| 1 | Holding Cost Raw Material Warehouse | IDR | 11,347.60 |
| 2 | Holding Cost Inventory WIP Dipping | IDR | 2,884.28 |
| 3 | Holding Cost Inventory WIP Lost Wax | IDR | 1,957.59 |
| 4 | Holding Cost Inventory WIP Blasting | IDR | 2,273.26 |
| 5 | Holding Cost Inventory WIP Drilling | IDR | 1,549.95 |
| 6 | Holding Cost Inventory WIP Pressure Testing | IDR | 2,997.73 |
| 7 | Holding Cost Inventory Finished Good | IDR | 1,860.66 |
| | Total | IDR | 24,871.07 |

Ploting the current cost integrated value stream map can be done based on the calculation of the value added cost and non value added cost. As discussed earlier, a cost integrated value stream map is created by adding a cost analysis, which integrates the cost line into the VSM as illustrated in Figure 2. After being analyzed using the 5W & 1H methods, the future cost integrated value stream map is designed (see Figure 3). Some recommended implementation plans include; for example, shipping raw materials that were done once a week is changed to 2 weeks. Delivery of raw materials is intended to prevent excessive accumulation of inventory of raw material; combining several processes such as combining drilling and welding processes; apply the Kanban system between shipping and final inspection, between painting and pressure testing, and between turning and blasting, as well as between sintering and lost wax. After comparing between current and future, the results of improvement can be seen in Table 9.

**Table 9**. Comparison results between current state and future state

| Metric | Baseline before | Baseline after | Reduce |
|---|---|---|---|
| Total value stream inventory | 241 pcs/day | 55 pcs/day | 186 pcs/day |
| Total processing lead time | 12.05 days | 2.75 days | 9.3 days |
| Total processing time | 15814.62 seconds/pc | 15634.62 seconds/pc | 180 seconds |
| Total non value added cost | IDR 24,871.07/pc | IDR 5,673.8/pc | IDR 7,178.83/pc |
| Total value added cost | IDR 438.203,67/pc | IDR 431.024,84/pc | IDR 19,197.27/pc |
| Uptime | 81.50% | 81.50% | 0 (No different) |
| Travel distance | 260 meters | 190 meters | 70 meters |

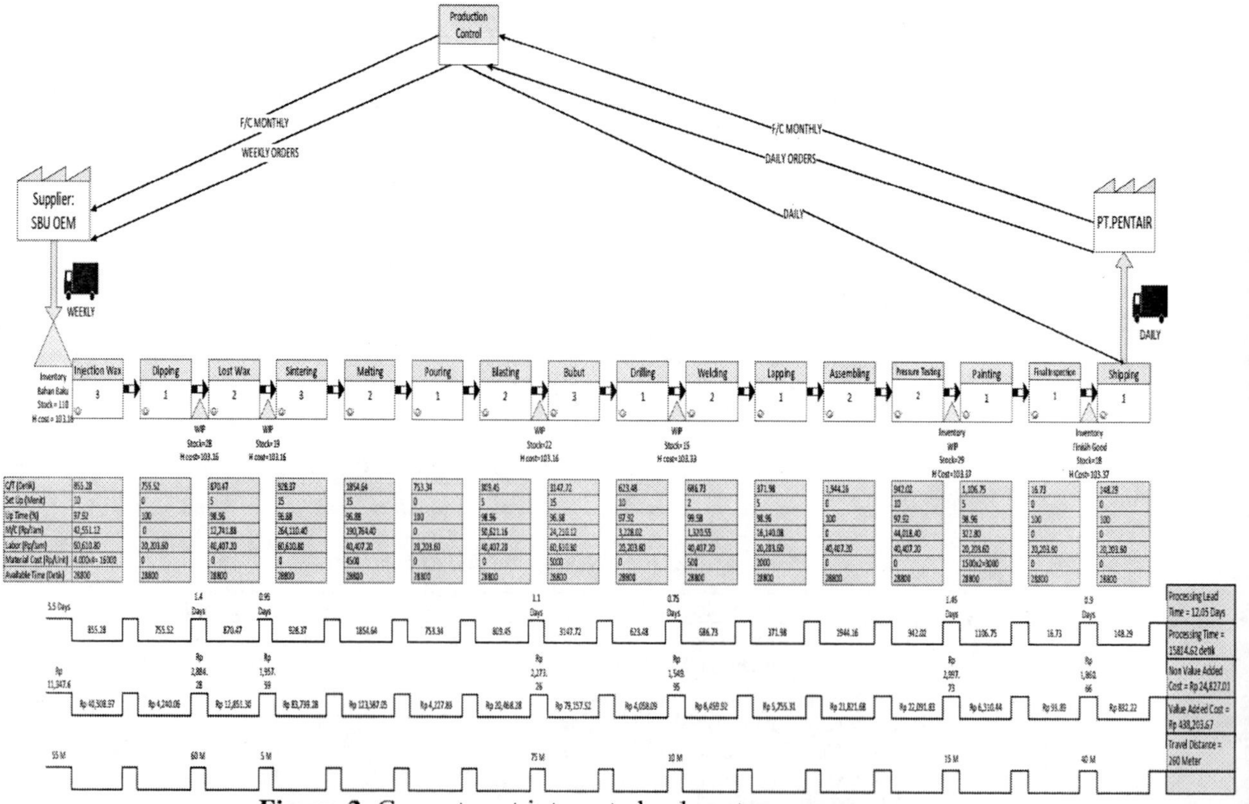


**Figure 2**. Current cost integrated value stream map

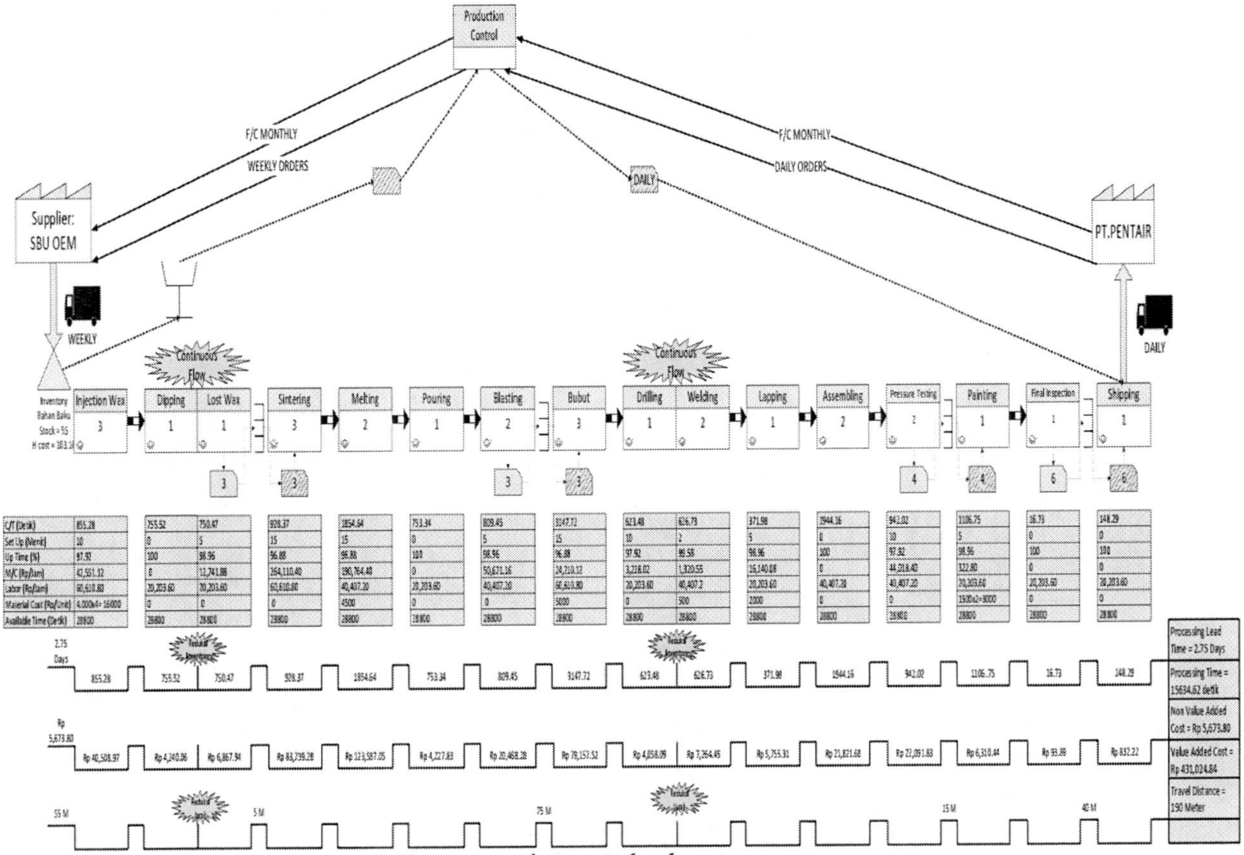


**Figure 3**. Future cost integrated value stream map

## 4. Conclusion

The cost approach integrated into the value stream map can help identify costs and activities that can be minimized or eliminated. This approach is considered more effective in helping management understand the activities or processes that trigger costs, as well as focus on continuous improvement. Further research is suggested to consider more accurate overhead costs

## 5. References

[1]. Kosasih, W and L Widodo 2018 Peningkatan Daya Saing Industri Kecil Menengah dengan Pendekatan Lean Manufacturing. *Jurnal Bakti Masyarakat Indonesia*, **1**(1), pp. 200-210.

[2]. Wilson, Lonnie 2010 *How to Implement Lean Manufacturing*. (USA: The McGraw-Hill Companies, Inc.)

[3]. Hines, P and D Taylor 2000 *Going Lean*. (United Kingdom: Lean Enterprise Research Center Cardiff).

[4]. Womack, J and D Jones 1996 *Lean Thinking*. (New York, NY: Simon and Schuster).

[5]. Rother, M, J Shook, J Womack, and D Jones 1999 *Learning to See: Value Stream Mapping to Add Value and Eliminate MUDA*. (Massachusetts, USA: The Lean Enterprise Institute).

[6]. Locher, Drew A 2008 *Value Stream Mapping for Lean Development*. (New York: Taylor & Francis Group).

[7]. Kosasih, W, et.al. 2019 Applying Value Stream Mapping Tools and Kanban System for Waste Identification and Reduction (Case Study: a Basic Chemical Company). *IOP Conference Series: Materials Science and Engineering* **528**, 012050, pp 1-8. doi:10.1088/1757-899X/528/1/012050

[8]. Firmansyah, F and Moses L Singgih 2015 Perancangan Lean Production System Dengan Pendekatan Cost Integrated Value Stream Mapping pada Divisi Kapal Niaga Studi Kasus PT PAL Indonesia. *Prosiding Seminar Nasional Manajemen Teknologi XXIII*, Institut Teknologi Sepuluh Nopember.
[9]. Abuthakeer, SS, PV Mohanram, and GM Kumar 2010 Activity Based Costing Value Stream Mapping. *International Journal of Lean Thinking*, **1**(2), pp. 51-64.
[10]. Garrison, R, E Noreen and P Brewer 2004 *Managerial Accounting*. 11th Edition. (Boston, MA: Mc-Graw Hill).
[11]. Samyrn, LM 2012 *Akuntansi Manajemen: Informasi Biaya untuk Mengendalikan Aktivitas Operasi & Investasi*. (Indonesia: Kencana Prenadamedia Group).

# Measurement of Feasibility and Risk Level on Modern Embroidery Kebaya Boutique Establishment in Jakarta

**R W Hanaa[1], and E Chumaidiyah[2*]**

Industrial Engineering, Telkom University, JL. Telekomunikasi 1 Bandung, Indonesia
[1]hanaarosyada@gmail.com, [2]endangcn@gmail.com
*Corresponding Author

**Abstract.** Business X is a creative business in Kebaya clothing, fabrics, and modern embroidered batik. The survey states that many medium business category turn into micro businesses due to the increasing number of imported textile products, resulting in a narrower market share. Under this condition, the company decided to create a Business X boutique considering that the products were not sold alone but at department stores that has a very high competition because of the large apparel industry continued to expand its business. In this paper, a feasibility analysis and the risk level measurement of the opening of this boutique are carried out. The market aspect begins with the distribution of questionnaires to obtain the demand and income projections. The technical aspects is for determining location using factor rating, layout and funding needs. Financial aspects estimate sales revenue and cashflow. The calculation results show NPV>0, PBP in 2,989 years, and IRR(40%)>MARR(10.99%), it can be concluded that the Business X boutique establishment is feasible. The risk is 21.7% and with a MARR (10.99%), the rate is 32.69%. Based on the rate value that is smaller than the IRR(40%) and NPV>0, then by observing the level of risk, this business is feasible.
**Keywords**: Feasibility Analysis, Risk Level, NPV, IRR, PBP

## 1. INTRODUCTION

Indonesia is a country with internationally recognized cultural diversity, and batik craft is one of it. On October 2, 2009, batik was designated by UNESCO as a humanitarian heritage for verbal and non-material culture, so the government designated October 2 as national batik day. With the establishment of national batik day, batik has become increasingly desirable to both local and international communities. Locals make batik an outfit that must be used in institutions or offices, schools, to attend important events such as weddings, graduations, and so on. In this regard, the need for batik cloth has increased, resulting in an increase of batik production in Indonesia each year. Table 1 shows the development of batik business units in Indonesia from 2011 to 2015.

**Table 1.** Development of Batik Business Unit in Indonesia in 2011-2015

| Year | Business Unit | Year | Business Unit |
|---|---|---|---|
| 2011 | 41.623 unit | 2014 | 46.365 unit |
| 2012 | 43.704 unit | 2015 | 47.755 unit |
| 2013 | 45.015 unit | | |

(Source: detikfinance, 2015)

Based on the Table 1, the development of batik business units in Indonesia increased by 14.7% from 2011 to 2015. This shows that batik is in great demand and needed by the local and international community. In addition, batik production has good business opportunities seen from the value of batik on Table 2 which continues to increase from year to year.

**Table 2.** Batik Value in 2011-2015

| Year | Business Unit |
|---|---|
| 2011 | Rp 1,909 trillion |
| 2012 | Rp 2,005 trillion |
| 2013 | Rp 2,065 trillion |
| 2014 | Rp 2,127 trillion |
| 2015 | Rp 2,191 trillion |

(Source: detikfinance, 2015)

The value of batik increased by 14.7% from 2011 to 2015. Besides the value of batik that continues to increase, the batik business opportunity can also be seen from the export value from year to year because people who have an interest in batik are not only from the local community but also from the international community. Table 3 shows the export value of batik increased by 14.7% from 2011 to 2015 shows.

**Table 3.** Batik Export Value in 2011-2015

| Year | Export Value |
|---|---|
| 2011 | Rp 43,961 trillion |
| 2012 | Rp 46,159 trillion |
| 2013 | Rp 47,543 trillion |
| 2014 | Rp 48,970 trillion |
| 2015 | Rp 50,439 trillion |

(Source: detikfinance, 2015)

Based on the data, with the increasing production of batik which shows the increasing number of competitors, it can be used as an opportunity to open a creative business based on batik clothing such as dresses, kebaya, and others. In addition, there is a competing threat to the clothing business, namely imported high-quality and low-priced products so that the textile business needs to have a strategy to improve the quality of the cloth and run an effective production to reduce production costs[1]. Running a business in the creative industry is currently preferred by entrepreneur in Indonesia. Creative products can attract more consumers because the uniqueness of these products makes it different from the products offered in the market in general.

Starting with a focus on attracting consumers around the Business X outlets, currently Business X consumers increasing along with marketing activities that has been carried out through exhibitions, fashion shows, and used as a wardrobe for news presenter on television. In addition, the technology in this digital age has provided new facilities and new ways for people to carry out various activities, from communicating, trading, creating products, looking for information, to find entertainment and so on[2]. Based on this situation, sales and demand for Business X products continues to increase every time. Business X products can now be obtained at one of the department stores in Jakarta.

According to Badan Pusat Statistik DKI Jakarta in 2015, from the various types of industries surveyed, the apparel industry had the largest quantity with a total of 281 companies or 21.24 percent of the total number of large and medium manufacturing industry companies. However, many small industries included in the medium category turned into micro businesses such as the small apparel industry in Indonesia last year with total 46,601 due to the increasing number of imported textile products, resulting in a narrower market share [3]. On the other hand, many large textile industries are expanding continuously.

Under these conditions, the company decided to create a Business X boutique considering that the products that produced so far were not sold in the Business X boutique but in department stores where the place has very high competition because the large apparel industry continues to expand its business. Therefore, it is necessary to conduct a feasibility analysis of the establishment of this boutique so decisions can be made whether the establishment of this boutique is feasible from the market aspects, technical aspects, and financial aspects or not.

## 2. METHODS

Figure 1 shows the concept of the feasibility analysis that conducted in this study. The feasibility analysis starts from the market aspect, continues with the technical aspects and the results of these aspects will be processed in the financial aspect and sensitivity analysis [4]. In this paper, we will discuss the risk level that will be measured to determine the feasibility of this business.

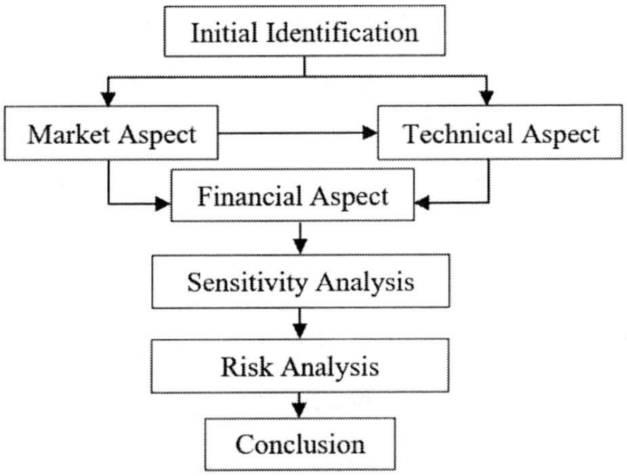


**Figure 1.** Conceptual Model

In initial identification stage, a goal is determined and the problem that will be examined in this paper is identified. Market aspects is an initial step to find out market opportunities related to consumer behaviour, prices, and market competition [5]. At this stage, an analysis of available markets, potential markets and target markets was carried out through questionnaires from the population of women aged 15 to 54 years from DKI Jakarta citizens with the sample determination using convenience sampling. The output of this analysis are demand and estimated income that will be used in data processing and financial aspects analysis.

Technical aspect consisting analysis of location, layout and the funding requirement. Determination of location using the factor rating method by assigning weights to factors in accordance with the order of importance of these factors in determining the location. The value of each factor on each alternative is multiplied by its weight and normalized so that the selected location is obtained. This aspect ends with the layout and investment fund identification.

Data processed on the financial aspect obtained from the market and technical aspects. The parameters that will be used to measure whether a business is feasible to run or not are NPV, IRR, and Payback Period. NPV is obtained by calculating the difference or change in value between the cash flow of the current investment and the cash flow of net income (cash flow) in the future [4]. IRR is a method of calculating interest rates that equates the present investment value with the value of future net cash receipts. Business is considered feasible if the IRR > MARR. IRR calculation is as follows.

$$IRR = i_1 \frac{NPV_1}{(NPV_1 - NPV_2)} (i_2 - i_1) \quad (1)$$

Pay Back Period (PBP) is a certain period of time that shows the cumulative cash flow flow equal to the amount of investment in the form of present value [6]. PBP calculation is as follows.

$$PBP = \frac{Investment\ Value\ (outlay)}{Net\ Cash\ Inflow\ (proceed)} x\ 1\ year \quad (2)$$

After the financial analysis, a sensitivity analysis is carried out to determine the business conditions if there is some undesirable changes occur. The analysis was carried out on NPV, IRR, and Payback Period which were obtained based on the changing condition. Furthermore, the risk level of the company is calculated based on the risks that can arise in this business so the feasibility of this business can be identified.

Research on the feasibility analysis of various types of businesses has been done before. One of the feasibility analysis studies is on bread business shows a positive NPV and IRR 67% so the business is feasible [7]. Research on risk measurement that has been done previously is on the barcoding project shows that the total risk is calculated by the value of each risk multiplied by its weight [8]. Other research shows that the feasibility and risk analysis of opening a branch of Gandul Rice Restaurant is feasible with the results showing the IRR 27.83%, the risk 6.26% and with MARR 14% the rate become 20.26%[9].

## 3. RESULT AND DISCUSSION

### *3.1 Demand Estimation*

Based on the results of the questionnaire, it shows potential market is 62% or 2,035,548 consumers, available markets is 54% of potential markets as many as 1,009,195 consumers, and the target market is 0.2% of available markets as many as 2,220 consumers. Demand projections for the next five years are obtained by multiplying this year's demand with an increase in the population of DKI Jakarta according to the BPS which is 1.02%. Demand projection from Enterprise X in 5 years is shown in Table 4.

**Table 4.** Demand Pojection

| Year | Demand |
|---|---|
| 2020 | 2220 |
| 2021 | 2243 |
| 2022 | 2266 |
| 2023 | 2289 |
| 2024 | 2312 |

Demand projection within the next 5 years for each product is obtained based on the sales percentage of Business X products within last four months. Based on these calculations, the demand projection for each product is obtained for the next five years in Table 5.

**Table 5.** Demand Projection for Each Products

| Product | Persentage | 2020 | 2021 | 2022 | 2023 | 2024 |
|---|---|---|---|---|---|---|
| Kebaya | 49% | 1079 | 1090 | 1101 | 1112 | 1124 |
| Dress Blouse | 19% | 415 | 419 | 424 | 428 | 432 |
| Fabric | 7% | 156 | 157 | 159 | 160 | 162 |
| Outer | 13% | 291 | 293 | 296 | 300 | 303 |
| Batik Skirt | 12% | 280 | 283 | 286 | 289 | 292 |
| Total | 100% | 2220 | 2243 | 2266 | 2289 | 2312 |

### *3.2 Determining Location and Layout*

Table 6 shows the result of the factor rating calculation to determine the location from three alternatives for opening the Business X boutique. The chosen location is on Jl. Prof. Dr. Soepomo with score 94.58.

**Table 6.** Factor Rating Result

| Factors | Wt | Jl. Panglima Polim | | Jl. Kemang Timur | | Jl. Prof. Dr. Soepomo | |
|---|---|---|---|---|---|---|---|
| | | Normalization | Wtd | Normalization | Wtd | Normalization | Wtd |
| Market Distance | 0.4 | 50.00 | 20.00 | 100.00 | 40.00 | 100.00 | 40.00 |
| Supplier Distance | 0.2 | 45.26 | 9.05 | 52.44 | 10.49 | 100.00 | 20.00 |
| Competitor Distance | 0.1 | 25.00 | 2.50 | 100.00 | 10.00 | 62.50 | 6.25 |
| Rent Price | 0.3 | 100 | 30.00 | 89.47 | 26.84 | 94.44 | 28.33 |
| Total | 1.00 | | 61.55 | | 87.33 | | 94.58 |

The building size in the chosen location is 5x10 meters. This boutique contained display room, warehouse, fitting room, and toilet. Based on the size and needs of the business, the layout is obtained as in Figure 2. One of the business processes that occur in this boutique as in Figure 3 is the process of serving customers. This process explains about the employees serving customers in buying the product.

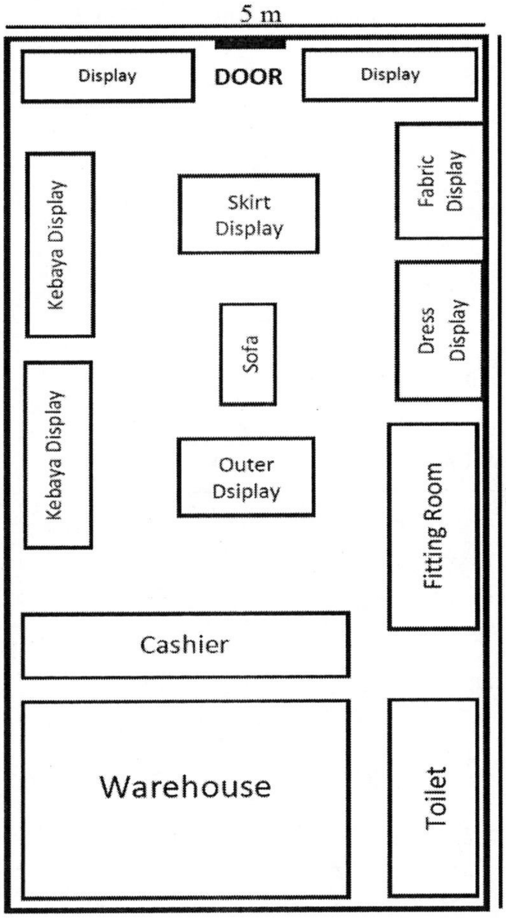


**Figure 2.** Layout

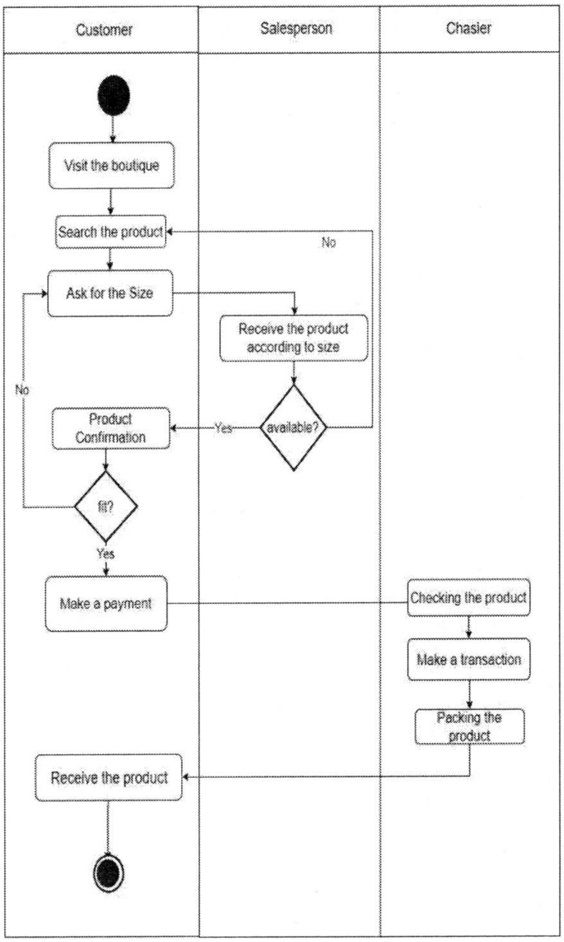


**Figure 3.** Serving Customers Process

IOP Conf. Series: Materials Science and Engineering **847** (2020) 012062 doi:10.1088/1757-899X/847/1/012062

### *3.3 Funding Requirement*
Funding requirements obtained based on the costs that has been calculated from this technical aspect are shown in Table 7.

**Table 7.** Funding Requirement

| Funding Requirement | | Total |
|---|---|---|
| Fixed Investment | Rp | 54,255,000 |
| Working capital | Rp | 189,268,686 |
| TOTAL | Rp | 243,523,686 |

Table 7 shows that fixed investment is Rp 54,255,000.00 and working capital for the first three months is 189,268,686.00. Investment was obtained from the cost of all equipment needed at the opening of this boutique. Working capital is obtained from total direct costs and indirect costs without depreciation for three months. With these fixed investment costs and working capital, a total funding requirement is Rp 243,523,686.00.

### *3.4 Income Estimation*
The income from this business is obtained from the selling price of the product multiplied by the demand of each product. The selling price will increase every year according to the credit value which is 3.01%. Based on the selling price, total revenue for the next five years is obtained as follows in Table 8.

**Table 8.** Income Estimation

| Year | Total Income |
|---|---|
| 2020 | Rp 866.378.676,00 |
| 2021 | Rp 902.609.991,00 |
| 2022 | Rp 940.356.473,00 |
| 2023 | Rp 979.681.485,00 |
| 2024 | Rp 1.020.651.039,00 |

### *3.5 Cashflow*
Table 9 shows the cash flow used to calculate NPV, IRR, and PBP to determine the feasibility of this business.

**Table 9.** Cashflow

| YEAR | 2019 | 2020 | 2021 | 2022 | 2023 | 2024 |
|---|---|---|---|---|---|---|
| Period | 0 | 1 | 2 | 3 | 4 | 5 |
| Cash Out (Rp) | 243,523,686 | | | | | |
| Cash In (Rp) | | 94,535,991 | 100,484,059 | 106,190,398 | 111,521,847 | 116,538,798 |
| Working Capital (Rp) | | | | | | 189,268,686 |
| Net Cash (Rp) | (243,523,686) | 94,535,991 | 100,484,059 | 106,190,398 | 111,521,847 | 305,807,484 |
| p/f factor (10.99%) | 1.0000 | 0.9010 | 0.8118 | 0.7314 | 0.6590 | 0.5937 |
| NPV (Rp) | (243,523,686) | 85,175,233 | 81,569,813 | 77,666,493 | 73,489,374 | 181,563,628 |
| NPV Cumulative (Rp) | (243,523,686) | (158,348,453) | (76,778,640) | Rp 887,853 | 74,377,226 | 255,940,855 |

### *3.6 Net Present Value*
Net Present Value is obtained from the initial investment of cash out at the beginning of year 0 minus cumulative income for the next five years. Based on the results, the cumulative NPV in the end of period is more than 0 (Rp 255,940,855.00) so it can be said that this business is feasible to run.

### *3.7 Internal Rate of Return*
Based on the calculation of cash flow, the IRR for five years is 40%, which means this business provides a profit of 40% for each year. In this study the MARR (Minimum Attractive Rate of Return) value used is 10.99%. The value of IRR obtained in this business is more than the MARR value so it can be said that this business is feasible.

### *3.8 Payback Period*
The payback period calculation is done to find out how long the business or project can return the investment. Payback period is calculated based on the cash flow that has been obtained previously. Based on calculations, the payback period of this business is 2,989 years which means this business takes 2,989 years to return the capital that was spent at the beginning.

### *3.9 Sensitivity Analysis*
This sensitivity analysis is used to measure the level of a condition or change that make this business unfeasible. Changes that may occur in this business are the increase in macloons costs, increase in labour costs, decrease in demand, and decrease in selling prices.

**Table 10.** Sensitivity

| | Percentage | NPV | IRR |
|---|---|---|---|
| Increase In Macloons | 24% | Rp 1,592,590 | 11.17% |
| | 25% | Rp (6,453,946) | 10.24% |
| | | 24.20% | |
| Increase In Labour Cost | 30% | Rp 2,353,735 | 11.27% |
| | 31% | Rp (4,058,123) | 10.51% |
| | | 30.37% | |
| Decrease In Demand | 19% | Rp 110,759,430 | 24.02% |
| | 20% | Rp (69,235,812) | 2.67% |
| | | 19.62% | |
| Decrease In Selling Price | 8% | Rp 1,362,326 | 11.15% |
| | 9% | Rp (22,806,066) | 8.30% |
| | | 8.06% | |

Based on Table 10, the increased of macloons cost sensitive by 24.20%, labor costs sensitive to an increase of 30.37%, demand is sensitive to a decrease of 19.62%, and the selling price was sensitive to a decrease of 8.06%.

### *3.10 Risk Analysis*
Risks in this business that can arise are the risk of decrease in demand, the risk of decrease in selling price, and the risk of equipment damage.

**Table 11.** Risk Measurement

| No | Risk | Likelihood (%) | Impact (Million Rp) | Risk (Rp) |
|---|---|---|---|---|
| 1 | Decrease in demand | 17 | 36 | 6120000 |
| 3 | Decrease in selling price | 4 | 9.5 | 380000 |
| 4 | Equipment Damage | 0.7 | 0.5 | 3500 |
| | Total | 21.7 | | |

Based on Table 11, the risk obtained from this business is 21.7%. Investment is considered feasible if the rate that has been added with MARR (10.99%) is smaller than IRR (40%). Based on calculations, the rate (32.69%) is smaller than IRR (40%) and the NPV is Rp 40,569,061.00. Based on the value of the rate (32.69%) which is smaller than the IRR (40%) and the NPV value> 0 then the business is feasible.

The NPV, IRR, and the risk rate result indicates that this business is feasible, then it is suggested that Business X can make this calculation as a consideration for opening the boutique.

## 4. CONCLUSION

Based on survey, historical sales data and competitors, the company is aiming for 0.2% of the available market so the target market is 2,220 customers in the first year. Demand continues to increase every year until the projected demand in 2024 is 2,312.

Fixed investment from equipment costs needed is Rp 54,255,000.00 while working capital is Rp 189,268,686.00 so the total funding requirement is Rp 243,523,686.00.

The results shows the NPV is Rp 255,940,855.00, payback period in 2,989 years and IRR of 40%. With the NPV value> 0, the payback period is less than the investment time, and the IRR is more than the MARR (10.99%), it can be said that the opening of the Business X boutique is feasible.

Based on the calculation, the cost of macloons sensitive at an increase of 24.20%, labour costs sensitive at an increase of 30.37%, demand sensitive at a decrease of 19.62%, and the selling price sensitive at a decrease of 8.06%.

Based on the calculation, the risk is equal to 21.7% and with a MARR value of 10.99%, the rate become 32.69% and the NPV value is Rp 40,569,061.00. Based on the smaller rate compared to IRR (40%) and NPV value> 0, this business is said to be feasible.

Suggestion for further research is to analyze the management and risk control that arise in the business.

## 5. REFERENCE

[1] Chumaidiyah E, Aurachman R and Sagita B H 2018 Strategy for Capability Development of Knitting Small Medium Enterprises Using SWOT Analysis. *International Conference on Industrial Engineering and Operations Management,* pp 1062-1069

[2] Chumaidiyah E, Aurachman R and Sagita B H 2019 Exploring the Internal and External Constraint of IT Business Startup in Bandung, Indonesia. *International Journal of Innovative Technology and Exploring Engineering,* pp 126-131

[3] Bisnis Indonesia 2016 *Industri Menengah Beralih Menjadi Mikro.* (Jakarta: Kementrian Perindustrian Republik Indonesia).

[4] Zativita F and Chumaidiyah E 2019 Feasibility analysis of Rumah Tempe Zanada Establishment in Bandung Using Net Present Value, Internal Rate of Return, and Payback Period. *IOP Conference Series: Materials Science and Engineering.* **505**, 012007. 10.1088/1757-899X/505/1/012007.

[5] Zuraida A, Chumaidiyah E and Tripiawan W 2019 Feasibility Analysis on the Development of Steel Sheet Zinc Plated and Galvalum Production Factory PT. S Steel. *IOP Conference Series: Materials Science and Engineering.* **528**. 012031. 10.1088/1757-899X/528/1/012031.

[6] Ibrahim H Y 2009 Studi Kelayakan Bisnis Edisi Revisi. Jakarta: Rineka Cipta.

[7] Purnamasari D and Hendrawan B 2013 Analisis Kelayakan Bisnis Usaha Roti Ceriwis sebagai Oleh-Oleh Khas Kota Batam. *Jurnal Akutansi, Ekonomi, dan Manajemen Bisnis,* **1**(1), pp. 83-87.

[8] Kusumawardhani A, Chumaidiyah E and Zulbetti R 2019 Operational Risk Analysis for Barcoding Project in PT. XYZ With Enterprise Risk Management Method. 10.2991/icoiese-18.2019.52.

[9] Ratnasari D and Chumaidiyah E 2014 Analisis Kelayakan Pembukaan Cabang Rumah Makan Nasi Gandul Pak Memet di Kota Semarang. *Jurnal Rekayasa Sistem & Industri,* **1**(2), pp. 55-61.

IOP Conf. Series: Materials Science and Engineering **847** (2020) 012063 doi:10.1088/1757-899X/847/1/012063

# The Effects of Sleep Quality on Vigilance and Driving Performance in a Train Simulator

**D Siswanto[1,2], H Iridiastadi[1,3] and K Muslim[1,4]**

[1]Institut Teknologi Bandung, Jl. Tamansari 64, Bandung, Indonesia
[2]hainiel@gmail.com, [3]hiridias@vt.edu, [4]kmuslim@ti.itb.ac.id

**Abstract.** Train driving activity requires high vigilance to prevent human errors that can lead to accidents. One factor mentioned to reduce vigilance is poor sleep quality. But so far, not many studies have proven the effect of sleep quality on vigilance. Previous researches did not clearly conclude this effect. In addition to being vigilant, sleep quality is also thought to effect driving performance because of the close relationship between vigilance and driving performance. Based on those backgrounds, this study aimed to prove the effects of sleep quality on vigilance and driving performance through experiments in a train simulator under monotonous condition, which in some studies is also proved to reduce vigilance. This study involved eight male subjects who experienced two good and poor sleep quality treatments. Through vigilance measurements with Psychomotor Vigilance Task (PVT) and electroencephalograph (EEG) using Multivariate Analysis of Variance (MANOVA) with significant value 0.05, it showed that there were effects of sleep quality on mean reaction times, minor lapses, mean relative band powers for theta, and speeding simultaneously (p-value = 0.015). The results of paired sample t-tests also showed the effects of monotonous driving conditions that affect mean reaction times (p-value = 0.01 on good sleep quality; p-value = 0.000 on poor sleep quality) and minor lapse (p-value = 0.000 on good sleep quality, p-value = 0.045 on poor sleep quality) at the beginning and the end of the experiment. From this study, it can be concluded that the quality of sleep affects vigilance and driving performance, so that poor sleep quality decreases vigilance.
**Keywords**: vigilance, sleep quality, monotonous, Psychomotor Vigilance Task, electroencephalograph

## 1. INTRODUCTION

Vigilance becomes an important factor in train driving that tends to be a repetitive task, tedious task with high risk but low level of control [1, 2]. It is described as a cognitive process related to a person's ability to maintain sustained attention for capturing signals over a prolonged period [3]. Vigilance is a combination of alertness and attention [4]. Vigilance decrement causes a slowdown in cognitive function and increases reaction time which can cause human errors while working [5]. Because train driving is considered to be a high-risk job [2], errors made while driving can lead to accidents. Train accidents have a more serious impact and incur very high costs even though their frequency is lower than road traffic accidents [6, 7]. Investigations on 23 train accidents in Australia found that 65% of accidents were caused by a slip in attention [8].

Driving is basically a vigilance task that requires psychomotor reactions and is very sensitive to fatigue, so vigilance decrement is often used as an indicator to detect fatigue [9]. Vigilance decrement is also expressed as a result of the influence of sleepiness [10]. Because driving is a picture of changes in vigilance, individuals who perform poorly on the vigilance test will also perform poorly when driving, especially in monotonous conditions [4]. Decreased alertness due to fatigue illustrates the reduced ability of individuals to drive because they are no longer able to maintain the sustained attention needed when driving [11, 12]. Low cognitive workload when driving a train and repetitive tasks lead to work monotony that can trigger a decrease in train driving awareness [13].

Sleep-related factors are often called as causes of fatigue which will reduce vigilance and work performance [14]. Prolonged wakefulness or sleep deprivation causes slowing of responses (errors of omission) and errors of response when there is no stimulus (errors of commission) [15]. Decreasing number of lapses (reaction time greater than 500 milliseconds) is also stated to be sensitive to the effect of sleep deprivation in healthy subjects [5]. Sleep deprivation is also stated to worsen cognitive performance such as vigilance, working memory, long-term memory and decision making [16]. In individuals who experience chronic sleep deprivation (sleep duration of 3 hours per night for 7 consecutive nights) there is also a decrease in mean reaction times and increased lapses [17].

Aside from sleep deprivation, poor sleep quality also affects performance when driving a train [18]. Sleep disorders or poor sleep quality is reported to be a cause of vigilance (sustained attention) decrement in 1266 train drivers in Belgium [2]. Investigations conducted on 30 train drivers at PT. Indonesian Railroad (KAI) DAOP II Bandung also showed 73.33% of drivers experienced poor sleep quality which made them sleepy while on duty [19]. A study on hospital nurses reported problems of reduced attention and increased errors when sleep quality deteriorated [20]. Poor sleep quality has also been associated with an increase in human errors committed by train drivers through a study of 4634 train drivers in South Korea [21]. Accumulation of exposure to the work environment with high demand but low levels of control (high strain) was associated with poor sleep quality and increased fatigue [1]. The characteristics of these high strains were found in train driving that has high responsibility and demands to be vigilant even in monotonous conditions or low level of control [2, 13]. So far, not many studies have shown an association between vigilance and sleep quality. Researches that have been done related to the association between vigilance and sleep quality have not given any conclusive results. One study proved that poor sleep quality decreases vigilance [22] while another study stated that vigilance alteration was not sensitive to sleep quality [23]. However, the results of the study [2] hinted at the importance of conducting studies on the effect of sleep quality on vigilance because it was reported to be the cause of decreased vigilance for most train drivers.

Poor driving performance due to fatigue closely related to accident risk [14]. This decrease in performance is associated with a decrease in vigilance as a sign of increased fatigue [27]. A decline in performance can also be seen from the failure or slowdown in responding to stimulus [28]. Furthermore, drivers who subjectively assess their quality of sleep as poorly, have a higher risk of an accident while driving [14].

The purpose of this study was to investigate the effects of sleep quality on vigilance and performance while driving a train simulator. The use of simulators is based on safety, cost, and convenience reasons [24, 25] and the ability to measure performance accurately [26]. From the results of this study, it was expected that the effects of sleep quality on vigilance and the relationship between vigilance alteration and train driving performance can be concluded.

## 2. METHODS

This study involved 8 subjects (male, 21-25 years), each undergoing two experimental treatments, namely good and poor sleep quality. The indicator used to assess sleep quality was the value of sleep efficiency. Sleep efficiency values greater than 85% ($SE \geq 85\%$) were categorized as good sleep quality, while sleep efficiency values less than 85% ($SE < 85\%$) were categorized as poor sleep quality [29]. Sleep efficiency was calculated as a comparison between total sleep time (TST) and the duration of the sleep episode (DSE) [30]. TST shows the length of time someone slept, while DSE shows the duration from the attempt to sleep (sleep onset latency) to the last wake without trying to fall asleep again (final awakening). In this study, DSE lasted 8 hours and TST measurements were performed with Fitbit Charge 2.

Psychomotor Vigilance Task (PVT) was used to measure alertness in this study. Vigilance test was performed with PC-PVT 2.0 with 5 minutes' duration [31]. PC-PVT was a development of PVT-192 which has become the gold standard. The advantages of PC-PVT were that it is easy to use, testing and analysis integrated in one system so that it minimizes the possibility of loss of data due to the process of transferring data from the device to the computer, the test results can be directly seen after testing, and simple data storage [31]. Parameters of PVT used as indicators of vigilance were mean reaction times and minor lapses (number of responses with reaction time greater than 500 milliseconds). PVT was carried out at the beginning and the end of the experiment with a train simulator. While driving a train simulator, vigilance is measured by electroencephalograph (EEG). The use of EEG to measure vigilance by reason of changes in brain waves can be an indicator of changes in vigilance [32]. The measured brain waves were theta (4-7 Hz) which is a sensitive measure to represent vigilance alteration because when power theta rises, vigilance decreases [33]. The EEG used in this study was Muse EEG 2, produced by InteraXon Inc., Toronto, Canada, and sold commercially (https://choosemuse.com).

The train simulator that was used was RailDriver Desktop Train Cab Controller (model RD-91-MDT-R) which was designed by PI Engineering USA with dimensions of 34 cm x 18 cm x 10 cm and weighs 2.5 kg. Each subject runs the train simulator for 120 minutes in a monotonous condition where the train did not stop from the first station to the end. Driving performance measured by the amount of speeding generated from the Train Simulator 2017. Speeding is the condition when the train's speed exceeds the allowable limit.

## 3. RESULTS AND DISCUSSIONS

Each subject experienced treatments of good and poor sleep quality. Good sleep quality is sleep continuity or uninterrupted sleep [34]. From this definition, for poor sleep quality, subjects were disturbed by waking up the subject every 2 hours. Sleep efficiency was calculated as a comparison between total sleep time (TST) and duration of the sleep episode (DSE) [30]. Sleep efficiency greater than 85% (SE ≥ 85%) categorized as good sleep quality and sleep efficiency less than 85% (SE < 85%) categorized as poor sleep quality [29]. The results of sleep efficiency for all subjects showed an average sleep efficiency of 91.82 ± 3.32% for good sleep quality and 74.74 ± 3.68% for poor sleep quality.

Vigilance measurement of each subject was carried out with PC-PVT 2.0 with a duration of 5 minutes at the beginning and at the end of the experiment. The 5-minute duration was chosen because it was not too short and has been validated to replace the PVT test duration for 10 minutes which is considered to be too long [35]. The results of the vigilance measurement with PC-PVT 2.0 can be seen in Table 1. Table 1 shows that mean reaction times and minor lapses were worse in poor sleep quality. Measurements at the end of the experiment (finish) also show a decrease in alertness that can be seen from the increase in mean reaction times and minor lapses. This indicates the hypothesis that there was a monotonous effect on alertness.

**Table 1.** Vigilance measurement results with PC-PVT 2.0

| Sleep Quality | Period | Reaction Times (min) | | Minor lapses | |
|---|---|---|---|---|---|
| | | *Mean* | *SD* | *Mean* | *SD* |
| Good | Start | 296.83 | 35.24 | 2.00 | 0.76 |
| | Finish | 355.25 | 27.20 | 3.88 | 1.13 |
| Poor | Start | 366.73 | 74.26 | 2.88 | 1.73 |
| | Finish | 389.27 | 57.09 | 5.75 | 1.83 |

Vigilance measurements were also carried out by the Muse EEG while subjects run the simulator. The Muse EEG produces absolute band powers data for theta (θ) which is then converted to relative band powers for theta [36]. Measurement of vigilance with the Muse EEG only done in the frontal lobe because vigilance was associated with the work of the frontal lobe [33]. Driving performance in a train simulator was measured by the amount of speeding, which is the violation of the speed limit. In the condition of good sleep quality, the average number of speeding was 12.75 ± 3.41. In poor sleep quality, the average number of speeding has almost doubled (23.88 ± 3.83). This indicates the effects of sleep quality on driving performance.

The effects of sleep quality on vigilance and driving performance were tested by Multivariate Analysis of Variance (MANOVA). The dependent variables involved were mean reaction times and minor lapses of the PC-PVT, mean relative band powers for theta, and the average number of speeding on the train simulator. The MANOVA test results can be seen in Table 2. The method used for the MANOVA test was Wilk's Lambda with a significance level (α) of 0.05. Testing was done with SPSS software. Based on the MANOVA test results, it was concluded that sleep quality simultaneously affected mean reaction times, minor lapses, the average of relative band powers for theta, and the average amount of speeding. The results of the EEG measurements while driving (Fig. 1) also showed the relative band powers for theta that were higher in poor sleep quality than in good sleep quality. This result was consistent with previous research which stated that when vigilance decreases, theta power increases [33].

While running the train simulator, subjects got monotonous driving conditions. To analyze the effects of monotonous driving conditions on vigilance, a paired sample t-test was performed on the results of the PC-PVT 2.0 test for the beginning and end of the experiment. Paired sample t-test results for mean reaction times and minor lapses with a significance level (α) of 0.05 can be seen in Table 3 through 6.

Paired sample t-test results showed that there were differences for mean reaction times and minor lapses at the beginning and the end of the driving experiment with the train simulator under monotonous conditions in all conditions of sleep quality. This showed the effects of monotonous conditions that cause vigilance decrement while driving. As can be seen from Table 1, there was vigilance decrement as indicated by increasing mean reaction times and minor lapses both in conditions of good and bad sleep quality.

**Table 2**. MANOVA test result

| **Multivariate**[a,b] | | | | | | |
|---|---|---|---|---|---|---|
| ***Within Subjects Effect*** | | ***Value*** | ***F*** | ***Hypothesis df*** | ***Error df*** | ***Sig.*** |
| Sleep Quality | Wilk's Lambda | 0.028 | 21.088[c] | 5.000 | 3.000 | 0.015 |

a. Design: Intercept
Within Subjects Design: Sleep Quality
b. Tests are based on average variables.
c. Exact statistic

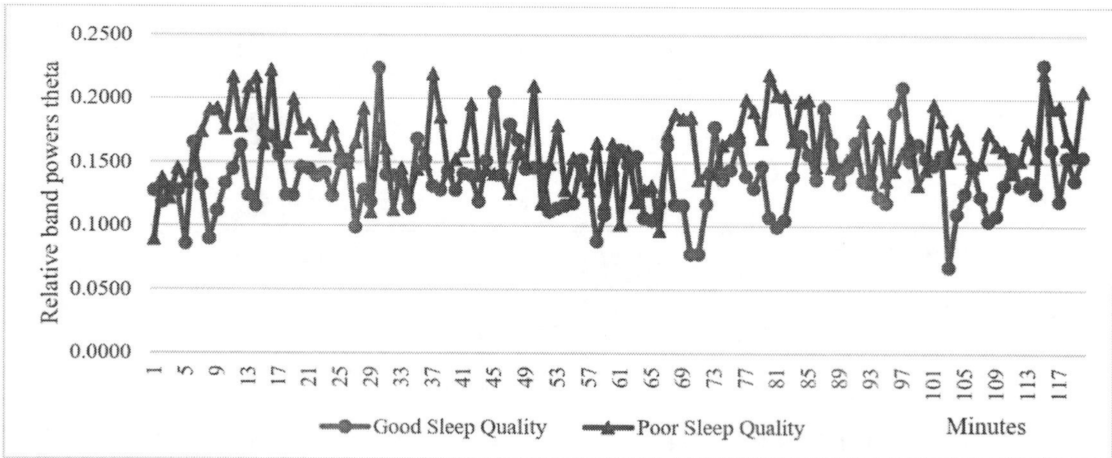


**Figure 1**. Relative band powers for theta while driving the train simulator

**Table 3**. Paired sample t-test for mean reaction times on good sleep quality

| **Paired Samples Test (Mean reaction times)** Paired Differences | | | | | | | | |
|---|---|---|---|---|---|---|---|---|
| | | | | 95% Confidence Interval of the Difference | | | | |
| | Mean | Std. Deviation | Std. Error Mean | Lower | Upper | t | df | Sig. (2-tailed) |
| Start-Finish (Good) | -33.157 | 26.927 | 9.520 | -55.668 | -10.646 | -3.483 | 7 | 0.010 |

**Table 4**. Paired sample t-test for mean reaction times on poor sleep quality

| **Paired Samples Test (Mean reaction times)** Paired Differences | | | | | | | | |
|---|---|---|---|---|---|---|---|---|
| | | | | 95% Confidence Interval of the Difference | | | | |
| | Mean | Std. Deviation | Std. Error Mean | Lower | Upper | t | df | Sig. (2-tailed) |
| Start-Finish (Poor) | -58.417 | 20.127 | 7.116 | -75.244 | -41.590 | -8.209 | 7 | 0.000 |

**Table 5**. Paired sample t-test for minor lapses on good sleep quality

| **Paired Samples Test (Minor lapses)** Paired Differences | | | | | | | | |
|---|---|---|---|---|---|---|---|---|
| | | | | 95% Confidence Interval of the Difference | | | | |
| | Mean | Std. Deviation | Std. Error Mean | Lower | Upper | t | df | Sig. (2-tailed) |
| Start-Finish (Good) | -3.375 | 1.408 | 0.498 | -4.552 | -2.198 | -6.780 | 7 | 0.000 |

**Table 6.** Paired sample t-test for minor lapses on poor sleep quality

| **Paired Samples Test (Minor lapses)** Paired Differences | | | | | | | | |
|---|---|---|---|---|---|---|---|---|
| | | | | 95% Confidence Interval of the Difference | | | | |
| | Mean | Std. Deviation | Std. Error Mean | Lower | Upper | t | df | Sig. (2-tailed) |
| Start-Finish (Poor) | -4.250 | 5.203 | 1.840 | -8.600 | 0.099 | -2.310 | 7 | 0.045 |

## 4. CONCLUSIONS

The results of the study concluded the effects of sleep quality on vigilance, as measured by PC-PVT 2.0 (mean reaction times and minor lapses) and EEG (relative band powers theta). Poor sleep quality worsens the vigilance needed when driving a train. Sleep quality also affects the driving performance (speeding) in the simulator. Monotonous driving condition also proved its effects shown by decreasing vigilance while driving. This is evidenced by a significant increase in mean reaction times and minor lapses from the beginning to the end of driving.

## 5. REFERENCES

[1] De Lange H, Kompier M A J, Taris T W, Geurts S A E, Beckers D G J, Houtman I L D and Bongers P M 2009 A hard day's night: a longitudinal study on the relationships among job demands and job control, sleep quality and fatigue. *J. Sleep Res.* **18,** 374-383.

[2] De Valck E, Smeekens L and Vantrappen L 2015 Periodic psychological examination of train driver' fitness in Belgium: deficit observed and efficacy of the screening procedure. *J. Occup. Environ. Med.* **0,** 1-8.

[3] Warm J 1984 *Sustained attention in human performance* (Wiley: Chichester).

[4] Thiffault P and Bergeron J 2003 Fatigue and individual differences in monotonous simulated driving. *Personality and Individual Differencess* **34,** 159-176.

[5] Mathis J and Hess C W 2009 Sleepiness and vigilance test. *Swiss Med. Wkly.* **139**(15-16) 214-219.

[6] Evans A W 2011 Fatal train accidents on Europe's railways: 1980–2009. *Accid. Anal. Prev.* **43,** 391–401.

[7] Stanton N A and Walker G H 2011 Exploring the psychological factors involved in the Ladbroke Grove rail accident. *Accid. Anal. Prev.* **43,** 1117–1127.

[8] Baysari M T, McIntosh A S and Wilson J R 2008 Understanding the human factors contribution to railway accidents and incidents in Australia. *Accid. Anal. Prev.* **40,** 1750–1757.

[9] Dinges D F and Mallis M M 1998 Managing fatigue by drowsiness detection: can technological promises be realized? in *Managing Fatigue in Transportation* (Ed. L. Hartley, Pergamon: Kidlington), 209-229.

[10] Dinges D F 1995 An overview of sleepiness and accidents. *J. Sleep Res.* **4,** 4-14.

[11] Thiffault P and Bergeron J 2003 Monotony of road environment and driver fatigue: a simulator study. *Accid. Anal. Prev.* **35,** 381-391.

[12] Chua E C P, Tan W Q, Yeo S C, Lau P, Lee I, Mien I H, Puvannendran K and Gooley J J 2012 Heart rate variability can be used to estimate sleepiness-related decrements in psychomotor vigilance during total sleep deprivation. *Sleep* **35,** 325-334.

[13] Dunn N and Williamson A 2012 Driving monotonous routes in a train simulator: the effect of task demand on driving performance and subjective experience. *Ergonomics* **55**(9), 997-1008.

[14] Williamson A, Lombardi D A, Folkard S, Stutts J, Courtney T K and Connor J L 2011 The link between fatigue and safety. *Accid. Anal. Prev.* **43,** 498-515.

[15] Doran S M, Van Dongen H P and Dinges D F 2001 Sustained attention performance during sleep deprivation: evidence of state instability. *Archives Italiennes de Biologie* **139,** 253-267.

[16] Alhola P and Polo-Kantola P 2007 Sleep deprivation: impact on cognitive performance. *Neuropsychiatric Disease and Treatment* **3**(5), 553-567.

[17] Belenky G, Wesensten N J, D R Thorne, Thomas M L, Sing H C, Redmond D P, Russo M B and Balkin T J 2003 Patterns of performance degradation and restoration during sleep restriction and subsequent recovery: a sleep dose-response study. *J. Sleep Res.***12,** 1–12.

[18] Pollard J 1991 *Issues in locomotive crew management and scheduling* (Federal Railroad Administration, US Departement of Transportation, Washington D.C).

[19] Siswanto D, Lestari V and Iridiastadi H 2017 Evaluation of machinist's fatigue at PT. Kereta Api Persero DAOP II Bandung. *International Journal of Technology* **3,** 932-941.

[20] Boughattas W, El Maalel O, Chikh R B, Maoua M, Houda K, Braham A, Rhif H, Chatti S, Debbabi F, Dogui M and Mrizak N 2014 Hospital night shift and its effects on the quality of sleep, the quality of life, and vigilance troubles among nurses. *International Journal of Clinical Medicine* **5,** 572-583.

[21] Jeon H J, Kim J H, Kim B N, Park S J, Fava M, Mischoulon D, Choi H, Roh S and Lee D 2014 Sleep quality, posttraumatic stress, depression, and human errors in train drivers: a population-based nationwide study in South Korea. *Sleep* **37**(12), 1969-1975.

[22] Batool-Anwar S, Kales S N, Patel S R, Varvarigou V, De Young P N and Malhotra A 2014 Obstructive sleep apnea and psychomotor vigilance task performance. *Nature and Science of Sleep* **6,** 65-71.

[23] Seitz V B 2014 *Using a sleep and activity monitor to operationalize fatigue risk management* (Thesis, Louisiana State University Department of Mechanical and Industrial Engineering).

[24] Reed M P and Green P A 1999 Comparison of driving performance on-road and in a low cost simulator using a concurrent telephone dialing task. *Ergonomics* **42,** 1015-1037.

[25] Sahami S and Sayed T 2013 How drivers adapt to drive in driving simulator, and what is the impact of practice scenario on the research? *Transportation Research Part F* **16,** 41-52.

[26] Gastaldi M, Rossi R and Gecchele G 2014 Effects of driver task related fatigue on driving performance. *Procedia - Social and Behavioral Sciences* **111,** 955 – 964.

[27] Lal S K L and Craig A 2001 A critical review of the psychophysiology of driver fatigue. *Biological Psychology* **55,** 173 – 194.

[28] Dorrian J, Roach G D, Fletcher A and Dawson D 2007 Simulated train driving: fatigue, self-awareness and cognitive disengagement. *Applied Ergonomics* **38,** 155-166.

[29] Sathyanarayana A, Joty S, Fernandez-Luque L, Ofli F, Srivastava J, Elmagarmid A, Arora T and Taheri S 2016 Sleep quality prediction from wearable data using deep learning. *JMIR mHealth and uHealth* **4**(4), 1-13.

[30] Reed D L and Sacco W P 2016 Measuring sleep efficiency: what should the denominator be? *Journal of Clinical Sleep Medicine* **12**(2), 263-266.

[31] Reifman J, Kumar K, Khitrov M Y, Liu J and Ramakrishnan S 2018 PC-PVT 2.0: an updated platform for psychomotor vigilance task testing, analysis, prediction, and visualization. *Journal of Neuroscience Methods* **304,** 39-45.

[32] Jap B T, Fischer P and Lal S 2011 *Assessing a potential electroencephalography based algorithm during a monotonous train driving task in train drivers* (Proceedings of the 6th International Conference on Broadband Communications & Biomedical Applications, November 21 - 24, 2011, The 6th International Conference on Broadband Communications & Biomedical Applications Melbourne, Australia), 139-144.

[33] Smit A S, Eling P A T M and Coenen A M L 2005 Vigilance: in search of sensitive measures. *NSWO* **16,** 141-147.

[34] Akerstedt T, Hume K, Minors D and Waterhouse J 1994 The meaning of good sleep: a longitudinal study of polysomnography and subjective sleep quality. *J. Sleep Res.* **3,** 152-158.

[35] Loh S, Lamond N, Dorrian J, Roach G and Dawson D 2004 The validity of psychomotor vigilance tasks of less than 10-minute duration. *Behavior Research Methods, Instruments, & Computers* **36**(2), 339-346.

[36] Wiechert G, Triff M, Liu Z, Yin Z, Zhao S, Zhong Z, Zhaou R and Lingras P 2016 *Identifying users and activities with cognitive signal processing from a wearable headband* (Proceeding of IEEE 15th International Conference on Cognitive Informatics and Cognitive Computing), 129-136.

IOP Conf. Series: Materials Science and Engineering **847** (2020) 012064 doi:10.1088/1757-899X/847/1/012064

# The NPD Process Design Canvas: Tool for NPD Process Creation

**M Iqbal[1,2,3] and A Suzianti [1,4]**

[1]Department of Industrial Engineering, Universitas Indonesia, Depok, Indonesia
[2]Industrial Engineering Study Program, Universitas Telkom, Bandung, Indonesia
[3]muhammad.iqbal830@ui.ac.id, [4]suzianti@eng.ui.ac.id

**Abstract.** New Product Development (NPD) is important for the growth and sustainability of companies. One of the most important factors in NPD is the management of the NPD process. NPD process is believed to be a crucial factor in NPD success. NPD process may vary between organizations. Companies must be able to implement the most suitable NPD process. Therefore, the design of NPD process is crucial. This article proposes a tool that can support organizations to design their NPD process. The tool is conceptualized based on the sequence of NPD design method. The tool is based on the form of 'space of elements' and called the NPD Process Design Canvas. The concept has been tested and several feedbacks identified. Some of the pros are its ability to capture the risks and elements of process design. Some of the cons are the variation of detail and inability to describe the existing process. Based on the feedbacks, improvements proposed are the addition of specific 'iterations' and 'reviews' analysis in the canvas, and the paradigm of the canvas as the integrator of several analyses carried out based on the sequence of NPD design method. Further research may focus on the testing of a more sophisticated canvas implementation and how the tool considers the position of companies' existing NPD process.
**Keywords**: New Product Development, Process, Design, Tools

## 1. Introduction

In a competitive business environment, company's ability to consistently deliver successful products is important. New product development (NPD) play important role in sustainability and growth of companies [1]. It is essential for a firm's business continuity, in almost all industries [2]. Considering its importance, NPD is expected to succeed. Prior studies related to the success factors of NPD show that NPD process is a crucial factor [3-6]. Therefore, it is beneficial for a company to have a well-designed, well-managed NPD process.

The product development process can be defined as a set of activities that enable the transformation of ideas into marketable products [7]. Focusing on the NPD process design and improvement is believed to have an important contribution to the success of NPD [3, 5, 8-10].

NPD process may differ from one organization to another, and deciding which NPD approach to be implemented is principal [11]. Unfortunately, not all organizations have a clear NPD process. It is found that in large organizations, the process tends to have a long decision-making stage, and in medium-small sized organizations, the process is lacking formal evaluation and monitoring [12]. Based on this, it is important for a company to be able to design their NPD process.

Designing the NPD process should consider several things, such as the risks in NPD [13] and the analysis of existing process implementation [11]. Unger and Eppinger [13] suggested a method that consists of identifying risks and assigning risks to a process that consists of stages, iterations, and reviews. It needs to take into consideration the specific NPD situation. This method is able to support organizations designing their NPD process [11], but an analytical tool to comprehensively analysed the whole NPD process design steps is needed. The focus of this article is to conceptualize the analytical tool that can be used to help organizations design their NPD process.

## 2. Methods

Based on the purpose, the method to design an analytical tool for the NPD process is divided into two main steps i.e. **Ideation** and **Concept Testing**, as shown in Figure 1.

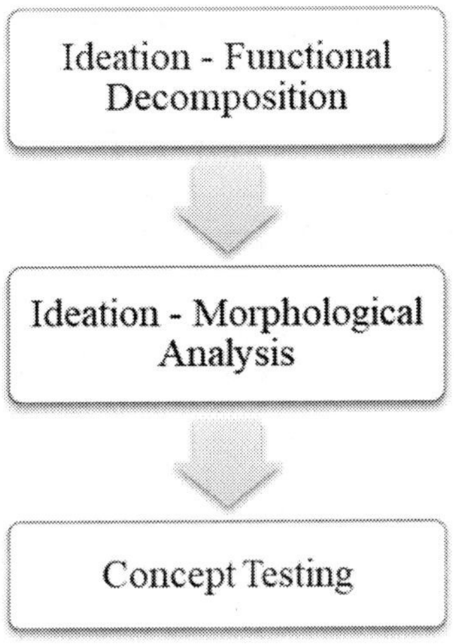


**Figure 1.** Design steps

To conceptualize the NPD process design tool, Ideation and Concept Testing will be carried out.

- **Ideation**. This is the phase where the tool concepts are generated. The result of the phase is the tool concept. Ideation is done using Functional and Morphological analysis. Functional decomposition can be beneficial for technical or non-technical design, and it can be done based on the sequence of user actions [14]. Morphological analysis can help to comprehensively manage ideas and integrate them into a solution [15]. Both techniques are used to help in capturing and managing design ideas.
- **Concept Testing**. This is the phase in which the concept is tested. Testing is done by applying the proposed concept. The respondents of the testing phase come from product development and innovation department from pharmaceutical, transportation & logistics, and building material companies. The prototype is given to respondents, and feedbacks are recorded. Based on the feedback, the early concept then improved.

**Table 1.** Research methodology

| | Steps | Data Collection and Analysis Techniques | Output |
|---|---|---|---|
| 1 | Ideation | Functional diagram | Overall function and Subfunctions |
| | | Morphological chart | Concept of the Analytical tool |
| 2 | Concept Testing | Physical model, Questionnaire | User feedback |

Functional Diagram is used to identify the function, Morphological Chart is used to generate concept, Physical Model is built and given to respondents, and Questionnaire is used to collect respondent's feedback, as shown in Table 1.

## 3. Results and Discussions

### *3.1. Ideation*

The analytical tool is based on the NPD process design with the sequence of [13]:

- Identifying the scope of NPD
- Risk documentation and prioritization
- Identification of loops, evaluation, or NPD phases
- Identification of planned iteration and process consolidation
- Identification of primary evaluation

While each of the steps may have specific methods, the NPD process design tool is conceptualized to support the observation of the whole process.

*3.1.1. Functional Decomposition.* "The first step to in decomposing the problem functionally is to represent it as a single black box operating on material, energy, and signal flows" [14]. While material and energy are relevant to physical technical products, 'signal flow', in this case 'information flow' seems to be the most appropriate approach to be used in the context of the NPD process design tool. The next step is to identify the subfunctions that will explain the overall function of the product. The decomposition is based on the researchers' understanding about the problem. The result of the functional decomposition is shown in Figure 2.

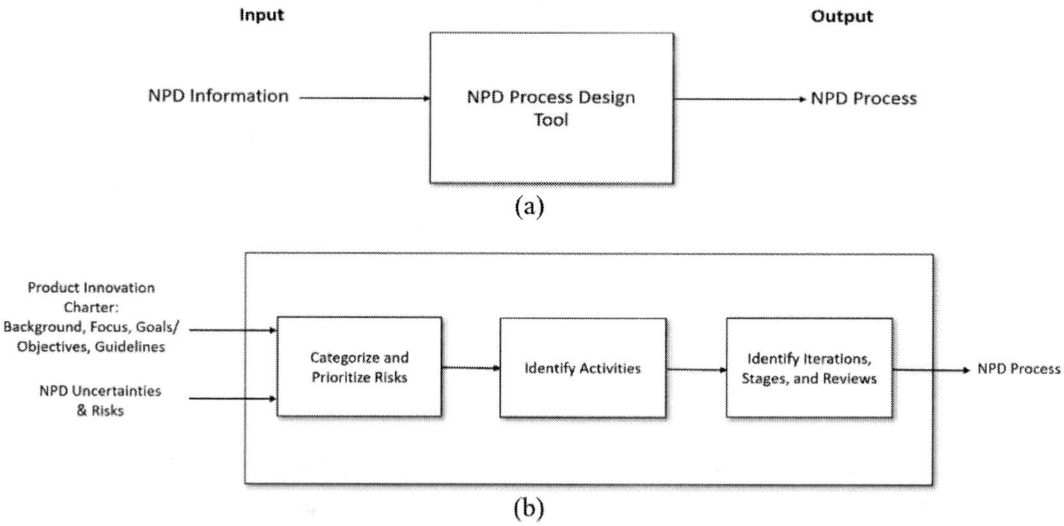


**Figure 2.** Functional decomposition: (a) The overall function (b) Subfunctions.

The primary function of the NPD process design is to get a specific NPD process relevant to the situations faced by the company. This specific condition is described in the Product Innovation Charter. The NPD process design sequence is based on the analysis risks, so it seems logical that the Product Innovation Charter and Risks are the inputs of the design. The subfunctions of the tool are referred to the sequence of the NPD process design. The subfunctions are stated as Categorize and Prioritize Risks, Identify Activities, and Identify Iterations, Stages, and Reviews as described in figure 2 (b).

IOP Conf. Series: Materials Science and Engineering **847** (2020) 012064 doi:10.1088/1757-899X/847/1/012064

*3.1.2. Morphological Analysis.* The next step is to find solutions based on the functional decomposition. For this purpose, a morphological chart is utilized. For relevant functions, solutions alternatives will be generated. These 'sub solutions' then analysed to get an intact design. The morphological chart is described in Table 2.

**Table 2.** Morphological Chart

| Functions | Options | | |
|---|---|---|---|
| Tool form | 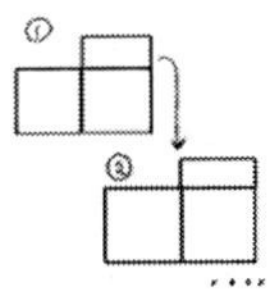<br>Sequence of Matrices | 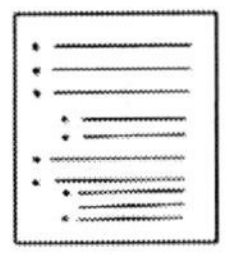<br>List of Description | 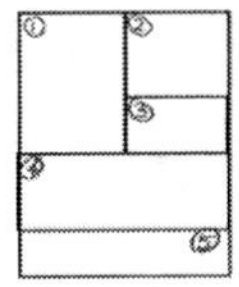<br>Space of Elements |
| Categorize and Prioritize Risks | 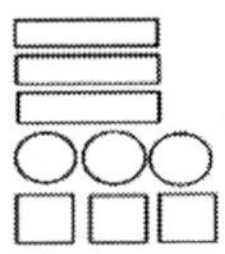<br>Arranged | 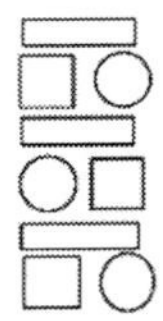<br>Not Arranged | |
| Identify Activities | 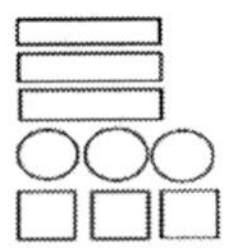<br>Arranged | 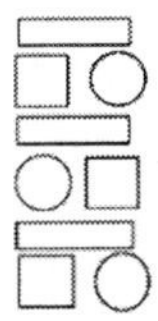<br>Not Arranged | |
| Identify Iterations, Stages, and Reviews | 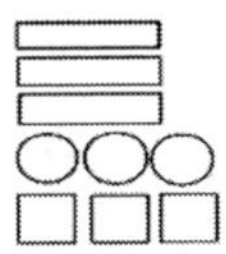<br>Arranged | 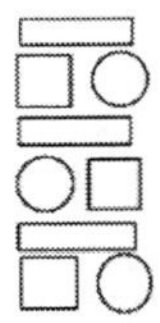<br>Not Arranged | |
| Process Finalization | 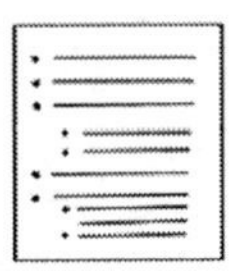<br>List of Description | 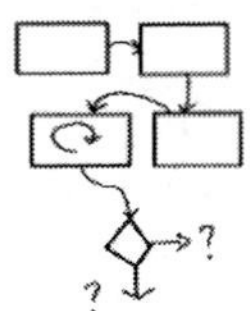<br>Sketches | 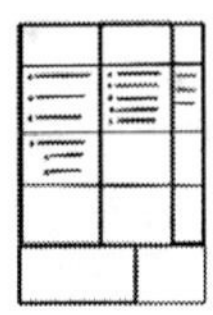<br>Tables |

The 'tool form' explores the possibility of the tool main form will be in the sequence of matrices represent the PDP process design sequence, in the list describing the sequence, or in the space with elements related to PDP process design. The risks, activities, iterations, stages, and reviews may be arranged or not arranged into defined categories. The process finalization describes on how to finalize the process design. Based on the combination of solution options, several tool concepts are generated. The concepts are represented in table 3.

**Table 3.** NPD Design Tool Concepts

| Concept | Combination | | | | | Concept Description |
|---|---|---|---|---|---|---|
| | Tool Form | Categorize and Prioritize Risks | Identify Activities | Identify Iterations, Stages, and Reviews | Process Finalization | |
| Concept 01: The NPD Process Design Matrices | Sequence of Matrices | Not arranged | Not arranged | Not arranged | Tables | The sequence will be identified using matrices, and connections between matrices is important as it showed the sequence progress. The specific arrangement is not needed as it can be identified on the matrices. The process is described in tables, as it is easier to 'translated' from matrices |
| Concept 02: The NPD Process Design Canvas | Space of Elements | Arranged | Arranged | Arranged | Sketches | As elements need to be elaborated and relations between elements need to be identified, it is beneficial to put all the elements in one space. Risks, activities, iterations, stages, and reviews need to be arranged since the 'empty space' does not beneficial to illustrate the connections between them. The process is finalized using sketches. |
| Concept 03: The NPD Process Design Manual | List of Description | Arranged | Arranged | Arranged | List of Description | A strict descriptive tool that describes the NPD process design sequence. Beneficial for inclusion of all-important information. Have risks of being too packed with textual descriptions. |

The 'best' concept then selected. The concept is chosen using the 'pros and cons', one of the concept selection approach based on the strengths and weaknesses of each concepts [14]. The concept with the design canvas approach (Concept 02) considered to be the best concept since it can elaborate on the connections between elements and can make the analysis performed not in a strictly linear approach. As example, risk can be re-identified as activities are listed. The 'space of elements approach' can support this process better compared to rigid matrices or textual description.

The sketches approach also could balance between the graphical and textual forms to describe a process. This concept is illustrated in Figure 3.

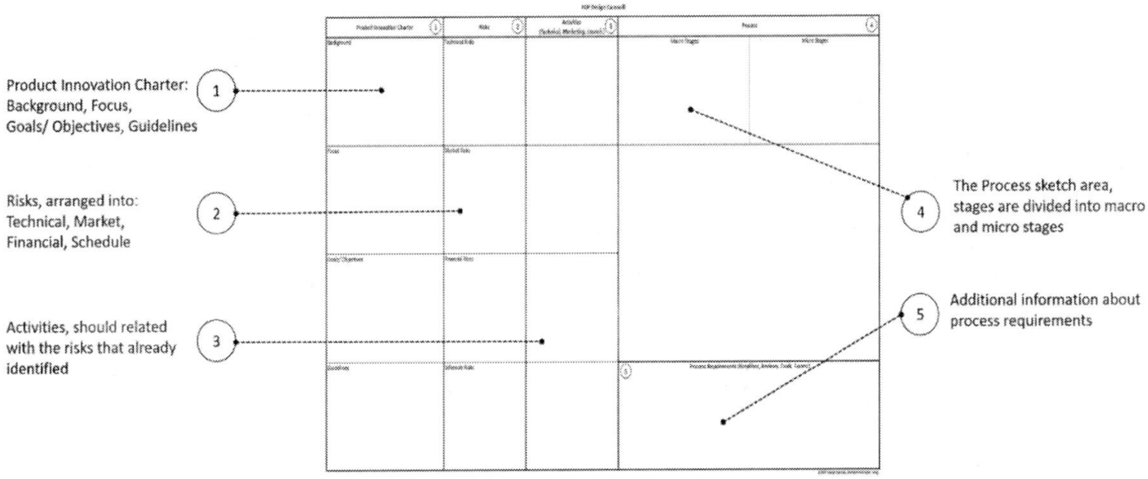


**Figure 3.** The NPD Process Design Canvas

The canvas is consisting of five main parts: **The Product Innovation Charter List, the Risks List, the Activities List, the Process Sketch Area**, and **the Additional Information Area**. The working sequence is started with the identification of project scope stated in the Product Innovation List. The next step is the identification of uncertainty and risks, and the identification of necessary activities to address those risks. The activities then arranged into a complete process and describe in graphical form (sketch). The process should describe the iterations, reviews, and stages. Additional information that cannot be explained from the process illustration then added into the **Additional Information Area**. The canvas may function as 'integrator' that captured information from analysis for each part.

### *3.2. Concept Testing*

The purpose of the concept testing is to collect the customer feedback for the NPD Process Design Canvas. The testing session is done using the **physical canvas (using printed paper) that is given to respondents who directly using the canvas to perform the sequences**. The respondents are from pharmaceutical manufacturers, transportation & logistics provider, and building material manufacturer. Respondents are representative of the innovation and product development functions at the companies. During the testing session, **feedback is collected as respondents asking questions and give comments**. At the end of the session, **respondents are asked to fill the questionnaire about the proposed concept**.

The result shows that risks are successfully identified. Risks of overdesign, trends, product similarity with the competitor, unmanaged design changes are several risks that must be faced by pharmaceutical companies. Limited market, different preferences in product design, and delayed product trial are main risks that must be solved by building material manufacturer. Maintenance, shifting customer behaviour, rapid technology changing, long approval stages are some risks that the transportation & logistic company must deal with.

NPD process also successfully designed. Although more detail must be completed for all the designed process, the process that is built based on the risk analysis has been done. Most of the processes are described in the gating system, explaining gates such as Idea Generation, Feasibility Study, Development, Testing, and Launching Preparation. **It is interesting that almost all process is described as a linear multi-stages sequence and no clear iterations and reviews that are stated.** It might be an indication that the canvas is not able to assist the conversion of activities into iterations, reviews, and gates.

The pros and cons of the concept also identified. The result is described in Table 4.

**Table 4.** Advantages and Disadvantages of The NPD Process Design Canvas

| Advantages | Disadvantages |
|---|---|
| Serve the needs of mapping the process comprehensively | Lack of product description |
| Risks can be studied in detail, enable risks anticipation | Elements have various detail; it is not easy to wrap it to a complete process |
| Elements in the canvas is adequate to design the NPD process | Unable to capture the information of existing NPD process |

Based on the result, several improvements for the canvas need to be done. The proposed improvements are

- ***The conversion of activities into process' iterations, reviews, and stages***. In the Process Sketch Area, additional boxes to identified iterations and reviews could be added under the Macro Stages and Micro Stages box. The illustration is as described in Figure 4.

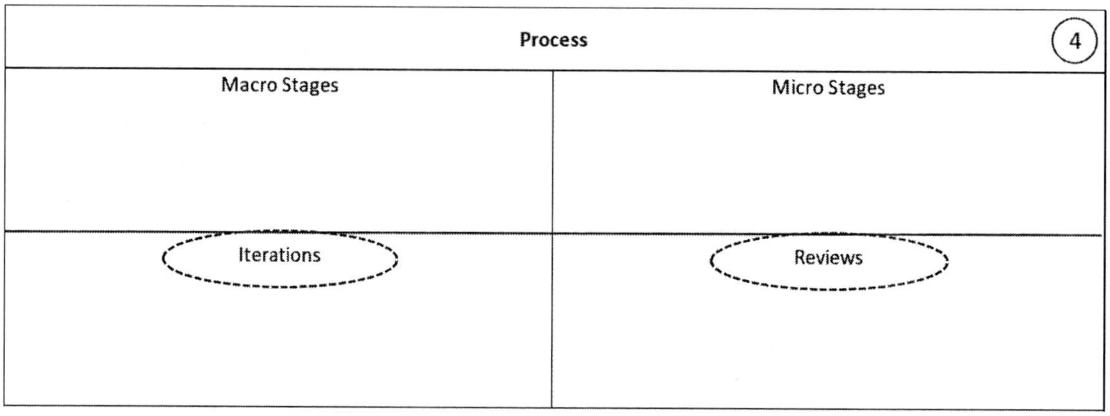


**Figure 4.** Additional 'iterations' and 'reviews' boxes in the Process Sketch Area

- ***NPD Process Design Canvas can be positioned as an integrator of several analyses***. All the sequences of NPD design [13] may need specific analysis. As an example, the process of risk prioritization may need some sort of rating method. The existing NPD process may be valuable information, and it may need a specific illustration space that cannot be fulfilled in the NPD design canvas. The result then can be included in the canvas. In that way, other tools are needed to comprehensively enable the design of the NPD process. This paradigm is important once the canvas will be implemented physically (such as in the whiteboard) or conceptually (such as in worksheet software or mobile apps).

## 4. Conclusion

The analytical tool that can be used to help organizations design their NPD process is conceptualized as the NPD Process Design Canvas. The tool concept is based on the 'space of elements' tool form, using 'sketches' to finalize the NPD process and provide arrangements of risks, activities, iterations, reviews, and stages. The concept has been tested and some improvements based on the feedbacks are the additional boxes for iterations and review identifications, and the position of NPD Process Design Canvas as an integrator of several analyses.

While this article explains the concept of the NPD Process Design Canvas, further testing needs to be done involving a more detailed analysis of risks, using a more sophisticated physical or software media. Other research also needed to address the issue about the position of organization's existing NPD process and how the NPD Process Design Canvas take it into consideration.

## 5. References

[1] R G Cooper 2013 New Products—What Separates the Winners from the Losers and What DrivesSuccess. *The PDMA Handbook of New Product Development.*

[2] R J Calantone, J B Schmidt, and X M Song 1996 Controllable Factors of New Product Success: A Cross-National Comparison. *Marketing Science,* **15**(4), pp. 341-358.

[3] H Florén, J Frishammar, V Parida, and J Wincent 2018 Critical success factors in early new product development: a review and a conceptual model. *International Entrepreneurship Management Journal,* **14**(2), pp. 411-427.

[4] H Evanschitzky, M Eisend, R J Calantone, and Y Jiang 2012 Success factors of product innovation: An updated meta-analysis. *Journal of product innovation management,* **29**, pp. 21-37.

[5] R G Cooper and E J Kleinschmidt 2007 Winning businesses in product development: The critical success factors. *Research-Technology Management,* **50**(3), pp. 52-66.

[6] H Ernst 2002 Success factors of new product development: a review of the empirical literature. *International journal of management reviews,* **4**(1), pp. 1-40.

[7] K B Kahn, S E Kay, R Slotegraaf, and S Uban 2013 *The PDMA handbook of new product development.* Wiley Online Library.

[8] M M Montoya-Weiss and R Calantone 1994 Determinants of new product performance: A review and meta-analysis," *Journal of product innovation management,* **11**(5), pp. 397-417.

[9] R G Cooper and E J Kleinschmidt 1995 Benchmarking the firm's critical success factors in new product development. *Journal of Product Innovation Management,* **12**(5), pp. 374-391.

[10] G S Lynn, K D Abel, W S Valentine, and R C Wright 1999 Key factors in increasing speed to market and improving new product success rates," *Industrial Marketing Management,* **28**(4), pp. 319-326.

[11] M Iqbal and A Suzianti 2019 Improvement of New Product Development Process by Evaluating the Existing Development Approach: Lesson Learned from Pharmaceutical and ICT Companies presented at the The 16th International Conference on Quality in Research, Padang, Indonesia.

[12] A Ghobadian and D Gallear 1997 TQM and organization size. *International journal of operations production management,* **17**(2), pp. 121-163.

[13] D Unger and S Eppinger 2011 Improving product development process design: a method for managing information flows, risks, and iterations. *Journal of Engineering Design,* **22**(10), pp. 689-699.

[14] K T Ulrich and S D Eppinger 2012 *Product Design and Development.* (New York: McGraw-Hill).

[15] A N Azammi, S Sapuan, M Ishak, and M T Sultan 2018 Conceptual design of automobile engine rubber mounting composite using TRIZ-Morphological chart-analytic network process technique," *Defence Technology,* **14**(4), pp. 268-277.

IOP Conf. Series: Materials Science and Engineering **847** (2020) 012065 doi:10.1088/1757-899X/847/1/012065

# EPR Approach for Better Waste Management System for Mobile Phone Design in Indonesia

**A B Junianto**[1,3], **D C Sugandha**[2,4]

[1] Universitas Agung Podomoro, APL Tower 5th floor, Grogol, West Jakarta, Indonesia
[2]Pelita Harapan University, Lippo Karawaci, Tangerang, Banten, Indonesia
[3]aloysius.baskoro@podomorouniversity.ac.id, [4]sugandhadewintacitra@gmail.com

**Abstract.** The influence of toxic electronic waste or electronic goods will occur if it is disposed of carelessly or managed in the wrong way. Open dumping and crude waste handling in landfill sites in Indonesia creating even more miserable situation (Damanhuri, 2010). Mobile phone or cell phone is considered as one of the fastest growing device in global society. More than 60% of global population have mobile phone, end even an individual may have more than one device. EPR could be taken as strategy in order to find better e-waste management of renewable and non-renewable resources specifically for Indonesia situation. EPR related action of WEEE in Indonesia is still uncertain especially for mobile phone product. It is observed that many used mobile phones are still disposed improperly by unauthorized party and may danger the people and environment. This study will observe how the mobile phone waste is handled by unorganized organization, how EPR is taken into account in mobile phone waste management system and what kind of solution that is based on EPR that could enhance environmentally save property of mobile phone design in the future, especially for Indonesia situation. It is important by considering Indonesian situation; the producers need to employ new design strategy not merely in order to meet regulation compliances but should extend the responsibility for better waste management friendly product design that is more suitable with Indonesia situation with minimum waste management infrastructure by involving End-of-Life Design Solution; better collection, separation, recyclability, reuse and ease of manufacture
**Keywords**: EPR, waste management, product design

## 1. Introduction

The problem of electronic waste or E-Waste can hazardous to human health and environment. This is due to the efect of dangerous toxic substances such as lead, mercury, chromium, cadmium, PBDE, and other heavy material if they are not managed properly. The influence of toxic electronic waste or electronic goods will occur if it is disposed of carelessly or managed in the wrong way. Open dumping and crude waste handling in landfill sites in Indonesia creating even more miserable situation (Damanhuri, 2010). Aboughaly and Gabbar (2019) in their study mentioned over 50 million metric tonnes of electronic-waste annualy across the world from abandoned equipment. Jakarta Provincial Environment Agency has processed more than one ton of electronic waste for the accumulation period from January to June 2019 (merdeka.com). Most of the waste were still being kept in large storage and the local government still need to collaborate with other third party for further recycling processes. However, the number of waste will increase significantly in near future due to the rapid development of technology and electronics industry that lead to the increasing consumption of electronic products which already becoming daily necessities.

Mobile phone or cell phone is considered as one of the fastest growing device in global society. More than 60% of global population have mobile phone, end even an individual may have more than one device. Heavy consumerism and tight competition among manufacturers also lead fast consumption by releasing new models each year. In Indonesia, the increasing numbers of electronic-waste of mobile phone should be taken into account by the manufacturing industry (producers), brand distributors and governments to draw attention to deliver better waste management system, and it is related to Extended Producers Responsibility (EPR) and Waste of Electrical and Electronic Equipment or WEEE (Börner and Hegger, 2018). EPR concept could be taken as strategy in order to find better e-waste management of renewable and non-renewable resources specifically for Indonesia's situation. EPR related action of WEEE in Indonesia is still uncertain especially for mobile phone industry. It is observed that many used mobile phones are still disposed improperly by unauthorized party and may danger the people and environment. This study will study how the mobile phone waste is handled by unorganized organization, how EPR is taken into account in mobile phone waste management system and what kind of solution that is based on EPR that could enhance environmentally save property of mobile phone design in the future, especially for Indonesia situation.

## 2. Method

This study employs qualitative method through three steps of processes:

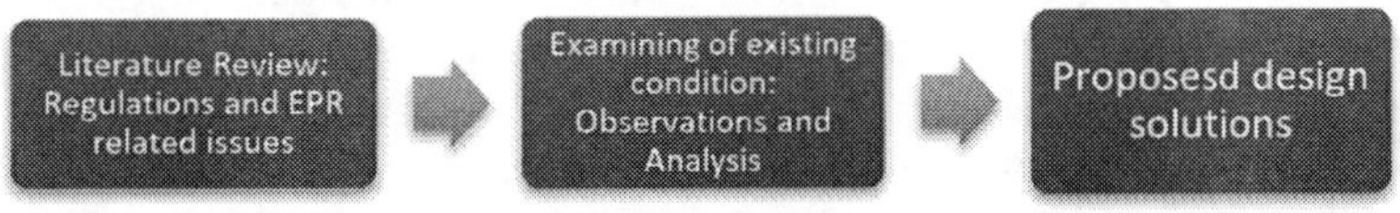


**Figure 1.** Study Method

This method is adapted from previous research by Börner and Hegger (2018) and Thorpe, et.al (2004) that build theoretical framework by taking E-Waste Governance (government regulation) and EPR strategy. Firstly, it will investigate WEEE related problems with Indonesian government regulations and EPR related issues by examining the literature of previous studies. Secondly, it would observe waste management problems of mobile phone concerning EPR cycle in society. By investigating the observed situation of unauthorized mobile phone service centre, to find some insights for proposed concept. Mobile phone components will be examined from the aspect of ease of dismantle through simple simulations with standard equipment. This is to see how difficult the component is disassembled and how it will effect on the EPR cycle. Thirdly, alternative design solutions of how EPR approach might be implemented in the design development of new products.

## 3. Results and Discussion

### *3.1. Regulations and EPR issues*

The Government of Indonesia Hazardous Waste Management Regulation: Act No.32/2009 concerning Environmental Protection and Management: Definition of Hazardous Waste Management (Article 1, point 23): Hazardous Waste Management shall be an activity covering the reduction, storage, collection, transportation, utilization, treatment and/or the pilling.

Damanhuri and Padmi (2009) explain some of the central issues regarding EPR of Law No. 18/2008 for example; every producer should indicate a label on their product how to proper handling of hazardous waste. Furthermore, the government regulations also imply how to manage hazardous waste as follows:

- The entire chains of waste transporting system shall apply for waste and recycling system since their first origin until final disposal;
- Selection of waste processing and dumping technologies that are safe and healthy, and conform with Indonesian situation;

- Open dumping and burning are forbidden and during five years after the passing of the law, open dumping would be completely banned; and prohibition to import waste into Indonesia territories, and to mix waste with dangerous wastes.

EPR or Extended Producer Responsibility or also called "Producer Takeback" (Thorpe, et.al. 2004) is the policies that should involve closed loop scenario in material management and should be thereof managed and processed by reusing or recycling the materials into new product or other packaging products with less energy consideration rather than producing new product from new material. This approach replaces traditional linear thinking and is based on the concept of supply chains manufacturing system, retailers and the consumer all share in end-of-life product management. In addition, recycling programs as part of the EPR, like existing now, cannot handle all of the hazardous waste. Thorpe, Kruszeska and McPerson (2004) mentioned that producers shall take bigger responsibility in recycling process from the product they had produced and design new product from the reused or recycled materials.

Chatterjee (2012) mentioned that there are two common practices in e-waste management system conducted by organized and unorganized sectors. The unorganized operators in India for example, according to Chatterjee are facing difficulty of save recycling technology. Furthermore, this practice needs to be reconsidered by providing alternative collecting mechanism by the unorganized operators. Therefore, unorganized operators will concentrate on collection, disassembly, segregation of e-waste, while, organized sector will concentrate on processing the PCBs to extract precious metals.

Thorpe et.al (2004) suggested that in order creating a strong, long-lasting Extended Producer Responsibility (EPR) program, the following elements should be included in the program design:

- Individual responsibility that is when a producer takes the end-of-life management of their own products, whereas they involve on cost sharing in the recycling process of their end-of-life products.
- Company can redesign their products to enhance durability, ease of repair, reuse, disassembly or recycling by their own initiative.
- Government mandate and regulations in the participation of *takeback* programs that will stop unorganized/illegal party from ruining the overall program.
- Reuse and recycling requirements: the number of minimum reuse and recycling materials as a major driver for design change need to be established. In addition, other method of waste processing with using large amount of energy should not be considered, this will imply to new directive of design to achieve full recover materials by employing reuse and recycling method.
- EPR programs need to consider for orphan (products still in used but whose producers no longer exist) and historic waste, as well as current end-of-life product waste. Years ago product designers did not design for reuse or recyclability. However, this waste must be dealt with. Designers or producers need to rethinking the possibility of the product become orphan waste. For example; Blackberry which no longer producing mobile phone, who is taking this responsibility?

### *3.2. End Of Life Phase and Design Implementation*

Huang, et.al. (2019) implied that there is strong correlation between durability and recyclability choices. They further explained that the more rigid recycling targets, it would lead to the increase recyclability and durability. So in terms of EPR, producers might need to find better recycling method to find valuable recycling materials in order to create more durable products as well as recyclability. Huang et.al. (2019) also stated that easier recycle of product design will increase recycling value at the end-of-life phase and durable product design will reduce the total volume has to recycle, and will lead to cost reduction as well.

The example of design implementation in end of life phase is conducted by Apple, Inc. Since 2017, Apple has been trying to use recycled and renewable raw materials. The raw material is designed so as not to produce too much waste and to have a long service life. When the product reaches the end of life phase, Apple ensures that the product returns back to the company using the trade in system, and the material will be recycled to become a new product. The strategy used is to create a circular supply chain

IOP Conf. Series: Materials Science and Engineering **847** (2020) 012065 doi:10.1088/1757-899X/847/1/012065

with strategies as follow; designing and building products and packaging Make a durable product by designing a durable hardware, providing repair services in various places, Collect products in the end of life phase through 'Apple Trade In' to extend the service life or utilize existing materials.

*3.3. Observations*

**Table 1.** Observation of Unauthorized Mobile Phone Service Center

| No. | Observation | Results | Pictures |
|---|---|---|---|
| 1. | Friday, 18 October 2019 – E-centre Supermall Karawaci **Christine Phone Cell** | • Parts / components that can still be used will be used for personal testing.<br>• Spare parts that are not used or outdated parts will be collected and thrown into the trash bin, because no people or institutions want to take or buy.<br>• There is no specific waste management system for e-waste components. |  |
| 2. | Friday, 26 October 2019 – E-centre Supermall Karawaci **Cinta Karya Cellular** | • Parts / components that can still be used will be stored inside.<br>• Parts that cannot be used anymore will be thrown to trash bin, because there are no people or institutions that want to take or buy.<br>• There is no specific waste management system for waste parts around the location.<br>• Phone Distributors do not care about broken cell phones or spare parts waste in the individual service center area. |  |
| 3. | Friday, 26 October 2019 – E-centre Supermall Karawaci **Permata Service** | • Parts or components that can be used will be stored<br>• Parts that are not used are only stored / for shop displays<br>• Most of the waste spare parts collected are LCD and batteries<br>• There are people who want to buy LCD parts that still work |  |
| 4. | Wednesday, 6 November 2019 - E-centre Supermall Karawaci<br><br>**Bagas Cellular** | • Spare parts that are still have function will be reused. Broken parts were sold to waste collectors.<br>• PCB parts can actually still be used, but because of the small number of requests, so inevitably they are discarded<br>• There is no special processing or management system in this service place.<br>• Theres is no special waste disposal site for electronic waste around, and there is no system / regulation that requires the safe sorting and disposal of electronic waste |  |
| 5. | Wednesday, 6 November 2019 - E-centre Supermall Karawaci<br><br>**King Service** | • Parts that cannot be used are immediately thrown in the trash.<br>• There is no spare parts waste management system. Cause: there is no special waste disposal site for electronic waste, and there is no system / regulation that requires the safe sorting and disposal of electronic waste.<br>• The owner feels there is no need for an unused component management system, because all this time when the component is disposed of, he feels safe |  |

IOP Conf. Series: Materials Science and Engineering **847** (2020) 012065 doi:10.1088/1757-899X/847/1/012065

*3.4. Analysis of EPR stages and current situation*

**Table 2.** Analysis of EPR stages, issues and current situation

| No | EPR stages | Regulations and issues | Current situation |
|---|---|---|---|
| 1. | Collection | The entire chains of waste transporting system shall apply for waste and recycling system since their first origin until final disposal.<br>Incentive mechanism for takeback program. | There is no spare parts waste management system. Cause: there is no special waste disposal site for electronic waste, and there is no system or rules that requires the safe sorting and disposal of electronic waste.<br>Spare parts that are not used or outdated parts will be collected and thrown into the trash bin, because no people or institutions want to take or buy.<br>There is no government mandate in terms of incentive mechanism for takeback program. |
| 2. | Processing/separation | Selection of waste processing and dumping technologies that are safe and healthy<br>In the case of everybody being incapable of treating Hazardous Waste individually the treatment shall be entrusted to other party. | Plenty of unorganized party still practicing unsafe process and there is no safety standard applied.<br>There are only 2 dry cell batteries collection and smelters in Central Java. |
| 3. | Recycling | Hazardous Waste Management shall be an activity covering the reduction, storage, collection, transportation, utilization, treatment and/or the pilling. | Most of existing Municipal Solid Waste management system in Indonesia, relies on the existence of landfill (open dumping) (Damanhuri and Padmi, 2009) |
| 4. | Reuse | Make recondition based on the product standard regulation.<br>Responsible for managing their waste from the production process. | Spare parts that are still have function will be reused. Broken parts were sold to waste collectors.<br>There is no control from the producer regarding standard regulation compliance. |
| 5. | Material extraction | Material extraction process should be conducted by employing environmental friendly process and not to consider other unsafe methods. | Most of the e-waste are disposed in open dumping and there is unclear situation regarding material extraction process from the e-waste especially from mobile phone e-waste.<br>Some illegal parties are practicing unsafe process of extracting, and mostly use incinerating process to eliminate the rest of unused parts. |
| 6. | Manufacturing | Producers are responsible to monitor the waste handing from collecting, processing, and recycling to manufacturing.<br>Producers are obligated to design environmentally safe product design. | All mobile phone producers are located outside Indonesia. |

*3.5. Analysis of mobile phone components*

Ease of dismantling/disassembly is one of the key points in EPR process (Thorpe et.al, 2004). Table 3 shows three different types mobile phones that were dismantled to simulate how difficult the dismantling process is undertaken by illegal party or personal user. Four major components were dismantled by using ordinary tools such as screwdriver, pliers, heat gun etc. Each component is given score range 5 (the most difficult) to 1 (the easiest). It is shown that iPhone has lower average score (2.75) that is easy to dismantle, and ASUS has highest score (4.75). It is indicated that in overall, the separation process is considered has medium to high ease of separation. For uni-body type of phone employs adhesive and make it harder to get dismantled. There are still plenty of mobile phone designs which do not provide ease of dismantling.

**Table 3.** Analysis of mobile phone dismantling process

| Mobile Phone Type | Battery pack | LCD screen | PCB | Housing structure | Score |
|---|---|---|---|---|---|
| **A. iPhone**  | 5 | 2 | 3 | 1 | **2.75** |
| **B. ASUS** 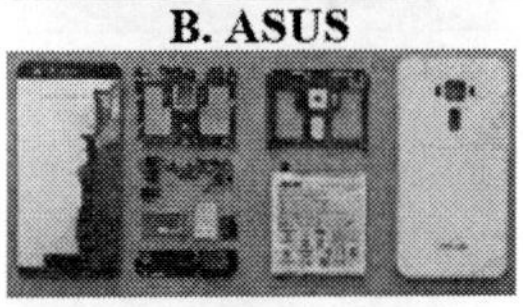 | 5 | 5 | 5 | 4 | **4.75** |
| **C. Samsung**  | 1 | 4 | 4 | 3 | **3** |

*3.6. The importance of mobile phone waste management*
The importance of mobile phone waste management is eminently needed for following reasons;

- The amount of mobile phone waste is bad for the environment. It is often neglected by the unorganized sector that will lead to environmental damage and depletion of natural resources of secondary material.
- Electronic waste containing precious metal base materials (some are dangerous) need to be processed properly and cannot just be reused just like that.
- Lack of effort in e-waste collection in Indonesia. The recycling technology used is unproper or lack of knowledge.
- Recycling electronic waste with proper handling will reduce the share of mining and therefore will reduce glass house effect emissions.
- Recycling with the use of appropriate systems and technologies, and safe for the environment reduce carbon footprint.

*3.7. Proposed Solution*
From the above discussions, then solution is proposed; to create mobile phone design concept that is easily recycled using the Extended Producer Responsibility system approach. This will involve users in mobile waste management and treatment systems in a closed-loop cycle. In terms of EPR, this study would like to propose End-of-Life phase design concept that is focusing on easy recycle product design (Huang, et.al 2019) which covering several issues; **collecting process, pre-treatment (separation), material recovering method, and safe disposal concept**. This end-of-phase concept is taken into account as a bottom up approach to see deeper problem of Indonesia situation related to EPR life cycle. The design concept implementation into practice might rethinking that each mobile phone components are easy to dismantled and replaced when its broken and could be marked so can be separated directly by the mobile phone service parties (authorized or unauthorized). This will include the service and system design that will relate to EPR application so that can comply with government regulation (i.e. Law No. 18/2008).

Table 4 below explains the variable design solutions that are associated with values in the EPR cycle. Score 1 is considered applicable that affect in all EPR cycles, and score 0 is not applied or is not related to EPR process. In the table it can be seen that the ease of dismantle gets the highest total score (5) so that it can be used as the main focus in the design of mobile phones that are capable of supporting electronic waste management. Followed by providing safe disposal system, minimizing usage of hazardous materials and takeback program.

**Table 4.** Design Solutions and EPR values

| Design solution variable | Better collecting | Better process/separation | Better recyclability | Better reuse material | Ease of manufacture | Score |
|---|---|---|---|---|---|---|
| Easy to dismantle components/parts | 1 | 1 | 1 | 1 | 1 | **5** |
| Provide safe disposal system | 1 | 0 | 0 | 1 | 1 | **3** |
| Minimizing usage of hazardous materials. | 0 | 1 | 1 | 0 | 1 | **3** |
| Incentive system in takeback program. | 1 | 0 | 0 | 1 | 0 | **3** |

## 4. Conclusion

The study shows that EPR has not well implemented in Indonesia especially in handling mobile phone waste issues. Unauthorized service and illegal party are still neglecting the safe treatment due to lack of awareness and technology. Hence, it is important by considering Indonesian situation; the producers need to employ new design strategy not merely in order to meet regulation compliances but should extend the responsibility for better waste management friendly product design that is more suitable with Indonesia situation with minimum waste management infrastructure by involving End-of-Life Design Solution; better collection, separation, recyclability, reuse and ease of manufacture. Further research might be needed to find the correlation between design variables and EPR values through statistical analysis.

## Acknowledgments

This paper is part of the author individual research and bachelor thesis research of product design by student Dewinta Citra Suganda from Product Design Department Pelita Harapan University.

## 5. References

[1] Aboughaly, M and Gabbar, HA 2019 Recent Technologies in Electronic-Waste Management. In *E-waste Recycling and Management* (pp. 63-80). Springer, Cham.
[2] Börner, L and Hegger, DL 2018 Toward design principles for sound e-waste governance: A research approach illustrated with the case of the Netherlands. *Resources, Conservation and Recycling*, **134**, pp.271-281.
[3] Chatterjee, S 2012 Sustainable electronic waste management and recycling process. *American Journal of Environmental Engineering*, **2**(1), pp.23-33.
[4] Damanhuri, E and Padmi, T 2009 Current situation of waste recycling in Indonesia. *3R Policies for Southeast and East Asia; Kojima, M., Damanhuri, E., Eds*, pp.23-52.
[5] Damanhuri, E 2010 Solid and hazardous waste management in Indonesia. *A paper. Bandung.*
[6] Huang, X, Atasu, A and Toktay, LB 2019 Design implications of extended producer responsibility for durable products. *Management Science*, **65**(6), pp.2573-2590.
[7] Kumar, A and Holuszko, M 2016 Electronic waste and existing processing routes: A Canadian perspective. *Resources*, **5**(4), p.35.

[8] Miah, MR, Saifuddoha, AM, Parvez, MS, Noor, A, Chakraborty, C and Tabassum, T 2013 Recycling of mobile phone waste. *Int. J. Sci. Eng. Res.*, **4**, pp.1813-1815.
[9] Thorpe, B, Kruszewska, I and McPherson, A 2004 Extended producer responsibility. *Clean Production Action*, **29**.
[10] Ylä-Mella, J, Pongrácz, E, Tanskanen, P and Keiski, RL 2007 March. Environmental Impact of Mobile Phones: Material Content. In *International Conference on Solid Waste Technology and Management* (s6), pp. 1-7.

# Vehicle Routing Problem with Heterogeneous Fleet, Split Delivery, Multiple Product, Multiple Trip, and Time Windows: A Case study in fuel distribution

**S M Nugroho[1], L Nafisah[2] , M S A Khannan[3], H Mastrisiswadi[4], and M N Ramdhani[5]**

[1,2,3,4,5]Industrial Engineering Department, Universitas Pembangunan Nasional Veteran Yogyakarta, Jl Babarsari No.2 Tambakbayan, Yogyakarta, Indonesia
[1]sm.nugroho@gmail.com, [2]laila@upnyk.ac.id, [3]shodiq@upnyk.ac.id, [4]mastrisiswadi@upnyk.ac.id, [5]mn.ramdhani@upnyk.ac.id

**Abstract.** The distribution of fuel in Pertamina co. Ltd Yogyakarta region is carried out using three types of vehicles, namely vehicles with a capacity of 32 kl, 24 kl, and 16 kl. In determining the distribution route, the operator chooses the gas station based on the first order and the type of vehicle determined based on the vehicle's ability to maneuver at the gas station. The determination of the route has not taken into account the time that affects the distribution, such as average speed, loading time, discharging time, and the time limit for completing the tour. So when in the high demand for fuel distribution, three gas stations are delayed from 29 gas stations in cluster I. This study analyzes the distribution of fuel in cluster 1 for Premium and Bio-Solar products. In determining the distribution route and distribution time using the VRP method by considering Heterogeneous Fleet, Split Delivery, Multiple Products, Multiple Trips, and Time Windows. The calculation is done using sequential insertion, where the selected customer is based on the smallest CT. The results of this study indicate that after processing, it was found that late customers were reduced to 1 in 3 late customers. Total mileage was reduced to 693.3 km from 701.5 km. The total tour completion time was reduced to 71,819 hours from 81,819 hours, and distribution costs increased to Rp. 20,754,119,875.00 from Rp. 18,511,210,625.00.
**Keywords**: Fuel Distribution, Vehicle Routing Problem, Sequential insertion, Heterogeneous Fleet-Split Delivery-Multiple Product-Multiple Trip-T-ime Windows.

## 1. Introduction

Fuel has become an essential commodity in people's lives, where fuel has become a source of energy used in society. Pertamina co. Ltd, as a producer and distributor of fuel in Indonesia, is always trying to meet consumer needs. Pertamina co. Ltd's Rewulu Depot is responsible for meeting fuel needs, especially in the Yogyakarta area. In determining the distribution route, the operator chooses the gas station based on the first order. The type of vehicle will be determined based on the vehicle's ability to maneuver at the gas station. If the vehicle load is still able to serve other gas stations, the next selected gas station is the nearest gas station based on a logical tree taking into account the vehicle's maneuverability at the gas station. The determination of the route above has not received the factors that influence the distribution time into consideration such as the average speed of the vehicle, the setup time of each location, the loading and discharging time, and the time limit for completing the tour. So when demand on Premium and Bio-Solar products is high, there are delays of 3 gas stations from 29 gas stations. Therefore, this study is directed to propose methods of determining distribution routes to reduce or eliminate delays in the distribution of fuel.

According to [1] the determination of distribution routes can be done using Vehicle Routing Problems (VRP). Previous research [2] about the Vehicle Routing Problem Model with Characteristics of Multiple Routes, Multiple Time Windows, Multiple Product, and Heterogeneous Fleet for a Single Depot. [2] examined the delivery of fuel distribution using ships.

This study aims to solve the vehicle routing problem that considers multiple trip variants, multiple product compartments, multiple time windows, split delivery, and heterogeneous fleets using truck tank fleets. This paper is a published version of [3]. A comparison of several similar studies can be seen in Table 1.

**Table 1.** Comparison between similar studies

| Author | Year | Topic | Multiple Fleet | Split Delivery | Multiple Product | Multiple Trip | Time Windows | Order Pickup | Method |
|---|---|---|---|---|---|---|---|---|---|
| Suprayogi& Priyandari [1] | 2008 | Vehicle Routing Problem | No | No | No | Yes | No | Yes | Sequential Insertion |
| Arvianto et al. [2] | 2014 | Vehicle Routing Problem with tanker freight | Yes | Yes | Yes | Yes | Yes | No | Sequential Insertion with Local Search |
| Windyarti [4] | 2019 | Vehicle Routing Problem | Yes | No | No | No | Yes | No | Mixed Integer Linear Programming Branch and Bound |
| Abadi et al. [5] | 2014 | Vehicle Routing Problem | No | No | No | Yes | Yes | No | Sequential Insertion |
| Park et al. [6] | 2016 | Inventory Routing Problem | Yes | Yes | No | No | Yes | No | Genetic Algorithm |
| Nugroho et al. (this study) | 2020 | Vehicle Routing Problem with a tanker truck | Yes | Yes | Yes | Yes | Yes | No | Sequential Insertion |

## 2. Methods

### *2.1. Vehicle routing problem*

According to [7], Vehicle Routing Problem or VRP is one of the most studied optimization problems related to the optimal route design for vehicles to use to serve a set of customers. Where customers can be served at time intervals. Certain vehicle travel schedules and needs to be determined. Here are four general objectives of VRP [8]:

1. Minimizing global transportation costs, related to distance and fixed costs related to vehicle use,
2. Minimizing the number of vehicles or drivers needed to serve consumer demand,
3. Balancing routes in terms of travel time and vehicle load, and

4. Minimize penalties due to unsatisfactory services to consumers, such as the inability to serve in full or late delivery.

*2.2. Several classes or variations of the main problems in VRP*

According to [8], there are several classes or variations of the main problems in VRP, namely:

1. Capacitated VRP (CVRP), is the simplest and most studied VRP class where the constraints are only in the form of limited vehicle capacity,
2. Distance Constrained VRP (DCVRP), is a VRP with constraints on the allowable distance travelled,
3. VRP with Time Windows (VRPTW), which is a VRP case where every customer has a service time limit,
4. VRP with Pickup and Delivery (VRPPD), is a VRP with mixed services, namely shipping and taking goods in one route,
5. VRP with Backhauls (VRPB), where the new collection can be done after all shipments are complete,
6. Split Delivery VRP (SDVRP), where consumers are served using different vehicles,
7. VRP with Multi Depot (MDVRP), where vendors use more than one depot to send consumers, and
8. Periodic VRP, where delivery is done within a certain period.

*2.3. Heterogeneous Fleet Vehicle Routing Problem Split Delivery Multiple Product Multiple Trip and Time Windows (HFVRPSDMPMTTW)*

According to [2], in determining the delivery route to achieve the goal of minimizing shipping costs, time, and distance, there are obstacles that must be resolved, namely:

1. Heterogeneous Fleet, the vehicle has specifications, compartments, and different types so that the capacity is also different, and each type of vehicle has a limited amount.
2. Multiple Trip is a vehicle capable of serving one or more customers in one delivery tour.
3. Time Windows, not only limited to depot and customer operating hours but also includes the time that affects the planning horizon such as setup time, loading time, and discharging time.
4. Multiple Products are types of products that are sent more than one.
5. Split Delivery is the delivery of products to customers that can be done directly with one shipment or several times of delivery.

Based on the characteristics above, the mathematical model of the Heterogeneous Fleet Vehicle Routing Problem Split Delivery Multiple Product Multiple Trip with Time Windows will be used as follows:

- Notation:
  - $i$ : location index; $i = 0$ is depot, $i = 1,2,\dots,N$ is customers
  - $t$ : tour index; $t = 1,2,3,\dots,NT$
  - $r$ : route index; $r = 1,2,3,\dots NR[t]$
  - $p$ : product index; $p = 0,1,\dots,,NP$
  - $k$ : position index; $k = 1,2,\dots,NL[t,r]$
  - $z$ : type of vehicles index; $z = 1,2,\dots,Z$
- Parameters:
  - $N$ : Number of $i$ set customers
  - $NP$ : Number of product type
  - $q_{L[t,z,r,k,p]}$ : Number of demands for $p$ product, $k$ location, $r$ route, $t$ tour, and $z$ vehicle
  - $Ws$ : Set up time
  - $LT$ : Loading time
  - $DT$ : Discharging time
  - $vz$ : Vehicle speed

| | | |
|---|---|---|
| $\tau_{[L[t,z,r,k]].[L[t,z,r,k+1]]}$ | : | Travel time between $k$ location to $k+1$, $r$ route, $t$ tour, $z$ vehicle |
| $Qz[p]$ | : | Compartment capacity for $p$ product in vehicle $z$ |
| $a_i^h$ | : | Lower bound for hole time windows $h$, $i$ locations |
| $b_i^h$ | : | Upper bound for hole time windows $h$, $i$ locations |
| $H$ | : | Horizon planning |

- Variables

| | | |
|---|---|---|
| $NV$ | : | Number of vehicles |
| $NTz$ | : | Number of tours by $z$ vehicle |
| $NR[t,z]$ | : | Number of routes in the $t$ tour by $z$ vehicle |
| $NL[t,z,r]$ | : | Number of locations on the $r$ route in the $t$ tour by $z$ vehicle |
| $L_{t,z,r,k}$ | : | Location at $k$ position, $r$ route, in $t$ tour by $z$ vehicle |
| $b_{Lt,z,r,k,p}$ | : | Load delivered on $r$ route, $t$ tour of $z$ vehicle for p product (unit in volume) |
| $y_{Lt,z,r,k,p}$ | : | The proportion of $p$ product shipments on $r$ route, $t$ tour, $z$ vehicle, and $k$ location |
| $Wp$ | : | Travel time |
| $Wt_{L,t,z,r,k}$ | : | Waiting time for vehicle in location at $k$ position, $r$ route, in $t$ tour by $z$ vehicle |
| $Jt_{L[t,z,r,k]}$ | : | Arrival at $k$ location, $t$ tour, $z$ vehicle, and $r$ route |
| $WLT$ | : | Loading time |
| $WDT$ | : | Discharging time |
| $Js_{L[t,z,r,NL[t,z,r]]}$ | : | Finished visit time at $NL[t,r]$ position on the $t$ tour, $z$ vehicle, and r route |
| $Jm_{L[t,z,r,NL[t,z,r]]}$ | : | Departure at $NL[t,r]$ position on the $t$ tour, $z$ vehicle, and $r$ route |
| $CT[t,z]$ | : | Completion time for $t$ tour by $z$ vehicle |
| $TCT$ | : | Total completion time |
| $RCT$ | : | Range of completion time |
| $TCD$ | : | Total distribution cost |
| $C_{SK}$ | : | Vehicle fixed cost |
| $C_{BB}$ | : | Fuel cost |
| $C_L$ | : | Loading/uinloading cost per unit |
| $C_G$ | : | Driver's cost for each visit |
| $C_M$ | : | Accommodation costs |
| $C_R$ | : | Retribution cost per vehicle per tour |

- Problem HFVRPSDMPMTTW can be formulated as follows:

$$Min\ NV = \sum NTz \tag{1}$$

$$Min\ TCT = \sum_{t=1}^{NT} CT[t,z] \tag{2}$$

$$Min\ RCT = Max\{CT_{[t,z]}\} - \min\{CT_{[t,z]}\} \tag{3}$$

$$Min\ TCD = \sum_{r=1}^{NR} \sum_{i=0}^{N} \sum_{j=1}^{N+1} C_{SKz} + C_g N_R + \sum_{r=1}^{NR} \sum_{z=0}^{Z} C_{BB} V_z W_p + 3C_M H N_v + 2\sum_{r=1}^{NR} \sum_{i=0}^{N} q_i C_l N_r + C_R N_V \tag{4}$$

subject to

$$L_{t,z,r,1} = L_{t,z,r,NL[t,r]} = 0 \tag{5}$$

$$L_{t,z,r,k} = i \tag{6}$$

$$NTz = NV \tag{7}$$

$$L_{t,z,r,k} = L_{t,z,r-1,NL[t,z,r-1]} = 0 \tag{8}$$

$$L_{t,z,r,k} \geq 1 \tag{9}$$

$$b_{Lt,z,r,P} \leq Q[p] \tag{10}$$

$$b_{Lt,z,r,P} \leq b_{Lt,z,r,P} + Q[p] \tag{11}$$

$$b_{Lt,z,r,P} \leq \sum_{k=2}^{NL[t,z,r]-1} q_{Lt,z,r,k,P} \cdot y_{L,t,z,r,k,P} \tag{12}$$

$$\sum_{t=1}^{NT} y_{L,t,z,r,k,P} = 1 \tag{13}$$

$$Q[p] = Q[p] - q[i * p] \tag{14}$$

$$\wedge_{h=1}^{mi}(a_i^h \leq Ti \leq b_i^h) \tag{15}$$

$$Wp = \tau_{[L[t,z,r,k]].[L[t,z,r,k+1]]} \tag{16}$$

$$WLT = \frac{bLt,z,r,k,p}{LT} \tag{17}$$

$$WDT = \frac{bLt,z,r,k,p}{DT} \tag{18}$$

$$CT_{[t]} = W_s \sum_{r=1}^{NR} \sum_{r=1}^{NL_{[t,z]}} L_{t,z,r,k} + (WDT + WLT) \sum_{r=1}^{NL_{[t,z]}} \sum_{i=1}^{n} b_{Lt,z,r,p} + \sum_{r=1}^{NR_{[t,z]}} \sum_{k=1}^{NL_{[t,z,r]}-1} \tau_{[L[t,z,r,k],[L[t,z,r,k+1]]]} + \sum_{r=1}^{NR_{[t,z]}} \sum_{k=2}^{NL_{[t,z,r]}-1} Wt_{Lt,z,,k} \tag{19}$$

$$CT_{[t]} = Js_{L[t,z,r,NL[t,z,r]]} \tag{20}$$

$$Jm_{L[t,z,r,NL[t,z,r]]} = JS_{l[t,z,r,NL[t,z,r]]} \tag{21}$$

$$Jm_{L[t,z,r,NL[t,z,r]]} = JS_{l[t,z,r,NL[t,z,r]]} + Wp \tag{22}$$

$$Wt_{L[t,z,r,k]} = \begin{cases} b_i^h - JtL[t,z,r,k], b_i^h \geq JtL[t,z,r,k] \\ 0, b_i^h \leq JtL[t,z,r,k] \end{cases} \tag{23}$$

$$CT_{[t,z]} \leq H \tag{24}$$

$$TCT = \sum_{t=1}^{NT} CT_{[t,z]} \tag{25}$$

The steps of forming the distribution route used to refer to the model used by [2] by modifying the departure step. The steps for forming a route are as follows:

Step 0:
a. Initiate the number of customers that have not been served and the number of tours, the number of vehicles, and the time of completion of the tour is 0 (not yet formed).

Step 1:
a. Establishment of the first tour and first route for the vehicle (ntz = ntz + 1) with the initial vehicle load (blt, z, r, p = 0).

Step 2:
a. Insert the customer that has not been served into the route that has been formed.
b. Set departure time (jml [t, z, r, nl [t, z, r]]) = 0. (first vehicle)
c. Set departure time (jml [t, z, r, nl [t, z, r]]) = setup time (ws) + loading time (wlt) of the previous vehicle. (2nd vehicle and so on)
d. Calculate loading time (wlt) using equation (17), discharging time (wdt) using equation (18) and calculate tour completion time (ct) based on equation (19) for each vehicle.
e. If ct> h, continue to step 8. If ct<h, select the customer who gives the shortest tour completion time and proceed to step 3.

Step 3:
a. Calculate the vehicle load (blt, z, r, P) based on Equation (12) and vehicle compartment (Q [p]) based on Equation (14).
b. If customer request q [i * p]> Q [p], then the request will be fulfilled by using another vehicle (split delivery).
c. Update the number of customers and proceed to step 4.

Step 4:
a. If all customers have been served, continue to step 9. Instead, continue to step 5.

Step 5:
a. Set the departure time (jml [t, z, r, NL [t, z, r]]) based on Equation (21)
b. Calculate loading time (WLT) using Equation (17), discharging time (WDT) using Equation (18) and calculate tour completion time (CT) based on Equation (19) for each vehicle.
c. If CT <H and q [i * p] <Q [p], proceed to step 6. If CT <H and q [i * p]> Q [p], continue to step 7. If CT> H, proceed to step 8.

Step 6:
a. Calculate the vehicle load (blt, z, r, P) based on Equation (12) and vehicle compartment (Q [p]) based on Equation (14).
b. If customer request q [i * p]> Q [p], then the request will be fulfilled using another vehicle (split delivery).
c. Update the number of customers and proceed to step 4.

Step 7:
a. Establish the next route (r = r + 1) for the vehicle (ntz = ntz + 1). Return to step 1.

Step 8:
a. Establish the next tour (t = t + 1) for the first vehicle route (ntz = ntz + 1). Return to step 1.

Step 9:
a. Set the number of vehicles based on Equation (7). Calculate the total CT based on Equation (2), calculate the time span based on Equation (3).

Step 10:
a. Calculate the TCT based on Equation (4)

## 3. Result and Discussion
Based on calculation using sequal insertion method, the route of each truck tanker, total distance and completion hour was solved and it can be summarized on Table 2 below.

**Table 2.** Route recapitulation of sequential insertion method

| No | Route | Vehicle Capacity (Kl) | Total Distance (Km) | Completion Time (Hour) |
|---|---|---|---|---|
| 1 | Depot-S18-S03-S20-S21 | 32 | 33.8 | 3.998 |
| 2 | Depot-S27-S29-44.552.18 | 32 | 26.3 | 3.661 |
| 3 | Depot-S24-S25-S02-S01 | 32 | 44 | 4.92 |
| 4 | Depot-S19-S04 | 32 | 23 | 3.796 |
| 5 | Depot-S04-S09-S21-S07 | 32 | 43 | 5.629 |
| 6 | Depot-S07-S10-S08 | 24 | 45 | 5.366 |
| 7 | Depot-S08-S17-S11-S15 | 32 | 83.3 | 7.371 |
| 8 | Depot-S26-S01 | 32 | 40.3 | 5.829 |
| 9 | Depot-44.552.06-S05-S13 | 24 | 82.6 | 7.407 |
| 10 | Depot-44.5561.12-S12 | 32 | 93.2 | 7.819 |
| 11 | Depot-S13-S23 | 24 | 88 | 7.826 |
| 12 | Depot-S14-S06 | 24 | 90.8 | 8.197 |

Table 2 shows the results of the route recapitulation using the insertion sequence method. The vehicle with the smallest capacity will result in faster tour completion time. However, a smaller vehicle capacity will require a greater number of vehicles to meet demand. As well as the start time of vehicle departure does not start simultaneously, causing the start time to be longer for the next vehicle. Therefore, larger capacity vehicles are preferred to meet customer demand. The following comparison of distribution routes using the sequential insertion method and the original can be seen in Table 3.

**Table 3.** Comparison of sequential insertion method with the original value (Pertamina co. LTD)

| No | Variable | Sequential Insertion Method | Original Value |
|---|---|---|---|
| 1 | Total Distance | 693.3 | 701.5 |
| 2 | Number of Vehicles | 12 | 13 |
| 3 | Total Completion Time | 71.819 | 81.819 |
| 4 | Completion Time Range | 4.536 | 6.252 |
| 5 | Total Distribution Cost | Rp. 20,754,119,875 | Rp. 18,511,210,625 |
| 6 | Total number of customers who are served late | 1 | 3 |

Based on the above comparison using the sequential insertion method produces a total distance of 693.3 km. It shows that the vehicle mileage using the sequential insertion method is smaller than the distance of the original, which is 701.5 km. The total number of customers who are served late using the sequential insertion method is 1. The total number of customers can be 0 if the total vehicle becomes 13, and there is one vehicle that does not have a maximum load of 16 kl capacity vehicles that only carry eight kl. However, this is better than before, namely that three customers have not been served. The sequential insertion method produces total tour completion time (TCT) and tour completion period (RCT) of 71,819 hours and 4,536 hours. It is caused by a smaller mileage, and a smaller number of vehicles (NV) compared to the original. The total distribution cost (TCD) using the sequential insertion method is Rp. 20,754,119,875.00. It shows that the cost of distribution using the sequential insertion method is greater than that of the original. It is due to the number of vehicles with a capacity of 32 kl more and the number of fixed costs for a 32 kl capacity vehicle of Rp 1,800,000,000.00.

## 4. Conclusion

We have solved the Vehicle Routing Problem in fuel distribution by using the sequential insertion method. The solution to the problem is done by selecting gas stations based on the smallest tour completion time. In this research, tour completion time is influenced by many factors, namely vehicle start time, setup time, loading time, discharging time, travel time, and waiting time. The case discussed in this study is the distribution of fuel at the Pertamina co. Ltd depot. Calculation results show that the total tour completion time is smaller than 71.819 hours from 81.819 hours. It is influenced by the selection of gas stations and the type of vehicle used so that the total distance and number of vehicles becomes less. Also, the number of customers served late decreased from 3 stations to 1 station. However, the sequential insertion method produces greater distribution costs due to the number of vehicles with a capacity of 32 kl has a larger fixed cost despite the larger capacity of 32 kl. Further research can be suggested as follows: (1) Need to consider other products and vehicle availability, (2) Need to consider customer goals in the block area, (3) Need to consider using a settlement method, (4) Using another method to solve the problem, like simulations and the Clark and Wright method, (5) Need to consider using metaheuristic, so it can be used for larger data, and route formation can be faster.

## 5. References

[1] Suprayogi and Priyandari Y 2008 Algoritma Sequential Insertion untuk Memecahkan Masalah Vehicle Routing Problem dengan Multi Trip, Time Windows dan Simultaneous Pickup Delivery. *J. Teknik Industri Universitas Sebelas Maret Surakarta,***7**(1), 88-96.

[2] Arvianto A, Setiawan A H, and Saptadi S 2014 Model Vehicle Routing Problem dengan karakteristik rute majemuk, Multiple Time Windows, Multiple Product dan Heterogenous Fleet dengan Depot Tunggal. *J. Teknik Industri Universitas Petra,* **16**(2), 85-96 *DOI:* https://doi.org/10.9744/jti.16.2.83-94

[3] Nugroho S M 2018 *Analisis Distribusi BBM Menggunakan Vehicle Routing Problem dengan Mempertimbangkan Heteregeneous Fleet, Split Delivery, Multiple Product, Multiple Trip, dan Time Windows* (studi kasus PT Pertamina Depot Rewulu Yogyakarta), undergraduate thesis unpublished, Industrial Engineering Department, Universitas Pembangunan Nasional Veteran Yogyakarta

[4] Windyatri W 2019 *Heterogeneous Vehicle Routing Problem with Multi Trips Untuk Penentuan Rute Distribusi BBM,* graduate thesis. Industrial Engineering Department, Universitas Gadjah Mada

[5] Abadi C, Susanty S, and Adianto H 2014 Penentuan Rute Kendaraan Distribusi Produk Roti menggunakan metode Nearest Neighbour dan Metode Sequential Insertion. *J. Teknik Industri Institit Teknologi Nasional (Itenas) Bandung,***01**(3), 152-163.

[6] Park Y B, Yoo J S, and Park H S 2016 A genetic algorithm for the vendor-managed inventory routing problem with lost sales. *Expert Systems With Applications* **53** pp 149-159

[7] Golden B L, Raghavan S, and Wasil E A 2008 *The Vehicle Routing Problem: Latest Advances and New Challenges*. Springer

[8] Toth P, and Vigo D 2002 *The Vehicle Routing Problem* SIAM. Philadelphia

# Analysis of socio-technical approach and Socio User Experience Network Analysis (SNA) to address objections ergonomic loom in the village of Lumban Suhi-Suhi

**Huda, L N[1], and Giffari, A F[2]**

[1]Department of Industrial Engineering Universitas Sumatera Utara, Kampus USU, Medan, Indonesia
[1]lnuruh@gmail.com

**Abstract.** Technological change is a good thing, but not necessarily be accepted by users. There are several factors that can cause rejection of the technology by a community. This paper aims to present a socio-technical analysis of rejection looms with ergonomic design in the village of Lumban Suhi-Suhi, Samosir Island and efforts should be made so that the tool can be accepted by the community. To achieve this objective, conducted a socio-technical approach using Socio Network Analysis (SNA) and the analysis of User Experience (UX). Data on the UX include effectiveness, efficiency, ease of use, usability, satisfaction, and attractiveness used to assess the performance of looms and looms ergonomic long as a technical comparison. While the relationship between the individual craftsmen were also collected to determine the most influential figures in the environment craftsmen as materials for SNA. The results showed that there are some gaps UX values between the two types of tools. Recommended strategy is to change the seat backrest ergonomic loom and perform social approach on three key figures in the community.
**Keywords**: User Experience, Socio Network Analysis, Ulos, Ergonomic Looms, Technology Rejection

## 1. Introduction

Ulos is one of the famous traditional cloths cultural heritage of Indonesia. One of the central region Ulos craftsmen were on the island. The loom used in this area is called Partonungan. This tool is a tool that has been used legacy passed down from ancestors to today's generation. The design has never been change trough the time passed. Based on academic preliminary research obtained that Partonungan looms cannot fulfil the standards of ergonomics because the weavers are using Partonungan in a bent position to cause a risk of Musculoskeletal Disorder (MSDS). On the other hand, Lumban Suhi-Suhi village weavers used to weave in a long time every day. If they are pursuing a target for the market (onan), weaver can weave for seven hours a day non-stop. The work posture of worker using Partonungan is shown in Figure 1.

**Figure 1.** Work position Villagers weave-Lumban Suhi-Suhi

Based on the above problems, through the activities of the Working Group Toba Lake Region Tourism and Sustainable Tourism (KK-PKDT & PB) produced a tool called "ergonomic loom". Improvements made is the addition of benches, footrests, wooden construction that is lighter, and the removal of wood to bind the seat and seat looms that are not tied to the loom. The purpose of this improvement is that the weaver avoid stooped posture, thereby reducing the risk of Musculoskeletal Disorders (MSDs) due weaving activities carried out for 5-7 hours per day. The design of the ergonomic loom shown in Figure 2.

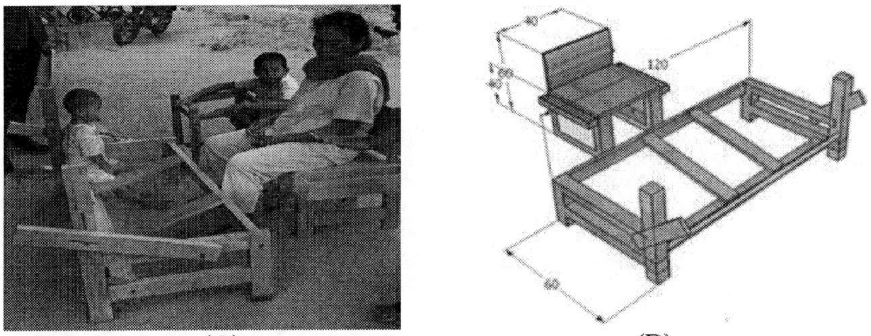

(A) (B)

**Figure 2.** Ergonomic Design of Weaving Tool (A) Actual Edition II
(B) Sketchup Design of Weaving

Ergonomic looms as research results have been produced and handed to the craftsmen. However, based on the reviews after the handover, this tool is no longer use by the weavers as expected. This condition indicating that the loom should be improved and so it can be re-used by the weavers. This paper analyses the factors that cause rejection of the use of the ergonomic loom and re-design solutions offer of ergonomic loom so ergonomic looms can be reusable.

Obstacles in applying the new technology is no longer a new thing in the world of industry. Several approaches have been carried-out to solve this problem. As performed Long, TB., Et al. [2] analyzed the situation of the producers and consumers of the barriers that lead to technological adaptations ever climate-smart agriculture in Europe. More research is about the difficulty of adaptation technologies decentralized renewable energy systems in Africa even though these technologies can improve living standards [3]. The study also reviewed the socio-technical aspects of society and its relationship with technology acceptance barriers. Both of these earlier studies, studied the habits of people to determine the source of the problems of adaptation technology barriers and assesses the technical side technology that is incompatible with the values of society. Follow the same logic, this study will also move from the public social analysis and studies to relate to the technical aspects of the technology, which will be assess by using User Experience parameters. After that recommendations will be generated to increase the adoptability of the proposed technology.

Based on preliminary research, many people refuse to use ergonomic loom because they wanted to maintain the tradition to use Partonungan looms that was inheritance in their family. However, not all families have such a legacy looms, only they still refused to use the new looms, influenced by fellow craftsmen. These conditions need to be analysed to see who are the key figures that can be a bridge to instil the idea that good ergonomic loom used. This will be done by using Social Network Analysis (SNA).

## 2. Methods

### *2.1. Data collection*

The data in this study is collected by questionnaires and direct observation in the Lumban Suhi-Suhi Village. There are 28 native artisan Involved in the study range from 10 to 46 years of experience. They are the expert of weaving from the area. This study sought to explore the barrier of the adoption of the ergonomics loom and to identify the technical aspect that could be enhanced to secure the adoption of the technology. This study was undertaken in Lumban Suhi-Suhi, Samosir Island, North Sumatra Province.

Data about the series of tasks done by artisan when weaving was collected as the subject of comparison in the technical use of each kind of loom. There are various questionnaires being used in this research, they are:

a. Single Ease Question (SEQ) questionnaire to see how satisfied respondents when using a traditional loom (Partonungan) and ergonomic loom.
b. *System Usability Scale* (SUS) questionnaires to see if traditional looms (Partonungan) and ergonomic loom are acceptable or not.
c. Kuesioner Atrakkdiff to gauge the views of respondents in terms of satisfaction with traditional loom (*Partonungan*) and ergonomic loom used.
d. Social Network Analysis (SNA) questionnaires aims to determine the interrelationship among weavers in order to discover which figures that may affect reception loom weavers in the community. Questionnaire form is open questionnaire with three questions that are used are:
   1) Among the crafts-men who had received the ergonomic loom, to whom do you learn to use the tool?
   2) When you have problems with the tool, to who do you comfortable to talk to?
   3) When you feel uncomfortable in using the ergonomic loom, to whom do you communicate to decide to use or not to use the loom?

In this paper, the evaluation of answering of question 1 will discuss more intensively.

### *2.2. Data Analysis*

Mainly two kind of analysis was being done. They are the social analysis and technical analysis. The technical approach will be carried out by the use of analysis User Interface. User Experience or UX commonly abbreviated as an overall experience of the user when using the product, system, or service. For the analysis of UX, the factors used in this study including Effectively which calculated by the success rate of each task and Efficiency which is calculated by the minute of doing the task, Ease of Use which is calculated using SEQ, Usability using SUS [5], and Satisfaction using Attrakkdiff Questionnaire. Social analysis performed using SNA analysis. According Binsar and Dana [4] *Social Network Analysis* is a technique for studying the relationship or social relations among members of a group. Mapping knowledge within the framework of Social Network Analysis can be visualized or be represented by a matrix and graph form. Social Network Analysis is a development method for understanding the relationships between actors in a network system. The SNA will be done using software Ucinet 6.0 to get the key person for each question by using eigenvector and degree centrality as the parameter.

After these analyses was conducted, then recommendations will be given according those analyses. Including which part of technical issues to improve and which artisan need to be prioritized in assimilating the frame of mind that the new loom is fine to be used.

## 3. Result and Discussion

### *3.1. Result*

The series of activities to be done in completing a piece of cloth Ulos, as shown in Table 1.

**Table 1.** Tasks in woving Cain Ulos

| *No.* | Task | No. | Task |
|---|---|---|---|
| 1 | Setting up the parts weave | 6 | Tidy up the thread, with Balobas |
| 2 | Installing Tundalan to waist | 7 | Cross main thread with Turakipahan |
| 3 | Tighten thread with Baliga | 8 | Cross yarn pattern with Turak kilat |
| 4 | Separating yarn with panggiunan | 9 | Shifts Balobas to above sidurukan |
| 5 | Set the fabric pattern with Lilis | 10 | Reconstitute looms after used |

Data in Table 1 represents the data used in the assessment of SEQ questionnaire for both types of looms. These task will be the technical value that determined the performance of each looms.

*3.1.1. User Interface (UX) Analysis.* UX analysis includes several dimensions, they are Effectivity and Efficiency, Ease of Use, Usability, and Satisfaction. Each dimension was rated for both types of looms to see performance comparison between the two. The difference in value between the two looms perspective is a value that would be ergonomic loom repair parameters.

Dimension Effectiveness rated to determine the level of success in performing the task [6], while the efficiency dimension is used to determine the time required to perform each task in Table 1 for each tool. Figure 3 shows the results of measurements of the dimensions of effectiveness.

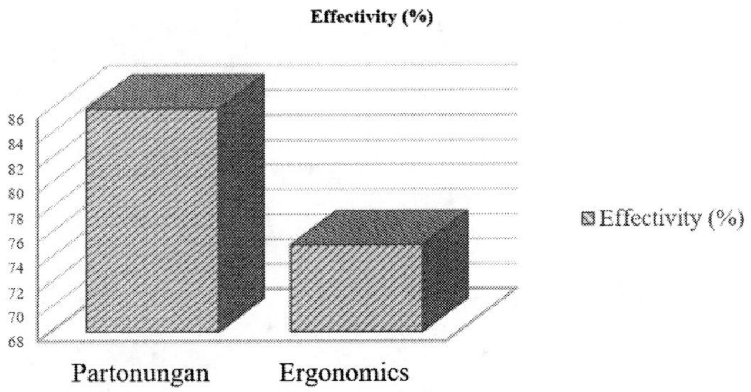


**Figure 3.** Effectivity value comparison between Partonungan and Ergonomics Loom

Table 2 shows the results in measurement efficiency of each task using both tools. The results are comparing.

**Table 2.** Efficiency value comparison between Partonungan and Ergonomics Loom

| Partonungan (Minute) | | Weaving Tool Ergonomics (Minute) | |
|---|---|---|---|
| *Task 1* | 168 | *Task 1* | 182 |
| *Task 2* | 55 | *Task 2* | 181 |
| *Task 3* | 144 | *Task 3* | 167 |
| *Task 4* | 153 | *Task 4* | 183 |
| *Task 5* | 347 | *Task 5* | 518 |
| *Task 6* | 151 | *Task 6* | 222 |
| *Task 7* | 45 | *Task 7* | 70 |
| *Task 8* | 41 | *Task 8* | 67 |
| *Task 9* | 44 | *Task 9* | 135 |
| *Task 10* | 46 | *Task 10* | 111 |
| **Total** | **1194** | **Total** | **1836** |
| **Average** | **119.4** | **Average** | **183.6** |

Based on Figure 3 and Table 2 above, the Partonungan looms are more effectively and efficiently used because it has fewer errors and faster to use. This can be caused by several factors, such as craftsmen has not been accustomed with a new engine, or the new machine has some bugs in its use.

Dimensions Ease of Use calculated knowing how easily respondents in using Partonungan looms and ergonomics looms during a series of weaving activities. Determination of Ease of Use questionnaires Single Ease Question (SEQ). Comparison of the results of calculating the dimensions of Ease of Use is shown in Figure 4.

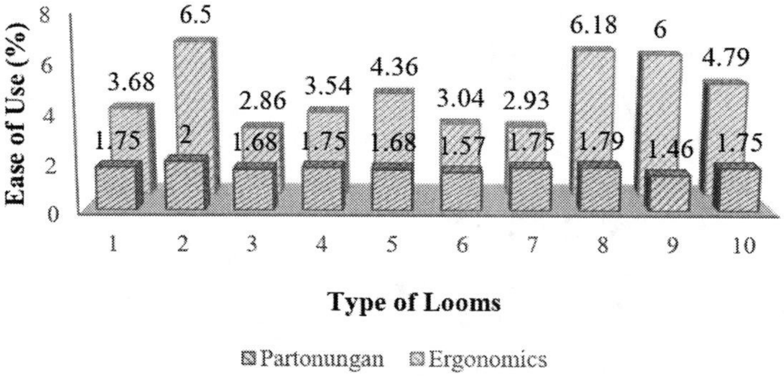


**Figure 4.** Comparisons of Weaving Tools Ease of Use Dimension between Ergonomic Looms and Partonungan

From the chart above we can conclude that the average weaver find it very difficult to perform a second task that is putting tundalan to their waist on ergonomic loom that obtain a value of 6.5 which means weaver very difficult to do the task. The smallest value obtained on the 3rd task that is tighten threads with baliga with a value of 2,86 which means it is easy to do.

Usability dimension analyzed by looking at the total value of the use of ergonomic loom and Partonungan based on questionnaires System Usability Scale (SUS). Based on the calculation Partonungan loom that get a score value of 82.86 were included into the excellent category and ergonomic loom score was 56.61 which included into good category. This shows that looms ergonomic considered quite beneficial in weaving, but more craftsmen find it easy to use the loom Partonungan.

Satisfaction dimension obtained from the result of questionnaires Attrakdiff recapitulation. The average satisfaction value of the Partonungan loom is at 40.07 which is higher than its value ergonomic loom which only valued at 29.21. Based on the questionnaire attrakdiff, it is found that the average value of each dimension of satisfaction from the quality of pragmatic, hedonic quality-identity, hedonic quality-simulation, and attractiveness, looms Partonungan has a higher value than the ergonomic loom. The lowest value was on pragmatic quality which is compatibility of tools functions to achieve its goals. This shows that the ergonomic looms less attractive because it can satisfy the expectations / the needs of the users in terms of its functionality.

*3.1.2. Socio using SNA.*SNA calculation results are grouped on the answers of the three questions on the questionnaire SNA (see section 2.1 (d)). The result analysis SNA diagram of question 1 using Ucinet 6.0program is shown in Figure 5.

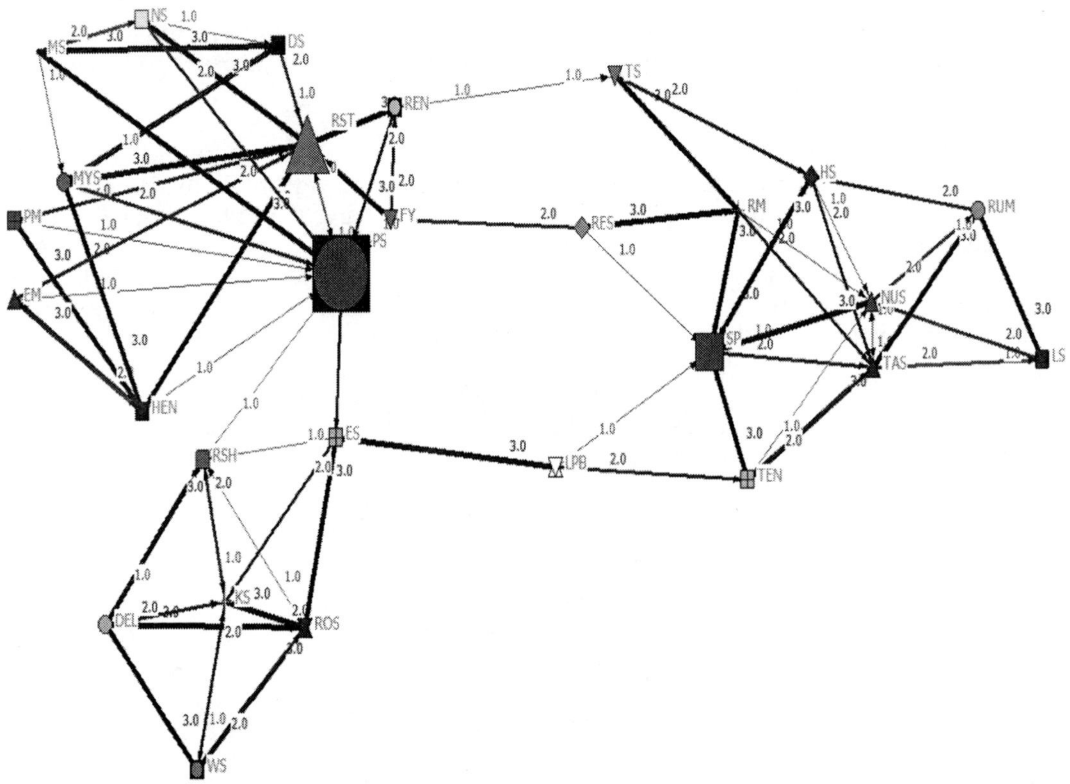


**Figure 5.** Relationship diagrams Question 1

Figure 5 shows the results of the first question is "Among the craftsmen who had received the ergonomic loom, to whom do you learn to use the tool?". The biggest green circle marks show the craftsmen who have the highest voting results based on questionnaires SNA. The quantitative indicators are the score of eigenvector and degree centrality. Figure 5 indicate that the craftsman with the initials RST is a figure who became a role model in the study of ergonomic loom. Recapitulation of workers with the highest score for the third SNA question are shown in Table 3.

**Table 3.** The most influential social craftsmen in the community based SNA Analysis

| No. | Question | Initials Name Craftsmen and Scores | |
|---|---|---|---|
| | | scores eigenvector | Score Degree centrality |
| 1 | Among the craftsmen who had received the ergonomic loom, to Whom do you learn to use the tool? | 1. RST = 0.429<br>2. PS = 0.387<br>3. MYS = 0.323 | 1. RST = 22.00<br>2. PS = 18.00<br>3. SP = 16.00 |
| 2 | When you have problems with the tool, to Whom do you comfortable to talk to? | 1. PS = 0.520<br>2. MYS = 0.335<br>3. RST = 0.291 | 1. PS = 25.00<br>2. RST = 19.00<br>3. TAS = 15.00 |
| 3 | When you have problems with the tool, to Whom do you comfortable to talk to? | 1. PS = 0,526<br>2. RST = 0.300<br>3. MSY = 0,296 | 1. PS = 25.00<br>2. RST = 17.00<br>3. NUS = 16.00 |

### *3.2. Discussion*
Based on this research, ergonomic loom has many shortcomings in its performance compared Partonungan looms. So it required some adjustments so that the loom can be accepted by the craftsmen. The strategy proposed for improvements include:

1. Technical factors

   Based on the results of the assessment several dimensions UX, Partonungan looms have a better performance than the ergonomic loom. The main problem of ergonomic looms located on the movement constraint in using a loom ergonomic for weaving, thereby reducing the sense of comfort and productivity craftsmen. Dimensions of Effectivity, Efficiency, and Ease of Use showed the lowest value in task 2which is to install tundalan to their waist. This is because the backrest on the seat which restrict the movement of the craftsmen's elbow. Hence the backrest of the chair are advised to be fixed so as not to disturb the movement of craftsmen in weaving.

**Figure 6.** Position ergonomic backrest on Ergonomic looms

2. Factors Social Analysis

   Each community has a different social dynamics depending on the circumstances of each. There are communities that tend to move in the interests of the group [7] and [8] and some are individual. Based on social factors, figures that may affect acceptance loom weavers in the use of the character ergonomic RST, PS, and MYS. These people are the most senior figures correspond to weavers' social culture community in the Lumban Suhi-Suhi Village. These characters are people who has the longest experience to work as a weaver and has the greatest social status in the community of craftsmen. This is in accordance with the customs of local culture, which is still very strong with traditional Batak culture. This trend indicates that the weavers' community will tend to follow the advice or direction of the three characters.

IOP Conf. Series: Materials Science and Engineering **847** (2020) 012067 doi:10.1088/1757-899X/847/1/012067

## 4. Conclusion

Transition of technology is concerning not only the potential benefits to be gained but also the adoptability of the technology to the users. In this study, the ergonomically designed loom is not well accepted by the craftsmen because they prefer to use their family legacy looms. This condition is not desirable, because the old designed could cause the MSDS for the craftsmen in the long term of use. The socio-technical approach is done to increase of the adoptability of the new technology. For the technical aspect, it is found that the old loom design is more effective and easy to use compare to the new design. The main cause is the back of the chair, which impede the movement of craftsmen at work. To facilitate the social approach, then there are three figures that should be prioritized is RST, PS, and MSY who has a big social influences over the weavers in this community.

## 5. References

[1] Allam, A H 2010 User Experience: Challenges and Opputunities. Journal Of Information Systems Research and Innovation, 28-36.

[2] L Nurul H, Nurlisa Ginting, Devi Azwar, and Indra 2018 *Proc. the 4th UPHEC* Yogyakarta: UPHEC 2018, **4,** 7-16.

[3] TB Long, Vincent Blok, Ingrid Coninx 2015 Barriers to the adoption and diffusion of technological innovations for climate-smart agriculture in Europe: evidence from the Netherlands, France, Switzerland and Italy *J. Cleaner Production* **112(1),** 9-21.

[4] Basil AS, Kirsten Martinus, Petra Tschakert, Raymond Willis 2018 Disruptive innovations and decentralized renewable energy systems in Africa: A socio-technical review *J. Energy Research &Social Science* **46,** 140-154.

[5] Agung DL dan Ratna DW 2011 Analisis potensi penyebaran informasi kesehatan melalui jejaring sosial (studi kasus pada 'forum jejaring peduli AIDS') *J. Buletin of Health System Research* **14(4),** 258-365.

[6] Wetzlinger, Werner & Auinger, Andreas & Dörflinger, Michael 2014 Comparing effectiveness, efficiency, ease of use, usability and user experience when using tablets and laptops *Lecture Notes in Computer Science* **8517**, 402-412.

[7] William Albert and Thomas Tullis 2013 *Measuring the User Experience 2nd Edition* (Wallltham: Elsevier Inc), 236-251.

[8] Anna Ramirez 2013 The influence of social networks on agricultural technology adoption *J. Procedia -Social behaviorial Sciences* **79,** 101-11.

[9] Julio Díaz-José, Roberto Rendón-Mede, Bram Govaerts, Jorge Aguilar-Ávila, Manrrubio Muñoz-Rodriguez 2015 Innovation Diffusion in Conservation Agriculture: A Network Approach *J. The European Journal of Development Research* **28,** 314-329.

[10] Barriers to the adoption and diffusion of technological innovations for climate-smart agriculture in Europe: evidence from the Netherlands, France, Switzerland and Italy.

# Determinants of Donation Behaviour on Flood Disasters in Indonesia

**D. S. Utomo[1], N. Paoprasert[2], and R. Yousuk[3]**

[1,2,3] Department of Industrial Engineering, Kasetsart University, 50 Ngam Wong Wan Road, Lat Yao, Chatuchak, Bangkok 10900, Thailand
[1]duthosuh.u@ku.th, [2] fengnpp@ku.ac.th, [3] fengryy@ku.ac.th

**Abstract.** This study used Structural Equation Modeling to apply Theories of Planned Behavior to flood disaster donation by adding attitude factors to charitable organizations, advertising, and flood victims. The proposed framework is tested to investigate the relationship between factors in forming intentions and in influencing behavior. This finding found that subjective norms, attitudes toward advertising and attitudes towards charitable organizations contributed to the intention of donations, while the flood victims and perceived behavioral control did not have a significant effect on donation behavior.
**Keywords**: Donation, Intention, Natural Disaster, Flood

## 1. INTRODUCTION

When natural disasters occur, many people feel empathy for the victims. Some decide to donate, while some just pray for them and refuse to take any action. Severe natural disasters may be neglected, especially if the media also ignores them. For example, the 2010 earthquake in Haiti that killed approximately 160,000 lives, raised $13 billion in donations during the early days of the disaster. However, in 2008, the Cyclone Nargis in Bangladesh and Myanmar that killed approximately 138,000 lives, raised only about $300 million [1].

Indonesia faces some natural threats including earthquakes, tsunamis, volcanic eruptions, floods, and droughts [2]. These sometimes occur repeatedly. Based on data from the Indonesian Disaster Management Agency in 2018, there were 5 major disasters that occurred in Lombok (earthquake), Palu (earthquake, tsunami, and liquefaction), Banten and Lampung (earthquake and tsunami). The total asset loss from the 2 major natural disasters (Lombok and Palu) that occurred in Indonesia was aapproximately $2,45 billion [3]. Flooding is a natural disaster that often occurs in Indonesia. In 2018 there were 679 incidents with around 1.45 million people affected and occurred in various cities and provinces [4]

There have been several studies conducted in the field of donations [5], [6], [7], [8] . Some of them used the theory of planned behaviour (TPB) to investigate the effect of each factor on the donation decision. TPB suggests three determinants of donation intention, namely attitudes, subjective norms and perceived behavioral control (PBC). The theory is based on consideration of people which is reasonable and uses information that might be possible, systematically. People determine the implications of their actions before they decide to do or not do certain behavior [9].

Knowles et. al. [7] studied the intention to contribute money by using perspectives from TPB that are revised and influenced by attitude, subjective norms, moral obligations, PBC, and past behavior. The results of their research indicate that intention was influenced by all factors examined except norms subjective.

The differences in the characteristics of each natural disaster that occured, such as the number of deaths, types of disasters, victims who need assistance, affect the amount of donations received [10]. This implies that the characteristics of disasters may affect decision making to donate. In our paper, the characteristic factors were included as factors that influenced the intention and behavior to donate to natural disasters.

Some studies on donation decisions used a variety of attitudes, such as attitudes towards advertising [11],[12],[13] attitudes helping others [14],[15],[16],[17],[13] attitudes towards charitable organizations [14],[13],[12],[18] and attitudes towards donations [18].

In the application of TPB on general donation behavior, attitude used must be clearly defined. Sometimes, it was found that attitudes toward helping others did not have a direct relationship with the intention to donate but through attitudes towards the organization and attitudes towards advertising [13]. Therefore, it was considered necessary to use all three attitudes in our study to investigate the factors that influence the intention of donating to natural disasters.

In this paper, we aim to propose a framework for testing the relationships of factors that influence donation behaviour towards natural flood disaster by adapting TPBs to add attitudes toward charitable organizations, attitudes towards advertising, and victims of flood disasters.

## 2. METHODS

As mentioned in the introduction, our proposed framework was established based mainly on the TPB. It is the theory that explains how behavioral intentions is influenced by attitudes, subjective norms and PBC [9].

Smith and McSweeney [8] used a revised TPB which consists of attitudes, norms (injunctive norms, descriptive and moral norms), PBC, and past results of behaviour to donate money to charitable organizations. Based on their results, attitude, PBC, injunctive norms, moral norms and past behaviour influenced intentions of charitable giving. However, descriptive norms do not contribute intention to donate.

The TPB model was extended, by Knowles et. Al. [7], used to confirm the variable relationship to intention to contribute. The results of the study show that attitudes, PBC, and past behaviour, moral norms predict the intention to contribute in the next time.

Mittelman and Méndez [18] investigated the extended TPB model of charity giving and the intention to donate to consider three different attitudes (charity, helping, and donation). Past behavior, moral norms, and PBC had a significant influence on the contribution intention; however, attitude toward charitable organization, and subjective norms were not significant.

Ranganathan and Henley [13] investigated religious factors in the donation decision-making process. The variable used in the model are religiosity and three attitudes (attitude towards helping others, attitude towards the advertisement, attitude towards charitable organizations) and behavioral intentions.

Evangelidis and Bergh [10] applied natural disaster characteristic variables from the International Disaster Database managed by CRED [19] as exogenous variables that affect donation decision. It was found that the decision on donations was based on number of deaths rather than the number of victims.

The proposed framework model can be seen in Fig.1. All latent variables are summarized in Table 1. In Figure 1, the relationships were linked based on several previous literatures. Tables 2-9 show the questions used as exogenous variables for each latent variable

**Table 1.** Research Variables

| Variable | References |
|---|---|
| Self-Reported Behavior (B) | [9],[20],[21],[7],[8] |
| PBC (P) | [9],[20],[21], [7], [8],[22] |
| Subjective Norm (S) | [9],[20],[21],[7] |
| Attitude Towards Advertisement (AA) | [11],[12], [13] |
| Attitude Towards Helping Others (AH) | [13],[14],[15],[16],[17], |
| Attitude Towards Charitable Organizations (AO) | [13], [14] |
| Donation Intention (I) | [9],[20],[21],[7],[8] |
| Natural Disaster Victims (V) | [10] |

**Table 2.** Indicators of Self-Reported Behavior

| Indicator | References |
|---|---|
| How often do you donate to past natural disasters? | [7],[8] |
| How many times have you donate to past natural disaster? | [7],[8] |
| How often do you contribute to natural disasters in the past two years? | [7],[8] |
| How many times have you donate to natural disasters in the past two years? | [7],[8] |

**Table 3**. Indicators of Perceived Behaviour Control

| Indicator | References |
|---|---|
| If I want to donate on natural disasters, then I can easily do it | [7],[8] |
| If in the future a natural disaster occurs, how much control do you have to donate to it | [7],[8] |
| Most of the factors that influenced me to contribute to natural disasters are myself | [7],[8] |

**Table 4**. Indicators of Subjective Norm

| Indicator | References |
|---|---|
| People who are close to me support me if I donate to natural disasters | [7] |
| What percentage of people who are important to you does not support you to contribute to natural disasters? | [7] |
| The opinion of most people who are important to me about donating | [7] |

**Table 5.** Indicators of Attitude Towards Advertisement

| Indicator | References |
|---|---|
| My reaction to the advertising of disaster donations | [23] |
| My attitude towards advertising donations of natural disasters | [23] |
| My feelings towards advertising about donations of natural disasters | [23] |

**Table 6**. Indicators of Attitude Towards Helping Others

| Indicator | References |
|---|---|
| Helping others is a must | [14] |
| For me, helping is important | [14] |
| People who need help must be helped | [14] |

**Table 7**. Indicators of Attitude Towards Charitable Organizations

| Indicator | References |
|---|---|

| | |
|---|---|
| Positive image on charities is must | [14] |
| Natural disaster donation organizations do useful things | [14] |
| Charity donation organizations are useful for helping those in need | [14] |

**Table 8.** Indicators of Behaviour Intention

| Indicator | References |
|---|---|
| I have a desire to donate money to natural disaster relief donation organizations in the next time | [7],[8] |
| How likely will you donate to natural disaster in the next time? | [7],[8] |
| I intend to contribute money to the natural disaster donation organization going forward | [7],[8] |

**Table 9**. Indicator of Natural Disaster Victims

| Indicator | References |
|---|---|
| Number of Dead and missing people | [10],[4] |
| Number of injured victims | [4] |
| Number of affected victims | [10],[4] |
| Number of Refugees | [4] |
| Number of damaged houses | [4] |

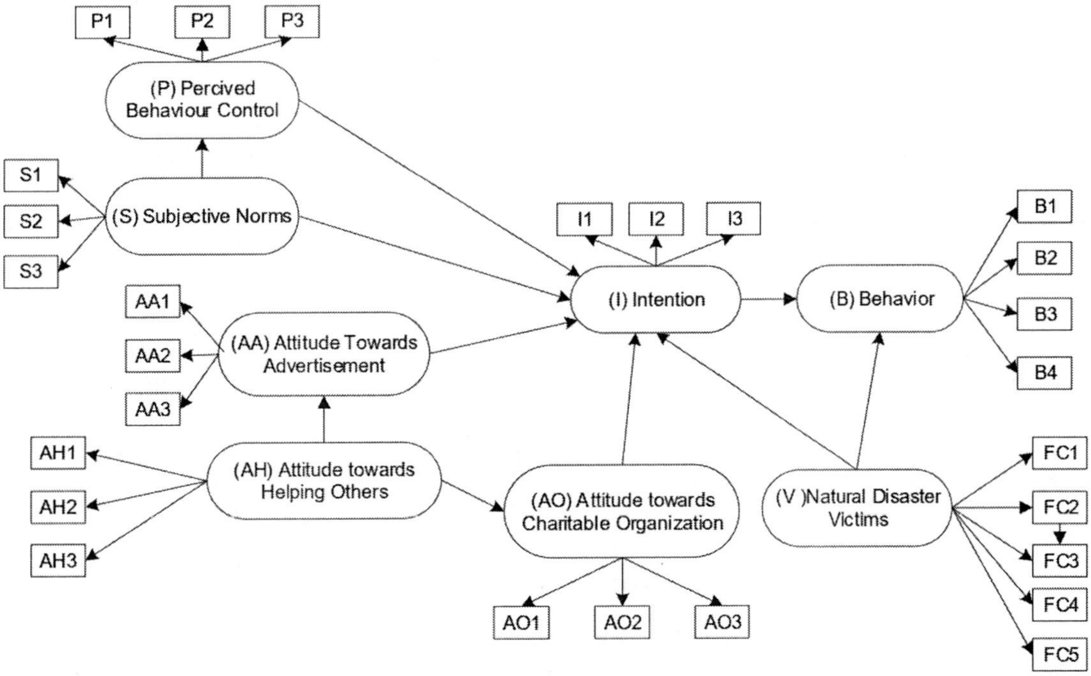


Figure 1. Path in Structural Model

Based on the basic TPB, PBC, subjective norms and attitudes influence the occurrence of intentions, and intentions also influence behavior [9]. In the proposed model, attitudes toward helping others have a significant relationship in influencing attitudes toward advertising and attitudes toward charitable organizations and then influencing intention [13]. Subjective norm has a significant relationship to PBC

[24], while victim of the natural disasters (the number of deaths and the number of homeless people) has a significant influence on the amount of donations given [10]. According to the Indonesian National Disaster Management Agency [4] , disaster victim variables measured by adopting several measures was found in.

The hypothesis was built based on the direction of the link in Figure 1. In this study, the number of samples taken by 200 respondents was taken randomly in Indonesia, 60% of the data came from the residents in the province of East Kalimantan and the rest were randomly answered by residents in other provinces. In this study a Confirmatory Factor Analysis (CFA) was conducted to evaluate the construct validity and to test the goodness of fit index of the measurement model. Structural Equation Modeling Analysis (SEM) was used to observe the relationships among latent variables in the proposed framework. Maximum likelihood with a robust standard error estimator was used to overcome data that were not normally distributed

## 3. RESULT AND DISCUSSION

In this study, testing the fit of the model using several criteria has been used by several researchers [25],[26],[27],[28]. The results can be seen in Table 10.

**Table 10**. The measurement model goodness of fit analysis results

| Goodness of Fit Criteria | Cut-off Value | Result | Conclusion |
|---|---|---|---|
| $\chi^2$ / df | < 2.00 | 0.115 | Fit |
| CFI | ≥ 0.95 | 1 | Fit |
| NNFI | > 0.90 | 1.030 | Fit |
| IFI | > 0.90 | 1.027 | Fit |
| RMSEA | < 0.05 | 0.000 | Fit |
| SRMR | <0.10 | 0.088 | Fit |

The measurement model was tested using the Confirmatory Factor Analysis (CFA) technique with measurements of Convergent Validity, namely Standardized Loading Factor (SLF), Construct Reliability (CR), and Average Variance Extracted (AVE). Table 11 shows the results of the three values for each indicator, namely SLF ≥ 0.5, CRCR ≥ 0.7, and AVE ≥ 0.5 meet the requirements, then the convergent validity can be achieved [25].

**Table 11**. Result of Reliability measurement

| Indicator | SLF | CR | AVE |
|---|---|---|---|
| B1, B2,B3,B4 | 0.67, 0.69,0.90, 0.85 | 0.9 | 0.6 |
| P1,P2,P3 | 0.56,0,74,0,78 | 0.7 | 0.5 |
| AA1, AA2, AA3 | 0.87,0,94,0,84 | 0.9 | 0.8 |
| AO1,AO2,AO3 | 0.80,0.96,0.84 | 0.9 | 0.8 |
| I1,I2,I3 | 0.89, 0.91,0.90 | 0.9 | 0.8 |
| S1, S2, S3 | 0.71, 0.64,0.76 | 0.7 | 0.5 |
| AH1, AH2, AH3 | 0.87, 0.92, 0.79 | 0.9 | 0.7 |
| FC1,FC2,FC3,FC4,FC5 | 0.86, 0.88,0.88,0.88,0.90 | 0.9 | 0.8 |

The hypotheses which describe the relationship between the variables depicted in Figure 1. A complete summary of the hypotheses can be seen in Table 12. The hypothesis is accepted using 95% confidence level so that the relationship with the value of $t_{0.05,199}$ > 1.97 can be accepted. The first obvious link shown in the results is Attitude Towards Helping Others (AH)→ Attitude Towards Charitable Organization (AO)→ Intention (I)→ Behaviour (B). The next link is Attitude Towards Helping Others (AH)→ Attitude Towards Advertisement (AA)→ Intention (I)→ Behaviour (B). Another link is Subjective Norms (S)→ Intention (I)→ Behaviour (B).

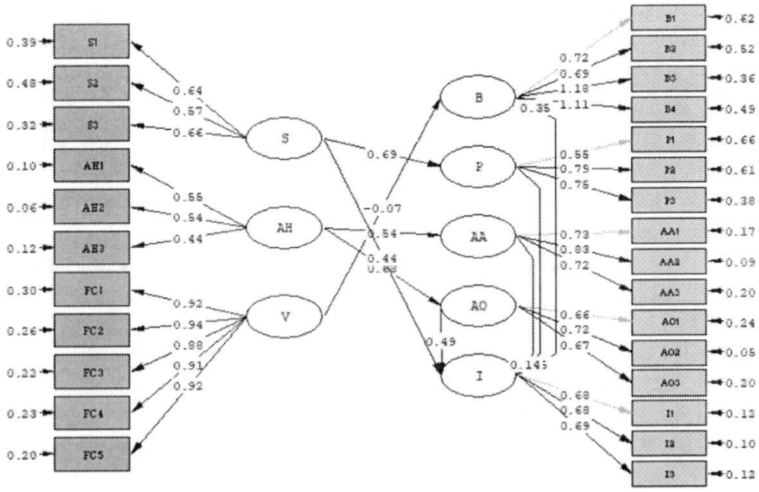


Figure 2: Path diagram of estimates

**Table 12**. Test of Research Hypotheses

| HP | Path | coefficient value | T value | Conclusion |
|---|---|---|---|---|
| H1 | Intention (I)→ Behaviour (B) | 0.34 | 5.61 | Accepted |
| | Natural Disaster Victims (V)→ Behaviour (B) | -0.07 | -0.99 | Rejected |
| | Subjective Norms (S) →Intention (I) | 0.44 | 2.03 | Accepted |
| H2 | Perceived Behaviour Control (P)→Intention (I) | -0.05 | -0.24 | Rejected |
| | Attitude Towards Advertisement (AA)→ Intention (I) | 0.14 | 2.14 | Accepted |
| H3 | Attitude Towards Charitable Organization (AO)→ Intention (I) | 0.49 | 6.35 | Accepted |
| H4 | Subjective Norms (S) → Perceived Behaviour Control (P) | 0.69 | 2.12 | Accepted |
| H5 | Attitude Towards Helping Others (AO)→ Attitude Towards Advertisement (AA) | 0.55 | 3.93 | Accepted |
| H6 | Attitude Towards Helping Others (AH)→ Attitude Towards Charitable Organization (AO) | 0.67 | 4.60 | Accepted |
| H7 | Subjective Norms (S)→ Intention (I)→ Behaviour (B) | 0.14 | 2.94 | Accepted |
| H8 | Perceived Behaviour Control (P)→ Intention (I)→ Behaviour (B) | -0.02 | -0.24 | Rejected |
| H9 | Attitude Towards Helping Others (AH)→ Attitude Towards Charitable Organization (AO)→ Intention (I)→ Behaviour (B) Attitude Towards Helping Others (AH)→ Attitude Towards Advertisement (AA)→ Intention (I)→ Behaviour (B) | 0.14 | 3.44 | Accepted |
| H10 | Attitude Towards Advertisement (AA)→ Intention (I)→ Behaviour (B) | 0.05 | 1.95 | Rejected |
| H11 | Attitude Towards Charitable Organization (AO)→ Intention (I)→ Behaviour (B) | 0.17 | 4.42 | Accepted |
| H12 | Attitude Towards Helping Others (AH)→ Attitude Towards Charitable Organization (AO)→ Intention (I) Attitude Towards Helping Others (AH)→ Attitude Towards Advertisement (AA)→ Intention (I) | 0.41 | 4.64 | Accepted |
| H13 | Subjective Norms (S)→ Perceived Behaviour Control (P)→ Intention (I) | -0.44 | -0.23 | Rejected |

## 4. CONCLUSION

In this study, factors that have been proven to significantly influence the intention to contribute are Subjective Norms, Attitudes Towards Advertising and Attitudes Towards Charitable Organizations. while Attitude Towards Helping Others has a significant indirect effect on intention through Attitude Towards Advertising and Attitude Towards Charitable Organizations. These factors besides influencing intention also influence Behaviour indirectly through intention.

Other factors, Control of Perception Behaviour does not have a significant influence on Intention to donate. Likewise, the perception of victims of natural disasters does not affect behaviour to donate, therefore in this study, it can be concluded when donating, people do not consider their perceptions of the condition of victims in flood disasters

Further research needs to be done to see or add other factors that can influence or strengthen existing factors in influencing the intention and behaviour to contribute to flood disasters.

## 5. REFERENCES

[1] K. Piper 2019 Why disaster relief is so hard , Tragedies from the Midwest to Mozambique leave us wanting to help. It's not easy. ed. https://www.vox.com/future perfect/2018/10/15/17927688/disasters-cyclone-idal-relief-mozambique.

[2] CFE-DM 2018 *Indonesia Disaster Management Reference Handbook* Center for Excellence in Disaster Management and Humanitarian Assistance ( CFE-DM).

[3] S.P. Nugroho 1999 Disaster Events During 2018, Thousand of Victims Died." https://www.bnpb.go.id/1999-kejadian-bencana-selama-tahun-2018-ribuan-korban-meninggal-dunia.

[4] BNBP 2018 [Online]. Available: http://bnpb.cloud/dibi/tabel1.

[5] M. Kashif, E. C. J. I. J. o. N. De Run, and V. S. Marketing 2015 Money donations intentions among Muslim donors: an extended theory of planned behavior model. **20**(1), pp. 84-96.

[6] M. Kashif, S. Sarifuddin, A. J. M. I. Hassan, and Planning 2015 Charity donation: intentions and behaviour. **33**(1), pp. 90-102.

[7] S. R. Knowles, M. K. Hyde, and K. M. J. J. o. A. S. P. White 2012 Predictors of young people's charitable intentions to donate money: An extended theory of planned behavior perspective. **42**(9), pp. 2096-2110.

[8] J. R. Smith, A. J. J. o. C. McSweeney, and A. S. Psychology 2007 Charitable giving: The effectiveness of a revised theory of planned behaviour model in predicting donating intentions and behaviour. **17**(5), pp. 363-386.

[9] I. Ajzen 1985 From intentions to actions: A theory of planned behavior, in *Action control*: Springer, pp. 11-39.

[10] I. Evangelidis and B. J. P. s. Van den Bergh 2013 The number of fatalities drives disaster aid: Increasing sensitivity to people in need. **24**(11), pp. 2226-2234.

[11] T. A. J. J. o. a. Shimp 1981 Attitude toward the ad as a mediator of consumer brand choice. **10**(2), pp. 9-48.

[12] C. P. Duncan and J. E. J. J. o. A. Nelson 1985 Effects of humor in a radio advertising experiment. **14**(2), pp. 33-64.

[13] S. K. Ranganathan, W. H. J. I. J. o. N. Henley, and V. S. Marketing 2008 Determinants of charitable donation intentions: a structural equation model. **13**(1), pp. 1-11.

[14] D. J. Webb, C. L. Green, and T. G. J. J. o. t. a. o. m. s. Brashear 2000 Development and validation of scales to measure attitudes influencing monetary donations to charitable organizations. **28**(2), pp. 299-309.

[15] F. F. Brunel and M. R. J. J. o. A. Nelson 2000 Explaining gendered responses to "help-self" and "help-others" charity ad appeals: The mediating role of world-views. **29**(3), pp. 15-28.

[16] J. A. Piliavin and H.-W. J. A. r. o. s. Charng 1990 Altruism: A review of recent theory and research. **16**(1), pp. 27-65.

[17] K.-P. Chiang 2003 Prediction of monetary donation: a path analysis. *Advances in Marketing* **138**.

[18] R. Mittelman, & Rojas-Méndez, J. 2018 Why Canadians give to charity: An extended theory of planned behaviour model. *International Review on Public and Nonprofit Marketing,* **15**(2), pp. 189-204.

[19] CERD. Centre for Research on the Epidemiology of Disasters. Emergency events database, www.emdat.be/glossary/9#affected

[20] I. Ajzen 1991 The theory of planned behavior," *Organizational behavior and human decision processes,* **50**(2), pp. 179-211.

[21] C. J. Armitage and M. Conner 2001 Efficacy of the theory of planned behaviour: A meta-analytic review. *British journal of social psychology.* **40**(4), pp. 471-499.

[22] S. J. C. p. Van der Linden 2011 Charitable intent: A moral or social construct? A revised theory of planned behavior model. **30**(4), pp. 355-374.

[23] M. B. Holbrook and R. J. J. o. c. r. Batra 1987 Assessing the role of emotions as mediators of consumer responses to advertising. **14**(3), pp. 404-420.

[24] M. Metawie, R. J. I. J. o. B. Mostafa, and S. Science 2015 Predictors of Egyptian university students 'charitable intentions: Application of the theory of planned behavior. **6**(8), pp. 204-215.

[25] J. Hair, W. Black, B. Babin, and R. Anderson 2009 Multivariate Data Analysis 7th Edition Pearson Prentice Hall. ed: JOUR.

[26] D. Hooper, J. Coughlan, and M. R. Mullen 2008 Structural equation modelling: Guidelines for determining model fit. *Electronic journal of business research methods,* **6**(1), pp. 53-60.

[27] L. t. Hu and P. M. Bentler 1999 Cutoff criteria for fit indexes in covariance structure analysis: Conventional criteria versus new alternatives. *Structural equation modeling: a multidisciplinary journal,* **6**(1), pp. 1-55.

[28] J. B. Chung and H.-K. Kim 2009 Competition, economic benefits, trust, and risk perception in siting a potentially hazardous facility. *Landscape and Urban Planning,* **91**(1), pp. 8-16.

IOP Conf. Series: Materials Science and Engineering **847** (2020) 012069 doi:10.1088/1757-899X/847/1/012069

# Designing the Smart Health Function towards Puskesmas (Citizen Health Centre) based on Smart City Concept

**Alif Miftahul Jannah, *Muharman Lubis, Rd. Rohmat Saedudin and Fritasya Dwiputri S**

Faculty of Industrial Engineering, Telkom University, Bandung, Indonesia
*muharmanlubis@telkomuniversity.ac.id

**Abstract**. According to Regional Regulation No.23 of 2014 Article 386 until 390 which requires all institutions to adopt architectural designs to bring up architectural innovations. A smart city is a concept of the application, development, and implementation of technology that is applied to a region (especially urban area) as a complex interaction between the various systems in it. One important development area to be developed is the health sector. In Indonesian constitution of article 28 H and 28 I that the state must guarantee the lives of all its citizens, include by providing appropriate services such as health services. This study aims to design an enterprise architecture in health functions within Citizen Health Centre management system based on the Smart City based on the TOGAF ADM framework, so that it is expected to help the city government work program in realizing smart city and business activities that currently run can be supported by information systems so as to increase satisfaction Bandung City communities. Based on the results of research on public health functions at the Bandung City Health Office obtained a design or blueprint that focuses on the preliminary phase, architecture vision, business architecture, system architecture, information technology, opportunities and solutions, and migration planning.
**Keywords:** enterprise architecture, TOGAF, information system, health function, migration planning

## 1. Introduction

There is an instruction from the president that requires all institutions to adopt architectural design with Presidential Regulation No. 85 2018 to bring up innovations in the delivery of architecture [1]. Local Government through innovation in Law No. 23 of 2014 concerning Regional Government. In Chapter XXI titled Regional Innovation. From Article 386 to Article 390 of Law 23/2014, it explains that in the context of improving the performance of the implementation of Regional Government, the Regional Government can innovate. Innovations as referred to are all forms of renewal in the administration of Regional Government. In formulating an innovation policy, the Regional Government refers to the principle of increasing efficiency, improving effectiveness, improving service quality, there is no conflict of interest, oriented to the public interest, carried out openly, fulfill the values of propriety, and the results cannot be accounted for their own sake. Many concepts are designed to create innovation both in terms of policies or even the laws created and in terms of technology, along with the development of technology and communication science, the concept of smart city is now emerging, these concepts develop by basing the application of information and communication technology in managing cities.

In Indonesia there are also several cities that have implemented smart cities. For example, the Makassar smart city that manages e-Kelurahan to support services in Karampuang Kelurahan Panakkukang, which belongs to a smart society that promotes local wisdom. As well as Surabaya City, as the city that won the title of Smart City Awards 2011 by winning three categories from four categories namely smart governance, smart living and smart environment. Likewise, Manado City implements smart city in the field of Tourism through the introduction of tourism potential (nature, culture, culinary, history) throughout Indonesia and the world through internet networks [2]. In applying the concept of smart city, there are several elements that need to be developed. Among the various fields that want to be developed for smart cities, one of the important fields that will be developed is the health sector, starting from the integration between health practitioners so that it leads to better and more efficient health services to the community. For this reason, it is necessary to build an information system that can help create a fast and efficient service within the scope of public services in the health sector. Thus, the management of puskesmas can be described as an effort to manage puskesmas in a sustainable manner, which is carried out with an integrated system and strategy, so that it can achieve its goals and targets in accordance with the plans and stages set and ultimately will prosper the urban population. This study aims to design an enterprise architecture on health functions in the smart city-based management system.

The step and procedure of implementation from the enterprise architecture can be used by organizations through adopting specific method or framework within the enterprise architecture. So with the enterprise architecture method, it is expected to be able to manage complex systems and be able to harmonize the business and IT that will be invested. The meaning of enterprise architecture is described as the key element that forms an organization where it is intended to achieve harmony between business processes and objectives. In designing enterprise architecture, a framework is needed so that modeling can be structured and appropriate. There are several types of framework that are commonly used in modeling enterprise architecture, including the Zachman framework, TOGAF (The Open Group Architecture Framework), FEAF (Federal Enterprise Architecture Framework), TEAF (Treasury Enterprise Architecture Framework), and others. Table 1 is a comparison of the four frameworks [3]. The first step that needs to be considered when implementing the TOGAF ADM is defining the preparations, namely by identifying the architectural context to be developed, the second is defining the strategy of architecture and determining the architectural parts to be designed, starting from business architecture, information system architecture, technology architecture, and determine the capabilities of the architecture to be designed and developed [4].

**Table 1**. Comparison of the four frameworks

| Criteria | Zachman | TOGAF | FEAF | Gartner |
|---|---|---|---|---|
| Taxonomy complete | 4 | 2 | 2 | 1 |
| Process completeness | 1 | 4 | 2 | 3 |
| Reference model guidance | 1 | 3 | 4 | 1 |
| Practice guidance | 1 | 2 | 2 | 4 |
| Maturity model | 1 | 1 | 3 | 2 |
| Business focus | 1 | 2 | 1 | 4 |
| Governance guidance | 1 | 2 | 3 | 3 |
| Partitioning guidance | 1 | 2 | 4 | 3 |
| Prescriptive catalog | 1 | 2 | 4 | 2 |
| Vendor neutrality | 2 | 4 | 3 | 1 |
| Information availability | 2 | 4 | 2 | 1 |
| Time to value | 1 | 3 | 1 | 4 |

The access gap is a phenomenon that has three different aspects, including global disparities in the differences in information and communication technology between countries, social differences in the differences in access to information and communications technology between different sectors of the country's society and democratic inequalities, which can also be understood in the name of the difference between those who use different digital means to engage in public life and those who do not [16]. The popular one is related to the digital Divide as a form and condition of the gap between people who have knowledge of digital technology with people who do not know at all. The factors causing the digital divide are the unequal provision of infrastructure, infrastructure in this case is the availability of access to roads, water, electricity, to the internet. Then, development has not been evenly distributed in each region. Meanwhile, the lack of cooperation between the government, the community, the private sector, academia and the lack of community awareness and differences in people's lifestyles are other implications. In addition, different regulations in each region can exacerbate the impact that occurs which leads to digital inequality in technical aspects, autonomy, skills and objectives.

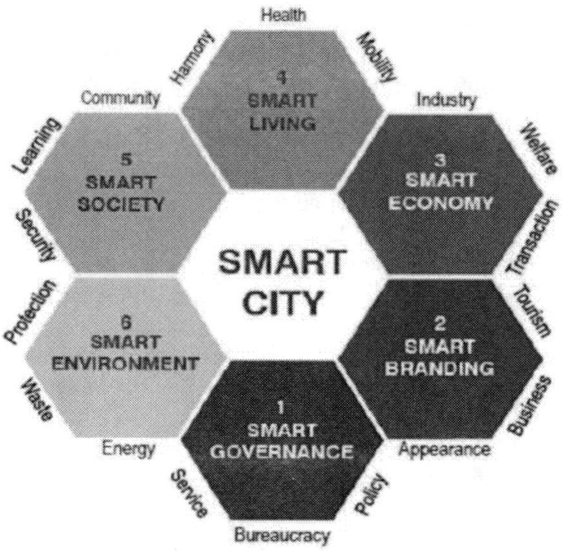


**Figure 1**. Smart City Dimensions

Smart city is a concept of applying, developing and implementing technology applied to an area (especially urban areas) as a complex interaction between various systems in it [3]. The purpose is to overcome the various characteristics of ecosystem innovation by all smart city ideas including being a green city, interconnected and integrated for all layers and forms of the city. Smart city planning uses a reference model to determine the concept of smart city layout and character. Dimensions of the concept of smart city explain the 6 (six) dimensions in the concept of smart city as the basis for the implementation of smart city which is then used in calculating the index of smart city 70 cities in Europe, the six dimensions include Smart Economy, Smart People, Smart Governance, Smart Mobility, Smart Environment and Smart Living [5]. Smart cities that provide a large amount of data and information in different dimensions make the complex process of impact assessment and management of major projects possible. Management and control of large projects as costly, long-term and large-scale developments are essential to their success. In addition, they can facilitate data management in smart cities, where they generally use new and innovative technologies, and can provide integrated data [6]. It also can be seen as specific geographical boundaries that have ability to control, monitor and manages resources and capital independently to be sustainable in term of growth and profit to its environment and citizen by preventing harmful effect, integrating the service management and optimizing the benefits [10][11]. Digital technology has great potential for improving efficiency and service. This is where smart cities can be deployed in a realistic way. However, many institutional conditions must be created to unleash this possibility especially considering the technological aspect of procedural through seller-initiated discourses on smart cities and at least at lower prices [12]. There are also many needs from the point of view of physical infrastructure, smart city, and digital technology that improve public services and better use of resources for the population, which reduces the impact on the environment [14]. Smart cities use ICT to smartly and efficiently use available resources while reducing costs and energy consumption, improving citizens' services and quality of life, and reducing their environmental footprint through develop innovation and a sustainable economy [15].

## 2. Methods

This study uses qualitative methods that are to give a clear picture and in accordance with facts in the field. The qualitative method is based more on the phenomenological nature of prioritizing appreciation. The purpose of this qualitative method is to look for the severity of a phenomenon or problem, develop the concept of sensitivity to the problem at hand. Data collection techniques that are often used are interviews where the researcher conducts a question and answer method with the Bandung City Health Office, regarding all actions relating to the public health function. In addition, researchers review the actions associated with the problematic issues raised. The results of these observations were directly recorded by the researcher and from the observation activities it can be seen the error or the process and the activity in the Bandung City Health Office.

## 3. Result and Discussion

In 1950 the Bandung City Health Service consisted of 10 Medical Centers and in 1972 it developed into 4 health centers consisting of: 1 Community Health Center, 18 Special Health Centers and then 18 Maternal and Child Health Centers and 6 maternity clinics. Based on Decree No. 50 of 1952 concerning implementation, namely the surrender as the Central Government on Health to regions in cities or cities. Personnel Management of the Health Service was gradually handed over to the local government of the Dati II Bandung Municipality and the status of the employees consisted of, Central Employees and Officers for Eradicating Smallpox and Eye Diseases. There are several problem identified such as There is often a delay in the process of receipt of a report by the Health Service which should be received every report in the month, inadequate communication information system becomes a hindering factor in the dissemination of health information and service quality improvement, Implementation of Standard Operational Procedure (SOP) has not been optimal resulting in low quality health services, the absence of health information service applications, to find out information about health such as room availability, ICU room availability and many more, the absence of mobile application to present relevant information and to create accessibility of interaction and lastly the service management is not optimal causing implementation programs that are less appropriate.

According to Presidential Regulation No. 95 of 2018 concerning Electronic-Based Government Systems (SPBE), which discussed SPBE components such as SPBE governance, SPBE management, SPBE services, and SPBE architecture. The SPBE principles that must be applied by government agencies based on Presidential Regulation No.95 of 2018 Article 2 are effectiveness, integration, continuity, efficiency, accountability, interoperability and security. The Health Information System road map is an indicative national health information system planning document for 2015-2019, which contains an overview of the current situation, direction and objectives to be achieved, implementation stage, targets of each stage, indicators of target achievement, financing, and organizing the implementation of the development and strengthening of the national health information system in the next five years in realizing an ideal health information system. Law No. 23 of 1992 concerning health, the President of the Republic of Indonesia has mandated health-oriented development through Healthy Indonesia 2010, which in essence undertakes development efforts need to put health policies in their implementation. In making efforts to develop and strengthen the central and regional governments, they have the authority to standardize, manage and develop national scale health information systems and facilitate the development of regional scale health information systems, then carry out management and development of provincial scale health information systems and management and development of health information systems within district or city scale.

IOP Conf. Series: Materials Science and Engineering **847** (2020) 012069 doi:10.1088/1757-899X/847/1/012069

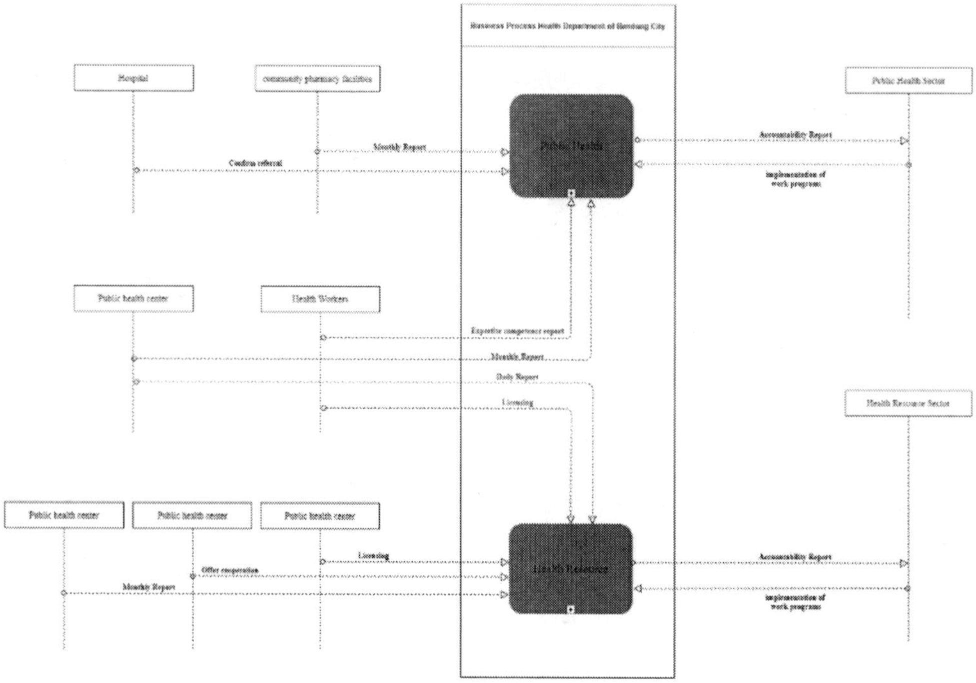


**Figure 2**. Existing Business Process of Bandung

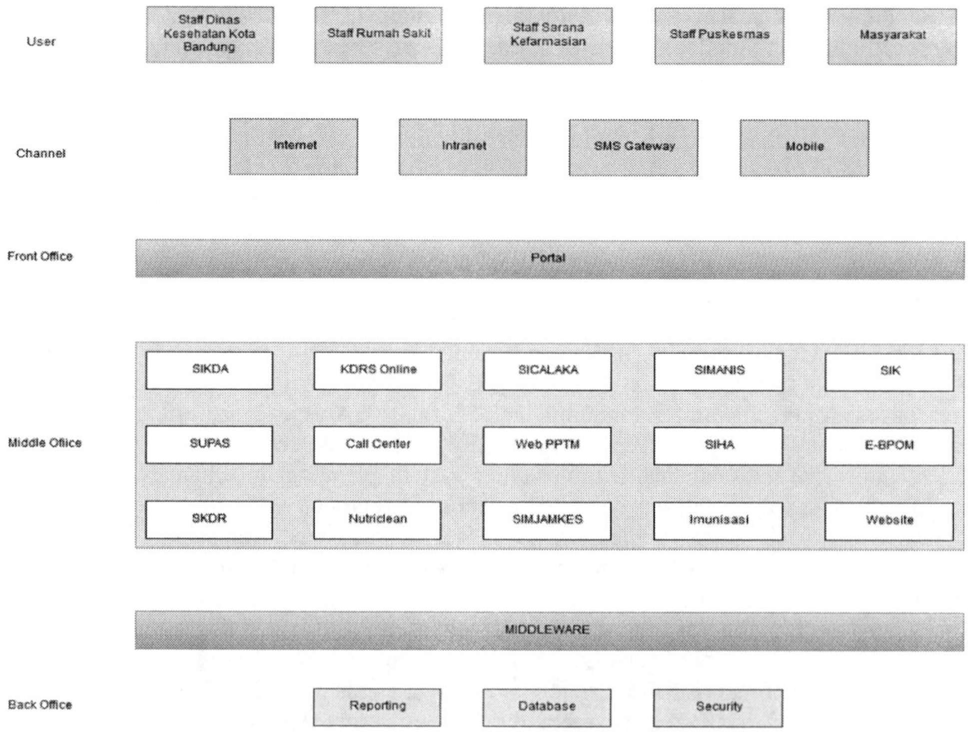


**Figure 3.** Solution Concept Diagram

Strategic issues that must be considered in efforts to develop, strengthen, and implement a health information system in the next five years include, among others, health information system policy and regulation, strengthening the coordination of health information systems, directed and measurable health information system planning, strengthening information system organizations health and standardization of health information systems. Meanwhile, the development of health information systems human resources, strengthening of ICT infrastructure in health care facilities, financing of health information systems requires insufficient funds and structuring transaction data in health care facilities to improve the quality and speed of service work processes and the availability and quality of data must also be serious attention. The government also needs to optimize data flow to increase availability, develop a health data bank and access to data sharing for the purpose of strengthening the use of information. Urban metabolism generally consists of product inputs and waste emissions with consistent negative externalities that exacerbate social and economic problems. Cities rely on many external resources, and the important fact is that you can always be a resource user. Promoting sustainability has been explained by encouraging natural capital stock. Other recent interpretations of urban sustainability are promoting a more centralized approach in which cities should respond to people's needs through sustainable solutions on social and economic dimensions [13].

## 4. Conclusions

Based on the results of the enterprise architecture design on the public health function, it can be concluded that the successful creation of smart city application in the enterprise architecture design is assessed from the smart city assessment indicator framework with CITYkeys namely integration, monitoring and evaluation and availability of government data. The results of the enterprise architecture design in this study produced a blueprint that can be used as a service reference for developing business services that are running. The results of the IT Roadmap design as a reference for smart cities in the development of information technology, namely the design of enterprise architecture in this study, use the TOGAF ADM method, with the main focus of research being the phase of business architecture, data architecture, application architecture, technology architecture, opportunities and solutions, and migration planning. Existing data, applications and technology available at the Bandung City Health Office are not yet fully in line with official requirements. So there needs to be improvement and development of data and applications in the Bandung City Health Office. Based on the analysis of technology architecture, the condition of the existing service technology does not meet several requirements, so it is necessary to develop and improve the target technology.

## 5. References

[1] Nilma. 2018 Analisis Cause Effect Mengenai Dampak dari Implementasi Bandung Smart City. *Faktor Exacta.*
[2] Pratama, IA. 2014 *Smart City Beserta Cloud Computing dan Teknologi Pendukung Lainnya.* (Bandung: Informatika).
[3] Supriana, I., and Watika 2011 Analisis perbandingan komponen dan karakteristik enterprise architecture framework. *Konferensi Nasional Sistem dan Informatika* 2011, 11-64.
[4] The Open Group. 2011 The Open Group Standard TOGAF Version 9.1.
[5] Giffinger R, Fertner C, Kramar H, Kalasek R, Milanovic NP and Meijers E. 2017 Smart Cities' Ranking of European Medium Size Citites. Center of Regional Science, Vienna UT, *White Paper.*
[6] Huovila A, Airaksinen M, Pinto-Seppa I, Piira K, Bosch P, Penttinen T, Neumann H-M and Kontinakis N. 2017 CITYkeys Smart City Performance Measurement System. *Int. J. for Housing Science and Its Applications*, 41(2), 113-125.
[7] Lubis M, Fauzi R and AR Lubis. 2018 Enterprise Application Integration for High School Students using Blended Learning System. *Proceedings ICIEE 2018, MATEC Web of Conference.*
[8] Lubis M, Fauzi R, Lubis AR and Fauzi R. 2018 Analysis of Project Integration on Smart Parking System in Telkom University. *Proceedings CITSM 2018, IEEE.*

[9] Lubis M, Fauzi R, Lubis AR and Fauzi R. 2018 A Case Study of Universitites Dormitory Residence Management System (DRMS) in Indonesia. *Proceedings CITSM 2018, IEEE.*
[10] Eremia M, Toma L and Sanduleac M. 2016 The Smart City Concept in the 21st Century. *Int. Conf. Interdisciplinarity in Engineering.*
[11] Ghosh P and Mahesh TR. 2015 Smart City: Concept and Challenges. *Int. J. on Advances in Engineering, Technology and Science,* **1**(1).
[12] Giron D. 2018 Innovative Governance of Large Urban Systems. *IGLUS Quaterly,* **4**(1), June 2018.
[13] Albino V, Berardi U and Dangelico RM. 2015 Smart Cities: Definitions, Dimensions, Performance and Initiative. *Journal of Urban Technology* **22**(1).
[14] Mohanty SP. 2016 Everything you wanted to know about smart cities. *IEEE Consumer Electronics Magazine,* **5**(3).
[15] Greco I and Cresta A. 2015 A Smart Planning for Smart City: The Concept of Smart City as an Opportunity to Re-think the Planning Models of the Contemporary. *Int. Conf. on Computational Science and Its Applications June 2015.*
[16] Srinuan C and Bohlin E. 2011 Understanding the digital divide: a literature survey and ways forward. *ECONSTOR* 2011.

# Scheduling and Allocation of Airport Service Manpower by considering Time and Work Constraints using M-MAPTWTC Method: a case study

**P Moengin[1] and F Darussalam[2]**

[1,2]Department of Industrial Engineering, Trisakti University, Jakarta, Indonesia
[1]parwadi@trisakti.ac.id
[2]fahridarussalam@gmail.com

**Abstract.** Airport Service workforce scheduling and allocation is important because the handling of ground handling jobs is also increasing. The system and allocation of workforce scheduling in the company that is currently being carried out is based on trials only (trial and error), this causes frequent shortages or excess labor. This can happen because in scheduling and allocating workforce has not considered the time span between the preparation time and flight time, as well as the work team constraints. To overcome the weaknesses of the system of scheduling and allocation of labor previously in this thesis, scheduling and allocation of labor is done by considering the time span between preparation time and flight time. The method used is using the m-MAPTWTC method. The results of a case study in a company using the m-MAPTWTC method show that the number of jobs done in the period 04.00 - 12.00 WIB as many as 80 flights can be done by 20 teams whereas previously the same number of jobs required 24 teams. With the m-MAPTWTC method it is also obtained the team allocation and work order for each job by each team.
**Keywords:** Scheduling and allocation of labor, job-teaming constraints, time span, m-MAPTWTC method.

## 1. Introduction

The growth of air transportation facilities users in Indonesia is so rapid every year, the average increase every year is 10%. Indonesia as an archipelagic country has limitations in the use of land transportation modes (cars and trains) and sea transportation modes causing the shift of consumers to use fast and cheap transportation facilities from land and sea transportation modes to air transportation modes. The rapid use of the aviation service industry in Indonesia is an opportunity for airlines. Airlines in Indonesia are competing to add consumers and strengthen their fleets and increase destinations. Three major airlines, namely Garuda Indonesia Group (Garuda Indonesia Airways and Citilink) with a total fleet of 200 aircraft, Lion Air (400 aircraft), and Air Asia (100 aircraft). In the flight system there are four interested parties, namely: Regulators (Government), Airport Managers (Angkasa Pura), Airlines (Airlines), and Ground Handling (PT. Gapura Angkasa and PT. JAS). PT. Gapura Angkasa is a company engaged in the field of Ground Handling (GH or Airport Services). The role of Ground Handling may not be seen directly for users of air transportation modes but for airlines it is very important because the punctuality of the aircraft to be ready to fly (ready take-off), passenger handling and cargo handling (letters, packages, etc.) is largely determined by the role of Ground Handling [1].

Business partners of PT. Gapura Angkasa namely domestic: Garuda Indonesia airlines and Citilink (Garuda Group), Batavia Air, Air Asia, Kalstar, Kartika Airlines, Lion Air and Wings Air (Lion Group), Sriwijaya Air, Sky Aviation International: China Airlines, Royal Brunei, Japan Airlines/All Nippon, Vietnam Airlines, Korean Airlines, Cathay Pacific, Airfrance (KLM). The problem faced by PT Gapura Angkasa is the scheduling and allocation of airport services workers in handling Garuda Indonesia aircraft. PT. Gapura Angkasa is currently not optimal in workforce scheduling and allocation system, where 80 flights are handled by 24 teams, and the allocation and scheduling of work teams is uneven with the number and growth of the Garuda Indonesia fleet. The purpose of this study is to provide suggestions regarding the scheduling and allocation of airport service workers.

Ground Handling (GH) comes from the word ground and handling. Ground means (on) land, in this case related to the airport. Handling means handling. In terms of (etymology) ground handling or ground service is an activity of an airline company which is related to the handling of passengers and their luggage, cargo, mail, auxiliary equipment for aircraft movement on the ground and the aircraft itself while at the airport, both for departure (departure) or arrival.

The scope of ground handling is in preflight services and post flight services. The scope of GH at the airport includes: handling passengers at the terminal, handling aircraft in the parking area called the Apron Area and cargo handling called Cargo Handling.

The Manpower Allocation Method with Time Windows and Job Teaming Constraints and a Limited number of teams was developed by Anders Dohn, Esben Kolind and Jens Clausen (Tecnical University of Denmark) [2]. The focus of research on this method (m-MAPTWTC) is carried out at ground handling companies at major airports in Europe regarding scheduling, focusing on the work team (group) of technicians involved in doing a job in this case the team from Ramp Handling, baggage handling, Catering Services, Fuel Services, Technical Services and Cleaning Services [3].

The m-MAPTWTC method is a refinement of the Manpower Allocation with Time Windows and Job Teaming Constraints (MAPTWTC) method developed first by Yanzhi Li, Andrew Lim, Brian Rodrigues Hongkong University in 2004 with a metaheuristic approach. The study of MAPTWTC, carried out in the port of Singapore, focuses on minimizing the number of workers needed in doing existing work [4].

## 2. Methods

The research begins by observing the main tasks and functions of the Ramp Coordinator in the field; during preparation to passenger embarkation and aircraft closing doors. From these observations note the time required to handle one aircraft (Loading - Unloading Baggage, Embarkation - Passenger Debarkation, Cleaning, and Other Preparations) ([4], [5]).

**Notation**

The notation used in the process of calculating this method are:

| | |
|---|---|
| K | = Number of available teams (K = 20) |
| V | = The set of all existing teams (K1, K2, K3, ..., K20) |
| Service center | = 0 |
| C | = The set of all tasks (A1, A2, ..., A80) |
| $a_i$ | = Preparation Time |
| $b_i$ | = *Actual Time Departure* (ATD) |
| $e_k$, $f_k$ | = start time and end time of each team (time span) |
| $t_{ij}$ | = Travel time from task -I to task-j |
| $a_i$, $b_i$, $e_k$, $f_k$, | = non-negative integers |
| $s_i$ | = integer variable and is defined as the start time of task -i |
| M | = a very large positive number |
| N | = the set of all tasks and service centers = $C \cup \{0\}$ |
| Decision variable: | |
| $x_{ijk}$ | = binary number |

$$x_{ijk} = \begin{cases} 1, \text{jif team k is doing task} - \text{I can directly do task} - \text{j} \\ 0, \text{if other} \end{cases}$$

**Model Manpower Allocation Problem with Time Windows and Job-Teaming Constraints (m-MAPTWTC)**

The formulation of the m-MAPTWTC model is as follows:

$$\max \sum_{k\in V} \sum_{i\in C} \sum_{j\in N} \chi_{ijk} \quad (1)$$

Subject to:

$$\sum_{k\in V} \sum_{j\in N} \chi_{ijk} \le r_i \qquad \forall i \in C \quad (2)$$

$$\sum_{j\in N} \chi_{0jk} = 1 \qquad \forall k \in V \quad (3)$$

$$\sum_{i\in N} \chi_{ihk} - \sum_{j\in N} \chi_{hjk} \ge 0 \qquad \forall h \in N, \forall k \in V \quad (4)$$

$$e_k + t_{0j} - M(1-\chi_{0jk}) \le s_j \qquad \forall j \in C,\ \forall k \in V \quad (5)$$

$$s_i + t_{i0} - M(1-\chi_{i0k}) \le f_k \qquad \forall i \in C,\ \forall k \in V \quad (6)$$

$$s_i + t_{ij} - M(1-\chi_{ijk}) \le s_j \qquad \forall i \in C,\ \forall j \in C, \forall k \in V \quad (7)$$

$$a_i \le s_i \le b_i \qquad \forall i \in C \quad (8)$$

$$\chi_{ijk} \in \{0.1\} \qquad \forall i \in N,\ \forall j \in N,\ \forall k \in V \quad (9)$$

$$s_i \in Z^+ \cup \{0\} \qquad \forall i \in C \quad (10)$$

The objective function (1) is to maximize the number of tasks assigned. An assignment is calculated multiples of time divided by team ($ri \ge 2$). Constraints / limitations (2) It is ensured that each job is carried out by a number of teams that are appropriate or may be lacking (only done by two teams), if some parts of the work are separate then the work is not done. Make sure each team starts work from the service center (3). Constraint (4) make sure the shift is not segmented (fragmented). Every work done by it must be completed. The next four constraints / constraints relate to time span. Ensure that each team must do its work during the available working hours (5) - (6). Check whether travel time is needed between one work and another (7). If the i-th plane is not visited, M large scalars make the related constraints non-binding. Constraint (8) make sure the job is within a certain time frame. Constraint (9) says that the decision variable is binary value. Constraint (10) is a non-negative integer ([6], [7]. [8]).

## 3. Results and Discussion

The following data is needed in data processing which includes the number of available teams, Task (aircraft handled), handling time, ATD time, windows time, start and finish time for each team, travel time

The following are the results of plotting the allocation of labor needed in one shift using the help of the lp Solve 5.5.2.0 Software and the computer specifications for Intel Core 2 Duo CPU (2 Ghz), Memory (3 GB). By using the collected data, the number of decision variables ($x_{ijk}$) is 128000, the number of constraints is 261060. The time required to execute the model is 4 hours.

**Table 1.** Work Allocation and Scheduling Table

| WORK ALLOCATION AND SCHEDULING TABLE | |
|---|---|
| **Work team (K)** | **Task (A) (Airlines)** |
| K1 | A2, A4, A5, A6, A7, A21, A41, A64, & A65 |
| K2 | A3, A4, A6, A7, A22, A42, A46, & A62 |
| K3 | A3, A5, A7, A23, A27, A43, & A63 |
| K4 | A6, A7, A24, A26, A44, & A66 |
| K5 | A4, A6, A25, A27, A45, & A64 |
| K6 | A3, A5, A6, A47, & A66 |
| K7 | A1, A4, A5, A7, A21, A22, A23, A27, A46, A47, A61, A62, A66, & A67 |
| K8 | A8, A28, A48, & A68 |
| K9 | A9, A29, A53, A54, & A69 |
| K10 | A10, A30, A49, A54, & A73 |
| K11 | A11, A33, & A74 |
| K12 | A34, A70, & A73 |
| K13 | A12, A31, A34, A54, & A73 |
| K14 | A8, A10, A11, A13, A14, A28, A32, A34, A50, A51, A52, A55, A69, A71, A72, & A64 |
| K15 | A17, A35, A75, & A77 |
| K16 | A16, A17, A40, A56, A60, A76, & A78 |
| K17 | A18, A40, A58, A60, & A77 |
| K18 | A19, A37, A57, A59, & A78 |
| K19 | A20, A39, A40, A60, & A79 |
| K20 | A15, A16, A18, A20, A36, A38, A56, A57, A58, A59, A60, A77, A79, & A80 |

The number of tasks completed amounted to 80 flights but only used 20 work teams. The previous workforce scheduling and allocation system requires 24 work teams to do 80 flights, with the team allocation and work order for each job by each team.

## 4. Conclusions

From these calculations conclusions can be drawn regarding the division of tasks of each team:

- Team K1 handled 9 aircraft, namely: 2, 4, 5, 6, 7, 21, 41, 64, and 65.
- Team K2 handled 8 aircraft, namely: 3, 4, 6, 7, 22, 42, 46 and 62

- Team K3 handles 7 aircraft, namely: 3, 5, 7, 23, 27, 43 and 63
- Team K4 handled 6 aircraft, namely: 6, 7, 24, 26, 44, and 66
- Team K5 handled 6 aircraft, namely: 4, 6, 25, 27, 45 and 64
- Team K6 handled 5 aircraft, namely: 3, 5, 6, 47 and 66
- Team K7 handled 14 aircraft, namely: 1, 4, 5, 7, 21, 22, 23, 27, 46, 47, 61, 62, 66 and 67
- Team K8 handled 4 aircraft, namely: 8, 28, 48 and 68
- Team K9 handled 5 aircraft, namely: 9, 29, 53, 54 and 69
- Team K10 handled 6 aircraft, namely: 10, 30, 49, 54, 72 and 74
- Team K11 handled 3 aircraft, namely: 11, 33 and 74
- Team K12 handled 3 aircraft, namely: 34, 70 and 73
- Team K13 handled 5 aircraft, namely: 12, 31, 34, 54 and 73
- Team K14 handled 16 aircraft, namely: 8, 10, 11, 13, 14, 28, 32, 34, 50, 51, 52, 55, 69, 71, 72 and 64
- The K15 team handled 4 aircraft, namely: 17, 35, 75 and 77
- The K16 team handled 7 aircraft, namely: 16, 17, 40, 56, 60, 76 and 78
- The K17 team handled 5 aircraft, namely: 18, 40, 58, 60 and 77
- The K18 team handled 5 aircraft, namely: 19, 37, 57, 59, and 78
- The K19 team handled 5 aircraft, namely: 20, 39, 40, 60 and 79
- The K20 team handled 15 aircraft namely: 15, 16, 18, 20, 36, 38, 56, 57, 58, 59, 60, 77, 79, and 80.

**Acknowledgment**
This research was funded by the Directorate of Research and Community Service, the Ministry of Research, Technology and Higher Education of the Republic of Indonesia in the 2019 budget year (Contract No. 9/AKM/MONOPNT/2019). For this reason, the researchers expressed their highest gratitude and appreciation.

## 5. References

[1] Abdul M, Suharto and Warpani, Eko P. 2009 *Ground Handling*, (Jakarta: PT. Raja Grafindo Persada).
[2] Clausen T. 2011 Airport Ground Staff Scheduling, Lyngby: *PhD thesis*, DTU Management Engineering Denmark.
[3] Dohn A. 2010 Rostering and Task Scheduling Applications in Manpower Planning, Lyngby: *PhD thesis*, DTU Management Engineering Denmark.
[4] Yanzhi L, Andrew L, and Brian R. 2004 Manpower Allocation with Time Windows and Job-teaming Constraints, *Wiley InterScience*, 302-311.
[5] Achterberg,T., T. Koch, and Martin A. 2005 Brancing rules revisited. *Operations Research Letters* **33**(1) 42-54.
[6] Moengin, P. 2010 Polynomial penalty method for solving linear programming problems, *IAENG International Journal of Applied Mathematics*, **10**(3)-06.
[7] Moengin, P. and Fitriana, R. 2015 Model of integrated production-inventory-distribution system: The case of billet steel manufacturing, *Lecture Notes in Engineering and Computer Science.*
[8] Moengin, P., Harahap, E.F., Adisuwiryo, S., and Fransiska, W.A. 2019 Integer Linear Programming Model and Algorithm to Integrate Heuristics Scheduling EDD, Inventory

Control and Distribution Problems in a Modular Production System, *IOP Conference Series: Materials Science and Engineering.*

# Investigating Supply Chain issues in the Food Processing Industry

**S Z Khan[1], MA Ashfaq[1], M U Awan[2], H URehman[2], S A Kamal[1], N T X Hoa[3] and M Shafiq[1,2*]**

[1]Department of Industrial Engineering, University of Engineering and Technology Taxila, Pakistan
[2]Institute of Quality and Technology Management, University of the Punjab, Lahore, Pakistan
[3]School of Economics and Management, Hanoi University of Science and Technology, Hanoi, Vietnam
*Corresponding author's email: shafiqaatir1@gmail.com

**Abstract**. With the recent increase in the complexity of food industries, numerous issues including rapid changes in taste and demand for food supplies are on the rise, particularly in developing countries. **Aim:** This research study was aimed at investigating supply chain management issues in food processing industries and influence of these factors on business performances. Methods: Research data from respondents was collected using online social platforms and Statistical Package for Social Sciences (SPSS) was employed to analyses the data. **Results**: Research findings indicated that factors including human resource management, technology, facility issues, and customer relations negatively influence business performance. **Conclusion**: Findings also reflect that human resource management issues represent a significant driver for improving performance of the food industries. This research study offers insight regarding barriers and drivers influencing the operations of food processing industries and what measures should be taken accordingly.
**Keywords**: Supply Chain Management Problems, Business Performance, Food Industry

## 1. Introduction

Pakistan is among one of the emerging economies in Asia [1],[2]. Nonetheless, the progress of supply chain operations lags far behind than numerous other Asian countries including India, Bangladesh, China and Japan owing to various problems [3]. Lack of advancement in the transport system and minimal utilization of modern technology in supply-chain management of food processing industries lead to significant economic loss in daily based business operation [4], [5].

According to (Patel and Deshpande, 2015), irregularities within food industries occur due to several logistical, social and economic factors aligned with the supply chain. Research studies have indicated how essential it becomes to increase productivity by improving the supply chain management issues[6].

Food Industries in Pakistan are not often well equipped with adequate supply chain operations. As the population of Pakistan continues to grow so are the problems in transportation and fulfilment of demand in food industries[7]. Instances of these problems include: diverse consumer tastes and requirements, the need for sustainable policies and the need to meet cold supply chain requisites in food transport [8]. The common existence of these difficulties front food processing industry with the competition to progress and evolve to meet requirements of customers [9, 10]. In spite of its weighty contributions, Pakistan's food processing industry is facing different problems hindering its economic growth. The food processing industry faces high frequencies of dynamic customer requirements and preferences effecting the possible progress and revenue generated by businesses. Despite the acknowledged benefits of the supply chain operations, its implementation has not been carried out on a significant scale particularly in the food industry. Therefore, to foster the rapid adoption of supply chain operations, it is vital to analyse the factors influencing its progress. The aim of this study is, therefore to inspect the problems related to supply chain management in the food processing industry of Pakistan and how these problems influence business performance. This research focuses on developing a theoretical model highlighting the key inhibitors of supply chain operations using a survey method. Based on the research findings and data analysis using SPSS, empirical validation of proposed factors has been carried out.

## 2. Literature Review

### *2.1. Supply chain management problems*

Supply chain management in the food processing industry primarily facilitates the effective movement of essential materials and information as well as the transport of the end product from distant factories to customers' local markets [11]. Supply chain management include logistics and transportation, operations management, materials and distribution management and marketing [12]. Research studies suggest that it is quite important to handle the problem of supply and demand correlation with all parties of business [13].

Now a days, customers are demanding more differentiated, higher quality products with higher levels of service at adequately low prices[14]. Therefore, supply chain management needs to be more effective in terms of ease and accuracy of the overall operation.

According to Hoffman, supply chain continues to struggle due to risk management issues caused by the external environmental factors as well as the factors threating chaos within the supply chain[15]. These risks can originate from customers, suppliers or the internal environment and consists of natural disasters, political and economic developments, changes to legislations and regulations, technological developments and the ability to respond to human behaviour[15, 16].The intense struggle for market share is a major issue within the supply chain. Another study [17], highlighted the absence of buyer-supplier trust and cooperation, supplier competencies, and interpersonal relationships as key barriers to supplier's performance.

### *2.2. Business performance*

In this era of electronic commerce and commercial globalization, business analytics is gradually becoming a central source of competitive advantage for enterprises[18]. Business performance can be seen as a company's ability of achieving set goals. Depending upon the end of business period results, business performance is a depiction of the level of work performed within the set out goals[19]. For a business to be successful, all owners or managers must ensure that the business operates as effectively and proficiently as possible. Nevertheless, improving the efficacy and proficiency of a business requires knowledge and foresight of the most critical elements of business operations and the preparedness to implement practices that enhance these crucial elements [20-22].

The improvement in supply chain performance is directly related to an individual business organization's performance which is operating in that supply chain [22].Successfully managing the supply chain performance results in improved communication, productivity, operational and strategic monitoring, problem solving and better decision making for businesses which ultimately results in a better supply chain management system[23, 24].

Based on the above discussion, it becomes clear that by improving supply chain management, business performance will also improve and contribute significantly to the economy. Therefore, it is important to identify which factors are inhibiting the growth in supply chain management of food processing industries and we can mitigate the effects of these issues in the food processing Industry of Pakistan. This research study has included several variables including Human Resource Management Problems, Technology Related Problems, Facility Related Problems, Customer Relation Management Problems, Supplier Relation Management Problems, Regulatory factors and Transportation and logistics problems for obtaining better insight into food processing industry. In light of the above discussion, this research study has presented a brief overview of factors which might influence the business performance of the food processing industry of Pakistan:

*2.2.1. Human Resource Management Problems (HRM).* HRM refers to the effective management of the people in an organization. It is a strategic approach that helps in providing a competitive advantage. The main purpose of HRM is to make sure that the firm achieves optimal success through people. Imposing proper recruitment and training methods, developing the abilities of employees and choosing the right person for a job can lead to improving the business's performance [25].

*2.2.2. Technology Related Problems (Tech).* Technology helps businesses to test new concepts, control market access opportunities and become more competitive. Businesses attract new demands and improve market activities and products due to technological advancements[26].Technological management creates new values, new products, new processes and improve workflows that enables business prosperity [27, 28].

*2.2.3. Facilities (Fac) and business performance.* A poor facility also represents a major issue affecting the performance of the business and constraining its success. Generally, business performance plummets due to the lack of state-of-the-art facilities and could cause delays to the sourcing and supply processes of a business [29]. Facilities having poor layout and design can lead to safety and health hazards for employees [9].

*2.2.4. Supplier relationship management (SRM) and business performance.* While outsourcing of logistics and manufacturing helps expand distribution channels, this in turn gives rise to more and more complex and dynamic supply chains. It has become more apparent to improve the supplier and buyer relationship to survive and thrive in the world of business[30].Procurement strategies such as cost reduction, predictable procurement and control over negotiations can be developed through a better supplier relationship management [31, 32].

*2.2.5. Customer relationship management (CRM) and business performance.* Effective management of customer relationships represents a major problem in business competitiveness. Businesses often demand adequate data about customers and their requirements[33].Customer relationship management and business performance essentially depend upon the effective and efficient utilization of knowledge resources. Therefore, the effectiveness of customer relationship management depends upon usefulness of the integration of the knowledge management process. Managers have to understand the importance of customer relationship management marketing strategy to sustain and gain edge over the competitors [34].

*2.2.6. Regulatory factors (RF) and business performance.* Legal factors influencing business include all regulatory and law indicators that can influence negatively or positively to the results obtained for market actions and decisions of the management of company operating in a particular country[35, 36]. Research studies have shown that regulatory factors affect the business performance negatively which is why questions relevant to these issues have been integrated in to the survey to obtain a brief idea as to whether they affect the business performance or not in context of a developing country [37, 38].

*2.2.7. Transportation and Logistics (Log) and business performance:* During the transportation of food products, these products must be placed in a closed container free of debris and odour as the presence of any contaminate could permeate packaging and ruin the goods being shipped [39]. Refrigerated vehicles are very essential for the transportation of perishable and freeze products [40].Consumers are becoming increasingly selective about food products and prefer to opt high-quality products. It doesn't take them much time to switch from one brand to another because of the lack of quality. Therefore it becomes important to investigate whether food processing industries incorporate sufficient measures to ensure quality of products [41].

In the light of literature review of above-mentioned factors and correlation of these factors with business performance, following research model is proposed in figure 1:

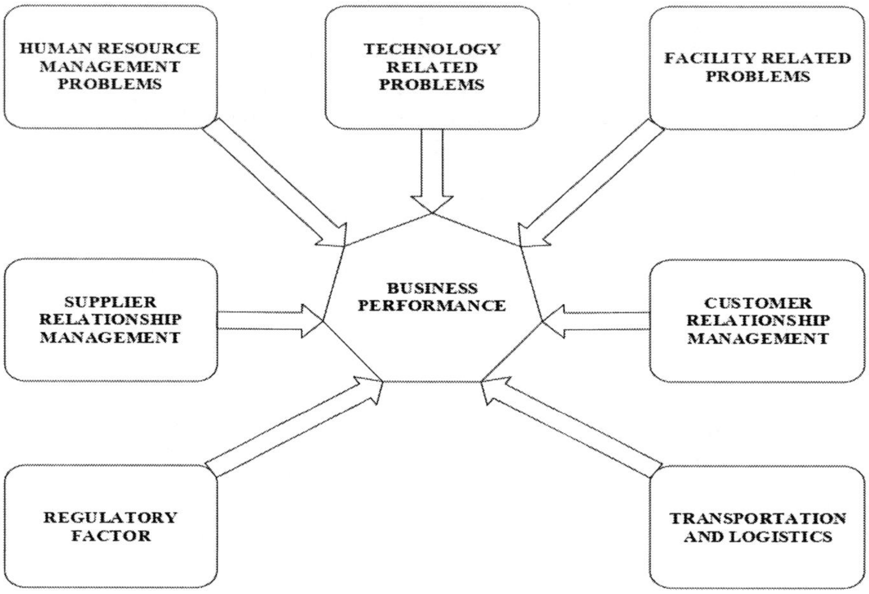


**Figure 1.** Research Model

In this research study, business performance is measured using return on investment, revenue growth, profit growth, customer satisfaction, and employee satisfaction as subjective measures which is in accordance with study conducted by [42].

## 3. Methods

Research data has been gathered from different food processing industries in Punjab and Khyber Pakhtunkhwa. The respondents were selected because of their extensive knowledge regarding supply chain operations in the food processing industry and its working environment.

### *3.1. Data Collection*

The sample for this research study consisted of 302 respondents including operators, supervisors, and managers. Data collection of self-administered questionnaires was completed in one month and a total 500 questionnaires were distributed. The questionnaires for this research was split into two sections. Section A was based upon the demographic details of respondents. The results of section A are presented in Table 1. From Table 1, it can be seen that majority of respondents were males. Highest percentage regarding employment of respondents was observed between 2-5 years. It can also be seen that majority of respondents held managerial positions. Based on responses obtained from these respondents, further assessment of collected data was carried out.

IOP Conf. Series: Materials Science and Engineering **847** (2020) 012071 doi:10.1088/1757-899X/847/1/012071

**Table 1.** Demographic Details of Respondents

| Variables | Categories | Total (N) | Percentage % |
|---|---|---|---|
| Gender | Male | 253 | 83.8 |
| | Female | 49 | 16.2 |
| Age | 18–25 | 101 | 33.4 |
| | 26–35 | 128 | 42.4 |
| | 36–45 | 59 | 19.5 |
| | 45 or above | 14 | 4.6 |
| Employment period | Under 2 years | 101 | 33.4 |
| | 2–5 years | 80 | 26.5 |
| | 6–10 years | 73 | 24.2 |
| | Above 10 years | 48 | 15.9 |
| Occupational position | Supervisor | 83 | 27.5 |
| | Manager | 95 | 31.5 |
| | Junior Manager | 94 | 31.1 |
| | Operator | 30 | 9.9 |

*3.2. Statistical analysis:*

Using the Pearson correlation coefficient, the association between supply chain management issues and their impact on business performance was assessed. The details are given in following sections.

*3.2.1. Reliability:*

To measure the reliability of the data, Cronbach's alpha was utilized. The measurement scales of reliability used in this study varied between .880 and 0.891 (above 0.7) as shown in Table 2 confirming the reliability of this model.

**Table 2.** Reliability of Research Model

| Factors | Items | Variance | Reliability (α) | Factors | Items | Variance | Reliability (α) |
|---|---|---|---|---|---|---|---|
| HRM | 3 | 12.250 | 0.891 | CRM | 5 | 11.810 | 0.875 |
| Tech | 3 | 11.065 | 0.884 | RI | 3 | 11.798 | 0.882 |
| Fac | 2 | 11.212 | 0.886 | Log | 4 | 11.434 | 0.873 |
| SRM | 3 | 11.744 | 0.88 | | | | |

*3.2.2. Validity*

Pearson correlation coefficients were used to measure convergent validity. From Table 3, the values of correlation among different factors confirm validity of the model.

**Table 3.** Convergent Validity among the factors

| | HRM | Tech | Fac | SRM | CRM | RI | Log | Bus_Perfor |
|---|---|---|---|---|---|---|---|---|
| HRM | 1 | | | | | | | |
| Tech | 0.508 | 1 | | | | | | |
| Fac | 0.484 | 0.547 | 1 | | | | | |
| SRM | 0.459 | 0.548 | 0.56 | 1 | | | | |
| CRM | 0.515 | 0.576 | 0.556 | 0.633 | 1 | | | |
| RI | 0.471 | 0.504 | 0.504 | 0.569 | 0.697 | 1 | | |
| Log | 0.55 | 0.621 | 0.584 | 0.637 | 0.678 | 0.602 | 1 | |
| Bus_Perfor | -0.423 | -0.3 | -0.073 | -0.036 | -0.059 | -0.15 | -0.179 | 1 |

*3.2.3. Regression Validity*

Predictive validity was determined using regression analysis. The Regression coefficient (R) highlights the correlation between predicted and observed business performance. In table 4, it can be seen that the value of R (0.759) reflects that the adopted research model adequately predicts business performance.

**Table 4**. Summary of Regression test

| R | R-Square | Adjusted R Square | Standard Error of s Estimate |
|---|---|---|---|
| 0.759 | 0.577 | 0.559 | 5.65177 |

Beta coefficient represents the relative strength of the factors among themselves. Beta coefficient is statistically significant if its p-value is smaller than 0.05[43]. The third column in Table 5 highlights that all the factors included in this research model are statistically significant.

**Table 5.** Results of Regression Analysis

| **Factors** | *Regression coefficient (B)* | *Standard Error* | *Regression coefficient(B)* | t-value | Significance level (p) |
|---|---|---|---|---|---|
| HRM | -0.116 | 0.072 | -0.118 | -1.608 | 0.039 |
| Tech | -0.078 | 0.063 | -0.098 | -1.227 | 0.021 |
| Fac | -0.056 | 0.063 | -0.069 | -0.886 | 0.016 |
| SRM | 0.2 | 0.08 | 0.021 | 0.249 | 0.049 |
| CRM | -0.067 | 0.098 | -0.065 | -0.684 | 0.029 |
| RI | -0.023 | 0.08 | -0.024 | -0.286 | 0.006 |
| Log | -0.114 | 0.087 | -0.12 | -1.305 | 0.004 |

## 4. Results and Discussion

This study highlights research model comprising of seven features including managements of human resources, technology, facility, supplier relationship, customer relationship, regulatory issues and logistics and how these features influence business performance in context of a food processing industry. Results shown in table 3 indicate that human resource management has a negative correlation with the business performance ($r = -0.423$) signifying that whenever problems regarding human resources arise, business performance of the firm decreases and vice versa. Table 5 also highlights that while predicting business development performance, human resource management problems are statistically significant ($\beta$=-0.118; $p = 0.039$). The negative value implies that HRM problems lead to reduce business performance which can be advocated with previous studies[44].

Results also suggest that a negative correlation exists between business performance and technology-related problems as shown in table 3. Analysis of the regression model also suggests that technology-related problems were statistically significant in predicting business performance ($\beta$ = -0.098, $p = 0.021$) as shown in table 5. These values suggest that the existence of technology-related problems negatively impact business performance in the food processing industry. This finding is consistent with other research study [45].

The facility-related problems also had a weak but negative correlation with the performance of the business ($r$ =-.073) indicating that unsuitable facility locations lead to a negative impact upon the performance of the business[46].Values displayed in table 5 ($\beta$ = -0.069; $p = 0.016$) show that these problems were also statistically significant in predicting the performance of the business.

The supplier relationship management problems had a very weak and negative relation with the performance of business ($\beta$ =-0.36). This suggests that problems associated to supplier relationship management does not significantly influence the business performance. These issues were also statistically significant in predicting business performance as shown in table 5 ($\beta$ = 0.021; $p = 0.049$).
Table 3 also highlights that a weak negative correlation exists between customer relationship management problems and business performance ($r = -0.059$). This reflects that any increase in problems related to customer relationship management, reduces the business performance. However, this reduction will only occur to a certain extent. In this analysis, customer relationship management problems proved to be significant in predicting business performance ($\beta$ = -0.065; $p = 0.029$).
It can also be seen that regulatory factors had a negative influence on the performance of the business ($r$ =-0.10). This suggests that when problems relevant to legal scenarios arise, it will affect the

performance of business but this effect might not be very strong. Regression analysis also reflected that legal issues were also statistically significant in predicting business performance ($\beta$ = -0.024; p = 0.006).It can also be seen that transportation issues also had a negative influence on the performance of the business (r = -0.179) as seen in table 3.

## 5. Conclusions

The research aimed to investigate the impact of supply chain management issues on business performance in Pakistan. This research has indicated that it is imperative to understand supply chain management problems and their origins in order to design approaches for avoiding and minimizing their negative effects on business performance. It is also highlighted in literature review of previous studies that although food processing industry is capable of making significant contribution to the economy of a country. Nevertheless, food processing industry in Pakistan faces multiple problems. Results obtained from Pearson correlation analysis exhibited that all of the seven factors exhibited correlation with the business performance indicating that increase in these problems, will reduce the business performance and vice versa. Furthermore, human resource management problems had a strong correlation in comparison to the other six factors. In comparison, supplier relationship management presented a weak correlation with the business performance. Results obtained from regression analysis also reflected that all the factors are found significant to predict business performance.

The results of the study have several implications. It highlights the importance of supply chain management and which directly influences the performance of the business in food processing industries. The overall organizing principle of this study is that business performance in food industry, among other things, can be improved by reducing the severity of supply chain management problems. It is also important to understand that there is currently a shortage of supply chain professionals in Pakistan. To mitigate this human resource management issue, it becomes important that competent supply chain professionals are recruited. Once they are hired, there should be regular trainings to develop their knowledge, skills and output. To solve technological issues, the rapid implementation of disruptive technologies is required. To be competitive in market regular maintenance of key technologies and the integration of technologies across all business functions should be made. ISO standards, sustainability or green supply chain management systems can be implemented to counter the problems associated with facilities management. The problem faced with facility management can be countered by practicing international standards and philosophies like ISO, sustainability and green SCM.

Some limitations are also outlined for this research study so that they can be addressed in the future. Data were collected from a sample size of 302 respondents located in the different regions of the provinces of Khyber Pakhtunkhwa and Punjab only in Pakistan. Future research studies should be focused on collecting data from other provinces of Pakistan based on face to face structured interviews.

## 6. References

[1] Chandio A, Yuansheng J, Rahman T, Khan M. 2015 Analysis of agricultural sub-sectors contribution growth rate in the agriculture GDP growth rate of Pakistan. *Int. Journal of Humanities and Social Science Invention*. **4**,101-5.

[2] Kashif U, Hong C, Naseem S. 2017 Comparison of Government Expenditure and Some Representative of Agricultural Product Prices and Their Impact on Agricultural Growth of Pakistan. *Russian Journal of Agricultural and Socio-Economic Sciences*. **63**,79-85.

[3] Rehman A, Jingdong L, Du Y. 2015 Last Five Years Pakistan Economic Growth Rate (GDP) And Its Comparison With China, India And Bangladesh. *Int. Journal of Scientific & Technology Research*. **4**,81-4.

[4] Hussain S, Ahmed W, Rabnawaz A, Jafar R, Akhtar H, Ju W, et al. 2015 Integration and Effective Supply Chain Management: A Review of Agriculture in Pakistan and China. *Journal of Economics and Sustainable Development*. **6**,56-61.
[5] Ziaullah M. Dairy 2014 Supply Chain Management and Critical an Investigations on Dairy Informal Channel Partners in Pakistan.
[6] Javaid M, Garg D, Ali A. 2012 Increasing sale, profit rate and productivity improvement using supply chain management: A case study. *Int. Journal of Appl. Engineering Research*. **7**,1351-7.
[7] Hong J, Zhang Y, Shi M. 2017 The impact of supply chain quality management practices and knowledge transfer on organisational performance: an empirical investigation from China. *Int. Journal of Logistics Research and Applications*. **21**,1-20.
[8] Raut R. 2019 Improvement in the Food Losses in Fruits and Vegetable Supply Chain - A Perspective of Cold Third-Party Logistics Approach. *Operations Research Perspectives*. **6**,100117.
[9] Shayan E, Ghotb F. 2019 *Supply Chain Management Problems*: A Review.
[10] Rashid K, Aslam H. 2012 Business excellence through total supply chain quality management. *Asian Journal on Quality*. **13**,309-24.
[11] Dharni K, Sharma R. 2015 Supply chain management in food processing sector: Experience from India. *Int. Journal of Logistics Systems and Management*. **21**,115.
[12] Shabbir M, Kassim N. 2018 Supply chain management drivers and sustainability of green initiatives in manufacturing enterprises: A case in Pakistan. *Int. Journal of Entrepreneurship* **22**.
[13] Chaudhuri A, Dukovska-Popovska I, Damgaard CM, Hvolby H-H, 2014 editors. Supply Uncertainty in Food Processing Supply Chain: Sources and Coping Strategies. Advances in Production Management Systems Innovative and Knowledge-Based Production Management in a Global-Local World; 2014 2014//; Berlin, Heidelberg: Springer Berlin Heidelberg.
[14] Ling C, Mansori S. 2018 The Effects of Product Quality on Customer Satisfaction and Loyalty: Evidence from Malaysian Engineering Industry. *Int. Journal of Industrial Marketing*. **3**,20-35.
[15] Hoffmann P, Schiele H, Krabbendam K. 2013 Uncertainty, supply risk management and their impact on performance. *Journal of Purchasing and Supply Management*. **19**(3),199-211.
[16] Andalib Ardakani D, Soltanmohammadi A. 2019 Investigating and analysing the factors affecting the development of sustainable supply chain model in the industrial sectors. *Corporate Social Responsibility and Environmental Management*. **26**(1),199-212.
[17] Stuart F, Verville J, Taskin N. 2012 Trust in buyer-supplier relationships: Supplier competency, interpersonal relationships and performance outcomes. *Journal of Enterprise Information Management*. **25**.
[18] Choughri R, Soubjaki M. 2017 The Future of Business Analytics for SMEs in MENA Region. *IOSR Journal of Business and Management (IOSR-JBM)*. **19**,86-92.
[19] Holbeche L. 2018 Organisational effectiveness and agility. *Journal of Organizational Effectiveness: People and Performance* **5**.
[20] Haggège M, Gauthier C, Rüling C-C. 2017 Business model performance: five key drivers. *Journal of Business Strategy*. **38**,6-15.
[21] Reddy. K JM, Rao. A N, L K. 2019 A review on supply chain performance measurement systems. *Procedia Manufacturing*. **30**,40-7.
[22] Bagher A. 2018 The effect of supply chain capabilities on performance of food companies. *Journal of Finance and Marketing*. **02**.
[23] Gunasekaran A, Subramanian N, Rahman S. 2017 Improving supply chain performance through management capabilities. *Production Planning & Control*. **28**(6-8),473-7.
[24] Ronoh R. 2016 Benefits of Supply Chain Management in the Manufacturing Sector. *Int. Journal of Sci. and Research (IJSR)*. **5**,1967-70.
[25] Dessler G. 2018 *Human resource management*. Sixteenth Edition. (New York: Pearson)

[26] Min S, Zacharia ZG, Smith CD. 2019 Defining Supply Chain Management: In the Past, Present, and Future. *Journal of Business Logistics*. **40**(1),44-55.
[27] Goldsby TJ, Zinn W. 2016 Technology Innovation and New Business Models: Can Logistics and Supply Chain Research Accelerate the Evolution? Journal of Business Logistics. **37**(2),80-1.
[28] Berekaa M. 2015 Nanotechnology in Food Industry; Advances in Food processing, Packaging and Food Safety. *Int. Journal of Current Microbiology and Applied Sci.s*. **4**,345-57.
[29] Fraser K. 2014 Facilities management: The strategic selection of a maintenance system. *Journal of Facilities Mgmnt*. **12**.
[30] Yunus EN, Kurniawan T. 2015 Revealing Unsuccessful Collaboration: A Case of Buyer-Supplier Relationship in the Pharmaceutical Industry. *Supply Chain Forum: An Int. Journal*. **16**(2),14-28.
[31] Lintukangas K. 2011 Supplier relationship management capability in global supply management. *Int. Journal of Procurement Mgmnt*. **4**.
[32] Dash A, Pothal LK, Tripathy S. 2018 Factors affecting Supplier Relationship Management: An AHP Approach. *IOP Conference Series: Materials Sci. and Engineering*. 390:012056.
[33] Smith A. 2019 Customer Relationships Management for Improved Productivity. *Int. Journal of Adult Vocational Education and Technology*. **10**,40-8.
[34] Srivastava SK, Chandra B, Sinha AP. 2018 Effectiveness of knowledge management on customer relationship management in hotel business performance. *Journal of Environmental Mgmnt and Tourism*. **9**,1277-88.
[35] Haislip J, Richardson V. 2017 The effect of Customer Relationship Management systems on firm performance. *Int. Journal of Accounting Information Systems*. **27**,16-29.
[36] Sitharam S, Hoque M. 2016 Factors affecting the performance of small and medium enterprises in KwaZulu-Natal, South Africa. *Problems and Perspectives in Mgmnt*. **14**,277-88 and 58.
[37] Olotu I, Adewale O, Olorunfemi M, Akande O, Landschoot S, Saeger S, et al. 2019 Occurrence of bacteria and endotoxins in fermented foods and beverages from Nigeria and South Africa. *Int. Journal of Food Microbiology*. **305**,108251.
[38] Anekwe R, Purity N-O, Chuka. 2019 Sustainability in the Nigerian Business Environment: Problems and Prospects. *The Int. Journal of Business and Management Research*. **3**,72-80.
[39] Ackerley N, Sertkaya A, Lange R. 2010 Food Transportation Safety: Characterizing Risks and Controls by Use of Expert Opinion. *Food Protection Trends*. **30**.
[40] Mercier S, Villeneuve S, Mondor M, Uysal I. 2017 Time-Temperature Management Along the Food Cold Chain: A Review of Recent Developments: Food preservation along the cold chain Comprehensive Reviews in Food Science and Food Safety. **16**.
[41] Anca V. 2019 Logistics and Supply Chain Management: An Overview. *Studies in Business and Economics*. **14**,209-15.
[42] Vij S, Bedi H. 2016 Are subjective business performance measures justified? *Int. Journal of Productivity and Performance Mgmnt*. **65**,603-21.
[43] Mansharamani H, Patil PS, Behere PB, Mansharamani D, Nagdive A. 2018 Psychiatric morbidity in children of alcoholic parents. *Ind Psychiatry J*. **27**(2),226-30.
[44] Van De Voorde K, Paauwe J, Van Veldhoven M. 2010 Predicting business unit performance using employee surveys: monitoring HRM-related changes. *Human Resource Mgmnt Journal*. **20**(1),44-63.
[45] Tavitiyaman P, Qu H, Zhang H. 2011 The Impact of Industry Force Factors on Resource Competitive Strategies and Hotel Performance. *Int. Journal of Hospitality Mgmnt - INT J HOSP MANAG*. **30**,648-57.
[46] Ko K-J, Park S-H. 2019 A Study on the Business Performance of Swimming Industry in Korea : Focusing on Location and Scale of Employment. *Korean Journal of Sports Sci.*. **28**,477-86.

# Development of Low-Cost Multi-Input Automated Storage and Retrieval System (AS/RS) for Educational Purposes

**Y Theopilus**[1,2], **S S Tjandra**[1,3] **and B Sagara**[1,4]

[1]Parahyangan Catholic University, 94th Ciumbuleuit Street, Bandung, Indonesia
[2]yansen_theopilus@unpar.ac.id, [3]sugih.sudharma@unpar.ac.id, [4]sagarabil@gmail.com

**Abstract.** Nowadays, learning about automation technology in higher education is important to produce graduates who are ready to work. However, in developing countries, this learning is often constrained by the problem of limited costs and places to procure practice tools. To overcome this problem, this research focuses on developing automation technology learning tools that are inexpensive and less space. The object is Automated Storage and Retrieval System (AS / RS) which is commonly used in various industries for material handling but is too expensive and large for learning. The AS / RS developed is an Arduino based AS / RS with multi-input modes that are pendant, coordinate, and RFID. The AS / RS is designed to facilitate students to learn the work system, electrical circuit, and programming algorithm. The research was held in several stages: problem identification, electrical hardware design, mechanical design, algorithm and programming design, prototyping and testing. The AS / RS prototype only costs $ 230 and requires 600 x 400 x 550 mm space. The testing was carried out on functionality, usability, and educational ability using task completion and essay post-test method. Based on the test, the three aspects of the designed AS / RS were very satisfying.
**Keywords**: Automation, Low Cost, Multi-input, Automated Storage and Retrieval System, Higher Education

## 1. Introduction

Entering the era of industrial revolution 4.0, companies in various industrial sectors try to apply various technologies to their processes and products to increase efficiency and effectiveness in their business. Industry 4.0 has changed the way we develop products and processes significantly, specifically in terms of design, process, operation, and service [1]. The technology which is used in various industrial sectors to improve efficiency and productivity is the automation of production or distribution system. Automation is a technology used to carry out various processes or procedures with little or no human assistance [2]. The application of automated processes in production or distribution system has various advantages, such as increasing labor productivity, improving product and process quality and reliability, reducing labor cost, improving worker safety, improving product quality, reducing lead time, accomplishing what cannot be done manually, and so many more. Although the application of automation within a company can cause an increase in the investment and maintenance costs, the significant positive impact caused provides more financial and non-financial benefits than the costs.

The rapid development of automation technology makes the world of education (especially higher education) should quickly adapt and educate the development of the technology to its students. This statement appears since the majority of graduates from higher education will work in companies that are most likely to utilize various technologies. If the higher education graduates are not prepared for the automation technology, they would have failed to meet the expectations of the workplaces because they do not have sufficient competence to work. Nowadays, higher education should be able to face the demands of the workplace in terms of automation and artificial intelligence, both understanding theory and how it is applied in real cases [3].

Learning theory and practice related to automation technology in various industrial sectors are very important, especially for several departments that are closely related such as industrial engineering, mechanical engineering, electrical engineering, chemical engineering, food engineering, computer science, logistics, and many more. However, the implementation of this learning should be supported by the ability and readiness of the lecturers, appropriate learning methods, and good facilities. The challenges that often arise in developing countries are very expensive learning tools and limited space. In some countries, especially developing countries, the procurement of good facilities and practice tools is often constrained by the problem of excessive costs and limited space. In some universities that have these limitations, students just can learn a variety of automation technologies from slideshows or learning video, without being able to practice more. This phenomenon makes its students cannot practice what they have learned. The impact of This phenomenon led to the tendency that they do not have the skills they should need in the workplace. In the long run, it will cause a big gap in the quality of graduates because the students of poor universities never have an adequate learning practice to prepare their students for the workplace demands.

To prepare engineering students to work in many industries, there are variety of automation technologies that need to be learned, such as rapid prototyping technology for product design, machine vision technology for inspection, automated storage and retrieval technology for material storage, flexible manufacturing system, cloud computing technology for data integration, and so many more. This research focuses on supporting the automated system learning for material handling and storage activities, where this activity is needed by so many kinds of industry. Automated Storage and Retrieval System (AS / RS) is an automated system of storing, retrieving, and moving material activity that is familiar in various industries. Material handling activity using AS / RS has so many advantages over the manual method using human, such as process acceleration, increased productivity, improved work health and safety, reduced labor costs, increased density and storage capacity, and increased storage security [4]. The movements that occur in the AS / RS system based on three axes of linear motion along the X-axis, Y-axis, and Z-axis. The movement along the X-axis is the AS / RS positioning movement in the storage rack column, the movement along the Y-axis is used to move the AS / RS to the appropriate storage rack line or level, and the movement along the Z-axis is used to position the fork against the specific rack in picking up or placing the material.

AS / RS is one of the automated systems in the industry that is commonly used for both small and large industries. Therefore, learning the AS / RS system is very important for many higher education students, especially engineering students who are closely related to the industry. However, AS / RS for industrial- scale are generally very large in size and very expensive, so that most universities have a limitation in the procurement of AS / RS in their laboratory to hone the practical ability of their students to operate and understand how AS / RS works. The production of most AS / RS at this time should be made custom to a few AS / RS sellers with very expensive price, relatively large size, and specific programming language that are difficult to learn. This phenomenon makes it difficult for universities to procure AS / RS for their students and makes it difficult for students to gain practical skills related to AS / RS mechanical system and programming.

Until now, some research has been carried out regarding the theoretical AS / RS exposure and AS / RS development over time [4], the AS / RS modeling and optimization to obtain optimal performances [5], the AS / RS design for various industrial sector [6], and many more. Unfortunately, no research developed the AS / RS system that is specifically intended for educational purposes to overcome the problems that have been described previously. Seeing the problems in the field of higher education in developing countries that arise and the absence of research to overcome these problems, this research aims to develop a low-cost multi-input Automated Storage and Retrieval System (AS / RS) for educational purposes. The AS / RS system uses multi-input so that the students can understand how AS / RS works with various possibilities of input. The AS / RS system also uses an Arduino Mega Microcontroller which uses an open-source programming platform using the C programming language to facilitate students in learning and practicing how to program and remake the AS / RS.

## 2. Method

On this research, the AS / RS system design has a specific purpose, that is for educational purposes. For making the development research more directed, we've made a mission statement containing the direction of the AS / RS system development. The mission statement is developed based on the educational problems on automation technology discussed in the previous chapter. The mission statement of AS / RS system development for educational purposes can be seen on Table 1.

**Table 1.** Mission Statement.

| Project Specification | Statement |
|---|---|
| Description | Low-cost multi-input Automated Storage and Retrieval System (AS/RS) to support students in higher education to learn about the operation system, mechanism, and programming of the AS/RS. |
| Benefit Preposition | • Low material and development costs.<br>• Using multi-input (pendant, coordinate, and RFID) so that students can learn more mechanism.<br>• Open-source programming using Arduino that is easy to learn the program and easy to remake.<br>• Modular hardware system that is easy to assemble and disassemble.<br>• Relatively small dimension so that it is easy to place. |
| Market | Primary : Higher Education, especially for engineering majors<br>Secondary : Industrial training center and high school |
| Scope and Assumption | • The AS/RS load capacity is small because the AS/RS is designed to focus on educational activity<br>• The input data command of AS/RS is retrieved from pendant, coordinate, and RFID.<br>• The AS / RS design uses a dedicated storage type.<br>• each rack or bay is only used to hold one type of component with one unit load (tote pan or other storage areas). |
| Stakeholders | Designer, Engineer, Manufacturer, Government, Distributor, Consumer, and User |

Based on the mission statement above, we developed a research method to produce an AS / RS system that is consistent with the mission statement. The research method can be seen on Figure 1.

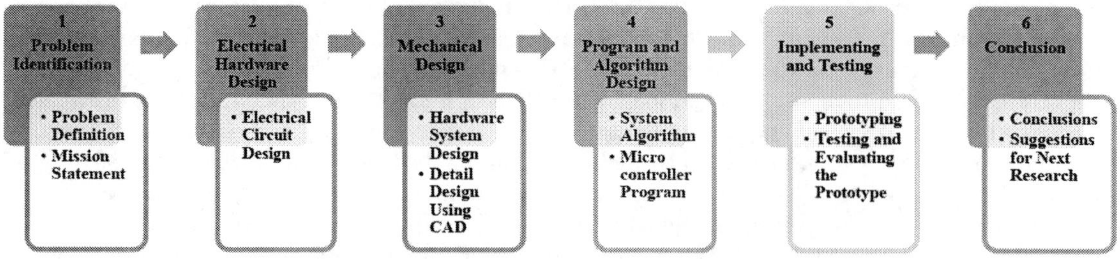


**Figure 1.** Research Methodology.

## 3. Result and Discussion

Based on the research methodology in the previous chapter, the AS / RS system is developed under the specified mission statement. The AS / RS type developed is a dedicated unit load AS / RS that uses an Arduino Mega 2560 Microcontroller with RAMPS 1.4 Shield Motor Driver that uses an open-source programming platform with C language that is easy to understand by the students. The development focuses on making AS / RS system that are inexpensive, have a lot of input modes, and have a simple program and electricity mechanism. If compared with the real AS / RS system in industrial scale, there are some simplifications made in the this AS / RS system because it is devoted to educational purposes, such as smaller dimensions, smaller capacity, smaller loads dimension, open-source programming platform, and more varied input data. A detailed explanation of the stages of developing the AS / RS system for educational purposes can be seen in the sections below.

### *3.1. Electrical Hardware Design*

The working system of the AS / RS developed was centered on the Arduino Mega microcontroller. All input components are connected to the microcontroller to be able to provide information to Arduino to be processed by the program algorithm in Arduino. However, because the number of pins on the Arduino Mega is limited and the need for a stepper driver to run all the stepper motors, all the inputs and outputs connected to the Arduino microcontroller through an intermediary shield namely Ramps 1.4 Shield. There is a power supply that is also connected to Ramps 1.4 Shield to provide voltage to the AS / RS circuit. The control system used in the AS / RS system is an open-loop control system because there is no feedback given from the motion mechanism and actuator. The AS / RS system electrical hardware design can be seen on Figure 2.

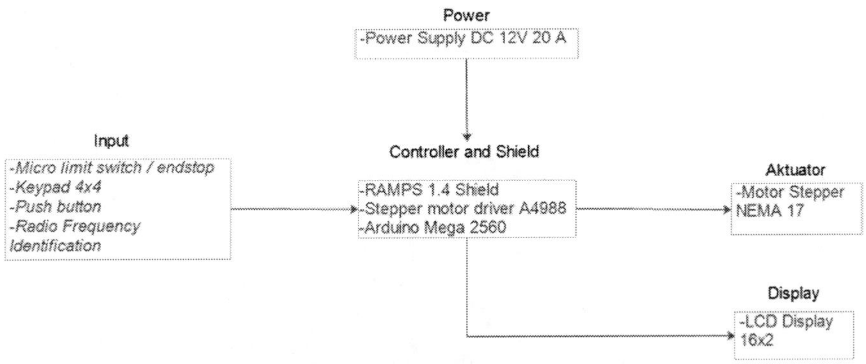


**Figure 2.** Electrical Hardware Scheme.

Arduino Mega 2560 is paired with Ramps 1.4 to facilitate the application of stepper motor in Arduino and also to increase the number of pins that can be used by other modules. Furthermore, all modules needed are installed such as NEMA 17 stepper motor, 16x2 LCD Display, 4x4 keypad, toggle switch, and RFID scanner. In the stepper motor module, the A4988 stepper driver should be installed first before attempting the motor in order to control the motor's performance. A heat sink is also installed in the driver to reduce the heat generated by a large and continuous supply of electrical voltage to drive the motor. In the LCD Display module, the installation uses a casing made by ABS plastic to protect and tidy up the arrangement of the circuit. A potentiometer is also needed in the installation of this module to adjust the intensity of the light of the display. Installation of the LCD display is made by soldering each slot with the jumper connector head. In the input module, the 4x4 keypad installation also uses ABS plastic casing to give the impression of neatness in the design. For other input modules such as push-buttons for pendant mode and RFID scanner for RFID mode, breadboard intermediaries are needed because the circuit is quite complicated. The complete AS / RS electrical hardware design can be seen on Figure 3.

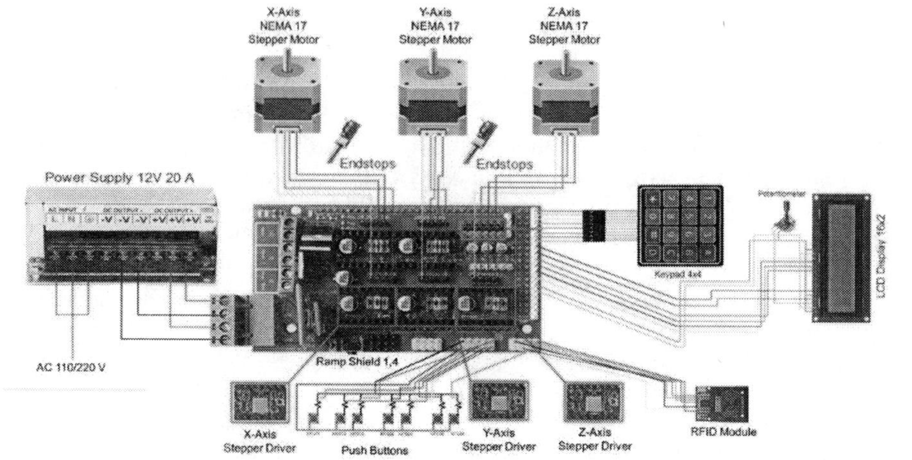


**Figure 3.** Electrical Hardware Design.

### *3.2. Mechanical Design*

The designed AS / RS utilizes the rotational motion of three motors into linear motion in the X, Y, and Z axes to be able to retrieve, move, and store material according to certain data inputs. The motion mechanism of the X-axis and Z-axis motors use a belt pulley system, whereas the Y-axis motion mechanism uses a lead screw system. The X-axis and Z-axis movements are designed to be able to move quickly so that the belt pulley mechanism is used. For the Y-axis, the movement should be more concerned with the level of precision and strength of the load lifting rather than speed, so that the lead screw mechanism is used. The AS / RS design has dimensions of 600 x 400 x 550 mm. With that size, the AS / RS is easy to carry and does not require much space. This machine requires 12V 20A of electric energy to run properly. The results of the CAD hardware AS / RS design can be seen on Figure 4.

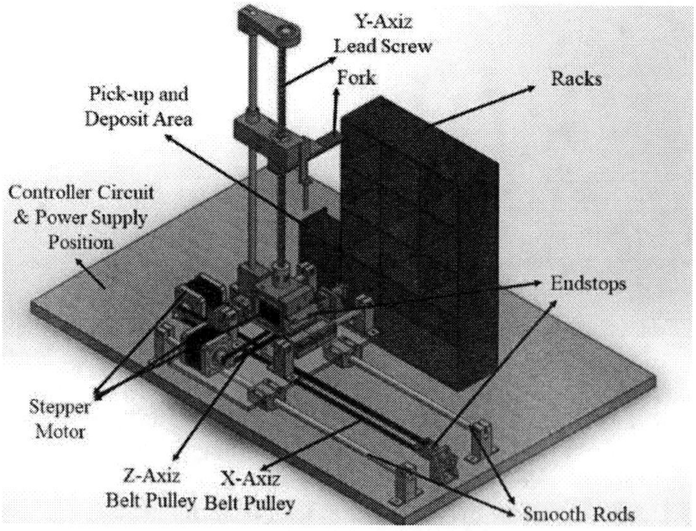


**Figure 4.** Hardware Design.

### *3.3. Software Design*

AS / RS Programming is designed using Arduino IDE software because the microcontroller used is Arduino Mega. The programming in the AS / RS design is divided into two major sections that are the

IOP Conf. Series: Materials Science and Engineering **847** (2020) 012072 doi:10.1088/1757-899X/847/1/012072

setup section and the main content section. The setup section (void setup) contains libraries, input and output declarations, and program default settings. The main content section contains the contents of the program itself (void loop) that are the program to run the function of pendant mode, coordinates mode, RFID mode, and auto home mode.

The beginning of the void setup is filled with all the libraries used. Library is a collection of programming code that can be used as needed. Library writing is done by writing #include at the beginning of writing. the void setup continues to define the step where all modules used need to be defined first based on the Ramp 1.4 Shield Structure. Each module of the AS / RS programming has a different way of defining steps.

In the void setup section, all declaration names are written in the same code: 'pinMode ()' followed by the name of the declaration used and its role (input or output). The motor acts as the output, while the push-button acts as an input and end-stop acts as a pull-up input. LCD initialization is done by writing the sentence 'lcd.begin ()' followed by the size of the LCD used. Code 'Serial.begin (9600)' is only used to activate the serial monitor for checking. Code 'SPI.begin ()' is used to activate the Serial Peripheral Interaction system and code 'rfid.init ()' is used to enable the type of RFID MFRC522 used. Motor X-axis, Y-axis, and Z-axis are also activated first by giving a digital 'LOW' value.

In the void loop section, before starting the operation program, the AS / RS opening title is displayed first on the LCD. After displaying the opening title on the LCD, the contents of the program operation are made. First of all, the user needs to choose the modes from the four modes provided, namely pendant mode, coordinate mode, RFID mode, and auto home mode. After the user selects a mode, it will enter into the chosen mode. The pendant void mode contains a program to run the fork from the push-button command. There are 7 push-buttons that are clockwise and counterclockwise on the three axes and one push-button as a stop. The coordinate void mode contains a program to run the fork according to the distance of the coordinate inputting input. The RFID void mode contains a program to run the machine in accordance with the number code of the card tag that is read by the RFID scanner. The auto home void mode contains a program to run all motors back to their default position (all coordinate axes are 0). The AS / RS program design algorithm can be seen on Figure 5.

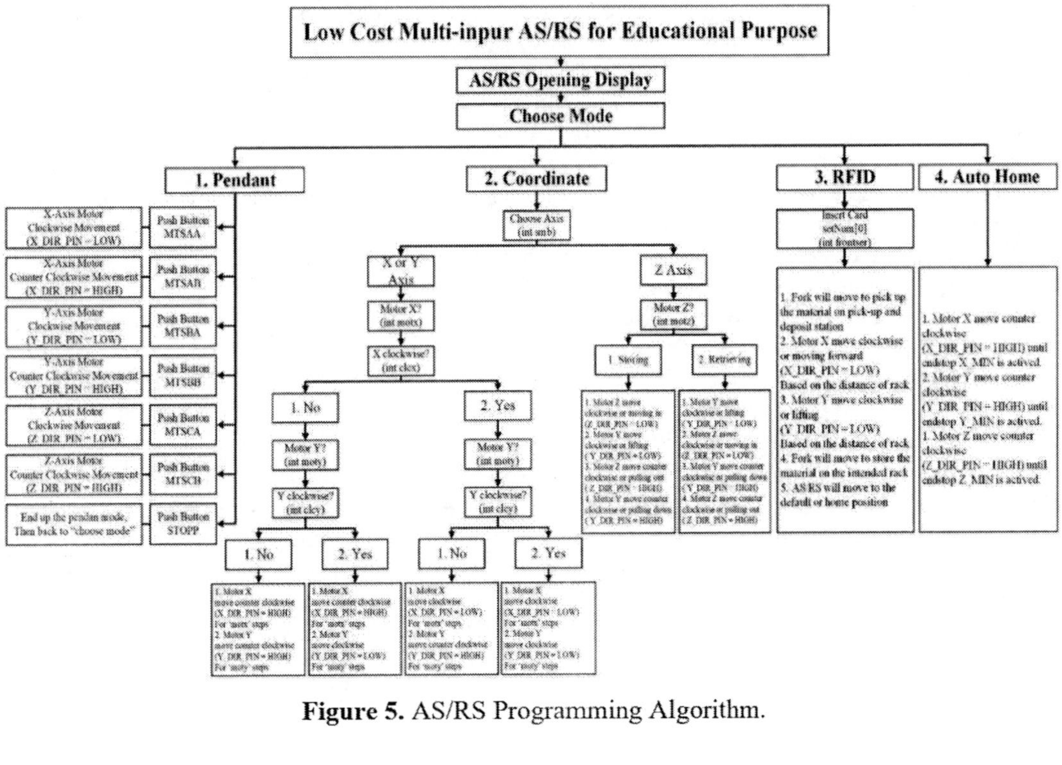


**Figure 5.** AS/RS Programming Algorithm.

*3.4. Implementation and Testing*

In the implementation stage, a final prototype of the designed AS / RS is made for testing and evaluating the function, the usability, and the ability to educate its users. The AS / RS design prototype for educational purposes can be seen on Figure 6.

**Figure 6.** AS/RS Prototype.

The cost of making the AS / RS prototype only costs around $ 230, which can be much reduced if it goes into the mass production. Thus, the costs required to produce this AS / RS will be much cheaper than buying an industrial AS / RS that has price up to thousands of dollar. In terms of space requirements, the AS / RS design only requires a small space, which is 600 x 400 x 550 mm, so that it will be easy for higher education to provide the space needed. Thus, the limited cost and space which is a problem for various educational institutions in learning the application of automation technology can be overcomed.

Most of the frames in the hardware, the shelves, and the Arduino box cases are using ABS plastic and manufactured using rapid prototyping technology (3D printing method). The AS / RS design is made open in terms of the appearance of the frame with the aim that the user can learn how the electronic circuits arranged can produce the motion mechanism that occurs. This system also makes it easier for users to learn the structure of the AS / RS to be remade by themselves. However, the Arduino circuit is wrapped in a modular ABS plastic box cases so that the position of the circuit is not easily changed and safer for the user. The microcontroller circuit and the program can be learned directly by the user and the program can be downloaded easily by the user through the Arduino device. Thus, the AS / RS electrical hardware design provides easy learning about the circuits and the AS / RS programming. The AS / RS can also improve user understanding of the motion mechanism and the algorithm of action using pendant mode, coordinate mode, and RFID mode as discussed in the previous section.

The prototype function testing was carried out by testing five potential users of the AS / RS, who are engineering students and lecturers in higher education, to carry out tasks related to the entire use of AS / RS design. Based on prior studies, five respondents are considered sufficient for a qualitative research [7]. The tasks given to the respondents were placing and retrieving materials in the unit load from and to 3 different shelves that have been determined by the researcher with pendant, coordinate, and RFID modes. Before the testing began, the respondents were given manual instructions that were made to facilitate users in operating the AS / RS. After carrying out these tasks, the respondents should have returned the fork position to the default position using the auto home menu. Based on the test conducted, all the tasks given can be carried out properly and correctly by all respondents. The AS / RS design performance also has 100% correct placement accuracy based on the user input. Thus, it can be concluded that the function and the ability of the designed AS / RS can be said to be good enough.

However, there are two notes based on tests conducted. First, the motor speed is considered too slow (X-axis speed = 89 mm/sec and Y-axis speed = 26 mm/sec). Second, some parts of the instruction manual are considered difficult to understand without images explanatory. Therefore, the motors would be replaced with NEMA 23 stepper motor that is stronger than NEMA 17 but still in accordance with their use, so that the speed of fork movement can increase up to twice.

Subsequent testing was to ensure that the designed AS / RS design successfully educates its users in terms of practices related to the working mechanism, the controller circuit, and the programming of AS / RS system. The test was held by evaluating the understanding of five industrial engineering students who have studied the AS / RS design for 120 minutes. The respondents were the students who had studied theories about automation technology in the manufacturing systems, computer programming with Java or C languages, basic electronics, and basic theories of AS / RS. The understanding evaluation was held by giving 15 essay questions on understanding the mechanism of AS / RS work and how to operate it, the AS / RS electronic circuit, and the Arduino programming. The answers of each respondent were assessed by a lecturer as well as a practitioner who is an expert in manufacturing, mechatronics, and Arduino programming. The assessment was done by giving a score of 0-100 for each respondent's answer, where a score of 0-50 is considered very unsatisfactory, a value of 50-60 is considered unsatisfactory, a value of 60-70 is considered quite satisfactory, a value of 70-80 is considered satisfactory, and a value of 80-100 is considered very satisfactory. Based on the evaluation of students's understanding, the average value of all respondents was 81.33. Therefore, it can be said that the AS / RS design has been able to educate students very satisfactorily. Based on the tests conducted, there are two things that need to be added to make the learning process more optimal. First, the students need the addition of a guideline book on AS / RS programming. Second, the students need a function of taking or storing material to several shelves at once in the designed AS / RS.

## 4. Conclusion

The AS / RS development research for educational purposes is one way to make the learning process that requires tools to be more effective and optimal, especially for some developing countries that often experience some financial and space constraints in developing their learning activities. Based on the research conducted, the designed AS / RS can be said to be successful in meeting its goal of facilitating the engineering students learning process about the AS / RS system with low procurement cost and minimum space. By using a similar methodology, subsequent research can also develop other learning practice tools with the same background and limitations. We all have already known that well-facilitated higher education can produce superior generations who can well compete in the future. Therefore, similar research can be developed with the same passion to develop the quality of humans in the industry 4.0 era which requires a good understanding of technology to well compete.

## 5. References

[1] Ślusarczyk B 2018 Industry 4.0 – Are We Ready? *Polish. J. Man. Stu.* **17** 232

[2] Groover M P 2001 *Automation, Production Systems and Computer-Integrated Manufacturing* (New Jersey: Prentice Hall)

[3] He Y, Liang L 2019 Application of Robotics in Higher Education in Industry 4.0 Era. *Uni. J. of Edu. Research.* **7** 1626

[4] Hu Y H, Huang S Y, Chen C, Hsu W J, Toh A C, Loh C K, and Song T 2005 Travel Time Analysis of a New Automated Storage and Retrieval System. *Comp. and. Op. Research.* **32** 1515

[5] Brezovnik S, Gotlih J, Balič J, Gotlih K, and Brezočnik M 2015 Optimization of an Automated Storage and Retrieval Systems by Swarm Intelligence. *Proc. Eng.* **100** 1309

[6] de Koster R M B M, Le-Duc T, and Yugang Y 2008 Optimal Storage Rack Design for A 3-Dimensional Compact AS/RS. *Int. J. of Prod. Research* **46** 1495

[7] Sim J, Saunders B, Waterfield J, Kingstone T 2018 Can Sample Size in Qualitative Research Be Determined A Priori? Int. J. of Soc. Research Method. **21** 619

# Workplace Design Process at Indonesian Manufacturing SMEs

**D Herwanto**[1,2,3] **and A Suzianti**[1,4]

[1]Industrial Engineering, Faculty of Engineering, Universitas Indonesia, Depok, Indonesia
[2]Industrial Engineering, Faculty of Engineering, Universitas Singaperbangsa Karawang, Karawang, Indonesia
[3]deneherwanto@yahoo.com, [4]suzianti@eng.ui.ac.id

**Abstract.** The workplace design process must not be ignored to get a good workplace. A good workplace can support occupational health and safety and comfort of workers while doing work so that it is expected to increase their work productivity. Unfortunately, the environmental conditions of workspaces in most SMEs (small and medium enterprises) in Indonesia do not meet the standards. Besides, production machinery, raw materials, and finished goods are placed irregularly which has the potential to inhibit material transfer. This study is aimed at obtaining an overview of the workplace design process carried out in Indonesian manufacturing SMEs as a first step to finding a suitable workplace design process model for manufacturing SMEs in Indonesia.
**Keywords**: Occupational Safety and Health, Productivity, Small and Medium Enterprises, Workplace, Workplace Design Process

## 1. Introduction

Workers need a good workplace to spend their best work. A good workplace can be earned through good workplace design.

Good workplace design should not be underestimated. According to Chim [1], good workplace design can reduce the rate of fatalities and accidents. Moreover, good workplace design is also able to support the working practices become more effective [2], as well as affecting the physical and psychological health, performance, satisfaction, comfort, safety, and productivity of employees [3, 4, 5, 6, 7]. Bakke [8] states that the design of the workplace is an important characteristic of the company. In some particular workplaces, workplace design can also become part of the company's business strategy [9].

The benefits of workplace design as mentioned above can be obtained through a good workplace design process. The process of workplace design is a complex work, requires a long process, and it involves many aspects in the birth of a workplace [10].

In reality, Launis et al [10] state that the practical experience in the industrial workplace design in several countries has shown results are not as expected. Santos [11] also showed that more than one-third of the workplace design should be repaired in the first year. It shows that the process of workplace design is very important to be taken seriously.

Given the importance of the process of workplace design to produce a good workplace design, some countries have published guidelines in the design of the workplace, such as Australia and Norway [8,

12]. The guide serves as a guideline in workplace design in an office to obtain an effective and efficient office workspace.

In connection with the workplace design in the manufacturing industry, some previous researchers have introduced workplace design process models for manufacturing industries [13, 14, 15, 16, 17, 18]. However, if we look further, models of the process of workplace design that was introduced by the previous researchers are more intended to be implemented in medium or large companies, not for small and medium enterprises (SMEs), while SMEs have a role strategic for the economy of a country, especially Indonesia.

Previous studies showed that small and medium enterprises (SMEs) have a very important role, even become a pillar in the country's economy [19, 20, 21, 22]. SMEs provide a substantial contribution, especially in developing countries, among which accounts for about 36% to 60% of GDP (Gross Domestic Product) and absorbs about 35% to 70% of the entire workforce [19, 20, 21].

SMEs' conditions in Indonesia, in general, are not much different from the condition of SMEs in other developing countries. The Statistics Indonesia data of 2016 showed that the number of MSMEs (micro, small, and medium enterprises) at least 99%, the contribution to GDP amounted to 62.57%, with employment amounted to 96.50% [23].

Despite having a considerable contribution, various studies indicate that the environmental conditions in the workplace of Indonesian SMEs are less good. The results of the literature review conducted by Hermawati et al [24] showed that most of the workspace in Indonesian SMEs is in stuffy and hot conditions. The results of Herwanto and Suzianti's literature search [25] also showed that in addition to the heat, workspace environment conditions of most of the manufacturing SMEs in Indonesia, such as noise, lighting, and other factors do not meet the standards set by the government. Moreover, Mufti et al [26] state that the layout of the work stations in Indonesian SMEs is less well thus increasing the risk of the safety and health of workers.

The workplace conditions in Indonesian SMEs as mentioned above can be caused by the process of workplace design which is not good. This study aimed to find out how the process of workplace design done on SMEs in Indonesia, especially SMEs engaged in the manufacturing industry. This study represents the initial phase of the design process to find a suitable workplace design process for manufacturing SMEs in Indonesia.

## 2. Methods

This study was conducted at three manufacturing SMEs located in Karawang, West Java, Indonesia. The three SMEs are engaged in the manufacturing of machine components, transportation equipment, and other components as well as a vendor for several major companies in Indonesia. The selection of the three SMEs is based on the consideration that Karawang is the largest industrial area in Southeast Asia with thousands of companies in it, whether small, medium, and large companies. Karawang also is one of the areas that has become the center of the development of machinery and transport equipment industry in Indonesia.

As mentioned above, this study aimed to find out how the process of workplace design done in the manufacturing SMEs in Indonesia so far, and how it has affected the workplace conditions of SMEs. This study represents the initial phase to find a suitable workplace design process model to be applied to manufacturing SMEs in Indonesia.

This study begins with the collection of data through interviews with the owners or managers of manufacturing SMEs in Indonesia which is the object of this study. Interviews were conducted to determine how the process of workplace design that had been done on SMEs that they manage. The interview is further illustrated in the form of a flow diagram of the process of workplace design.

In addition to interviews, data collection is also done through direct observation and measurement. Direct observation was conducted to determine the condition of the layout of the workplace of the SMEs concerned. As for knowing the environmental conditions in the workplace, then the direct measurement of the lighting and noise was carried out using a lux meter and noise meter.

The results of measurements of lighting and noise will then be compared to standard lighting and noise in the workspace accordance with the Regulation of the Minister of Manpower No. 5 of 2018 concerning Occupational Safety and Health of the Work Environment to determine whether environmental conditions (lighting and noise) in the workspace is already compliant or not yet.

Finally, an analysis is carried out to see the effect of the workplace design process on the layout and environmental conditions in the workplace on manufacturing SMEs.

## 3. Result and Discussion

### *3.1 Workplace Design Process at SME A*

SME A is one of the small manufacturing industries in Indonesia, located in the eastern part of Karawang. This SME produces products in the form of various machine components and other production auxiliary equipment. The products produced by this SME has been used by many companies in Indonesia.

Based on interviews with the owner of SME A, the workplace design process in SME A can be described as shown in Figure 1.

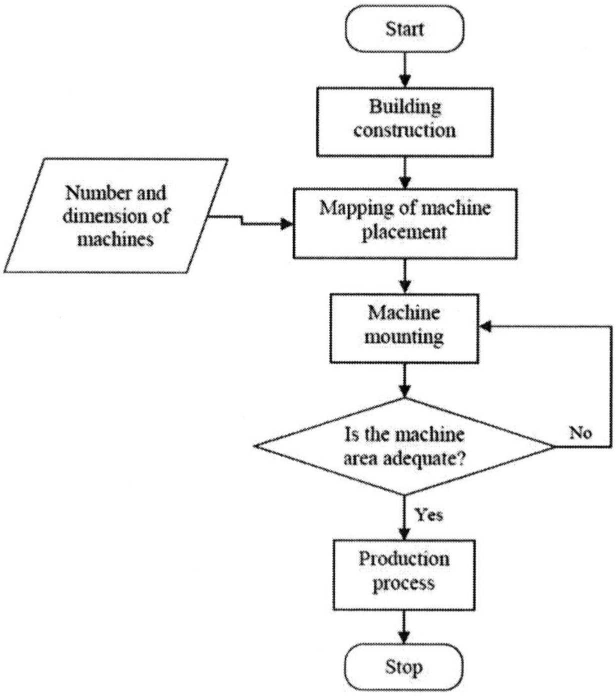


**Figure 1**. The process of workplace design at SME A

The process of workplace design in SME A begins with the construction of the workplace building. After building up the workplace, further mapping engine placement. This mapping is done by considering the number and dimensions of the machine, as well as space for the movement of workers, regardless of the type of the machine. The machines owned by SME A include five lathes with various dimensions, two milling machines, milling & drilling machines, cutting machines, bending machines, and workbench.

The next process is the installation of machines in the work area or production space based on the results of the mapping that has been done before with some adjustments according to the conditions of the existing workplace. Once the installation of the machine is finished as desired, the production process can then be carried out.

The results of the field observations on SME A show that the layout of machines and materials in the workplace area (production room) is irregular and not sequentially. This resulted in material handling and the movement of workers to be blocked.

The results of the measurement of the workplace environment condition indicate that the level of lighting in the production room of SME A varies about 40 to 600 lux. According to the Minister of Manpower Regulation No. 5 of 2018 on Occupational Safety and Health of the Work Environment, the work carried out in the SME A requires illumination with an intensity of 300 lux. This means that the lighting in some work areas (production rooms) in SME A is below the standard, while in some other areas that are above the standard.

Noise in the workplace in the SME A when all the machines in operation are above 85 dB, which means above the specified standard. Noise in the workplace area is further aggravated because the workplace of SME A is located on the edge of the highway. Another problem that occurs in SME A is a slippery floor due to splashes of coolant from the lathe to the floor so it is quite dangerous for workers.

*3.2 Workplace Design Process at SME B*

Just like SME A, SME B is also an SME engaged in the manufacturing industry. The products produced by SME B include the part of the body of a bus, the piano part, and others. In the production process, SME B uses seven types of machines, namely bending machines, cutting machines, stamping machines, milling machines, grinding machines, drilling machines, and welding machines.

According to the results of interviews with the owner of SME B, the process of workplace design on SME B can be described as shown in Figure 2.

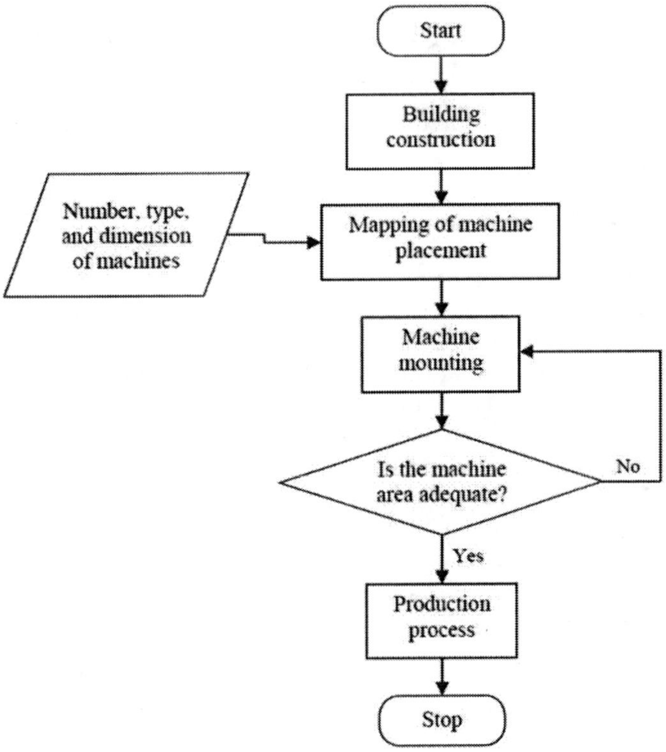


**Figure 2.** The process of workplace design at SME B

IOP Conf. Series: Materials Science and Engineering **847** (2020) 012073 doi:10.1088/1757-899X/847/1/012073

The process of workplace design on SME B begins with the construction of the building. The building used for the workplace in SME B is a rental building in a business district. The next step is mapping the location of the placement of machines in the building that has been established. The mapping is done by drawing on paper by taking into account the number, type, and dimensions of the machine. Having obtained the mapping as needed, then machines are installed based on mapping that has been done before until all the existing machines could be placed in the production room. The production process can be run after all the machines are in the area of production space.

One of the problems seen in SME B is machine placement that does not take machine characteristics into account. An example is the placement of cutting machines, stamping machines and bending machines that are positioned in the same area, resulting in a very high noise when the three machines operate together. The risk of noise exposure that can be experienced by these workers is higher because the working hours at SME B can reach 12 hours per day.

In addition to high noise, the level of lighting at the workplace (production room) of SME B is under 300 lux. According to the Minister of Manpower Regulation No. 5 of 2018 on Occupational Safety and Health of the Work Environment, work processes on SME B require a lighting intensity of 300 lux. Thus, the level of lighting in the workplace of SME B is below specified standards.

Another problem is the machine layout is irregular, causing the transfer of material from one machine to another to be blocked.

### *3.3 Workplace Design Process at SME C*

Slightly different from the two previous SMEs, SME C was originally a trading company located in Jakarta. Due to the high demand from customers, in 2011 SME C opened a business in manufacturing in the Karawang region. At the beginning of the establishment of manufacturing enterprises, SME C makes various products according to customers' demand with the job order system.

In 2017, SME C is selected by one of the major automotive component producing companies as a subcontractor for the gasket product with a mass-production system. For the gasket production process, based on the input or the request of the company giving the gasket product orders, SME C changed the layout of the workplace in the production room. Currently, besides producing gaskets, SME C is also still producing various engine components according to customer demand.

The process of workplace design on SME C as the result of an interview with the manager of SME C is shown in Figure 3.

As with the two previous SMEs, the process of workplace design on SME C begins with the construction of the workplace building first. After building up the workplace, the next step is to install the machine by considering the number and dimensions of existing machines. The installation of this machine was done without any mapping beforehand. Machine placement is deemed complete if all the machines have been placed and can be accommodated within the workspace. Under certain conditions, SME C receives input from customers to change machine placement.

The problems seen in the working space of SME C production are raw materials, production machines, and storage of finished goods that are placed in the same area, thus disturbing the material removal process. Besides, the distance between one machine and another in a sequence the working process is too far, causing the workers to experience fatigue at the time of the removal of material.

In addition to the things mentioned above, the noise condition in the production room of SME C exceeds the noise level standards which have been determined following the Regulation of the Minister of Manpower No. 5 of 2018 on Occupational Safety and Health of Work Environment. The lighting in this workspace uses only a few lights that are mounted on the side of the production room. This resulted in the lighting levels on production machines are under specified lighting standards. If such a thing is allowed, it will cause visual impairment in workers so that labor can lead to errors.

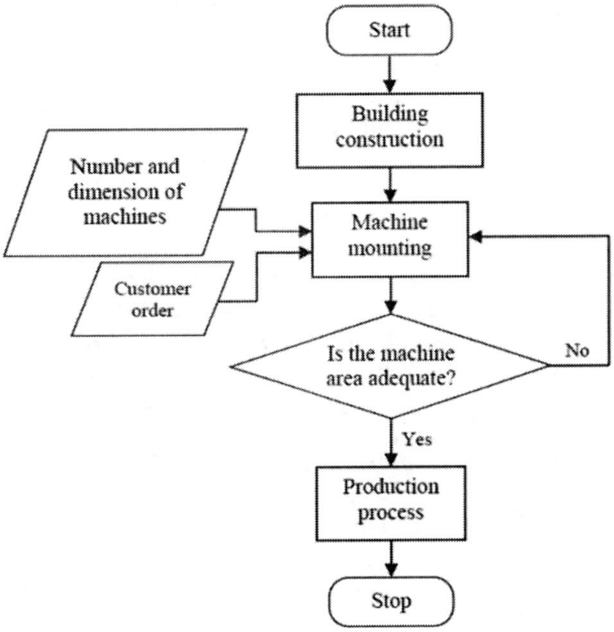


**Figure 3**. The process of workplace design at SME C

*3.4 Analysis of Workplace Design Process at Indonesian Manufacturing SMEs*

Based on the description above, it can be seen that the process of workplace design on the three SMEs have many similarities. The three SMEs started the process of workplace design on the construction of the workplace building. The workplace buildings currently used by SMEs are in part ready-made buildings and the SMEs concerned only buy or rent the building. However, some SMEs build their workplaces from scratch.

Factors considered by the management of SMEs when workplace design only the number and dimensions of the machines, and their placement in the workspace only considers "fit or not fit" of the machines in the workspace, regardless of the sequence of work and other aspects. This study shows the results corresponding to Mufti et al [26] which states that the majority of SMEs in Indonesia do not pay attention to the layout of facilities and workplaces when designing their workplaces.

In general, it can be said that the process of workplace design in the three SMEs has not paid attention to good procedures and important factors in designing workplaces. This resulted in some problems, such as the distance between machines of the successive working process too far and impede the removal of material, the floor is slippery due to splashes of coolant, and so on. Besides, noise and lighting in the workspace of SMEs do not meet the standards as a result of SME managers do not pay attention to these two factors in the process of workplace design.

The problems mentioned above may ultimately result in interference to safety and health and lower labor productivity.

Although this study shows quite clear results, this study still has at least two weaknesses. The first weakness is the number of manufacturing SMEs involved as an object in this study only three so that the results of this study can not be generalized to the entire manufacturing SMEs in Indonesia. The second weakness is this study has not considered the opinion of the workers in the workplace environment, it is not known how far the level of comfort of workers to their workplace environment.

To overcome these weaknesses, further studies need to be developed with the involvement of more manufacturing SMEs and consider the opinions of workers about the comfort of their workplace so that the results can describe the process of workplace design in the manufacturing SMEs in Indonesia and their effect on working conditions and comfort of workers in the workplace.

## 4. Conclusion

In general, the workplace design process carried out in manufacturing SMEs in Indonesia has not yet paid attention to good procedures. This resulted in the emergence of some problems in the workplace, which in turn can reduce productivity, safety, and occupational health.

## 5. References

[1] Chim J M Y 2018 6Ws in Ergonomics Workplace Design. *Proc. of the 20th Congress of the International Ergonomics Association (IEA 2018)* (Switzerland: Springer Nature) p 1282
[2] Tu'inukuafe A 2016 Future Space. *NZBusiness Magazine* **February 2016** 44
[3] Al Horr Y, Arif M, Kaushik A, Mazroei A, Katafygiotou M and Elsarrag E 2016 Occupant productivity and office indoor environment quality: A review of the literature. *Building and Environment* **105** 369
[4] Bangwal D and Tiwari P 2019 Workplace environment, employee satisfaction and intent to stay. *Int. J. of Contemporary Hospitality Management* **31** 268
[5] Candido C, Thomas L, Haddad S, Zhang F, Mackey M and Ye W 2018 Designing activity-based workspaces: satisfaction, productivity and physical activity. *Building Research & Information* **47** 275
[6] Li X, Han S, Gül M and Al-Hussein M 2019 Automated post-3D visualization ergonomic analysis system for rapid workplace design in modular construction. *Automation in Construction* **98** 160
[7] Nag P K 2018 Spatial and Behavioural Attributes in Office Design. *Office Buildings* (Singapore: Springer Singapore) p 29
[8] Bakke J W 2007 *The Nordic Workplace Design for Knowledge Work.* (Oslo: Nordic Innovation Centre)
[9] Chan J K, Beckman S L and Lawrence P G 2007 Workplace Design: A New Managerial Imperative. *California Management Review* **49** 6
[10] Launis M, Vuori M and Lehtela J 1996 Who is the workplace designer? - Towards a collaborative mode of action. *International Journal of Industrial Ergonomics* **17** 331
[11] Silva e Santos M 2012 The PhOCoe Model – ergonomic pattern mapping in participatory design processes. *Work* **41** 2643
[12] Department of Finance 2014 *Flexible and Efficient Workplace Design Guidance.* (Parkes: Commonwealth of Australia)
[13] Awad R, Fechter M and van Heerden J 2017 Integrated risk assessment and safety consideration during design of HRC workplaces. *Proc. of 22nd IEEE Int. Conf. on Emerging Technologies and Factory Automation (ETFA)*
[14] Battini D, Faccio M, Persona A and Sgarbossa F 2011 New methodological framework to improve productivity and ergonomics in assembly system design. *Int. J. of Industrial Ergonomics* **41** 30
[15] Caputo F, Greco A, Fera M and Macchiaroli R 2019 Digital twins to enhance the integration of ergonomics in the workplace design. *Int. J. of Industrial Ergonomics* **71** 20
[16] Harari Y, Bechar A, Raschke U and Riemer R 2017 Automated Simulation-Based Workplace Design That Considers Ergonomics and Productivity. *Int. J. of Simulation Modelling* **16** 5
[17] Mateus J C, Claeys D, Limère V, Cottyn J and Aghezzaf E H 2019 A structured methodology for the design of a human-robot collaborative assembly workplace. *The Int. J. of Advanced Manufacturing Technology* **102** 2663
[18] Del Rio Vilas D, Longo F and Monteil N R 2012 A general framework for the manufacturing workstation design optimization: a combined ergonomic and operational approach. *SIMULATION* **89** 306

[19] Gyensare M A, Anku-Tsede O, Boakye K O and Twumasi E 2018 Occupational Health and Safety and Employee Engagement: Evidence from the SMEs Sector in Ghana. *Advances in Safety Management and Human Factors* (Switzerland: Springer International Publishing) p 202
[20] Sinclair R C, Cunningham T R and Schulte P A 2013 A Model for Occupational Safety and Health Intervention Diffusion to Small Businesses. *American J. of Industrial Medicine* **56** 1442
[21] Surienty L 2018 OSH Implementation in SMEs in Malaysia: The Role of Management Practices and Legislation. *Proc. of the 20th Congress of the International Ergonomics Association (IEA 2018)* (Switzerland: Springer Nature) p 650
[22] Unnikrishnan S, Iqbal R, Singh A and Nimkar I M 2015 Safety Management Practices in Small and Medium Enterprises in India. *Safety and Health at Work* **6** 46
[23] Muharram A 2018 *Evaluasi, Pelaksanaan dan Rencana Kebijakan dan Program Kementerian Koperasi dan SME.* (Jakarta: Kementerian Koperasi dan SME)
[24] Hermawati S, Lawson G and Sutarto A P 2014 Mapping ergonomics application to improve SMEs working condition in industrially developing countries: a critical review. *Ergonomics* **57** 1771
[25] Herwanto D and Suzianti A 2019 An Environmental Ergonomics Review of Small Medium Enterprises Workplace Condition in Indonesia. *Proc. of 2nd Asia-Pacific Conference on Research in Industrial and Systems Engineering (APCoRISE)*
[26] Mufti D, Ikhsan A and Putri T M 2019 Workplace Ergonomic Risk Assessment Toward Small-Scale Household Business. *IOP Conf. Series: Materials Science and Engineering* **528** 012013

# A System Performance Analysis of Ship to Shore Operation Considering Crane Availabilities using Simulation Approach

**BD Aqsha[1,2], N Siswanto[1,3], Suparno[1,4], and M I Tama[1,5]**

[1]Institut Teknologi Sepuluh Nopember, Surabaya, Indonesia
[2]Bayudwiaqsha@gmail.com, [3]siswanto@ie.its.ac.id, [4]suparno@ie.its.ac.id,
[5]indratama25@gmail.com

**Abstract.** Indonesia which consists of many islands has difficulty in establishing national distribution logistics. Because of the sea traffic expansion in Eastern part of Indonesia, the existing port is limited and cannot handle such a problem. To accommodate this problem, new port is built in Surabaya, East Java, Indonesia. However, the new port usually faces to handle an unlimited demand with limited resources. This paper evaluates the port productivity of container loading and unloading activities. Equipments used in container loading and unloading at this port include cranes and other equipments. Because of many complexities in the observed system, discrete event simulation is used to evaluate the terminal operation performance. The simulation results show that STS crane quantities and formations for loading and unloading activities are affecting the new port productivity.
**Keywords**: *Container Terminal; Discrete Event Simulation; Port Logistics; Ship to Shore Crane.*

## 1. Introduction

Indonesia as one of the largest archipelago states in the world needs many harbors ports, and sea terminals to develop national commodities distributions. Production competition from national to international markets, internal distribution efficiency, and integrated national economy are affected by seaport system performance. The distribution system using sea transportation must be controlled effectively and efficiently. However, many Indonesian ports are lack of efficient and not managed well. According to Indonesian Statistics Bureau, unloading and loading activities are increasing from year to year in domestic and international ports in Indonesia [1]. These activities are the main business process in a seaport. If these activities are held, many queue events will rise in incoming and outgoing vessels. To handle such a problem, shore to crane operation builds and one of the main tools in the operation is the crane availabilities. However, crane reliability sometimes makes the unloading and loading process hold up. Some of the researches consider the routing decision for those vessels in minimizing the cost and the route [2]. These situations can decrease system performance in port terminals [3]. The case study of this research is in the new seaport terminal in Surabaya, East Java, Indonesia. This port is a new port built with the eco-green port concept. The port is using environmentally-friendly equipment to run their operations. However, this port also has an increase in container flow that has to be loaded and unloaded. Unfortunately, to cope with this problem, many pieces of equipment should be adjusted. Adjusting equipments in seaport terminals is a complex decision because many seaport operations can be affected, from economical until the technical perspective.

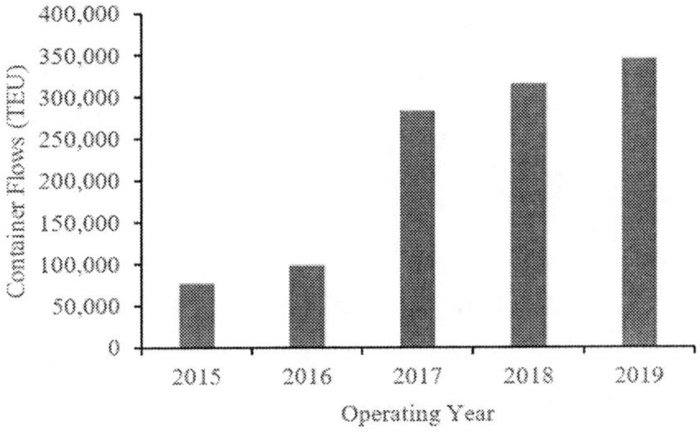


**Figure 1.** Graphic of Container Flows in the Case Study Port Terminal

Port terminal logistic research is done by many academics and researchers. This problem is so complex that using a mathematical model needs to be done with careful planning. That is why these kinds of problems usually solved using a simulation approach to cope with the uncertainty. The focus of the research goal should be pointed to evaluate the unloading and loading activities by using discrete event simulation (DES) to model the problems [4]. This is difference with other research which is focus on analysing crane availabilities [5]. The other research use mathematical model due to without stochastic settings to understand the behaviour the crane strength in a ship to shore operation [3]. Then, another analytical research study a new port development with limited capacity to handle the increaseing demand [6]. In the port development, some of the port should manage their environment to lower their pollution [7]. From this perspective, there is a limited study about crane availabilities in a ship to shore operations at a new port that performs an eco-green labelled port. This kind of novelty is used in this research.

The steps to conduct this research is (1) to make a model system and simulation of loading and unloading considering crane availabilities, then (2) to determine how many equipment should be prepared to find the best performances in the activities. Thus, this research objective is to develop a simulation model of unloading and loading activities considering ship to shore crane and testing the additional equipments needed to improve the system performance. The novelty of this research is the implementation of crane decision to ship to shore operation in an uncertain environment at new port labelled as an eco-green port. For the managerial implication, the simulation can give an insight analysis of the significant effects of adding more equipments in a ship to shore crane to port terminal performance. This research does not analysis financial and economic aspects and let these as our further research. The research assumptions are (1) the service operational activity is working normally, (2) the berthing time is the same, (3) there are no break down and maintenance activity in the crane, (4) the unloading rate is constant, and finally (5) the ship delivery time is in stochastic settings.

## 2. Method

### *2.1. Model Development*

This first step of this research is conducting survey in the port to collect secondary data. The survey is done togather information about the main activity, facility and equipment, container service procedure, operational performance, and the problems in container operations. After the information is obtained, the model is developed using Simulation Software in the form of a block flow diagram. Then the entities, activities, and the resource of the model is filled up in the software. The following is a general description of container loading and unloading activities. The system variables are shown in Table 1 as below:

IOP Conf. Series: Materials Science and Engineering **847** (2020) 012074 doi:10.1088/1757-899X/847/1/012074

**Table 1.** System Variables

| Decision Variables | Response Variables | State Variables |
|---|---|---|
| *STS Crane Quantities* | *Service Level* | *Crane Busy & Idle Status* |
| *STS Crane Formation* | *Utility Level* | *Port Busy and Idle Status* |

### *2.2. Model Verification and Validation*
After the model is built, the verification phase begins for testing the model justification and reliability according to researcher observation and from other related resources. Then the model needs to be validated by using hypotheses testing of two samples variances in T-test with significance level 95%. If the model is not verified or validated, then the model will be reformulated. The hypothesis is shown below:

$H_0$: The model simulation data variance is equal to the real data variance
$H_1$: The model simulation data variance is not equal to the real data variance.

### *2.3. Scenario Testing*
After the model is verified and validated, the model is simulated using five alternative scenarios. The alternatives are compared to the existing scenario for their system performances as seen in Table 2. The used response variables for comparison are the service and utility level of the port.

**Table 2.** System Scenarios of STS Crane Quantities and Formations.

| No | Scenario | Formation | Information |
|---|---|---|---|
| *1* | *Existing* | *2-1-1-1* | • *Two STS Crane for one vessel*<br>• *Other vessels served by one STS Crane* |
| *2* | *1$^{st}$* | *1-1-1-1-1* | • *Each vessel served to be one STS Crane* |
| *3* | *2$^{nd}$* | *2-2-1* | • *One vessel is served by one STS Crane*<br>• *Other vessels served by two STS Crane* |
| *4* | *3$^{rd}$* | *3-1-1* | • *Three STS Crane for one vessel*<br>• *Other vessels served by one STS Crane* |
| *5* | *4$^{th}$* | *3-2* | • *Three STS Crane for one vessel*<br>• *Other vessel served by two STS Crane* |
| *6* | *5th* | *4-1* | • *Four STS Crane for one vessel*<br>• *Other vessel served by one STS Crane* |

## 3. Results and Discussions
Container service in the domestic region has a 450-meter port length for berthing operations. Container vessels can dock at the same time in the dock if the total length of the docked vessel does not exceed the total length of the pier. Also, the loading and unloading activities can be performed if there is still STS equipment available. The stochastic parameters of the model ship are the arrival time, the vessel length overall (LOA), the vessel gross register tonnage, the number of unloaded containers, and the number of loaded containers. Modelling is developed by changing the number of STS equipment in serving

container loading and unloading vessels as in Table 2. Ships that dock at the pier will be regulated by the amount of equipment to be used in container loading and unloading activities. With a total domestic dock length of 450 m, the number of STS equipment at the domestic pier is 5 units and total trucks available are 50 units. The model flow is stated in the Figure 3 below.

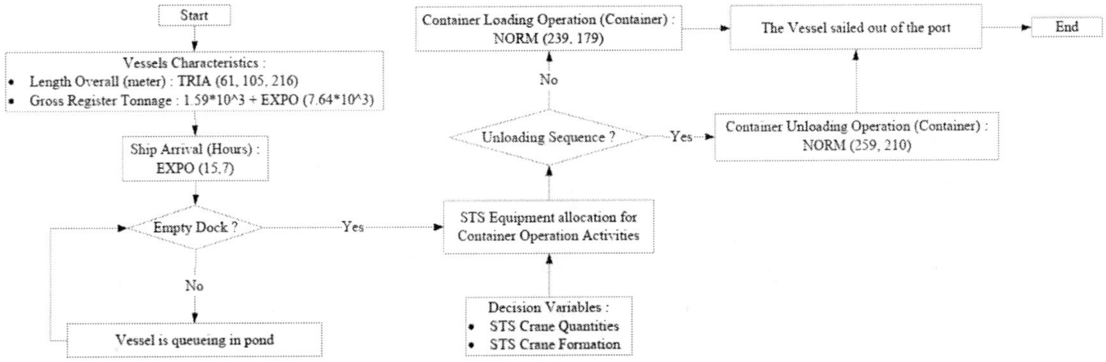


**Figure 2.** The System Model Flow

The model limitations is only reviewed the problem in container loading and unloading activities in vessels using ship to shore cranes. The model assumptions is that no machine failures in the cranes, so the STS cranes capacities are the same to load/unload 25-30 containers per hours. After the model is developed using the software, the model is verified and validated. The research process is continued to the validation test to make sure that the model simulation fits with the real system. For the simulation result based on an existing condition is shown in the Table 3 below.

**Table 3.** Simulation Result for Validation.

| Month Simulation | Incoming Vessels | Unloading and Loading Activities (Hours) | |
|---|---|---|---|
| | | Effective Time | Total Time |
| *July* | *51* | *832.22* | *1057.36* |
| *August* | *59* | *997.99* | *1264.07* |
| *September* | *45* | *694.65* | *879.61* |
| *October* | *43* | *645.56* | *817.92* |
| *November* | *48* | *760.71* | *961.93* |
| *December* | *50* | *745.72* | *979.06* |

The simulation result is compared to the real existing data as in Table 3. The validation is using a T-test statistic with a significance level of 95%. The parameters used for validation are the number of incoming vessels, the effective time, and the port total time. The calculated t-value is compared to the upper and lower limit of the two tail of T-table. If the simulation is inside the limit, then the null hypothesis is accepted, and the model simulation is considered as a representation of the real system. The validation test result is shown in the Table 4 below.

**Table 4.** Validation Test Result.

| Validated Parameters | Calculated T-Value | T Critical two-tail | Condition |
|---|---|---|---|
| *Incoming Vessels* | *1.004720719* | *2.570581836* | *Accept H0* |
| *Effective Time* | *-0.078268079* | *2.570581836* | *Accept H0* |
| *Total Time* | *0.244462296* | *2.570581836* | *Accept H0* |

*Significance Level 95%

After the model is validated, this research has a model of simulation that imitates the real system. Models of several scenarios are ready to be developed to obtain simulation results which have the most positive impacts on operational activities. These activities include the number of docked ships, the length of loading and unloading activities at the docks and the utilization of equipments. Three indicators, known as response variables, are chosen: the number of serviced vessels, the average effective time, and the utilization percentage.

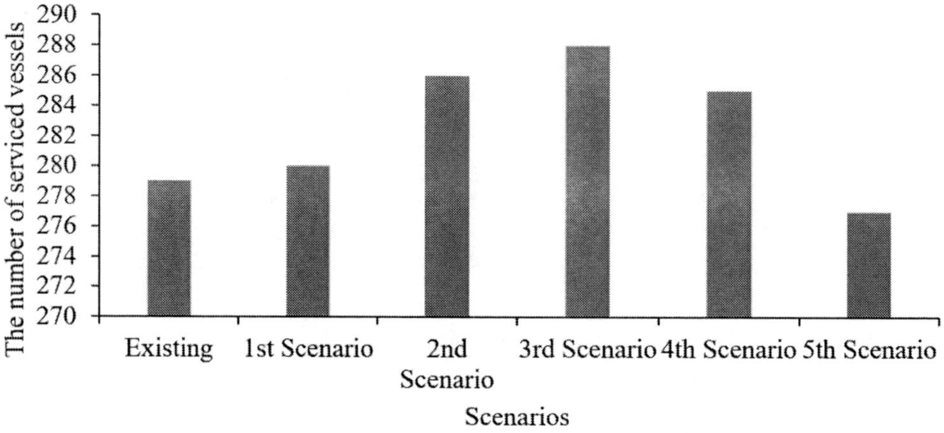


**Figure 3.** The Number of Serviced Vessels Chart

The simulation is conducted with the lenght of six months simulation time with the result that the highest vessel can be serviced is on the 3rd scenario for 288 vessels. However, the lowest serviced vessels can be utilized is 277 vessels with the 5th scenario. The high serviced vessel quantity cannot be strong evidence for evaluating port productivity. It still needs more response variables to conclude which scenario produce the most productivity.

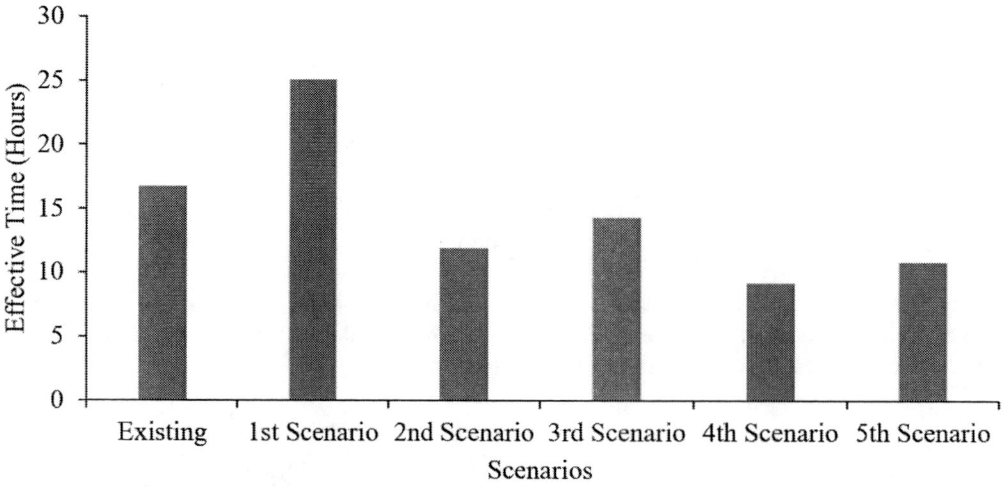


**Figure 4.** Chart of Effective Time

The effective time of the unloading and loading activities is significant to the port system performance. Effective time is a time for unloading and loading activities without any disruptions. The effective time is calculated from the first time the container is unloaded to the last time the container is loaded to the ship. The highest effective time is reached from the 4th scenario with 9.19 hours. The lowest effective time is the result from the first alternative scenario with 25.09 hours.

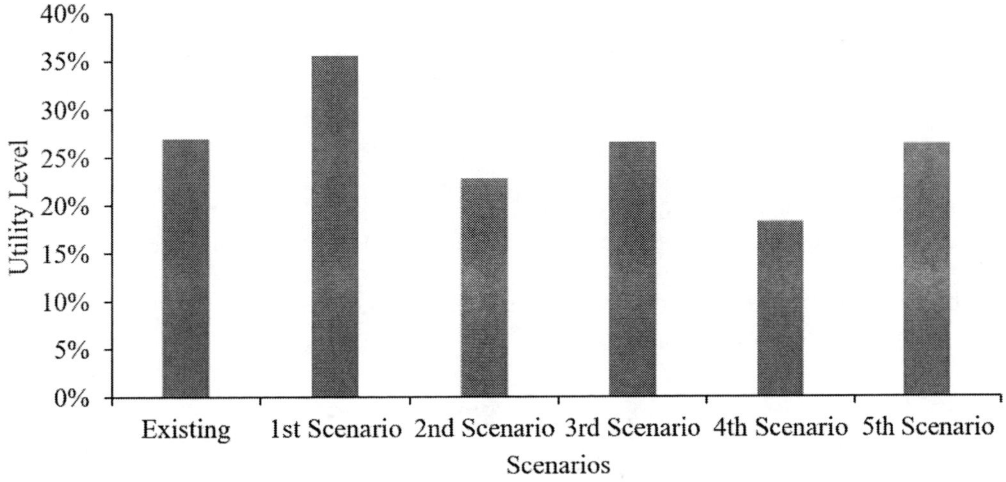


**Figure 5.** Chart of Utility Level

The utilization level in the port terminal system is an equipment indicator of handling unloading and loading containers in the docks. The higher the value of the utilization indicates that the equipment is all used. From the simulation results, it can be seen that the highest utilization is reached from the first scenario. Higher utilization can make the operation time longer. However, the 4th scenario has the lowest utilization value. Thus, this scenario has the fastest time of operation.

## 4. Conclusion

Port development in Indonesia is one of the fundamental factors to improve national logistics performance. To support this event, many ports are being developed, one of them is the case study port. This kind of port is performing an eco-green labelled to minimize carbon emission. However, this new port's development is having a difficult problem, with limited resources and tools. So, the port must handle the demand increase of container processing from incoming vessels. A discrete event model and simulation are used in this research to solve the problem. The existing model shows that the port will be experienced limitations of the resources in the next few years with the options that the available cranes and the crane formation must be decided. Several simulation scenarios are conducted that show crane assignment decision is having a significant relationship to port system performance in a ship to shore operations. The best system performance has ever been recorded from the simulation is using the 4th scenario with 3 STS crane for one vessel and 2 STS crane for other vessels. With this configuration, the effective time is 9.19 hours and the utilization equipment is 18.21%.

The research is having some limitations to the problems. The paper focus only on the number of cranes availabilities for the loading and unloading operations. The research could be more interesting if there are reliability analyses for the use of cranes and the trucks. Besides determining the number of trucks, scheduling and routing decision can be added to make the system more complex.

## 5. References

[1] Badan Pusat Statistik 2017 *Statistik Transportasi Laut.* (Jakarta: Badan Pusat Statistik).
[2] Siswanto, N., Wiratno, S.E., Rusdiansyah, A., and Sarker. R. 2019 Maritime Inventory Routing Problem With Multiple Time Windows. *J. Ind. Manag. Optim.*, vol. 15(3), pp. 1185–1211.

[3] Tang, G., Shi, C., Wang, Y., and Hu, X. 2019 Strength Analysis of the Main Structural Component in Ship-to-Shore Cranes under Dynamic Load. *IEEE Access*, 7, pp. 23959–23966.
[4] Siswanto, N., Kurniawati, U., Latiffianti, E., Rusdiansyah, A., and Sarker. R. 2018 A Simulation study of sea transport based fertilizer product considering disruptive supply and congestion problems. *Asian Journal of Shipping and Logistics*, 34(4.), pp. 269–278.
[5] Ursavas, E. 2015 Priority control of berth allocation problem in container terminals. *Ann. Oper. Res.*, pp. 1–20.
[6] Kotachi, M., and Msakni, M. 2016 A Discrete Event Simulation for The Logistics of Hamad's Container Terminal of Qatar. in *Proceedings of the 2016 Winter Simulation Conference*, pp. 2262–2271.
[7] Souf-Aljen, A. S., Maimun, A., Rahimuddin, and Zairie, N. 2016 Port capacity forecasting and the impact of the dredging works on port sea operations using discrete event simulation. *J. Teknol.*, vol. 78, no. 9–4, pp. 31–40, 2016.
[8] Yuan, Z., 2016 A Brief Literature Review on Ship Management in Maritime Transportation. Brussels.
[9] Chopra, S., and P. Meindl. 2007 Supply Chain Management. Strategy, planning & operation. *Das Summa Summ. des Manag.*, pp. 265–275.

# Comparison Analysis of Tajuk's Onion Production Efficiency (*Allium Ascalonicum*) in Rejoso Sub-District – Nganjuk

**D R Ulansari[1,4] and I N Pujawan[1,3]**

[1]Institut Teknologi Sepuluh Nopember, Jalan HOS Cokroaminoto 12, Surabaya, Indonesia
[2]dewirekno140@gmail.com, [3]pujawan@ie.its.ac.id

**Abstract**. Nganjuk is one of production area for Tajuk's onion. This research aimed to compare the results of technical and allocation efficiency from Tajuk's onion production in rejoso sub district, whether there are any differences or not. In addition, it will be used by the researcher to determine the factors which cause inefficiency and find sorts of way to fix the problems. The method used in this study was DEA and would be continued by conducting comparative test. The results in this study showed that there was difference between technical effeciency and allocation effeciency of Tajuk's onion production in Rejoso Sub-District on June-September, 2019, showed that W value on Mann Whitney test met 67,94 percent, contained sig value 0,000. The three main factors which caused technical inefficiency were pesticide, water flow, and the land area. Meanwhile, allocative inefficiency is caused by certain main factors; pesticide, seeds, land area, and working hour. The pesticide usage can be dealt by conducting studies and comparative study among farmers, remesured the land area of sewer and flowerpot along with plants gap; making new irrigation; using stored up seeds only before the harvest time, and allocating the numbers of labors that arranged by land capacity.
**Keywords**: Onion, Data Envelopment Analysis, Efficiency, Production.

## 1. Introduction

Onion is a seasonal plant whereas everyone demands for this comodity everyday. The demands for onion will consistently increase as the people's necessities keep increasing, along with the growth numbers of citizen populations. The following is the Projection Results of Onion Consumption in Indonesia in 2016-2020, which is explained on Table 1.

The onion consumption has increased from year to year. Looking at the projection on the national onion consumption in 2016-2020, conducted by the Agricultural Data and Information System Center, illustrates that the consumption of onion in 2016 to 2018 will consistently increase (eventhough the calculation from 2018 to 2019 have decreased). Whereas, the national consumption is obtained from consumption per capita multiplied by the total population calculated by BPS. The national onion consumption in 2020 is calculated to reach 755,687 tons with an average growth of 2.03% per year. Through this information, therefore it can be concluded that national consumption until to the next calculations will increase and thus it is necessary to make a strategy for managing the onion supply, in order to be competent in meeting the national consumption needs, including in minimizing onion import activity.

**Table 1.** The Projection Results of Onion Consumption in Indonesia in 2016-2020

| Year | Consumption (kg/cap/year) | Development (%) | Population (000 person) | Development (%) | National Consumption (Ton) | Development (%) |
|---|---|---|---|---|---|---|
| 2015 | 2,71 | | 252.165 | | 684.124 | |
| 2016 | 2,73 | 0,74 | 255.462 | 1,31 | 698.178 | 2,05 |
| 2017 | 2,73 | 0,04 | 258.705 | 1,27 | 707.299 | 1,31 |
| 2018 | 2,84 | 3,84 | 261.891 | 1,23 | 743.509 | 5,12 |
| 2019 | 2,79 | -1,73 | 265.015 | 1,19 | 739.392 | -0,55 |
| 2020 | 2,82 | 1,08 | 267.974 | 1,12 | 755.687 | 2,20 |
| Mean (%/year) | | 0,79 | | 1,22 | | 2,03 |

Source: Outlook onion Center for Data and Information Systems Ministry of Agriculture, 2016

Nganjuk, as one of the onion production area in East Java, has locations as onion's planting and growth area in several districts. One of them is Rejoso Sub-District with one of their superior onions so called the Nganjuk Original Local Plant (Tajuk's Onion). One effort that can be used to measure the productivity is to measure the productive efficiency. *Data Envelopment Analysis* (DEA) method is the most popular method of measuring efficiency. DEA method was first developed by Charles Cooper and Rhodes through his paper which has been published by the European Journal of Operation Research in 1978 (Falhanawati, 2013: 23). Agricultural research using the DEA model and yet closely related to onion research has been done by previous researchers, including a research of the difference of onion production efficiency based on its types by Gebremariam et al. (2019), Anik et al. (2017) who puts efficiency measurement points on natural resource use, and Ray et al. (2016) who researches the driving factors of onion cultivation's production and efficiency. In this research, the results of DEA testing for technical efficiency and allocative efficiency will be compared using different tests with a view to knowing whether there are differences in efficiency or not. This is due to the fact that the allocation of inputs in technical terms is obviously different, that can be seen from the real use of production factors (with technical units, such as kg and liters) and in the form of funding. This is also considering that the funding allocation from one farmer and another farmer is different, although in terms of technique they are the same but have different prices. Therefore, the researchers feel the urge to conduct a comparative test, which hopes to contribute in realocating the use of proper inputs both technically and allocatively funding.

Based on the introduction above, and knowing that Rejoso district in Nganjuk as the location of Tajuk's Onion cultivate area, thus this research aims to find out:

1. Is there any level difference on technical efficiency and allocative effeciency of Tajuk's onion in Rejoso district-nganjuk?
2. What factors do influence the technical efficiency and allocative inefficiency of Tajuk's onion in Rejoso district-nganjuk?

## 2. Research methods

### *2.1 Research Approach*

This research elaborated the effeciency of onion production by using quantitative approach through non parametric DEA method by CSR approach. The quantitative approach intregrated hypothesis testing by measured data and would be discovered how a variable's effect towards another variable, therefore it would make a generalized conclusion.

### *2.2 Research Scope And Limitation*

This research was conducted in Nganjuk, rejoso sub-district. The sample was a part of populations' characteristic and quantity (Anshori and Sri, 2009:94). This research's sampling was conducted by two stages; *purposive sampling* and *quota sampling*. Sugiyono (2007:68) explained that purposive

IOP Conf. Series: Materials Science and Engineering **847** (2020) 012075 doi:10.1088/1757-899X/847/1/012075

sampling is the determining technique sample by certain consideration. The purposive sampling in this research had criteria as follows:

- Farmers who had/rented a land to farming with more than 0,700 *Ha* large, based on RDKK data of Rejoso district, Nganjuk. In working the agricultural land, the farmers who worked on field more than 0,700 *Ha* large, employed another labors other than their family; and
- Farmers over than 20 years old.

And to meet the purposive sampling criteria above, the research continued by conducting quota sampling. According to Sugiyono (2007:67) quota sampling is a technique to determine the population sample which has certain characteristics up to the wanted amount. The terms from quota sampling in this research were:

1. Choosing the farmers who had/rented a land with more than 200 *rho* (the unit determined by the Nganjuk community, 1 rho = 14 m2) or equivalent to 2.800 m2.
2. Taking 30 samples of farmers in each village that met the criteria above.
   Based on the determined sample above, the villages sample in this research were, Mungkung Village, Ngadiboyo Village, Setren Village, Rejoso Village, Sidokare Village, Gempol Village, Sukorejo Village, and Mojorembum Village.

*2.3 The definitions of Variable*

This research used six variable input and one variable output. The used data in this research was racio data. Because, there were two types of efficiency counting, the operational definitions of each variables was showed on Table 2.

**Table 2.** Operational variables and definitions according to research objective

| Variable's Name | Symbols | Units | Information |
|---|---|---|---|
| First Purpose | | | |
| Production Amount | Y | *Kg* | The amount of shallots/onion varieties produced by farmers in one harvest season. |
| Fertilizer | $X_1$ | *Kg* | The amount of fertilizer used to produce Red Onion Varieties in one harvest season. |
| Pesticide | $X_2$ | *Liter* | The amount of pesticide used to produce Red Onion Varieties in one harvest season. |
| Seeds | $X_3$ | *Kg* | The amount of seeds used to produce Red Onion Varieties in one harvest season. |
| Water flow | $X_4$ | *Liter* | The amount of water flow used to produce Red Onion Varieties in one harvest season. |
| Working Hours | $X_5$ | Hour | The amount of working hour used to produce Red Onion Varieties in one harvest season. |
| Land Areas | $X_6$ | $M^2$ | The amount of land areas used to produce Red Onion Varieties in one harvest season. |
| Second Purpose | | | |
| Production Sales | Y | Rupiah | Farmer's income from the sale of Onion produced in one harvest season. |
| Fertilizer | $X_1$ | Rupiah | Fertilizer costs paid by farmers to produce onion in one harvest season. |
| Pesticide | $X_2$ | Rupiah | Pesticide costs paid by farmers to produce Onions in one harvest season. |
| Seeds | $X_3$ | Rupiah | Seeds costs paid by farmers to produce Onions Varieties in one harvest season. |
| Water flow | $X_4$ | Rupiah | Water flow costs paid by farmers to produce Onions Varieties in one harvest season. |
| Working Hours | $X_5$ | Rupiah | Working hours costs paid by farmers to produce Onions Varieties in one harvest season. |
| Land Areas | $X_6$ | Rupiah | Land areas costs paid by farmers to produce Onions Varieties in one harvest season. |

*2.4 Analysis Technique*
The technique used in this study was the DEA model (*data cross section*). The analysis model in this study was used to analyze the obtained data in order to discover to what extent the efficiency values of each sample in each *Decision Making Unit* (DMU). *Decision Making Unit* (DMU) in this research was sample or farmers.

*2.4.1 Descriptive Statisticss.* In descriptive statistics was presented and was explained from the *mean*, lowest value (minimum), and highest value (maximum), from the input and output variables, and the standard deviation of each input variable were fertilizer, pesticides, seeds, water flow, working hours, and land area and pest repeller; as well as the output variable, which was the total production of Tajuk's Onion.

*2.4.2 Data Envelopment Analysis.* The steps that needed to be taken to analyze the DEA were as follows: determined DMU (farmers), inputs and outputs; imported the data and performing the optimization methods; output maximization and CRS assumptions; executed the efficiency table for the target to see the efficiency of each village as well as from each input and output; and interpreted all of the results. Pre-determined input and output variables would be analyzed using the CRS model to determine *Overall Technical Efficiency*. The measurement of efficiency values assuming the CRS would be based on formula 1 and 2. The measurement results would produce an efficiency value scale.

$$Max\ h_k = \frac{\sum_{r=1}^{s} u_r y_{rj}}{\sum_{i=1}^{m} u_i x_{ij}} \tag{1}$$

$$\sum_{j=1}^{n} u_j y_{rj} \geq y_{rk} \tag{2}$$

*2.4.3 Independent Sample Mann-Whitney Test.* Comparative test was conducted to answer the hypothesis that had been set in this research. The first step was conducted a Kolmogorov-Smirnov data normality test to discover whether the data was normally distributed or not (Siregar, 2014: 152). The purpose of the Mann-Whitney Test model wasto determine whether there is a significant difference between the technical efficiency and the allocative efficiency of the Tajuk's Onion in the Rejoso Sub-district, Nganjuk , in the Harvest Season of 2019. This study was classified as a large sample, where $n_1$ or $n_2 \geq 20$ and both sample groups had standard deviations or was obtained from the same variant. Because of that, the significance test for the null hypothesis test used the criticism price Z in the Normal Table.

## 3. Result and discussion

*3.1 Statistical description*
This researh used six variable inputs, along with one variable output, which will be reviewed from two aspects; namely technical aspect and allocative aspect. Furthermore, descriptions from each variables are presented and are therefore described on Table 3.

Based on the Table 3, it can be described further in the following points:

- The data from fertilizer variable, seed variable, working hour, land area, both of technical and allocative, the *mean* value were greater than the standard of deviation. Since the mean > deviation, then those variables data werenot extensively expanded.
- The data from pesticide variable and water flow, both of technical and allocative, the mean value were lower than the standard of deviation. Since the *mean* value < standard of deviation, then those variables data were extensively expanded.
- The variable data of total product, technically, were not greatly expanded since the *mean* value > standard of deviation, yet allocatively expanded largely since the *mean* value > standard of deviation.

**Table 3**. Descriptions of research variables

| Variables | Information | Mean | Min | Max | Std. Deviation |
|---|---|---|---|---|---|
| Fertilizer | T: (*Kg*) | 830,78 | 112 | 6800 | 690,27 |
| (input) | A: (Rp) | 2.379.274 | 175.000 | 8.745.000 | 1.571.353 |
| Pesticide | T: (*Ml*) | 12184,93 | 700 | 110000 | 14444,5 |
| (input) | A: (Rp) | 4.101.012,5 | 45.000 | 35.493.000 | 4.593.852,47 |
| Seeds | T: (*Kg*) | 573,53 | 200 | 2500 | 282,33 |
| (input) | A: (Rp) | 12.231.827 | 3.940.000 | 50.000.000 | 6.094.924,1 |
| Water flow | T: (*l*) | 11478,45 | 1943 | 96446 | 14905,72 |
| (input) | A: (Rp) | 1.974.551 | 50.000 | 21.325.000 | 2.143.725 |
| Working Hours | T: (Hour) | 778,37 | 364 | 1794 | 289,33 |
| (input) | A: (Rp) | 7.608.404,2 | 960.000 | 35.500.000 | 4.841.728,9 |
| Land Areas | T: ($m^2$) | 4403,12 | 2800 | 16800 | 2072,2 |
| (input) | A: (Rp) | 2.974.313,7 | 9.250 | 19.000.000 | 2.888.470,1 |
| Total Production | T: (*Kg*) | 12185,13 | 6000 | 110000 | 10323,22 |
| (output) | A: (Rp) | 51.095.958,3 | 7.000.000 | 640.000.000 | 54.549.568,2 |

[a]N: 240 farmers; [b]T: Technique; [c]A: Allocative

*3.2 Analysis of The Calculation Result in Technical and Allocative Efficiency Values*

Productive efficiency was a method from several relative capability from certain companies in order to use *input* for producing *output* on the certain level (Nurhapsa et.al, 2017:17). Soekartawi (1993) also described exactly the same thing. Meanwhile, the farmers were stated to have attained the allocative efficiency if they got great profit. In this part, the researcher would discuss further the results of efficiency values from each villages, both technically and allocatively. It turned out that DMU-DMU had technical efficiency than allocative efficiency. There was a summary from the discussion result of technical and allocative efficiency in each villages on Table 4 as follows:

**Table 4**. Results of calculation of technical efficiency and allocative efficiency

| No | Villages | Technical Efficiency | | | | Allocative Efficiency | | | |
|---|---|---|---|---|---|---|---|---|---|
| | | E = 1 | dlm % | E < 0,5 | dlm % | E = 1 | dlm % | E < 0,5 | dlm % |
| 1 | Mungkung | 4/30 | 13,33 | 8/30 | 26,67 | 8/30 | 26,67 | 4/30 | 13,33 |
| 2 | Ngadiboyo | 4/30 | 13,33 | 0/30 | 0 | 10/30 | 33,33 | 7/30 | 23,33 |
| 3 | Setren | 19/30 | 63,33 | 0/30 | 0 | 10/30 | 33,33 | 1/30 | 3,33 |
| 4 | Rejoso | 10/30 | 33,33 | 0/30 | 0 | 5/30 | 16,67 | 21/30 | 70 |
| 5 | Sidokare | 14/30 | 46,67 | 0/30 | 0 | 6/30 | 20 | 5/30 | 16,67 |
| 6 | Gempol | 15/30 | 50 | 0/30 | 0 | 12/30 | 40 | 2/30 | 6,67 |
| 7 | Sukorejo | 13/30 | 43,33 | 0/30 | 0 | 5/30 | 16,67 | 24/30 | 80 |
| 8 | Mojorembun | 18/30 | 60 | 0/30 | 0 | 9/30 | 30 | 4/30 | 13,33 |
| **Total** | | **97/240** | **40,42** | **8/240** | **3,33** | **65/240** | **27,08** | **68/240** | **28,33** |

Source: Processing Results with OSDEA

Based on the summary of the results of Table 4, in the technical calculation, the highest efficiency value from the eight villages observed in this study was in Setren Village (63.33%), followed by Mojorembun Village (60%), Gempol Village (50%), Sidokare Village (46 , 67%), Sukorejo Village (43.33%), Rejoso Village (33.33%), Mungkung Village (13.33%) and Ngadiboyo (13.33%). If they were accumulated, the total number of DMUs that had been efficient was 97 out of 240 observational samples or equivalent to 40.42 percent in the calculation of technical efficiency. There were only two villages that had a total efficiency value of more than 50 percent, namely Setren and Mojorembun.

The allocative calculations in eight villages turned out to be different from the technical calculations. The total results of efficiency in eight villages presented 27.08 percent with the sequence and details of Gempol Village (40%), Ngadiboyo Village (33.33%), Setren Village (33.33%), Mojorembun Village (30%), Mungkung Village (26.67%), Sidokare Village (20%), Rejoso Village

(16.67%), and Sukorejo Village (16.67%). That was, still farther from expected. This certainly would give influence in productivity unless the efficiency can be managed properly.

Furthermore, the calculation results of efficiency values less than 0.5 even the allocative calculation enhanced and turned into 28.33 percent (68/240). Results that were more than two digits were in Sukorejo Village (24) and Rejoso Village (21). All villages needed to pay attention to the allocation of inputs so that there were no waste so that efficiency could be achieved and could be even more productive. On the calculation results in Table 4 showed that efficiency value still farther than the target (80 percent farmers had efficiency) both from technical and allocative efficiency.

### *3.3 Model Analysis and Hypothesis Testing*

*3.3.1 Normality Test.* The normality test is the basis for determining what statistical models are best for conducting hypothesis testing. In this study the normality test was carried out with the Kolmogorov-Smirnov test. Since $P_{value}<P_{table}$, thus the decision of this test was received $H_0$ and refused $H_1$, this meant that the data was not normally distributed. Because the data groups were not normally distributed, then the analysis model used was the non-parametric independent sample Mann-Whitney Test model.

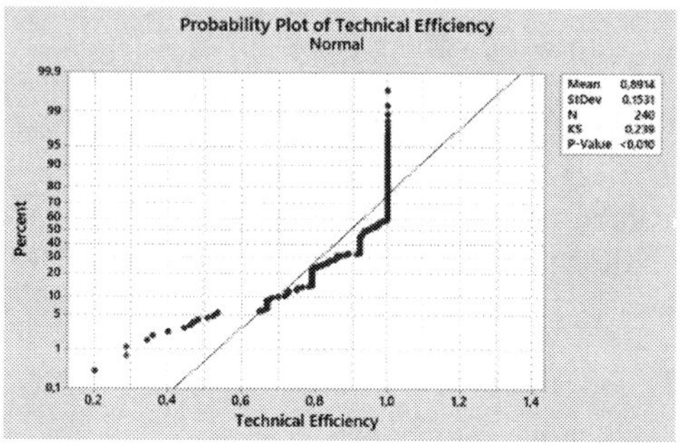


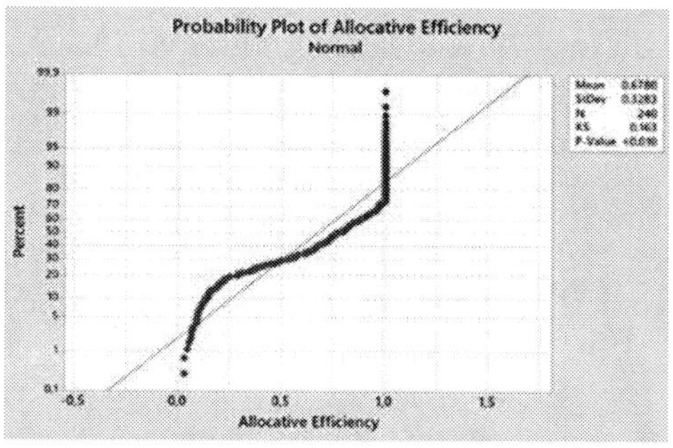


**Figure 1** and **Figure 2**: Efficiency and Allocative Efficiency Normality Test Results

*3.3.2 Independent Sample Mann-Whitney Test.* The Mann-Whitney Test results showed that for the efficiency value the CRS assumption had W value of 67938.00 or 67.94 percent with a significance value of 0,000 or less than 0.05. Since the value of $W_{value}<W_{table}$, then the decision of this test was to reject H0 and accept H1, which means that there were differences in technical efficiency and allocative production of Tajuk's Onion in Rejoso Sub-district in the June-September 2019 harvest season.

*3.3.3 Analysis Causes of Technical Inefficiency and Allocative Production of Tajuk's Onion in Rejoso Sub-district.* This research used DEA method with the assumption of CRS and output oriented approach, thus the slack variable results would show how much the excess input was used. The over input allocation with the same output, can cause the inefficiency.The slack variable calculation was conducted using OSDEA and summarized on Table 5.

On Table 5 explained that what dominates the cause of technical inefficiencies in Mungkung and Ngadiboyo Villages were the land area, meanwhile in Setren, Rejoso and Sidokare Villages were Pesticides, while in Gempol, Sukorejo, and Mojorembun Villages were water flow. Table 5 also explained that what dominates the causes of allocative inefficiency in Mungkung, Rejoso, and Gempol Villages were Pesticides, while in Ngadiboyo and Sukorejo Villages were Seeds and in Setren village was Working Hours. The cause of allocative inefficiency in the villages of Sidokare and Mojorembun was Land Area.

**Table 5**. Slack variable that causes inefficiency

| No. | Village | *Main Cause of Technical Inefficiency Input* | *Main Cause of Allocative Inefficiency Input* |
|---|---|---|---|
| 1 | Mungkung | Land Areas | Pesticide |
| 2 | Ngadiboyo | Land Areas | Seeds |
| 3 | Setren | Pesticide | Working Hours |
| 4 | Rejoso | Pesticide | Pesticide |
| 5 | Sidokare | Pesticide | Land Areas |
| 6 | Gempol | Water Flow | Pesticide |
| 7 | Sukorejo | Water Flow | Seeds |
| 8 | Mojorembun | Water Flow | Land Areas |

This result was supported by Yadav (2015: 193) which stated that doing unnecessary work efficiently is not fully productive and it would be more appropriate to do things efficiently in output oriented (rather than input oriented) to achieve results. Whereas, in order to achieve the efficiency, Curristine (2007: 9) explains that it can be obtained by increasing the scale of operations. In line with this, Sheth (2002) in Roghanian et al. (2012) in his research also explained that productivity can be achieved through a combination of effectiveness and efficiency.

## 4. Conclusions and suggestions

After conducting research at that location and processing it according to the research method, the researcher gets some discussion of the results that can be concluded, as follows:

1. There was a difference on the technical efficiency level and allocative production level of the Tajuk's onion in the Rejoso Sub-district in the June-September 2019 harvest season.
2. Factors that influenced the efficiency of the technique based on its slack calculation were pesticides, land area, and water flow. While the factors that influence the allocative efficiency in the slack calculation were pesticides, land area, seeds, and working hours.

Based on the results of research and empirical studies, the following authors submit some input / suggestions that are expected to be usefu, as follows:

1. Based on the calculation results of the technical and allocative efficiency, it is recommended that farmers who have not meet the optimum efficiency to follow the amount of allocation of the farmers' used input that has been stated as efficient.
2. Factors causing inefficiencies that occur in both types of technical and allocative calculations are land area and pesticides. The problems on land area can be minimized by rearranging the size of the ditches and flowerpot, as well as plant gap, therefore the land of production can be optimal because the spacing will affect the land itself. As for pesticides, it would be better for farmers in one village to do comparative study among farmers to discover which types of pesticides are effective and have realistic prices. Thus, it is necessary to reduce costs yet people's needs are still fulfilled. Furthermore, water flow input factors also influence the technical inefficiencies. This can be minimized by creating new irrigation channels for villages that do not yet have irrigation channels. Because, from the price facet, irrigation channels are cheaper when compared to diesel (especially renting it).
3. The seeds problem, as one of the factors causing the allocative inefficiency can be minimized by using seeds that have been stockpiled by the farmers themselves (leaving some previous harvests) and not buying additional seeds near the harvest season. Because, when it is close to harvest season, the price of seeds will be more expensive. The problem of working hours can be overcome by allocating the amount of labors that is adjusted to the land capacity.
4. The last suggestion is for the local government to pay attention to the primary production for the crops productivity, especially Onion. This is related to the provision of irrigation facilities which will influence for the farmers. In addition, the regulations are needed related to the sale of onion

IOP Conf. Series: Materials Science and Engineering **847** (2020) 012075 doi:10.1088/1757-899X/847/1/012075

commodities thus that local farmers do not experience decreasing selling prices as a result of the soaring commodities from other regions. Besides, the government is expected to be able to determine the ceiling price and floor price limits and controls the prices of commodity sales in the marketplace, so that the selling prices of the main producers (farmers) to consumers (last) do not experience great gap.

Thus, this research still limited to:

1. The technique of sampling on the second phase, after determining the main sample which uses purposive sampling, is quota sampling
2. The methode analysis which uses *non-parametric approach* is DEA. Therefore, it is not capable in identifying the external factors or environmental variable that cause inefficiency.

## 5. References

[1] Anik, Asif R, MD. Abdus S, and Sanzidur R 2017 Drivers of Production and Efficiency of Onion Cultivation in Bangladesh. *Bulgarian Journal of Agricultural Science*, **23**(1), 34-41.

[2] Anshori, Muslich and Sri I 2009 *Metodologi Penelitian Kuantitatif.* 1st edition. (Surabaya: Pusat Penerbitan dan Percetakan UNAIR, AUP).

[3] Curristine, Teresa, Zsuzsanna L, and Isabelle J 2007 Improving Public Sector Efficiency: Challenges and Opportunities. *OECD Journal on Budgeting*, **7**(1), 1-41.

[4] Falhanawati, Yudnina 2013 Analisis Tingkat Efisiensi Perbandingan Perbankan Syariah dengan Menggunakan Medote Data Envelopment Analysis (DEA*). Final Project* (unpublished). (Jakarta: Fakultas Ekonomi dan Bisnis UIN).

[5] Gebremariam, Haregu G, Leake G, and Assefa H Tekle 2019 Efficiency of Male and Female as Irrigated Onion Growers. *International Journal of Vegetable Science,* (January).

[6] Outlook Bawang Merah 2016 *Pusat Data dan Sistem Informasi Pertanian Kementerian Pertanian.* ISSN: 1907-1507.

[7] Ray, Jui, Debabrata P, Arabinda M, Ranjan C, and Debashis S 2016 A Case Study on Resource-use Efficiency in Onion Cultivation in West Bengal, India. *Journal of Economics Affairs*, **61**(2), 321-326.

[8] Roghanian, Parastoo, Amran R, and Hamed G 2012 Productivity Through Effectiveness and Efficiency in The Banking Industry. *Procedia – Social and Behavioral Sciences,* **40**, 550-556.

[9] Sadono, Endiarjati Dewandaru 2015 Penerapan Fungsi Produksi Cobb-Douglas dalam Menganalisis Technical Efficiency Petani Bawang Merah di Desa Buahan, Kecamatan Kintamani, Kabupaten Bangli, pada Musim Panen 2015. *Final Project* (unpublished). (Yogyakarta: Fakultas Ekonomi dan Bisnis Universitas Gadjah Mada).

[10] Sugiyono 2007 *Statistika untuk Penelitian.* 12th edition. (Jawa Barat: Anggota Ikatan Penerbit Indonesia IKAPI).

[11] Yadav, Pooja, and Col. S Marwah 2015 The Concept of Productivity. *International Journal of Engineering and Technical Research (IJETR)*, **3,** 192-196, ISSN: 2321-0869.

# Assessment of Safety Culture Maturity Level in Production Area of a Steel Manufacturer

**A Sudiarno**[1,2], **A A C Sudarni**[1,3]

[1]Department of Industrial Engineering Institut Teknologi Sepuluh Nopember, ITS, Surabaya, Indonesia
[2]adithya.sudiarno@gmail.com, [3]adiekastikaclara@gmail.com

**Abstract.** Safety culture is widely known to have the ability to encapsulate all factors including perception, psychology, and attitude. This ability serves as an important role for organizations in occupational safety and health. Further, this ability is used as an effort to reduce the risk of accidents. One of the efforts made is to measure the level of maturity of the safety culture in the steel manufacturer production area by involving the variables of safety culture, namely commitment, leadership, responsibility, competence, engagement & involvement, information & communication, risk, and organizational learning. This study aims to measure the level of safety culture maturity in the activities production area of steel manufacturers. Data collection was carried out by distributing questionnaires with a purposive sample of 107 workers. Two expert judgments are involved as determinants of the safety culture variable weights using the AHP (Analytical Hierarchy Process) method. The results showed that the level of safety culture maturity in the production area of steel manufacturer was 3,64 , which falls in the proactive category. The priority for improvement of safety culture variables starts from the leadership to the information & communication.
**Keywords**: Analytical hierarchy process, Occupational safety and health, Safety culture, Safety culture maturity level

## 1. Introduction

Steel manufacturing is an industry where safe working procedures are important, as workers face many risks due to the nature of the job [1]. One of the steel manufacturing in Indonesia is a steel manufacturer in the Sidoarjo district which is engaged in the steel smelting industry. Steel manufacturer is a steel smelting company that was founded in 1976. The raw material used in the production process is scrap which is then processed into steel billets and steel wire rods. The production process at the steel manufacturer consists of SMS (Steel Melting Shop) and RML (Rolling Mill). Every implementation of steel production activities at steel manufacturers, especially in each SMS and RML production area has the potential to cause danger resulting in work accidents.

Work accidents that occur in the manufacturing industry, especially in smelting furnaces can pose various undesirable risks. This is because the work environment in the manufacturing industry has a work environment that creates many risks such as the nature of the work, hot and noisy environmental conditions, as well as heavy routine work tasks and the risk of accidents both injury and burns [1]. Research conducted by Holmes explains that accidents are defined as any unplanned event that results in injury or ill health of people, or damage or loss to property, plant, materials or the environment or a loss of a business opportunity [2]. Problems faced by this research, there are accident data that occurred in the steel manufacturer production area during the 2014 - 2018 period revealed that there was an increase in the number of work accidents in 2018 if we were compared to the previous 4 years with 10 case. Work accidents experienced by steel manufacturer workers in 2018 have severe a higher rate of severe work accidents than in previous years. Severe work accidents experienced by fractures in the ankles, middle and index fingers and stab wounds on the right foot caused by being exposed to the workpiece. According to [5] with the enhancement occurrence of work accidents, a work safety culture is needed as a safety that is felt, assessed, and prioritized in an organization.

An organization has a varied safety culture and must have a strong commitment [4] so that the safety culture can affect safety [1]. Just like Borwn say in [1], there have been several previous studies on safety in manufacturing that have focused on behavior, attitudes, climate, or culture with the results showing that culture can form the basis for unsafe attitudes and behavior. Safety culture can be widely known for its ability to encapsulate all factors including perception, psychology, attitude and managerial. This is what makes one an important component of organizational culture that discusses individual safety, safety performance, and several things that are prioritized by safety organizations [5], [6]. The development of the safety culture model concept has been carried out by several researchers including [7]. Safety culture in an organization is assessed by using three aspects of indicators namely personal aspects, behavioral aspects, and situation aspects, while [8] safety culture is assessed by the P2T model namely people, procedures, and technology models. [9] Argues that modeling in safety culture is multi-dimensional, there is no consensus regarding the dimensions used, depending on the safety culture model it uses. The research that was done uses nine components or dimensions, namely: leadership, communication, organizational goals and values, supportive environment, responsibility, learning, trust in people and systems, resilience, engagement.

As did previous research, while in the research that will be conducted by using eight variables of safety culture, namely; commitment, leadership, responsibility, engagement and involvement, risk, competence, information and communication, and organizational learning. The variables used in the study are used as a tool to measure the maturity level of safety culture in steel manufacturer companies, especially in the company's production area. One measure that can be used is safety culture maturity level. Safety culture maturity is a model that assesses the level of understanding of safety culture at every level of the organization in the company. The assessment of safety culture maturity for this steel manufacturer is assessed using five levels namely; basic, reactive, planned, proactive and resilient [10]. This study aims to measure the level of safety culture maturity in the activities production area of a steel manufacturers to describe the stages of improvement of safety culture. Furthermore, the results of measuring safety culture maturity in the study serve as recommendations for improving the safety culture in the production area of the steel manufacturers.

## 2. Methods

This research involves eight safety culture variables consisting of commitment (C), leadership (L), responsibility (R), competence (CO), engagement & involvement (EI), information & communication (IC), risk (RI) variables and organizational learning (OL) as a model of safety culture research [11]. Safety culture modeling with these variables is used in the assessment of the maturity level of safety culture in the production area of the steel manufacturers. The level of maturity assessment was carried out in three stages, including:

1. Assessment of the average from eight safety culture variables based on the results of the development of a safety culture questionnaire with a scale of 1-5 which was distributed to 107 workers from 224 workers in the production area of the steel manufacturer. The number of samples obtained using a purposive sample by means of inclusion and exclusion criteria. The results of the development of safety culture questionnaire have a total of 80 questions in which 40 questions are from the variable commitment, leadership, information & communication, and organizational learning (each 10 questions variable), 9 questions from the responsibility variable,

7 questions from the engagement & involvement variable, and 24 questions from risk and competence variables (each variable is 12 questions).

2. Assessment of the weights of each safety culture variable by using the AHP (Analytical Hierarchy Process) method with Expert Choice V11 software tools. This assessment was completed by experts with the rating scale 1-9 listed in table 1.

**Table 1**. Weighing Scale of Safety Culture Variables

| Scale | Definition |
|---|---|
| 1 | Equally important |
| 3 | Variable A is slightly more important than variable B |
| 5 | Variable A is more important than variable B |
| 7 | Variable A is more important than variable B |
| 9 | Variable A is absolutely more important than variable B |
| 2,4,6,8 | Value between the two closest numbers |

3. Then an assessment of safety culture maturity level is carried out in the production area from the product of the average value and weight of each variable safety culture. The assessment of safety culture maturity level is assessed using five levels namely; basic, reactive, planned, proactive and resilient.

## 3. Result and Discussion

### *3.1 The Average Assessment of Safety Culture Variables*

Steel manufacturer is a company engaged in the steel smelting industry located in Sidoarjo district. Research conducted on 107 workers in the steel manufacturer production area, namely SMS and RML, obtained a value average on each safety culture variable based on a questionnaire that had been distributed. The safety culture variables used in this research are commitment (C), leadership (L), responsibility (R), competence (CO), engagement & involvement (EI), information & communication (IC), risk (RI) and organizational learning (OL). The average assessment of safety culture variables is the first assessment of safety culture maturity level. The results of the assessment are shown in table 2.

**Table 2.** Average Value Variables of Safety Culture

| No | Department | Variables of Safety Culture | | | | | | | |
|---|---|---|---|---|---|---|---|---|---|
| | | C | L | R | EI | RI | CO | IC | OL |
| 1 | Department 1 | 4,13 | 3,87 | 3,99 | 3,94 | 3,93 | 3,79 | 4,03 | 4,06 |
| 2 | Department 2 | 3,58 | 3,49 | 3,74 | 3,57 | 3,55 | 3,60 | 3,65 | 3,55 |
| 3 | Department 3 | 3,64 | 3,61 | 3,59 | 3, 47 | 3,45 | 3,49 | 3,43 | 3,53 |
| 4 | Department 4 | 3,85 | 3,78 | 3,92 | 3,94 | 3,84 | 3,73 | 3,87 | 3,83 |
| 5 | Department 5 | 3,48 | 3,35 | 3,51 | 3,73 | 3,68 | 4,03 | 4,09 | 3,96 |
| 6 | Department 6 | 3,45 | 3,74 | 3,73 | 3,94 | 3,78 | 3,68 | 3,98 | 3,70 |
| 7 | Department 7 | 3,46 | 3,50 | 3,62 | 3,73 | 3,83 | 3,73 | 3,66 | 3,68 |
| 8 | Production Area Steel Manufacturer | 3,65 | 3,62 | 3,73 | 3,76 | 3,72 | 3,72 | 3,81 | 3,76 |

Based on table 2 regarding the average assessment of safety culture variables, the results were obtained that the smallest value obtained by the leadership variable was 3,35 in department 5. While for the largest value in the assessment of the average safety culture variable was obtained on the commitment variable which was 4,13 in department 1. So that the average grade results - The average safety culture variable in all production areas obtained the smallest value in the leadership variable of 3,62 while the largest value in the information and communication variable was 3,81.

### *3.2. The Weighs Assessment Safety Culture Variables*

The next assessment is to assess the safety culture variable weights using the AHP (Analytical Hierarchy Process) method with the use of expert choice v11 software. Analytical Hierarchy Process is a framework for a person in making complex problem decisions, giving an assessment based on knowledge, experience, and feelings using paired comparisons, and to obtain a set of priorities that are considered as a reasonable solution in solving problems. The assessment of the safety culture variable weights was conducted by 2 experts who were considered to have knowledge and experience in the field of occupational safety and health, especially in the area of production of PT. X. The weighting assessment of 2 experts can use the participant table by selecting the combined column in expert choice v11 software so that the results of weighting the safety culture variables are shown in Table 3.

**Table 3.** Safety Culture Variable Weight Values

| No | Variable | Expert 1 | Expert 2 | Combined |
|---|---|---|---|---|
| 1 | Commitment | 0,125 | 0,116 | 0,122 |
| 2 | Leadership | 0,137 | 0,129 | 0,144 |
| 3 | Responsibility | 0,192 | 0,119 | 0,160 |
| 4 | Engagement & involvement | 0,106 | 0,052 | 0,077 |
| 5 | Risk | 0,069 | 0,234 | 0,138 |
| 6 | Competence | 0,136 | 0,095 | 0,114 |
| 7 | Information & Communication | 0,113 | 0,190 | 0,148 |
| 8 | *Organizational Learning* | 0,121 | 0,066 | 0,097 |
| | Total | 1 | 1 | 1 |

### *3.3. Assessment of Safety Culture Maturity Level*

This assessment is needed to determine the level of safety culture owned by the steel manufacturer production area. There are five levels of safety culture level according to research Foster's namely; basic, reactive, planned, proactive and resilient. The safety culture maturity level assessment is formulated using equation 1 on each safety culture variable so that the results are shown in Table 4.

$$Safety\ Culture\ Maturity\ Leve = Average\ x\ Weight \quad (1)$$

**Table 4.** Safety Culture Maturity Level

| No | Department | Level Safety Culture Variable Level | | | | | | | |
|---|---|---|---|---|---|---|---|---|---|
| | | C | L | R | EI | RI | CO | IC | OL |
| 1 | Department 1 | 0,50 | 0,56 | 0,64 | 0,30 | 0,54 | 0,43 | 0,60 | 0,39 |
| 2 | Department 2 | 0,44 | 0,50 | 0,60 | 0,27 | 0,49 | 0,41 | 0,54 | 0,34 |
| 3 | Department 3 | 0,44 | 0,52 | 0,57 | 0,27 | 0,48 | 0,40 | 0,51 | 0,34 |
| 4 | Department 4 | 0,47 | 0,54 | 0,63 | 0,30 | 0,53 | 0,43 | 0,57 | 0,37 |
| 5 | Department 5 | 0,42 | 0,48 | 0,56 | 0,29 | 0,51 | 0,46 | 0,61 | 0,38 |
| 6 | Department 6 | 0,42 | 0,54 | 0,60 | 0,30 | 0,52 | 0,42 | 0,59 | 0,36 |
| 7 | Department 7 | 0,42 | 0,50 | 0,58 | 0,29 | 0,53 | 0,43 | 0,54 | 0,36 |
| 8 | Production Area PT,X | 3,11 | 3,64 | 4,18 | 2,02 | 3,60 | 2,98 | 3,96 | 2,54 |
| Safety Culture Maturity Level | | 3,25 | | | | | | | |

Based on table 4, the safety culture maturity level assessment in the steel manufacturer production area is 3.25 where the value is included in the planned category. The results obtained from table 4 are a numerical data value using the AHP method so that the resulting value has an uncertain value, meaning that the value is presented in order of scale and criteria so that it has more subjective properties. Evaluation with criteria that do not have certainty and subjective assessment requires a concept fuzzy. One of the concepts fuzzy that can be used TFN (Triangular Fuzzy Number). According to [12] the TFN (Triangular Fuzzy Number) method is a method special class of fuzzy numbers whose members are defined by three main numbers expressed as (l, m, r) with namely the lowest value, middle value, and highest value. The use of the TFN method is illustrated by the safety culture maturity level scale contained in table 5 and the results of the TFN for safety culture maturity level in the steel manufacturer production area are in table 6

**Table 5.** Safety Culture Maturity Level with The TFN Method

| Scale *Safety Culture Maturity* | Scale TFN Scale | | | Information |
|---|---|---|---|---|
| | $TFN_l$ | $TFN_m$ | $TFN_r$ | |
| 1 | 1 | 1 | 2 | *basic* |
| 2 | 1 | 2 | 3 | *reactive* |
| 3 | 2 | 3 | 4 | *planned* |
| 4 | 3 | 4 | 5 | *proactive* |
| 5 | 4 | 5 | 5 | *resilient* |

**Table 6.** Assessment Results Safety Culture Maturity Level

| No | Variable | $TFN_l$ | $TFN_m$ | $TFN_r$ | Average |
|---|---|---|---|---|---|
| 1 | Commitment | 2,676 | 3,663 | 4,554 | 3,615 |
| 2 | Leadership | 2,622 | 3,608 | 4,471 | 3,547 |
| 3 | Responsibility | 2774 | 3763 | 4557 | 3665 |
| 4 | Engagement and involvement | 2,756 | 3,744 | 4,590 | 3,673 |
| 5 | Risk | 2,714 | 3,699 | 4,574 | 3,644 |
| 6 | Competence | 2709 | 3,699 | 4,551 | 3,630 |
| 7 | Information & Communication | 2,803 | 3,793 | 4,602 | 3,702 |
| 8 | Organizational Learning | 2,728 | 3,721 | 4,580 | 3,654 |
| | Safety Culture Maturity Level | 2,723 | 3,711 | 4,560 | 3,641 |

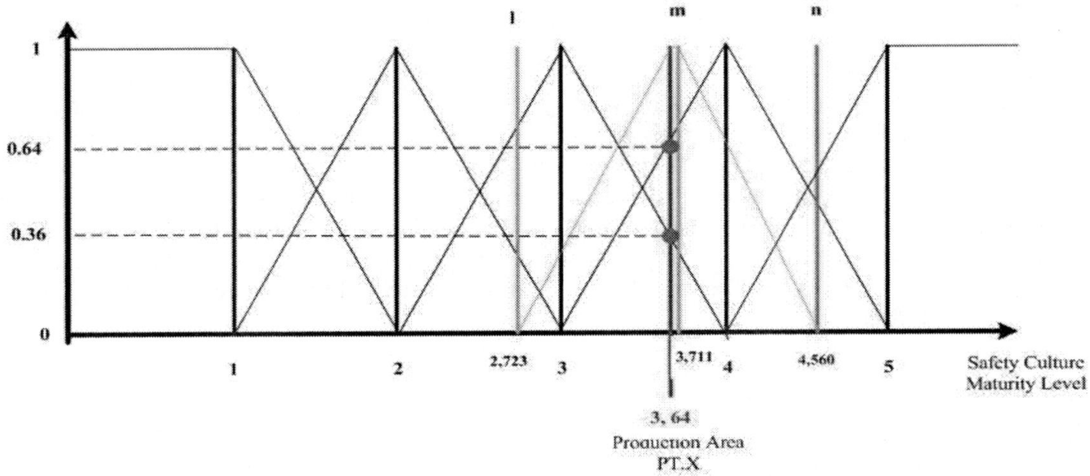


**Figure 1.** Triangular Curve Membership Chart of Safety Culture Maturity Level

IOP Conf. Series: Materials Science and Engineering **847** (2020) 012076 doi:10.1088/1757-899X/847/1/012076

Based on table 6 the assessment of the level of safety culture variables at the safety culture maturity level of the steel manufacturer production is a variable leadership 3,547 and the highest value is owned by the information & communication variable with a value of 3.705. Termination of the proactive category owned by production area steel manufacturer by using a triangular curve membership chart. From this research, the safety culture maturity level rating of the steel manufacturer production area is 3.64 where the value is included in (or belongs to) the proactive category. For this value of 3.64 we can see from Fig 1 that the corresponding score obtained for proactive value is 0.64, and its corresponding score for planned value is 0.36.

Hudson explains that the proactive category at the level of safety culture maturity is that there is improved performance, something that is not expected to be considered a challenge [13]. The involvement of workers seems to have begun to take the initiative and doesn't carry out too orders from the leadership regarding safety factors. The concept of Hudson's safety culture maturity model proves to be a useful tool for the organization in assessing its safety culture. But the model proposed by Foster has a difference where the concept of the model created has a direct relationship to the standards of the company's own management with safety culture categories that have also been established. So in the proactive category proposed by [10], there is a target setting where senior managers are involved in determining safety and in overseeing managers are responsible for the results of safety activities that are made to meet and on target. In this case, the proactive categorization in the steel manufacturer production area is obtained through observations that the manager of the production area department and the manager of occupational safety and health are jointly responsible for the safety culture with proof that every month a supervision is held in the form of safety patrol, providing safety-related training and occupational health, there are standard operating procedures in every production activity, as well as integrated safety audits in accordance with research from Foster. However, the facts that occur in the plant there are still some workers sometimes still violate the rules of OSH which are caused by the lack of strict action from the manager of the production area department. This needs to be considered given that the improvement of safety culture still needs to be done in improving the safety culture in the production area of steel manufacturers to achieve resilient levels.

Recommendations for improvements given in the production area of steel manufacturers can be given as follows such as role models on leadership variables, conducting audits to maintain company commitment in safety culture, as well as on competency variables always checking all production equipment. To guarantee workers from work accidents, the recommendation on the risk variable is to provide a warning alarm that indicates a process failure or natural disaster. Furthermore, recommendations given to organizational learning variables regarding safety culture are discussions between the SHE Department and the production area department regarding safety, and as a form of responsibility for safety culture, safety patrols should be carried out. The involvement and participation of workers to improve safety culture in the production area of steel manufacturers is very important, therefore, a reward and punishment system is needed. In order for workers to understand the culture of workplace safety, the recommendation for information and communication variables is to conduct safety education directly to workers. For this research, the priorities for improving safety culture must also be applied more to the variable leadership because it has the lowest safety culture maturity value when compared to other variables. The flow of improvement in the safety culture in the steel manufacturer production area can be suggested starting from the variables of leadership, commitment, competence, risk, organizational learning, responsibility, engagement & involvement, and information & communication.

## 4. Conclusion

The level of safety culture maturity in the production area of steel manufacturer has a value of 3,64 where the value is included in the proactive category. The lowest level of safety culture variable assessment on the safety culture maturity level of steel manufacturer production area is the variable leadership 3,562 and the highest value is owned by the information & communication variable with a value of 3,725. The priority of improving safety culture in the production area of steel manufacturer production is on the leadership variable because it has the lowest value in the assessment of safety culture maturity level. So that the flow of safety culture improvement to reach the generative category in the production area of steel manufacturer starts from leadership, commitment, competence, risk, organizational learning, responsibility, engagement & involvement, and information & communication.

## 5. References

[1] Nordlof, H., 2015. Safety Culture and Reasons for Risk Tahing at A Large Steel Manufacturing Company Investigating The Worker Perspective. *Safety Science,* **73**, pp. 126-135

[2] Tarik, B., and Adil, H.A., 2018. Occupational Health and Safety in The Moroccan Construction Sites: Preliminary Diagnosis. *International Journal of Metrology and Quality Engineering,* **9(6)**, pp. 1-9.

[3] Corrigan, S., Kay, A. and Ryan, M., 2018. Human Factors and Safety Culture: Challenges and Opportunities for the Port Environment. *Safety science.*

[4] Hopkins, A., 2006. Studying Organisational Cultures and Their Effects on Safety. *Safety Science,* **44(10),** pp. 875-889.

[5] Choudhry, R.M., Fang, D. and Mohamed Sherif., 2007. The Nature of Safety Culture: A Survey of The State of The Art. *Safety Science,* 45(10), pp. 993-1012.

[6] Machfudiyanto, R. A., Latief, Y., Arifuddin, R. and Yogiswara, Y., 2017. Identification of Safety Culture Dimensions Based on The Implementation of OSH Management System in Construction Company. *Procedia Engineering,* **171**, pp. 405-412.

[7] Cooper, M. D., 2000. Towards a Model of Safety Culture. *Safety Science,* **2(36)**, pp. 111-136.

[8] Reniers, G. L., Cremer, K. and Buytaert, J., 2011. Continuously and Simultaneously Optimizing an Organization's Safety and Security Culture and Climate: The Improvement Diamond for Excellence Achievement and Leadership in Safety & Security (IDEAL S&S) Model. *J. Cleaner Pod,* **11(19),** pp. 1239-1249.

[9] Lingard, H., Zhang, R., Harley, J. and Blismas, N., 2014. *Health and Safety Culture, Construction Work Health and Safety Project, Centre for Construction Work Health and Safety Research.* Melbourne: RMIT

[10] Foster , P. and Hoult, S., 2013. *The Safety Journey: Using a Safety Maturity Model for Safety.* Minerals, **3**, pp. 59-72

[11] Hermawan, E., 2018. Structural Equation Modek Dimensions Of Safety Culture In Operational And Maintenance Services of Coal Fired Power Plant. *Thesis Ph.D.,* Surabaya: Institut Teknologi Sepuluh November.

[12] Yang, H., 2008. The Evaluation For Cooperative Partner Selection Based on TFN-AHP. *Fifth International Conference on Fuzzy Systems and Knowledge Discovery,* pp. 203-207.

[13] Hudson, P., 2007. Implementing A Safety Culture In A Major Multi-National. *Safety Science,* **47**, pp. 697-722.

# ITSM Analysis using ITIL V3 in Service Operation in PT.Inovasi Tjaraka Buana

**Muharman Lubis*, Rizky Cherthio Annisyah and Lyvia Winiyanti L**

Telkom University, Jalan Telekomunikasi No. 1, Bandung, 40257, Indonesia
*muharmanlubis@telkomuniversity.ac.id

**Abstract.** PT. Inovasi Tjaraka Buana is a company in ISP (Internet Service Provider) sector that uses information technology to sustain activities the company. At this time the company is experiencing scaling-up, which is a drastic increase in the number of user of internet services, and coverage areas that make companies have to develop for various aspects of the company. The problem is that companies cannot at this time balancing service operations when handling an incident that occurs with drastic addition of the number of users and the amount of coverage area. To solve the problem, it is necessary to implement Information Technology Service Management (ITSM). The method used in this research is data collection by interview and observation techniques. This research will produce Incident Management Flow and Problem Management Flow. ITIL, which is a framework that illustrates best practices that focus on managing IT services, IT development and operations, which can help companies to overcome problems, so the company can apply it in order to balance service operations when handling incidents that occur.
**Keywords**: ITSM, ITIL, incident management flow, problem management flow.

## 1. Introduction

The application of information technology that continues to develop very quickly has a major impact on an organization [1] almost all company organizations implement IT because it is believed to help improve the effectiveness and efficiency of business process companies, including companies engaged in services [2]. PT. Inovasi Tjaraka Buana is a company in ISP (Internet Service Provider) sector that uses information technology to sustain activities the company. At this time the company is experiencing scaling-up, which is a drastic increase in the number of user of internet services, and coverage areas that make companies have to develop for various aspects of the company. The problem is that companies cannot at this time balancing service operations when handling an incident that occurs with drastic addition of the number of users and the amount of coverage area. One observation was supported by evidence in the form of many complaints forms complaints by users to the company[3] to solve the problem, it is necessary to implement information management service technology. The IT Service Management (ITSM) framework has been widely used to increase the value of the company's service to stakeholders [4]. ITSM is a method for managing an IT system within an organization, focusing on the perspective of customer IT services within the organization. ITSM focuses on IT-related activities and efforts to provide a framework for building relationships between IT managers and IT customers. The submitted framework includes ITIL [5]. ITIL provides a systematic approach to seeing and evaluating integrated systems in terms of people, technology, processes and organizations. The process of identifying the process, design, implementation and evaluation that exists within the ITIL framework can be used as a tool or tool to look at more existing problems and make it easier to propose recommendations accordingly. In some IT companies ITIL is often used as a tool to obtain quality standards such as ISO 9000 and ISO / IEC 20000 [6]. In this research an ITSM analysis will be done using the v3 item by focusing on service operations [1]. The results of this analysis will produce incident management flow and problem management flow at PT Inovasi Tjaraka Buana.

Management of Information Technology Services or better known as ITSM is a management method information technology systems that are centred on the perspective of consumers of IT services on business company [7]. ITSM can defined as 'a set of processes who work together to make sure directly the quality of IT services, according to service level agreed by customer'[8].Competition among companies in now has become very orientation towards IT, thus causing companies to have to have various quality standards for improve business processes. ITSM (IT Service Management) is a source of practical guidance to be used as an agar quality standard the creation of process improvements in the company [9]. An effective application of ITSM is done by integrating three main elements, namely people, processes and technology, into a well-designed system, which is based on best industry practice [10] Modern companies are hardly experiencing increasing competition The need to use information technology to increase competitiveness can be avoided. BSM (Business Service Management) is a compiled solution (Best Practice) The IT of the world, known as the IT Infrastructure Library (IT-IL). Accordingly The general information technology infrastructure library (IT-IL) defines how technology can work Procedures (processes) and individuals (human resources) can help in simplifying business processes and efficiency Different areas of the company [10]. ITIL (Information Technology Infrastructure Library) is a framework that describes best practices IT management service. ITIL provides a framework For IT management and quality improvement IT services provided [5]. What are ITIL best practices and their benefits?

**Table 1.** Benefits of ITIL and Compared

| ITIL, ISO 20000, and COBIT Compared | | | |
|---|---|---|---|
| Parameter | ITIL | ISO 20000 | COBIT |
| Ownership | A service management framework owned by Axelos | A service management standard from ISO (Geneva) | An IT governance framework from ISACA |
| Implementation | As a framework, it can be adopted and adapted to suit IT organizations needs | As a standard, it has to be implemented in spirit and principles by IT organization | As a framework, it can be adopted and adapted to suit IT organizations needs |
| Certificate | ITIL certificate awarded to individuals only; can't be awarded to an organization | ISO 20000 certificate awarded to organizations and individuals as assessor, implementer, etc. | COBIT certificate awarded to individuals only; can't be awarded to an organization |
| Scope/Coverage | ITIL is a framework of best practices for service management and is complementary to ISO 20000 | The ISO 20000 standard is complementary to ITIL | The COBIT framework has more scope coverage compared to ITIL |
| Flexibility | ITIL is flexible; only required practices for an organization can be implemented | In order to prove compliance with ISO 20000, organizations must implement all standard requirements | COBIT is flexible; only IT governance needed for an organization can be implemented |
| Benefits of Certification | The certificate helps individuals as a knowledgebase in service | The certificate helps an organization to improve its services, | The certificate helps individuals in their careers for performing IT governance roles and eventually, |

**Table 1.** Benefits of ITIL and Compared

| ITIL, ISO 20000, and COBIT Compared | | | |
|---|---|---|---|
| Parameter | ITIL | ISO 20000 | COBIT |
| | management and eventually, the organization for efficient management of IT services | demonstrates reliability and high quality of service | the organization for increased customer satisfaction |
| Validity Period | The individual certificate is valid for the specified version in the certificate | The organization's certificate must be renewed every 3 years, with surveillance audits to be conducted on a yearly basis | The individual certificate is valid for life for the specified version in the certificate |
| Synergy | Adopting the ITIL framework helps an organization comply with the ISO 20000 standard | An organization that has ISO 20000 can easily adopt ITIL practices | As a framework with more scope, it helps an organization to adopt ISO 20000 or ITIL practices with reduced efforts |
| Miscellaneous | ITIL is widely implemented by organization selling IT services, system integrators, etc. For their clients business having IT as a backbone | ISO 20000 is widely adopted by organizations that are in the IT consultancy business, or equivalent for their own organization | COBIT is widely implemented by organizations that have an IT department, but that are NOT in the IT consultancy business, e.g., banking, insurance, etc. |

**Figure 1.** Best Practices of ITIL

From the table above we can see the advantages of using ITIL and how it compares to ITIL, ISO 2000, and COBIT. Besides that ITIL framework is very appropriate to be used as a guide in developing a governance because it is best practice and has a detailed library to develop steps in the procedure [11], and can be seen using other ITIL best practices [12][13][14]. Actually, ITIL is built into five main components in the ITIL Service Lifecycle: service strategy, service design, service transition, Continual Service Improvement, and service operation [1]. Service Operation is a lifecycle stage that covers all daily operational activities managing IT services. Inside are various guidelines on how to manage IT services efficiently and effectively. Service Operation has 5 processes and 4 functions, namely: Event Management, Incident Management, Problem Management, Access Management, and Request Fulfilment, Service Desk, Technical Management, Application Management, and IT Operations Management [15]. The following are some of the processes carried out in Service Operation [16] such as event, incident problem, request fulfilment and access management. In some province, there are times when media committees have not been established yet or allocated according to the regulations. At the same time, prison imprisonment and fine relief cannot provide a deterrent to perpetrators of crime, possibly in connection with managing and providing services in an electronic environment. As a result, human resources, technical infrastructure, public participation, supervision, and socialization are important factors that increase awareness and satisfaction of this organizational commitment [17][18]. One of the simplest ways to calculate availability depends on two numbers. The period in which the service must be available during the reporting period. This is the agreed service time (AST). Measure the downtime (DT) during the period, then take the agreed downtime and convert it to a percentage.

$$Availability = \frac{AST-DT}{AST} \times 100\% \qquad (1)$$

**Table 2.** Percentage Degradation of Service, Outage Duration and Maximum Frequency

| IT Function That Is Not Available | % Degradation of Service | Outage duration | Maximum frequency |
|---|---|---|---|
| Receiving email | 100% | Up to 2 minutes | 2 events per hour |
| | | | 5 events per day |
| | | | 10 events per week |
| Reading public folders | 50% | 2 minutes to 30 minutes | 2 events per week |
| | | | 6 events per quarter |
| Updating public folders | 10% | 30 minutes to 4 hours | 4 events per year |
| Accessing shared calendars | 30% | 4 hours to 8 hours | 1 event per year |
| Updating shared calendars | 10% | | |

Most incidents do not result in complete termination of service for all users. Some users might not be affected while others might not have services. In the most extreme cases, there may be one user with a damaged PC who does not have access to services. It might be considered a loss of 100% service, but that would make IT an entirely unrealistic destination and not a fair measure of availability. On the other hand, it can be chosen to have the service available while others can access it. However, it doesn't take much imagination to find out how customers feel if a marked service is available when many people don't have access to it.

One way to measure its impact is to calculate the percentage of minutes lost by the user. To do this:

1. Calculate Potential User Minutes. This is the total number of users multiplied by the duration of their work.
2. Calculate User Outage Minutes. This is the total number of users who cannot work multiplied by the time they cannot work.

$$Availability = \frac{PotentialUserMinutes-UserOutageMinutes}{PotentialUserMinutes} \times 100\% \qquad (2)$$

## 2. Methods

The method used in this research is data collection by interview and observation techniques [19]. Data collection is one of the important stages in a process research [20]. Observations were carried out by observing the ongoing conditions at the organization's location. This condition can be in the form of media that supports the processes related to incident management, problem management that occurs at PT Inovasi Tjaraka Buana. Interviews conducted directly to management informants and staff related to data and information needed by researchers. This research expect to produce an incident management and problem management flow.

## 3. Result and Discussion

### *3.1. Incident Management Flow*

Management that ensures that service services in an organization can be returned to its original condition as quickly as possible and minimize the negative impacts that may be caused [10]. Having a structured workflow ensures that incidents are handled effectively. This optimizes problem resolution for customers and businesses. The incident management flow of PT. Inovasi Tjaraka Buana. Where is the initial stage when complaints about incidents that enter via WhatsApp or Line, then incidents will be grouped according to priority, after the incident is grouped, the incident will be handled if the incident cannot be handled then it will contact the technician, if the incident in a low priority level will be resolved use the service desk and the incident has been resolved. If the incident cannot be resolved using a service desk then contact the technician if the incident has experienced problem management and has not yet been resolved, the technician goes to the field until the problem can be resolved and the incident can be said to be finished.

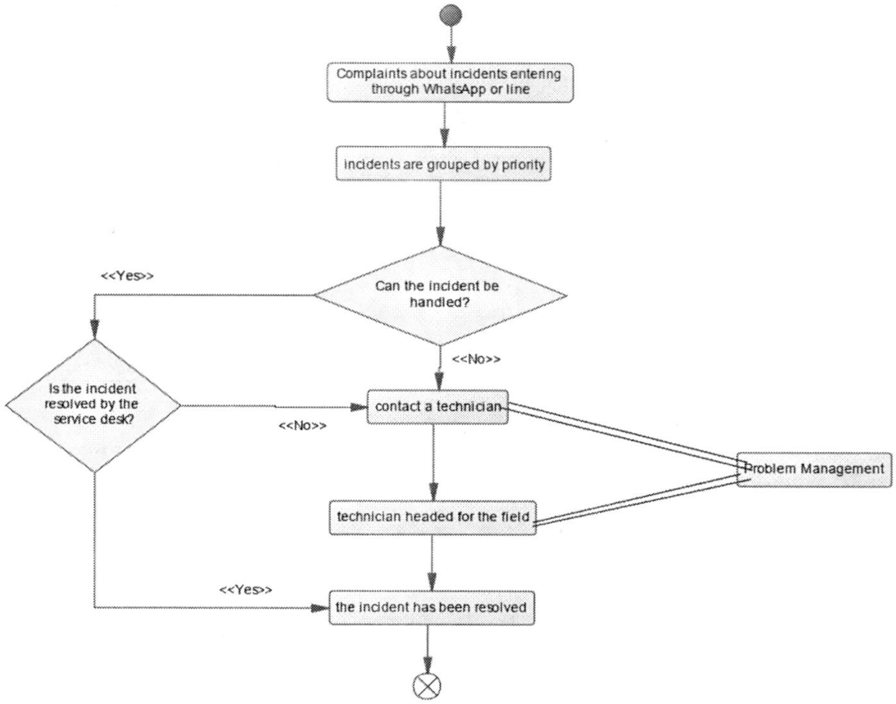


**Figure 2.** Incident Management Flow

### *3.2. Problem Management Flow*

Information has dramatically shifted to a particular paradigm over the past decade as an individual's prerequisite for the presentation and creation of quality and participation in social interaction and communication. Government efforts to provide legal assurance in disclosure should be appreciated and supported. As such, there are several requirements to support execution and maintain intent, such as commitment leadership, bureaucratic culture, and public awareness. A society that wants to adopt openness as a value of inclusive importance will not only allow its citizens a wide range of individual freedom of expression, but will go one step further and actually open the process intended by public demand. Information is a fundamental requirement for the development of each individual and social environment as an important part of the digital ecosystem in education systems and national defense. To minimize adverse effects on business caused by incidents and problems caused by errors in infrastructure [10]. Problem management is an identification process Problems that occur directly. The purpose is to prevent The problem of the accident and the result eliminate recurring incidents and reduce impact incidents that cannot be prevented [5]. Furthermore, capturing all relevant information helps the team to find the root cause of the problem which is an important part of ITIL problem management. Management problem PT. Inovasi Tjaraka Buana. When there is a new problem that occurs there will be a warning management problem alert after that the problem will be grouped in accordance with priorities and the technician will go to the field to solve the problem if the problem has been resolved, the technician will record the errors that occur and the problem has been resolved.

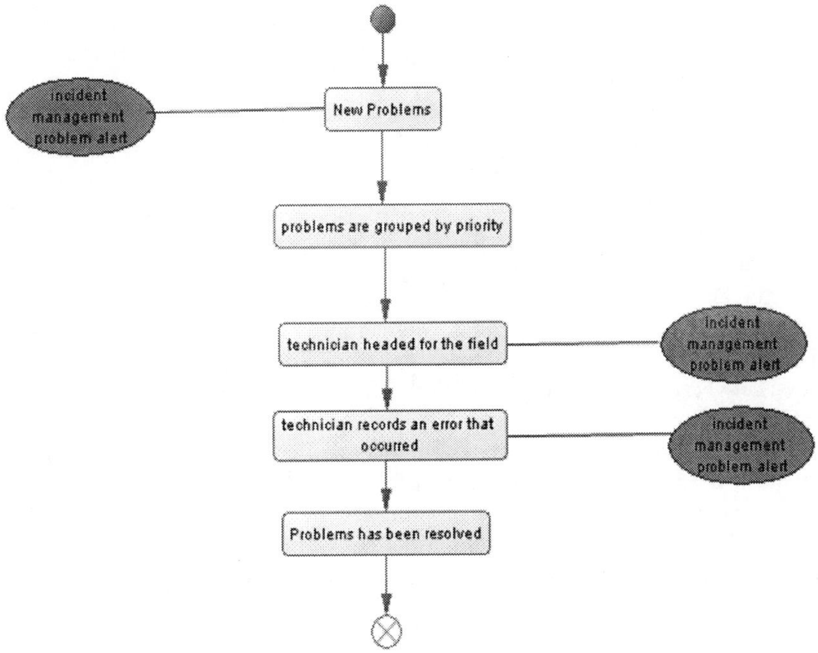


**Figure 3.** Problem Management Flow

## 4. Conclusion

Based on research that has been done about ITSM Analysis using ITIL V3 in Service Operations in PT. Inovasi Tjaraka Buana, then has a picture that operating services that have well-managed and appropriate services, will provide significant benefits and benefits for the continuity of business activities in the company. ITIL, which is a framework that illustrates best practices that focus on managing IT services, IT development and operations can help companies to overcome problems, so the company can apply it in order to balance service operations when handling incidents that occur. The author suggests that companies must have incident management flow and problem management flow so that when incidents repeatedly occur the company can immediately handle quickly and recorded.

## 5. References

[1] S Hanief and IW Jefriana 2018 Framework Itil V3 Domain Service Operation Dalam Analisis Pengelolaan Teknologi Blended Learning. *J. Teknol. Inf. dan Komput.* **4**(1), pp. 59–65.

[2] AM Arifin and R Hanafi 2015 Analisis Dan Perancangan Itsm Domain Service Operation Pada Layanan Akademik Institut Pemerintahan Dalam Negeri (Ipdn) Dengan Menggunakan Framework Itil Versi 3 Analysis and Design ITSM Service Operation Domain on Academic Services of Institut Pemerinta **2**(2), pp. 5798–5803.

[3] RD Putra, E Darwiyanto, G Agung and A Wisudiawan 2016 Audit Teknologi Informasi Dengan Menggunakan Framework COBIT 5 Domain DSS (Deliver, Service and Support) pada PT. Inovasti Tjaraka Buana," *eProceedings Eng.*, **3**(1), pp. 930–937.

[4] HT. Sukmana, LK Wardhani, S Khairunnisa, KO Lee and R Wati 2019 ITSM software ranking for small medium enterprises based on ITIL V3 quick win criteria using fuzzy SIR method," *Adv. Sci. Technol. Eng. Syst.*, **4**(2), pp. 288–298.

[5] AF Wijaya, MNN Sitokdana and P Hapsari 2019 Analysis of Information Technology Services

Management using the ITIL V3 Domain Service Operation Framework on SIMDA (Case Study: Boyolali Regency Inspectorate). **100**(41), pp. 126–131.

[6] YP Handoko 2017 Pemanfaatan ITIL v3 untuk Mengatasi Masalah Layanan TI pada Sistem Terintegrasi di Perguruan Tinggi. *Semin. Nas. Sains, Tek. Komput. dan Rekayasa*, **3**(1), pp. 1–10.

[7] S Romadini, A Fajar and S Iqbal 2018 Perancangan sistem manajamen layanan teknologi informasi pada layanan reseller dan dropship bandros menggunakan iso 20000-1 : 2011 area general requirements dan design and transition of new or changed services (Studi Kasus : CV kabita informatika ) DESIG. **5**(2), pp. 3428–3435.

[8] F. Salim 2017 Pengaruh Information Technology Service Management (ITSM) terhadap Kinerja Industri Perbankan. (November), p. 5.

[9] B Uddin 2014 Evaluasi penerapan manajemen layanan ti menggunakan kerangka kerja it infrastructure library (ITIL) sub domain service desk, incident management dan problem management. *Jurnal TEDC* **8**(2).

[10] L Magdalena and K Kunci 2011 Analisis Problem Management pada IT Helpdesk dengan implementasi ITSM dan SLA (Studi Kasus: Citigroup Indonesia). *J. Digit*, **1**(2), pp. 97–112.

[11] TP Silitonga, A Holil and N Ali 2010 Program Manajemen Helpdesk Dan Dukungan Ti Berdasarkan Framework ITIL V3 (Studi Kasus Pada Biro Teknologi Informasi BPK-RI). SemnasIF 2010.

[12] R Wijaya and RV Hari 2016 Perancangan Tata Kelola Service Desk Menggunakan Kerangka Kerja Information Technology Infrastructure Library (ITIL) V.3 Pada PT Berlian Jasa Terminal Indonesia. *Pros. Semin. Nas. Manaj. Teknol. XXV*, pp. 1–8.

[13] R Wijesinghe, H Scheepers and S Mcloughlin 2015 Defining the optimal level of business benefits within IS/IT projects: Insights from benefit identification practices adopted in an IT Service Management (ITSM) project. *ACIS 2015 Proc. - 26th Australas. Conf. Inf. Syst.*

[14] M Bahtiar and AR Perdanakusuma 2018 Evaluasi Tingkat Kematangan Sistem Informasi Manajemen Rumah Sakit Menggunakan ITIL (Information Technology Structure Library) Versi 3 (Studi Pada : Rumah Sakit Umum Universitas Muhammadiyah Malang ). *J. Pengemb. Teknol. Inf. dan Ilmu Komput. Univ. Brawijaya*, **2**(11), pp. 4525–4531.

[15] L Ramadani 2016 Perancangan Tata Kelola Manajemen Layanan Teknologi Informasi Menggunakan ITIL V3 Domain Service Design Di Pemerintahan Kota Bandung Designing Information Technology Service Management Governance Using Itil V3 Service Design Domain in Bandung City. **3**(2), pp. 3397–3402.

[16] E Wahyu and T Darmaningrat 2018 Pimpinan Redaksi Faizal Mahananto Dewan Redaksi Tata Pelaksana Usaha Achmad Syaiful Susanto Rini Ekowati Sekretariat. **08**(01).

[17] M. Lubis, TF Kusumasari and L Hakim 2018 The Indonesia Public Information Disclosure Act (UU-KIP): Its Challenges and Responses. *Int. J. of Electrical and Computer Engineering*, **8**(1), pp. 94–103.

[18] M. Lubis and FA Maulana 2010 Information and Electronic Law Effectiveness (UU-ITE) in Indonesia. *Proc. IEEE CITSM* 2010.

[19] N. Iin 2010 Dasar Teori Perbaikan Analisis Jabatan Menggunakan Metode Observasi dan Proses Hierarki Analitik Di PT. Leading Garment Industries. *e-Proceeding Eng.* **6**(1), pp. 5–45.

[20] I. Poetra and H. Mulyono 2019 Analisis dan Perancangan Sistem Informasi Layanan Pelanggan PDAM Tirta Mayang Kota Jambi Berbasis SMS. *J. Manaj. Sist. Inf.* **4**(2), pp. 92–101.

IOP Conf. Series: Materials Science and Engineering **847** (2020) 012078 doi:10.1088/1757-899X/847/1/012078

# Design and Analysis of Electrical Ergonomic Bionic Grip Wrench

**A A Ramadhan[1], F A Putra[2], H Wirawan[3] and T R Sahroni[4]**

[1,2,3,4]Industrial Engineering Department, BINUS Graduate Program – Master of Industrial Engineering, Bina Nusantara University 11480, Jakarta, Indonesia.
[1]andi.ramadhan@binus.ac.id,[2]fajri.putra@binus.ac.id,[3]harries.wirawan@binus.ac.id; [4]taufik@binus.edu

**Abstract.** Increasing demand for the use of bionic grip locks as well as customers and markets is increasing day by day, requiring improved design of technical tools to increase efficiency and effectiveness. Supporting tools design needs to be modified for the tools capability improvement and to provide new ways of working. The concept of innovative design can be applied to tools designing for up living the work way more productive in order to boost the manufacturing and service performances, meeting the mounted up customer demand. Taking the wrench product design as an example of observation and improvement initiative opportunity, this article discovers the chance of the mechanical and machinery products innovation, with the project methodology presented in applying the theories of product redesign from a manual conventional wrench to an advance design and puts forward the thought of the wrench innovative engineering design, principles and the process to electrical and mechanical mechanism. As a result, the proposed design scheme is evaluated, which indicates that the optimized design approach can be applied to a manual wrench design.
**Keywords**: wrench, product, redesign, machinery, tools, improvement

## 1. Introduction

Machinery tools industry supports an important role in providing the engineering equipment for the national economy contribution. Mechanical product level is one of the absolute parameter of the country modern manufacturing practices. Nation's industry, agriculture, construction, transportation, national defence and the advancement of science and technology are related with the development of mechanical products, which mechanical engineering design essence is innovation. Pay attention to the machinery products innovative design is a fundamental way to enhance the competitive advantage of the products itself. Economic attraction and market globalization asks for the improvisation of product design and manufacture to meet nowadays requirement which needs the way wore efficient way of working.

Recent customers make product decisions based on subjective factors such as convenience, ease of use, and pricing. People spend their money on what feels right and are often unable to provide an explanation of their reason. While there is no simple formula to design a product that can be exactly successful in the market, inclusion of the needs of the target population in the design process is an essential part of that process. The design of the product is believed to be one of the most important factors influencing the success of the product in a market, the interaction of the market environment and the new product strategy and execution, includes the product itself (e.g. product design, product advantages) [1]. This is the primary reason that motivates our current research on user preferences for improved engineering tools product.

A wrench (or spanner) is a tool used to provide grip and mechanical advantage in applying torque to turn objects usually rotary fasteners, such as nuts and bolts or keep them from turning [2]. In British English, spanner is the standard term. The most common shapes are called open-ended spanner and ring spanner. The term wrench refers to various types of adjustable spanner. In American English, wrench is the standard term. The very most common shapes are called open-end wrench and box-end wrench [3]. In American English, spanner refers to a specialized wrench with a series of pins or tabs around the circumference (these pins or tabs fit into the holes or notches cut into the object to be turned.) In American commerce, such a wrench may be called a spanner wrench to distinguish it from the British sense of spanner. Higher quality wrenches are typically made from chromium-vanadium alloy tool steels and are often drop forged [4].

Quantitative data was gathered, for conventional product design exists these days. On the total research, design, development, tooling, and launch marketing costs. Each project was then classified qualitatively according to the main type of design expertise involved-product design (e.g., furniture, textiles, engineering tools), engineering design, engineering plus industrial design (e.g., electronics plus ergonomics), and graphic design. The project outcomes were also assessed qualitatively and quantitatively: first, whether the project was implemented (put into production), then its financial results, and finally indirect benefits such as learning design-management skills [5].

This paper describes the necessity of new wrench product in order to utilize the bio-mechanic mechanism and ergonomic aspect. Furthermore, the new design focusses on design of grip mechanism to minimize the required human-energy during the process.

## 2. Method

In this stage, the overall process begins with 3D mechanical drawing of conventional wrench, 3D mechanical drawing of non-electrical bionic grip wrench, and design the 2D & 3D mechanical design of electrical bionic grip wrench. Using the Inventor program, we make the stress analysis of the nuts gripper due to critical part design analysis. Figure 1 shows the flowchart of the development stage of the research.

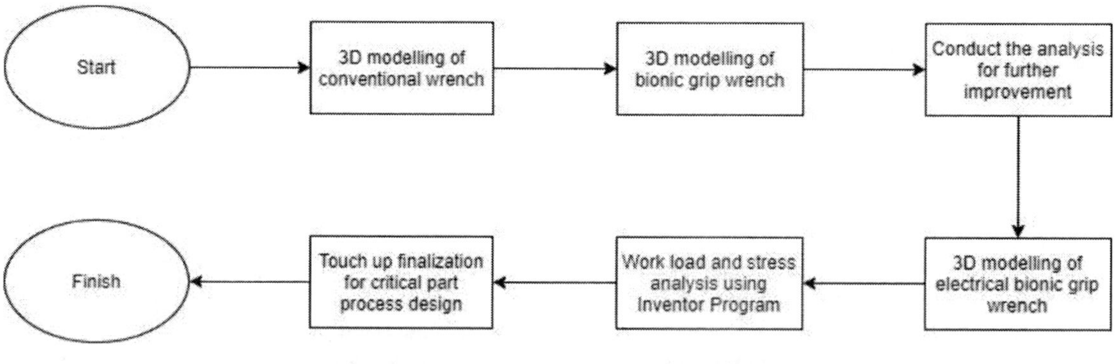


**Figure 1. Development Stage Flowchart**

We observe the current design of conventional wrench before the improvement or product redesign was implemented to the current one. Then, we did some research regarding the updated wrench product design that exists in the market. Thought it still has another room for improvement, we try to think the concept of rotary motor that could grip the nuts with its bionic grip, then fasten the nuts with rotary force, resulted by the motor and the power source. We used the Inventor program to do the design and make some stress analysis as a result came from the nuts grip and fastening. This step will consider the manufacture-ability of this product redesign and decide the most suitable material used.

After the product redesign done, we analyze the 3D modelling with stress and force experienced by the product. From this analysis, we would like to know and understand that the redesigned product could eliminate manual human effort and capable enough of receiving the feedback of stress and force generated by the materials.

## 3. Result and Discussion

The result of this development research will generate the new product design that improves the bionic grip wrench with its ability with electrically rotary motor. The stress analysis generated by Inventor somehow will show the force as a result of nuts grip and fastening of each six-sides grip. The idea was obtained by the observation of the authors to minimize the effort generated by human. As the importance of time and motion study, the new design is proposed and to be called as electrical bionic grip wrench, performing the rotary force to fasten the nuts after conduct the grip step and it is available to do the fastening activities for following types of nuts : flange bolts, hex bolts, hexagon nuts, and lock nuts. Figure 2 shows the current design conventional wrench that needs manual positioning and manual force driven by the operator.

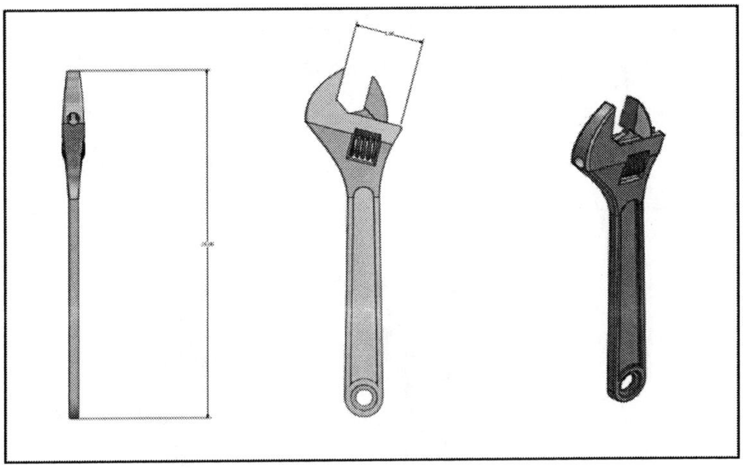

**Figure 2.** Conventional Wrench Design

In this stage, authors observe the updated design of bionic grip wrench, that exists in the market. This product design reduces the processing time to do the manual positioning for the nuts and bolts, as it has several sides on the material surface that needs to be fit before do the fastening. Figures 3 shows the design of bionic grip wrench that successfully reduce the lead time of tool manual positioning.

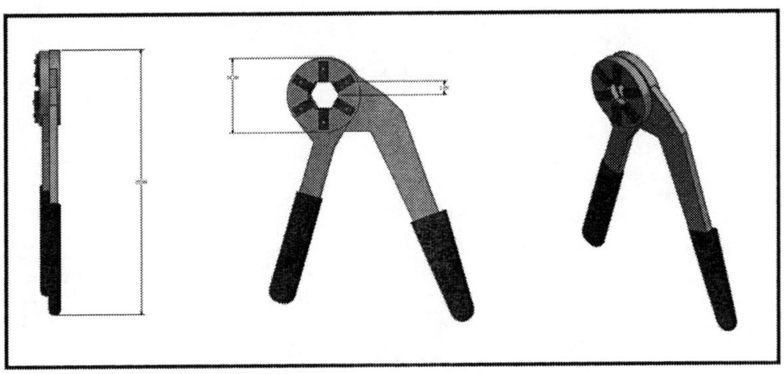

**Figure 3.** Bionic Grip Wrench Design

Coninuing to the next stage of development, the authors observe the chance and possibility to reduce the waste of the activity. Found and discovered that, after the bionic arm grips the nuts or the bolts, it is still requires the human effort to do the bolt fastening by rotating the tools, manually. Hence, the authors initiate the new design that implementing the motor to be able performing rotary force after material grips been done. It reduces the human manual effort by utilizing electrical power that generates rotary force of the wrench. Figure 4 shows the proposed design to eliminate the problems stated above.

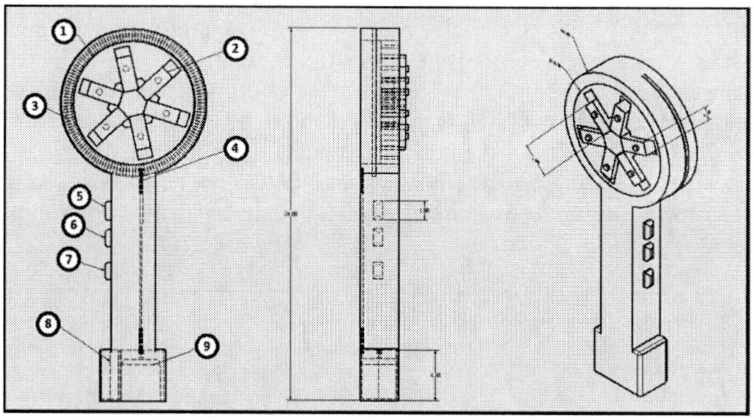


**Figure 4.** 2D Electrical Bionic Grip Wrench Design

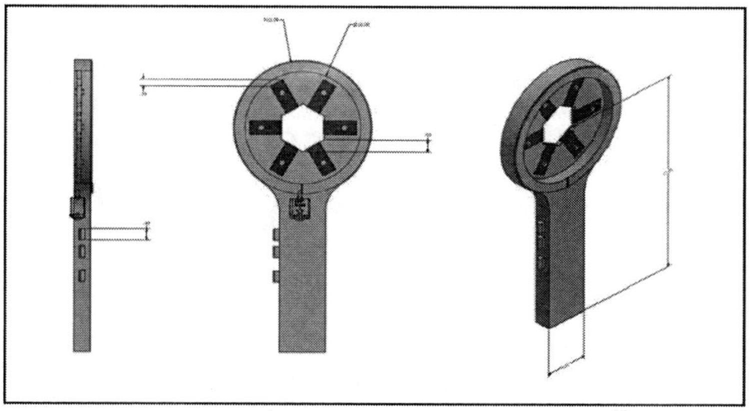

**Figure 5.** 3D Electrical Bionic Grip Wrench Design

**Table 1.** Product's Part Name

| No | Parts Name |
|---|---|
| 1 | Main Disk |
| 2 | Lock Teeth |
| 3 | Gear Disk |
| 4 | Revolve Lever |
| 5 | Fasteners Teeth Button |
| 6 | Looseners Teeth Button |
| 7 | Revolving Disc Button |
| 8 | Lithium Ion Battery |
| 9 | Dynamo bosch GSB 1080 Li |

The inventor program shows proposed bionic grip's teeth material is steel on yield strength 30,022.8 pounds per square inch (psi), means the maximum stress that can be applied along material. The ultimate tensile strength shows 50,038 psi that means the maximum force before the material fractures. Table 2 shows the summary of the product's performance parameter

**Table 2.** Product's Performance Parameter

| Name | Minimum | Maximum |
|---|---|---|
| Volume | 170184 mm^3 | |
| Mass | 0.848246 lbmass | |
| Von Mises Stress | 0.00000000103897 MPa | 0.9676116 MPa |
| 1st Principal Stress | -0.266194 MPa | 0.826204 MPa |
| 3rd Principal Stress | -1.252 MPa | 0.118056 MPa |
| Displacement | 0 mm | 0.0000815169 mm |
| Safety Factor | 15 ul | 15 ul |
| Stress XX | -0.436512 MPa | 0.164633 MPa |
| Stress XY | -0295959 MPa | 0.313883 MPa |
| Stress XZ | -0.321135 MPa | 0.357921 MPa |
| Stress YY | -1.12067 MPa | 0.824607 MPa |
| Stress YZ | -0.514684 MPa | 0.512301 Mpa |
| Stress ZZ | -0.985721 MPa | 0.73289 MPa |

For the bionic grip arms section, Figure 6 shows the stress analysis performed by Inventor program. Analyzing the result, the product's maximum stress is 0.9761 MPA with safety factor is 15. For displacement analysis, the maximum value is 8.152 mm.

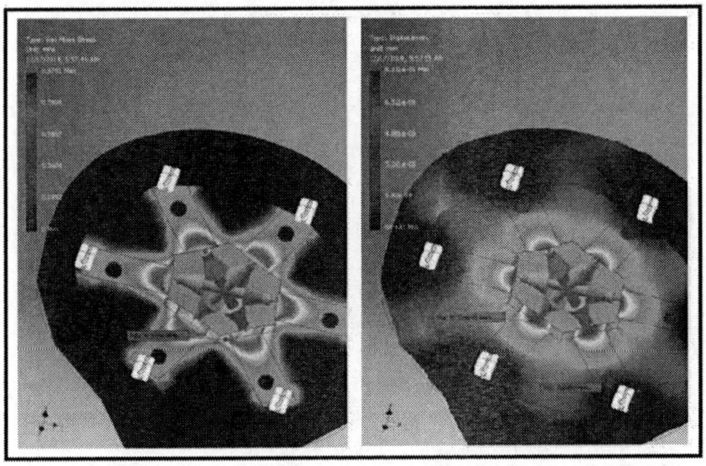

**Figure 6.** Product's Maximum Stress and Displacement Analysis

XX, XY, and XZ section stress are shown on below Figure 6, with the value of maximum stress of 0.1646 MPa for XX section, 0.3139 MPa for XY section, and 0.3579 MPa for XZ section.

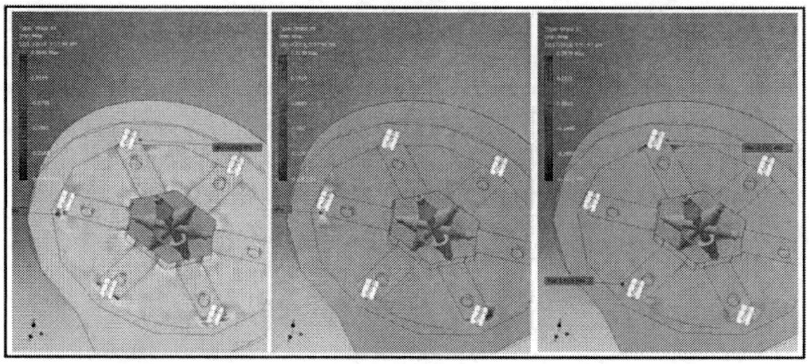

**Figure 7.** XX, XY, XZ Section Stress Analysis

YY, YZ, and ZZ section stress are shown on below Figure 8, with the value of maximum stress of 0.824 MPa for YY section, 0.5123 MPa for YZ section, and 0.7329 MPa for ZZ section.

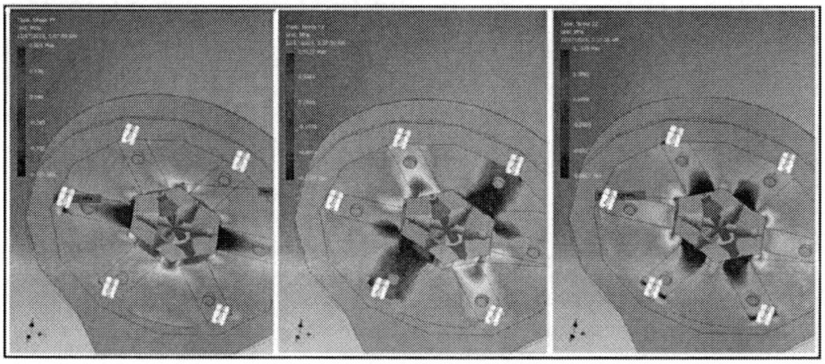

**Figure 8.** YY, YZ, ZZ Section Stress Analysis

## 4. Conclusion

Design, 3D modeling, and stress and force analysis are performed using the Inventor program. The results are proposed to eliminate the human manual effort that contributes to the longer processing time needed compared to the manual grip position, the rotating force carried out by the human motor, and the design of manufacturability. Stress analysis is carried out mainly on the side of the bionic handle which directly contributes the main rotating force to tighten the nuts and bolts, the maximum force that applies to the locking gear is $\bar{x} = 0.4843$, the design can be used on regular nuts with a size of 3-6 cm. This type of battery uses Lithium Ion with a charging input of 110V = 240V AC, 1.5 A. To provide power to the dynamo Bosch GSB 1080 Li that could produce 45 nm.

## 5. References


[1] Cooper R G, and Kleinschmidt E J 1987 New products: what separates winners from losers? *Journal of product innovation management* **4**(170).

[2] Dheraaj S, Ashish S, Gaurav V, Manasi C, Amit G 2002 Innovative Design of Multi Jaw Wrench and Its Analysis. *International Conference on Manufacturing Excellence.*

[3] Lu X S and Zhou H 2000 Pneumatic automation system optimal design. *Shanghai: Shanghai Science and Technology Literature Press.*

[4] Brij N Agrawal and Max F Platzer 2018 *Standard Handbook for Aerospace Engineers* (Beijing: McGraw-Hill Education) p 128.

[5] Roy R 1994 Can the benefits of good design be quantified *Design Management Journal* **5**(9).

# Risk Assessment User Interface Design's by using Object Orientation Programming Approximation in Wooden Toys Industry

**M Mutiasari[1], W N Tanjung MT[1], N Parwati[1], A T Purwandari[1], and U W Islamiah [1]**

[1]Industrial Engineering Department, Al Azhar Indonesia University, Jakarta 12110, Indonesia
[1]miamutiasari30@gmail.com, [2]widya@uai.ac.id, [3]niken.parwati@gmail.com, [4]aprilia_prilia@yahoo.com , [5]uswatunislamiah10@gmail.com.

**Abstract** - Supply chain is a group of companies that work together to create and deliver products to end users [7]. Whereas supply chain risk is defined as an imbalance between demand and supply. Supply chain disruptions can cause various problems such as length of waiting time, out of stock, inability to meet customer demand, and rising costs [1]. If supply chain risk occurs, it will certainly cause material and non-material losses. To anticipate this loss, proper supply chain risk management is essential. In this study, the author developed [11] research entitled "Risk Management Analysis Using FMECA and ANP Methods in the Supply Chain of Wooden Toy Industry", IOP Conference Series: Materials Science and Engineering, 2019 "The development carried out by the author is" Risk Assessment User Interface Design's by using Object Orientation Programming Approximation in Wooden Toys Industry" aims to build information systems so that users more easily determine the greatest risk in the company and can minimize the impact of risks that may occur so that it is more effective and efficient . There are 6 risk factors with 25 risk variables from the data questionnaire results [11] and the biggest risk factor is eviroment with marcoeconomic variable risk of 518,778.
**Keywords**: supply chain, risk management, Failure Mode Effects and Criticality Analysis (FMECA), User Interface, Object Oriented programming.

## 1. INTRODUCTION

Companies in the small and medium industry sector are companies that are vulnerable to supply chain risks. So that the need for SCRM, SCRM is the implementation of strategies to manage every day through continuous risk assessment with the aim of reducing vulnerability or risk [8]. Therefore, with the development of existing computer technology, especially in business and information systems can be designed to minimize this loss. The information system is designed based on object oriented FMECA method. This information system serves to determine priority risks that occur in a company, so that the risks that exist in the company can be minimized and can facilitate the user in determining risks effectively and efficiently.

## 2. LITERATURE REVIEW

### *2.1 Supply Chain Risk Management*

In improving supply chain efficiency and performance, one way is to estimate and manage the possible risks that can arise and disrupt the system so that the company can avoid negative adverse impacts on the company. Supply Chain Risk Management (SCRM) is a potential source of risk and implements appropriate strategies through coordinated among members who need supply chains, to reduce supply chain needs [12]. According to [5] risk assessment covers the stages of risk identification that aims to identify risks that can affect the achievement of organizational goals.

### *2.2 Failure Mode Effects and Criticality Analysis (FMECA)*

Risk identification and determination of risks that have a major effect on company performance can be assessed using the FMECA method. FMECA is an evolutionary method from FMEA which consists of two separate analyzes, namely FMEA and Criticality Analysis (CA). The FMEA (Failure Mode Effects Analysis) method is a method used to identify potential failure modes, determine their effect on product or system operations, and identify actions to reduce these failures. While FMEA must be completed before conducting criticality analysis [6]. By doing CA means providing additional benefits by showing a quantitative ranking of the system and / or subsystem failure mode [4]. Criticality analysis is performed using a statistical approach that is the frequency distribution to determine classes and intervals only. Frequency distribution table is the arrangement of data in a table that has been classified according to certain classes or certain categories [9]. In FMECA, a failure or risk assessment is represented in a value named Risk Priority Number (RPN). RPN is an assessment resulting from the multiplication of three factors and is a value that is an indication of the seriousness of a potential failure or risk, if the RPN value is higher the higher the level of seriousness of a potential failure or risk. These factors are severity, occurrence, and detection (SOD). According to [1] AHP for weighting risk sources and risk variables and FMECA to determine the priority level of risk.

### *2.3 User Interface*

Designing the interface is the most important part of designing the system. an interface must be simple, an interface must be complete, and an interface must have fast performance. The interface describes a collection of objects and operations that can be used to manipulate objects [2]. In the interface development process, the focus must be on the interface elements and the objects that the user sees and uses, rather than the capabilities of a program. The process that in detail illustrates how interface design and development is seen in the picture above.

### *2.4. Object Oriented System Design Approach*

Object-oriented system design approach is a new approach technique in seeing problems and systems (software systems, information systems, or other systems) [2]. This approach views the system to be developed as a collection of objects in the real world. When abstracting and modeling objects, the data and processes owned by the object will be encapsulated (wrapped) into a single unit. In software engineering, the concept of object-oriented approach can be applied at the analysis, design, programming, and software testing stages. There are various techniques that can be used at each of these stages, with certain modeling rules and aids. Use case diagram: This diagram uses the set of use-cases and actors (a special type of class). This diagram is very important for organizing and modeling a system needed by the user. Activity Diagram: This diagram illustrates the various activities flow in the system designed, how each flow starts, the decisions that might occur, and how they end [10]. Sequence Diagram: A sequence diagram is an interaction that is determined at the time a message is sent at a certain time.

## 3. METHODOLOGY

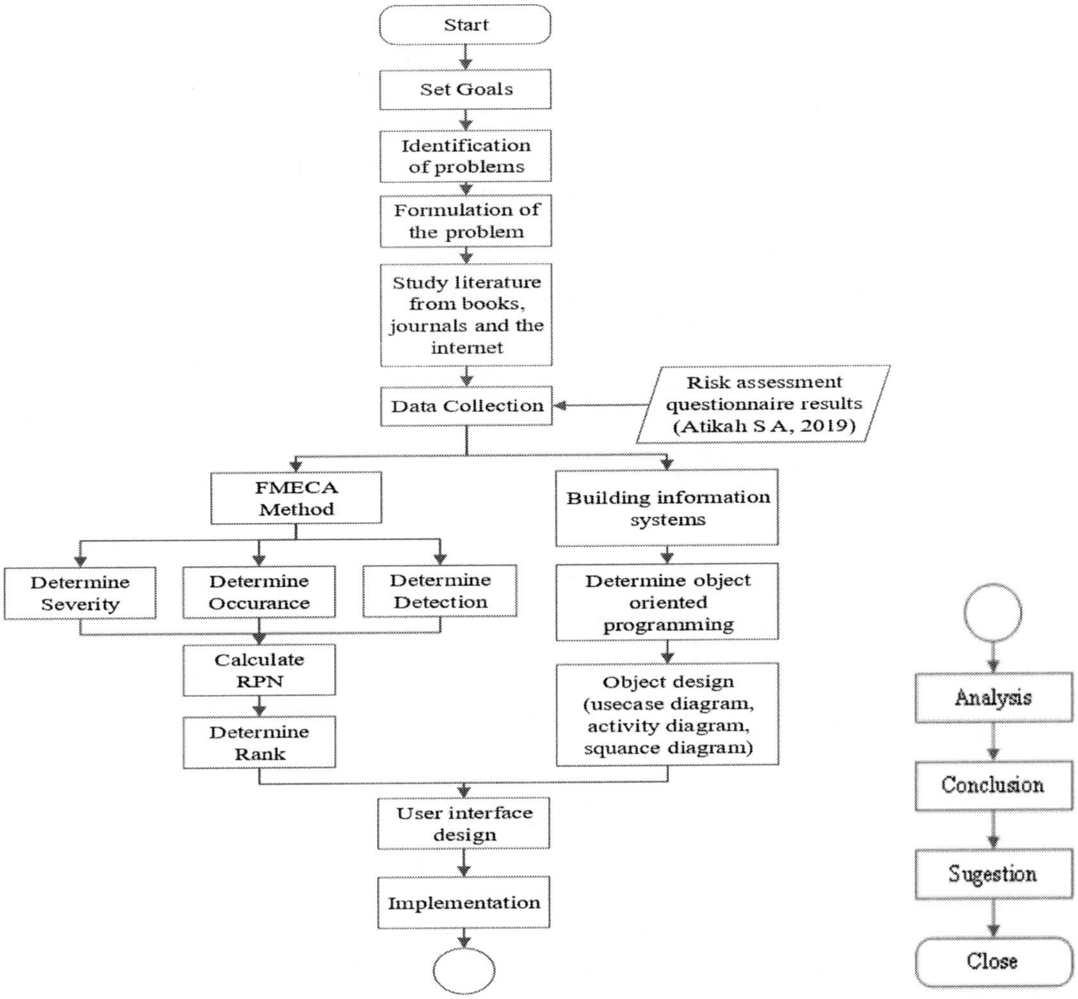


**Figure 1.** Research Flowchart

Based on the flowchart in Figure 1, the research flow is divided into 3 stages. The first step is to set goals, identify problems, formulate problems, study literature, and collect data. In the second stage, it is divided into two parts, namely the measurement of the magnitude of risk in the supply chain of the Wooden Toy Industry using the FMECA method. Data is taken from previous research [11] and building information systems, determining object oriented, designing object oriented, then combined to design user interface design and implementation. The third stage is to analyze the results of calculations and draw conclusions from the results of the analysis that has been made, so that the formulation of the problem and the research objectives that have been set can be answered and closed.

## 4. RESULT AND DISCUSSION

### *4.1. Supply Chain Risk Identification Based on Literature and Interviews*

Identification of risks in the supply chain is carried out based on a literature study in the form of a reference journal according to [3] regarding risk categories in the supply chain lines that are generally found in companies. The categories are arranged in the form of sources of risk or *risk events* which are

IOP Conf. Series: Materials Science and Engineering **847** (2020) 012079 doi:10.1088/1757-899X/847/1/012079

risk events and risk variables or *risk agents* which are risk agents from the results of *risk event* classification.

**Table 1**. Supply Chain Risk Factors and Variables

| ***Risk Event*** | ***Risk Agent*** |
|---|---|
| Demand Risk | Competitor Moves<br>Delays in Delivery to Customers<br>*Forecast Error*<br>Market Saturation |
| Environment Risk | Macroeconomic Uncertainty<br>Natural Disasters<br>Policy Uncertainty<br>Social, Culture & Politic Uncertainty |
| Financial Risk | Cost/Price Risks<br>Exchange Rate Risk |
| Information Risk | Breakdown of IT Infrastructure<br>Distorted Information<br>Inadequate Information Security<br>Information Delay<br>*Wrong Choice of Communication* |
| Operational Risk | Capacity Inflexibility<br>Design Changes<br>Disruption in Production<br>Inventory Risks<br>Variability in Production Process |
| Supply Risk | Dependency on Single Supplier<br>Inflexibility of Supplier<br>Poor Delivery Performance<br>Supplier Poor Quality<br>Supplier Bankruptcy |

The risk classification is carried out based on the activities carried out by the company in the supply chain channel. Then do a compilation based on sub-risks or in this study referred to as risk events. Risks are classified as risk agents. Agent risk is the risk that causes a risk event to occur. Based on the results of secondary data collection from previous studies conducted by [11]. risks contained in Table 1, 25 risk agents were obtained from 6 risk events.

**Table 2.** Results of the Risk Assessment Questionnaire [11]

| Risk Factor | Risk Variable | P1 | | | P2 | | | P3 | | |
|---|---|---|---|---|---|---|---|---|---|---|
| | | S | O | D | S | O | D | S | O | D |
| Demand | Competitor moves | 5 | 10 | 1 | 5 | 5 | 3 | 1 | 1 | 1 |
| | Delays in delivery to customers | 8 | 8 | 1 | 1 | 3 | 3 | 5 | 5 | 2 |
| | Forcast errors | 5 | 8 | 5 | 5 | 7 | 5 | 1 | 8 | 2 |
| | Market saturation | 7 | 2 | 2 | 5 | 5 | 3 | 5 | 3 | 8 |
| Environment | Macroeconomic uncertainty | 10 | 8 | 5 | 9 | 7 | 8 | 10 | 8 | 8 |
| | Natural disasters | 8 | 2 | 6 | 3 | 1 | 8 | 8 | 8 | 10 |
| | Policy uncertainty | 10 | 8 | 5 | 5 | 5 | 8 | 3 | 2 | 2 |
| | Social uncertainty | 10 | 8 | 5 | 3 | 4 | 3 | 5 | 2 | 4 |
| Financial | Business risk | 5 | 5 | 5 | 6 | 7 | 7 | 8 | 10 | 5 |
| | Cost/price risk | 8 | 1 | 2 | 5 | 6 | 1 | 1 | 2 | 4 |
| Information | Exchange rate risk | 10 | 1 | 2 | 3 | 2 | 2 | 3 | 5 | 2 |
| | Breakdown of IT infrastructure | 6 | 6 | 5 | 6 | 3 | 2 | 3 | 2 | 1 |
| | Distorted information | 5 | 4 | 8 | 5 | 4 | 4 | 2 | 2 | 1 |
| | Inadequate information security | 8 | 5 | 6 | 6 | 5 | 3 | 2 | 3 | 1 |
| | Information delay | 5 | 5 | 5 | 5 | 2 | 2 | 8 | 5 | 2 |
| Operational | *Wrong choice of communication* | 6 | 7 | 5 | 7 | 4 | 3 | 8 | 7 | 3 |
| | Capacity inflexibility | 5 | 5 | 1 | 6 | 4 | 3 | 8 | 10 | 1 |
| | Design changes | 8 | 5 | 5 | 6 | 5 | 3 | 8 | 5 | 2 |
| | Disruption in production | 6 | 6 | 4 | 3 | 3 | 3 | 5 | 8 | 2 |
| | Inventory risk | 4 | 5 | 5 | 4 | 7 | 4 | 1 | 1 | 1 |
| Supply | Dependency on single supplier | 5 | 2 | 3 | 6 | 5 | 3 | 5 | 3 | 1 |
| | Inflexibility of supplier | 2 | 2 | 2 | 6 | 6 | 3 | 8 | 8 | 3 |
| | Poor delivery performance | 7 | 1 | 1 | 8 | 6 | 4 | 1 | 2 | 2 |
| | Supplier poor quality | 8 | 1 | 1 | 7 | 5 | 4 | 5 | 8 | 5 |
| | Supplier bankruptcy | 10 | 1 | 1 | 5 | 3 | 2 | 3 | 8 | 2 |

Based on Table 2, it can be seen that each expert has a different assessment. Assessment uses a *Lickerd* scale (ordinal) which ranges from 1 to 10 to make it easier for experts to do the assessment. The *severity of the* assessment given is related to how severe the impact of the risk is if it occurs. And on the *detection* assessment given is related to how easily the risk can be detected if it occurs.

## *4.2. Data Processing and Analysis*

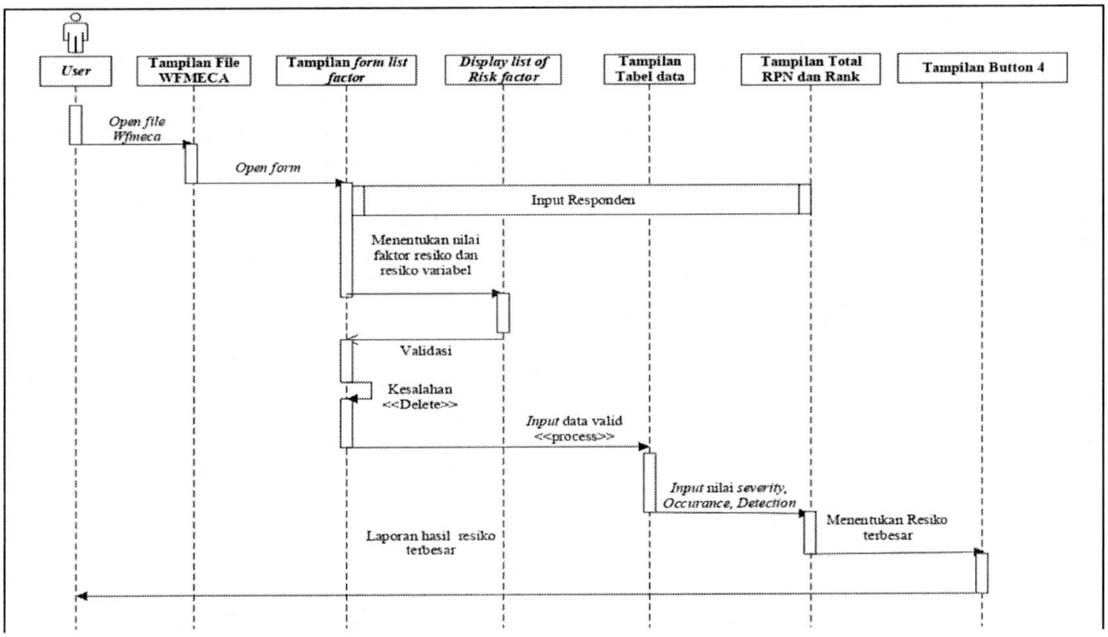


**Figure 2.** Squance Diagram Object Oriented

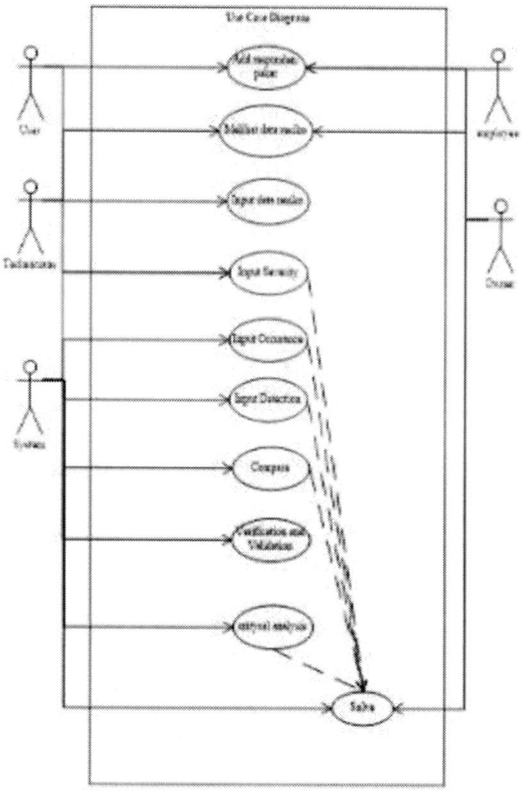

**Figure 3.** Use case diagram user interface OOP

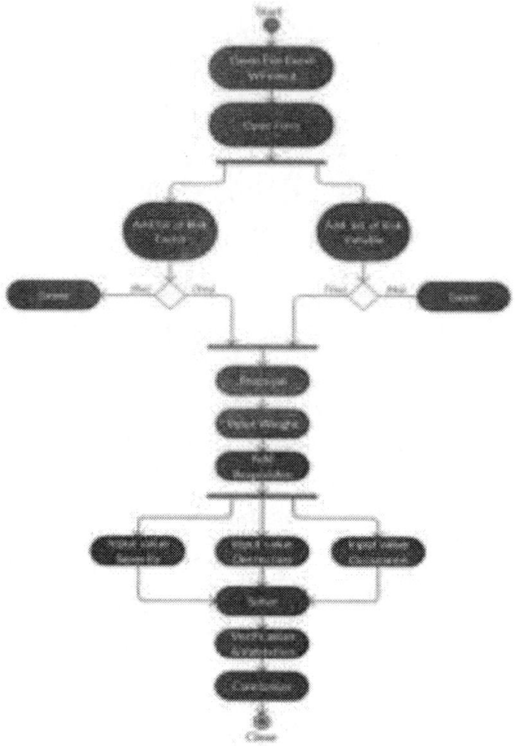

**Figure 4**. Activity Diagram Object Oriented

From use case diagrams, sequence diagrams and activity diagrams explain the interactions between interrelated objects and are arranged in a time sequence based on the activities carried out.

### *4.3. Data Processing Using Methods*

In the activities of the supply chain of the Wooden Toy Industry, there are 25 risk variables which are grouped into 6 risk factors namely demand, environment, finance, information, operations and supply. To identify and determine the magnitude of the risks and risks that often occur in supply chain activities, in this study data processing was performed using the method (FMECA) [11].

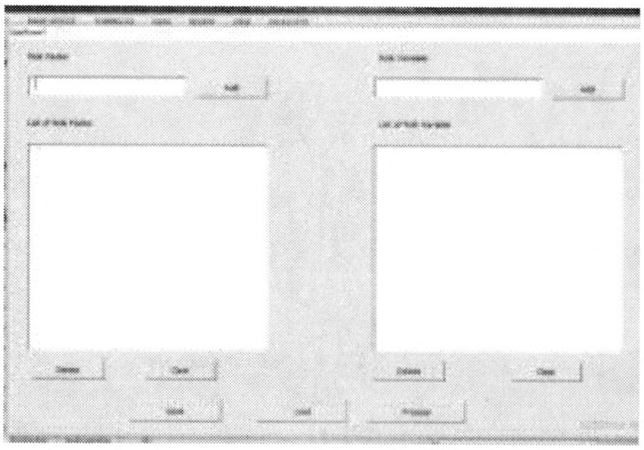

**Figure 5.** Display user interface on Ms. Excel

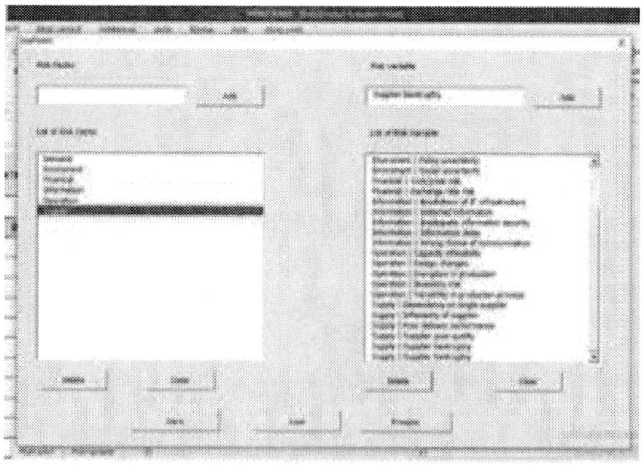

**Figure 6.** Display list Factor risk and List variable risk

| Risk Factor | Weight | Risk Variable | S | O | D | RPN | Rank |
|---|---|---|---|---|---|---|---|
| Demand | 0.22 | competitor moves | 3 | 4 | 2 | 32.5926 | 24 |
| Demand | | Delays in delivery to customers | 3 | 5 | 2 | 49.7778 | 17 |
| Demand | | Forcast errors | 8 | 4 | 4 | 112.444 | 7 |
| Demand | | Market saturation | 3 | 6 | 4 | 81.8519 | 10 |
| Environment | 0.082 | Macroeconomic uncertainty | 8 | 10 | 7 | 518.778 | 1 |
| Environment | | Natural disasters | 4 | 6 | 8 | 188.778 | 3 |
| Environment | | Policy uncertainty | 5 | 6 | 5 | 150 | 5 |
| Environment | | Social uncertainty | 5 | 6 | 4 | 111 | 8 |
| Financial | 0.234 | Cost/price risk | 7 | 6 | 6 | 283.105 | 2 |
| Financial | | Exchange rate risk | 3 | 5 | 2 | 32.6667 | 23 |
| Information | 0.094 | Breakdown of IT infrastructure | 3 | 5 | 2 | 28.4444 | 25 |
| Information | | Distorted information | 4 | 5 | 3 | 48.8889 | 18 |
| Information | | Inadequate information security | 3 | 4 | 4 | 57.7778 | 16 |
| Information | | Information delay | 4 | 5 | 3 | 77.037 | 12 |
| Information | | Wrong choice of communication | 4 | 6 | 3 | 72 | 14 |
| Operation | 0.179 | Capacity inflexibility | 6 | 7 | 4 | 154 | 4 |
| Operation | | Design changes | 6 | 6 | 2 | 66.8519 | 15 |
| Operation | | Disruption in production | 5 | 7 | 3 | 122.222 | 6 |
| Operation | | Inventory risk | 6 | 5 | 3 | 79.3333 | 11 |
| Operation | | Variability in production process | 4 | 3 | 3 | 43.3333 | 19 |
| Supply | 0.191 | Dependency on single supplier | 3 | 5 | 2 | 41.4815 | 20 |
| Supply | | Inflexibility of supplier | 5 | 5 | 3 | 75.8519 | 13 |
| Supply | | Poor delivery performance | 3 | 5 | 2 | 37.3333 | 22 |
| Supply | | Supplier poor quality | 5 | 7 | 3 | 103.704 | 9 |
| Supply | | Supplier bankruptcy | 4 | 6 | 2 | 40 | 21 |

**Figure 7.** Display severity, occurance, detection

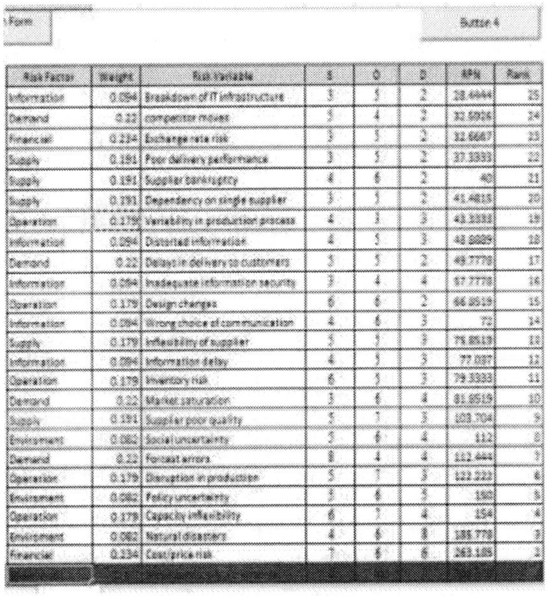


| Risk Factor | Weight | Risk Variable | S | O | D | RPN | Rank |
|---|---|---|---|---|---|---|---|
| Information | 0.094 | Breakdown of IT infrastructure | 3 | 5 | 2 | 28.4444 | 25 |
| Demand | 0.22 | competitor moves | 3 | 4 | 2 | 32.5926 | 24 |
| Financial | 0.234 | Exchange rate risk | 3 | 5 | 2 | 32.6667 | 23 |
| Supply | 0.191 | Poor delivery performance | 3 | 5 | 2 | 37.3333 | 22 |
| Supply | 0.191 | Supplier bankruptcy | 4 | 6 | 2 | 40 | 21 |
| Supply | 0.191 | Dependency on single supplier | 3 | 5 | 2 | 41.4815 | 20 |
| Operation | 0.179 | Variability in production process | 4 | 3 | 3 | 43.3333 | 19 |
| Information | 0.094 | Distorted information | 4 | 5 | 3 | 48.8889 | 18 |
| Demand | 0.22 | Delays in delivery to customers | 3 | 5 | 2 | 49.7778 | 17 |
| Information | 0.094 | Inadequate information security | 3 | 4 | 4 | 57.7778 | 16 |
| Operation | 0.179 | Design changes | 6 | 6 | 2 | 66.8519 | 15 |
| Information | 0.094 | Wrong choice of communication | 4 | 6 | 3 | 72 | 14 |
| Supply | 0.179 | Inflexibility of supplier | 5 | 5 | 3 | 75.8519 | 13 |
| Information | 0.094 | Information delay | 4 | 5 | 3 | 77.037 | 12 |
| Operation | 0.179 | Inventory risk | 6 | 5 | 3 | 79.3333 | 11 |
| Demand | 0.22 | Market saturation | 3 | 6 | 4 | 81.8519 | 10 |
| Supply | 0.191 | Supplier poor quality | 5 | 7 | 3 | 103.704 | 9 |
| Environment | 0.082 | Social uncertainty | 5 | 6 | 4 | 112 | 8 |
| Demand | 0.22 | Forcast errors | 8 | 4 | 4 | 112.444 | 7 |
| Operation | 0.179 | Disruption in production | 5 | 7 | 3 | 122.222 | 6 |
| Environment | 0.082 | Policy uncertainty | 5 | 6 | 5 | 150 | 5 |
| Operation | 0.179 | Capacity inflexibility | 6 | 7 | 4 | 154 | 4 |
| Environment | 0.082 | Natural disasters | 4 | 6 | 8 | 185.778 | 3 |
| Financial | 0.234 | Cost/price risk | 7 | 6 | 6 | 283.105 | 2 |
| [illegible] | [illegible] | [illegible] | [illegible] | [illegible] | [illegible] | [illegible] | [illegible] |

**Figure 8**. Display result

Figure 5 explains the appearance of the user interface on the excel macros that have been designed, then in Figure 6 input the risk agent data and risk variables in the list of factors, then click Save and Process, then the table will come out like Figure 7. From Figure 8 Displays the results of clicks from button 4 which displays the rank results to find out the highest risk. From the data generated RPN highest amounted to 518 778 with a level of critical extremely high, in part *risk factor and Environment* and *Risk Variable Macroeconomic uncertainty*. It is causing a risk not to be accepted it must immediately do mitigation.

## 5. CONCLUSIONS AND RECOMMENDATIONS

The prototype design of the object-oriented programming industry's toy-based supply chain risk management system can be completed using the FMECA method on Macro Excel, making it easier for users to determine the greatest risk in the company in order to minimize risk. There are 25 risk variables and risk variable classification into six risk factors, namely demand, environment, finance, information, operations, and supply. Based on data processing using the Macro Excel user interface with the Failure Mode, Effects and Criticality Analysis (FMECA) methods, The results show that the risk factors identified in the environment (environment) with macroeconomic uncertainty (total economic uncertainty) the highest of 518,778.

## ACKNOWLEDGMENTS

The author would like to thank you because this work was supported in part by the Al Azhar University LP2M Indonesia

## 6. REFERENCES

[1] Bahauddin, Achmad and Minata, Arina 2015 *Analisis dan Strategi Penanganan Risiko Supply Chain Pada PT. Batik Banten Indonesia Menggunakan AHP dan FMECA*. Universitas Ageng Tirtayasa, Cilegon.
[2] Grady Booch1991 *Object-Oriented Analysis and Design with Application*. Benjamin/Cummings.

[3] Jafarnejad, A, Ebrahimi, M, Ali Abbaszadeh, M and Abtahi, SM 2014 Risk Management in Supply Chain using Consistent Fuzzy Preference Relations. *International Journal of Academic Research in Business and Social Sciences*, **4**(1).
[4] Lipol, Lefayet S, Haq, J 2011 *Risk Analysis Method: FMEA/FMECA in The Organizations*. (Boras: University of Boras).
[5] Mahadewi Lufina 2017 *Proses Manajemen Resiko*. Retrievied from : https://swa.co.id/swa/my-article/proses-manajemen-risiko (accessed on 28 october 2019)
[6] Paolo, Berno 2010 *A Tool for The Integration of FMECA and Diagnostic Analysis*. (Italia: Politechnic Milan).
[7] Sumajouw, MD and Sompie, BF 2014 Manajemen Risiko pada Perusahaan Jasa Pelaksanaan Konstruksi di Provinsi Papua (Studi Kasus di Kabupaten Sarmi). *Jurnal Ilmiah Media Engineering* **4**(2), 109-118.
[8] Schlegel, Gregory L, Trent, Robert J 2014 *Supply Chain Risk Management: An Emerging Discipline*. CRC Press Taylor & Francis Group, Boca Raton.
[9] Siagian, D and Sugiarto 2006 *Metode Statistika*. (Jakarta: Gramedia Pustaka Utama)
[10] Suryadi, L 2015 Analisis dan Perancangan Sistem Informasi Pendistribusian Gas Lpg Bersubsidi dengan Metodologi Object Oriented Studi Kasus PT. XYZ. In *Seminar Nasional Informatika (SEMNASIF)* **1**(5).
[11] Tanjung WN, SA Atikah, et al. 2019 Risk Management Analysis Using FMECA and ANP Methods in the Supply Chain of Wooden Toy Industry. *IOP Conference Series: Materials Science and Engineering*.
[12] Zaroni.2015. *Manajemen Risiko Rantai Pasok dalam Model SCOR*. Retrieved from Supply Chain Indonesia:http://supplychainindonesia.com/new/manajemen-risiko-rantai-pasok-dalam-model-scor/ (accessed on January 31, 2019)

IOP Conf. Series: Materials Science and Engineering **847** (2020) 012080 doi:10.1088/1757-899X/847/1/012080

# Supply Chain Risk Evaluation Information System Design using Fuzzy House of Risk in Wooden Toy Industry

**U W Islamiah[1], W Nurchayanty Tanjung[1], N Parwati[1], A Tri Purwandari[1] and M Mutiasari[1]**

[1]Industrial Engineering Department, Al Azhar Indonesia University, Jakarta 12110, Indonesia
[1]uswatunislamiah10@gmail.com, [2]widya@uai.ac.id, [3]niken.parwati@gmail.com, [4]aprilia_prilia@yahoo.com, [5]miamutiasari30@gmail.com

**Abstract.** In the continuity of business processes, companies cannot avoid the possibility of disruptions in their supply chains. This disorder is referred to as risk. The problem often faced by companies today is a poor internal control system in the production process in the supply chain channel. Therefore it is necessary to propose solutions in the form of new procedures that meet all the criteria of good internal control in the company. To minimize risk events, analysis and design of supply chain risk information systems has been made using a DFD based approach. In this study, the observation of risk in the supply chain path of one of the wooden toy industries, namely CV. Atham Toys. Aims to minimize the risks that occur by creating a DFD-based information system to identify priority risks that occur in the company. Based on observations, there are 25 risk agents that can cause 6 risk events. Then the priority risk chosen using the Fuzzy House of Risk method is price or cost risk with an ARPj value of 499.
**Keywords**: supply chain risk management, fuzzy, house of risk, information systems

## 1. Introduction

A company's business processes are inseparable from the risks that can cause harm to the company. Risk is something that leads to uncertainty over the occurrence of an event during a certain time interval where the event causes loss [1]. For this reason, with the development of existing computer technology, especially in business, information systems can be created to minimize this loss. The information system is designed based on data flow diagrams (DFD) with the fuzzy house of risk (FHOR) approach. This information system is designed to determine priority risks that occur in a company, so that the risks that exist in the company can be minimized or even eliminated.

## 2. Literature Review

### *2.1. Risk Management*

Risk is the probability of an event that results in a loss when the event occurs during a certain period. The effect of risk can be measured by multiplying the frequency of events and the impact of those events. Risks can arise from any event, but can be managed based on organizational needs. The approach to managing risk is called risk management. Generally used to avoid, reduce, transfer, divide or accept these risks [2].

### *2.2. Supply Chain Risk Management*

SCRM is the identification of potential sources of risk and implementation of appropriate strategies through a coordinated approach between members of the supply chain with the purpose to reduce the vulnerability of the chain of supply. Supply chain risk management seriously pays attention to sources of risk along the supply chain and how to deal with or minimize these risks [3].

### *2.3. Fuzzy*

Some aspects in the real world always or usually are outside the mathematical model and are inexact. This uncertainty is the basic concept of the emergence of fuzzy logic [4]. The key idea of fuzzy theory is that an element has a degree of membership (membership degree) in an uncertain state. The membership function shows the membership value of an element in a set. The membership value of an element ranges from 0 and 1 [5].

### *2.4. House of Risk*

This House of Risk (HOR) is a modification of FMEA (Failure Mode and Effects Analysis) and the House of Quality (HOQ) model to prioritize which sources of risk are first selected to take the most effective actions to reduce risk [6]. In the house of risk approach, the calculation of the value of ARPj (Aggregate Risk Potential of risk agent j) can be calculated by the formula [7]:

$$ARP\,j = Oj \sum Si\ Rj \tag{1}$$

Information:

ARPj : Aggregate Risk Potential
Oj : Occurance
Si : Severity
Rij : Relationship between Risk Event and Risk Agent

### *2.5. Data Flow Diagram*

Data flow diagrams are a logical model of data or processes created to illustrate where the data originates from, and where the data goes out of the system, where the data is stored, what processes produce the data, and interactions between the stored data, and the processes imposed on that data [8].

## 3. Methodology

The research flowchart can be seen in Figure 1.

## 4. Result and Discussion

### *4.1. Data Collection*

The data needed in this study is qualitative judgment data. This assessment comes from a questionnaire that was asked to experts related to the assessment of risks in the company's supply chain lines.

*4.1.1. Supply Chain Risk Identification Based on Literature and Interviews.* Risk identification is carried out based on reference journals according to (Jafarnejad, et al., 2014) about risk categories in supply chain lines that are generally found in companies and then adjusted to company conditions through discussions with company experts [9]. The risk classification can be seen in Table 1.

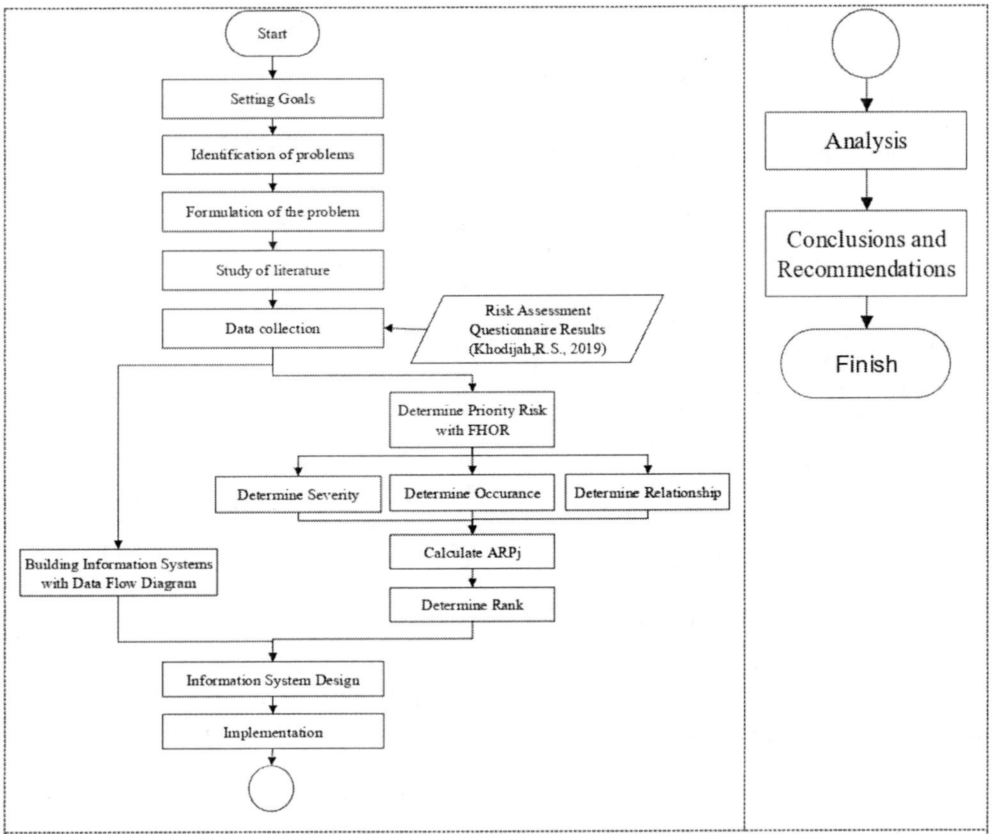


**Figure 1.** Research Flowchart

**Table 1.** Risk Classification

| Risk Event | Risk Agent | Risk Event | Risk Agent |
|---|---|---|---|
| Demand Risk | Competitor Moves | Operational Risk | Capacity Inflexibility |
| | Delays in Delivery to Customers | | Design Changes |
| | *Forecast Error* | | Disruption in Production |
| | Market Saturation | | Inventory Risks |
| Environment Risk | Macroeconomic Uncertainty | | Variability in Production Process |
| | Natural Disasters | Supply Risk | Dependency on Single Supplier |
| | Policy Uncertainty | | Inflexibility of Supplier |
| | Social, Culture & Politic Uncertainty | | Poor Delivery Performance |
| Financial Risk | Cost/Price Risks | | Supplier Poor Quality |
| | Exchange Rate Risk | | Supplier Bankruptcy |
| Information Risk | Breakdown of IT Infrastructure | | |
| | Distorted Information | | |
| | Inadequate Information Security | | |
| | Information Delay | | |
| | *Wrong Choice of Communication* | | |

The risk classification is carried out based on the activities carried out by the company in the supply chain line. Then do the compilation based on sub-risk or in this study referred to as a *risk event* or risk events. The risks are classified as *risk agents*. Risk agents are risks that cause risk events to occur. Based on the results of risk collection contained in Table 1, 25 risk agents were obtained from 6 risk events.

**Table 2.** Results of the Risk Assessment Questionnaire [10]

| Risk Event | Risk Agent | Mr. Thamrin | | | Mrs. Hera | | | Mrs. Weni | | |
|---|---|---|---|---|---|---|---|---|---|---|
| | | S | O | R | S | O | R | S | O | R |
| Demand Risk | Competitor Moves | 7 | 10 | 1 | 6 | 5 | 3 | 4 | 1 | 1 |
| | Delays in Delivery to Customers | 7 | 8 | 9 | 6 | 3 | 9 | 4 | 5 | 9 |
| | *Forecast Error* | 7 | 8 | 3 | 6 | 7 | 1 | 4 | 8 | 3 |
| | Market Saturation | 7 | 2 | 9 | 6 | 5 | 3 | 4 | 3 | 9 |
| Environment Risk | Macroeconomic Uncertainty | 7 | 8 | 1 | 3 | 7 | 3 | 7 | 8 | 9 |
| | Natural Disasters | 7 | 2 | 9 | 3 | 1 | 9 | 7 | 8 | 3 |
| | Policy Uncertainty | 7 | 8 | 3 | 3 | 5 | 1 | 7 | 2 | 1 |
| | Social, Culture & Politic Uncertainty | 7 | 8 | 3 | 3 | 4 | 3 | 7 | 2 | 1 |
| Financial Risk | Cost/Price Risks | 7 | 5 | 9 | 7 | 7 | 9 | 5 | 10 | 9 |
| | Exchange Rate Risk | 7 | 1 | 3 | 7 | 6 | 3 | 5 | 2 | 3 |
| Information Risk | Breakdown of IT Infrastructure | 5 | 1 | 9 | 7 | 2 | 1 | 4 | 5 | 3 |
| | Distorted Information | 5 | 6 | 3 | 7 | 3 | 1 | 4 | 2 | 3 |
| | Inadequate Information Security | 5 | 4 | 3 | 7 | 4 | 1 | 4 | 2 | 3 |
| | Information Delay | 5 | 5 | 3 | 7 | 5 | 3 | 4 | 3 | 3 |
| | *Wrong Choice of Communication* | 5 | 5 | 3 | 7 | 2 | 1 | 4 | 5 | 1 |
| Operational Risk | Capacity Inflexibility | 7 | 7 | 3 | 5 | 4 | 3 | 7 | 7 | 3 |
| | Design Changes | 7 | 5 | 9 | 5 | 4 | 3 | 7 | 10 | 9 |
| | Disruption in Production | 7 | 5 | 9 | 5 | 5 | 9 | 7 | 5 | 9 |
| | Inventory Risks | 7 | 6 | 3 | 5 | 3 | 1 | 7 | 8 | 1 |
| | Variability in Production Process | 7 | 5 | 3 | 5 | 7 | 1 | 7 | 1 | 1 |
| Supply Risk | Dependency on Single Supplier | 7 | 2 | 3 | 7 | 5 | 3 | 5 | 3 | 3 |
| | Inflexibility of Supplier | 7 | 2 | 3 | 7 | 6 | 3 | 5 | 8 | 3 |
| | Poor Delivery Performance | 7 | 1 | 9 | 7 | 6 | 3 | 5 | 2 | 9 |
| | Supplier Poor Quality | 7 | 1 | 9 | 7 | 5 | 9 | 5 | 8 | 9 |
| | Supplier Bankruptcy | 7 | 1 | 9 | 7 | 3 | 9 | 5 | 8 | 9 |

### *4.2. Data Processing*

#### *4.2.1. Designing Data Flow Diagrams*

Through interviews and conducting a study of the division of labor in the supply chain activities CV. Atham Toys, it can be identified several actors who are directly related to the system. The following is a grouping of the actors identified in the risk management information system in the supply chain.

Can be seen from the context diagram (Figure 2), that there are 4 actors who play a role in the risk management information system in the supply chain CV. Atham Toys namely admin, employees, technicians and managers. At this level, the risk management information system carried out by admins, employees, and technicians gets feedback.

In *Data Flow Diagram* Level 1, there are 3 activities carried out namely risk identification, questionnaire data input and risk reports that occur. For risk identification activities, there is one actor who has a role, that is admin and has a *data store* that is a risk system. In the questionnaire data input activity the actors in charge are employees and there is a *data store*, the questionnaire system. Actor technicians act as submit data on SOR values that have been inputted by employees. While the last activity is a risk report that occurs in a company. Actors who play a role in this activity are company managers.

IOP Conf. Series: Materials Science and Engineering **847** (2020) 012080 doi:10.1088/1757-899X/847/1/012080

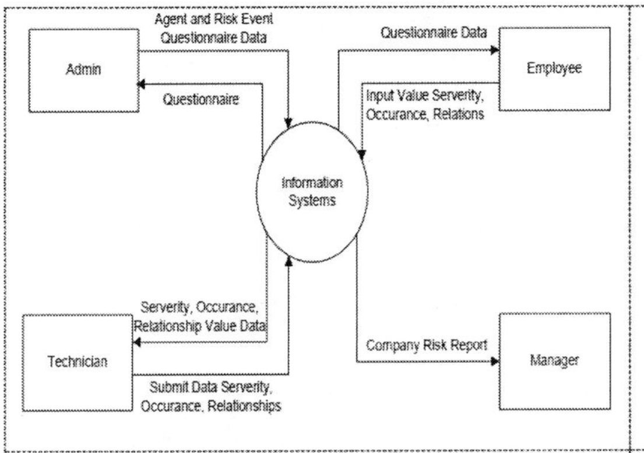


**Figure 2.** Context Diagram of Risk Management Information Systems

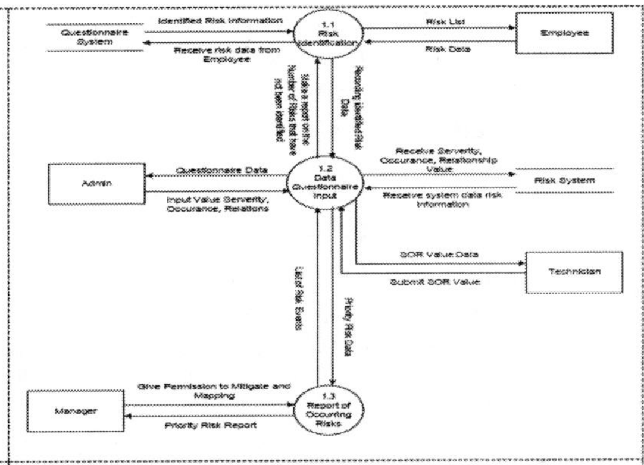


**Figure 3.** Level 1 DFD Risk Management Information Systems

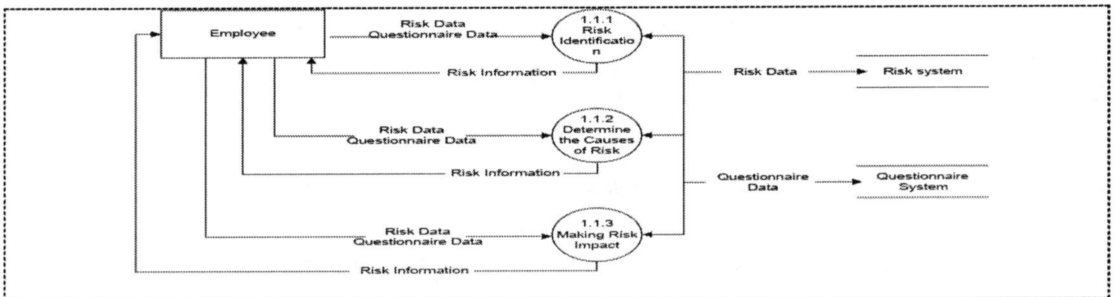


**Figure 4.** Level 2 DFD Risk Identification of Risk Management Information Systems

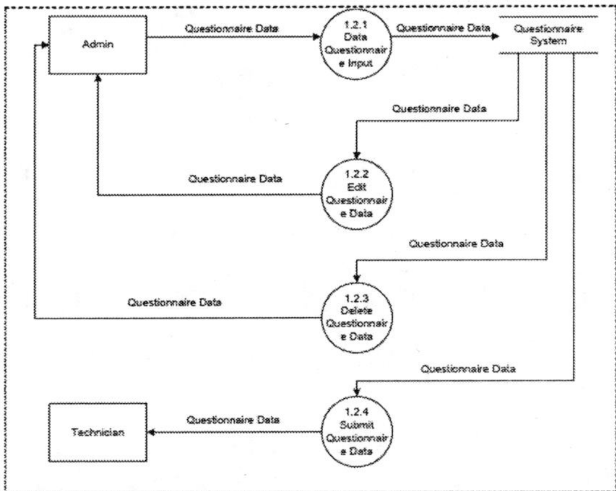


**Figure 5.** Level 2 DFD Data Input Questionnaire Risk Management Information System

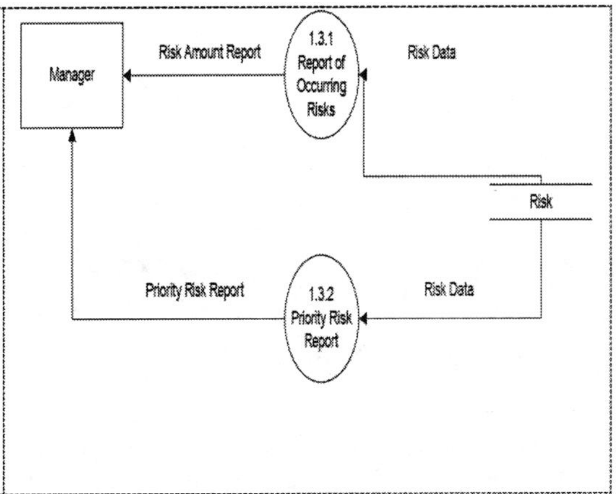


**Figure 6.** Level 2 DFD Risk Management Information System Risk Report

*Data Flow Diagram* (DFD) level 2 is an activity in detail or detail of each activity. In DFD level 2 risk identification, the activities carried out are risk identification, determining the cause of risk, and making risk impact. The actor in charge is the admin and there are 2 *data stores,* namely the risk system and the questionnaire system. At DFD level 2 input questionnaire data, the activities carried out are input questionnaire data, edit questionnaire data, delete questionnaire data, and submit questionnaire data. Actors in charge are employees and technicians and there is one *data store*, the questionnaire system. In DFD level 2 risk reports, the activities carried out are risk reports that occur and priority risk reports. The acting actor is the manager of the company and there is 1 *data store,* namely risk.

*4.2.2. Data Processing Using Methods FHOR*
When the user enters or opens Microsoft Excel for the first time, the initial program will appear as shown in Figure 7. On the program's start screen, there is a Button 1 button that functions to continue the program. If you want to continue the program, it will click the Button 1 button and if you want to exit the program then click the x button in the upper right corner.

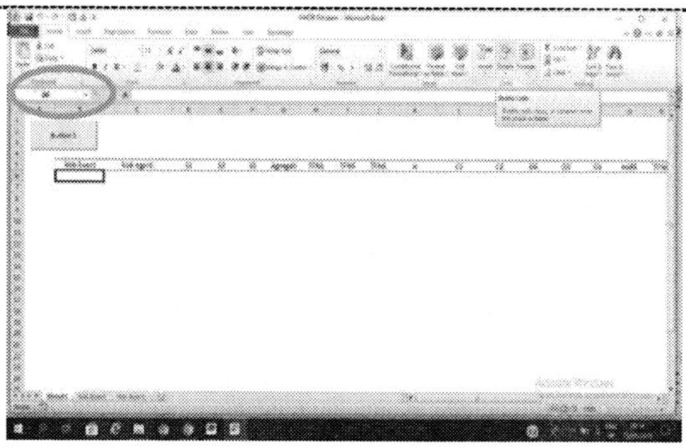

**Figure 7.** Initial Display of Information Systems Using the FHOR Method

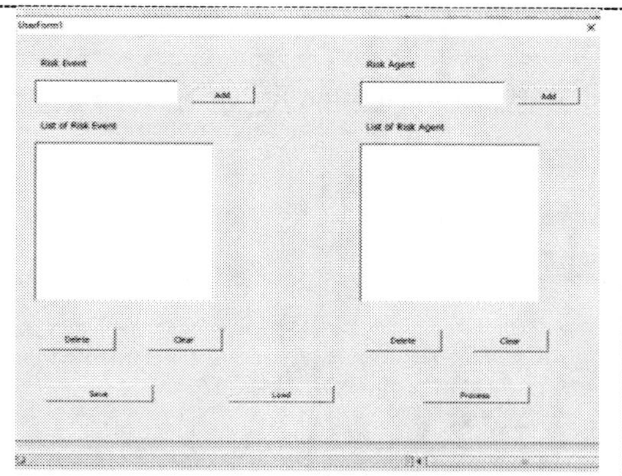


**Figure 8.** *Risk Event* and *Agent List Form*

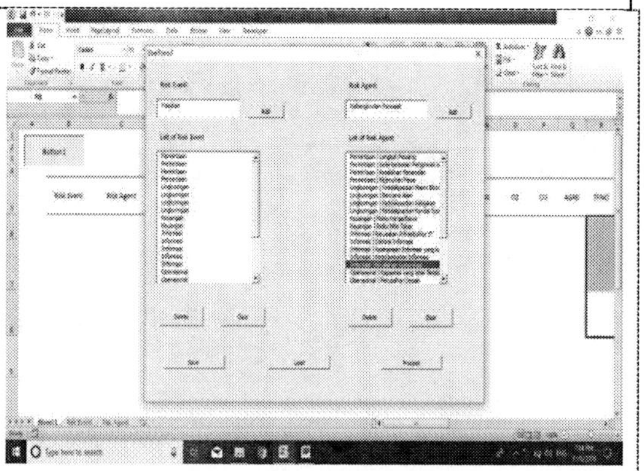

**Figure 9.** *Input Event Risk* and *Agent Risk*

After clicking the *Commond Button* 1 *button*, a *risk event* and *agent risk list form* will appear. In this *form*, the *user is* expected to input the *agent's risk* and *risk events* that occur in a company. In the case of a risk and risk agent list form there is a clear button that functions to delete all the lists that have been entered, the delete button to delete one of the lists that have been entered, the save button to save the list, and the process button that is useful for processing the list. After resolving the company's risk events and risk agents, the user is expected to press the process button so that it appears as shown in Figure 10.

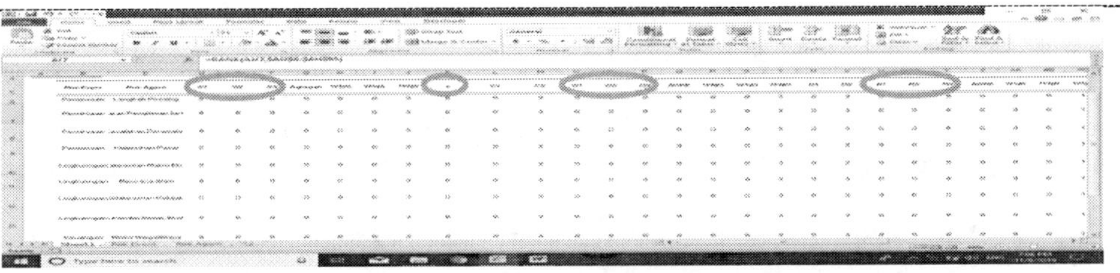

**Figure 10.** Display After Pressing the *Process* Button

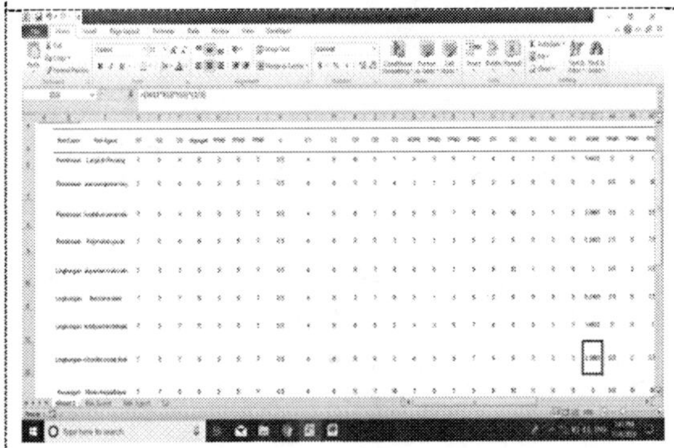

**Figure 11.** Input SOR Value

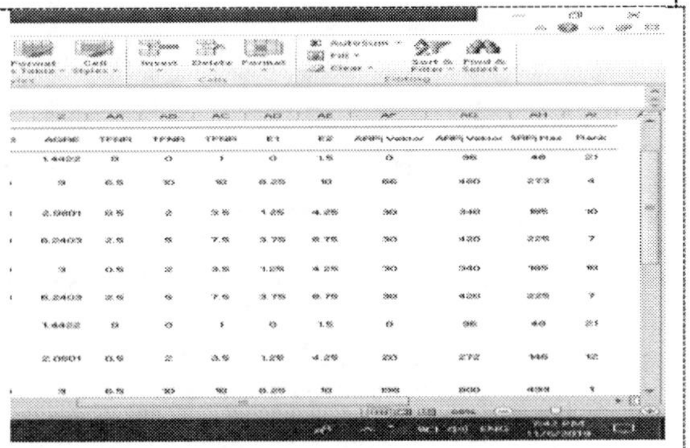

**Figure 12.** *Solution for* FHOR Method

The next step is to enter the value of the results of the questionnaire that was distributed to 3 experts. The values entered are alpha value, serverity value, event value, and relationship value for each expert. After entering the serverite values, events, and relationships of each expert, aggregation values will automatically be obtained as shown in Figure 11.

After entering all the values, the ARPj value for each risk agent and risk event will be obtained as shown in Figure 12. From the results of this ARPj a rating process will be carried out which will be ranked from the value that has the largest ARPj. The biggest ARPj value is the risk that must be prioritized and is expected to mitigate this risk so as not to cause huge losses to the company.

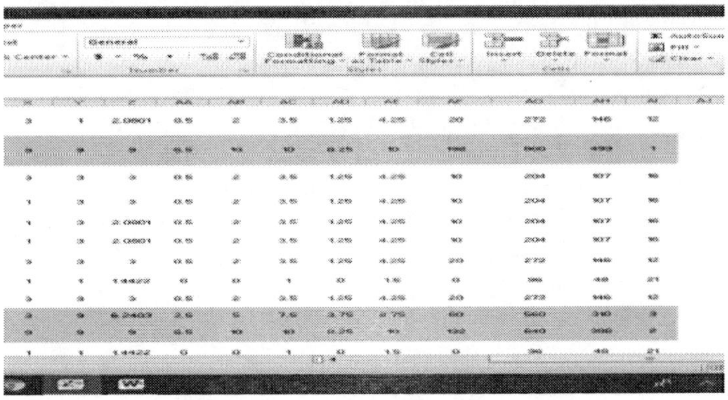

**Figure 13.** Company Priority Risk

## 5. Conclusion

Based on data collection and analysis results of data processing that has been done, it was found that the actors involved in the design of information systems consist of 4 actors, namely admin, employees, technicians, and managers. In DFD consists of 2 levels, where each level has a certain activity. On the CV. Atham Toys also found several risks that may and often occur. These risks are classified based on *risk events* and *risk agents*. Then there are 6 risk events with 25 risk agents causing these risk events. Then the results of data processing with the FHOR (*Fuzzy House of Risk*) information system obtained the risk agent with the largest ARP namely A9 or the price or cost risk agent with an ARP value of 499. Then the risk agent that must be prioritized and considered urgent in the supply chain line CV. Atham Toys is a price or cost risk agent.

**Acknowledgments**
This work was supported in part by LP2M Al Azhar Indonesia University

## 6. References

[1] Sumajouw, MD and Sompie, BF 2014 Manajemen Risiko pada Perusahaan Jasa Pelaksanaan Konstruksi di Provinsi Papua (Studi Kasus di Kabupaten Sarmi). *Jurnal Ilmiah Media Engineering* **4**(2), 109-118.
[2] Nurlailah Badariah, DS 2013 Analisa Supply Chain Risk Management Berdasarkan Metode Failure Mode and Effects Analysis (FMEA). *Jurnal Teknik Industri*, 110-11.
[3] Kamalahmadi, M and Parast, M 2016 A Review of The Literature on The Principles of Enterprise and Supply Chain Resilience: Major Findings and Directions for Future Research. *International Journal of Production Economics*, **171**, 116-133.
[4] Marimin, Umano, M, Hatono, I and Tamura, H 1998 Linguistic Labels for Expressing Fuzzy Preference Relations in Fuzzy Group Decision Making. *IEEE Transactions on Systems, Man, and Cybernetics, Part B: Cybernetics*, 205-218.
[5] Azmiyati, S and Hidayat, S 2017 Pengukuran Kinerja Rantai Pasok pada PT. Louserindo Megah Permai Menggunakan Model SCOR dan FAHP. *Jurnal Al-Azhar Indonesia Seri Sains dan Teknologi*, **3**(4), 163-170.
[6] Nasution, S, Arkeman, Y, Soewardi, K, and Djatna, T 2014 Identifikasi dan evaluasi risiko menggunakan fuzzy FMEA pada rantai pasok agroindustri udang. *Journal of Industrial Research (Jurnal Riset Industri)*, **8**(2).
[7] Pujawan, IN, and Geraldin, LH 2009 House of Risk: A Model for Proactive Supply Chain Risk Management. *Business Process Management Journal*, 953-967.
[8] Afyenni, R 2014 Perancangan data flow diagram untuk Sistem informasi sekolah (studi kasus pada sma pembangunan Laboratorium UNP). *Jurnal Teknoif*, **2**(1).
[9] Jafarnejad, A, Ebrahimi, M, Ali Abbaszadeh, M, and Abtahi, SM 2014 Risk Management in Supply Chain using Consistent Fuzzy Preference Relations. *International Journal of Academic Research in Business and Social Sciences*, **4**(1).
[10] Tanjung, WN, Khodijah, RS, Hidayat, S, Ripmiatin, E, Atikah, SA, and Asti, SS 2019 Supply Chain Risk Management on Wooden Toys Industries by using House of Risk (HOR) and Analytical Network Process (ANP) Method. In *IOP Conference Series: Materials Science and Engineering* **528**(1), 012086.

# ERP Implementation in Crisis Management: A Case Study of Government-Owned Electricity Company

**Muharman Lubis[1,2], Exa Parmita[1] and Lyvia Winiyanti Lumingkewas[1,3]**

[1]Telkom University, Jalan Telekomunikasi No. 1, Bandung, 40257
[2]muharmanlubis@telkomuniversity.ac.id, [3]lwinyanti2@gmail.com

**Abstract**. Enterprise Resource Planning plays an outstanding role in integrating business functions and facilitating the exchange of information and communication. Through the ERP system integration, significant speed and productivity gains become the objective that the company wants to achieve directly and comprehensively. One of the companies in the public service sector that potentially implements ERP is the state-owned company in the electricity sector. Unfortunately, until now there are still problems that often occur such as mass power outages, electricity supply crises, and other technical and nontechnical problems. By adopting ERP for crisis management, the company can categorize and meet the time to resolve (TTR) incidents and problems that are set to analyze access management to determine the level of security of the company's location, so the company can determine the equipment in each location including access rights. This research will identify the implementation of ERP that is suitable for company crisis management implementation.
**Keywords**: ERP, crisis management, incidents, problems.

## 1. Introduction

Nowadays, Information technology has become a critically important part of all industries and activities. Few areas of industrial, commercial and service activities require information technology to achieve their goals [1]. It has been applied to increase productivity and help achieve quality, time and satisfaction standards for consumers and employees, which has aimed to obtain a specific long or short-term goals in the business world through a series of systems consisting of information systems and supporting infrastructure. The excellent solution today is the Enterprise Resource Planning or better known as ERP. ERP can integrate all existing operations into the functional area of the company, for example between departments and between different locations [2]. With this system integration, data previously obtained from different systems to combine into one system in a standard format. So there are no practical differences between jobs, between departments and between different sites [3]. One of the companies in public service sector that potentially implements ERP is the state-owned company in the electricity sector that ensures everyone in Indonesia access electricity. Companies that have high risks such as electricity companies must be able to handle problems precisely and quickly to always provide the best service for customers. Unfortunately, until now there are still problems that often occur such as mass power outages, electricity supply crises and other technical and nontechnical problems. To solve the problem, the company needs IT breakthroughs such as integrated crisis management by adopting ERP that can be adjusted to IT Service Management standards such as ITIL [4]. By adopting ERP for crisis management, the company can categorize and meet the time to resolve (TTR) incidents and problems that are set, analyze access management to determine the level of security of the company's location, so the company can determine the equipment in each location including access rights. This research will identify the implementation of ERP that is suitable for company crisis management implementation.

## 2. Literature Review

The main objective of implementing ERP is to solve old system problems, provide a single and integrated technology platform thus helping companies gain a competitive advantage and thus compete globally [5]. In particular, ERP plays an outstanding role in integrating business functions and facilitating the exchange of information and communication. Through the ERP system integration, significant speed and productivity gains can be achieved [6]. ERP systems can be a useful tool for companies to build a robust infrastructure for information systems and to enable management to make better decisions based on accurate and timely information. Moreover, this system improves product quality and customer response and increases the exchange of information and the quality of information between different departments within the company. It also exceeds business limits for suppliers, customers, and other partners in the supply chain [7]. Ultimately, this will improve overall business performance, helping to gain a competitive advantage in the global economy and increase long-term profitability. But on the other hand, ERP system implementation is very expensive and time-consuming [5]. The reason for this is the fact that many companies become ERP systems and limit the market share of ERP systems. Many leading companies in the world have successfully implemented ERP systems such as Cisco costs for operational has been lowered and sales increased considerably since implemented ERP. Chevron Texaco has improved supply chain management by implementing an ERP system. Earns $ 100 million net annual profit. The IBM R/3 app was one of the most successful. Another successful ERP implementation, Indian power distribution utilities/companies (known as Discoms) faced a financial crisis in 2003 amid monopolistic activities. Some Discoms's do not accept a minimum return (ROR). The electricity sector has a large deficit due to the deteriorating financial situation. To overcome this problem, the Indian government adopted ERP. The main purpose of using an ERP solution is to improve operations and efficiency, reduce costs and improve customer service, as well as adhere to standard settings and reduce ATC losses [8]. However, there have been some significant failures. Dell believed that the ERP system could not handle its sales volume and suffered a financial loss of approximately $200,000 when the ERP system was designed for two years [9]. It becomes the essential reasons why DELL cancels the ERP project and lost $115 million. Another symbolic case was FoxMeyer. The pharmaceutical giant incurred a loss of $100 million and failed after implementation failure.

## 3. Research Methodology

This paper uses observation, informal interviews, and secondary data as sources of information. Observations were conducted by observing the existing process of the incident and problem management in the organization. Meanwhile, informal interviews were conducted with management, staffs and users of the state-owned company in the electricity sector. In addition, the process of scheduling becomes a consideration for the actual session, which informants should have proper experiences and knowledge in terms of in electricity field and was intended to emphasize relevant electricity issues and challenges that may be appropriate to crisis management. Secondary data sources were the result of a review of publications on the ERP implementation in the industry and crisis management, a search engine that indexes references and summaries in the broad areas of electricity. After the data is collected, then modelling the existing crisis management business processes, from the existing business processes then analyzed and implementing ERP as crisis management.

## 4. Result and Analysis

As a result, the incident management system has two basic principles that company should considered wisely and carefully. First, the local-aware structure must be flexible enough to easily scale, especially as additional resources outside of the jurisdiction, and to add to the level of demand imposed by escalating events. Secondly, it should use proper tools and mechanism in day-to-day emergencies with good coordination among team members to forms the basis for an expanded structure for dealing with disasters. Many incident management systems have been devised over the years and their use remains intermittent to predict worst-case scenarios, while the resources and strategies of many agencies and jurisdictions to manage large incidents have been reported to be improved through several phases especially if the requirement asked to [12]. The following is a list of major incident cases in state-owned company in the electricity sector:

**Table 1.** Major Incident

| Major Incident | | | |
|---|---|---|---|
| Case | System Failure | System failure is the inability of a system or component to perform the functions required in accordance with its specifications | Missed payments, double charged customer, error work flow, 404 not found, timed out. |
| | Virus Attack | A virus is a program or application that can spread into other programs or documents that aim to steal data and damage data or documents | Broken operating system, slow response, corrupted file, damage process |
| | Force Majeure | Force Majeure is an event or situation that occurs outside the authority of the parties concerned | Fire and earthquake |

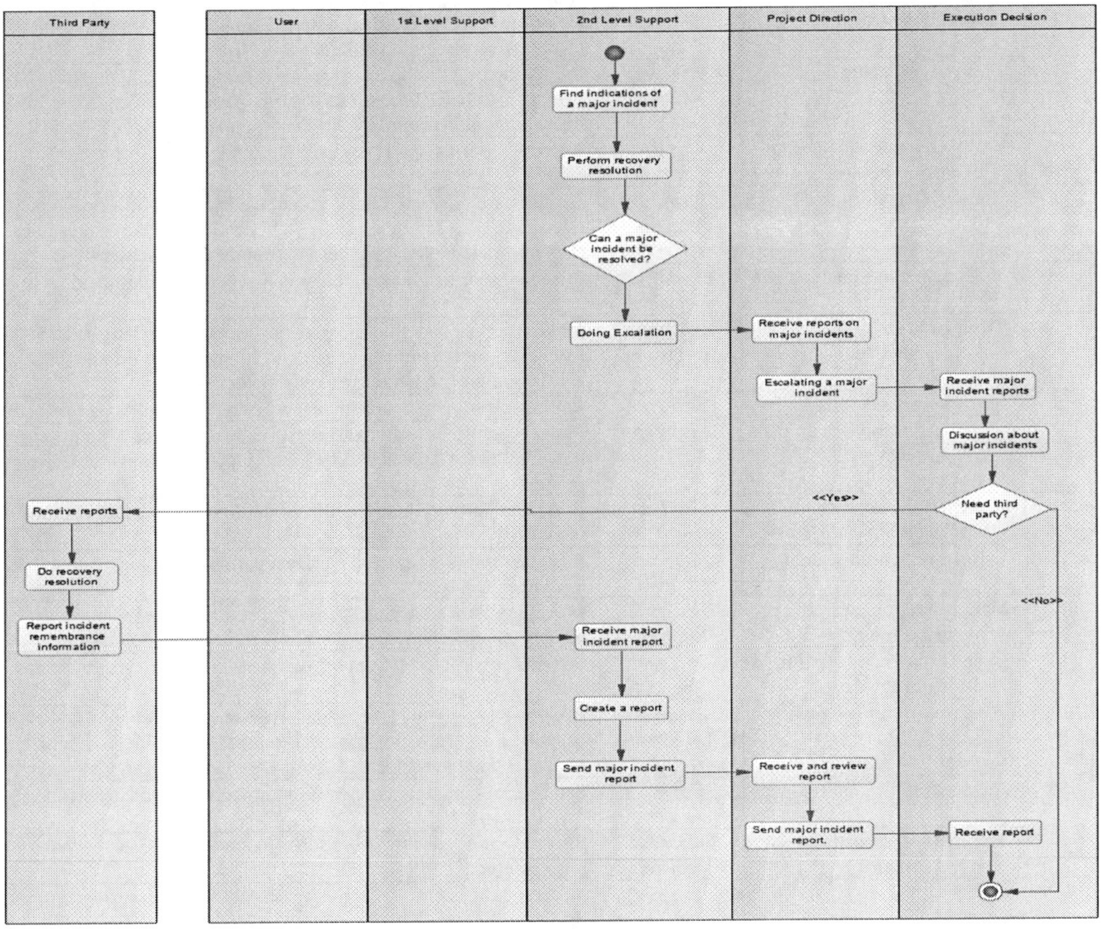


**Figure 1.** Crisis Management Business Process Existing

Nowadays, the development of information technology offers many new opportunities so that the exchange of data, information and knowledge can be obtained respectively even when the government is constrained by distance, time and place to provide a means to improve performance. Therefore, incident management systems need to be aware that small mistakes maybe exposed to competitors or illegal people in order to take advantage of the opportunity to profit from the trouble that has occurred [11]. Freedom of information in terms of availability and accessibility requires consideration of the context, purpose, and interests of the community to promote community understanding and promote citizen support. Actually, the challenge of implementing a particular regulation is related to an attempt to develop commitment and consistency. On the other hand, large gaps between policies, plans, and realizations can degrade implementation. It became more complicated when bureaucracy tended to maintain the status quo rather than develop public transparency and the ability to access the necessary information [10]. The concept of incident command has existed in law enforcement and fire brigade for many years, and has emerged as a collection of organizational rules designed to meet the needs of field commanders and operations forces at the forefront of emergency response [12].

**Table 2.** Crisis Management Business Process Existing Explanation

| No | Activity | Authority | Explanation |
|---|---|---|---|
| 1 | Find indications of major incidents | 2nd Level Support | Report incidents to the Helpdesk via email, phone, self-service and walk in |
| 2 | Perform recovery resolution | 2nd Level Support | Helpdesk receives incident reports from users via email, phone, chat, self-service, and walk in |
| 3 | Escalate | 2nd Level Support | Escalating by IT support to service management and deputy managers to hold discussions regarding major incident solutions that cannot be resolved by 2nd level support |
| 4 | Receive reports regarding major incidents | Project Direction | Receive reports from IT support regarding major incidents |
| 5 | Escalating major incidents | Project Direction | Escalating by deputy managers and management of service management to senior managers and heads of STI divisions related to major incidents that cannot be resolved by IT support |
| 6 | Receive major incident reports | Executive Decision | Receive reports of major incidents that cannot be resolved |
| 7 | Discussion about the major incident | Executive Decision | Make discussions with senior managers, STI division heads, deputy managers, service management, and IT support to find the best solution related to handling major incidents. |
| 8 | Receive reports | Third Party | Receive reports on major incidents that IT support cannot resolve |
| 9 | Perform resolution and recovery | Third Party | Perform resolution and recovery of major incidents. resolution and recovery that is expected to be major incidents can be resolved and returned to normal |
| 10 | Report information on handling major incidents | Third Party | Provide reports to IT support that handling major incidents can be resolved |
| 11 | Receive major incident reports | 2nd Level Support | Receive reports from third parties regarding major incidents that have been resolved |
| 12 | Making a report on handling major incidents | 2nd Level Support | Making reports from third parties related to major incidents that have been resolved |
| 13 | Send reports on handling major incidents | 2nd Level Support | Sending reports to IT Support that handling major incidents can be resolved |
| 14 | Receive and review reports | Project Direction | Receive and review reports that have been made by the 2nd level of support as an evaluation material so that the problem of major incidents that have successfully resolved does not happen again |
| 15 | Send major incident reports | Project Direction | Send major completed incident reports to senior managers and STI division heads |
| 16 | Receive reports and review reports | Executive Decision | Receive and review reports related to major incidents that have been resolved |

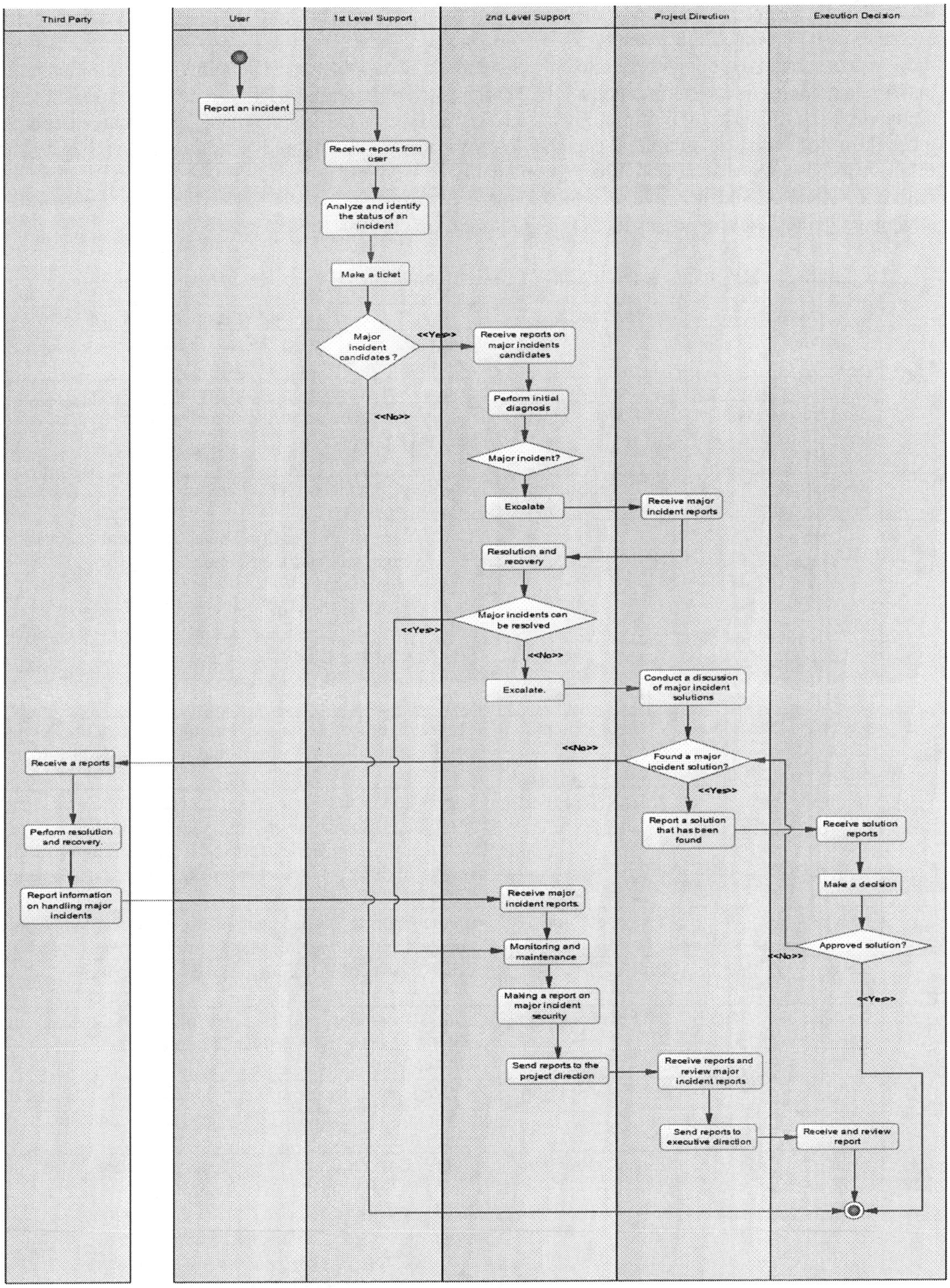


**Figure 2.** ERP Management Crisis of Major Candidates Service Incidents

Companies are realizing that in order to survive in a global business environment, organizations need to improve not only the efficiency of their organizations, but also their entire supply chains. This is because today's competition extends not only between companies, but also into the supply chain. For these reasons, many companies need to stay current and invest heavily in developing and implementing better technologies and systems, such as ERP systems [5]. Inherently, ERP systems are a useful tool for companies to build a strong information systems infrastructure and enable managers to make better and more informed decisions on time. It can also improve product quality and customer responsiveness, and improve information sharing and information quality between various departments within the company. Ultimately, this should improve overall business performance and help you achieve a competitive advantage in the global economy and increase long-term profitability [9].

**Table 3.** ERP Management Crisis of Major Candidates Service Incidents Explanation

| No. | Activity | Authority | Information |
|---|---|---|---|
| 1 | Reporting an incident | User | Report an incident to the Service Desk.<br>Reporting tools that are via ERP System |
| 2 | Receive reports | 1st Level Support | Service Desk receives incident reports from users. |
| 3 | Analyse and identify the status of an incident | 1st Level Support | Identify the status of an incident based on reports, tools, users or IT Support |
| 4 | Recording of incident tickets | 1st Level Support | Recording incident reports includes the following information:<br>a. a reporter's identity (name, organization, location, email address, telephone);<br>b. report source<br>c. a description of the signs / symptoms of the incident, time of occurrence and supporting evidence files.<br>Once an incident has been recorded, the status will change to on progress |
| 5 | Receive candidate reports on major reports | 2nd Level Support | Receive reports of major incident incidents by the Service Desk for immediate resolution and recovery by IT Support |
| 6 | Initial diagnosis | 2nd Level Support | Check whether there have been previous similar incidents / complaints (known errors) in the log (Knowledge Base). Preliminary diagnosis is expected to provide input for the overall handling of incidents and if possible can provide solutions to incidents |
| 7 | Escalate | 2nd Level Support | Provide reports regarding major incidents to Deputy Managers and Processing Management Services |
| 8 | Receive major incident reports | Project Direction | receive reports of major incidents by IT Support |
| 9 | Resolution and Recovery | 2nd Level Support | Perform resolution and recovery of major incidents. the resolution and recovery expected of major incidents can be resolved without the need to escalate |
| 10 | Escalate | 2nd Level Support | Provide reports that major incidents cannot be resolved by IT Support to Deputy Managers and Management of Service Management. |
| 11 | Conduct a discussion of major incident solutions | Project Direction | Hold discussions with the Deputy Manager with the support team to find the best solution related to handling major incidents.<br>Matters that need to be discussed are:<br>- The first way to handle major incidents<br>- the impact of handling it<br>- how long will it take to handle a major incident |

**Table 3.** ERP Management Crisis of Major Candidates Service Incidents Explanation

| No. | Activity | Authority | Information |
|---|---|---|---|
| 12 | Do the solutions that have been found | Project Direction | Provide reports on the discussion of solutions that have been found by IT Support, Deputy Management, and Processing Management Services related to major incidents that cannot be resolved by IT Support. |
| 13 | Receive a solution report | Executive Decision | Receive reports on major incident solutions that have been found by IT Support, Deputy Managers, and Processing Management Services. |
| 14 | Make decisions | Executive Decision | Make decisions by senior managers and STI division heads regarding solutions to handling major incidents. |
| 15 | Receive reports | Third Party | Receive reports about major incidents that IT Support cannot resolve |
| 16 | Make a resolution and recovery | Third Party | Perform resolution and recovery of major incidents. Resolution and recovery that are expected to be major incidents can be resolved |
| 17 | Report information on handling major incidents | Third Party | provide reports to IT Support that handling major incidents can be resolved |
| 18 | Receive major incident reports | 2nd Level Support | Receive reports from Third Party regarding major incidents that have been resolved |
| 19 | Monitoring and maintenance | 2nd Level Support | Monitoring and maintaining services related to major incidents that have been resolved so that services remain stable |
| 20 | Make a report on handling major incidents | 2nd Level Support | make a report regarding major incidents that have been resolved |
| 21 | Send reports to project directions | 2nd Level Support | Sending major incident reports that have been resolved to the Deputy Manager and Management of Service Management |
| 22 | Receive reports and review major incident reports | Project Direction | Submitting and reviewing reports that have been made by 2nd level of support as evaluation material so that the major issues of successful incidents at Resolved do not recur. |
| 23 | send report to executive decision | Project Direction | Sending major incident reports that have been resolved to the Senior Manager and STI Division Head |
| 24 | Receive report and review report | Executive Decision | Receive and review reports related to major incidents that have been resolved |

## 5. Conclusion

ERP systems can be a useful tool for companies to build a robust infrastructure for information systems and to enable management to make better decisions based on accurate and timely information. The main difference from existing crisis management and targeting after using an ERP system is the existence of process automation and integrating internal and external information flow within the organization that makes the process of handling incidents and problems more effective and efficient. In addition, each incident and problem can be categorized to be resolved according to a predetermined time to resolve so the company can resolve the problem and even prevent it to maintaining the quality of service to customers.

## 6. References

[1] S Saberi, 2016 An Incident Management System for Debt Collection in Virtual Banking," in *(IJACSA) International Journal of Advanced Computer Science and Applications*, 7(12).

[2] MA Abd Elmonem, ES Nasr and MH Geith, 2016 Benefits and challenges of cloud ERP systems – A systematic literature review, in *Future Computing and Informatics Journal*, **1**(2), pp. 1–9.

IOP Conf. Series: Materials Science and Engineering **847** (2020) 012081 doi:10.1088/1757-899X/847/1/012081

[3] PP Dewi, N Luh and P Asriani, 2019 Analisis Faktor-Faktor Kesuksesan Penerapan Enterprise Resource Planning (ERP) Pada Perusahaan Pengguna ERP Wilayah Bali, **5**(1), pp. 39–48.
[4] M Sarnovsky and J Surma, 2018 Predictive Models for Support of Incident Management Process in It Service Management, in *Acta Electrotechnica et Informatica*, **18**(1), pp. 57–62.
[5] AS Shatat, 2015 Critical success factors in enterprise resource planning system implementation: An exploratory study in Oman, *The Electronic Journal of Inf. Sys. Evaluation*, **18**(1), p. 36-45.
[6] N Karia, 2016 Enterprise Resource Planning (ERP). Systems in the Egyptian Higher Education Institutions: Benefits, Challenges and Issues, in *International Conference on Industrial Engineering and Operations Management*, pp. 1935–1943.
[7] A Badewi, 2018 ERP benefits capability framework : orchestration theory perspective.
[8] T Bhattacharya and P Chellasamy, 2016 An analysis of ERP security issues in ERP implementation process of Indian power distribution companies (Discoms), **2**(7), pp. 34–38.
[9] A Jayakody, EMMN Samaranayake, NNWG. Punchihewa and BVR. Jeewantha, 2019 Impact of Critical Success Factors on ERP Implementation: Typical Organizations in Sri Lankan Context, *Global Journal of Comp. Sci.* **19**(1).
[10] M. Lubis, TF Kusumasari and L Hakim, 2018 The Indonesia Public Information Disclosure Act (UU-KIP): Its Challenges and Responses, *Int. J. of Electrical and Computer Engineering*, **8**(1), pp. 94–103.
[11] M. Lubis and FA Maulana, 2010 Information and Electronic Law Effectiveness (UU-ITE) in Indonesia, *Proc. IEEE CITSM 2010.*
[12] RW Perry, 2003 Incident Management Systems in Disaster Management, *Disaster Prevention and Management* **12**(5): 405-412.
[13] M. Lubis and FA Maulana, 2010 Information and Electronic Law Effectiveness (UU-ITE) in Indonesia, *Proc. IEEE CITSM.*

# Discrete Event Simulation Modelling for Classifying the Container Yard Availability Considering Dock Unloading Activity

**M I Tama[1,2], N Siswanto[1,3], Suparno[1,4], and BD Aqsha[1,5]**

[1]Institut Teknologi Sepuluh Nopember, Surabaya, Indonesia
[2]indratama25@gmail.com, [3]siswanto@ie.its.ac.id, [4]suparno@ie.its.ac.id,
[5]bayudwiaqsha@gmail.com

**Abstract.** The development of systems and technology in the industry especially in the supply chain sector currently has a significant influence on the speed and accuracy along the supply chain. This is one of the reasons to build a container loading and unloading service company located in Surabaya. Terminal Teluk Lamong is the one and only Indonesian company that applies the green port concept with renewable loading and unloading equipment technology. One of the unique and prominent renewable concepts is the docking station, which aims to speed up unloading activities. Because the condition of container unloading activities is more common, this research focuses on container unloading activity only. Currently, the port has five yards available with 30 docking gates on the unloading side. The problem is the yard occupancy ratio (YOR) in unloading activity which has a low percentage value (27%). Even though the percentage value should be around 60%, therefore the value of yard utility didn't optimal yet. A discrete-event simulation (DES) aims to help classify which yards are suitable for unloading activities and the rest for loading activities. The model verification is obtained "no error" and model validation is less than 5%. For all scenarios conclude that the best scenario is to use three-yard containers with six gates of docking for each yard.
**Keywords**: Discrete-event Simulation, Docking Station, Container Yard.

## 1. Introduction

The development of systems and technology in the industry especially in the supply chain sector currently has a significant influence on the speed and accuracy along the supply chain. Throughout the supply chain the influence of various facilities has an important role in the value of the supply chain itself (value chain), one of which is warehouse. In the modern industrial warehouse used not only as a storage area, but as a place of consolidation of products to be distributed. This consolidation functions as a collection of products from several suppliers that will be distributed directly to various destinations, namely consumers. Consolidation does not only occur in the manufacturing industry but also in the service industry. Terminal Teluk Lamong is a new container loading and unloading service company in Surabaya. Founded in 2013-2014 with renewable loading and unloading equipment technology. However, based on the performance of service operations in container yards, the yard occupancy ratio (YOR) is relatively low (under 30%). Whereas the flow of containers in terminal from 2015 to 2018 continues to increase (Figure 1). The loading ports have their own daily demand rates, which vary from one month to another due to the seasonal cycle [3]. Unlike in [4] the flow of

IOP Conf. Series: Materials Science and Engineering **847** (2020) 012082 doi:10.1088/1757-899X/847/1/012082

containers in the terminal do not have any time windows. They may be loaded and unloaded any time because of complete facilities.

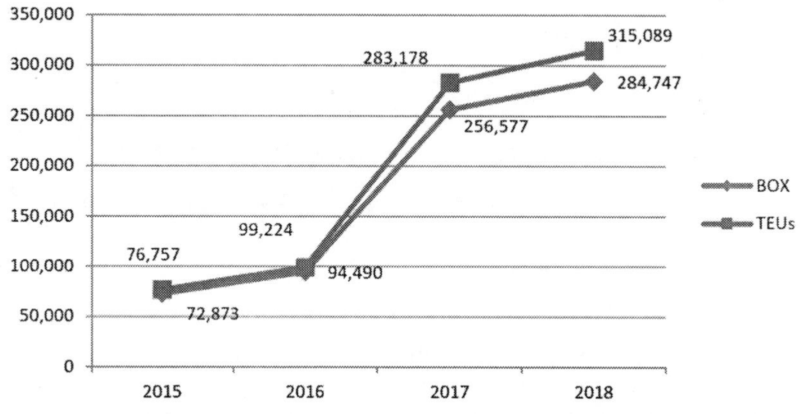


**Figure 1.** Domestic Container Flow (2015-2018)

This indicates the need for improved loading and unloading performance, one of which is the performance in the container yard itself. The purpose of this study is to create a loading and unloading service system model that uses the docking concept and determine the number of docking gates and the number of container yards used to accommodate containers coming from the dock. Operating performance in CY (container yard) is usually measured by the length of time it receives and delivery time, when the container enters the waterside until it exits to the landside. Currently, the port has five yards available with 30 docking gates. Therefore at the operating level, uncertainty is very high, such as ship arrival, truck arrival, box volume to be transported. As a result, adjustments are often made to the original layout. To improve efficiency, planning in the field must take into account uncertainties that occur during the operational level [2]. The problem is in the utility yard in unloading activity or yard occupancy ratio (YOR) which has a low percentage value. Even though the percentage value should be around 60%, therefore value of yard utility didn't optimal yet (Table 1). Level of stacking field usage or YOR containers, are the ratio of the number of container stacking field usage calculated in 1 box container per day or m² per day with capacity stacking available.

**Table 1.** Yard Occupancy Ratio In 2018

| Month | Unloading Service Time (Hour) | Total Box | Settles (Day) |
|---|---|---|---|
| January | 725 | 11748 | 42 |
| February | 712 | 13608 | 55 |
| March | 745 | 10308 | 35 |
| April | 758 | 12334 | 43 |
| May | 745 | 9627 | 43 |
| June | 708 | 6462 | 55 |
| July | 780 | 11492 | 43 |
| August | 725 | 11310 | 49 |
| September | 685 | 14366 | 42 |
| October | 743 | 14486 | 47 |
| November | 715 | 13632 | 47 |
| December | 739 | 14310 | 47 |
| | **Total** | **143683** | **548** |
| | **Yard Occupancy Ratio (%)** | **27,32** | |

## 2. Methods

For those matter we take the most lots in one month and simulate it with discrete-event simulation. In discrete event simulation (DES), the operation of a system is represented as a chronological sequence of events. Every event happens instantly in time and marks a change in status in the system. Discrete event simulation is an experimental approach that is often used; allows a high level of detail to be modelled because assumptions about buffer space, processing time distribution, or priority delivery can be modelled [1]. According to [5], creating a discrete event simulation model using the simulation software can analyze several material problems in the terminal. The problems faced are cargo transportation, and handling engine performance, also faces several problems regarding stockpiling of raw materials and production of materials.

For assumptions in this study include the number of docking gates used for simulation, there are 6 gates, no damage to loaded containers, no disruption in the dock area (ship/vessel) during the loading-unloading container process, and do not consider berths. In addition, the purpose of implementing docking only refers to the increase in the percentage of YOR (yard occupancy ratio). At this stage describes the problem conditions in container yards (CY). This is to make it easier to get the identification of problems that occur in container yards (CY). The following is a scheme of activities and conditions in the company (Figure 2).

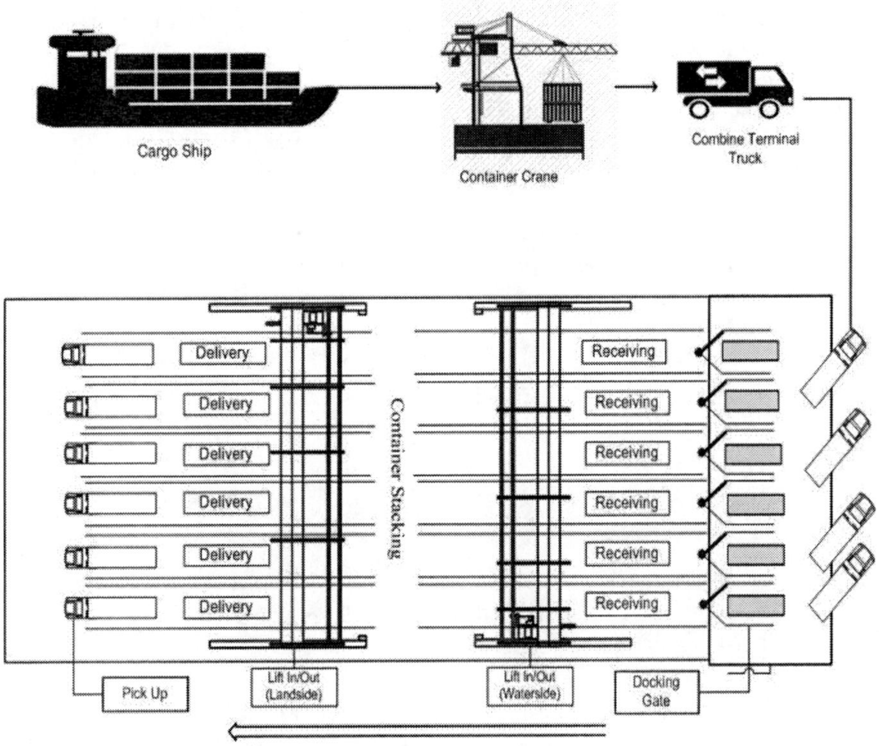


**Figure 2.** Unloading Activity In Docking Station

In the simulation stage, modelling must be done first before making the simulation model. In making models must draw a real system in the company. Before making a simulation model it is better to make a flowchart to help drawings of the actual system. Flowcharts describe the process flow to facilitate the making of simulations in software. The flowchart model will be created in simulation software to create a real simulation. The results of the model created must be able to interpret the actual results that exist in the company. Conceptual models in existing condition can be interpreted in flow logic diagram or flowchart. There are three sub-models for describe existing condition in systems. The first conceptual sub-model for cargo arrival can be seen in the figure 3.

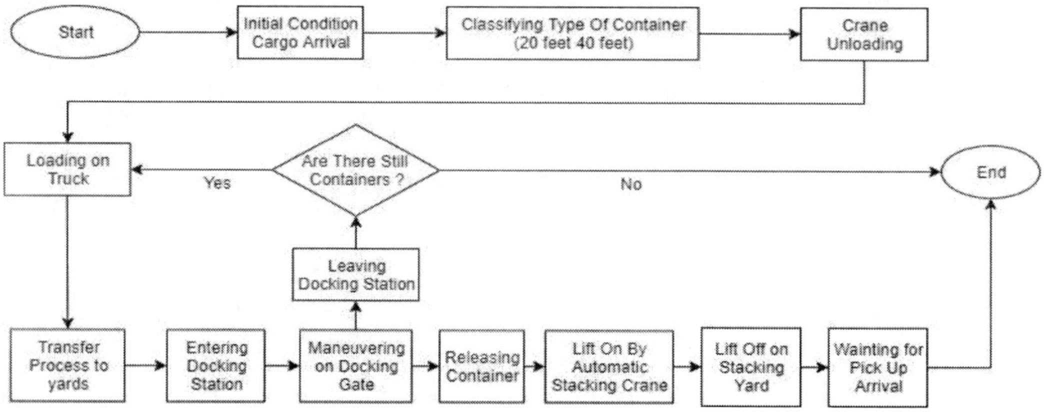


**Figure 3.** Flowchart for Cargo Arrival

The scheme above (Figure 3) explains the arrival of the container or the ship's cargo will enter the dock according to the data fitting that has been done, then the container will be separated based on the type, namely 20 feet and 40 feet. Furthermore, containers enter the pier and will be processed by crane or can be called ship to shore (STS) with working time based on the existing distribution. There are 5 STS that work for each container arrival. Then the container will be loaded into combine terminal truck (CTT). This CTT has a capacity of 40 feet for one load and if the container of 20 feet type is entered it will first be batched 2 times. CTT will go directly to container yard (CY) for the stacking process. This CTT has a speed specification of 20 km / h which will be converted to a model of approximately 333 meters / minute, and has a distance from the pier to container yard as far as 1.51 km or 1515 meters. The second sub-model for choosing yard considering the yard's capacity can be seen in figure 4.

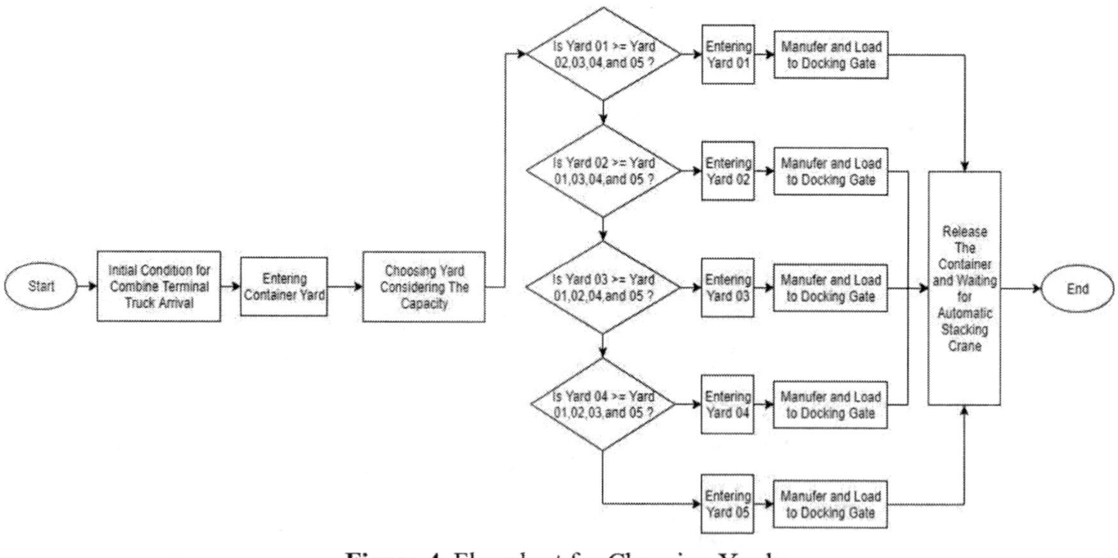


**Figure 4.** Flowchart for Choosing Yard

The scheme above (Figure 4) explains CTT enters the yard, before entering the yard CTT have to choose the biggest ammount yard's capacity. After that condition, CTT must maneuver to place containers into the docking gate. This manoeuvring process requires the fastest time of 4-5 minutes and the longest until 10 minutes. After the CTT has finished unloading the docking station, the CTT will return to the dock for the next unloading process and if the number of containers runs out the CTT will automatically stop at the docking station (buffer area). Containers included in the docking station will be separated for the purpose of separating 20 feet container types. Furthermore, automatic stacking cranes (ASC) will pick up containers at a speed of 60 meters / minute with a distance of 250 meters to be stacked into a stacking field created with an assign module for capacity updates. ASC is only able to process or carry only one container box. After that the container which is piled up will wait for pickup truck for the delivery process. The last sub-model for delivery process can be seen in figure 5.

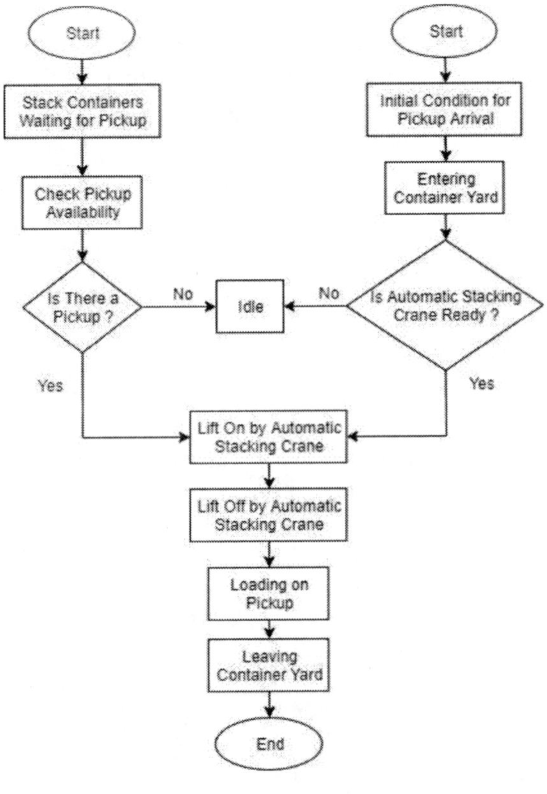


**Figure 5.** Flowchart for Pickup Arrival

Figure 5 explain the pickups that come are based on the deposition time for each container that is already in the stacking yard. The minimum deposit time for each container is 4 minutes and the longest is up to 14-15 minutes. When a pickup arrives, it triggers a signal for container pickup and yard capacity will be updated to lose the container. The decide module is used to check the available capacity in each yard, lost containers based on the highest stacking capacity in the yard which will be reduced by picking up containers in the yard. The outgoing containers are then processed again by the same automatic stacking crane (ASC) with the same specifications for transportation at the docking station at a speed of 60 meters / minute and a distance of 250 meters. After being processed by ASC the container will be automatically exited or disposed.

IOP Conf. Series: Materials Science and Engineering **847** (2020) 012082 doi:10.1088/1757-899X/847/1/012082

## 3. Results and Discussion

After making the model in the simulation software, then the results can be verified whether an error occurs to ensure the model runs as desired. Model verification which interprets the results of "*no error*". The next step is to validate the model. This validation aims to see the suitability of the actual model with the simulation model that has been made. This validation compares the number of boxes, service time, and yard occupancy ratio. After the model has been verified, the model needs to be validated to see whether the model matches the real system in the field. This validation process involves output parameters such as the number of container boxes entered, service time, and the percentage of yard occupancy ratio (YOR). The actual output that has been calculated and the simulation output that comes out will be tested by t-Test: Paired Two Sample for Means in Excel which aims to see whether the t-stat value exceeds or less than the t-critical value of two tails. The validation results can be seen in (Table 2).

**Table 2.** Results for Validation Test

| Validation Parameter | t-stat | t-critical two tail | | Conclusion |
|---|---|---|---|---|
| | | Lower limit | Upper limit | |
| Total Box | 0,177 | -2,069 | 2,069 | Valid |
| Service Time | -0,102 | -2,069 | 2,069 | Valid |
| Yard Utility | -2,241 | -2,776 | 2,776 | Valid |

Making improvement scenarios aims to provide priority choices with different parameters. In the repair simulation model, each scenario has a number of different docking gates and a number of yards. There is a combination of the number of docking with the number of yards used. The combination of docking there are 4 types and the number of yards there are 3 types, further creating 12 existing scenarios and for each scenario will be replicated 3 times to take the average results of the scenario. For more details scenario and the results can be seen in (Table 3).

**Table 3.** Scenario Test and Results

| Scenario | Yard | Dock | Total Box | Service Time (Hours) | Utility (%) | |
|---|---|---|---|---|---|---|
| | | | | | Dock | Yard |
| A1 | 3 | 3 | 17411 | 1567,67 | 0,17 | 39,42 |
| A2 | 3 | 4 | 15832 | 1483,33 | 0,11 | 37,05 |
| A3 | 3 | 5 | 14538 | 1456,00 | 0,09 | 36,47 |
| A4 | 3 | 6 | 13317 | 1335,67 | 0,06 | 35,30 |
| B1 | 4 | 3 | 16600 | 1512,33 | 0,12 | 34,31 |
| B2 | 4 | 4 | 14533 | 1439,00 | 0,08 | 32,06 |
| B3 | 4 | 5 | 15079 | 1457,33 | 0,07 | 32,47 |
| B4 | 4 | 6 | 13460 | 1398,00 | 0,05 | 31,33 |
| C1 | 5 | 3 | 16974 | 1570,00 | 0,10 | 29,04 |
| C2 | 5 | 4 | 13688 | 1421,67 | 0,06 | 27,05 |
| C3 | 5 | 5 | 14609 | 1448,00 | 0,05 | 27,37 |
| C4 | 5 | 6 | 15857 | 1493,67 | 0,05 | 26,51 |

Scenario selection based on utility parameters and service time, therefore interpreted each parameter to compare which scenario is best. The parameters ranging from total box, service time, docking utility, and yard occupancy ratio that have been tested also have a difference from the output value. The results of each parameter interpretation can be explained as follows:

### *3.1 Scenario of Total Box*

In (Figure 6) the addition of a docking gate to each yard will affect the arrival of the container box to be processed. As more yards are added, container reception will decrease. That is because the length of the selection of yards based on existing capacity. Besides the arrival of the container box will also affect the service time or unloading work time. If more and more total container boxes arrive, the loading time will also be longer.

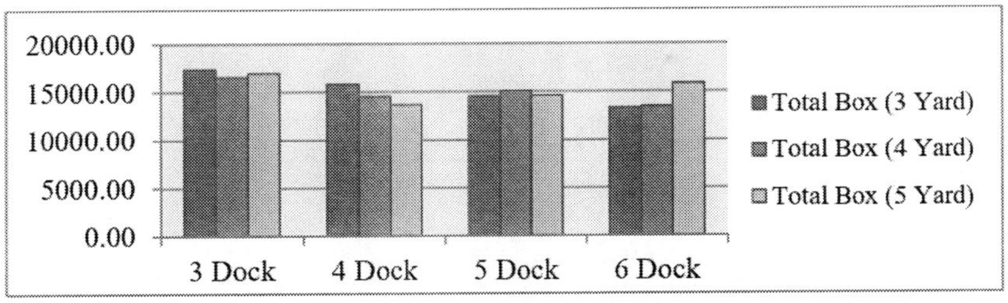


**Figure 6.** Total Box Comparison Chart

### *3.2 Scenario of Service Time*

In (Figure 7), the graph for the addition of docking gates and yards affect the length of unloading work (hours). The more docking gates and the number of yards that will be used, the service time for loading activities will decrease. No less significant difference is also influenced by the arrival of container boxes as well. The same example also applies to the use of 6 docking gates with 5 yards, where service time increases due to the arrival of container boxes also increase under these conditions.

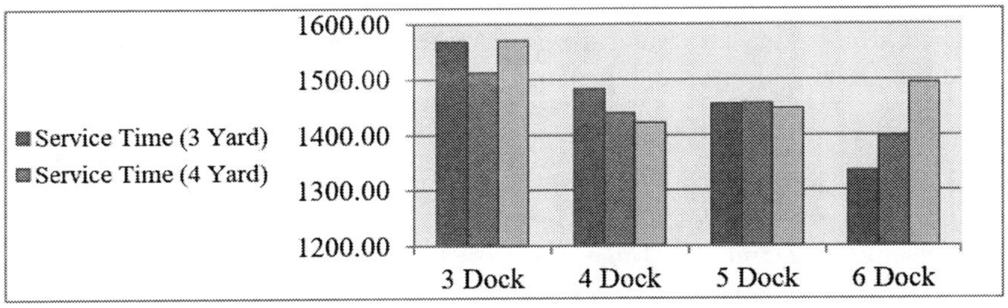


**Figure 7.** Service Time Comparison Chart

### 3.3 *Scenario of Dock Utility*

In (Figure 8), the picture shows the use of docking gate facilities will affect the percentage value of the facility as well. The less docking gate facilities are used, the higher the percentage of utility will be. That is because the docking gate facility often accepts incoming boxes. In addition, container box receipts also influence the percentage value of utilities where more box containers come, the value of docking gate utilities also increases.

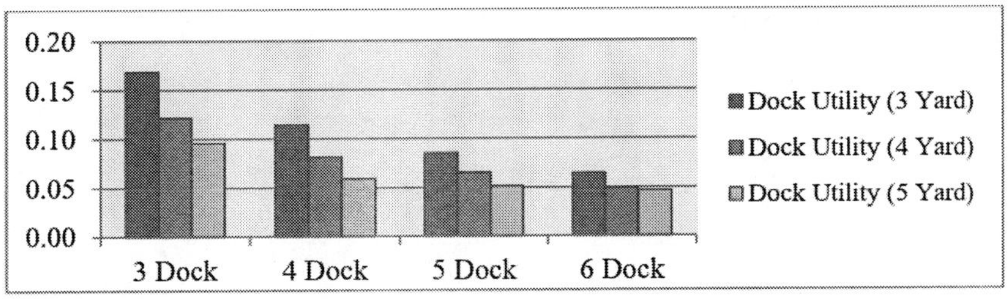


**Figure 8.** Dock Utility Comparison Chart

### 3.4 *Scenario of Yard Occupancy Ratio*

In (Figure 9), based on the yard occupancy ratio graph at below, it can be concluded that the more yards used, the value of utility percentages will decrease. The optimal use of yards based on the graph above or only on the use of 3 yards has the highest utility value.

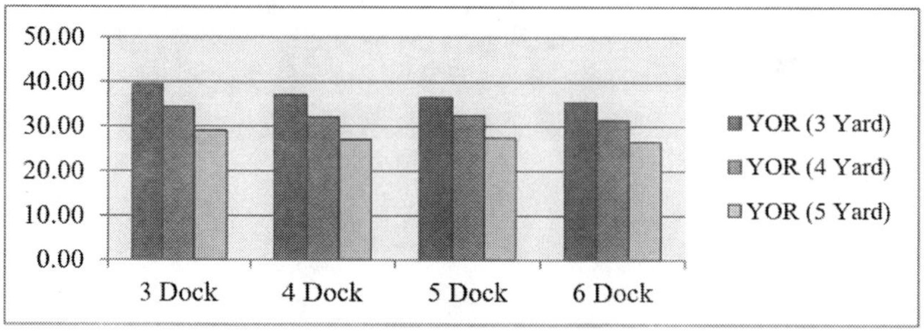


**Figure 9.** Yard Occupancy Ratio Comparison Chart

## 4. Conclussion

From the scenario results obtained for each addition of the docking gate will affect the service of container box reception and loading time. The calculation result is obtained the optimal number of uses by using 4 docking gate facilities in each yard. Based on the results of the scenario that has been made, the use of yards does not have to be all required yards. Scenario A4 has a high yard utility with low service time. This explains the use of yards which used to be 5 will be reduced to 3 yards with the use of docking gates amounting to 6. The consideration of selecting the above scenario is based on 2 important parameters, namely utility and service time. Although Scenario A1 has the highest utility, consideration of service time results has suboptimal results. So scenario A4 has the best output compared to scenario A1. Here describes the work time (service time) will affect the waiting time or service time for the process of receiving containers.

## 5. References

[1] Dewa, M, and Chidzuu. L., 2013. *Managing Bottlenecks In Manual Automobile Assembly Systems Using Discrete Event Simulation. The South African Journal of Industrial Engineering,* **24(2)**, pp. 155-166.

[2] Nurminarsih, S., Rusdiansyah A., and Maulin, M.P., 2018. *Space-Sharing Strategy for Building Dynamic Container Yard Storage Considering Uncertainty on Number of Incoming Containers. Journal Industrial Engineering,* **19(2)**, pp. 67-74.

[3] Siswanto, N., Kurniawati, U., Latiffianti, E., Rusdiansyah, A., and Sarker. R., 2018. *A Simulation study of sea transport based fertilizer product considering disruptive supply and congestion problems. Asian Journal of Shipping and Logistics*, **34(4)**, pp. 269–278.

[4] Siswanto, N., Wiratno, S.E., Rusdiansyah, A., and Sarker. R., 2019. *Maritime inventory routing problem with multiple time windows, J. Ind. Manag. Optim.*, **15(3)**, pp. 1185–1211.

[5] Yuan, Z., Peilin, Z., and Chuanbo, Y., 2010. *Research on the Simulation of Industry Port Raw Material Terminal. International Conference on Computer Design and Applications (ICCDA),* **5(1)**, pp. 108-133.

# Lean Startup Approach on Product Design and Manufacture Facility Planning in Uncertain Business Climate

**W S Dewobroto[1] and I A Marie[2]**

[1]Entrepreneurship Departement, Podomoro University, Jakarta, Indonesia
[2]Industrial Engineering Departement, Trisakti University, Jakarta, Indonesia
[1]wisnu.dewobroto@podomorouniversity.ac.id, [2] iveline.annemarie@trisakti.ac.id

Abstract– Successful companies can provide the products that suit to consumer desire. There are two approaches in the product manufacturing process which are Product Development Process and Lean Startup. This study aims to demonstrate the conception of Lean Startup and its application in product design particularly in uncertain business climate. After comparing the two methods, Lean Startup approach is appropriate since the validation process in each product development process will certainly increase the possibility of products that suit to the consumer desires and reduce the time needed starting from product design to the market. In order to support Lean concept in the whole processes, there are two techniques that might be applied to support lean operational system, which are the formation of one worker multiple machine (OWMM) cells and Group Technology (GT) cells.
**Keywords**: Facility Planning, Flexible Business Model, Lean Startup, Product Design

## 1. Introduction

A company is categorized as successful if it performs the ability to provide products in accordance with customer and consumer desire. In sustainable design process, there is a process known as Product Development Process that focuses on the needs of users or consumers. As business conditions continue to change, the Lean Startup approach adopted from the manufacturing approach has evolved and now it is applied to every process in management and business, even new products. In terms of product manufacturing, the Lean Startup approach emphasizes the process of building a business from a product, and there is no fundamental detail regarding the product technical sides.

In the development of engineering and business science, many things have evolved following consumer needs, national and global industries are adapting with current development. The development of manufacturing system (extended manufacturing system) directs the actors in the manufacturing system to move forward, make continuous improvements, adjust to the technological advancement and the needs of modern industrial societies, and see the company as part of the other companies' supply chain.

This article will mainly discuss about the product design and manufacturing process by comparing the two approaches; Product Development Process and Lean Startup. Then, the discussion continues with the argumentation of the most suitable method, especially in uncertain business conditions, followed by an impact on the design of facility. Because it is known that product design is influenced by aesthetic, functional, and material considerations and impacts on the manufacturing process to meet the values needed by consumers. Facility planning will determine how the activities of production facilities in an industry can be arranged in such way so that it can support the efforts to achieve the main objectives effectively and efficiently.

## 2. Research Methodology
This article aims to explain Lean Startup method and its application in Product Design. Then, facility planning techniques will be obtained to support the Lean Startup approach in the Product Design process. The theory references obtained become the basis and main tool in this study.

## 3. Result and Discussion
Product design is the whole activities of product manufacturing started from observing market potential to producing, selling, and delivering the products to customers. The good product design characteristics are as following:

1. Prioritizing the quality measured from the products that suit to the consumer needs and availability to buy the products.
2. Product manufacturing costs that meet the product **value**. Product manufacturing costs will affect to the selling price and product profit.
3. Calculating a product quality dimensions which include [2]: Performance, Durability, Conformance to Specification, Features, Reliability, Aesthetics, and Perceived Quality.

In the process of gaining a good product, there will be two approaches discussed; Product Development Process and Lean Startup.

### *3.1 Product Development Process*
Product development process is generally divided into some phases:

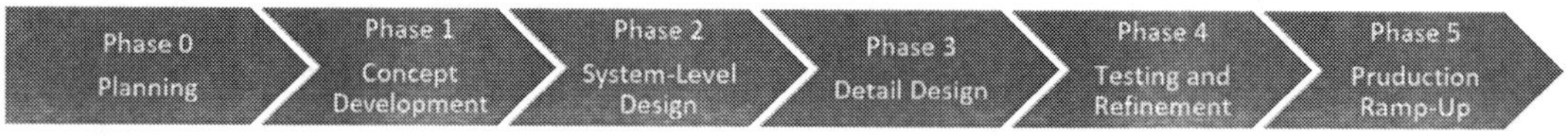


Figure 1. Product Development Process [12]

The process begins with a planning phase related to technology advancement and advanced research activities. The output of planning phase is project mission statement, which is the input needed to start the concept development phase and a guide for each development. The completion of product development process is the product launching sent to the customers later.

Product development process consists of some significant activities, namely:

1. Identification of consumer need. This activity is designated to understand the consumer needs and communicate them effectively to the development team.
2. Target specification. Specification is the translation of consumer needs into technical requirements.
3. Concept drafting. The product concept is the illustration or estimation of technology, working principles, and product form.
4. Concept selection. Concept selection is an activity in which various concepts are analysed successively, then eliminated to identify the most promising concepts.
5. Concept testing. One or more concepts are tested to find out whether the needs of consumers have been recognized, estimate the market potential of the product, and identify some weaknesses that must be fixed during the next development process.
6. Final specification decision. The target specifications set at the beginning of the process are reviewed after the process is selected and tested.
7. Project planning. At the end of this concept development, the team formulates a detailed development schedule, sets strategies to minimize development time, and identifies resources used to finalize the project.

8. Economic analysis. Economic analysis is used to ensure the continuation of the overall development program and to solve specific bargaining, for example between manufacturing costs and development costs.
9. Analysis of competing products. An understanding of the competing products is important for successful new product positioning and can be a rich source of ideas for product design and production processes.
10. Prototype making and modelling. Each stage in the concept development process involves many forms of models and prototypes. This includes, among others, proof of concept models that will assist the development team in demonstrating the feasibility of the model shown to the consumers to evaluate their ergonomics and style, while the worksheet model is for technical choices.

### 3.2. *Lean Startup*

In principle, the Lean Startup process trains to reduce waste by increasing the contact frequency with customers, to the extent that testing can be done, invalid market assumptions can be avoided as early as possible, and finally a valid product can be acquired. The initial validation phase begins with the hypothesis of who the prospective customers are, what they need, how many, and so on. The consumer hypothesis is then validated directly to the potential customers to determine whether the product is suitable to what is desired [1].

In a business, Lean Startup is the method that minimizes risk, especially by relying on iteration (repetitive steps) from product to market to find quality feedback as quickly as possible and as often as possible from the market [1].

Another important thing in the Lean Startup method is Validated Learning which can be defined as the process of discovering relevant facts related to the design of a product. This process is initiated by using assumptions / hypotheses which are later examined by field experiment to obtain responses from the prospective customers whether the hypothesis is correct.

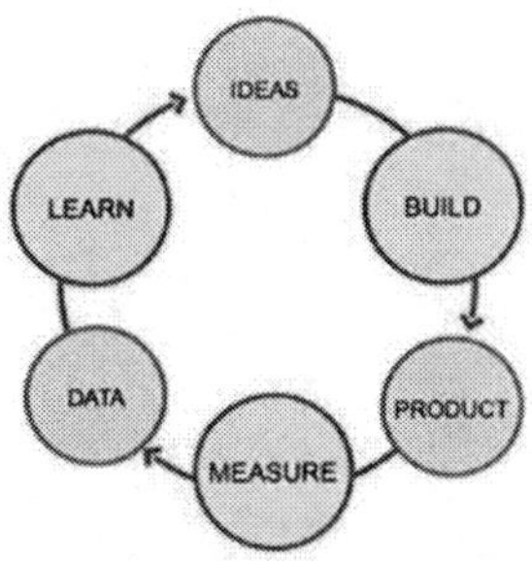


Figure 2. Lean Startup main process

The main cycle or process in Lean Startup is Build Measure and Learn. The purpose of Build-Measure-Learn is to provide the facts needed in Validate Learning. The process of Build-Measure-Learn is represented by these three repetitive phases as demonstrated below:

- Build: Create products based on the main hypotheses that have been examined with Validated Learning. The first product made is Minimum Viable Product ("MVP"). MVP is a minimum product that only has core features to further examine some hypotheses that have not been validated.
- Measure: Collect the responses, suggestions, input, and feedback from MVP users and measure the results obtained with the aim of acquiring knowledge related to the tested hypothesis.
- Learn: Make conclusions from the Measure process results to find the validation of tested hypothesis. The results from Learn stage are further used as the basis for product or strategy improvements or changes.

The conclusion drawn from this process also decides whether the Startup Founder must "Persevere" or "Pivot". Persevere is continuing the development process with the same strategy or product, while Pivot is making changes or replacing some parts or the whole strategy or product.

### *3.3. The Comparison of Product Development and Lean Startup*

From the discussion of the two methods in product design process, the main similarities and differences in the process can be found below:

- **Focus on Consumers**. In term of consumer orientation, both methods have similarities. Users or consumers are important and must be ensured that the process includes consumer need identification in which it is hidden and unspoken like the explicit ones. Later, the identification becomes the basis for compiling product specifications and developed by the development team.
- **Design Process**. When compared to the two methods, the product development process looks meticulous and is done in more static context [12]. The process is started from market analysis, resources needed, and ended with prototypes. The iteration or validation process is accomplished after System-Level Design.

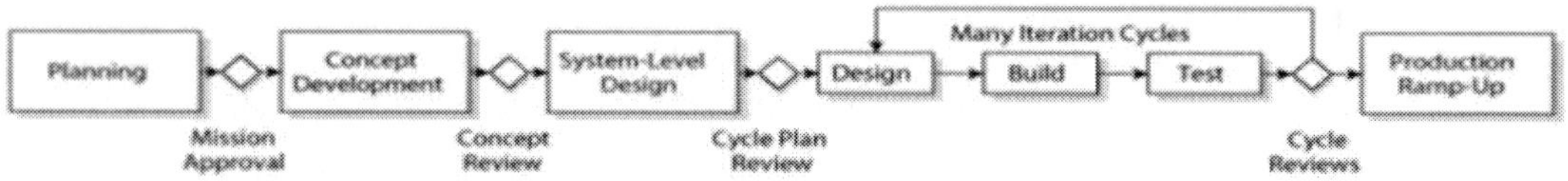


Figure 3. Product Development Process

Meanwhile, Lean Startup process is executed by the time of planning and forming product Value as shown in Figure 4. The iteration is performed to the final stage quickly.

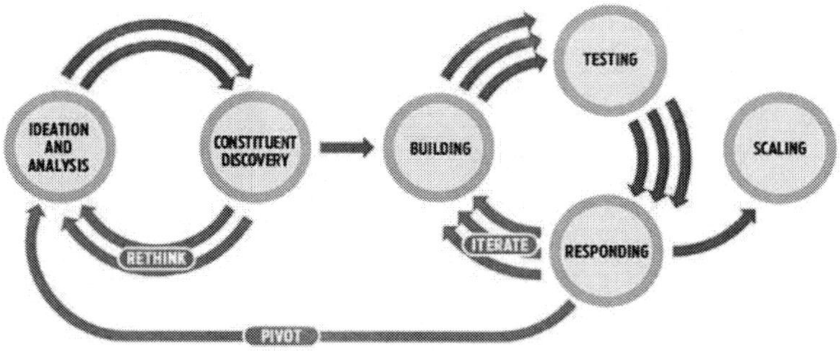


Figure 4. Iteration Process in Lean Startup

Changes (Pivot) can occur even during the planning stage to the end hence a valid product will be certainly obtained at the end of the process. In other words, Lean Startup performs the process of development and implementation in parallel, while in Product Development Process, the implementation is executed after the design process is complete. When this is applied to companies, especially in developing new products, it will certainly increase the possibility of products that suit to consumer desires and also reduce the time needed from product design to the market. But good communication is necessary in every part of the company, especially between designers and product development teams who may have different perspectives and thinking.

### *3.4. An Uncertain Business Climate*

A successful company is indicated by its ability to control all forms of uncertainty into competitive advantages. It is undeniable that many companies expect a deterministic environment, such as constant level of demand, delivery time from fixed suppliers, low price volatility of raw materials, small

variability in production processes, and so forth. However, uncertainty is inevitable and the company must live with it. Companies must have provisions to deal with uncertain conditions.
Uncertainty in the business world can occur because:

- Consumer demand rapidly changes both in volume and design variety
- Design and manufacturing capabilities of suppliers are always changing
- The use of new and various technology platform in industries
- Competitor behaviour

In this case, flexibility turns out to be important and interpreted as the company's ability to overcome the overall uncertainties precisely and quickly [3]. In this case, there are three components that must be assured to create a more-resistant-to-uncertainty company.

1. **Product Development Flexibility**. Companies must have the competence and capability in developing flexible products. In a highly competitive market characterized by increasingly shorter product life cycles, the company must be flexible in modifying product designs or even producing new products altogether. In this case, a fast and valid product development process is needed. From this, it can be said that the application of Lean Startup approach is appropriate for product development process when business conditions are full of uncertainties. With the validation in each process, it can be assured that each process has valid results or output.

   It is known before the Learn Startup process, Value Proposition needs to be made as a basis for validation of each product design process. Value Proposition is a method used in deciding how far the product or service offered has a high value according to its target customers (Osterwalder and Pigneur, 2010). In other words, it is about how a company can offer products or services that are different from its competitors. Value Proposition has two focuses, customer segments and value.

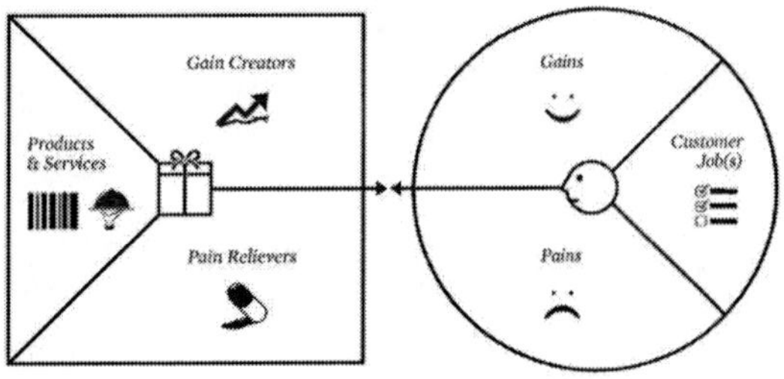


Figure 5. ***Value Proposition Canvas***

   Value Proposition Canvas is a tool that will describe Customer Profile and Value map in detail. On the Customer profile, market segments can be specifically seen and thus the target market and the needs or expectations of target market will be more detailed. Customer Profile consists of three parts including customer gain, customer pain, and customer jobs.

   Besides, Value Map illustrates in detail the value of a product from the consumer's point of view. Value Map carries three parts, which are products and services, pain relievers, and gains creators.

   The Value Proposition Canvas shows the details of the compatibility between two sides (Consumer and Value) so that the pre-launch product holds a value that meets the consumer desires and finds out solutions to the consumers' problems. The three adjustments are Problem-Solution Fit, Product-Market Fit, and Business Model Fit.

2. **Manufacturing Flexibility**. Companies are required to be flexible in producing various products (Volume flexibility) and several types of products at once (mix flexibility). The challenge companies must face is the fact that flexibility in producing goods is not something easy. A company is said to be flexible if it has a flat average cost curve, meaning the companies can produce the products in different quantities with the same average cost. Nevertheless, not as it seems, producing a number of products per month will be cheaper than producing a number of fluctuating items. Flexibility is not unlimited. The flexibility to produce more products, to a certain extent, can be

achieved through a number of ways, such as machine flexibility, labour flexibility, routing flexibility, and material handling flexibility.
The challenges faced by the manufacturing system are not merely about consumer demand change, equipment design development, and global competition, but now also increasingly need to take into account the lack of future energy resources and environmental responsibility. Manufacturing companies are demanded to have high flexibility to deal with the current phenomena. Lean Startup as an approach in product design, followed by the application of lean operating systems and lean thinking processes in every operational activity of business organizations or companies, has become a necessity to continuously excel in global competition.

3. **Supply chain flexibility.** In a strong supply chain, materials and information run smoothly without obstacles, thus all forms of waste can be eliminated. Besides the aspect of efficiency, flexibility in the supply chain is required to meet all consumer demands on time. Changes in products, services, customers, markets, and business practices potentially direct the changes in industrial supply chain network facilities. In supply chain management approach, the company tries to regulate the flow of information via supply chain to obtain synchronization level so that the company can better respond to customer needs while minimizing costs. The facility must assist business organizations to achieve supply chain excellence with the objective of increasing ROA, increasing customer satisfaction, reducing costs, and performing integrated supply chains. All facilities in the supply chain must be characterized with flexibility, modularity, upgradability, adaptability, and selective operability [11].

From the three components mentioned above, companies today must have a flexible business model. In other words, in addition to flexible product development, a flexible business model is needed to address the business climate and uncertain product demand [6]. The intended business model is those which concern on the network power pertaining products, manufacturing, and supply chain.

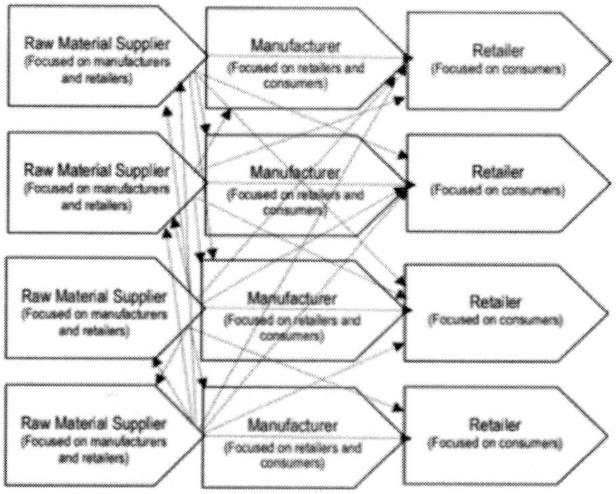


Figure 6. Flexible Business Model (Network focused approach)

*3.5. Facility Design Techniques to support Flexibility in Product Development, Manufacturing, and Supply Chain.*

Facing the changing and developing era, shorter cycle, rapid information technology advancement, and consumer trends disruption will affect the facility design layout on a production floor. Facilities Planning is the process of analysing, conceptualizing, designing, and actualizing a system for the manufacture of goods and services, generally represented as floor plan, or physical facility arrangement which includes: equipment, land, buildings and other means to optimize relations between implementers, the flow of goods, the flow of information, and the procedures needed to achieve goals efficiently, economically, and safely.

Facilities Planning scholarship continues to grow with the addition of quantitative approaches and the principle optimization which are also supported by the availability of layout and simulation software packages used to acquire facility layout design results that can optimize industrial activities. The facility planning process will be strongly influenced by the strategic business plan and concepts, techniques and technology to be the factor of consideration in manufacturing strategy and assembly.

It is known that the outputs from product design stages, design process, and design schedules become the input in the design of a facility. Product design decisions, processes, schedules and facilities are made in a sequential and mutually influential manner. A clear vision is needed regarding what to do and how to do it (including concepts, techniques and technological considerations). In most cases, changes in product design, design process, and design schedule of manufacturing facilities will require modifications in layout, control, and storage facility.

Related to product design activities, a facility planner must be able to anticipate the level of uncertainty that will occur, consider the mission of planned facility, implement specific activities, and direct related activities. By considering the type of product that will be produced by the facility, the business philosophy has to consider facilities and related external factors, such as the economy, availability of labour, the level of competition, and other factors that may cause changes. If it is decided that the facilities can accommodate changes related to the use of facility and its mission, then a design with a high degree of flexibility and a very general area are needed. Conversely, if the product has high level of trust, the facility can be designed to optimize the production [11]

As the need for more competitive design facilities among manufacturing companies becomes stricter and borderless (global), every business organization must be able to create various strategic efforts to respond to customer value needs by reducing waste and making creative and innovative efforts to get smooth and efficient flow of material and information in the value chain. On the other hand, customer needs are increasingly varied and unique, while manufacturing equipment is increasingly automated and has integrated architectural design. It will be a significant challenge for manufacturing companies to adjust their facility design to these changes.

There are two techniques that might be applied to support lean operational systems, namely the formation of one worker multiple machine (OWMM) cells and Group Technology (GT) cells.

Figure 7. One Worker Multiple Machine (OWMM) Layout

Figure 7. illustrates five OWMM cell machines used by 1 operator in the middle (U-shape is the most common form). Operators work by moving around in a circular area. In this layout, different parts or products can be made by changing the engine setup.

Furthermore, the facility layout based on product group or product family or group technology layout is a layout based on the grouping of products or components that are made, in this case the grouping is not based on the similarity of the final product types but it is grouped based on the processing steps, shapes, machines, or equipment used. In brief, the technology layout group is a combination of fixed and process layout. Group technology is a method for managing similar entities (components, processes, tools, etc.) to reduce manufacturing complexity by dividing manufacturing systems into sub-manufacturing cells.

## 4. Conclusion and Suggestion for Further Research

From the comparison results between two approaches in product design, it can be concluded that:

- Both Product Development Process approaches Lean Startup approach equally orient to consumers as product users.
- To become a successful company that is able to control all forms of uncertainty to continue having a competitive advantage, it requires a flexible, fast, and valid product development process. Hence, Lean Startup approach is appropriate for the product development process when business conditions are full of uncertainties. Involving validation in each process, valid results or output will be certainly acquired.
- Lean Startup as an approach in product design, followed by the application of lean operating systems and lean thinking processes in every operational activity of business organizations or companies has become a necessity that must be done to be able to continue to excel in competition. There are two techniques that might be applied to support lean operational systems, which are the formation of one worker multiple machine (OWMM) cells and Group Technology (GT) cells.

## 5. References

[1] Blank, Steve and Bob Dorf 2012 *The Startup Owner's Manual.* (California: K&S Ranch, Inc.)
[2] Chen, J 2015 The Role of Reward System in Product Innovations: an Examination of New Product Development Projects, *Project Management Journal.*
[3] Febransyah, A 2009 *Menikmati Ketidakpastian – Tahapan Kritis dalam mendesain kesuksesan,* (Jakarta: PT Gramedia Pustaka Utama).
[4] Henne, C 2012 The Industrial Designer's role in Applying Lean Startup principles in IT Startups, *Journal of Product Design*, Norwegian University of Science and Technology.
[5] Krajewski, L.J., Ritzman, L.P., and Malhotra, M.K. 2013 *Operation Management: Processes and Supply Chains*, 10th Ed., Pearson Education Limited.
[6] Katy J. Mason, Steganos Mouzas 2012 *Flexible Business Models*. European Journal of Marketing.
[7] Lean Production for Competitive Advantage 2011 A Comprehensive Guide to Lean Methodologies and Management Practices. John Nicholas. Taylor and Francis Group. USA.
[8] Osterwalder, Alexander and Pigneur, Yves 2010 *Business Model Generation: A Handbook for Visionaries, Game Changers, and Challengers. New Jersey: John Wiley & Sons, Inc.*
[9] Osterwalder, Alexander, Pigneur, Yves, et al. 2014 *Value Proposition Design. New Jersey: John Wiley & Sons, Inc.*
[10] Ries, Eric. 2011. The Lean Startup: How Today's Entrepreneurs Use Continuous Innovation to Create Radically Successful Businesses. The New York Times Bestseller.
[11] Tompkins, J.A., White, J.A., Bozer, Y.A., and Tanchoco, J.M.A. 2003 *Facilities Planning*. 3rd Ed., John Wiley & Sons.
[12] Ulrich, K and Eppinger D, 2012 *Product Design and Development,* McGraw Hill.

IOP Conf. Series: Materials Science and Engineering **847** (2020) 012084 doi:10.1088/1757-899X/847/1/012084

# Design of Mobile and Integrated Tyre Repair Tools for Motorcycle

**T Eben Haezar[1,2], W Samdan[1,3], A Yudha[1,4] and S Taufik Roni[1,5]**

[1]Industrial Engineering Department, BINUS Graduate Program – Master of Industrial Engineering, Bina Nusantara University, Jakarta, Indonesia
[2]eben.tobing@binus.ac.id,[3]samdan.wibawa@binus.ac.id,
[4]yudha.aprilianto@binus.ac.id, [5]taufik@binus.edu

**Abstract-** Tyres are a very important component of a motorcycle. Not only vehicle driving components, tyres also play a role in safety, passenger comfort, fuel consumption efficiency, endurance components of car legs and so forth. So, the condition of the tyre must receive special attention from the owner of a motorized vehicle. When riding on a motorcycle, tyre leakage is one of the problems that is often encountered. Not only because of sharp objects, tyres can also leak due to several other reasons. If problems are not immediately sought, leaks in tyres can cause loss of control, endangering the safety of motorists and other road users. And the place and timing of leakage events that cannot be predicted are often a scourge for motorists. After a leak, motorists, especially motorcycle riders, are very dependent on the existence of patched tyre entrepreneurs. With helps from IoT application based in advance, the design of this tool will be an opportunity for tyre patch entrepreneurs to be able to "Pick Up the Ball" to every customer / driver who has a leak. The design of this tool is quite concise and integrated thru Programmable Logic Control (PLC) with each main component making easier for entrepreneurs for patching tyres and can reduce downtime repair / replacement of the tyre.
**Keywords:** Tyres, Automatic Controller, Market opportunity, Motorcycle, Repair.

## 1. INTRODUCTION

### *1.1 Market Analysis*

The Indonesian motorcycles industry which providing primary private transport sector for people and goods, is currently world's top three for the last two decades and six time bigger in size compared to any corelating industries. The industry hits its peak approximately 10 years ago with sales record at 8 million units that year. In 2017 the industry experienced progressive declining sales and hits the lowest record ever with only 5.8 million units sold. In 2018, this industry has showed a positive trend with 8.4% rise in sales with over 6 million units sold at the end of the year, fueled by a quite positive economic trend [1].

Based on DKI Jakarta Transportation Statistics data for 2018, passenger cars recorded the highest growth of 6.48% per year in the 2012-2016 period. In 2012 the number of passenger cars in Jakarta was 2.74 million units while in 2016 it increased to 3.52 million units. If it is assumed that the growth of passenger cars is still the same, the number of passenger cars in Jakarta in 2017 will reach 3.75 million units and 2018 will be 3.99 million units. Based on DKI Jakarta Transportation Statistics data for 2018, passenger cars recorded the highest growth of 6.48% per year in the 2012-2016 period. In 2012 the number of passenger cars in Jakarta was 2.74 million units while in 2016 it increased to 3.52 million units. If it is assumed that the growth of passenger cars is still the same, the number of passenger cars in Jakarta in 2017 will reach 3.75 million units and 2018 will be 3.99 million units [2].

On the Length of Road perspective, the highway as a link between one region and another is one of the important infrastructures in land transportation. The road connects the production centers with the marketing area. Economic mobility relies heavily on the reliability and level of service of the road transport network. At present and in the future road infrastructure development is increasingly colored by aspects of regional development. This means that infrastructure development will be increasingly demanded to be able to support the movement of people, goods, and services within the framework of regional development perspective [3].

**Table 1.** Growth Motorcycle / Passenger Vehicle in DKI Jakarta until 2016

| Type of Vehicles | 2012 | 2013 | 2014 | 2015 | 2016 | Growth per year (%) |
|---|---|---|---|---|---|---|
| Motorclycle | 10.825.973 | 11.949.280 | 13.084.372 | 13.989.590 | 13.310.672 | 5,30 |
| Passenger Vehicle | 2.742.414 | 3.010.403 | 3.266.009 | 3.469.168 | 3.525.925 | 6,48 |
| Heavy Vehicle | 561.918 | 619.027 | 673.661 | 706.014 | 689.561 | 5,25 |
| Bus | 358.895 | 360.223 | 362.066 | 363.483 | 338.730 | -1,44 |
| Special Vehicle | 129.113 | 133.936 | 137.859 | 139.801 | 141.516 | 2,32 |
| **Total** | **14.618.313** | **16.072.869** | **17.523.967** | **18.668.056** | **18.006.404** | **5,35** |

### *1.2 Problem in Field*

Tyre patching tool is one type of appropriate tools used by user in assisting the tyre filling process in a motorcycle. This tool consists of 4 main components, namely electric compressors, electric heating tyres, power generators and Human Machine Interface (HMI). The four components work in an integrated manner with each other so that it facilitates how to use and accelerates the time in tyre repair activities. Generally, currently the tyre patching equipment in the workshop still uses conventional tyre patching tools. So that the time of the tyre filling process cannot be ascertained, adjusting to the condition of the patching tools found in the workshop [4]. In the tyre filling process, the results of the tyre filling are stated to be good if the patching material and inner tyre are tightly integrated. This is possible if the tyre patch can determine the temperature and length of time patching. Therefore, to get the perfect tyre filling results, it is necessary to have a component that can regulate the temperature of the filling followed by the time of the filling process [5]. Small reciprocating compressors, as applied in the automotive industry for pneumatic suspension, vertically adjustable seats and tyre repair devices are produced in large quantities and so the effort for business companies to find cost-detailed solutions is very high. Especially the avoidance of oversized designs is an important point. To meet this demand without a cost explosion in field tests, simulation tools should be introduced [6]. In order to create opportunities for the operation of on-site power generation, new and well distributed small-scale of generators should present tendency to produce smaller power systems.[7].

With the imbalance in the volume of motorized vehicles with a lot and the length of the road that can be passed by the motorized vehicle, the chances of a tyre leak occurring in become very large, for example in DKI Jakarta. This phenomenon is an opportunity for tyre patching businesses. Most of the tyre patch workers use the "Wait for the Ball" principle, but the principle is inversely proportional to the expectations of motorized motorists. Tyre leaks when driving is often a scourge for motorists, the uncertainty in the distance where tyre patches and the time and energy spent looking for them is an opportunity to design compact and practical tyre patches and can be developed with online request-based applications. make it easy for businesses to meet customers and change the principle of "Wait for the Ball" to "Pick Up the Ball" and be an answer to the fear of motorists when experiencing a tyre leak. This condition creates business opportunities for the existence of the mobile and integrated tyre repair

## 2. METHODS

The design of this tool is done by carrying out several steps. The step of this design starts with the Reference Theory and Problem Identification by raising some statistics from the growth of the number of motorized vehicles and the growth of the number of roads available in Indonesia to show the business scale and identifying cases that often occur in motor vehicle users, especially the occurrence of leaks tyres on motorized vehicles. With the above considerations, this tool is designed through a control philosophy based on real activities carried out by conventional tyre patch entrepreneurs.

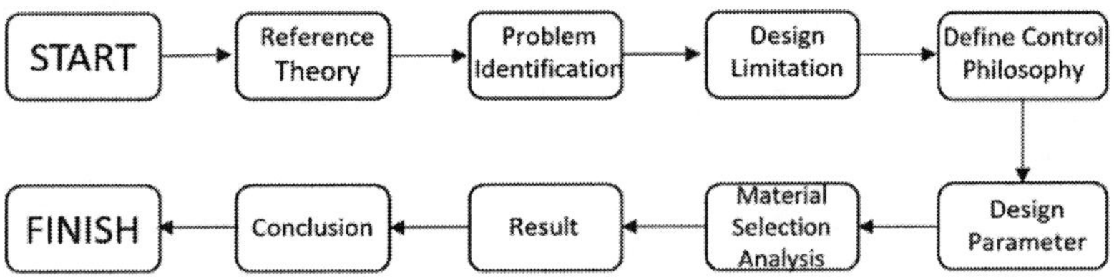


**Figure 1.** Methodology Diagram

### *2.1 Control Philosophy*

This tool is designed using PLC and uses Human Machine Interface (HMI). PLC and HMI are used to integrate these 3 major components, namely the Heating Element, Air Compressor and Power Generator Set. This tool is also equipped with transmitters as safety devices when operating. This tool can be used in 2 modes, Automatic Mode and Manual Mode. Manual Mode can turn on and off the Compressor and Heater individually.

*2.1.1 Automatic Mode.* In this mode, the device is designed to run automatically between Heating Element (HE01) and Air Compressor (AC02). Both tools will get electric power from the Diesel Engine Generator (DEG03). The entyre process of this tool will be ordered via HMI (HMI04). The tyre will first be dismantled by a technician and look for the hole that caused the leak. After the tyre has been dismantled and the hole has been found, the tyre is ready to be patched. By pressing the Automatic Push Button (PB00), HMI04 will command HE01 to heat the element. The heating time of the element takes 3 minutes and the length of the filling process on the heating element is 3 minutes. After 3 minutes of patching, HMI04 will order HE01 to stop. After the tyre has been completely patched, the technician will immediately replace the tyre on the vehicle. A moment HE01 stops, then HMI04 will give a message that is accompanied by a buzzer sound to be able to technician order HMI04 to run AC02. The compressor will automatically turn OFF if the message is ignored by the technician for 15 minutes. Compressors need 1-2 minutes to fill 1 tyre of a motorcycle tyre. HMI04 will order AC02 to stop when the desired tyre pressure has been read by a Pressure Transmitter (PT02).

*2.1.2 Manual Mode.* In this mode it can be used according to the actual conditions required by a tyre. Technicians can turn HE01 and AC02 on or off by pressing the Push Button Heater (PB01) and Push Button Compressor (PB02). The process in this mode, Safety Devices and Regulator Switch continues to function.

IOP Conf. Series: Materials Science and Engineering **847** (2020) 012084 doi:10.1088/1757-899X/847/1/012084

*2.1.3 Safety Devices.* This tool is designed with a Safety Device equipped with a transmitter sensor. In addition, there is a Timer Command from HMI04 as a switch function on this tool, the Temperature Transmitter on the Heating Element (TT01) also functions as a Switch and as an indicator on the HMI. TT01 also provides warning protection if the temperature of the element reaches 75 °C and provides shutdown protection that can turn off automatically HE01 when the temperature of the element reaches 80 °C. Pressure Transmitter on the Air Compressor (PT02) functions as a switch when the pressure in the tyre reaches set pressure which are desired. In addition, PT02 also provides warning protection when the pressure in the tyre reaches 35 psig and provides shutdown protection that can automatically turn off AC02 when the pressure in the tyre reaches 40 psig. The Level Transmitter in DEG03 (LT03) serves as a warning and shutdown protection when the fuel level of the device touches the specified threshold. All Analog Signals from each transmitter will be sent to HMI04.

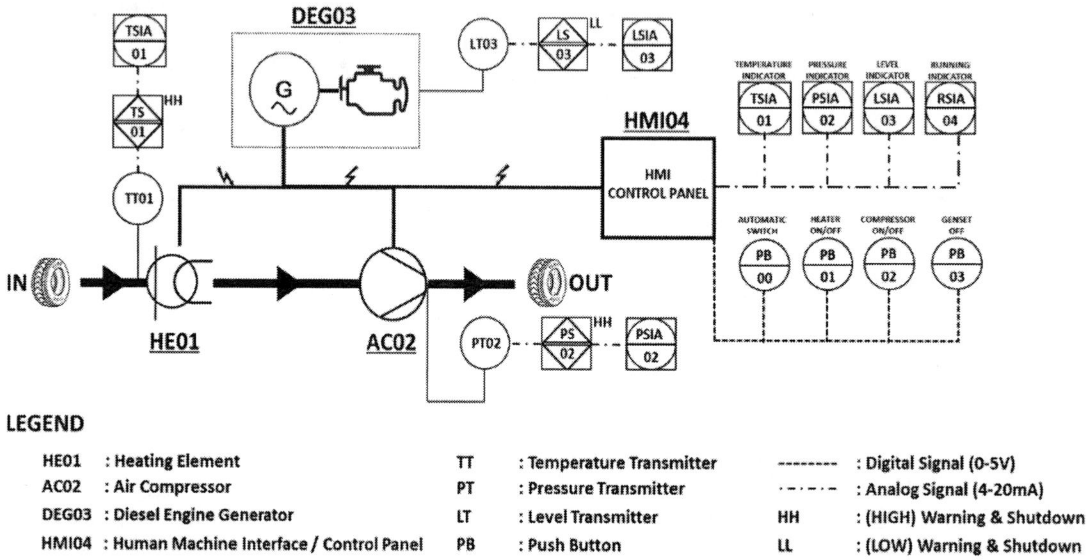


**Figure 2.** Process Flow Diagram

## *2.2 Main Component*

*2.2.1 Heating Element.* Electric heating is a process where heat energy is produced through electrical energy conversion. An electric heater is an electrical device that converts an electric current into heat. The heating element that works on the principle of Joule heating, inside every electric heater is an electrical resistor. Electric current that passing through the resistor will convert the electrical energy into heat energy. The example of using electric heating is on electric tyre system. The heating plate is part of a heat system that encounters the tyres. This plate is used in patching process as a medium of heat transfer from heater to tyre.

The design of a tyre patch equipment system that is sourced from electricity is designed to be environmentally friendly, because it does not cause air pollution, such as the system of tyre patches generally. The design of the heating system of the tyre patch tool consists of several major components and supporting components. The main component consists of a voltmeter that serves to know the large voltage used in the filling process [8]. Amperemeter is used to determine the current strength in the series of tyre patches [9]. The Rise of Thermo-Electric effect will be produced by the combination of 2 different conductors in thermocouple as the tools to measure the temperature. [10].

In the process of working system of electric tyre patches will be obtained a data that can be used to determine the rate of heat transfer on the heating plate and tyres based on the formula of conduction heat transfer [11]. The heat transfer rate Data (Q) on the heater plate is obtained from the Iron heat conductivity (K), the area of the cross section (A) and the plate thickness (L) of the heater plate as well as the temperature difference between T1 danT2. So that in the Q counting on the heating plate applies the formula (1) [12] as follows:

$$Q = k.\,A.\,\frac{T_1 - T_2}{L} \qquad (1)$$

Description:

Q = Heat transfer rate (Watt)
k = Iron heat conductivity (W/m. °C)
A = Area of the heating plate section ($m^2$)
T1 = Heater temperature (°C)
T2 = Heating plate temperature (°C)
L = Thick heating plate (m)

Acquisition of Q data on tyres is an advanced calculation of the Q calculation on the heater plate. The difference in Q calculation on tyres based on K and ΔT. Where K is used is K of tyres and ΔT comes from the temperature difference between T2 and T3. So in the Q counting on tyres apply formula (2) [13],[15] follows:

$$Q = k.\,A.\,\frac{T_2 - T_3}{L} \qquad (2)$$

Description:

Q = Heat transfer rate (Watt)
K = Rubber heat conductivity (W/m. °C)
A = Area of the heating plate section ($m^2$)
T2 = Heating plate temperature (°C)
T1 = Tyre temperature (°C)
L = Tyre thickness (m)

Power (P) is obtained from the magnitude of the voltage and the current strong into the system. In this case the voltage (V) of electricity entered into the system amounted to 220 volts, as per the power provided by PLN. For the current strong (I) obtained from the heater capability to transmit the amount of electrical current. So in the Q counting based on the voltage and current apply the formula [12] follows:

$$P = V.\,I.\,Cos\,Ø \qquad (3)$$

Description:

Q = Power (Watt)
V = Voltage (Volt)
I = Current (Aampere)
Cos Ø = Power Factor

*2.2.2 Air Compressor.* Air Compressor used is reciprocating type. A reciprocating compressor also known as piston compressor is a positive-displacement compressor type by using forces of pistons driven through a crankshaft to deliver pressurized gasses. The intake gas enters the suction filters into inlet manifold, enters the chamber cylinder and later compressed by piston driven in a reciprocating motion via a crankshaft, and then discharged [14].

Bicycle form is the simplest reciprocating type. Dynamic piston will suction the air into the cylinder and then compressed it. If only one piston used is called a single-acting compressor. If both pistons on top and undersides are used, the compressor is double acting. Relationship of the absolute pressure on Outlet and Inlet nozzle is called Pressure Ratio. Accordingly, Compressor Chamber draws in air at atmospheric pressure (1 bar(a) and compresses it into 7 bar working pressure at a pressure ratio of (7 + 1)/1 = 8 [14].

At Suction Stage, Stroke volume is a volumetric capacity of the cylinder. Clearance volume is a volumetric capacity below inlet and outlet valve and should be above the piston. Gap between those volumetric capacities will produce pressurized free air delivery. p/V diagram and the actual diagram show the practical characteristic of volumetric compressor. For Reciprocating Compressor, there is always a gap between piston and cylinder, valves are never completely sealed. Valves cannot 100% full open and close then minimum delay will always be occurred, which causes a pressure drop when the air suction and compressed through the chambers. Because of this design, the gas will also have temperature rises when flowing into the cylinder.
Compression work with isothermal compression:

$$W = ln(\frac{P2}{P1}).\ (P_1V_1) \quad (4)$$

Compression work with isentropic compression:

$$W = \frac{K}{K-1}.\ (P_2V_2 - P_1V_1) \quad (5)$$

W = compression work (J)
$P_1$ = initial pressure (Pa)
$V_1$ = initial volume (m3)
$P_2$ = final pressure (Pa)
K = isentropic exponent: K ≈ 1,3 – 1,4

These relations show that more work is required for isentropic compression than for isothermal compression [14].

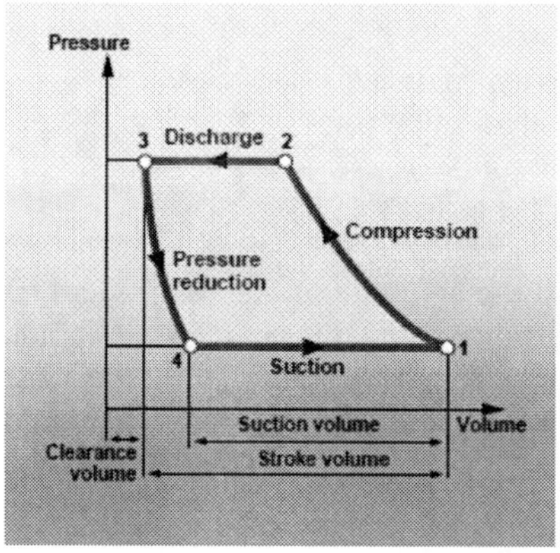


**Figure 3.** P/V Diagram

*2.2.3 Diesel Engine Power Generator.* Genset (generator set) is a device that functions to produce electricity. The set is equipped with a combination of two different devices, namely engine and generator or alternator. Engine will work as a rotating device while the generator or alternator as a power generation device. The engine can be a diesel engine or gasoline engine, while the generator or alternator is a copper coil or coil consisting of a stator (static coil) and rotor (rotating coil).

In a simple method it can be explained that the engine rotates the rotor on the generator so that a magnetic field arises on the generator stator coil, the magnetic field arising on the stator and interacting with the rotating rotor will produce an electric current. The electric current generated by the generator will have a voltage difference between the two poles of the generator so that when connected to a load it will produce electric power [16]

*2.2.4 Programmable Logic Controller/PLC.* A programmable logic controller (PLC) or programmable controller is an digital computerized control has been ruggedized and adapted for the control for any kind of manufacturing processes, such as assembly lines, or robotic devices, Packaged

System or any activity that requires high reliability control and ease of programming and process fault diagnosis for safety reason. [17]

To emulate the functions of electro-mechanical relay is the basic function of a programmable controller. Variable inputs are given a unique code, and specific PLC instructions can test whether the input states of activity or inactivity. A logical "AND" function performed by relay contacts will not allowing current to pass unless all contacts are checked. A series of "EXAMINE IF ON" instructions will energize its output storage bit if all the input bits are on, meanwhile a parallel set of instructions will perform a logical "OR" function as well. On the wiring diagram of electro-mechanical relay, a group of contacts controlling one coil is called a "rung" of a "ladder diagram ", this concept is also used to describe PLC logic.[18]. Some models of PLC limit the number of series and parallel instructions in one "rung" of logic. The output of each rung sets or clears a storage bit, that can be associated with a physical output address or an "internal coil" with no physical connection.[19] These internal coils can be used as a common element in multiple separate rungs. Unlike physical relays, there is no limit for number of inputs, outputs or internal coils referenced in a PLC program.

## 3. RESULT AND DISCUSSION

### *3.1 Main Component Selection*

The selection process of several main tools such as the Heating Element, Air Compressor and Diesel Engine Power Generator above are based on the control philosophy described earlier. The control philosophy of this tool also has an important role in determining the type of PLC will be used. In order to achieve target time in filling the tyre within 10 minutes using working temperature 70$^{o}$C, the design requires supply power minimum 650 watts, 8 bar(g) working discharge pressure and minimum 110 Liters per minutes (lpm) free air delivery capacity. Brands currently available on market are considered to use as tools.

**Table 2.** Main Component Selection

| Heating Element (HE01) | | Air Compressor (AC02) | | Diesel Engine Power Generator (DEG03) | | PLC (HMI04) | |
|---|---|---|---|---|---|---|---|
| Brand | : Local/Concept Designed | Brand | : Greeloy | Brand | : Yamaha | Brand | : Schneider |
| Power Required | : 150 Watt | Type | : GA-61/9 | Type | : EF 1000FW | Type | : TM3D18 |
| Voltage | : 220/50 | COO | : China | COO | : Japan | Voltage | : 120/50 |
| Design Temp | : 90 DegC | Power Required | : 500 Watt | Capacity | : Max 800 Watt | | |
| Working Temp | : 70 DegC | Voltage | : 220/50 | LxWxH(mm) | : 400x300x420 | | |
| Capacity | : 2-3 Minutes/tire | Working Pressure | : 8 barg | Weight (kg) | : 28 | | |
| LxWxH(mm) | : 300x150x200 | Capacity | : 118 Lpm | | | | |
| Weight (kg) | : 10 | Vessel Capacity | : 9 L | | | | |
| | | LxWxH(mm) | : 440x440x590 | | | | |
| | | Weight(Kg) | : 27 | | | | |

### *3.2 Product Designed Drawing*

The tool is designed to be compact and easy handling for the end user. Design is divided into compartments for Air Compressor, Heating Element, Diesel Engine Generator and tools. Design is limited on height by 1,000 mm to ease the tyre patching handling during transportation. Diesel Engine Generator and Compressor are located on one compartment to maintain the centre of gravity (COG). Design has maximum 65 kgs in weight.

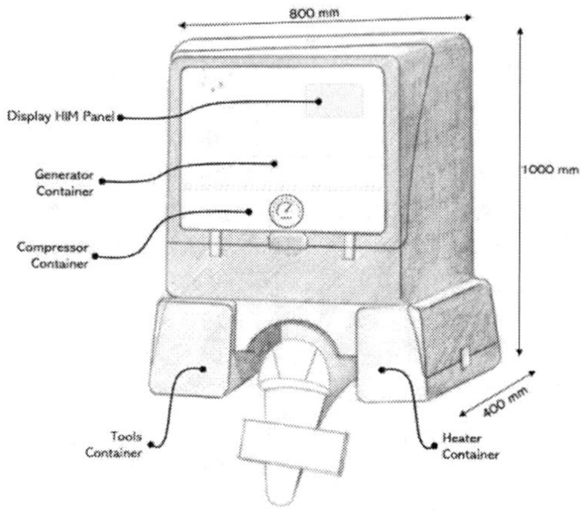


**Figure 4.** Design Container Drawing

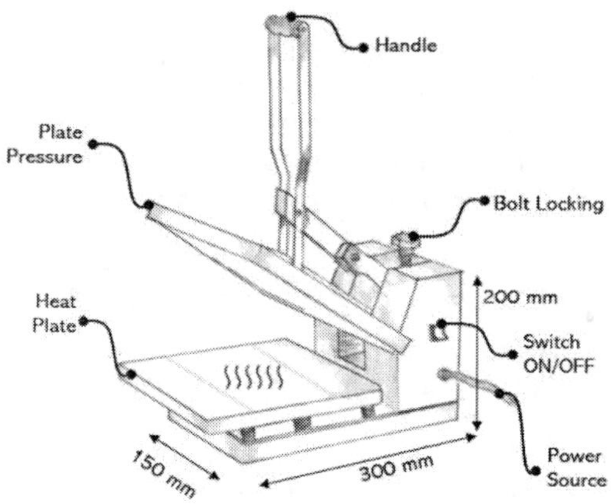


**Figure 5.** Design Heating Element

## 4. CONCLUSION

The concept design of this tool can make things easier for patching tyre businesses to fulfil the needs of consumers. In the other side, customers / motorbike rider's problems when a tyre leak occurs will be quickly solved. Later in the future, this market can be combined with online / IoT based applications and Artificial Intelligence to be able to know the whereabouts of consumers. It is concluded that this product could be useful for motorcycle in order to prevent tyre puncture during transportation.

## 5. REFERENCES

[1] MotorcyclesData 2019. *https://motorcyclesdata.com/2019/12/15/indonesia-motorcycles/*

[2] Biro Pusat Statistik Provinsi DKI Jakarta 2018 Statistik Transportasi DKI Jakarta 2018 *https://jakarta.bps.go.id* pp 7-14

[3] Widowati H 2019. *https://databoks.katadata.co.id/datapublish/2019/08/02/berapa-jumlah-kendaraan-di-dki-jakarta*

[4] Benidiktus A 2012 Perancangan Alat Press Tambal Ban Elektrik dengan Pengatur Waktu untuk Ban Dalam Sepeda Motor. *http://repository.upnyk.ac.id/4016/* (Yogyakarta: UPN "Veteran" Yogyakarta) pp 1-2

[5] Andrizal 2003. *Pembuatan Alat Tambal Ban Dalam (Benen) Kendaraan Semi Otomatis* (Padang: Lembaga Penelitian Universitas Negeri Padang) pp 15-16

[6] Landschuetzer C and Andreas W 2010 Simulation Potentialities For a Reciprocating Compressor. *Proceedings in Manufacturing Systems* Vol 5 No 2 pp 1-2

[7] Quinonez-Varela G and A Cruden 2007 Development of a Small-Scale Generator set model for Local Network Voltage and Frequency Stability Analysis. *IEEE Transactions on Energy Conversion* Vol 22 No 2 pp 1-2

[8] Kay B.B and Harmon J.L 1969 Twelve functions in a new digital meter. *Hewlett-Packard Journal* 20(7) pp 2-13

[9] Wulandari D.C and Wildian 2014. Rancang bangun ammeter DC tipe nondestructive berbasis mikrokontroler ATmega8535 dengan sensor efek hall ACS712. *Jurnal Fisika Unand* 3(2) pp 121-127

[10] Bull K 2008 Thermocouples: Basic function and sources of error thermistors and thermocouples: Matching the tool to the task in thermal validation. *The Journal of Validation Technology* 14(2) pp 73-76

[11] Zhengyu W, Xiaoyang Y, Guanghui L and Guangyu T 2014. Online cutting temperature measurement system based thermocouple. *International Journal of Control and Automation* 7(10) pp 355-364

[12] Cengel Y.A 2003. *Heat transfer: A practical approach* (2nd ed.) (Boston: Mc Graw-Hill). *http:// www.ncbi.nlm.nih.gov/pmc/articles/PMC341276/pdf/nar00305-0003.pdf.*
[13] Ortuño M, Márquez A, Gallego S, Neipp C and Beléndez A 2011. An experiment in heat conduction using hollow cylinders. *European Journal of Physics* 32(4) pp 1065-1075
[14] Atlas Copco Airpower N.V 2015. *Compressed Air Manual Book 8th Edition* pp 20-22
[15] Ashari A, Mufarida A and Efan A 2016. Analisis perpindahan panas pada alat tambal ban elektrik. *J-Proteksion, Jurnal Kajian Ilmiah dan Teknologi Teknik Mesin* 1(1) pp 37-42
[16] Aysegul A and Mehmet K 2009. Exergetic and thermoeconomic analyses of diesel engine powered Cogeneration. *Applied Thermal Engineering* 29 (2009) pp 234–241
[17] Liping G 2009. Design Projects in a Programmable Logic Controller (PLC) Course in Electrical Engineering Technology. *Technology Interface Journal*
[18] MM. Lashin 2014. Different applications of programmableogic controller (PLC). *International Journal of Computer Science, Engineering and Information Technology (IJCSEIT)* Vol 4
[19] Zulfakar A and Mohamed K.H 2009. Modeling of a Ladder Logic Processor for High Performance Programmable Logic Controller. *Third Asia International Conference on Modelling & Simulation*

# Optimasation of Mechanical Cassava Peeling System Parameters

**I M Gana[1,3], A A Shhu[1,4] and A Gbabo[2,5]**

[1]Department of Agricultural and Bioenvironmental Engineering, School of Engineering Technology, Federal Polytechnic Bida, Niger State, Nigeria
[2]Department of Agricultural and Bioresources Engineering, Federal University of Technology Minna, Niger State, Nigeria
[3]ganaibro74@yahoo.com, [4]shhuaa@yahoo.com, [5]agidids@yahoo.com

**Abstract.** This study focused on investigations of effects of mechanical parameters (peeling speed, cutter length) and handling parameter (cassava tuber length) of a cassava peeling machines on the machine output (peeling and recovery efficiencies) with the view of optimizing the parameters. The serious issues of existing techniques of peeling cassava are moderately low peeling and recovery efficiencies because of the irregular shape and size of cassava tubers. The results of the trial of the machine utilizing the cassava tubers revealed that all the parameters have significant effects on the peeling and recovery efficiencies of the machine. The cutter length had a more significant impact on the peeling and recovery efficiencies. Maximum values of 83.5 % and 97.2% for peeling and recovery efficiencies respectively, with an attractive quality of 0.864 were achieved from peeling cassava tuber of 50 mm length with cutter length of 2 mm at a speed of 328 rpm. The investigation's discoveries give the standard machine input parameters which are equipped for improving quality, peeling and recovery efficiencies of a mechanical cassava peeling system.
**Keywords**: cassava, cutter length, machine, optimum, peeling

## 1. Introduction

Cassava is presently a significant crop for both local utilization and export promotion. It has been reported to have various advantages, for example, well being, monetary and as methods for accomplishing food availability [1]. Among the most important unit processes involved in cassava processing are peeling, grating, boiling, drying, milling, sieving, extrusion and frying [2]. Cassava peeling is and has always been a major challenge for design engineers engaged in manufacturing machines and equipment for cassava processing [3]. This is because of its unusual shape, dimension, peel thickness, and even the weight of the same variety [4]; [5]. Presently, cassava peeling operation is accomplished using a manual and mechanical technique. As reported by [6], the manual technique comprised the use of a knife that is, tedious and time-consuming. The mechanical approach, however, entails the use of the machinery and equipment in cassava peeling. Generally, the efficiency of a cassava peeler depends on the peeling mechanism, which could be in the form of perforated drums or a blade, peeling speed or the cassava tubers, length, size, volume and of the cassava tuber variety. Some of the recent research work on mechanical cassava peeling techniques, their peeling and recovery efficiencies are as follows; [7] designed a cassava peeling machine with an abrasive drum as the peeling mechanism. The peeling and recovery efficiencies of the machine were 65% and 95% respectively. The major limitation of this machine is its time consumption. A machine with an external peeling mechanism that produced cassava peeling and recovery efficiencies of 60% and 94%, respectively was developed by [5]. The drawback of this machine was the high losses and grating of cassava tubers. An automated cassava peeler was designed by [8]. The peeling and recovery efficiencies of this machine were 48% and 86% respectively. This machine needed expertise and could not absolutely peel smaller tuber sizes. A peeler that operates based on Lathe machine principle was developed by [9], the machine had a peeling efficiency of 65% and recovery efficiency of 88%. The major limitation of this machine is its higher mechanical damage. [10] developed an automated cassava peeler with knife-edge peeling mechanism. The peeling and recovery efficiencies of the machine were 83% and 58% respectively. The major problem of the machine is it could not peel tiny sizes of tubers and has a significant loss. A double action, self- fed cassava peeling machine with peeling and recovery efficiencies of 80% and 92%, respectively, was developed by [10]. This design had a complex working mechanism, high peel retention and requires hand trimming of tubers and skill operator. Generally, all the shortcomings of these machine centered on the size of peeling mechanism, speed of peeling and the span of the straight portion of a tuber. Hence, to address poor product consistency, relatively low efficiency in peeling and recovery of mechanical peeling system, there is needed to optimize the size of the peeling blade, speed of peeling and the length of the cassava tuber. The objective of this study is to determine the effects of peeling blade, speed of peeling and the length of the cassava tuber on efficiency in peeling and recovery. Also, to established optimum parameters that would yield the best machine output.

## 2. Methods

*2.1 Experimental setup and procedure*

A central composite rotatable design of response surface methodology (rsm) made up of three factors varied at five levels was used in this study [11]. The three variables are: peeling speed and cutter size (mechanical variables of the machine) and tuber size (processing variables). The size of the cutter was varied by 0.12, 0.15, 0.2, 0.25 and 0.28cm. The size of the tuber ranged as 1.6, 5, 10 15, 18.4 cm. The peeling speed was varied according to 1232 rpm, 1300 rpm, 1400 rpm, 1500 rpm and 1568 rpm. The test was performed using the template matrix indicated in Table 1. The machine's power switch was on with no load, to test the machine's functionality. The tuber was then fed into the system via the hopper. As the shaft rotates the circular disc at base of the machine creates centrifugal force, which swung the tuber to the cylinder's wall where the tuber's outer skin was peeled. Water was allowed to flow into the machine and this aid in separating the peeled skin from the flesh tubers. The water and peeled skin flowed out of the machine through the lower chute while the flesh cassava tubers are discharged out of the machine through the upper chute

*2.2 Testing of the machine*

A set of empirical test was conducted to examine the influence of cutter size, peeling speed, and tuber size on peeling and recovery efficiency of a mechanical peeler [2].

*2.3. Statistical analysis*

Analysis of variance (ANOVA) was performed to estimate the influence of major variables and their potential correlation effect on the responses.

*2.4 Machine performance.*

The performance of the machine was determined by the peeling and recovery efficiency.

*2.4.1 Peeling Efficiency.* This is the percentage of the tuber skin peels by the machine to the total mass of the tuber's skin. It was calculated as reported by [2]

$$P_E = \frac{Mm_t}{M_{mt}+M_{ht}} \times 100 \quad (1)$$

Where$P_E$is the peeling efficiency (%), $M_{mt}$is a mass of the tuber skin peel by the machine (kg),$M_{ht}$ is a mass of the peel removed by hand after machine peeling (kg).

*2.4.2. Recovery efficiency.* This is percentage mass of the flesh tuber after peeling to the total mass of the tuber less the mass of the tuber skin. It was calculated as reported [2]

$$R_E = \frac{M_{ft}}{M_{cb}-M_{sb}} \times 100 \quad (2)$$

Where,$R_E$is the recovery efficiency (%),$M_{ft}$is mass of the flesh tuber after peeling (kg), $M_{cb}$is mass of the cassava tuber (kg), $M_{sb}$ is mass of the tuber skin (kg).

## 3. Results and discussion

The independent parameters strongly influence cassava peeling process are the cutter size, peeling speed and tuber size. An experiment was conducted to peel cassava tubers by varying and combining the different levels of the independent variables and the responses were measured after each trial. The results obtained are as follows:

### *3.1. Effects of independent variables on peeling efficiency*

The result of the statistical analysis of variance (ANOVA) of the peeling efficiency presented in table 1, showed that the model and all the independent variables were significant with F - value of 41.073 and P-value probability > F less than 0.0500 respectively. The cutter size had a more positive effect on the peeling performance with a coefficient of estimation of 11.7. The size of the tuber had a significant negative effect on the peeling efficiency with a coefficient of estimation of -7.593[13]. The R-square value of 0.9753 was very close to 1 as recommended by [1].

**Table 1.** Regression analysis of response of peeling efficiency

| Source | Coefficient Estimate | Standard Error | F Value | p-value Prob > F | R-Squared | |
|---|---|---|---|---|---|---|
| Model | 80.153 | 1.326 | 41.073 | < 0.0001 | 0.9737 | Significant |
| A-Cutter length | 11.701 | 0.880 | 176.699 | < 0.0001 | | |
| B-Speed of Cutting | 7.248 | 0.880 | 67.801 | < 0.0001 | | |
| C-Cassava Length | -7.593 | 0.880 | 74.403 | < 0.0001 | | |
| AB | 2.863 | 1.150 | 6.199 | 0.0320 | | |
| AC | -0.211 | 1.150 | 0.033 | 0.8579 | | |
| BC | 3.336 | 1.150 | 8.414 | 0.0158 | | |
| $A^2$ | -4.899 | 0.856 | 32.695 | 0.0002 | | |
| $B^2$ | -0.165 | 0.856 | 0.037 | 0.8504 | | |
| $C^2$ | -1.993 | 0.856 | 5.413 | 0.0423 | | |
| Lack of Fit | | 5 | 4.264 | 0.0687 | | not significant |

*3.1.1. Effects of tuber and cutter sizes on the peeling efficiency.* Figures 1 and 2 show the response surface and contour plot for the influence of tuber and cutter lengths and peeling output, respectively. From figure 1, peeling the cassava tubers with cutter size of 2 mm, it was observed that the peeling efficiency remains nearly steady with a value of 84% as the cassava tuber size enlarged from 15 mm to 50 mm, and then decreased significantly to 62% with more increase in the cassava tuber size. This may

IOP Conf. Series: Materials Science and Engineering **847** (2020) 012085 doi:10.1088/1757-899X/847/1/012085

be as a result of uneven shape of the cassava tubers which makes it difficult to peel the skin as the length of the cassava increased.

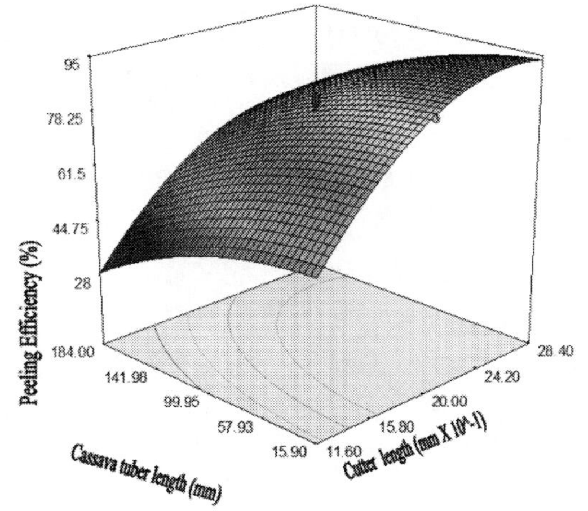


**Figure 1.** Response Surface for Influence of Tuber Size and Cutter Length on Peeling Efficiency

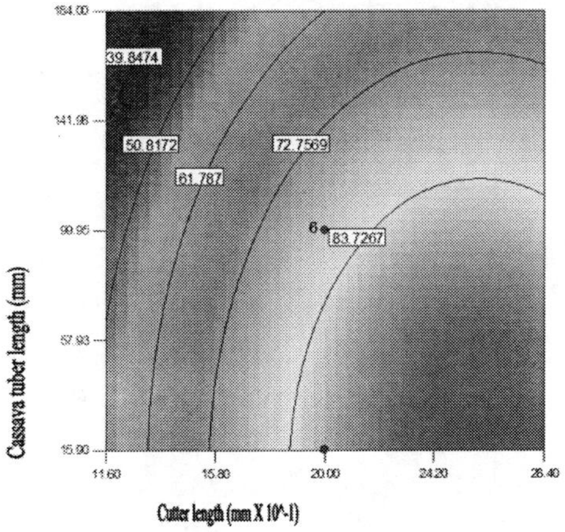


**Figure 2.** Contour Plot for Influence of Tuber Size and Cutter Length on Peeling Efficiency

### *3.2. Effects of independent variables of recovery efficiency*

The result of the statistical analysis of variance (ANOVA) of the recovery efficiency presented in table 2, showed that the model and all the independent variables were significant with F - value of 48.6 and P-value probability > F less than 0.0500 respectively. The cutter size had a greater effect on the recovery performance, with an approximate coefficient of -7.89, followed by the peeling speed with value of -5.82 and then the tuber size with a value of -4.6. All the three factors had significant negative effect on the recovery efficiency [11]. As suggested by [1], the R- squared value of 0.9777 was very close to 1.

*3.2.1. Effects of cassava size and cutter size on recovery efficiency.* Figures 3 and 4, respectively, display the response surface and contour plot for the influence of cassava size and cutter size on the recovery performance. The recovery performance improved from 85% to 94% using a cutter size of 2 mm as the size of the cassava increased from 20 mm to 50 mm and then decreased to 70% as the tuber size increased to 187 mm. This could be as results of cutting away some part of the cassava flesh together with the skin layer as a result of uneven shape and size of the tuber. This was agreed with the report from an earlier study [13] in which cassava recovery of a mechanical peeler was found to be low because of loss of more useful flesh. However, the recovery remains almost unchanged with a value of 94% as the cutter size increases from 1.2 mm to 2 mm and then decreases substantially to 73% with a further increase in the cutter size to 2.8 mm. This could be as a result of penetration of the blade deep into the cassava flesh beyond the skin layer, which resulted in cutting away some useful flesh together with the outer layer.

### *3.3 The optimisation of the machine parameters*

The optimisation of the parameters was carried using a numerical technique with the goals of maximizing the efficiencies of peeling and recovery. The result of optimum and the predicted values of the parameters are presented in figures 5 to 10. The desirability (quality) of values of all the dependent and independent variables is presented in figure 11, while the ramp of the optimisation is shown in figure 12. From figure 5, using constant values of cutter and tuber sizes of 207 mm and 50 mm, respectively, the peeling efficiency, increased from 78% to 95% as the speed increases from 266 rpm to 434 rpm. But this high speed had a negative effect on the recovery efficiency, as can be seen in figure 8. Therefore, optimum speed of 328 rpm that best favoured the responses was selected. This gave a predicted value of peeling efficiency of 85.32%.

**Table 2.** Regression analysis of response of the recovery efficiency

| Source | Coefficient Estimate | Standard Error | F Value | p-value Prob > F | R-Squared | |
|---|---|---|---|---|---|---|
| Model | 90.089 | 0.888 | 48.616 | < 0.0001 | 0.9777 | Significant |
| A-Cutter length | -7.889 | 0.589 | 178.929 | < 0.0001 | | |
| B-Speed of Cutting | -5.815 | 0.589 | 97.220 | < 0.0001 | | |
| C-Cassava Length | -4.593 | 0.589 | 60.654 | < 0.0001 | | |
| AB | -2.132 | 0.770 | 7.658 | 0.0199 | | |
| AC | -1.087 | 0.770 | 1.991 | 0.1885 | | |
| BC | -0.252 | 0.770 | 0.107 | 0.7499 | | |
| A^2 | -4.930 | 0.574 | 73.748 | < 0.0001 | | |
| B^2 | -2.699 | 0.574 | 22.108 | 0.0008 | | |
| C^2 | 0.169 | 0.574 | 0.087 | 0.7739 | | |
| Lack of Fit | | | 0.735 | 0.6280 | | not significant |

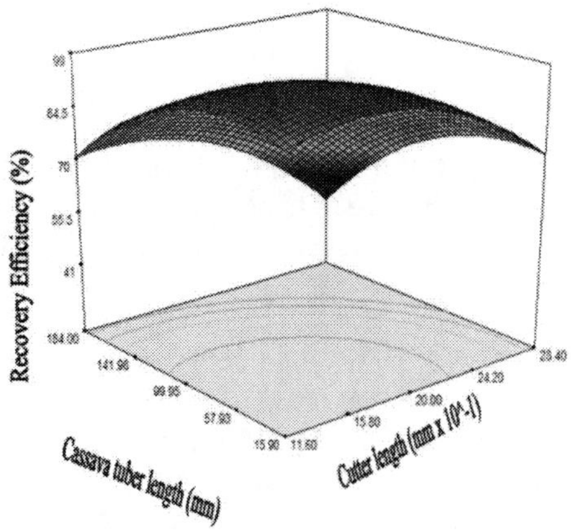


**Figure 3.** Response Surface for Influence of Tuber Size and Cutter Length on Recovery Efficiency

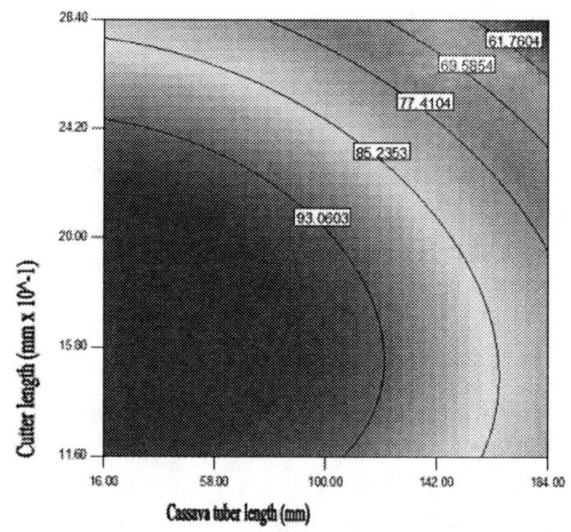


**Figure 4.** Contour Plot for Influence of Tuber Size and Cutter Length on Recovery Efficiency

With constant values of peeling speed and tuber size of 327.54 rpm and 50 mm, respectively (figure 6), the peeling efficiency, increased from 68% to 93% as the cutter size increases from 1.5 mm to 2.5 mm. However the bigger cutter size had a negative effect on the recovery efficiency, as can be seen in figure 9. This is as a result of increase in depth of penetration of the cutter into the cassava tuber that resulted in peeling off more of the tuber flesh as the cassava skin thickness ranges from 1.5 mm to 2 mm. This agreed with the outcome of an earlier study [1], where segregation of materials by a size reduction system decreased with a decrease in cutter surface area. Therefore the optimum cutter size of 2.07 mm that best favoured the two responses was selected and this gave a predicted value of peeling efficiency of 85.32%. From figure 7, using constant values of cutter size of 2.07 mm and peeling speed of 327.54 rpm, the peeling efficiency slightly decreased from 87.5% to 85.32% as the tuber size increases from 16 mm to 50 mm and then drastically reduced to 57% with further increased with tuber length to 184 mm. Considering the recovery efficiency, as can be seen in figure 10 an optimum tuber size 50 mm that best favoured the responses was selected. This gave a predicted value of peeling efficiency of 85.32%. With constant values of cutter and tuber sizes of 2.07 mm and 50 mm, respectively as shown in figure 8, the recovery efficiency remains unchanged with a value of 95.92% as the peeling speed increased from 266 rpm to 327.32 rpm and then decreased to 74% with further increase in peeling speed to 434 rpm. Therefore, optimum speed of 327.32 rpm that best favoured the responses was selected. This gave a predicted value of recovery efficiency of 95.92%.

From figure 9, using a constant value of peeling speed and tuber size of 327.54 rpm and 50 mm, respectively, the recovery efficiency remains unchanged with a value of 95.92% as the cutter size increases from 1.5 mm to 2.0 mm, afterward decreased to 85% with further increase in cutter size to 2.5 mm. Value of cutter size below 2 mm had a negative effect on the peeling efficiency as shown in figure 6. This is because the tuber skin thickness ranged from 1.5 to 2 mm. The decreased on the efficiency when peeling with cutter size above 2 mm could be as a result of deep penetration of the cutter into the cassava tuber that resulted in peeling off more of the tuber. Therefore the optimum cutter size of 2.07 mm that best favoured the two responses was selected and this gave a predicted value of peeling efficiency of 95.92%. From figure 10, using constant values of cutter size of 2.07 mm and peeling speed of 327.54 rpm, the recovery efficiency slightly decreased from 97.5% to 94.32% as the tuber size increases from 16 mm to 50 mm and then drastically reduced to 82% with further increased in tuber length to 184 mm. Considering the peeling efficiency (figure 7) an optimum tuber size of 50 mm that best favoured the responses was selected. This gave a predicted value of peeling efficiency of 95.92%. From figure 11, maximizing the efficiencies of peeling and recovery gave a quality value of 0.777 for peeling efficiency, 0.9641 for recovery efficiency and combined quality value of 0.864. The ramp of the optimization shown in figure 12 gave the summary of optimum values of variables; cutter length of 2.07 mm, speed of 328 rpm, and cassava length of 50 mm, which produced peeling efficiency of 85.4% and recovery efficiency of 98.04 %.

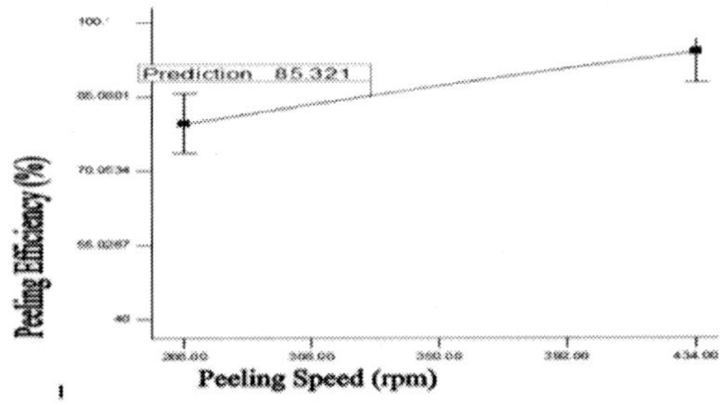


**Figure 5.**Optimum and Predicted Values of the Peeling Speed with respect to Peeling Efficiency

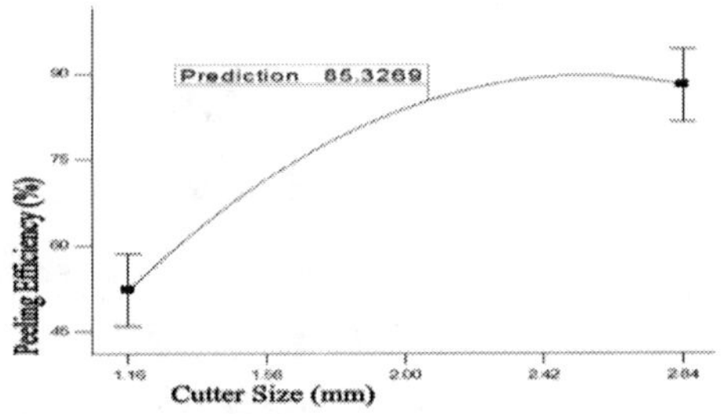


**Figure 6.**Optimum and Predicted Values of the Cutter Size with respect to Peeling Efficiency

IOP Conf. Series: Materials Science and Engineering **847** (2020) 012085 doi:10.1088/1757-899X/847/1/012085

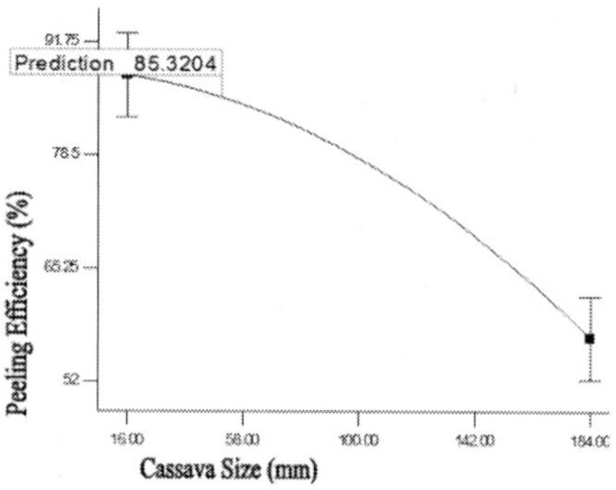


**Figure 7.** Optimum and Predicted Values of the Cassava Size with respect to Peeling Efficiency

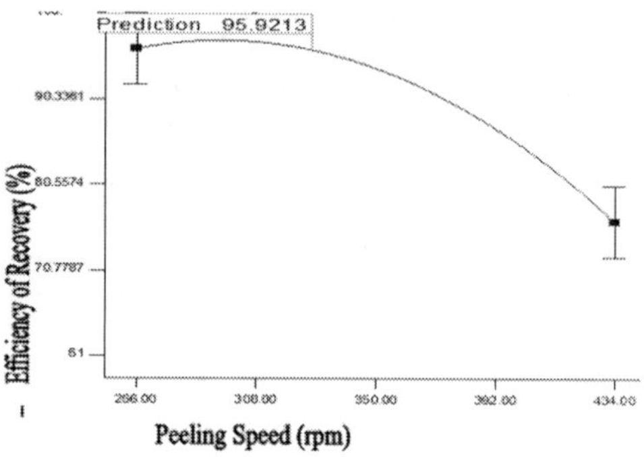


**Figure 8.** Optimum and Predicted Values of the Peeling Speed with respect to Recovery Efficiency

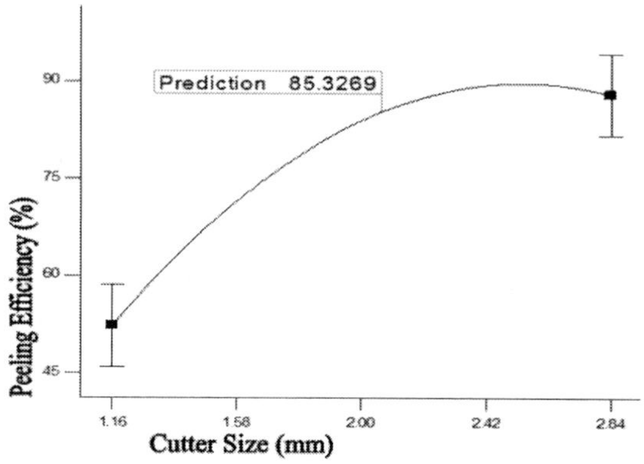


**Figure 9.** Optimum and Predicted Values of the Cutter Size with respect to Recovery Efficiency

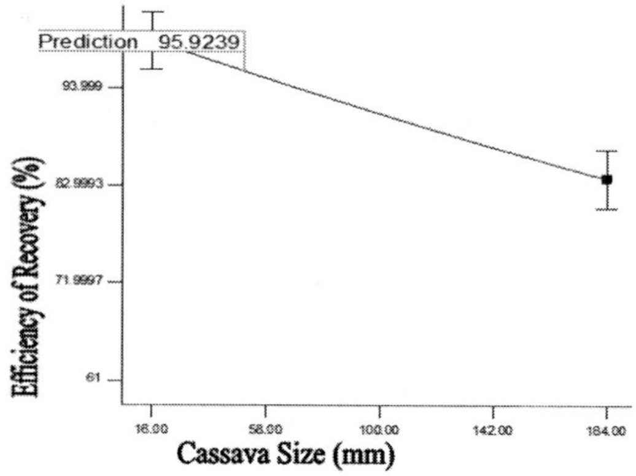


Figure **10.** Optimum and Predicted Values of the Cassava Size with respect to Recovery Efficiency

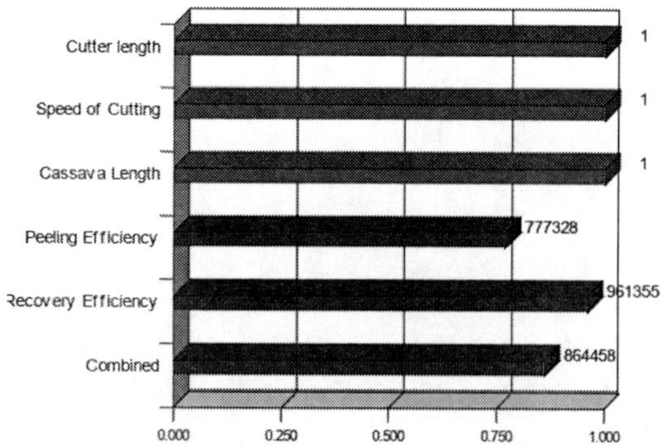


**Figure 11.** Result of desirability of the individual and combine responses

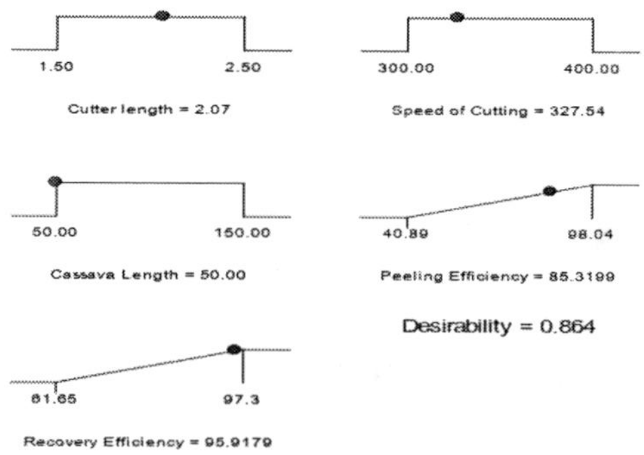


**Figure 12. Ramp** for optimization of the independent variables

## 4. Conclusion

The cutter size, peeling speed and tuber size strongly influence the cassava peeling and recovery efficiency. The cutter size had a more positive effect on the recovery efficiency, followed by the peeling speed while the size of the tuber had a significant negative effect on the peeling efficiency. On the other hand all the three factors had significant negative effect on the recovery efficiency. The optimisation of the parameters gave optimum values of variables; cutter length of 2.07 mm, speed of 328 rpm, and cassava length of 50 mm, which produced peeling efficiency of 85.4% and recovery efficiency of 98.04 %. The result of this investigation give the standard machine input parameters which are equipped for improving quality, peeling and recovery efficiencies of a mechanical cassava peeling system and this will increase production and utilization.

## 5. References

[1] Onyenwoke C A and Simonyan K J 2014 Cassava Post-Harvest Processing and Storage in Nigeria: A Review *African Journal of Agricultural Research* **9**(53), pp 3853-3863

[2] Jimoh M O, Olukunle O J, Manuwa S I, and Amumeji O T 2014 Theoretical analysis of tuber movement during mechanical peeling of cassava *Journal of Mechanical and Civil Engineering* **6**(I), pp 27-36

[3] Alhassan E A, Ijabo O J and Afolabi E O 2018 Development of Cassava Peeling Machine using an Abrasive Mechanism *JPE* **21**(1), pp 61-65

[4] Adetan D A, Adekoya L O, and Aluko O B 2003 Characterization of some Properties of Cassava Root Tubers *Journal of Food Engineering* **59**, pp 349-353

[5] Oluwole K O O and Adio M A 2013 Design and Construction of a Batch Cassava Peeling Machine *Journal of Mechanical Engineering and Automation* **3**(1), pp 16-21.

[6] Hassan A B 2012 Design and Fabrication of a Cassava Peeling Machine. *Journal of Engineering (IOSRJEN). ISSN: 2250-3021* **2**(6), pp 01-08

[7] Ukenna R U and Okechukwu V J 2014 Development and Modification of Cassava Peeling and Washing Machine *Agricultural Engineering Department, B. Eng. Thesis. Federal University of Technology Owerri (FUTO), Nigeria*

[8] Olukunle O J and Akinnuli B O 2013 Theory of an Automated Cassava Peeling System. *International Journal of Engineering and Technology (IJEIT)* **2**, pp 2277-3754

[9] AbdulkadirBH2012 Design and Fabrication of a Cassava Peeling Machine. *Journal of Engineering* **2**(6) pp 01-08.

[10] Olukunle O J, Ogunlowo A S and Sanni L 2010 The Search for Effective Cassava Peeler *The West Indian Journal* **32**(1 and 2), pp 42-47.

[11] Aworanti O A, Agarry A O and Ajani A O 2013 Statistical Optimization of Process Variables for Biodiesel Production from Waste Cooking Oil Using Heterogeneous Base Catalyst *British Biotechnology Journal* **3**(2), pp 116-132

[12] Gbabo A, Gana I M and Solomon M D 2012 Effect of Blade Types on the Blending Efficiency and Milk Consistency of a Grain Drinks *Academic Research International* **12**(3), pp 2223-9944.

[13] Egbeocha C C, Asoegwu S N and Okereke N A A 2016 A Review on Performance of Cassava Peeling Machines in Nigeria Futo *Journal Series (FUTOJNLS)* **2**(1), pp 140 – 168

# Influence of moisture dependent physical properties of fluted pumpkin vital to development it's processing equipments

**J J Shiru[1,2] and I M Gana[1,3]**

[1]Department of Agricultural & Bio-environmental Engineering, Federal Polytechnic Bida, Niger State, Nigeria
[2]jonathanshiru@yahoo.com, [3]ganaibro74@yahoo.com

**Abstract.** Knowledge of physical and mechanical properties of seeds is an important factor for designing agricultural machines and equipment. A fluted pumpkin is a very important vegetable crop as well known for its high nutritional, medicinal and economic values, but the data on its physical and mechanical properties is still lacking. The aim of this study is to determine the effect of moisture content on the physical (basic dimensions, porosity, bulk and true densities, coefficient of friction, repose angle and sphericity) and mechanical properties of fluted pumpkin seed. These properties were evaluated as a function of moisture content at 45, 50, 55 and 60%. The results revealed that length, width and thickness increased with increasing moisture content. Similarly, an average mass of the seed, volume, surface area, sphericity, angle of repose all increased with increasing moisture content. Bulk density decreased as the moisture content increases. The parameters obtained from this study could be used as a reference data for the design of fluted pumpkin seed handling and processing equipment.
**Keywords**: fluted pumpkin, physical, mechanical, properties.

## 1. Introduction

The fluted pumpkin is cultivated in the rain forest zones of Nigeria and other western African countries [1]. As reported by [2] and [3], the economic value, health and nutritional benefits of fluted pumpkin have contributed in its more cultivation and consumption in Nigeria in many other parts of the world. The understanding of physical properties of agricultural materials among other engineering properties is essential to the development of various machine and equipment [4]. As it help in numerous design consideration such as size and capacity of the equipment, shape and angle of machine part and power requirement determination among others. In order to optimize the design of an Agricultural machine, basic engineering knowledge of the crops is important and it is needed by fabricators, engineers, food processors and scientists [5]. Among these engineering properties, physical and mechanical property data are essential for the design of a specific machine or for the analysis of product behaviour in under certain operational conditions [6]. For example, crop grading and packaging is dependent of physical characteristics such as size and shape of the crop. The axial dimensions of the agricultural product determine the size, while the diameter of the product determines the shape [5]. Bulk density is a major factor considered in the construction of storage, transport and separation facilities and machines, as well as in the determination of capacity of machines and equipment [5]. [7] suggested that the performance and efficiency of agricultural machine solemnly dependent on correlation between moisture and physical properties of the material to be processed by the machines. As the optimal efficiency of the machines could only be achieved with a specific range of moisture contents. In the prediction of seed pressure on walls of machines and equipment, frictional properties such as static angle and coefficient friction are important. As well coefficient of friction is critical in the design of storage bins, hoppers, screw conveyors, forage harvesters and threshers [7].

Generally, the development of fluted pumpkin processing equipment is hindered by insufficient data available on its physical properties. Hence, the investigation on effects of change in moisture content level on physical properties of fluted pumpkin seed will help in developing various equipments for its processing. The aim of this study is to investigate the influence of moisture content level (45, 50, 55 and 60% wet basis) used in processing fluted pumpkin seed on a number of physical properties such as length, width, thickness, sphericity, surface area, volume, true and bulk densities, porosity, angle of repose and coefficient of friction that would aid in the design of numerous equipments for processing.

## 2. Methods

The fluted pumpkin seeds (*Telforia occidentalis*) varieties locally recognized as "Ugu" used in this study were purchased from farmer's farm in Jima in Doko district area of Niger State. The seeds were cleaned and sorted before divided into samples [5]. The samples were then conditioned to obtain four different levels of moisture content of 45%, 50%, 55% and 60% wet basis by adding pre-determined volume clean water as reported by [2] and [5]. These moisture levels represent the processing and handling moisture content range of the seed. The moisture content of the seed was determined as reported by [1] by drying the ground seed samples in an vacuum oven set at 105 0C for average 18 hours.

To calculate the average size of the seed, 50 seeds were chosen at random from each of the four moisture contents. The three main longitudinal dimensions including length A, breadth B, and thickness C, were calculated using a micrometer screw gauge with an accuracy of 0.001 mm [6]; [8]. Seed sphericity and surface area were determined using the expression reported by [6]; [9] and are shown in equation (1) and (2) respectively

$$S_p = \frac{(A \times B \times C)^{0.333}}{A} \quad (1)$$

$$S_{as} = \pi \mathrm{De}^2 \quad (2)$$

Where $S_p$ is the sphericity of the seed; A is the linear length of the seed (mm); B is the breadth of the seed (mm); C is the thickness of the seed (mm), $S_{as}$ is surface area of seed ($cm^2$); De is the geometric mean diameter (cm)

The mass of the seed was calculated using an electronic balance with accuracy of 0.001g [8]. At each level of the moisture content, the true density is the ratio of mass of a single seed to the volume of that seed and was calculated using the water displacement method [5]. The bulk density was calculated as reported [5] filling a cylindrical container of 30 cm by 20 cm with the seeds. The container volume was determined, after the mass of the seed was weighed. The buck density was then calculated using the expression in equation (3)

$$\rho_S = \frac{M_{PS}}{V_{ct}} \quad (3)$$

Where, $\rho_s$ is the bulk density of a given volume of the seed ($kg/m^3$), $M_{PS}$ is the mass of the given volume of the seed (kg), $V_{ct}$ is the volume of the container used ($m^3$)

The porosity was calculated in terms of true density and bulk density using the expression in equation (4) as stated by [5]

$$\mathrm{P} = \frac{\rho_T - \rho_S}{\rho_T} \times 100 \quad (4)$$

Where the P is is the porosity of the seed in percentage, $\rho_T$ is the true density of the seed ($kg/m^3$), $\rho_s$ is the bulk density of a given volume of the seed ($kg/m^3$)

The seeds were poured into a hollow container of 30 cm and 20 cm height and diameter respectively. The container was filled and slowly raised until a conical pile of seeds was obtained free of the seeds. The angle of repose was determined by the expressing in equation [5] as reported by [9].

$$\theta = \tan^{-1}\frac{2H}{D} \quad (5)$$

Where $\theta$ the angle of repose (o), H is is the cone height (cm), D is the cone diameter (cm)

The coefficient of static friction of the seeds against glass, plywood and mild steel surfaces were determined. A hollow cylinder of height and diameter of 30 cm and 20 cm respectively was used to determine the coefficient. The cylinder was placed on the surface and then filled with the seeds. The material full of the seeds was slowly lifted with a screw tool until it began slipping. From a graduated scale the angle of tilt was read, and the slope of this angle was taken as the static friction coefficient [10].

## 3. Results and Discussion

### *3.1 Seed dimensions*

The results obtained at the highest moisture level of the seed are similar to those reported by [11] and [12] for sweet maize and maize (Swam 1) respectively. All the dimensions of the seeds decreased with the decrease in moisture level. The seeds enlarge in length, breadth and thickness. The correlation between the axial dimensions and seed moisture level is shown in Table 1. The results point out that increase in size is dependant of moisture content. Within the range of analyzed moistures the seed increased in length by 2.45% (3.67-3.76 cm), likewise it breadth and thickness by 3.16% (3.16-3.26 cm) and 11.18% (1.61-1.79 cm) respectively. The interaction existing between axial dimensions and moisture levels for the seed were found to be linear and can be expressed as shown in Equation (6) to Equation (8) and Figure 1 to Figure 3:

$$S_a = 0.062M + 33.87, R^2 = 0.896 \quad (6)$$

$$S_b = 0.066M + 28.76, R^2 = 0.985 \quad (7)$$

$$S_c = 0.112M + 11.22, R^2 = 0.933 \quad (8)$$

where: $S_a$ is the linear length of the seed, $S_b$ is the linear breadth of the seed, $S_c$ is the linear thickness of the seed, M is the moisture content (%)

The expansion is more linear in breadth of the seed coefficient of correlation value of 0.985. The seeds dimension is very vital feature regarded in determination of sieve apertures and other features in machine design [12]. This in lined with the report of [13], where grain size was found to be a significant factor considered in the design of sieve size in agricultural processing machine design. Also from Table 2, the results of Analysis of variance (ANOVA) of the seed linear dimensions at the the study moisture ranged indicated that increased in moisture content have no significant effects on the linear dimension. Therefore this may not affect design data within the study moisture ranged.

**Table 1.** The values of Physical Properties of Fluted Pumpkin Seeds at the Study Moisture Levels (wb)

| S/N | Charateristics | Moisture Content Levels | | | | |
|---|---|---|---|---|---|---|
| | | Unit | 45% | 50% | 55% | 60% |
| 1 | Length of seed | (mm) | 36.7±16.6 | 36.9 ±16.2 | 37.3±15.4 | 37.6±11.7 |
| 2 | Breadth of seed | (mm) | 31.6±12.87 | 32.2±11.0 | 32.5±9.2 | 32.6±11.7 |
| 3 | Thickness | (mm) | 16.1±6.2 | 17.1±6.5 | 17.3±6.6 | 17.9±8.3 |
| 4 | Mass of seed | (g) | 12.004±7.2 | 16.6±10.9 | 16.34±15.1 | 16.97±11.5 |
| 5 | Bulk density | ($g/cm^3$) | 0.5127 | 0.5066 | 0.5008 | 0.4594 |
| 6 | Surface area | ($cm^2$) | 256±167 | 259±145 | 263±159 | 269±150 |
| 7 | Sphericity | | 0.730±0.135 | 0.739±0.151 | 0.739±0.239 | 0.742±0.151 |
| 8 | Angle of repose | (°) | 29.55 | 30.07 | 32.28 | 41.08 |
| 9 | | | | Static coefficient of friction | | |
| | Plywood | | 0.6249 | 0.7536 | 0.9325 | 1.1917 |
| | Mild steel | | 0.5498 | 0.6249 | 0.7265 | 0.9325 |
| | Glass | | 0.4500 | 0.5295 | 0.6249 | 0.7265 |

**Table 2.** Result of Analysis of variance of linear dimension

| Term | DOF | Sum Sqr | Mean Sqr | F Value | Prob>F |
|---|---|---|---|---|---|
| A-Moisture content | 3 | 1.049167 | 0.35 | | |
| B-Dimension | 2 | 875.1517 | 437.58 | | |
| AB | 6 | 1.288333 | 0.21 | | |
| Residuals | 0 | 0 | | | |

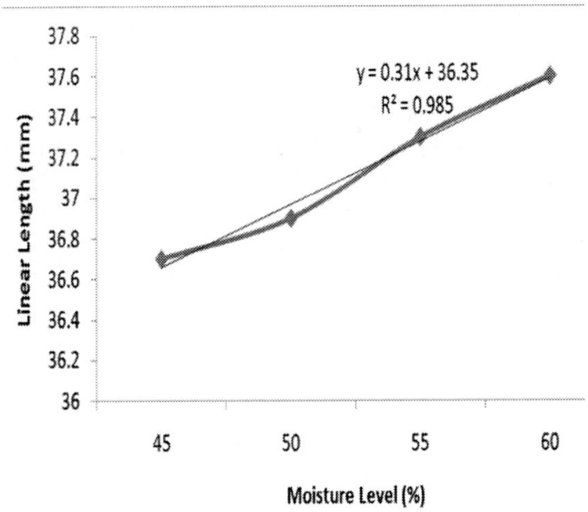


Figure 1. Relationship between linear length and moisture level

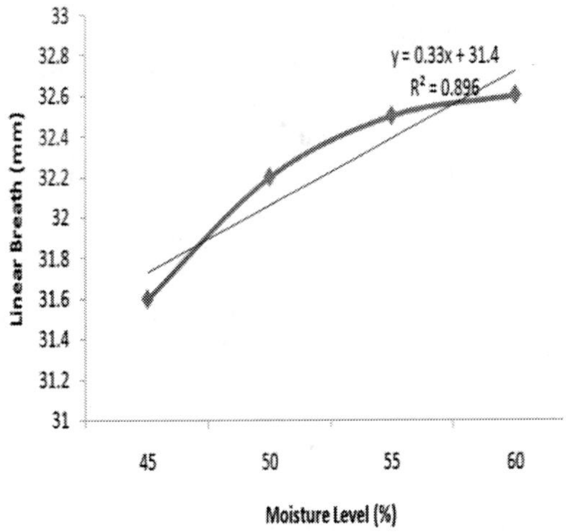


**Figure 2.** Relationship between linear breadth and moisture level

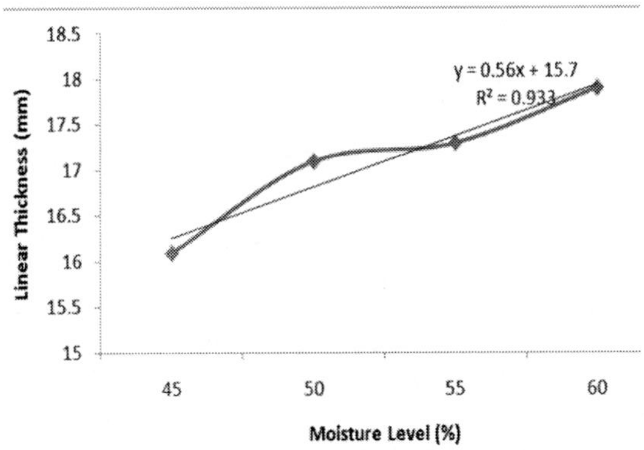


**Figure 3.** Relationship between linear thickness and moisture level

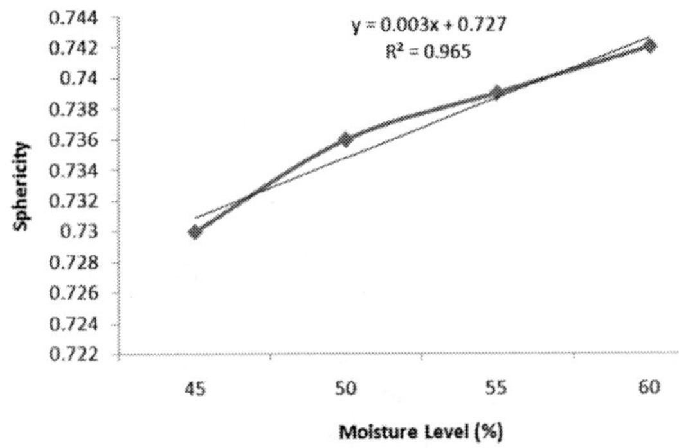


**Figure 4.** Relationship between sphericity and moisture level

From Figure 4, the sphericity of seeds decreased with decrease in moisture level and the relationship between moisture level and sphericity can be represented by the regression equation (9) and coefficient of correlation value of 0.965 indicated that the relationship is more of linear.

$$S_{sp} = 0.0001M + 0.695, \quad R^2 = 0.965 \tag{9}$$

where: $S_{sp}$ is the sphericity of the seed, M is the moisture content (%)

From Figure 8, the surface area of seeds decreased with decrease in moisture level and the relationship between moisture content and surface area can be represented by the regression Equation (10) and coefficient of correlation value of 0.975 indicated that the relationship is more of linear.

$$S_{as} = 4.3M + 251, \quad R^2 = 0.975 \tag{10}$$

where: $S_{as}$ is the surface area of the seed, M is the moisture content (%)

From Table 1 the mass of the fluted pumpkin seed varied from 12.0 g to 16.97 g. At the higher moisture level of 60 percent, the maximum values of 16.97 and 12.0 g were obtained from the higher moisture level of 60 percent and the lower level of 45 percent. It suggests an improvement in mass 41.42% with a rise in moisture level from 45 per cent to 60 per cent (wb). This factor is vital in designing sorting and grading equipment as well as in calculation of power requirement of an agricultural machine [11]. A linear relationship existed between the mass of the seed and moisture content level as shown in Equation (11) and Figure 5:

$$S_M = 0.303M - 0.451, \quad R^2 = 0.704 \tag{11}$$

where: $S_M$ is the mass of the seed (kg), M is the moisture content (%)

As shown in Table 1 the bulk density values increased with decreased moisture level for all levels measured (Figure 6). The moisture content correlation was linear as shown in Equation (12).The finding is consistent with previous research by [11] on sweet maize seed. The increase in bulk density with decrease in moisture level could be because of lower rate of volume expansion is relative to weight gain. This is a vital variable which affects the efficiency and capacity of agricultural machine and its power requirement. For example to process five hundred grammes of this seed at the study moisture levels with equipment with a cylindrical chamber putting into considering the bulk density obtained at each of the level. As shown in Table 3, there are no any significant differences in volume of the seed and the dimension of the container at the various moisture levels. This is an indication that change in the study moisture level does not affect the volume of the seed to be process.

$$S_{bd} = -0.003M + 0.668, \quad R^2 = 0.785 \tag{12}$$

where: $S_{bd}$ is the bulk density of the seed (kg/m$^3$), M is the moisture content (%)

**Table 3.** Relationship between moisture level and some design parameter

| Parameter | **45%** | **50%** | **55%** | **60%** |
|---|---|---|---|---|
| Mass of the seed (g) | 500 | 500 | 500 | 500 |
| Bulk density (g/cm$^3$) | 0.5127 | 0.5066 | 0.5008 | 0.4594 |
| Volume of the seed (cm$^3$) | 975 | 986.97 | 998.4 | 1088 |
| Height of the container (cm) | 10 | 10 | 10 | 10 |
| Radius of the container (cm) | 5.567 | 5.607 | 5.63 | 5.88 |

The angle of repose of the seed decreased linearly with decrease in moisture content from 41.08° to 29.55°. Figure 7 suggests an decrease -39 % in angle of repose with decrease in moisture level from 60 per cent to 45 per cent (wet basis). This is because as the seed absorbed more moisture it becomes heavier and therefore sticks to surface of the material. [9] reported similar finding. This parameter is a vital factor that affects flow of material in processing machines. The relationship can be expressed by the Equation (13) and Figure 7:

$$S_{ag} = 0.303M - 0.451, \quad R^2 = 0.704 \tag{13}$$

where: $S_{ag}$ is the angle of repose of the seed (kg/m$^3$), M is the moisture content (%)

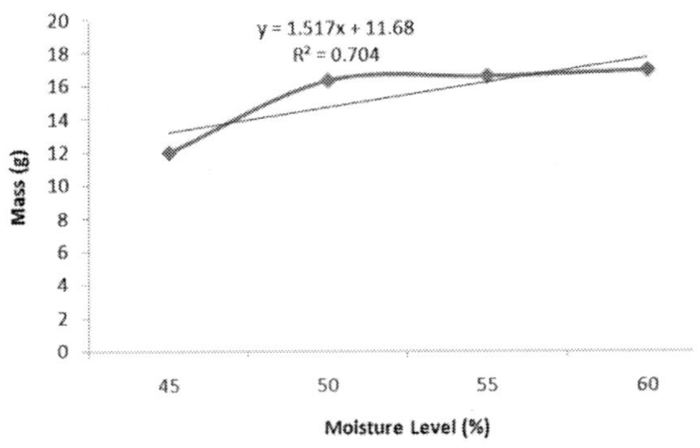


**Figure 5.** Relationship between mass of the seed and moisture level

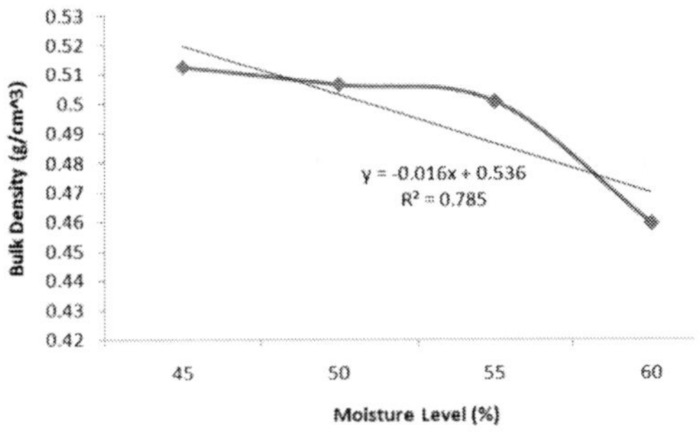


**Figure 6.** Relationship between bulk density and moisture level

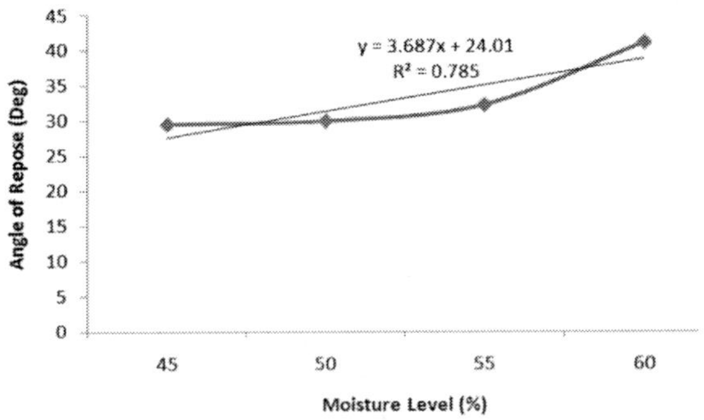


**Figure 7.** Relationship between angle of repose and moisture level

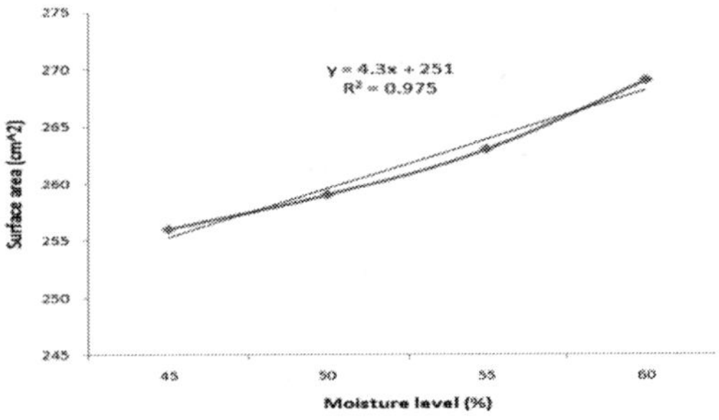


**Figure 8.** Relationship between surface area and moisture level

### *3.2 Static coefficient of friction*

The static friction coefficient of the seeds on glass, mild steel and plywood surfaces decreased linearly with decreasing moisture level. Concrete has the highest static friction coefficient followed by plastic. From all the levels of moisture content, the values were lowest on glass surfaces followed by mid steel and then the highest value from plywood. This is because the glass surface was smoother than the other material under the analysis. Similar trends have been published [5]; [9] on sweet corn. This aspect is crucial in the design of hopper and discharge outlets for the processing machine [5]. The moisture content correlation was linear as shown in Equation (14) to Equation (16) and Figure 8. Also from Table 4, the results of Analysis of variance (ANOVA) of the static coefficient at the study moisture ranged indicated that increased in moisture level of the seed and material surfaces have significant effects on the static coefficient of friction of the seed with moisture level having more effects than the material surface. The moisture level of 45% had coefficient of estimate of -0.18, 50% had value of -0.086 and 55% had value of 0.040. The material surfaces of plywood and mild steel have coefficient of estimate of 0.3 and 0.013 respectively. Therefore, if a machine part that is being influenced by this factor is constructed considering a particular level of moisture content say 45% that part of machine may not function well processing the seed with higher or lower level of moisture compared to the design moisture. Also, if an angle of the machine part was constructed considering particular material surface say plywood, this part may not work well if the surface is replaced with different material. Consequently, the level of moisture and the material surface are very vital factors to consider in the fabrication of equipment for processing fluted pumpkin seed.

$$S_{PW} = 0.187M + 0.405, \quad R^2 = 0.976 \tag{14}$$

$$S_M = 0.125M + 0.396, \quad R^2 = 0.944 \tag{15}$$

$$S_G = 0.092M + 0.351, \quad R^2 = 0.997 \tag{16}$$

where: $S_{PW}$ is the static coefficient of the seed on plywood, $S_M$ is the static coefficient of the seed on mild steel, $S_G$ is the static coefficient of the seed on glass, M is the moisture content (%)

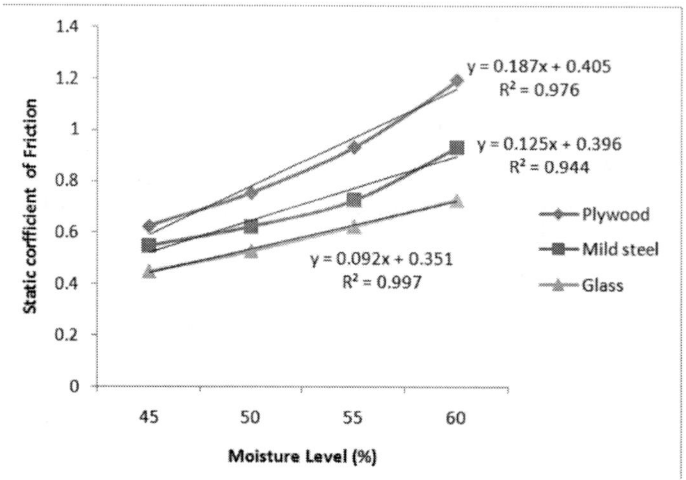


**Figure 9.** Relationship between static coefficient of friction and moisture level

**Table 4.** Result of Analysis of static coefficient of friction

| Source | Sum of Squares | Df | Mean Square | F Value | p-value Prob > F | |
|---|---|---|---|---|---|---|
| Model | 0.40223 | 5 | 0.080446 | 68.82366 | < 0.0001 | Significant |
| A-Moissture content | 0.277039 | 3 | 0.092346 | 79.00456 | < 0.0001 | |
| B-Type of Surface | 0.125192 | 2 | 0.062596 | 53.5523 | 0.0001 | |
| Residual | 0.007013 | 6 | 0.001169 | | | |
| Cor Total | 0.409244 | 11 | | | | |

## 4. Conclusion

Within the study moisture level of 45-60% (w.b), the linear dimensions of the seeds insignificantly increased with the increase in moisture. The mass, volume, sphericity, angle of repose and the static coefficient of friction all decreased with a decrease in the seed moisture level. While the bulk density increased with decrease in moisture level. The implication of the findings is that for any characteristics that is been influenced significantly by the moisture change, the data obtained from one level of moisture of the seed with respect to that characteristic cannot be used to fabricate a machine that will process seed at different moisture level. This also applies to the type of material surface. But for characteristics that is not been influenced significantly by the moisture change, data obtained from one level of moisture of the seed with respect to that characteristic can be used to fabricate a machine that will process seed at different moisture level within the study range of moisture.

## 5. References

[1] Odiaka N I 2008 Diversity and production methods of fluted pumpkin (*Telfairia occidentalis*); Experience with Vegetable Farmers in Markurdi, Nigeria *African Journal of Biotechnology* **7**(8) pp 944 – 954

[2] Ehiagbonare O A 2008 Preparation and Characterization of Activated Carbon Derived from Fluted Pumpkin Stem Waste *Res J Chen sci* **1**(3) pp 10 – 17.

[3] Ajayi S A 2004 Conservation status of *Telfairia spp*. In Sub-Saharan Africa. *In regional Workshop on Plant Genetic Resource and Food Security in West and Central Africa, IITA, Ibradan* April pp 26 – 30

[4] Aremu D O, Ogunlowo Q O, Babajide N A and Makinde O O 2016 Selected engineering properties of fluted pumpkin (*Telfairia occidentalis*) seeds *Journal of Multidisciplinary Engineering Science and Technology* **3** (4) pp 4594 - 4598

[5] Gana I M 2016 Development and Optimisation of Functional Parameters of an Automated Grain Drinks Processing Machine (Unpublished PhD. Engineering Thesis Federal University of Technology, Minna)

[6] Singh K.K and Sahay K M 2003 Unit Operation of Agricultural Processing Physical Properties Second Revised Edition, Vikas Publishing House Pvt. Ltd India

[7] Mohammed R S and Reza A 2010 The Moisture Content Effect on Some Physical and Mechanical Properties of Corn (Sc.704) *Journal of Agricultural Science* **2** (4) pp 125 - 134

[8] Simonya K J, El-Okene A M and Yijep Y D 2007 Some Physical Properties of Samaru Sorghum 17 Grains *Agricultural Engineering International: The CIGR Journal Manuscript Fp 07 008* vol IX

[9] Zubairu U B, Aimrun W, and Amin M S 2014 Determination of some Physical Properties of common Malaysian Rice MR219 Seeds *Australian Journal of Crop Science* **8** (3) pp 332-337.

[10] Uchenna S.H (2006) Studies of selected Physiochemical Properties of Fluted Pumpkin (*Telfairia occidentalis*) Seed Oil and Tropical Almond (*Terminalia catalpa l.*) Seed Oil *Pakistan Journal of Nutrition* **5** (4) pp 304 – 306.

[11] Karababa E and Coskuner Y 2007 Moisture Dependent Physical Properties of Dry Sweet Corn Kernels *International Journal of Food Prop* **10** pp 549 – 560.

[12] Sobukola O P, Kajihausa O E, Onwuka V I, and Esan T P 2013 Physical Properties of High Quality Maize (Swam 1 Variety) Seeds (Zey mays) as Affected by Moisture Levels *African Journal of Food Science* **7** (1) pp 1 – 8.

[13] Abdullahi M H 2011 Determination of Some Physical Properties of nut meg (*Mynstical fragrans)* Seed Research *Journal of Applied Services Engineering and Technology* **2** (7) pp 667 – 672

[14] Chukwu O and Orhevba B A 2011 Determination of Selected Engineering Properties of Soya Beans (*Glycine max*) related to Design of Processing Machine *Journal of Agricultural Food Technology* **1** (6) pp 68-72.

IOP Conf. Series: Materials Science and Engineering **847** (2020) 012087 doi:10.1088/1757-899X/847/1/012087

# Workload and Fatigue Assessment on Air Traffic Controller

**V Triyanti[1,2], H A Azis[1], H Iridiastadi[1] and Yassierli[1]**

[1]Institut Teknologi Bandung, Jl. Ganesa 10, Bandung, Indonesia
[2]Atma Jaya Catholic University of Indonesia, Jl. Jend Sudirman 51, Jakarta, Indonesia
vivi.triyanti@atmajaya.ac.id

**Abstract.** Workload experienced during certain work periods may cause fatigue. In Air Traffic Control work, a high air traffic rate may increase the workload and fatigue, especially during rush hour and may lead to risky situations. This study intended to estimate the value of the mental workload and fatigue of an Air Traffic Controller (ATCo) in airport X during their work period. The NASA-TLX and Samn Perelli Scale were used to quantify mental workload and fatigue. The questionnaires were distributed to ATCo and 257 questionnaires were obtained. Results showed the value of the mental load at airport X tended to be higher than other airports (average 66.53), especially for mental demand, time pressure, and effort. The value of fatigue before and after work also increased (average 1,749 to 3,506) included in the safe category. However, there was no convincing evidence that differences in mental load values affect the value of fatigue. These results indicated that although the workload of ATCO was relatively high, the ATCo in Airport X still can handle it and did not cause excessive fatigue.
**Keywords**: Air Traffic Controller, fatigue, workload, Samn Perelli scale, NASA-TLX

## 1. Introduction

Task demands related to cognitive work will lead to complexity in thinking systems. The accumulation of all task demands and complexities that occurred in the human thinking system may cause workload [1], especially mental workload. Mental workload is the amount of attention allocated to carry out an activity [2]. The impact of high mental workloads (or even moderate) experienced in a long duration was the occurrence of fatigue, especially mental fatigue [3,4,5]. Researches showed that the main trigger for mental fatigue was time-on-task (TOT) [5,6,7]. For example, mental demands experienced for 2 hours continuously will cause mental fatigue and potentially slow down the work (increase reaction time) [6].

Mental fatigue will cause several negative impacts on one's performance and motivation. The fatigue can reduce the level of alertness, concentration, speed, and accuracy in doing work [3, 4]. Mental fatigue was worsened when someone also experiences drowsiness [8]. A person who experienced fatigue due to lack of sleep will experience decreased ability in information processing, memory, problem-solving, and decision making [9]. The potency for mental fatigue increased when work involved stress, such as time pressure [10] or pressure due to high work risks [11].

Air Traffic Control is a service task whose duty is to prevent conflicts between aircraft. Its main tasks are to regulate air traffic and to provide information and support for pilots [12]. To prevent collisions, continuous vigilance is needed [1]. Constant vigilance is needed to ensure that all aircraft in the jurisdiction are always safe. The safe criteria are when aircraft separated at a safe distance and can reach their destination safely, according to the International Civil Aviation Organization (ICAO) rules [13].

The task demands related to the complexity of air traffic, airspace, and operational traffic conditions [14] cause mental workload on ATC officers.

Some previous studies have estimated the amount of mental workload on ATCo [2,12,14]. However, those studies had not yet explained the relationship of the workload with fatigue experienced by the ATCo. Besides, these studies have not determined whether the workload experienced by ATC officers at the airport was classified as high or not, compared to the mental workload experienced by similar professions (ATC) at other airports.

This study was aimed to estimate the mental workload and fatigue conditions of ATC officers in a real work system. The ATC work was chosen because it has a combination of three things, namely the potential to experience a high mental workload, the need for constant vigilance, and high work risks. High task demands have the potential to cause mental workloads and fatigue at ATC officers, which in turn can reduce performance and cause the risk of aircraft-related incidents [5].

By using ATC as the object of study, the objective of this study was to estimate the value of mental load and fatigue for ATC officers. Through this study, the expected output was the estimation of mental load and fatigue conditions in real work situations in ATC. Based on this goal, the hypotheses developed were 1) ATC is a job with a high mental workload and 2) There is an increase in fatigue after carrying out their duties. A field survey was conducted on the ATCo to collect data related to this hypothesis, on the actual ATC working conditions. The survey focused on values and factors that influence mental workload and fatigue. This survey did not discuss performance and its relation to mental workloads and fatigue.

## 2. Methods

### *2.1. Participant and Apparatus*

The survey was conducted in May 2019. There are two ATC units were examined, namely ACC and APP. In one shift, numbers of questionnaires were randomly distributed to officers in both units with the help of supervisors. If there were difficulties, the researcher assisted the ATCo in filling them. After the questionnaire was filled in, the researcher ensured that all questions have been filled out completely and following the provisions before the questionnaire was collected. During each shift, the assessment of workload was conducted only one time. However, controllers must assess the level of fatigue they experienced two times, which were before and after doing work. From 3 days of research, 263 questionnaires had been collected, but only 257 data can be processed. The six questionnaires were disused because they were filled out by supervisors or field managers. Supervisors or managers should not fill the questionnaires because they were not directly in charge of managing the traffic, but only supervised the work of ATCo. This research used a simple random sampling design. Samples are taken randomly, without regard to differences in conditions that exist in the population [15].

### *2.2. Workload Assessment*

In this study, the unweighted NASA-TLX approach was used to estimate the workload. NASA TLX is a multi-aspect approach that can also be used to estimate the value of mental demand and physical demand. The approach was chosen because it proved to be closely correlated with several other subjective load assessment methods, such as SWAT and WP [16]. NASA – TLX also has relatively good validity, reliability, and sensitivity values [17]. Also, this scale was widely used to assess mental workloads in the field of Aviation [9].

NASA-TLX consists of 6 subscales: Mental Demand (MD), Physical Demand (PD), Time Pressure (Temporal Demand / TD), Frustration (Frustration), Effort (Effort / EF), and Performance (PE). Twenty equal intervals - marked between labels at each end from 'low' to 'high' or 'good' to 'bad' - were used to rank on these dimensions, producing a score between 0 and 100. An average of the six-dimensional workloads represented the overall workload.

IOP Conf. Series: Materials Science and Engineering **847** (2020) 012087 doi:10.1088/1757-899X/847/1/012087

*2.3 Fatigue Assessment*

The Samn-Perelli Crew Status Check (referred to as Samn-Perelli scale) was a method for measuring mental fatigue that was originally developed for military air transport operations [18]. It had been widely used in flight studies [9,19]. Consistent with the definition of ICAO fatigue [13], this scale ranking was directly related to the likelihood of performance degradation on a scale of 1 to 7. For interpretation of this score, categorization can be used [20]. This criterion was based on operational experience in military airlift operations.

**Table 1** Interpretation of scores from the Samn-Perelli Scale [20]

| Score | Interpretation |
|---|---|
| 1-3 | There is likely no performance decrease related to fatigue that was expected |
| 4 | Performance degradation is possible but not a significant factor |
| 5 | Moderate fatigue. Decreased performance related to fatigue may occur. Execution of tasks is permitted but not recommended unless urgent |
| 6 | Severe fatigue. Decreased performance related to fatigue may occur. The execution of tasks was not recommended. |
| 7 | Severe fatigue. Performance must be disrupted. The execution of tasks was not recommended. Flight safety is in danger |

## 3. Result and Discussion

The results are presented sequentially for each hypothesis.

*3.1. Hypothesis 1: ATC is a work with a high mental workload*

The mental workload was calculated using the unweighted NASA-TLX approach on a scale of 1-100. To estimate the category of mental workloads, a percentile approach was used. Twenty-four studies had been conducted on ATCo to determine the percentile limit of mental load [21] (Table 2). This value was then compared to the data collected in the current study. It appears that the 25th, 50th, and 75th percentiles of this study are far greater than those produced in the 24 ATC studies. This result can also be interpreted that the mental workload felt by the ATCo at the study site was classified as high.

**Table 2** Percentile of ATCo's workload value of this research compared to literature [21]

| Percentile | Workload value | | |
|---|---|---|---|
| | Average of 24 ATC | Rounded | This research |
| P-25% | 42.81 | 43 | 55 |
| P-50% | 52.44 | 52 | 65 |
| P-75% | 68.32 | 68 | 74 |

Research from [22] was used to interpret each mental load value in this study. The research suggested that mental load with values <50 means "acceptable", 50-70 means "within limits", and mental load > 70 as "sometimes acceptable in certain situations." Based on the previous studies, a justification for mental load categories for ATC tasks was carried out in the medium, high and very high categories (Table 3).

**Table 3** Categorization of NASA-TLX value of ATCo

| Justification | Category | Interpretation |
|---|---|---|
| < 43 – 52 | Medium | Acceptable |
| 52 – 68 | High | On borderline |
| > 68 | Very high | Only can be accepted in a certain situation |

With justification in Table 3, values 52 - 68 indicated a high mental workload. Based on the assessment of mental load, the total score of mental workloads was 66.53, which was included in the high category. Mental demands, physical demands, and time pressure are aspects of workload that are directly felt by workers. On the other hand, performance, effort, and feeling of frustration are aspects of workload that are felt as impacts [16]. From the scores in Table 4, it is shown that the highest workload value is Effort. However, all aspects of workload cause a relatively high workload, especially in aspects of mental demands and time pressure. However, with the very high value of mental demands, time pressure, and effort, the ATCo felt that they had done the job with good performance, so the feeling of frustration felt was also relatively lower.

**Table 4** NASA-TLX Workload assessment of ATCo

| Aspects | Score | SD | Category |
|---|---|---|---|
| Mental demand | 75 | 3.99 | Very high |
| Physical demand | 66 | 4.50 | High |
| Time pressure | 74 | 3.85 | Very high |
| Performance | 53 | 3.30 | High |
| Effort | 81 | 3.58 | Very high |
| Frustation | 50 | 5.69 | Medium |
| Average | 66.53 | 2.90 | High |

*3.2. Hypothesis 2.1: There is an increase in fatigue during his work period*

Mental fatigue was assessed using the Samn-Perelli scale with a value of 1-7; the higher means more tired [13]. From Table 5, it can be seen that the average value of fatigue before work is 1,749, which is included in the category "Very excited, responsive, but not at its peak." Values below 3 indicate that the officer is relatively tired. A value of 4 indicates slight fatigue [20]. The value of fatigue after work, on average, is 3,506, which belongs to the category "A little tired, less than fresh". For a score of 4, a decrease in performance may occur but not a significant factor [20]. This value condition indicates that the maximum working time setting of 2 hours as implemented by the ATC organizer has been effective enough to prevent excessive fatigue.

Based on data processing, it is known that the potential for fatigue continues to occur because all data showed an increase in the value of fatigue before and after work. With a paired t-test with a significance level of 0.05, it was concluded that fatigue before and after work was significantly different ($p$-value = 0.0000 and T value = 23.22), with a higher level of fatigue after work (average difference of 1.7569). However, the average value of fatigue both before and after work is still at a reasonable level, because it is only still slightly above 3 with a relatively small standard deviation.

**Table 5** Fatigue value before and after work

| Group | Average | Variance | SD |
|---|---|---|---|
| Before work | 1.749 | 0.874 | 0.935 |
| After work | 3.506 | 1.330 | 1.153 |

*3.3. Discussion*

The main objective of this study was to assess workload and fatigue experienced by Air Traffic Controller. The overall results of the study indicate that officers at Airport X ATC have a tendency to have a high workload during their work shifts. When viewed from the traffic rate, the actual average traffic during 2019 decreased considerably compared to the previous year (Figure 1). In 2018, the average number of flights per month is relatively higher because at that time flight fares have not been adjusted. In 2019 there was an increase in flight rates [23,24], which caused a decrease in the number of passengers and the number of flights. Referring to this data, there are two possibilities for the data obtained. The first possibility, the officer still does not feel a reduction in workload associated with a reduction in traffic. This is possible if there is an adjustment in the number or extent of sectors that must be handled by an officer. This adjustment could cause the mental burden felt by officers to remain the same, despite the reduction in air traffic. The second possibility is that the workload in the previous conditions is very high.

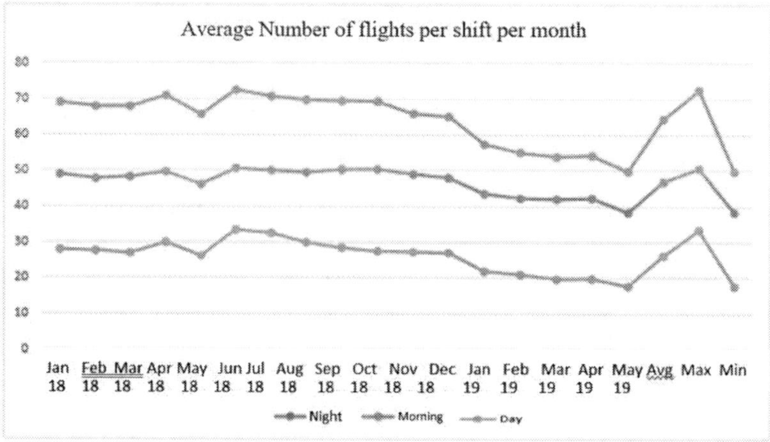


**Figure 1.** Average Number of flights per shift per month

The second possibility is that the workload in the previous condition is indeed higher. This is indicated by the average number of flights per hour which is close to the maximum value, which is 80 aircraft per hour [25]. This value mainly occurs in the morning shift. From this data, it can also be seen that high traffic occurs in the morning shift (06.00-12.00) and night (20.00-10.00). These two shifts are likely the second most optimal time as departure and return times from business destinations.

Based on the results, we may also analyze the relationship between the mental workload and fatigue. There is a possibility that a higher workload may cause a higher level of fatigue. The mental load perceived by ATCo was classified into the medium, high and very high, according to Table 3. Based on these categories, the level of fatigue was compared for each pair. By using two independent sample T-tests, it is concluded that there was no strong evidence that differences in mental workload level have a significant effect on fatigue value (Table 6). However, if examined further, we can see in all workload conditions, the fatigue tend to increase after work. The increasing value tends to be higher in higher workload. It means that higher workload tends to cause a higher level of fatigue, although the increase is not significant.

Although we can have explanations of workload conditions in Airport X, the research still has many limitations. There are a lot of factors that can both workload and fatigue that have not been explored yet in this manuscript. The first factor is the ATC work schedule. ATC is a job that must be run by 24 hours continuously without stopping. Work shift and amount of work time can have an impact on the risk of fatigue. For example, working when workers are biologically programmed to sleep (which can interfere with the worker's body clock) and working for long periods can cause fatigue. In addition to work-related factors, individual and environmental factors that are not related to work can also affect fatigue. The quantity and quality of sleep, health, age, sex, emotions, and social conditions are one example of this factor. Analyzing all these factors. The analysis of factors that can influence the mental burden and fatigue of ATC officers can be an interesting discussion for future research.

**Table 6** Output *two samples independent t-test* of workload (low, medium, high) to fatigue difference

| Hari | Count | Mental fatigue | |
|---|---|---|---|
| | | Average | SD |
| Medium | 34 | 1.676 | 1.389 |
| High | 102 | 1.765 | 0.913 |
| Very high | 121 | 1.793 | 1.715 |
| Comparison | | *p*-value | *T*-value |
| Medium – High | | 0.662 | 0.438 |
| Medium – very high | | 0.778 | 0.282 |
| High- Very high | | 0.393 | 0.856 |

## 4. Conclusion

Air Traffic Control is a job that requires the ability to remain vigilant and make decisions continuously throughout its working period. These task demands cause relatively high workloads. At Airport X, the workload felt by officers was also relatively high, especially in aspects of effort load, mental demand, and frustration. The total workload value is even higher than the average ATC work at other airports. However, the high workload does not cause excessive fatigue in ATCo. Fatigue before and after work significantly increased, but still within the limits of reasonableness. Furthermore, this increase in fatigue is related to an increase in mental load, but no evidence proved that the relationship is significant. For further research, it is advisable to investigate further the factors associated with mental burden and fatigue, both factors related to or not related to work.

## 5. References

[1] Hilburn B 2004 Cognitive complexity in air traffic control – A literature review, *EEC Note 04/04*, Eurocontrol.

[2] Ranchet M, Morgan J C, Akinwuntan A E, and Devos H 2017 Cognitive workload across the spectrum of cognitive impairments: A systematic review of physiological measure *Neuroscience and Biobehavioral Reviews* **80** 516-37

[3] Mehta R K and Parasuraman R 2014 Approach effects of mental fatigue on the development of physical fatigue: A neuroergonomic *Human Factors: The Journal of the Human Factors and Ergonomics Society* **56** 4 645-56

[4] Cutsem, J V Marcora S, De Pauw K, Bailey S, Meeusen R, and Roelands R 2017 The effects of mental fatigue on physical performance: a systematic review *Sports Medicine* (Switzerland: Springer International)

[5] Dasari D, Shou G and Ding L 2017 ICA-Derived EEG Correlates to mental fatigue, effort, and workload in a realistically simulated Air Traffic Control task *Frontiers in Neuroscience* 11 297 1-12

[6] Lorist M M, Boksem M A S, Ridderinkhof K R 2005 Impaired cognitive control and reduced cingulate activity during mental fatigue, *Cognitive Brain Research,* **24** 2 199-205

[7] Cazzoli D, Antoniades C A, Kennard C, Nyffeler T, Bassetti C L, and Müri R M 2014 Eye movements discriminate fatigue due to chronotypical factors and time spent on task - a double dissociation *PLoS One* **9** 1 e87146 1-5

[8] Work Safe Victoria 2008 *Fatigue Prevention in the workplace* (New Zealand: Work Safe Victoria)

[9] Gawron V J, Kaminski M A, Serber M L, Payton, Hadjimichael M, Jarrott W M, Estes S L, Neal T A 2011 Human Performance and Fatigue Research for Controllers Revised, MTR Mitre Technical Report, Federal Aviation Administration

[10] Cheng S Y and Hsu H T 2011 Mental fatigue measurement using EEG *in* Nota G, ed. *Risk Management Trends* InTechOpen 203-228

IOP Conf. Series: Materials Science and Engineering **847** (2020) 012087 doi:10.1088/1757-899X/847/1/012087


[11] Loft S, Sanderson P, Neal A, and Mooij M 2007 Modeling and Predicting Mental Workload in En Route Air Traffic Control: Critical Review and Broader Implications *Human Factors* **49** 3 376–99

[12] Hilburn B 2004 Cognitive complexity in air traffic control – A literature review, *EEC Note 04/04*, Eurocontrol

[13] International Civil Aviation Organization (ICAO), International Federation of Air Traffic Controllers Associations (IFATCA), and Civil Air Navigation Services Organization (CANSO) 2016 *Fatigue Management Guide for Air Traffic Service Providers, first edition* International Civil Aviation Organization

[14] Tobaruela G, Schuster W, Maajumdar A, Ochieng W Y, Martinez L and Hendrickx 2014 A method to estimate Air Traffic Controller mental workload based on traffic clearances *Journal of Air Transport Management* **39** 59-71

[15] Malhotra, N.K. 2009 Marketing Research: An Applied Orientation 6th Edition (UK:Pearson)

[16] Hart S G and Staveland L E 1988 Development of NASA-TLX (Task Load Index): Results of empirical and theoretical research. in P. A. Hancock and N. Meshkati (Eds.) *Human mental workload* (139-183) (Amsterdam: North Holland Press)

[17] Hoonakker P, Carayon P, Gurses A, Brown R, McGuire K, Khunlertkit A, Walker J M 2011 Measuring workload of ICU nurses with a Questionnaire survey: the NASA Task Load Index (TLX), *IIE Trans Healthc Syst Eng.* **1**, 2, 131–143.

[18] Samn S and Perelli L. 1982 *Estimating Aircrew Fatigue: A Technique with Application to Airlift Operations (SAM-TR82-21)* (Brooks Air Force Base, TX: U.S. Air Force)

[19] Signal T L, Gander P H, Anderson H. and Brash S 2009 Scheduled napping as a countermeasure to sleepiness in air traffic controllers. *Journal of Sleep Research* **18** 11-19

[20] Chang Y-H, Yang H-H, Hsu W-J 2019 Effects of work shifts on fatigue levels of Air Traffic Controllers *Journal of Air Transport Management* **76**, 1–9

[21] Grier R A 2015 How high is high? A meta-analysis of NASA-TLX global workload scores *Proceedings of the Human Factors and Ergonomics Society 59th Annual Meeting*, Los Angeles, California

[22] Eiterheim M H and Fernandes A 2016 The NASA Task Load Index for rating workload acceptability *Human Factors and User Needs in Transport, Control, and the Workplace - HFES - Europe Annual Meeting 2016*, Prague

[23] Samadi B K 2019 *Peraturan Menteri Perhubungan Nomor 20 Tahun 2019 tentang Tata Cara dan Formulasi Perhitungan Tarif Batas Atas Penumpang Pelayanan Kelas Ekonomi Angkutan Udara Niaga Berjadwal Dalam Negeri* Jakarta: Kementrian Perhubungan Indonesia

[24] Samadi B K 2019 *Keputusan Menteri Nomor 72 Tahun 2019 tentang Tarif Batas Atas Penumpang Pelayanan Kelas Ekonomi Angkutan Udara Niaga Berjadwal Dalam Negeri* Jakarta: Kementrian Perhubungan Indonesia

[25] Samadi B K 2017 Instruksi *Menteri Perhubungan Nomor 16 Tahun 2017 tentang Perataan Distribusi Jadwal Penerbangan Dan Slot Time Di Bandar Udara Internasional Soekarno - Hatta Tangerang Untuk Peningkatan Pelayanan Dan Keselamatan Penerbangan* Jakarta: Kementrian Perhubungan Indonesia

IOP Conf. Series: Materials Science and Engineering **847** (2020) 012088 doi:10.1088/1757-899X/847/1/012088

# Solar Panel-based Wireless Battery Charging System using Fuzzy Control Method

**Alfarid Hendro Yuwono[1], Muhammad Rivai[2], and Tri Arief Sardjono[3]**

Department of Electrical Engineering, Institut Teknologi Sepuluh Nopember, Surabaya, Indonesia.
[1]alfarid.man@gmail.com, [2]muhammad_rivai@ee.its.ac.id, [3]sardjonotriarief@bme.its.ac.id.com

**Abstract.** Rapid technological development leads to the growth of electronic devices powered by batteries. This requires a simple, fast and safe battery charging system. In this study, we have designed and realized a wireless battery charging system supplied by solar panel using the fuzzy control method. This system is equipped with solar tracking to produce optimal electric power. Furthermore, fuzzy control system is applied to determine the amount of charging voltage by taking into account the temperature of the battery. The experimental results show that the solar tracking system can improve the performance of solar panel. The system always detects the temperature of the battery and will reduce the voltage value during the charging process as the temperature increases. This battery charging system will also automatically decrease the charging voltage when the battery voltage approaches the maximum value. This system is expected to be used as a fast battery charging system without affecting the battery life.
**Keywords**: Fuzzy control method, Solar panel, Wireless battery charging.

## 1. Introduction

Currently, technological developments have an impact on the increasing number of portable electronic devices that are supplied with batteries such as cellphones, computers, and vehicles. This requires a simple battery charging system such as the wireless power transfer method. This method transfers energy of the electromagnetic field using inductive coupling. The energy is sent through the coil to an electrical device which can then be used for battery charging [1]. Research on the wireless power transfer is being developed that can be applied in various fields. In the medical field, wireless power transfer is used to supply implantable electronic devices to a patient's body [2]. The automotive industry has also begun to develop wireless power transfer systems to charge electric cars [3], [4]. One of the problems regarding wireless power transfer systems is the length of time required to charge an electronic device [5]. This is because the power transferred by the system tends to be constant regardless of the condition of the power available on the charged battery. Improper charging methods can affect the battery life.

Nowadays solar panels are considered as an alternative solution for a source of electricity that is low cost and does not produce pollution. Solar panels are still rarely used in both communities and industrial areas. The solar tracking system can be applied in order the solar panels can produce optimal electrical energy [6]. Among the control methods, fuzzy logic is often used widely in various fields such as the solar tracking system [7], robot navigation [8], [9], and electrostatic air filtration system [10]. This is because the fuzzy control method is able to apply to non-linear elements without having to find an exact mathematical model [11]. In this study, we have designed and realized a wireless battery charging system supplied by solar panel using the fuzzy control method. This system is expected to produce optimal solar energy and fast charging time without affecting the battery life.

## 2. Methods

The overall block diagram of a solar panel-based wireless battery charging system is depicted in figure 1. In general, the system consists of three main parts including solar tracking, wireless battery charging, and fuzzy control method. The Arduino microcontrollers are employed to measure the sensor signal, to determine the amount of transferred energy, and to control the battery charging system.

### *2.1. The solar tracking*

The solar tracking system is applied in order the solar panel can follow the movement of the sun which keeps the surface always perpendicular to the sunlight as shown in figure 2(a). This system involves a 100 WP solar panel, a servo motor, the Arduino Nano microcontroller, and two Light Dependent Resistor (LDR) sensors. The LDR resistance decreases with increasing light intensity and vice versa. The change in resistance will be represented by a change in voltage in a voltage divider circuit. This will give action to the servo motor to change the direction of the solar panel so that the two sensors receive similar energy indicating the surface of the solar panel is perpendicular to sunlight. The electrical energy produced by solar panels is stored in a battery or accumulator for supplies in the absence of sunlight either when covered by clouds or at night. The solar tracking algorithm is illustrated in figure 2(b).

### *2.2. The wireless battery charging*

The working principle of wireless power transfer is to convert DC signals into sinusoidal AC signals which are then sent to the primary coil and emitted in the form of magnetic flux so that it can be received by the secondary coil [12], as shown in figure 3. The sine wave applied to the coils can provide higher power transfer efficiency [13]. The rectifier and low pass filter circuits are employed to ensure the battery receives a DC signal for charging purposes. In this experiment, we set the distance between the coil is about 1 cm with the series-parallel (SP) topology. The transmitter inverter is provided by the AD9833 Programmable Waveform Generator, X9C103 Digitally Controlled Potentiometer, and TDA7386 Amplifier modules. The AD9833 module can be programmed to produce sine, triangle, and square wave signals with a frequency range of 0 MHz to 12.5 MHz. This device has the function of turning off the power when the device is not in use to minimize electrical power consumption. The X9C103 module has 100 wiper tap points which the position can be stored in non-volatile memory. The TDA7386 module has a high output power capability of 4 x 49 W / 4 Ω load with high cut-off frequencies up to 200 kHz.

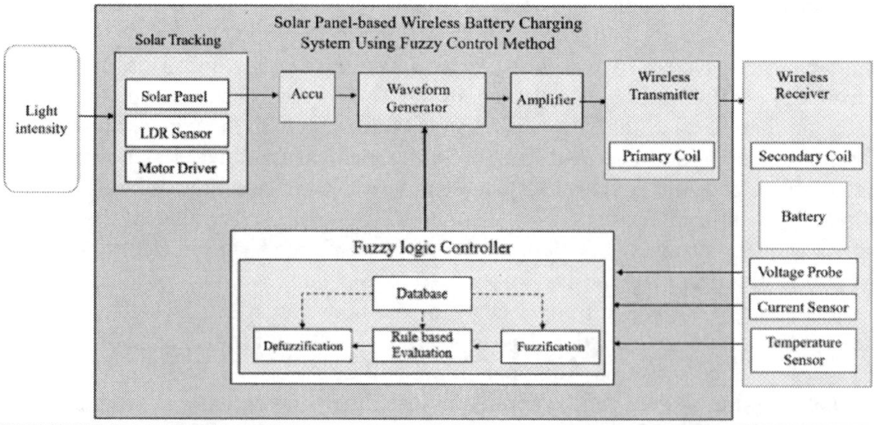


**Figure 1.** The overall block diagram of a solar panel-based wireless battery charging system.

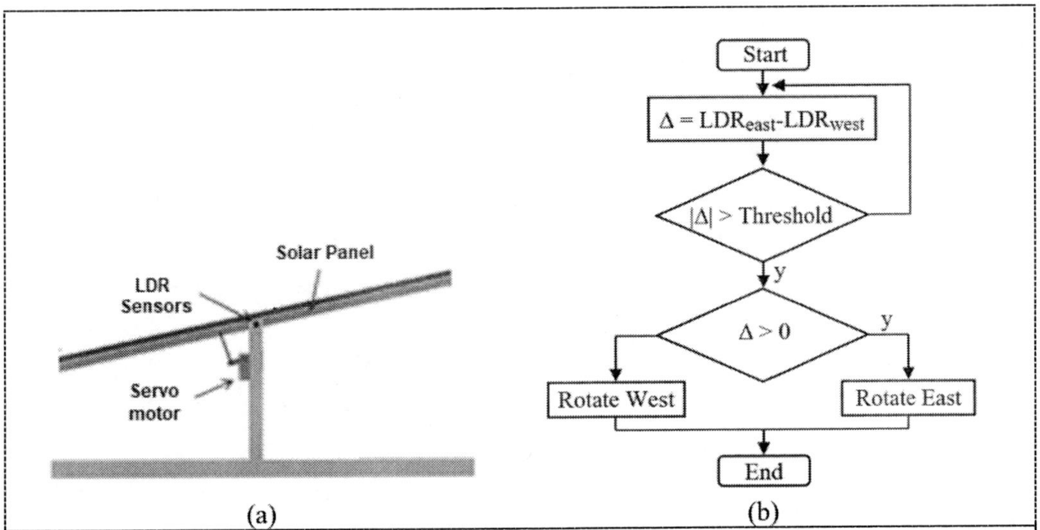


**Figure 2.** The solar tracking system: (a) mechanical construction, and (b) solar tracking algorithm.

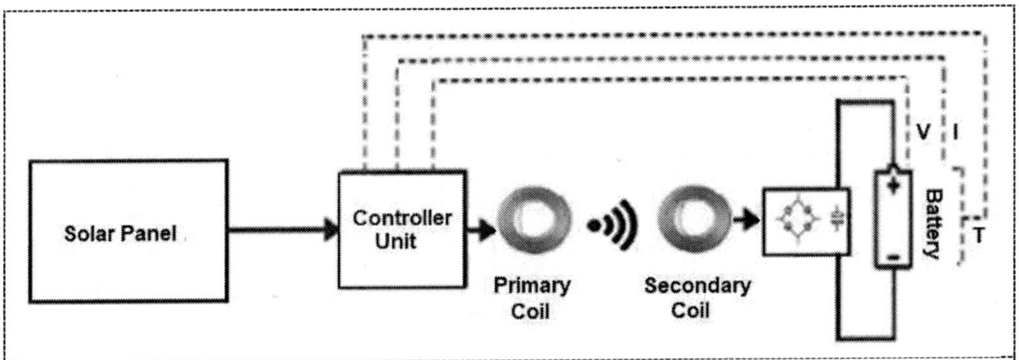


**Figure 3.** The wireless battery charging.

### *2.3. The fuzzy control method*

Fuzzy logic is a mathematical means that represents decision making based on imprecise and non-numeric information. In this method, the variables can assume linguistic values. Fuzzy logic consists of three stages, namely fuzzification, inference, and defuzzification depicted in figure 4. Fuzzification is the process of converting crisp input into a fuzzy input [14]. In this study, there are two input membership functions, namely the electric current (I) and the temperature (T) of the battery as shown in figure 5(a). Fuzzy control system is used to determine the amount of charging voltage by taking into account the temperature of the battery. When the battery temperature is low, the charging voltage will be high so that the battery fills up quickly. Moreover, when the temperature of the battery begins to rise, the charging voltage will be adjusted so that it does not damage the battery. The analog signals from the current and temperature sensors are converted to digital values using a 10-bit Analog-to-digital converter (ADC) embedded in the Arduino Uno microcontroller.
Inference consists of IF-THEN rules expressed as:

$$if\ X_1\ is\ A_1\ and\ ...\ and\ X_n\ is\ A_n\ then\ Y\ is\ B \quad (1)$$

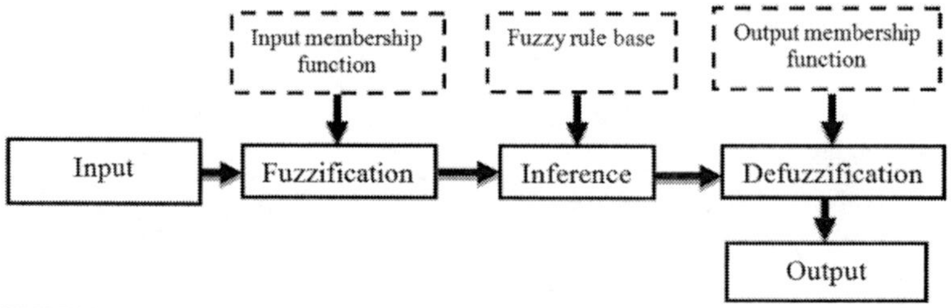


**Figure 4.** The block diagram of fuzzy logic.

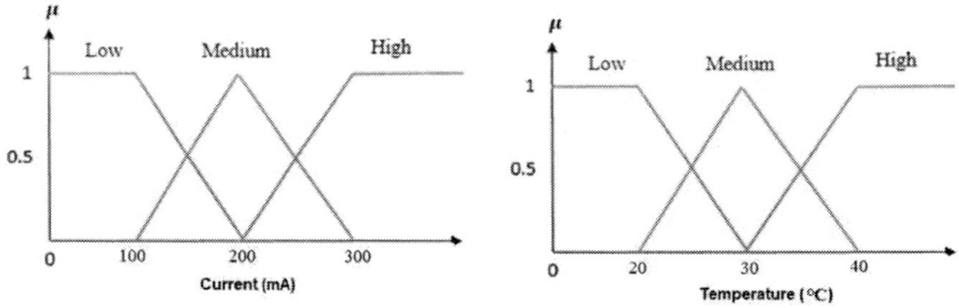


(a)

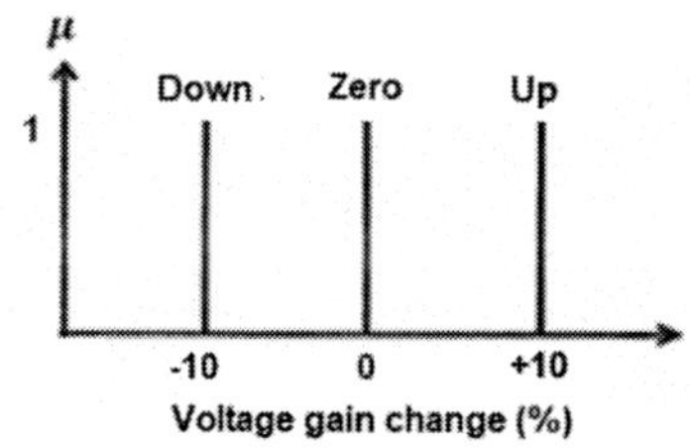


| T \ I | $I_L$ | $I_M$ | $I_H$ |
|---|---|---|---|
| $T_L$ | Up | Up | Zero |
| $T_M$ | Up | Zero | Down |
| $T_H$ | Zero | Down | Down |

(b) (c)

**Figure 5.** The Fuzzy Interference System: (a) input membership functions, (b) output membership functions, and (c) rule evaluation.

Meanwhile, the implication function of the rules can be declared as:

$$\mu_B(y) = \max[min[\mu_{A1}(input(i)), \mu_{A2}(input(j)), \ldots]] \tag{2}$$

The fuzzy output membership functions, and the rule evaluation are shown in figure 5(b) and figure 5(c), respectively. Defuzzification is the process of converting fuzzy output to crisp output [15]. The defuzzification method used in this study is the Center of Area (CoA) which is expressed as follows:

$$Z_0 = \sum_{i=1}^{n} \frac{\mu(Z_i).Z_i}{\mu(Z_i)} \tag{3}$$

## 3. Result and Discussion

The solar tracking system used in the experiments is shown in figure 6(a). In the first experiment, we compare the output power of solar panel with and without tracking system recorded every hour. The result of this experiment is shown in figure 6(b). In the morning and evening, the solar panel equipped with the tracking system produces higher electrical power. The solar panel equipped with solar tracking provides a higher average electric power of 44%. This shows that the solar tracking system can improve the performance of solar panel.

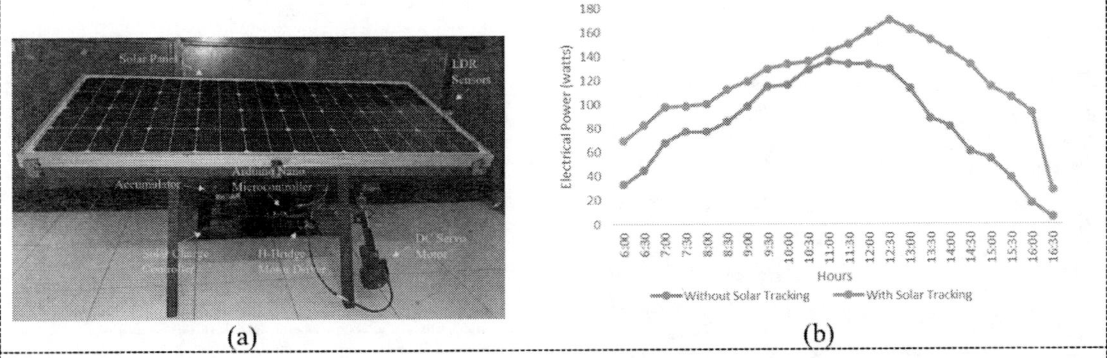


**Figure 6.** (a) The solar tracking system used in the experiments, and (b) the output power of the solar panel with and without tracking system.

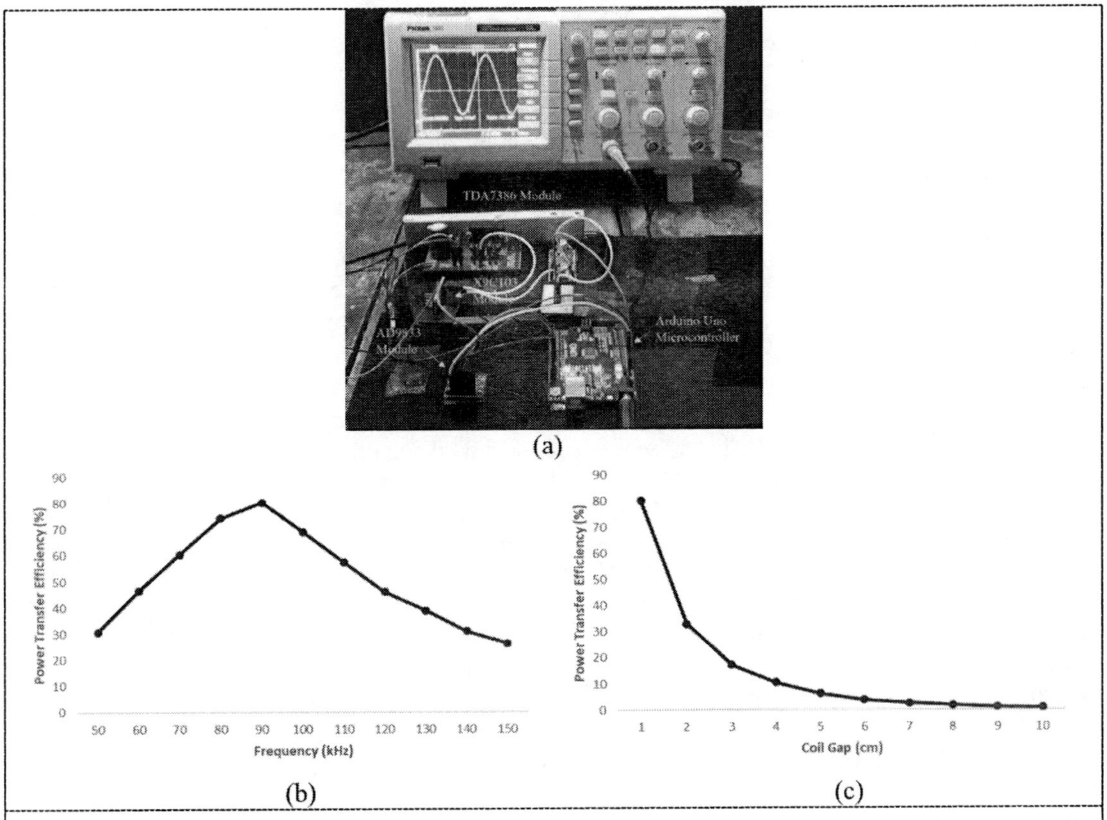


**Figure 7.** (a) Sinusoidal waveform generator, (b) frequency response of wireless power transfer, and (c) the power transfer as a function of the coil gap.

The next experiment is wireless power transfer. Several measurements are carried out using sinusoidal signals with frequencies between 50 kHz and 150 kHz as shown in figure 7(a). This signal is fed to the primary coil, then the power is received in the secondary coil. The results of this experiment indicate that the optimal value for wireless power transfer is a sine wave with a frequency of 90 kHz

IOP Conf. Series: Materials Science and Engineering **847** (2020) 012088 doi:10.1088/1757-899X/847/1/012088

shown in figure 7(b). This system can transfer power with an efficiency of 80% at a distance between the coils of 1 cm as shown in figure 7(c).

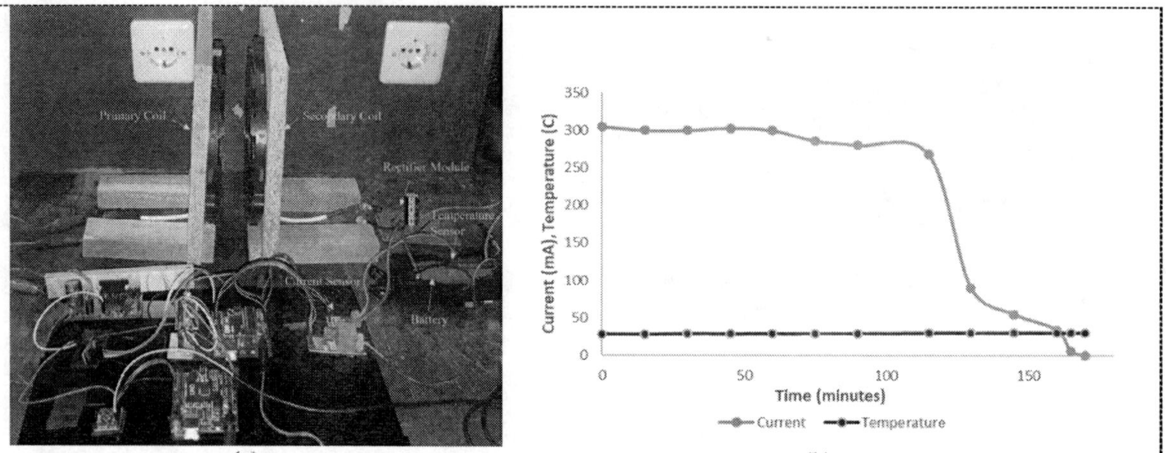


(a) (b)

**Figure 8.** (a) Wireless battery charging system, and (b) the charge time.

The last experiment is wireless battery charging system using fuzzy control method as shown in figure 8(a). The experimental results show that the initial charging voltage is set to the maximum. The temperature tends to increase during the battery charging process. The system always detects the temperature of the battery and will reduce the voltage value during the charging process as the temperature increases. Therefore, the battery can avoid damage due to overheating. This battery charging system will also automatically decrease the charging voltage when the battery voltage approaches the maximum value indicated by a decreasing current, as shown in figure 8(b). This system is expected to be used as a fast battery charging system without affecting the battery life.

## 4. Conclusion

In this study, we have designed and realized a wireless battery charging system supplied by solar panel using the fuzzy control method. This system is equipped with solar tracking to produce optimal electric power. Furthermore, fuzzy control system is applied to determine the amount of charging voltage by taking into account the temperature of the battery. The experimental results show that the solar panel equipped with solar tracking provides a higher average electric power of 44%. The optimal value for wireless power transfer is a sine wave with a frequency of 90 kHz. This system can transfer power with an efficiency of 80% with a distance between the coils of 1 cm. The system always detects the temperature of the battery and will reduce the voltage value during the charging process as the temperature increases. This battery charging system will also automatically decrease the charging voltage when the battery voltage approaches the maximum value. For future works, sensor signals as the control feedback will be sent via a wireless connection. The buck-boost converter will be employed to increase the system efficiency.

## Acknowledgments

We would like to thank LPPM Institut Teknologi Sepuluh Nopember, Surabaya, and Kementerian Riset, Teknologi dan Pendidikan Tinggi Republik Indonesia for financial aid support.

## 5. References

[1] Bomber A, and Rosa L, 2006. Wireless Power Transmission: An Obscure History, Possibly a Bright Future. *Phys* 464: *Appl. Opt.*, pp. 1–15.

[2] Jiang H, Zhang J, Lan D, Chao, Liou S, Shahnasser H, Fechter R, Hirose S, Harrison M, Roy S, 2013. A Low-Frequency Versatile Wireless Power Transfer Technology for Biomedical Implants. *IEEE Trans. Biomed. Circuits Syst*, **7**, pp. 526–535

[3] Yashchenko V, Turgaliev V, Kozlov D, Vendik I, and Katsay A, 2017. Adaptive impedance-matching network for wireless power transfer system with off-center receiver, *Electromagnetics Research Symposium*, pp. 2185–2189

[4] Samanta S, Rathore AK, and Thrimawithana DJ, 2017. Analysis and Design of Current-Fed Half-Bridge (C)(LC)–( LC) Resonant Topology for Inductive Wireless Power Transfer Application. *IEEE Trans. Ind. Appl.*, **53**, pp. 3917–3926

[5] Mou X, and Sun H, 2017. Analysis of multiple segmented transmitters design in dynamic wireless power transfer for electric vehicles charging. *Electron. Lett*, **53**, pp. 941–943

[6] Nayak SR, Pradhan CR, Ali SM, and Sabat R, 2012. Application of Solar Energy Using Artificial Neural Network and Particle Swarm Optimization. *International Journal of Advances in Engineering & Technology*, pp. 550-560

[7] Guo L, Han J, and Otieno AW, 2013. Design and Simulation of a Sun Tracking Solar Power System. *ASEE Annual Conference & Exposition*, pp.1-6

[8] Rivai M, Rendyansyah, and Purwanto D, 2015. Implementation of Fuzzy Logic Control in Robot Arm for Searching Location of Gas Leak. *International Seminar on Intelligent Technology and Its Application*, pp 69-74

[9] Putra R, Rivai M, and Irfansyah AN, 2018. Unmanned Surface Vehicle Navigation Based on Gas Sensors and Fuzzy Logic Control to Localize Gas Source. *Journal of Physics: Conference Series*, **1201**, pp.1-8

[10] Meivita DN, Rivai M, and Irfansyah AN, 2018. Development of an Electrostatic Air Filtration System Using Fuzzy Logic Control. *International Journal on Advanced Science, Engineering and Information Technology*, **8**, pp. 1284-1289

[11] Asadi H, Kaboli A, Mohammadi A, and Oladazimi M, 2011. Fuzzy Logic Control Technique in Li-Ion Battery Charger. *International Conference on Electrical, Electronics and Civil Engineering*, pp. 179-183

[12] Yashchenko V, Turgaliev V, Kozlov D, Vendik I, and Katsay A, 2017. Adaptive impedance-matching network for wireless power transfer system with off-center receiver. *Electromagnetics Research Symposium - spring*, pp. 2185–2189

[13] Sarwar MB, Nallagownden P, Baharudin Z, and Muthuvalu MS, 2016. Review of Different Techniques used for Wireless Transmission of Electrical Energy. *MATEC Web of Conferences*, **38**, pp. 1-7.

[14] Sugiarto K, Rivai M, Irfansyah AN, 2019. Control of Livestock Waste Odors Using Gas Sensors and Fuzzy Logic. *International Conference on Information & Communication Technology and System*, pp. 81-86

[15] Rendyansyah, Rivai M, and Purwanto D, 2019. Olfactory Arm Mobile Robot for Object Inspection Based on Fuzzy Logic and Support Vector Machine. *Journal of Physics: Conf. Series*, **1196**, pp.1-9

IOP Conf. Series: Materials Science and Engineering **847** (2020) 012089 doi:10.1088/1757-899X/847/1/012089

# The Application of Soft System Methodology to design the Conceptual Model for Intelligent Supply Chain Model of Natural Fibre Agroindustry

**N Nurhasanah[1,2], Machfud[1]*, D Mangunwidjaja[1], and M Romli[1]**

[1]Agro-industrial Engineering Department, Faculty of Agricultural Technology, IPB University, Indonesia
[2]Industrial Engineering Department, Faculty of Science and Technology, Universitas Al Azhar Indonesia
[1,2]nunung_nurhasanah@apps.ipb.ac.id, *machfud21@gmail.com
*Corresponding email

**Abstract.** Soft system methodology in designing an intelligent supply chain is defined as a system-based method that is holistically constructed without reduction, according to representation from the real world, coming from stakeholders interacting one to another to gain value added and improve profits through data training and data saving on cloud presented from human to system of real-world database. This study aimed to apply a soft system methodology approach in designing an intelligent supply chain model of natural fibre agroindustry. A conceptual model was successfully developed and produced eight activities which were compared though assessment criteria of efficiency, efficacy, and effectiveness. The eight activities mentioned are productivity improvement, data mining to predict demand and stock, development of a collaborative planning forecasting and replenishment model, development of an intelligent decision support system, digital platform construction, value added improvement, enhancement of efficiency and response to buyers, as well as improvement in supply chain performance. This study has not yet carried out the stages of taking action to improve the problem situation, and will be carried out in further study.
**Keyword:** Collaboration, intelligent decision support system, performance, value added, data mining

## 1. Introduction

Indonesia shows the potential for plant fibre processing industries. However, their market is not able to compete with Malaysia, and some other Asean kenaf fibre (KF) producing countries [1]. Kenaf (*Hibiscus cannabicus L.*) is a fibre plant which potentially contributes to economical and environmental benefits [2]. However, the plantation area is currently increasingly shrinking to 3,344 Ha in East Java [2]. Its largest producers worldwide include India, Bangladesh, and USA at 970,000 ton [3]. FAO claimed that Indonesia is the third largest kenaf importing country [4], [5]. Improvement in kenaf value added (VA) during transformation process to produce innovated products deals with lack of raw material supply [2]. In order to ensure VA improvement in the supply chain network, availability of KF is supposed to be maintained. This is achieved through an availability system managed with information sharing in the form of collaboration [6] in an intelligent supply chain (ISC) model.

Over 86 titles of scientific relevant paper studied, there were 31 titles related to supply chain, 55 titles related to ISC. From of all these articles, no article has been found uses SSM in developing ISC models for natural fiber cases [6]. Therefore, this research would specifically review soft system methodology (SSM) on an ISC model of natural fibre.

SSM is an approach to organize problems that are unstructured, complex, and difficult to define [7] [8] [9] [10] [11] [12] [13]; which contains interaction between components of humans and technology [14], and assisting to comprehend perception in the minds of people whom are involved in the situation [12]. SSM uses systems thinking in the research action cycle [12], dividing perspective on the real world and existing systems, the model is considered a learning tool rather than a tool for predicting, and an organized search for situations [15]. It is performed through Focus Group Discussion (FGD) [16]. SSM consists of human activities as it involves a number of stakeholders with different point of views, interests, and understandings [17]. It offers a systemic framework according to logical steps [18] that produce a process analysis. Issues of technology and decision maker are independent, with a number of conflicting stakeholder-related world's view and objectives [10]. Some research found to implement SSM in power plant construction [14], health industries [7], retailer [15], government offices for community service [8], construction [12], coffee agroindustry [19], wood-processing industries [16], tuna fishery businesses [9], education [20] [11], batik industry [17], *high techonology* industry [21], and textile industry [13]. It is concluded that SSM is a systemic approach which requires in-depth discussion, such as Forum grup discussion (FGD), in depicting complex and unstructured problems in order to be transformed into real-world data as database from an ISC model of natural fibre agroindustry (ISCMNA). In this approach, behaviour of the real-world system is modelled in system thinking into a conceptual model through CATWOE elements.

Collaborative planning, forecasting and replenishment (CPFR) is a business process that manages demand uncertainty, point of sale data, promotion and replenishment plans [22], a strategy that combines the intelligence of several trading partners in planning and fulfilling customer demand [23]. Stated by [24] CPFR as a business that combines the efforts of business partners in planning and meeting customer demand with the aim of increasing efficiency in the entire supply chain.

## 2. Method

Method employed in this study was according to that designed by Checkland [25]. Table 1 presents the research stages of this study.

**Table 1.** Research stages

| No | SSM | Description | Solution by this research |
|---|---|---|---|
| 1 | Situation of unstructured problems | Describing roles of stakeholders and aspects related to ISCMNA in the real world | Observing and discussing with stakeholders |
| 2 | Expression of problem situation | Designing rich picture to express the problems, as well as collecting data and information | |
| 3 | Formulation of root definition from system activities | Constructing definition on expression of problem situation related to ISCMNA in the real world and on system thinking | Formulating ISCMNA root definition model by identifying CATWOE elements |
| 4 | Conceptual model | Depicting real ISCMNA problems derived from root definition | Formulating a conceptual model |
| 5 | Comparison | Comparing design of ISCMNA model on system thinking in the real world according to efficiency, efficacy, and effectiveness (3E) | Comparing the conceptual model with the real world, according to database, using 3E principal |
| 6 | Recommendation | Developing ideas of ISCMNA design in order to be applied in the real world | Determining an ISCMNA model design |

Figure 1 shows the research framework. Identification of unstructured problem would be conducted based on interview and direct observation using two categories. The output was database that would be stored for comparison between real-world behaviour and real-world system thinking.

## 3. Result and discussion

*3.1. The complex, problematic and unstructured problem of supply chain of natural fibre agroindustry*
Review on complex, problematic, and unstructured problems was conducted based on field observation and FGD, starting at Balittas as a seed supplier; continued at Agriculture Agency of Garut Regency; kenaf fibre agroindustry (KFA) of Cibiuk Sub-district Garut Regency as cultivation and KF producing area; Cikutra of Cibeunying Kidul Bandung City as a producing and marketing area of woven fabric; and Indonesian tree community (ITC) Office in Bandung City as a Non Government Organization.

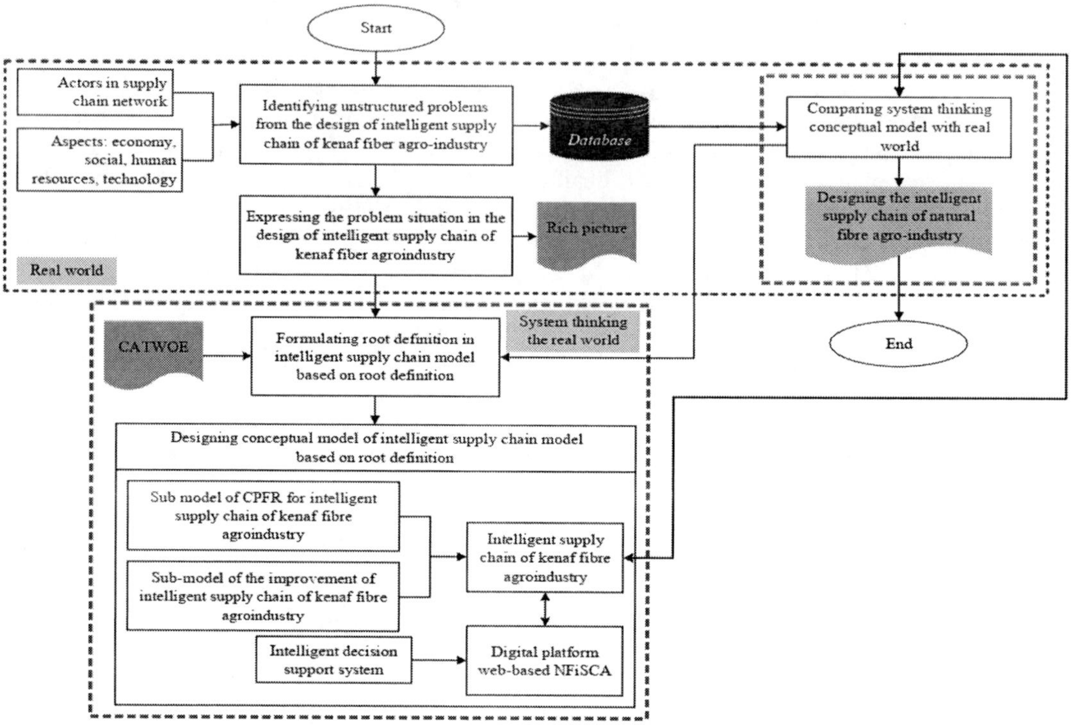


**Figure 1.** Research framework.

Kenaf fibre supply chain (KFSC) involves a number of actors in its network, i.e. KFA in upstream network, woven fabric manufacturing industry (WFMI) in middle-stream network, textile and textile product (TTP) industries in downstream network, kenaf farm workers, retting workers, weaving craftsmen, Balittas, investors, transportation service, seed suppliers, banks, exhibitors, and buyers, which include human resources, technology, economy, and social.

International price for dried KF is considerably promising for KFA, however this price is not applicable in purchase and sale agreement as there is monopoly power from foreign-owned company in East Java. Accordingly, KFA was unwilling to continue the agreement, considering that it was detrimental for kenaf farm workers and retting workers. This resulted in non-continuous business operation and non-optimal response to buyers. KFA has not gained VA of KF derived from core stem, that is further processed into core chip and kenaf core powder [26], as marketing in upstream supply chain network was not assured, although demand for these two products have existed in free market. This occurred since there was not a stock system managed well by KFA. Therefore, model design of ISCMNA is considered important to improve performance of KFA in responding as well as behaving efficiently and transparently during purchase and sale.

Low fibre productivity in local free market and high fibre price in global free market led to difficulty for Innatex and buyers to obtain high quality KF. WFMI in the middle network was not assured with availability of dried fibre raw material. Accordingly, they could not deal with urgent demands as buyers have to wait for the availability of raw material. Their incapability to respond to buyers properly may be prevented by enhancing collaboration in implementing an intelligent-based platform digital technology of Natural Fibre Supply Chain Analytics (NFiSCA) [8] between fibre traders and woven fabric buyers. High import of kenaf dried fibre [6] [7] became a global competition challenge for KFA, WFMI, and TTP actors along the supply chain network. In addition, monopoly power of KF production impacts on difficulty of kenaf industry actors to develop and compete. The presence of an intelligence-based Internet of Things (IoT) system which allows optimum response and facilitation of collaboration between demand and stock along the supply chain will assist improvement in performance and competitiveness of agroindustry.

In relation to technological aspect, problem addressed is manual harvesting. In addition, KFA has not employed digital device of information technology. Employment of an intelligent system-based information technology digital device is able to enhance sensitivity and collaboration between KFA and buyers at local and global levels. Collaboration which promote sales rate of dried KF will increase VA of KF. Collaboration is used to identify availability of fibre to deal with unexpected demand from buyers. Currently, KFA has not owned a system which is able to record stock and demand rate of fibre. The presence of an intelligent system in managing stock and demand system will facilitate improvement in VA. Besides of an intelligent system, each actor in the supply chain definitely needs to act optimal according to their respective tasks and functions.

WFMI developed 55 industrial houses in Pekalongan. These industries provide processing service of fibre into woven fabric in traditional manner. They actively follow exhibition events, such as *Ina Craft* and *Market Craft*. The exhibitions facilitate Innatex to come across with foreign buyers as they have not owned marketing human resources to promote and publish the products. Innatex is not able to use digital facility, rather relying on orders from phone and social media channel. Furthermore, they have not initiated to invest on loom machine, rather handing over the work to a traditional woven production houses, leading to an increased income for the production houses. The weaving process is performed manually, without machine that involves electricity and fuel, while dyeing process is carried out with manual dyeing aid. Low quality of KF will result in increased production costs for involvement of scouring process. Footwear industries which become an illustration of TTP industries in downstream network produce 200 pair of shoes per month through a handmade process, which requires more time. However, the price is relatively high, targeted for upper and middle classes. These products rely on digital marketing for their export to Canada, France, Sweden, England, Hongkong, Philippines, Singapore, and Malaysia. In addition to using online facility, these products are marketed directly through offline shops and exhibitions for IDR355,000-IDR1,800,000 per pair. As is the case for WFMI, these industries use exhibition events to introduce their products to foreign buyers. Since these products are still produced in a limited number, orders are frequently made from offline buyers. However, more buyers come from online media such as website, Instagram, and Facebook. They have been assisted with skilled marketing human resources who apply IoT-based devices, and therefore there is only minimum use of machine. Industries are able to minimize the use of glue and synthetic dye, and replace them with natural yarn and dye.

Social aspects in downstream include consumer's understanding towards an environmentally friendly concept. Consumers are actors who gain benefits from this transformation system. Introducing these products to middle and upper as well as premium consumer classes are easier as compared to lower consumer class. Thus, the challenge faced is educating consumers about environmentally friendly shoes products being offered.

### 3.2. *Expression of problem situation of supply chain of natural fibre agroindustry*

FAO [4], [5] recorded a gap between consumption and production of natural fibre by 90.38% [27]. According to expression of problem situation, it is obvious that a synergic collaboration was not built

IOP Conf. Series: Materials Science and Engineering **847** (2020) 012089 doi:10.1088/1757-899X/847/1/012089

among supply chain network. This is noticed from weakness shown by KFA, i.e. low responsiveness to consumers, since an ISC model and a digital device to facilitate the collaboration process have not been implemented. Currently, agroindustry has not been able to optimally respond to overseas orders. Each of stakeholders should detect incompatibility in the process and products produced. Monitoring process to detect incompatibility can be performed by establishing a system that is able to recognize and control occurrence of incompatibility by intelligent decision making along the supply chain network.

There were 4 problems related to kenaf agroindustry development that were identified through literature study, i.e. unfulfillment of domestic demand for kenaf natural fibre [4], [5]; high import level of kenaf natural fibre [4], [5]; unstable price of international kenaf natural fibre; uncertainty of kenaf natural fibre fulfilment to industries, giving a impact on stock costs and shipment schedule distructive; as well as in-optimum utilization of available land [28]. Problems identified in upstream supply chain network include absence of marketing network to penetrate global market, un-implemented international price for KF, and unavailable access to free market, except for *PT XYZ*. In middle-stream network, there are still unassured availability of raw materials, long processing steps, and in-optimum response to global market. Meanwhile, problems identified in downstream supply chain include unassured fulfilment of demand from buyers due to seasonal raw materials. Based on observation of real-world situation, decisions regarding demand-based production and availability level taken by ASK have not been through a precise calculation and certain approaches.

Accordingly, an ISC model to overcome complex demand and availability problems in the real world should be developed. Supply chain flow along upstream to downstream network has been established, however decisions taken in every supply chain flow currently did not employ an ISC which is able to assure effectiveness and responsiveness of supply chain actors. In addition, this model is essential in promoting intelligent system-based decision making which is able to prevent issues of demand and stock rapidly, accurately, and dynamically. The rich picture diagram has shown in Figure 2.

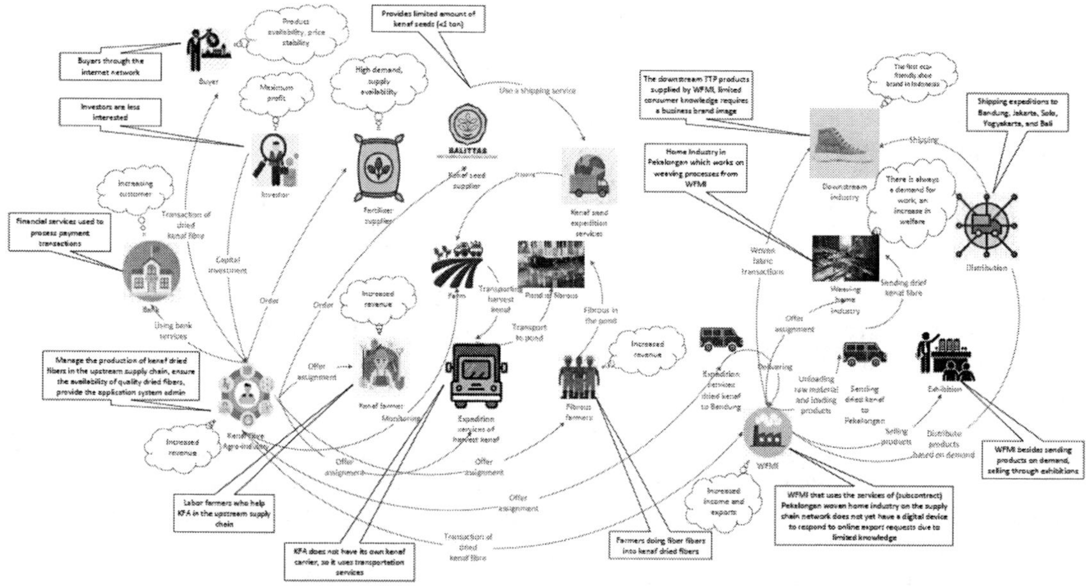


**Figure 2.** Rich picture of intelligent supply chain model of natural fibre agroindustry

*3.3. Root definitions* related to the situation of the supply chain of natural fiber agroindustry

Formulation of core definition from description of CATWOE elements is a concept which represents terms as presented in Table 2, and has been developed by [13].

**Table 2.** CATWOE elements and their descriptions

| CATWOE Elements | Results of Description |
|---|---|
| *Client* | Woven fabric-manufacturing industries; woven fabric-processing industries; buyers |
| *Actor* | Kenaf farm workers; kenaf retting workers; KFA; WFMI & TTP industries; weaving craftsmen |
| *Transformation* | (1) Process of kenaf cultivation to produce dried KF; (2) Stock planning system of woven fabric and TTP products; and (3) NFiSCA web-based digital platform information system |
| *Weltanschauung* | (1) Improvement in VA of KF will enhance welfare of kenaf farm workers, retting workers, and KFA in the upstream supply chain network; (2) Improvement in efficiency and response to buyers by assuring stock of woven fabric and textile products in the supply chain network; and (3) Improvement in supply chain performance (SCP) by implementing an ISC model |
| *Owner* | Kenaf fibre agroindustry |
| *Environment* | (1) Low productivity of KF in Indonesia; (2) Wide distribution of low quality seeds in the market; (3) Low assurance of marketing; (4) Industry 4.0 force to the actors to produce highly competitive products; and (5) Government policy to encourage potential excellence of Indonesian KF in order to create competitiveness of local products that are able to play a role in global market |

Establishment of root definition according to elaboration of CATWOE elements in [29] by means of an *O* system; operated by *A*; in *T* problem situation; to satisfy *C*; by *W*; in *E*. Results of root definition are as follows: "An agroindustrial system of KF; which is run by kenaf farm workers, retting workers, KFA, WFMI and processing industries, as well as weaving craftsmen; through kenaf processing, stock planning system, and information system of NFiSCA web-based digital platform; in order to meet demand from woven fabric manufacturing industries, and buyers; to improve VA, efficiency, and response towards consumers, and performance through an ISC model; on conditions of low kenaf productivity in Indonesia, wide distribution of low quality seeds in the market, low assurance of marketing, demand for industry 4.0, and in-optimum government policy in encouraging potential excellence of Indonesian KF in global competitiveness".

### *3.4. Conceptual model of intelligent supply chain model of natural fiber agroindustry*

Conceptual model presented in Figure 3, was established based on root definition which produces important activities in the model design of ISCMNA. Activities referred are improvement in kenaf productivity, construction of NFiSCA digital platform, data mining for prediction of demand and stock, VA improvement, development of an ISC model, improvement in efficiency and response to buyers, and improvement in SCP. These eight activities would create two models developed in an ISCMNA model design. The first sub-model, CPFR, would be developed to achieve an objective of value addition activities, as well as improvement in efficiency and response to consumers. The second sub-model, performance improvement, would be developed to achieve an objective of improvement in SCP. These two sub-models are expected to be integrated in the ISCMNA model in order to support decisions developed by NFiSCA web-based digital platform. The digital platform is developed through an IDSS, which defines management of database and model.

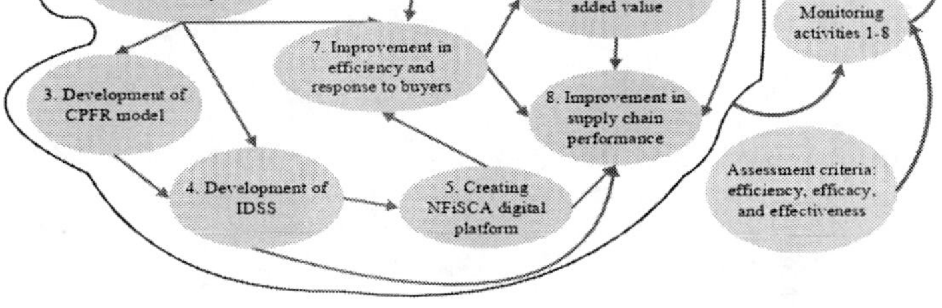


**Figure 3**. *Conceptual model of intelligent supply chain model of natural fiber agroindustry*

### *3.5. Identification of gap analysis between ideal situations and the real world*

Table 3 presents a gap analysis as adopted from table [13]. Gap analysis is done by describing the activities in the conceptual model, current practice, and the contribution of the model that will be implemented.

**Table 3.** Gap analysis of conceptual model and problem situation

| Activity | Current practice | Contribution of model |
|---|---|---|
| Improvement in kenaf productivity | Purchase of seeds has been performed from certified suppliers | • Scheduling planting time to predict harvesting time |
| Data mining to predict demand and stock | ASK and Innatex have not carried out prediction of dried KF demand & stock | • Predicting demand and stock of dried KF |
| Development of CPFR model | The current supply chain has not utilized CPFR model | • Identifying the supply chain business process<br>• Analyzing the supply chain network<br>• Developing CPFR model |
| Development of IDSS | The current supply chain has not utilized IDSS | • Developing IDSS |
| Development of NFiSCA digital platform | Unemployment of information system along supply chain to facilitate business process | • Developing NFiSCA digital platform<br>• Carrying out user acceptance test on KFA and Innatex |
| VA improvement | • KFA has not optimally conducted improvement in kenaf fibre VA<br>• WFMI and downstream industries have created variation in products<br>• Collaboration for information sharing of product demand and stock has not existed | • Publishing KF-based products through NFiSCA digital platform |
| Improvement in efficiency and response to buyers | Not being able to optimally respond to buyers | • Developing a collaboration system through information sharing facilitated with NFiSCA digital platform |
| Improvement in SCP | Performance evaluation along supply chain network is never performed | • Presenting performance of each stakeholder in real time<br>• Presenting solution to improve performance |

Table 4 shows assessment criteria based on 3E of ISCMNA model design from eight activities elaborated in the conceptual model.

**Table 4.** Assessment criteria of intelligent supply chain model based on 3E

| Activity | Efficiency | Efficacy | Effectiveness |
|---|---|---|---|
| Productivity improvement | Efficiency of production costs will impact on increased productivity | Improvement in productivity of supply chain and agroindustry revenue | Performance improvement is affected by a productivity improvement value |
| Data mining to predict demand and stock | Efficiency of stock costs | Increasing VA | Success percentage of prediction will improve SCP |
| Development of CPFR model | Efficiency of production and inventory costs | Improving SCP and agroindustry revenue | Success percentage of development of this model will improve SCP |
| Development of IDSS | Efficiency of supply chain of KFA | Improving SCP and agroindustry revenue | Success percentage of development of intelligent decision support systems |
| NFiSCA digital platform development | Being able to reduce production costs through time efficiency in processes of collaboration, marketing, publication, and performance monitoring | Improving performance of supply chain and agroindustry performance | Success percentage of the platform and readiness for implementation will improve SCP |
| VA improvement | Efficiency of production time will improve productivity and revenue | Improving SCP and agroindustry revenue | VA of products becomes a success value of agroindustry SCP |
| Improvement in efficiency and response to buyers | Time efficiency will improve SCP and revenue | Improving SCP | High efficiency and responding success become a success value of agroindustry SCP |

**Table 4.** Assessment criteria of intelligent supply chain model based on 3E

| Activity | Efficiency | Efficacy | Effectiveness |
|---|---|---|---|
| Improvement in SCP | Efficiency of time and costs will be achieved by performance improvement | This activity impacts on improvement in KFA competitiveness | Value of performance improvement becomes a success indicator of agroindustry development |

Results of conceptual model in real-world system thinking has been evaluated with all data contained in real-world database. Deserved and desirable changes for an ISCMNA model were established, which consist of eight activities, i.e. improvement in kenaf productivity; data mining to predict demand and stock; development of CPFR model; development of IDSS; construction of NFiCSA digital platform; improvement in VA; improvement in efficiency and response to buyers; and improvement in SCP.

## 4. Conclusion

The conceptual model was developed and produced eight activities which were assessed for each activity with the criteria of 3E. The conceptual model produces eight activities, namely improvement in kenaf productivity; data mining to predict demand and stock; development of CPFR model; development of IDSS; construction of NFiCSA digital platform; improvement in VA; improvement in efficiency and response to buyers; and improvement in SCP. The limitation of this study is that the SSM stage has only been carried out until stage 6, while the seventh stage, which is to take corrective action to the problem, has not been carried out in this study. Future reaserch of this study is developing the detail of the models, the models are: CPFR, developing IDSS, and creating NFiSCA digital platform. This study recommended that the government should provide assistance and issue policies to support the development of kenaf fiber that has the potential to be favored into competitive products that support Indonesia making 4.0.

**Acknowledgment**

A gratitude is expressed to Ministry of Research and Higher Education for support given to this work through BPPDN scholarship, as well as to Universitas Al Azhar Indonesia.

## 5. References

[1] Subagyo A, 2012 Strategi pengembangan industri serat alam daun nanas untuk meningkatkan taraf hidup masyarakat *Inov. dan Kewirausahaan* **1**, 1 p. 23–28.

[2] Ardani H K, 2013, Pengembangan serat kenaf (Hibiscus cannabinus L.) sebagai filler komposit bermatriks polimer (abs) pada aplikasi helm, IPB.

[3] Adekomaya O Jamiru T Sadiku R and Huan Z, 2016 A review on the sustainability of natural fiber in matrix reinforcement - A practical perspective *J. Reinf. Plast. Compos.* **35**, 1 p. 3–7.

[4] Food and agriculture, 2018, Food and agriculture organization statistical. [Online]. Available: http://www.fao.org/faostat/en/#data/QC. [Accessed: 07-Nov-2018].

[5] [FAO] Food and Agriculture Organization, 2016 Jute, kenaf, sisal, abaca, coir and allied fibres - Statistics December 2015 *FAO* p. 1–40.

[6] Nurhasanah N Machfud Mangunwidjaja D and Romli M, 2019 A Literature Review on The Design of Intelligent Supply Chain for Natural Fibre Agroindustry *Int. J. supply Chain Manag.* X p. 1–18.

[7] Berge L and Dahl F A, 2011 Using soft systems methodology as a precursor for an emergency department simulation model *OR Insight* **24**, 3 p. 168–189.

[8] Pollack J, 2009 Multimethodology in series and parallel : strategic planning using hard and soft OR p. 156–167.

[9] Nurani T W Wahyuningrum P I Wisudo S H Gigentika S and Arhatin R E, 2018 Model designs of Indonesian tuna fishery management in the Indian Ocean (FMA 573) using soft system methodology approach *Egypt. J. Aquat. Res.* **44**, 2 p. 139–144.

[10] Antunes C H Dias L Dantas G Mathias J and Zamboni L, 2016 An application of Soft Systems Methodology in the evaluation of policies and incentive actions to promote technological innovations in the electricity sector *Energy Procedia* **106**, 2 p. 258–278.
[11] Soemartono T, 2014 Reconstruction of Education Policy in Jembrana Bali , Best Practices of Creative and Innovative Leadership using Soft Systems Methodology based Action Research *Procedia - Soc. Behav. Sci.* **115**, Iicies 2013 p. 269–282.
[12] Maqsood T Finegan A D and Walker D H T, Five case studies applying Soft Systems Methodology to Knowledge Management.
[13] Ngai E W T *et al.*, 2012 Development of the conceptual model of energy and utility management in textile processing: A soft systems approach *Intern. J. Prod. Econ.* **135**, 2 p. 607–617.
[14] Panchanan Beheraa R P M & A P, 2015 Understanding Construction Supply Chain Management *Prod. Plan. Control Manag. Oper.* July p. 20.
[15] Winter M, 2006 Problem structuring in project management: an application of soft systems methodology *J. Oper. Res. Soc.* **5**, 7 p. 802–812.
[16] Soetara A Machfud Affandi M J and Maulana A, 2018 The design on Conceptual Model for Continuation of Lean Manufacturing (LM) Implementation in Indonesia Wood Processing Factory using Soft System Methodology *Int. J. Adv. Sci. Eng. Inf. Technol.* **8**, 4 p. 1302–1306.
[17] Novani S Sarjono U and Hermawan P, 2014 An Application of Soft System Methodology in Batik Industrial Cluster Solo by using Service System Science Perspective *Procedia - Soc. Behav. Sci.* **115**, Iicies 2013 p. 324–331.
[18] Novani S and Mayangsari L, 2017 Soft Systems Agent-Based Methodology : Multi-methods Approach Between Soft Systems Methodology and Agent-Based Modeling Soft System Agent Based Methodology : Multi Methods Approach between Soft System Methodology and Agent Based Modeling June 2019.
[19] Fadhil R Maarif M S Bantacut T and Hermawan A, 2018 Situational analysis and intervention strategy for Gayo coffee agroindustry institution in Indonesia *J. Food, Agric. Environ.* **16**, 1 p. 31–40.
[20] Mehregan M R Hosseinzadeh M and Kazemi A, 2012 An application of Soft System Methodology *Procedia Soc. Behav. Sci.* **41** p. 426–433.
[21] Liu W B Meng W Mingers J Tang N and Wang W, 2012 Developing a performance management system using soft systems methodology : A Chinese case study *Eur. J. Oper. Res.* **223**, 2 p. 529–540.
[22] Singhry H B and Abd Rahman A, 2018 Enhancing supply chain performance through collaborative planning, forecasting, and replenishment *Bus. Process Manag. J.* **24**, 4 p. 965–984.
[23] Panahifar F Byrne P J and Heavey C, 2015 A hybrid approach to the study of CPFR implementation enablers *Prod. Plan. Control* **26**, 13 p. 1090–1109.
[24] Hidayat A, 2015, Kolaborasi Perencanaan Peramalan dan Pengisian Kembali Persediaan (CPFR), *Supply Chain Indonesia*. [Online]. Available: http://supplychainindonesia.com/new/kolaborasi-perencanaan-peramalan-dan-pengisian-kembali-persediaan-cprf/. [Accessed: 10-Nov-2018].
[25] Checkland P, 2000 *Systems thinking, system practise* Reprinted Chichester: John Wiley & sons.
[26] Nurhasanah N Machfud M Mangunwidjaja D and Romli M, 2019 Value added of kenaf fibre in natural fibre agroindustry supply chain network in *International conference of innovation in technology and management for sustainable agroindustry*.
[27] Nurhasanah N Machfud M Mangunwidjaja D and Romli M, 2019 A conceptual framework on The design of intelligent supply chain for natural fibre agroindustry in *International Conference on Industrial, Mechanical, Electrical and Chemical Engineering 2019* p. 8.
[28] Miyagawa and Tranggono, 2012, Kebutuhan serat kenaf sebagai bahan baku industri PT TBINA, *Balittas*, p. 54–57.
[29] Wilson B, 2001 *Soft system methodology* Chichester: John Wiley & sons.

IOP Conf. Series: Materials Science and Engineering **847** (2020) 012090 doi:10.1088/1757-899X/847/1/012090

# Analysing the cause of idle time in loading and unloading operation at Indonesian international port container terminal: Port of Tanjung Priok case study

**F Suprata[1] , C Natalia[2] , and A Sugioko[3]**

[1]Innovation System Laboratory, Department of Industrial Engineering, Atma Jaya Catholic University of Indonesia, South Tangerang, Indonesia
[2]Statistics Industry and Decision Making Laboratory, Department of Industrial Engineering, Atma Jaya Catholic University of Indonesia, South Tangerang, Indonesia
[3]Production System Laboratory, Department of Industrial Engineering, Atma Jaya Catholic University of Indonesia, South Tangerang, Indonesia
[1]ferdian.suprata@atmajaya.ac.id, [2]chrisnatalia@atmajaya.ac.id, and
[3]andre.sugioko@atmajaya.ac.id

**Abstract.** Prior study shows that major Indonesian ports still need some improvements to achieve a better logistics performance. This study aims to analyze which variables can influence the present of idle time during loading and unloading process at Container Terminal Operation 2 Tanjung Priok Port Jakarta. The Tanjung Priok Port is chosen since the port is designed as the hub port in the national sea transport system of the country where it handles more than 50% Indonesia's trans-shipment cargo traffic. This study uses service blueprint, multiple regression analysis, and one-way anova method. The findings show that at the existing Tanjung Priok Port Container Terminal has 17 (seventeen) failure spots, which are categorized into 6 (six) variables namely head truck, computer system, weather, stevedoring worker, equipment failure, and operator activities, that cause the idle time at the port service. Nevertheless, based on the multiple regression analysis, only 5 (five) variables have a significant influence on the cause of idle time excluding weather variable. Finally, with one-way anova test it is found that operator activities at the gate-in process is the most significant influence to the idle-time at the port service.
**Keywords**: Container, Idle Time, Service Blueprint, Regression Analysis, Tanjung Priok Port

## 1. Introduction

The global economy growth has an effect to the global international trade where ports plays important gateways to support global international trade. In many countries, ports can be seen as major accelerator of local economic development in the rapid global competitive market [1]. Historically since the early stages of industrialization, ports were designed to support local, regional and global economies activities. One may assume that there is a correlation between ports and the economy growth of a country. Ports are thus having influence to the economic development of countries on local, national, and international level [1].

As the world largest archipelago country and located between the indian and the pacific oceans, Indonesia can become a global maritime axis. Indonesian President Joko Widodo, Jokowi, in late 2014 presented the global maritime axis concept to rejuvenate country's identity as a maritime nation [2].

Prior study showed that key objective of this global maritime axis concept was to enhance inter-island connectivity and upgrading major port infrastructure in Indonesia. By issuing Presidential Regulation No. 26/2012 on Blueprint of National Logistics System Development (SISLOGNAS), Indonesian Government showed its commitment to promote the improvement of national competitiveness as well as to encourage the implementation of the Masterplan of Acceleration and Expansion of Indonesia's Economic Development (MP3EI) from period 2011 to 2025. All of these governances were developed to become a guidance for relevant stakeholders [3].

Indonesia has a potential to become a host of international hub port. Nevertheless, the productivity and capacity at major Indonesia ports are currently not able to accomodate the increasing demand in the flow of goods for both domestic and international. Major improvements in Indonesian Ports are needed to be implemented so that it can support efficient and effective maritime logistics and connect all islands in Indonesia. In fact, delivering goods from Padang to Jakarta costs more than three times compare with shipping the same container from Jakarta to Singapore [4]. Study done by the World Bank in 2015 argued that the deficiency in maritime logistics, operations, and infrastructure may result in poor connectivity between islands, undermining trade, and less competitiveness. The study also showed that the operation of most Indonesia ports leading to inefficiencies that would have a direct influence to the high logistics costs for business and consumers. The logistics costs in Indonesia equal to 24 percent of Gross Domestic Product (GDP) which can slowdown country's economic growth [5]. In the latest World Logistics Performance Index (LPI) 2018, Indonesia ranked at 46th place in the world. Compare with other South East Asia Countries, Indonesia ranked below Singapore (7th Place), Thailand (32nd Place), Vietnam (39th Place), and Malaysia (41st Place) [6].

In relation to port competitiveness, it is realized that the main problem of Indonesian ports comprises 3 (three) main issues including the absent of international hub ports, low productivity and capacity of ports, and poor port performance services [7]. Therefore, this study was conducted to analyze the cause of inefficient performance ports in Indonesia. Previous study showed that idle time at the port service leveraged to the high cost of Terminal Handling Charge (THC) [8]. However, different with the previous one, this study will specify which variables may cause an idle time during the loading and unloading container at Container Terminal Operation 2 Tanjung Priok Port Jakarta. The Tanjung Priok Port is choosen as a case study since the port is the largest port in Indonesia with more than 50 (fifty) percent container in Indonesia access via this port. The objectives of this study are to identify which variables can cause the idle time, to analyze which dominant variables that have a significant influence to idle time, and to recommend port operator for reducing the idle time during loading and unloading container. The scope of this study focuses at Container Operational Terminal 2 Tanjung Priok Port for loading and unloading domestic container during period October 2016 until December 2016.

### *1.1 Port Service Performance Indicator*

Port can be defined as a place at the edge of an ocean, river, or lake that serve the need to transport passengers, animals and goods. Roa et al. in 2013 [9] defined port as an area which was connected to sea, ocean or river and was considered as entities, while Flere in 1967 [10] defined port as a place which was provided facilities of terminals and services for ships to deliver goods and/or passengers. Nevertheless, This study will follow the port definition as it was defined according to National Law No. 21 of 1992 (UU21/1992) and Government Regulation (PP) No. 69 Year 2001: *"A port is a place which consists of land and surrounding specified border waters, is a place for ships to anchor and harbor, passengers to embark and disembark and/or goods loading and unloading, is equipped with navigation safety facilities and port activities support, and also a place for various intermodal transportation activities".*

Nowadays, many scholars are conducting researches about port performance measures. Study done by Loke et al in 2015 [8] on the delay factors of container transshipment in major Malaysian ports, showed that the main factor to the delay was the idle time during loading and unloading container process. Comparable with major Malaysian ports, Tanjung Priok port has alike challenge dealing with inefficient container operating activities. As the matter of fact, the average dwelling time at Tanjung

Priok Port is 5.5 days which is the longest in Asia while in some items it can take up to a month to clear custom and excise [11]. In today's dynamic competitive market, various determinants can affect port performance including the local market character, organizational and physical capacity, system integration in logistics, maritime accessibility, competition, dock equipment and parking field, delivery service and inter-connection to the hinterland areas [12]. As a consequence, in current port operating services, there may also valid and useful to include effectiveness and efficiency aspects in measuring port performance which is usually associated with efficiency in operational activities [13].

With regard to the port service performance indicators, the Ministry of Transportation of the Republic of Indonesia through the Directorate General of Sea Transportation has issued the Decree of the Directoral General of Sea Transportation No. UM.002/38/18/DJPL-11 year 2011 about Standard Port Operational Service Performance [14]. Inside the document the indicator of port service performance was defined including: Waiting Time (WT), Approach Time (AT), Effective Time per Bert Time (ET:BT), Work Productivity (Loading/Unloading) (Ton/Aisle/Hours), Work Productivity (Box/Crane/Hours), Bert Occupancy Ratio (BOR), Shed Occupancy Ratio (SOR), and Yard Occupancy Ratio (YOR). Correspondingly, this study uses the mentioned eight indicators of port performance with the case of Tanjung Priok port. The performance of Tanjung Priok port for period 2011 to 2015 is given to the following table 1.

**Table 1.** Performance of Tanjung Priok Port, Source: [15]

| Year | WT (Hours) | AT (Hours) | ET/BT (%) | T/A/H | B/C/H | BOR (%) | SOR (%) | YOR (%) |
|---|---|---|---|---|---|---|---|---|
| 2011 | 1.00 | 1.00 | 79.42 | 44.07 | 15.83 | 54.62 | 40.51 | 42.59 |
| 2012 | 1.00 | 1.00 | 80.50 | 55.96 | 16.10 | 56.73 | 31.40 | 44.80 |
| 2013 | 1.00 | 1.00 | 82.00 | 72.23 | 15.58 | 50.25 | 34.33 | 45.37 |
| 2014 | 0.18 | 0.91 | 76.75 | 70.67 | 18.42 | 49.24 | 30.69 | 40.41 |
| 2015 | 1.00 | 1.00 | 64.98 | 75.19 | 22.36 | 36.12 | 15.23 | 36.21 |

Folowing the decree of the Directoral General of Sea Transportation No. UM. 002/38/18/DJPL-11 year 2011, the performance score scale of Tanjung Priok Port was categorized into three score levels, namely 1 (Poor), 2 (Fair), and 3 (Good). Table 2 shows the port performance scale of Tanjung Priok Port during 2011 – 2015. the ship service performance is formulated as follows: Bert Time (BT) = Bert Working Time (BWT) + Idle Time (IT). Hence, BWT = Effective Time (ET) + Not Operation Time (NOT). From table 2, it is clear that WT and ET/BT have fair performance indicator while work productivity (Box/Crane/Hours) has poor score performance.

**Table 2.** Performance Scale of Tanjung Priok Port, Source: [15]

| Year | WT (Hours) | AT (Hours) | ET/BT (%) | T/A/H | B/C/H | BOR (%) | SOR (%) | YOR (%) |
|---|---|---|---|---|---|---|---|---|
| 2011 | 2 | 3 | 2 | 3 | 1 | 3 | 3 | 3 |
| 2012 | 2 | 3 | 2 | 3 | 1 | 3 | 3 | 3 |
| 2013 | 2 | 3 | 3 | 3 | 1 | 3 | 3 | 3 |
| 2014 | 3 | 3 | 2 | 3 | 1 | 3 | 3 | 3 |
| 2015 | 2 | 3 | 1 | 3 | 2 | 3 | 3 | 3 |

### *1.2 Idle Time*

Looking at the formulation given in the previous paragraph, idle time has an impact to overall container operational process. Dirgahayu in 1999 [16] defined the idle time was a time which was not used properly during bert working time. Ferry Setiawan, Trimaijon, and Ferry Fatnanta (2016) [17] explained that factors which influenced the idle time also include waiting the head truck arrival, the operator arrival, equipment failure, ship maintenance, and weather. While PT Pelabuhan Indonesia (2000) argued that idle time was a time outside break time which was not used during loading and unloading process. Other factors which cause idle time include to the following [16]: late to start working, leave early from work, waiting for the truck arrival, waiting for equipment maintenance, shipping to shore system, waiting for loading and unloading process, and late document submission.

## 2. Methods

This study uses both primary and secondary data. First, as a primary data, this study uses participatory observation during 3 (three) months period from October 2016 until December 2016 and followed by an interview with the Gate Coordinator at PT. Pelabuhan Tanjung Priok. The participatory observation was focused on getting some insights during the loading and unloading container process and it was mainly taking place in South Section Terminal 2 Tanjung Priok Port. The data was gathered to find which factors can influence the idle time as well as to identify the failure points of each phases at service blueprint framework. The population of this study monitored 53 (fifty three) container operational shipping samples. The observation phase was resulted in time stamp records of each operational activities during loading and unloading container process. Following the observation phase, the interview with with Gate Coordinator at PT. Pelabuhan Tanjung Priok Container Terminal Operation 2 was performed. The main goal of this interview was to validate all data recorded during the observation phase. Second, this study uses literature reviews such as histories data from January till December 2016, academic papers, reports, and jurnals as a secondary data to gather various information needed in the implementation of study.

## 3. Result and Discussion

This study uses service blueprint, multiple regression analysis, and one-way anova method. The data analysis is done by using SPSS software to perform hypothesis testing. First, The service blueprint is used as a framework which allows company to explore all the issues in creating or managing a service. The process of designing a blueprint encompasses to the following issues [18]: *Identifying process,* this step is mapping the processes that construct the service. *Isolating fail points*, identification of fail points and the design of fail-safe process are essential. By analyzing the fail point at the design stage, one can reduce the service failures. At Tanjung Priok Container Terminal Operational Port, the process included to the following steps: gate-in, yard operation, bert operation, and gate out. Actions of each steps will be separated into front-stage operator and backstage operator. The front-page operator is a person who is directly contacted with the customer while the backstage is a person who is not having direct contact to the customer. Each step will result in source of documents. In the following table 3 each idle variable is explained:

**Table 3.** Factors which influence Idle Time

| Idle Variable | Attributes | Sources | Description |
|---|---|---|---|
| Head Truck | Container arrive at the break time | Arwinas Dirgahayu (1999) [16] | Delay due to the availability of head truck |
| Computer System | Limitation on the availability of computer system | Araia Weldeselassie Simon (1999) [19] | Most operation still use manual instead of automatization using computer system |
| | Limitation on the availability of computer system | Araia Weldeselassie Simon (1999)[19] | Most operation still use manual instead of automatization using computer system |
| Weather Factor | Rain | Harmaini Wibowo (2010) [20] | Delay due to hard rain |
| | Hard wind | Ferry Setiawan, Trimaijon, Ferry Fatnanta (2016) [17] | Delay due to hard wind |
| Stevedoring Worker | Less worker | Ferry Setiawan, Trimaijon, Ferry Fatnanta (2016) [17] | Delay due to availability of stevedoring workers |

IOP Conf. Series: Materials Science and Engineering **847** (2020) 012090 doi:10.1088/1757-899X/847/1/012090

**Table 3.** Factors which influence Idle Time

| Idle Variable | Attributes | Sources | Description |
|---|---|---|---|
| Equipment Failure | Waiting for stevedoring | Ferry Setiawan, Trimaijon, Ferry Fatnanta (2016) [17] | Delay due to the indiscipline stevedoring |
| | Equipment maintenance | Arwinas Dirgahayu (1999) [16] | Equipment failure during loading and unloading process |
| | Limitation of equipment | Arwinas Dirgahayu (1999) [16] | Delay due to Limitation on tools or equipment |
| | Equipment Quality | Harmaini Wibowo (2010) [20] | Delay due to low-quality equipment |
| Operator Activities | Waiting for operator arrival | Ferry Setiawan, Trimaijon, Ferry Fatnanta (2016) [17] | Delay due to the indiscipline operators |

Next, the following table 4 describes container operating process at Terminal II Container Operational Port including the fail points (F) or waiting point (W):

**Table 4.** Service Blueprint Result

| Process | Activities | Fail Point / Waiting | Variable Category | Front Stage | Person in charge | Back Stage |
|---|---|---|---|---|---|---|
| Gate In (GI) | Inspection | | | Permitting Container to be checked | Security 1 | |
| | | | | Requesting Receiving Card (R/C) e-service from the driver | Security 1 | Delivering R/C to Operator GI 1 |
| | | | | Controlling Physical Container | Security 1<br>Security 2 | Security report the result to Operator GI 1 |
| | | F1.1 | Computer System | Receiving R/C from Security 1 | Operator GI 1 | Documenting the physical result in R/C |
| | | F1.2 | Equipment Failure | Permitting Container to weighbridge | Security 1 | Delivering R/C to Operator GI 2 |
| | Weight Calculation | F1.3 | Equipment Failure | Calculating Total Container Weight | Operator GI 1 | Delivering the net weight to Operator GI 3 |
| | | F1.4 | Computer System | Delivering Seal Number, Type, Size, Voyage Name to R/C | Operator GI 2 | |
| | | | | Separating First Page R/C to GI and Second Page R/C to Yard Operation | Operator GI 2 | |
| | Container Movement Strip (CMS) Allocation | F1.5 | Operator Activity, Equipment Failure | CMS Allocation | Operator GI 3 | Printing CMS for Driver |
| | Delivering Print out CMS | | | Delivering Second Page to Driver for submitting to Yard Operation | Operator GI 2 | Delivering Print Out CMS to Operator GI 2 |
| | | | | Delivering CMS to driver for showing the location | Operator GI 2 | |

IOP Conf. Series: Materials Science and Engineering **847** (2020) 012090 doi:10.1088/1757-899X/847/1/012090

**Table 4.** Service Blueprint Result

| Process | Activities | Fail Point / Waiting | Variable Category | Front Stage | Person in charge | Back Stage |
|---|---|---|---|---|---|---|
| Yard Operation (YO) | | F2.1 | Operator Activity | Requesting R/C | Operator YO | Controlling the container location |
| | | F2.2 | Operator Activity | Reading Location above the ship for setting | Operator YO | Confirming Location Availability |
| | | F2.3 | Equipment Failure, Weather | Delivering Location above the ship for Operator Rubber Tyred Gantry (RTG) | Operator YO | |
| | | F2.4 | Equipment Failure, Weather | Operating RTG for container movement | Operator RTG | |
| | | | | Receiving R/C from Operator | Admin YO | |
| | | F2.5 | Operator Activity | Documenting Location above ship with Single Stack | Admin YO | |
| | | | | Documenting the loading and unloading process with Hand Held | Operator Documenta tion | Writing Seal Number with Tally Sheet |
| Bert Operation (BO) | | F3.1 W3.1 | Operator Activity | Documenting the loading and unloading process with Hand Held | Operator Documenta tion | Writing Seal Number with Tally Sheet |
| | | F3.2 W3.2 | Equipment Failure | Writing Seal Number with Tally Sheet | Operator Documenta tion | Mechanic Team Standby |
| | | F3.3 W3.3 | Stevedoring Worker, Weather, Equipment Failure, Head truck | Giving the instruction to move the container from ship | Stevedorin g Worker | |
| | | F3.4 W3.4 | Stevedoring Worker, Weather, Equipment Failure | Listening the instruction for container location | Stevedorin g Worker | |
| | | F3.5 W3.5 | Weather, Equipment Failure | Operating HMC for container movement | Operator HMC | |
| | | | | Permitting container to the gate out | Operator | |
| Gate Out (GO) | | | | Requesting R/C e-service from driver | Operator GO 1 | |
| | | F4.1 W4.1 | Operator Activity | Entering data R/C for Delivering Container Letter | Operator GO 1 | Comparing the data with delivering container letter |
| | | F4.2 W4.2 | Equipment Failure | Printing out Delivering Container Letter | Operator GO 1 | |
| | | | | Permitting Container to leave | Operator GO 1 | |

From table 4, it is clear that there are 17 (seventeen) Fail Points (F) and 7 (seven) Wait Points (W). Concurrently, the fail point's variables are categorized into the following 6 (six) variables, namely head truck, computer system, operator activities, stevedoring worker, weather, and equipment failure. From the mentioned fail points, all variables will be tested by using multiple regression. This multiple regression analysis is used as a technique for creating a straight-line equation to make estimates [21]. The result of this multiple regression is formulated to the following: Idle Time = 0.374 + 0.804 X1 (Head_Truck) + 1.069 X2 (Computer_System) + 0.077 X3 (Weather) + 1.754 X4 (Stevedoring_Worker) + 1.045 X5 (Equipment_Failure) + 2.157 X6 (Operator_Activities). The 0.374 Constanta means that the ship has 0.374 hour idle time during loading and unloading container process. Variable weather has sig value 0.876 > 0.05, accordingly this variable weather can be excluded from the new formulation. The new regression model is thus given to the following formulation: Idle Time = 0.374 + 0.804 X1 (Head_Truck) + 1.069 X2 (Computer_System) + 1.754 X4(Stevedoring_Worker) + 1.045 X5 (Equipment_Failure) + 2.157 X6 (Operator_Activities). As operator_activities variable has the largest coefficient, 2.157, this operator_activities variable will be analyzed further with one-way ANOVA test. Therefore, one can find: in which step of operator activities is the most significant influence to the delay of loading and unloading container process.

To perform one-way ANOVA, extra container operation shipping record is needed. According to SLOVIN's formula with 5% margin error, the population must meet minimal 53 (fifty three) container operational shipping samples to test. From this one-way ANOVA test, it was resulted that the F-Score = 19.930 > the Critical F-Value = 3, therefore one can argue that there is a different influence among operators in each loading and unloading container activities. Next, ANOVA Post Hoc test was performed by using Tukey's Honestly Significant Difference (HSD) post hoc test which finally this Tukey's HSD test resulted that the operator activities variable at gate in phase was the most significant influence to the idle time during loading and unloading container process. The following recommendations are made to improve the port service performance:

- Each phases can be targeted with the amount of successful loading and unloading container and the target finished container processing time. Adding extra facilities so that the process can be faster;
- Bureaucracy and regulation should be less complex especially with regards to the permit. The Port Management should reduce corruption in the authorize port area;
- The payment system for the stevedoring worker should be changed from lump sum wages to the daily wages. Each stevedoring worker can be paid based on hour works and the amount of work done.

This study also has a few limitation. First, it is associated with the validity on the variables that cause the idle time during loading and unloading container process. Through the research process, it was discovered that many variables may also include social, environment, economic factors which may have influence to the delay of loading and unloading process. Second, this study should include extra interview with other gate coordinator and port operators in the fields. Accordingly, comparison with other terminals can be considered to the future study.

## 4. Conclusion

Based on the analysis done from this study, the following conclusion can be deducted: Fail points are causing the idle time during loading and unloading container process which can be categorized into 6 (six) variables, namely head truck, computer system, weather, stevedoring worker, equipment failure, and operator activities. Weather Variables has the least significant influence to the idle time, it can thus be excluded to the main causes of idle time. This is mainly because the sig value of this weather variable, 0.876 > 0.05. Further, it is found that the most influence to the delay on the container process is at the gate in stage on the operator activities. The main reason is due to indiscipline from the operators with regards to the working time. At the gate in phase, the person in charge is not always stand-by which can lead to the delay of loading and unloading container process. Human factor is thus play an important role to the port performance in overall.

## 5. References

[1] Jouili, T. A. and Allouche, M. A., 2016. Impacts of Seaport Investment on Economic Growth. *Promet Traffic&Transportation,* **28**(4), pp. 365-370.

[2] Sambhi, N., 2015. Jokowi's 'Global Maritime Axis':Smooth Sailing or Rocky Seas Ahead? *JSTOR,* **11**(2), pp. 39-56.

[3] Chairuddin, I., 2015. An overview of Indonesian logistic performance and the Sislognas 2012. *Jurnal Manajemen Bisnis Transportasi dan Logistik ,* **1**(3), pp. 353 - 367.

[4] The World Bank, 2014. *The World Bank News and Broadcast.* [Online] Available at: http://go.worldbank.org/Z6VE3IDAF0 [Accessed 14 January 2020].

[5] The World Bank, 2015. *The World Bank News.* [Online] Available at: https://www.worldbank.org/en/news/feature/2015/05/26/the-tale-of-two-ports-in-indonesia [Accessed 13 January 2020].

[6] The World Bank, 2018. *International LPI.* [Online] Available at: https://lpi.worldbank.org/international/global [Accessed 13 January 2020].

[7] Republic of Indonesia, 2012. *Indonesian Presidential Regulation about blueprint of national logistics system development (SISLOGNAS),* Jakarta: s.n.

[8] Loke, K. B., Saharuddin, A. H., Ibrahim, A. R. and Rizal, I., 2005. *The study on the delay factor of container transhipment.* Selangor, Kolej Universiti Sains dan Teknologi Malaysia.

[9] Roa, I, Peña, Y, Amante, B and Goretti, M 2013, 'Ports: definition and study of types, sizes, and business models', *Journal of Industrial Engineering and Management,* **6**(4), pp. 1055–1064. DOI: http://dx.doi.org/10.3926/jiem.770.

[10] Flere, WA 1967, Port Economics, Ward & Foxlow.

[11] Container Management, 2020. *News.* [Online] Available at: https://containermag.com/2015/06/22/president-sack-officials-responsible-tanjung-priok-delays/ [Accessed 16 January 2020].

[12] J, T. and Heng, W., 2005. Port privatization, efficiency and competitiveness: some empirical evidence from container ports (terminals). *Transportation Research Part A: Policy and Practice,* **39**(5), pp. 405 - 424.

[13] Brooks, M. R. and Pallis, T., 2013. Considering the perspectives of port users. *Port Technology International* **60**, pp. 27 - 28 .

[14] Ministry of Transportation Republic of Indonesia , 2012. *Directorate General of Sea Transportion.* [Online] Available at: http://www.dephub.go.id/post/read/standar-kinerja-pelayanan-operasional-pelabuhan-di-audit-6-bulan-sekali-10626?language=en[Accessed 16 January 2020].

[15] Kamil, I., Plamonia, M., Hasan, A. and Alias, B., 2018. Can Supply Chain Connectivity Effect on the Relationship between Port Performance and Economic Growth? *Supply Chain Management,* **7**(6), pp. 462 - 475.

[16] Dirgahayu, C. A., 1999. *Petunjuk Penangan Kapal dan Barang di Pelabuhan ,* Jakarta: PT. Pelabuhan Indonesia II.

[17] Setiawan, F., Trimaijon and Fatnanta, F., 2016. *Identifikasi Faktor-Faktor Yang Mempengaruhi Nilai Idle Time (IT) DI Pelabuhan Dumai (Dermaga A),* Pekanbaru: FT Universitas Riau.

[18] Shostack, G. L., 1984. Designing Services That Deliver. *Operation Management ,* 62, No. 1(Harvard Business Review), pp. 133 - 139.

[19] Simon, Araia Weldeselassie, Analysis of the problems of productivity and performance of the Port of Assab [Eritrea] 1999. World Maritime University Dissertations. 100.

[20] Wibowo,H. 2010. *Analisis Faktor-Faktor Yang Mempengaruhi Waktu Tunggu Kapal Di Pelabuhan Tanjung Emas Semarang.* Universitas Diponegoro, Semarang.

[21] Sekaran, U. and Bougie, R., 2009. *Research Methods for Business A Skill- Building Approach.* 5$^{th}$ ed. West Sussex: Wiley.

IOP Conf. Series: Materials Science and Engineering **847** (2020) 012091 doi:10.1088/1757-899X/847/1/012091

# Combination of Value Stream Mapping and House of Risk Methods to Eliminate Waste in Productivity Enhancement in Production Area of Fertilizer Company

**F Suryanti[2,3], A Sudiarno[1,4], , A A C Sudarni[3,5]**

[1,2]Department of Industrial Engineering Institut Teknologi Sepuluh Nopember, ITS, Surabaya, Indonesia
[3]fitsuryanti@gmail.com, [4]adithya.sudiarno@gmail.com, [5]adiekastikaclara@gmail.com

**Abstract.** Success in ensuring agricultural production through fertilizer production is the key to strengthening national food security. PT X is a domestic company producing ZA fertilizer. In this study, the endeavor to implement lean manufacturing with a nine wastes (E-DOWNTIME) approach to eliminate waste, so productivity can be increased by relying on a combination of VSM and HOR methods to identify wastes to propose remedies to the risks arising from waste on the ZA production floor, through FGD. This research focuses on risks to waste each production process flow according to the business process adopted. The results showed a waste of waiting, defects, excess processing, and EHS on the ZA production floor. 28 risk events were identified and 28 risk agents were identified. After two alternative repairs were selected namely replacement of the drum filter shaft and replacement of the turbine rotor, the duration of the NVA was 79 minutes reduced. Corrosion, aging, and abrasion machines / tools are the selected risk agents, representing that operating machines have shown a decline in performance due to age, so that a gradual machine update is necessary.
**Keywords**: VSM, HOR, Lean manufacturing, E-DOWNTIME, Productivity

## 1. Introduction

Indonesia is famous as an agricultural country. The performance of the national agricultural industry contributes to the growth of the agricultural sector which increases every year. Success in ensuring agricultural production is the key to strengthening national food security. Fulfillment of quality fertilizer production as a result of market demand that continues to increase from year to year. High-quality, subsidized fertilizer, stable prices and easily accessible can bring benefits for producers to consumers. The great responsibility of meeting fertilizer needs in order to strengthen national food security while at the same time striving for fertilizer exports as a commodity that benefits the country, is carried out by strengthening the national fertilizer industry.

PT X is a fertilizer producing company. PT X is not only known to produce fertilizer products, but also non-fertilizer products. PT X has a vision to become a producer of fertilizers and other chemical products that are highly competitive and the products most demanded by consumers. To realize this vision requires commitment from management to the production line and is proven by optimal performance that will produce high productivity. Like the concrete efforts that have been carried out by many other manufacturing industries that hold fast to the company's commitment and continue to strive to achieve its vision and mission. However, company data shows that there are problems on the production floor that have the potential to cause waste. There are several waste classified into nine types namely Environmental, Health, and Safety (EHS), Defect, Overproduction, Waiting, Non-utilized / Underutilized Talent, Transportation, Over Inventory, Motion, and Excess Processing or abbreviated as E-DOWNTIME [1]. Thus, the application of methods from many scientific fields is needed in order to eliminate the waste immediately.

The fields of lean manufacturing and safety did not escape the focus of the industry to be developed to adapt to real conditions. The use of lean manufacturing has been proven to be widely applied and is the most effective method in the manufacturing industry throughout the world in an effort to improve the operational efficiency of the company through the success of increasing yields with fewer resources compared to traditional manufacturing systems. Lean manufacturing is a system that aims to do production with minimum labor, minimum production area, minimum resource consumption, minimum inventory, minimum defective goods, short production time, and minimizing consumer dissatisfaction [2]. This success is related to the company's ability to identify and eliminate waste or waste in its manufacturing floor manufacturing system in order to be more responsive to customer demand while maintaining the quality of the products produced.

From a lean perspective, all types of waste found throughout the value stream process, which transforms from input to output must be eliminated in order to increase product value and subsequently increase customer value [3]. Value Stream Mapping (VSM) is one of the lean manufacturing methods to describe the conditions of the production process flow. Not only that, VSM can find various types of waste and try to eliminate it [4]. In VSM we can know the cycle time and lead time of the company's production process. For the process of making goods, value streams include suppliers of raw materials, manufacturing and assembling of goods, as well as distribution networks to users of the goods [3].

HOR is a method for identifying and calculating risks that must be borne by the company to find priority corrective actions that are appropriate to be implemented with reference to the company's ability. The HOR model can allow companies to choose a set of risk agents to treat and then prioritize proactive actions, to reduce the overall impact of the risk events caused by the risk agent [5]. The business process involved in preparing HOR is useful in guarding the identification of each risk that occurs in all activities in the business process that is used in a coherent manner. The method used to identify risks to HOR is the development of the Supply Chain Order Reference (SCOR) which consists of plan, source, make, deliver, and return. The risk identification method using SCOR which is generally implemented in the supply chain field is considered to have weaknesses, in addition to its implementation limitations that are only in the supply chain field, this method cannot accommodate changes management aspects [6].

## 2. Methods

In this study data processing was performed using a combination of two methods namely Value Stream Mapping (VSM) and (HOR), while the data collection to compile VSM (Current State Mapping and Future State Mapping) and HOR is done by interview and Focus Group Discussion (FGD) and by involving four section heads who are experts in the field of production.

1. Mapping the production flow with VSM in order to describe the production process from upstream to downstream, so as to make it easier to identify the waste and its causes that occur in each production flow of ZA products. Waste that has been identified is a risk event which will then be assessed by the HOR method. The whole process was carried out through interview techniques and FGDs.
2. Risk assessment of waste that occurs in each production flow of ZA products is carried out using the HOR phase 1 method, which begins by determining the risk event or risk event. Waste that occurs in the production process flow that has been mapped previously using the VSM method, is a risk event that will be assessed based on the level of severity and determined the cause or risk agent. After the risk agent is determined, it is then assessed based on the level of occurrence, and determines the correlation value between the risk event and the risk agent.

IOP Conf. Series: Materials Science and Engineering **847** (2020) 012091 doi:10.1088/1757-899X/847/1/012091

3. The next stage after conducting a risk assessment of waste using HOR method phase 1, the selected risk agent results based on ranking are used as input in designing alternative improvements with the HOR method phase 2. HOR phase 2 is used to prioritize preventive actions that companies must take to maximizing cost effectiveness in managing selected risk agents in HOR phase 1, thus taking into account the value of Total Effectiveness of Action (TEk) and Effectiveness to Difficulty Ratio (ETDk).
4. After identifying risks to choosing an alternative improvement with HOR, a production flow mapping is carried out using future state mapping after the implementation of selected preventive actions on the production floor of ZA products. In the future state mapping, it can be seen the difference in terms of time reduced by the results of risk mitigation efforts.

## 3. Result and Discussion

*3.1Mapping Production Flow With Current State Mapping*

Mapping in the form of current state mapping stated on Figure 1, mapping the production process flow of ZA products from carbonation, reaction, filtration, neutralization, evaporation and crystallization to drying and cooling. In the mapping, waste or waste identified and its causes are identified as risk events and risk agents, each of 28 risk events and 28 risks agent. The overall risk event and risk agent according to the E-DOWNTIME approach, are classified as waiting, defect, excess processing, and EHS waste. The process of producing ZA products requires time in a single production.

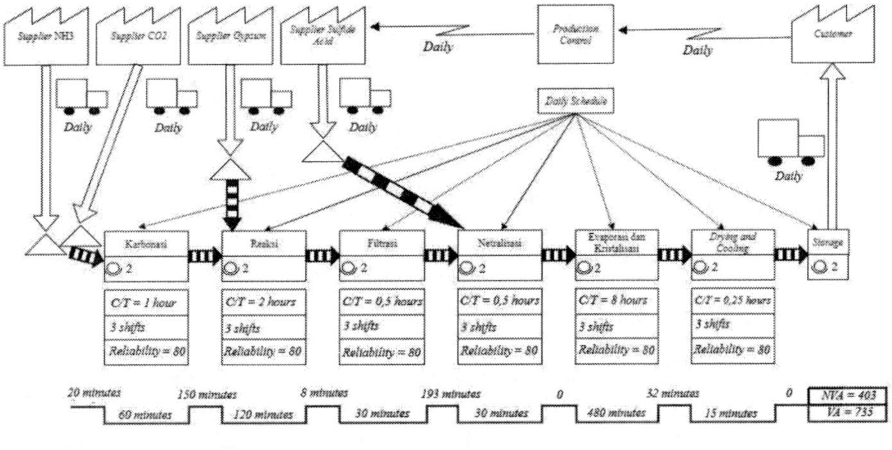


**Figure 1** Current State Mapping of ZA Product

*3.2 Risk Assessment of Waste With HOR Phase 1*

According to the Supply Chain Operation Reference (SCOR) business process including plan, source, make, deliver, and return, there are 26 risk events with a composition of 1 risk event on the plan, 7 risk events on source, 16 risk events on make, 1 risk event on deliveries, and 1 risk event on returns. The most risk event is in make with a percentage of 64%. Whereas the severity that occurred was included in criteria 3 and 4. Criteria 3 became the most severe severity with 15 risk events or 54%. After identifying the risk event, a risk agent or risk agent is determined. There are a number of 28 risk agents, with the distribution of occurrence or the likelihood of occurrence in criteria 1, 2, 3, and 4. Criteria 3 is the most common occurrence criteria with a total of 15 risk agents or with a percentage of 53%. In ZA products, selected risk agents that will be continued on HOR 2 are aging machines / tools, corrosion machines / tools, ship arrival is unpredictable and abrasion machines / tools with notations respectively A8, A9, A3, and A11. The cumulative percentage of ARP values A8, A9, A3, and A11 are 8.5%, 14.6%, 19.9% and 25.2% respectively, based on Table 1 and Figure 2.

**Table 1** Calculation of The Selected ARP Value

| Risk Agent | | Rank | ARP | Cumulative ARP | % ARP | Cumulative % ARP |
|---|---|---|---|---|---|---|
| Aging machines / tools | A8 | 1 | 252 | 252 | 8,5% | 8,5% |
| Machine / tool corrosion | A9 | 2 | 180 | 432 | 6,1% | 14,6% |
| The arrival of the ship is unpredictable | A3 | 3 | 156 | 588 | 5,3% | 19,9% |
| Abrasion machine / tool | A11 | 4 | 156 | 744 | 5,3% | 25,2% |

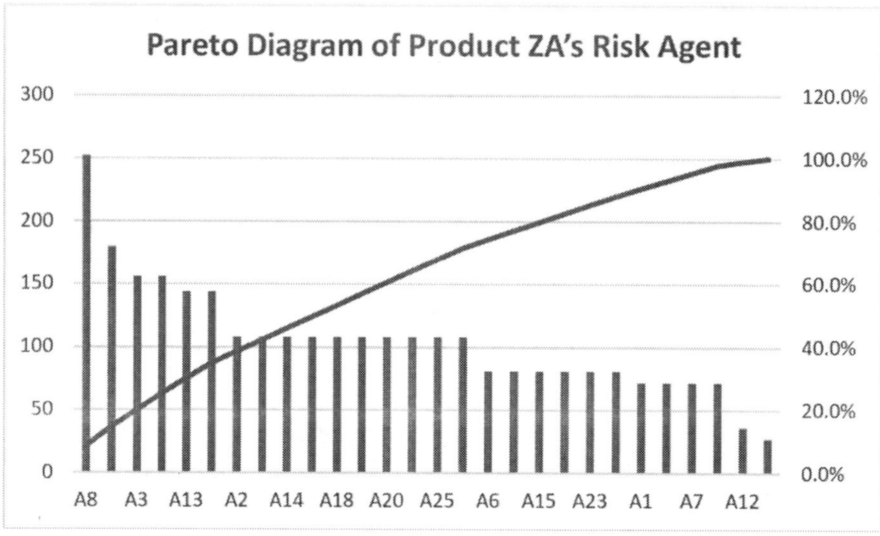


**Figure 2** Pareto Diagram of ZA Product's Risk Agent

*3.3 Alternative Design Improvement With HOR Phase 2*

The results of the FGD agreed on the existence of 8 preventive actions as mitigation actions for selected risk agents, which were notated as PA1 to PA8. Preventive action has a correlation scale of 3 and 9 which means that preventive action has a moderate and large role in preventing the occurrence of risk agents and then the total effectiveness of action (TEk) value is carried out. The total effectiveness of action is formulated using equation 1. The acquisition of the total value of effectiveness of action (TEk) is further divided by the level of difficulty with a scale of 4 and 5 which means that the application of preventive action requires considerable and difficult resources, based on equation 2. This distribution results in the value of effectiveness to difficulty ratio (ETDk) which is then processed using the Pareto technique to obtain selected preventive actions, namely the replacement of the drum filter and the replacement of the turbine rotor, which are notated as PA3 and PA2, respectively. The cumulative percentage of PA3 and PA2 ETDk values are 15.8% and 28.5%, respectively, based on Table 2 and Figure 3.

$$TE_k = \sum_j \mathbf{ARP}_j E_{jk} \quad (1)$$
$$ETD_k = TE_k / D_k \quad (2)$$

**Table 2** Calculation of The Selected ETDk Value

| Preventive Action | | Rank | ETDk | Cumulative ETDk | % ETDk | Cumulative % ETDk |
|---|---|---|---|---|---|---|
| Replacement drum shaft filter | PA3 | 1 | 567 | 567 | 15,8% | 15,8% |
| Turbine rotor replacement | PA2 | 2 | 453,6 | 1020,6 | 12,7% | 28,5% |

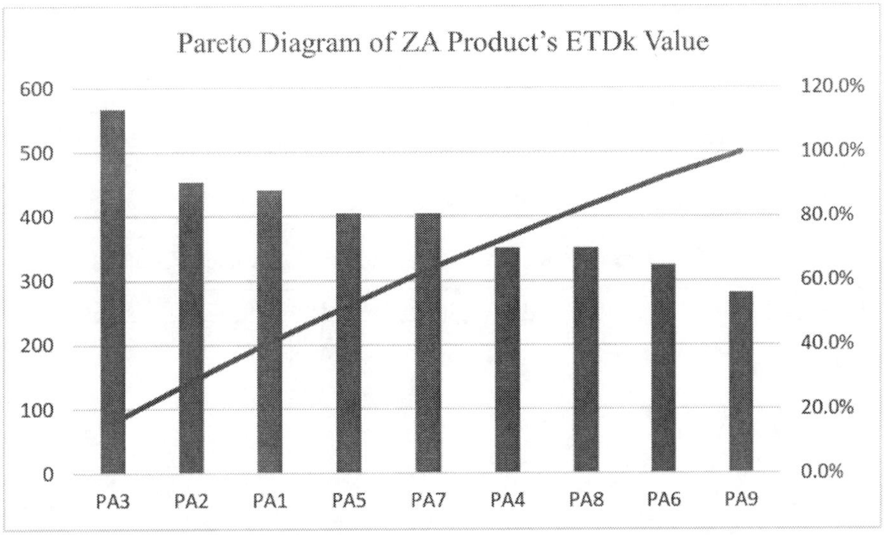


**Figure 3** Pareto Diagram of ZA Product's ETDk Value

*3.4 Mapping Production Flow With Future State Mapping*

Mapping of production flow with future state mapping of ZA products in Figure 4, is done after the selection of alternative improvements to the waste or waste that occurs on the production floor. The duration listed is an estimate of the time that occurred after the two alternative improvements selected, namely the replacement of the drum filter and the replacement of the turbine rotor successfully applied to the carbonation and filtration process. Information regarding the duration of value added (VA), non-value added (NVA) and non-value added (NVA) is reduced by 79 minutes shown in Table 3.

IOP Conf. Series: Materials Science and Engineering **847** (2020) 012091 doi:10.1088/1757-899X/847/1/012091

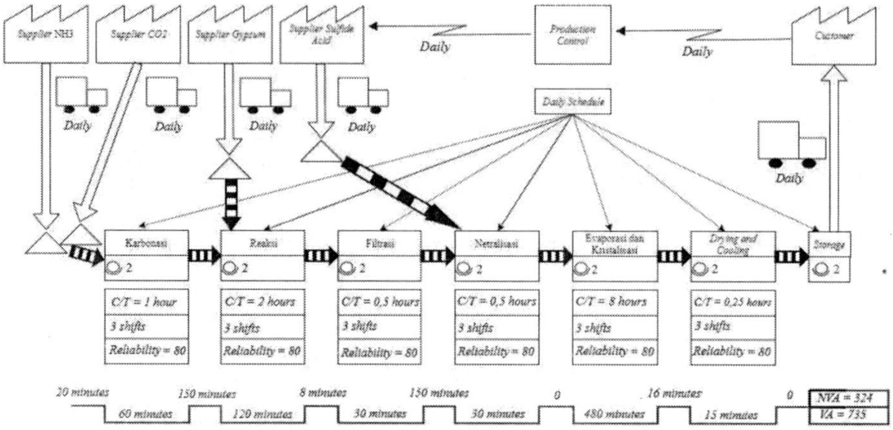


**Figure 4** Future State Mapping of ZA Product

**Table 3** Total Duration of VA, NVA, and NVA Reduced Products ZA

| Production Flow | VA (minutes) | NVA (minutes) | NVA Reduced (minutes) |
|---|---|---|---|
| Carbonation | 60 | 150 | 20 |
| Reaction | 120 | 8 | 0 |
| Filtration | 30 | 150 | 43 |
| Neutralization | 30 | 0 | 0 |
| Evaporation and Crystallization | 480 | 16 | 0 |
| Drying and Cooling | 15 | 0 | 0 |

In ZA products, the most dominating risk agent with an cumulative amount of 8.5% ARP is aging machines / tools. The aging machine / tool is related to the lifetime of the machine / tool. This happens because, there are still many machines / tools that have not been replaced with new units, so that a gradual replacement of machines / tools is needed. Recommendations for improvement for further research is to develop methods in the field of lean service and apply quantitative methods.

## 4. Conclusion

Current state mapping of the production process flow of ZA products from carbonation, reaction, filtration, neutralization, evaporation and crystallization to drying and cooling. In this mapping, waste or waste identified and their causes are identified as risk events and risk agents, each of 28 risk events and 28 risk agents. The overall risk event and risk agent for ZA products according to the E-DOWNTIME approach are classified as waiting, defect, excess processing, and EHS waste. The purpose of HOR phase 1 is to obtain a selected risk agent or cause of a risk event. On the production floor of ZA products, aging machines / tools and corrosion machines / tools are the selected risk agents denoted by A8 and A9, and the cumulative percentage of ARP values are 8.5% and 14.6% respectively. The purpose of HOR phase 2 is to obtain selected preventive actions. On the production floor of ZA products, there are 8 alternatives for preventive action improvement alternatives. Then two selected preventive actions are produced, namely the replacement of the drum filter and the replacement of the turbine rotor notated as PA3 and PA2 with a cumulative percentage of ETDk value of 15.8% and 28.5%. After two alternative improvements were chosen, namely the replacement of the drum filter and the replacement of the turbine rotor successfully applied to the carbonation and filtration process, eliminating waste and reducing NVA by 79 minutes. Corrosion machines / tools, aging machines / tools, and abrasion machines / tools are the selected risk agents for ZA products, representing that operating machines have shown a decline in performance due to age, so a gradual renewal of the machine is needed.

IOP Conf. Series: Materials Science and Engineering **847** (2020) 012091 doi:10.1088/1757-899X/847/1/012091

## 5. Reference

[1] L. Stack, 2010 Lean Process and DOWNTIME, USA: The Productivity Pro.
[2] J. P. Womack, D. T. Jones and D. Roos, 1991 The Machine That Change the World: The Story of Lean Production, New York: Harper Perennial.
[3] V. Gaspersz, 2007 Lean Six Sigma for Manufacturing and Services Industries, Jakarta: PT Gramedia Pustaka Utama.
[4] N. A. Wahab, 2013 A conceptual Model of Lean Manufacturing Dimensions, *Procedia Technology* **11**.
[5] I. N. Pujawan and L. H. Geraldin, 2009 House Of Risk: A Model For Proactive Supply Chain Risk Management, *Business Process Management Journal*.
[6] S. H. Huan, S. K. Sheoran and G. Wang, 2004 A Review and analysis of supply chain operations reference (SCOR) model, *Supply Chain management*.

IOP Conf. Series: Materials Science and Engineering **847** (2020) 012092 doi:10.1088/1757-899X/847/1/012092

# Measurement of Physiological and Psychological Workloads of Mechanical Department Operator PT. XYZ

**B Cahyadi[1,2], A S Maryanti[1,3], G A Timang[1,4]**

[1]Department of Industrial Engineering, Faculty of Engineering, Pancasila University, Jakarta, Indonesia
[2]bambang.cahyadi@univpancasila.ac.id, [3]amandasalsabila98@gmail.com, [4]gitatimang20@gnail.com

**Abstract.** PT. XYZ is an electrical machinery repair services company. The amount and density of operator activity can affect worker fatigue. This research conducted to calculate the level of physiological and psychological workloads and ideals rest periods for operators in the Mechanical Department of PT. XYZ. The method to calculate physiological workload is direct measurement by measuring heartbeat, whereas the method to calculate psychological workload is NASA-TLX Method. The results show that the most severe physiological workload experienced by Mr. Nur who is a medium-sized lathe operator with a value of energy consumption of 8.497 Kcal/minute before a break and 7.602 Kcal/minute after a break, the ideal rest time for Mr. Nur is 112 minutes with work duration from 07.00-12.00 WIB and 66 minutes with work duration from 13.00-16.00 WIB, while Mr. Supardi and Mr. Didik requires a rest time of 103 minutes and 38 minutes with work duration from 07.00-12.00 WIB and the level of psychological workload for the 12 operators are high and very high. Therefore the company can minimize the level of physiological and psychological workload by calculating the workload of the operator in a balanced manner.
**Keywords:** physiological and psychological workload, energy expenditure, energy consumption, ideal rest time, NASA-TLX.

## 1. INTRODUCTION

The development of industry today especially machinery progressing very rapidly, it is because in every production activity is never separated from the use of tools or machines as operational support. The machines will operate in accordance with a properly when supported by the operating standards and the correct treatment to prevent damage to either the light damage or major damage.

As required maintenance for machinery production, also affect the increase in service repairs and installation of machinery and equipment. This is evidenced in the growth of the industry is the highest production increased repair and installation of machinery and equipment, amounted to 13.29% [1]. Each company in the production activities need to pay attention to the machine or tool used remains in good condition. In order to keep the machine in good condition, it needs regular maintenance and service as well as the corrective action on the part that was damaged or had trouble shooting, therefore it is necessary for the continuation of engine repair services production process.

PT. XYZ is an electrical machinery repair services company. The company has three departments, one of them is Department of Mechanics. In this department, there are 12 operators where each

IOP Conf. Series: Materials Science and Engineering **847** (2020) 012092 doi:10.1088/1757-899X/847/1/012092

operator doing a different job or more than one job in accordance with the settlement of repair items. Repair items that do require a different time compatible with the problem and dealing between the company and the customer. Many operators and density of activity may impact on worker fatigue.

Based on the results of the questionnaire fatigue, there are three criteria for fatigue experienced by operators such as workload, duration of action and the physical working environment. Out of 12 operators who filled out a questionnaire, 8 operators agreed that fatigue is caused by the criteria of workload, 8 operators agreed that fatigue is caused by the criteria of duration and 5 operators agreed that the fatigue is caused by environmental criteria physical labor.

Based on the information above, this research will focus on the criteria of workload and duration of action in view of these two criteria is fatigue often experienced by the operator. Referring to the results of these studies it is necessary to measure the physiological workload by using direct measurement methods and psychological workload using NASA-TLX method based on the parameters to be measured against the operator so that work can be adapted to the capabilities of the operator, and to compare whether the level of physical fatigue matches the level of mental fatigue of the worker.

## 2. METHODS

### *2.1. Workload*

Workload is an effort that must be issued by a person to meet the "demand" of the work, the intended workload is the size (portion) of the limited capacity of the operator that required to do a job [2].

### *2.2. Duration Of Work*

The duration is the span of time or the length of time, therefore the duration/hours of work in Indonesia was regulated by Undang-Undang No.13 of 2003 about Manpower [4]. For several companies, working hours, rest periods, and overtime included in the Collective Labor Agreement (CLA). This is according with Clause 77, Verse 1, of UU No.13/2003, which requires each employer to implement the provisions of working hours.

### *2.3. Physiological Workload*

The psychological workload (mental workload) is defined as the evaluation of the hose operator maturity (current capacity is being motivated by the existing workload) when doing a job (metacontroller activity) to achieve specific goals [12]. Some examples of activities that are dominated by mental activity is like a surgical operation, assembly carefully, looking at the target at the time of shooting, see micro-sized object through a microscope and others [3]. Mental burden has a fairly high correlation to the mistakes made (error) or mental workload has a correlation with the level of performance [13].

Working physiological (physical) is a work that requires physical energy of human muscle as a source of energy (power). Physical work is often referred to as "manual operation" which is the work performance will entirely depend on a man who serves as a source of energy (power) or job control. In terms of the physical work, the energy consumption are the main factors and benchmarks used as a determinant of weight/lightness of the physical work [5].

Factors that affecting the workload is ratings physical workload can be done by two methods objectively, is direct assessment methods and indirect assessment methods [6].

Direct measurement method is to measure the energy expended (energy expenditure) through the intake of oxygen during the work. The more weight the more work energy expended [7]. The amount of energy expended for a job can be measured by calculating the heart rate and demographic factors. Research on the measurement of energy one of which is expressed Keytel [8]:

$$EE = -20.4022 + (0.4472\ HR) - (0.1263\ w) + (0.074\ A) \quad (2.1)$$

Where:
EE= Energy expenditure (kcal/min)

HR= Heart rate (beats/min)
w = Weight (kg)
A = Age (years)

With a linear relationship between heart rate and energy absorption, one can generally use heartbeat to build a "heavy" work. In Table 1 lists the severity of work with energetic demands and blood circulation [9].

**Table 1.** Occupational Classification Based on Energy Expenditure and Heart Rate

| Classification Profession | Total Energy Expenditure | | Heart Rate (beats/ minute) |
|---|---|---|---|
| | (KJ/ minute) | (Kcal/ minute) | |
| Light | 10 | 2.5 | ≤ 90 |
| Moderate | 20 | 5 | 90-100 |
| Weight | 30 | 7.5 | 100-120 |
| Very Heavy | 40 | 10 | 120-140 |
| Extreme Weight | 50 | 12.5 | 140-160 |

After doing the calculations above, we can calculate the energy consumption (K) using the following formula equation [10]:

$$K = Et - Ei \tag{2.2}$$

Where:
K = Energy Consumption (kcal / min)
Et = Energy expenditure on certain working time (kcal/min)
Ei = Energy expenditure at a time before work (kcal/min)

*2.4. Work Break Cycle*
Humans are unable to maintain the level of physical activity for a longer period. They need a break periodically to recover from the effects of the work performed. Throughout the years, various studies offer suggestions for the rest period permitted ie [11]:

$$RT = 0 \quad \text{for } K < S \tag{2.3}$$

$$RT = \frac{\left(\frac{K}{S} - 1\right) \times 100 + \frac{T(K - S)}{K - BM}}{2} \quad \text{for } S \leq K < 2S \tag{2.4}$$

$$RT = \frac{T(K - S)}{K - BM} \times 1.11 \quad \text{for } K \geq 2S \tag{2.5}$$

Where:
RT = Rest required (minutes)
K = Expenditure of energy to work (Energy expenditure) or consumption energy (kcal/min)
S = Recommended energy expenditure average (4 kcal/min for women, 5 kcal / min for men)
T = Expected total work time (minute)
BM = Basal metabolism, is the energy required by the body in a state of rest total both body and spirit (BM for women = 1.4 and BM for male = 1.7)

*2.5. Measurement Method of NASA-TLX*
NASA-TLX (National Aeronautics and Space Administration-Task Load Index). This method is a questionnaire developed by the emergence of subjective measurement needs an easier but more sensitive to the measurement of workload. NASA-TLX method is a multidimensional rating procedure, which divides the workload on the basis of the average loading of six dimensions, namely

IOP Conf. Series: Materials Science and Engineering **847** (2020) 012092 doi:10.1088/1757-899X/847/1/012092

the Mental Demand, Physical Demand, Temporal Demand, Effort, Own Performance, and Frustation Level. NASA-TLX is divided into two phases, a comparison of each scale (Paired Comparison) and giving value to the job (Event Scoring).

## 3. RESULTS AND DISCUSSION

### *3.1 Measurement Of Physiological Workload Experienced By The Operator Of The Mechanical Department PT. XYZ*

Data processing carried out in this study is to calculate the physical workload by operators in the Department of Mechanical PT. XYZ. In the data processing, calculating physical workload experienced by operators in the Department of Mechanical PT. XYZ, which consists of several stages:

1. Take measurements of heart rate to 12 mechanical operators by using Polar M200.
2. Asking for general information such as weight and age of the mechanical operators where these data as variables to be taken into the calculation.
3. Calculate the energy expidenture obtained from the heart rate measurements, body weight and age of the mechanical operators.
4. Calculate energy consumption based on reducing energy expidenture after work and before work with.
5. Classifying the types of work 12 operators with energy consumption has been calculated.

**Table 2**. The Results Of The Calculation Of Energy Expideture, Energy Consumption and Workload Categories Of 12 Mechanical Operators

| No. | Operator Name | Time Measurement | *Energy Expenditure* (Kcal/min) | Energy Consumption (Kcal/min) | Workload Category |
|---|---|---|---|---|---|
| 1 | Mr. Jumali | Before Work (07.00 WIB) | 19.768 | -14.758 | - |
| | | After Work (12.00 WIB) | 5.010 | | |
| | | Before Work (13.00 WIB) | 17.085 | -10.286 | - |
| | | After Work (16.00 WIB) | 6,799 | | |
| 2 | Mr. Satimin | Before Work (07.00 WIB) | 8.964 | 2.236 | Light |
| | | After Work (12.00 WIB) | 11,200 | | |
| | | Before Work (13.00 WIB) | 14.331 | -4.472 | - |
| | | After Work (16.00 WIB) | 9.859 | | |
| 3 | Mr. Johanes | Before Work (07.00 WIB) | 6.714 | 1,342 | Light |
| | | After Work (12.00 WIB) | 8.056 | | |
| | | Before Work (13.00 WIB) | 11,633 | 5.814 | Moderate |
| | | After Work (16.00 WIB) | 17.477 | | |
| 4 | Mr. Suryanto | Before Work (07.00 WIB) | 11.303 | -2.683 | - |
| | | After Work (12.00 WIB) | 8.620 | | |
| | | Before Work (13.00 WIB) | 9.514 | 1,789 | Light |
| | | After Work (16.00 WIB) | 11.303 | | |
| 5 | Mr. Nur | Before Work (07.00 WIB) | 5.052 | 8.497 | Weight |
| | | After Work (12.00 WIB) | 13.549 | | |
| | | Before Work (13.00 WIB) | 13.102 | 7.602 | Weight |
| | | After Work (16.00 WIB) | 20.704 | | |
| 6 | Mr. Supardi | Before Work (07.00 WIB) | 5.705 | 8.050 | Weight |
| | | After Work (12.00 WIB) | 13.755 | | |
| | | Before Work (13.00 WIB) | 18.227 | 2.236 | Light |
| | | After Work (16.00 WIB) | 20.463 | | |
| 7 | Mr. Kinan | Before Work (07.00 WIB) | 11.954 | 1,342 | Light |
| | | After Work (12.00 WIB) | 13.296 | | |
| | | Before Work (13.00 WIB) | 8.824 | -3.578 | - |
| | | After Work (16.00 WIB) | 5.246 | | |
| 8 | Mr. Pras | Before Work (07.00 WIB) | 12.078 | -3.130 | - |
| | | After Work (12.00 WIB) | 8.948 | | |
| | | Before Work (13.00 WIB) | 15.209 | -4.919 | - |
| | | After Work (16.00 WIB) | 10.289 | | |

IOP Conf. Series: Materials Science and Engineering **847** (2020) 012092 doi:10.1088/1757-899X/847/1/012092

**Table 2**. The Results Of The Calculation Of Energy Expideture, Energy Consumption and Workload Categories Of 12 Mechanical Operators

| No. | Operator Name | Time Measurement | *Energy Expenditure* (Kcal/min) | Energy Consumption (Kcal/min) | Workload Category |
|---|---|---|---|---|---|
| 9 | Mr. Sabar | Before Work (07.00 WIB) | 15.950 | -5.814 | - |
| | | After Work (12.00 WIB) | 10.136 | | |
| | | Before Work (13.00 WIB) | 22.658 | -7.155 | - |
| | | After Work (16.00 WIB) | 15.503 | | |
| 10 | Mr. Royyan | Before Work (07.00 WIB) | 10.321 | -4.472 | - |
| | | After Work (12.00 WIB) | 5.849 | | |
| | | Before Work (13.00 WIB) | 14.346 | 0,447 | Light |
| | | After Work (16.00 WIB) | 14.793 | | |
| 11 | Mr. Didik | Before Work (13.00 WIB) | 16.971 | -0.894 | - |
| | | After Work (16.00 WIB) | 16.076 | | |
| | | Before Work (07.00 WIB) | 10.263 | 5.814 | Moderate |
| | | After Work (12.00 WIB) | 16.076 | | |
| 12 | Mr. Novianto | Before Work (07.00 WIB) | 2,426 | -1.342 | - |
| | | After Work (12.00 WIB) | 1,085 | | |
| | | Before Work (13.00 WIB) | 2,873 | 0,447 | Light |
| | | After Work (16.00 WIB) | 3.321 | | |

*3.2 Calculate The Ideal Rest Time for Operators in The Mechanical Department PT. XYZ*

After knowing the results of the Energy Consumption (K) is obtained, and then calculated the ideal rest time (RT) for the operator. In the calculation of rest time by using equations 2.3 and 2.4, this is in accordance with the provisions of $K < S$ dan $S \leq K < 2S$. Then for Work Expected Total Time (T) at 07.00-12.00 WIB taken for 5 hours or 300 minutes and at 13.00-16.00 WIB for 3 hours or 180 minutes because total hours worked when the operator. In the Table 3 is the calculation of ideal rest time obtained for each operator.

**Table 3.** Ideal Resting Time Calculation Results

| No. | Operator Name | Measurement Time | RT(min) | Information |
|---|---|---|---|---|
| 1 | Mr. Jumali | Before Work (07.00 WIB) | - | Not Defined |
| | | After Work (12.00 WIB) | | |
| | | Before Work (13.00 WIB) | - | |
| | | After Work (16.00 WIB) | | |
| 2 | Mr. Satimin | Before Work (07.00 WIB) | 0 | Enough Rest Time |
| | | After Work (12.00 WIB) | | |
| | | Before Work (13.00 WIB) | - | Not Defined |
| | | After Work (16.00 WIB) | | |
| 3 | Mr. Johannis | Before Work (07.00 WIB) | 0 | Enough Rest Time |
| | | After Work (12.00 WIB) | | |
| | | Before Work (13.00 WIB) | 26 | Require additional rest time of 26 minutes |
| | | After Work (16.00 WIB) | | |
| 4 | Mr. Suryanto | Before Work (07.00 WIB) | - | Not Defined |
| | | After Work (12.00 WIB) | | |
| | | Before Work (13.00 WIB) | 0 | Enough Rest Time |
| | | After Work (16.00 WIB) | | |
| 5 | Mr. Nur | Before Work (07.00 WIB) | 112 | Require additional rest time of 112 minutes |
| | | After Work (12.00 WIB) | | |
| | | Before Work (13.00 WIB) | 66 | Require additional rest time of 66 minutes |
| | | After Work (16.00 WIB) | | |
| 6 | Mr. Supardi | Before Work (07.00 WIB) | 103 | Require additional rest time of 103 minutes |
| | | After Work (12.00 WIB) | | |
| | | Before Work (13.00 WIB) | 0 | Enough rest time |
| | | After Work (16.00 WIB) | | |

**Table 3.** Ideal Resting Time Calculation Results

| No. | Operator Name | Measurement Time | RT(min) | Information |
|---|---|---|---|---|
| 7 | Mr. Kinan | Before Work (07.00 WIB) | 0 | Enough rest time |
| | | After Work (12.00 WIB) | | |
| | | Before Work (13.00 WIB) | - | Not Defined |
| | | After Work (16.00 WIB) | | |
| 8 | Mr. Pras | Before Work (07.00 WIB) | - | |
| | | After Work (12.00 WIB) | | |
| | | Before Work (13.00 WIB) | - | |
| | | After Work (16.00 WIB) | | |
| 9 | Mr. Sabar | Before Work (07.00 WIB) | - | |
| | | After Work (12.00 WIB) | | |
| | | Before Work (13.00 WIB) | - | |
| | | After Work (16.00 WIB) | | |
| 10 | Mr. Royyan | Before Work (07.00 WIB) | - | |
| | | After Work (12.00 WIB) | | |
| | | Before Work (13.00 WIB) | 0 | Enough rest time |
| | | After Work (16.00 WIB) | | |
| 11 | Mr. Didik | Before Work (07.00 WIB) | 38 | Require additional rest time of 38 minutes |
| | | After Work (12.00 WIB) | | |
| | | Before Work (13.00 WIB) | - | Not Defined |
| | | After Work (16.00 WIB) | | |
| 12 | Mr. Novianto | Before Work (07.00 WIB) | - | |
| | | After Work (12.00 WIB) | 0 | |
| | | Before Work (13.00 WIB) | | Enough rest time |
| | | After Work (16.00 WIB) | | |

### *3.3 Measurement Of Psychological Workload Experienced By The Operator Of The Mechanical Department PT. XYZ*

Counting mental workload experienced by operators in the Department of Mechanical PT. XYZ, which consists of several stages:

1. Counting the number of tally of any indicators that have been selected by the operator which serves as a weight.
2. Calculating the value of the product.
3. Calculating Weighted Workload (WWL).
4. Calculate the average WWL.
5. Interpretation score to determine the workload generated category.

Data processing is performed using the NASA-TLX method, the following is the result of mental workload calculation for each operator.

**Table 4.** Results of Product Value and WWL

| No. | Operator Name | Product Value | | | | | | WWL |
|---|---|---|---|---|---|---|---|---|
| | | MD | PD | TD | OP | FR | EF | |
| 1 | Mr. Jumali | 160 | 450 | 240 | 70 | 0 | 360 | 1280 |
| 2 | Mr. Satimin | 100 | 165 | 50 | 180 | 50 | 300 | 845 |
| 3 | Mr. Johannis | 210 | 240 | 120 | 180 | 100 | 240 | 1090 |
| 4 | Mr. Suryanto | 140 | 100 | 400 | 0 | 20 | 360 | 1020 |
| 5 | Mr. Nur | 240 | 400 | 0 | 160 | 70 | 320 | 1190 |
| 6 | Mr. Supardi | 285 | 160 | 375 | 85 | 70 | 270 | 1245 |
| 7 | Mr. Kinan | 80 | 0 | 340 | 360 | 320 | 150 | 1250 |
| 8 | Mr. Pras | 0 | 80 | 350 | 320 | 80 | 240 | 1070 |
| 9 | Mr. Sabar | 0 | 80 | 350 | 320 | 80 | 240 | 1070 |
| 10 | Mr. Royyan | 55 | 325 | 120 | 195 | 0 | 150 | 845 |
| 11 | Mr. Didik | 60 | 140 | 400 | 240 | 40 | 140 | 1020 |
| 12 | Mr. Novianto | 240 | 400 | 0 | 160 | 70 | 320 | 1190 |

**Table 5**. Mental Workload Score of The Mechanical Department Operator PT. XYZ

| No. | Operator Name | Score Of WWL | Workload Category |
|---|---|---|---|
| 1 | Mr. Jumali | 85.333 | Very high |
| 2 | Mr. Satimin | 56.333 | High |
| 3 | Mr. Johannis | 72.667 | High |
| 4 | Mr. Suryanto | 68 | High |
| 5 | Mr. Nur | 79.333 | High |
| 6 | Mr. Supardi | 83 | Very high |
| 7 | Mr. Kinan | 83.333 | Very high |
| 8 | Mr. Pras | 71.333 | High |
| 9 | Mr. Sabar | 71.333 | High |
| 10 | Mr. Royyan | 56.333 | High |
| 11 | Mr. Didik | 68 | High |
| 12 | Mr. Novianto | 79.333 | High |

In this study, the result of the level of physical work load felt by the operator will affect the level of fatigue, ideal rest time is required to restore the energy that has been taken out by the operator while performing the work and the level of workload that operator mentally experienced, is very important because in addition to the physical workload, mental workload can affect the level of operator fatigue.

The conclusions of this research are formulated to answer the problem is to know how big physical workload and mental workload accepted by the operator during the work, how the ideal breaks time for operators.

Further studies are needed on the factors that affect physiological and psychological workloads, especially the physical work environment, the relationship between the two workloads and observation not only for the operators of the Mechanical Department but also for another Departments, to know that there are comparisons of which Department operators are most heavy workload, and it is necessary to measure the operator's work productivity before and after adding the rest period so the effect of the rest time can be seen on the productivity produced.

## 4 CONCLUSIONS

Based on the results of data processing and analysis has been done, it can be concluded as follows.

1. The level of physiological workload heaviest suffered by Mr. Nur before the break with a KE of 8.497 kcal/minutes and 7.602 kcal/minutes after the break when operating a medium-sized lathes. While other operators produce KE varying between light and undefined.
2. Work breaks are ideal for operators who categorized the heavy workload that Mr. Nur requires 112 minutes to break from 07.00-12.00 WIB and 66 minutes from 13.00-16.00 WIB. Then, while working from 07.00-12.00 WIB, Mr. Supardi takes 103 minutes for break and Mr. Didik takes 38 minutes for break. While the other operators has already sufficient so it does not need to increase the rest time.
3. The level of psychological work load experienced by all operators belong to the criteria of high and very high with the highest indicator is Effort (EF) with a value of 2770.

## 5. REFERENCES

[1] Admin, 2019 *Manufacturing Industrial Production Growth fourth quarter 2018*. the Central Bureau of Statistics, Jakarta.

[2] R. A. M. Puteri and Z. N. K. Sukarna, 2017 Workload Analysis Method Using CVL and NASA-TLX at PT. ABC, *Spectrum of Industry*, XV(2), p. 212.

[3] P. Fithri and W. F. Anisa, 2017 Measurement of Psychological and Physiological Workload Workers in Textile Industry, *Journal of Industrial System Optimization*, XVI(2), pp. 120-130.

[4] Admin, 2003 Law of the Republic of Indonesia Number 13 of 2003 on Labor. pp. 1-50.

[5] S. Wignjosoebroto, 2003 *Ergonomics, Motion and Time Studies.* 3rd ed., (Surabaya: Widya Guna), p. 272.

[6] T. Sholichul and L. Sudiajen, 2004 *Ergonomics for Occupational Health, Safety and Productivity*, (Surakarta: UNIBA Press).

[7] P. Suma'mur, 1982 *Ergonomics For Work Productivity* (Jakarta: Yayasan Karya Swabhawa).

[8] L. Keytel, J. Goedecke, T. Noakes, H. Hiiloskorpi, Laukkanen R. and L. V. D. M. E. Lamber 2005 Prediction of energy expenditure from heart rate monitoring during submaximal exercise, *Journal of Sports Sciences* XXIII(3).

[9] K. H. E. Kroemer, H. J. Kroemer and K. E. Kroemer-Elbert, 2010 *Engineering Physiology*, 4th ed., (Heidelberg: Springer).

[10] S. Widodo, 2008 Determination of the Old Breaks Based on Workload Using Physiological Approach (Case Study: Oil Mill White Wood Krai), Department of Industrial Engineering, Universitas Muhammadiyah Surakarta, Surakarta.

[11] B. M. Pulat, 1992 *Fundamentals of Industrial Ergonomics*, (USA: Waveland Press, Inc.)

[12] P. A. Hancock and N. Meshkati, 1988 "Human Mental Workload", Amsterdam: North-Holland.

[13] R. Asdyanti, 2011 "Analysis of Mental Workload Relationship with Employee Performance Category Management Department Contract Business Unit at Chevron Indonesia".

IOP Conf. Series: Materials Science and Engineering **847** (2020) 012093 doi:10.1088/1757-899X/847/1/012093

# Clustering on Small-scale Food Manufacturing Industry in West Jakarta: A Fuzzy Analytical Hierarchy Process Approach

**S Sulistiandi[1], B Marpaung[1] and O Sunardi[1*]**

[1]Department of Industrial Engineering, Faculty of Engineering and Computer Science, Krida Wacana Christian University, Jakarta 11470, Indonesia
[*]oki.sunardi@ukrida.ac.id

**Abstract**. The small-scale food manufacturing industry has become the most dominant sector in Jakarta, in terms of number and manpower absorption. However, they are challenged with slow growth rates, indicated by their inability to meet national demands. The Government believes that the small-scale food manufacturing industry in Jakarta was formed without comprehensive planning, and yet, they are spread throughout Jakarta. This study aims to cluster the small-scale food manufacturing industry, which was focused on the West Jakarta area as a pilot project. By employing the Fuzzy Analytical Hierarchy Process (Fuzzy AHP), five criteria were considered during the assessment: availability of land, suppliers, facilities and infrastructure, labor, and markets. As a result, an area clustering map was proposed.
**Keywords:** small-scale industry, food manufacturing, clustering, fuzzy AHP.

## 1. INTRODUCTION

The small-scale food manufacturing industry is people-intensive. As a people-intensive industry, the quest for efficiency becomes a major concern. Small-scale industry competitiveness can be reflected by several indicators: the ability of the industry to create and produce high-quality products, the capacity of the industry to export high-quality products, and the propensity of the enterprises (SMEs) to sustain in the global economics [1]. The propensity of the small-scale industry to sustain in the current economics is considered as the most crucial indicator for most of the growing Asian countries.

The Ministry of Industry noted the investment in the manufacturing industry sector continues to grow significantly. In 2014, capital investment reached Rp195.74 trillion, then rose to Rp222.3 trillion in 2018. This increased investment boosted employment up to 18.25 million people in 2018, which contributed 14.72 percent of the total workforce national. The Government is targeting, throughout 2019 manufacturing industry growth can reach 5.4 percent. Subsectors that is expected to grow high include the food manufacturing industry [2].

Moreover, the Ministry of Trade of the Republic of Indonesia records that the total trade of Indonesian Processed Foods to Europe in 2018 was USD 981.9 million (surplus of USD 277.5 million). To date, 5 (five) main Indonesian processed food products are exported to Europe include Prep or Pres Fish (HS 1604), Crustaceans (HS 1605), Tobacco (HS 2401), Fruits (HS 2008), and Coconut (HS 2008) HS 0801). While the 3 (three) main destination countries for Indonesia's processed food exports are the Netherlands (USD 142.6 million), Germany (USD 98.9 million), and the UK (USD 78.2 million).

Several countries in Europe have been identified as potential markets for Indonesian Processed Food products including Spain (snacks), Germany (tea), Denmark (frozen seafood), and the European Union (coffee) [3].

The competitiveness of the Indonesian food industry, however, is highly contributed by medium and large enterprises. The lack of competitiveness for the small-scale industry is caused by several conditions: location, the government supports, human resources, and strategic partnership [4]. For the Indonesian context, the basic possible barrier is because of the lack of location planning. Thus far, small industries grow naturally, grouping by type of product. Industrial areas like this are known as industrial clusters. In general, industrial clusters are not formed based on certain considerations. According to industry service data, for example, the Capital City of Jakarta has several small-scale industrial clusters such as tofu/ tempeh, apparel, household appliance, *mukena*, and so forth. The cluster that has the most number of business units is the tofu/*tempe* cluster, consisted of 987 business units located in West Jakarta.

Thus, this paper aims to develop a clustering decision with a special focus on the small-scale food manufacturing industry in West Jakarta area.

### *1.1. Industrial Clustering*

Clusters are formed because of the interconnectedness of institutions with similar needs and complement each other. Besides, clustering helps entrepreneurs to make decisions when there is an opportunity to do business [5]. Industrial clustering is quite beneficial for small and medium enterprises because each of the similar industries in one cluster can work together with each other, making it more effective and efficient in the production process [6]. Industrial clustering is one way to improve industrial competitiveness and can improve relations between similar entrepreneurs by doing business or cooperating, who have similar needs for capabilities, technology, and infrastructure [7].

The previous study recommends industrial development in strategic locations to building a good and competitive trade market [4, 8]. However, strategic location alone is not enough, industrial clustering at strategic locations is something that should be considered towards a competitive manufacturing industry. Previous studies implied several indicators to measure the potential competitiveness of manufacturing industry: skilful employee sourcing [4], suppliers' specialization [4, 7], facilities and infrastructure [6], production layout [6], quality of the product [6, 9], and delivery cost [9].

### *1.2. Criteria in Clustering Decision*

There are several considerations in locating the appropriate industrial cluster. Some criteria should be considered [8, 10], such as characteristics and size of the population, availability of labor, proximity to production sources, promotion, economic base, compatibility with facilities, competitive situation, and convenience of a store location.

Another study used seven criteria in choosing the location of the processing industry: raw materials, land conditions, labor, marketing, spatial planning, supporting facilities, and adequate infrastructure [11]. For this study, five criteria were analyzed: land availability, infrastructures, labor source, suppliers, and market.

## 2. METHODS

Five criteria were analyzed using the Fuzzy Analytical Hierarchy Process (Fuzzy AHP). Analytical Hierarchy Process (AHP) is a method of Multi-Criteria Decision Making (MCDM) developed by Saaty in the early 1970s. AHP is an analytical method used to create a problem-based model that has no structure and can be used to solve quantitative problems, and as well as problems that require opinions [12, 13].

Besides, AHP can also be used to solve problems in complex situations. AHP is a general theory of measurement used to find the ratio scale, both from discrete and continuous pair comparisons. These comparisons can be taken from actual measurements or a baseline scale that reflects the strength of feelings and relative preferences [12]. Further, fuzzy logic will deliver better results compared to

IOP Conf. Series: Materials Science and Engineering **847** (2020) 012093 doi:10.1088/1757-899X/847/1/012093

conventional theories in general, because fuzzy logic develops existing theories by minimizing deficiencies in previous theories, such as not being able to provide optimal results when research is subjective. Fuzzy logic has a role to minimize uncertainty. The uses of a triangular fuzzy number are one approach to minimize uncertainty in the AHP scale in the form of values.

For this study, three experts were involved as informants. The experts consist of a government official who has the experience and knowledge in a small industry, an industrial consultant who used to dealing with regional planning and development, and a higher education expert who has the concentration of research related to the industry, economic development, and regional development. Experts were given the freedom to assess the existing criteria with five priority weights of the AHP fuzzy method: EI (Equal Importance), MI (Moderate Importance), SI (Strong Importance), VSI (Very Strong Importance), EMI (Extremely More Importance) [13]. The assessment questionnaire was filled by filling in the available fields with a checkmark (V). The following table represents the fuzzy importance rating scale used in the questionnaire.

**Table 1.** Priority assessment scale

| Priority scale | Notes | Membership function |
|---|---|---|
| 9 | Extremely more importance (EMI) | (8,9,10) |
| 7 | Very strong importance (VSI) | (6,7,8) |
| 5 | Strong importance (SI) | (4,5,6) |
| 3 | Moderate importance (MI) | (2,3,4) |
| 1 | Equal importance (EI) | (1,1,2) |

In processing data using the AHP method, it is necessary to calculate the consistency test, where the consistency test is useful to determine whether the pairwise comparisons produced are consistent or not. The measurement of consistency from a pairwise comparison matrix is based on the largest eigenvalue [12].

## 3. RESULT AND DISCUSSION

The assessment of the importance of each criterion was carried out by experts selected as informants in this study. In the questionnaire distributed several sub-districts became indicators, namely as follows, KL (Land Availability), SP (Facilities and Infrastructure), PM (Suppliers), TK (Labour), and PS (Market). The results obtained by filling out the questionnaire were comparisons between criteria. All questionnaire results can be arranged in the form of pairwise comparison matrices containing fuzzy number scales.

In assessing the relative importance of two elements, a reciprocal axiom applies, meaning that if the element in column one is given a value 4 times more important than the element in column two, then the element in column two must be equal to 1/4 times more important than the element in column one. Scale numbers that have been made on the paired matrix were then converted to the TFN (Triangular Fuzzy Number) scale. Fuzzy numbers are usually shown in the form of three numbers, i.e. (l, m, u). These parameters represent the smallest possible values, the most probable values, and the largest values that represent fuzzy problems.

Before defuzzifying the assessment matrix, the assessment matrix of respondent one, respondent two, and respondent three were combined into one assessment matrix, using the following formula: Let A1 = (l1, u1); A2 = (l2, u 2); A3 = (l3, u3). Then the combined assessment matrix is formulated as follows:

$$A_g(l,u) = \sqrt[3]{(l_1 * l_2 * l_3),(u_1 * u_2 * u_3)} \quad (1)$$

The combined matrix of criteria assessment is shown in Table 2.

**Table 2.** Criteria Assessment Combination Matrix

| | KL | | SP | | PM | | TK | | PS | |
|---|---|---|---|---|---|---|---|---|---|---|
| KL | 1 | 1.5 | 0.77 | 1.06 | 5.09 | 6.10 | 0.31 | 0.42 | 0.30 | 0.37 |
| SP | 1.09 | 1.51 | 1 | 1.5 | 3.44 | 4.51 | 0.43 | 0.51 | 0.33 | 0.40 |
| PM | 0.17 | 0.20 | 0.23 | 0.30 | 1 | 1.5 | 0.17 | 0.20 | 0.12 | 0.14 |
| TK | 2.77 | 3.68 | 1.98 | 2.33 | 5.09 | 6.10 | 1 | 1.5 | 0.24 | 0.30 |
| PS | 2.76 | 3.35 | 2.53 | 3.10 | 7.11 | 8.11 | 3.81 | 4.75 | 1 | 1.5 |

Defuzzification used in this study aims to convert fuzzy values into crisp values. The calculation of crisp values for the confidence level of $\alpha = 0.5$ is in Table 3.

**Table 3.** Combined Crisp Matrix

| | **KL** | **SP** | **PM** | **TK** | **PS** |
|---|---|---|---|---|---|
| **KL** | 1 | 0.91 | 0.37 | 0.36 | 0.33 |
| **SP** | 1.30 | 1 | 0.47 | 0.42 | 0.36 |
| **PM** | 0.18 | 0.26 | 1 | 0.16 | 0.13 |
| **TK** | 3.23 | 2.15 | 1.25 | 1 | 0.27 |
| **PS** | 3.06 | 2.81 | 4.28 | 2.87 | 1 |
| **TOTAL** | 8.77 | 7.14 | 7.36 | 4.81 | 2.10 |

C.R.'s calculation results obtained was 0.00. Therefore, it can be concluded that the paired matrix is consistent, and the results of the weighting of the criteria contained in the matrix can be used for further data processing. Table 4 shows the eigenvector calculation.

**Table 4.** Eigenvector calculation

| | **KL** | **SP** | **PM** | **TK** | **PS** | **weight** | **Rank** |
|---|---|---|---|---|---|---|---|
| **KL** | 0.11 | 0.13 | 0.05 | 0.08 | 0.16 | 0.11 | **4** |
| **SP** | 0.15 | 0.14 | 0.06 | 0.09 | 0.17 | 0.12 | **3** |
| **PM** | 0.02 | 0.04 | 0.14 | 0.03 | 0.06 | 0.06 | **5** |
| **TK** | 0.37 | 0.30 | 0.17 | 0.21 | 0.13 | 0.24 | **2** |
| **PS** | 0.35 | 0.39 | 0.58 | 0.60 | 0.48 | 0.48 | **1** |

There were five fuzzy weighing scales used, namely STA (strongly not submitted), TA (not submitted), CA (simply submitted), A (submitted), SA (strongly proposed). Each weighing scale has different membership functions, such as STA (1,1,2), TA (2,3,4), CA (4,5,6), A (6,7,8), SA (8,9,10). The results of the performance evaluation of the type of industry in each district can be seen in Table 5.

**Table 5.** Performance evaluation based on industry types

| District | Tofu | Cake & Bread | Chili Sauce | Pudding | Wet Noodle | Dry Noodle | Seasoning | Jelly |
|---|---|---|---|---|---|---|---|---|
| Cengkareng | 56.24 | 76.66 | 64.36 | 100.00 | 59.55 | 29.75 | 26.10 | 51.27 |
| Grogol Pet. | 56.24 | 77.44 | 82.70 | 43.11 | 67.67 | 30.50 | 70.97 | 30.99 |
| Kalideres | 100.00 | 79.50 | 63.76 | 94.35 | 83.97 | 50.00 | 51.27 | 51.91 |
| Kebon Jeruk | 34.36 | 73.43 | 100.00 | 65.70 | 83.77 | 54.21 | 47.51 | 81.94 |
| Kembangan | 34.36 | 56.25 | 96.48 | 58.27 | 100.00 | 47.09 | 100.00 | 54.81 |
| Palmerah | 14.48 | 100.00 | 73.76 | 65.70 | 82.93 | 100.00 | 75.68 | 100.00 |
| Taman Sari | 14.48 | 66.17 | 49.25 | 27.09 | 41.33 | 25.56 | 34.70 | 11.94 |
| Tambora | 14.48 | 50.72 | 33.74 | 16.50 | 21.32 | 19.36 | 14.67 | 33.38 |

By combining with Simplex LP method, it is expected to find the optimal solution. The industry clustering optimization formulation is stated as follows:

$$\textbf{Max } Z = 56{,}24\ x_{11}+76{,}66\ x_{12}+64{,}36\ x_{13}+100\ x_{14}+59{,}55\ x_{15}+29{,}75\ x_{16}+26{,}10\ x_{17}+51{,}27\ x_{18}+56{,}24\ x_{21}+77{,}44\ x_{22}+82{,}70\ x_{23}+43{,}11\ x_{24}+67{,}67\ x_{25}+20{,}50\ x_{25}+20{,}50\ x_{26}+70{,}97\ x_{27}+30{,}99\ x_{28}+100\ x_{31}+79{,}50\ x_{32}+63{,}76\ x_{33}+94{,}35\ x_{34}+83{,}97\ x_{35}+50\ x_{36}+51{,}27\ x_{37}+51{,}91\ x_{38}+34{,}36\ x_{41}+73{,}43\ x_{42}+100\ x_{43}+65{,}70\ x_{44}+83{,}77\ x_{45}+54{,}21\ x_{46}+47{,}51\ x_{47}+81{,}94\ x_{48}+34{,}36\ x_{51}+56{,}25\ x_{52}+96{,}48\ x_{53}+58{,}27\ x_{54}+100\ x_{55}+47{,}09\ x_{56}+100\ x_{57}+54{,}81\ x_{58}+14{,}48\ x_{61}+100\ x_{62}+73{,}76\ x_{63}+65{,}70\ x_{64}+82{,}93\ x_{65}+100\ x_{66}+75{,}68\ x_{67}+100\ x_{68}+14{,}48\ x_{71}+66{,}17\ x_{72}+49{,}25\ x_{73}+27{,}09\ x_{74}+41{,}33\ x_{75}+25{,}56\ x_{76}+34{,}70\ x_{77}+11{,}94\ x_{78}+14{,}48\ x_{81}+50{,}72\ x_{82}+33{,}74\ x_{83}+16{,}50\ x_{84}+21{,}32\ x_{85}+19{,}36\ x_{86}+14{,}67\ x_{87}+33{,}38\ x_{88}$$

**Constraints:**

$$X_{11}+x_{12}+x_{13}+x_{14}+x_{15}+x_{16}+x_{17}+x_{18} = 3$$
$$X_{21}+x_{22}+x_{23}+x_{24}+x_{25}+x_{26}+x_{27}+x_{28} = 2$$
$$X_{31}+x_{32}+x_{33}+x_{34}+x_{35}+x_{36}+x_{37}+x_{38} = 987$$
$$X_{41}+x_{42}+x_{43}+x_{44}+x_{45}+x_{46}+x_{47}+x_{48} = 1$$
$$X_{51}+x_{52}+x_{53}+x_{54}+x_{55}+x_{56}+x_{57}+x_{58} = 2$$
$$X_{61}+x_{62}+x_{63}+x_{64}+x_{65}+x_{66}+x_{67}+x_{68} = 0$$
$$X_{71}+x_{72}+x_{73}+x_{74}+x_{75}+x_{76}+x_{77}+x_{78} = 0$$
$$X_{81}+x_{82}+x_{83}+x_{84}+x_{85}+x_{86}+x_{87}+x_{88} = 26$$

From the results of optimization using the SOLVER menu on MS-Excel, the overall performance score has increased from before clustering. The overall performance score before clustering was 99901, and after clustering the overall performance score was 100177. Figure 1 represents the proposed clustering for small-scale food manufacturing industries in the West Jakarta area. Area and Industry Clusters were named in Bahasa Indonesia and the picture was manually constructed based on performance evaluation scores.

Figure 1 can be explained as follows. An area which obtained a score of 100 shows a "strong" capability to serve as an industry cluster. Kalideres District, for example, was found very suitable for tofu industry. The lower the score, the less likely an area to be developed as an industrial cluster. In this case, Taman Sari District was identified as one area which was not suitable for almost all types of industries due to several possible reasons: less availability of land and high density of housing.

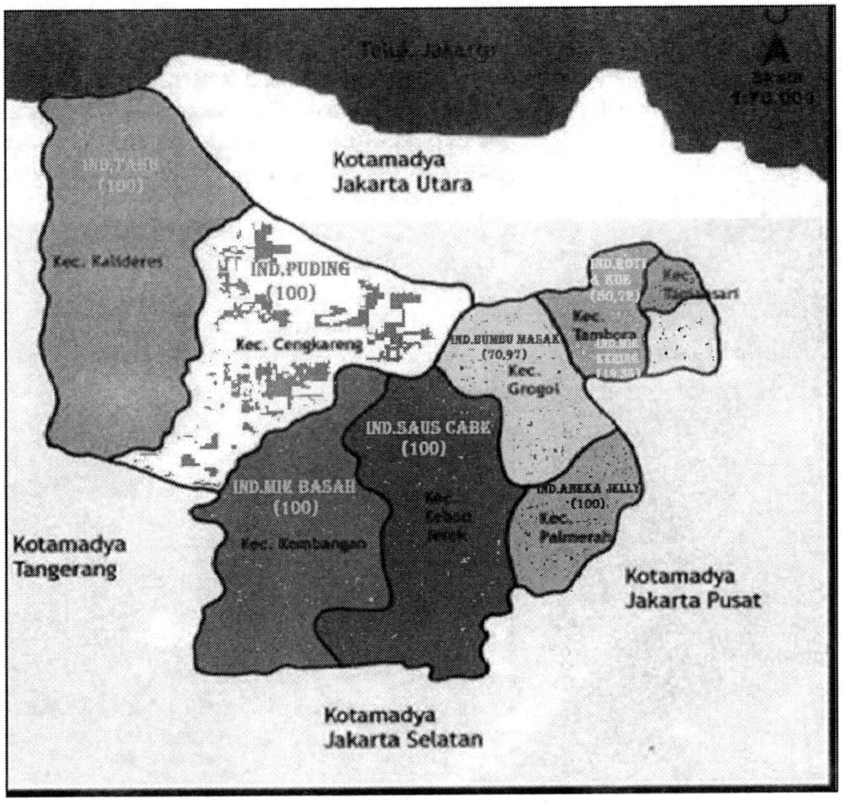


**Figure 1.** Proposed clusters

## 4. CONCLUSION

The fuzzy AHP approach showed that the "market" shown the greatest importance, followed by "labor", "facilities and infrastructure", "land availability", and "suppliers". The distribution of clusters was obtained as follows: tofu industry in Kalideres District, pudding industry in Cengkareng District, wet noodle industry in Kembangan District, chili sauce industry in Kebon Jeruk District, seasoning industry in Grogol District, various jelly industries in Palmerah District, bread & cake industry and the dried noodle industry in Tambora District. There is one sub-district which is not proposed at all to form an industrial cluster (Taman Sari District), due to the following conditions: the availability of land is quite minimal, and the density of housing is high.

This study could help the provincial government to better reallocating its industrial clusters. However, the outcome of this study could be more enriched. Three experts involved in this study were from government official, academician, and industrial consultant. Further study should incorporate views from business owners, since they would provide richer perspectives.

## 5. REFERENCES

[1] World Economic Forum, 2017 *The inclusive growth and development report 2017*, January 2017, accessed on 06 March 2019 from http://weforum.org.

[2] The Ministry of Industry, 2019 *Investasi manufaktur diyakini semakin moncer seusai Pemilu*, April 2019, accessed on 07 March 2019 from http://kemenperin.go.id.

[3] The Ministry of Foreign Affairs, 2019 *American and European Market*, May 2019, accessed on 06 June 2019 from https://pasaramerop.kemlu.go.id.

[4] N.U. Handayani, H. Santoso, and A.I. Pratama, 2012 Faktor-faktor yang mempengaruhi daya saing klaster mebel di Kabupaten Jepara, *Jurnal Teknik Industri*, **13**(1), pp. 22–30.
[5] M. Momeni, M. Mohseni, and M. Soofi, 2015 Clustering stock market companies via K-means algorithm, *Journal of Business and Management*, **59**(3), pp. 234-273.
[6] M. Najib, H. Yagi, and A. Kiminami, 2011 Competitiveness of Indonesian Small and Medium Food Processing Industry: Does the location matter? *International Journal of Business and Management*, **6**(9), pp. 57-67.
[7] S. Lewis, M. Sundar, J. Jiang, and S. Lu, 2009 Cluster analysis method for identification of failures in a manufacturing environment, *Journal of Manufacturing Technology Management*, **56**(2), pp. 243-265.
[8] M. Kirankabes and M. Arik, 2014 Industrial clustering approach in regional development, *Journal of Applied Business and Economics*, **16**(3), pp. 257-273.
[9] C. Boja, 2011 Clusters models, factors and characteristics, *International Journal of Economic Practices and Theories*, **1**(1), pp. 34-43.
[10] S. Zuliarni and R. T. Hidayat, 2013 Analisis pertimbangan pebisnis restoran kelas kecil di lingkungan kampus Universitas Riau dalam pemilihan lokasi usaha, *Jurnal Aplikasi Bisnis*, **3**(2), pp. 100-119.
[11] N. G. Candy and A. Pamungkas, 2013 Penentuan alternatif lokasi industri pengolahan sorgum di Kabupaten Lamongan, *Jurnal Teknik ITS*, **2**(2), pp. 211-214.
[12] M. Javanbarg, C. Scawthorn, J. Kiyono, and B. Shahbodaghkhan, 2012 Fuzzy AHP-based multicriteria decision making system using Particle Swarm optimization, *Journal of Manufacturing Technology Management*, **54**(2), pp. 245-274.
[13] T. Hartanto, M. Ginting, and O. Sunardi, 2019 A Fuzzy Analytic Hierarchy Process approach for determining the criteria success factors of MRT parts' e-procurement: the case of Jakarta MRT project, *IOP Conference Series: Materials Science and Engineering*, **528**(1), 012001.

IOP Conf. Series: Materials Science and Engineering **847** (2020) 012094 doi:10.1088/1757-899X/847/1/012094

# Management of Plastic Waste Recycling by Value Stream Mapping

**Tiena Gustina Amran**[1,2,3]

[1]University of Trisakti, Jl. Kyai Tapa No. 1, Jakarta, Indonesia
[2]tiena@trisakti.ac.id, [3]amran.tiena@gmail.com

**Abstract.** This study explains two recycling process category with plastic pellets and fuel as main products. Fuel can be utilized by vehicles and plastic pellets has economical value. Reduction of waste in plastic waste recycling process was conducted using value stream mapping approach. The research of plastic recycling was done through literature review and direct survey. In first stage, several analyses were made: plastic waste processing category, Value Added Time (VAT) and Non Value Added Time (NVAT), also process time per work station. As the result, current stream mapping was obtained. In the second analysis stage, comparison between two processing category revealed that the highest dissipation came from motion waste, waiting time waste, and movement waste. Big picture mapping was used to improve the quality of waste recycling process based on the procedures that has been established Fore mentioned dissipation can be reduced by adding the number of operator equipped with safety gear in washing and drying.
**Keywords:** Plastic Waste, Fuel, Plastic Category, Value Steam Mapping

## 1. Introduction

Plastic as one of the most massive raw material production is a raw material that is relatively cheap, durable, and versatile. This impacts the creation of plastic products that has been beneficial for the society in economics, occupational, and other activities related to quality of life. On the other hand, plastic can impact negatively such as ecological damage. Therefore, various ways are done to decrease the amount of plastic garbage in Indonesia. One of them is to recycled plastics into other product or material. Recycling means is the utilization opportunity of plastics from the waste of consumption in accordance to the form, criteria, and type of the recycling moreover, the priority of plastic waste recycling is to make recycled products that is resembled the original products so its quality does not decrease, for example pellet. So far there's many factories based in Indonesia that can process plastic waste into pellet [1]. The production has been exported abroad. Four categories of plastic recycling includes: (1) Recycling output known as virgin material or product's raw material product, (2) Recycling output has no similar characteristic with the original product, (3) Recycling output is in the form of fuel or chemicals such as plastic scraps which become urban waste, (4) Recycling output is in the Form Of Energy, Not Materials [2]. There are various tools and techniques to implement lean principles to industry. Among the various tools of lean, VSM is one of the most important tools in lean manufacture to apply value stream mapping (VSM) for enabling leanness in an Indian camshaft manufacturing organisation. Then Vinodh [3] to explore the practical feasibility of deploying value stream mapping (VSM) for enabling lean manufacturing.

Miya and Ngacho [4] stated that VSM is one of the techniques tool of lean thinking philosophy that can be used to analyze the current state of a process, materials and information flow necessary to bring out products or services to the customer. VSM tool is concerned with finding the best techniques and solutions to reduce or eliminate waste in a process and identify opportunities for improvement. The effective technique of value stream mapping can be used to reduce or eliminate waste in processes of solid waste management, mitigate challenges of SWM and identify opportunities for improvement.

**Tabel 1.** Previous Research in Value Stream Mapping

| No | Author | VSM | Results |
|---|---|---|---|
| 1 | Guilherme Luz Tortorella, Lucila M. S. Campos and Glauco Garcia Martins Pereira da Silva [5] | VSM, | Considering the five studied RCs, resulting in an increase of 22% in employment opportunities. |
| 2 | Rubiano O. Oscar; Peña M. Claudia C.; Paz R. Juan C. [6] | VSM, | The results confirm some advantages of the proposed VSM methodology as a practical redesign tool of waste and materials flows in the value stream. |
| 3 | Anne Miya, Christopher Ngacho [4] | VSM, | Policies and legislation are very necessary to plan, design and implement solid waste management operations, however with no enforcement the policies are toothless. |
| 4 | Sri Hartini, Udisubakti Ciptomulyono, Maria Anityasari, Sriyanto, Darminto Pudjotomo [7] | VSM, Sustainable concept, Sus-VSM, | The result is that the developed tools can analyze problems in economic, environment, and social aspects as the basic of improvement. |
| 5 | Putri Citra Marifa, Feny Yuliana Andriani, Sri Indrawati, Anggita Noviyanti Parmasari, Hardiyanti Budiman, and Atika Kamilia [8] | VSM, VALSAT, QFM | Based on the result of the research, there are seven types of production waste: overproduction (9,62%), inventory (17,3%), defect (23,08%), motion (9,62%), transportation (9,62%), Over processing (9,62%) and waiting (21,15%). |
| 6 | Guilherme Luz Tortorella, Flávio Sanson Fogliatto, Michel Anzanello, Giuliano Almeida Marodin, Mayara Garcia & Rafael Reis Esteves [9] | VSM, | It is recommended to integrate VSM to other analytical tools able to handle system variation, presenting the dynamics between system components and validating the future state before its real implementation takes place. Computer modelling and simulation may fulfil these requirements and are proposed as future research development towards a more holistic approach to lean system analysis. |

## 2. Methods

The methods consist of five steps. Step one, from this research is to conduct preliminary study regarding plastic factories existing in Indonesia, then continued with factory visiting and chose the two factories for the study case that will be analyzed in this research. Step two, to categorize raw materials and process of plastic waste processing mechanically with two indicators that is plastic recycling [1]. Step three, identify the existing waste processing that is existing today. In this step, it is explained more detail regarding physical and non-physical resource management of both factories. Step four, conduct analysis Value Stream Mapping in both cases with these steps:

- Creating Big Picture Mapping
- Calculating activities time (Value Added & Non Value Added)
- Making and spreading questionnaire
- Conducting VALSAT analysis
- Analysing wastefulness in accordance to the 8 waste rules in Lean Manufacturing

Step five, is to analyse the best practice and give recommendation for plastic waste processing management in accordance to the analysis data in Step Three and Step Four. Research Location: an example of reprocessing pellet into pellet can be seen in the ongoing process in MBG, where pellet is reprocessed into broom handle. AJP factory does not conduct the process.

Mechanical recycling process of plastic explains by Fisher in Gabriel [1] proves that the material made of plastic can be recycled into raw material with three categories, and also a quartener recycled product that does not produce material as an output, but an energy instead. The three mentioned categories can be seen in Figure 1.

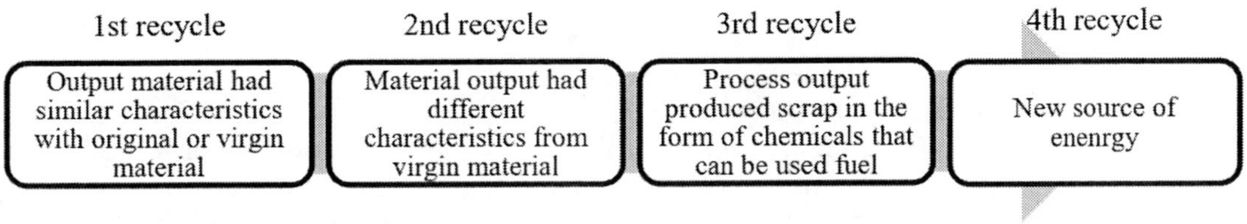


**Figure 1.** Categories of Material Produces from Recycled Plastic Waste

## 3. Result and Discussion

*Value Stream Mapping* (VSM) shows details of material flow, operational lead time, yield, uptime, human resource needed, frequencies of delivery, inventories, inventories amount, batch size, setup time, process time, and holistic efficiency of process. Through this method, general analysis will be conducted against the waste from the process of the plastic waste treatment/production in MBG and AJP with a few steps such as, sorting, weighing, grinding, cleansing, drying, packaging, and or reprocess the plastic waste into goods. During the process of production with unsophisticated tools and little man power, there has been a few wastes in both factory. One example is when the grinding process. There is motion waste which is the repeating movements of hands and feet to push raw material into the machine until the raw material is run out or the grinding process is done. Through visits to both factories, it is known that the outline of the plastic waste processing mechanism in both factories is quite similar with the process that is defined by Manrich and Santos. Below is the identification results:

**Table 2.** Plastic Waste Processing Mechanism in Factory Category 1 and 3.

| Steps of Mechanical Recycling | Case study: Factory Category 1 | Case study: Factory Category 3 |
|---|---|---|
| Stage 1: Identification, separation, and sorting plastic raw material | Manually will be sorted by 10 female worker, seperated by waste types, cleansed with water and put in a plastic sack of 1 ton | There is no specific sorting process. There is not much plastic that is not uniform because it has been sorted by suppliers since the delivery of the raw materials. |
| Stage 2: Grinding | One male worker will put raw materials into grinding pipes. After the raw materials is put into the tip of the pipe, worker will press it with foot so the raw material enters the 0.5 m diameter grinding pipes | There are 2-3 male worker put the uniform PE or PP raw material gradually into the grinding machine/ When putting the raw materials, worker will separate dirt or a not uniform plastic aside. The grinding machine capacity is up to 1 ton. |
| Stage 3: Cleansing, with or without additional cleansing product | The grinded plastic in each colors enters the gravity pipes. Gravity pipes is connected to the water channel that consist of cleansing soap and alum. The waste grindin result will be taken by cleansing water flow through pipes and enters the shelter section to be put into the plastic filter manually. The larger grinded product that is larger than the diameters of the | The grinded plastic falls into the hole below the machine and pushed by conveyor that connects the grinding machine with the cleansing pool. The cleansing pool is filled with water that is constantly moved with a waterwheel so plastic will be pushed forward. The capacity of the cleansing pool is up to 1 ton. |

IOP Conf. Series: Materials Science and Engineering **847** (2020) 012094 doi:10.1088/1757-899X/847/1/012094

**Table 2.** Plastic Waste Processing Mechanism in Factory Category 1 and 3.

| Steps of Mechanical Recycling | Case study: Factory Category 1 | Case study: Factory Category 3 |
|---|---|---|
| | filter holes is put into clean water, while the smaller one is outcasted to reach the final disposal channel. | |
| Stage 4: Drying | The clean plastic waste will be sheltered in the filter and put into the drying machine. The machine capacity is up to 3 drying baskets (approximately 7 kg). The machine will be closed and drying process will take for 20-25 minutes. | The grinded plastic is carried away by the pool water into the conveyor that bring the grinded plastic into the drying machine. The drying cubicle is 1x 1.5 m big with 2 m height. There are 2 drying cubicle that is sometimes supervised by 1 person to ensure that the dry plastic has been sucked by vacuum. |
| Stage 5: Silos | - | The grinded plastic that has been dried is sucked by a long vacuum through the storage room near the printing machine. One operator will help the dry grinded plastic to enter the machine by a little push with a wood. |
| Stage 6: Agglutination | - | High temperature machines will process the plastic into solid chunks |
| Stage 7: Extrusion | - | Plastic chunks is printed into plastic noodles |
| Step 8: Granulation | - | Plastic noodles is distributed to cold water and cut in the form of small seed pieces |

Analysis of Stream Mapping after conducting Value Added Tools (VAT) and Non Value Added Tools (NVAT) for waste management

*3.1. Case Study: Factory category 1 (MBG).*

Identification of Plastic Waste Category that is used by MBG: Plastic waste categories used by MBG can be seen in Table 3. below,

**Table 3.** Identification of Plastic Waste Category in Factory Category 1 (MBG).

| | Waste Management | Output | Cause of Material Degradation |
|---|---|---|---|
| Mechanical Recycling Process | Residual value medium: all brands of bottle packaged drink (transparent, blue, green), soya sauce bottle, sauce, etc. | Recycled plastic pellets | Processing of plastic waste into grinded plastic pellets |
| | Residual value medium: all brands of glass packaged drink | | Processing of plastic waste into plastic pellets |
| | Residual value low: infusion bottle waste, drinking gallon cap waste | | Processing of plastic waste into alternative fuel |

**Table 4.** Analysis of VAT and NVAT For Process in Factory Category 1 (MBG).

| Process Type | *Cycle Time* (s) VAT | *Cycle Time* (s) NVAT | *Lead Time* (s) |
|---|---|---|---|
| Transporting | | | |
| Raw material of plastic waste is transported to the chopping machine | | 180 | |
| Grinding | | | |
| Raw material into the tool | 1980 | | |
| Cleansing | 1260 | | |
| (a) Grinded material is washed with alum water | | | |
| (b) Grinded material is stored in the filter and moved into a clean water basin | | | |
| (c) Grinded material is washed with clean water manually | | | |
| (d) Grinded material is stored in a filter and put into drying machine | | | |
| (e) Separation of objects carried by the grinded materials | | | |
| (f) Waste is distributed to the back of the factory | | | |
| Drying | 1286 | | |
| (a) Grinded material is dried | | | |
| (b) Grinded material is dried in the sun | | | |
| (c) *Quality Control* of the grinded material | | | |
| Packing | | | |
| (a) *Flakes* are put into sacks | 382 | | |
| (b) Products are exported to Hongkong & China | | | |
| **Total (s)** | **4908** | **180** | **5088** |

Resume of process time per work station Factory category 1 (minutes).

| Transporting | Chopping | Cleansing | Drying | Packing | Leadtime |
|---|---|---|---|---|---|
| 3 minutes | 33 minutes | 21 minutes | 21 minutes | 6 minutes | 84.8 minutes |

Resume shows that Lead Time for 10 kg of raw materials are 84.8 minutes. It means that every once in 1 hour and 28.8 minutes, a 10 kg plastic waste raw material are completely processed. Lead time for 10 kg per unit can be approached by above calculation. After knowing activities that includes in VA or NVA, a big picture mapping (Figure 2) can be shown as below

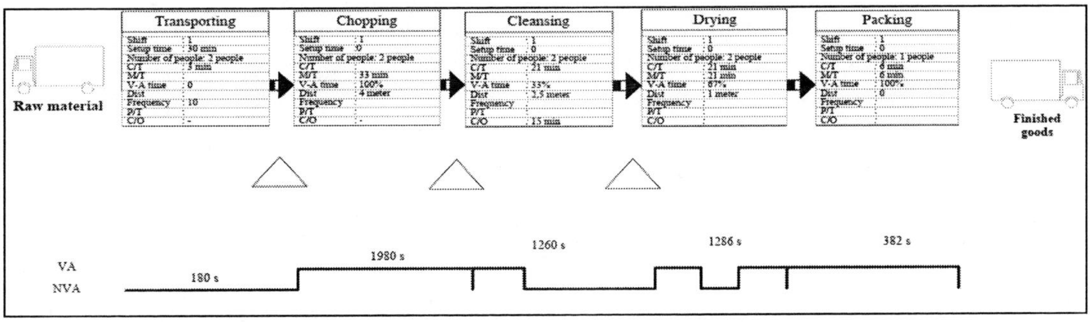


**Figure 2.** Value Stream Mapping in Category 1 Thermoplastic Recycling System

Analysis: Through observation, the most dominant waste is unnecessary motion, waiting, and excessive transportation. By doing re-crosschecking towards types of the most dominant waste regarding to the waste management executor (owner) by using 7 waste questionnaire by interview. Below is the recapitulation of 7 waste for Factory Category 1 – MBG

**Table 5.** The Difference in Analysis of The Most Dominant Waste MBG (Questionnare).

| | Over production | Waiting | Exceessive transportation | Inapproptiate processing | Unnecessary inventory | Unnecessary motion | Defect | |
|---|---|---|---|---|---|---|---|---|
| Score | 1 | 1 | 4 | 2 | 3 | 3 | 0 | |
| Total value | | | | | | | | 14 |

Therefore, the comparisons of the most dominant waste type based on researcher and owner is as below,

**Table 6.** Comparison of Dominat Waste Between Owner and Researcher

| | | | |
|---|---|---|---|
| Researcher | Unnecessary Motion | Waiting | Excessive Transportation |
| Owner | Unnecessary Motion | Unnecessary Inventory | Excessive Transportation |
| Root cause | Bad workplace organization and layout, inconsistent work method, and machines with bad design | Waiting:<br>The length of duration in product changing and inconsistent work method<br>Unnecessary inventory:<br>Unbalanced workflow, bad tools, inaccurate production forecast, incapable vendors, batch size that no big enough and the length of duration in product changing | Bad factory layout, no team coordination, bad factory maintenance, not suitable factory organization, and too many warehouses that is too far. |

There are differences between identification according to researcher's observation and owner's questionnaire result which are in waiting and unnecessary inventory aspects. Both aspects will become an evaluation for integrated plastic waste management. By using Seven Value Stream Mapping Tools, it is obtained a VALSAT recapitulation as below:

**Table 7.** VALSAT Recapitulation MBG.

| Waste | Process Activity Mapping | Supply Chain Response Matrix | Production Variety Funnel | Quality Filter Mapping | Demand Amplification Mapping | Decision Point Analysis | Physical Structure |
|---|---|---|---|---|---|---|---|
| Over Production | 1 | 2 | | 1 | 3 | 3 | |
| Waiting | 9 | 9 | 1 | | 3 | 3 | |
| Transport | 36 | | | | | | 1 |
| Inappropriate processing | 18 | | 6 | 2 | | 2 | |
| Unnecessary Inventory | 9 | 27 | 9 | | 27 | 9 | 3 |
| Unnecessary Motion | 27 | 3 | | | | | |
| Defect | 0 | | | 9 | | | |
| Total | 100 (44,84%) | 41 (18,38%) | 16 (7.17%) | 12 (5.38%) | 33 (14.79%) | 17 (7.62%) | 4 (1.79%) |
| Ranking | 1 | 2 | 5 | 6 | 3 | 4 | 7 |

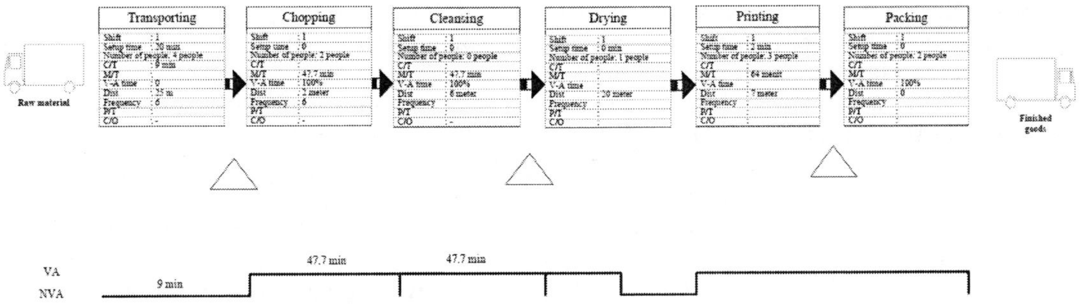


**Figure 3**. Value Stream Mapping in Category 3 Thermoplastic Recycling System

*3.2. Case study: Factory Category 3.*

Waste management 2: Below is the result of 7 waste questionnaire recapitulation for Factory Category 3. Based on 7 Waste Questionnaire evaluation, it can be understood that by lean there are no significant waste for AJP. There is also a difference in the most dominant waste evaluation between researcher and owner.

**Table 8.** Difference of Analysis Result in Most Dominant Waste Factory Category 3 (AJP).

| Recapitulation | Over production | Waiting | Exceessive transportation | Inapproptiate processing | Unnecessary inventory | Unnecessary motion | Defect |
|---|---|---|---|---|---|---|---|
| Score | 0 | 1 | 0 | 0 | 0 | 1 | 1 |
| Total value | | | | | | | 3 |

| | | | |
|---|---|---|---|
| Researcher | Unnecessary Motion | Waiting | Excessive Transportation |
| Owner | Unnecessary Motion | Waiting | Defect |
| Root cause | Bad workplace organization and layout, inconsistent work method, and machines with bad design | The length of duration in product changing and inconsistent work method | Excessive transportation: Bad factory layout, no team coordination, bad factory maintenance, not suitable factory organization, and too many warehouses that is too far.<br>Defect: Error in process, ineffective training, and no standard procedures. |

**Table 9.** Recapitulation of VALSAT Calculation for Factory Category 3.

| Waste | Process Activity Mapping | Supply Chain Response Matrix | Production Variety Funnel | Quality Filter Mapping | Demand Amplification Mapping | Decision Point Analysis | Physical Structure |
|---|---|---|---|---|---|---|---|
| Over Production | 0 | 0 | | 0 | 0 | 0 | |
| Waiting | 9 | 9 | 1 | | 3 | 3 | |
| Transport | 0 | | | | | | 0 |
| Inappropriate processing | 0 | | 0 | 0 | | 0 | |
| Unnecessary Inventory | 0 | 0 | 0 | | 0 | 0 | 0 |
| Unnecessary Motion | 9 | 1 | | | | | |
| Defect | 1 | | | 9 | | | |
| Total | 19 (42.2%) | 10 (22.22%) | 1 (2.22%) | 9 (20%) | 3 (0.03%) | 3 (0.03%) | 0 (0%) |
| Ranking | 1 | 2 | 5 | 3 | 4 | 4 | 6 |

IOP Conf. Series: Materials Science and Engineering **847** (2020) 012094 doi:10.1088/1757-899X/847/1/012094

## 4. Conclusion

In the Big mapping of plastic waste management process flow factory Category 1, the main waste that happens in managing plastic pellets are motion and waiting. The problem in transportation can't be deeply reviewed because the factory layout is prioritized for raw material hoarding and the grinding process is conducted in series. Improvement recommendation in waste that happens in the production process of plastic pellets are to add new tools whether for cleanser or dryer and to improve machines flow so that it follow the gravity direction. The plastic waste management in factory category 3, the main waste happens in unecessary motion, waiting, excessive transportation and defect. Improvements can be done by putting raw materials in accordance to the capacity of machines periodically so that there is no delay in waste management process that is linear to the management cost. Improvement of management process in factory layouts is that it needs a raw material excess flow that later will be processed into oil fuel and plastic pellets in parallel. This is supposed to save space, cost, energy and can increase the revenue of plastic waste entrepreneur.

**Acknowledgements**

The author thanks to Universitas Trisakti, the institution which is research sponsored and supported to attend the research and seminar.

## 5. References

[1] Gabriel DS. 2016 How to Increase Plastic Waste Acceptance for Mechanical Recycling: An Introduction to Material Value Conservation and its Phenomenon. Key Engineering Materials [Internet]. Trans Tech Publications, Ltd.; Aug;705:362–7.

[2] Dipak Kumar S and Pranav S. Plastic to oil. IOSR Journal of Mechanical and Civil Engineering. IOSR-JMCE. e-ISSN: 2278-1684. p-ISSN: 2320-334X. pp 446-48. www.iosrjournals.org

[3] Vinodh S, Arvind KR, Somanaathan M. 2010 Application of value stream mapping in an Indian camshaft manufacturing organisation. Journal of Manufacturing Technology Management [Internet]. Emerald; Sep 12;21(7):888–900

[4] Miya A, Ngacho C. 2017 An Assessment of Significance of Value Stream Mapping to Mitigate Challenges Related to Solid Waste Management: Insights from Literature. Mediterranean Journal of Social Sciences [Internet]. Richtmann Publishing; Sep 1;8(5):41–50.

[5] Tortella GL, Campos LMS, Pereira da Silva GGM. Improving productivity of solid waste recycling centres through lean implementation: a comparative analysis based on value stream mapping. pp 1-10.

[6] Rubiano O. Oscar; Peña M. Claudia C.; Paz R. Juan C. 2015 Solid Waste Management Applying Value Stream Mapping. A Small and Medium Sized Enterprise Case. Universidad del Valle, Facultad de Ingeniería, Escuela de Estadística. Feb; 17:43-50.

[7] Hartini S, Ciptomulyono U, Anityasari M, Sriyanto, Pudjotomo D. 2018 Sustainable-value stream mapping to evaluate sustainability performance: case study in an Indonesian furniture company. Ma'mun S, Tamura H, Purnomo MRA, editors. MATEC Web of Conferences [Internet]. EDP Sciences;154:01055.

[8] Marifa PC, Andriani FY, Indrawati S, Parmasari AN, Budiman H, Kamilia A. 2018 Production waste analysis using value stream mapping and waste assessment model in a handwritten batik industry. Ma'mun S, Tamura H, Purnomo MRA, editors. MATEC Web of Conferences [Internet]. EDP Sciences;154:01076.

[9] Tortorella GL, Fogliatto FS, Anzanello M, Marodin GA, Garcia M, Reis Esteves R. 2016 Making the value flow: application of value stream mapping in a Brazilian public healthcare

organisation. Total Quality Management & Business Excellence [Internet]. Informa UK Limited; Mar 4;28(13-14):1544–58.

[10] Gabriel DS. 2016 Value Chain Upgrading Scheme of Thermoplastic Recycling Manufacturing System: A Product Quality Perspective. Key Engineering Materials [Internet]. Trans Tech Publications, Ltd.; Aug;705.

[11] Intifada GS and Witantyo. 2012 Minimasi Waste (Pemborosan) Menggunakan Value Stream Analysis Tool Untuk Meningkatkan Efisiensi Waktu Produksi. *Jurnal Teknik POMITS,* 1:1-6.

IOP Conf. Series: Materials Science and Engineering **847** (2020) 012095 doi:10.1088/1757-899X/847/1/012095

# Assessing supply chain practices and how they are perceived to impact performance of firms in Sierra Leone: A Case Study in a telecommunication company *(Sierratel)*

**Hassan Andrew Fornah[1], and I Nyoman Pujawan[2]**

[1,2]Institut Teknologi Sepuluh Nopember, Surabaya, Indonesia
hassanfofie@gmail.com, pujawan@gmail.com

**Abstract.** As competition is no longer between organizations but among supply chain players, Supply chain practices have become a major strategy employed by many businesses to gained competitive advantage and improving organizational performance. The research aimed at assessing Supply Chain Practices and how they are perceived to impact the performance of firms in Sierra Leone a case study in a Telecommunication Company (Sierratel). Two paradigms of the supply chain were developed for this research. Supply chain enablers (Organization, People, Technology, Plan, Source, Deliver, Information Sharing) and supply chain non-financial performance (Flexibility, Responsiveness, Reliability and Agility). Self-administered questionnaires were sent to thirty-four respondents through google form using a simple stratified random sampling technique. Descriptive statistics is applied to analyse the data through SPSS. The findings showed differences in various supply chain practices adopted and implemented by the company. People, Technology, Delivery, Responsiveness, Reliability and Agility did not create much impact on the company's performance. Further analysis also revealed that supply chain processes as one of the practices among the others adopted and implemented are almost in its right footpath as compared to strategy and performance but continuous improvement of supply chain practices will attain a sustainable competitive advantage and hence firm's performance.
**Keywords**: supply chain practices, supply chain strategy, enablers, performance, competitive advantage, continuous improvement

## 1. Introduction

Nowadays, organization has realized that they only control and command a small portion of the value chain in their operation and the opportunities for creating an atmosphere of continuous improvement is very much imminent. Theses realization by supply chain (SC)practitioners demands a strategy and processes for securing competitive advantage. Supply chain management (SCM) is a new phenomenon in the service industry and may play a significant role towards the successful achievement of any organization. Supply chain management practices (SCMP) are set of standard events designed and undertaken by organizations to enhance the effective management of supply chain. SC practices have served as an engine that accelerated economic growth for businesses. SCM has perceived as a strategic response to the global challenges ahead from the SC players due to the over dependencies of the inflow and outflow of information, products and services among SC members[1]. Studies by [2], stated that SC is a whole set of activities involved in taking and managing the product from point of production to point of consumption. Much interest and attention have been seen over the recent years especially in the field of operations management research. Development of new ideas such as outsourcing, decreased in inventories, Just- in-Time Philosophy, cooperation and collaboration, and strategic partnerships within the SC [3].

### *1.1 Background*

SC practices have become paramount for most business and therefore examining the relationship between organizational performance and SC performance has prompted managers to seek better knowledge and understanding of SC. This has prompted many organizations recently to adopt and integrates different supply chain practices like e- collaboration within the supply chain network.[4],SCM is an integrated management approach dealing with interdependent network of suppliers, manufacturers, distributors, and retailers to facilitates the smooth flows of goods, services, and information from suppliers to end customers with minimum cost while sustaining required service levels[5]. According to [6], SC manages supply of relations as well as managing a network of interconnected businesses with the aim of providing the required product/service to the end users with satisfaction. The number of competitors in the market is increasing and expanding gradually and organizations has to re-establish their operations to respond to the changing market dynamics to produce goods and services of high quality that differentiates from others through efficient and effective management of the SC [7]. Over the years the nature of SC has changed to the extent that organizations no longer compete against each other on the basis of quality as it was practiced in the nineties [8] As competition in the 1990s continues to intensified in the markets. Improving customer service level in terms of delivering the product or service to the right place, at the right time and at lowest cost became the challenges associated. To be more competitive organizations should begin to think ahead to improve efficiencies within their entire SC network. Understanding SC strategies as well as implementing effective SC practices is an essential prerequisite to global competition and profitability [9],[10]. However,[11], mentioned that, to survive in tight competition, continuous improvement of SC practices should be adopted and integrated by companies as it is largely believed by many practitioners that, better SC practices will lead to better performance. Because organizations no longer compete between themselves but among supply chain the current trend of SC practices has become a future path for scoring competitive advantage and improved organizational performance[12]. The new source of business competition lies outside the walls of the firm. It's determined by how effectively these organizations link their operations with the SC partners

### *1.2 Problem Statement*

SC practices are designed to reduce cost, if not to eradicate the high incidence of wastages among the supply chain networks. SCM itself is compounded by a network of key players both at the strategic and operational levels. This problem sometimes leads to poor coordination and fall in performance standards. Survey has revealed that a high percentage of companies in Sierra Leone are still prone to traditional practices and lack modern SC practices that would improve performance. Such practices as it is observed had little or no effect in increasing the performance of companies especially SIERRATEL to meet global competitiveness. Many studies on supply chain management have been carried out but majority of them focused on processing and manufacturing sectors. Supply chain management within the services sector, especially in the telecommunication sector, has not been fully addressed by many researchers. This is justified by [13], that, SCM practices, in general, is not fully matured in businesses rendering services and conclusions reached are untenable. Studies by [14] have found varied impacts of SCM on organizational performance. Some of the findings include but not limited to; improved performance measurement, continuous improvement of organizational practices. Supply chain practices require effective collaboration, cooperation and strategic partnership at all levels of management both internal or external to attained organizational performance of which presently this has not been adequately showcased in the service sector of Sierra Leone with emphasis placed on the Sierrtel.

The critical challenges mostly faced by the company includes the adoption and integration of effective SC practices, Poor Information and Communication Technology (ICT) and others. The company should be able to quickly identify the best SCM practices to respond to changes in the market.

Information technology is a pathway for connectivity and integration in doing businesses thereby linking different supply chain players to a whole operation in an efficient and effective way. Technology is often seen as an enabler in SC due to its substantial reduction in paperwork, improve communication and reduce supply chain cycle times if appropriately implemented [15].There are numerous complaints from customers on the poor quality connection lines and response time causing dissatisfaction among customers. Major concerns raised by customers is that the company lacks a comprehensive strategy to improve performance despite it enormous potential for growth. To ensure successful growth, the company's SC should eliminate quality defects through the implementation of a robust and continuous improvement strategy to increase the level of service with minimum cost. It required management to invest on advanced IT systems, recruit trained and qualified people to handle current technology.

### *1.3 Objectives of the Study*

This study includes the following: (1) To assess the SCPs and how they are perceived to impact performance. (2). To provide a framework for effective and excellent supply chain practices., (3) To identify supply chain strategies that will enable companies to achieve superior performance.

## 2. Methods

### *2.1 Research Design*

The study is designed into qualitative and quantitative in nature. To describe some aspect of the objectives and to further understand the relationship between the variables of the study, a research framework was designed to form the basic construct of the research and can be seen below in Figure 1

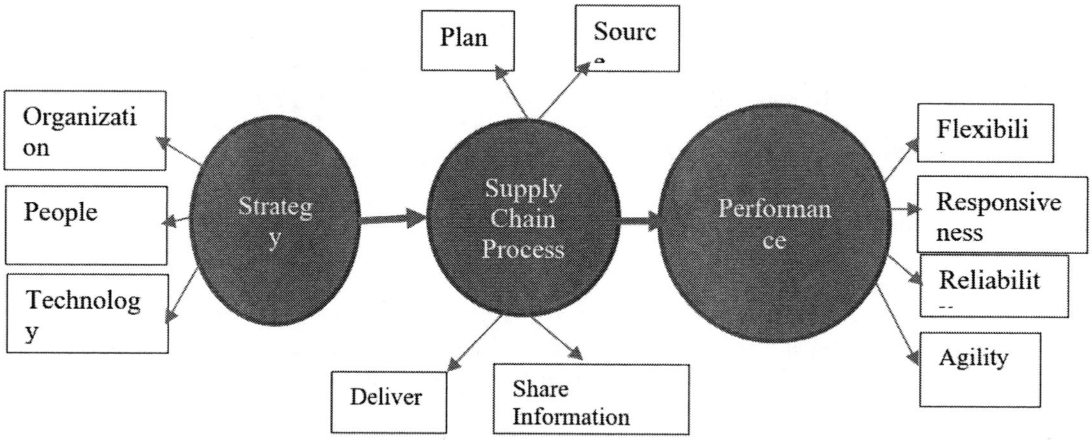


**Figure 1** Expanded Model View

### *2.2 Population and Sample Size*

The researcher identified a special population of interest to investigate a specific issue whose findings is used to generalized the whole population. [16], define the target population as a population consisting of a specific group of subjects to whom the researcher plans to generalize their findings. The population of the study includes all directors, managers, procurement officers, supervisors and warehouse officers of the company. A sample size of thirty-four (34) respondents was drawn from the sample frame using simple stratified random sampling to promote the needs for efficiency and representativeness. No fixed percent of the target population is used as a representative sample of this study as this is justified by [16]. It may depend on the nature of the population of interest or the data to be collected.

*2.3 Research Instrument*

The instruments used in the research are questionnaire and personal interview. To ensure convenient comparison and analysis of the variables under investigation, the use of statistical tools such as tabulation, charts and the transformation of the absolute data value into percentages is also adopted to examine the magnitude of the SCMP and its impact on performance. Finally, an interview is justified because a high percentage of the respondents are not familiar with SCMP to respond to the items in the questionnaires whilst at the same time rephrasing the questions to avoid ambiguity.

*2.4 Sources of Data*

The research utilizes both primary and secondary data as the main sources of the research data.

*2.5 Data Collection Procedure*

The data was collected through a questionnaire. The questionnaire is administered using google form platform method. To ensure effective data collection, the questionnaire was designed in an open and closed-ended question in order to enable the respondent to answer the questions independently. To avoid missing out very important details from the interview, the researcher also made phone calls to asked questions and then record the responses by means of note-taking. The questionnaires was presented into blocks and marked on a Likert scale of 1 to 5. The scale is labelled with Strongly Agree [5 ] to Strongly disagree [1]. Finally, email was first sent to respondents at an earlier period explaining the aims and objectives of the study and requesting their participation. Also, similar mails and phone calls was done repeatedly to reminding them to respond to the survey.

*2.6 Pilot testing of the research instrument*

A Pre-testing of the research instrument was done prior to the main study on a group of respondents. Mugenda and Mugenda,(2003), as cited in [17] stated that a sample size of 1- 10% of the sample frame is a suitable frame to engage in a pilot testing. Three (3) respondents were used for the pilot testing to determine the suitability, appropriateness, and clarity of the questionnaire in addressing the variables under investigation to determine the reliability of the instrument.

*2.7 Reliability and Validity of the Study*

The greater the degree of consistency and stability in an instrument the greater its reliability.[18],states that "validity refers to an empirical measure adequately reflects the real meaning of the concept under consideration". Mugenda & Mugenda (2003) as applied to [17] hold that, the accuracy of data to be collected largely depends on the data collection instruments in terms of validity and reliability achieved by having objective questions included in the questionnaire. The questionnaire was tested for both internal consistency and stability of items in the survey and to prove how the items measuring a concept hang together as a set. [19], stated that Cronbach's Alpha is a reliability coefficient that indicates how well the items in a set are positively correlated to one another. The closer the alpha value to 1, the higher the internal consistency but for the purpose of this study, Cronbach Alpha coefficient of 0.5 and above is accepted and suitable for the questionnaire. Sekaran (2003) as cited in [20], further claims that a study is only reliable when another researcher using the same procedure and studying the same phenomenon arrives at similar conclusions

*2.8 Data analysis and presentations*

[19] asserts that there are three objectives in data analysis; getting a feel for the data, testing the goodness of the data, and Hypotheses testing to answering the research question, Establishing the goodness of data lends credibility to all subsequent analysis and findings because it measures the reliability and the validity of the study. Thus, the study used the Statistical Package for Social Sciences (SPSS) version 23 and Ms – Excel to analyze the data. Three techniques of data analysis test were used in the study such as descriptive statistics, reliability and correlation analysis.

## 3. Result and Discussion

### *3.1 Supply Chain Strategy*

The research uses three paradigms as strategy to assess the supply chain practices of the research as shown below in Table 1

**Table 1.** Supply Chain Strategy

| | N | Range | Min | Max | Sum | Mean | Std. Dev. | Variance |
|---|---|---|---|---|---|---|---|---|
| Organization | 34 | 3.80 | 1.20 | 5.00 | 136.90 | 4.0265 | .70681 | .500 |
| People | 34 | 2.70 | 2.30 | 5.00 | 132.90 | 3.9088 | .67211 | .452 |
| Technology | 34 | 2.70 | 2.30 | 5.00 | 132.50 | 3.8971 | .54688 | .299 |
| Valid N (listwise) | 34 | | | | | | | |

From the data presented in Table 1 above, it shows that "organization" as a supply chain strategy score the highest mean and standard deviation of (4.0265) and (0.70681) respectively, followed by people and technology. 34% (12) of the respondents of the company strongly agree that the company assign responsibilities to employees with set targets as well as the organization has a well-defined policy that is periodically reviewed. Also, 33% (11) respondents agree that employees are not involved in some decision-making process and the company do not have sophisticated IT systems for their operations. Meanwhile, the same 34% (12) employees agree that the company has a focus and visionary leadership as well as a centralized and decentralized work policy but disagree with the fact that the company has a shorter span of control. More so, 66% of the respondents disagree that the organization invests heavily in IT infrastructure and their IT system enables cross-enterprise connectivity. They also failed to agree that, the organization provides training capacity for their employees and the right people are assigned in the right positions.

### *3.2 Supply Chain Processes*

Four constructs were used by the researcher to assess the various supply chain processes. The mean and standard deviation were the main measures used by the researcher to analyse the results. A mean of 4 and above indicates that the respondents agree or strongly agree while a mean less than 2 indicates strongly disagree or disagree and a mean of 3 shows that the employees are neutral in their opinion. The findings of this study are presented in Tables below.

**Table 2** Supply chain Process

| | N | Range | Min. | Max. | Sum | Mean | Std. Dev. | Var. |
|---|---|---|---|---|---|---|---|---|
| Plan | 34 | 2.30 | 2.70 | 5.00 | 138.40 | 4.0706 | .57342 | .329 |
| Source | 34 | 2.10 | 2.90 | 5.00 | 140.60 | 4.1353 | .48548 | .236 |
| Deliver | 34 | 2.40 | 2.30 | 4.70 | 129.60 | 3.8118 | .59583 | .355 |
| Information Sharing | 34 | 2.00 | 3.00 | 5.00 | 137.40 | 4.0412 | .47934 | .230 |
| Valid N (listwise) | 34 | | | | | | | |

The results presented in Table 2 above clearly shown that the sourcing policy adopted by the company has the highest mean score of (4.1353) interns of its supply chain processes as compared to plan (4.0706), information sharing (4.0412) and deliver (3.8118). This means that since source has the highest mean, the respondents agree that the company have good supplier relationship, an inventory plan for supplier; regularly update their suppliers with their performance records; supplier has high level of reliability and manage supply risk systematically. Also, the planning process of the company yield a mean of (4.0706) This means that the company have a strategic plan which has a timeline for implementation, do a collaborative planning with suppliers, plan for capacity requirement and finally, periodical reviewed of the strategic plan for improvement. With regards to the information sharing

IOP Conf. Series: Materials Science and Engineering **847** (2020) 012095 doi:10.1088/1757-899X/847/1/012095

process of the company, it has a mean score of (4.0412). This indicates that the respondents agreed with the view that, the company inform their suppliers in advance of any changing needs or events that may affect their relationship, share information of core business processes. Meanwhile, the mean score of the delivery process (3.8118) fall below the average which means that the respondents disagreed with the delivery processes of the company. For instance, they disagreed that the company's distribution centres are closer to the customers, delivery schedules are shorter than competitors; make provisions for potential delivery disruptions such as breakdowns and natural disaster.

### *3.3 Supply chain Performance*

The researcher used non-financial performance measurements to assess the performance of the company. This includes flexibility, responsiveness, reliability and agility are shown in the Table below.

**Table 3** Supply Chain Non- Financial Performance

| | N | Range | Min | Max | Sum | Mean | Std. Dev. | Var. |
|---|---|---|---|---|---|---|---|---|
| Flexibility | 34 | 2.20 | 2.80 | 5.00 | 136.10 | 4.0029 | .61026 | .372 |
| Responsiveness | 34 | 2.30 | 2.50 | 4.80 | 133.30 | 3.9206 | .66596 | .444 |
| Reliability | 34 | 3.60 | 1.40 | 5.00 | 135.60 | 3.9882 | .73020 | .533 |
| Agility | 34 | 3.20 | 1.60 | 4.80 | 126.00 | 3.7059 | .73069 | .534 |
| Valid N (listwise) | 34 | | | | | | | |

From the Table 3, there is empirical evidence showing that flexibility have an average mean of (4.0028) which indicates that, respondents agreed with the fact that flexibility has a direct effect on the performance of the company. Most of the respondents interviewed express their opinion that the company has the ability to change planned delivery schedules as well as quickly adapt to changes in demand in the market. Meanwhile, the company's responsiveness practices have a mean of 3.9206 indicating that the company could not respond to customers' request promptly; response time is not always available on a 24 hours bases; and customers complaints are not handled with the attention needed. Both reliability and agility also showed a mean of (3.9882) and (3.7059) respectively.

### *3.4 Summary of Supply Chain Practices*

**Table 4** Summary of Supply Chain Practices

| | N | Range | Min | Max | Sum | Mean | Std. Dev. | Var. |
|---|---|---|---|---|---|---|---|---|
| Strategy | 34 | 2.70 | 2.20 | 4.90 | 133.90 | 3.9441 | .53315 | .284 |
| Process | 34 | 1.60 | 2.90 | 4.50 | 136.70 | 4.0147 | .42411 | .180 |
| Performance | 34 | 2.50 | 2.30 | 4.80 | 133.10 | 3.9044 | .62140 | .386 |
| Valid N (listwise) | 34 | | | | | | | |

This section sought to give an overall summary of the findings made from the different supply chain practices adopted and implemented by the company for the period under reviewed.
Table 4 shows a summary analysis of the different supply chain practices adopted and implemented by the company for the period under review. Statistical data clearly indicates that the company is only able to achieve an average mean of 4.0206 for it supply chain processes while the mean scores for strategy (3.9382) and performance (3.9147) practices fall below average, meaning that the two practices did not contribute much impact on the performance of the company. Finally, it can be concluded that the higher/ stronger the company's supply chain practices rating the better the company's overall performance and improvement it will be. This is inconsistent with the studies done by [8] and [12]

### *3.5 Correlation*
A Pearson correlation was used to show the direction, strength and significance of the bivariate relationships between strategy, processes and performance. The (r) between 1.0 which indicates a positive relationship and (r) -1 indicate a negative correlation

**Table 5 Correlation between Strategy, Process and Performance**

| | | SC Strategy | SC Process | SC Performance |
|---|---|---|---|---|
| SC Strategy | Pearson Correlation | 1 | .653** | .653** |
| | Sig. (2-Tailed) | | .000 | .000 |
| | N | 34 | 34 | 34 |
| SC Process | Pearson Correlation | .653** | 1 | .834** |
| | Sig. (2-Tailed) | .000 | | .000 |
| | N | 34 | 34 | 34 |
| SC Performance | Pearson Correlation | .653** | .834** | 1 |
| | Sig. (2-Tailed) | .000 | .000 | |
| | N | 34 | 34 | 34 |

**. Correlation is significant at the 0.01 level (2-tailed).

According to Table 5, strategy shows a positive relationship with process and performance. The path coefficient of strategy was 0.00 and is statistically significant ($p < 0.05$) A correlation of 0.653 is moderately a positive correlation with process and performance and that there is evidence to say that a correlation exists in the population. Also, there is a strong direct relationship between process and performance (0.834) and statistically significant ($p < 0.05$). This means, there is enough evidence to justify that, a positive relationship exist between process and performance ($r= 0.834$) indicated that the level of process accounted for a large portion of (83.4%) of the variability in strategy.

## 4. Conclusion
From the research carried -out, it can be concluded that the company's supply chain strategy is below average to fully achieve its strategic objectives. This is due to the company's failure to invest heavily on IT infrastructure. The technology used is not sophisticated to meet the current trend of the market making business operations more difficult. There is inadequate training capacity for personnel and lack of expertise in the supply chain. People are not assigned in their rightful position. The results from the findings also shows that the SC process scored the highest mean and standard deviation than strategy and performance. This is as a result of (plan, source and information sharing) scored a mean of 4 and above as compared to strategy and performance with only one indicator above four for each. The company's sourcing policy is in the right path with the current practices, as well as the good supplier's relationship and high level of reliability. The performance indicator became the least among the supply chain practices of the company. The analysis of the result showed that only flexibility had an impact on the performance. The company's responsiveness practices are below average meaning that, the company could not respond to customers' requests promptly and response time is not always available on a twenty-four (24) hours bases. Also, customer complaints are not handled with the attention needed. Both reliability and agility shown a mean that is below average respectively.

### *4.1 Recommendations*
The following recommendations where drawn from the research for management of the company, other organizations and future academic researchers wanting to undertake a similar venture.
Firms have to design a clear and comprehensive strategy that will embrace supply chain best practices. This is so because the strategy shows the direction or path the business is going in terms of efficiency and responsiveness. The company should invest heavily in IT infrastructure and installed sophisticated systems to avoid the problem of poor network connectivity nationwide. Training programs should be

developed and implemented more frequently for performance and continuous improvement. The company should also have the willingness to share vital information or knowledge of core business processes with its strategic partners. Performance has to be measured to meets and adjust to the rapid changes in the market. Finally, to ensure sustainability and continuous improvement of performance, all three levels of management should collaborate through a closed-loop system.

## 5. References


[1] L. Xu and B. M. Beamon, 2006 Supply chain coordination and cooperation mechanisms: An attribute-based approach, *J. Supply Chain Manag.*, **42**(1), pp. 4–12.

[2] S. Sindhu and A. Panghal, 2016 Robust retail supply chains - the driving practices, *Int. J. Adv. Oper. Manag.*, **8**(1), p. 64.

[3] D. Kern, R. Moser, E. Hartmann, and M. Moder, 2012 Supply risk management: model development and empirical analysis, *Int. J. Phys. Distrib. Logist. Manag.*, **42**(1), pp. 60–82.

[4] P. Hove-Sibanda and R. I. D. Pooe, 2018 Enhancing supply chain performance through supply chain practices, *J. Transp. Supply Chain Manag.*,**12(**September), pp. 0–13.

[5] L. Avelar-Sosa, J. L. García-Alcaraz, and J. P. Castrellón-Torres, 2014 The effects of some risk factors in the supply chains performance: A case of study, *J. Appl. Res. Technol.*, **12**(5), pp. 958–968.

[6] C. M. Harland, 2005 Supply Chain Management: Relationships, Chains and Networks, *Br. J. Manag.*, **7**(s1), pp. S63–S80.

[7] J. R. Stock and S. L. Boyer, 2009 Developing a consensus definition of supply chain management: A qualitative study, *Int. J. Phys. Distrib. Logist. Manag.*, **39**(8), pp. 690–711.

[8] S. E. Fawcett, P. Osterhaus, G. M. Magnan, J. C. Brau, and M. W. McCarter, 2007 Information sharing and supply chain performance: The role of connectivity and willingness, *Supply Chain Manag.*, **12**(5), pp. 358–368.

[9] D. Power and S. Rahman, 2001 Critical Success Factors in Agile Supply Chain Management. *International Journal of Physical Distribution & Logistics Management Article information*.

[10] S. Li, S. S. Rao, T. S. Ragu-nathan, and B. Ragu-nathan, 2005 Development and validation of a measurement instrument for studying supply chain management practices, **23**, pp. 618–641.

[11] I. N. Pujawan, 2008 SCAT: Supply Chain Assessment Tool toward Excellence, *Proc. 9th Asia Pasific Ind. Eng. Manag. Syst. Conf. 2008*, no. August 2015, pp. 580–585.

[12] I. Musa and I. Nyoman Pujawan, 2018 The relationship among the resiliency practices in supply chain, financial performance, and competitive advantage in manufacturing firms in Indonesia and Sierra Leone," *IOP Conf. Ser. Mater. Sci. Eng.*, **337**(1).

[13] T. K. Hong and S. Zailani, 2011 Service supply chain practices from the perspective of Malaysian tourism industry, *IEEE Int. Conf. Ind. Eng. Eng. Manag.*, no. February, pp. 539–543.

[14] M. Blowfield and C. Dolan, 2010 Outsourcing governance: Fairtrade's message for C21 global governance, *Corp. Gov.*, **10**(4), pp. 484–499.

[15] R. B. Handfield and C. Bechtel, 2002 The role of trust and relationship structure in improving supply chain responsiveness, **31**, pp. 367–382.

[16] J. W. Best and J. V Kahn, 2006 *Welcome to Research in Education*, Tenth Educ. Pearso.

[17] P. R. O. Amos Sila Mwangangi, Rukia Atikiya, Joyce Nzulwa, 2017 Effect of Community Related CRS on Performance of Manufacturing Firms in Kenya, **9**(24), no. August, pp. 68–79.

[18] E. Babbie, 2011 *The Basics of Social Research*, Fifth. (USA: Wadsworth).

[19] U. Sekaran, 2000 *Research Methodds For Business Fourth Edition*.

[20] L. O. Oyewobi, B. Suleiman, and A. Muhammad-Jamil, 2012 Job Satisfaction and Job Commitment: A Study of Quantity Surveyors in Nigerian Public Service, *Int. J. Bus. Manag.*, **7**(5), pp. 179–192.

IOP Conf. Series: Materials Science and Engineering **847** (2020) 012096 doi:10.1088/1757-899X/847/1/012096

# Investigating the impact of supply chain management on the performance of manufacturing industries in Sierra Leone: case study of Sierra Leone bottling company (SLBC)

**Alimamy Kamara[1] and I Nyoman Pujawan[2]**

[1,2]Institut Teknologi Sepuluh Nopember, Surabaya, Indonesia
alimamykamara58@gmail.com, pujawan@gmail.com

**Abstract.** Supply Chain (SC) is a series or network of companies who work collectively to make and deliver products and services to end-users. SC has become an important way of gaining competitive advantage and a way of improving companies' performance. The impact of supply chain management (SCM) on the performance of manufacturing industries is very important in gaining competitive advantage. A weakness in SCM has a negative impact on the profit and can seriously affect the production and delivery of products. This research conceptualizes and develops four main dimensions to improve the performance of SLBC on Strategy and design, enabling infrastructure, process, and performance. To investigate the impact of SCM on the performance, this research employed descriptive data which was tested base on the available data through the use of software packaging for social scientists (SPSS) and Minitab. From the results, Process was found to have a strong correlation with performance (0.821), enabling infrastructure and performance (0.596) which shows that there is a cordial relationship between the variables. Also, performance has the highest mean (4.0760) which indicate it output factors have a very great impact on profitability, productivity and the entire performance. The company faced several setbacks in achieving its goals.
**Keywords**: Supply Chain Management, Supply Chain Impact, Performance Measurement, Customer Relationship Management, Implementation Strategies.

## 1. Introduction

Supply Chain has to do with the taking and managing of a product(s) which begin from the point of production on to the point of consumption. The process from the production to the consumption has to do with the upstream and downstream. SC nowadays play a key role in gaining competitive advantage over other competitors and a way of improving companies' performance as competition is very high among companies. [1] started that since competition is no more among organizations, for companies to gain advantage and improve organizational performance over other competitors, they must create value chain through supply chain integration. [2] mentioned that the demand of customers in terms of how they ask for products is now becoming more sophisticated it is calling for individualized goods and services. It is for this reason that organizations are trying to ensure that they increase their revenue, and at the same time, they reduce cost. A major element of effective SCM includes both upstream collaboration and downstream integration of the partners of the firm's and the customers in an efficient and effective manner [3]. The study specifically covers problems that are connected with the systems of supply chain management and how best they will minimize expenses to a level that maximizing profit and investigating the impact it has on the entire industry. The Manufacturers and Suppliers are different units that are financially independent to each other, but they pursue their own motive in order to maximize their returns [1].

The supply chain management system is the Centre for any organization more especially for manufacturing companies. Manufacturing companies need a proper management system. According to [4] A holistic view of supply chain systems can deliver an insights on how the participants supply chain operate independently and collectively towards attaining a real-time, global supply chain. In the process of trying to attain the goal of SCM system, each organization in the network has its own plan and operates independently from the others. [2] Stated how supply chain management has become very important in order to gain competitive advantage. As a result of this, people from academics and industries are now showing great concern in SCM with many models and techniques which are extremely useful in order to maximize the market with more efficiency and effectiveness.

*1.1 Background*

Supply chain is now becoming a very important concept in a comprehensive analysis. It is a link that connects the different elements of a manufacturing process that start from the stage of production up to the stage of delivery. In recent times, SCM has become well-known, and it is shown by marked increases in conferences, professional development programs, practitioner and academic publications [5]. SCM focuses at the management function that purely deals with the management of goods, works and services from the original producer, through some transformation processes in a firm in order to add value to the ultimate consumer and backflow of remains and empties together with the related information [6]. The motive of SCM is to see how best companies can incorporate activities within and across organizations for providing customer value. As also mentioned by [5], [2], supply chain management contains the preparation of all activities involved in sourcing and procurement, conversion, and involved in acquiring, manufacturing, issuing, and selling the products or services to the end customers. According to [7] they explained that "Supply chain management includes coordination and collaboration with channel partners, which can be suppliers, intermediaries, third-party service providers, and customers which integrates supply and demand management within and across companies".

In order to make sure there is a great turnout in long term performance of SCM and other industries, SCM has a key role to play in order to have a massive improvement within an organization. One major important role of supply chain management is that it increases competitiveness in order to satisfy customers. In these current days, SCM has become an integral part of a business and it is essential to any company's success and customer satisfaction. In order to have a better improvement, sometimes supply chain managers involves in going beyond the suppliers that interface with the organization. For this to be actualized, this may include removing cost and increasing quality.

*1.2 Problem Statement*

There are several manufacturing companies that are faced with many challenges which normally have an adverse effect on the output of the firm's performance [8]. The problem of how best companies should ensure they minimize their cost and satisfy the need of their customers on time has existed for a very long time. It is of essence for companies to know the real problem before thinking about how to solve the problem. [9] mentioned that the "idea of a problem statement, a problem-solving method, and the application of the problem-solving method to a problem are given detailed formulations that are purely based on the type of the problem". Any weaknesses in the process of transforming a product(s) can seriously disturb the manufacture and distribution of products to final consumers. One of the major problems relating to researchers is to identify the reasons why companies tend not to implement SCM, notwithstanding its advantages and its driving forces [10]. The weaknesses that might have occurred in a production process may have adverse impacts on the overall outcome of the company. There has always been a shortage of the company's products during festive seasons. In recent years, there is very high competition with other beverage producing companies which had prompted the SLBC to revisit their packaging system as other beverage manufacturing companies are using plastic bottles that are disposable after use. There are also problems with lead time, transportation facilities and holding of raw materials as well as finished products.

Management of the company should ensure that they do not have shortage of products particular during the festive season, and this can be done by ensuring they follow the company supply chain procedures by having all materials needed early. Also, strict measures should be taken in order to solve the packaging problem of the company by making sure that they gain a competitive advantage over other competitors in the market This can be solved by using plastic bottles which are disposable and not returnable. Finally, the problem of lead time, transportation facilities and the holding of raw materials can be solved through early preparation and having sufficient vehicles to transport the product. The company should ensure they have an intermediary that exists between the company and the suppliers by making sure that the company receives its supplies at the appropriate place, time and quantity.

### *1.3 Objectives of the Study*

The reason for SCM increasing is to compete at the SC levels, organizations must ensure they incorporate strategies. The objective of this study is to examine the impact of Supply Chain Management on the SC performance of SLBC. However, in order to ascertain its motive, key objectives had been investigated which includes:

- To evaluate the supply chain management of the SLBC in order to make rapid and drastic improvement in its supply chain process/procedures
- To evaluate the importance of SCM on the overall performance of SLBC.
- To evaluate the challenges/ problems faced by SLBC while adopting supply chain management

## 2. Methods

This part focuses on the approaches used to carry out this research. It describes the methodology used in collecting and analysing the data for this research. This study assumes a cross-sectional survey design that uses both quantitative and qualitative approaches. A questionnaire was obtained to gather information from the employees asking them to indicate their level of agreement on the questions asked by each variable. The questions were designed on the review of the literature and from the author.

### *2.1 Research Design*

[11], has mentioned that research design is a set of conceptual procedures in which research is conducted and analysed in order to measure the variables specified, it gives a clear picture about how the analysis of the data design would look like. The study adopted descriptive statistics. The methodology used in this study is vital for any research work to be carried out. [12] stated that "A research design is the arrangement of conditions for collection and analysis of data in a manner that aims to combine relevance to the research purpose with economy in procedure". A cross-section study was used. The use of research design is very useful because it helps the smooth sailing of the several research processes, by ensuring the research is well-organized as a way of obtaining the highest information with a minimal expenditure of effort and time [12].

### *2.2 Sample Size and Techniques*

A sample is a small number of people that are normally selected for proper observation and for further analysis [13]. The study used stratified random sampling to members who are in the network of manufacturing and distribution of the operations of the SLBC. The Sample was drawn from SLBC Limited and focused on all departments dealing with Supply Chain management issues. According to [12], a sample design has to do with a certain way of getting sample output from a particular population. 30 respondents out of 100 were drowned from the sample frame through the use of stratified sampling. [12], stated that a representative sample could be 30% of the targeted population. The sample included 3 members of SLBC from the Administration and HR department, 5 members of staff from the Logistics department, 5 members from the Marketing/commercial department, 6 members from the production/Technical department, 5 members from the supervision department, 3 managers, and 3 members from the Finance department.

*2.3 Collection of Data*

The process of collecting data take effect after a problem(s) and a plan have been identified [12]. They are known to be un-interpreted materials on which a decision is to be created and rely on facts which may include anything known to be true or exist [11]. This paper develops and administers an online questionnaire that investigates the performance of the manufacturing company through Google from a well-represented sample of members from the SLBC. The questionnaire was divided into various sections. In collecting the data, both secondary and primary data were used.

*2.4 Data Analysis Procedure*

In order to show a convincing conclusion in analysing the data, this paper shows different statistical procedures where a test was done base on the available data. Data analysis which involves the systematic process of establishing a relationship between and among data collected through various instruments used. The study thus used descriptive statistics mainly mean, variance and standard deviation to summarize the response. According to [14], descriptive statistics offer simple summaries about the sample that include the statistical measures of central tendency and dispersion. This was used to analyse the objective and to integrate both qualitative and quantitative techniques in the data analysis [14]. A reliability test was done, and the analysis was done through a software packaging for Social Scientists (SPSS) version 23 and Minitab 17.

## 3. Results and Discussions

The process of a company in understanding the need of their customers help the company to deliver high good products or service to their customers [15] This paper sought to establish the different SCM process on the impact of SLBC that had been adopted by different institutions. Table 1 shows the summarized output on the impact of SCM improvement. It evaluates the SC processes that help to determine the weak links and identify possible improvement ways in the areas of strategy and design, enabling infrastructure, process, and performance. The study of supply chain that concerns manufacturing industries are done with the help of certain key variables like, lack of sophisticated information system to help the IT department, and lack of support among members in the SC [3]. The discussion regarding the rapid improvement of the company SC, it is clearly shown from the output that the average mean score obtained from the four variables were rated high which concluded that the SCM of the SLBC is effective and efficient. It has also proven that there is a relationship between the factors and performance. The problem of packaging, storage, lead time, outdated machinery, having one production centre are issues that need key intention in order to gain competition in the market and maximize profit.

*3.1 The Improvement process implementation of SCM for the variables*

This research sought to establish the different SCM process on the impact of SLBC that had been adopted by different institutions. The respondents were asked to rate the variables base on their level of agreement working with the company with different statements on a scale of 1-5. The outputs of the responses are provided in table 1.

**Table 1**: Mean of Strategy, Enabling Infrastructure, Process, and Performance.

| | N | Range | Minimum | Maximum | Mean | Std. Deviation | Variance |
|---|---|---|---|---|---|---|---|
| Strategy and Design | 30 | 1.60 | 3.00 | 4.60 | 4.0733 | .40166 | .161 |
| Enabling Infrastructure | 30 | 1.85 | 2.75 | 4.60 | 4.0170 | .41836 | .175 |
| Process | 30 | .86 | 3.64 | 4.50 | 4.0570 | .20785 | .043 |
| Performance | 30 | 1.19 | 3.33 | 4.52 | 4.0760 | .27535 | .076 |
| Valid N (listwise) | 30 | | | | | | |

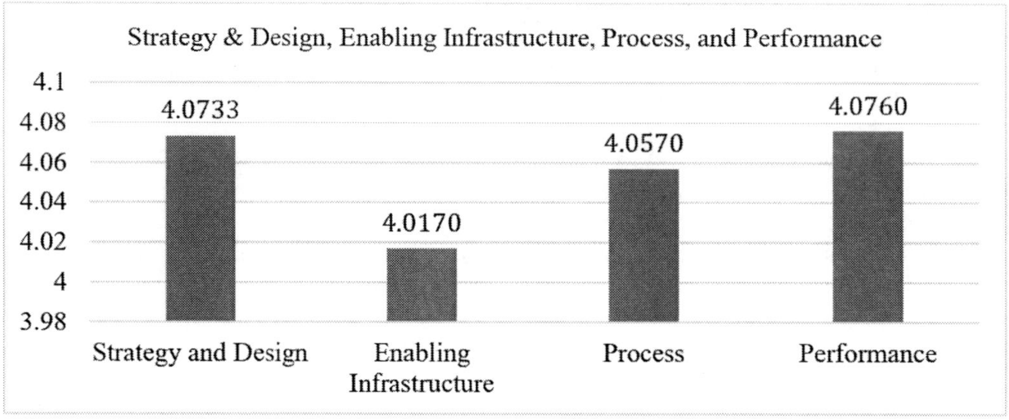


**Figure 1**: Strategy & Design, Enabling Infrastructure, Process, and Performance

The descriptive statistics in Table 1 and Figure 1 was obtained from the first objective of the research that tries to evaluate the SCM of the company that aim at it improvement for Strategy and design, enabling infrastructure, Process, and Performance that shows the way in which the variables have been adopted either to a great extent or to a moderate extent. A variable(s) can be adapted to a great extent if (mean lies between 3.51 and 5) and for a moderate extent the mean must lie between 2.51 and 3.5. From the results, Strategy and design show an output of 4.0733 with a standard deviation and variance values 0.40166 and 0.161 respectively. Enabling infrastructure has a mean of 4.0170 which is the smallest when compared to the others with a minimum value of 2.75 and a maximum value of 4.60, Process has a range of 0.86, mean value of 4.0570 and a minimum value of 3.64 and a maximum of 4.50. Performance has the highest output of 4.0760. It has a variance and a standard deviation of 0.076 and 0.2753 respectively it also has a range of 1.19 with a minimum and a maximum value of 3.33 and 4.52 respectively. The results clearly show how each variable has a role in attaining the success factor in determining the SC performance of the study [17] These findings show that the SCM improvement process implementation has been adopted and implemented to a great extent, where there is no moderate extent from the results. It doesn't mean that the SLBC did not have grounds to cover in terms of improving the company SC.

*3.2 The important and relationship of SCM on the performance of SLBC*

In an extent to know the relationship that exists among the following variables, a correlation analysis tests were done in order to know the relationship between two continuous variables in terms of how strong the relationship is, and in what direction the relationship goes. The strength of the relationship lies between -1 to +1. In order to know whether the relationship is strong or weak as shown in table 2.

IOP Conf. Series: Materials Science and Engineering **847** (2020) 012096 doi:10.1088/1757-899X/847/1/012096

**Table 2**: Correlation of the variables

| | | Strategy & Design | Enabling Infrastructure | Process | Performance |
|---|---|---|---|---|---|
| Strategy and Design | Pearson Correlation | 1 | .618** | .203 | .140 |
| | Sig. (2-tailed) | | .000 | .283 | .461 |
| | N | 30 | 30 | 30 | 30 |
| Enabling Infrastructure | Pearson Correlation | .618** | 1 | .500** | .596** |
| | Sig. (2-tailed) | .000 | | .005 | .001 |
| | N | 30 | 30 | 30 | 30 |
| Process | Pearson Correlation | .203 | .500** | 1 | .821** |
| | Sig. (2-tailed) | .283 | .005 | | .000 |
| | N | 30 | 30 | 30 | 30 |
| Performance | Pearson Correlation | .140 | .596** | .821** | 1 |
| | Sig. (2-tailed) | .461 | .001 | .000 | |
| | N | 30 | 30 | 30 | 30 |

Strategy and design plus enabling infrastructure have a positive correction coefficient of 0.618 which shows that the correlation between them is strong. The output shows that the correlation is significant at a level of 5% because the p-value for strategy design and enabling infrastructure is less than the alpha value. The correlation coefficient for process and performance is 0.821 positive which means that the relationship between process and performance has positive properties. The strength of the correlation is also strong because the P-value is < alpha, then it implies that the correlation between process and performance is statistically significant at 5% level.

The correlation between Enabling infrastructure and process is positive 0.500. The strength of the correlation 0.500, in this case, it can be said to be moderate. Since the P-value (0.005) is less than 0.05, it indicates that the correlation is significant at 5% level. Also, the correlation coefficient that exists between Enabling infrastructure and performance is positive 0.596. This result shows that there is a moderate linear relation between enabling infrastructure and performance. Since the P-value between enabling infrastructure and performance is less than the alpha value of 0.05, it indicates that the correlation is significant.

A Pearson correlation was used to assess the predictive relationship between the following factors: strategy and design, enabling infrastructure, process, and performance with the overall performance of the company. Finding show showed a positive correlation among the factors toward the performance of the company. Process was found to have a very strong correlation with performance (0.821), enabling infrastructure and performance (0.596) as it is shown in Table 4.12. The output clearly shows that the factors have a strong impact on the aspect of productivity, profitability and the overall performance.

The variables were supported by sub-variables. Enabling infrastructure comprises of (ICT, human resource, and organizational structure), process (CRM, warehouse management, delivery, planning, and sourcing) while performance (reliability, responsiveness, and flexibility).

## 4. Conclusion

### *4.1 Conclusions*
The implementation of effective and efficient SCM can significantly improve organizational performance through cost reduction [15]; [18]; [19]. Based on the topic under review which problem statement was developed with a specific objective set out upon which a particular methodology was drowned, data collected were presented, analysed and interpreted. From the above findings, the researcher has come to the following conclusions: Firstly, the SCM systems have a very great impact on the aspect of productivity, profitability and the overall performance of the operations of SLBC. Secondly, it was concluded that strategy design and enabling infrastructure, plus process and performance have a strong positive relationship with the SLBC while process and performance have a moderate relationship with enabling infrastructure. Thirdly, the researcher concluded that the SLBC faced some challenges in their bid to implement SCM best practices. These include the failure on the part of top management officials to give deserved attention to the functions of the suppliers and distributors, delays in the clearing of materials at the quay, poor technology, old and outdated equipment and machinery, lack of branch production centre, long lead time, and inadequate number of distributors. For any SCM to be more effective, it must require managing relational exchange with other supply chain institutions [20].

### *4.2 Recommendation*
In order to improve performance, profitability, and the productivity of SLBC, the following recommendations have been put forward:

- The company should ensure they try to decentralize its production process by extending production branch in the provinces
- Management of the company should replace the old and outdated equipment and machinery, and also introducing technological innovations in the company production process.
- The company need to build more outlet throughout the country, where customers can easily have access to the company products.
- In order to avoid supply management challenges arising from the suppliers in future contracts, management of SLBC should consider a supplier who has demonstrated reliability in their previous contract by fulfilling its obligation of the contract.
- From the output of the correlation, out of four variables, only two were found to have a strong and positive relationship. Management should ensure that there is a cordial relationship that exists between the manufacturer (buyer) and the supplier, manufacturer and distributors, and they should have an intermediary that exists between the company and its suppliers by ensuring that they receive their supplies on time.
- From the output, the SLBC should strengthen the company SCM in order to continue improving the performance of the company

## 5. References
[1] C. Wu, 2008 *"Knowledge creation in a supply chain,"* (Supply Chain Manag). vol. 13, no. 3, pp. 241–250.
[2] E. Hofmann and S. Knébel 2016 "*Supply Chain Differentiation: Background , Concept and Examples,*".
[3] Kumar C G and Nambirajan T 2013 "*Supply Chain Management Concerns in Manufacturing* (Industries," IUP J. Suppl*y)* ..., vol. X, no. 4, pp. 69–83.
[4] Crandall C, Richard E and Crandall William R, 2010 "*Principles of Supply Chain Management.* (USA: Taylor & Francis Group)
[5] K. Burgess, P. J. Singh, and Koroglu R 2006 "*Supply chain management: A structured literature review and implications for future research,*" (Int. J. Oper. Prod. Manag*).*, vol. 26, no. 7, pp. 703–729,

[6] M. Amukanga 2018 "*Factors Influencing Supply Chain Management Performance Among Sugar Companies In Kakamega County*, A Case Study of Mumias Sugar Company," (*IOSR J.* Bus. Manag)., vol. 20, no. 6, pp. 89–95,

[7] Queiroz M. M, Telles R, Bonilla S H, and Telles R 2019 "*Blockchain and supply chain management integration*: a systematic review of the literature,".

[8] Thi Duc Nguyen N 2019 "*Experiences in Supply Chain Quality Management Practices in Vietnam*: Studying on Steel Manufacturing Company A and Packaging Manufacturing Company B," (Int. J. Bus. Manag)., vol. 14, no. 7, p. 171.

[9] Annamalai N, Kamaruddin S, Abdul Azid I and Yeoh T 2013 "*Importance of problem statement in solving industry problems,*" (Appl. Mech. Mater)., vol. 421, no. February 2016, pp. 857–863.

[10] W. H. Wan Mahmood W H *et al 2013* "*Manufacturing performance in green supply chain management,*" (World Appl. Sci. J)., vol. 21, no. SPECIAL ISSUE2, pp. 76–84.

[11] Mohajan H K 2018 "*Qualitative Research Methodology in Social Sciences and Related Subjects,*" (J. Econ. Dev. Environ. People), vol. 7, no. 1, p. 23.

[12] Kothari C R 2004 *In Reply: Behaviour Therapy*, Second Edi., (New Age International) (P) Ltd. vol. 112, no. 483.

[13] Best J W and Kahn J V 2006 "*Welcome to Research in Education*" Tenth.

[14] Hassani G, Hossein, Ghodsi, Mansoureh and Howell 2010 "*A note on standard deviation and error,*" (Res. tn Educ)., vol. 29.

[15] Reichhart A and Holweg M 2007 "*Creating the customer-responsive supply chain*: A reconciliation of concepts," (Int. J. Oper. Prod. Manag), vol. 27, no. 11, pp. 1144–1172.

[16] Tan K C 2001 "*A framework of supply chain management literature,*" (Eur. J. Purch. Supply Manag), vol. 7, no. 1, pp. 39–48.

[17] Rozar N M, Mahmood W H W, Ibrahim A and Razik M A 2015 "*A Study of Success Factors in Green Supply Chain Management in Manufacturing Industries in Malaysia,*" (J. Econ. Bus. Manag), vol. 3, no. 2, pp. 287–291.

[18] Kopczak L R and Johnson M E "MIT SMR, Spring, Kopczak L R and M. E. Johnson M E 2003, *The Supply-Chain Management Effect. Reprint* ... Page 1 of 2 Article Abstract:" vol. 44, no. 3, pp. 1–2.

[19] Naslund D and Williamson S 2010 "*What is Management in Supply Chain Management*-A Critical Review of Definitions, Frameworks and Terminology.," (J. Manag. Policy Pract)., vol. 11, no. 4, pp. 11–28.

[20] Esper T L, Clifford Defee C and Mentzer J T 2010 "*A framework of supply chain orientation,*" (Int. J. Logist. Manag), vol. 21, no. 2, pp. 161–179.

IOP Conf. Series: Materials Science and Engineering **847** (2020) 012097 doi:10.1088/1757-899X/847/1/012097

# Determination of Standard Time and Output Production of Spring Frame Mattress Components using Work Sampling Method

**Taufiqur Rachman[1], Iphov Kumala Sriwana[2], and Stefanny Liyanawati Zalukhu[3]**

Industrial Engineering Department, Esa Unggul University, Jl. Arjuna Utara No.9, Kebon Jeruk, Jakarta Barat, Indonesia
[1]taufiqur.rahman@esaunggul.ac.id, [2]iphov.kumala@esaunggul.ac.id,
[3]stefannyliyanawati@gmail.com

**Abstract.** PT.CMAP which is a company that manufactures mattresses with production process starts from making foam (foaming), mattress covers, spring frames and finishing. Based on observations, obtained that the process of spring frame has the lowest percentage of daily production results. The purpose of this research is to determine the percentage of productive activities, cycle time, normal time, standard time and daily output of the spring frame mattress production process at PT.CMAP. The method used in this research is work sampling method that begins with preliminary sampling, data uniformity test, data adequacy test, calculating cycle time, calculating normal time by including performance factors, calculating standard time by including allowance factor and calculating daily output. The results obtained from this research that the standard time for the process of spring frame mattress components at PT.CMAP which consists of the spring round process is 5.04 minutes with the daily output is 90 pcs, the semi-finished spring frame process is 10.99 minute with the daily output is 41 pcs, the list frame process is 14.81 minutes with the daily output is 31 pcs and spring frame shooting CL process is 13.40 minutes with the daily output is 34 pcs.
**Keywords**: work sampling, sampling, cycle time, normal time, standard time

## 1. INTRODUCTION

In the current situation of business competition, productivity is one of the factors that support the company to grow and survive. Productivity cannot be realized suddenly, but through gradual and continuous improvement, which must always be done continuously, in order to create a good habit in increasing work productivity.[1]

PT.CMAP which is a company that manufactures mattresses with production process starts from making foam (foaming), mattress covers, spring frames and finishing. Based on observations obtained that the percentage of the daily production results for the making foam (foaming) process is 100%. For the mattress cover process is consisting of quilting process is 94%, cutting process is 115%, and sewing process is 101%. For the spring frame process is consisting of spring round process is 91%, list frame process is 68%, semi-finished frame process is 48%, and spring frame shooting CL process is 47%. For the finishing process which consists of assembling is 100%, sewing corner process is 97% and packing process is 107%. Based on these data it can be seen that the spring frame process has the smallest percentage and must get attention to be immediately improved so that productivity can increase.

Work sampling method is very suitable for use in observing work that is non-repetitive and has a relatively long time. Basically the implementation steps are quite simple, by observing work activities for randomly taken time intervals of one or more machines or operators and then recording whether they are working or idle. Work sampling is very useful in industry, especially in the manufacture of quality products. The number of observations that must be carried out in work sampling activities is influenced by two factors, are confidence level and degree of accuracy. Generally, the work sampling method can be used for several things, among others: [5, 7]

1) Measure the Delay Ratio of a number of machines, operators/employees or other work facilities.
2) Determine the Performance Level of someone in working time based on the times when the person works or idle, especially for manual work.
3) Determine a standard time for a work operation process.

Work sampling method is one method that can be used to overcome these problems which is a statistical sampling technique based on sampling theory. This method can estimate a certain amount, for example the proportion of productive activities through sampling.

Based on the problems that occur at PT.CMAP, the purpose of this research is to determine the percentage of productive activities, cycle time, normal time and standard time and daily output in the production process of spring frames mattress at PT.CMAP.

## 2. METHODS

This research used work sampling method to determine the standard time of the production process of skeletal components in PT. CMAP which consists of 4 processes, among others spring round process, semi-finished spring frames process, list frame process, and spring frame shooting CL process, so that it can know the amount of daily production output. There are several stages in this research, among others:

1) Preliminary sampling
   In this step a number of observations are made on the work activities of the observed operators to find out the best work system and find out the time interval taken at random. In this step, productive percentages can determine using formula: [6]

$$\bar{p} = (\Sigma X_i / \Sigma n_i) \times 100 \quad (1)$$

   Where $X_i$ is percentage of productive period i and $n_i$ is number of observations of the period i.
2) Data uniformity test
   Data uniformity test aims to determine whether the data taken is within the control limits or not between the upper control limit (UCL/BKA) and the lower control limit (LCL/BKB), with the equation used is: [6]

$$BKA = \bar{p} + 3\sqrt{\bar{p}\,(1 - \bar{p})/n} \quad \text{and} \quad BKB = \bar{p} - 3\sqrt{\bar{p}\,(1 - \bar{p})/n} \quad (2)$$

3) Data adequacy test
   The number of observations needed for the level of accuracy and confidence that has been determined, known through the formula: [6]

$$N' = K^2\,(1 - \bar{p})/S^2 \times \bar{p} \quad (3)$$

   Where $\bar{p}$ is the productive percentage, N' is amount of data required, K is the amount of the constant depends on the confidence level (this research used 90%) and S is the degree of accuracy in decimal numbers (this research used 10% ≈ 0.1).
4) Calculate cycle times (WS)
   In the work sampling method, in order to determine the cycle time it must first be known the time for 1 output and the productive percentage of each process obtained from the preliminary sampling. For productive percentages use formula (1), while for time for 1 output uses the following formula. [2, 3, 4]

IOP Conf. Series: Materials Science and Engineering **847** (2020) 012097 doi:10.1088/1757-899X/847/1/012097

$$\text{Time for 1 output} = \sum \text{Output time} / \sum \text{Output} \tag{4}$$

After time for 1 output and productive percentage are known, the cycle time can be determined using the following formula. [2, 3, 4]

$$\text{WS} = \text{The productive percentage} \times \text{Time for 1 output} \tag{5}$$

5) Calculate normal time (WN)
To be able to determine the normal time, performance factors must first be determined, which in this research uses the Westinghouse method that directs the assessment of four factors that are considered to determine fairness or irregularities in work, namely skills, effort, working conditions, and consistency. Determination of the performance value using the westinghouse method performance table contained in Sutalaksana, dkk. (6). After the performance value is obtained, the performance factor value (P) can be calculated using the following formula.

$$\text{P} = 1 + \sum \text{Westinghouse performance value} \tag{6}$$

After the performance factor value is obtained, the normal time can be calculated using the following formula. [6]

$$\text{WN} = \text{WS} \times \text{P} \tag{7}$$

6) Calculate standard time (WB)
To be able to determine the standard time, it must first be determined the allowance factor value are given for three things namely personal needs, eliminating fatigue and inevitable obstacles. Determination of the allowance factor value (A) uses the allowance table contained in Sutalaksana, dkk. (6). After the allowance factor value is known, the standard time can be calculated by the following formula. [6]

$$\text{WB} = \text{WN} + \text{A(WN)} \tag{8}$$

## 3. RESULT AND DISCUSSION

In each production process, there are regulations that must be obeyed, especially in terms of achieving the daily production targets set by the company. Table 1 is a mattress daily production data in PT.CMAP.

**Table 1**. Mattress daily production data

| No | Component | Process | Target | Daily Production Result | | | | | Total | % Achieving |
|---|---|---|---|---|---|---|---|---|---|---|
| | | | | 1 | 2 | 3 | 4 | 5 | | |
| 1 | Foaming | Sponge beam | 13 | 13 | 13 | 13 | 13 | 13 | 65 | 100 |
| | | Rolling foam | 12 | 12 | 12 | 12 | 12 | 12 | 60 | 100 |
| 2 | Mattress cover | Quilting | 220 | 189 | 210 | 206 | 218 | 210 | 1033 | 94 |
| | | Cutting | 80 | 80 | 85 | 100 | 120 | 86 | 471 | 118 |
| | | Sewing | 35 | 36 | 35 | 35 | 36 | 36 | 178 | 102 |
| 3 | Spring frame | Spring round | 110 | 107 | 96 | 109 | 102 | 97 | 511 | 93 |
| | | Semi-finished spring frame | 40 | 25 | 15 | 25 | 20 | 15 | 100 | 50 |
| | | List frame | 25 | 18 | 17 | 16 | 17 | 18 | 86 | 69 |
| | | Spring frame CL shooting | 40 | 25 | 20 | 18 | 15 | 20 | 98 | 49 |
| 4 | Finishing | Assembling | 25 | 25 | 25 | 25 | 25 | 25 | 125 | 100 |
| | | Corner sewing | 25 | 24 | 24 | 25 | 25 | 23 | 121 | 97 |
| | | Packing | 70 | 70 | 80 | 77 | 80 | 73 | 380 | 109 |

From Table 1 it can be seen that in the components of spring frame process have a low achieving value, which will then be the focus of this research.

### *3.1. Preliminary Sampling Data*

Preliminary sampling is carried out in each process of manufacture spring frame components with the number of observations as much as 4 times with different times in each process. For spring round process carried out for 30 minutes, for semi-finished spring frames process is carried out for 30 minutes, for list frame process is carried out for 120 minutes, and for spring frame shooting CL process carried out for 120 minutes. In table 2 can be seen the results of preliminary sampling for all processes in the manufacture of spring frame components at PT. CMAP.

**Table 2**. Observation data in the components of spring frame process

| Process | Activities | Observation Number | | | | Total |
|---|---|---|---|---|---|---|
| | | 1 | 2 | 3 | 4 | |
| Spring Round Process | Productive | 56 | 55 | 53 | 57 | 221 |
| | Non-productive | 4 | 4 | 7 | 3 | 18 |
| | Inevitable | 0 | 1 | 0 | 0 | 1 |
| | Total | 60 | 60 | 60 | 60 | 240 |
| | % Productive | 93.3 | 91.7 | 88.3 | 95.0 | |
| | Output | 8 | 7 | 7 | 6 | 28 |
| Semi-finished Spring Frames Process | Productive | 152 | 146 | 148 | 146 | 592 |
| | Non-productive | 3 | 11 | 10 | 11 | 35 |
| | Inevitable | 3 | 1 | 0 | 1 | 5 |
| | Total | 158 | 158 | 158 | 158 | 632 |
| | % Productive | 96.2 | 92.4 | 93.7 | 92.4 | |
| | Output | 4 | 3 | 4 | 3 | 14 |
| List Frame Process | Productive | 40 | 39 | 41 | 38 | 158 |
| | Non-productive | 11 | 12 | 10 | 13 | 46 |
| | Inevitable | 0 | 0 | 0 | 0 | 0 |
| | Total | 51 | 51 | 51 | 51 | 204 |
| | % Productive | 78.4 | 76.5 | 80.4 | 74.5 | |
| | Output | 8 | 8 | 8 | 8 | 32 |
| Spring Frame Shooting CL Process | Productive | 38 | 37 | 38 | 36 | 149 |
| | Non-productive | 2 | 2 | 4 | 5 | 13 |
| | Inevitable | 3 | 4 | 1 | 2 | 10 |
| | Total | 43 | 43 | 43 | 43 | 72 |
| | % Productive | 88.4 | 86.0 | 88.4 | 83.7 | |
| | Output | 12 | 11 | 12 | 10 | 45 |

### *3.2. Data Uniformity Test*

Data uniformity test is conducted to determine whether the data obtained from preliminary observations are uniform and do not exceed the upper control limit (UCL/BKA) and the lower control limit (LCL/BKB) that have been determined. Following are the calculation of BKA and BKB for the manufacturing spring round process which use equation (1) dan (2).

$$\bar{p} = \frac{221}{240} \times 100\% = 92.1\%$$

$$BKA = 92.1 + 3\sqrt{\frac{92.1\,(100 - 92.1)}{240}} = 97.3\% \quad \text{and} \quad BKB = 92.1 - 3\sqrt{\frac{92.1\,(100 - 92.1)}{240}} = 86.9\%$$

For the results of other process calculations can be seen in Table 3.

**Table 3.** Result of Data Uniformity Test

| Process | $\bar{p}$ | BKA | BKB | Result |
|---|---|---|---|---|
| Spring round | 92.1 | 97.3 | 86.9 | Uniform |
| Semi-finished spring frame | 93.7 | 96.6 | 90.8 | Uniform |
| List frame | 77.5 | 86.2 | 68.7 | Uniform |
| Spring frame CL shooting | 86.6 | 94.4 | 78.8 | Uniform |

Based on Table 3 it can be seen that the average percentage of productive activity ($\bar{p}$) for the whole process is in the range of BKA and BKB, so it can be concluded that the data of all processes are uniform and can be continued to the next stage.

### *3.3. Data Adequacy Test*

To test the adequacy of the data, the confidence level used is 90% with a value of 1,645 and the level of accuracy used is 10% (0.10). The following is the calculation of the data sufficiency test for the manufacturing spring round process which use equation (3).

$$N' = \frac{1.645^2\,(100 - 92.1)}{0.10^2 \times 92.1} = 23.26 \approx 24$$

The calculation results of the data adequacy test for the manufacturing spring round process obtained an N' value of 23.26 (equivalent to 24 data) which means that the amount of data needed to be said to be sufficient is 24 data. With the number of preliminary observational data (N) for the manufacturing spring round process of 240 data, it can be concluded that the data is adequate because N'<N. For the results of data adequacy tests for other process can be seen in Table 4.

**Table 4.** Result of Data Adequacy Test

| Process | N | N' | Result |
|---|---|---|---|
| Spring round | 240 | $23.26 \approx 24$ | Adequate |
| Semi-finished spring frame | 204 | $18.28 \approx 19$ | Adequate |
| List frame | 632 | $78.78 \approx 79$ | Adequate |
| Spring frame CL shooting | 172 | $41.77 \approx 42$ | Adequate |

Based on Table 4 it can be seen that the amount of observational data (N) for the whole process is greater than the calculation of the amount of data needed (N'), so it can be concluded that the data of the entire process is sufficient and can be continued to the next stage.

### *3.4. Calculation of Cycle Times (WS)*

Cycle time is the time needed for each time the manufacturing process. The following is the calculation of the cycle time (WS) for the manufacturing spring round process which use equation (4) dan (5).

$$\text{Time for 1 output} = \frac{30+30+30+30}{8+7+7+6} = 4.29 \text{ minute} \quad \text{and} \quad \text{The productive percentage} = 92.1\%$$

$$WS = 0.921 \times 4.29 = 3.95 \text{ minute}$$

For the results of other process calculations can be seen in Table 5.

**Table 5.** Cycle Time Calculation Results

| Process | Time for 1 Output (minute) | Cycle Time/WS (minute) |
|---|---|---|
| Spring round | 4.29 | 3.95 |
| Semi-finished spring frame | 8.57 | 8.03 |
| List frame | 15.00 | 11.62 |
| Spring frame CL shooting | 10.67 | 9.24 |

### *3.5. Calculation of Normal Time (WN)*

For the normal time calculation, an performance factor is needed because in the time measurements made on observations there are some irregularities that occur. There are many factors that influence these irregularities, so that performance factors are needed. The method used to determine the performance factor is the Westinghouse method, which consists of four factors, namely skills, effort, working conditions, and consistency. Table 6 is the value of the performance factor of all the processes that exist in the manufacture of spring frame components.

**Table 6**. Performance Factor Value

| Process | Performance Factor | Class | Symbol | Performance | Total |
|---|---|---|---|---|---|
| Spring round | Skills | Good | C2 | 0.03 | 0.06 |
| | Effort | Good | C1 | 0.05 | |
| | Working conditions | Fair | E | −0.03 | |
| | Consistency | Good | C | 0.01 | |
| Semi-finished spring frame | Skills | Good | C1 | 0.06 | 0.09 |
| | Effort | Good | C2 | 0.02 | |
| | Working conditions | Average | D | 0.00 | |
| | Consistency | Good | C | 0.01 | |
| List frame | Skills | Good | C1 | 0.06 | 0.08 |
| | Effort | Good | C1 | 0.05 | |
| | Working conditions | Fair | E | −0.03 | |
| | Consistency | Average | D | 0.00 | |
| Spring frame CL shooting | Skills | Good | C1 | 0.06 | 0.12 |
| | Effort | Excellent | B2 | 0.08 | |
| | Working conditions | Fair | E | −0.03 | |
| | Consistency | Good | C | 0.01 | |

After the performance factor value is determined, the normal time calculation can be done. The following is the calculation of the normal time for the manufacturing spring round process which use equation (6) dan (7).

$P = 1 + 0.06 = 1.06$ And $WN = 3.95 \text{ minute} \times 1.06 = 4.18 \text{ minute}$

For the results of other process calculations can be seen in Table 7.

**Table 7**. Normal Time Calculation Results

| Process | Performance Factor (P) | Normal Time/WN (minute) |
|---|---|---|
| Spring round | 1.06 | 4.18 |
| Semi-finished spring frame | 1.09 | 8.75 |
| List frame | 1.08 | 12.55 |
| Spring frame CL shooting | 1.12 | 10.35 |

### *3.6. Calculation of Standard Time (WB)*

For the calculation of the standard time needed allowance factor which is the time required by the operator for the needs of resting after a cycle of product execution. The allowance factor is also useful in delaying the activity carried out which is caused by several factors. The value of allowance factors for all processes in the manufacture of spring frame components can be seen in Table 8.

After the allowance factors for each process are determined, the standard time calculation can be done with the detailed calculation for the spring round process which use equation (8) as follows.

$WB = 4.18 + 20.30\%(4.18) = 5.03 \text{ minute}$

For the results of other process calculations can be seen in Table 9.

**Table 8**. Allowance Factor Value

| Process | Allowance Types | Allowance (%) | Total (%) |
|---|---|---|---|
| Spring round | Personal | 2.00 | 20.30 |
| | Fatigue | 18.00 | |
| | Inevitable | 0.30 | |
| Semi-finished spring frame | Personal | 2.00 | 24.83 |
| | Fatigue | 18.00 | |
| | Inevitable | 4.83 | |
| List frame | Personal | 2.00 | 18.00 |
| | Fatigue | 16.00 | |
| | Inevitable | 0.00 | |
| Spring frame CL shooting | Personal | 2.00 | 28.50 |
| | Fatigue | 24.00 | |
| | Inevitable | 2.50 | |

**Table 9**. Standard Time Calculation Results

| Process | Allowance Factor/A (%) | Standard Time/WB (minute) |
|---|---|---|
| Spring round | 20.30 | 5.03 |
| Semi-finished spring frame | 10.99 | 10.92 |
| List frame | 14.81 | 14.81 |
| Spring frame CL shooting | 13.40 | 13.30 |

*3.7. Output Calculation*

From the calculation of standard time which is the time needed to work on 1 set of frame components per, it can be determined the output generated from each process for each day with a total working hours of 450 minutes. Calculation of output per day for the manufacturing process per round is as follows.

$$\text{Output} = \frac{\text{Total work hours}}{\text{Standard time}} = \frac{450}{5.03} = 89.29 \approx 90 \text{ pcs}$$

In Table 10, the daily output is displayed for each process of spring frame components.

**Table 10**. Output of Each Process in the Spring Frame Manufacturing

| Process | Standard Time/WB (minute) | Output (pcs) | Target |
|---|---|---|---|
| Spring round | 5.03 | 90 | 110 |
| Semi-finished spring frame | 10.92 | 41 | 40 |
| List frame | 14.81 | 31 | 25 |
| Spring frame CL shooting | 13.30 | 34 | 40 |

From Table 10 which is the daily output in each process of spring frame components, it can be seen that there are two processes in the production of spring frame components which cannot meet the targets of PT. CMAP, namely spring round process and spring frame CL shooting process.

## 4. CONCLUSION

Based on the low achievement of the target of the production of spring frame components compared with the other components production in the manufacture of mattresses at PT. CMAP, obtained several conclusions, including:

1) Percentage of productive activities for the process of spring frame mattress components at PT.CMAP consisting of spring round process is 92.1%, semi-finished spring frame process is 93.7%, list frame process is 77.5% and spring frame CL shooting process is 86.6%.

2) The cycle time for spring frame mattress at PT.CMAP which consists of spring round process is 3.95 minutes, semi-finished spring frame process is 8.03 minutes, frame lists process is 11.62 minutes and spring frame CL shooting process is 9.24 minutes.
3) The normal time for spring frame mattress at PT. CMAP which consists of spring round process is 4.18 minutes, semi-finished spring frame process is 8.75 minutes, frame lists process is 12.55 minutes and spring frame CL shooting process is 10.35 minutes.
4) The standard time for spring frame mattress at PT. CMAP which consists of spring round process is 5.03 minutes, semi-finished spring frame process is 10.92 minutes, frame lists process is 14.81 minutes and spring frame CL shooting process is 13.30 minutes.
5) The daily output of spring round processes is is 90 pcs, semi-finished spring frame process is 41 pcs, frame lists process is 31 pcs and spring frame CL shooting process is 34 pcs.
6) PT. CMAP must review the targets set in the production process of spring frame component, because based on the calculation of standard time and daily output there are two processes that cannot meet the set targets.

This research only uses four preliminary sampling for each process, so that for continuous research it can use a larger preliminary sampling so that the results obtained can be better.

## 5. REFERENCES

[1] Andhana, M. R. and Prastawa, H. Analisis Produktivitas Pekerja Dengan Metode Work Sampling Pada Filling Shed I Produk Premium PT.Pertamina Tbbm Semarang Group. 2019. *Ind. Eng. Online J.* **7**. 1–8

[2] Andi, Wibowo, D.K. and Prasetya, A., 2004. Analisa Produktifitas Pekerja dengan Metode Work Sampling: Studi Kasus Pada Proyek X dan Y. Civil Engineering Dimension

[3] Jono. 2015. Pengukuran Beban Kerja Tenaga Kerja Dengan Metode Work Sampling (Studi Kasus di PT. XY Yogyakarta)

[4] Kurniawan, Hadi and Budiawan,Wiwik. Analisa Produktivitas Pekerja Dengan Metode Sampling Kerja Pada Bagging Section PT. Chandra Asri Petrochemical Tbk. Program Studi Teknik Industri, Fakultas Teknik, Universitas Diponegoro Semarang

[5] Fajrah, Nofriani. Uswansyaf, M. Gamma. Budiman, Arif. Mahata, Raisa Batura Endo. 2019. Sampling Pekerjaan Stasiun Kerja Pemotongan Pada Pabrik Tahu Pak Buyung, Jurusana Teknik Industri, Fakultas Teknik Universitas Andalas

[6] Sutalaksana, dkk. 2006. *Teknik Perancangan Sistem Kerja*. ITB. Bandung

[7] Rachman, Taufiqur. 2013. Penggunaan Metode Work Sampling Untuk Menghitung Waktu Baku Dan Kapasitas Produksi Karungan Soap Chip Di PT. SA. *Jurnal Inovisi*. **9**. 48–60

[8] Wignjoseobroto, Sritomo. 1989. *Teknik Tata Cara dan Pengukuran Kerja*, PT Guna Widya. Jakarta

# AUTHOR INDEX


Adisuwiryo, S ........ 419
Ahmad ........ 458
Akbar, Wydzka Tasha Aulia ........ 373
Alhilman, Judi ........ 144, 152
Alhilman, J. ........ 176
Amarilies, Harummi Sekar ........ 48
Ambarita, R ........ 168
Amran, Tiena Gustina ........ 727
Andra, D Jovie ........ 9
Andriyan, Tri ........ 313
Andriyani, Neng Resi ........ 466
Annisyah, Rizky Cherthio ........ 596
Anussornnitisarn, P. ........ 200
Aqsha, BD ........ 574, 633
Ardinia, D ........ 24
Arifin, F ........ 320
Ariyanti, Silvi ........ 442
Ariyon, Muhammad ........ 233
Arta, A K J ........ 427
Arumsari ........ 246
Ashfaq, MA ........ 549
Atmaji, Fransiskus Tatas Dwi ........ 152
Atmaji, FTD ........ 176
Aulia, Zulhilmi Naufal ........ 287
Awan, M U ........ 549
Azis, H A ........ 674
Bachtiar, Marsellinus ........ 450
Bon, Abdul Talib ........ 342
Cahyadi, B ........ 712
Chen, Tai-Jung ........ 225
Chiang, Chin-Hsin ........ 225
Christata, Bellachintya Reira ........ 88
Chumaidiyah, Endang ........ 373
Chumaidiyah, E ........ 481
Darmawan, I ........ 95
Darmawan, M ........ 40
Darussalam, F ........ 543
Daryanto, Yosef ........ 88
Dewi, D R S ........ 434
Dewi, Sarastya ........ 152
Dewobroto, W S ........ 641
Dias, S P ........ 119
Dimyati, Tjutju Tarliah ........ 136
Divia, G A ........ 419
Doaly, C O ........ 427, 473
Dwiputri, S Fritasya ........ 536
Ekananda, Hilman Ardian ........ 358
Fauzia, SW ........ 72
Fauziyah, Salma Defina ........ 295
Febrianti, Erna ........ 350
Fithri, P ........ 9
Fitriana, Rina ........ 80, 295, 313
Fornah, Hassan Andrew ........ 736
Gana, I M ........ 658, 666
Gbabo, A ........ 658
Giffari, A F ........ 520
Giffari, F ........ 262
Ginting, A P ........ 119
Gozali, L ........ 1
Habibie, Muhammad Fadhil ........ 144
Haezar, T Eben ........ 649
Halim, Alfred Darius ........ 127
Hanaa, R W ........ 481
Hartono, N. ........ 208
Hendra, Franka ........ 168
Herwanto, D ........ 566
Hidayah, N Y ........ 40
Hidayat, S ........ 72
Hidayat, T P ........ 411
Hoa, N T X ........ 549
Huda, L N ........ 520
Husen, Novan Pizary ........ 365
Ibrahim, N. G. ........ 254
Ikatrinasari, Zulfa Fitri ........ 381
Iqbal, M ........ 496
Iridiastadi, H ........ 489, 674
Islamiah, U W ........ 609, 617
Ismail, A H ........ 208
Jannah, Alif Miftahul ........ 536
Jason, Andreas ........ 442
Jou, Y. T. ........ 388
Junianto, A B ........ 504
Kamal, S A ........ 549
Kamara, Alimamy ........ 744
Kamaralo, MK ........ 176
Khan, S Z ........ 549
Khannan, M S A ........ 512
Kosasih, Wilson ........ 127, 458, 473
Kristiana, S P D ........ 403
Kristina, H J ........ 458
Kristiyani, Ika Murti ........ 88
Kurniati, N ........ 278
Larasati, Dea Prameswari ........ 80
Lee, Yu-Ching ........ 225
Lilajati, M A ........ 403
Lim, N ........ 1

Limlawan, V ........ 200
Lubis, Muharman ........ 536, 596, 625
Lumingkewas, Lyvia Winiyanti ........ 625
Machfud ........ 688
Magdalena, R ........ 119, 403
Mangunwidjaja, D ........ 688
Marie, I A ........ 320, 641
Marpaung, B ........ 720
Maryanti, A S ........ 712
Mastrisiswadi, H ........ 512
Moengin, P ........ 543
Mufidah, Ilma ........ 48, 358
Mukti, B C ........ 327
Muqimuddin ........ 217
Muslim, K ........ 489
Mutiasari, M ........ 609, 617
Nadlifatin, Reny ........ 48
Nafisah, L ........ 512
Nasution, S.R. ........ 1
Natalia, C ........ 411, 697
Ngadono, Teguh Sri ........ 381
Nugroho, S M ........ 512
Nurdina, R ........ 103
Nurhasanah, N ........ 72, 688
Nurrahman, Ahmad Arif ........ 365
Oktavia, C W ........ 403, 411
Omran, A ........ 303
Ong, A K S ........ 388
Paoprasert, N. ........ 528
Parmita, Exa ........ 625
Parwati, N ........ 56, 103, 609, 617
Pasaribu, Kristina ........ 458
Pham, D T ........ 208
Pittayachawan, S ........ 434
Prasetyani, R ........ 64, 327
Prasetyawan, Y. ........ 254, 262
Pratama, A T ........ 32
Prianggara, F ........ 32
Pujawan, I Nyoman ........ 736, 744
Pujawan, I N ........ 581
Purwandari, A Tri ........ 617
Purwandari, A T ........ 56, 103, 609
Purwaningsih, Sri Retno ........ 192
Putera, B S A ........ 262
Putra, F A ........ 603
Putranto, K Eko ........ 168
Rachman, Taufiqur ........ 752
Rahayu, Mira ........ 350, 358
Ramadhan, A A ........ 603
Ramdhani, M N ........ 512
Ratanadewi, Steffi ........ 450
Ratnamirah, A ........ 56
Redi, A A N Perwira ........ 48
Rendra, Meldi ........ 373
Reza, Refiandi ........ 233
Rifqi, J N ........ 246
Rimantho, D ........ 24, 111
Rivai, Muhammad ........ 681
Robielos, R A C ........ 388
Romli, M ........ 688
Roni, S Taufik ........ 649
Rudolf, DI ........ 396
Rukmana, Otong ........ 365
Saedudin, Rd. Rohmat ........ 536
Sagara, B ........ 558
Sahroni, T R ........ 603
Salomon, L L ........ 427
Salomon, Lithrone Laricha ........ 127
Samdan, W ........ 649
Saputra, Fajar ........ 335
Saragih, Johnson ........ 80, 295, 313
Saraswati, T ........ 32
Sardjono, Tri Arief ........ 681
Sasongko, A ........ 320
Savitri, T ........ 32
Septiana, Nuzul Fatma ........ 160
Septiani, W ........ 419
Setiawan, Aldo ........ 233
Setyawati, E P ........ 111
Sevani, Nina ........ 335
Shabara ........ 473
Shafiq, M ........ 549
Shhu, A A ........ 658
Shiru, J J ........ 666
Sholihin, M Y M ........ 327
Silalahi, Frans Arianto ........ 350
Singgih, M. L. ........ 217
Siswanto, D ........ 489
Siswanto, N ........ 574, 633
Siti, A ........ 64
Soenandi, Iwan Aang ........ 335
Soesanto, R P ........ 95
Sriwana, Iphov Kumala ........ 270, 752
Sudarni, A A C ........ 589, 705
Sudiarno, A ........ 589, 705
Sugandha, D C ........ 504
Sugioko, A ........ 697
Sukarno, Iwan ........ 160
Sulistiandi, S ........ 720
Sunardi, O ........ 720
Sundari, A S ........ 111
Suparno ........ 184, 574, 633
Suprata, F ........ 697
Suprayitno, H ........ 184

Supriyono ........................................................................ 168
Surbakti, Feliks P Sejahtera ................................................... 17
Suryanti, F ........................................................................ 705
Susanto, B S P .................................................................... 278
Suwandi, Arief..................................................................... 287
Suzianti, A ................................................................ 496, 566
Syafrizal, M ........................................................................ 40
Syauqillah, Nadya ............................................................... 270
Tait, E ................................................................................ 434
Tama, M I .................................................................. 574, 633
Tanjung, W Nurchayanty ...................................................... 617
Tanjung, W N ............................................................... 56, 103
Tanjung MT, W N ............................................................... 609
Taufik ................................................................................. 9
Theopilus, Y ....................................................................... 558
Timang, G A ................................................................ 64, 712
Tjandra, S S ....................................................................... 558
Tjhong, S G......................................................................... 411
Tripiawan, Wawan................................................................ 144
Tripiawan, W ...................................................................... 95
Triyanti, V .................................................................. 396, 674
Ulansari, D R ...................................................................... 581
Umaiyah ............................................................................ 184
URehman, H ....................................................................... 549
Utomo, D. S. ....................................................................... 528
Wahyukaton......................................................................... 239
Wee, H M ........................................................................... 388
Wee, Hui Ming .................................................................... 192
Widodo, E........................................................................... 184
Widodo, Lamto ............................................................ 442, 458
Widodo, L ........................................................................... 1
Winiyanti, L Lyvia................................................................. 596
Wirawan, H......................................................................... 603
Wirdianto, E......................................................................... 9
Yahya, SHS ........................................................................ 303
Yassierli.............................................................................. 674
Yogaswara, Yogi .................................................................. 466
Yousuk, R. .......................................................................... 528
Yu, James .......................................................................... 192
Yudha, A............................................................................. 649
Yuwono, Alfarid Hendro ....................................................... 681
Zalukhu, Stefanny Liyanawati ............................................... 752
Zeybek, S ........................................................................... 208